Straight Lines

Slope: $m = \dfrac{y_2 - y_1}{x_2 - x_1}$

Point-slope form: $y - y_1 = m(x - x_1)$

Slope-intercept form: $y = mx + b$

Horizontal line: $y = b$

Vertical line: $x = a$

If $m_1 = m_2$, then $L_1 \parallel L_2$

If $m_1 = -\dfrac{1}{m_2}$, then $L_1 \perp L_2$

Distance: $d = \sqrt{(x_2 - x_1)^2 + (y_2 - y_1)^2}$

Midpoint: $x_m = \dfrac{x_1 + x_2}{2}$; $y_m = \dfrac{y_1 + y_2}{2}$

Quadratic Formula

If $ax^2 + bx + c = 0$ $(a \neq 0)$,

then $x = \dfrac{-b \pm \sqrt{b^2 - 4ac}}{2a}$

Properties of Logarithms

$x = a^y$ is equivalent to $y = \log_a x$, $a > 0$

1. $\log_a (M \cdot N) = \log_a M + \log_a N$

2. $\log_a \left(\dfrac{M}{N}\right) = \log_a M - \log_a N$

3. $\log_a M^n = n \log_a M$

4. $\log_a \sqrt[n]{M} = \dfrac{1}{n} \cdot \log_a M$

5. $\log_a 1 = 0$

6. $\log_a \dfrac{1}{M} = -\log_a M$

7. $\log_a a^x = x$

8. $a^{\log_a x} = x$

9. $\log x = \log_{10} x$

10. $\ln x = \log_e x$

Technical Mathematics

Technical Mathematics

SECOND EDITION

Dale Ewen
Parkland Community College

Joan S. Gary
Parkland Community College

James E. Trefzger
Parkland Community College

Upper Saddle River, New Jersey
Columbus, Ohio

Library of Congress Cataloging-in-Publication Data

Ewen, Dale
Technical mathematics / Dale Ewen, Joan S. Gary, James E. Trefzger.—2nd ed.
p. cm.
Includes index.
ISBN 0-13-048810-0
1. Mathematics. I. Gary, Joan S. II. Trefzger, James E. III. Title.

QA39.3.E95 2005
510—dc22

2004003378

Editor in Chief: Stephen Helba
Senior Acquisitions Editor: Gary Bauer
Editorial Assistant: Natasha Holden
Development Editor: Michelle Churma
Production Editor: Louise N. Sette
Production Supervision: *The GTS Companies*/York, PA Campus
Design Coordinator: Diane Ernsberger
Cover Designer: Eric Davis
Production Manager: Pat Tonneman
Marketing Manager: Leigh Ann Sims

This book was set in Times Roman by *The GTS Companies*/York, PA Campus. It was printed and bound by Courier Kendallville, Inc. The cover was printed by The Lehigh Press, Inc.

Pearson Education Ltd.
Pearson Education Singapore Pte. Ltd.
Pearson Education Canada, Ltd.
Pearson Education—Japan
Pearson Education Australia Pty. Limited
Pearson Education North Asia Ltd.
Pearson Educación de Mexico, S.A. de C.V.
Pearson Education Malaysia Pte. Ltd.

10 9 8 7
ISBN 0-13-048810-0

Preface

Technical Mathematics, Second Edition (formerly *Mathematics for Technical Education*), provides the mathematics skills for students considering a career in a technical or engineering technology program and emphasizes the mathematical concepts that students need to be successful.

The text covers the following major areas: fundamental concepts and measurement; fundamental algebraic concepts; exponential and logarithmic functions; right-triangle trigonometry, the trigonometric functions, and trigonometric formulas and identities; complex numbers; matrices; polynomial and rational functions; statistics for process control; and analytic geometry.

Key and New Features

- Clear explanations supported by detailed and well-illustrated examples.
- More than 5700 exercises.
- Important formulas and principles highlighted within the text.
- Detailed descriptions of real-world applications of mathematical concepts.
- Calculator story boards integrated throughout the text to illustrate step-by-step calculator operations.
- Chapter summary and review sections help students prepare for quizzes and examinations.
- Development of the metric system in Appendix B.
- Two appendices of instructions for using graphing calculators—one for a basic calculator (TI-83 Plus or TI-84 Plus) and another for an advanced calculator (TI-89).
- **New! Chapter 2: Review of Geometry** now expanded into a complete chapter and covered early in the text.
- **New! Nine Extended Applications in Technology Features**—placed at the end of selected chapters.
- **New! Electronic Study Guide on CD: Free in Book!** This unique tutorial tool contains the entire book on CD along with question banks for each section objective covered in the text. As students work through the self-study questions, the quizzes are immediately graded and they can access the text on CD to review concepts that they have not understood. The Electronic Study Guide helps students master the material and prepares them for taking classroom tests.
- **New! Companion Website** Self-grading practice quizzes help students prepare for tests at www.prenhall.com/ewen.

- **New! Online Graphing Calculator Help** for popular TI, HP, Casio, GFX, and Sharp Calculators—available at the book companion website.
- **New! Student Solutions Manual** This helpful study tool includes worked solutions to all odd-numbered exercises in the text. Order online at www.prenhall.com (0-13-110286-9).
- **Instructor's Manual** contains solutions for selected odd-numbered exercises, answers for even-numbered exercises, and sample chapter tests and answers.

Illustration of Some Key Features

Numerous Examples Since many students learn by example, a large number of detailed and well-illustrated examples including key steps are used throughout the text. See page 286 for examples.

Abundant Exercises To reinforce key concepts for students, we have provided a large variety of well-illustrated exercises. See page 140 for examples.

Extended Applications in Technology Ten more in-depth presentations of special applications in technology are included at the end of selected chapters to provide student interest and class discussion. See page 345 for an example.

Calculator Story Boards Actual calculator screens are used to show students the sequence of step-by-step operations, as shown on page 530.

To the Faculty

The topics have been arranged with the assistance of faculty who teach in a variety of technical programs. However, we have also allowed for many other compatible arrangements. The topics are presented in an intuitive manner, with technical applications integrated throughout whenever possible. The large number of detailed examples and exercises are features that students and faculty alike find essential.

The text is written at a language level and a mathematics level that are cognizant of and beneficial to most students in technical programs. We assume that students have a mathematics background that includes one year of high school algebra or its equivalent and some geometry. The introductory chapters are written so that students who are deficient in some topics may also be successful. The material in this book should be completed in two semesters or three quarters and serves as a foundation for more advanced work in mathematics. A companion book, *Technical Mathematics with Calculus,* contains these same 20 chapters and continues with calculus for those engineering technology and other programs that require a development of practical calculus. *Technical Calculus* is also available for those who prefer a comprehensive two-volume alternative.

Chapters 1 through 3 provide the basic skills that are needed early in almost any technical program. Chapters 4 through 9 complete the basic algebraic foundation, and Chapters 10 through 14 include the trigonometry necessary for the technologies. Chapters 15 though 18 include some advanced topics needed for some programs. Chapter 19 addresses basic statistics. Chapter 20 (analytic geometry) completes a comprehensive mathematics background needed in many programs; some programs include this chapter at the end of the first year, while other programs include this chapter at the beginning of an introductory calculus course.

We have included Appendix C on the basic graphing calculator (TI-83 Plus or TI-84 Plus) and Appendix D on the advanced graphing calculator (TI-89) so that faculty have the option of which, if any, graphing calculator to use in their course. Graphing calculator uses are integrated into some of the examples in the text.

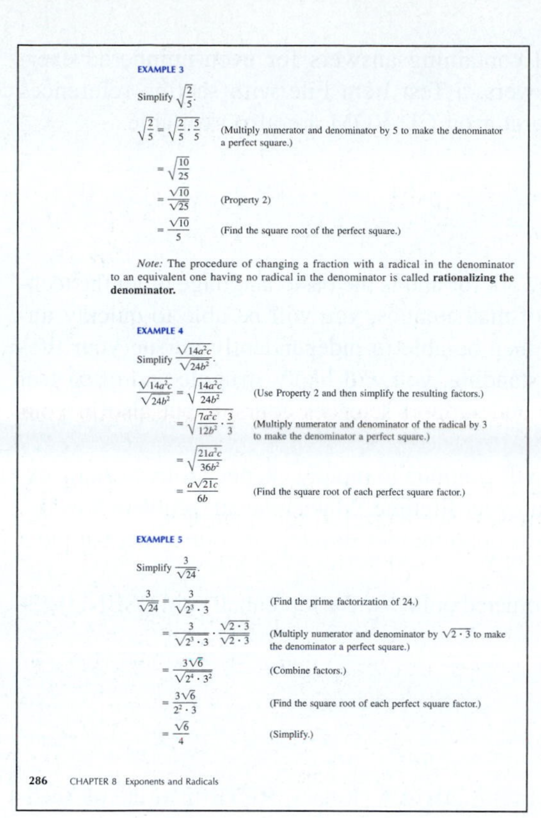

EXAMPLE 3

Simplify $\sqrt{\frac{2}{5}}$.

$$\sqrt{\frac{2}{5}} = \sqrt{\frac{2}{5}\cdot\frac{5}{5}}$$ (Multiply numerator and denominator by 5 to make the denominator a perfect square.)

$$= \sqrt{\frac{10}{25}}$$

$$= \frac{\sqrt{10}}{\sqrt{25}}$$ (Property 2)

$$= \frac{\sqrt{10}}{5}$$ (Find the square root of the perfect square.)

Note: The procedure of changing a fraction with a radical in the denominator to an equivalent one having no radical in the denominator is called **rationalizing the denominator.**

EXAMPLE 4

Simplify $\frac{\sqrt{14a^2c}}{\sqrt{24b^2}}$.

$$\frac{\sqrt{14a^2c}}{\sqrt{24b^2}} = \sqrt{\frac{14a^2c}{24b^2}}$$ (Use Property 2 and then simplify the resulting factors.)

$$= \sqrt{\frac{7a^2c}{12b^2}\cdot\frac{3}{3}}$$ (Multiply numerator and denominator of the radical by 3 to make the denominator a perfect square.)

$$= \sqrt{\frac{21a^2c}{36b^2}}$$

$$= \frac{a\sqrt{21c}}{6b}$$ (Find the square root of each perfect square factor.)

EXAMPLE 5

Simplify $\frac{3}{\sqrt{24}}$.

$$\frac{3}{\sqrt{24}} = \frac{3}{\sqrt{2^3\cdot 3}}$$ (Find the prime factorization of 24.)

$$= \frac{3}{\sqrt{2^3\cdot 3}}\cdot\frac{\sqrt{2\cdot 3}}{\sqrt{2\cdot 3}}$$ (Multiply numerator and denominator by $\sqrt{2\cdot 3}$ to make the denominator a perfect square.)

$$= \frac{3\sqrt{6}}{\sqrt{2^4\cdot 3^2}}$$ (Combine factors.)

$$= \frac{3\sqrt{6}}{2^2\cdot 3}$$ (Find the square root of each perfect square factor.)

$$= \frac{\sqrt{6}}{4}$$ (Simplify.)

286 CHAPTER 8 Exponents and Radicals

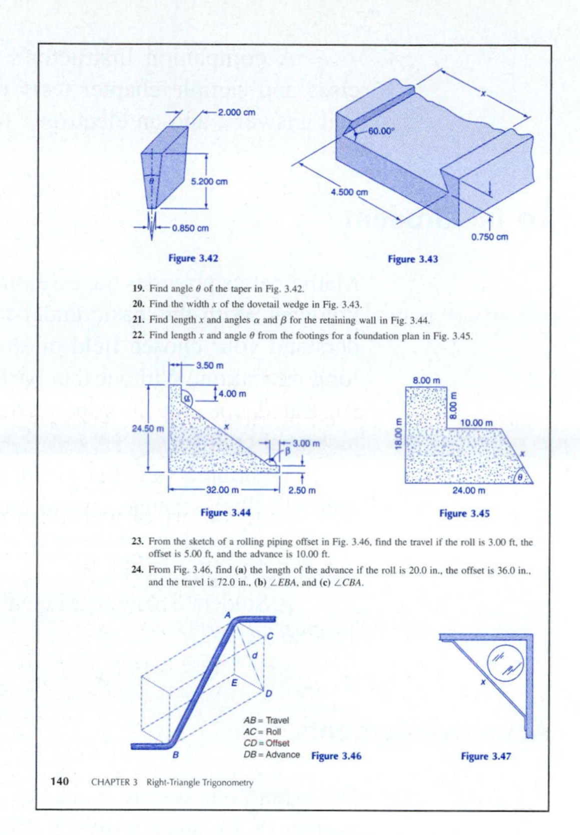

Figure 3.42

Figure 3.43

19. Find angle θ of the taper in Fig. 3.42.
20. Find the width x of the dovetail wedge in Fig. 3.43.
21. Find length x and angles α and β for the retaining wall in Fig. 3.44.
22. Find length x and angle θ from the footings for a foundation plan in Fig. 3.45.

Figure 3.44

Figure 3.45

23. From the sketch of a rolling piping offset in Fig. 3.46, find the travel if the roll is 3.00 ft, the offset is 5.00 ft, and the advance is 10.00 ft.
24. From Fig. 3.46, find **(a)** the length of the advance if the roll is 20.0 in., the offset is 36.0 in., and the travel is 72.0 in., **(b)** $\angle EBA$, and **(c)** $\angle CBA$.

Figure 3.46

Figure 3.47

140 CHAPTER 3 Right-Triangle Trigonometry

CHAPTER **9**

SPECIAL NASA APPLICATION

The Space Shuttle Landing, Part II*

Part I appears at the end of Chapter 4. Reading that Introduction will be helpful to this application.

As the Shuttle comes out of its communication blackout, it travels a circular path around an imaginary cone that will line it up with the center line of the runway. Coming out of its turn, the Space Shuttle should be at an altitude of 13,365 ft, have a speed of 424 mi/h, and be 7.5 mi (horizontal distance) from the runway. It is now 86 s to touchdown. The nose is down so that the Space Shuttle can descend steeply to a point 7500 ft from the runway threshold, where its altitude should be 1750 ft. The vehicle then enters a transitional phase. The Shuttle's nose is raised as it heads for a position where its altitude is 131 ft and its distance from the runway threshold is 2650 ft. The Shuttle is now 17 s to touchdown. From here the Space Shuttle enters the final phase, aiming at a point 2200 ft down the runway. See Fig. 4.33. From point (39600, 13365) to point (7500, 1750), we can use a linear function (model) to describe the path; its glide slope is 19.9°. We can also use a linear function between the last two points (2650, 131) and (−2200, 0); its glide slope is 1.5°. The path from a steep glide slope to a shallow one is not a linear function. See Fig. 9.15.

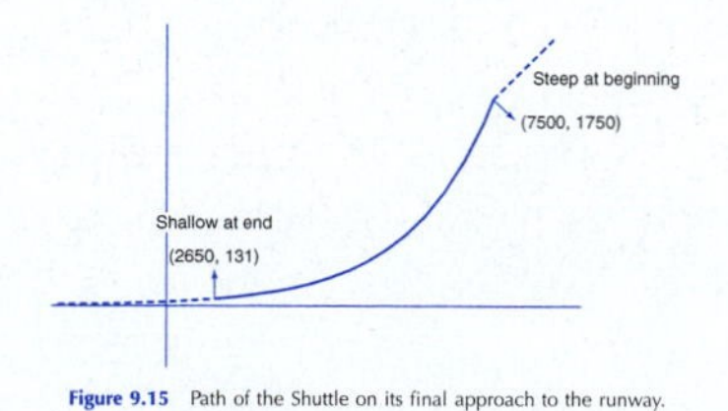

Figure 9.15 Path of the Shuttle on its final approach to the runway.

*From NASA–AMATYC–NSF Project Mathematics Explorations II, grant principals John S. Pazdar, Patricia L. Hirschy, and Peter A. Wursthorn; copyright Capital Community College, 2000.

345

a point on its graph. If the remainder had been 19, you would have known that (7, 19) was a point on the graph of the polynomial (and that $x - 7$ was *not* one of its factors). Unfortunately, the use of polynomial division to test potential factors and evaluate points easily becomes very tedious, especially in practical work. Instead, consider the method illustrated in Example 3.

1. Use a graphing calculator to locate the x-intercepts (or zeros) of the polynomial.
2. Use the Factor Theorem to infer a corresponding factor from each x-intercept (or zero).
3. Use the calculator's **TRACE** feature to evaluate other points that you may wish to investigate.

EXAMPLE 3

Graph the polynomial $f(x) = 2x^4 + x^3 - 6x^2 + x + 2$ to locate its x-intercepts (its zeros).

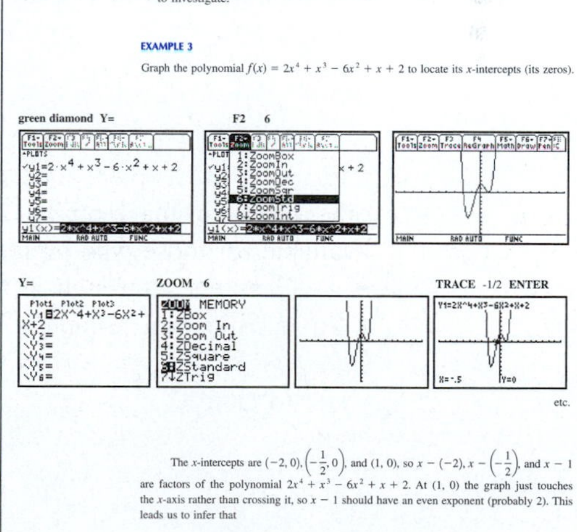

The x-intercepts are $(-2, 0)$, $\left(-\frac{1}{2}, 0\right)$, and $(1, 0)$, so $x - (-2)$, $x - \left(-\frac{1}{2}\right)$, and $x - 1$ are factors of the polynomial $2x^4 + x^3 - 6x^2 + x + 2$. At (1, 0) the graph just touches the x-axis rather than crossing it, so $x - 1$ should have an even exponent (probably 2). This leads us to infer that

$$2x^4 + x^3 - 6x^2 + x + 2 = 2(x + 2)(x + \tfrac{1}{2})(x - 1)^2$$

Note that the leading coefficient (in this case, 2) must be included as a constant factor to reproduce the original polynomial in factored form. Some people prefer to multiply $2\left(x + \frac{1}{2}\right) = 2x + 1$ giving

$$2x^4 + x^3 - 6x^2 + x + 2 = (x + 2)(2x + 1)(x - 1)^2$$

530 CHAPTER 16 Polynomials of Higher Degree

A companion Instructor's Manual containing answers for even-numbered exercises and sample chapter tests with answers, a Test Item File with section references and answers, and an electronic test generator on CD-ROM are also available.

To the Student

Mathematics provides the essential framework for and is the basic language of all the technologies. With this basic understanding of mathematics, you will be able to quickly understand your chosen field of study and then be able to independently pursue your lifelong education. Without this basic understanding, you will likely struggle and often feel frustrated not only in your mathematics and support sciences courses, but also in your technical courses.

Technology and the world of work will continue to rapidly change. Your working career will likely change several times during your lifetime. Mathematical, problem-solving, and critical-thinking skills will be crucial as opportunities develop in your career path in a rapidly changing world.

A Student Solutions Manual can be ordered online at www.prenhall.com (ISBN: 0-13-110286-9).

Acknowledgments

We extend our sincere thanks to our reviewers: David A. Kelsey, RETS Institute of Technology (KY), and Carolyn S. Rieffel, Louisiana Technical College–LEFC.

We sincerely thank Patricia L. Hirschy, John S. Pazdar, and Peter A. Wursthorn for allowing us to use portions of their NASA–AMATYC–NSF projects.

We also give a special thanks to our Prentice Hall editor, Gary Bauer; our Prentice Hall associate editor, Michelle Churma; our Prentice Hall production editor, Louise Sette; to project manager, Penny Walker at *The GTS Companies*/York Campus; our copyeditor, Philip Koplin; and to Joyce Ewen for her excellent proofing assistance.

If anyone wishes to correspond with us regarding suggestions, criticisms, questions, or errors, please contact Dale Ewen through Prentice Hall.

Dale Ewen
Joan S. Gary
James E. Trefzger

Contents

7 Quadratic Equations 261

8 Exponents and Radicals 278

9 Exponentials and Logarithms 304

10 Trigonometric Functions 347

11 Oblique Triangles and Vectors 373

12 Graphing the Trigonometric Functions 421

13 Trigonometric Formulas and Identities 445

14 Complex Numbers 476

15 Matrices 498

16 Polynomials of Higher Degree 526

17 Inequalities and Absolute Value 542

18 Progressions and the Binomial Theorem 561

19 Basic Statistics 574

20 Analytic Geometry 608

1 Fundamental Concepts

INTRODUCTION

In a steel mill, the temperature of molten steel, the size of the ingots, the time it takes a batch of molten steel to be poured, and the electronic circuitry that keeps the assembly lines moving—all involve measurements and calculations using those measurements.

In this chapter, we introduce many of the basic mathematical terms and methods of calculating that are required in electronics, construction, and manufacturing plant operations. We also need to learn about ratios, proportions, and variation as well as how to perform operations on algebraic expressions. In addition to these topics, this chapter lays the foundation for working with formulas, exponents, and radicals.

Objectives

- Add, subtract, multiply, and divide signed numbers.
- Work with numbers in scientific notation.
- Know the difference between the accuracy and the precision of a measurement.
- Perform arithmetic operations with measurements.
- Apply the rules for order of operations.
- Name the number of terms in and the degree of a polynomial.
- Identify the numerical coefficient of a term.
- Use the rules of exponents for positive integral exponents.
- Simplify polynomial expressions.
- Add, subtract, multiply, and divide polynomials.
- Evaluate formulas.

1.1 THE REAL NUMBER SYSTEM

The **positive integers** are the counting numbers; that is, 1, 2, 3, ... Note that the positive integers form an infinite set. The **negative integers** may be defined as the set of opposites of the positive integers; that is, $-1, -2, -3, \ldots$. **Zero** is the dividing point between the positive integers and the negative integers and is neither positive nor negative. The set of **integers** consists of the positive integers, the negative integers, and zero.

The **rational numbers** are those numbers that can be represented as the ratio of two integers, such as $\frac{3}{4}$, $\frac{-7}{5}$, and $\frac{5}{1}$. The **irrational numbers** are those numbers that cannot be represented as the ratio of two integers, such as $\sqrt{3}$, $\sqrt[3]{16}$, and π.

The set of **real numbers** is the set consisting of the rational numbers and the irrational numbers. (See Fig. 1.1.) With respect to the real number line as in Fig. 1.2, we say there is a one-to-one correspondence between the real numbers and the points on the number line; that is, for each real number there is a corresponding point on the number line, and for each point on the number line there is a corresponding real number. As a result, we say the number line is dense, or "filled."

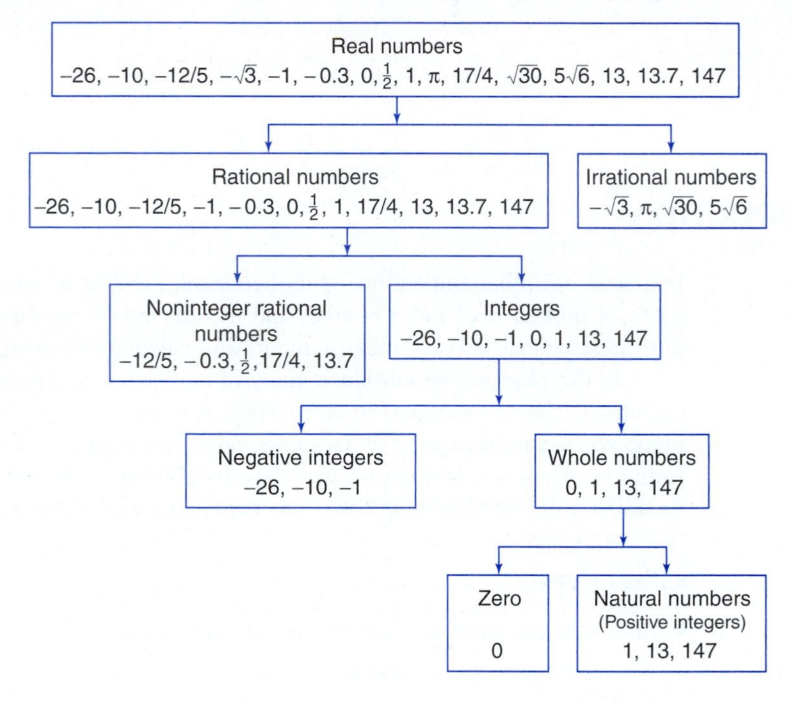

Figure 1.1 Real numbers.

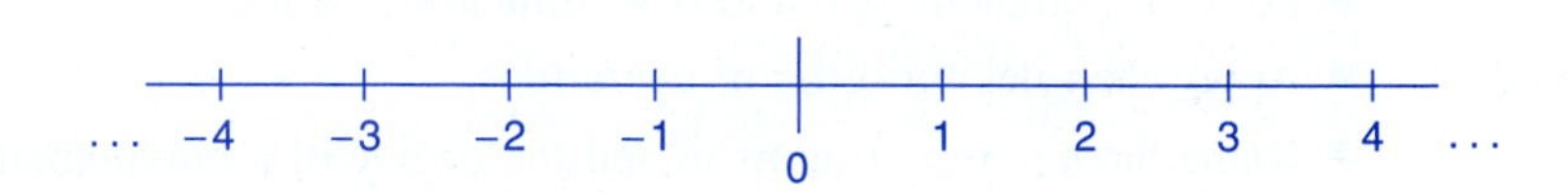

Figure 1.2 The real number line.

The set of complex numbers is introduced in Chapter 14.

A **prime number** is defined as a positive integer greater than one that is evenly divisible* only by itself and one. The first ten prime numbers are 2, 3, 5, 7, 11, 13, 17, 19, 23, and 29.

The following are some properties of real numbers:

1. $a + b = b + a$ — Commutative property of addition

2. $ab = ba$ — Commutative property of multiplication

3. $(a + b) + c = a + (b + c)$ — Associative property of addition

4. $(ab)c = a(bc)$ — Associative property of multiplication

*By "evenly divisible" we mean that the remainder is zero after the indicated division is completed.

5. $a(b + c) = ab + ac$ — Distributive property of multiplication over addition

6. $a + (-a) = 0$ — Additive inverse or negative property

7. $a \cdot \frac{1}{a} = 1\ (a \neq 0)$ — Multiplicative inverse or reciprocal property

8. $a + 0 = a$ — Identity element of addition

9. $a \cdot 1 = a$ — Identity element of multiplication

Operations with Signed Numbers

The rules for operations with signed numbers are often stated in terms of their absolute values. Therefore, we shall first define absolute value.

The **absolute value** of a real number n, written $|n|$, is defined as

$|n| = n$ if $n \geq 0$ ($\geq$ means greater than or equal to.)

$|n| = -n$ if $n < 0$ ($<$ means less than.)

EXAMPLE 1

Find the absolute value of each number.

(a) $|+7| = 7$

(b) $|-3| = -(-3) = 3$

(c) $|0| = 0$

ADDING TWO SIGNED NUMBERS

1. If the numbers have the *same* signs, add their absolute values and use the common sign before the sum.
2. If the numbers have *different* signs, find the difference of their absolute values. To this result attach the sign of the number whose absolute value is larger.

EXAMPLE 2

Add each pair of signed numbers.

(a) $(+7) + (+2) = +9$

(b) $(-3) + (-2) = -5$

(c) $(+6) + (-4) = +2$

(d) $(-8) + (+1) = -7$

ADDING THREE OR MORE SIGNED NUMBERS

1. Find the sum of the positive numbers.
2. Find the sum of the negative numbers.
3. Add the resulting positive sum and negative sum.

EXAMPLE 3

Add $(+2) + (-3) + (-7) + (+5) + (-9) + (+3) + (-6)$.

Step 1: $(+2) + (+5) + (+3) = +10$

Step 2: $(-3) + (-7) + (-9) + (-6) = -25$

Step 3: -15

SUBTRACTING SIGNED NUMBERS

To subtract signed numbers, change the sign of the subtrahend (number being subtracted), and add the resulting signed numbers: $a - b = a + (-b)$.

EXAMPLE 4

Subtract.

(a) $(-6) - (+2) = (-6) + (-2) = -8$
(b) $(+5) - (+9) = (+5) + (-9) = -4$
(c) $(-3) - (-8) = (-3) + (+8) = +5$
(d) $(+9) - (-2) = (+9) + (+2) = +11$
(e) $(-3) - (-7) - (+5) = (-3) + (+7) + (-5) = -1$

MULTIPLYING (OR DIVIDING) TWO SIGNED NUMBERS

1. If the numbers have the *same* signs, find the product (or quotient) of their absolute values and place a positive sign before the result.
2. If the numbers have *different* signs, find the product (or quotient) of their absolute values and place a negative sign before the result.

EXAMPLE 5

Multiply.

(a) $(+3)(+2) = +6$
(b) $(-8)(-6) = +48$
(c) $(+7)(-9) = -63$
(d) $(-5)(+4) = -20$

EXAMPLE 6

Divide.

(a) $(+8) \div (+2) = +4$
(b) $(-20) \div (-4) = +5$
(c) $(-36) \div (+6) = -6$
(d) $(+26) \div (-2) = -13$

MULTIPLYING AND/OR DIVIDING THREE OR MORE SIGNED NUMBERS

Multiply and/or divide the absolute values of the numbers. Then place a negative sign before the result if there is an odd number of negative numbers, or place a positive sign if there is an even number of negative numbers.

EXAMPLE 7

(a) Multiply.

$$(+2)(-4)(+5)(-1)(-3) = -120$$

(Note that there are three negative signs.)

(b) Simplify.

$$\frac{(+3)(-4)(+1)}{(-2)(-5)(-6)} = +\frac{1}{5}$$

(Note that there are four negative signs.)

Exercises 1.1

Find each absolute value.

1. $|-15|$ **2.** $|+9|$ **3.** $|+11|$ **4.** $|-2|$

5. $|-8|$ **6.** $|+1|$ **7.** $|3-5|$ **8.** $|1-(-6)|$

Perform the indicated operations and simplify.

9. $(-3)+(-6)$ **10.** $(-5)+(+6)$ **11.** $(+7)+(-9)$

12. $(-2)+(-4)$ **13.** $(+6)+(+4)$ **14.** $(+4)+(-10)$

15. $(-5)+(+8)$ **16.** $(-8)+(-5)$ **17.** $(-\frac{1}{3})+(-\frac{3}{4})$

18. $(+\frac{1}{2})+(-\frac{5}{16})$ **19.** $(-3\frac{4}{9})+(+4\frac{5}{12})$ **20.** $(-1\frac{5}{8})+(-2\frac{1}{12})$

21. $(-13)+(+3)+(-7)+(+6)+(-2)$

22. $(+5)+(-1)+(-9)+(-6)+(+4)$

23. $(-2)+(-3)+(+6)+(-4)+(+1)+(-5)$

24. $(+1)+(-7)+(-6)+(+4)+(+11)+(-3)+(-4)$

25. $(+4)-(+9)$ **26.** $(+8)-(-2)$ **27.** $(-2)-(-8)$

28. $(-11)-(+6)$ **29.** $(+7)-(-2)$ **30.** $(-2)-(-5)$

31. $(-6)-(+8)$ **32.** $(+6)-(+8)$ **33.** $4-7$

34. $14-20$ **35.** $-12.5-3.2$ **36.** $-0.75-(-1.45)$

37. $(-\frac{2}{3})-(-\frac{1}{4})$ **38.** $(+\frac{2}{5})-(-\frac{1}{4})$ **39.** $(+\frac{1}{12})-(+1\frac{2}{3})$

40. $(-1\frac{3}{8})-(+2\frac{5}{12})$ **41.** $(+3)-(-6)-(+9)-(-6)$

42. $(-7)-(+6)-(+8)-(-2)-(-4)$ **43.** $(+3)+(-4)-(-7)-(+2)+(+7)$

44. $(-2)-(-5)+(+4)-(+6)+(-9)$ **45.** $8+9-16+4-5-6+1$

46. $5-7-6+2-3+19$ **47.** $9-7+4+3-8-6-6+2$

48. $-4+6-7-5+6-8-1$ **49.** $(-5)(-6)$

50. $(-2)(+6)$ **51.** $(+2)(-12)$

52. $(-5)(-7)$ **53.** $(-8)\div(-2)$ **54.** $(-16)\div(+8)$

55. $\dfrac{+54}{-30}$ **56.** $\dfrac{-8}{-28}$ **57.** $(-\frac{2}{3})(-\frac{9}{16})$

58. $(-1\frac{1}{4})(+3\frac{1}{5})$ **59.** $(+1\frac{3}{8})\div(-1\frac{5}{6})$ **60.** $(-1\frac{1}{8})\div(-3\frac{3}{8})$

61. $\dfrac{-\frac{7}{16}}{\frac{21}{32}}$ **62.** $\dfrac{2\frac{1}{2}}{-1\frac{3}{4}}$ **63.** $(+2)(-7)(-6)(-1)$

64. $(-6)(+2)(+4)(-5)$ **65.** $(-2)(+3)(-1)(+5)(+2)$ **66.** $(-3)(+5)(-8)(+2)(-1)$

67. $\dfrac{(-6)(+4)(-7)}{(-12)(-2)}$ **68.** $\dfrac{(-9)(-4)}{(+6)(-10)}$

69. $\dfrac{(+36)(-18)(+5)}{(-4)(-8)(+30)}$ **70.** $\dfrac{(+6)(-15)(-24)}{(-12)(+6)(+20)(-3)}$

71. The temperature at 6:00 A.M. is $-15°$. By noon the temperature increases by $35°$. Find the temperature at noon.

72. During March, Mary's small business had an income of \$1875 and expenses of \$2055. Find her monthly profit.

73. The temperature at 2:00 P.M. is $18°$. The temperature at 11:00 P.M. is $-12°$. Find the difference in temperatures.

74. Bill, a diver, is 120 ft below the surface of the Pacific Ocean. Heather is directly above Bill in a balloon that is 260 ft above the Pacific Ocean. Find the distance between Bill and Heather.

75. A diver is 65 ft below the surface of the Atlantic Ocean. A second diver is 25 ft below the first diver. A third diver is 15 ft above the second diver. Find the depth of the third diver.

1.2 ZERO AND ORDER OF OPERATIONS

The following properties of zero with respect to addition, subtraction, and multiplication are well known and widely used:

1. $a + 0 = a$
2. $a - 0 = a$
3. $a \cdot 0 = 0$
4. If $a \cdot b = 0$, then either $a = 0$ or $b = 0$.

Observe that the number zero must be treated as a very special case in the real number system when it comes to division.

From the definition of division, we have

$$\frac{a}{b} = q \quad \text{if and only if} \quad a = b \cdot q$$

For example,

$$\frac{18}{3} = 6 \quad \text{if and only if} \quad 18 = 3 \cdot 6$$

First, if $a = 0$ and $b \neq 0$, what is the nature of the quotient q? By substituting in the preceding definition, we have

$$\frac{0}{b} = q \quad \text{only if} \quad 0 = b \cdot q$$

Since the product of b and q is 0 and $b \neq 0$, then $q = 0$. That is, $\frac{0}{b} = 0\ (b \neq 0)$.

On the other hand, if $a \neq 0$ and $b = 0$, what is the nature of q? By substituting, we have

$$\frac{a}{0} = q \quad \text{only if} \quad a = 0 \cdot q$$

The product $0 \cdot q$ equals 0 leads to a contradiction since $a \neq 0$; that is, there is no real number q which, when multiplied by 0, gives a number a that is not 0. Hence, we say that no quotient exists and division by zero is *meaningless.*

Finally, if $a = 0$ and $b = 0$, what is the nature of q? Substituting yields

$$\frac{0}{0} = q \quad \text{only if} \quad 0 = 0 \cdot q$$

Note that any real number value of q satisfies $0 = 0 \cdot q$; that is, there is no unique value of q that can be determined. Hence, we say that no unique quotient exists and that the form $\frac{0}{0}$ is *indeterminate.*

EXAMPLE 1

Perform the indicated operations where possible.

(a) $8 \cdot 0$ (b) $\frac{0}{5}$ (c) $\frac{12}{0}$ (d) $\frac{0}{0}$

(a) $8 \cdot 0 = 0$

(b) $\frac{0}{5} = 0$

(c) $\frac{12}{0}$ is meaningless.

(d) $\frac{0}{0}$ is indeterminate.

EXAMPLE 2

Find the values of x that make $\frac{4x + 3}{(5x - 1)(2x + 4)}$ meaningless.

This fraction is meaningless for those values of x for which *only* the denominator is zero. Thus, set each factor in the denominator equal to zero and solve for x.

$$\begin{aligned} 5x - 1 &= 0 & 2x + 4 &= 0 \\ 5x &= 1 & 2x &= -4 \\ x &= \frac{1}{5} & x &= -2 \end{aligned}$$

A fraction is indeterminate for those values of x for which *both* the numerator and the denominator are zero.

The operations with zero are summarized as follows.

OPERATIONS WITH ZERO

1. $a + 0 = a$
2. $a - 0 = a$
3. $a \cdot 0 = 0$
4. If $a \cdot b = 0$, then either $a = 0$ or $b = 0$.
5. $\frac{0}{b} = 0 \quad (b \neq 0)$.
6. $\frac{a}{0}$ is meaningless $\quad (a \neq 0)$.
7. $\frac{0}{0}$ is indeterminate.

When we treat a group of numbers or quantities as a unit, we use the following grouping symbols: parentheses ( ), brackets [ ], braces { }, and the fraction bar —.

What is the value of $8 - 3 \cdot 2$? Is it 10? Is it 2? Some other number? It is very important that each arithmetic operation have only one result and that we each perform the exact same operations on a given computation or problem. As a result, the following order-of-operations convention is followed by everyone.

ORDER-OF-OPERATIONS CONVENTION

1. Do all operations within any grouping symbols, beginning with the innermost pair if grouping symbols are contained within each other.
2. Then do all multiplications and divisions in the order in which they occur from left to right.
3. Finally, do all additions and subtractions in the order in which they occur from left to right.

EXAMPLE 3

Evaluate.

$$\begin{aligned} 2 - \{6 + [(2 - 7) + (4 + 6)] - 7\} + 1 &= 2 - \{6 + [-5 + 10] - 7\} + 1 \\ &= 2 - \{6 + 5 - 7\} + 1 \\ &= 2 - 4 + 1 \\ &= -1 \end{aligned}$$

EXAMPLE 4

Evaluate.

Step 1: $6(2 + 5) - 33 \div 11 - 2 \cdot 6 = 6(7) - 33 \div 11 - 2 \cdot 6$

Step 2: $= 42 - 3 - 12$

Step 3: $= 27$

EXAMPLE 5

Evaluate.

Step 1: $\dfrac{-3(4 - 6) - 18 \div 3 \cdot 2}{18 \cdot 2 \div 4 + 1} = \dfrac{-3(-2) - 18 \div 3 \cdot 2}{18 \cdot 2 \div 4 + 1}$

Step 2: $= \dfrac{6 - 12}{9 + 1}$

Step 3: $= \dfrac{-6}{10}$

$= -\dfrac{3}{5}$

Now let us return to the earlier problem of finding the value of $8 - 3 \cdot 2$. According to the preceding convention, $8 - 3 \cdot 2 = 8 - 6 = 2$.

Exercises 1.2

Perform the indicated operations where possible.

1. $0 - 3$
2. $-4 + 0$
3. $6 \cdot 0$
4. $(0 - 3)0$
5. $\dfrac{9 \cdot 0}{3}$
6. $\dfrac{8 + (-8)}{5}$
7. $\dfrac{13}{6 - 6}$
8. $\dfrac{5 \cdot 0}{18 - 18}$
9. $\dfrac{7 + (-7)}{6 - 6(3 + 2)}$
10. $\dfrac{2 - 2}{2 - |-8| + 6}$
11. $\dfrac{5 \cdot 0}{-6 - (-6)}$
12. $\dfrac{6 \cdot 15 \cdot 8 \cdot 10}{18 \cdot 0 \cdot 16}$

Find the values of x that make each fraction meaningless.

13. $\dfrac{5x}{x - 2}$
14. $\dfrac{3x - 7}{2x + 1}$
15. $\dfrac{2 - x}{x(3x + 4)}$
16. $\dfrac{4 + 3x}{2x(x + 1)}$
17. $\dfrac{5x + 1}{(x + 1)(2x - 1)}$
18. $\dfrac{3}{x(x + 1)(x - 1)}$

Find the values of x that make each fraction indeterminate.

19. $\dfrac{6x^2}{2x}$
20. $\dfrac{2 - x}{(2x - 7)(x - 2)}$
21. $\dfrac{12x - 10}{(3 + x)(5 - 6x)}$
22. $\dfrac{(1 - 2x)(3x - 7)}{5x(2x - 1)}$
23. $\dfrac{(2x + 1)(x - 3)}{(6 - 2x)(2x + 1)}$
24. $\dfrac{(3x - 6)(1 - x)}{(x - 1)(2x - 6)}$

Evaluate.

25. $15 + 2 \cdot 4$
26. $20 - 8 \cdot 2$
27. $3 - 8(5 - 2)$
28. $(6 + 2)3 - 5$
29. $12 + 4 \div 2$
30. $18 \div 6 - 2$

31. $6 - 7(4 + 1)$

32. $4 + 2(9 - 6)$

33. $5 - [4 + 6(2 - 7) - 3]$

34. $19 + \{6 - [5 + 2(8 + 2)] - 8\} + 1$

35. $26 \cdot 2 \div 13 + 4 - 7 \cdot 2$

36. $36 \div 9 \cdot 3 - 6 \cdot 4 + 1$

37. $3(3 - 7) \div 6 - 2$

38. $24 \div 6 - 1 + 4 \cdot 3$

39. $6 \cdot 8 \div 2 \cdot 72 \div 24 + 4$

40. $12 \cdot 9 \div 18 \cdot 64 \div 8 + 2$

41. $18 \div 6 \cdot 24 \div 4 \div 6$

42. $4 \cdot 9 \div 3 \div 6 \cdot 9 \cdot 2 \div 12$

43. $7 + 6(3 + 2) - 6 - 5(4 - 2)$

44. $5 - 3(7 - 2) - 5 \cdot 2 - 2(4 - 7)$

45. $\dfrac{4 - 7(6 - 2)}{9 \div 3 + 7}$

46. $\dfrac{16 \cdot 2 \div 8(2 - 3)}{5 - 2 \cdot 3}$

47. $\dfrac{5 \cdot 12 \div 6 \cdot 2 - (-4)}{6 - 2(5 + 4)}$

48. $\dfrac{48 \div 4 \div 3 \cdot 6 - 3}{4 + 2(6 - 2)}$

1.3 SCIENTIFIC NOTATION AND POWERS OF 10

In technical work it is often necessary to refer to very large and very small numbers. **Scientific notation** is a method of writing such numbers while avoiding the writing of many zeros. When a number is written in scientific notation, it is expressed as a product of a decimal between 1 and 10 and a power of 10.

First, let's review positive powers of 10.

$$
\begin{aligned}
10^1 &= 10 &&= 10\\
10^2 &= 10 \cdot 10 &&= 100\\
10^3 &= 10 \cdot 10 \cdot 10 &&= 1000\\
10^4 &= 10 \cdot 10 \cdot 10 \cdot 10 &&= 10{,}000\\
&\vdots && \vdots\\
10^n &= \underbrace{10 \cdot 10 \cdot 10 \cdot \cdots \cdot 10}_{n \text{ factors}} &&= \underbrace{1000 \cdots 0}_{n \text{ zeros}}
\end{aligned}
$$

For example, if $n = 5$, $10^5 = 100{,}000$; if $n = 7$, $10^7 = 10{,}000{,}000$.

Recall the following properties for negative powers of 10:

$$
\begin{aligned}
10^{-1} &= \frac{1}{10} &&= \frac{1}{10} = 0.1\\
10^{-2} &= \frac{1}{10 \cdot 10} &&= \frac{1}{10^2} = 0.01\\
10^{-3} &= \frac{1}{10 \cdot 10 \cdot 10} &&= \frac{1}{10^3} = 0.001\\
10^{-4} &= \frac{1}{10 \cdot 10 \cdot 10 \cdot 10} &&= \frac{1}{10^4} = 0.0001\\
&\vdots && \vdots\\
10^{-n} &= \frac{1}{\underbrace{10 \cdot 10 \cdot 10 \cdot \cdots \cdot 10}_{n \text{ factors}}} &&= \frac{1}{10^n} = \underbrace{0.000 \cdots 0}_{(n-1) \text{ zeros}}1
\end{aligned}
$$

For example, if $n = 6$, $10^{-6} = 0.000001$; if $n = 9$, $10^{-9} = 0.000000001$.

The laws of exponents involving powers of 10 are shown in the following boxes. A more detailed study of the general laws of exponents will be given later.

LAW 1

$$10^m \cdot 10^n = 10^{m+n}$$

EXAMPLE 1

Multiply.

(a) $10^3 \cdot 10^2 = 10^{3+2} = 10^5$

(b) $10^6 \cdot 10^{-2} = 10^{6+(-2)} = 10^4$

(c) $10^{-3} \cdot 10^{-4} = 10^{(-3)+(-4)} = 10^{-7}$

LAW 2

$$\frac{10^m}{10^n} = 10^{m-n}$$

EXAMPLE 2

Divide.

(a) $\dfrac{10^8}{10^2} = 10^{8-2} = 10^6$

(b) $10^{-2} \div 10^5 = 10^{(-2)-5} = 10^{-7}$

(c) $\dfrac{10^6}{10^{-3}} = 10^{6-(-3)} = 10^9$

LAW 3

$$(10^m)^n = 10^{mn}$$

EXAMPLE 3

Raise each power of 10 to the indicated power.

(a) $(10^2)^3 = 10^{(2)(3)} = 10^6$

(b) $(10^{-4})^2 = 10^{(-4)(2)} = 10^{-8}$

(c) $(10^{-3})^{-2} = 10^{(-3)(-2)} = 10^6$

LAW 4

$$10^{-n} = \frac{1}{10^n} \quad \text{and} \quad \frac{1}{10^{-n}} = 10^n$$

EXAMPLE 4

Rewrite each power of 10 using positive exponents.

(a) $10^{-3} = \dfrac{1}{10^3}$

(b) $\dfrac{1}{10^{-4}} = 10^4$

Note that the second part of Law 4 can be shown by using the first part; that is,

$$\frac{1}{10^{-n}} = \frac{1}{\frac{1}{10^n}} = 1 \div \frac{1}{10^n} = 1 \times \frac{10^n}{1} = 10^n$$

The zero power of 10 is one; that is, $10^0 = 1$. To show this, we use the substitution principle, which states that if $a = b$ and $a = c$, then $b = c$.

$$\frac{10^n}{10^n} = 1 \qquad \text{(Any number other than zero divided by itself equals one.)}$$

and

$$\frac{10^n}{10^n} = 10^{n-n} \qquad \text{(Law 2)}$$

$$= 10^0$$

Therefore,

$$10^0 = 1 \qquad \text{(Substitution)}$$

LAW 5

$$10^0 = 1$$

With the preceding properties of powers of 10 in mind, let's now write numbers in scientific notation.

SCIENTIFIC NOTATION

A number is expressed in scientific notation as a product of a decimal between 1 and 10 and a power of 10; that is, $N \times 10^m$ where $1 \leq N < 10$ and m is an integer.

Note: A number in scientific notation has *one* nonzero digit to the left of the decimal point.

A number greater than 10 is expressed in scientific notation as a product of a decimal between 1 and 10 and a *positive* power of 10.

EXAMPLE 5

Write each number greater than 10 in scientific notation.

(a) $2380 = 2.38 \times 10^3$
(b) $52{,}600 = 5.26 \times 10^4$
(c) $31{,}000{,}000 = 3.1 \times 10^7$
(d) $681.4 = 6.814 \times 10^2$
(e) $32.765 = 3.2765 \times 10^1$

A number between 0 and 1 is expressed in scientific notation as a product of a decimal between 1 and 10 and a *negative* power of 10.

EXAMPLE 6

Write each number less than 1 in scientific notation.

(a) $0.0617 = 6.17 \times 10^{-2}$
(b) $0.00024 = 2.4 \times 10^{-4}$
(c) $0.63 = 6.3 \times 10^{-1}$
(d) $0.0000004 = 4 \times 10^{-7}$

A number between 1 and 10 is expressed in scientific notation as a product of a decimal between 1 and 10 and the *zero* power of 10.

EXAMPLE 7

Write each number between 1 and 10 in scientific notation.

(a) $8.23 = 8.23 \times 10^0$
(b) $1.04 = 1.04 \times 10^0$

CHANGING A NUMBER FROM DECIMAL FORM TO SCIENTIFIC NOTATION

1. Move the decimal point to a position immediately after the first nonzero digit reading from left to right.
2. If the decimal point is moved to the *left,* the exponent of 10 is the number of places that the decimal point has been moved.
3. If the decimal point is moved to the *right,* the exponent of 10 is the negative of the number of places that the decimal point has been moved.
4. If the original position of the decimal point is after the first nonzero digit, the exponent of 10 is zero.

CHANGING A NUMBER FROM SCIENTIFIC NOTATION TO DECIMAL FORM

1. Multiply the decimal part by the power of 10 by moving the decimal point *to the right* the same number of decimal places as indicated by the power of 10 if it is *positive.*
2. Multiply the decimal part by the power of 10 by moving the decimal point *to the left* the same number of decimal places as indicated by the power of 10 if it is *negative.*

Supply zeros as needed.

EXAMPLE 8

Write each number in decimal form.

(a) $3.45 \times 10^2 = 345$
(b) $1.06 \times 10^5 = 106{,}000$
(c) $2.77 \times 10^{-2} = 0.0277$
(d) $8.15 \times 10^{-6} = 0.00000815$
(e) $4.92 \times 10^0 = 4.92$

Note: In this text we use numerous examples to illustrate how to use calculators that use algebraic logic. In Appendix C we give detailed instructions for those students and/or classes that prefer to use a basic graphing calculator, the TI-83 or TI-84 Plus. In Appendix D we give detailed instructions for those students and/or classes that prefer to use an advanced graphing calculator, the TI-89.

Numbers expressed in scientific notation can be entered into many calculators. The results may then also be given in scientific notation.

EXAMPLE 9

Find $\dfrac{3.24 \times 10^{-5}}{7.2 \times 10^{-12}}$ and write the result in scientific notation.

3.24 **EE** -5/7.2 **EE** -12 **ENTER**

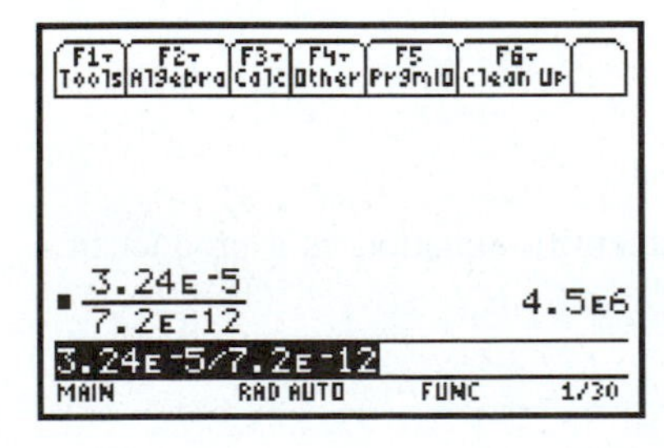

MODE **2nd QUIT**

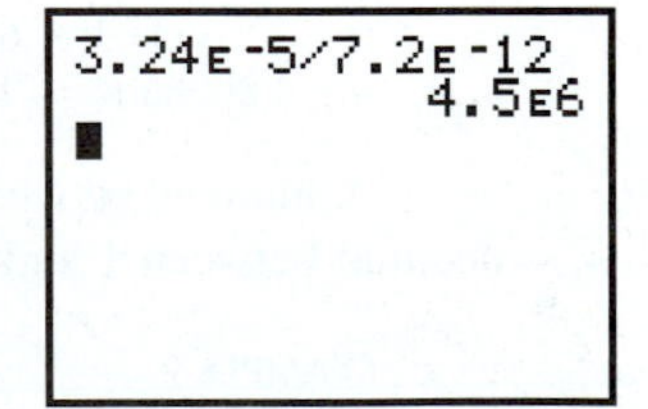

3.24 **2nd EE** -5/7.2 **2nd EE** -12 **ENTER**

Thus, the quotient is 4.5×10^6.

Note: In this text two graphing calculator models are illustrated. The Texas Instruments TI-89 screens are easily recognizable because of the menu bar across the top (F1 Tools, F2 Algebra, etc.). The screen images without a menu bar are from a TI-83 Plus. In almost all cases, they will be identical to those of a regular TI-83 calculator.

EXAMPLE 10

Find the value of $\dfrac{(-6.3 \times 10^4)(-5.07 \times 10^{-9})(8.11 \times 10^{-6})}{(5.63 \times 10^{12})(-1.84 \times 10^7)}$ and write the result in scientific notation, rounded to three significant digits.

-6.3 **EE** 4*-5.07 **EE** -9*8.11 **EE** -6/(5.63 **EE** 12*-1.84 **EE** 7) **ENTER**

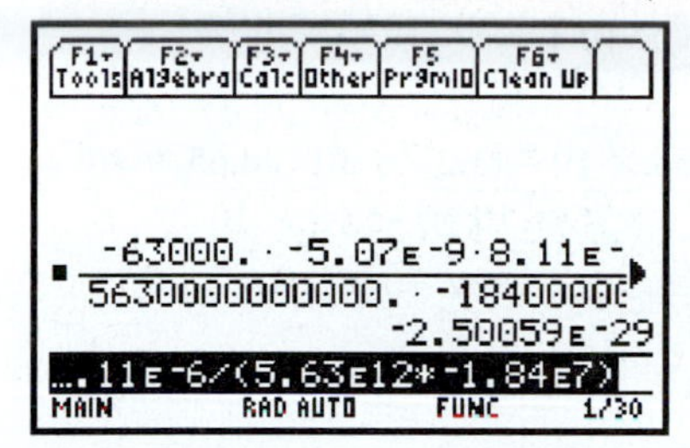


```
-6.3E4*-5.07E-9*
8.11E-6/(5.63E12
*-1.84E7)
  -2.5005938E-29
```

-6.3 **2nd EE** 4*-5.07 **2nd EE** -9*8.11 **2nd EE** -6/(5.63 **2nd EE** 12*-1.84 **2nd EE** 7) **ENTER**

Thus, the result rounded to three significant digits is -2.50×10^{-29}.

Perhaps a quick review of significant digits is in order. All nonzero digits are significant. Zeros between significant digits are significant; for example, 809 m has three significant digits. Zeros to the right of a significant digit *and* a decimal point are significant; for example, 52.30 m has four significant digits. When a measurement is written, any zero with a bar over it is significant. A more detailed discussion is given in Section 1.4.

Exercises 1.3

Perform the indicated operations using the laws of exponents. Express each result as a power of 10.

1. $10^3 \cdot 10^8$ **2.** $10^4 \div 10^{-5}$ **3.** $(10^3)^3$

4. $10^{-4} \cdot 10^{-6}$ **5.** $(10^{-4})^2$ **6.** $\dfrac{1}{10^{-6}}$

7. $10^{-2} \div 10^{-5}$ **8.** $(10^3)^{-2}$ **9.** $\dfrac{1}{10^{-4}}$

10. $(10^{-2})^3$ **11.** $\dfrac{10^{-2} \cdot 10^{-7} \cdot 10^{-3} \cdot 10^0}{10^3 \cdot 10^4 \cdot 10^{-2}}$ **12.** $\dfrac{10^3 \cdot 10^{-5} \cdot 10^6}{10^{-2} \cdot 10^{-7} \cdot 10^5}$

13. $(10^{-2} \cdot 10^3 \cdot 10^{-5})^{-4}$ **14.** $(10^2 \cdot 10^{-5} \cdot 10^9)^3$ **15.** $\left(\dfrac{10^2 \cdot 10^{-4}}{10^{-3} \cdot 10^6}\right)^3$

16. $\left(\dfrac{10^2 \cdot 10^{-5} \cdot 10^{-3}}{10^{-4} \cdot 10^6}\right)^5$ **17.** $\left(\dfrac{10^{-3} \cdot 10^5 \cdot 10^{-7}}{10^2 \cdot 10^{-1} \cdot 10^{-5}}\right)^{-4}$ **18.** $\left(\dfrac{10^5 \cdot 10^{-2}}{10^{-6} \cdot 10^3 \cdot 10^{-4}}\right)^{-3}$

Write each number in scientific notation.

19. 2070 **20.** 60,000 **21.** 0.091

22. 0.00001264 **23.** 5.61 **24.** 370,000

25. 8,500,000 **26.** 6700 **27.** 0.000006

28. 11.7 **29.** 10,060 **30.** 0.000102

Write each number in decimal form.

31. 1.27×10^2 **32.** 1.105×10^{-4} **33.** 6.14×10^{-5}
34. 2.96×10^0 **35.** 9.24×10^6 **36.** 7.72×10^{-1}
37. 6.96×10^{-9} **38.** 3.14×10^7 **39.** 9.66×10^0
40. 3.18×10^{-3} **41.** 5.03×10^4 **42.** 1.19×10^{-8}

Find each value. Round each result to three significant digits written in scientific notation.

43. $(6.43 \times 10^8)(5.16 \times 10^{10})$
44. $(4.16 \times 10^{-5})(3.45 \times 10^{-7})$
45. $(1.456 \times 10^{12})(-4.69 \times 10^{-18})$
46. $(-5.93 \times 10^9)(7.055 \times 10^{-12})$
47. $(7.46 \times 10^8) \div (8.92 \times 10^{18})$
48. $(1.38 \times 10^{-6}) \div (4.324 \times 10^6)$
49. $\dfrac{-6.19 \times 10^{12}}{7.755 \times 10^{-8}}$
50. $\dfrac{1.685 \times 10^{10}}{1.42 \times 10^{24}}$
51. $\dfrac{(5.26 \times 10^{-8})(8.45 \times 10^6)}{(-6.142 \times 10^9)(1.056 \times 10^{-12})}$
52. $\dfrac{(-2.35 \times 10^{-9})(1.25 \times 10^{11})(4.65 \times 10^{17})}{(8.75 \times 10^{23})(-5.95 \times 10^{-6})}$
53. $\dfrac{(4.68 \times 10^{-15})(5.19 \times 10^{-7})}{(-7.27 \times 10^{-16})(4.045 \times 10^{-8})(1.68 \times 10^{24})}$
54. $\dfrac{(3.86 \times 10^5)(5.15 \times 10^{-9})(1.91 \times 10^8)}{(2.34 \times 10^{-4})(1.35 \times 10^6)(9.05 \times 10^{10})}$

1.4 MEASUREMENT

Measurement can be defined as the comparison of a quantity with a standard unit. Centuries ago, parts of the human body were used as standards of measurement. However, such standards were neither uniform nor acceptable to all. Later, each of the various standards was defined. A fourteenth-century legal definition of the standard English inch was: The length of three barley corns, round and dry, taken from the center of the ear, and laid end to end. To complicate matters, many countries introduced or defined their own standards, which were not consistent with those of other countries. In 1670 Gabriel Mouton, a Frenchman, recognized the need for a simple, uniform, worldwide measurement system, and he proposed a decimal system that was the basis for the modern metric system. By the 1800s, metric standards were adopted worldwide. In 1960 the modern metric system was adopted; it is identified in all languages by the abbreviation SI (for Système International d'Unités—the international system of units of measurement, written in French).*

Approximate Versus Exact Numbers

Up to this time in your study of mathematics, all numbers and all measurements have probably been treated as exact numbers. An **exact number** is a number that has been determined as a result of counting, such as 24 students enrolled in this class, or that has been defined in some way, such as 1 hour = 60 minutes or 1 inch = 2.54 cm, a conversion definition agreed to by the world governments' bureaus of standards. The treatment of the addition, subtraction, multiplication, and division of exact numbers normally is the emphasis, or main content, of grade-school mathematics.

However, nearly all data of a technical nature involve **approximate numbers;** that is, they have been determined as a result of some measurement process—some direct, as with a ruler, and some indirect, as with a surveying transit. Before studying how to

*See Appendix B for a detailed introduction to the metric system.

perform the calculations with approximate numbers (measurements), we first must determine the "correctness" of an approximate number. First of all, we realize that no measurement can be found exactly. The length of the cover of this book can be found using many instruments. The better the measuring device used, the better is the measurement.

A measurement may be expressed in terms of its accuracy or its precision.

Accuracy and Significant Digits

The **accuracy** of a measurement refers to the number of digits, called **significant digits,** which indicate the number of units we are reasonably sure of having counted when making a measurement. The greater the number of significant digits given in a measurement, the better is the accuracy, and vice versa.

EXAMPLE 1

The average distance between the moon and the earth is 239,000 miles. This measurement indicates measuring 239 thousands of miles, and its accuracy is indicated by three significant digits.

EXAMPLE 2

A measurement of 0.035 cm indicates measuring 35 thousandths of a centimetre; its accuracy is indicated by two significant digits.

EXAMPLE 3

A measurement of 0.0200 mg indicates measuring 200 ten-thousandths of a milligram; its accuracy is indicated by three significant digits.

Notice that a zero is sometimes significant and sometimes not. To clarify this, we give the following rules for significant digits.

RULES FOR SIGNIFICANT DIGITS

1. All nonzero digits are significant; for example, 356.4 m has four significant digits (this measurement indicates 3564 tenths of metres).
2. All zeros between significant digits are significant; for example, 406.02 km has five significant digits (this measurement indicates 40,602 hundredths of kilometres).
3. A zero in a number greater than one which is specially tagged, such as by a bar above it, is significant; for example, $13\bar{0},000$ km has three significant digits (this measurement indicates $13\bar{0}$ thousands of kilometres).
4. All zeros to the right of a significant digit *and* a decimal point are significant; for example, 36.10 cm has four significant digits (this measurement indicates $361\bar{0}$ hundredths of centimetres).
5. Zeros to the right in a whole number measurement that are not tagged are *not* significant; for example, 2300 m has two significant digits (23 hundreds of metres).
6. Zeros to the left in a measurement less than one are *not* significant; for example, 0.00252 m has three significant digits (252 hundred-thousandths of a metre).

Precision

The **precision** of a measurement refers to the smallest unit with which a measurement is made; that is, the position of the last significant digit.

EXAMPLE 4

The precision of the measurement 239,000 mi is 1000 mi. (The position of the last significant digit is in the thousands place.)

EXAMPLE 5

The precision of the measurement 0.035 cm is 0.001 cm. (The position of the last significant digit is in the thousandths place.)

EXAMPLE 6

The precision of the measurement 0.0200 mg is 0.0001 mg. (The position of the last significant digit is in the ten-thousandths place.)

Unfortunately, many people use *accuracy* and *precision* interchangeably. A measurement of 0.0006 cm has good precision and poor accuracy when compared with the measurement 368.0 cm, which has much better accuracy (one versus four significant digits) and poorer precision (0.0001 cm versus 0.1 cm).

EXAMPLE 7

Find the accuracy and precision of each of the following measurements.

	Measurement	*Accuracy (significant digits)*	*Precision*
(a)	3463 ft	4	1 ft
(b)	3005 mi	4	1 mi
(c)	10,809 kg	5	1 kg
(d)	36,000 tons	2	1000 tons
(e)	$88\bar{0}0$ mi	3	10 mi
(f)	1,349,000 km	4	1000 km
(g)	$600\bar{0}$ m	4	1 m
(h)	0.00632 kg	3	0.00001 kg
(i)	0.0401 m	3	0.0001 m
(j)	0.0060 g	2	0.0001 g
(k)	14.20 m	4	0.01 m
(l)	30.00 cm	4	0.01 cm
(m)	100.060 g	6	0.001 g

Exercises 1.4

Find the accuracy (the number of significant digits) of each measurement.

1. 205 in. **2.** 14.7 m **3.** 60.0 cm

4. 35,000 ft **5.** 6.010 km **6.** $10,\bar{0}00$ mi

7. $16\bar{0}0$ Ω **8.** 120 V **9.** 0.060 g

10. 0.0250 A **11.** $20\bar{0}$ mm **12.** 205,000 Ω

Find the precision of each measurement.

13. 3.6 cm
14. 7.0 m
15. 16.00 cm
16. 4.100 mi
17. $16\bar{0}$ mm
18. 304,000 km
19. 6.00 m
20. 360 V
21. $3\bar{0}00\ \Omega$
22. 0.050 km
23. 0.0040 A
24. 63.500 g

*In each set of measurements, find the measurement that is **(a)** the most accurate and **(b)** the most precise.*

25. 15.2 m; 0.023 m; 0.06 m
26. 256 ft; 400 ft; 270 ft
27. 0.642 cm; 0.82 cm; 14.02 cm
28. 6.2 m; 4.7 m; 3.0 m
29. 0.0270 A; 0.035 A; 0.00060 A; 0.055 A
30. 164.00 km; 5.60 km; 4.000 km; 0.05 km
31. 305,000 Ω; 38,000 Ω; $4\bar{0}0{,}000\ \Omega$; 80,000 Ω
32. 1,300,000 V; 35,000 V; $60{,}\bar{0}00$ V; 20,000 V

*In each set of measurements, find the measurement that is **(a)** the least accurate and **(b)** the least precise.*

33. 13.2 m; 0.057 m; 0.08 m
34. 372 yd; 300 yd; 560 yd
35. 16.8 km; 0.52 km; 15.05 km
36. 6.5 kg; 460 kg; 0.075 kg
37. 0.0370 A; 0.030 A; 0.00009 A; 0.41 A
38. 284.0 mi; 6.35 mi; 7.000 mi; 0.05 mi
39. 205,000 Ω; 43,000 Ω; $6\bar{0}0{,}000\ \Omega$; 500,000 Ω
40. 1,400,000 V; 27,000 V; $50{,}\bar{0}00$ V; 30,000 V

1.5 OPERATIONS WITH MEASUREMENTS

If someone measured the length of one of two parts of a shaft with a micrometer calibrated in 0.01 mm as 12.27 mm and another person measured the second part with a ruler calibrated in mm as 23 mm, would the total length be 35.27 mm? Note that the sum 35.27 mm indicates a precision of 0.01 mm. The precision of the ruler is 1 mm, which means that the measurement 23 mm with the ruler could actually be anywhere between 22.50 mm and 23.50 mm using the micrometer (which has a precision of 0.01 mm). That is, any measurement between 22.50 and 23.50 can be read only as 23 mm using the ruler. Of course, this means that the tenths and hundredths digits in the sum 35.27 mm are actually meaningless. In other words, *the sum or difference of measurements can be no more precise than the least precise measurement.*

ADDING OR SUBTRACTING MEASUREMENTS

1. Make certain that all of the measurements are expressed in the same unit. If they are not, change them all to the same unit.
2. Add or subtract.
3. Then round the result to the same precision as the least precise measurement.

EXAMPLE 1

Add the measurements 1250 cm, 1562 mm, 2.963 m, and 9.71 m.

First, convert all measurements to the same unit, such as metres.

$$1250 \text{ cm} = 12.5 \text{ m}$$
$$1562 \text{ mm} = 1.562 \text{ m}$$

Next, add.

$$\begin{array}{r} 12.5\ \ \ \text{m} \\ 1.562\ \text{m} \\ 2.963\ \text{m} \\ \underline{9.71\ \ \text{m}} \\ 26.735\ \text{m} \rightarrow 26.7\ \text{m} \end{array}$$

Then round this sum to the same precision as the least precise measurement, which is 12.5 m. Thus, the sum is 26.7 m.

EXAMPLE 2

Subtract the measurements: 2567 g − 1.60 kg.

First, convert all measurements to the same unit, such as grams.

$$1.60 \text{ kg} = 16\bar{0}0 \text{ g}$$

Be careful not to change the number of significant digits. Next, subtract.

$$\begin{array}{r} 2567\ \text{g} \\ \underline{16\bar{0}0\ \text{g}} \\ 967\ \text{g} \rightarrow 970\ \text{g} \end{array}$$

Then round this difference to the same precision as the least precise measurement, which is $16\bar{0}0$ g. Thus, the difference is 970 g.

Suppose that you need to find the area of a rectangular room that measures 11.4 m by 15.6 m. If you multiply the numbers 11.4 and 15.6, the product 177.84 implies an accuracy of five significant digits. But note that each of the original measurements contains only three significant digits. To rectify this inconsistency, we say that *the product or quotient of measurements can be no more accurate than the least accurate measurement.*

MULTIPLYING OR DIVIDING MEASUREMENTS

1. First multiply or divide the measurements as given.
2. Then round the result to the same number of significant digits as the measurement with the least number of significant digits.

EXAMPLE 3

Multiply the measurements: 11.4 m × 15.6 m.

$$11.4 \text{ m} \times 15.6 \text{ m} = 177.84 \text{ m}^2$$

Round this product to three significant digits, which is the accuracy of the least accurate measurement (which is the accuracy of each measurement in this example). That is,

$$11.4 \text{ m} \times 15.6 \text{ m} = 178 \text{ m}^2$$

EXAMPLE 4

Divide the measurements: 78,000 m^2 ÷ 654 m.

$$78{,}000 \text{ m}^2 \div 654 \text{ m} = 119.26606 \text{ m}$$

Round this quotient to two significant digits, which is the accuracy of the least accurate measurement (78,000 m^2). That is,

$$78{,}000 \text{ m}^2 \div 654 \text{ m} = 120 \text{ m}$$

EXAMPLE 5

Use the rules for multiplication and division of measurements to evaluate the following:

$$\frac{(25.0 \text{ kg})(14 \text{ m/s})}{0.104 \text{ m}}$$

First, multiply and divide the numbers (3365.384615 . . .), and round the result to two significant digits, which is the accuracy of the least accurate measurement (14 m/s). Then multiply and divide the units as you did the numbers.

$$\frac{\text{kg (m/s)}}{\text{m}} = \frac{\text{kg}}{\text{s}}$$

Then

$$\frac{(25.0 \text{ kg})(14 \text{ m/s})}{0.104 \text{ m}} = 3400 \text{ kg/s}$$

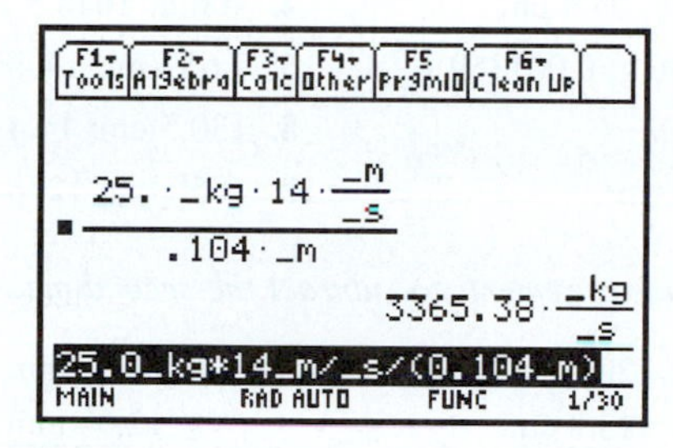

EXAMPLE 6

The TI-89 calculator can be used to convert units from one system to another, either by performing unit calculations as shown in the previous example, or by simply requesting the conversion as shown in the following frame. The underscore _ is used to indicate units (to distinguish them from variable names) and is obtained by pressing **green diamond** then the **MODE** key. Pressing **2nd** then the **MODE** key accesses the small arrow, which acts as a conversion symbol.

If a cable weighs 1.3 lb/ft, find its weight in g/cm. (The rounded answer is 19 g/cm.)

Note that the TI-89 symbol for grams is _gm (the symbol _g is used to represent the gravitational constant on the TI-89).

Note: When we multiply or divide measurements, the units do not need to be the same. The units must be the same when we add or subtract measurements. Also, the units are multiplied and/or divided in the same manner as the corresponding numbers. Any power or root of a measurement should be rounded to the same accuracy as the given measurement.

Obviously, such calculations with measurements should be done with a calculator. When no calculator is available, you may round the original measurements or any intermediate results to one more digit than the accuracy or precision required in the final result.

If both exact numbers and approximate numbers (measurements) occur in the same calculation, only the approximate numbers are used to determine the accuracy or precision of the result.

There are even more sophisticated methods for dealing with the calculations of measurements. The method we use, and indeed if we should even follow any given procedure, depends on the number of measurements and the sophistication needed for a particular situation.

The procedures for operations with measurements shown here are based on methods followed and presented by the American Society for Testing and Materials (ASTM).

Exercises 1.5

Use the rules for addition of measurements to find the sum of each set of measurements.

1. 15.7 in.; 6.4 in.

2. 178 m; 33.7 m; 61 m

3. 45.6 cm; 13.41 cm; 1.407 cm; 24.4 cm

4. 406 g; 1648.5 g; 39.74 g; 68.1 g

5. 1.0443 g; 0.00134 g; 0.08986 g; 0.001359 g

6. 7.639 mi; 14.48 mi; 1.004 mi; 0.68 mi

7. 14 V; 1.005 V; 0.018 V; 3.5 V

8. 130.5 cm; 14.4 cm; 1.457 m

9. 10.505 cm; 9.35 mm; 13.65 cm

10. 1850 cm; 1276 mm; 2.816 m; 4.02 m

Use the rules for subtraction of measurements to subtract the second measurement from the first.

11. 16.3 cm
12.4 cm

12. 120.2 cm
13.8 cm

13. 15.02 mm
12.6 mm

14. 162 mm
15.3 cm

15. 16.61 oz
11.372 oz

16. 94.1 g
32.74 g

17. 6.000 in.
2.004 in.

18. 0.54861 in.
0.234 in.

19. Four pieces of metal of thickness 0.149 in., 0.407 in., 1.028 in., and 0.77 in. are to be bolted together. What is the total thickness of the four pieces?

20. Five pieces of metal of thickness 2.47 mm, 10.4 mm, 3.70 mm, 1.445 mm, and 8.300 mm are clamped together. What is the total thickness of the five pieces?

21. Find the current going through R_5 in the circuit in Fig. 1.3. *Hint:* $I_1 + I_2 + I_3 = I_4 + I_5$.

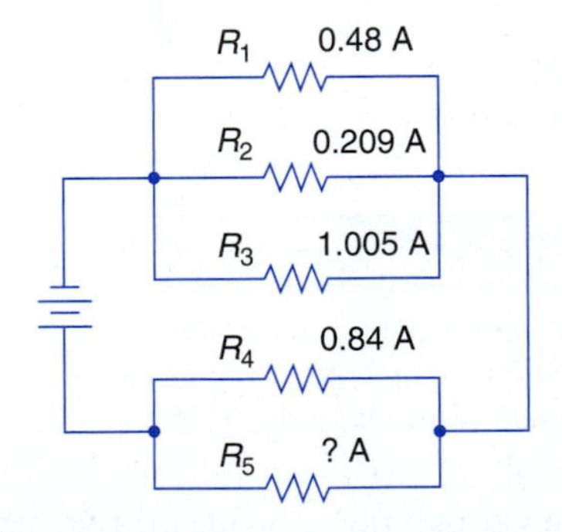

Figure 1.3

22. Find the sum of the following resistances: 15 Ω, 120 Ω, 6.5 Ω, 0.025 Ω, and 2375 Ω.

Use the rules for multiplication and division of measurements to evaluate each of the following.

23. (17.7 m)(48.2 m)

24. (540 cm)(28.0 cm)

25. (4.6 in.)(0.0285 in.)

26. (8.2 km)(6.75 km)

27. (34.2 cm)(26.1 cm)(28.9 cm)

28. (0.065 m)(0.0282 m)(0.0375 m)

29. $19.4\ \text{m}^3 \div 9.3\ \text{m}^2$

30. $4300\ \text{V} \div 14.5\ \text{A}$

31. $\dfrac{490\ \text{cm}}{6.73\ \text{s}^2}$

32. $\dfrac{5.03\ \text{km}}{4.7\ \text{s}}$

33. $\dfrac{0.447\ \text{N}}{(1.43\ \text{m})(4.0\ \text{m})}$

34. $\dfrac{(120\ \text{V})^2}{50.0\ \Omega}$

35. $\dfrac{(4\bar{0}\ \text{kg})(3.0\ \text{m/s})^2}{5.50\ \text{m}}$

36. $\dfrac{190\ \text{g}}{(3.4\ \text{cm})(1.6\ \text{cm})(8.4\ \text{cm})}$

37. Find the area of a rectangle measured as 6.2 cm by 17.5 cm ($A = lw$).

38. The formula for the volume of a rectangular solid is $V = lwh$, where l = length, w = width, and h = height. Find the volume of a rectangular solid when $l = 12.4$ ft, $w = 9.6$ ft, and $h = 5.4$ ft.

39. Find the volume of a cube with each edge 8.50 cm long ($V = e^3$, where e is the length of each edge).

40. The formula $s = 4.90t^2$ gives the distance s in metres a body falls in a given time t. Find the distance a ball falls in 2.6 s.

41. Given K.E. $= \frac{1}{2}mv^2$, where $m = 2.37 \times 10^6$ kg, and $v = 10.4$ m/s. Find K.E.

42. A formula for finding the horsepower of an engine is $p = \dfrac{d^2n}{2.50}$, where d is the diameter of each cylinder in inches and n is the number of cylinders. What is the horsepower of an 8-cylinder engine if each cylinder has a diameter of 3.00 in.? (*Note:* Eight is an exact number. The number of significant digits in an exact number has no bearing on the number of significant digits in the product or quotient.)

43. Six pieces of metal, each of thickness 2.08 mm, are fitted together. What is the total thickness of the six pieces?

44. Find the volume of a cylinder having a radius of 6.1 m and a height of 8.3 m. The formula for the volume of a cylinder is $V = \pi r^2 h$.

45. In 1970 in the United States 4,2$\bar{0}$0,000,000 bushels (bu) of corn were harvested from 66,800,000 acres. In 2000 there were 10,200,000,000 bu harvested from 73,100,000 acres. What were the yields in bushels per acre, and what was the increase in yield?

1.6 ALGEBRAIC EXPRESSIONS

An **algebraic expression** is a combination of finite sums, differences, products, quotients, roots, and powers of numbers and of letters representing numbers. Some examples of algebraic expressions are

$$6x - 7y + 3 \qquad \frac{5a^2bc}{9a - 1} \qquad (6x + 1)^2 \qquad 3a\sqrt{6a + 4}$$

In an expression a **variable** quantity may be represented by a letter. This letter may be replaced by any number from a given replacement set, such as the set of real numbers. A **constant** may also be represented by a letter. This letter may be replaced by only one number in a given situation.

A **term** is an expression or part of an expression involving only the product of numbers or letters. Terms may be connected by plus or minus signs, which indicate addition or subtraction of the terms, respectively.

Terms may have two or more **factors** connected by signs indicating multiplication. The term $7xyz$ has four factors, namely, 7, x, y, and z. The **coefficient** of a factor (or factors) is the product of the remaining factors. In the term $7xyz$, the coefficient of xyz is 7;

the coefficient of $7z$ is xy; the coefficient of $7y$ is xz. The **numerical coefficient** of $12xy^2$ is 12.

The algebraic expression a^4 indicates that the letter a is to be used as a factor four times. We say that a^4 is the *fourth power* of a; it may also be written $a \cdot a \cdot a \cdot a$. The factor that is expressed as a power is called the **base.** The number that indicates the number of times the base is to be used as a factor is the **exponent.** For example, in the seventh power of x, written x^7, x is the base and 7 is the exponent.

A *polynomial in one variable, x,* is a special type of algebraic expression defined as follows:

$$a_nx^n + a_{n-1}x^{n-1} + a_{n-2}x^{n-2} + \cdots + a_2x^2 + a_1x + a_0$$

where n is a positive integer and the coefficients $a_n, a_{n-1}, a_{n-2}, \ldots, a_2, a_1, a_0$ are real numbers.

When an algebraic expression contains only one term, it is called a **monomial.** A **binomial** is an algebraic expression containing exactly two terms. A **trinomial** is an algebraic expression containing exactly three terms. More generally, an algebraic expression with two or more terms is called a **multinomial.**

EXAMPLE 1

The following polynomials in one variable are examples of monomials, binomials, and trinomials.

Monomials (one term)	*Binomials (two terms)*	*Trinomials (three terms)*
$3a$	$3y^2 + 4y$	$8x^2 + 4x + 9$
$7z^2$	$5a^3 - 6$	$6b^3 - 4b^2 + 7b$
$6m^3$	$8w^2 + 5w^3$	$5 + a - a^4$

The **degree of a monomial in one variable** is the same as the exponent of the variable.

EXAMPLE 2

Find the degree of each monomial.

(a) $4x^2$ has degree 2.
(b) $-16x^5$ has degree 5.
(c) $3y^4$ has degree 4.
(d) 8 has degree 0.

Note: A constant has degree 0.

The **degree of a polynomial** is the same as the highest-degree monomial in the polynomial.

EXAMPLE 3

Find the degree of each polynomial.

(a) $3x^2 - 4x + 7$ has degree 2.
(b) $2y^6 - 7y^4 + 4y^3 - 8y + 10$ has degree 6.
(c) $5y^8 - 9y^7 - y^5 + 4y^3 - y$ has degree 8.

A polynomial is in **decreasing order** if each term is of some degree less than the preceding term. For example, the following polynomial is in decreasing order:

$$5x^6 - 4x^4 + 5x^3 - 9x^2 - 6x + 1$$

A polynomial is in **increasing order** if each term is of some degree larger than the preceding term. For example, the following polynomial is in increasing order:

$$4 - 7x + 3x^2 - x^3 + 5x^6$$

The **degree of a monomial in more than one variable** equals the sum of the exponents of its variables.

EXAMPLE 4

Find the degree of each monomial.
(a) $3x^2y^2$ has degree 4. (b) $-2ab^4$ has degree 5.
(c) $12x^2y^3z$ has degree 6.

EXAMPLE 5

Find the degree of each polynomial.
(a) $3x^2y^3 - 4xy^2 + xy$ has degree 5.
(b) $4abc - 6a^2bc^3 - 10bd$ has degree 6.
(c) $2w^4x^2y - 7w^3x^3y^2 + 13w^2x^4y^2 + wx^3y$ has degree 8.

To add and subtract algebraic expressions, combine like terms. **Like terms** have identical letters and powers of letters.

EXAMPLE 6

Add $5x^2 + 7x - 4$ and $3x - x^2 - 5$.

$$\begin{aligned}(5x^2 + 7x - 4) + (3x - x^2 - 5) &= (5x^2 - x^2) + (7x + 3x) + (-4 - 5)\\ &= 4x^2 + 10x - 9\end{aligned}$$

You may prefer to arrange the expressions so that the like terms appear in the same vertical column and then add.

$$\begin{array}{r} 5x^2 + 7x - 4 \\ \underline{-x^2 + 3x - 5} \\ 4x^2 + 10x - 9 \end{array} \quad \text{(Add like terms.)}$$

To subtract one multinomial from another, use the subtraction principle.

$$a - b = a + (-b)$$

That is, to subtract, add the opposite of each quantity being subtracted.

EXAMPLE 7

Subtract $6x^2 - 5x + 4$ from $-3x^2 + 4x + 7$.

$$\begin{aligned}(-3x^2 + 4x + 7) - (6x^2 - 5x + 4) &= -3x^2 + 4x + 7 - 6x^2 + 5x - 4\\ &= -9x^2 + 9x + 3\end{aligned}$$

REMOVING GROUPING SYMBOLS

1. To remove grouping symbols preceded by only a plus sign, remove the grouping symbols and leave the sign of each term unchanged within the grouping symbols.
2. To remove grouping symbols preceded by only a minus sign, remove the grouping symbols and change the sign of each term within the grouping symbols.
3. When sets of grouping symbols are contained within each other, remove each set starting with the innermost set and finishing with the outermost set.

EXAMPLE 8

Perform the indicated operations and simplify.

$$\begin{aligned}&(3x^2 - 5x + 4) - (6x^2 - 6x + 1) + (2x^2 - 4x)\\&= 3x^2 - 5x + 4 - 6x^2 + 6x - 1 + 2x^2 - 4x\\&= -x^2 - 3x + 3\end{aligned}$$

EXAMPLE 9

Perform the indicated operations and simplify.

$$\begin{aligned}&-[3x + (x - y) - (3y + 2x)] - [-(x - 5y) + (y - x)]\\&= -[3x + x - y - 3y - 2x] - [-x + 5y + y - x]\\&= -[2x - 4y] - [-2x + 6y]\\&= -2x + 4y + 2x - 6y\\&= -2y\end{aligned}$$

How do powers affect the order of operations that we discussed in Section 1.2? Since raising a number to a power is repetitive multiplication, we simplify all powers before multiplications and divisions. Therefore, when exponent operations are included, the order of operations is as follows.

ORDER OF OPERATIONS

1. Do all operations within any grouping symbols, beginning with the innermost pair if grouping symbols are contained within each other.
2. Simplify all powers.
3. Do all multiplications and divisions in the order in which they occur from left to right.
4. Do all additions and subtractions in the order in which they occur from left to right.

To evaluate an algebraic expression, substitute the given numerical values in place of the letters and follow the order of operations as given in the preceding box.

EXAMPLE 10

Evaluate $4a^3b + 5b + 2[6 - b]^2$ when $a = -2$ and $b = -3$.

First, substitute as follows:

$$4(-2)^3(-3) + 5(-3) + 2[6 - (-3)]^2$$

$= 4(-2)^3(-3) + 5(-3) + 2[9]^2$	(Do the subtraction within the brackets first.)
$= 4(-8)(-3) + 5(-3) + 2(81)$	(Then simplify all powers.)
$= 96 - 15 + 162$	(Do all multiplications from left to right.)
$= 243$	(Do all additions and subtractions from left to right.)

EXAMPLE 11

Evaluate $\dfrac{3x^2 + 8y}{4z^3}$ when $x = 4$, $y = -5$, and $z = 2$.

First, substitute as follows:

$$\frac{3(4)^2 + 8(-5)}{4(2)^3}$$

$$= \frac{3(16) + 8(-5)}{4(8)} \quad \text{(Simplify the powers.)}$$

$$= \frac{48 - 40}{32} \quad \text{(Do all multiplications.)}$$

$$= \frac{8}{32} \quad \text{(Subtract. } \textit{Note:} \text{ Treat the fraction bar here as a pair of grouping symbols.)}$$

$$= \frac{1}{4} \quad \text{(Divide.)}$$

Exercises 1.6

Classify each expression as a monomial, a binomial, or a trinomial.

1. $3x + 4$ **2.** $5a^2$ **3.** $8x + 2x^2 - x^3$

4. $5x^2 + 6x$ **5.** $-3xy$ **6.** $7x^3 + 4x^2 + x$

7. $x^2 + x$ **8.** $1 - x^2$ **9.** $3ab^2 - 4a^2b - 5a^2b^2$

10. $5x^2 - 5y^2 + 5z^2$ **11.** $8ab - 3a^2$ **12.** $4 - x^3y^3$

Find the degree of each monomial.

13. $5a^2$ **14.** $-6x^3$ **15.** $-7x^4$

16. $10x^5$ **17.** $3y^{10}$ **18.** 4

19. $4x^2y^3$ **20.** $6xy$ **21.** $-2a^2b^3c$

22. $19p^2q^3r^5$ **23.** $10ab^2c$ **24.** $-16xy^3z^4$

Write each polynomial in decreasing order and find its degree.

25. $3x^2 + 2 + 5x$ **26.** $1 - 8x^2 + x^3 - x^5$

27. $5x^2 + 9x^8 - 5x^4 + 6x^3$ **28.** $8x^5 - 5x^4 - 2x^6 - x$

29. $3y^3 + 5 - 3y + 4y^5$ **30.** $7 - 3z + 4z^3 - 6z^2$

Write each polynomial in increasing order and find its degree.

31. $2x^3 - 3x^4 + 4x$ **32.** $a^3 - 1 + a^2 - a$

33. $5c^3 - 8c^5 + c - 7 + 3c^4$ **34.** $6x^3 - 4x^5 - x + 7x^4 + 2$

35. $5y^3 - 6y^6 + 2 - 8y^4 + 2y$ **36.** $7y^2 + 4 - 2y^5 - 6y + y^8$

Find the degree of each polynomial.

37. $8a^2b - 4ab^2 + 6ab$ **38.** $5xy^4 - 14x^2y + 7$

39. $6x^3y - 5x^2y^2 + 6x^5 - 2$ **40.** $6ab^4 - 4a^2b^2 + 6b^3 - 4$

41. $3x^2y^4z - 4xy^3z^4 + 9x^3y^2z^3$

42. $16x^3yz + 12x^4yz - 4x^5y^2 + 3$

43. $a^3b^4c + 4a^2b^2c^2 - 6a^9 - 3b^2$

44. $2x^5y^5 + 3x^2y^3z - 5y^6 + z^8$

Perform the indicated operations and simplify.

45. $(3x^2 - 4x + 8) + (6x^2 - x - 3)$
46. $(-4x^2 + 3x - 2) + (x^2 - x + 4)$
47. $(5x^2 + 7x - 9) + (-x^2 - 6x + 4)$
48. $(3x^2 + 2x - 8) + (9x^2 - 8x + 2)$
49. $(-3x^2 - 5x - 3) - (5x^2 - 2x - 7)$
50. $(-6x^2 + 2x - 4) - (-8x^2 - 7)$
51. $(-6x^2 + 7) - (4x^2 + x - 3)$
52. $(5x^2 + 2x) - (3x^2 - 5x + 2)$
53. $(3x^2 + 4x - 4) + (-x^2 - x + 2) - (-2x^2 + 2x + 8)$
54. $(-x^2 + 7x - 9) + (10x^2 - 11x + 4) - (-12x^2 - 15x + 3)$
55. $(-4x^2 + 6x - 2) - (5x^2 + 7x - 4) + (5x^2 - 6x + 1)$
56. $(5x^2 - 3x + 9) - (-2x^2 - 3x + 4) - (4x^2 - 2x - 1)$
57. $(3x^2 - 1 + 2x) - (9x^2 + 3 - 9x) - (3x - 4 - 2x^2)$
58. $(4 - 3x - x^2) - (3x^2 - 1 - 4x) + (4x^2 + 3 - x)$
59. $(5x^2 - 12x - 1) - (11x^2 + 4) + (4x + 7) - (3x - 2)$
60. $(3x^2 + 5) - (6x - 7) - (3x^2 + 6x - 13) + (5x - 9)$
61. $(3x^3 + 5x - 2) + (6x^2 - 10x + 1) - (4x^2 - 1) - (-5x^3 - 3)$
62. $(x^2 + 1) - (x^3 - 1) - (x^2 - x + 1) + (1 - x^3)$
63. $(3x^2 + 2x - 1) - (1 - 5x) - (3x^2 + x) + (3x - 4x^2) - (6x^2 + x^3)$
64. $(5x + 4) - (6x^2 + 4x - 7) - (6 - 5x + x^2) + (3x - 7x^2) + (5x^2 - 14) - (-3x)$
65. $-(x - 3y) - [(x + 2y) + (3y - 2x)]$
66. $[(5x + 3y) + (-2x - 2y)] - [-(x + 6y) - (-3x + 2y)]$
67. $-\{(5x + 3y) - (2x + 5y) - [3x + (4y + x)]\}$
68. $-[-(5x - 6y) - (-3x + 2y) + 6x - (4y - x)]$

Evaluate each expression when $a = -2$, $b = 3$, $c = -1$, and $d = 1$.

69. $a - b$

70. $a + 2c$

71. $3a - 2c$

72. $4d - 3bc$

73. $3a^2b^3c^2$

74. $4a^3b^4c^2d^3$

75. $(-b + 3cd)^3$

76. $(a + 6c)^2$

77. $b - a(c + d)$

78. $\dfrac{2a + 5b}{3b - 2c}$

79. $\dfrac{12a + 6b^2}{9c + 10a}$

80. $\dfrac{4a^3 - 5b}{3c^2 + d}$

81. $(4a^2bc)^3$

82. $(-a^2cd)^4$

83. $\dfrac{5}{ab} - \dfrac{6a}{5b} - \dfrac{cd}{b}$

84. $\left(\dfrac{4a^2 + 3b}{b - a}\right)^2$

1.7 EXPONENTS AND RADICALS

In Section 1.3 we studied the laws of exponents involving powers of 10. There are similar, but more general, laws of exponents for any base, given as follows:

LAWS OF EXPONENTS

In each of the following, a and b are real numbers and m and n are positive integers.

1. $a^m \cdot a^n = a^{m+n}$

2. (a) $\dfrac{a^m}{a^n} = a^{m-n} \quad (m > n \quad \text{and} \quad a \neq 0)$

(b) $\dfrac{a^m}{a^n} = \dfrac{1}{a^{n-m}} \quad (m < n \quad \text{and} \quad a \neq 0)$

3. $(a^m)^n = a^{mn}$

4. $(ab)^n = a^n b^n$

5. $\left(\dfrac{a}{b}\right)^n = \dfrac{a^n}{b^n} \quad (b \neq 0)$

The following examples illustrate the laws of exponents.

EXAMPLE 1

$$x^2 \cdot x^3 = x^{2+3} = x^5 \qquad \text{(Law 1)}$$

This result can also be shown as follows:

$$x^2 \cdot x^3 = (x \cdot x)(x \cdot x \cdot x) = x^5$$

EXAMPLE 2

$$\frac{x^7}{x^4} = x^{7-4} = x^3 \qquad [\text{Law 2(a)}]$$

This result can also be shown as follows:

$$\frac{x^7}{x^4} = \frac{\cancel{x} \cdot \cancel{x} \cdot \cancel{x} \cdot \cancel{x} \cdot x \cdot x \cdot x}{\cancel{x} \cdot \cancel{x} \cdot \cancel{x} \cdot \cancel{x}} = x^3$$

EXAMPLE 3

$$\frac{x^2}{x^5} = \frac{1}{x^{5-2}} = \frac{1}{x^3} \qquad [\text{Law 2(b)}]$$

This result can also be shown as follows:

$$\frac{x^2}{x^5} = \frac{\cancel{x} \cdot \cancel{x}}{\cancel{x} \cdot \cancel{x} \cdot x \cdot x \cdot x} = \frac{1}{x^3}$$

EXAMPLE 4

$$(x^3)^4 = x^{(3)(4)} = x^{12} \qquad \text{(Law 3)}$$

This result can also be shown as follows:

$$(x^3)^4 = x^3 \cdot x^3 \cdot x^3 \cdot x^3 = x^{12}$$

EXAMPLE 5

$$\begin{aligned} (3y^3)^4 &= (3)^4(y^3)^4 && \text{(Law 4)} \\ &= 81y^{12} && \text{(Law 3)} \end{aligned}$$

This result can also be shown as follows:

$$(3y^3)^4 = (3y^3)(3y^3)(3y^3)(3y^3) = 81y^{12}$$

EXAMPLE 6

$$\begin{aligned}(-x^2)^4 &= (-1)^4(x^2)^4 && \text{(Law 4)}\\ &= 1 \cdot x^8 && \text{(Law 3)}\\ &= x^8\end{aligned}$$

EXAMPLE 7

$$\left(\frac{x^4}{y^3}\right)^2 = \frac{x^8}{y^6} \qquad \text{(Laws 5 and 3)}$$

This result can also be shown as follows:

$$\left(\frac{x^4}{y^3}\right)^2 = \left(\frac{x^4}{y^3}\right)\left(\frac{x^4}{y^3}\right) = \frac{x^8}{y^6}$$

The next general law of exponents is given as follows:

$$a^0 = 1, \qquad \text{where } a \text{ is any real number and } a \neq 0$$

To show that this law for zero exponents is valid, we use the same reasoning as in Section 1.3 by using the substitution principle: If $a = b$ and $a = c$, then $b = c$.

$$\frac{a^n}{a^n} = 1 \qquad \text{(Any number other than zero divided by itself equals one.)}$$

and

$$\begin{aligned}\frac{a^n}{a^n} &= a^{n-n} && \text{(Law 2, where } m = n\text{)}\\ &= a^0\end{aligned}$$

Therefore,

$$a^0 = 1 \qquad \text{(substitution)}$$

The inverse process of raising a number to a power is called **finding the root of a number.** Square roots and cube roots are the roots most often used in technical problems. The nth root of a number a is written $\sqrt[n]{a}$, where n is the **index,** a is the **radicand,** and the symbol $\sqrt{}$ is called a **radical sign.**

The square root of a number a, written $\sqrt{a}$, is that number which, when multiplied by itself, is a. The index for square root is 2, but it is not usually written. For example, $\sqrt{36} = 6$ because $(6)(6) = 36$; however, $(-6)(-6) = 36$ is also true. Therefore, it seems that

$$\sqrt{36} = 6 \quad \text{and} \quad \sqrt{36} = -6$$

From the principle of substitution we know that

$$\begin{aligned}\text{if} \quad p = q \quad &\text{and} \quad p = r\\ \text{then} \quad q = r\end{aligned}$$

Applying this principle to $\sqrt{36} = 6$ and $\sqrt{36} = -6$, then we must accept the statement that $6 = -6$, which we know is false. Therefore, one of the assumptions must be false. The quantity $\sqrt{36}$ is a real number, and each real number has only one value. Mathematicians have agreed that the quantity $\sqrt{36}$ is a positive number and must have a positive value. As a result, we say that $\sqrt{36} = 6$ is true and $\sqrt{36} = -6$ is false.

What is the square root of a negative number, such as -49? That is, what real number squared equals -49? The product $(7)(7) = +49$ and the product $(-7)(-7) = +49$; hence, there is no real number whose square is -49. In fact, there is no real number whose square is negative. For this reason, the square root of a negative number is undefined within the set of real numbers.

As a result of the preceding discussions, we can now define the square root of a *nonnegative* number a, written $\sqrt{a}$, as that nonnegative number which, when multiplied by itself, is a.

EXAMPLE 8

Simplify each square root.

(a) $\sqrt{81} = 9$ (b) $\sqrt{100} = 10$

(c) $\sqrt{0} = 0$ (d) $\sqrt{25} = 5$

(e) $\sqrt{8^2} = 8$ (f) $\sqrt{10^8} = 10^4$

(g) $\sqrt{-4}$ is not a real number. (h) $-\sqrt{16} = -(+4) = -4$

The cube root of a number a, written $\sqrt[3]{a}$, is that number which, when multiplied by itself three times, is a. For example, $\sqrt[3]{8} = 2$ because $2 \cdot 2 \cdot 2 = 8$.

EXAMPLE 9

Simplify each cube root.

(a) $\sqrt[3]{27} = 3$ because $(3)(3)(3) = 27$

(b) $\sqrt[3]{64} = 4$ because $(4)(4)(4) = 64$

(c) $\sqrt[3]{8} = 2$ because $(2)(2)(2) = 8$

(d) $\sqrt[3]{-125} = -5$ because $(-5)(-5)(-5) = -125$

(e) $\sqrt[3]{-27} = -3$ because $(-3)(-3)(-3) = -27$

(f) $-\sqrt[3]{-8} = -(-2) = 2$

Note: A negative quantity under the radical sign does not present a problem for cube roots.

Roots of numbers, or radicals, are discussed in more detail in Chapter 8, but one property of radicals needs some discussion now:

$$\sqrt{ab} = \sqrt{a}\,\sqrt{b} \quad \text{where } a > 0 \text{ and } b > 0$$

That is, *the square root of a product of positive numbers equals the product of the square roots of its factors.* This property is used to simplify radicals when either a or b is a perfect square. A **perfect square** is the square of a rational number. The first ten positive integral perfect squares are 1, 4, 9, 16, 25, 36, 49, 64, 81, and 100.

Examples 10 through 12 illustrate how to simplify square roots.

EXAMPLE 10

Simplify $\sqrt{18}$.

$\sqrt{18} = \sqrt{9 \cdot 2}$ (Find the largest perfect square factor of 18.)

$= \sqrt{9}\,\sqrt{2}$ $(\sqrt{ab} = \sqrt{a}\,\sqrt{b})$

$= 3\sqrt{2}$ (Find the square root of the perfect square.)

EXAMPLE 11

Simplify $\sqrt{48}$.

$\sqrt{48} = \sqrt{16 \cdot 3}$ (Find the largest perfect square factor of 48.)

$= \sqrt{16}\,\sqrt{3}$ $(\sqrt{ab} = \sqrt{a}\,\sqrt{b})$

$= 4\sqrt{3}$ (Find the square root of the perfect square.)

EXAMPLE 12

Simplify $\sqrt{360}$.

$$\begin{aligned}\sqrt{360} &= \sqrt{36 \cdot 10} && \text{(Find the largest perfect square factor of 360.)}\\ &= \sqrt{36}\,\sqrt{10} && (\sqrt{ab} = \sqrt{a}\,\sqrt{b})\\ &= 6\sqrt{10} && \text{(Find the square root of the perfect square.)}\end{aligned}$$

The square root of a number is simplified when the number under the radical contains no perfect square factors. To find the decimal value of the square root of a number, use the square root key on your calculator.

For order-of-operation purposes, treat a radical as a grouping symbol. That is, do all operations under the radical before finding the root, as in the following example.

EXAMPLE 13

Simplify $\sqrt{6^2 + 18}$.

$$\begin{aligned}\sqrt{6^2 + 18} &= \sqrt{36 + 18} && \text{(Find the power.)}\\ &= \sqrt{54} && \text{(Add.)}\\ &= \sqrt{9 \cdot 6} && \text{(Find the largest perfect square factor of 54.)}\\ &= \sqrt{9}\,\sqrt{6} && \\ &= 3\sqrt{6} && \text{(Find the square root of the perfect square.)}\end{aligned}$$

Pressing **ENTER** on the TI-89 home screen will simplify numerical square roots. To obtain a decimal approximation instead, press **green diamond ENTER** (≈).

2nd multiplication sign 6^2+18) **ENTER**

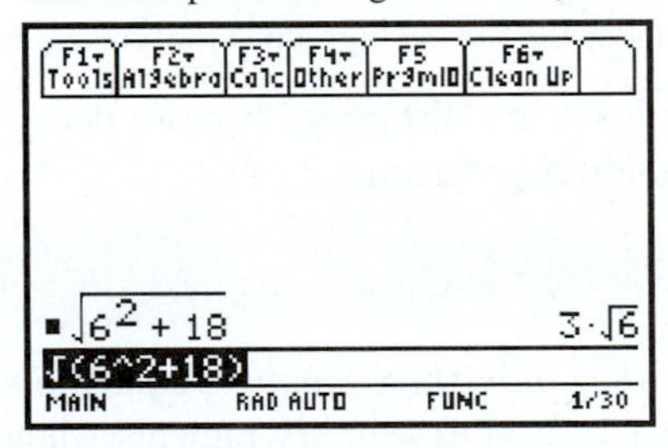

green diamond ENTER

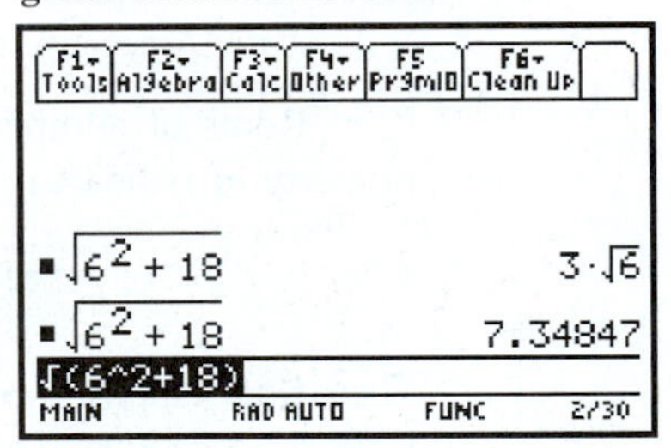

Exercises 1.7

Using the laws of exponents, perform the indicated operations and simplify.

1. $x^5 \cdot x^7$
2. y^3y^6
3. $(3a^2)(4a^3)$
4. $5y^3 \cdot 6y^4$
5. $\dfrac{m^9}{m^3}$
6. $\dfrac{c^{12}}{c^3}$
7. $\dfrac{x^2}{x^6}$
8. $\dfrac{y^2}{y^8}$
9. $\dfrac{12x^8}{4x^4}$
10. $\dfrac{36x^{10}}{2x^2}$
11. $\dfrac{15x^2}{3x^5}$
12. $\dfrac{5y^6}{35y^8}$
13. $(a^2)^3$
14. $(x^5)^2$
15. $(c^4)^4$
16. $(b^7)^6$
17. $(9a)^2$
18. $(4m^2)^3$
19. $(2x^2)^5$
20. $(3c^6)^4$
21. $\left(\dfrac{3}{4}\right)^2$
22. $\left(\dfrac{a}{2}\right)^3$
23. $\left(\dfrac{2}{a^3}\right)^4$
24. $\left(\dfrac{x^2}{y}\right)^5$
25. 4^0
26. $(3x)^0$
27. $3x^0$
28. $7(x^2)^0$
29. $(-3x)^2$
30. $(-2x^2)^5$
31. $(-t^3)^4$
32. $(-s^2)^5$

33. $(-a^2)^3$ 34. $(-c^3)^6$ 35. $(-2a^2b)^3$ 36. $(-a^2b^3)^4$

37. $(3x^2y^3)^2$ 38. $(-5a^2b^4)^2$ 39. $(-3x^3y^4z)^3$ 40. $(2a^2b^3c^4)^3$

41. $\left(\frac{2x^2}{3y^3}\right)^2$ 42. $\left(\frac{-3x^4}{4y^2}\right)^3$ 43. $\left(\frac{-4x}{3y^2}\right)^2$ 44. $\left(\frac{5xy^2}{7z^3}\right)^2$

45. $\left(\frac{-1}{6y^3}\right)^2$ 46. $\left(\frac{-2}{3x^3}\right)^3$

Simplify each root. In Exercises 47 through 60, the result is an integer.

47. $\sqrt{4}$ 48. $\sqrt{16}$ 49. $\sqrt{64}$ 50. $\sqrt{49}$

51. $\sqrt{121}$ 52. $\sqrt{144}$ 53. $\sqrt{5^{16}}$ 54. $\sqrt{10^6}$

55. $\sqrt[3]{125}$ 56. $\sqrt[3]{343}$ 57. $\sqrt[3]{-216}$ 58. $\sqrt[3]{-64}$

59. $\sqrt[3]{512}$ 60. $\sqrt[3]{1000}$ 61. $\sqrt{45}$ 62. $\sqrt{12}$

63. $\sqrt{50}$ 64. $\sqrt{80}$ 65. $\sqrt{72}$ 66. $\sqrt{75}$

67. $\sqrt{4^2 + 32}$ 68. $\sqrt{12^2 - 44}$ 69. $\sqrt{3 \cdot 4^2 - 4 \cdot 2^2}$ 70. $\sqrt{3^4 + 5 \cdot 2^3}$

Evaluate and round to three significant digits.

71. $\sqrt{329}$ 72. $\sqrt{492}$ 73. $\sqrt{2596}$ 74. $\sqrt{87{,}500}$

75. $\sqrt{0.00472}$ 76. $\sqrt{0.924}$ 77. $\sqrt{16 + 36}$ 78. $\sqrt{81 - 49}$

79. $\sqrt{5^2 + 8^2}$ 80. $\sqrt{9^2 - 9 \cdot 2^2}$

81. $\sqrt{(2.73 \times 10^4)^2 + (1.00 \times 10^5)^2}$ 82. $\sqrt{(3.45 \times 10^{-3})^2 + (6.85 \times 10^{-4})^2}$

83. $\sqrt{(115)^2 + (15.5 - 84.6)^2}$ 84. $\dfrac{1}{2\pi\sqrt{(1.23 \times 10^{-5})(4.45 \times 10^{-12})}}$

1.8 MULTIPLICATION OF ALGEBRAIC EXPRESSIONS

To multiply monomials, multiply their numerical coefficients and multiply each set of like letter factors using the laws of exponents.

EXAMPLE 1

Multiply: $(-3a^2b^3)(5ab^2)$.

$$(-3a^2b^3)(5ab^2) = (-3 \cdot 5)(a^2 \cdot a)(b^3 \cdot b^2) = -15a^3b^5$$

EXAMPLE 2

Multiply: $(2x^2yz^2)(4x^3z)(6xy^2z)$.

$$(2x^2yz^2)(4x^3z)(6xy^2z) = (2 \cdot 4 \cdot 6)(x^2 \cdot x^3 \cdot x)(y \cdot y^2)(z^2 \cdot z \cdot z) = 48x^6y^3z^4$$

To multiply a multinomial by a monomial, multiply each term of the multinomial by the monomial using the distributive property:

$$a(b + c) = ab + ac$$

EXAMPLE 3

Multiply: $5a(6b + 3c)$.

$$5a(6b + 3c) = (5a)(6b) + (5a)(3c) = 30ab + 15ac$$

EXAMPLE 4

Multiply: $4x(3x^2 - 2x + 5)$.

$$\begin{aligned} 4x(3x^2 - 2x + 5) &= (4x)(3x^2) + (4x)(-2x) + (4x)(5) \\ &= 12x^3 - 8x^2 + 20x \end{aligned}$$

EXAMPLE 5

Multiply: $6a^2b(-2a^3b^2 + 3a^2b - b + 1)$.

$$\begin{aligned} &6a^2b(-2a^3b^2 + 3a^2b - b + 1) \\ &\quad = (6a^2b)(-2a^3b^2) + (6a^2b)(3a^2b) + (6a^2b)(-b) + (6a^2b)(1) \\ &\quad = -12a^5b^3 + 18a^4b^2 - 6a^2b^2 + 6a^2b \end{aligned}$$

In general, to multiply two multinomials, multiply each term of the first multinomial by each term of the second and simplify. *Note:* The method is most similar to the method of multiplying whole numbers.

EXAMPLE 6

Multiply: $(5x + 4)(3x - 6)$.

$$\begin{array}{rrl} & 5x + 4 & \\ & \underline{3x - 6} & \\ & -30x - 24 & \leftarrow \; -6(5x + 4) \\ \underline{15x^2 + 12x} & & \leftarrow \; 3x(5x + 4) \\ 15x^2 - 18x - 24 & & \end{array}$$

EXAMPLE 7

Multiply: $(3x^2 + 4x - 7)(2x^3 - x^2 - 2)$.

$$\begin{array}{rl} 3x^2 + 4x - 7 & \\ \underline{2x^3 - x^2 - 2} & \\ -6x^2 - 8x + 14 & \leftarrow \; -2(3x^2 + 4x - 7) \\ -3x^4 - 4x^3 + 7x^2 \qquad\quad & \leftarrow \; -x^2(3x^2 + 4x - 7) \\ \underline{6x^5 + 8x^4 - 14x^3 \qquad\qquad\qquad\quad} & \leftarrow \; 2x^3(3x^2 + 4x - 7) \\ 6x^5 + 5x^4 - 18x^3 + x^2 - 8x + 14 & \end{array}$$

Next, we discuss a special method for mentally finding the product of two binomials, a special product that occurs again and again in our work. This method is called the **FOIL method.** The initials FOIL (First, Outer, Inner, Last) are used to help keep track of the order of multiplying the terms of the two binomials as illustrated in the following examples.

EXAMPLE 8

Multiply: $(2x + 3)(4x - 5)$.

Outer product

$(2x + 3)(4x - 5)$

Inner product

F: product of *First* terms of the binomials: $(2x)(4x) = 8x^2$

O: *Outer* product: $(2x)(-5) = -10x$

I: *Inner* product: $(3)(4x) = 12x$ } sum $= 2x$

L: product of *Last* terms of the binomials: $(3)(-5) = -15$

$$(2x + 3)(4x - 5) = 8x^2 + 2x - 15$$

EXAMPLE 9

Multiply: $(4x - 2)(5x - 6)$.

Outer product: $(4x - 2)(5x - 6)$ — outer terms $4x$ and -6

Inner product: inner terms -2 and $5x$

$$
\begin{aligned}
&F: \quad (4x)(5x) = && 20x^2 \\
&O: \quad (4x)(-6) = -24x \\
&I: \quad (-2)(5x) = -10x \quad \Big\} \text{ sum } = && -34x \\
&L: \quad (-2)(-6) = && 12 \\
&(4x - 2)(5x - 6) = && 20x^2 - 34x + 12
\end{aligned}
$$

You should do these steps mentally and write only the final result.

EXAMPLE 10

Multiply: $(2x + 3)(5x + 1)$.

$$(2x + 3)(5x + 1) = 10x^2 + 17x + 3$$

EXAMPLE 11

Multiply: $(6x + 4)(2x - 3)$.

$$(6x + 4)(2x - 3) = 12x^2 - 10x - 12$$

EXAMPLE 12

Multiply: $(5x^2 - 6)(4x^2 - 1)$.

$$(5x^2 - 6)(4x^2 - 1) = 20x^4 - 29x^2 + 6$$

EXAMPLE 13

Multiply: $(3x + 2)(3x - 2)$.

$$(3x + 2)(3x - 2) = 9x^2 - 4$$

To find the power of a binomial, first rewrite the binomial as a product, and then use the most appropriate multiplication method.

EXAMPLE 14

Multiply: $(x - 3)^2$.

$$(x - 3)^2 = (x - 3)(x - 3) = x^2 - 6x + 9$$

EXAMPLE 15

Multiply: $(3x - 2)^3$.

$$
\begin{aligned}
(3x - 2)^3 &= (3x - 2)(3x - 2)(3x - 2) \\
&= (9x^2 - 12x + 4)(3x - 2)
\end{aligned}
$$

Use FOIL to find the product of the first two binomials. Then find the final product as follows:

$$
\begin{array}{r}
9x^2 - 12x + 4 \\
3x - 2 \\
\hline
-18x^2 + 24x - 8 \\
27x^3 - 36x^2 + 12x \quad \\
\hline
27x^3 - 54x^2 + 36x - 8
\end{array}
$$

The TI-89 uses the **expand** command (press **F2** then **3**) to perform polynomial multiplication and to simplify using the properties of exponents.

F2 3

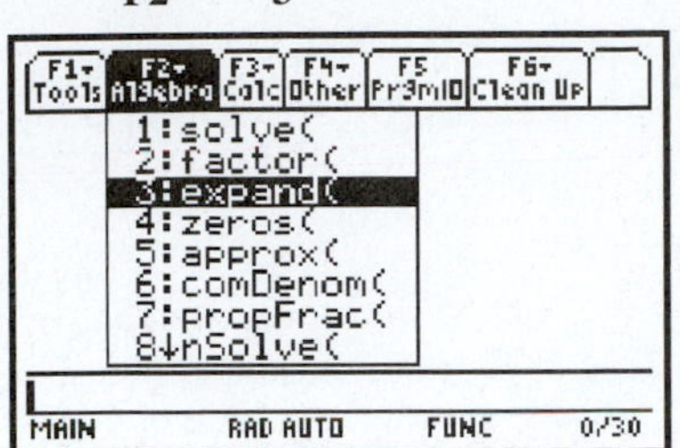

(3x-2)^3) **ENTER**

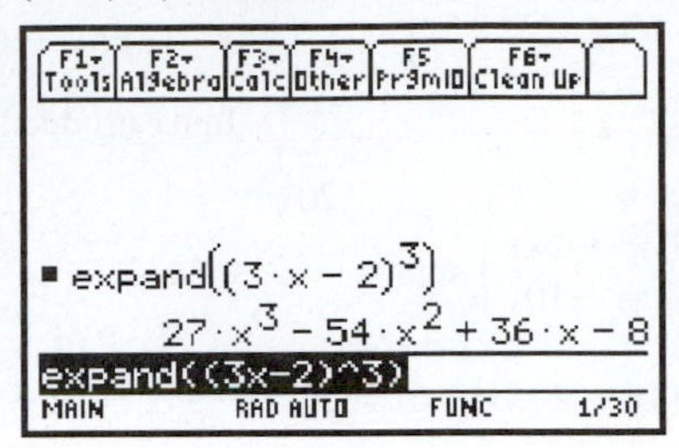

F2 3 (2x^3*y^2)^4/x^2) **ENTER**

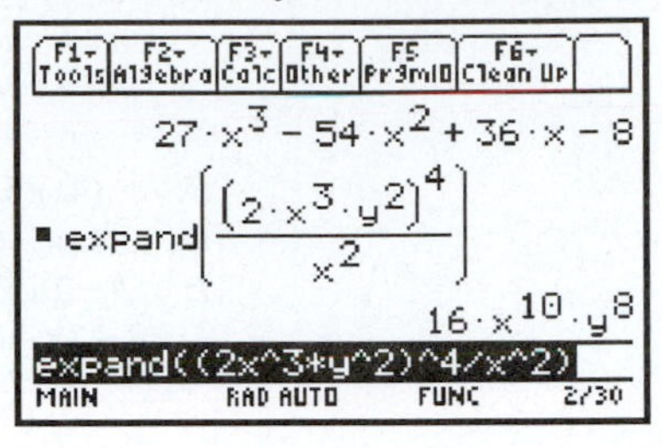

Exercises 1.8

Find each product.

1. $(4x^2)(8x^3)$
2. $(-6x^3)(12x^7)$
3. $(-4a^2b)(6a^3b^2)$
4. $(-4x^2y)(-7x^4y^2)$
5. $(12a^2bc^3)(-4ac^2)$
6. $(-2x^2y^2z)(7x^2yz^4)$
7. $(-3a^2b^4c)(2ab^2c^5)(-4ab^3)$
8. $(3x^2yz^5)(5x^3y^2)(-2y^3z^2)$
9. $3a(4a - 7b)$
10. $-5c(a + 2b - 3c)$
11. $3x(2x^2 + 4x - 5)$
12. $5x(-3x^2 - x + 7)$
13. $-5x^2(3x^2 - 5x + 8)$
14. $-3x^3(4x^2 + 6x - 2)$
15. $6ab^3(4a^2b - 8a^3b^4)$
16. $4a^2b(-2a^2b + 6ab^3)$
17. $-3a^2b^4(-a^4b^3 + 3ab - b^2)$
18. $-8a^3b^2(-2ab^3 - 3ab + a^2)$
19. $(3x - 7)(2x + 5)$
20. $(6x - 1)(5x - 3)$
21. $(6x + 3)(8x + 5)$
22. $(5x + 7)(3x - 2)$
23. $(3x + 4)(3x - 4)$
24. $(5x + 9)(5x - 9)$
25. $(4x + 1)(6x - 1)$
26. $(5x + 2)(3x + 4)$
27. $(3x - 7)(2x - 3)$
28. $(6x - 3)(x + 6)$
29. $(3x + 8y)(6x + 4y)$
30. $(2x + 7y)(9x - y)$
31. $(5s - 9t)(8s + 2t)$
32. $(4a - 5b)(7a - 10b)$
33. $(-3x + 4)(5x + 6)$
34. $(-11x + 2)(-10x - 3)$
35. $(3x^2 - 1)(2x^2 + 7)$
36. $(5x^2 + 4)(6x^2 + 9)$
37. $(5x^2 - 6)(6x^2 - 5)$
38. $(3x^2 + 4)(4x^2 - 5)$
39. $(2x + 5)(3x^2 + 4x - 3)$
40. $(3x - 4)(2x^2 - 6x - 4)$
41. $(x^2 + x + 2)(x^2 - x + 3)$
42. $(3x^2 + 8x - 2)(5x^2 + x - 7)$
43. $(x + y - 7)(x - y + 4)$
44. $(x - 3y + 7)(2x - y - 3)$
45. $(3x^2 + 2x - 6)(5x^2 - 4x - 1)$
46. $(4x^2 - 6x + 4)(-2x^2 - x + 5)$
47. $(3x^2 + 5x + 2)(4x^2 - 3)$
48. $(3x^2 - 5x - 4)(2x^2 - 6x)$
49. $(2x - 5)^2$
50. $(4x + 3)^2$
51. $(3x + 8)^2$
52. $(6x - 4)^2$
53. $(-5x + 2)^2$
54. $(-3x^2 - 4)^2$
55. $(3x - 4)^2(x^2 - 2x + 1)$
56. $(2x + 5)^2(x^2 + 3x - 1)$
57. $(2x - 1)^3$
58. $(4x + 3)^3$
59. $(2a + 5b)^3$
60. $(2 - 4x)^3$

1.9 DIVISION OF ALGEBRAIC EXPRESSIONS

To divide monomials, divide their numerical coefficients and divide each set of like letter factors using the laws of exponents.

EXAMPLE 1

Divide: $\dfrac{24x^2y^5}{3xy^3}$.

$$\frac{24x^2y^5}{3xy^3} = \frac{24}{3} \cdot \frac{x^2}{x} \cdot \frac{y^5}{y^3} \quad \text{(Divide the numerical coefficients and each set of like letter factors.)}$$

$$= 8xy^2 \quad \text{(Subtract exponents.)}$$

EXAMPLE 2

Divide: $\dfrac{18a^3b^6}{-8a^5b^2}$.

$$\frac{18a^3b^6}{-8a^5b^2} = \frac{18}{-8} \cdot \frac{a^3}{a^5} \cdot \frac{b^6}{b^2} \quad \text{(Divide the numerical coefficients and each set of like letter factors.)}$$

$$= -\frac{9}{4} \cdot \frac{1}{a^2} \cdot b^4 \quad \text{(Subtract exponents.)}$$

$$= -\frac{9b^4}{4a^2}$$

An alternate method for this division of monomials involves cancellation as follows:

$$\frac{18a^3b^6}{-8a^5b^2} = \frac{\overset{9}{\cancel{18}}\ \overset{1}{\cancel{a^3}}\overset{b^4}{\cancel{b^6}}}{\underset{-4}{\cancel{-8}}\ \underset{a^2}{\cancel{a^5}}\underset{1}{\cancel{b^2}}} = -\frac{9b^4}{4a^2}$$

To divide a multinomial by a monomial, divide each term in the multinomial by the monomial and simplify.

EXAMPLE 3

Divide: $\dfrac{15x^2 - 6x}{3x}$.

$$\frac{15x^2 - 6x}{3x} = \frac{15x^2}{3x} - \frac{6x}{3x} = 5x - 2 \quad \text{(Divide each term in the multinomial by } 3x \text{ and simplify.)}$$

EXAMPLE 4

Divide: $\dfrac{6a^3 - 10a^2 + 4a}{2a^2}$.

$$\frac{6a^3 - 10a^2 + 4a}{2a^2} = \frac{6a^3}{2a^2} - \frac{10a^2}{2a^2} + \frac{4a}{2a^2} \quad \text{(Divide each term in the multinomial by } 2a^2 \text{ and simplify.)}$$

$$= 3a - 5 + \frac{2}{a}$$

DIVIDING ONE MULTINOMIAL (DIVIDEND) BY A SECOND MULTINOMIAL (DIVISOR)

1. Arrange each multinomial in descending powers of one of the variables.
2. To find the first term of the quotient, divide the first term of the dividend by the first term of the divisor.
3. Multiply the divisor by the first term of the quotient and subtract the product from the dividend.
4. Continue this procedure of dividing the first term of each remainder by the first term of the divisor until the final remainder is zero or until the remainder is of degree less than the divisor.

EXAMPLE 5

Divide: $3x^3 + 2x^2 - 7x + 2$ by $x + 2$.

$$\begin{array}{rrl}
 & 3x^2 - 4x + 1 & \text{(quotient)} \\
\text{(divisor)} \quad x + 2) & \overline{3x^3 + 2x^2 - 7x + 2} & \text{(dividend)} \\
 & \underline{3x^3 + 6x^2 } & \\
 & -4x^2 - 7x & \\
 & \underline{-4x^2 - 8x } & \\
 & x + 2 & \\
 & \underline{x + 2} & \\
 & 0 & \text{(remainder)}
\end{array}$$

That is,

$$\frac{3x^3 + 2x^2 - 7x + 2}{x + 2} = 3x^2 - 4x + 1$$

EXAMPLE 6

Divide: $\dfrac{2x^3 + 11x^2 + 10}{2x + 3}$.

$$\begin{array}{rr}
 & x^2 + 4x - 6 \\
2x + 3) & \overline{2x^3 + 11x^2 + 10} \\
 & \underline{2x^3 + 3x^2 } \\
 & 8x^2 \\
 & \underline{8x^2 + 12x } \\
 & -12x + 10 \\
 & \underline{-12x - 18} \\
 & 28
\end{array}$$

Since 28 is the remainder (which is of degree less than the divisor, $2x + 3$), it is usually written in the form of a quotient, $\dfrac{28}{2x + 3}$.

That is,

$$\frac{2x^3 + 11x^2 + 10}{2x + 3} = x^2 + 4x - 6 + \frac{28}{2x + 3}$$

The TI-89 uses the **propFrac** command (press **F2** then **7**) to perform polynomial division. Note that the remainder term will always appear first, followed by the quotient. This is the opposite of the customary order.

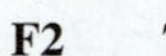

(2x^3+11x^2+10)/(2x+3)) **ENTER**

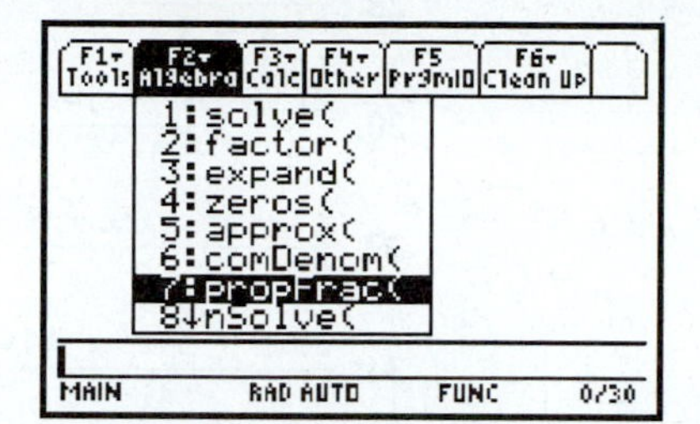

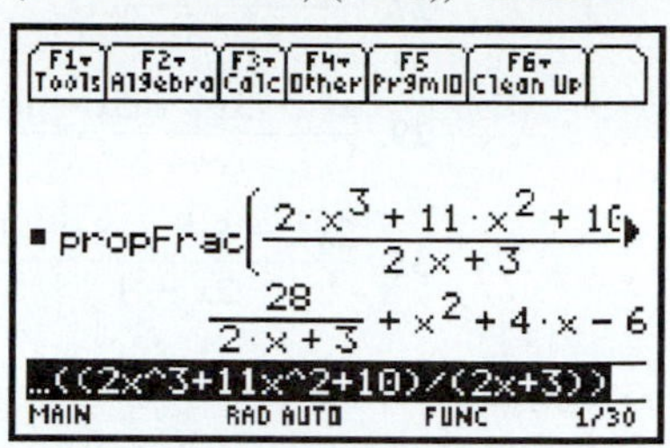

EXAMPLE 7

Divide: $\dfrac{10x^4 - 11x^3 - 15x^2 + 28x - 12}{2x^2 + x - 3}$.

$$
\begin{array}{r}
5x^2 - 8x + 4 \\
2x^2 + x - 3 \overline{)\,10x^4 - 11x^3 - 15x^2 + 28x - 12}\\
\underline{10x^4 + 5x^3 - 15x^2} \\
-16x^3 + 28x \\
\underline{-16x^3 - 8x^2 + 24x} \\
8x^2 + 4x - 12\\
\underline{8x^2 + 4x - 12}
\end{array}
$$

Thus,

$$\frac{10x^4 - 11x^3 - 15x^2 + 28x - 12}{2x^2 + x - 3} = 5x^2 - 8x + 4$$

Exercises 1.9

Find each quotient.

1. $\dfrac{24x^2}{4x}$
2. $\dfrac{-18x^2y^3}{3xy}$
3. $\dfrac{36a^3b^5}{-9a^2b^2}$
4. $\dfrac{-48a^5b^3}{-12a^2b}$
5. $\dfrac{45x^6y}{72x^3y^2}$
6. $\dfrac{32a^5b^2}{28a^8b^4}$
7. $\dfrac{-25x^3}{20x^5y^2}$
8. $\dfrac{40a^2b^4}{-15b^5}$
9. $\dfrac{3a(2b^2)^2}{(12ab)^2}$
10. $\dfrac{(3x^2y)(4x^2)^2}{xy(2x^3)^3}$
11. $\dfrac{(4st^2)^2(3t^2)^3}{t^2(9st)^2}$
12. $\dfrac{(3a^2)^2(ab^2)^2(2b)}{4a(3ab)^2(b^2)^3}$
13. $\dfrac{24x^2 - 16x + 8}{8}$
14. $\dfrac{36x^2 + 18x + 12}{6}$
15. $\dfrac{15x^5 - 20x^4 + 10x^2}{5x}$
16. $\dfrac{27x^3 - 33x^2 - 21x}{3x}$
17. $\dfrac{-28x^5 + 35x^4 - 49x^3}{7x^3}$
18. $\dfrac{20x^4 - 25x^3 + 10x^2 - 15x}{5x^2}$
19. $\dfrac{-64x^7 + 48x^5 + 36x^3 + 24x}{8x^4}$
20. $\dfrac{27x^8 - 18x^5 + 36x^4 - 12x}{9x^3}$
21. $\dfrac{4a^2b - 6a^2b^2 + 8ab}{2ab}$
22. $\dfrac{8a^2b^3 + 12ab^4 - 16a^3b^2}{8a^2b}$
23. $\dfrac{9m^2n^2 + 12mn^3 - 15m^3}{-3mn}$
24. $\dfrac{25p^2q^3 - 40p^4 + 20pq^2}{-10p^2q^2}$
25. $\dfrac{224x^4y^2z^5 - 168x^3y^3z^4 - 112xy^4z^2}{28xy^2z^2}$
26. $\dfrac{175m^2n^3p + 125mnp^4 - 225n^2p^3}{-75mn^2p^3}$

27. $\dfrac{2x^2 + x - 15}{x + 3}$

28. $\dfrac{x^2 - 2x - 3}{x - 3}$

29. $\dfrac{x^3 - 3x^2 + 5x - 6}{x - 2}$

30. $\dfrac{2x^3 + 5x^2 + 14}{x + 3}$

31. $\dfrac{2x^3 - 5x^2 + 8x + 1}{2x - 1}$

32. $\dfrac{x^3 - x^2 - 2x + 8}{x + 2}$

33. $\dfrac{6x^3 + 31x^2 + 26x - 15}{3x + 5}$

34. $\dfrac{20x^3 - 54x^2 + 44x - 12}{4x - 6}$

35. $\dfrac{-16x^3 + 37x - 24}{4x - 3}$

36. $\dfrac{25x^3 - 56x + 52}{5x - 4}$

37. $\dfrac{9x^4 - 3x^3 - 2x^2 + 12x + 4}{3x + 1}$

38. $\dfrac{8x^4 - 12x^3 + 10x^2 - 19x}{2x - 3}$

39. $\dfrac{2x^3 + x^4 - 10 + 11x - 6x^2}{x^2 - x + 2}$

40. $\dfrac{x^6 + x^4 - x^3 + x^2 + 1}{x^2 + x + 1}$

41. $\dfrac{x^3 - 64}{x + 4}$

42. $\dfrac{x^4 + 4}{x^2 - 2x + 2}$

43. $\dfrac{8x^3 + 1}{2x + 1}$

44. $\dfrac{16a^4 + 1}{2a - 1}$

1.10 LINEAR EQUATIONS

An **equation** is a statement that two algebraic expressions are equal. In particular, a **linear equation** in one variable is an equation in which each term containing that variable is of the first degree. To **solve** an equation means to find what number or numbers can replace the variable to make the equation a true statement.

EXAMPLE 1

Solve $2x + 4 = 6x - 8$.

If we let $x = 2$, then

$$2(2) + 4 = 6(2) - 8$$

which is a *false* statement. Hence, 2 *is not* a solution. But if we let $x = 3$, then

$$2(3) + 4 = 6(3) - 8$$

which is a *true* statement. Hence, 3 *is* a solution (sometimes called a **root**).

FOUR BASIC PROPERTIES USED TO SOLVE EQUATIONS

1. If the same quantity is added to each side of an equation, the resulting equation is equivalent to the original equation.*
2. If the same quantity is subtracted from each side of an equation, the resulting equation is equivalent to the original equation.
3. If each side of an equation is multiplied by the same (nonzero) quantity, the resulting equation is equivalent to the original equation.
4. If each side of an equation is divided by the same (nonzero) quantity, the resulting equation is equivalent to the original equation.

*Two equations are equivalent when they have the same solutions.

EXAMPLE 2

Solve $x - 6 = 11$.

$$
\begin{aligned}
x - 6 &= 11 \\
x - 6 + 6 &= 11 + 6 \qquad \text{(Property 1)} \\
x &= 17
\end{aligned}
$$

EXAMPLE 3

Solve $x + 4 = 1$.

$$
\begin{aligned}
x + 4 &= 1 \\
x + 4 - 4 &= 1 - 4 \qquad \text{(Property 2)} \\
x &= -3
\end{aligned}
$$

EXAMPLE 4

Solve $\frac{x}{4} = 9$.

$$
\begin{aligned}
\frac{x}{4} &= 9 \\
\left(\frac{x}{4}\right)4 &= (9)4 \qquad \text{(Property 3)} \\
x &= 36
\end{aligned}
$$

EXAMPLE 5

Solve $3x = 15$.

$$
\begin{aligned}
3x &= 15 \\
\frac{3x}{3} &= \frac{15}{3} \qquad \text{(Property 4)} \\
x &= 5
\end{aligned}
$$

SOLVING A FIRST-DEGREE EQUATION IN ONE UNKNOWN OR VARIABLE

1. Eliminate any fractions by multiplying each side by the lowest common denominator of all fractions in the equation.
2. Remove any grouping symbols.
3. Combine like terms on each side of the equation.
4. Isolate all the unknown terms on one side of the equation and all other terms on the other side of the equation.
5. Again combine like terms where possible.
6. Divide each side by the coefficient of the unknown.
7. Check your solution by substituting it in the original equation.

EXAMPLE 6

Solve $3x + 4 = 28$.

$$3x + 4 = 28$$
$$3x = 24 \quad \text{(Subtract 4 from each side.)}$$
$$x = 8 \quad \text{(Divide each side by 3.)}$$

Check:

$$3(8) + 4 = 28 \quad \text{(Substitute } x = 8.\text{)}$$
$$28 = 28$$

EXAMPLE 7

Solve $\frac{1}{3}x = \frac{1}{4}x - 2$.

$$\frac{1}{3}x = \frac{1}{4}x - 2 \quad \text{[Multiply each side by the lowest common denominator (L.C.D.), 12.]}$$
$$4x = 3x - 24$$
$$x = -24 \quad \text{(Subtract } 3x \text{ from each side.)}$$

Check:

$$\frac{1}{3}(-24) = \frac{1}{4}(-24) - 2 \quad \text{(Substitute } x = -24.\text{)}$$
$$-8 = -8$$

EXAMPLE 8

Solve $4(x + 1) = 6 - (3x - 12)$.

$$4(x + 1) = 6 - (3x - 12)$$
$$4x + 4 = 6 - 3x + 12 \quad \text{(Remove parentheses.)}$$
$$4x + 4 = -3x + 18 \quad \text{(Combine like terms.)}$$
$$7x + 4 = 18 \quad \text{(Add } 3x \text{ to each side.)}$$
$$7x = 14 \quad \text{(Subtract 4 from each side.)}$$
$$x = 2 \quad \text{(Divide each side by 7.)}$$

Check:

$$4(2 + 1) = 6 - (3 \cdot 2 - 12) \quad \text{(Substitute } x = 2.\text{)}$$
$$12 = 12$$

EXAMPLE 9

Solve the following:

$$\frac{2(6y + 1)}{5} - 3(y + 1) = \frac{3 - y}{4} - 3$$
$$8(6y + 1) - 60(y + 1) = 5(3 - y) - 60 \quad \text{(Multiply each side by the L.C.D., 20.)}$$
$$48y + 8 - 60y - 60 = 15 - 5y - 60 \quad \text{(Remove parentheses.)}$$
$$-12y - 52 = -5y - 45 \quad \text{(Combine like terms.)}$$
$$-7y - 52 = -45 \quad \text{(Add } 5y \text{ to each side.)}$$
$$-7y = 7 \quad \text{(Add 52 to each side.)}$$
$$y = -1 \quad \text{(Divide each side by } -7.\text{)}$$

Check:

$$\frac{2[6(-1) + 1]}{5} - 3(-1 + 1) = \frac{3 - (-1)}{4} - 3 \quad \text{(Substitute } y = -1.)$$

$$-2 - 0 = 1 - 3$$

$$-2 = -2$$

EXAMPLE 10

Solve $8.24(3.6x + 18.6) = 246.7$.

$$8.24(3.6x + 18.6) = 246.7$$

$$(8.24)(3.6x) + (8.24)(18.6) = 246.7 \quad \text{(Remove parentheses.)}$$

$$(8.24)(3.6x) = 246.7 - (8.24)(18.6) \quad \text{[Subtract (8.24)(18.6) from each side.]}$$

$$x = \frac{246.7 - (8.24)(18.6)}{(8.24)(3.6)} \quad \text{[Divide each side by (8.24)(3.6).]}$$

$$x = 3.15 \quad \text{(rounded to three significant digits)}$$

Note: When you check a root that has been rounded, there will often be an acceptable small difference on each side of the equation.

Exercises 1.10

Solve each equation.

1. $x + 9 = 7$
2. $x - 4 = 12$
3. $x - 6 = 10$
4. $x + 3 = -4$
5. $5x = 35$
6. $\frac{x}{4} = 5$
7. $\frac{x}{8} = -9$
8. $6x = 72$
9. $4x - 6 = 14$
10. $7x + 8 = 36$
11. $5y + 6 = 16$
12. $2x - 4 = 10$
13. $8 - 3x = 14$
14. $7 - 2y = 3$
15. $3x - 2 = 5x + 8$
16. $4x + 9 = 7x - 15$
17. $8 - 6x = 5x + 25$
18. $1 - 7x = 16 - 2x$
19. $3(x - 5) = 27$
20. $-4(3x + 2) = 14$
21. $4(y + 2) = 30 - (y - 3)$
22. $2(x - 3) = 4 + (x - 14)$
23. $6(3x + 1) + (2x + 2) = 5x$
24. $4(2y - 3) - (3y + 7) = 6$
25. $16(x + 3) = 7(x - 5) - 9(x + 4) - 7$
26. $5 - 2(x - 5) = -3(x + 4)$
27. $4(2x - 1) - 2x = 3 - 5(2x - 5)$
28. $6x - 2(4x - 2) - (x - 4) = x + 4$
29. $1.5x - 3.8 + 0.4(2x - 5) = 8$
30. $3.6(2x + 3) - 2.4(5x - 2) = 0.6(4.5 - 2x) - 1.4(x - 10)$
31. $-[x - (4x + 5)] = 2 + 3(2x + 3)$
32. $10y - (4 - y) = 2[-3 - (2 + y) - 12]$
33. $\frac{2}{3}y - 4 = \frac{3}{4}y - 5$
34. $\frac{3}{5}x + 2 = \frac{1}{3}x + 10$
35. $2 - \frac{3}{7}x = \frac{x}{4} - \frac{15}{2}$
36. $\frac{1}{2}y + \frac{1}{6} = \frac{2}{3}y + \frac{1}{9}$
37. $3\left(\frac{2}{5}x - 7\right) = 4\left(\frac{1}{3}x - 5\right)$
38. $-2\left(\frac{1}{2}x + 6\right) = 7\left(5 - \frac{1}{2}x\right) + 8$
39. $\frac{3(y + 2)}{4} - \frac{y - 3}{2} = \frac{3}{4}y$
40. $\frac{5(x - 4)}{6} + \frac{x - 2}{12} = \frac{2x}{3} + 1$

41. $\dfrac{4(x-3)}{9} - \dfrac{2x}{3} = 1$

42. $\dfrac{2}{3}\left(x - \dfrac{1}{2}\right) - \left(x + \dfrac{1}{4}\right) = \dfrac{3}{2}x$

43. $\dfrac{3}{5}\left(\dfrac{2}{3}y + \dfrac{1}{5}\right) = \dfrac{2}{3}\left(\dfrac{3}{4}y - \dfrac{1}{3}\right)$

44. $\dfrac{1}{4}\left(\dfrac{4}{5}x - 3\right) = -\dfrac{1}{2}\left(\dfrac{2}{5}x + \dfrac{3}{2}\right)$

Solve each equation. Round your calculator results to three significant digits.

45. $18.7x + 253 = 28.6x$

46. $6.19x - 39.4 = 82.4$

47. $4.12x + 6.18 = 12.6x - 3.6$

48. $0.96 + 4.08x = 6.21 - 1.33x$

49. $6.3 - 0.4(9.36 - 2x) = 1.2x$

50. $4.5x + 3.2 = 0.6(5x + 14.5)$

51. $2.76(1.81x + 59.2) + 16.7 = 763$

52. $29.2(2.4x - 6.5) = 10(3.4 - x)$

53. $2\pi(2.50 \times 10^6)L = 2590$

54. $\dfrac{V_1}{2.62 \times 10^4} = \dfrac{5.63 \times 10^5}{4.16 \times 10^7}$

55. $2.50 \times 10^{-3} = \dfrac{(9.00 \times 10^9)(8.50 \times 10^{-7})q}{(0.15)^2}$

56. $1.59 \times 10^{-3} = \dfrac{1}{2\pi f(1.55 \times 10^{-6})}$

1.11 FORMULAS

A **formula** is an equation, usually expressed in letters, that shows the relationship between quantities.

EXAMPLE 1

The formula $P = VI$ states that the electric power P equals the product of the voltage drop V and the current I.

EXAMPLE 2

The formula $p = \dfrac{F}{A}$ states that the pressure p equals the quotient of the force F and the area, A.

EXAMPLE 3

The formula $F = \dfrac{mv^2}{r}$ states that the centripetal force F of a rotating body equals the product of the mass m and the square of the velocity v divided by r, the radius of the circular path.

SOLVING A FORMULA

To **solve a formula for a given letter** means to isolate the given letter on one side of the equation and express it in terms of all the remaining letters by having all the other letters and numbers appear on the opposite side of the equation. We solve a formula using the same principles that we use in solving any equation.

EXAMPLE 4

Solve $P = VI$ for V.

$$P = VI$$

$$\frac{P}{I} = V \qquad \text{(Divide each side by } I.\text{)}$$

Note that $\frac{P}{I} = V$ and $V = \frac{P}{I}$ are equivalent equations.

EXAMPLE 5

Solve $p = \frac{F}{A}$ for F and then for A.

First solve for F.

$$p = \frac{F}{A}$$

$$pA = F \qquad \text{(Multiply each side by } A.\text{)}$$

Now solve for A.

$$p = \frac{F}{A}$$

$$pA = F \qquad \text{(Multiply each side by } A.\text{)}$$

$$A = \frac{F}{p} \qquad \text{(Divide each side by } p.\text{)}$$

EXAMPLE 6

Solve $F = \frac{mv^2}{r}$ for v^2.

$$F = \frac{mv^2}{r}$$

$$Fr = mv^2 \qquad \text{(Multiply each side by } r.\text{)}$$

$$\frac{Fr}{m} = v^2 \qquad \text{(Divide each side by } m.\text{)}$$

The TI-89's home screen provides an excellent work area to solve a formula for a given letter. Enter the original formula, then just describe the operations to be performed on both sides of the equation.

Type the formula and press **ENTER.**

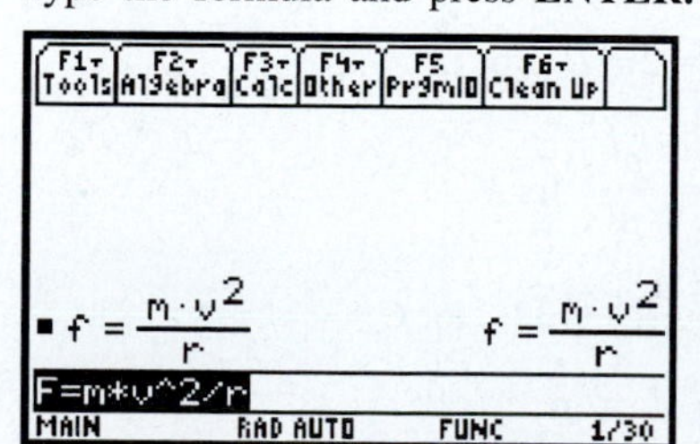

The uppercase **F** is interpreted as **f.** Remember to interrupt variable names with a multiplication sign.

multiplication sign **alpha R ENTER**

Multiply both sides by **r.** Note **ans(1)** appears automatically when you start a line with multiplication.

division sign **alpha M ENTER**

Divide both sides by **m.** Again, **ans(1)** appears automatically and refers to the preceding answer.

EXAMPLE 7

Solve $V = E - Ir$ for r.

One way:

$$V = E - Ir$$
$$V - E = -Ir \qquad \text{(Subtract } E \text{ from each side.)}$$
$$\frac{V - E}{-I} = r \qquad \text{(Divide each side by } -I.\text{)}$$

Alternate way:

$$V = E - Ir$$
$$V + Ir = E \qquad \text{(Add } Ir \text{ to each side.)}$$
$$Ir = E - V \qquad \text{(Subtract } V \text{ from each side.)}$$
$$r = \frac{E - V}{I} \qquad \text{(Divide each side by } I.\text{)}$$

Note that the two results are equivalent. Take the first result, $\frac{V - E}{-I}$, and multiply numerator and denominator by -1.

$$\left(\frac{V - E}{-I}\right)\left(\frac{-1}{-1}\right) = \frac{-V + E}{I} = \frac{E - V}{I}$$

Exercises 1.11

Solve each formula for the given letter. These formulas are in common use in various technical and scientific areas.

1. $W = JQ$ for J

2. $v = f\lambda$ for λ

3. $R_T = R_1 + R_2 + R_3$ for R_2

4. $V = IZ$ for I

5. $E = IR$ for R

6. $C_T = C_1 + C_2 + C_3 + C_4$ for C_3

7. $C = \frac{Q}{V}$ for Q

8. $I = \frac{Q}{t}$ for t

9. $V = \frac{W}{Q}$ for Q

10. $E = mc^2$ for m

11. $Q = \frac{I^2Rt}{J}$ for R

12. $R = \rho\frac{L}{A}$ for A

13. $P = \frac{(\text{O.D.})}{N + 2}$ for N

14. $\text{hp} = \frac{D^2N}{2.5}$ for N

15. $R = \frac{kL}{D^2}$ for L

16. $F = \frac{wa}{g}$ for g

17. $R = \frac{\pi}{2P}$ for P

18. $X_L = 2\pi fL$ for f

19. $\frac{V}{V'} = \frac{T}{T'}$ for T

20. $\frac{V}{V'} = \frac{p'}{p}$ for V'

21. $C = \frac{5}{9}(F - 32)$ for F

22. $F = \frac{9}{5}C + 32$ for C

23. $\frac{I_s}{I_p} = \frac{N_p}{N_s}$ for N_s

24. $\frac{V_1P_1}{T_1} = \frac{V_2P_2}{T_2}$ for T_1

25. $\frac{\Delta d}{d} = 1.22\frac{\lambda}{a}$ for a

26. $\frac{I_1}{I_2} = \frac{r_2^2}{r_1^2}$ for I_2

27. $l = n\frac{\lambda}{2}$ for λ

28. $l = (2n + 1)\frac{\lambda}{4}$ for λ

29. $\Delta L = \alpha L(T - T_0)$ for T

30. $A = ab + \frac{d}{2}(a + c)$ for d

31. $f' = f\left(\frac{v + v_0}{v - v_s}\right)$ for v

32. $f' = f\left(\frac{v + v_0}{v - v_s}\right)$ for v_s

33. $F = \frac{q_1 q_2}{4\pi\epsilon_0 r^2}$ for q_1

34. $\alpha = \frac{\Delta\rho}{\rho\Delta T}$ for ΔT

35. $E - IR - \frac{q}{C} = 0$ for R

36. $X_C = \frac{1}{2\pi fC}$ for f

37. $\frac{1}{R_T} = \frac{1}{R_1} + \frac{1}{R_2}$ for R_2

38. $\frac{1}{C_T} = \frac{1}{C_1} + \frac{1}{C_2}$ for C_T

39. $\frac{1}{R_T} = \frac{1}{R_1} + \frac{1}{R_2} + \frac{1}{R_3}$ for R_T

40. $\frac{1}{C_T} = \frac{1}{C_1} + \frac{1}{C_2} + \frac{1}{C_3}$ for C_2

41. $\frac{1}{f} = \frac{1}{s_0} + \frac{1}{s_i}$ for s_0

42. $\frac{1}{f} = \frac{1}{s_0} + \frac{1}{s_i}$ for f

43. $\frac{1}{f} = (n - 1)\left(\frac{1}{R'} - \frac{1}{R''}\right)$ for n

44. $\frac{1}{f} = (n - 1)\left(\frac{1}{R'} - \frac{1}{R''}\right)$ for R''

45. $eV = hf - \phi$ for f

46. $hf = -13.6\left(\frac{1}{m^2} - \frac{1}{n^2}\right)$ for m

47. $I_2 = \frac{Z_3}{Z_2 + Z_3}I_T$ for Z_2

48. $I_3 = \frac{Z_2}{Z_2 + Z_3}I_T$ for Z_2

49. $R_A = \frac{R_1 R_3}{R_1 + R_2 + R_3}$ for R_1

50. $R_B = \frac{R_1 R_2}{R_1 + R_2 + R_3}$ for R_3

1.12 SUBSTITUTION OF DATA INTO FORMULAS

Problem solving is basic to all technical fields. An important part of problem solving involves analyzing the given data, finding a formula that relates the given quantities with the unknown quantity, substituting the given data into this formula, and finding the unknown quantity. Working with formulas in this way is one of the most important skills that you will learn.

USING A FORMULA TO SOLVE A PROBLEM WHERE ALL BUT THE UNKNOWN QUANTITY IS GIVEN

1. Solve the formula for the unknown quantity.
2. Substitute each known quantity with its units.
3. Use the order-of-operations procedures to find the numerical quantity and to simplify the units.

This is the most useful method when you use a calculator, which we assume you will. We shall use this method most often in this text.

EXAMPLE 1

Given the formula $A = bh$, $A = 150\text{ m}^2$, and $b = 25$ m. Find h.

First solve for h.

$$A = bh$$

$$\frac{A}{b} = \frac{bh}{b} \qquad \text{(Divide each side by } b.\text{)}$$

$$\frac{A}{b} = h$$

Then substitute the data.

$$h = \frac{A}{b}$$

$$h = \frac{150\text{ m}^2}{25\text{ m}}$$

$$= 6.0\text{ m}$$

Note: Follow the rules for calculations with measurements in this section.

EXAMPLE 2

Given the formula $P = 2l + 2w$, $P = 712$ cm, and $w = 128$ cm. Find l.

First, solve for l.

$$P = 2l + 2w$$

$$P - 2w = 2l + 2w - 2w \qquad \text{(Subtract } 2w \text{ from each side.)}$$

$$P - 2w = 2l$$

$$\frac{P - 2w}{2} = \frac{2l}{2} \qquad \text{(Divide each side by 2.)}$$

$$\frac{P - 2w}{2} = l \qquad \left(\text{or } l = \frac{P}{2} - w\right)$$

Then, substitute the data.

$$l = \frac{P - 2w}{2}$$

$$l = \frac{712\text{ cm} - 2(128\text{ cm})}{2}$$

$$= \frac{712\text{ cm} - 256\text{ cm}}{2}$$

$$= \frac{456\text{ cm}}{2} = 228\text{ cm}$$

EXAMPLE 3

Given the formula $V = \frac{1}{2}lw(D + d)$, $V = 54.85\text{ in}^3$, $l = 4.50$ in., $w = 3.65$ in., and $d = 3.25$ in. Find D.

First, solve for D.

$$V = \frac{1}{2}lw(D + d)$$

$$2V = lw(D + d) \qquad \text{(Multiply each side by 2.)}$$

$$2V = lwD + lwd \qquad \text{(Remove parentheses.)}$$

$$2V - lwd = lwD \qquad \text{(Subtract } lwd \text{ from each side.)}$$

$$\frac{2V - lwd}{lw} = D \qquad \text{(Divide each side by } lw.\text{)}$$

$$D = \frac{2V}{lw} - d$$

Then, substitute the data.

$$D = \frac{2(54.85 \text{ in}^3)}{(4.50 \text{ in.})(3.65 \text{ in.})} - 3.25 \text{ in.}$$
$$= 6.68 \text{ in.} - 3.25 \text{ in.}$$
$$= 3.43 \text{ in.}$$

EXAMPLE 4

Given $R = \frac{\rho L}{A}$, $R = 3.25\ \Omega$, $\rho = 1.72 \times 10^{-6}\ \Omega$ cm, and $A = 0.0250$ cm^2. Find L.

$$R = \frac{\rho L}{A}$$
$$RA = \rho L \qquad \text{(Multiply each side by } A.)$$
$$L = \frac{RA}{\rho} \qquad \text{(Divide each side by } \rho.)$$
$$L = \frac{(3.25\ \Omega)(0.0250 \text{ cm}^2)}{1.72 \times 10^{-6}\ \Omega \text{ cm}} \qquad \text{(Substitute the data.)}$$
$$= 47{,}200 \text{ cm} \quad \text{or} \quad 472 \text{ m}$$

EXAMPLE 5

If a satellite orbits the earth in a circular path $50\bar{0}$ mi above the earth's surface, find the acceleration due to the pull of gravity at that height.

The formula to use is $g = \frac{GM}{s^2}$, where g is the acceleration due to the force or pull of gravity; G is a constant, $6.67 \times 10^{-11} \frac{\text{m}^3}{\text{kg s}^2}$; M is the mass of the earth, 5.96×10^{24} kg; and s is the distance from the center of the earth to the satellite.

The average radius of the earth is 3960 mi; therefore, the average radius of the orbit is 3960 mi + $50\bar{0}$ mi = 4460 mi or 7.18×10^6 m.

Substituting, we have

$$g = \frac{GM}{s^2}$$
$$g = \frac{\left(6.67 \times 10^{-11} \frac{\text{m}^3}{\text{kg s}^2}\right)(5.96 \times 10^{24} \text{ kg})}{(7.18 \times 10^6 \text{ m})^2}$$
$$= 7.71 \text{ m/s}^2 \text{ or } 25.3 \text{ ft/s}^2$$

(For comparison purposes, the value of g on the earth's surface is 9.80 m/s^2, or 32.2 ft/s^2.)

Formulas involving reciprocals are often used in electronics and physics. We next consider an alternate method for substituting data into such formulas and solving for a specified letter using a calculator. First, solve for the reciprocal of the specified variable. Then substitute the given data and follow the calculator screens as shown in the following example.

EXAMPLE 6

Given the formula $\frac{1}{R} = \frac{1}{R_1} + \frac{1}{R_2} + \frac{1}{R_3}$, $R = 1.5\ \Omega$, $R_1 = 3.0\ \Omega$, $R_3 = 12\ \Omega$. Find R_2.

First, solve the formula for the reciprocal of R_2.

$$\frac{1}{R} = \frac{1}{R_1} + \frac{1}{R_2} + \frac{1}{R_3}$$
$$\frac{1}{R_2} = \frac{1}{R} - \frac{1}{R_1} - \frac{1}{R_3}$$

Then, substitute the data.

$$\frac{1}{R_2} = \frac{1}{1.5\ \Omega} - \frac{1}{3.0\ \Omega} - \frac{1}{12\ \Omega}$$

Next, use your calculator as follows:

1/1.5-1/3.0-1/12 **ENTER** 1/ **2nd ANS ENTER**

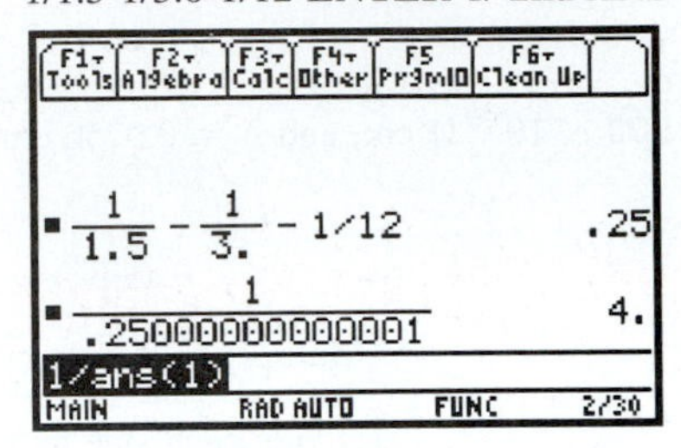

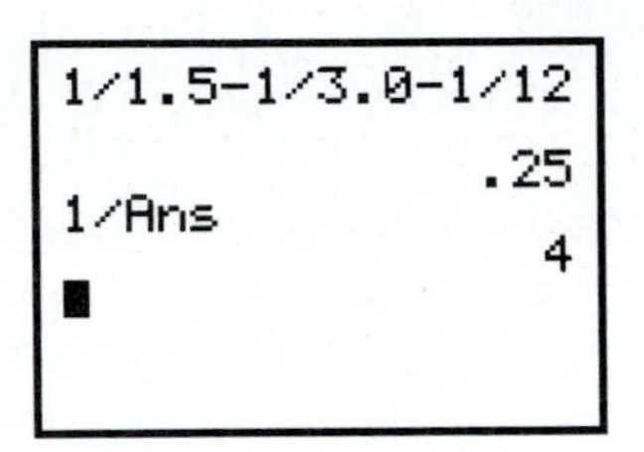

1/1.5-1/3.0-1/12 **ENTER** 1/ **2nd ANS ENTER**

Thus, $R_2 = 4.0\ \Omega$.

Exercises 1.12

Solve for the given letter. Then substitute the data to find the value of the given letter. Follow the rules for calculating with measurements.

Formula	*Data*	*Find*
1. $A = bh$	$A = 24.0\ \text{in}^2$, $h = 6.00$ in.	b
2. $C = \pi d$	$C = 358$ cm	d
3. $A = \frac{1}{2}bh$	$A = 144\ \text{m}^2$, $b = 8.00$ m	h
4. $A = 2\pi rh$	$A = 1450\ \text{ft}^2$, $r = 16.0$ ft	h
5. $V = \pi r^2 h$	$V = 1950\ \text{m}^3$, $r = 12.6$ m	h
6. $P = 2(a + b)$	$P = 248$ in., $b = 45$ in.	a
7. $V = lwh$	$V = 2.50 \times 10^4\ \text{ft}^3$, $l = 44.5$ ft, $h = 19.7$ ft	w
8. $A = \left(\frac{a + b}{2}\right)h$	$A = 205.2\ \text{m}^2$, $a = 16.50$ m, $b = 19.50$ m	h
9. $C = \frac{5}{9}(F - 32°)$	$C = 55°$	F
10. $(\Delta L) = \alpha L(T - T_0)$	$\Delta L = 0.025$ m, $\alpha = 1.3 \times 10^{-5}/°\text{C}$ $L = 25.000$ m, $T_0 = 25°\text{C}$	T

11. Given $Q = mc\Delta T$, where $m = 50\bar{0}$ g, $c = 0.214$ cal/g °C, and $\Delta T = 40.0°\text{C}$. Find Q.

12. Given $E = \dfrac{wv^2}{2g}$, where $w = 2.8 \times 10^4$ N, $v = 2.5 \times 10^4$ m/s, and $g = 9.80\ \text{m/s}^2$. Find E.

13. Given $X_L = 2\pi fL$, where $X_L = 75.0\ \Omega$ and $f = 60.0$ Hz. Find L.

14. Given $I = \dfrac{E}{R_L + r}$, where $I = 2.00$ A, $R_L = 28.0\ \Omega$, and $r = 0.500\ \Omega$. Find E.

15. Given $P\ (\text{in hp}) = \dfrac{(\text{force in lb})(\text{distance in ft})}{\left(550\dfrac{\text{ft-lb}}{\text{s}}\right)(\text{time in s})}$ hp. What horsepower engine is required to hoist $8\bar{0}$ tons of coal per hour from a mine shaft $2\bar{0}0$ ft deep?

16. Given $\dfrac{V_P}{V_s} = \dfrac{N_P}{N_s}$, where $V_P = 12\bar{0}$ V, $V_s = 3{,}5\bar{0}0$ V, and $N_s = 12{,}\bar{0}00$ turns. Find N_p.

17. Given $\dfrac{P_1}{P_2} = \dfrac{V_2}{V_1}$, where $P_1 = 270$ kPa, $V_1 = 750$ cm^3, and $P_2 = 210$ kPa. Find V_2.

18. Given $X_C = \dfrac{1}{2\pi fC}$, where $X_C = 10\bar{0}0\ \Omega$ and $C = 2.65 \times 10^{-6}$ F. Find f.

19. Given $\dfrac{\Delta d}{d} = 1.22\dfrac{\lambda}{a}$, where $\Delta d = 1.75$ mm, $\lambda = 5.50 \times 10^{-7}$ m, and $a = 2.00 \times 10^{-3}$ m. Find d.

20. Given $F = k\dfrac{q_1 q_2}{r^2}$, where $F = -2.5 \times 10^{-2}$ N, $k = 9.00 \times 10^9$ N m^2/C^2, $q_1 = 8.0 \times 10^{-7}$ C, and $r = 0.15$ m. Find q_2.

21. Given $\dfrac{V_1 P_1}{T_1} = \dfrac{V_2 P_2}{T_2}$, where $V_1 = 125$ m^3, $P_1 = 13.5$ MPa, $T_1 = 295$ K, $P_2 = 18.4$ MPa, and $T_2 = 305$ K. Find V_2.

22. Given $\dfrac{F_1}{F_2} = \dfrac{r_1^2}{r_2^2}$, where $F_2 = 60\bar{0}0$ N, $r_2 = 12.0$ cm, and $r_1 = 2.50$ cm. Find F_1.

23. Given $f = \dfrac{nv}{2l}$, where $f = 384$ Hz $= 384$/s, $n = 1$, and $v = 348$ m/s. Find l (in cm).

24. Given $f = \dfrac{(2n + 1)v}{4l}$, where $f = 384$ Hz $= 384$/s, $n = 2$, and $l = 112$ cm. Find v (in m/s).

25. Given $f' = f\left(\dfrac{v - v_0}{v + v_s}\right)$, where $f = 425$ Hz $= 425$/s, $v = 343$ m/s, $v_0 = 0$, and $v_s = 25.0$ m/s. Find f'.

26. Given $f' = f\left(\dfrac{v + v_0}{v + v_s}\right)$, where $f' = 438$ Hz $= 438$/s, $v = 343$ m/s, $v_0 = 35.0$ m/s, and $v_s = 15.0$ m/s. Find f.

27. Given $\dfrac{1}{R} = \dfrac{1}{R_1} + \dfrac{1}{R_2}$, where $R = 60.0\ \Omega$ and $R_2 = 24\bar{0}\ \Omega$. Find R_1.

28. Given $\dfrac{1}{Z} = \dfrac{1}{Z_1} + \dfrac{1}{Z_2} + \dfrac{1}{Z_3}$, where $Z_1 = 25.0\ \Omega$, $Z_2 = 15.0\ \Omega$, and $Z_3 = 50.0\ \Omega$. Find Z.

29. Given $\dfrac{1}{R} = \dfrac{1}{R_1} + \dfrac{1}{R_2} + \dfrac{1}{R_3}$, where $R_1 = 3\bar{0}\ \Omega$, $R_2 = 4\bar{0}\ \Omega$, and $R_3 = 5\bar{0}\ \Omega$. Find R.

30. Given $\dfrac{1}{R} = \dfrac{1}{R_1} + \dfrac{1}{R_2} + \dfrac{1}{R_3}$, where $R = 15\ \Omega$, $R_1 = 3\bar{0}\ \Omega$, and $R_3 = 6\bar{0}\ \Omega$. Find R_2.

31. Given $\dfrac{1}{f} = \dfrac{1}{s_o} + \dfrac{1}{s_i}$, where $f = 15$ cm and $s_o = 2\bar{0}$ cm. Find s_i.

32. Given $\dfrac{1}{f} = (n - 1)\left(\dfrac{1}{R'} - \dfrac{1}{R''}\right)$, where $R' = 2\bar{0}$ cm, $R'' = 4\bar{0}$ cm, and $n = 1.50$. Find f.

33. Given $s = v_i t + \dfrac{1}{2}at^2$, where $s = 40.0$ m, $t = 3.70$ s, and $a = -6.00$ m/s^2. Find v_i.

34. Given $v_f^2 = v_i^2 + 2as$, where $v_f = 0$, $v_i = 25.0$ m/s, and $a = -7.50$ m/s^2. Find s.

35. Given $I_1 = \dfrac{R_2}{R_1 + R_2} I_2$, where $R_1 = 5.0\ \Omega$, $R_2 = 25.0\ \Omega$, and $I_1 = 0.125$ mA. Find I_2.

36. Given $R_A = \dfrac{R_1 R_3}{R_1 + R_2 + R_3}$, where $R_A = 2.00\ \Omega$, $R_1 = 4.00\ \Omega$, and $R_2 = 2.00\ \Omega$. Find R_3.

1.13 APPLICATIONS INVOLVING LINEAR EQUATIONS

Many technical problems can be expressed mathematically as linear equations. Solving such a problem then becomes a problem of solving a linear equation. What follows is a suggested outline for the solution to this type of problem. The most difficult task is that of "translating" the stated problem into mathematical terms. Because of the numerous ways of describing any given problem, there is no simple process that can be given for making the mathematical translation. The ability to make an accurate translation will come from experience. We cannot overemphasize the need to work out numerous problems to gain this experience.

SOLVING APPLICATION PROBLEMS

Step 1: Read the problem carefully at least twice.

(a) The first time you should read straight through from beginning to end. Do not stop this time to think about setting up an equation. You are only after a general impression at this point.

(b) Now read through *completely* for the second time. This time begin to think ahead to the next steps.

Step 2: If possible, draw a picture or diagram. This will often help you to visualize the possible mathematical relationships needed in order to write the equation.

Step 3: Choose a symbol to represent the unknown quantity (the quantity not given as a specific value) in the problem. Be sure to label this symbol, the unknown variable, to indicate what it represents.

Step 4: Write an equation that expresses the information given in the problem. To obtain this equation, look for information that will express two quantities that can be set equal to each other. Do not rush through this step. Try expressing your equation in words to see if it corresponds to the original problem.

Step 5: Solve for the unknown variable.

Step 6: Check your solution in the equation from Step 4. If it does not check, you have made a mathematical error. You will need to repeat Step 5. Also check your solution in the original verbal problem for reasonableness.

Note: Do not use rules for calculating with measurements in this section.

EXAMPLE 1

A piece of lumber 7 ft 7 in. long is to be sawed into nine equal pieces. If the loss per cut is $\frac{1}{8}$ in., how long will each piece be?

After reading the problem, draw a diagram (see Fig. 1.4).

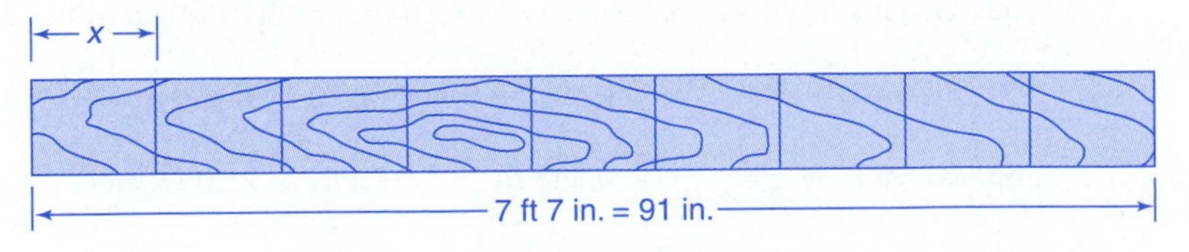

Figure 1.4

Let x = the length of each sawed piece

$\frac{1}{8}$ = the length of each cut

Then,

$9x$ = the total length of the 9 pieces

$8\left(\frac{1}{8}\right) = 1 =$ the total length of the 8 cuts (*Note:* 8 cuts will be needed.)

Since the total of the lengths of the pieces and the cuts equals the total length of the original board, we have the equation

$$
\begin{aligned}
9x + 1 &= 91 \\
9x &= 90 && \text{(Subtract 1 from each side.)} \\
x &= 10 && \text{(Divide each side by 9.)}
\end{aligned}
$$

Check:

$$
\begin{aligned}
9 \text{ pieces } 10 \text{ in. each} &= 90 \text{ in.} \\
8 \text{ cuts } \tfrac{1}{8} \text{ in. each} &= \underline{\ 1 \text{ in.}\ } \\
\text{Total} &= 91 \text{ in.} = 7 \text{ ft } 7 \text{ in.}
\end{aligned}
$$

EXAMPLE 2

Bill wants to use 160 m of fence to enclose a rectangular portion of a lot. He wants the length to be 20 m longer than the width. Find the dimensions of the fenced portion.

Let x = width

$x + 20$ = length

160 = perimeter

Draw a diagram (see Fig. 1.5).

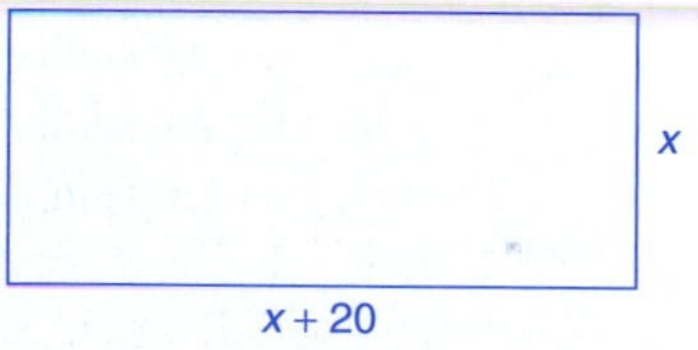

x

$x + 20$

Figure 1.5

The perimeter of a rectangle is given by the formula

$$
\begin{aligned}
2l + 2w &= P \\
2(x + 20) + 2x &= 160 && \text{(Substitute.)} \\
2x + 40 + 2x &= 160 && \text{(Remove parentheses.)} \\
4x + 40 &= 160 && \text{(Combine like terms.)} \\
4x &= 120 && \text{(Subtract 40 from each side.)} \\
x &= 30 && \text{(Divide each side by 4.)} \\
x + 20 &= 50
\end{aligned}
$$

Thus, the width is 30 m and the length is 50 m.

Check:

$$
\begin{aligned}
2l + 2w &= 2(50 \text{ m}) + 2(30 \text{ m}) \\
&= 100 \text{ m} + 60 \text{ m} \\
&= 160 \text{ m}
\end{aligned}
$$

which is the amount of fence used.

EXAMPLE 3

Helen bought 50 acres of land for $111,000. Part of the land cost $1500 per acre and the rest cost $3500 per acre. How much land was sold at each rate?

Let x = the amount of land sold at $1500 per acre

$50 - x$ = the amount of land sold at $3500 per acre

Therefore,

$$\begin{aligned} 1500x + 3500(50 - x) &= 111{,}000 \\ 1500x + 175{,}000 - 3500x &= 111{,}000 \\ -2000x + 175{,}000 &= 111{,}000 \\ -2000x &= -64{,}000 \\ x &= 32 \\ 50 - x &= 18 \end{aligned}$$

Thus,

32 acres sold at $1500 per acre

18 acres sold at $3500 per acre

Check:

32 acres at $1500 per acre = $ 48,000

18 acres at $3500 per acre = $ 63,000

Total = $111,000

EXAMPLE 4

The cruising air speed of a small plane is 200 km/h. The wind is blowing from the north at 60 km/h. How far north can the pilot expect to fly and still return within 3 h?

Let 200 km/h = speed of plane in still air

60 km/h = speed of wind

140 km/h = speed of plane traveling north

260 km/h = speed of plane traveling south

x = elapsed time traveling north in hours

$3 - x$ = elapsed time traveling south in hours

The basic formula is $d = rt$ (distance = rate × time). We can make a chart of the information.

	d	r	t
Traveling north	$140x$	140	x
Traveling south	$260(3 - x)$ ↑ $d = rt$	260	$3 - x$

Here the distances are the same; that is,

$$\begin{aligned} 140x &= 260(3 - x) \\ 140x &= 780 - 260x \\ 400x &= 780 \\ x &= 1.95 \text{ h} \end{aligned}$$

which is the time traveling north.

The distance traveled north is then

$$\begin{aligned} d &= 140x \\ &= (140 \text{ km/h})(1.95 \text{ h}) \\ &= 273 \text{ km} \end{aligned}$$

We leave the check to you.

EXAMPLE 5

A 12-qt cooling system is checked and found to be filled with a solution that is 40% antifreeze. The desired strength of the solution is 60% antifreeze. How many quarts of solution need to be drained and replaced with pure antifreeze to reach the desired strength?

The amount of 40% antifreeze solution that needs to be drained equals the amount of pure antifreeze that must be added in order to increase the concentration of the mixture to 60% antifreeze. Let us denote this amount by x. This relationship can be shown by the diagram in Fig. 1.6.

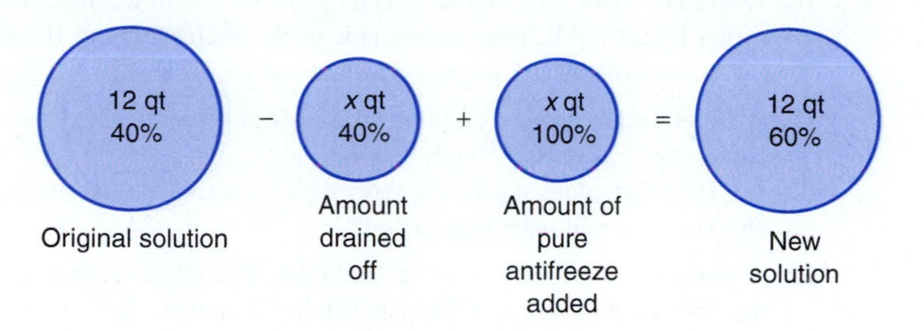

Figure 1.6

Write an equation in terms of antifreeze.

$$\begin{aligned} (0.40)(12) - (0.40)(x) + (1.00)(x) &= (0.60)(12) \\ 4.8 \quad - \quad 0.4x \quad + \quad x \quad &= \quad 7.2 \\ 4.8 + 0.60x &= 7.2 \\ 0.60x &= 2.4 \\ x &= 4 \text{ qt} \end{aligned}$$

That is, 4 qt of the old solution needs to be drained off and replaced with 4 qt of pure antifreeze. The check is left to you.

Exercises 1.13

Solve the following application problems. (Do not use the rules for calculating with measurements.)

1. Separate \$216 into two parts so that one part is three times the other.
2. The difference between two numbers is 6. Their sum is 30. Find the numbers.
3. A set of eight built-in bookshelves is to be constructed in a room with a floor-to-ceiling clearance of 9 ft 6 in. If each shelf is $\frac{3}{4}$ in. thick and equal spacing is desired between shelves, what is the space between the shelves? (Assume there is no shelf against the ceiling and no shelf on the floor.)

4. A deck railing 23 ft 8 in. long is to be made with eight evenly spaced redwood posts 4 in. by 4 in. (Posts are located on the ends.) What is the distance between each post?

5. The length of a rectangle is 4 m longer than its width. The perimeter is 56 m. Find the length and the width of the rectangle.

6. The perimeter of an isosceles triangle is 122 cm. Its base is 4 cm shorter than one of its equal sides. Find lengths of the sides of the triangle.

7. Forty acres of land were purchased for $20,400. The land facing the highway cost $650 per acre. The remainder cost $450 per acre. How much land was sold at $650 per acre? At $450 per acre?

8. A man deposited $6800 into two different savings accounts. One account earns interest at 8% and the other at $10\frac{1}{2}$%. The total interest earned from both accounts at the end of one year was $669. Find the amount deposited at each rate.

9. A person earned $1440 per week for Smith Contracting Co. He quit this job to work for Jones Contracting Co. for $1660 per week. It cost him $4840 in moving expenses to relocate near his new place of employment. How many weeks will it take his increase in salary to cover the cost of his moving expenses?

10. A woman wishes to enclose a rectangular yard with a fence. She plans to use the entire back of her house (70 ft long) as one side of the enclosed yard. If she has 50 yd of fencing, find the dimensions of the largest yard that can be enclosed.

11. A freight train leaves a station and travels at 45 mi/h. Three hours later a passenger train leaves and travels at 75 mi/h.
    (a) How long does it take the passenger train to overtake the freight train?
    (b) How far will each train travel?

12. A plane has a cruising speed of 270 mi/h. The wind velocity is 30 mi/h.
    (a) What is the speed of the plane flying with the wind?
    (b) What is the speed of the plane flying against the wind?
    (c) How far can the plane fly with the wind and return in 9 h?

13. Two planes are 760 mi apart, leave at the same time, fly toward each other, and meet in 4 h. If their speeds differ by 20 mi/h, find the speed of each plane.

14. A plane averaging 120 mi/h and an express train averaging 60 mi/h depart at the same time from the same city headed for the same destination. If the train arrives 45 min later than the plane, how far did they each travel?

15. A man has two piles of solder (a mixture of tin and lead) available. One pile is 30% tin and the other is 70% tin. How much of each must be used to make a 20-lb mixture that is 45% tin?

16. A bar of metal contains 10% silver. Another bar contains 15% silver. How many kilograms of each must be taken to make a 10-kg bar containing 12% silver?

17. A 12-qt cooling system is checked and found to be filled with 25% antifreeze. The desired strength is 50% antifreeze. How many quarts need to be drained and replaced with pure antifreeze to reach the desired strength?

18. In testing a hybrid engine, engineers are trying various mixtures of gasoline and methanol. How much of a 90% gasoline mixture and a 75% gasoline mixture would be needed for 600 L of an 85% gasoline mixture?

19. The sum of three electric currents is 4.65 A. The greatest current is four times the least. The third current is 0.75 A greater than the least. Find the three currents.

20. A test solution of gasohol contains 60 L of gasoline and 15 L of alcohol. How much pure alcohol must be added so that the resulting solution is 25% alcohol?

21. Enclose a rectangular plot with 280 m of fencing so that the length is twice the width and the area is divided into two equal parts. Find the length and the width of the lot.

22. Lee is in charge of a fireworks display for a Fourth of July celebration. She notices that she hears the explosion 2.40 s after she sees the display in the sky. If sound travels at 331 m/s, how far is she from the aerial display?

23. The sum of three resistances connected in series in a circuit is 540 Ω. The second resistance is 15 Ω greater than the first, and the third is 60 Ω greater than the second. Find the size of each resistance.

24. One of three angles is 5° more than the smallest angle. The third angle is three times the middle angle. The sum of the three angles is 120°. Find the measure of each angle.

1.14 RATIO AND PROPORTION

A **ratio** is the quotient of two numbers or quantities. The ratio $\frac{30\text{ m}}{40\text{ m}}$ compares two linear (length) measurements. This ratio may also be represented by a division sign (30 m ÷ 40 m), by a colon (30 m : 40 m), or by the word *to* (30 m to 40 m). Ratios should be expressed in simplest form; that is,

$$\frac{30\text{ m}}{40\text{ m}} = \frac{3}{4}$$

Note that the ratio of two like quantities has no units.

In technical applications ratios are sometimes used to compare unlike quantities (a ratio of unlike quantities is often called a **rate**). For example, if we travel 300 mi in 6 h, our average speed can be expressed by the ratio $\frac{300\text{ mi}}{6\text{ h}}$, or $50\frac{\text{mi}}{\text{h}}$. Other examples of ratios include pressure, which may be defined as the ratio of force per unit area; mass density, which may be defined as the ratio of mass per unit volume; and power, which may be defined as the ratio of work per unit time.

EXAMPLE 1

Find the ratio of 90 cm to 72 cm.

$$\frac{90\text{ cm}}{72\text{ cm}} = \frac{5}{4}$$

EXAMPLE 2

Find the ratio of 24 yd : 12 ft.

First, change to a common unit, when possible.

$$\frac{24\text{ yd}}{12\text{ ft}} = \frac{24\text{ yd}}{4\text{ yd}} = \frac{6}{1}$$

or

$$\frac{24\text{ yd}}{12\text{ ft}} = \frac{72\text{ ft}}{12\text{ ft}} = \frac{6}{1}$$

EXAMPLE 3

The mechanical advantage (*MA*) of a simple machine such as a pulley system may be defined as the following ratio:

$$MA = \frac{\text{resistance force}}{\text{effort force}}$$

If 130 N of effort lifts a 6500-N load, the *MA* is

$$\frac{6500\text{ N}}{130\text{ N}} = \frac{50}{1}$$

EXAMPLE 4

The mechanical advantage (*MA*) of a hydraulic press may be calculated by the ratio

$$MA = \frac{r_l^2}{r_s^2}$$

where

$$r_l = \text{radius of the larger or resistance piston}$$
$$r_s = \text{radius of the smaller or effort piston}$$

Find the *MA* of a hydraulic press that has a large piston of radius 10 in. and a small piston of radius 2 in.

$$MA = \frac{r_l^2}{r_s^2} = \frac{(10\text{ in.})^2}{(2\text{ in.})^2} = \frac{100\text{ in}^2}{4\text{ in}^2} = \frac{25}{1}$$

EXAMPLE 5

Mass density is defined as the ratio of mass per unit volume. Find the mass density of a piece of rock that has a volume of 320 cm^3 and a mass of 880 g.

$$\text{Mass density} = \frac{\text{mass}}{\text{volume}} = \frac{880\text{ g}}{320\text{ cm}^3} = 2.75\text{ g/cm}^3$$

A **proportion** is a statement that two ratios are equal. The proportion $\frac{a}{b} = \frac{c}{d}$ may also be written as $a:b = c:d$. The first and fourth terms, a and d, are called the **extremes;** the second and third terms, b and c, are called the **means.**

In any proportion the product of the means equals the product of the extremes; that is, if

$$\frac{a}{b} = \frac{c}{d}$$

then

$$bc = ad$$

We can use this principle to solve proportions.

EXAMPLE 6

Solve the proportion $\frac{x}{12} = \frac{5}{6}$.

$$\frac{x}{12} = \frac{5}{6}$$

$6x = 60$ (The product of the means equals the product of the extremes.)

$x = 10$ (Divide each side by 6.)

EXAMPLE 7

Solve the proportion $\dfrac{x}{21 - x} = \dfrac{3}{4}$.

$$\frac{x}{21 - x} = \frac{3}{4}$$

$$4x = 3(21 - x) \quad \text{(The product of the means equals the product of the extremes.)}$$

$$4x = 63 - 3x \quad \text{(Remove parentheses.)}$$

$$7x = 63 \quad \text{(Add } 3x \text{ to each side.)}$$

$$x = 9 \quad \text{(Divide each side by 7.)}$$

EXAMPLE 8

Cut a board 14 ft long into two pieces whose lengths are in the ratio 2 : 5. Find the lengths of the two pieces.

Let

$$x = \text{length of one piece}$$

$$14 - x = \text{length of the other piece}$$

$$\frac{x}{14 - x} = \frac{2}{5}$$

$$2(14 - x) = 5x$$

$$28 - 2x = 5x$$

$$28 = 7x$$

$$4 = x$$

Therefore, $x = 4$ ft is the length of one piece and $14 - x = 10$ ft is the length of the other.

EXAMPLE 9

A car travels 275 mi using 11 gal of gasoline. How far can the car travel on 35 gal?

This problem can be solved using a proportion as follows:

$$\frac{275 \text{ mi}}{11 \text{ gal}} = \frac{x \text{ mi}}{35 \text{ gal}}$$

$$11x = (275)(35)$$

$$x = \frac{(275)(35)}{11}$$

$$x = 875 \text{ mi}$$

Exercises 1.14

Express each ratio as a fraction in simplest form.

1. 36 mi : 8 mi

2. $\dfrac{240 \text{ ft}}{12 \text{ yd}}$

3. 2500 m to 25 km

4. $\dfrac{36 \text{ h}}{45 \text{ min}}$

5. $12 \text{ ft}^2 : 24 \text{ in}^2$

6. 18 kg to 360 g

7. $\dfrac{1.75 \text{ m}^2}{35 \text{ cm}^2}$

8. $12 \text{ ft}^3 : 24 \text{ in}^3$

9. $\dfrac{(28 \text{ in.})^2}{(7 \text{ in.})^2}$

10. $\dfrac{(8 \text{ cm})^3}{(2 \text{ cm})^3}$

11. Find the mechanical advantage of a pulley system in which it takes 250 lb of effort to lift 7500 lb of load.

12. Find the mechanical advantage of a hydraulic press that has pistons of radii 36 cm and 4 cm (see Example 4).

13. The scale of a given map is $\frac{3}{4}$ in. to 2500 ft. Express this ratio in simplest form.

14. Pressure may be defined as the ratio of force to area.
    (a) Find the pressure when a force of 170 lb is exerted on an area of $4\frac{1}{4}$ in^2.
    (b) Find the pressure (in kPa) when a force of 72 N is exerted on an area of 4 cm^2 [1 pascal (Pa) = 1 N/m^2].

15. A flywheel has 72 teeth and a starter drive gear has 18 teeth. Find the ratio of flywheel teeth to drive-gear teeth.

16. A transformer has 200 turns in the primary coil and 18,000 turns in the secondary coil. Find the ratio of secondary turns to primary turns.

17. A 1850-ft^2 house sells for $80,475. Find the ratio of cost to area (price per square foot).

18. If 27 ft^3 of cement are needed to make 144 ft^3 of concrete, find the ratio of concrete to cement.

19. A 350-gal spray tank can cover 14 acres. Find the rate of application in gallons per acre.

20. A bearing bronze mix includes 192 lb of copper and 30 lb of lead. What is the ratio of copper to lead?

21. What is the alternator-to-engine-drive ratio if the alternator turns at 1125 rpm when the engine is idling at 500 rpm?

22. If 35 gal of oil flow through a feeder pipe in 10 min, express the rate of flow in gallons per minute.

23. A flywheel has 72 teeth and a starter drive gear has 15 teeth. Find the ratio of flywheel teeth to drive-gear teeth.

24. A certain transformer has a voltage of 18 V in the primary circuit and 5850 V in the secondary circuit. Find the ratio of the primary voltage to the secondary voltage.

25. If the total yield from a 45-acre field is 6075 bu, express the yield in bushels per acre.

26. The ratio of the voltage drops across two resistors connected in series equals the ratio of their resistances. Find the ratio of the voltage drops across a 960-Ω resistor and a 400-Ω resistor.

27. Steel can be worked into a lathe at a cutting speed of 25 ft/min. Stainless steel can be worked at 15 ft/min. What is the ratio of the cutting speed of steel to the cutting speed of stainless steel?

28. The power gain of an amplifier is the ratio $\dfrac{\text{power output}}{\text{power input}}$. Find the power gain of an amplifier whose output is 18 W and input is 0.72 W.

Solve each proportion for x.

29. $\dfrac{x}{9} = \dfrac{81}{27}$

30. $\dfrac{25}{96} = \dfrac{75}{x}$

31. $\dfrac{64}{2x} = \dfrac{16}{84}$

32. $\dfrac{1}{75} = \dfrac{5x}{125}$

33. $\dfrac{x}{5 - x} = \dfrac{2}{3}$

34. $\dfrac{x}{8 - x} = \dfrac{1}{3}$

35. $\dfrac{x}{28 - x} = \dfrac{3}{4}$

36. $\dfrac{a}{b} = \dfrac{c}{x}$

37. $\dfrac{m}{x} = \dfrac{n}{p}$

38. $\dfrac{x}{m} = \dfrac{p}{q}$

39. $\dfrac{x}{2a} = \dfrac{4}{a}$

40. $\dfrac{x}{-a} = \dfrac{a}{b}$

Solve each proportion using a calculator. Round each result to three significant digits.

41. $\dfrac{30.2}{276} = \dfrac{85.6}{x}$ **42.** $\dfrac{284}{7.8} = \dfrac{x}{13.1}$ **43.** $\dfrac{32.5}{115} = \dfrac{2450}{x}$

44. $\dfrac{x}{563.2} = \dfrac{34.5}{244}$ **45.** $\dfrac{x}{0.477} = \dfrac{2.75}{16.1}$ **46.** $\dfrac{47.9}{x} = \dfrac{0.355}{0.0115}$

47. If 2000 bolts cost $240, what is the cost of 750 bolts? Of 2800 bolts?

48. If $\frac{1}{4}$ in. on a map represents 25 mi, what distance is represented by $2\frac{1}{8}$ in.?

49. If a man receives $182.40 for 30 h of work, how much will he receive for 36 h of work?

50. A car travels 54 km on 4.5 L of gasoline.
(a) How far can the car travel on 40 L?
(b) How much gasoline would be used on a 3600-km trip?
(c) Find the car's gas "mileage" in kilometres per litre.

51. A fuel pump delivers 45 mL of fuel in 540 strokes. How many strokes are needed to use 75 mL of fuel?

52. If $2\frac{3}{4}$ ft^3 of sand make 8 ft^3 of concrete, how much sand is needed to make 120 ft^3 of concrete?

53. If a builder sells a 1500-ft^2 home for $94,500, find the price of a 2100-ft^2 home of similar quality. *Note:* Assume that the price per square foot remains constant.

54. If 885 bricks are used in constructing a wall 15 ft long, how many bricks are needed for a similar wall 35 ft long?

55. The sum of the length and the width of a proposed feed lot is 1260 ft. Find the length and the width if their ratio must be 5 : 2.

56. Divide $450 in wages between two people so that the money is in the ratio of 4 : 5.

57. On an assembly line 68,340 parts need to be divided into two groups in the ratio of $\frac{5}{12}$. How many parts are in each group?

58. The ratio of the length and the width of a rectangular field is 5 : 6. Find the dimensions of the field if its perimeter is 4400 m.

59. Brass is an alloy of copper and zinc in the ratio of 3 : 2. How much copper and how much zinc are contained in 2500 lb of brass?

60. If 12 machines can produce 27,000 bolts in 6 h, how many bolts can be produced in 10 h? How many machines would it take to increase the production to 45,000 bolts in 6 h?

61. The owner of a building assessed at $45,000 is billed $638.40 for property taxes. How much should the owner of a similar building next door, assessed at $78,000, be billed for property taxes?

62. How much wettable powder do you put in a 300-gal spray tank if 20 gal of water and 3 lb of chemical pesticide are applied per acre?

63. A farmer uses 150 lb of insecticide on a 40-acre field. How many pounds are needed for a 180-acre field at the same rate of application?

64. If 100 bu of corn per acre remove 90 lb of nitrogen, phosphorus, and potassium (N, P, and K, respectively), how many pounds of N, P, and K are removed by a yield of 135 bu per acre?

65. If a farmer has a total yield of 38,400 bu of corn from a 320-acre farm, what total yield should be expected from a similar 520-acre farm?

66. A copper wire 750 ft long has a resistance of 1.563 Ω. How long is a copper wire of the same cross-sectional area whose resistance is 2.605 Ω? *Note:* The resistance of these wires is proportional to their length.

67. If the voltage drop across a 28-Ω resistor is 52 V, what is the voltage drop across a 63-Ω resistor that is in series with the first one? *Note:* Resistors in series have voltage drops proportional to their resistances.

68. If the ratio of secondary turns to primary turns in a given transformer is 35 to 4, how many secondary turns are there if the primary coil has 136 turns?

69. An 8-V automotive coil has 250 turns of wire in the primary circuit. If the secondary voltage is 15,000 V, how many secondary turns are in the coil? *Note:* The ratio of secondary voltage to primary voltage equals the ratio of secondary turns to primary turns.

70. The clutch linkage on a vehicle has an overall advantage of 24 : 1. If the pressure plate applies a force of 504 lb, how much force must the driver apply to release the clutch?

71. The clutch linkage on a vehicle has an overall advantage of 30 : 1. If the driver applies a force of 15 N to the clutch, how much force is applied to the pressure plate?

72. A certain concrete dry mix is composed of 1 part cement, $2\frac{1}{2}$ parts sand, and 4 parts gravel by volume. Assume that 60 ft^3 of gravel are used.
(a) How much cement is needed?
(b) How much sand is needed?

73. A triangle whose sides are in the ratio of 3 : 4 : 5 is a right triangle, as shown in Fig. 1.7. If the length of the hypotenuse is 85 m, what are the lengths of the legs?

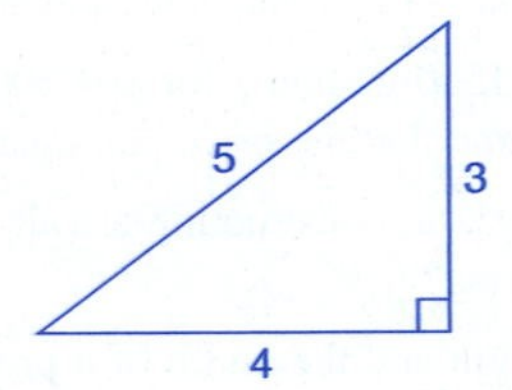

Figure 1.7

74. Assume that 75 m^3 of the concrete mix in Exercise 72 are needed for a job.
(a) How much cement is needed?
(b) How much sand is needed?
(c) How much gravel is needed?

75. You have 360 ft of fence. What size lot in the shape of a 3 : 4 : 5 right triangle can be fenced using all the fence?

1.15 VARIATION

Often scientific laws and technical principles are given in terms of **variation.** To help you understand and use this idea and terminology effectively in your other courses, we now develop the basic concepts of variation.

DIRECT VARIATION

If two quantities, y and x, change and their ratio remains constant $\left(\frac{y}{x} = k\right)$, the quantities **vary directly,** or y is **directly proportional** to x. In general, this relationship is written in the form $y = kx$, where k is the **proportionality constant.**

EXAMPLE 1

Consider the following data:

y	4	10	6	20	240
x	6	15	9	30	360

Note that y varies directly with x because the ratio $\frac{y}{x}$ is always $\frac{2}{3}$. Also note that $y = \frac{2}{3}x$, where $k = \frac{2}{3}$.

In general, if y varies directly as x (or $y = kx$) and x *increases,* then y also *increases.* Similarly, if x *decreases,* then y also *decreases.*

EXAMPLE 2

The circumference C of a circle varies directly with the radius r. This relation is written $C = kr$, where $k = 2\pi$, the proportionality constant.

EXAMPLE 3

The weight w of a body varies directly with its mass m. This relation is written $w = km$, where $k = g$, the acceleration of gravity, which is the proportionality constant.

EXAMPLE 4

In a given wire the electric resistance R varies directly with the wire's length L. This relation is written $R = kL$, where k is the proportionality constant.

EXAMPLE 5

Charles's law states that if the pressure on a gas is constant, its volume V varies directly with its absolute temperature T. This relation is written $V = kT$, where k is the proportionality constant.

INVERSE VARIATION

If two quantities, y and x, change and their product remains constant ($yx = k$), the quantities **vary inversely,** or y is **inversely proportional** to x. In general, this relation is written $y = \frac{k}{x}$, where k is called the proportionality constant.

EXAMPLE 6

Consider the following data:

y	6	2	12	3	96
x	8	24	4	16	$\frac{1}{2}$

Note that y varies inversely with x because the product is always 48. Also note that $y = \frac{48}{x}$, where $k = 48$.

In general, if y varies inversely as x (or $y = k/x$) and x *increases,* then y *decreases.* Similarly, if x *decreases,* then y *increases.*

EXAMPLE 7

Boyle's law states that if the temperature of a gas is constant, its volume V varies inversely with its pressure p. This relation is written $V = \frac{k}{p}$, where k is the proportionality constant.

EXAMPLE 8

The rate r at which an automobile covers a distance of 200 mi varies inversely with the time t. This relation is written $r = \frac{k}{t}$ or $r = \frac{200}{t}$, where $k = 200$ is the proportionality constant.

For many relationships, one quantity varies directly or inversely with a power of the other.

EXAMPLE 9

The area A of a circle varies directly with the square of its radius r. This relation is written $A = kr^2$, where $k = \pi$, which is the proportionality constant.

EXAMPLE 10

In wire of a given length, the electric resistance R varies inversely with the square of its diameter D. This relation is written $R = \frac{k}{D^2}$, where k is the proportionality constant.

JOINT VARIATION

One quantity **varies jointly** with two or more quantities when it varies directly with the product of these quantities. In general, this relation is written $y = kxz$, where k is the proportionality constant.

EXAMPLE 11

Coulomb's law for magnetism states that the force F between two magnetic poles varies jointly with the strengths s_1 and s_2 of the poles and inversely with the square of their distance apart d. This relation is written $F = \frac{ks_1s_2}{d^2}$, where k is the proportionality constant.

Once we are able to express a given variation sentence as an equation, the next step is to find the value of k, the proportionality constant. This value of k can then be used for all the different sets of data in a given problem or situation.

EXAMPLE 12

Suppose that y varies directly with the square of x and that $x = 2$ when $y = 16$. Find y when $x = 3$ and when $x = \frac{1}{2}$.

First write the variation equation:

$$y = kx^2$$

Then, to find k, substitute the values for x and y.

$$16 = k2^2$$
$$16 = 4k$$
$$4 = k$$

Therefore,

$$y = 4x^2$$

For $x = 3$,

$$y = 4(3)^2 = 36$$

For $x = \frac{1}{2}$,

$$y = 4\left(\frac{1}{2}\right)^2 = 1$$

EXAMPLE 13

At a given temperature the electric resistance R of a wire varies directly with its length L and inversely with the square of its diameter D. The resistance of 20.0 m of copper wire of diameter 0.81 mm is 0.67 Ω. What is the resistance of 40.0 m of copper wire 1.20 mm in diameter?

First write the variation equation:

$$R = \frac{kL}{D^2}$$

Then find k.

$$0.67\ \Omega = \frac{k(20.0\text{ m})}{(0.81\text{ mm})^2}$$

$$k = 0.022\frac{\Omega\text{ mm}^2}{\text{m}}$$

Therefore,

$$R = \frac{0.022L}{D^2}$$

For $L = 40.0$ m and $D = 1.20$ mm,

$$R = \frac{\left(0.022\frac{\Omega\text{ mm}^2}{\text{m}}\right)(40.0\text{ m})}{(1.20\text{ mm})^2} = 0.61\ \Omega$$

Exercises 1.15

For each set of data, determine whether y varies directly with x or inversely with x. Also find k.

1.

y	6	15	54	1.5
x	8	20	72	2

2.

y	4	2	$\frac{1}{2}$	5
x	5	10	40	4

3.

y	$\frac{3}{4}$	$\frac{1}{4}$	$\frac{13}{20}$	$\frac{9}{16}$
x	2	6	$\frac{60}{26}$	$\frac{8}{3}$

4.

y	14	21	$\frac{7}{3}$	$\frac{14}{5}$
x	16	28	4	$\frac{16}{5}$

5.

y	12	18	0.75	1.5
x	20	30	2.5	5

6.

y	2	$\frac{1}{3}$	18	$\frac{2}{11}$
x	7	$\frac{7}{6}$	63	$\frac{7}{11}$

Write a variation equation for each.

7. y varies directly with z.

8. p varies inversely with q.

9. a varies jointly with b and c.

10. m varies directly with the square of n.

11. r varies directly with s and inversely with the square root of t.

12. d varies jointly with e and the cube of f.

13. f varies jointly with g and h and inversely with the square of j.

14. m varies directly with the square root of n and inversely with the cube of p.

First find k; then find the given quantity for Exercises 15 through 28.

15. y varies directly with x; $y = 8$ when $x = 24$. Find y when $x = 36$.

16. m varies directly with n; $m = 198$ when $n = 22$. Find m when $n = 35$.

17. y varies inversely with x; $y = 9$ when $x = 6$. Find y when $x = 18$.

18. d varies inversely with e; $d = \frac{4}{5}$ when $e = \frac{9}{16}$. Find d when $e = \frac{5}{3}$.

19. y varies directly with the square root of x; $y = 24$ when $x = 16$. Find y when $x = 36$.

20. y varies directly with the square of x; $y = 216$ when $x = 3$. Find y when $x = 6$.

21. y varies jointly with s and t; $y = 7.2$ when $s = 36$ and $t = 0.8$. Find y when $s = 52$ and $t = 1.5$.

22. y varies jointly with x and the square of z; $y = 150$ when $x = 3$ and $z = 5$. Find y when $x = 12$ and $z = 8$.

23. p varies directly with q and inversely with the square of r; $p = 40$ when $q = 20$ and $r = 4$. Find p when $q = 24$ and $r = 6$.

24. m varies inversely with n and the square root of p; $m = 18$ when $n = 2$ and $p = 36$. Find m when $n = 9$ and $p = 64$.

25. Newton's law of universal gravitation states that the gravitational force F between two objects varies jointly with their masses m_1 and m_2 and inversely with the square of the distance r between the centers of mass of the two objects. If $F = 1.98 \times 10^{20}$ N when $m_1 = 5.97 \times 10^{24}$ kg, $m_2 = 7.35 \times 10^{22}$ kg, and $r = 3.84 \times 10^{8}$ m, find F when $m_1 = 5.97 \times 10^{24}$ kg, $m_2 = 1.90 \times 10^{27}$ kg, and $r = 7.95 \times 10^{11}$ m.

26. The amount of illumination E on a surface varies directly with the intensity I of the source and inversely with the square of the distance r from the source. If the illumination is 8.07 lux and the intensity is 405 lumens when the source is 2.00 m away, find the illumination when the intensity is 1250 lumens and the source is 0.250 m away.

27. Use Charles's law in Example 5 to find the volume of oxygen when the absolute temperature is $3\bar{0}0$ K if the volume is $15\bar{0}0$ cm^3 when the temperature is $25\bar{0}$ K.

28. Use Boyle's law in Example 7 to find the volume of nitrogen when the pressure is 25.0 lb/in^2 if the volume is $35\bar{0}0$ ft^3 when the pressure is 15.0 lb/in^2.

29. Charles's law and Boyle's law combined state that the volume V of a gas varies directly with its absolute temperature T and inversely with its pressure p. If the volume of acetylene is $15{,}\bar{0}00$ ft^3 when the absolute temperature is $40\bar{0}$°R and its pressure is 20.0 lb/in^2, find its volume when its temperature is 575°R and its pressure is 50.0 lb/in^2.

30. The electric power P generated by current varies jointly with the voltage drop V and the current I. If 6.00 A of current in a $22\bar{0}$-V circuit generate 1320 W of power, how much power do 8.00 A of current in a 115-V circuit generate?

31. The electric power P used in a circuit varies directly with the square of the voltage drop V and inversely with the resistance R. If 180 W of power in a circuit are used by a $9\bar{0}$-V drop with a resistance of 45 Ω, find the power used when the voltage drop is 120 V and the resistance is $3\bar{0}$ Ω.

32. The power P of an engine varies directly with the square of the radius r of its piston. If a $4\bar{0}$-hp engine has a piston of radius 2.0 in., what power does an engine with a piston of radius 4.0 in. have?

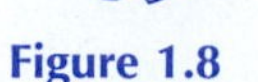

Figure 1.8

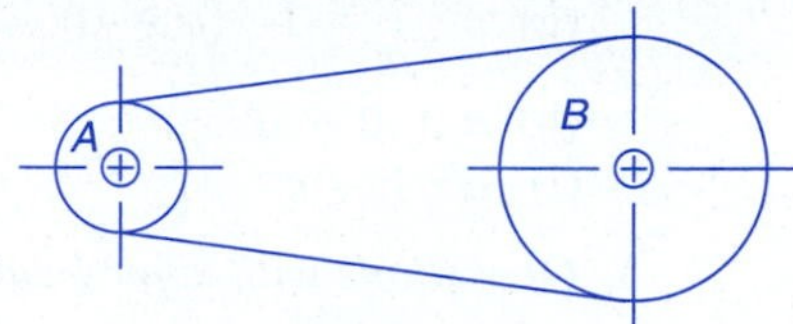

Figure 1.9

33. For two rotating gears as in Fig. 1.8, the number of teeth t for either gear is inversely proportional to the number of revolutions n that the gear makes per unit of time. A gear with 50 teeth rotating at $40\overline{0}$ rpm (revolutions per minute) turns a second gear at 125 rpm. How many teeth does the second gear have?

34. A gear with 75 teeth rotates at 32 rpm and turns a second gear with 25 teeth. How fast does the second gear rotate?

35. For two rotating pulleys connected with a belt as in Fig. 1.9, the diameter d of either pulley varies inversely with the number of revolutions n that the pulley makes per unit of time. A small pulley, 25 cm in diameter and rotating at 72 rpm, rotates a large pulley that is 75 cm in diameter. What is the speed of the large pulley?

36. One pulley 22 cm in diameter revolves at 680 rpm. A second pulley revolves at 440 rpm. What is its diameter?

37. One pulley 15 in. in diameter revolves at 150 rpm. A second pulley is 45 in. in diameter.
- **(a)** What is the speed of the second pulley?
- **(b)** If the speed of the first pulley doubles, what is the speed of the second pulley?

38. The number of workers needed to complete a particular job is inversely proportional to the number of hours that they work. If 12 electricians can complete a job in 72.0 h, how long will it take 8 electricians to complete the same job? Assume that each person works at the same rate no matter how many persons are assigned to the job.

CHAPTER 1 SUMMARY

1. *Basic terms:*
- (a) *Positive integers:* 1, 2, 3, ...
- (b) *Negative integers:* $-1, -2, -3, \ldots$.
- (c) *Integers:* $\ldots, -3, -2, -1, 0, 1, 2, 3, \ldots$.
- (d) *Rational numbers:* Numbers that can be represented as the ratio of two integers.
- (e) *Irrational numbers:* Numbers that cannot be represented as the ratio of two integers.
- (f) *Real numbers:* Set of numbers consisting of the rational numbers and the irrational numbers.
- (g) *Prime number:* A positive integer greater than one that is evenly divisible only by itself and one.
- (h) *Absolute value:* The absolute value of a real number n, written $|n|$, is defined as $|n| = n$ if $n \geq 0$ or $|n| = -n$ if $n < 0$.

2. *Basic properties of real numbers:*

(a) $a + b = b + a$	Commutative property of addition
(b) $ab = ba$	Commutative property of multiplication
(c) $(a + b) + c = a + (b + c)$	Associative property of addition
(d) $(ab)c = a(bc)$	Associative property of multiplication
(e) $a(b + c) = ab + ac$	Distributive property of multiplication over addition

(f) $a + (-a) = 0$ — Additive inverse or negative property

(g) $a \cdot \frac{1}{a} = 1 \quad (a \neq 0)$ — Multiplicative inverse or reciprocal property

(h) $a + 0 = a$ — Identity element of addition

(i) $a \cdot 1 = a$ — Identity element of multiplication

3. *Operations with signed numbers:* See Section 1.1.

4. *Operations with zero:*

(a) $a + 0 = a$

(b) $a - 0 = a$

(c) $a \cdot 0 = 0$

(d) If $a \cdot b = 0$, then either $a = 0$ or $b = 0$.

(e) $\frac{0}{b} = 0 \quad (b \neq 0)$

(f) $\frac{a}{0}$ is meaningless $\quad (a \neq 0)$

(g) $\frac{0}{0}$ is indeterminate

5. *Laws of exponents for powers of 10:*

(a) $10^m \cdot 10^n = 10^{m+n}$

(b) $\frac{10^m}{10^n} = 10^{m-n}$

(c) $(10^m)^n = 10^{mn}$

(d) $10^{-n} = \frac{1}{10^n}$ and $\frac{1}{10^{-n}} = 10^n$

(e) $10^0 = 1$

6. *Scientific notation:* A number is expressed in scientific notation as a product of a decimal between 1 and 10 and a power of 10 in the form $N \times 10^m$ where $1 \leq N < 10$ and m is an integer.

7. *Accuracy:* The accuracy of a measurement refers to the number of significant digits it contains. The number of significant digits indicates the number of units we are reasonably sure of having counted when making the measurement.

8. *Rules for significant digits:*

(a) All nonzero digits are significant.

(b) All zeros between significant digits are significant.

(c) A zero in a number greater than one which is specially tagged, such as by a bar above it, is significant.

(d) All zeros to the right of a significant digit *and* a decimal point are significant.

(e) Zeros to the right in a whole number measurement that are not tagged are *not* significant.

(f) Zeros to the left in a measurement less than one are *not* significant.

9. *Precision:* The precision of a measurement refers to the smallest unit with which the measurement is made, that is, the position of the last significant digit.

10. *To add or subtract measurements:*

(a) Make certain that all of the measurements are expressed in the same unit. If they are not, change them all to the same unit.

(b) Add or subtract.

(c) Then round the result to the same precision as the least precise measurement.

11. *To multiply and/or divide measurements:*

(a) First multiply and/or divide the measurements as given.

(b) Then round the result to the same number of significant digits as the measurement with the least number of significant digits.

12. An *algebraic expression* is a combination of finite sums, differences, products, quotients, roots, and powers of numbers and of letters representing numbers.

13. A *polynomial in one variable* is defined as follows:

$$a_nx^n + a_{n-1}x^{n-1} + a_{n-2}x^{n-2} + \cdots + a_2x^2 + a_1x + a_0$$

where n is a positive integer and the coefficients $a_n, a_{n-1}, a_{n-2}, \ldots, a_2, a_1, a_0$ are real numbers.

14. *Basic terms:*

(a) *Monomial:* algebraic expression with one term.

(b) *Binomial:* algebraic expression with two terms.

(c) *Trinomial:* algebraic expression with three terms.

(d) *Multinomial:* algebraic expression with two or more terms.

15. *Degree:*

(a) The degree of a monomial in one variable is the same as the exponent of the variable.

(b) The degree of a monomial in more than one variable equals the sum of the exponents of its variables.

(c) The degree of a polynomial is the same as the highest-degree monomial in the polynomial.

(d) A polynomial is in decreasing order if each term is of some degree less than the preceding term.

(e) A polynomial is in increasing order if each term is of some degree larger than the preceding term.

16. *Removing grouping symbols:*

(a) To remove grouping symbols preceded by only a plus sign, remove the grouping symbols and leave the sign of each term unchanged within the grouping symbols.

(b) To remove grouping symbols preceded by only a minus sign, remove the grouping symbols and change the sign of each term within the grouping symbols.

(c) When sets of grouping symbols are contained within each other, remove each set starting with the innermost set and finishing with the outermost set.

17. *Order of operations:*

(a) Do all operations within any grouping symbols, beginning with the innermost pair if grouping symbols are contained within each other.

(b) Simplify all powers.

(c) Do all multiplications and divisions in the order in which they occur from left to right.

(d) Do all additions and subtractions in the order in which they occur from left to right.

18. To evaluate an algebraic expression, substitute the given numerical values in place of the letters, and follow the order of operations.

19. *Laws of exponents:* In each of the following, a and b are real numbers and m and n are positive integers.

(a) $a^m \cdot a^n = a^{m+n}$

(b) (i) $\dfrac{a^m}{a^n} = a^{m-n} \quad (m > n \quad \text{and} \quad a \neq 0)$

(ii) $\dfrac{a^m}{a^n} = \dfrac{1}{a^{n-m}} \quad (m < n \quad \text{and} \quad a \neq 0)$

(c) $(a^m)^n = a^{mn}$
(d) $(ab)^n = a^n b^n$
(e) $\left(\frac{a}{b}\right)^n = \frac{a^n}{b^n} \quad (b \neq 0)$
(f) $a^0 = 1 \quad (a \neq 0)$

20. *Roots:*

(a) The square root of a number, $\sqrt{a}$, is that nonnegative number which, when multiplied by itself, is a.
(b) The cube root of a number, $\sqrt[3]{a}$, is that number which, when multiplied by itself three times, is a.
(c) The nth root of a number a is written $\sqrt[n]{a}$, where n is the index, a is the radicand, and $\sqrt{\ }$ is called the radical sign.
(d) The square root property states that $\sqrt{ab} = \sqrt{a}\sqrt{b}$, where $a > 0$ and $b > 0$.
(e) The square root of a number is simplified when the number under the radical contains no perfect square factors.

21. *Multiplication of algebraic expressions:*

(a) To multiply monomials, multiply their numerical coefficients and multiply each set of like letters using the laws of exponents.
(b) To multiply a multinomial by a monomial, multiply each term of the multinomial by the monomial.
(c) To multiply two multinomials, multiply each term of the first multinomial by each term of the second and simplify.
(d) Use the FOIL method to multiply two binomials mentally.

22. *Division of algebraic expressions:*

(a) To divide monomials, divide their numerical coefficients and divide each set of like letters using the laws of exponents.
(b) To divide a multinomial by a monomial, divide each term of the multinomial by the monomial.
(c) To divide one multinomial by a second multinominal:

Arrange each multinomial in descending powers of one of the variables.

To find the first term of the quotient, divide the first term of the dividend by the first term of the divisor.

Multiply the divisor by the first term of the quotient and subtract the product from the dividend.

Continue this procedure of dividing the first term of each remainder by the first term of the divisor until the final remainder is zero or until the remainder is of degree less than the divisor.

23. *Equations:*

(a) An *equation* is a statement that two algebraic expressions are equal.
(b) A *linear equation in one variable* is an equation in which each term containing that variable is of first degree.
(c) *Four basic properties used to solve equations:*

If the same quantity is added to each side of an equation, the resulting equation is equivalent to the original equation.

If the same quantity is subtracted from each side of an equation, the resulting equation is equivalent to the original equation.

If each side of an equation is multiplied by the same (nonzero) quantity, the resulting equation is equivalent to the original equation.

If each side of an equation is divided by the same (nonzero) quantity, the resulting equation is equivalent to the original equation.

(d) *To solve a first-degree equation in one variable:*

Eliminate any fractions by multiplying each side by the lowest common denominator of all fractions in the equation.

Remove any grouping symbols.

Combine like terms on each side of the equation.

Isolate all the unknown terms on one side of the equation and all other terms on the other side of the equation.

Again combine like terms where possible.

Divide each side by the coefficient of the unknown.

Check your solution by substituting it in the original equation.

24. A *formula* is an equation, usually expressed in letters, that shows the relationship between quantities.

25. To *solve a formula for a given letter* means to isolate the given letter on one side of the equation and express it in terms of all the remaining letters by having all the other letters and numbers appear on the opposite side of the equation. We solve a formula using the same principles that we use in solving any equation.

26. *To use a formula to solve a problem where all but the unknown quantity is given:*

Solve the formula for the unknown quantity.

Substitute each known quantity with its units.

Use the order-of-operations procedures to find the numerical quantity and to simplify the units.

27. *To solve application problems:*

Step 1: Read the problem carefully at least twice.

(a) The first time you should read straight through from beginning to end. Do not stop this time to think about setting up an equation. You are only after a general impression at this point.

(b) Now read through *completely* for the second time. This time begin to think ahead to the next steps.

Step 2: If possible, draw a picture or diagram. This will often help you to visualize the possible mathematical relationships needed in order to write the equation.

Step 3: Choose a symbol to represent the unknown quantity (the quantity not given as a specific value) in the problem. Be sure to label this symbol, the unknown variable, to indicate what it represents.

Step 4: Write an equation that expresses the information given in the problem. To obtain this equation, look for information that will express two quantities that can be set equal to each other. Do not rush through this step. Try expressing your equation in words to see if it corresponds to the original problem.

Step 5: Solve for the unknown variable.

Step 6: Check your solution in the equation from Step 4. If it does not check, you have made a mathematical error. You will need to repeat Step 5. Also check your solution in the original verbal problem for reasonableness.

28. A *ratio* is the quotient of two numbers or quantities.

29. A *proportion* is a statement that two ratios are equal.

30. In any proportion, the product of the means equals the product of the extremes.

31. *Variation:*

(a) *Direct:* If two quantities, y and x, change and their ratio remains constant $\left(\frac{y}{x} = k\right)$, the quantities *vary directly,* or y is *directly proportional* to x. In general, this relation is written in the form $y = kx$, where k is the *proportionality constant.*

(b) *Inverse:* If two quantities, y and x, change and their product remains constant ($yx = k$), the quantities *vary inversely,* or y is *inversely proportional* to x. In general, this relation is written $y = \frac{k}{x}$, where k is again called the proportionality constant.

(c) *Joint:* One quantity *varies jointly* with two or more quantities when it varies directly with the product of these quantities.

CHAPTER 1 REVIEW

Perform the indicated operations and simplify.

1. $(-3) + (+6) + (-8)$

2. $(+4) - (+7) + (-3) - (-6)$

3. $(-3)(+5)(-7)(-2)$

4. $\frac{(+6)(-12)(+4)}{(+9)(-8)(-2)}$

5. $8 - 3 \cdot 4$

6. $4 + 10 \div 2$

7. $6 - (8 + 4) \div 6$

8. $5 + 2(6 - 9)$

9. $48 \div 8 \cdot 3 + 2 \cdot 5 - 17$

10. $3 \cdot 4 \div 6 - 8 + 6(-3)$

11. $\frac{3 \cdot 5 + 6(4 \div 2)}{8 \div 4 \cdot 8 - 3}$

12. $\frac{18 \div 9 \cdot 2 - 3}{5 - (2 - 3)}$

13. $\frac{4 \cdot 0}{8}$

14. $\frac{-5 - (15 - 20)}{9 \cdot 0 + (-4 - (-4))}$

15. $\frac{5 - 6(2 + 4)}{-16 - (-16)}$

16. $(10^2)^4$

17. $\frac{10^{-2}}{10^{-6}}$

18. $10^3 \cdot 10^4$

19. Write 3,420,000 in scientific notation.

20. Write 5.61×10^{-4} in decimal form.

Find each value. Round each result to three significant digits written in scientific notation.

21. $(8.54 \times 10^7)(4.97 \times 10^{-14})$

22. $\frac{1.85 \times 10^{12}}{6.17 \times 10^{-18}}$

*For each measurement, find its **(a)** accuracy and **(b)** precision.*

23. 307 m

24. 0.050 A

25. $12\bar{0},000$ V

Use the rules for addition of measurements to find the sum of each set of measurements.

26. 19.80 L; 14.4 L; 6.000 L; 17.431 L

27. 12.600 cm; 10.40 mm; 16.75 cm; 7.005 m

Use the rules for multiplication and division of measurements to find the value of each of the following.

28. (18.5 m)(21.6 m)

29. $\frac{49.7 \text{ m}^3}{16.0 \text{ m}^2}$

30. $\frac{680 \text{ lb}}{(14.5 \text{ in.})(18.6 \text{ in.})}$

31. Classify $4x^3 - 6x^2 + 5x$ as a monomial, binomial, or trinomial, and give its degree.

Perform the indicated operations and simplify.

32. $(3x^2 + 5x - 7) + (13x^2 - x - 14)$

33. $(-5x^2 - 2x - 3) + (12x^2 + x - 8)$

34. $(-4x^2 - 3x + 1) - (-5x^2 + 2x - 1)$

35. $(7x^2 + x - 5) - (2x^2 - 3x - 5)$

36. $-[(2a - 3b) - (-4a + b) + (-a - 4b)]$

37. $-3[-(2a - b) - (a - b) + (6a + 5b)]$

38. Evaluate $\dfrac{3a^2b - 6b}{2b + 1}$ for $a = -2$ and $b = -3$.

Perform the indicated operations and simplify.

39. $y^5 \cdot y^4$

40. $\dfrac{b^{12}}{b^3}$

41. $(3x^2)(-5x^4)$

42. $\dfrac{18m^6}{3m^2}$

43. $(a^3)^4$

44. $(5a^3)^2$

45. $\left(\dfrac{y}{x^2}\right)^3$

46. $(5x^2)^0$

47. $(-s^3)^3$

48. $(x^5 \cdot x^4)^2$

49. $(2a^3b^2)^3$

50. $\left(\dfrac{-3x^6}{6x^2}\right)^2$

51. $\left(\dfrac{-4x^8}{2x^2}\right)^3$

Simplify each root.

52. $\sqrt{49}$

53. $\sqrt[3]{27}$

54. $\sqrt{63}$

55. $\sqrt{108}$

Find each square root rounded to three significant digits.

56. $\sqrt{3147}$

57. $\sqrt{0.0205}$

Find each product.

58. $(-3a^2bc^3)(8a^3bc)$

59. $5x(2x - 6y)$

60. $-3x^2(4x^3 - 3x^2 - x + 4)$

61. $3a^2b(-5a^2b^3 + 3a^3 - 5b^2)$

62. $(2x + 7)(3x - 4)$

63. $(5x - 3)(6x - 9)$

64. $(4x + 6)(5x + 2)$

65. $(2x - 3)(3x + 2)$

66. $(8x - 5)^2$

67. $(x + y - 7)(2x - y + 3)$

Find each quotient.

68. $\dfrac{81x^2y^3z}{27x^2yz^2}$

69. $\dfrac{(4a)^2(5a^2b)^2}{50a^3b^2}$

70. $\dfrac{48x^2 - 24x + 15}{3x}$

71. $\dfrac{15x^3 - 25x + 35}{5x^2}$

72. $\dfrac{18m^2n^5 + 24mn - 8m^6n^2}{6m^2n^3}$

73. $\dfrac{6x^2 - x - 12}{2x - 3}$

74. $\dfrac{3x^3 + 5x - 3}{x + 1}$

75. $\dfrac{2x^3 - 5x^2 + 8x + 7}{2x - 1}$

Solve each equation.

76. $6x + 7 = -17$

77. $8 - 3x = 12 - 2x$

78. $3(x + 8) = 2(x - 3)$

79. $3(2x - 1) - 2(3x - 1) = 3x + 1$

80. $\dfrac{1}{4}x + 8 = \dfrac{3}{4}x - \dfrac{7}{2}$

81. $\dfrac{2}{3}x - \dfrac{4}{5} = \dfrac{1}{5}x + \dfrac{4}{3}$

82. $\dfrac{5(x - 2)}{6} - \dfrac{3x}{2} = \dfrac{5}{3}x$

83. $\dfrac{3}{4}(2x - 3) - \dfrac{2}{3}(6x - 2) = x$

84. Solve $24.6x - 45.2 = 0.4(39.2x + 82.5)$ and round the result to three significant digits.

Solve each formula for the indicated letter.

85. $S = \frac{\pi D}{12}$ for D

86. $C = \frac{Q}{V}$ for V

87. $v = v_0 - gt$ for t

88. $(\Delta V) = \beta V(T - T_0)$ for T_0.

89. Given $v = v_0 + at$, $v = 18\bar{0}$ m/s, $a = 9.80$ m/s^2, and $t = 10.0$ s. Find v_0.

90. Given $Z = \sqrt{R^2 + X_L^2}$, $R = 40.0\ \Omega$, and $X_L = 37.7\ \Omega$. Find Z.

91. Given $\frac{1}{R} = \frac{1}{R_1} + \frac{1}{R_2} + \frac{1}{R_3}$, $R = 4.5\ \Omega$, $R_1 = 12.5\ \Omega$, and $R_2 = 8.5\ \Omega$. Find R_3 rounded to three significant digits.

92. A mechanic needs to insert shims for a total thickness of 0.090 in. If there is already an 0.018-in. shim in place, how many 0.0040-in. shims must be inserted?

93. A rectangle is 5 m longer than it is wide. Its perimeter is 90 m. Find its dimensions.

94. A bar of metal contains 15% silver. Another bar contains 20% silver. How many ounces of each must be taken to make a 30-oz bar containing 18% silver?

95. A car is traveling at 55 mi/h. A police officer is 10 mi behind traveling at 75 mi/h and is dispatched to intercept this car. How long will it take the police officer to overtake the car?

Express each ratio in simplest form.

96. $\frac{360 \text{ ft}}{441 \text{ ft}}$

97. $\frac{600 \text{ mm}}{1.5 \text{ m}}$

98. $\frac{(2.8 \text{ in.})^2}{(1.2 \text{ in.})^2}$

Solve each proportion for x.

99. $\frac{13}{24} = \frac{x}{96}$

100. $\frac{39}{12} = \frac{3x}{48}$

101. $\frac{x}{60 - x} = \frac{2}{3}$

102. $\frac{a}{6 + x} = \frac{b}{a}$

103. $\frac{x}{c} = \frac{b}{a}$

104. $\frac{270}{15} = \frac{x}{3.15}$

105. If 8000 people use 40,000 gal of water per day, how much water will 20,000 people use?

106. Separate 2370 parts into two groups in the ratio of 3 : 7. How many parts are in each group?

In Exercises 107 through 110, write the variation equation.

107. y varies directly with the square root of z.

108. y varies jointly with v and the square of u.

109. y varies directly with p and inversely with q.

110. y varies jointly with m and n and inversely with the square of p.

111. y varies jointly with x and z; $y = 576$ when $x = 24$ and $z = 8$. Find y when $x = 36$ and $z = 48$.

112. y varies directly with the square of p and inversely with the square root of q; $y = 6$ when $p = 4$ and $q = 16$. Find y when $p = 8$ and $q = 4$.

113. Hooke's law states that within the limits of perfect elasticity, the force F needed to stretch a spring varies directly with the distance d the spring is stretched. If 40.0 N stretch a spring 1.5 m, what force is needed to stretch the spring 6.0 m?

114. The centrifugal force F of a body moving in a circular path varies jointly with its weight W and the square of its velocity v. If 19.2 lb of centrifugal force are produced by a 20.0-lb weight moving at 16.0 ft/s, what force is produced by a 50.0-lb weight moving at 10.0 ft/s?

115. Coulomb's law for electric charges states that the force F between two point charges varies jointly with the charges q_1 and q_2 and inversely with the square of the distance d between them. Two charges of $+20\ \mu$C (microcoulombs) and $+12\ \mu$C that are 6.0 cm apart produce a repulsive force of $6\bar{0}0$ N. What attractive force is produced by two charges of $-36\ \mu$C and $+12\ \mu$C that are 12 cm apart?

CHAPTER 1

SPECIAL NASA APPLICATION

Spare Parts for the Space Shuttle*

Many Shuttle parts are not repairable. They must be replaced with new parts when they wear out. For instance, a Shuttle has two types of tires: two nose landing gear tires in front and four larger, main landing gear tires behind them. When the Shuttle lands, the four main landing gear tires hit the ground first, followed by the two nose landing gear tires. For this reason the nose landing gear tires at the front of the Shuttle receive less wear and can be used twice before being replaced, while the main landing gear tires are replaced after each flight.

It is, of course, necessary to have nonrepairable items on hand when they are needed. How do you make sure that you have a replacement when a bulb burns out in your house? You probably keep a few new bulbs on hand and purchase more when very few are left. NASA's logistics engineers do exactly this for each nonrepairable item. They determine how many of each item to keep in stock, referred to as the minimum stock level, and how many to order, referred to as the order quantity.

Main landing gear tires are nonrepairable items with a life of one landing. Each Shuttle has four main landing gear tires in two dual-wheel configurations. The tires are inflated with gaseous nitrogen to a pressure of 315 lb/in^2. The maximum allowable load per main landing gear tire is 123,000 lb, and they are rated at 258 mi/h. The tires cost $5560 each in May 1997 and are made especially for the Shuttles.

We now turn to the problem of determining how many main landing gear tires to keep in stock and how many to order so that a sufficient, but not excessive, number of new tires is available when needed.

PART A CALCULATING THE MINIMUM STOCK LEVEL (MSL)

The minimum stock level (MSL) is the number of items that should be kept in stock. When the number of items in stock declines to the MSL, it is time to order more of those items. The MSL is determined by the number of items needed for each flight (the "issue rate"), the number of flights per year, the amount of lead time needed for the order (to account for things like delivery time), and a built-in "safety pad" of 12 months of stock.

*From NASA–AMATYC–NSF Project Mathematics Explorations I, grant principals John S. Pazdar, Peter A. Wursthorn, and Patricia L. Hirschy; copyright Capital Community College, 1999.

The "safety pad" accounts for the possibility of a problem with the order or a higher demand for the item than had been anticipated.

$$\text{MSL} = \left(\frac{\text{average issue rate}}{\text{flight}} \times \frac{\text{flights}}{\text{year}} \times \frac{1 \text{ year}}{12 \text{ months}}\right) \times (\text{lead time} + \text{safety pad of 12 months})$$

1. What effect does the factor "1 year/12 months" have on the equation?
2. For the main landing gear tires, what is the issue rate (that is, how many main landing gear tires are used on each flight)?
3. As of May 1997, the number of flights per year is eight and the lead time for ordering the main landing gear tires is 9 months. Fill in the following blanks and calculate the minimum stock level:

$$\text{MSL} = \left(\frac{\_\_\text{ tires}}{\text{flight}} \times \frac{\_\_\text{ flights}}{\text{year}} \times \frac{1 \text{ year}}{12 \text{ months}}\right) \times (\_\_\text{ months} + 12 \text{ months})$$
$$= \_\_\text{ tires}$$

4. Write a sentence to explain what your answer means.

PART B CALCULATING THE ORDER QUANTITY (OQ)

The logistics engineers use the minimum stock level to decide whether to order more parts. If the number of items on hand (the number of serviceable parts, or "Ser.") is greater than the MSL, then no items need to be ordered. However, if the number of serviceable parts is less than or equal to the MSL, the engineers must order more parts. The order quantity (OQ) tells the logistics engineers how many parts to order. The order quantity depends on the issue rate, the number of flights per year, the support period required (the amount of time the quantity of items ordered is expected to last, in this case 12 months), the number of items already ordered (the "due-in quantity"), and the number of items already on hand. Keep in mind that items are ordered when the number of serviceable parts on hand (Ser.) is less than or equal to the MSL. If Ser. exceeds the MSL, there is a sufficient number of items in stock and the order quantity is zero (no order required).

$$\text{OQ} = \frac{\text{average issue rate}}{\text{flight}} \times \frac{\text{flights}}{\text{year}} \times \frac{1 \text{ year}}{12 \text{ months}} \times (\text{support period required})$$
$$+ \text{MSL} - \text{due-in quantity} - \text{Ser.} \qquad \text{if MSL} \geq \text{Ser.}$$

$$\text{OQ} = 0 \qquad \text{if MSL} < \text{Ser.}$$

1. For the main lan\ding gear tires
   (a) Average issue rate / flight = ______
   (b) Flights / year = ______
   (c) Support period required = ______
   (d) Minimum stock level (MSL) = ______
   (e) Due-in quantity = 0 tires
   (f) Ser. = 54 tires
   (g) $\text{OQ} = \left(\frac{\_\_\text{ tires}}{\text{flight}} \times \frac{\_\_\text{ flights}}{\text{year}} \times \frac{1 \text{ year}}{12 \text{ months}}\right) \times 12 \text{ months}$
   $+ \_\_\text{ tires} - \_\_\text{ tires} - \_\_\text{ tires}$
   $= \_\_\text{ tires}$
2. Write a sentence to explain what your answer means.

CHAPTER 1

APPLICATION

Precision-Machined Computer Disks

Imagine a 747 airplane flying $\frac{1}{32}$ in. above the ground. This is the equivalent of the head of a disk drive flying 60 mi/h over a computer disk at a height of 1 to 2 μin. With such a low glide height, the disk must be extremely smooth and flat. Thickness is also an issue, since up to 10 disks may be stacked in one hard drive, with a head on the top and bottom of each disk. When disks are stacked, spacer rings are used to separate them. The disks need to be of uniform thickness so that the spacer rings can be uniform for ease of assembly.

Several computer component manufacturers produce aluminum disk substrates for the magnetic thin-film disks used in computer hard disk drives. The manufacturing process begins with operators measuring the thickness of the unmachined disks and sorting them into six to eight categories, which differ by $\frac{1}{10,000}$ in. The next step is chemical etching, which is done for a specified time determined by the initial thickness of the disks. The chemicals remove the tough oxide surface layer, making the disks easier to grind, and they reduce the thickness variation to $\pm\frac{2}{10,000}$ in., so that after the chemical bath the disks have four categories of thickness.

Disks have a hole in the center and an outside diameter of 95 mm for desktop computers or 65 mm for laptops. After the etching, the diameters are adjusted and the outer and inner edges are beveled at an angle specified by the customer. The beveling aids in the automated handling of the disks during later processing. The beveled edge tolerance is $\pm\frac{20}{10,000}$ in., while the inner and outer diameter tolerances are $\pm\frac{4}{10,000}$ in. and $\pm\frac{10}{10,000}$ in., respectively. The machine that bevels the edges holds a disk in place by a vacuum, which causes the disk to lose its flatness; the disks are then moved through an oven, since heat reestablishes the flatness.

Finally, the disks are ground smooth and flat to a uniform thickness of 0.0307 in. with a tolerance of $\pm\frac{3}{10,000}$ in. The grinding technology is so important to the process that some companies custom fabricate their own grinding stones.

Flatness is measured by subtracting the lowest point on the disk from the highest point. Usually the specifications require that the maximum difference be 200 or 300 μin. Smoothness or surface finish has two measurements: roughness average (RA) and roughness total (RT). RA is the average of the many peaks and valleys as measured with a contact stylus. The RA is usually between 0.4 and 0.5 μin. but must be less than or equal to 0.8 μin. RT is the difference of the highest peak and the lowest valley (i.e., the worst case). It is usually 4 to 5 μin. but must be less than or equal to 8 μin.

Because thickness, flatness, and smoothness are all critical to the finished disk substrate, manufacturers use advanced test equipment to monitor disks at each production stage.

This description of the procedure for manufacturing computer disk substrates illustrates the importance of accuracy and precision in measurements. Almost all manufacturing processes and technical applications use these concepts.

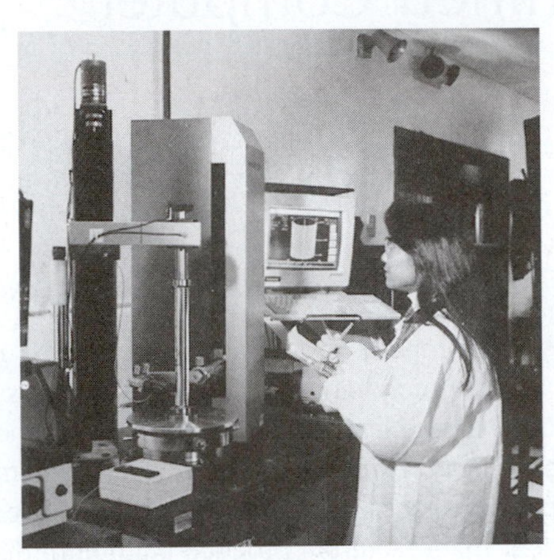

A laboratory technician uses video monitors to conduct a quality control check.

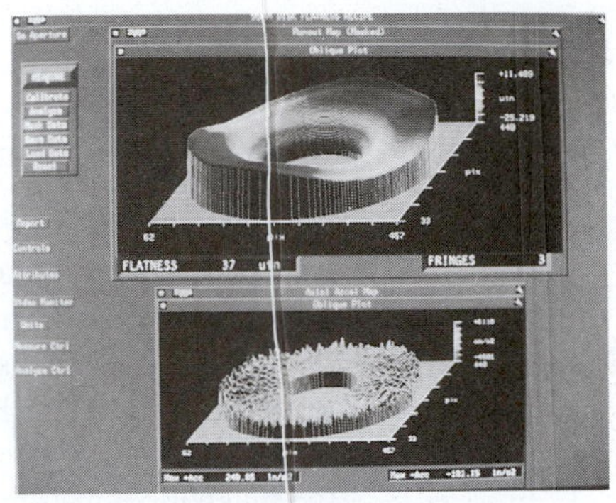

This laser monitor displays the surface finish of a disk substrate.

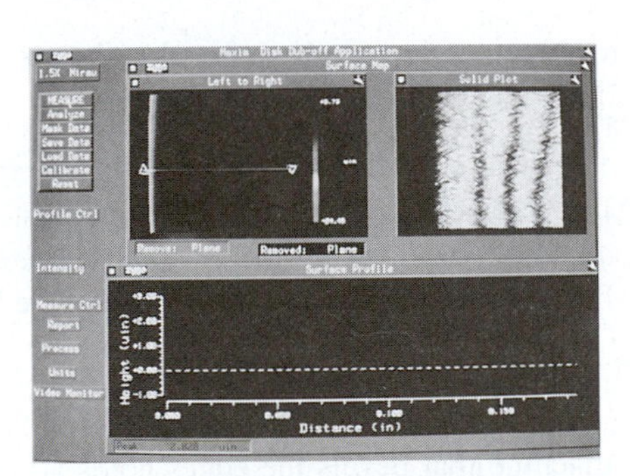

A computer displays a graph of a cross section of a disk surface, where the height is measured in microinches.

Substrates are used in computer laser printers.

(Photos courtesy of Cerion Technologies, Inc. Champaign, IL.)

2
Review of Geometry

INTRODUCTION

Lines, angles, area, and volume have many applications in all areas of technology. A basic knowledge of the fundamentals of geometry is necessary to solve technical problems as well as to understand mathematical discussions and developments. We shall briefly review the most basic, often-used geometric terms, relationships, formulas, and theorems.

Objectives

- Apply the basic definitions and relationships for angles, lines, and geometric figures to geometric and technical applications.
- Find the area and the perimeter of triangles and quadrilaterals.
- Find the area and the circumference of circles.
- Find the volume, the lateral surface area, and the total surface area of geometric solids.

2.1 ANGLES AND LINES

Some basic fundamentals of geometry must be understood to solve many technical applications and to follow some of the mathematical developments and discussions in this text, in technical support courses, and in many on-the-job training programs.

An **angle** is formed by two lines with one common point. The common point is called the **vertex** of the angle. The parts of the lines that form the angle are called the **sides** of the angle. An angle is often designated by a number, a single letter, or three letters; for example, in Fig. 2.1 the angle is referred to as $\angle 1$, $\angle B$, or $\angle ABC$. The middle letter of the three letters must be at the vertex.

The measure of an angle is the amount of rotation needed to make one side of the angle coincide with the other. One of the most common units for the measure of an angle is the degree (°); one degree is 1/360 of one complete revolution, that is, 360° = one revolution.

Angles may be classified by their degree measure. A **right angle** is an angle with a measure of 90°. In a sketch or a diagram, a right angle is often noted by placing ⊥ or ⊥ in the angle as shown in Fig. 2.2(a). An **acute angle** is an angle with a measure

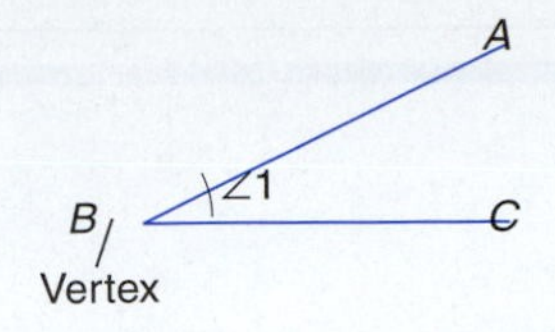

Figure 2.1 Basic parts of an angle.

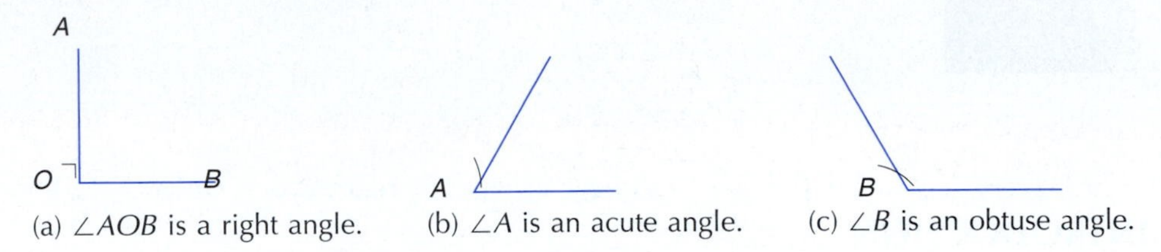

(a) $\angle AOB$ is a right angle. (b) $\angle A$ is an acute angle. (c) $\angle B$ is an obtuse angle.

Figure 2.2

less than 90°. An **obtuse angle** is an angle with a measure greater than 90° but less than 180°. A **straight angle** is an angle with a measure of 180°. In summary:

Name of angle	*Measure of angle*
Right angle	90°
Acute angle	Between 0° and 90°
Obtuse angle	Between 90° and 180°
Straight angle	180°

Let's first study some geometric relationships of angles and lines in the same plane. When two different lines **intersect,** they have only one point in common (see Fig. 2.3).

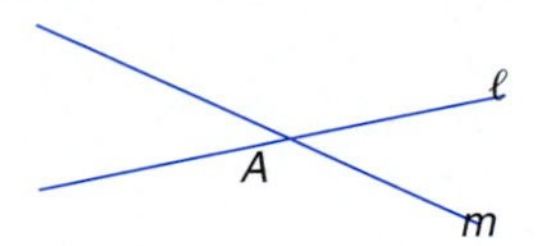

Figure 2.3 Lines ℓ and m intersect at point A.

Two lines in the same plane are **parallel** (||) if they do not intersect even when extended (see Fig. 2.4).

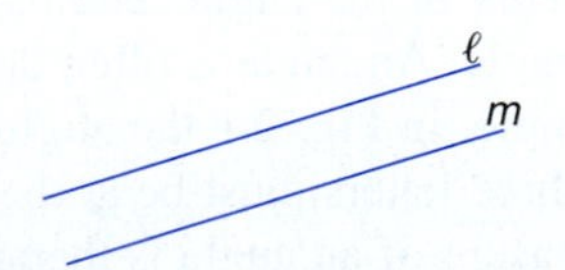

Figure 2.4 Lines ℓ and m are parallel.

Two angles are **adjacent** if they have a common vertex and a common side between, with no other interior points. In Fig. 2.5, $\angle 1$ and $\angle 2$ are adjacent angles because both have a common vertex B and a common side BD between them.

Two lines in the same plane are **perpendicular** ($\perp$) if they intersect and form equal adjacent angles. Each of the adjacent angles is a right angle. In Fig. 2.6, $AB \perp CD$ because $\angle 1$ and $\angle 2$ are equal adjacent (right) angles.

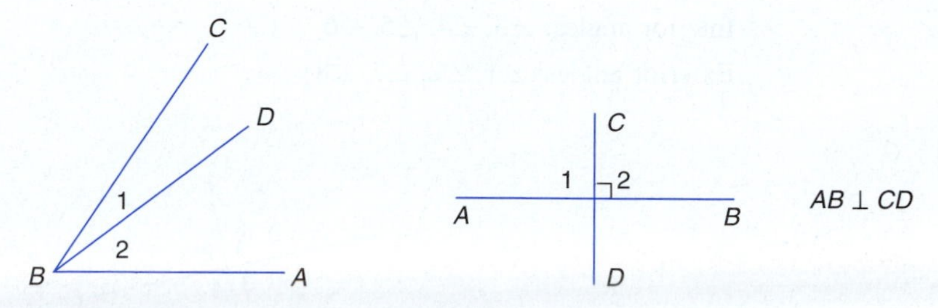

Figure 2.5 Adjacent angles. **Figure 2.6** Perpendicular lines ($AB \perp CD$).

Complementary angles are two angles the sum of whose measures is 90°. In Fig. 2.7(a), $\angle A$ and $\angle B$ are complementary: $64° + 26° = 90°$. In Fig. 2.7(b), $\angle LMN$ and $\angle NMP$ are complementary: $55° + 35° = 90°$.

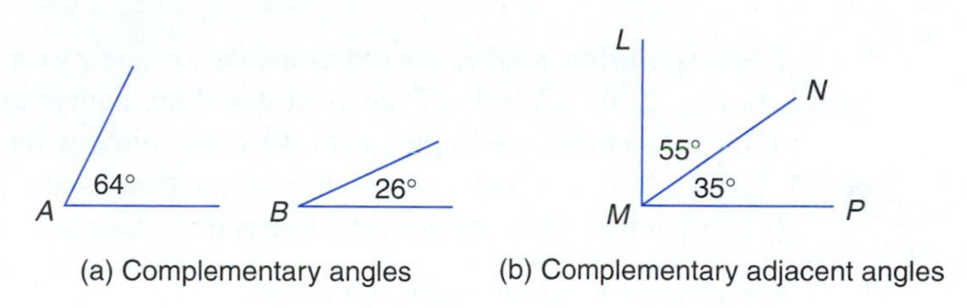

Figure 2.7

Supplementary angles are two angles the sum of whose measures is 180°. In Fig. 2.8(a), $\angle C$ and $\angle D$ are supplementary: $60° + 120° = 180°$. Two adjacent angles with their exterior sides in a straight line are *supplementary*. In Fig. 2.8(b), $\angle 1$ and $\angle 2$ are supplementary: $\angle 1 + \angle 2 = 180°$.

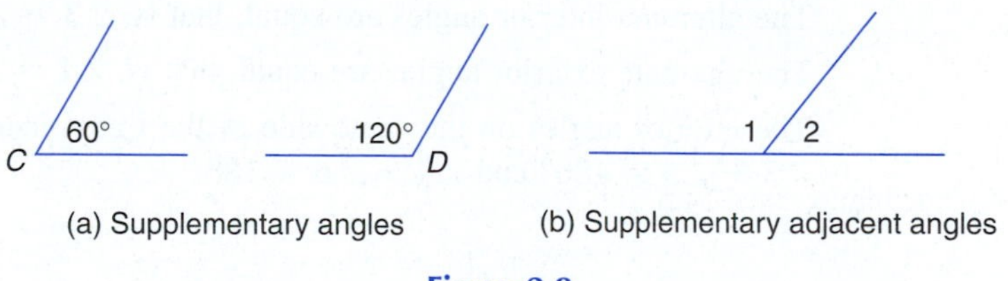

Figure 2.8

Two **vertical angles** are the opposite angles formed by two intersecting lines. In Fig. 2.9, $\angle 1$ and $\angle 3$ are vertical angles, as are $\angle 2$ and $\angle 4$.

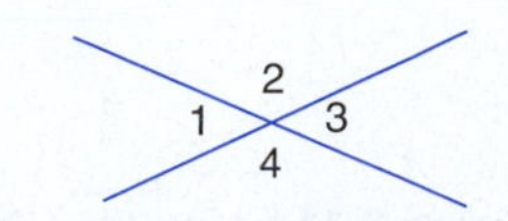

Figure 2.9 Vertical angles.

A **transversal** is a line that intersects two or more lines in different points in the same plane (Fig. 2.10). **Interior angles** are angles formed inside lines ℓ and m by the transversal. **Exterior angles** are angles formed outside lines ℓ and m by the transversal.

Interior angles: $\angle 3, \angle 4, \angle 5, \angle 6$

Exterior angles: $\angle 1, \angle 2, \angle 7, \angle 8$

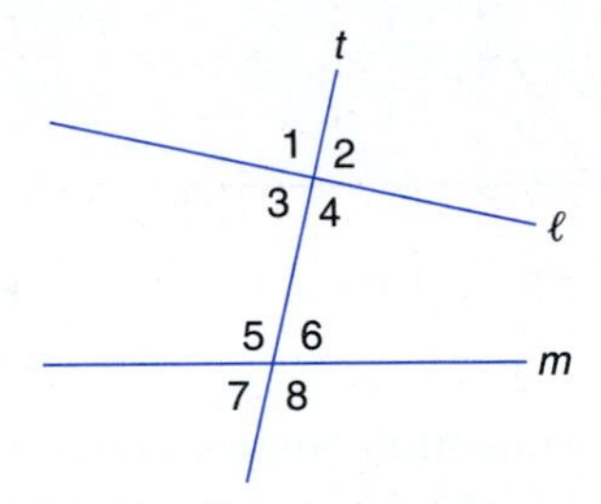

Figure 2.10 Line t is a transversal of lines ℓ and m.

Corresponding angles are exterior-interior angles on the same side of the transversal. In Fig. 2.10, $\angle 3$ and $\angle 7$ are corresponding angles, as are $\angle 2$ and $\angle 6$.

Alternate angles are angles with different vertices on opposite sides of the transversal. In Fig. 2.10, $\angle 1$ and $\angle 6$ are alternate angles, as are $\angle 1$ and $\angle 8$.

If two parallel lines are cut by a transversal, then

- the *corresponding* angles are *equal.*
- the *alternate-interior* angles are *equal.*
- the *alternate-exterior* angles are *equal.*
- the *interior* angles on the same side of the transversal are *supplementary.*

In Fig. 2.11, lines ℓ and m are parallel and t is a transversal.

The corresponding angles are equal, that is, $\angle 1 = \angle 5$, $\angle 2 = \angle 6$, $\angle 3 = \angle 7$, and $\angle 4 = \angle 8$.

The alternate-interior angles are equal, that is, $\angle 3 = \angle 6$ and $\angle 4 = \angle 5$.

The alternate-exterior angles are equal, that is, $\angle 1 = \angle 8$ and $\angle 2 = \angle 7$.

The interior angles on the same side of the transversal are supplementary, that is, $\angle 3 + \angle 5 = 180°$ and $\angle 4 + \angle 6 = 180°$.

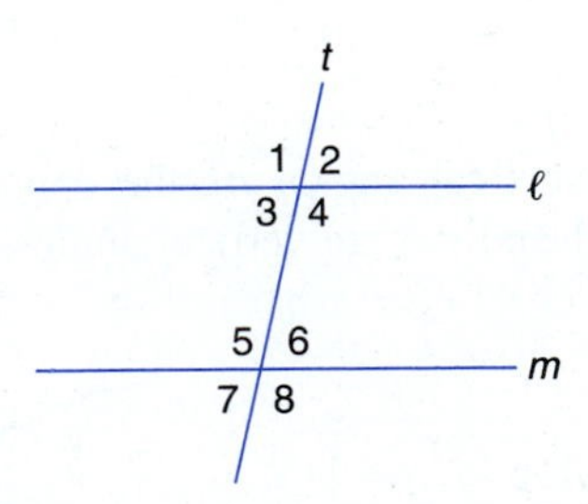

Figure 2.11 Line t is a transversal of parallel lines ℓ and m.

EXAMPLE 1

In Fig. 2.12, lines ℓ and m are parallel and t is a transversal. If $\angle 2 = 70°$, find the measure of $\angle 5$.

There are several ways of finding $\angle 5$. We will show two ways.

Method 1: Since $\angle 2$ and $\angle 4$ form a straight angle, they are supplementary. So, $\angle 4 = 180° - \angle 2 = 180° - 70° = 110°$.
Since $\angle 4$ and $\angle 5$ are alternate-interior angles, they are equal. So, $\angle 4 = \angle 5 = 110°$.

Method 2: Since $\angle 2$ and $\angle 3$ are vertical angles, they are equal. So, $\angle 2 = \angle 3 = 70°$. Since $\angle 3$ and $\angle 5$ are interior angles on the same side of the transversal, they are supplementary. So, $\angle 5 = 180° - \angle 3 = 180° - 70° = 110°$.

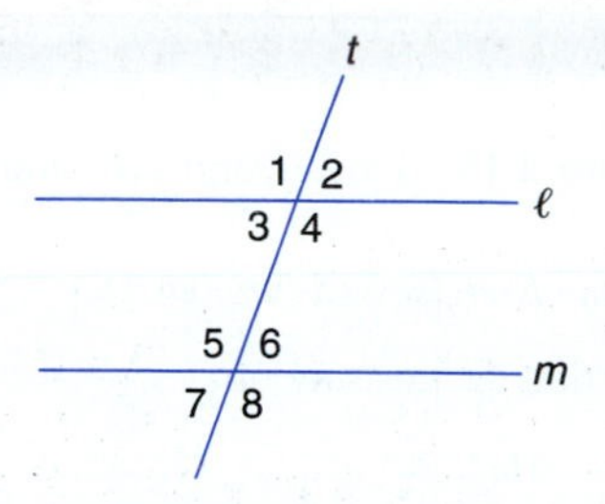

Figure 2.12 Line t is a transversal of parallel lines ℓ and m.

A *straight line* is infinitely long. A *line segment* is a specific portion of a line (see Fig. 2.13).

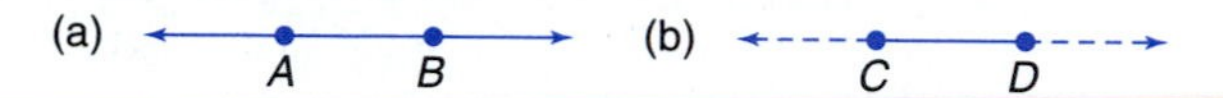

Figure 2.13 (a) Line AB is infinitely long and contains points A and B. (b) Line segment CD is the portion of the line that contains points C and D and all of the points between C and D.

Exercises 2.1

Use Fig. 2.14 for Exercises 1 through 8.

1. Name two acute angles.
2. Name two obtuse angles.
3. Name two right angles.
4. Name two acute vertical angles.
5. Name two obtuse vertical angles.
6. Name two equal angles that form a straight angle.

If $\angle 1 = 125°$, find the measure of

7. $\angle 2$.
8. $\angle 4$.

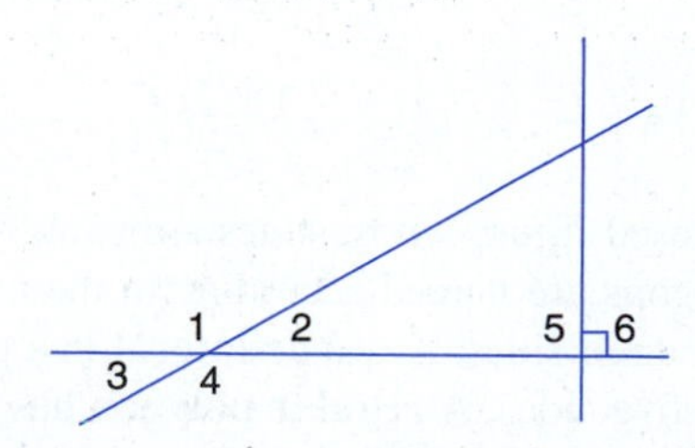

Figure 2.14

Use Fig. 2.15 for Exercises 9 through 14.

9. Name three pairs of complementary angles.

10. Name two pairs of supplementary angles.

If $\angle 5 = 50°$, find the measure of

11. $\angle 6$.

12. $\angle 7$.

13. $\angle 2$.

14. $\angle 4$.

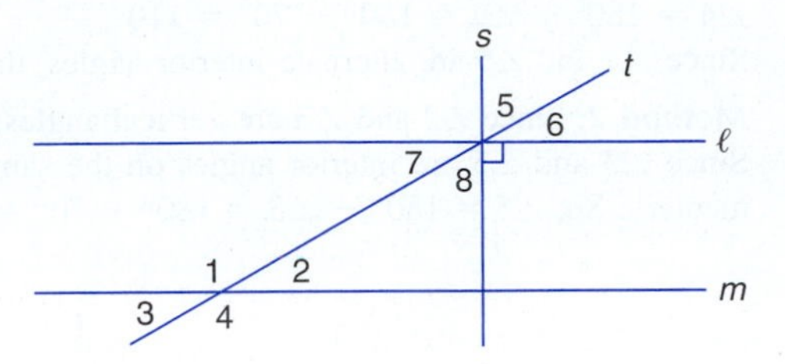

Figure 2.15 Lines s and t are transversals of parallel lines ℓ and m.

Use Fig. 2.16 for Exercises 15 through 18.

If $\angle 1 = 130°$, find the measure of

15. $\angle 4$.

16. $\angle 8$.

17. $\angle 5$.

18. $\angle 7$.

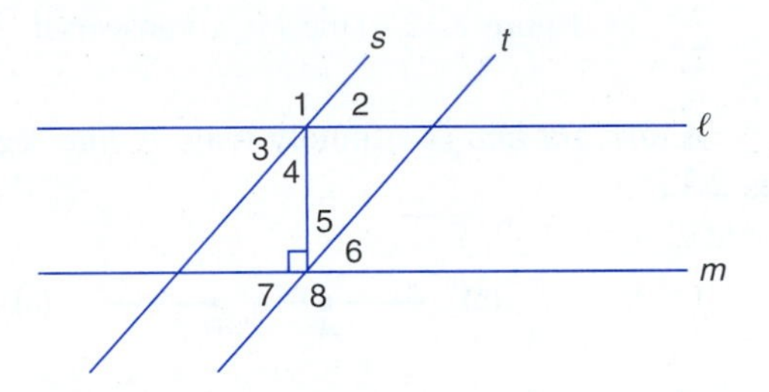

Figure 2.16 Parallel lines s and t are transversals of parallel lines ℓ and m.

19. A plumber needs to connect two parallel pipes as shown in Fig. 2.17. Find $\angle 1$.

20. A truss is shown in Fig. 2.18. Given $\angle C = 55°$, $\angle CAD = 80°$, and $\angle ADB = 85°$, find the measure of $\angle BDE$.

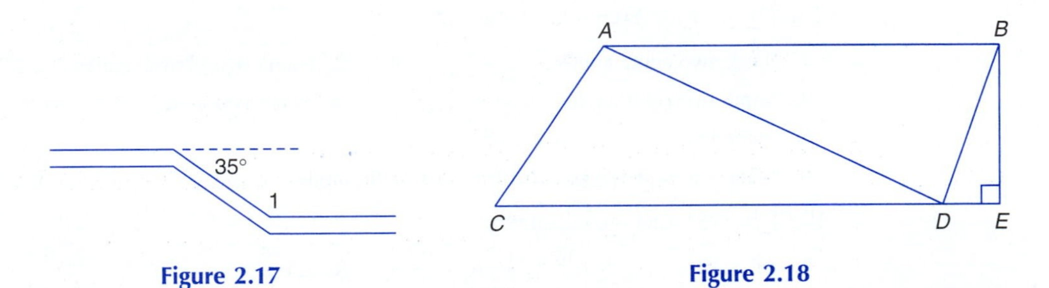

Figure 2.17

Figure 2.18

2.2 TRIANGLES

A **polygon** is a closed figure whose sides are straight-line segments. A polygon is shown in Fig. 2.19. Polygons are named according to the number of sides they have. A **triangle** is a polygon with three sides. A **quadrilateral** is a polygon with four sides. A **pentagon** is a polygon with five sides. A **regular polygon** has all of its sides equal and all of its interior angles equal. (See Fig. 2.20.)

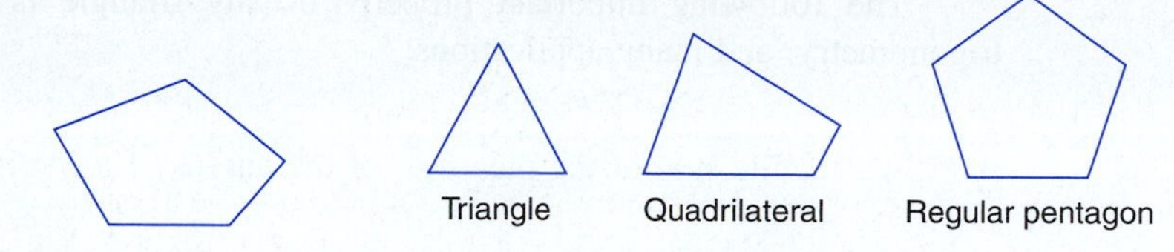

Figure 2.19 Polygon **Figure 2.20**

Polygons with 6 to 10 sides are named as follows.

Number of sides	*Name of polygon*
6	Hexagon
7	Heptagon
8	Octagon
9	Nonagon
10	Decagon

Triangles are often classified in two ways:

1. By the number of equal sides
2. By the measures of the angles of the triangle

An **equilateral** triangle is a triangle in which all three sides are equal. All three angles are also equal (60°). An **isosceles** triangle is a triangle in which two sides are equal. The angles opposite the equal sides are also equal. A **scalene** triangle is a triangle in which no two sides are equal. No angles are equal either. Figure 2.21 shows triangles classified according to their sides.

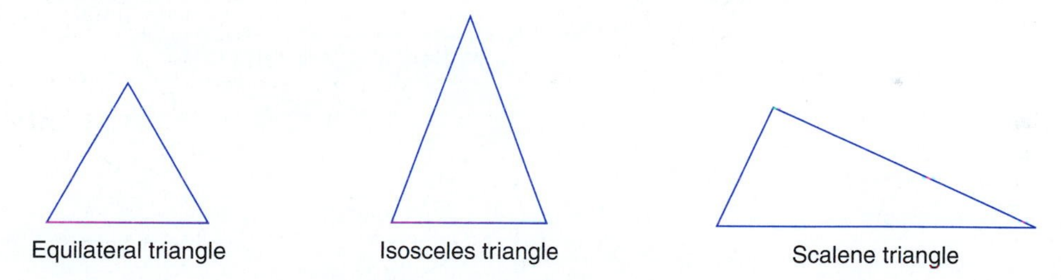

Figure 2.21 Triangles classified by

Triangles may also be classified or named in terms of the measures of their angles. An **acute triangle** is a triangle in which all three angles are acute. An **obtuse triangle** is a triangle with an obtuse angle. A **right triangle** is a triangle with a right angle. Figure 2.22 shows triangles classified according to their angles. An **oblique triangle** is a triangle that does not contain a right angle.

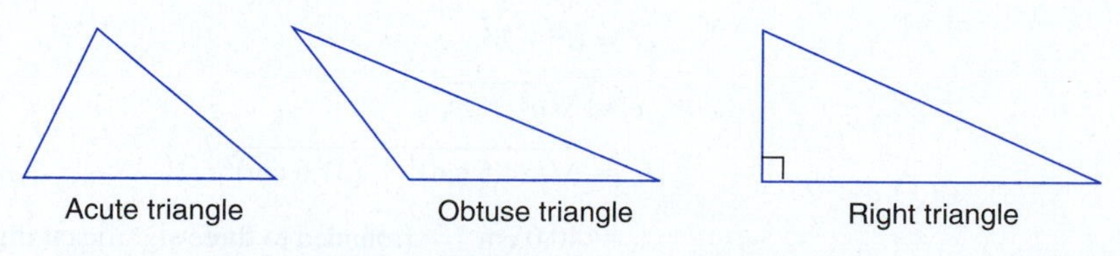

Figure 2.22 Triangles classified by

The following important property of any triangle is often used in geometry, trigonometry, and many applications:

The sum of the measures of the angles of any triangle is 180°.

This property may be shown using Fig. 2.23. First, draw any triangle ABC. Then, draw a line through vertex B and parallel to side AC. Note that

$$\angle 1 + \angle B + \angle 2 = 180° \qquad \text{(a straight angle)}$$

Then, note that the following two sets of alternate interior angles are equal; $\angle A = \angle 1$ and $\angle C = \angle 2$. Substituting these quantities in the equation above, we have

$$\angle A + \angle B + \angle C = 180°$$

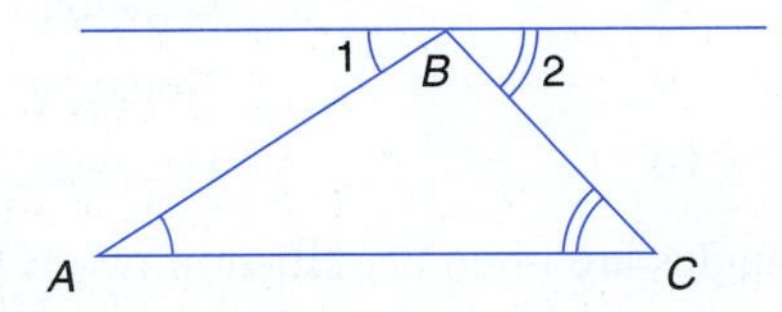

Figure 2.23 $\triangle ABC$ with a line through vertex B and parallel to side AC.

The lengths of the sides of a triangle are often represented by the lowercase letters a, b, and c. In a right triangle, the side opposite the right angle is called the **hypotenuse**, c, and the other two sides are called **legs**, a and b, as shown in Fig. 2.24. The **Pythagorean theorem** relates the lengths of the sides of any right triangle as follows: *The square of the hypotenuse of a right triangle is equal to the sum of the squares of the two legs.*

PYTHAGOREAN THEOREM

$$c^2 = a^2 + b^2$$

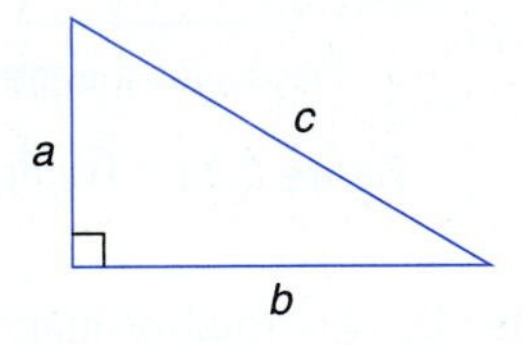

Figure 2.24

EXAMPLE 1

Find the length of the hypotenuse of the triangle in Fig. 2.25.

Substitute $a = 24.5$ cm and $b = 31.6$ cm in the formula as follows:

$$c^2 = a^2 + b^2$$

$$c = \sqrt{a^2 + b^2}$$

$$c = \sqrt{(24.5 \text{ cm})^2 + (31.6 \text{ cm})^2}$$

$$= 40.0 \text{ cm} \qquad \text{(rounded to three significant digits)}$$

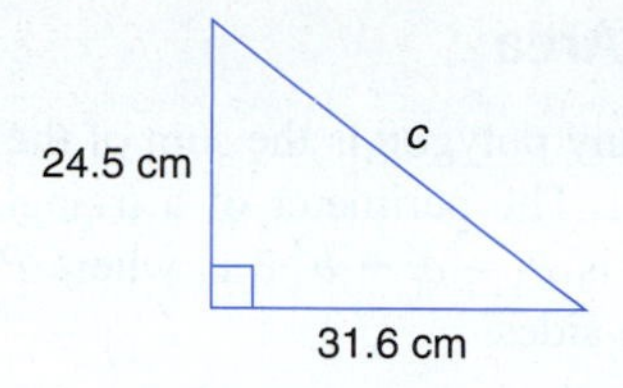

Figure 2.25

EXAMPLE 2

Find the length of the unknown side in Fig. 2.26.

Substitute $c = 55.0$ cm and $b = 50.0$ cm in the formula as follows:

$$c^2 = a^2 + b^2$$

$$a^2 = c^2 - b^2 \qquad \text{(Solve for } a^2.\text{)}$$

$$a = \sqrt{c^2 - b^2}$$

$$a = \sqrt{(55.0 \text{ cm})^2 - (50.0 \text{ cm})^2}$$

$$a = 22.9 \text{ cm}$$

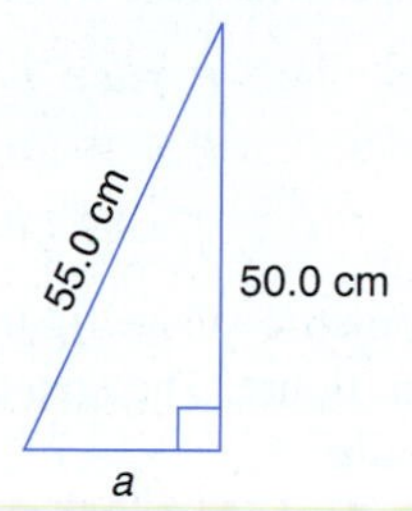

Figure 2.26

A **median** of a triangle is a line segment joining any vertex to the *midpoint* of the opposite side. If $AD = DB$ in Fig. 2.27(a), then CD is a median. If all three medians of a triangle are drawn, they intersect at a single point called the *centroid* of the triangle.

An **altitude** (height) of a triangle is the *perpendicular* line segment from any vertex to the opposite side (or the opposite side extended). If $CD \perp AB$ as in Fig. 2.27(b), then CD is an altitude. If all three altitudes of a triangle are drawn, they intersect at a single point called the *orthocenter* of the triangle.

An **angle bisector** of a triangle is a line segment that bisects any angle and intersects the opposite side. If $\angle 1 = \angle 2$ in Fig. 2.27(c), then CD is an angle bisector. If all three angle bisectors of a triangle are drawn, they intersect at a single point called the *circumcenter* of the circle.

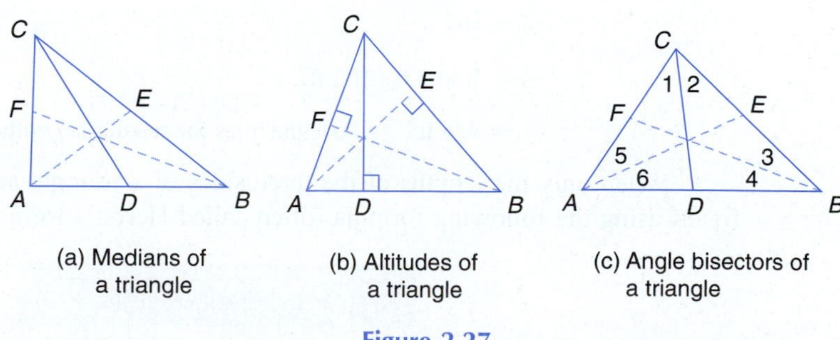

Figure 2.27

Perimeter and Area

The **perimeter** of any polygon is the sum of the lengths of the sides, or the total distance around the polygon. The perimeter of a triangle is the sum of the lengths of the three sides. The formula is $P = a + b + c$, where P is the perimeter and a, b, and c are the lengths of the three sides.

EXAMPLE 3

Find the perimeter of the triangle in Fig. 2.28.

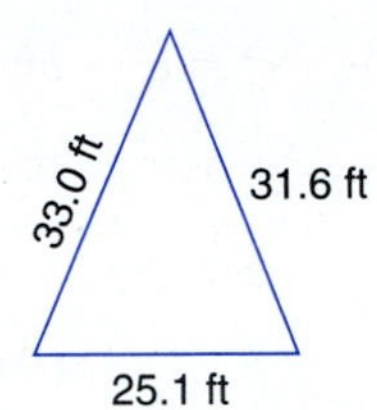

Figure 2.28

Using the perimeter formula, we have

$$P = a + b + c$$
$$P = 25.1 \text{ ft} + 31.6 \text{ ft} + 33.0 \text{ ft}$$
$$= 89.7 \text{ ft}$$

The **area** of a geometric plane figure is the number of square units of measure enclosed by the geometric figure. The area of a triangle with base b and height h is found using the following formula:

$$A = \tfrac{1}{2}bh$$

EXAMPLE 4

Find the area of the triangle in Fig. 2.29.

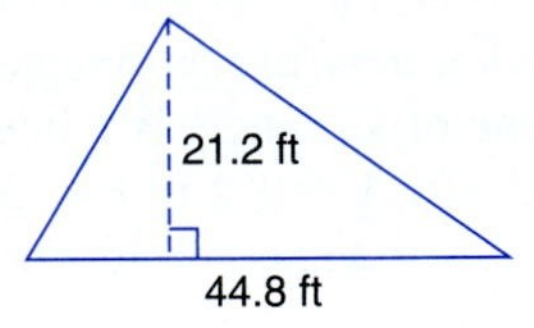

Figure 2.29

Using the area formula, we have

$$A = \tfrac{1}{2}bh$$
$$A = \tfrac{1}{2}(44.8 \text{ ft})(21.2 \text{ ft})$$
$$= 475 \text{ ft}^2 \quad \text{(Use the rules for calculating with measurements.)}$$

When only the lengths of the three sides of a triangle are known, its area may be found using the following formula (often called Heron's formula)

$$A = \sqrt{s(s - a)(s - b)(s - c)}$$

where a, b, and c are the lengths of the three sides of the triangle and $s = \tfrac{1}{2}(a + b + c)$.

EXAMPLE 5

Find the area of the triangle in Fig. 2.30.

Using Heron's formula, we have

$$s = \tfrac{1}{2}(a + b + c)$$

$$s = \tfrac{1}{2}(215 \text{ m} + 244 \text{ m} + 147 \text{ m})$$

$$= 303 \text{ m}$$

Then,

$$A = \sqrt{s(s - a)(s - b)(s - c)}$$

$$A = \sqrt{(303 \text{ m})(303 \text{ m} - 215 \text{ m})(303 \text{ m} - 244 \text{ m})(303 \text{ m} - 147 \text{ m})}$$

$$= 15{,}700 \text{ m}^2 \qquad \text{(rounded to three significant digits)}$$

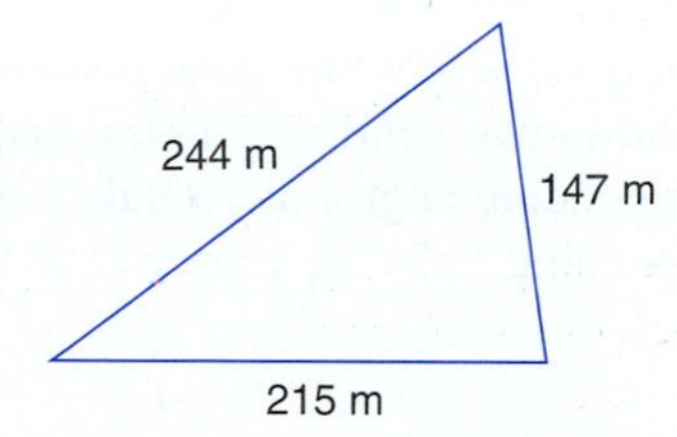

Figure 2.30

EXAMPLE 6

Find the area of the triangular metal plate in Fig. 2.31.

$$A = \tfrac{1}{2}bh$$

$$A = \tfrac{1}{2}(30.5 \text{ in.})(28.0 \text{ in.})$$

$$= 427 \text{ in}^2$$

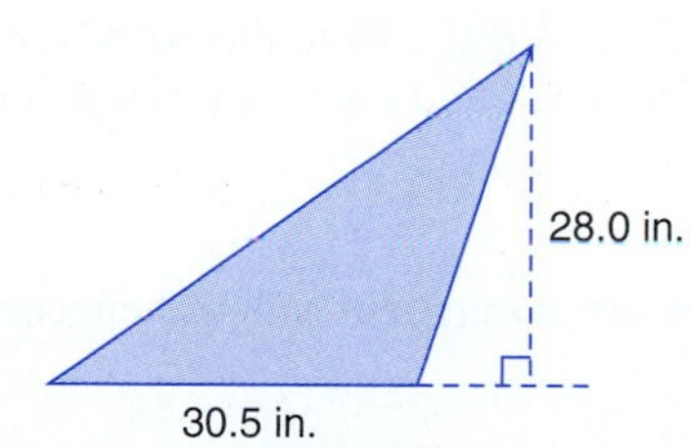

Figure 2.31

EXAMPLE 7

Find the area of the triangular plot of land in Fig. 2.32.

Using Heron's formula, we have

$$s = \tfrac{1}{2}(a + b + c)$$

$$s = \tfrac{1}{2}(1250 \text{ ft} + 1850 \text{ ft} + 3060 \text{ ft})$$

$$= 3080 \text{ ft}$$

Then,

$$A = \sqrt{s(s - a)(s - b)(s - c)}$$

$$A = \sqrt{(3080 \text{ ft})(3080 \text{ ft} - 1250 \text{ ft})(3080 \text{ ft} - 1850 \text{ ft})(3080 \text{ ft} - 3060 \text{ ft})}$$

$$= 372{,}000 \text{ ft}^2 \qquad \text{(rounded to three significant digits)}$$

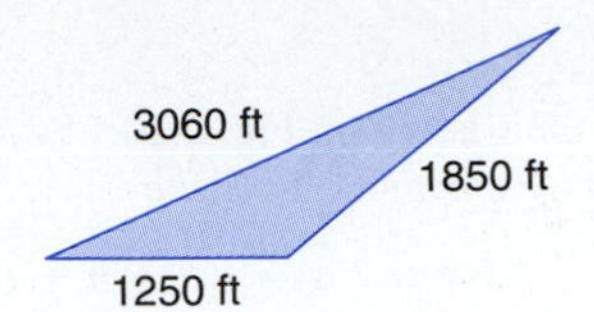

Figure 2.32

Similar Triangles

Two triangles with the same shape are called **similar triangles.** The conditions for similar triangles are:

1. The measures of their corresponding angles are equal.
2. The lengths of their corresponding sides are proportional.

If one of these conditions is met, the other condition is also met.

Figure 2.33 shows two similar triangles, that is, $\triangle ABC$ is similar ($\sim$) to $\triangle A'B'C'$. Note that the corresponding angles are equal: $\angle A = \angle A'$, $\angle B = \angle B'$, and $\angle C = \angle C'$ and that the corresponding sides are proportional (by a factor of 2).

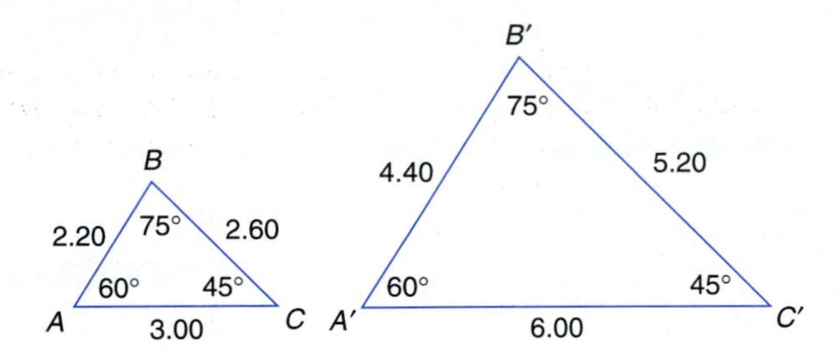

Figure 2.33 $\triangle ABC \sim \triangle A'B'C'$. Note the equal corresponding angles, $\angle A = \angle A'$, $\angle B = \angle B'$, and $\angle C = \angle C'$, and the proportional corresponding sides (factor of 2).

Two triangles are **congruent** when their corresponding angles and corresponding sides are equal.

EXAMPLE 8

A ramp is to be built so that it reaches a height of 4.50 ft over a horizontal distance of 12.00 ft. Two braces equidistant from the ends and each other are used to support the ramp as shown in Fig. 2.34. Find the height of the braces.

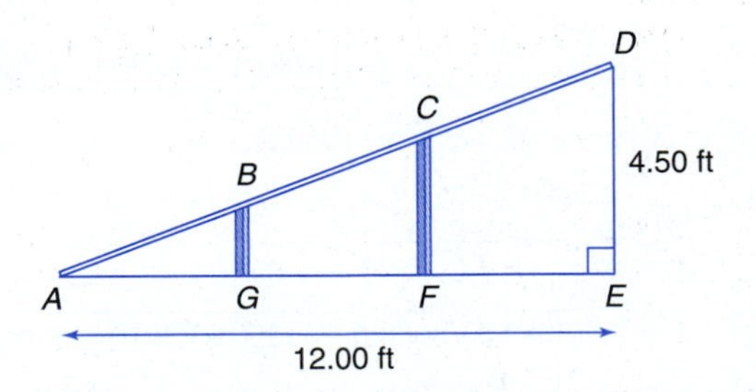

Figure 2.34

The braces will be placed 12.00 ft/3 or 4.00 ft apart. Next, note that $\triangle ABG \sim \triangle ADE$ because their corresponding angles are equal. Thus, the lengths of their corresponding sides are proportional.

$$\frac{AG}{AE} = \frac{BG}{DE}$$

$$\frac{4.00 \text{ ft}}{12.00 \text{ ft}} = \frac{BG}{4.50 \text{ ft}}$$

$$BG = \frac{(4.00 \text{ ft})(4.50 \text{ ft})}{12.00 \text{ ft}}$$

$$BG = 1.50 \text{ ft}$$

Next, note that $\triangle ACF \sim \triangle ADE$ because their corresponding angles are equal. Thus, the lengths of their corresponding sides are proportional.

$$\frac{AF}{AE} = \frac{CF}{DE}$$

$$\frac{8.00 \text{ ft}}{12.00 \text{ ft}} = \frac{CF}{4.50 \text{ ft}}$$

$$CF = \frac{(8.00 \text{ ft})(4.50 \text{ ft})}{12.00 \text{ ft}}$$

$$CF = 3.00 \text{ ft}$$

Exercises 2.2

(Use the rules for calculating with measurements.)

Find the length of the unknown side in each right triangle.

1. 6.00 cm; 8.00 cm; c

2. 24.0 in.; 11.0 in.; c

3. b; 2250 m; 3640 m

4. 295 ft; a; 182 ft

*Find **(a)** the area and **(b)** the perimeter of each triangle.*

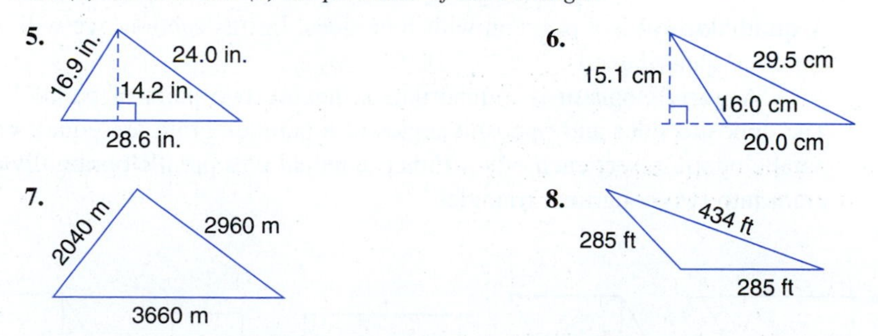

9. Find the area and the perimeter of an equilateral triangle with sides 124 cm long.

10. Find the area and the perimeter of an isosceles triangle with two sides 45.6 in. long and the third side 18.0 in. long.

11. Find the area and the perimeter of an isosceles right triangle with legs 275 ft long.

12. Can you draw an equilateral right triangle? If so, describe the triangle. If you cannot, why not?
13. Describe the sets of medians, altitudes, and angle bisectors in an equilateral triangle.
14. Are congruent triangles similar? Why or why not?
15. If the measures of two angles of a triangle are 45.7° and 65.4°, what is the measure of the third angle?
16. If the measure of one angle of a right triangle is 35.75°, what are the measures of the other two angles?
17. Two pieces of steel are welded to form a right angle. If their lengths are 6.00 ft and 9.50 ft, respectively, what is the distance between the two unwelded ends?
18. A helicopter flies due east for 65.3 mi, then due north for 21.5 mi. How far is the helicopter from the beginning point?
19. A tower is 275 ft tall. If a guy wire is attached 45 ft from the top and anchored in the ground 60.0 ft from the base of the tower, how long is the guy wire?
20. A piece of 6.00-in.-diameter round metal stock is to be milled into a square piece of stock with the largest dimensions possible. What will be the length of each side of the square piece of stock?
21. A person measures the lengths of fencing along a triangular lot as follows: 925 ft, 624 ft, and 835 ft. What is the area of the field?
22. Find the area of a triangular piece of metal whose sides measure 42.1 cm, 28.5 cm, and 38.0 cm.
23. A tree casts a shadow 45 ft long when a person 6.0 ft tall casts a shadow 4.0 ft long. How tall is the tree?
24. In Fig. 2.35, $AB \parallel DE$. Find the length of **(a)** DE and **(b)** BC.

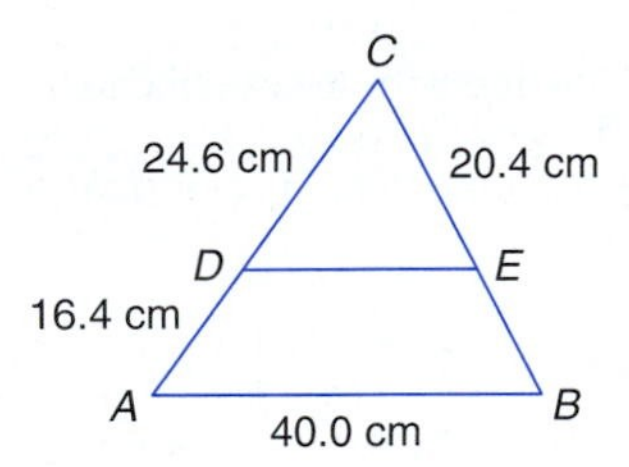

Figure 2.35

2.3 QUADRILATERALS

A **quadrilateral** is a polygon with four sides. In this section, we will consider the most common quadrilaterals.

A **parallelogram** is a quadrilateral having two pairs of parallel sides (Fig. 2.36). The opposite sides and opposite angles of a parallelogram are equal. The diagonals of a parallelogram bisect each other. Either diagonal of a parallelogram divides the parallelogram into two congruent triangles.

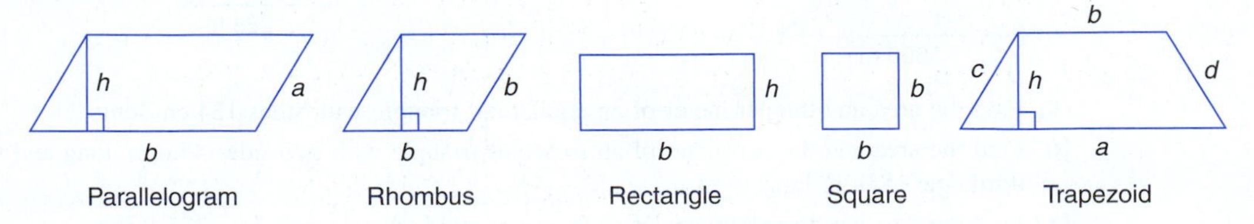

Figure 2.36

A **rhombus** is a parallelogram with equal sides (Fig. 2.36).

A **rectangle** is a parallelogram with right angles (Fig. 2.36). The longer side is often called the *length* and the shorter side is called the *width*. The diagonals of a rectangle are equal.

A **square** is a rectangle with equal sides (Fig. 2.36).

A **trapezoid** is a quadrilateral with only one pair of parallel sides (Fig. 2.36). The parallel sides are called *bases*.

SUMMARY OF FORMULAS FOR AREA AND PERIMETER OF QUADRILATERALS

Quadrilateral	*Area*	*Perimeter*
Parallelogram	$A = bh$	$P = 2(a + b)$
Rhombus	$A = bh$	$P = 4b$
Rectangle	$A = bh$	$P = 2(b + h)$
Square	$A = b^2$	$P = 4b$
Trapezoid	$A = \frac{1}{2}h(a + b)$	$P = a + b + c + d$

Note: Follow the rules for working with measurements in this chapter.

EXAMPLE 1

Find the area and the perimeter of the parallelogram in Fig. 2.37.

The formula for the area of a parallelogram is

$$A = bh$$
$$A = (48.0 \text{ cm})(24.0 \text{ cm})$$
$$= 1150 \text{ cm}^2 \quad \text{(rounded to three significant digits)}$$

The perimeter is

$$P = 2(a + b)$$
$$P = 2(30.0 \text{ cm} + 48.0 \text{ cm})$$
$$= 156.0 \text{ cm}$$

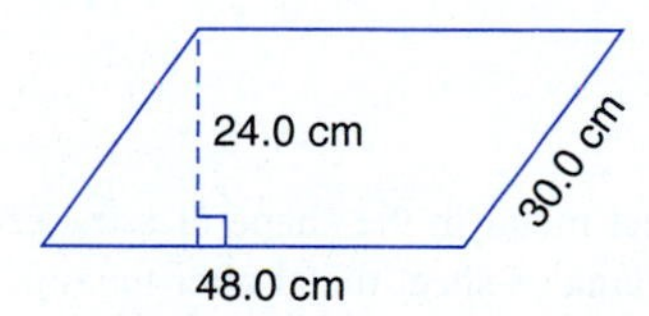

Figure 2.37

EXAMPLE 2

Find the length of the diagonal braces for the gate in Fig. 2.38.

Since a diagonal brace is the hypotenuse of a right triangle, substitute $a = 48.0$ in. and $b = 72.0$ in. in the Pythagorean formula as follows:

$$c^2 = a^2 + b^2$$
$$c = \sqrt{a^2 + b^2}$$
$$c = \sqrt{(48.0 \text{ in.})^2 + (72.0 \text{ in.})^2}$$
$$= 86.5 \text{ in.}$$

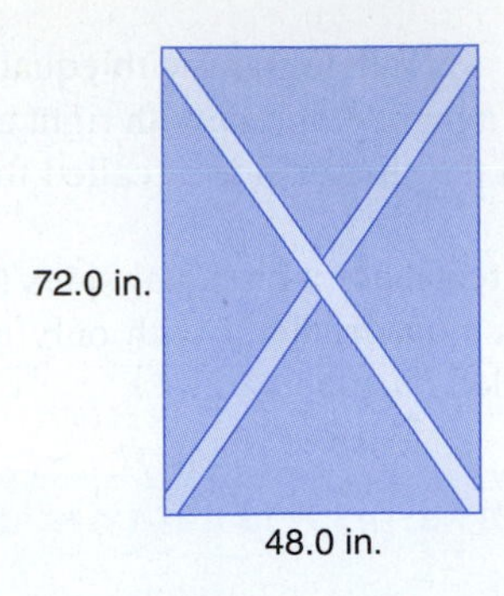

Figure 2.38

EXAMPLE 3

A vacant rectangular lot measures 124.6 ft by 95.5 ft. It must be fenced and then fertilized. (a) If fencing is installed at \$6.75 per running foot, find the cost of fencing the lot. (b) If one bag of fertilizer covers 2500 ft^2 and costs \$13.95, find the cost to fertilize the grass in the lot.

(a) The length of fencing needed is the same as the perimeter of the lot. The perimeter is

$$P = 2(b + h)$$
$$P = 2(124.6 \text{ ft} + 95.5 \text{ ft})$$
$$= 440.2 \text{ ft}$$

Thus, the cost is (440.2 ft) (\$6.75/ft) = \$2971.35.
(b) First, find the area.

$$A = bh$$
$$A = (124.6 \text{ ft})(95.5 \text{ ft})$$
$$= 11{,}900 \text{ ft}^2$$

The amount of fertilizer needed is found by dividing the total area by the area covered by one bag:

$$\frac{11{,}900 \text{ ft}^2}{2500 \text{ ft}^2} = 4.76 \text{ bags, or } 5 \text{ bags}$$

The cost is (5 bags)(\$13.95/bag) = \$69.75.

EXAMPLE 4

A piece of sheet metal in the shape of a trapezoid has a square hole in it as shown in Fig. 2.39. Find the area of sheet metal after the hole has been removed.

First, let's find the area of the trapezoid.

$$A = \tfrac{1}{2}h(a + b)$$
$$A = \tfrac{1}{2}(12.0 \text{ in.})(20.0 \text{ in.} + 24.0 \text{ in.})$$
$$= 264 \text{ in}^2$$

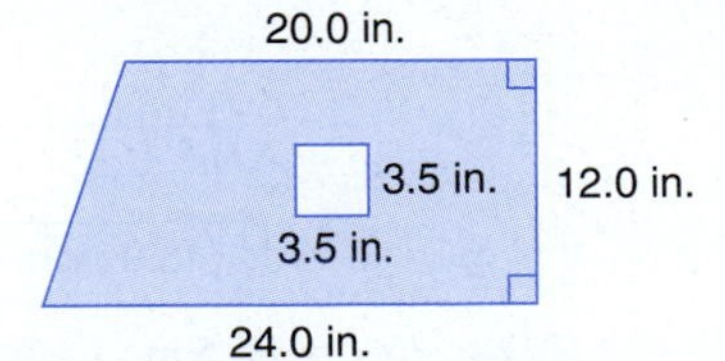

Figure 2.39

Next, find the area of the square hole.

$$A = b^2$$
$$A = (3.5 \text{ in.})^2$$
$$= 12 \text{ in}^2 \quad \text{(rounded to two significant digits)}$$

Then, the area of sheet metal is $264 \text{ in}^2 - 12 \text{ in}^2 = 252 \text{ in}^2$.

Exercises 2.3

Find (a) the perimeter and (b) the area of each quadrilateral.

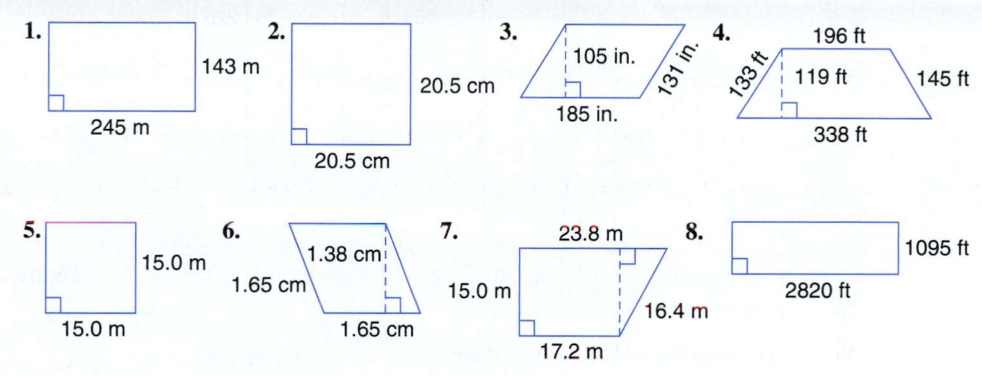

9. The area of a parallelogram is 556 m^2. Find its base if its height is 33.2 m.
10. Find the area of a rectangle with length 235 ft and width 104 ft.
11. The area of a trapezoid is 3350 cm^2, its height is 36.0 cm, and one of the bases is 45.0 cm long. Find the length of the other base.
12. Find the length of the side of a square ranch whose area is 37.5 mi^2.
13. A rectangular piece of sheet metal has an area of 7560 in^2. Find its length if its width is 79.2 in.
14. Find the amount of sheathing needed for the roof shown in Fig. 2.40. If one square = 100 ft^2, how many squares of shingles need to be purchased?
15. Estimate the cost of painting the exterior of the building in Fig. 2.40. The total area of the openings not to be painted is 375 ft^2. The cost is \$0.90/ft^2.

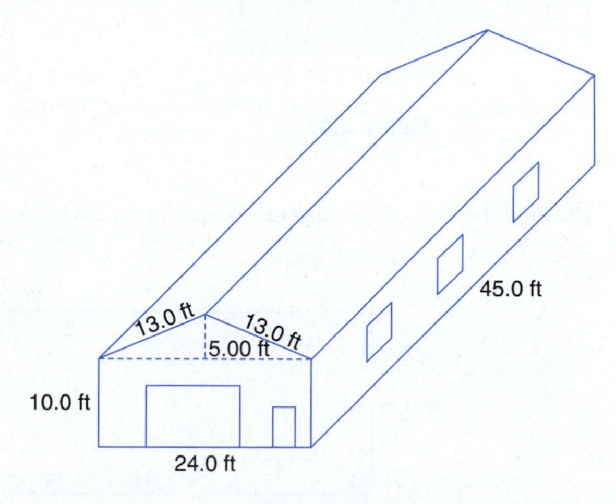

Figure 2.40

16. A rectangular building measures 145 ft by 204 ft. **(a)** Find the area of a sidewalk 3.00 ft wide that completely surrounds the building. **(b)** Find the perimeter of the outer edge of the sidewalk.

17. Find the floor space of the stores in the building shown in Fig. 2.41.

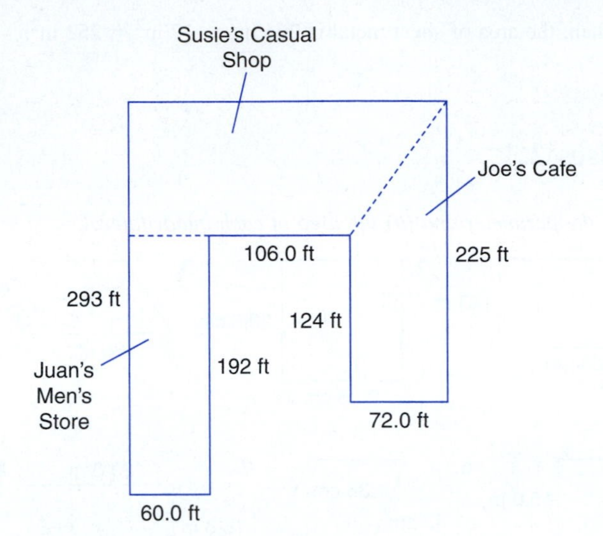

Figure 2.41

18. Find the shaded area of the diagram shown in Fig. 2.42.

19. Find the area of the diagram shown in Fig. 2.43.

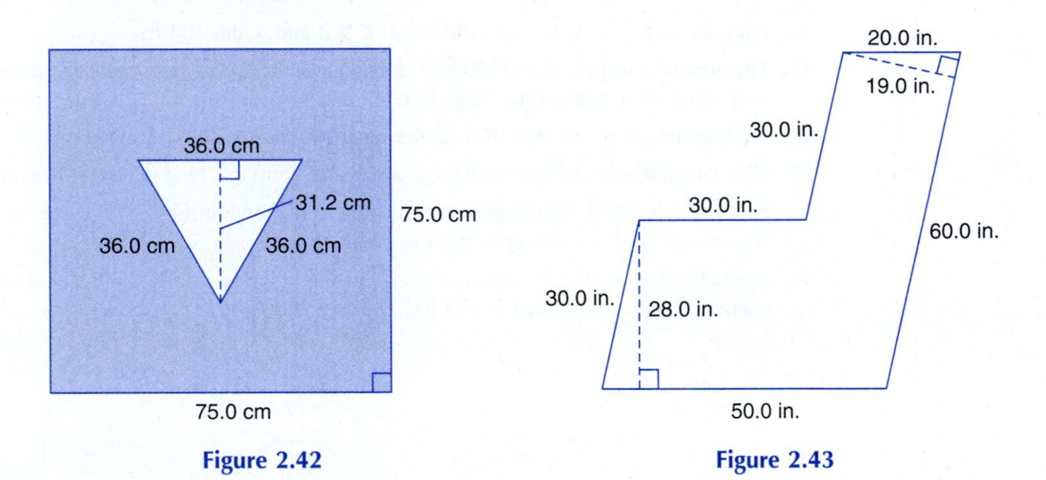

Figure 2.42

Figure 2.43

20. Find the area of the diagram shown in Fig. 2.44.

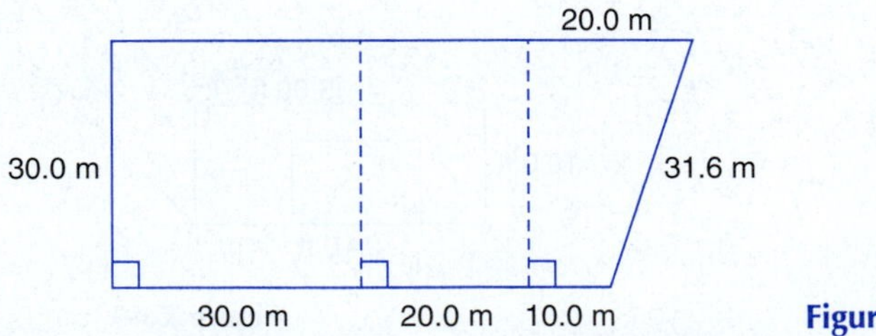

Figure 2.44

2.4 CIRCLES

A **circle** is the set of all points on a curve equidistant from a given point called the *center*. There are several key terms related to circles that you need to know. A *radius* is a line segment joining the center and any point on the circle. A *chord* is a line segment joining any two points on the circle. A *diameter* is a chord passing through the center whose length is twice the radius. A *tangent* is a line that intersects a circle at only one point. A *secant* is a line that intersects a circle in two points. See Fig. 2.45 for an illustration of these key terms.

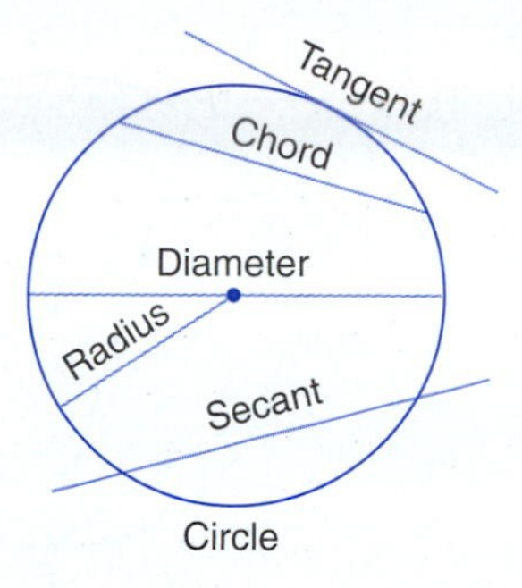

Figure 2.45

Circumference and Area of a Circle

The **circumference** of a circle is the distance around the circle. The ratio of the circumference of any circle to the length of its diameter is a constant called π (the Greek letter pi). The irrational number π cannot be written exactly as a decimal; two often-used decimal approximations are 3.14 and 3.1416. When solving problems with π, use the π key on your calculator.

The following formulas are used to find the circumference and the area of a circle, where C is the circumference and A is the area of a circle; d is the length of the diameter and r is the length of the radius.

Circumference	*Area*
$C = 2\pi r$	$A = \pi r^2$
$C = \pi d$	$A = \dfrac{\pi d^2}{4}$

EXAMPLE 1

Find the circumference and the area of a circle if the length of its radius is 24.0 cm.

First, find the circumference.

$$C = 2\pi r$$
$$C = 2\pi(24.0 \text{ cm})$$
$$= 151 \text{ cm} \qquad \text{(rounded to three significant digits)}$$

The area is

$$A = \pi r^2$$
$$A = \pi(24.0 \text{ cm})^2$$
$$= 1810 \text{ cm}^2$$

Circular Arcs

An angle formed between two radii with vertex at the center of a circle is a *central angle*. One important relationship to remember is *the sum of the measures of all the central angles of any circle is 360°*.

An *inscribed angle* is an angle with vertex on the circle and whose sides are chords. An *intercepted arc* is that part of the circle between two sides of a central angle or of an inscribed angle. In Fig. 2.46, C is the center of the circle, $\angle ACB$ is a central angle, and $\angle DEF$ is an inscribed angle. $\overset{\frown}{AB}$ is the intercepted arc of the central angle $\angle ACB$, and $\overset{\frown}{DF}$ is the intercepted arc of the inscribed angle $\angle DEF$.

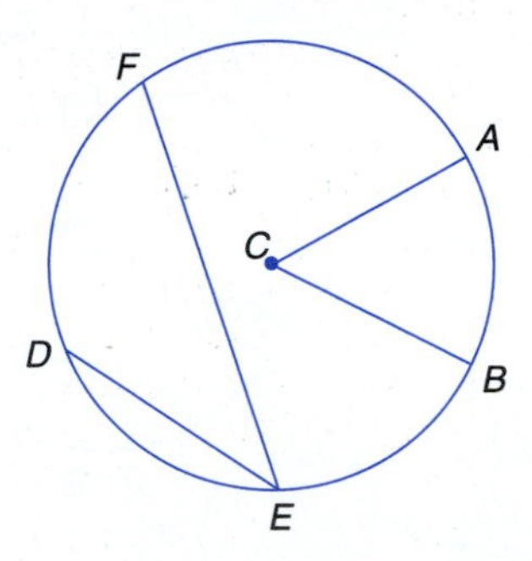

∠ACB is a central angle. ∠DEF is an inscribed angle.

Figure 2.46

The measure of an arc of a circle is equal to the measure of the central angle of that arc (Fig. 2.47). The measure of an inscribed angle in a circle is equal to one-half the measure of its intercepted arc.

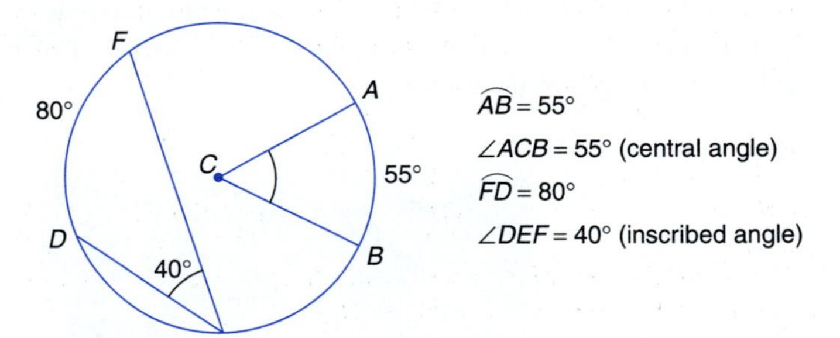

Figure 2.47

The measure of an angle formed by two intersecting chords is equal in number to one-half the sum of the measures (the average) of the intercepted arcs (Fig. 2.48).

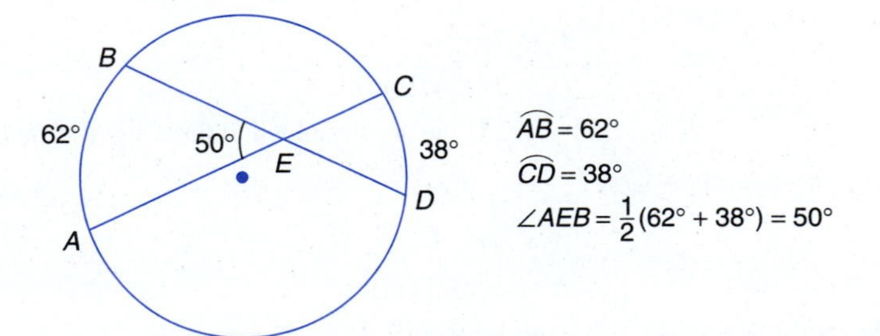

Figure 2.48

Other Relationships

A line tangent to a circle is perpendicular to the radius at the point of tangency (Fig. 2.49).

A diameter that is perpendicular to a chord bisects the chord (Fig. 2.50).

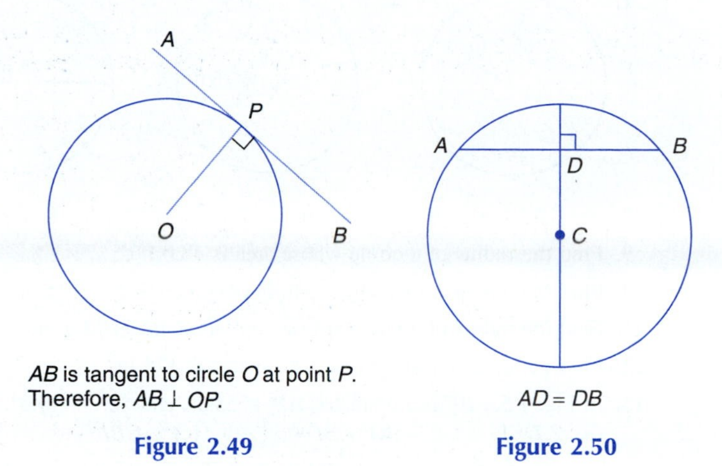

AB is tangent to circle O at point P. Therefore, $AB \perp OP$.

Figure 2.49

$AD = DB$

Figure 2.50

A *semicircle* is one-half of a circle. An angle inscribed in a semicircle is a right angle (Fig. 2.51).

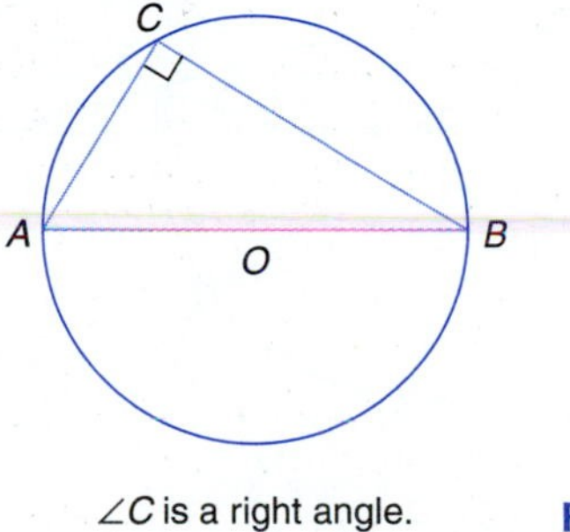

$\angle C$ is a right angle.

Figure 2.51

Two tangent line segments to a circle drawn from a point outside the circle are equal in length. The line segment drawn from the center of the circle to this point outside the circle bisects the angle formed by the tangents (see Fig. 2.52).

Sectors and segments of a circle are discussed in Section 10.4.

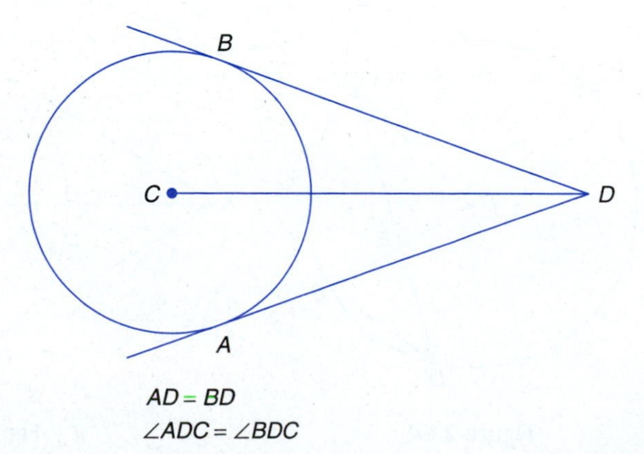

$AD = BD$

$\angle ADC = \angle BDC$

Figure 2.52

Exercises 2.4

Find (**a**) *the circumference and* (**b**) *the area of each circle in Exercises 1 through 8.*

1. 65.2 m **2.** 6.75 ft **3.** 125 mi **4.** 1150 km

5. $r = 10.5$ cm **6.** $d = 24.0$ ft **7.** $d = 2240$ m **8.** $r = 175$ km

9. Find the radius of a circle whose area is 48.0 ft^2.

10. Find the diameter of a circle whose circumference is 245 cm.

11. Find the radius of a circle whose circumference is 73.5 in.

12. Find the diameter of a circle whose area is 175 m^2.

13. In Fig. 2.53, BF is a diameter, $\widehat{AB} = 58.0°$, and $\widehat{DF} = 40.0°$. Find the measure of each angle.
(a) $\angle DEF$ **(b)** $\angle BDA$ **(c)** $\angle CBF$ **(d)** $\angle BDF$

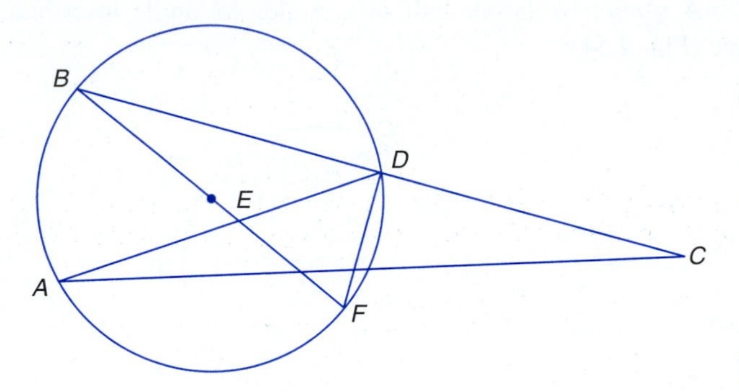

Figure 2.53

14. In Fig. 2.54, $\widehat{AB} = 66.0°$ and $AC \parallel BD$. Find the measure of each angle.
(a) $\angle ACB$ **(b)** $\angle BDA$ **(c)** $\angle CAE$ **(d)** $\angle AEB$

15. In Fig. 2.55, $\angle BAC = 30.0°$. Find
(a) $\widehat{BF}$ **(b)** $\widehat{BD}$ **(c)** $\widehat{EF}$ **(d)** $\angle ACB$ **(e)** $\angle ADE$

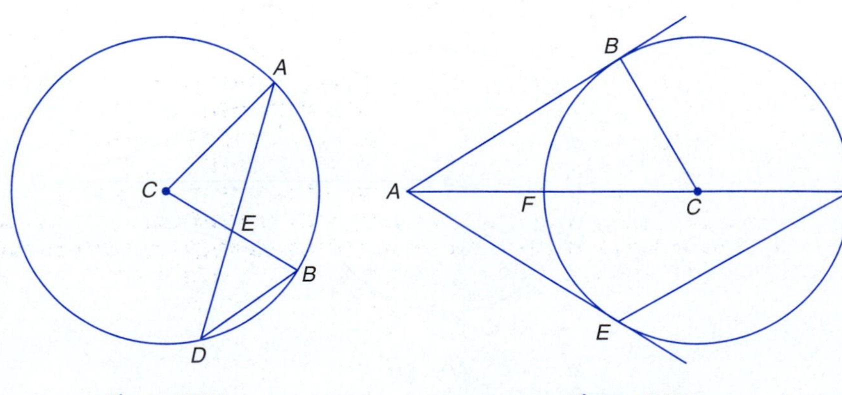

Figure 2.54 **Figure 2.55**

16. In Fig. 2.55, $\angle BAC = 30.0°$, $r = 20.0$ cm, and $AB = 34.6$ cm. Find
(a) AC **(b)** AF **(c)** AE **(d)** AD

17. In Fig. 2.56, $AB = 68.2$ cm and $AC = 74.0$ cm. Find the radius of the circle.

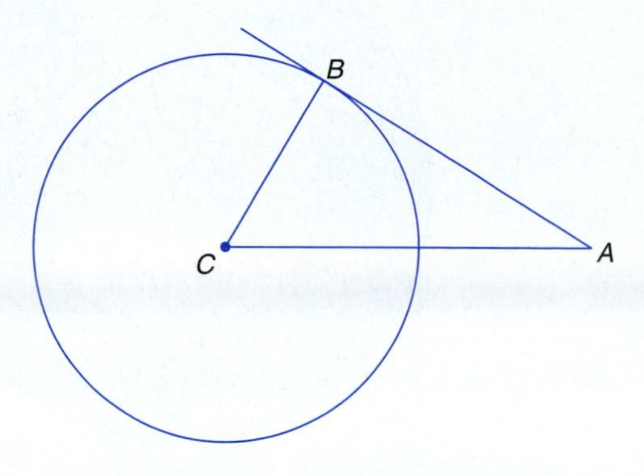

Figure 2.56

18. In Fig. 2.57, $\angle ACB = 70.0°$ and the radius of the circle is 10.0 cm. Find
(a) $\angle ABC$ **(b)** $\angle ABD$ **(c)** AB if $BE = 24.3$ cm

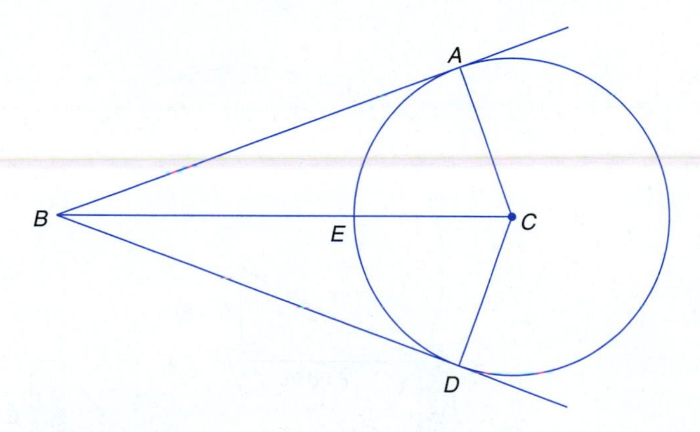

Figure 2.57

19. Boyd needs to punch six equally spaced holes in a circular arrangement. Find the measure of the central angle between two adjacent holes.

20. Find **(a)** the area and **(b)** the perimeter in Fig. 2.58.

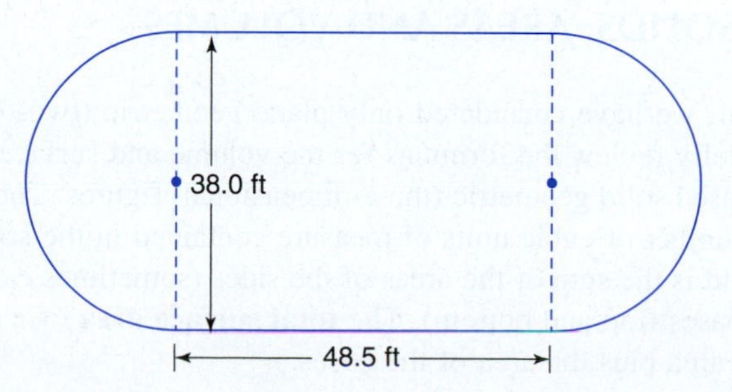

Figure 2.58

21. Find the shaded area in Fig. 2.59 if the length of the outer radius is 36.0 ft and the length of the inner radius is 20.0 ft.

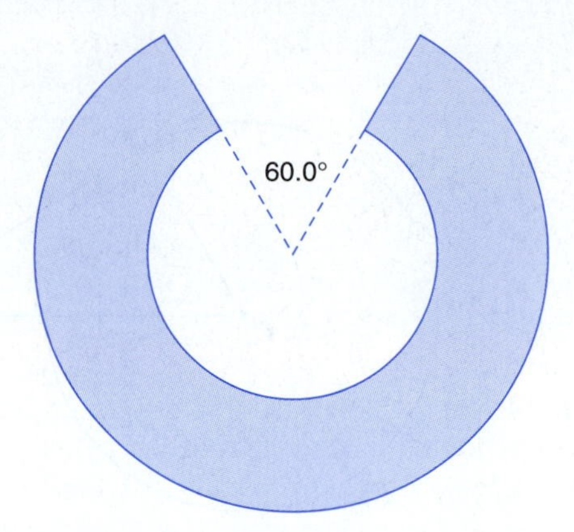

Figure 2.59

22. Find the diameter of the circular duct that will fit between two ceiling joists 16.0 in. apart and touch the ceiling as shown in Fig. 2.60.

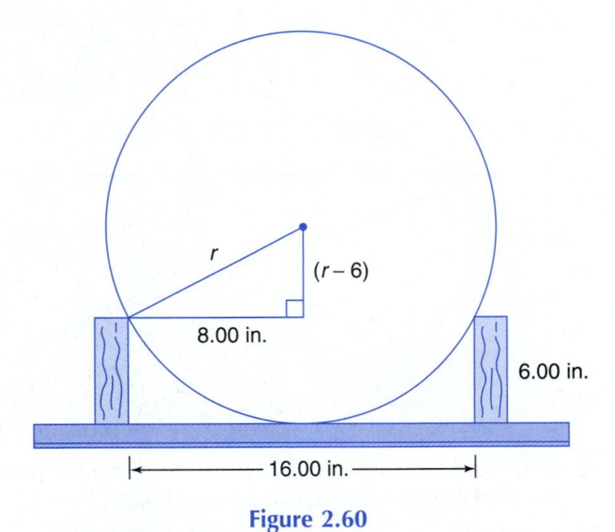

Figure 2.60

2.5 GEOMETRIC SOLIDS: AREAS AND VOLUMES

Thus far, we have considered only plane geometric (two-dimensional) figures. We will now briefly review the formulas for the volume and surface area of the most basic, commonly used solid geometric (three-dimensional) figures. The **volume** of a geometric solid is the number of cubic units of measure contained in the solid. The **lateral surface area** of a solid is the sum of the areas of the sides (sometimes called faces) excluding the area of the bases (top and bottom). The **total surface area** of a solid is the sum of the lateral surface area plus the area of the bases.

A **prism** is a solid whose sides are parallelograms and whose bases are a pair of polygons that are parallel and have the same size and the same shape. A *right prism* has rectangular lateral faces, which are therefore perpendicular to the bases. A *rectangular solid* is a right prism whose bases and lateral faces are rectangles.

A **right circular cylinder** is a solid formed by rotating a rectangle about one of its sides. The parallel bases are circles, the radius of the cylinder is the radius of either base, and the height of the cylinder is the distance between the parallel bases.

A **pyramid** is a solid whose base is a polygon and whose lateral faces are triangles with a common vertex, called the *apex*. A pyramid is often named after its base; for example, a triangular pyramid has a base that is a triangle, a square pyramid has a base that is a square, and so on.

A **right circular cone** is a solid formed by rotating a right triangle about one of its legs. The base is a circle, the radius of the cone is the radius of the circular base, the height of the cone is the length of the leg of the right triangle about which it is rotated, and the slant height is the length of the hypotenuse.

A **sphere** is a solid formed by rotating a circle about its diameter. The radius of the sphere is the distance from the center to any point on the surface of the sphere.

Figure 2.61 shows each of these geometric solids along with the corresponding formulas for its volume and lateral surface area. We use B as the area of the base, r as the length of the radius, and h as the height of the geometric solid.

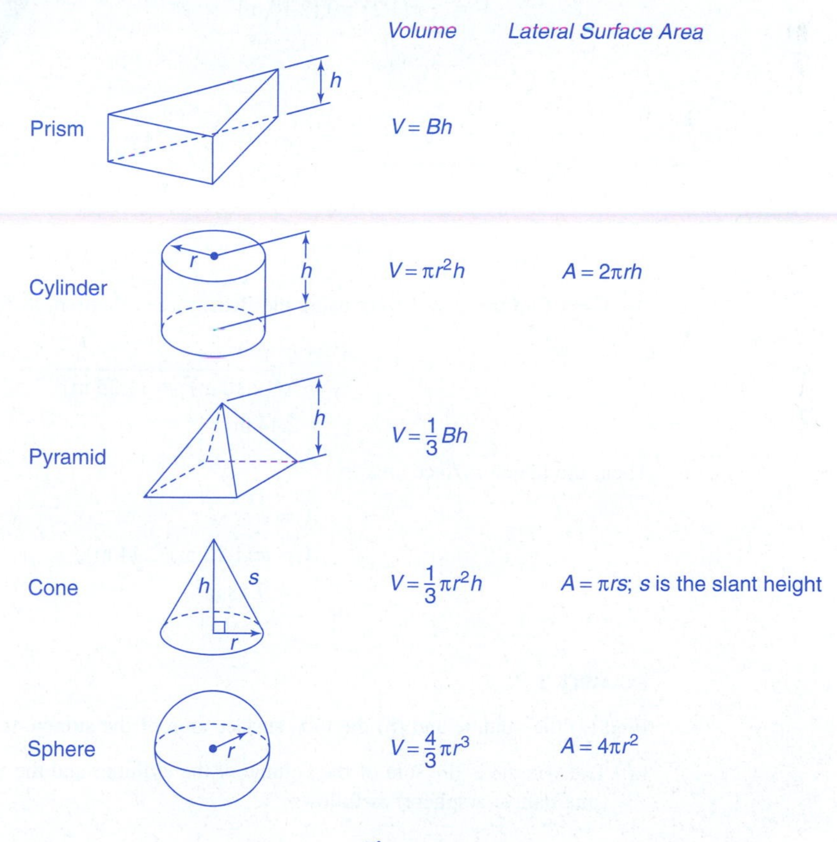

Figure 2.61

EXAMPLE 1

How many cubic yards of concrete are needed for a driveway 18.0 ft wide, 33.0 ft long, and 4.00 in. thick?

Note that this driveway is a rectangular prism, whose volume is found by the formula

$$V = Bh = \ell wh$$

$$V = (33.0\text{ ft})(18.0\text{ ft})(4.00\text{ in.}) \times \frac{1\text{ ft}}{12\text{ in.}}$$

$$= 198\text{ ft}^3 \times \left(\frac{1\text{ yd}}{3\text{ ft}}\right)^3$$

$$= 7.33\text{ yd}^3$$

EXAMPLE 2

Find (a) the volume and (b) the lateral surface area of the right circular cone feed storage bin shown in Fig. 2.62.

(a) We have

$$V = \frac{1}{3}\pi r^2 h$$

$$V = \frac{1}{3}\pi(1.25\text{ m})^2(2.10\text{ m})$$

$$= 3.44\text{ m}^3$$

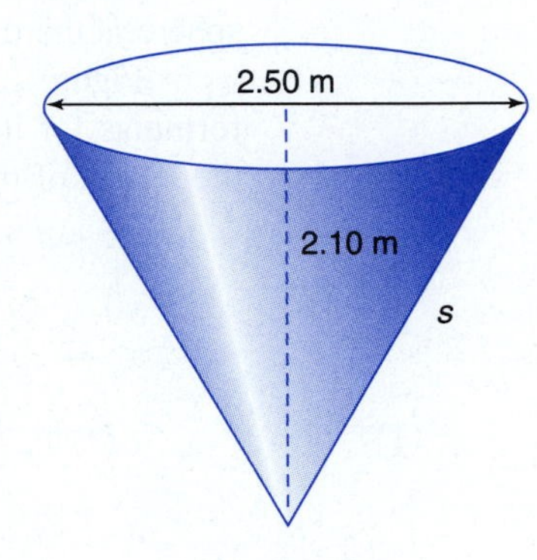

Figure 2.62

(b) First, find the slant height using the Pythagorean theorem as follows:

$$s^2 = h^2 + r^2$$

$$s = \sqrt{(2.10\text{ m})^2 + (1.25\text{ m})^2}$$

$$= 2.44\text{ m}$$

Then, the lateral surface area is

$$A = \pi rs$$

$$A = \pi(1.25\text{ m})(2.44\text{ m})$$

$$= 9.58\text{ m}^2$$

EXAMPLE 3

Find (a) the volume and (b) the total surface area of the storage tank in Fig. 2.63.

(a) The volume is the sum of the volume of the cylinder and the volume of the hemisphere (one-half of a sphere) as follows:

$$V = \pi r^2 h + \frac{1}{2}\left(\frac{4}{3}\pi r^3\right)$$

$$V = \pi(10.3\text{ cm})^2(28.5\text{ cm}) + \frac{2}{3}\pi(10.3\text{ cm})^3$$

$$= 11{,}800\text{ cm}^3$$

(b) The total surface area is the sum of the areas of the circular end of the cylinder, the lateral surface area of the cylinder, and the hemisphere as follows:

$$A = \pi r^2 + 2\pi rh + \tfrac{1}{2}(4\pi r^2)$$
$$A = \pi(10.3 \text{ cm})^2 + 2\pi(10.3 \text{ cm})(28.5 \text{ cm}) + 2\pi(10.3 \text{ cm})^2$$
$$= 2840 \text{ cm}^2$$

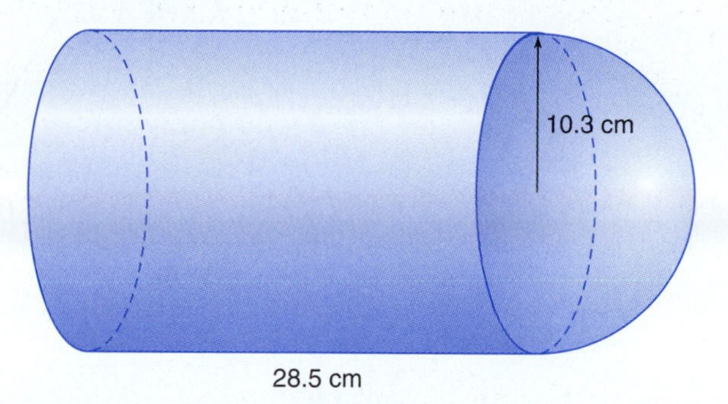

Figure 2.63

Exercises 2.5

1. Find **(a)** the volume, **(b)** the lateral surface area, and **(c)** the total surface area of the rectangular solid in Fig. 2.64.

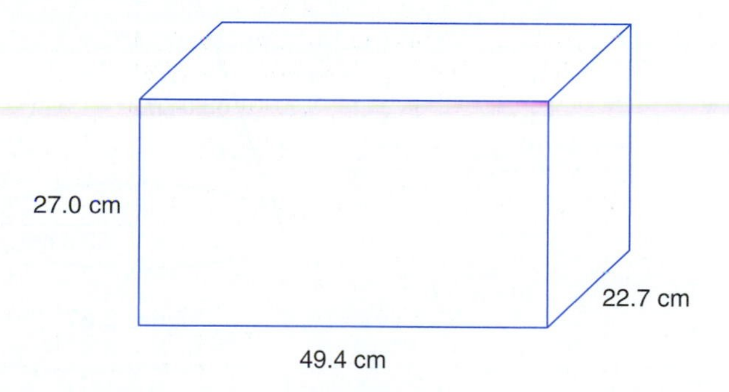

Figure 2.64

2. Find **(a)** the volume, **(b)** the lateral surface area, and **(c)** the total surface area of the right circular cylinder in Fig. 2.65.

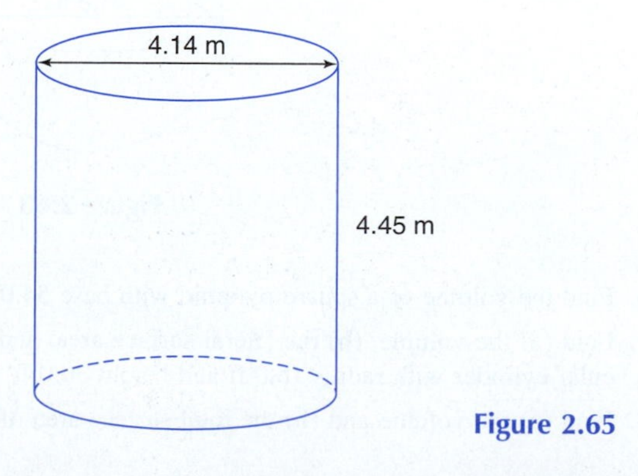

Figure 2.65

3. Find the volume of the rectangular pyramid in Fig. 2.66.

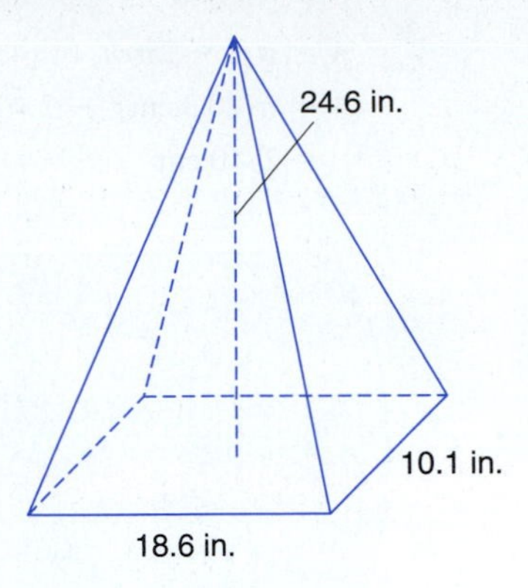

Figure 2.66

4. Find **(a)** the volume, **(b)** the lateral surface area, and **(c)** the total surface area of the right circular cone in Fig. 2.67.

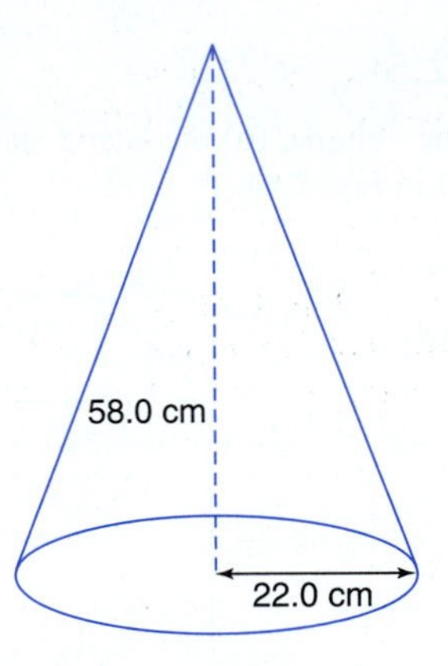

Figure 2.67

5. Find **(a)** the volume and **(b)** the total surface area of the sphere in Fig. 2.68.

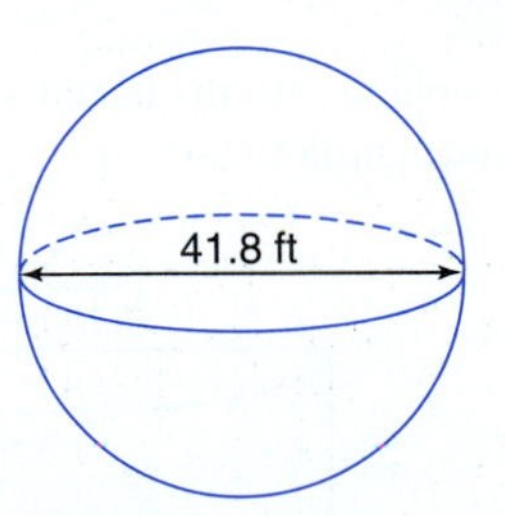

Figure 2.68

6. Find the volume of a square pyramid with base 54.0 cm on a side and 30.0 cm high.
7. Find **(a)** the volume, **(b)** the lateral surface area, and **(c)** the total surface area of a right circular cylinder with radius 15.0 ft and height 30.0 ft.
8. Find **(a)** the volume and **(b)** the total surface area of a sphere with radius 75.3 cm.

9. Find **(a)** the volume, **(b)** the lateral surface area, and **(c)** the total surface area of a right circular cone with radius 12.5 ft and slant height 18.4 ft.

10. Find **(a)** the volume, **(b)** the lateral surface area, and **(c)** the total surface area of a cube 5.00 cm on a side.

11. A rectangular lead sleeve is shown in Fig. 2.69. **(a)** How many cubic inches of lead are contained in this sleeve? **(b)** Find its weight if lead weighs 708 lb/ft^3.

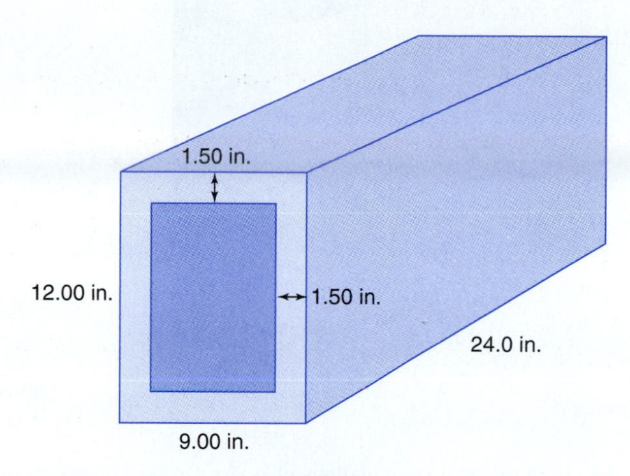

Figure 2.69

12. **(a)** How many square feet of sheet metal are needed to form the trough whose ends are semicircles as shown in Fig. 2.70? **(b)** How many cubic feet of liquid will the trough hold?

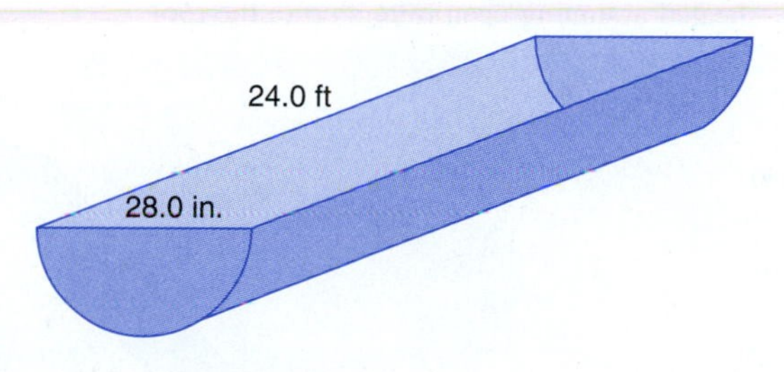

Figure 2.70

13. The circular building in Fig. 2.71 is made of metal weighing 17.8 lb/ft^2. **(a)** What is the weight of the top? **(b)** What is the weight of the top and sides? The bottom consists of concrete. **(c)** Find the total volume capacity of the building.

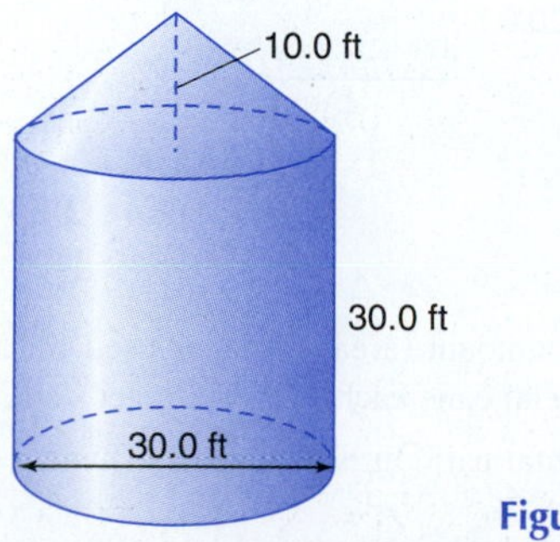

Figure 2.71

14. Find **(a)** the volume of the cylindrical silo with a hemispherical top shown in Fig. 2.72. **(b)** Find the area of the silo that would be painted.

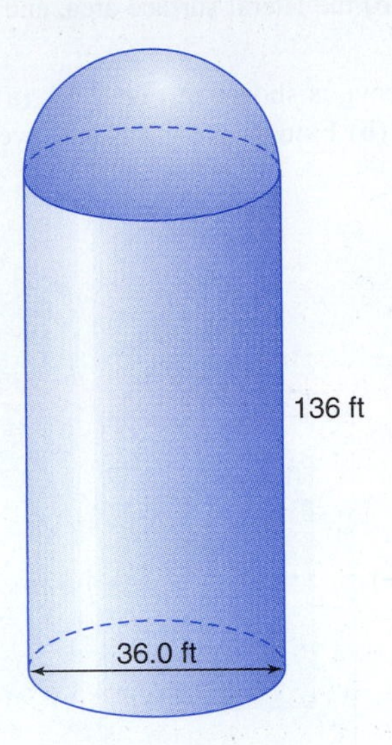

Figure 2.72

15. Find **(a)** how many gallons of water can be stored in a spherical water tank 60.0 ft in diameter (7.48 gal are contained in 1 ft^3), **(b)** the surface area of the water tank, and **(c)** how many gallons of paint would be necessary to paint the tank if 1 gal covers 125 ft^2.
16. For the building in Fig. 2.73, find **(a)** the area of the four sides to be painted assuming no windows, **(b)** the area of the roof to be covered with shingles, **(c)** the storage capacity of the building assuming materials may be stored up to the gutter line, and **(d)** the volume to be heated assuming open rafters up to the roof.

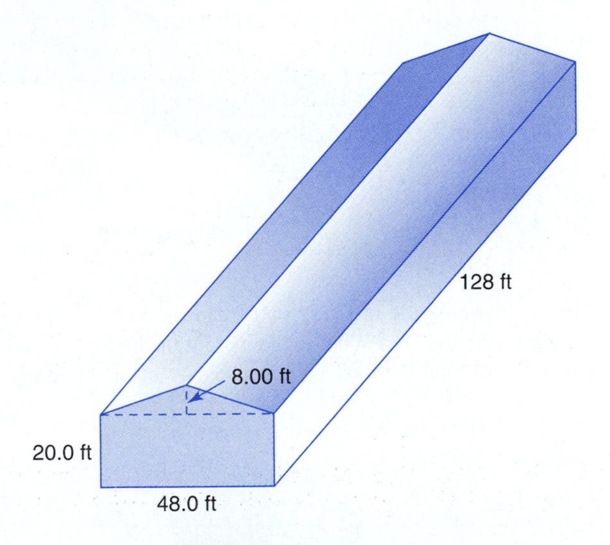

Figure 2.73

17. Find the total amount (area) of paper used for labels that completely cover the sides of 1750 cylindrical metal cans each with diameter 7.40 cm and height 10.0 cm.
18. An experimental balloon is designed to have a diameter of 8.75 m. How much material is needed?

19. Find the volume of gravel stored in a conical pile whose circumference measures 256 ft and whose slant height measures 42 ft. If gravel weighs 3400 lb/yd^3, how many 22-ton truckloads are needed to move the gravel?

20. Sheet metal is used to fabricate a tank whose ends are isosceles trapezoids as shown in Fig. 2.74. **(a)** How many cubic feet of liquid will it hold when it is full? **(b)** How many cubic feet of liquid will it hold when the liquid level is one-half the depth of the tank?

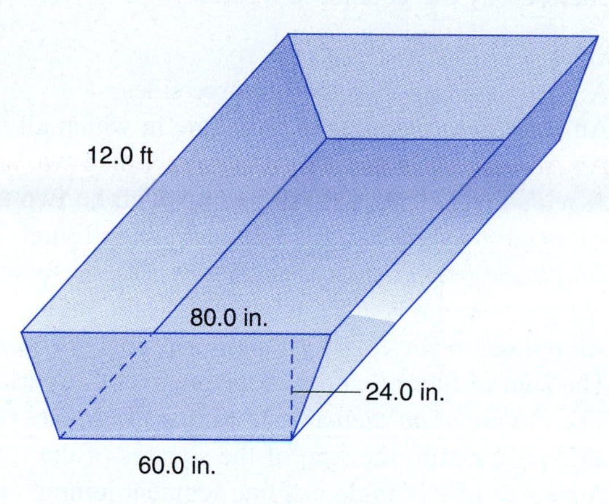

Figure 2.74

CHAPTER 2 SUMMARY

1. *Angles and lines*
   (a) An *angle* is formed by two lines with one common point, called the *vertex*. The parts of the lines that form an angle are called the *sides* of the angle.
   (b) A *right angle* is an angle with a measure of 90°. An *acute angle* is an angle with a measure less than 90°. An *obtuse angle* is an angle with a measure greater than 90° but less than 180°.
   (c) Two lines *intersect* if they have one point in common. Two lines are *parallel* ($\|$) if they do not intersect even when extended. Two lines in the same plane are *perpendicular* ($\perp$) if they intersect and form equal adjacent angles.
   (d) Two angles are *adjacent* if they have a common vertex and a common side between them. Two angles are *complementary* if the sum of their measures is 90°. Two angles are *supplementary* if the sum of their measures is 180°. Two *vertical angles* are the opposite angles formed by two intersecting lines.
   (e) A *transversal* is a line that intersects two or more lines in different points in the same plane. *Interior angles* are angles formed inside the lines by the transversal. *Exterior angles* are angles formed outside the lines by the transversal. *Corresponding angles* are exterior-interior angles on the same side of the transversal. *Alternate angles* are angles with different vertices on opposite sides of the transversal.
   (f) If two parallel lines are cut by a transversal, then
   - the corresponding angles are equal.
   - the alternate-interior angles are equal.
   - the alternate-exterior angles are equal.
   - the interior angles on the same side of the transversal are supplementary.

2. *Polygons*

(a) A *polygon* is a closed figure whose sides are straight-line segments. Polygons are named according to the number of sides they have.
(b) A *regular* polygon has all of its sides equal and all of its interior angles equal.
(c) The *perimeter* of a polygon is the sum of the lengths of the sides.
(d) The *area* of a geometric plane figure is the number of square units of measure enclosed by the geometric figure.

3. *Triangles*

(a) A *triangle* is a polygon with three sides.
(b) An *equilateral triangle* is a triangle in which all three sides are equal.
(c) An *isosceles triangle* is a triangle in which two sides are equal.
(d) A *scalene triangle* is a triangle in which no two sides are equal.
(e) An *acute triangle* is a triangle in which all three angles are acute.
(f) An *obtuse triangle* is a triangle with an obtuse angle.
(g) A *right triangle* is a triangle with a right angle.
(h) An *oblique triangle* is a triangle that does not contain a right angle.
(i) The sum of the measures of the angles of any triangle is 180°.
(j) The *Pythagorean theorem* states that the square of the hypotenuse of a right triangle is equal to the sum of the squares of the two legs, or $c^2 = a^2 + b^2$.
(k) A *median* of a triangle is a line segment joining any vertex to the *midpoint* of the opposite side.
(l) An *altitude* (height) of a triangle is the *perpendicular* line segment from any vertex to the opposite side (or the opposite side extended).
(m) An *angle bisector* of a triangle is a line segment that bisects any angle and intersects the opposite side.
(n) Two triangles are *similar* when they have the same shape. The measures of their corresponding angles are equal, and the lengths of their corresponding sides are proportional.
(o) Two triangles are *congruent* when their corresponding angles and corresponding sides are equal.

4. *Quadrilaterals*

(a) A *quadrilateral* is a polygon with four sides.
(b) A *parallelogram* is a quadrilateral having two pairs of parallel sides.
(c) A *rhombus* is a parallelogram with equal sides.
(d) A *rectangle* is a parallelogram with right angles.
(e) A *square* is a rectangle with equal sides.
(f) A *trapezoid* is a quadrilateral with only one pair of parallel sides, called the bases.

5. *Circles*

(a) A *circle* is the set of all points on a curve equidistant from a given point called the center.
(b) A *radius* of a circle is a line segment joining the center and any point on the circle.
(c) A *chord* is a line segment joining any two points on the circle.
(d) A *diameter* is a chord passing through the center of the circle whose length is twice the radius.
(e) A *tangent* is a line that intersects a circle at only one point.
(f) A *secant* is a line that intersects a circle in two points.
(g) The *circumference* of a circle is the distance around the circle.

(h) A line tangent to a circle is perpendicular to the radius at the point of tangency.
(i) A diameter that is perpendicular to a chord bisects the chord.
(j) A *semicircle* is one-half of a circle. An angle inscribed in a semicircle is a right angle.
(k) Two tangent line segments to a circle drawn from the same point outside the circle are equal in length. The line segment drawn from the center of the circle to this point outside the circle bisects the angle formed by the tangents.

6. *Circular Arcs*

(a) A *central angle* is an angle formed between two radii with vertex at the center of a circle.
(b) The sum of the measures of all the central angles of *any* circle is 360°.
(c) An *inscribed angle* is an angle with vertex on the circle and whose sides are chords.
(d) An *intercepted arc* is that part of the circle between two sides of a central angle or of an inscribed angle.
(e) The measure of an arc of a circle is equal to the measure of the central angle of that arc.
(f) The measure of an inscribed angle of a circle is equal to one-half the measure of its intercepted arc.
(g) The measure of an angle formed by two intersecting chords is equal in number to one-half the sum of the measures (the average) of the intercepted arcs.

7. *Geometric Solids*

(a) The *volume* of a geometric solid is the number of cubic units of measure contained in the solid.
(b) The *lateral surface area* of a solid is the sum of the areas of the sides, or faces, excluding the area of the bases.
(c) The *total surface area* of a solid is the sum of the lateral surface area plus the area of the bases.
(d) A *prism* is a solid whose sides are parallelograms and whose bases are a pair of polygons that are parallel and have the same size and shape. A *right prism* has rectangular lateral faces, which are therefore perpendicular to the bases.
(e) A *rectangular solid* is a right prism whose bases and lateral faces are rectangles.
(f) A *right circular cylinder* is a solid formed by rotating a rectangle about one of its sides.
(g) A *pyramid* is a solid whose base is a polygon and whose lateral faces are triangles with a common vertex, called the apex.
(h) A *right circular cone* is a solid formed by rotating a right triangle about one of its legs.
(i) A *sphere* is a solid formed by rotating a circle about its diameter.

8. *Formulas from Chapter 2:*

	Area	Perimeter
Triangle	$A = \frac{1}{2}bh$	$P = a + b + c$
Rectangle	$A = bh$	$P = 2(b + h)$
Square	$A = b^2$	$P = 4b$
Parallelogram	$A = bh$	$P = 2(a + b)$
Rhombus	$A = bh$	$P = 4b$
Trapezoid	$A = \frac{1}{2}h(a + b)$	$P = a + b + c + d$
Circle	$A = \pi r^2$	$C = 2\pi r$ or $C = \pi d$

	Volume	Lateral surface area
Prism	$V = Bh$	
Cylinder	$V = \pi r^2 h$	$A = 2\pi rh$
Pyramid	$V = \frac{1}{3}Bh$	
Cone	$V = \frac{1}{3}\pi r^2 h$	$A = \pi rs$
Sphere	$V = \frac{4}{3}\pi r^3$	$A = 4\pi r^2$

CHAPTER 2 REVIEW

1. In Fig. 2.75, two parallel lines are cut by a transversal. If $\angle 5 = 30°$, find the measure of each numbered angle.

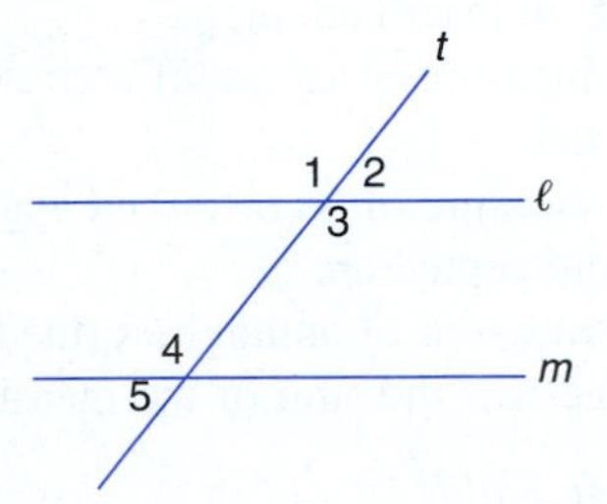

Figure 2.75

2. Name pairs of numbered angles in Fig. 2.75 that are **(a)** vertical angles, **(b)** alternate-interior angles, **(c)** corresponding angles, and **(d)** supplementary angles.

3. Name the polygon that has **(a)** three sides, **(b)** four sides, **(c)** five sides, **(d)** six sides, and **(e)** eight sides.

*Find **(a)** the area and **(b)** the perimeter of each polygon.*

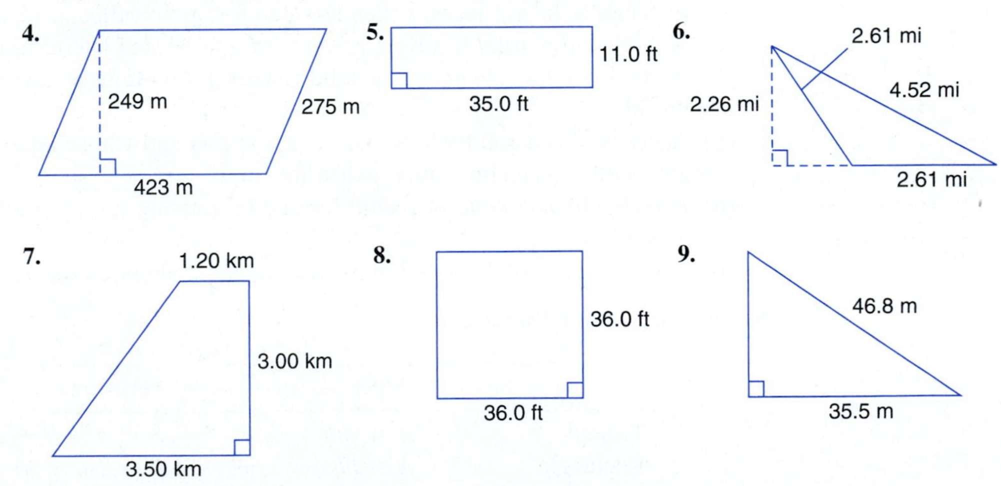

10. Find **(a)** the area and **(b)** the circumference of a circle with radius 14.5 m.

11. Find the hypotenuse of a right triangle with legs 126 ft and 214 ft.

12. The area of a rectangle is 2440 m^2 and its length is 76.2 m. Find its width.

13. Find the diameter of a circle whose area is 1230 in^2.

14. Find **(a)** the area and **(b)** the perimeter of an equilateral triangle with one side 12.0 cm long.

15. Find **(a)** the volume, **(b)** the lateral surface area, and **(c)** the total surface area of a rectangular solid 24.0 ft wide by 36.0 ft long by 18.0 ft high.

16. Find **(a)** the volume, **(b)** the lateral surface area, and **(c)** the total surface area of a right circular cylinder 45.0 m high and 9.50 m in diameter.

17. Find **(a)** the volume, **(b)** the lateral surface area, and **(c)** the total surface area of an inverted right circular cone 41.6 in. high and 22.0 in. in diameter.

18. Find the volume of a rectangular pyramid whose base measures 28.0 cm by 44.0 cm and whose height is 16.5 cm.

19. Find **(a)** the volume and **(b)** the total surface area of a sphere whose diameter is 32.0 m.

20. Find **(a)** the volume, **(b)** the lateral surface area, and **(c)** the total surface area of the solid in Fig. 2.76.

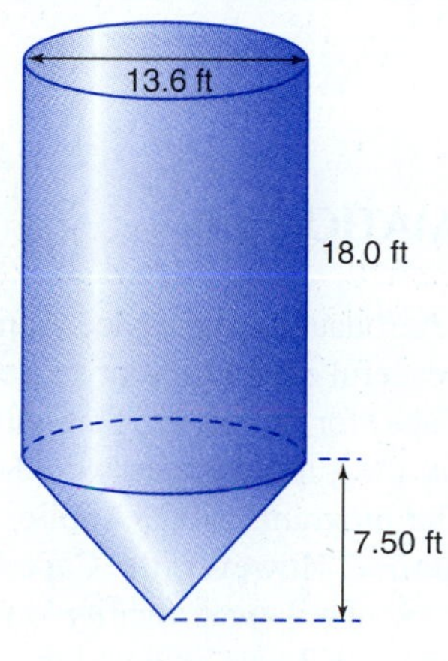

Figure 2.76

CHAPTER 2

SPECIAL NASA APPLICATION

Window-to-Wall Ratio*

BACKGROUND INFORMATION

The National Aeronautics and Space Administration (NASA) was established in October 1958 for the peaceful exploration of space. NASA adapted facilities at the Cape Canaveral Air Station (Cape) for its early rocket launches. In May 1961, President John F. Kennedy announced that the United States would send men to the Moon and back by the end of the decade. The program, called Apollo, would require the largest rocket ever built, the 363-ft-tall Saturn V. However, the Cape facilities were inadequate for launching the Saturn V rockets. A new launch facility was built north of the Cape Canaveral Air Station and in November 1963 was renamed the John F. Kennedy Space Center (KSC). Over the years, facilities at both Kennedy Space Center and the Cape continued to serve both manned and unmanned space activities.

Two factors have led NASA to study energy use at the Cape.

First, many Cape Canaveral Air Station facilities are old and not designed for their current uses. The facilities at the Cape were built as airplane hangars, service and repair facilities, and barracks. Today the same buildings are used for activities and equipment as diverse as electronic monitoring and control units for launches, assembly facilities and clean rooms for preparation of payloads to be delivered to space, and biological experiments that test food and oxygen sources for long-term space travel. The facilities lack most or all modern energy-saving features or equipment.

Second, the federal government has mandated energy savings through the Energy Policy Act of 1992 (EO 12902). This act requires energy reductions in federal buildings by 2005 and permits contracts with private companies to install energy-saving equipment. An energy savings performance contract (ESPC) specifies cost-saving measures to be performed, projects energy reductions, and provides for the government to pay the private contractor with savings from reduced energy costs in the future.

WINDOW-TO-WALL RATIO

The window-to-wall ratio of a building is important since a higher ratio means there will be more heat transfer between inside and outside. This, in turn, will indicate a higher energy use.

*From NASA–AMATYC–NSF Project Mathematics Explorations II, grant principals John S. Pazdar, Patricia L. Hirschy, and Peter A. Wursthorn; copyright Capital Community College, 2000.

Exterior windows and walls have one side facing inside the building and the other facing outside the building. We will find the areas of each and form the ratio. The following chart corresponds to the front wall of Home Outline Z in Figure 2.77. In Home Outline Z, 1 mm represents 0.25 ft.

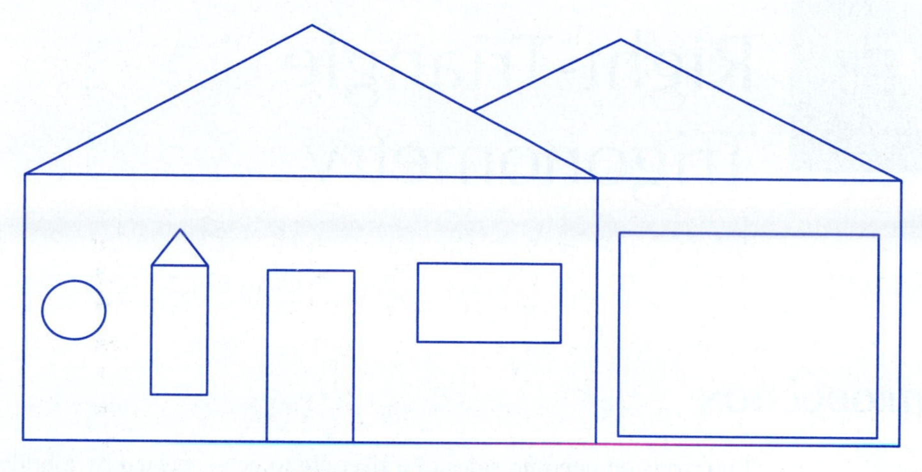

Figure 2.77 Home Outline Z.

Example of Window and Wall Information

Home Outline Z	Window dimension	Window area	Exterior wall dimensions	Exterior wall area
Window 1	Circle: radius = 4 mm	$\pi \cdot 16$ or 50.3 mm^2	35 mm by 76 mm	2660 mm^2
Window 2 (window with triangular top)	Triangle: $b = 8$ mm, $h = 5$ mm Rectangle: 8 mm by 17 mm	156 mm^2	Same wall	Same wall
Window 3	Rectangle: 10 mm by 19 mm	190 mm^2	Same wall	Same wall
Door	Rectangle: 12 mm by 23 mm	—	—	—
Total	—	396.3 mm^2	—	2660 mm^2

The dimensions of the windows and exterior walls have been measured to the nearest millimetre. The attic and garage walls were not included. Divide the window area by the wall area to find the window-to-wall ratio (WWR).

$$\text{WWR} = \frac{396.3\text{ mm}^2}{2600\text{ mm}^2} \times 100\% = (0.149) \times (100\%) = 14.9\%$$

1. When the WWR was formed, what happened to the units (mm^2) that were attached to the wall and window areas?
2. If we had measured in inches instead of millimetres, how would the resulting WWR have changed?
3. Why is it not necessary to know the true dimensions of a given building to determine the WWR?

3 Right-Triangle Trigonometry

INTRODUCTION

Two roads on opposite sides of a river are to be connected by a bridge. Because of rock formations, the roads are not directly opposite each other, and the least expensive way to connect them is to build the bridge at an angle of 75.0° with the near side of the river. If the width of the river is 89.1 m, what must be the length of the bridge?

To solve this problem we need to use a trigonometric function. In this chapter we learn how to find a side or an angle of a right triangle given other pieces of information. We use the given information and our knowledge of trigonometric functions to solve problems that are not solvable using only algebra. (The preceding problem is number 7 in Exercises 3.4.)

Objectives

- Understand the degree/minute/second and radian measures of an angle.
- Know the Pythagorean theorem.
- Know the ratio definitions of the trigonometric functions.
- Know the values of the trigonometric functions for key angles.
- Use a calculator to evaluate trigonometric functions.
- Solve right triangles.

3.1 THE TRIGONOMETRIC RATIOS

Trigonometry is concerned with the measurement of the parts of a triangle—sides and angles. It is a basic tool used in the development of mathematics and various fields of technology. Hipparchus and Ptolemy created this branch of mathematics in the second century B.C. Its earliest applications were in astronomy, surveying, and navigation. Trigonometry continues to play a crucial role in today's problems of engineering, science, and technology. It is used, for example, to explain wave phenomena in physics, such as sound and electricity. As we shall soon see, trigonometry is based on working with certain ratios.

*Some geometry topics are repeated in this chapter for those who skipped Chapter 2. If you covered Chapter 2, omit them here.

The generation or formation of an **angle** is the result of rotating a line segment about a fixed point from one position to another. To be more specific, consider a fixed point O and the line segment OP (see Fig. 3.1). Before the line segment is rotated about point O, its initial (beginning) position, OP, is called the **initial side** of the angle θ (theta) as shown. The final position of the line segment, OQ, is called the **terminal side** of the angle. The fixed point O is called the **vertex** of the angle.

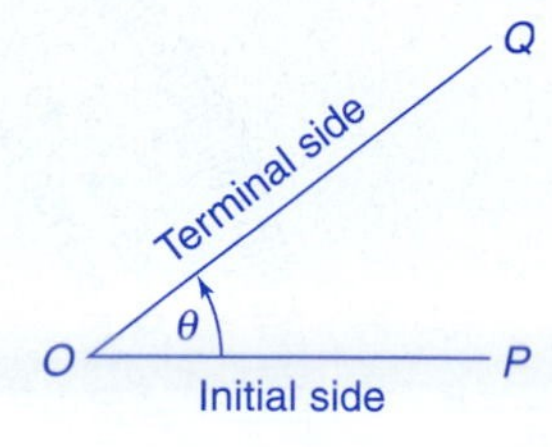

Figure 3.1 Angle θ.

A question now arises: How much is the rotation; that is, how large is the angle? Angles can be measured using any of three basic units of measure: revolutions, degrees, and radians.

One revolution (rev) [see Fig. 3.2(a)] corresponds to one complete rotation of the initial side. When an object, such as a gear, makes several rotations, its angle displacement is usually measured in revolutions. This is a common unit in industry.

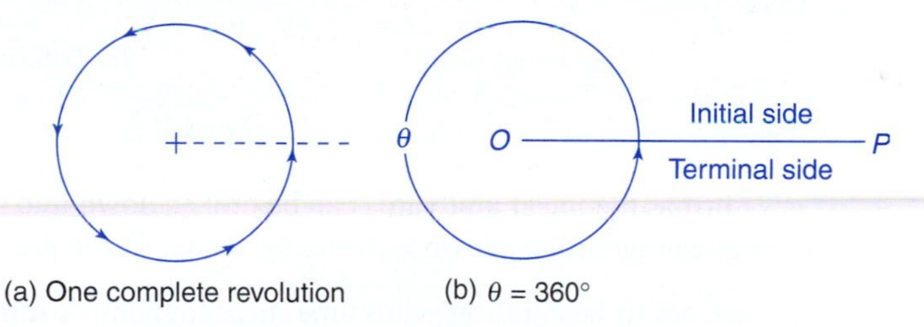

Figure 3.2 One complete revolution = 360°.

A **degree** is $\frac{1}{360}$ of one complete rotation of a line about a point. That is, if we rotate the line OP until it reaches its original position, we will have rotated an angle of 360 degrees, as shown in Fig. 3.2(b).

$$1 \text{ complete rotation} = 360 \text{ degrees}$$
$$1 \text{ degree} = \frac{1}{360} \text{ of a complete rotation}$$

Thus, we measure angles by measuring how many degrees the terminal side of the angle has been rotated from the initial side. The symbol ° is used to denote degrees.

EXAMPLE 1

Construct an angle of $\frac{1}{6}$ of a rotation.

An angle generated by $\frac{1}{6}$ of a rotation has a measure of $\frac{1}{6} \times 360° = 60°$ as shown in Fig. 3.3.

The protractor (see Fig. 3.4) is an instrument marked in degrees that is commonly used to measure angles.

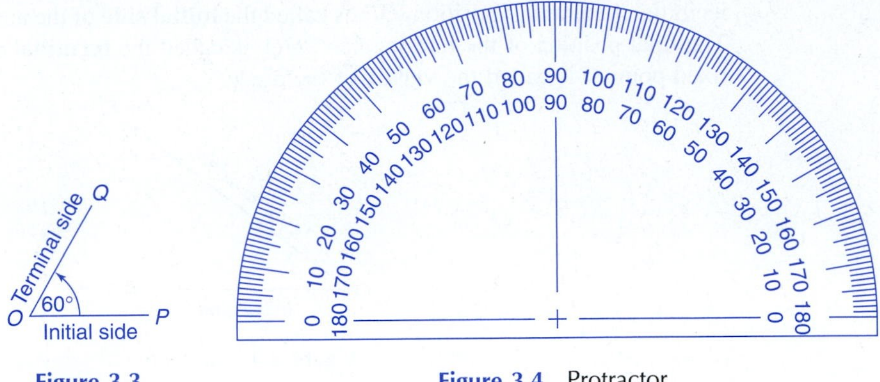

Figure 3.3

Figure 3.4 Protractor.

An **acute angle** is an angle whose measure is less than 90° [see Fig. 3.5(a)]. An **obtuse angle** is an angle whose measure is more than 90° but less than 180° [see Fig. 3.5(b)].

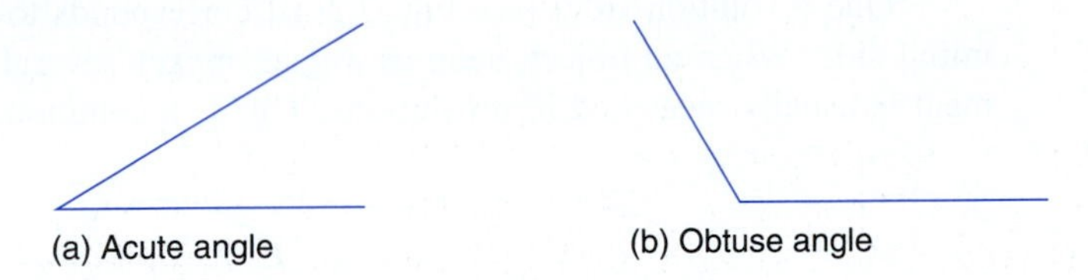

Figure 3.5

Just as the linear unit yards can be broken down into smaller units of feet and inches, degrees can be broken down into smaller units. These are the minute and the second, and they are not to be confused with time measurement. A **minute** in trigonometry is $\frac{1}{60}$ of a degree. A degree is divided into 60 equal parts called minutes. The symbol $'$ is used to denote minutes.

$$1' = \frac{1^\circ}{60}$$

$$1^\circ = 60'$$

A **second** is defined to be $\frac{1}{60}$ of a minute. The symbol $''$ is used to denote seconds.

$$1'' = \frac{1'}{60}$$

$$1' = 60''$$

$$1^\circ = 3600''$$

EXAMPLE 2

Change $63^\circ 12'$ to degrees.

Since $12' = \dfrac{12^\circ}{60}$,

$$63°12' = 63\frac{12}{60}°$$

$$= 63\frac{1}{5}°$$

$$= 63.2°$$

EXAMPLE 3

Change 78°15′45″ to degrees.

First

$$15' = \frac{15°}{60} = 0.25°$$

and

$$45'' = \frac{45'}{60} = 0.75'$$

$$= \frac{0.75°}{60} = 0.0125°$$

Then

$$78°15'45'' = 78° + 0.25° + 0.0125°$$
$$= 78.2625°$$

78 **2nd |** 15 **2nd =** 45 **2nd 1 2nd MATH 2 9 ENTER**

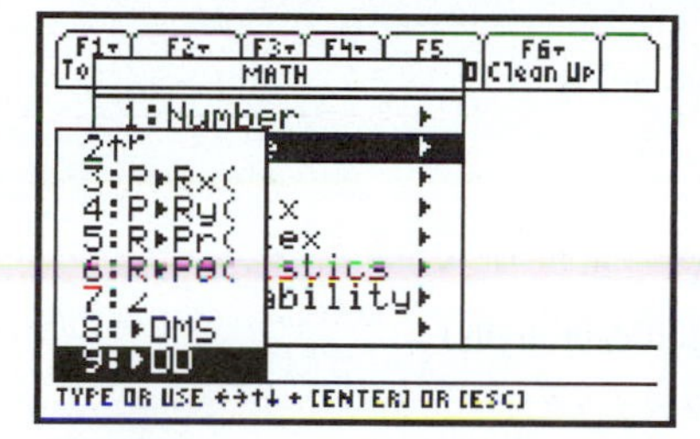

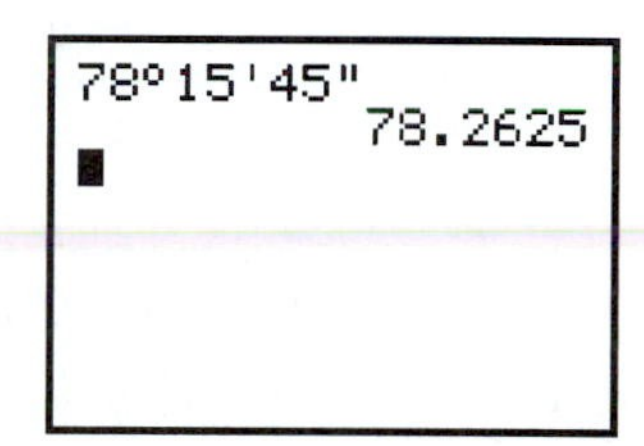

78 **2nd ANGLE 1** 15 **2nd ANGLE 2** 45 **ALPHA + ENTER**

EXAMPLE 4

Change 141.38° to degrees and minutes.

141.38 **2nd | 2nd MATH 2 8 ENTER**

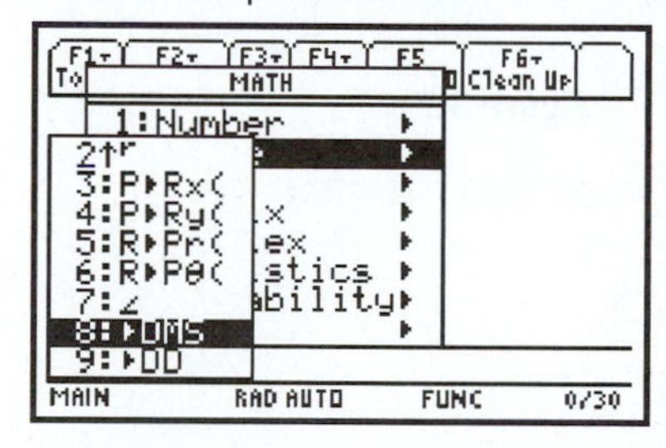

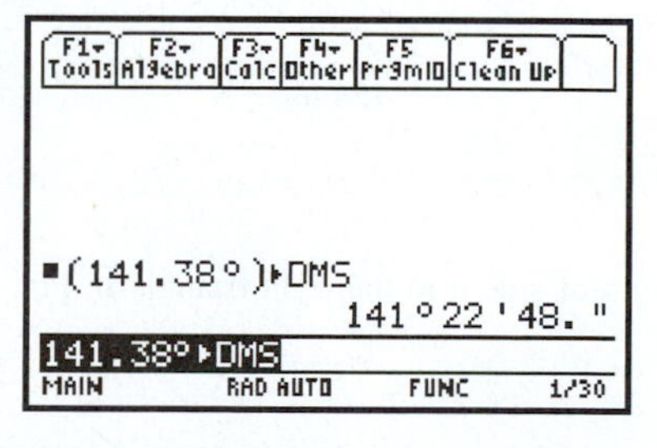

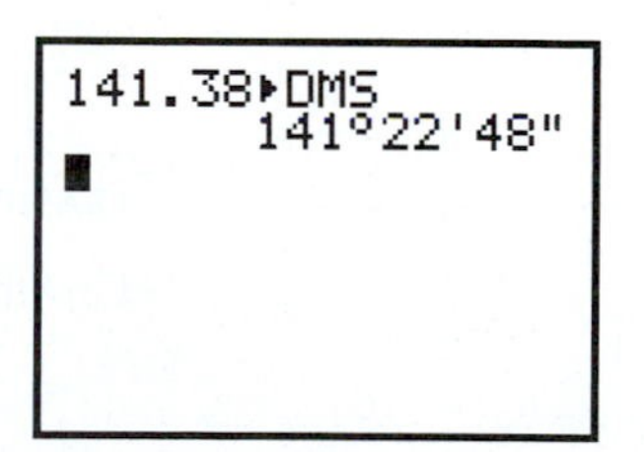

141.38 **2nd ANGLE 4 ENTER**

Thus, the result is 141°23′ (to the nearest minute).

The radian unit is discussed in Chapter 10.

A right triangle has one right angle, two acute angles, a hypotenuse, and two legs. As shown in Fig. 3.6, the right angle is usually labeled with the capital letter C. The vertices of the two acute angles are labeled with the capital letters A and B. The hypotenuse is usually labeled with the lowercase letter c. The legs are the sides opposite the acute angles. The

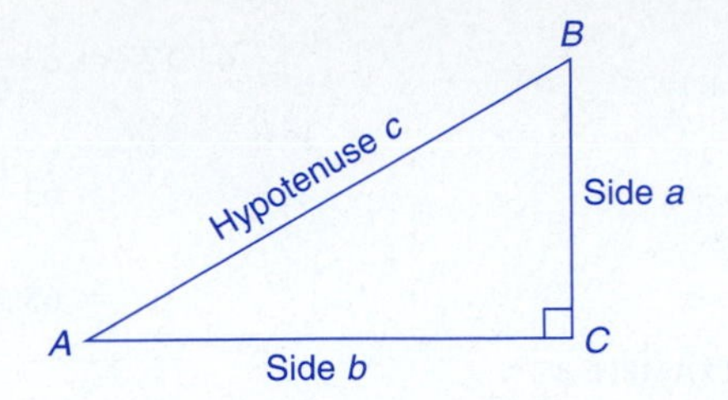

Figure 3.6 Common labels of angles and sides of a right triangle.

leg (side) opposite angle A is labeled a, and the leg opposite angle B is labeled b. Note that each side of the triangle is labeled with the lowercase of the letter of the angle opposite it.

The **Pythagorean theorem** gives the relationship among the sides of a right triangle.

PYTHAGOREAN THEOREM

$$c^2 = a^2 + b^2 \quad \text{or} \quad c = \sqrt{a^2 + b^2}$$

In words, in a right triangle, the square of the length of the hypotenuse is equal to the sum of the squares of the lengths of the two legs.

EXAMPLE 5

Find the length of the hypotenuse in the right triangle in Fig. 3.7.

Using the Pythagorean theorem, we obtain

$$c = \sqrt{a^2 + b^2}$$
$$c = \sqrt{(115 \text{ m})^2 + (184 \text{ m})^2}$$
$$= 217 \text{ m} \qquad \text{(to three significant digits)}$$

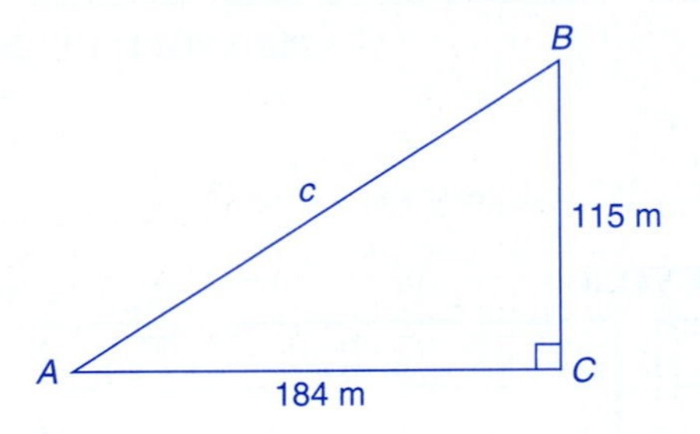

Figure 3.7

EXAMPLE 6

Find the length of side a in the right triangle in Fig. 3.8.

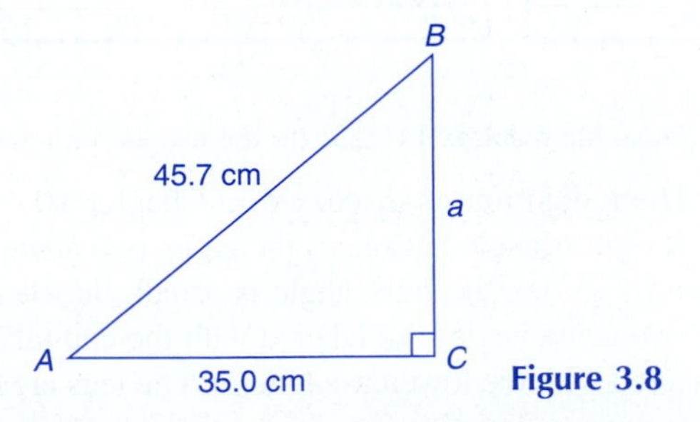

Figure 3.8

Use the Pythagorean theorem,

$$c^2 = a^2 + b^2$$

and solve for a.

$$\begin{aligned} a^2 &= c^2 - b^2 \\ a &= \sqrt{c^2 - b^2} \\ a &= \sqrt{(45.7 \text{ cm})^2 - (35.0 \text{ cm})^2} \\ &= 29.4 \text{ cm} \qquad \text{(to three significant digits)} \end{aligned}$$

The following six trigonometric ratios express the relationships between an acute angle of a right triangle and the lengths of two of its sides (see Fig. 3.9):

The **sine** of angle A, abbreviated $\sin A$, is equal to the ratio of the length of the side opposite angle A, a, to the length of the hypotenuse, c.

The **cosine** of angle A, abbreviated $\cos A$, is the ratio of the length of the side adjacent to angle A, b, to the length of the hypotenuse, c.

The **tangent** of angle A, abbreviated $\tan A$, is equal to the ratio of the length of the side opposite angle A, a, to the length of the side adjacent to angle A, b.

The **cotangent** of angle A, abbreviated $\cot A$, is equal to the ratio of the length of the side adjacent to angle A, b, to the length of the side opposite angle A, a.

The **secant** of angle A, abbreviated $\sec A$, is equal to the ratio of the length of the hypotenuse, c, to the length of the side adjacent to angle A, b.

The **cosecant** of angle A, abbreviated $\csc A$, is equal to the ratio of the length of the hypotenuse, c, to the length of the side opposite angle A, a.

$$\sin A = \frac{\text{side opposite } A}{\text{hypotenuse}} = \frac{a}{c} \quad \text{or} \quad \sin B = \frac{b}{c}$$

$$\cos A = \frac{\text{side adjacent to } A}{\text{hypotenuse}} = \frac{b}{c} \quad \text{or} \quad \cos B = \frac{a}{c}$$

$$\tan A = \frac{\text{side opposite } A}{\text{side adjacent to } A} = \frac{a}{b} \quad \text{or} \quad \tan B = \frac{b}{a}$$

$$\cot A = \frac{\text{side adjacent to } A}{\text{side opposite } A} = \frac{b}{a} \quad \text{or} \quad \cot B = \frac{a}{b}$$

$$\sec A = \frac{\text{hypotenuse}}{\text{side adjacent to } A} = \frac{c}{b} \quad \text{or} \quad \sec B = \frac{c}{a}$$

$$\csc A = \frac{\text{hypotenuse}}{\text{side opposite } A} = \frac{c}{a} \quad \text{or} \quad \csc B = \frac{c}{b}$$

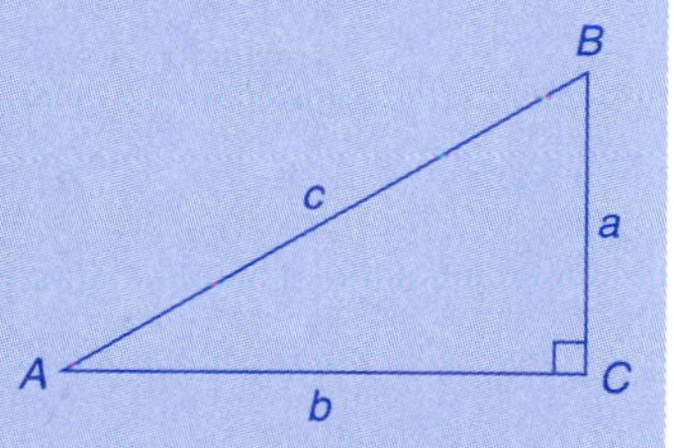

Figure 3.9

We shall discuss the trigonometric ratios of nonacute angles in Chapter 10.

EXAMPLE 7

Find the six trigonometric ratios of angle A in the right triangle in Fig. 3.10.

Using the Pythagorean theorem, we have

$$\begin{aligned} c &= \sqrt{(3.00 \text{ m})^2 + (4.00 \text{ m})^2} \\ &= 5.00 \text{ m} \end{aligned}$$

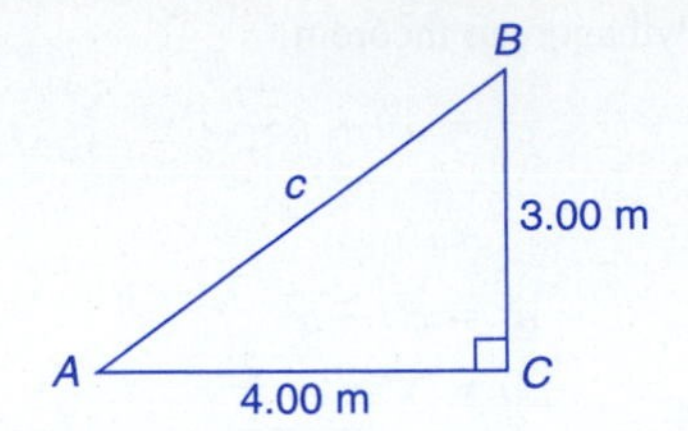

Figure 3.10

Use the definitions as follows:

$$\sin A = \frac{\text{side opposite } A}{\text{hypotenuse}} = \frac{3.00 \text{ m}}{5.00 \text{ m}} = 0.600$$

$$\cos A = \frac{\text{side adjacent to } A}{\text{hypotenuse}} = \frac{4.00 \text{ m}}{5.00 \text{ m}} = 0.800$$

$$\tan A = \frac{\text{side opposite } A}{\text{side adjacent to } A} = \frac{3.00 \text{ m}}{4.00 \text{ m}} = 0.750$$

$$\cot A = \frac{\text{side adjacent to } A}{\text{side opposite } A} = \frac{4.00 \text{ m}}{3.00 \text{ m}} = 1.33$$

$$\sec A = \frac{\text{hypotenuse}}{\text{side adjacent to } A} = \frac{5.00 \text{ m}}{4.00 \text{ m}} = 1.25$$

$$\csc A = \frac{\text{hypotenuse}}{\text{side opposite } A} = \frac{5.00 \text{ m}}{3.00 \text{ m}} = 1.67$$

Let us further examine this relationship between an acute angle of a right triangle and the lengths of two of its sides. Draw any two right triangles with a common acute angle A, as shown in Fig. 3.11. Note that $\triangle ABC$ and $\triangle ADE$ are similar. Two triangles are similar when two angles of one triangle have the same measure as two corresponding angles of the second. In Fig. 3.11, angle A is a common angle to both $\triangle ABC$ and $\triangle ADE$; angle ACB and angle AED are both right angles and therefore are of equal measure. From geometry, we know that the corresponding sides of similar triangles are proportional. This means that $\frac{a}{b} = \frac{c}{d}$. Thus, no matter how points B and D are chosen, the numerical ratios $\frac{a}{b}$ and $\frac{c}{d}$, which are used to define $\tan A$, are the same.

Similar arguments using Fig. 3.11 can be given to show that the other trigonometric ratios do not depend upon the choice of the points in determining the defining ratios. Angle A is the only determining factor.

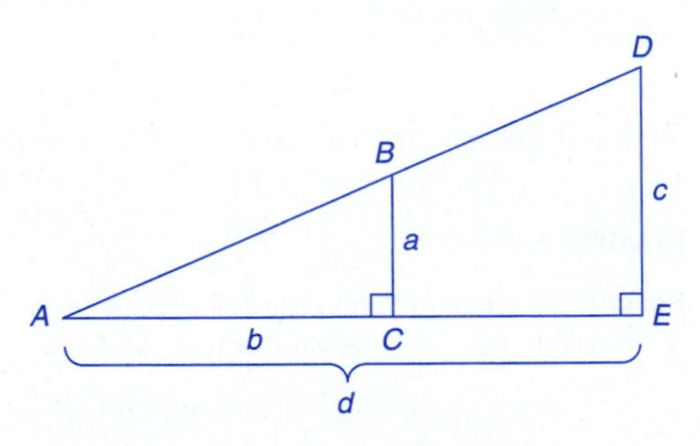

Figure 3.11 Two right triangles with common acute angle A.

EXAMPLE 8

Find the six trigonometric ratios of angle A (a) using $\triangle ABC$ and (b) using $\triangle ADC$ in Fig. 3.12.

(a) First find the length of hypotenuse AB in $\triangle ABC$.

$$AB = \sqrt{3^2 + 4^2} = \sqrt{9 + 16} = \sqrt{25} = 5$$

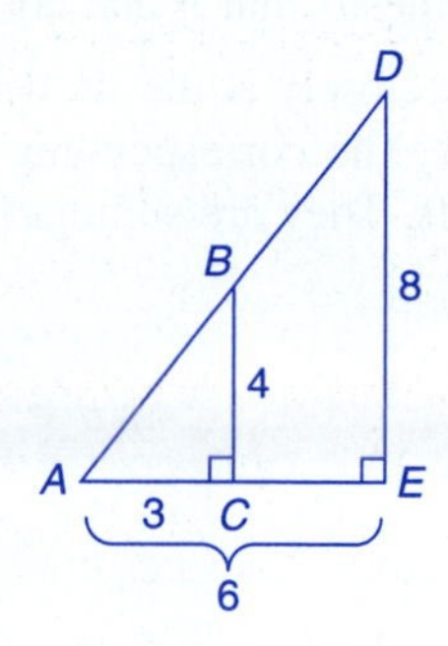

Figure 3.12

Then, using the definitions, we have

$$\sin A = \frac{\text{side opposite } A}{\text{hypotenuse}} = \frac{4}{5}$$

$$\cos A = \frac{\text{side adjacent to } A}{\text{hypotenuse}} = \frac{3}{5}$$

$$\tan A = \frac{\text{side opposite } A}{\text{side adjacent to } A} = \frac{4}{3}$$

$$\cot A = \frac{\text{side adjacent to } A}{\text{side opposite } A} = \frac{3}{4}$$

$$\sec A = \frac{\text{hypotenuse}}{\text{side adjacent to } A} = \frac{5}{3}$$

$$\csc A = \frac{\text{hypotenuse}}{\text{side opposite } A} = \frac{5}{4}$$

(b) Find the length of the hypotenuse AD in $\triangle ADE$.

$$AD = \sqrt{6^2 + 8^2} = \sqrt{36 + 64} = \sqrt{100} = 10$$

Then

$$\sin A = \frac{\text{side opposite } A}{\text{hypotenuse}} = \frac{8}{10} = \frac{4}{5}$$

$$\cos A = \frac{\text{side adjacent to } A}{\text{hypotenuse}} = \frac{6}{10} = \frac{3}{5}$$

$$\tan A = \frac{\text{side opposite } A}{\text{side adjacent to } A} = \frac{8}{6} = \frac{4}{3}$$

$$\cot A = \frac{\text{side adjacent to } A}{\text{side opposite } A} = \frac{6}{8} = \frac{3}{4}$$

$$\sec A = \frac{\text{hypotenuse}}{\text{side adjacent to } A} = \frac{10}{6} = \frac{5}{3}$$

$$\csc A = \frac{\text{hypotenuse}}{\text{side opposite } A} = \frac{10}{8} = \frac{5}{4}$$

Note that any other right triangle with the same angle A would give the same six ratios.

Look closely at each of the six ratios in Examples 7 and 8. Do you see each of the following relationships?

1. The values of sin A and csc A are reciprocals of each other.

2. The values of cos A and sec A are reciprocals of each other.

3. The values of tan A and cot A are reciprocals of each other.

Now, look closely at the six definitions. Note that this same reciprocal relationship exists there, too. The corresponding pairs of reciprocals are called **reciprocal trigonometric functions.** They are summarized in the following box.

$$\sin\theta = \frac{1}{\csc\theta} \qquad \csc\theta = \frac{1}{\sin\theta}$$

$$\cos\theta = \frac{1}{\sec\theta} \qquad \sec\theta = \frac{1}{\cos\theta}$$

$$\tan\theta = \frac{1}{\cot\theta} \qquad \cot\theta = \frac{1}{\tan\theta}$$

Exercises 3.1

Use a protractor to draw each angle.

1. 35° **2.** 126° **3.** 240° **4.** 333°

Change each angle to degrees.

5. 15′ **6.** 6′ **7.** 120′ **8.** 47′

Change each angle to minutes.

9. $\frac{1}{2}^\circ$ **10.** $\frac{2}{3}^\circ$ **11.** 0.4° **12.** 0.7°

Change each angle to degrees.

13. 37°12′ **14.** 142°30′ **15.** 75°47′ **16.** 120°11′

Change each angle to degrees and minutes.

17. $69\frac{1}{3}^\circ$ **18.** 183.5° **19.** 23.3° **20.** $7\frac{2}{5}^\circ$

Change each angle to degrees.

21. 34°24′15″ **22.** 65°27′36″ **23.** 19°18′27″ **24.** 135°48′9″

Change each angle to degrees, minutes, and seconds.

25. 18.21° **26.** 35.84° **27.** 8.925° **28.** 29.608°

Use △ABC in Fig. 3.13 for Exercises 29 through 60.

29. The side opposite angle A is ____.

30. The side opposite angle B is ____.

31. The hypotenuse is ____.

32. The side adjacent to angle A is ____.

33. The side adjacent to angle B is ____.

34. The angle opposite side a is ____.

35. The angle opposite side b is ____.

36. The angle opposite side c is ____.

37. The angle adjacent to side a is ____.

38. The angle adjacent to side b is ____.

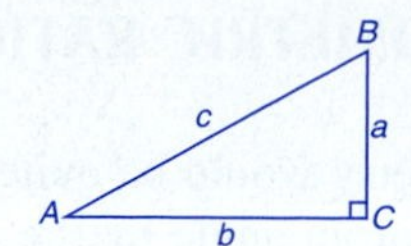

Figure 3.13

Find the length of the third side of each right triangle, rounded to three significant digits.

39. $a = 5.00$ cm, $b = 12.0$ cm

40. $a = 13.5$ ft, $b = 18.5$ ft

41. $a = 115$ mi, $c = 208$ mi

42. $a = 46.7$ km, $c = 75.6$ km

43. $b = 377$ yd, $c = 506$ yd

44. $b = 1450$ mi, $c = 2960$ mi

45. $a = 35.7$ m, $b = 16.8$ m

46. $a = 105$ m, $c = 537$ m

47. $a = 2.25$ cm, $c = 3.75$ cm

48. $b = 155$ mi, $c = 208$ mi

Find the six trigonometric ratios of angle A rounded to three significant digits.

49. $a = 5.00$ cm, $b = 12.0$ cm

50. $a = 13.5$ ft, $b = 18.5$ ft

51. $a = 335$ m, $c = 685$ m

52. $a = 19.8$ km, $c = 40.5$ km

53. $b = 3.00$ km, $c = 6.00$ km

54. $b = 239$ mi, $c = 307$ mi

Find the six trigonometric ratios of angle B rounded to three significant digits.

55. $a = 5.00$ cm, $b = 12.0$ cm

56. $a = 13.5$ ft, $b = 18.5$ ft

57. $a = 4.60$ m, $c = 9.25$ m

58. $a = 1.62$ km, $c = 4.05$ km

59. $b = 4.50$ ft, $c = 9.00$ ft

60. $b = 27.5$ in., $c = 51.2$ in.

61. Find the six trigonometric ratios of angle A **(a)** using $\triangle ABC$ and **(b)** using $\triangle ADE$ in Fig. 3.14. Round each result to three significant digits.

62. Find the six trigonometric ratios of angle B **(a)** using $\triangle BAC$ and **(b)** using $\triangle BFG$ in Fig. 3.15. Round each result to three significant digits.

63. Show that $\tan A = \dfrac{\sin A}{\cos A}$ by using the defining ratios. (*Hint:* The right-hand expression is a complex fraction that reduces to tan A.)

64. Given that $\sin A = 0.8387$ and $\cos A = 0.5446$, compute the remaining four trigonometric functions of A. Round each result to four significant digits.

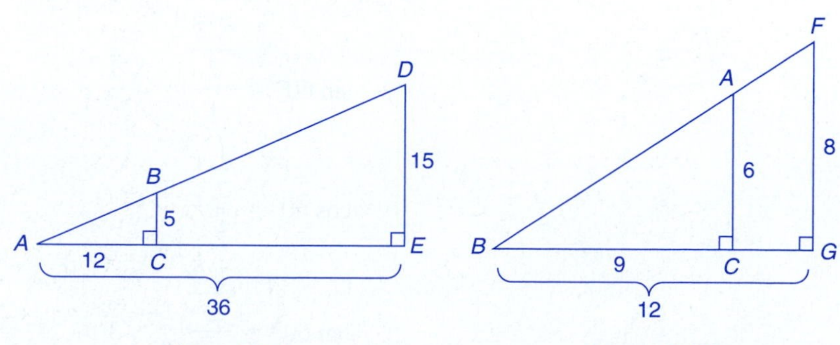

Figure 3.14

Figure 3.15

3.2 VALUES OF THE TRIGONOMETRIC RATIOS

Our work with trigonometry would be quite limited if we always had to rely on computing trigonometric ratios of an angle by the methods used thus far. In practice we are often given the measurement of the angle in degrees or radians and need the value of one of the trigonometric ratios of this angle. In this text we shall present trigonometry using a calculator. Before learning to use a calculator we see how these trigonometric values have been obtained for certain angles.

Let us first consider the angle of measure 60°. For this purpose we construct an equilateral triangle as shown in Fig. 3.16. From geometry we know that each angle of triangle OPQ must be 60°.

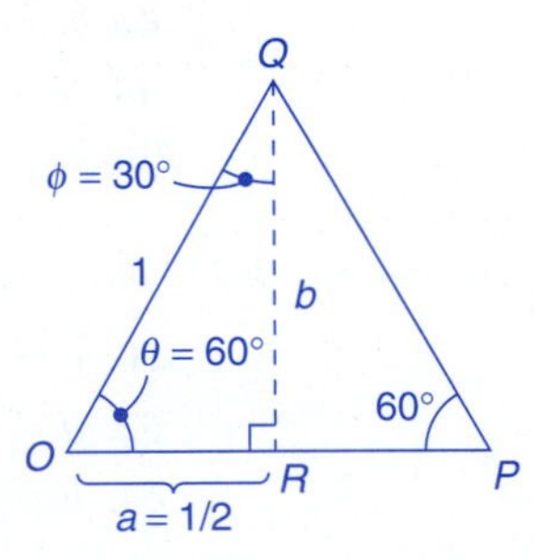

Figure 3.16 Equilateral triangle.

We shall let each side be of length 1. By construction, $OR = a = \frac{1}{2}$, which is the side adjacent to angle θ in right triangle ORQ. The opposite side, $QR = b$, can be computed by using the Pythagorean theorem.

$$1^2 = a^2 + b^2$$

$$1 = \left(\frac{1}{2}\right)^2 + b^2$$

$$1 - \frac{1}{4} = b^2$$

$$b^2 = \frac{3}{4}$$

$$b = \sqrt{\frac{3}{4}} = \frac{\sqrt{3}}{2}$$

Using the trigonometric definitions and $\triangle ORQ$, we have

$$\sin 60° = \frac{\frac{\sqrt{3}}{2}}{1} = \frac{\sqrt{3}}{2}$$

$$\cos 60° = \frac{\frac{1}{2}}{1} = \frac{1}{2}$$

$$\tan 60° = \frac{\frac{\sqrt{3}}{2}}{\frac{1}{2}} = \sqrt{3}$$

$$\cot 60° = \frac{\frac{1}{2}}{\frac{\sqrt{3}}{2}} = \frac{1}{\sqrt{3}} = \frac{\sqrt{3}}{3}$$

$$\sec 60° = \frac{1}{\frac{1}{2}} = 2$$

$$\csc 60° = \frac{1}{\frac{\sqrt{3}}{2}} = \frac{2}{\sqrt{3}} = \frac{2\sqrt{3}}{3}$$

Since $\phi = 30°$ and side $b = \frac{\sqrt{3}}{2}$ in Fig. 3.16, we can compute the trigonometric ratios of 30°. Side a is the side opposite angle ϕ, and side b is the adjacent side. The same right triangle, $\triangle ORQ$, has been redrawn in Fig. 3.17 for convenience in viewing angle ϕ.

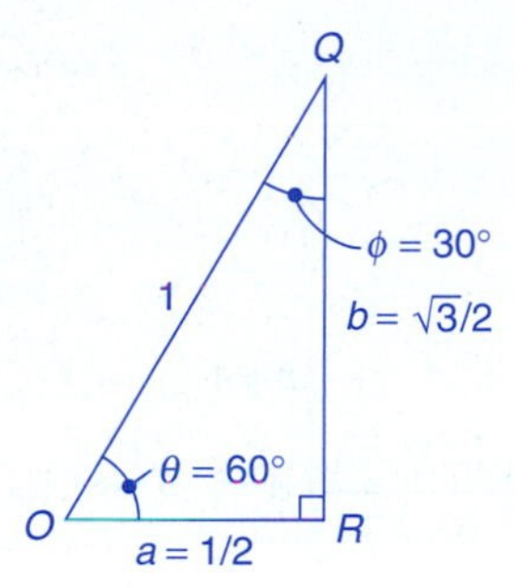

Figure 3.17 Triangle *ORQ* from Figure 3.16.

Again using the trigonometric definitions and $\triangle ORQ$, we have

$$\sin 30° = \frac{\frac{1}{2}}{1} = \frac{1}{2}$$

$$\cos 30° = \frac{\frac{\sqrt{3}}{2}}{1} = \frac{\sqrt{3}}{2}$$

$$\tan 30° = \frac{\frac{1}{2}}{\frac{\sqrt{3}}{2}} = \frac{1}{\sqrt{3}} = \frac{\sqrt{3}}{3}$$

$$\cot 30° = \frac{\frac{\sqrt{3}}{2}}{\frac{1}{2}} = \sqrt{3}$$

$$\sec 30° = \frac{1}{\frac{\sqrt{3}}{2}} = \frac{2}{\sqrt{3}} = \frac{2\sqrt{3}}{3}$$

$$\csc 30° = \frac{1}{\frac{1}{2}} = 2$$

Two acute angles are **complementary** if their sum is 90°. Thus, angles of 30° and 60° are complementary angles. In fact, it is always true that the two acute angles of any right triangle are complementary since their sum must be 90°.

Observe that we have found

$$\sin 30° = \frac{1}{2} = \cos 60°$$

$$\sin 60° = \frac{\sqrt{3}}{2} = \cos 30°$$

In fact, this is true of any two complementary angles θ and ϕ.

$$\sin \theta = \cos \phi$$

From this relationship we can interpret the "co" in cosine to mean that it is the sine of the *co*mplementary angle. The sine and cosine are said to be **cofunctions.** Similar statements can be made for tangent and cotangent as well as for secant and cosecant.

We shall use Fig. 3.18 to evaluate the trigonometric ratios of 45°.

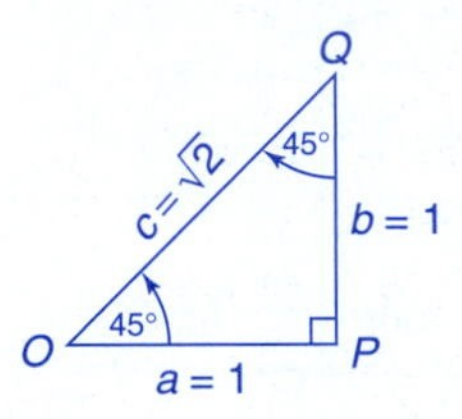

Figure 3.18 Isosceles right triangle.

Since $\triangle OPQ$ is isosceles, the adjacent side and opposite side are equal. If we let $a = b = 1$, then $c = \sqrt{1^2 + 1^2} = \sqrt{2}$. Therefore,

$$\sin 45° = \frac{1}{\sqrt{2}} = \frac{\sqrt{2}}{2}$$

$$\cos 45° = \frac{1}{\sqrt{2}} = \frac{\sqrt{2}}{2}$$

$$\tan 45° = \frac{1}{1} = 1$$

$$\cot 45° = \frac{1}{1} = 1$$

$$\sec 45° = \frac{\sqrt{2}}{1} = \sqrt{2}$$

$$\csc 45° = \frac{\sqrt{2}}{1} = \sqrt{2}$$

However, we shall need to know the values of trigonometric ratios of angles besides 30°, 45°, and 60°. For this purpose we shall use calculators. Most calculators have buttons to evaluate the sine, cosine, and tangent of an angle.

EXAMPLE 1

Find sin 26° and cos 36.75° rounded to four significant digits.

MODE down arrows right arrow **2 ENTER 2nd SIN** 26) **green diamond ENTER 2nd COS** 36.75) **ENTER**

MODE 2nd QUIT

SIN 26) **ENTER COS** 36.75) **ENTER**

That is, sin 26° = 0.4384 and cos 36.75° = 0.8013 rounded to four significant digits.

If a calculator does not have buttons for cotangent, secant, and cosecant, how can you evaluate these functions? Recall that

$$\cot\theta = \frac{1}{\tan\theta} \qquad \sec\theta = \frac{1}{\cos\theta} \qquad \text{and} \qquad \csc\theta = \frac{1}{\sin\theta}$$

These are called the *reciprocal trigonometric functions.*

EXAMPLE 2

Find sec 19° and cot 27.5° rounded to four significant digits.

1/ **2nd COS** 19) **green diamond ENTER** 1/ **2nd TAN** 27.5) **ENTER**

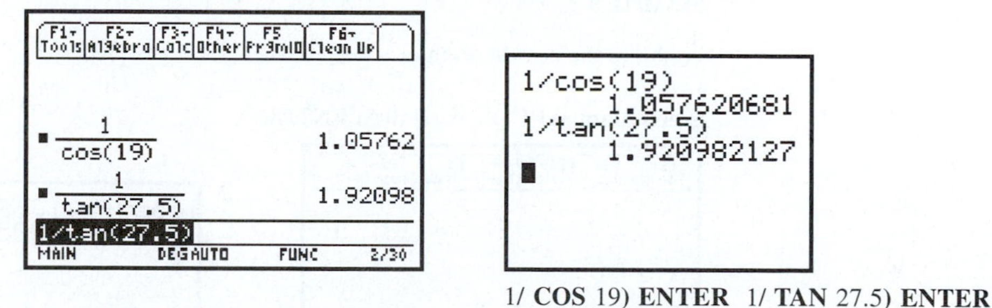

1/ **COS** 19) **ENTER** 1/ **TAN** 27.5) **ENTER**

That is, sec 19° = 1.058 and cot 27.5° = 1.921 rounded to four significant digits.

A calculator may also be used to find the angle when the value of a trigonometric ratio is given. The procedure is shown by the following examples. We shall first limit our angle θ to $0° \le \theta \le 90°$ in this chapter.

EXAMPLE 3

Find θ to the nearest tenth of a degree when $\sin \theta = 0.4321$.

green diamond SIN^{-1} .4321) **ENTER**

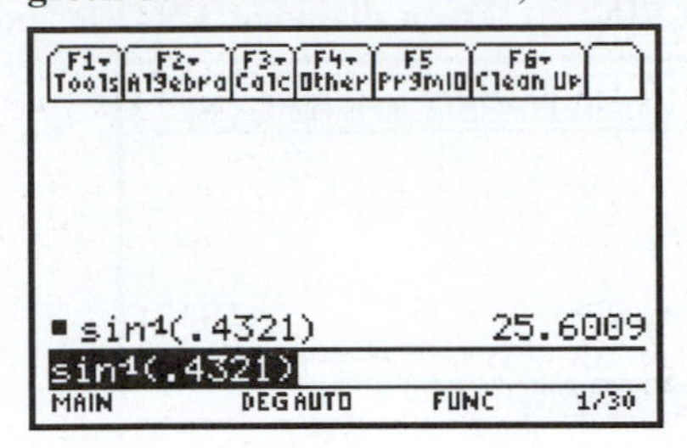

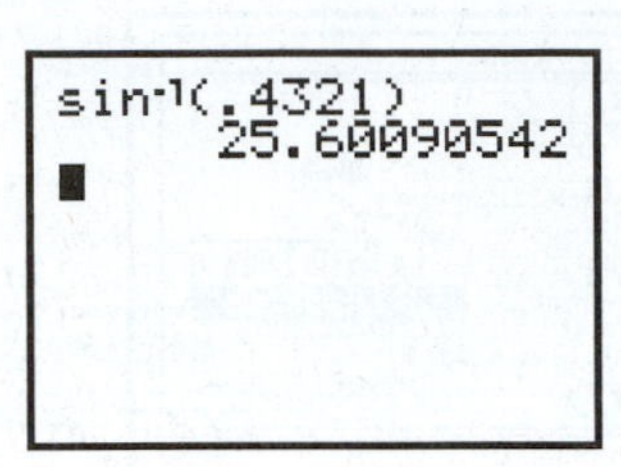

2nd SIN^{-1} .4321) **ENTER**

Thus, $\theta = 25.6°$ to the nearest tenth of a degree.

EXAMPLE 4

Find θ to the nearest tenth of a degree when $\cos \theta = 0.6046$ and when $\tan \theta = 2.584$.

green diamond COS^{-1} .6046) **ENTER green diamond TAN^{-1}** 2.584) **ENTER**

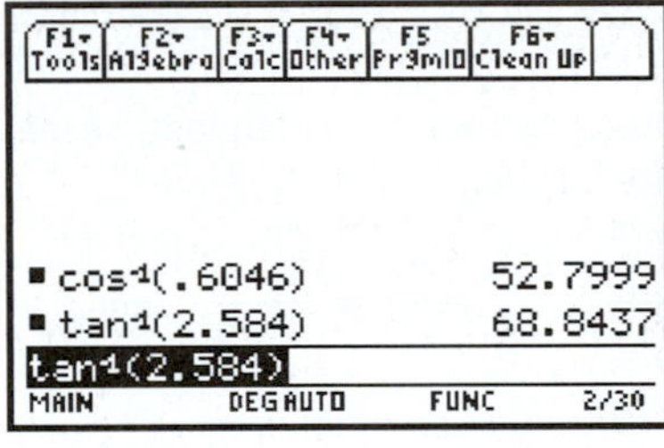

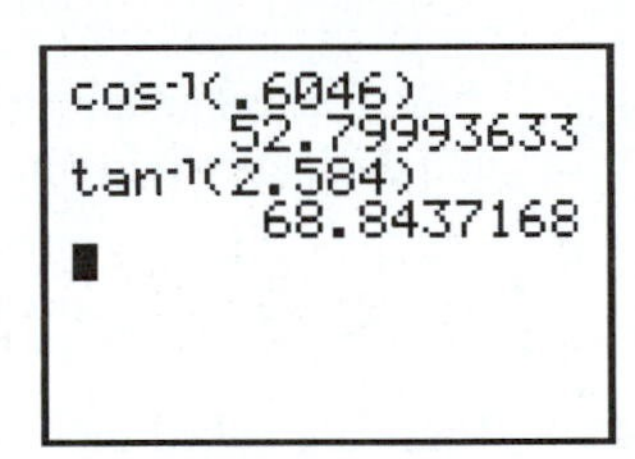

2nd COS^{-1} .6046) **ENTER 2nd TAN^{-1}** 2.584) **ENTER**

Thus, $\theta = 52.8°$ and $\theta = 68.8°$, respectively, to the nearest tenth of a degree.

EXAMPLE 5

Find θ to the nearest tenth of a degree when $\sec \theta = 1.365$.

green diamond COS^{-1} 1/1.365) **ENTER**

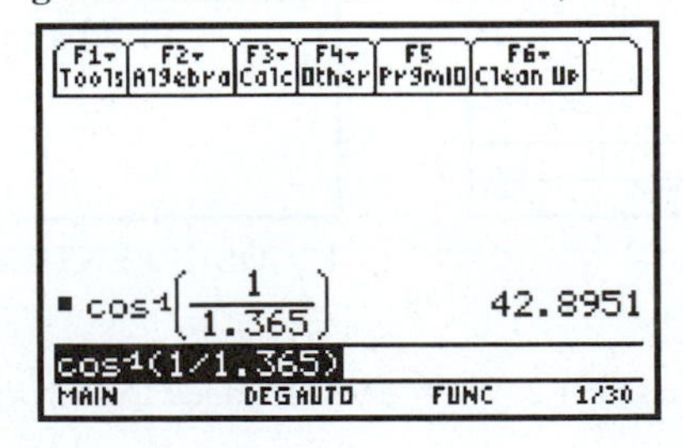

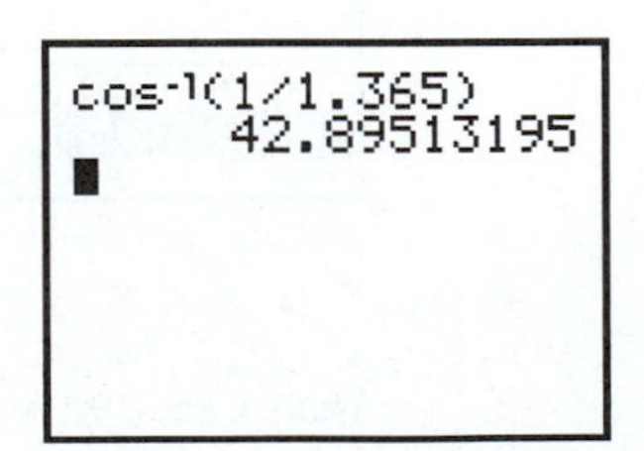

2nd COS^{-1} 1/1.365) **ENTER**

Thus, $\theta = 42.9°$ to the nearest tenth of a degree.

EXAMPLE 6

Find tan 58°16′24″ rounded to five significant digits.

2nd TAN 58 **2nd** | 16 **2nd** = 24 **2nd 1 ENTER**

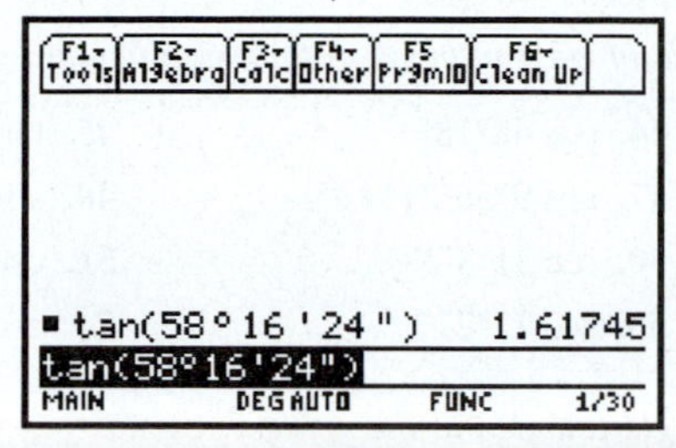

TAN 58 **2nd ANGLE 1** 16 **2nd ANGLE 2** 24 **ALPHA +) ENTER**

Thus, tan 58°16′24″ = 1.6175 rounded to five significant digits.

EXAMPLE 7

Find θ to the nearest second when $\sin\theta = 0.2587$.

green diamond SIN^{-1} .2587) **2nd MATH 2 8 ENTER**

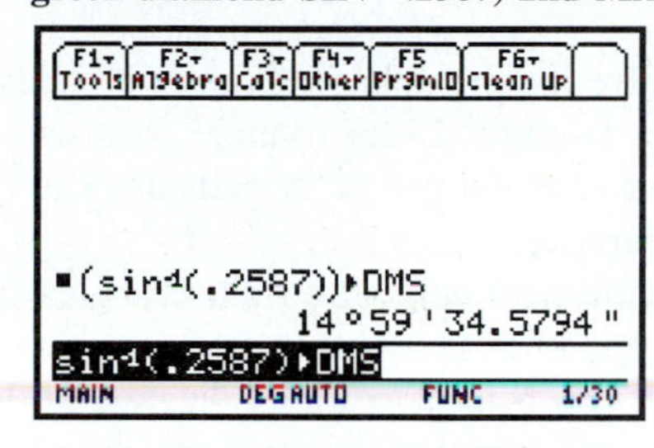

2nd SIN^{-1} .2587) **2nd ANGLE 4 ENTER**

Thus, θ = 14°59′35″ rounded to the nearest second.

Exercises 3.2

Use a calculator to find the value of each rounded to four significant digits.

1. sin 18.5° **2.** cos 27.6° **3.** tan 41.4°
4. sin 13.6° **5.** cos 77.2° **6.** tan 87.1°
7. sec 34.7° **8.** csc 80.5° **9.** cot 34.0°
10. sec 19.0° **11.** csc 49.8° **12.** cot 74.1°
13. sin 46.72° **14.** cos 19.51° **15.** tan 73.8035°
16. csc 34.9625° **17.** sec 8.3751° **18.** cot 16.3795°

Use a calculator to find each angle rounded to the nearest tenth of a degree.

19. $\sin\theta = 0.4305$ **20.** $\cos\theta = 0.7771$ **21.** $\tan\theta = 0.4684$
22. $\sin\theta = 0.2096$ **23.** $\cos\theta = 0.1463$ **24.** $\tan\theta = 1.357$
25. $\tan\theta = 3.214$ **26.** $\cos\theta = 0.5402$ **27.** $\sin\theta = 0.1986$
28. $\cot\theta = 3.270$ **29.** $\sec\theta = 2.363$ **30.** $\csc\theta = 5.662$
31. $\cot\theta = 0.5862$ **32.** $\sec\theta = 3.341$ **33.** $\csc\theta = 2.221$
34. $\csc\theta = 1.333$ **35.** $\sec\theta = 6.005$ **36.** $\cot\theta = 8.307$

Use a calculator to find each angle rounded to the nearest hundredth of a degree.

37. $\cos\theta = 0.4836$ **38.** $\sin\theta = 0.1920$ **39.** $\cot\theta = 1.5392$

40. $\tan\theta = 2.5575$ **41.** $\csc\theta = 2.4075$ **42.** $\sec\theta = 1.2566$

Use a calculator to find the value of each rounded to four significant digits.

43. $\sin 36°24'$ **44.** $\cos 48°18'$ **45.** $\tan 52°43'38''$

46. $\tan 17°35'52''$ **47.** $\cos 9°56'21''$ **48.** $\sin 21°8'46''$

49. $\cot 36°15'44''$ **50.** $\sec 31°8'27''$ **51.** $\csc 84°35'53''$

52. $\cot 51°40'11''$ **53.** $\sec 72°27''$ **54.** $\csc 7°52''$

Use a calculator to find each angle rounded to the nearest second.

55. $\sin\theta = 0.8556$ **56.** $\cos\theta = 0.2749$ **57.** $\tan\theta = 6.2662$

58. $\tan\theta = 0.3254$ **59.** $\cos\theta = 0.5966$ **60.** $\sin\theta = 0.2694$

61. $\cot\theta = 0.8678$ **62.** $\sec\theta = 3.3424$ **63.** $\csc\theta = 2.3770$

64. $\cot\theta = 4.5065$ **65.** $\sec\theta = 1.1678$ **66.** $\csc\theta = 8.1407$

3.3 SOLVING RIGHT TRIANGLES

Many technical problems involve finding unknown values of sides or angles of triangles. We call this process **solving a triangle.** Every triangle consists of six parts: three sides and three angles. If at least one side and two other parts are known, it is possible to find the values of the other three parts.

In this chapter we are concerned only with right triangles. Since one angle is 90°, we need to know only a side and one other part to find the other three parts. When you are solving triangles, it is often very helpful to draw a diagram of the triangle to be solved. For this purpose we shall follow customary practice and label the angles and sides as shown in Fig. 3.19.

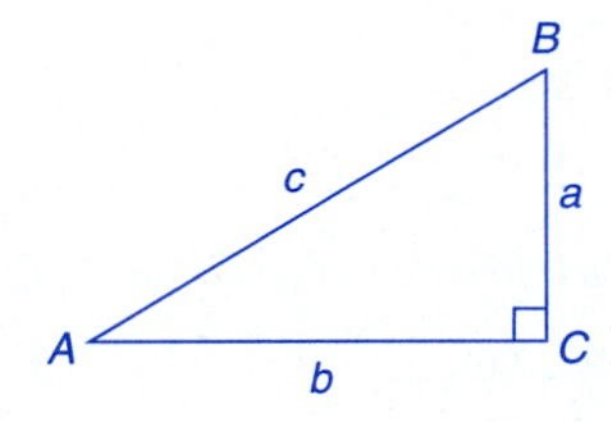

Figure 3.19 General right triangle *ABC*.

From the Pythagorean theorem, we know that

$$c^2 = a^2 + b^2 \quad \text{or} \quad c = \sqrt{a^2 + b^2}$$

Recall that the two acute angles of a right triangle are complementary. That is,

$$A + B = 90°$$
$$A = 90° - B$$
$$B = 90° - A$$

Thus, once we know the value of one acute angle, we can find the value of the other. We now solve right triangles. Using examples, we shall see how to find the unknown parts.

When calculations with measurements involve a trigonometric ratio, we use the following rule for significant digits:

Angles expressed to the nearest:	*Lengths of sides of a triangle will contain:*
1°	Two significant digits
0.1° or 1′	Three significant digits
0.01° or 1″	Four significant digits

EXAMPLE 1

Find side c given side $a = 6.00$ cm and side $b = 8.00$ cm.

First, draw a triangle as in Fig. 3.20. Then

$$c = \sqrt{a^2 + b^2}$$
$$c = \sqrt{(6.00 \text{ cm})^2 + (8.00 \text{ cm})^2}$$
$$= \sqrt{36.0 \text{ cm}^2 + 64.0 \text{ cm}^2}$$
$$= \sqrt{100.00 \text{ cm}^2}$$
$$= 10.0 \text{ cm}$$

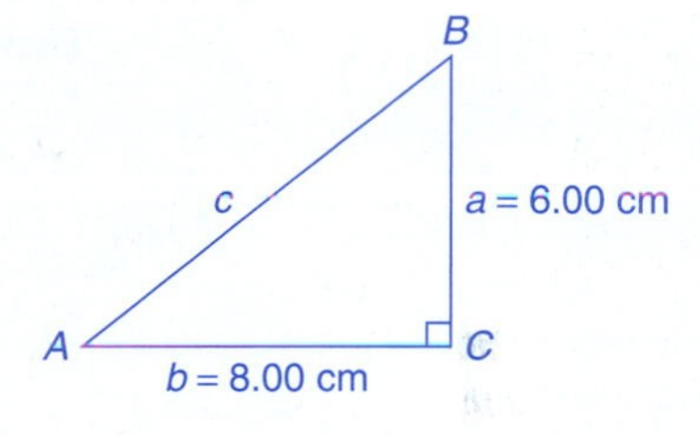

Figure 3.20

EXAMPLE 2

Find side a given side $b = 6.00$ m and side $c = 11.0$ m.

First, draw a triangle as in Fig. 3.21. Since

$$c^2 = a^2 + b^2$$

we have

$$a^2 = c^2 - b^2$$
$$a = \sqrt{c^2 - b^2}$$
$$a = \sqrt{(11.0 \text{ m})^2 - (6.00 \text{ m})^2}$$
$$= 9.22 \text{ m}$$

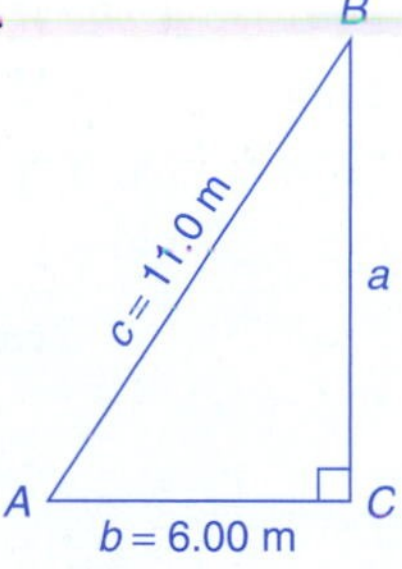

Figure 3.21

These two examples did not require the use of a trigonometric ratio. However, such use will be necessary in the following examples.

EXAMPLE 3

Find angle B to the nearest tenth of a degree given $b = 7.00$ m and $c = 9.00$ m.

Draw a triangle as in Fig. 3.22. From the definition of sin B,

$$\sin B = \frac{\text{side opposite } B}{\text{hypotenuse}}$$
$$= \frac{7.00 \text{ m}}{9.00 \text{ m}} = 0.7778$$
$$B = 51.1° \quad \text{(to the nearest tenth of a degree)}$$

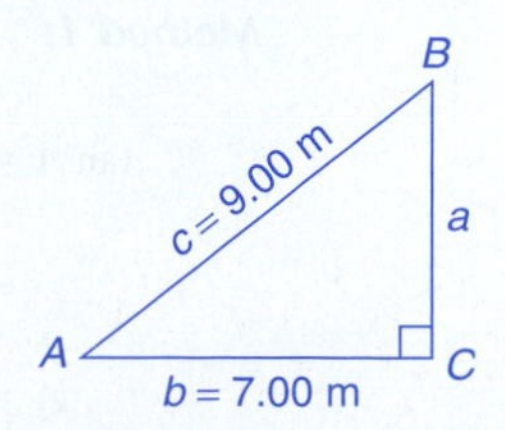

Figure 3.22

This angle is found using a calculator as follows:

green diamond SIN^{-1} 7/9) green diamond ENTER

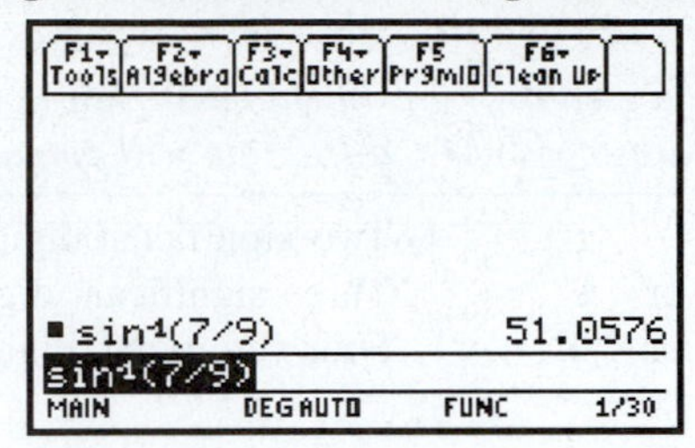

2nd SIN^{-1} 7/9) ENTER

So, $B = 51.1°$ rounded to the nearest tenth of a degree.

EXAMPLE 4

Find angle A to the nearest hundredth of a degree given $b = 4.250$ km and $c = 9.750$ km.

Draw a triangle as in Fig. 3.23. In this case we have

$$\cos A = \frac{\text{side adjacent to } A}{\text{hypotenuse}}$$

$$= \frac{4.250 \text{ km}}{9.750 \text{ km}} = 0.4359$$

$A = 64.16°$ (to the nearest hundredth of a degree)

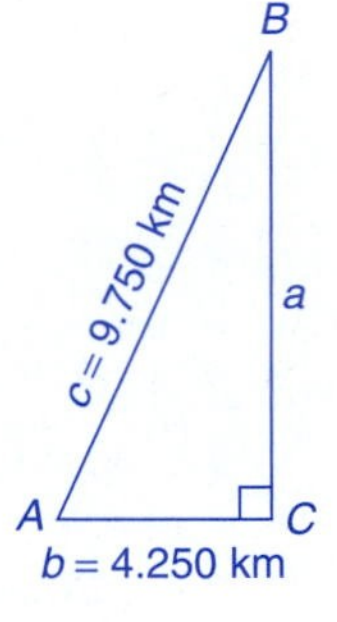

Figure 3.23

EXAMPLE 5

Find side a given $c = 12.00$ mi and $B = 24.00°$.

Draw a triangle as in Fig. 3.24.

$$\cos B = \frac{\text{side adjacent to } B}{\text{hypotenuse}}$$

$$\cos 24.00° = \frac{a}{12.00 \text{ mi}}$$

$$a = (12.00 \text{ mi})(\cos 24.00°)$$

$$= (12.00 \text{ mi})(0.9135)$$

$$= 10.96 \text{ mi}$$

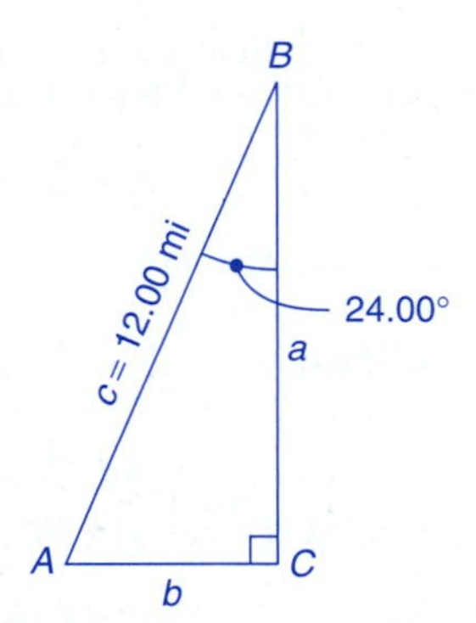

Figure 3.24

In the next example we solve a right triangle completely; that is, we find all sides and angles not given or known.

EXAMPLE 6

Solve the right triangle given $a = 6.00$ ft and $b = 4.00$ ft.

Draw a triangle as in Fig. 3.25.

Method 1:

$$\tan A = \frac{\text{side opposite } A}{\text{side adjacent to } A}$$

$$= \frac{6.00 \text{ ft}}{4.00 \text{ ft}} = 1.500$$

$A = 56.3°$ (to the nearest tenth of a degree)

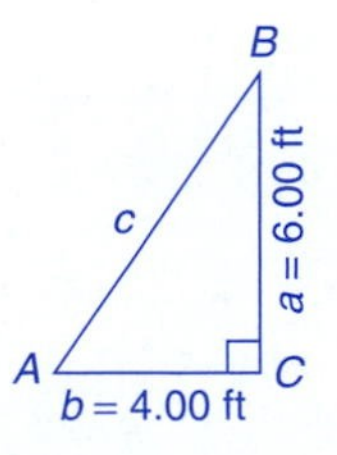

Figure 3.25

To find B we let

$$B = 90° - A$$
$$B = 90° - 56.3°$$
$$= 33.7°$$

Finally,

$$c = \sqrt{a^2 + b^2}$$
$$c = \sqrt{(4.00 \text{ ft})^2 + (6.00 \text{ ft})^2}$$
$$= \sqrt{52.0 \text{ ft}^2} = 7.21 \text{ ft}$$

Method 2: We could have found $c = 7.21$ ft first. Then

$$\sin A = \frac{\text{side opposite } A}{\text{hypotenuse}}$$
$$= \frac{6.00 \text{ ft}}{7.21 \text{ ft}} = 0.8322$$
$$A = 56.3°$$

Angle B would be found in the same way as in Method 1.

A triangle is not determined if only the angles are known. If $A = 60°$ and $B = 30°$, then we know that the triangle with sides $a = \frac{1}{2}$, $b = \frac{\sqrt{3}}{2}$, and $c = 1$ is a possible solution. But so is the triangle with sides $a = 1$, $b = \sqrt{3}$, and $c = 2$. In fact, there are infinitely many other solutions as well. Thus, unless the value of a side is given, we are unable to determine a specific triangle.

Note: While all six trigonometric ratios may be used to solve a right triangle, we usually restrict our choices here to sine, cosine, and tangent because these buttons appear on calculators. In later chapters the other three trigonometric ratios will become more important and will receive more attention.

Exercises 3.3

Solve each right triangle. Assume the standard labeling, as shown in Fig. 3.26.

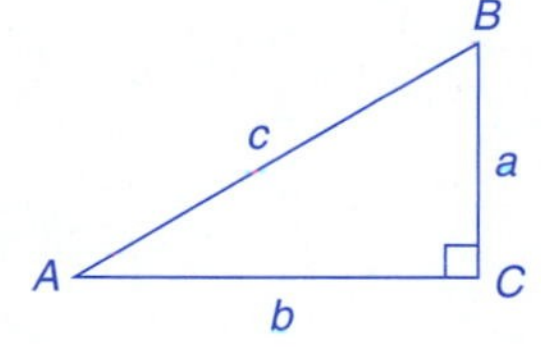

Figure 3.26

In Exercises 1 through 10, find each side to three significant digits and each angle to the nearest tenth of a degree.

1. $a = 4.00$ ft, $b = 8.00$ ft

2. $a = 7.00$ m, $B = 43.2°$

3. $c = 21.0$ cm, $A = 27.3°$

4. $a = 5.00$ in., $c = 9.00$ in.

5. $b = 7.50$ m, $c = 13.4$ m

6. $a = 9.20$ cm, $A = 72.4°$

7. $a = 12.4$ mi, $b = 7.70$ mi

8. $c = 9.40$ km, $B = 17.3°$

9. $b = 25\bar{0}$ km, $B = 37.0°$

10. $a = 12\bar{0}$ yd, $b = 95\bar{0}$ yd

In Exercises 11 through 18, find each side to four significant digits and each angle to the nearest hundredth of a degree.

11. $a = 14.21$ cm, $c = 37.42$ cm

12. $c = 7.300$ m, $A = 49.35°$

13. $a = 6755$ mi, $A = 68.75°$

14. $b = 13{,}530$ km, $c = 25{,}550$ km

15. $c = 45.32$ m, $B = 15.80°$

16. $a = 500\bar{0}$ ft, $B = 25.00°$

17. $b = 2572$ ft, $c = 4615$ ft

18. $a = 3.512$ mi, $c = 5.205$ mi

In Exercises 19 through 26, find each side to two significant digits and each angle to the nearest degree.

19. $b = 1500$ mi, $c = 3500$ mi

20. $a = 15$ ft, $A = 3\bar{0}°$

21. $c = 45$ m, $B = 5\bar{0}°$

22. $a = 36$ ft, $b = 16$ ft

23. $a = 140$ ft, $A = 37°$

24. $b = 3700$ m, $A = 59°$

25. $a = 3.5$ mi, $B = 22°$

26. $a = 0.36$ mi, $c = 0.44$ mi

In Exercises 27 through 34, find each side to four significant digits and each angle to the nearest second.

27. $a = 1753$ m, $B = 37°41'30''$

28. $a = 28{,}570$ ft, $b = 37{,}550$ ft

29. $a = 495.5$ ft, $c = 617.0$ ft

30. $c = 5.632$ km, $A = 18°6'45''$

31. $a = 37.52$ m, $A = 58°11'25''$

32. $b = 7753$ ft, $c = 8455$ ft

33. $c = 6752$ ft, $B = 27°5'16''$

34. $b = 55.60$ km, $B = 75°7'8''$

3.4 APPLICATIONS OF THE RIGHT TRIANGLE

Many applications, both technical and nontechnical, are solved using the trigonometric definitions and right triangles. The following examples illustrate some of these basic applications.

EXAMPLE 1

A mine shaft extends down at an angle of 5°. How long will the shaft have to be to reach a vein of ore which is 65 ft directly below the surface?

First, draw a triangle as in Fig. 3.27.

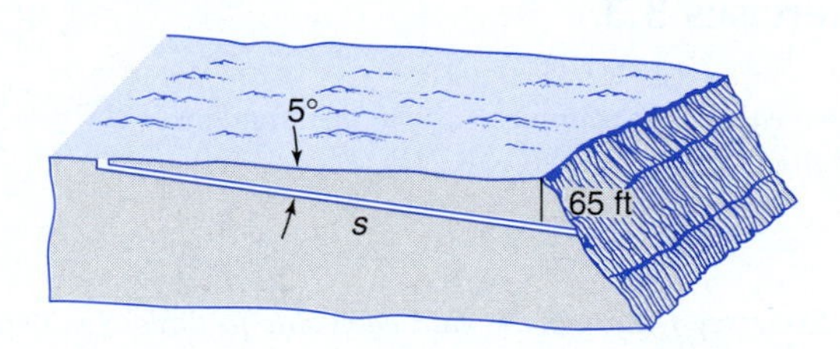

Figure 3.27

$$\sin A = \frac{\text{side opposite } A}{\text{hypotenuse}}$$

$$\sin 5° = \frac{65 \text{ ft}}{s}$$

$$s(\sin 5°) = 65 \text{ ft} \quad \text{(Multiply each side by } s\text{.)}$$

$$s = \frac{65 \text{ ft}}{\sin 5°} \quad \text{(Divide each side by } \sin 5°\text{.)}$$

$$= 750 \text{ ft} \quad \text{(Round to two significant digits.)}$$

EXAMPLE 2

On one grade a railroad track rises 1.00 m for each 42.0 m of track. What angle does the track make with the level ground?

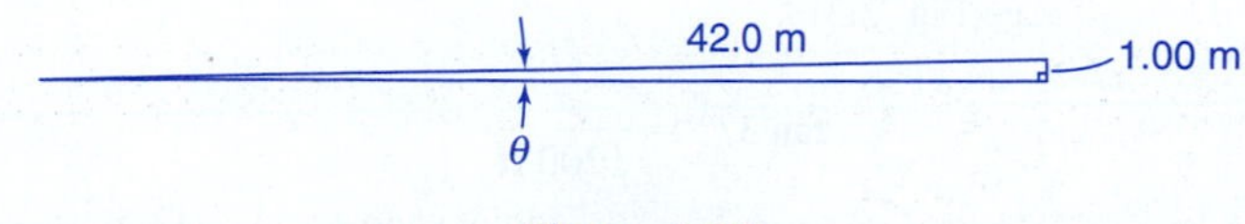

Figure 3.28

From Fig. 3.28,

$$\sin\theta = \frac{\text{side opposite } \theta}{\text{hypotenuse}}$$

$$\sin\theta = \frac{1.00\text{ m}}{42.0\text{ m}} = 0.0238$$

$$\theta = 1.4^\circ \qquad \text{(to the nearest tenth of a degree)}$$

The **angle of depression** is the angle between the horizontal and the line of sight to an object that is *below* the horizontal. The **angle of elevation** is the angle between the horizontal and the line of sight to an object that is *above* the horizontal. In Fig. 3.29, α is the angle of depression for an observer in the helicopter sighting down to the building on the ground and β is the angle of elevation for an observer in the building sighting up to the helicopter. (*Note:* $\alpha = \beta$.)

EXAMPLE 3

A horizontal distance of 225 ft is measured from the base of a vertical cliff. The angle of elevation from this distance measures 67°. Find the height of the cliff.

First, draw a sketch as in Fig. 3.30. Then

$$\tan 67^\circ = \frac{h}{225\text{ ft}}$$

$$h = (\tan 67^\circ)(225\text{ ft}) \qquad \text{(Multiply each side by 225 ft.)}$$

$$= 530\text{ ft}$$

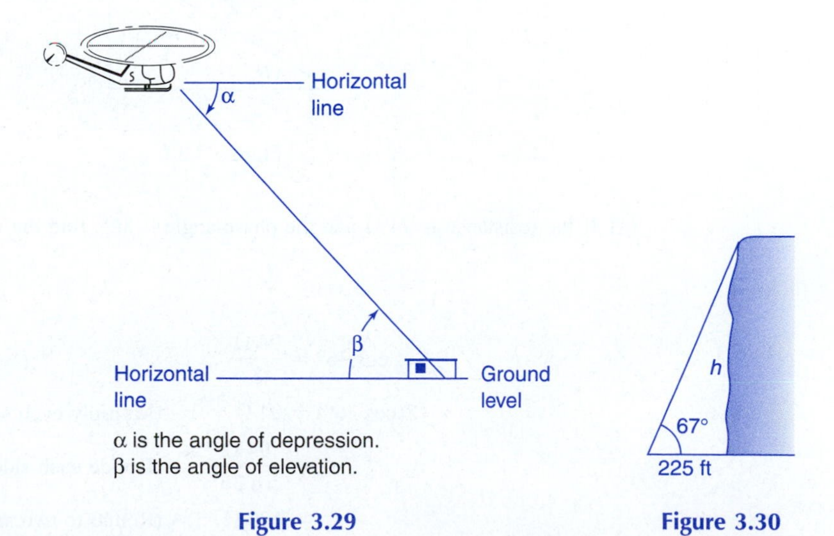

Figure 3.29

Figure 3.30

EXAMPLE 4

Carla is piloting a helicopter in the wilderness at 1200 ft above the ground searching for a downed plane. As Carla spots it, she measures its angle of depression as 53°. She also spots a road whose angle of depression is 15°. Find the distance of the downed plane to the road (see Fig. 3.31).

$$\tan 37° = \frac{x}{1200 \text{ ft}}$$

$$x = (\tan 37°)(1200 \text{ ft}) \quad \text{(Multiply each side by 1200 ft.)}$$

$$= 9\bar{0}0 \text{ ft}$$

$$\tan 75° = \frac{y}{1200 \text{ ft}}$$

$$y = (\tan 75°)(1200 \text{ ft})$$

$$= 4500 \text{ ft}$$

The distance from the downed plane to the road is $4500 \text{ ft} - 9\bar{0}0 \text{ ft} = 3600 \text{ ft}$.

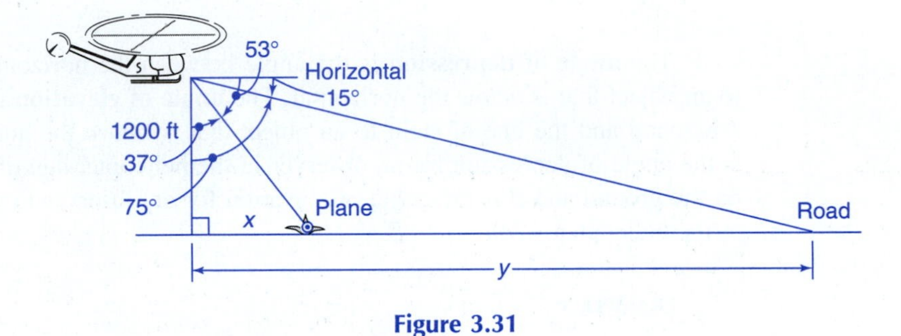

Figure 3.31

EXAMPLE 5

In ac (alternating current) circuits the relationship between the impedance Z, the resistance R, and the phase angle ϕ is shown by the right triangle in Fig. 3.32.

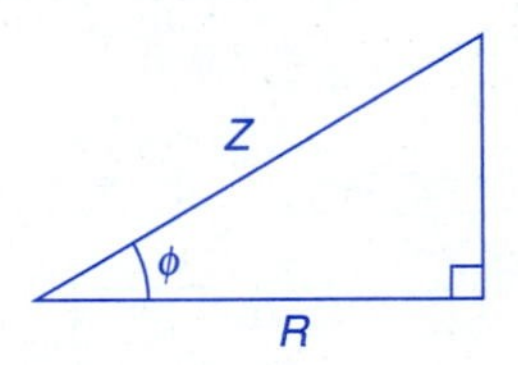

Figure 3.32

(a) If the resistance is 94 Ω and the phase angle is 36°, find the impedance.

$$\cos \phi = \frac{R}{Z}$$

$$\cos 36° = \frac{94\ \Omega}{Z}$$

$$Z(\cos 36°) = 94\ \Omega \quad \text{(Multiply each side by } Z\text{.)}$$

$$Z = \frac{94\ \Omega}{\cos 36°} \quad \text{(Divide each side by } \cos 36°\text{.)}$$

$$= 120\ \Omega \quad \text{(Round to two significant digits.)}$$

(b) If the resistance is 145 Ω and the impedance is 210 Ω, what is the phase angle?

$$\cos \phi = \frac{R}{Z}$$

$$\cos \phi = \frac{145\ \Omega}{210\ \Omega} = 0.6905$$

$$\phi = 46° \qquad \text{(Round to the nearest degree.)}$$

Exercises 3.4

1. A tower casts a shadow 42 m long when the angle of elevation of the sun is 51°. Find the height of the tower.
2. Find the width W of the river in Fig. 3.33.

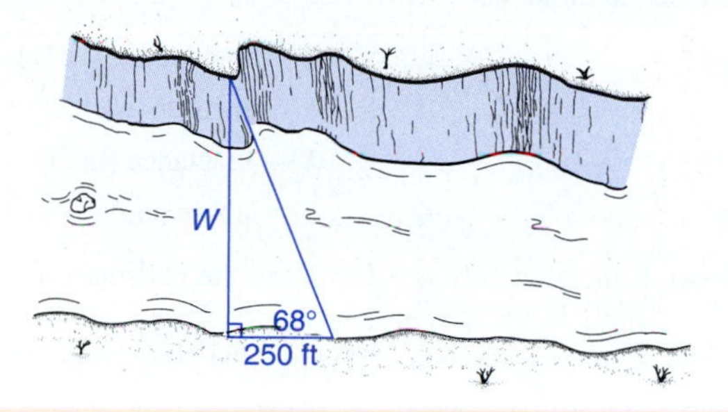

Figure 3.33

3. A roadbed rises 250 ft for each 3600 ft of road. Find the angle of elevation of the roadbed.
4. When directly over one town, the pilot of a plane notices that the angle of depression of another town is 11°. If the altimeter registers 8400 ft, what is the ground distance (in mi) between the two towns?
5. A smokestack is $19\overline{0}$ ft high. Find the length of a guy wire which must be fastened to the stack 25 ft from the top and which makes an angle of 40.0° with the ground.
6. A railroad track has an angle of elevation of 1.0°. What is the difference in altitudes of two points on the track which are **(a)** 1.00 mi apart and **(b)** 1.00 km apart?
7. A bridge is to be built across a river at an angle of 75.0° with the near side of the river. If the width of the river is 89.1 m, what must be the length of the bridge?
8. If the span of the bridge in Exercise 7 is to be 16.0 m above the roadway and the angle of elevation of the approach is to be 5.0°, how long will the approach be?
9. The cliff shown in Fig. 3.34 has a dangerous boulder on its face which may fall on the roadway below. At a point 275 ft from the base of the cliff, the angle of elevation to the boulder is 42.0° and the angle of elevation to the top of the cliff is 62.0°. How far down from the top will a scaffold have to be lowered to reach the boulder?
10. The pathway for a mineshaft is described as follows: The first shaft extends north and down at an angle of 3.5° for 225 ft to a vertical shaft which is 125 ft long. A third shaft then extends north and down at an angle of 6.8° for 175 ft. What is the total depth below ground level at the end of the third shaft? Also, what is the net horizontal distance from the ground opening to the end of the third shaft?

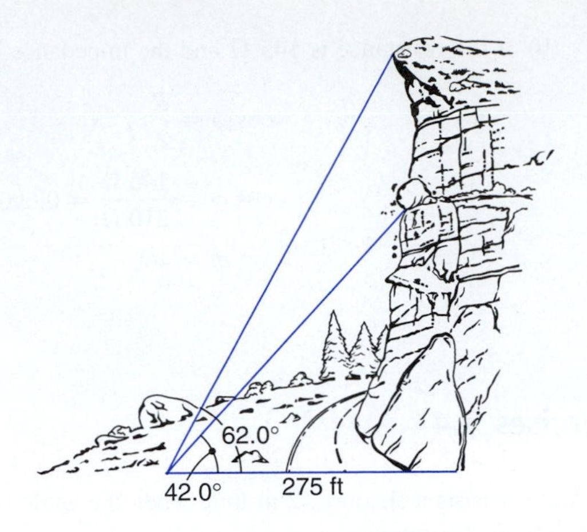

Figure 3.34

11. The right triangle in Fig. 3.35 shows the relationship among impedance, resistance, and reactance in an ac circuit, where

$$Z = \text{impedance (in } \Omega)$$
$$R = \text{resistance (in } \Omega)$$
$$X = \text{reactance (in } \Omega)$$
$$\phi = \text{phase angle}$$

(a) If the reactance is 82.6 Ω and the resistance is 112 Ω, find the impedance and the phase angle to the nearest tenth of a degree.
(b) If the resistance is 250 Ω and the phase angle is 23°, find the impedance and the reactance.

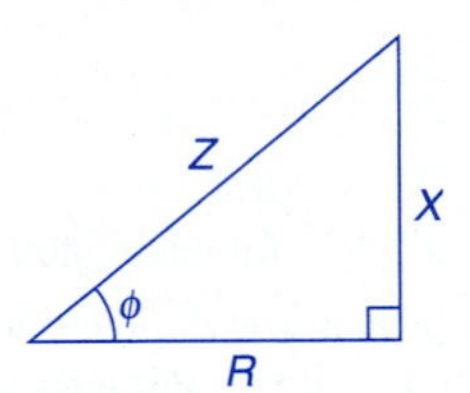

Figure 3.35 Impedance–resistance relationship.

12. A machinist needs to drill five holes in a circular plate. The centers of the holes must be 8.00 cm from the center of the plate and be spaced in a circular pattern equidistant from each other. How far apart (straight-line distance) will the centers of the adjacent holes be placed?

13. The corner of the metal plate in Fig. 3.36 is cut off. Find length x and angle α.

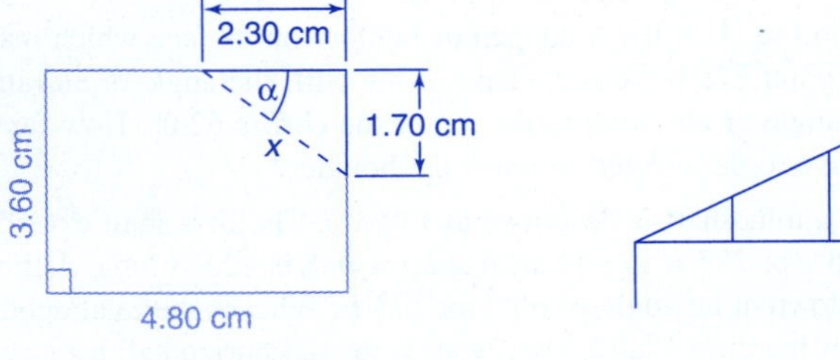

Figure 3.36

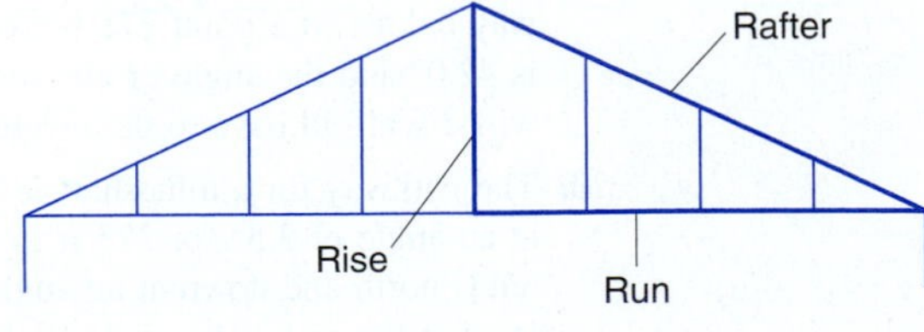

Figure 3.37

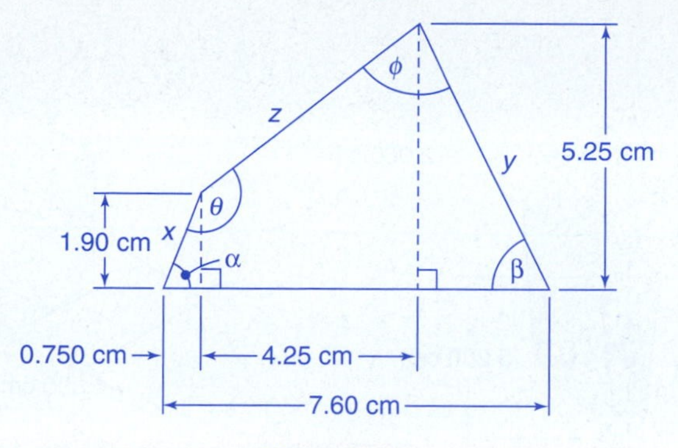

Figure 3.38

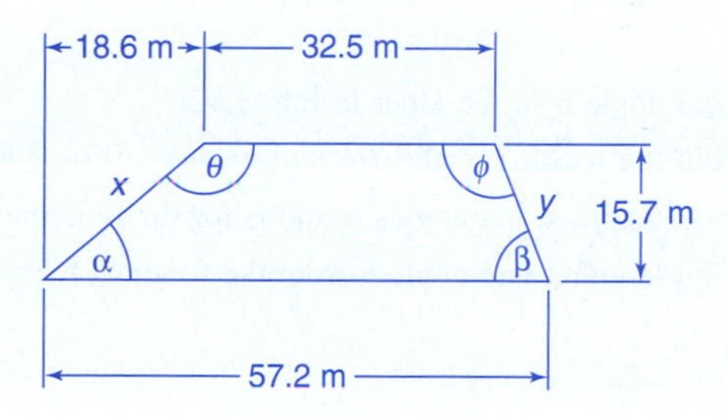

Figure 3.39

14. The pitch of a roof is the ratio of the rise to the run of a rafter (Fig. 3.37). If a roof has a rise of 6.50 ft and a run of 16.0 ft, find **(a)** the angle between the rafters and the horizontal, and **(b)** the length of the rafters.

15. Find lengths x, y, and z and angles α, β, θ, and ϕ in Fig. 3.38.

16. In the trapezoid in Fig. 3.39, find lengths x and y and angles α, β, θ, and ϕ.

17. Figure 3.40 is a schematic for a thread. Find length x.

18. Find length x on the bolt in Fig. 3.41.

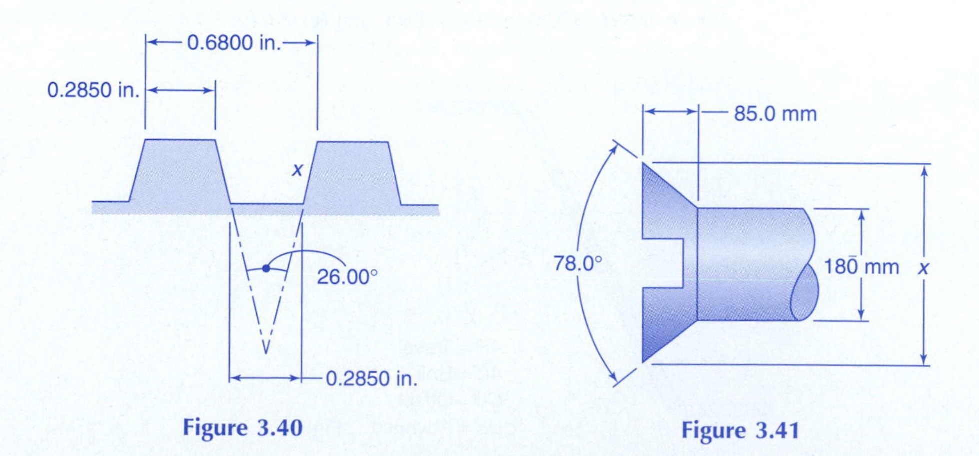

Figure 3.40

Figure 3.41

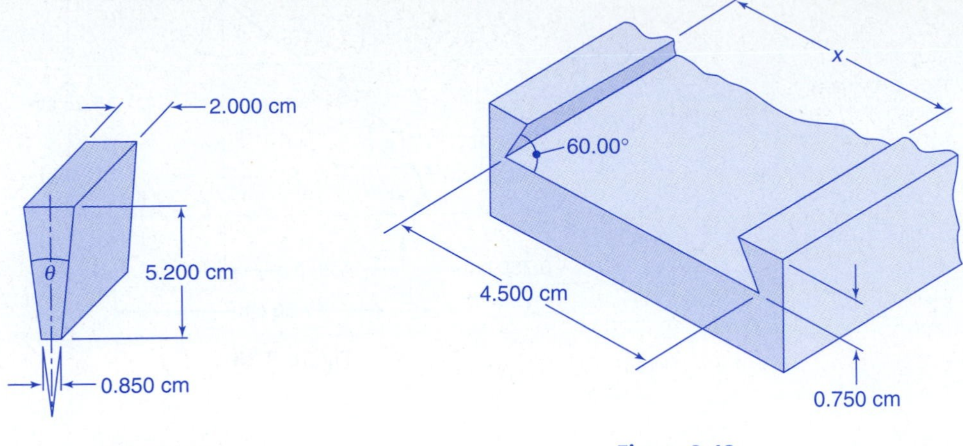

Figure 3.42

Figure 3.43

19. Find angle θ of the taper in Fig. 3.42.

20. Find the width x of the dovetail wedge in Fig. 3.43.

21. Find length x and angles α and β for the retaining wall in Fig. 3.44.

22. Find length x and angle θ from the footings for a foundation plan in Fig. 3.45.

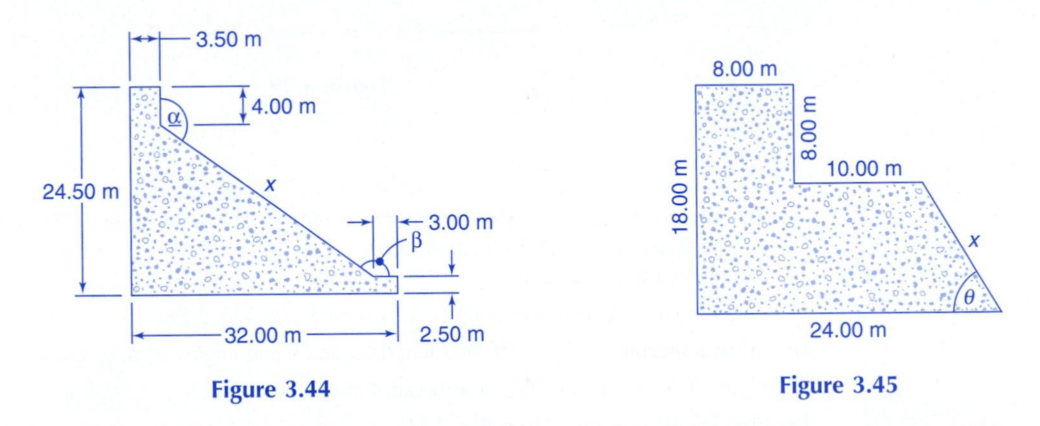

Figure 3.44

Figure 3.45

23. From the sketch of a rolling piping offset in Fig. 3.46, find the travel if the roll is 3.00 ft, the offset is 5.00 ft, and the advance is 10.00 ft.

24. From Fig. 3.46, find **(a)** the length of the advance if the roll is 20.0 in., the offset is 36.0 in., and the travel is 72.0 in., **(b)** $\angle EBA$, and **(c)** $\angle CBA$.

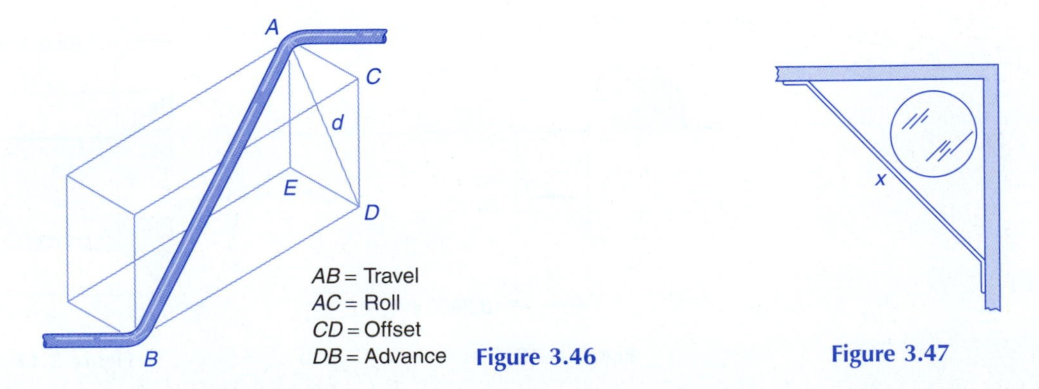

Figure 3.46

Figure 3.47

25. A swimming pool with a sloped bottom measures 48.0 ft long and 18.0 ft wide. The depth of the water is 12.00 ft at the deep end and 4.00 ft at the shallow end. Find the angle of the sloped bottom with the horizontal.

26. A roof is in the shape of an isosceles triangle. If the rafters are 18.0 ft long and the width of the building is 30.0 ft, what angle do the rafters make with the horizontal? Assume no overhang.

27. A cylindrical tank of diameter 16.0 ft is placed in a corner as in Fig. 3.47. Using 45° fittings and assuming 1.0 ft of clearance between walls, pipe, and tank, what is the length x of the pipe that cuts across the corner?

28. Find the radius of the circle in Fig. 3.48.

29. Find the radius of the circle in Fig. 3.49.

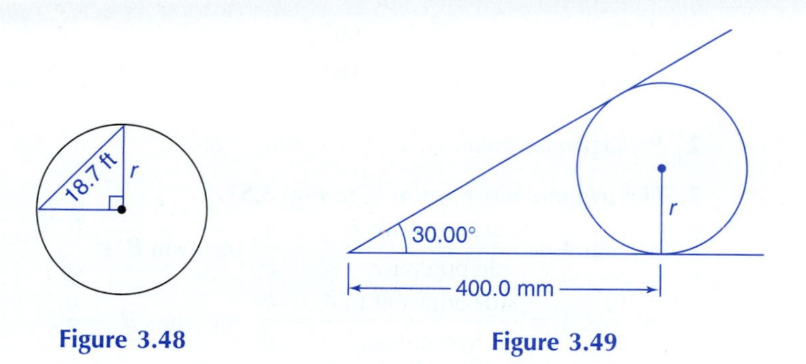

Figure 3.48 Figure 3.49

30. Find the radius of the circle in Fig. 3.50.

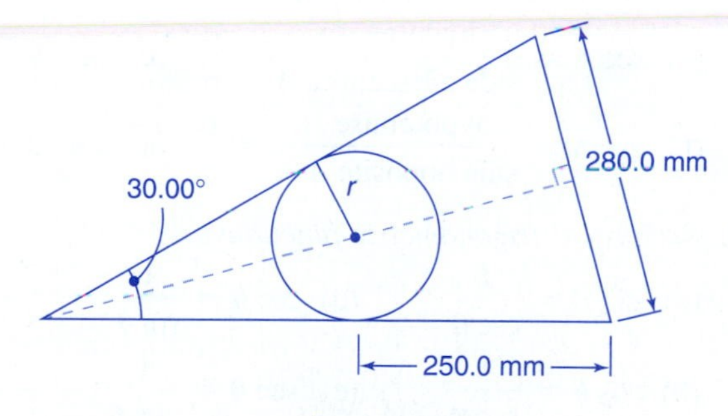

Figure 3.50

31. Find the radius of the circle in Fig. 3.51.

32. Find the radius of the smaller circle in Fig. 3.52.

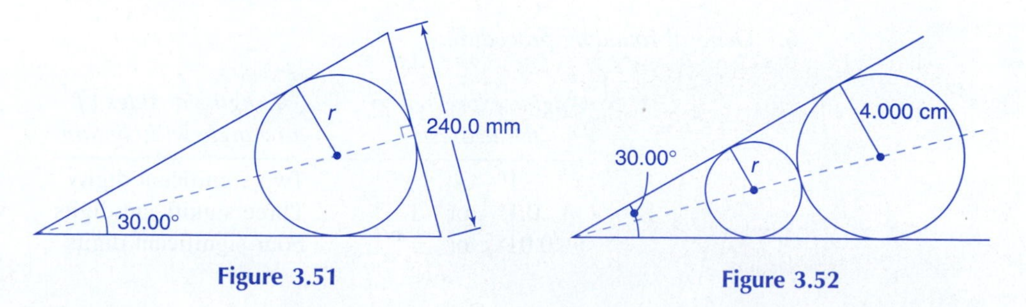

Figure 3.51 Figure 3.52

33. At a horizontal distance of 125.5 ft from the base of a tower, the angle of elevation to its top is 71°24′30″. Find the height of the tower.

34. In right triangle ABC, the hypotenuse $AB = 18.34$ in. and $BC = 10.15$ in. Find angle A to the nearest second and length AC.

35. A line joins a vertex of a square to the midpoint of an opposite side and forms a right triangle. Find the two acute angles rounded to the nearest second.

CHAPTER 3 SUMMARY

1. One complete rotation = 360°

$$1° = 60'$$
$$1' = 60''$$

2. *Pythagorean theorem:* $c^2 = a^2 + b^2$ or $c = \sqrt{a^2 + b^2}$ (see Fig. 3.53).

3. *The trigonometric ratios* (see Fig. 3.53):

(a) $\sin A = \dfrac{\text{side opposite } A}{\text{hypotenuse}} = \dfrac{a}{c}$ or $\sin B = \dfrac{b}{c}$

(b) $\cos A = \dfrac{\text{side adjacent to } A}{\text{hypotenuse}} = \dfrac{b}{c}$ or $\cos B = \dfrac{a}{c}$

(c) $\tan A = \dfrac{\text{side opposite } A}{\text{side adjacent to } A} = \dfrac{a}{b}$ or $\tan B = \dfrac{b}{a}$

(d) $\cot A = \dfrac{\text{side adjacent to } A}{\text{side opposite } A} = \dfrac{b}{a}$ or $\cot B = \dfrac{a}{b}$

(e) $\sec A = \dfrac{\text{hypotenuse}}{\text{side adjacent to } A} = \dfrac{c}{b}$ or $\sec B = \dfrac{c}{a}$

(f) $\csc A = \dfrac{\text{hypotenuse}}{\text{side opposite } A} = \dfrac{c}{a}$ or $\csc B = \dfrac{c}{b}$

Figure 3.53

4. *Reciprocal trigonometric functions:*

(a) $\sin\theta = \dfrac{1}{\csc\theta}$ (d) $\csc\theta = \dfrac{1}{\sin\theta}$

(b) $\cos\theta = \dfrac{1}{\sec\theta}$ (e) $\sec\theta = \dfrac{1}{\cos\theta}$

(c) $\tan\theta = \dfrac{1}{\cot\theta}$ (f) $\cot\theta = \dfrac{1}{\tan\theta}$

5. The two acute angles of a triangle are complementary; that is,

$$A + B = 90°$$

6. *General rounding procedure:*

Angles expressed to the nearest:	*Lengths of sides of a triangle will contain:*
1°	Two significant digits
0.1° or 1′	Three significant digits
0.01° or 1″	Four significant digits

CHAPTER 3 REVIEW

Change each angle to degrees.

1. 129°30′ **2.** 76°12′

Change each angle to degrees and minutes.

3. $35\frac{2}{3}°$ **4.** 314.3°

5. Change 16°27′45″ to degrees.

6. Change 38.405° to degrees, minutes, and seconds.

Find the length of the third side of each right triangle rounded to three significant digits.

7. $a = 16.0$ m, $c = 36.0$ m **8.** $a = 18.7$ mi, $b = 25.5$ mi

9. Find the six trigonometric ratios of angle A rounded to three significant digits when $b = 127$ cm and $c = 235$ cm.

Use a calculator to find the value of each rounded to four significant digits.

10. cos 14.6° **11.** sin 51.7° **12.** tan 29.5°

13. sec 16.7° **14.** cot 29.1° **15.** csc 79.2°

Use a calculator to find each angle rounded to the nearest tenth of a degree.

16. $\sin\theta = 0.6075$ **17.** $\cos\theta = 0.3522$ **18.** $\tan\theta = 1.2345$

19. $\sec\theta = 1.3290$ **20.** $\cot\theta = 0.9220$ **21.** $\csc\theta = 1.2222$

Use a calculator to find the value of each rounded to four significant digits.

22. sin 41°37′55″ **23.** tan 75°9′27″ **24.** sec 34°14′35″

Use a calculator to find each angle rounded to the nearest second.

25. $\cos\theta = 0.4470$ **26.** $\tan\theta = 0.2408$ **27.** $\csc\theta = 3.4525$

Solve each right triangle.

28. $a = 7.00$ m, $b = 9.50$ m **29.** $b = 15.75$ cm, $B = 36.50°$

30. $c = 1700$ km, $A = 2\bar{0}°$ **31.** $a = 245.7$ m, $A = 35°14'32''$

32. A ranger at the top of a fire tower observes the angle of depression to a fire to be 3°. If the tower is 250 ft tall, what is the ground distance to the fire from the tower?

33. A tower which is known to be 175 feet high is sighted from the ground. The angle of elevation is found to be 2°. How far away is the tower?

34. A roadbed rises 175 ft for each $41\bar{0}0$ ft of road. Find the angle of elevation of the roadbed.

35. Find lengths x and y and angles α and β in the trapezoid in Fig. 3.54.

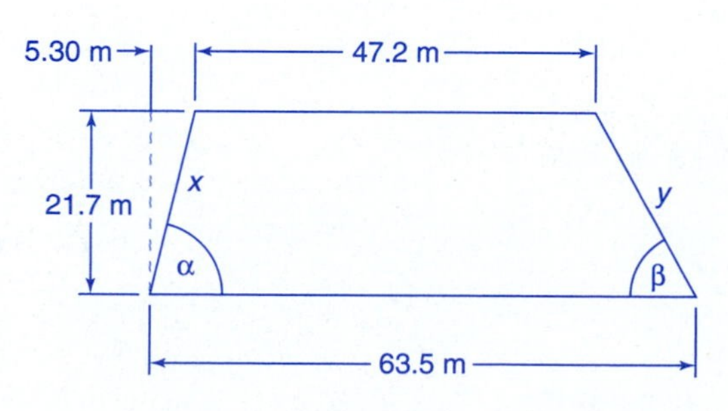

Figure 3.54

36. A roof is to be built to cover a building 56.0 ft wide as in Fig. 3.55. If the slope of the roof is to be 20.0°, to what height x should the roof rise at the center?

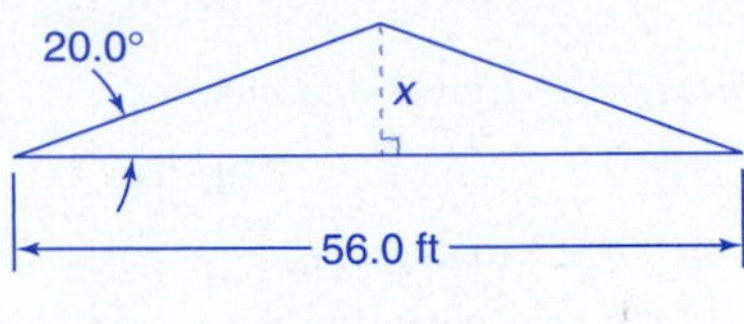

Figure 3.55

37. **(a)** If the reactance in an ac circuit is 75 Ω and the resistance is 42 Ω, find the impedance and the phase angle (see Exercise 11 of Section 3.4).

(b) If the reactance is 94 Ω and the phase angle is 47°, find the resistance and the impedance.

38. Find the distances x and y across the corners of a hex-bolt if the distance across the flats is 2.50 cm as shown in Fig. 3.56.

39. A ship leaves port and travels 275 mi at an angle of 24.5° east of north. It then travels 125 mi due east. Find the distance of the ship from the port.

40. Find length x in Fig. 3.57.

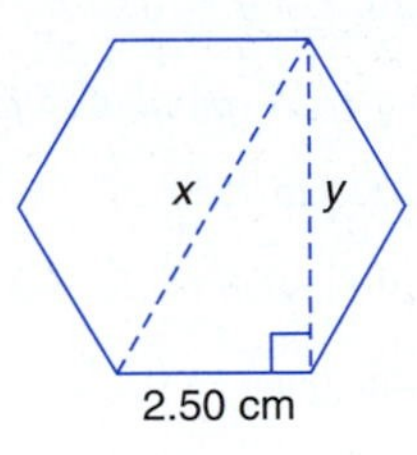

Figure 3.56

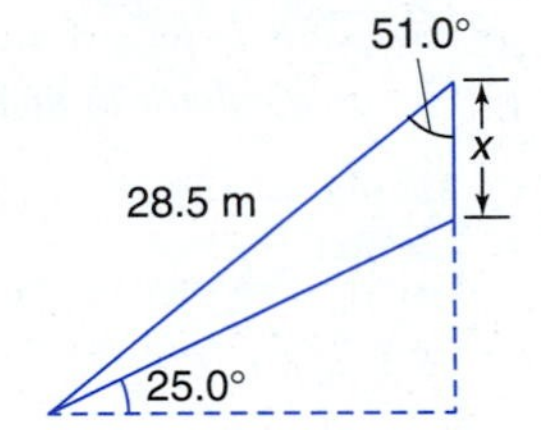

Figure 3.57

4 Equations and Their Graphs

INTRODUCTION

The path of a golf ball in flight, the relationship between the strength of a beam and its length, and the cost of manufacturing a particular item can all be described using the concept of a function. In this chapter we study functions and relate their equations to a graph. This graphic representation, which connects algebra and geometry, is extremely helpful in solving problems.

Objectives

- Identify relations that are functions.
- Determine the domain and the range of a function.
- Use functional notation.
- Solve equations graphically.
- Find the slope of a line.
- Find the equation of a line given appropriate information.
- Apply the characteristics of parallel and perpendicular lines.
- Apply the distance and midpoint formulas.

4.1 FUNCTIONS

In common usage, a relation means that two or more things have something in common. We say that a brother and a sister are related because they have the same parents, or that a person's career potential is related to his or her education and work experience.

In mathematics a **relation** is defined as a set of ordered pairs of numbers in the form (x, y). Sometimes an equation, a rule, a data chart, or some other type of description is given that states the relationship between x and y. In an ordered pair the first element or variable, called the **independent variable,** may be represented by any letter, but x is normally used. The second element or variable is normally represented by the letter y and is called the **dependent variable** because its value depends on the particular choice of the independent variable.

All of the numbers that can be used as the first element of an ordered pair or as replacements for the independent variable of a given relation form a set of numbers called the **domain.** The domain is often referred to as the set of all x's. We can think of these x-values as "inputs." The **range** of a relation is the set of numbers that can be used as the second element of an ordered pair or as replacements for the dependent variable. The range is often referred to as the set of all y's. We can think of these y-values as "outputs."

EXAMPLE 1

Given the relation described in ordered pair form $A = \{(1, 2), (3, 5), (7, 9), (6, 3)\}$, find its domain and its range.

The domain is the set of first elements: $\{1, 3, 6, 7\}$. The range is the set of second elements: $\{2, 3, 5, 9\}$.

Note: Braces { } are normally used to group elements of sets.

EXAMPLE 2

Given the relation in equation form $y = x^2$, find its domain and its range.

The domain is the set of possible replacements for the independent variable x. Note that there are no restrictions on the numbers that you may substitute for x. That is, we may replace x by any real number. We say that the domain is the set of real numbers.

After each replacement of x, there is no possible way that we can obtain a negative value for y because the square of any real number is always positive or zero. Thus, the range is the set of nonnegative real numbers, or $y \geq 0$.

EXAMPLE 3

Find the domain and the range of the relation $y = \sqrt{x - 4}$.

Note that no value of x less than 4 may be used because the square root of any negative number is not a real number. Thus, the domain is the set of real numbers greater than or equal to 4, or $x \geq 4$.

After each possible x-replacement, the square root of the resulting value is never negative, so the range is $y \geq 0$.

FUNCTION

A **function** is a special relation: a set of ordered pairs in which no two distinct ordered pairs have the same first element.

In equation form, a relation is a function when for each possible value of the first or independent variable, there is only one corresponding value of the second or dependent variable. In brief, for a relation to be a function, each value of x must correspond to one, and only one, value of y.

In computer or calculator terms, the domain of a function is the set of possible "inputs" and its range is the set of possible "outputs" which can result from the computation that the function describes. To satisfy the definition of a function, a computation must give a predictable output (one and only one y-value) for a given input (an x-value). Put another way, in a function, an input must unambiguously *determine* what the output will be.

EXAMPLE 4

Is the relation $B = \{(3, 2), (6, 7), (5, 3), (1, 1), (3, 7)\}$ a function? Find its domain and its range.

B is not a function because it contains two different ordered pairs that have the same first element: (3, 2) and (3, 7). In other words, the fact that both 2 and 7 correspond to 3 causes the relation B not to be a function. The domain of B is $\{1, 3, 5, 6\}$. The range of B is $\{1, 2, 3, 7\}$.

Does the set $A = \{(1, 2), (3, 5), (7, 9), (6, 3)\}$ from Example 1 describe a function? Yes, because no two ordered pairs have the same first element.

EXAMPLE 5

Is the relation $x = y^2$ a function? Find its domain and its range.

Can we find two ordered pairs that have the same first element? Yes, for example, (9, 3) and (9, −3) as well as (16, 4) and (16, −4) and many others. Therefore, $x = y^2$ is not a function because for at least one x-value, there corresponds more than one y-value.

To find the domain, note that each x-value is the square of a real number and can never be negative. Thus, the domain is $x \geq 0$.

There are no restrictions on replacements for y; therefore, the range is the set of all real numbers.

Consider the relations in Examples 2 and 3. Are they functions? Note that in the relation $y = x^2$, for each value of x there is only one corresponding value of y. For example, (2, 4), (−2, 4), (3, 9), (−3, 9), (4, 16), (−4, 16), and so forth. Therefore, $y = x^2$ is a function.

Example 3 was the relation $y = \sqrt{x - 4}$. Here we find that for each x-value there corresponds only one y-value; for example, (5, 1), (8, 2), $(10, \sqrt{6})$, ... Therefore, $y = \sqrt{x - 4}$ is a function.

In summary, a function is a relationship between two sets of numbers, the domain and the range, which relates each number, x, in the domain to one and only one number, y, in the range.

Next let's consider the following, more intuitive function: On a summer vacation trip, you are driving 65 mi/h using cruise control on an interstate highway. You want to relate the distance and the time you are traveling. First, we know that distance equals rate times time, or $d = rt$. Also, since $r = 65$, we have the relation $d = 65t$. As you drive along, you begin to think how far you can drive in 1 h:

$$d = 65t = 65(1) = 65 \text{ mi}$$

How far can you drive in 3 h?

$$d = 65t = 65(3) = 195 \text{ mi}$$

Is this relation a function? Yes, because for each value of t, there is one and only one value of d; that is, during each driving time period, there is one and only one distance traveled. What are the domain and the range? First, note that $t \geq 0$ and $d \geq 0$. While there are no theoretical upper limits on t and d, the practical limits depend on the amount of time and the distance that you want to travel.

Functional Notation

To say that y is a function of x means that for each value of x from the domain of the function, we can find exactly one value of y from the range. This statement is said so often that we have developed the following notation, called **functional notation,** to write that y is a function of x:

$$y = f(x)$$

with $f(x)$ read "f of x." *Note:* $f(x)$ does **not** mean f times x.

In each of the following equations, y can be replaced by $f(x)$, and the resulting equation is written in functional notation.

Equation	*Functional notation form*
$y = 3x - 4$	$f(x) = 3x - 4$
$y = 5x^2 - 8x + 7$	$f(x) = 5x^2 - 8x + 7$
$y = \sqrt{6 - 2x}$	$f(x) = \sqrt{6 - 2x}$

Functional notation can be used to simplify statements. For example, find the value of $y = 3x^2 + 5x - 6$ for $x = 2$. Using substitution, we replace x with 2 as follows:

$$y = 3x^2 + 5x - 6$$
$$y = 3(2)^2 + 5(2) - 6 = 16$$

The statement "Find the value of $y = 3x^2 + 5x - 6$ for $x = 2$" may be abbreviated using functional notation as follows:

$$\text{given } f(x) = 3x^2 + 5x - 6, \text{ find } f(2)$$

EXAMPLE 6

Given the function $f(x) = 5x - 4$, find each of the following:

(a) $f(0)$

Replace x with 0 as follows:

$$f(0) = 5(0) - 4 = 0 - 4 = -4$$

(b) $f(7)$

Replace x with 7 as follows:

$$f(7) = 5(7) - 4 = 35 - 4 = 31$$

A function is usually named by a specific letter, such as $f(x)$, where f names the function. Other letters, such as g in $g(x)$ and h in $h(x)$, are often used to represent or name functions.

EXAMPLE 7

Given the function $g(x) = \sqrt{x + 4} + 3x^2$, find each of the following:

(a) $g(5)$

Replace x with 5 as follows:

$$g(5) = \sqrt{5 + 4} + 3(5)^2 = 3 + 75 = 78$$

(b) $g(-3)$

Replace x with -3 as follows:

$$g(-3) = \sqrt{-3 + 4} + 3(-3)^2 = 1 + 27 = 28$$

(c) $g(-10)$

Replace x with -10 as follows:

$$g(-10) = \sqrt{-10 + 4} + 3(-10)^2 = \sqrt{-6} + 300$$

which is not a real number because $\sqrt{-6}$ is not a real number. Another way of responding to Part (c) is to say, "Since -10 is not in the domain of $g(x)$, $g(-10)$ has no real value."

Letters may also be used with functional notation as illustrated by the following example.

EXAMPLE 8

Given the function $f(x) = x^2 - 4x$, find each of the following:

(a) $f(a)$

Replace x with a as follows:

$$f(a) = a^2 - 4a$$

(b) $f(3c^2)$

Replace x with $3c^2$ as follows:

$$f(3c^2) = (3c^2)^2 - 4(3c^2) = 9c^4 - 12c^2$$

(c) $f(a + 5)$

Replace x with $a + 5$ as follows:

$$\begin{aligned} f(a + 5) &= (a + 5)^2 - 4(a + 5) \\ &= a^2 + 10a + 25 - 4a - 20 \\ &= a^2 + 6a + 5 \end{aligned}$$

Using a calculator, we have

2nd CUSTOM F2 6

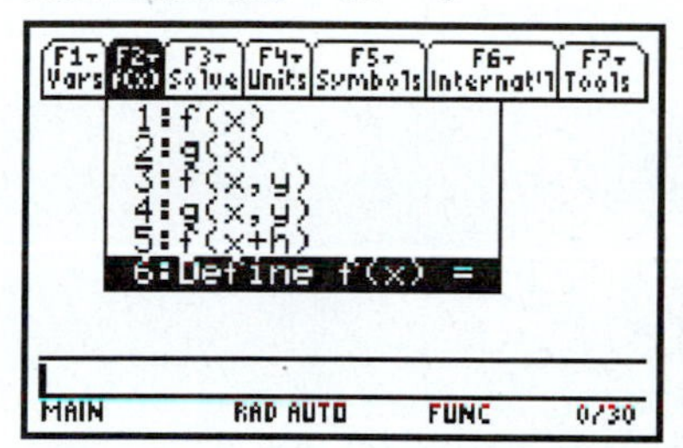

x^2-4x **ENTER**

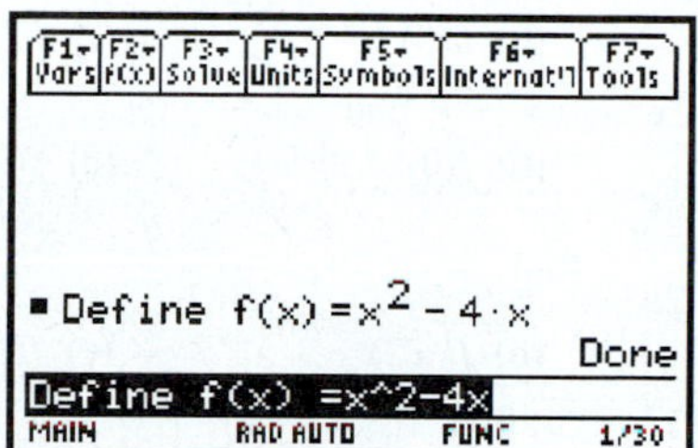

(**2nd CUSTOM** restores standard menus.)

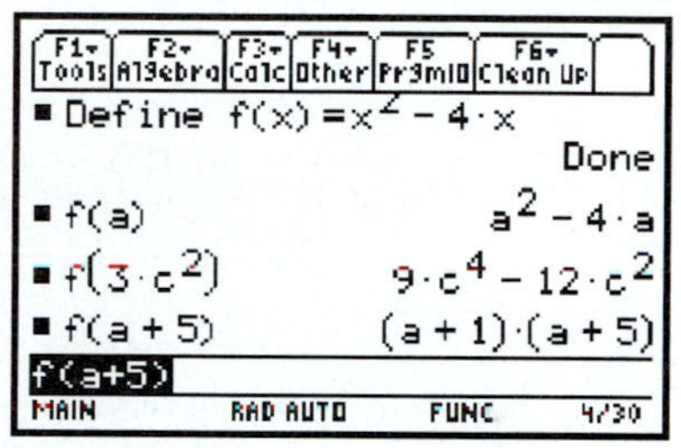

Note that the last answer is factored. Use **expand** to multiply if you wish.

Other letters, such as t in $f(t)$ and r in $f(r)$, are used in applications to name independent variables.

EXAMPLE 9

Given the function $f(t) = 0.50t + 5.4$, find each of the following:

(a) $f(3.2)$

Replace t with 3.2 as follows:

$$\begin{aligned} f(t) &= 0.50t + 5.4 \\ f(3.2) &= 0.50(3.2) + 5.4 \\ &= 1.6 + 5.4 \\ &= 7.0 \end{aligned}$$

(b) $f(t_0)$

Replace t with t_0 as follows:

$$\begin{aligned} f(t) &= 0.50t + 5.4 \\ f(t_0) &= 0.50t_0 + 5.4 \end{aligned}$$

Exercises 4.1

Determine whether or not each relation is a function. Write its domain and its range.

1. $A = \{(2, 4), (3, 7), (9, 2)\}$

2. $B = \{(5, 2), (3, 3), (1, 2)\}$

3. $C = \{(2, 5), (7, 3), (2, 1), (1, 3)\}$

4. $D = \{(0, 2), (5, -1), (2, 7), (5, 1)\}$

5. $E = \{(3, 2), (5, 2), (2, 2), (-2, 2)\}$

6. $F = \{(3, 4), (3, -4), (-3, -4), (-3, 4)\}$

7. $y = 2x + 5$

8. $y = -3x$

9. $y = x^2 + 1$

10. $y = 2x^2 - 3$

11. $x = y^2 - 2$

12. $x = 3y^2 + 4$

13. $y = \sqrt{x + 3}$

14. $y = \sqrt{3 - 6x}$

15. $y = 6 + \sqrt{2x - 8}$

16. $y = 16 - \sqrt{x + 5}$

17. Given the function $f(x) = 8x - 12$, find
(a) $f(4)$ **(b)** $f(0)$ **(c)** $f(-2)$

18. Given the function $g(x) = 20 - 4x$, find
(a) $g(6)$ **(b)** $g(0)$ **(c)** $g(-3)$

19. Given $g(x) = 10x + 15$, find
(a) $g(2)$ **(b)** $g(0)$ **(c)** $g(-4)$

20. Given $f(x) = x^2 - 4$, find
(a) $f(6)$ **(b)** $f(0)$ **(c)** $f(-6)$

21. Given $h(x) = 3x^2 + 4x$, find
(a) $h(5)$ **(b)** $h(0)$ **(c)** $h(-2)$

22. Given $f(x) = -2x^2 + 6x - 7$, find
(a) $f(3)$ **(b)** $f(0)$ **(c)** $f(-1)$

23. Given $f(t) = \dfrac{5 - t^2}{2t}$, find
(a) $f(1)$ **(b)** $f(-3)$ **(c)** $f(0)$

24. Given $g(t) = \sqrt{21 - 5t}$, find
(a) $g(1)$ **(b)** $g(-3)$ **(c)** $g(2)$ **(d)** $g(8)$

25. Given $f(x) = 6x + 8$, find
(a) $f(a)$ **(b)** $f(4a)$ **(c)** $f(c^2)$

26. Given $g(x) = 8x^2 - 7x$, find
(a) $g(z)$ **(b)** $g(2y)$ **(c)** $g(3t^2)$

27. Given $h(x) = 4x^2 - 12x$, find
(a) $h(x + 2)$ **(b)** $h(x - 3)$ **(c)** $h(2x + 1)$

28. Given $f(y) = y^2 - 3y + 6$, find
(a) $f(y - 1)$ **(b)** $f(y^2 + 1)$ **(c)** $f(1 - 4y)$

29. Given $f(x) = 3x - 1$ and $g(x) = x^2 - 6x + 1$, find
(a) $f(x) + g(x)$ **(b)** $f(x) - g(x)$ **(c)** $[f(x)][g(x)]$ **(d)** $f(x + h)$

30. Given $f(t) = 5 - 2t + t^2$ and $g(t) = t^2 - 4t + 4$, find
(a) $f(t) + g(t)$ **(b)** $g(t) - f(t)$ **(c)** $[f(t)][g(t)]$ **(d)** $g(t + h)$

Find the domain of each function.

31. $f(x) = \dfrac{3x + 4}{x - 2}$

32. $f(t) = \dfrac{8}{6t + 3}$

33. $g(t) = \dfrac{2t + 4t^2}{(t - 6)(t + 3)}$

34. $g(x) = \dfrac{3x - 10}{x^2 + 4}$

35. $f(x) = \dfrac{12}{\sqrt{15 - 3x}}$

36. $g(t) = \dfrac{9}{\sqrt{5t + 20}}$

4.2 GRAPHING EQUATIONS

Consider a plane in which two number lines intersect at right angles. Let the point of intersection be the zero point of each line and call it the **origin.** Each line is called an **axis.** The horizontal number line is usually called the ***x*-axis** and the vertical line is usually called the ***y*-axis.** On each axis the same scale (unit length) is preferred but not always possible in all applications. Such a system is called the **rectangular coordinate system,** or the **Cartesian coordinate system.** (The name *Cartesian* is after René Descartes, the

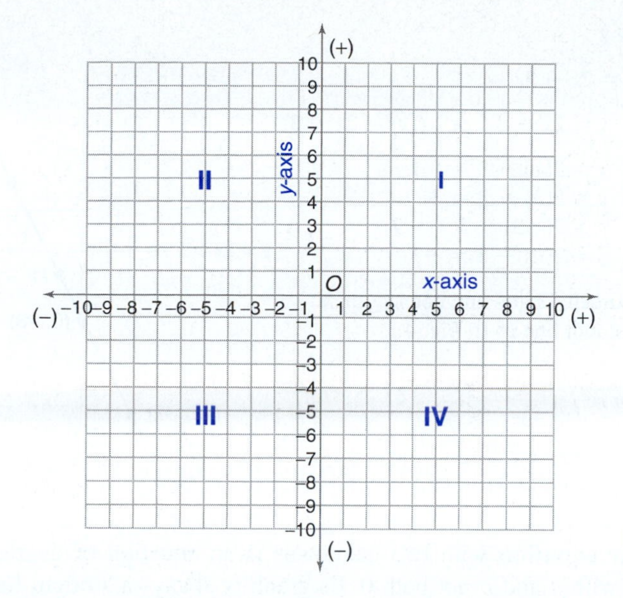

Figure 4.1 The rectangular coordinate system.

seventeenth-century French mathematician who first conceived this idea of combining algebra and geometry together in such a way that each could aid the study of the other.) The plane is divided by the axes into four regions called **quadrants.** The quadrants are numbered as shown in Fig. 4.1.

In the plane there is a point that corresponds to each ordered pair of real numbers (x, y). Likewise, there is an ordered pair (x, y) that corresponds to each point in the plane. Together x and y are called the **coordinates** of the point; x is called the **abscissa** and y is called the **ordinate.** This relationship is called a **one-to-one correspondence.** The location, or position, of a point in the plane corresponding to a given ordered pair is found by first counting right or left from O (origin) the number of units along the x-axis indicated by the first number of the ordered pair (right if positive, left if negative). Then from this point reached on the x-axis, count up or down the number of units indicated by the second number of the ordered pair (up if positive, down if negative).

EXAMPLE 1

Plot the point corresponding to each ordered pair in the coordinate plane:

$A(3, 1)$ $B(2, -3)$ $C(-4, -2)$ $D(-3, 0)$ $E(-6, 2)$ $F(0, 2)$ (See Fig. 4.2.)

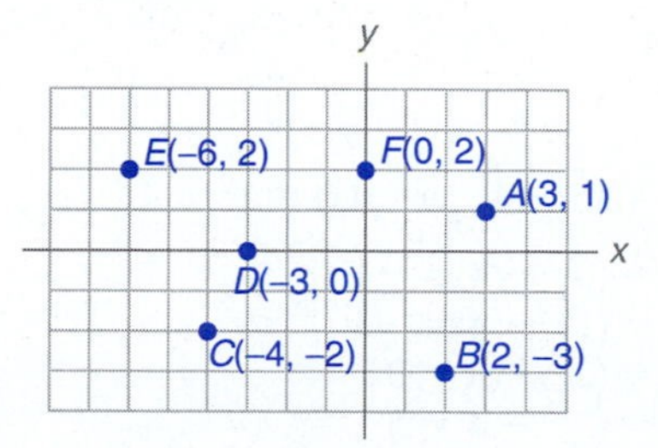

Figure 4.2

To graph equations we plot a sample of ordered pairs and connect them with a smooth curve. To obtain the sample, we need to generate ordered pairs from a given equation. One way to generate these ordered pairs is by randomly choosing a value for x, replacing this value for x in the equation, and solving for y.

EXAMPLE 2

Graph $y = 2x - 3$.

x	y	$y = 2x - 3$ or $f(x) = 2x - 3$
1	-1	$y = 2(1) - 3 = -1$
3	3	$y = 2(3) - 3 = 3$
-2	-7	$y = 2(-2) - 3 = -7$
0	-3	$y = 2(0) - 3 = -3$

Plot the ordered pairs and connect them with a smooth line as in Fig. 4.3.

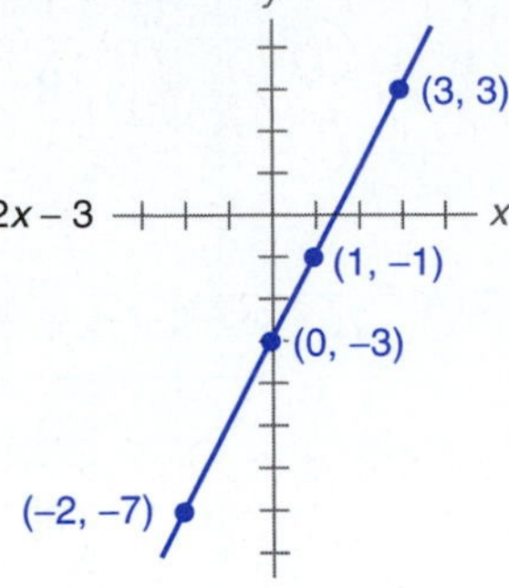

Figure 4.3

A **linear equation** with two unknowns is an equation of degree one in the form $ax + by = c$ with a and b not both 0. Its graph is always a straight line. Therefore, two ordered pairs are sufficient to graph a linear function, since two points determine a straight line. However, finding a third point provides good insurance against a careless error.

EXAMPLE 3

Graph $y = -3x + 5$.

x	y	$y = -3x + 5$ or $g(x) = -3x + 5$
0	5	$y = -3(0) + 5 = 5$
2	-1	$y = -3(2) + 5 = -1$
-1	8	$y = -3(-1) + 5 = 8$

See Fig. 4.4.

Figure 4.4

The graph of an equation that is not linear is usually a curve of some kind and hence requires several points to sketch a smooth curve.

EXAMPLE 4

Graph $y = x^2 - 4$.

x	y	$y = x^2 - 4$ or $f(x) = x^2 - 4$
0	-4	$y = (0)^2 - 4 = -4$
1	-3	$y = (1)^2 - 4 = -3$
2	0	$y = (2)^2 - 4 = 0$
3	5	$y = (3)^2 - 4 = 5$
-1	-3	$y = (-1)^2 - 4 = -3$
-2	0	$y = (-2)^2 - 4 = 0$
-3	5	$y = (-3)^2 - 4 = 5$

See Fig. 4.5.

Figure 4.5

You should be able to write, graph, evaluate, and simplify functions in both the "$y =$" and "$f(x) =$" forms. The "$y =$" form is commonly used in defining a function on a graphing calculator and in writing the equation of a graph in terms of x and y. The functional notation form "$f(x) =$ " is especially important in calculus and is also used in physics and engineering formulas to emphasize that a computation is, indeed, a function (that one quantity depends on the other). Either form may be used interchangeably.

EXAMPLE 5

Graph $y = 2x^2 + x - 5$.

x	y	$y = 2x^2 + x - 5$
0	−5	$y = 2(0)^2 + (0) - 5 = -5$
1	−2	$y = 2(1)^2 + (1) - 5 = -2$
2	5	$y = 2(2)^2 + (2) - 5 = 5$
−1	−4	$y = 2(-1)^2 + (-1) - 5 = -4$
−2	1	$y = 2(-2)^2 + (-2) - 5 = 1$
−3	10	$y = 2(-3)^2 + (-3) - 5 = 10$

See Fig. 4.6.

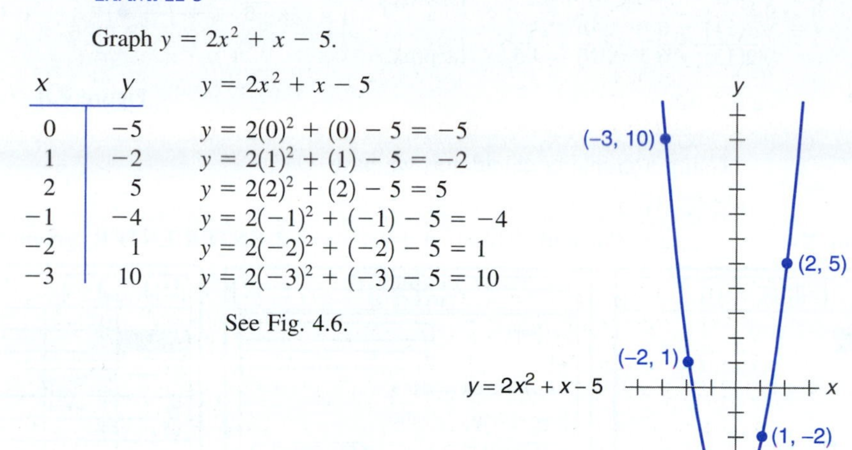

Figure 4.6

For a more complicated function, more ordered pairs are usually required to obtain a smooth curve. It may also be necessary to change the scale of the graph in order to plot enough ordered pairs to obtain a smooth curve. To change the scale means to enlarge or reduce the unit length on the axes according to a specified ratio. This ratio is chosen on the basis of fitting the necessary values in a given space allowed for the graph.

EXAMPLE 6

Graph $y = x^3 + 4x^2 - x - 4$.

x	y	$y = x^3 + 4x^2 - x - 4$
0	−4	$y = (0)^3 + 4(0)^2 - (0) - 4 = -4$
1	0	$y = (1)^3 + 4(1)^2 - (1) - 4 = 0$
2	18	$y = (2)^3 + 4(2)^2 - (2) - 4 = 18$
3	56	$y = (3)^3 + 4(3)^2 - (3) - 4 = 56$
−1	0	$y = (-1)^3 + 4(-1)^2 - (-1) - 4 = 0$
−2	6	$y = (-2)^3 + 4(-2)^2 - (-2) - 4 = 6$
−3	8	$y = (-3)^3 + 4(-3)^2 - (-3) - 4 = 8$
−4	0	$y = (-4)^3 + 4(-4)^2 - (-4) - 4 = 0$
−5	−24	$y = (-5)^3 + 4(-5)^2 - (-5) - 4 = -24$

See Fig. 4.7.

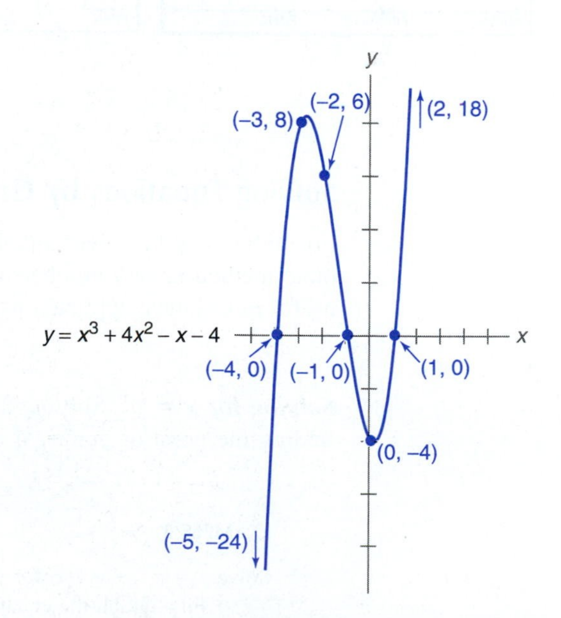

Figure 4.7

EXAMPLE 7

Graph $y = \sqrt{2x - 6}$.

x	y	$y = \sqrt{2x-6}$	
3	0	$y = \sqrt{2(3) - 6} = \sqrt{0} = 0$	
5	2	$y = \sqrt{2(5) - 6} = \sqrt{4} = 2$	
7	2.8	$y = \sqrt{2(7) - 6} = \sqrt{8} = 2.8$	(approx.)
8	3.2	$y = \sqrt{2(8) - 6} = \sqrt{10} = 3.2$	(approx.)
11	4	$y = \sqrt{2(11) - 6} = \sqrt{16} = 4$	
13	4.5	$y = \sqrt{2(13) - 6} = \sqrt{20} = 4.5$	(approx.)

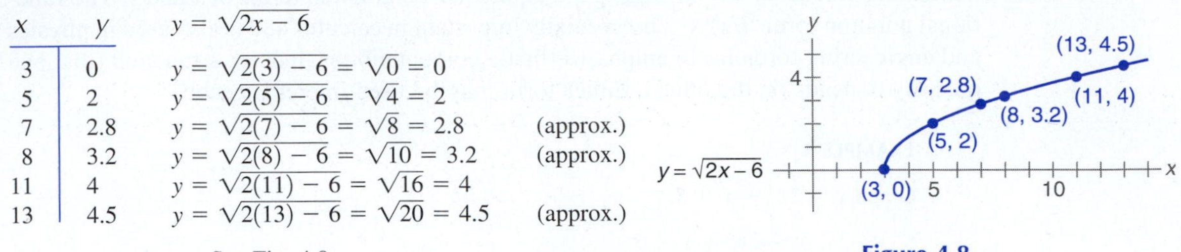

See Fig. 4.8.

Figure 4.8

Using a graphing calculator, we have

green diamond Y=

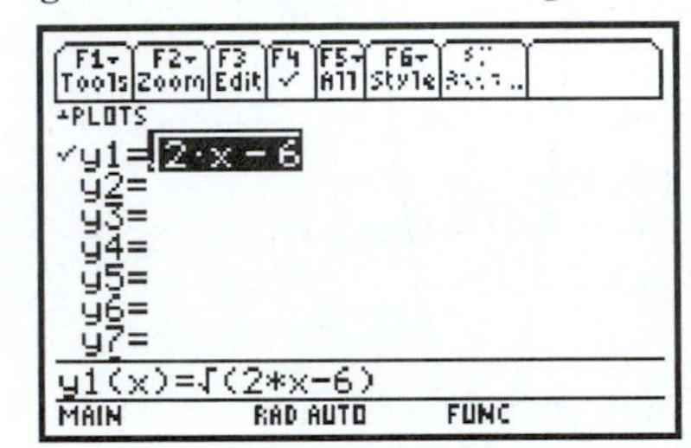

green diamond Tblset 3 down arrow **2 ENTER ENTER**

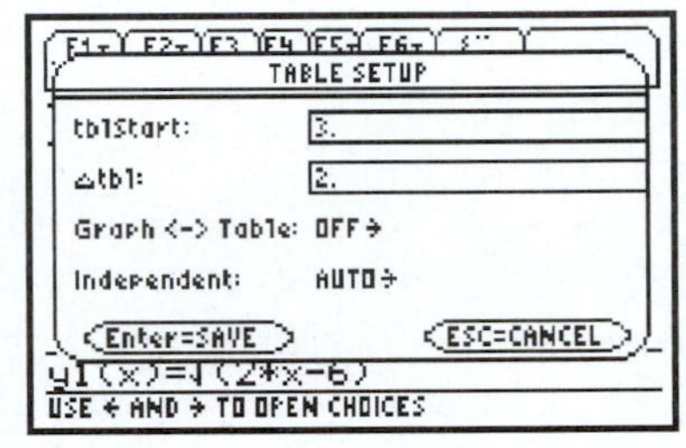

green diamond TABLE

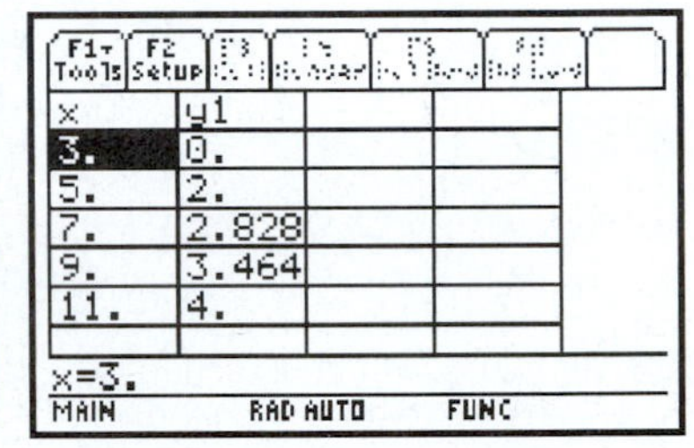

green diamond WINDOW

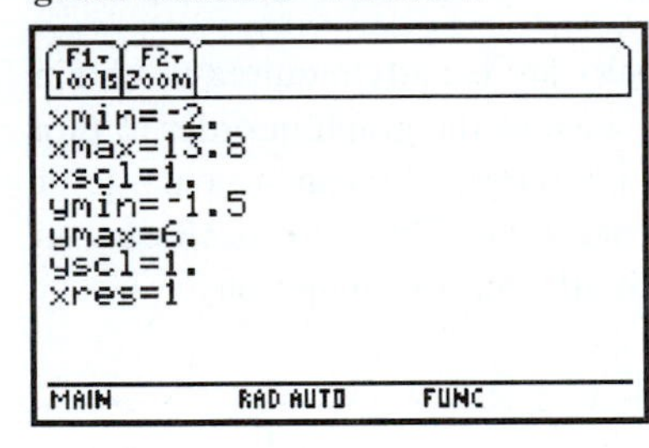

green diamond GRAPH

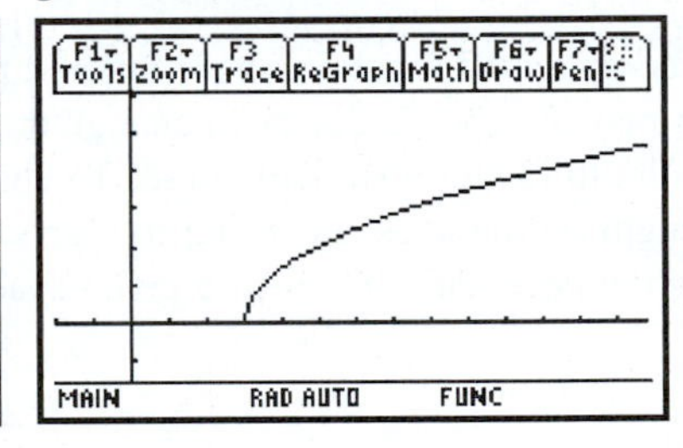

F3 (Trace) 8 ENTER

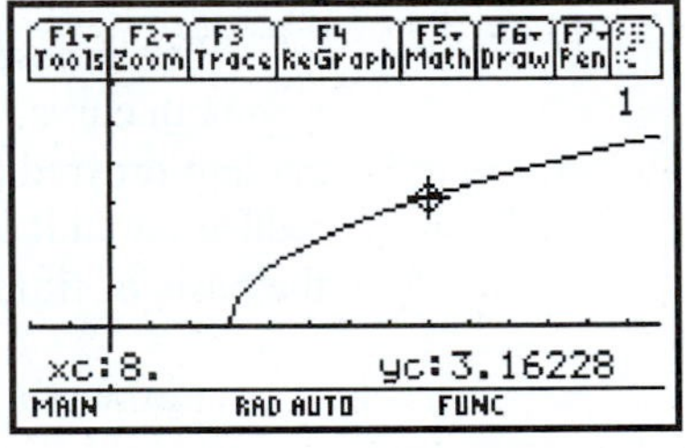

Specific function values can be calculated on the **Trace** screen.

Solving Equations by Graphing

Equations may be solved graphically. This method is particularly useful when an algebraic method is very cumbersome, cannot be recalled, or does not exist; it is especially useful in technical applications.

Solving for $y = 0$ Solving the equation $y = x^2 - x - 6$ for $y = 0$ graphically means finding the point or points, if any, where the graph crosses the line $y = 0$ (the x-axis).

EXAMPLE 8

Solve $y = x^2 - x - 6$ for $y = 0$ graphically.

First, graph the equation $y = x^2 - x - 6$ (see Fig. 4.9).

x	y
1	−6
2	−4
3	0
4	6
0	−6
−1	−4
−2	0
−3	6

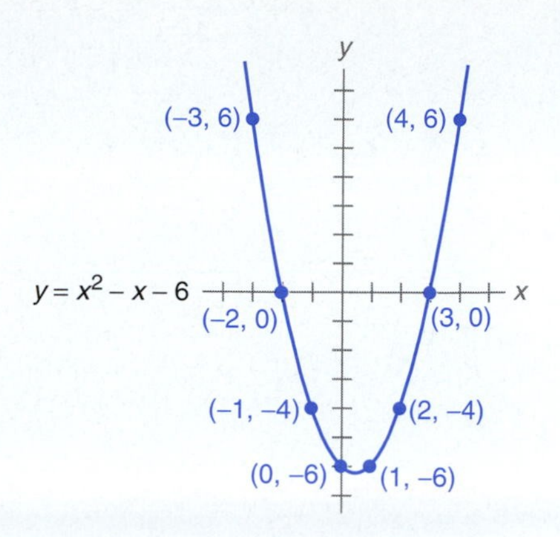

Figure 4.9

Then, note the values of x where the curve crosses the x-axis: $x = -2$ and $x = 3$. Therefore, from the graph the solutions of $y = x^2 - x - 6$ for $y = 0$ are $x = -2$ and $x = 3$.

Sometimes the curve crosses the x-axis between the unit marks on the x-axis. In this case we must estimate as closely as possible the point of intersection of the curve and the x-axis. If a particular problem requires greater accuracy, we can scale the graph to allow a more accurate estimation.

EXAMPLE 9

Solve $y = x^2 + 2x - 4$ for $y = 0$ graphically.

First, graph the equation $y = x^2 + 2x - 4$ (see Fig. 4.10).

x	y
0	−4
1	−1
2	4
−1	−5
−2	−4
−3	−1
−4	4

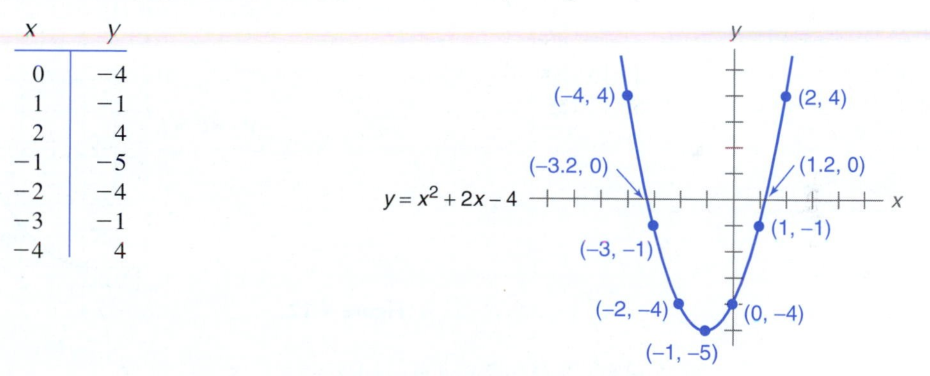

Figure 4.10

The values of x where the curve crosses the x-axis are approximately 1.2 and −3.2. Therefore, the approximate solutions of $y = x^2 + 2x - 4$ for $y = 0$ are $x = 1.2$ and $x = -3.2$.

Solving for $y = k$

EXAMPLE 10

Solve the equation from Example 9, $y = x^2 + 2x - 4$, for $y = 4$ and $y = -3$.

First, find the values of x where the curve crosses the line $y = 4$. From the graph in Fig. 4.11, the x-values are 2 and −4. Therefore, solving for $y = 4$, we find that the solutions of $y = x^2 + 2x - 4$ are $x = 2$ and $x = -4$.

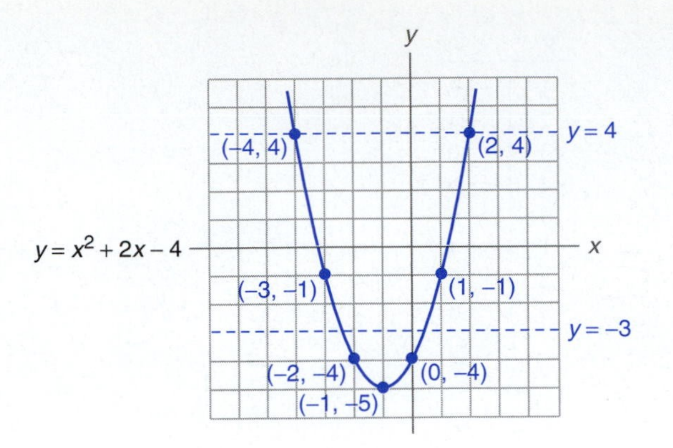

Figure 4.11

Next, find the values of x where the curve crosses the line $y = -3$. From the graph in Fig. 4.11, the approximate x-values are 0.4 and -2.4. That is, for $y = -3$, the solutions of $y = x^2 + 2x - 4$ are approximately $x = 0.4$ and $x = -2.4$.

Note that $y = x^2 + 2x - 4$ has no solutions for $y = -7$.

EXAMPLE 11

The voltage V in volts in a given circuit varies with time t in milliseconds according to the equation $V = 6t^2 + t$. Solve for t when $V = 0$, 70, and 120.

t	V
0	0
1	7
2	26
3	57
4	100
5	155

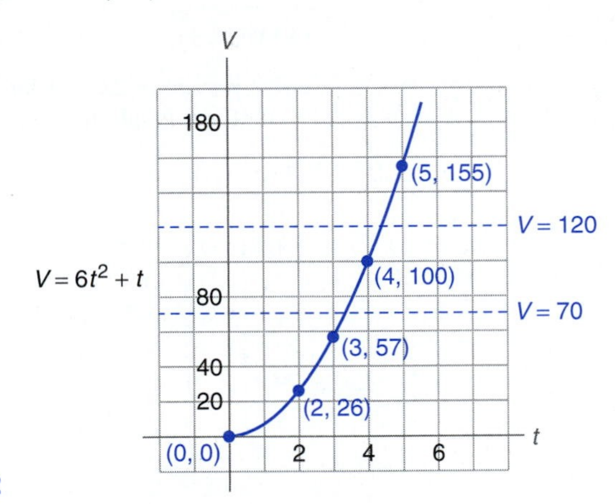

Figure 4.12

In Fig. 4.12, we used the scales

$$t\text{: } 1 \text{ square} = 1 \text{ ms}$$
$$V\text{: } 1 \text{ square} = 20 \text{ V}$$

From the graph,

$$\text{at } V = 0 \text{ V}, \quad t = 0 \text{ ms}$$
$$\text{at } V = 70 \text{ V}, \quad t = 3.3 \text{ ms}$$
$$\text{at } V = 120 \text{ V}, \quad t = 4.4 \text{ ms}$$

Negative values of time t are not meaningful in this example.

EXAMPLE 12

The work w done in a circuit varies with time t according to the equation $w = 8t^2 + 4t$. Solve for t when $w = 60$, 120, and 250.

t	w
0	0
1	12
2	40
3	84
4	144
5	220
6	312

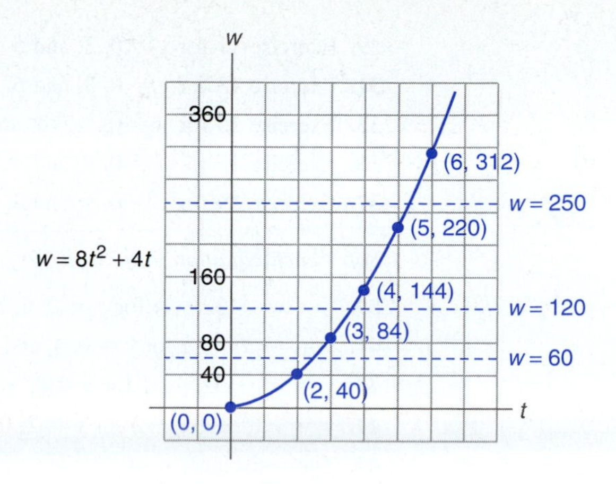

Figure 4.13

In Fig. 4.13, we used the scales

$$t\text{: } 1 \text{ square} = 1 \text{ unit}$$
$$w\text{: } 1 \text{ square} = 40 \text{ units}$$

From the graph,

$$\text{at } w = 60, \quad t = 2.5$$
$$\text{at } w = 120, \quad t = 3.6$$
$$\text{at } w = 250, \quad t = 5.4$$

Exercises 4.2

1. Plot the point corresponding to each ordered pair in the coordinate plane.

$A(2, 5)$ $B(-3, -6)$ $C(-4, 0)$ $D(2, -1)$
$E(3, 4)$ $F(0, 3)$ $G(0, -3)$ $H(5, 0)$
$I(-4, -1)$ $J(-2, 2)$ $K(3, -5)$ $L(-3, 4)$

2. Plot the point corresponding to each ordered pair in the coordinate plane.

$P_1(3, 2)$ $P_2(-2, 5)$ $P_3(-1, -6)$ $P_4(4, -7)$
$P_5(0, 4)$ $P_6(-5, 0)$ $P_7(-4, 4)$ $P_8(6, 0)$
$P_9(-8, 5)$ $P_{10}(7, -2)$ $P_{11}(-5, -3)$ $P_{12}(0, -8)$

Graph each equation.

3. $y = 2x + 1$
4. $y = 3x - 4$
5. $-2x - 3y = 6$
6. $2y = -4x - 3$
7. $y = x^2 - 9$
8. $y = x^2 + x - 6$
9. $y = x^2 - 5x + 4$
10. $y = x^2 + 3$
11. $y = 2x^2 + 3x - 2$
12. $y = -x^2 + 2x + 4$
13. $y = x^2 + 2x$
14. $y = x^2 - 4x$
15. $y = -2x^2 + 4x$
16. $y = -\frac{1}{4}x^2 - \frac{3}{2}x + 2$
17. $y = x^3 - x^2 - 10x + 8$
18. $y = x^3 - 4x^2 + x + 6$
19. $y = x^3 + 2x^2 - 7x + 4$
20. $y = x^3 - 8x - 3$
21. $y = \sqrt{x + 4}$
22. $y = \sqrt{3x - 12}$
23. $y = \sqrt{12 - 6x}$
24. $y = \sqrt{3 - x}$

Solve each equation graphically for the given values.

25. Exercise 7 for $y = 0, -5$, and 2.
26. Exercise 8 for $y = 0, 6$, and -3.
27. Exercise 9 for $y = 0, 2$, and -4.
28. Exercise 10 for $y = 0, 4$, and 6.

29. Exercise 11 for $y = 0$, 3, and 5.

30. Exercise 12 for $y = 0$, 4, and -2.

31. Exercise 13 for $y = 0$, 3, and 6.

32. Exercise 14 for $y = 0$, -2, and 3.

33. Exercise 15 for $y = 0$, 5, -4, and $-1\frac{1}{2}$.

34. Exercise 16 for $y = 0$, -1, and 1.5.

35. Exercise 17 for $y = 0$, 2, and -2.

36. Exercise 18 for $y = 0$, 2, and 8.

37. Exercise 19 for $y = 0$, 4, and 8.

38. Exercise 20 for $y = 0$, 2, and -3.

Solve each equation graphically for the given values.

39. $y = x^2 + 3x - 4$ for $y = 0$, 6, and -2.

40. $y = 2x^2 - 5x - 3$ for $y = 2$, 0, and -3.

41. $y = -\frac{1}{2}x^2 + 2$ for $y = 0$, 4, and -4.

42. $y = -\frac{1}{4}x^2 + x$ for $y = 0$, $\frac{1}{2}$, and -4.

43. $y = x^3 - 3x^2 + 1$ for $y = 0$, -2, and -0.5.

44. $y = -x^3 + 3x + 2$ for $y = 2$, 0, and 3.

45. The resistance, r, of a resistor in a circuit of constant current varies with time, t in ms, according to the equation $r = 10t^2 + 20$. Solve for t when $r = 90\ \Omega$, $180\ \Omega$, and $320\ \Omega$.

46. An object dropped from an airplane 2500 m above the ground falls according to the equation $h = 2500 - 4.95t^2$, where h is the height in metres above the ground and t is the time in seconds. Find the times for the object to fall to a height of 2000 m, 1200 m, and 600 m above the ground. Also find the time it takes to hit the ground.

47. The energy dissipated (work lost) w by a resistor varies with the time t in milliseconds (ms) according to the equation $w = 5t^2 + 6t$. Solve for t when $w = 2$, 4, and 10.

48. The resistance r in ohms (Ω) in a given circuit varies with time t in milliseconds according to the equation $r = 10 + \sqrt{t}$. Find t when $r = 14.1\ \Omega$, $14.3\ \Omega$, and $14.7\ \Omega$. (*Hint:* Choose a suitable scale for the graph and graph only the part you need.)

49. A given inductor carries a current expressed by the equation $i = t^3 - 15$, where i is the current in amperes and t is the time in seconds. Find t when i is 5 A and 15 A.

50. The charge q in coulombs (C) in a given circuit varies with the time t in microseconds (μs) according to the equation $q = t^2 - \dfrac{t^3}{3}$. Find t when q is 5 C and 10 C.

51. A machinist needs to drill four holes 2.00 in. apart in a straight line in a metal plate as shown in Fig. 4.14. The first hole is placed at the origin, and the line forms an angle of 36.0° with the vertical axis. Find the coordinates of the other three holes.

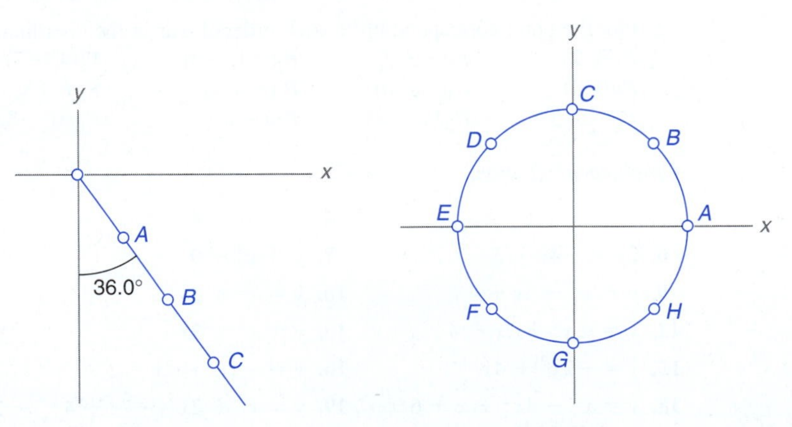

Figure 4.14

Figure 4.15

52. A machinist often uses a coordinate system to drill holes by placing the origin at the most convenient location. A bolt circle is the circle formed by completing an arc through the centers of the bolt holes in a piece of metal. Find the coordinates of the centers of eight equally spaced, $\dfrac{1}{4}$-in. holes on a bolt circle of radius 4.00 in. as shown in Fig. 4.15.

4.3 THE STRAIGHT LINE

Analytic geometry is the study of the relationships between algebra and geometry. The concepts of analytic geometry provide us with ways of algebraically analyzing a geometrical problem. Likewise, with these concepts we can often solve an algebraic problem by viewing it geometrically.

We now develop several basic relations between equations and their graphs. The **slope** of a nonvertical line is the ratio of the difference of the y-coordinates of any two points on the line to the difference of their x-coordinates when the differences are taken in the same order (see Fig. 4.16).

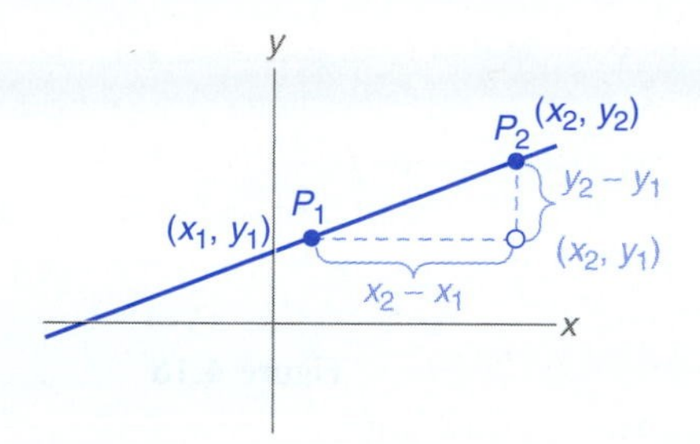

Figure 4.16 Slope of line through P_1 and P_2.

SLOPE OF A LINE

If $P_1(x_1, y_1)$ and $P_2(x_2, y_2)$ represent any two points on a straight line, then the slope m of the line is

$$m = \frac{y_2 - y_1}{x_2 - x_1}$$

EXAMPLE 1

Find the slope of the line passing through $(-2, 1)$ and $(3, 5)$.

If we let $x_1 = -2$, $y_1 = 1$, $x_2 = 3$, and $y_2 = 5$ as in Fig. 4.17, then

$$m = \frac{y_2 - y_1}{x_2 - x_1} = \frac{5 - 1}{3 - (-2)} = \frac{4}{5}$$

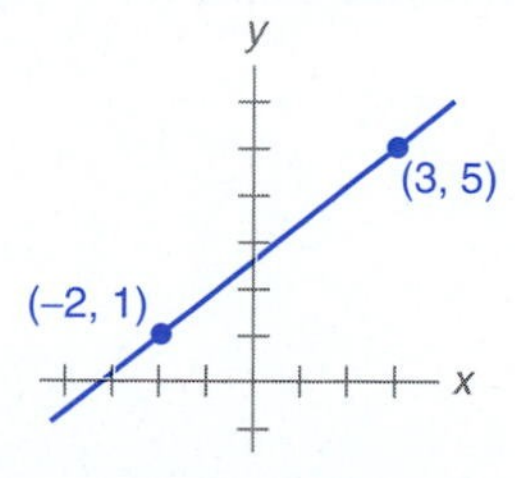

Figure 4.17

Note that if we reverse the order of taking the differences of the coordinates, the result is the same.

$$\frac{y_1 - y_2}{x_1 - x_2} = \frac{1 - 5}{-2 - 3} = \frac{-4}{-5} = \frac{4}{5} = m$$

EXAMPLE 2

Find the slope of the line passing through $(-2, 4)$ and $(6, -6)$.

If we let $x_1 = -2$, $y_1 = 4$, $x_2 = 6$, and $y_2 = -6$ as in Fig. 4.18, then

$$m = \frac{y_2 - y_1}{x_2 - x_1} = \frac{-6 - 4}{6 - (-2)} = \frac{-10}{8} = -\frac{5}{4}$$

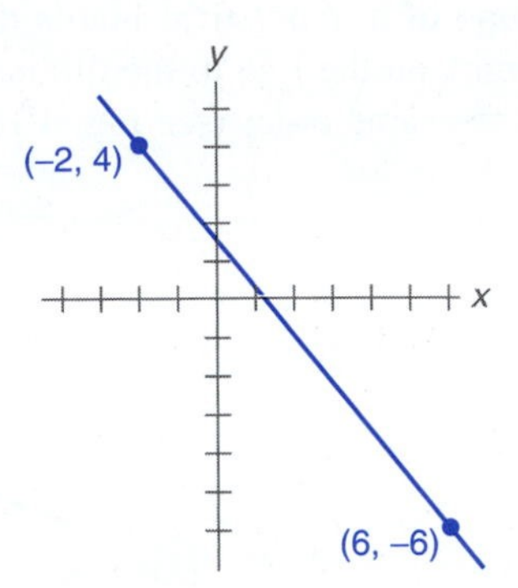

Figure 4.18

Note that in Example 1, the line slopes upward from left to right, while in Example 2, the line slopes downward. In general, we have the following:

1. If a line has positive slope, then the line slopes upward from left to right ("rises").
2. If a line has negative slope, then the line slopes downward from left to right ("falls").
3. If the line has zero slope, then the line is horizontal ("flat").
4. If the line is vertical, then the line has undefined slope because $x_1 = x_2$, or $x_2 - x_1 = 0$. In this case, the ratio $\dfrac{y_2 - y_1}{x_2 - x_1}$ is undefined because division by zero is undefined.

We can use these facts to assist us in graphing a line if we know the slope of the line and one point P on the line. The line can be sketched by drawing a line through the given point P and a point Q which is plotted by moving one unit to the right of P, then moving vertically m units. That is, a point moving along a line will move vertically an amount equal to m, the slope, for every unit move to the right as in Fig. 4.19.

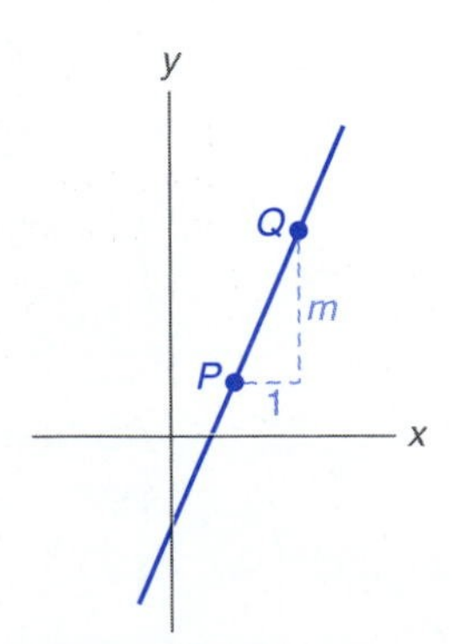

Figure 4.19 The slope m corresponds to the vertical change for each horizontal change of 1.

EXAMPLE 3

Graph a line with slope -2 that passes through the point (1, 3).

Since the slope is -2, points on the line drop 2 units for every unit move to the right. The line passes through (1, 3) and (2, 1) as in Fig. 4.20.

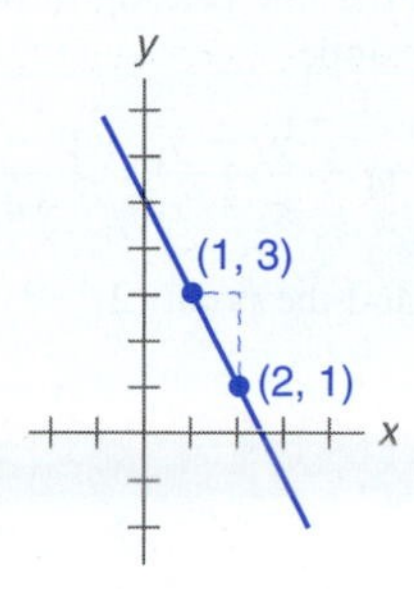

Figure 4.20

Knowing the slope and one point on the line will also determine the equation of the straight line. Let m be the slope of a given nonvertical straight line and let (x_1, y_1) be the coordinates of a point on this line. If (x, y) is any other point on the line as in Fig. 4.21, then we have

$$\frac{y - y_1}{x - x_1} = m$$

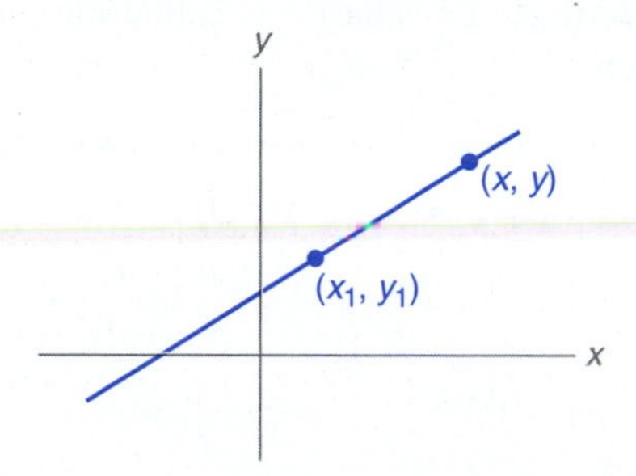

Figure 4.21

By multiplying each side of the equation by $(x - x_1)$, we obtain the following result:

POINT-SLOPE FORM OF A STRAIGHT LINE

If m is the slope and (x_1, y_1) is any point on a nonvertical straight line, its equation is

$$y - y_1 = m(x - x_1)$$

EXAMPLE 4

Find the equation of the line with slope 3 that passes through the point $(-1, 2)$.

Here $m = 3$, $x_1 = -1$, and $y_1 = 2$. Using the point-slope form, we have

$$\begin{aligned} y - y_1 &= m(x - x_1) \\ y - 2 &= 3[x - (-1)] \\ y - 2 &= 3x + 3 \\ y &= 3x + 5 \end{aligned}$$

The point-slope form can also be used to find the equation of a straight line that passes through two points.

EXAMPLE 5

Find the equation of the line passing through the points (2, −3) and (−4, 9).

First, find the slope.

$$m = \frac{y_2 - y_1}{x_2 - x_1} = \frac{9 - (-3)}{-4 - 2} = \frac{9 + 3}{-6} = \frac{12}{-6} = -2$$

Substitute $m = -2$ and the point (2, −3) in the point-slope form.

$$\begin{aligned} y - y_1 &= m(x - x_1) \\ y - (-3) &= -2(x - 2) \\ y + 3 &= -2x + 4 \\ 2x + y - 1 &= 0 \end{aligned}$$

Note: We could have used the other point (−4, 9) in the point-slope form to obtain the equation

$$y - 9 = -2[x - (-4)]$$

which also simplifies to

$$2x + y - 1 = 0$$

A nonvertical line will intersect the y-axis at some point in the form $(0, b)$ as in Fig. 4.22. This ordinate (y-value) b is called the **y-intercept** of the line. If the slope of the line is m, then

$$\begin{aligned} y - y_1 &= m(x - x_1) \\ y - b &= m(x - 0) \\ y - b &= mx \\ y &= mx + b \end{aligned}$$

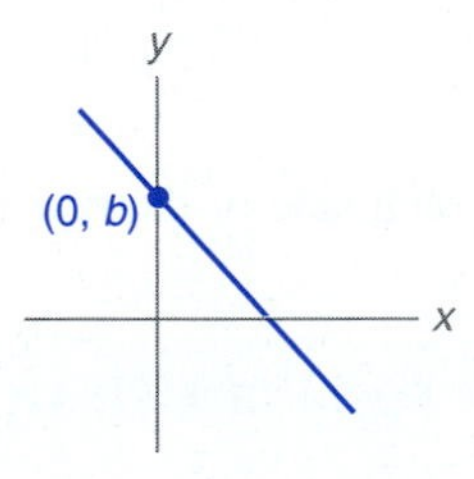

Figure 4.22 b is the y-coordinate of the point where the line crosses the y-axis.

SLOPE-INTERCEPT FORM OF A STRAIGHT LINE

If m is the slope and $(0, b)$ is the y-intercept of a nonvertical straight line, its equation is

$$y = mx + b$$

EXAMPLE 6

Find the equation of the line with slope $\frac{1}{2}$ that crosses the y-axis at $b = -3$.

Using the slope-intercept form, we have

$$y = mx + b$$
$$y = \frac{1}{2}x + (-3)$$
$$y = \frac{1}{2}x - 3$$

or

$$x - 2y - 6 = 0$$

A line parallel to the x-axis has slope $m = 0$ (see Fig. 4.23). Its equation is

$$y = mx + b$$
$$y = (0)x + b$$
$$y = b$$

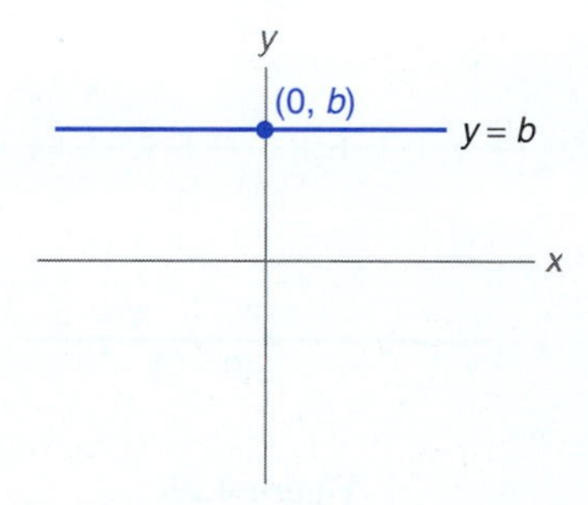

Figure 4.23 Horizontal line.

EQUATION OF A HORIZONTAL LINE

If a horizontal line passes through the point (a, b), its equation is

$$y = b$$

EXAMPLE 7

Find the equation of the line parallel to and 3 units above the x-axis.

The equation is $y = 3$.

By writing the equation of a nonvertical straight line in the slope-intercept form, we can quickly determine the line's slope and a point on the line (the point where it crosses the y-axis).

EXAMPLE 8

Find the slope and the y-intercept of $3y - x + 6 = 0$. Graph the line.

Write the equation in slope-intercept form; that is, solve for y.

$$3y - x + 6 = 0$$
$$3y = x - 6$$
$$y = \frac{x}{3} - 2$$
$$y = \left(\frac{1}{3}\right)x + (-2)$$

So $m = \frac{1}{3}$ and $b = -2$ (see Fig. 4.24).

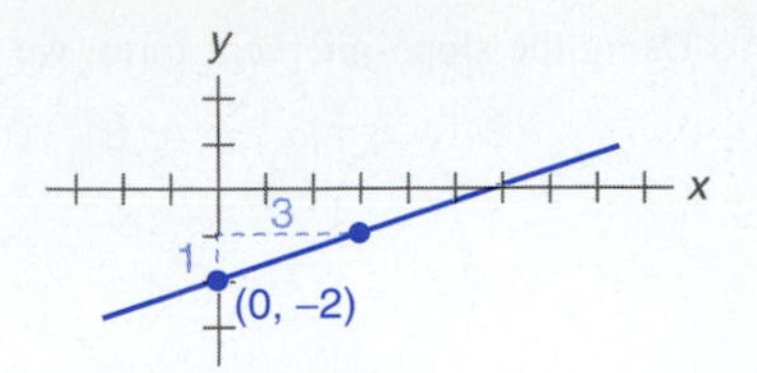

Figure 4.24

EXAMPLE 9

Describe and graph the line whose equation is $y = -5$.

This is a line parallel to and 5 units below the x-axis (see Fig. 4.25).

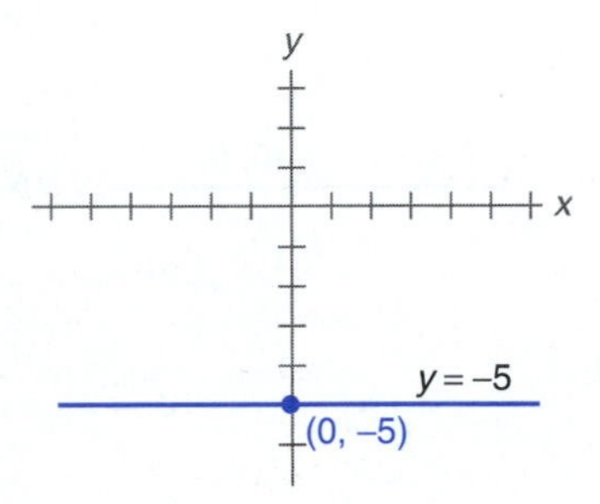

Figure 4.25

If a line is vertical, then we cannot use any of these equations since the line has undefined slope. However, note that in this case, as shown in Fig. 4.26, the line crosses the x-axis at some point in the form $(a, 0)$. All points on the line have the same x-coordinate as the point $(a, 0)$. This characterizes the line, giving us the following equation:

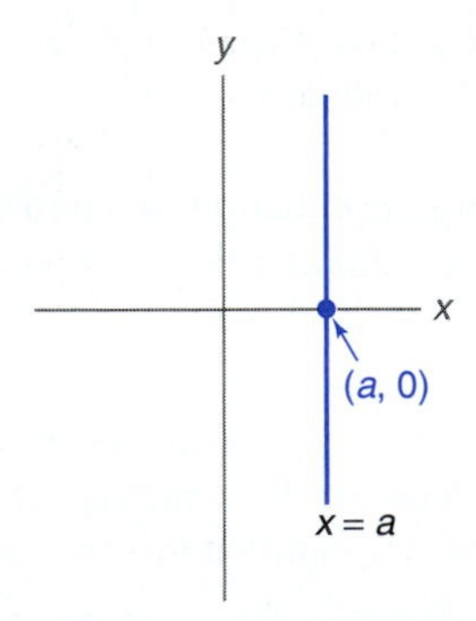

Figure 4.26 Vertical line.

EQUATION OF A VERTICAL LINE

If a vertical line passes through the point (a, b), its equation is

$$x = a$$

EXAMPLE 10

Describe and graph the line whose equation is $x = 2$.

This is a line perpendicular to the x-axis that crosses the x-axis at the point (2, 0) (see Fig. 4.27).

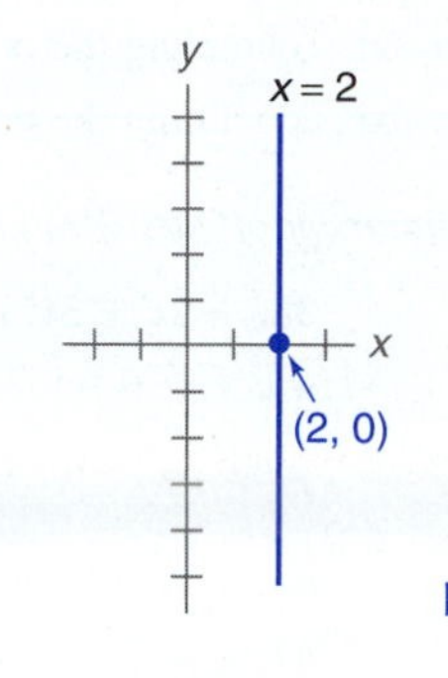

Figure 4.27

EXAMPLE 11

Write the equation of the line perpendicular to the x-axis that crosses the x-axis at the point (−3, 0).

The equation is $x = -3$.

Note: All the equations presented in this section can be put in the form

$$Ax + By + C = 0 \quad \text{with } A \text{ and } B \text{ not both } 0.$$

This is known as the **general form** of the equation of a line.

Any equivalent form of the equation of a line is acceptable. The specific form used usually depends on the application. The most common forms are the point-slope form, $y - y_1 = m(x - x_1)$; the slope-intercept form, $y = mx + b$; and the general form, $Ax + By + C = 0$.

Exercises 4.3

Find the slope of each line passing through the given points.

1. (4, 2), (3, 1)
2. (−3, 2), (−1, −2)
3. (4, −5), (2, 3)
4. (−6, −4), (5, −3)
5. (−3, 2), (6, 2)
6. (4, −7), (4, 3)
7. (5, 7), (−3, 2)
8. (−3, 6), (−1, 3)

Graph each line passing through the given point with the given slope.

9. (2, −1), $m = 2$
10. (0, 1), $m = -3$
11. (−3, −2), $m = \frac{1}{2}$
12. (4, 4), $m = -\frac{1}{3}$
13. (4, 0), $m = -2$
14. (−3, 1), $m = 4$
15. (0, −3), $m = -\frac{3}{4}$
16. (5, −2), $m = \frac{3}{2}$

Find the equation of the line with the given properties.

17. Passes through (−2, 8) with slope −3.
18. Passes through (3, −5) with slope 2.
19. Passes through (−3, −4) with slope $\frac{1}{2}$.
20. Passes through (6, −7) with slope $-\frac{3}{4}$.
21. Passes through (−2, 7) and (1, 4).
22. Passes through (1, 6) and (4, −3).
23. Passes through (6, −8) and (−4, −3).
24. Passes through (−2, 2) and (7, −1).
25. Crosses the y-axis at −2 with slope −5.
26. Crosses the y-axis at 8 with slope $\frac{1}{3}$.
27. Has y-intercept 7 and slope 2.
28. Has y-intercept −4 and slope $-\frac{3}{4}$.
29. Parallel to and 5 units above the x-axis.
30. Parallel to and 2 units below the x-axis.

31. Perpendicular to the x-axis and crosses the x-axis at $(-2, 0)$.

32. Perpendicular to the x-axis and crosses the x-axis at $(5, 0)$.

33. Parallel to the x-axis containing the point $(2, -3)$.

34. Parallel to the y-axis containing the point $(-5, -4)$.

35. Perpendicular to the x-axis containing the point $(-7, 9)$.

36. Perpendicular to the y-axis containing the point $(4, 6)$.

Find the slope and the y-intercept of each straight line.

37. $x + 4y = 12$ **38.** $-2x + 3y + 9 = 0$ **39.** $4x - 2y + 14 = 0$

40. $3x - 6y = 0$ **41.** $y = 6$ **42.** $x = -4$

Graph each equation.

43. $y = 3x - 2$ **44.** $y = -2x + 5$ **45.** $5x - 2y + 4 = 0$

46. $4x + 3y + 6 = 0$ **47.** $x = 7$ **48.** $x = -2$

49. $y = -3$ **50.** $y = 2$ **51.** $6x + 8y = 24$

52. $3x - 5y = 30$ **53.** $x - 3y = -12$ **54.** $x + 6y = 8$

55. A metal rod is 43.0 cm long at temperature $-15.0°C$ and 43.2 cm long at 55.0°C. These data can be listed in (x, y) form as $(-15.0, 43.0)$ and $(55.0, 43.2)$. Find the slope (as a simplified fraction) of the straight line passing through these two points.

4.4 PARALLEL AND PERPENDICULAR LINES

PARALLEL LINES

Two lines are parallel if either one of the following conditions holds:

1. They are both perpendicular to the x-axis [see Fig. 4.28(a)].
2. They both have the same slope [see Fig. 4.28(b)]. That is, if the equations of the lines are

$$L_1: \quad y = m_1x + b_1 \quad \text{and} \quad L_2: \quad y = m_2x + b_2$$

then

$$m_1 = m_2$$

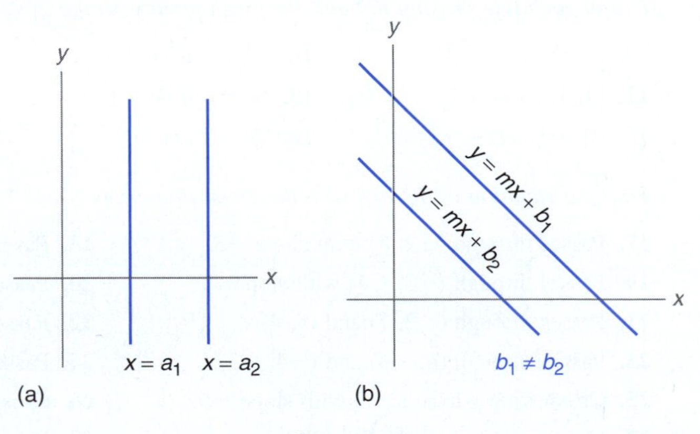

Figure 4.28

PERPENDICULAR LINES

Two lines are perpendicular if either one of the following conditions holds:

1. One line is vertical with equation $x = a$ and the other is horizontal with equation $y = b$.
2. Neither is vertical and the slope of one line is the negative reciprocal of the other. That is, if the equations of the lines are

$$L_1: \quad y = m_1x + b_1 \quad \text{and} \quad L_2: \quad y = m_2x + b_2$$

then

$$m_1 = -\frac{1}{m_2}$$

To show this second relationship for perpendicular lines, consider the triangle in Fig. 4.29, where L_1 is perpendicular to L_2. Let

$(c, 0)$ represent the point P
$(d, 0)$ represent the point R
(e, f) represent the point Q

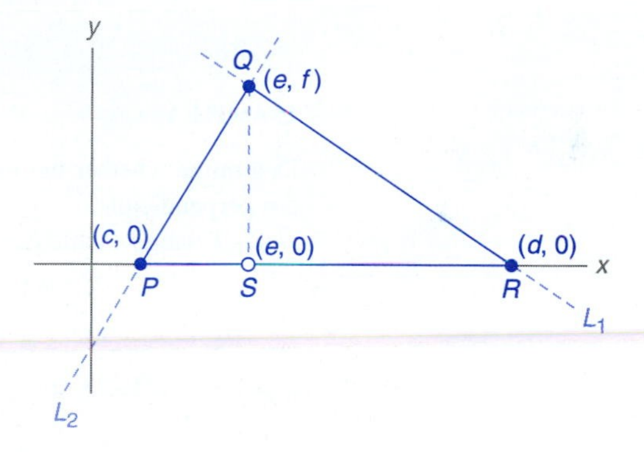

Figure 4.29

Draw QS perpendicular to the x-axis. Then S must be represented by $(e, 0)$.

Triangles PSQ and QSR in Fig. 4.29 are similar. (Note that angle PQS equals angle QRS.) From geometry we know that

$$\frac{PS}{QS} = \frac{QS}{SR} \tag{1}$$

In this case,

$PS = e - c$	(the distance from c to e on the x-axis)
$QS = f$	(the distance from 0 to f on the y-axis)
$SR = d - e$	(the distance from e to d on the x-axis)

Substituting these values in Equation (1), we have

$$\frac{e - c}{f} = \frac{f}{d - e}$$

Multiplying each side of the equation by $(d - e)f$ gives

$$f^2 = (d - e)(e - c) \tag{2}$$

Compute slopes m_1 and m_2 as follows:

$$m_1 = \frac{f - 0}{e - d} = \frac{f}{e - d}$$

$$m_2 = \frac{f - 0}{e - c} = \frac{f}{e - c}$$

$$(m_1)(m_2) = \frac{f}{e - d} \cdot \frac{f}{e - c} = \frac{f^2}{(e - d)(e - c)}$$

Substituting from Equation (2), we have

$$(m_1)(m_2) = \frac{(d - e)(e - c)}{(e - d)(e - c)} = \frac{d - e}{e - d} = -\left(\frac{e - d}{e - d}\right) = -1$$

or

$$(m_1)(m_2) = -1$$

Dividing each side of this equation by m_2, we have

$$m_1 = \frac{-1}{m_2}$$

EXAMPLE 1

Determine whether the lines given by the equations $3y + 6x - 5 = 0$ and $2y - x + 7 = 0$ are perpendicular.

Change each equation into slope-intercept form; that is, solve for y.

$$y = -2x + \frac{5}{3} \qquad \text{(Slope is } -2.)$$

and

$$y = \frac{1}{2}x - \frac{7}{2} \qquad \left(\text{Slope is } \frac{1}{2}.\right)$$

Since

$$-2 = \frac{-1}{\frac{1}{2}} \qquad \left(-2 \text{ is the negative reciprocal of } \frac{1}{2}.\right)$$

the lines are perpendicular.

EXAMPLE 2

Find the equation of the line through $(-3, 2)$ and perpendicular to $2y - 3x + 5 = 0$.

We can find the slope of the desired line by finding the negative reciprocal of the slope of the given line. First find the slope of the line $2y - 3x + 5 = 0$. Writing this equation in slope-intercept form, we have

$$y = \frac{3}{2}x - \frac{5}{2}$$

The slope of this line is $m = \frac{3}{2}$. The slope of the line perpendicular to this line is then equal to $-\frac{2}{3}$, the negative reciprocal of $\frac{3}{2}$. Now using the point-slope form, we have

$$y - y_1 = m(x - x_1)$$

$$y - 2 = -\frac{2}{3}[x - (-3)]$$

$$y - 2 = -\frac{2}{3}(x + 3)$$

or

$$2x + 3y = 0$$

EXAMPLE 3

Find the equation of the line through $(2, -5)$ and parallel to $3x + y = 7$.

First, find the slope of the given line by solving its equation for y.

$$y = -3x + 7$$

Its slope is -3. Any line parallel to this line has the same slope. Now, write the equation of the line with slope -3 passing through $(2, -5)$.

$$y - y_1 = m(x - x_1)$$

$$y - (-5) = -3(x - 2)$$

$$y + 5 = -3x + 6$$

$$y = -3x + 1 \quad \text{or} \quad 3x + y = 1$$

EXAMPLE 4

Judging the slopes of lines using a graphing calculator can be misleading unless a "square" viewing window is chosen. In particular, perpendicular lines don't look like they are intersecting at a 90° angle unless incremental changes along the x- and y-axes have the same meaning. The **ZoomSqr** feature will square up the current viewing window by choosing the *smaller* unit and using it on both axes (the effect is to zoom *out* to square the viewing window). As the last frame shows, **ZoomDec** is a square viewing window to begin with (each pixel on the x- and y-axes represents 0.1 units).

Graph $y = 3x - 4$ and $y = -\frac{1}{3}x - 2$.

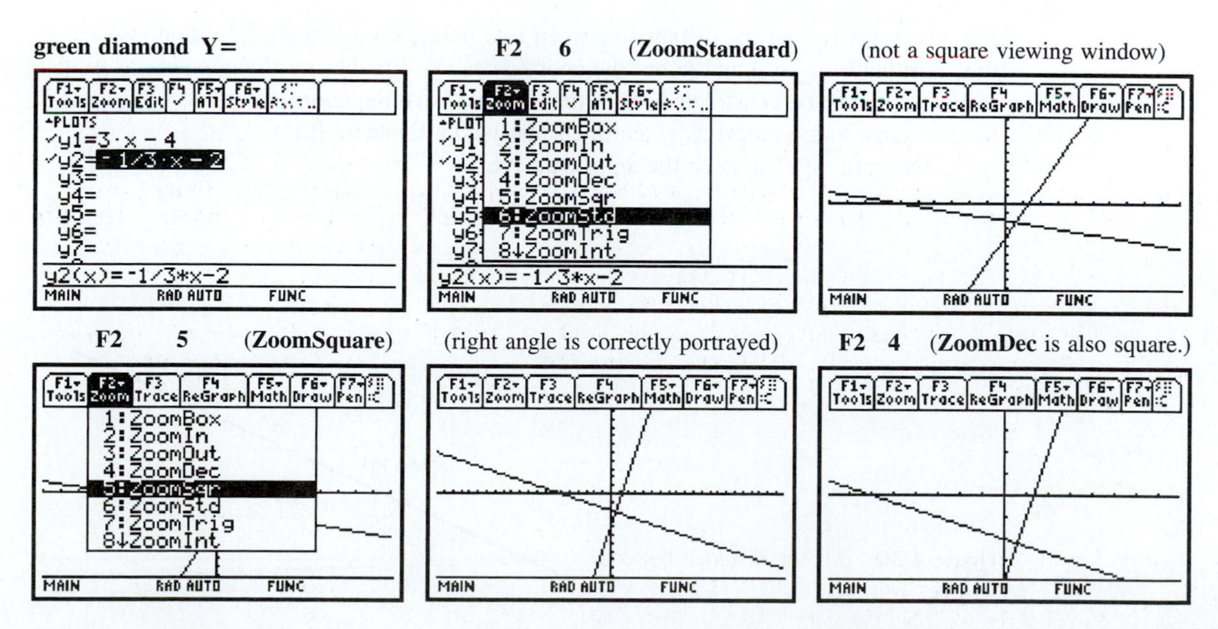

Exercises 4.4

Determine whether each given pair of equations represents lines that are parallel, perpendicular, or neither.

1. $x + 3y - 7 = 0; -3x + y + 2 = 0$
2. $x + 2y - 11 = 0; x + 2y + 4 = 0$
3. $-x + 4y + 7 = 0; x + 4y - 5 = 0$
4. $2x + 7y + 4 = 0; 7x - 2y - 5 = 0$
5. $y - 5x + 13 = 0; y - 5x + 9 = 0$
6. $-3x + 9y + 22 = 0; x + 3y - 17 = 0$

Find the equation of the line that satisfies each set of conditions.

7. Passes through $(-1, 5)$ and is parallel to $-2x + y + 13 = 0$.
8. Passes through $(2, -2)$ and is perpendicular to $3x - 2y - 14 = 0$.
9. Passes through $(-7, 4)$ and is perpendicular to $5y = x$.
10. Passes through $(2, -10)$ and is parallel to $2x + 3y - 7 = 0$.
11. Passes through the origin and is parallel to $3x - 4y = 12$.
12. Passes through the origin and is perpendicular to $4x + 5y = 17$.
13. Has x-intercept 6 and is perpendicular to $4x + 6y = 9$.
14. Has y-intercept -2 and is parallel to $6x - 4y = 11$.
15. Has y-intercept 8 and is parallel to $y = 2$.
16. Has x-intercept -4 and is perpendicular to $y = 6$.
17. Has x-intercept 7 and is parallel to $x = -4$.
18. Has y-intercept -9 and is perpendicular to $x = 5$.
19. The vertices of a quadrilateral are $A(-2, 3)$, $B(2, 2)$, $C(9, 6)$, and $D(5, 7)$.
    (a) Is the quadrilateral a parallelogram? Why or why not?
    (b) Is the quadrilateral a rectangle? Why or why not?
20. The vertices of a quadrilateral are $A(-4, 1)$, $B(0, -2)$, $C(6, 6)$, and $D(2, 9)$.
    (a) Is the quadrilateral a parallelogram? Why or why not?
    (b) Is the quadrilateral a rectangle? Why or why not?

4.5 THE DISTANCE AND MIDPOINT FORMULAS

We now wish to find the distance between two points on a straight line. Suppose P has the coordinates (x_1, y_1) and Q has the coordinates (x_2, y_2). Then a triangle similar to that in Fig. 4.30 can be constructed. Note that R must have the coordinates (x_2, y_1). (Point R has the same x-coordinate as Q and the same y-coordinate as P.)

Using the Pythagorean theorem, we have

$$PQ^2 = PR^2 + QR^2 \tag{3}$$

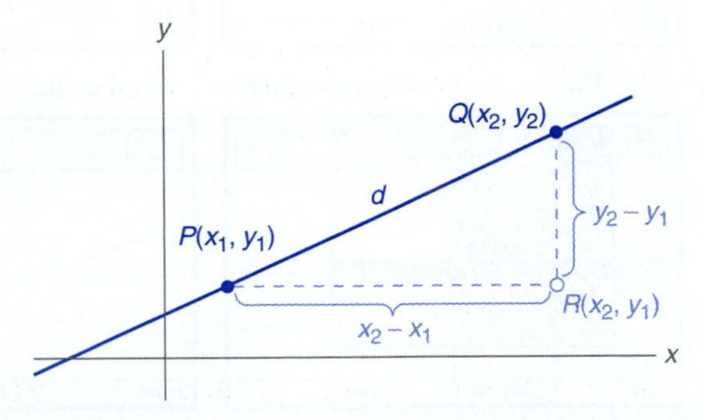

Figure 4.30 d is the distance between points P and Q.

Observe that

$$PR = x_2 - x_1 \qquad \text{(the horizontal distance between } x_1 \text{ and } x_2 \text{ on the } x\text{-axis)}$$
$$QR = y_2 - y_1 \qquad \text{(the vertical distance between } y_1 \text{ and } y_2 \text{ on the } y\text{-axis)}$$

Substituting these values for PR and QR in Equation (3) gives

$$PQ^2 = (x_2 - x_1)^2 + (y_2 - y_1)^2$$

DISTANCE FORMULA

The distance between two points $P(x_1, y_1)$ and $Q(x_2, y_2)$ is given by the formula

$$d = PQ = \sqrt{(x_2 - x_1)^2 + (y_2 - y_1)^2}$$

EXAMPLE 1

Find the distance, d, between (3, 4) and (−2, 7).

$$d = \sqrt{(x_2 - x_1)^2 + (y_2 - y_1)^2}$$
$$d = \sqrt{(-2 - 3)^2 + (7 - 4)^2}$$
$$= \sqrt{(-5)^2 + (3)^2} = \sqrt{25 + 9} = \sqrt{34}$$

Note that we can reverse the order of (3, 4) and (−2, 7) in the formula for computing d without affecting the result.

$$d = \sqrt{(3 - (-2))^2 + (4 - 7)^2}$$
$$= \sqrt{(5)^2 + (-3)^2} = \sqrt{25 + 9} = \sqrt{34}$$

MIDPOINT FORMULA

The coordinates of point $Q(x_m, y_m)$ which is midway between two points $P(x_1, y_1)$ and $R(x_2, y_2)$ are given by

$$x_m = \frac{x_1 + x_2}{2} \qquad y_m = \frac{y_1 + y_2}{2}$$

Figure 4.31 illustrates the midpoint formula. First look at points P, Q, and R. Triangles PSQ and QTR are congruent. This means that

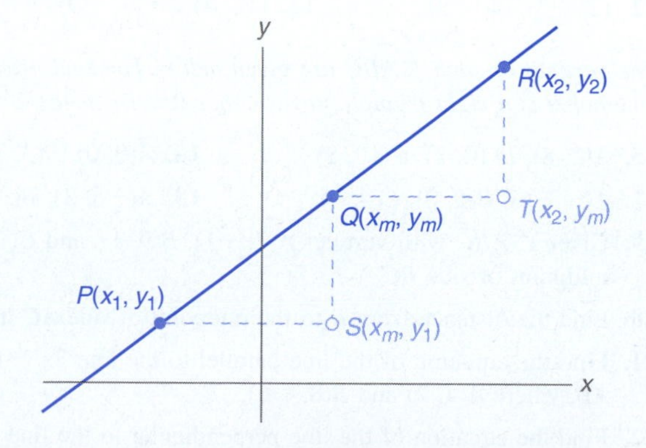

Figure 4.31 Point Q is the midpoint between points P and R.

$$PS = QT$$

Since

$$PS = x_m - x_1$$

and

$$QT = x_2 - x_m$$

then

$$x_m - x_1 = x_2 - x_m$$

or

$$2x_m - x_1 = x_2$$
$$2x_m = x_1 + x_2$$
$$x_m = \frac{x_1 + x_2}{2}$$

The formula for y_m is found in the same manner.

EXAMPLE 2

Find the point midway between $(2, -3)$ and $(-4, 6)$.

$$x_m = \frac{x_1 + x_2}{2} = \frac{2 + (-4)}{2} = \frac{-2}{2} = -1$$
$$y_m = \frac{y_1 + y_2}{2} = \frac{-3 + 6}{2} = \frac{3}{2}$$

The midpoint is $\left(-1, \frac{3}{2}\right)$.

Exercises 4.5

Find the distance between each pair of points.

1. $(4, -7), (-5, 5)$
2. $(4, 3), (-2, -1)$
3. $(3, -2), (10, -2)$
4. $(6, -2), (6, 4)$
5. $(5, -2), (1, 2)$
6. $(2, -3), (-1, 1)$
7. $(3, -5), (3, 2)$
8. $(2, -4), (6, -4)$

Find the coordinates of the point midway between each pair of points.

9. $(2, 3), (5, 7)$
10. $(0, 5), (2, -4)$
11. $(3, -2), (0, 0)$
12. $(2, -3), (4, -3)$
13. $(11, 4), (-11, -9)$
14. $(4, 10), (-6, -8)$

The vertices of each $\triangle ABC$ are given below. For each triangle, find **(a)** *the perimeter,* **(b)** *whether it is a right triangle,* **(c)** *whether it is isosceles, and* **(d)** *its area if it is a right triangle.*

15. $A(2, 8), B(10, 2), C(10, 8)$
16. $A(0, 0), B(3, 3), C(3, -3)$
17. $A(-3, 6), B(5, 0), C(4, 9)$
18. $A(-6, 3), B(-3, 7), C(1, 4)$
19. Given $\triangle ABC$ with vertices $A(7, -1)$, $B(9, 1)$, and $C(-3, 5)$, find the distance from A to the midpoint of side BC.
20. Find the distance from B to the midpoint of side AC in Exercise 19.
21. Find the equation of the line parallel to the line $3x - 6y = 10$ and through the midpoint of AB, where $A(4, 2)$ and $B(8, -6)$.
22. Find the equation of the line perpendicular to the line $2x + 5y = 12$ and through the midpoint of AB, where $A(-3, -4)$ and $B(7, -8)$.

23. Find the equation of the line perpendicular to the line $4x + 8y = 16$ and through the midpoint of AB, where $A(-8, 12)$ and $B(6, 10)$.

24. Find the equation of the line parallel to the line $5x - 6y = 30$ and through the midpoint of AB, where $A(3, 11)$ and $B(7, -5)$.

In Exercises 25 through 28, start with a graph and then use the distance formula and the slopes to confirm the given geometric figure.

25. Show that the figure $ABCD$ with vertices $A(-2, 2)$, $B(1, 3)$, $C(2, 0)$, and $D(-1, -1)$ is a rectangle.

26. Show that the figure $ABCD$ with vertices $A(2, 6)$, $B(7, 2)$, $C(8, 5)$, and $D(3, 9)$ is a parallelogram.

27. Show that the figure $ABCD$ with vertices $A(-12, 8)$, $B(3, 2)$, $C(5, 7)$, $D(-5, 11)$ is a trapezoid with one right angle.

28. Show that if the coordinates of the vertices of a triangle are (a, b), $(a + c, b)$, and $(a + c, b + c)$, the triangle is a right triangle.

CHAPTER 4 SUMMARY

1. *Basic terms:*
   (a) *Relation:* Set of ordered pairs, usually in the form (x, y).
   (b) *Independent variable:* First element of an ordered pair, usually x.
   (c) *Dependent variable:* Second element of an ordered pair, usually y.
   (d) *Domain:* Set of all first elements of the ordered pairs in a relation, or set of all x's.
   (e) *Range:* Set of all second elements of the ordered pairs in a relation, or set of all y's.
   (f) *Function:* Set of ordered pairs in which no two distinct ordered pairs have the same first element.
   (g) *Functional notation:* To write an equation in variables x and y in functional notation, solve for y and replace y with $f(x)$.
   (h) *Linear equation with two unknowns:* An equation of degree one in the form $ax + by = c$ where a and b are not both 0.

2. *Slope of a line:* If $P_1(x_1, y_1)$ and $P_2(x_2, y_2)$ represent any two points on a line, then the slope m of the line is

$$m = \frac{y_2 - y_1}{x_2 - x_1}$$

   (a) If a line has positive slope, the line slopes upward from left to right.
   (b) If a line has negative slope, the line slopes downward from left to right.
   (c) If a line has zero slope, the line is horizontal.
   (d) If a line has undefined slope, the line is vertical.

3. *Point-slope form of a line:* If m is the slope and $P_1(x_1, y_1)$ is any point on a nonvertical line, its equation is

$$y - y_1 = m(x - x_1)$$

4. *Slope-intercept form of a line:* If m is the slope and $(0, b)$ is the y-intercept of a nonvertical line, its equation is

$$y = mx + b$$

5. *Equation of a horizontal line:* If a horizontal line passes through the point (a, b), its equation is

$$y = b$$

6. *Equation of a vertical line:* If a vertical line passes through the point (a, b), its equation is

$$x = a$$

7. *General form of the equation of a straight line:*

$$Ax + By + C = 0 \quad \text{where } A \text{ and } B \text{ are not both } 0$$

8. *Parallel lines:* Two lines are parallel if either one of the following conditions holds:
   (a) They are both perpendicular to the x-axis.
   (b) They both have the same slope. That is, if the equations of the lines are

$$L_1: \quad y = m_1x + b_1 \quad \text{and} \quad L_2: \quad y = m_2x + b_2$$

then

$$m_1 = m_2$$

9. *Perpendicular lines:* Two lines are perpendicular if either one of the following conditions holds:
   (a) One line is vertical with equation $x = a$ and the other is horizontal with equation $y = b$.
   (b) Neither is vertical and the slope of one line is the negative reciprocal of the other. That is, if the equations of the lines are

$$L_1: \quad y = m_1x + b_1 \quad \text{and} \quad L_2: \quad y = m_2x + b_2$$

then

$$m_1 = -\frac{1}{m_2}$$

10. *Distance formula:* The distance between two points $P(x_1, y_1)$ and $Q(x_2, y_2)$ is given by the formula

$$d = PQ = \sqrt{(x_2 - x_1)^2 + (y_2 - y_1)^2}$$

11. *Midpoint formula:* The coordinates of the point $Q(x_m, y_m)$ which is midway between two points $P(x_1, y_1)$ and $R(x_2, y_2)$ are given by

$$x_m = \frac{x_1 + x_2}{2} \quad \text{and} \quad y_m = \frac{y_1 + y_2}{2}$$

CHAPTER 4 REVIEW

Determine whether or not each relation is a function. Write its domain and its range.

1. $A = \{(2, 3), (3, 4), (4, 5), (5, 6)\}$

2. $B = \{(2, 6), (6, 4), (2, 1), (4, 3)\}$

3. $y = -4x + 3$

4. $y = x^2 - 5$

5. $x = y^2 + 4$

6. $y = \sqrt{4 - 8x}$

7. Given $f(x) = 5x + 14$, find
   (a) $f(2)$ **(b)** $f(0)$ **(c)** $f(-4)$

8. Given $g(t) = 3t^2 + 5t - 12$, find
   (a) $g(2)$ **(b)** $g(0)$ **(c)** $g(-5)$

9. Given $h(x) = \dfrac{4x^2 - 3x}{2\sqrt{x - 1}}$, find
   (a) $h(2)$ **(b)** $h(5)$ **(c)** $h(-15)$ **(d)** $h(1)$

10. Given $g(x) = x^2 - 6x + 4$, find
    (a) $g(a)$ **(b)** $g(2x)$ **(c)** $g(z - 2)$

Graph each equation.

11. $y = 4x + 5$
12. $y = x^2 + 4$
13. $y = x^2 + 2x - 8$
14. $y = 2x^2 + x - 6$
15. $y = -x^2 - x + 4$
16. $y = \sqrt{2x}$
17. $y = \sqrt{-2 - 4x}$
18. $y = x^3 - 6x$

Solve each graphically.

19. Exercise 12 for $y = 5$, 7, and 2.
20. Exercise 13 for $y = 0$, -2, and 3.
21. Exercise 15 for $y = 2$, 0, and -2.
22. Exercise 18 for $y = 0$, 2, and -3.
23. The current i in a given circuit varies with the time t according to $i = 2t^2$. Find t when $i = 2$, 6, and 8.
24. A capacitor receives a voltage V where $V = 4t^3 + t$ where t is in seconds. Find t when V is 40 and 60.

Use the points $(3, -4)$ and $(-6, -2)$ in Exercises 25 through 27.

25. Find the slope of the line through the two points.
26. Find the distance between the two points.
27. Find the coordinates of the point midway between the two points.

Find the equation of the line that satisfies each condition in Exercises 28 through 31.

28. Passes through $(4, 7)$ and $(6, -4)$.
29. Passes through $(-3, 1)$ with slope $\frac{2}{3}$.
30. Crosses the y-axis at -3 with slope $-\frac{1}{3}$.
31. Is parallel to and 3 units to the left of the y-axis.
32. Find the slope and the y-intercept of $3x - 2y - 6 = 0$.
33. Graph $3x - 4y = 12$.

Using the slope-intercept form of each line, determine whether each given pair of equations represents lines that are parallel, perpendicular, or neither.

34. $x - 2y + 3 = 0$, $8x + 4y - 9 = 0$
35. $2x - 3y + 4 = 0$, $-8x + 12y = 16$
36. $3x - 2y + 5 = 0$, $2x - 3y + 9 = 0$
37. $x = 2$, $y = -3$
38. $x = 4$, $x = 7$
39. Find the equation of the line parallel to the line $2x - y + 4 = 0$ that passes through the point $(5, 2)$.
40. Find the equation of the line perpendicular to the line $3x + 5y - 6 = 0$ that passes through the point $(-4, 0)$.

CHAPTER 4
SPECIAL NASA APPLICATION
The Space Shuttle Landing, Part I*

INTRODUCTION

Space Shuttle missions have become a familiar topic on the evening news and undoubtedly the fiery launch of a Shuttle from the Kennedy Space Center is a marvel to behold. However, the final and most critical phases of a Space Shuttle mission are deorbit, reentry into the atmosphere, and landing. This application focuses on the final landing approach of the Shuttle, but first we give a brief description of the events from deorbit until final approach.

The Space Shuttle has no power and essentially acts as a glider, although it is much heavier and has less of an aerodynamic shape than an actual glider (it has been referred to as a "flying brick"). It has only one chance to land since, without engines, it cannot climb and try another approach. The process begins half a world away from the landing runway, when the Space Shuttle is traveling 200 mi above the ground at a speed of over 17,000 mi/h. It is now about 60 min to touchdown. During the deorbit burn, the Space Shuttle travels tail first and loses some speed and altitude. Once the burn is complete, the Space Shuttle position is reversed, its nose is raised, and the atmospheric entry begins. It is now about 31 min to touchdown. During this phase, there is a tremendous heat buildup around the Shuttle and portions of the vehicle's exterior reach 2800°F (you have probably heard about the tiles used on the surface of the vehicle to protect it at this critical time). The heat strips electrons from the air around the Space Shuttle, enveloping it in a sheath of ionized air that blocks all communication with the ground for about 12 min. During this interval the pilot performs several banking maneuvers called roll reversals or S-turns to control descent. When the Space Shuttle comes out of the communications blackout, its speed is about 8275 mi/h and 12 min remain to touchdown. It is now committed to a particular landing site and must begin the final approach with enough altitude and speed to reach the touch-down point. At this point the vehicle travels a circular path around an imaginary cone that will line it up with the center line of the runway. Once the Shuttle comes out of this turn, it is ready for its final approach to the runway.

*From NASA–AMATYC–NSF Project Mathematics Explorations II, grant principals John S. Pazdar, Patricia L. Hirschy, and Peter A. Wursthorn; copyright Capital Community College, 2000.

FINAL APPROACH TO THE RUNWAY

Coming out of its turn, the Shuttle should be at an altitude of 13,365 ft, have a speed of 424 mi/h, and be 7.5 mi (horizontal distance) from the runway. It is now 86 s to touch-down. The nose is down so that the Space Shuttle can descend steeply to a point 7500 ft from the runway threshold, where its altitude should be 1750 ft. The vehicle then enters a transitional phase. The Shuttle's nose is raised as it heads for a position where its altitude is 131 ft and its distance from the runway threshold is 2650 ft. The Shuttle is now 17 s to touch-down. From here the Space Shuttle enters the final phase, aiming at a point 2200 ft down the runway.

To examine the *path* of the Shuttle on its final approach to the runway, we establish a coordinate system with the runway threshold as the origin; the vertical axis is altitude in feet, the horizontal axis is distance from the runway threshold in feet, and the touchdown point, P, is to the left of the y-axis (see Fig. 4.32).

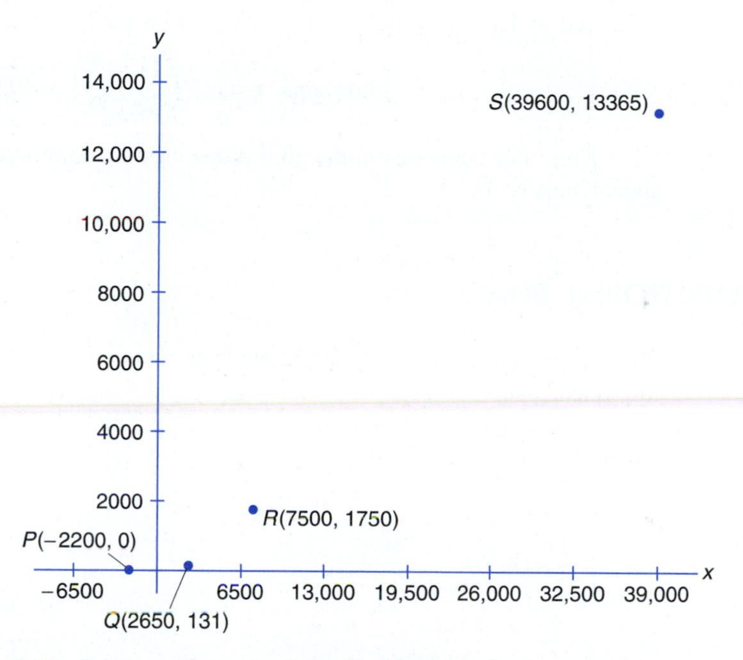

Figure 4.32 Points on the path of the Shuttle on its final approach to the runway.

Notice that the data points increase from left to right; the farther the Shuttle is from the touch-down point, the higher it is above ground. Relative to the coordinate system, the Shuttle travels from right to left and moves down.

THE FIRST PHASE

The first phase is the initial steep descent of the Shuttle, which takes it from the first point, S, where the altitude is 13,365 ft to the second point, R, where the altitude is 1750 ft. Refer to the previous description of the final approach to the runway. This phase is best modeled as a *linear function.*

The slope of the line segment containing the first two data points is

$$\frac{13{,}365 - 1750}{39{,}600 - 7500} = \frac{11{,}615}{32{,}100} = 0.36184$$

The slope indicates how much the Shuttle descends for each foot that it nears the touch-down point. For every foot (measured horizontally) that the Shuttle is closer to the runway, its altitude decreases by about 0.36 ft.

Using the points (39,600, 13,365) and (7500, 1750), we write the equation of the line which describes this linear descent.

$$y = 0.36184x - 963.78505$$

This function has a domain of $7500 \leq x \leq 39{,}600$. This means we can use the function to predict the altitude of the Shuttle when its horizontal distance from the runway threshold is in this domain. For example, if the Shuttle's horizontal distance from the runway threshold is 10,000 ft, then the altitude is

$$0.36184(10{,}000) - 963.78505 \approx 2655 \text{ ft}$$

The slope of this line segment gives the steepness of descent of the Shuttle during this phase, but NASA flight engineers prefer to use *glide slope*. Glide slope is the angle, measured in degrees, which the line makes with the horizontal. Thus,

$$\text{glide slope} = \tan^{-1}\left(\frac{11{,}615}{32{,}100}\right) = 19.9°$$

Note: For comparison, the glide slope used by commercial aircraft when landing is approximately 3°.

THE TRANSITIONAL PHASE

The transitional phase takes the Shuttle from its steep linear descent path to its final shallow linear path (see Fig. 4.33). During this time, the vehicle shifts from nose down to nose up. This phase begins at the point where the altitude is 1750 ft and ends at the point where the altitude is 131 ft.

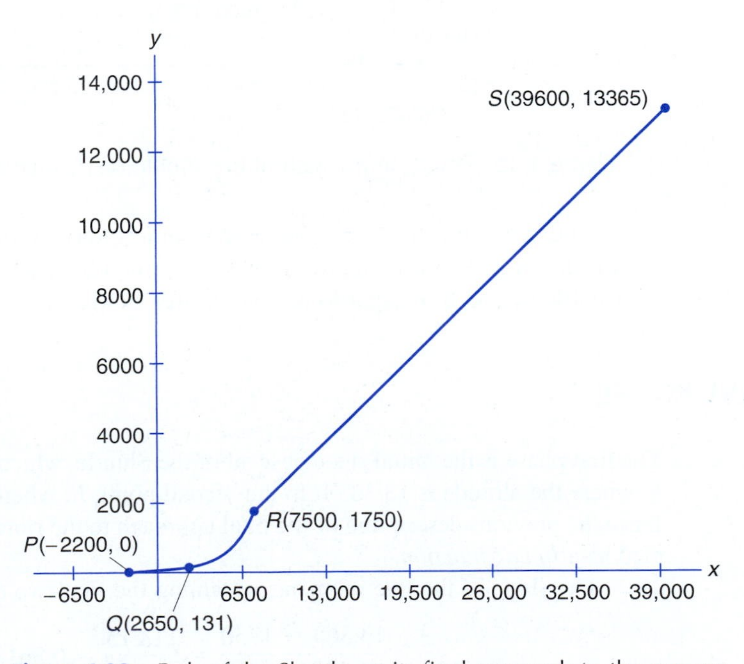

Figure 4.33 Path of the Shuttle on its final approach to the runway.

After the transitional phase, the Shuttle travels from an altitude of 131 ft to 0 ft. The equation of the line containing the points (2650, 131) and (−2200, 0) is

$$y = 0.02701x + 59.42268$$

Notice that the slope of this equation is 0.02701, with a glide slope of $\tan^{-1}\left(\frac{131}{4850}\right) = 1.5°$. The transition from a steep glide slope of 19.9° to a shallow one of 1.5° cannot be accomplished using a straight line. We need a different function model for the transitional phase. See Chapter 9, Special NASA Application, for a development of this nonlinear model.

5
Factoring and Algebraic Fractions

INTRODUCTION

The properties that allow us to change the form of algebraic expressions without changing their values or their characteristics are very important. Expressions often need a different format so that we can simplify them or enter them into a calculator or a computer.

In this chapter we will study factoring algebraic expressions, simplifying algebraic fractions, and solving equations with algebraic fractions.

Objectives

- In an algebraic expression, find the greatest common factor.
- Identify and factor the difference of two squares.
- Identify and factor the square of a binomial.
- Factor trinomials.
- Factor using the method of grouping.
- Identify and factor the sum or difference of two cubes.
- Add, subtract, multiply, and divide algebraic fractions.
- Solve equations containing algebraic fractions.

5.1 SPECIAL PRODUCTS

In Chapter 1 we introduced certain fundamental algebraic concepts and operations. Before proceeding, we must develop some additional algebraic techniques that are necessary in the development of later topics.

Certain types of algebraic products occur so often that we must be able to recognize them on sight. We must also be able to do the multiplications quickly and mentally. To illustrate the first two types of special algebraic products, we give the general form of each followed by two examples.

$$a(x + y + z) = ax + ay + az$$

EXAMPLE 1

$$3x(x^2 + 2x - 4) = 3x^3 + 6x^2 - 12x$$

EXAMPLE 2

$$4a^2b^3(7a^4b^3 - 5a^2b^3 - ab + 3a^2) = 28a^6b^6 - 20a^4b^6 - 4a^3b^4 + 12a^4b^3$$

$$(x + y)(x - y) = x^2 - y^2$$

EXAMPLE 3

$$(3a + 4)(3a - 4) = 9a^2 - 16$$

EXAMPLE 4

$$(6a^2 + 5xy^3)(6a^2 - 5xy^3) = 36a^4 - 25x^2y^6$$

There are two general forms of the *square of a binomial*.

$$(x + y)^2 = x^2 + 2xy + y^2$$

EXAMPLE 5

$$(2x + 3y)^2 = 4x^2 + 12xy + 9y^2$$

EXAMPLE 6

$$(5ab^2 + 4c)^2 = 25a^2b^4 + 40ab^2c + 16c^2$$

$$(x - y)^2 = x^2 - 2xy + y^2$$

EXAMPLE 7

$$(3x - 4)^2 = 9x^2 - 24x + 16$$

EXAMPLE 8

$$(2a^2 - 5y^3)^2 = 4a^4 - 20a^2y^3 + 25y^6$$

The *product of two binomials* may be found mentally using the FOIL method discussed in Section 1.8.

EXAMPLE 9

$$(x + 4)(x + 6) = x^2 + 10x + 24$$

EXAMPLE 10

$$(x + 3)(x - 5) = x^2 - 2x - 15$$

EXAMPLE 11

$$(2x + 3)(5x + 4) = 10x^2 + 23x + 12$$

EXAMPLE 12

$$(3x - 4)(-2x + 7) = -6x^2 + 29x - 28$$

The general form of the *square of a trinomial* is as follows:

$$(x + y + z)^2 = x^2 + y^2 + z^2 + 2xy + 2xz + 2yz$$

EXAMPLE 13

$$(4x + y + 3z)^2 = 16x^2 + y^2 + 9z^2 + 8xy + 24xz + 6yz$$

The *square of a trinomial* may also be found using grouping and the square of a binomial as follows for Example 13:

$$\begin{aligned}(4x + y + 3z)^2 &= [(4x + y) + 3z]^2 \\ &= (4x + y)^2 + 2(4x + y)(3z) + (3z)^2 \\ &= 16x^2 + 8xy + y^2 + 24xz + 6yz + 9z^2\end{aligned}$$

EXAMPLE 14

$$(3a - b^2 - 2c^3)^2 = 9a^2 + b^4 + 4c^6 - 6ab^2 - 12ac^3 + 4b^2c^3$$

or

$$\begin{aligned}(3a - b^2 - 2c^3)^2 &= [(3a - b^2) - 2c^3]^2 \\ &= (3a - b^2)^2 - 2(3a - b^2)(2c^3) + (2c^3)^2 \\ &= 9a^2 - 6ab^2 + b^4 - 12ac^3 + 4b^2c^3 + 4c^6\end{aligned}$$

Or, by calculator

F2 **3**

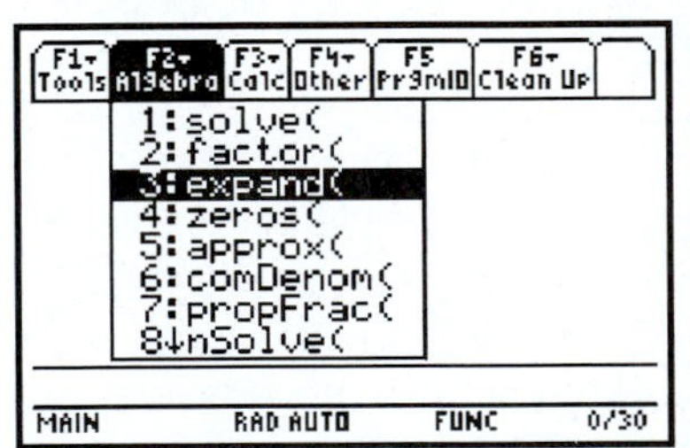

(3a-b^2-2c^3)^2) **ENTER**

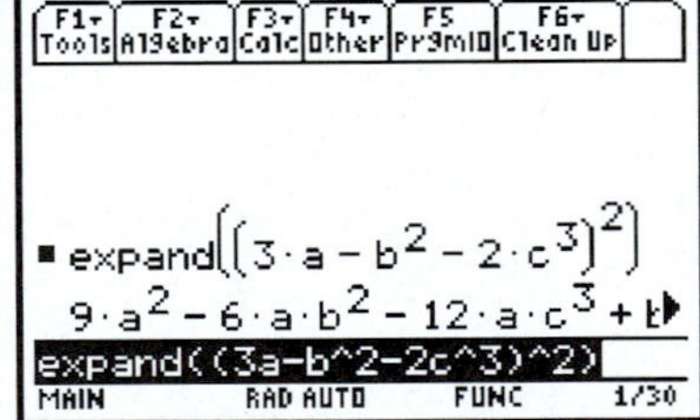

up arrow **2nd** right arrow

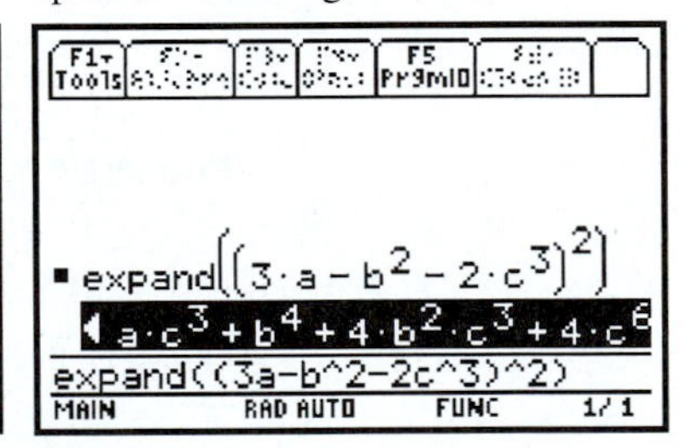

There are two general forms of the *cube of a binomial.*

$$(x + y)^3 = x^3 + 3x^2y + 3xy^2 + y^3$$

EXAMPLE 15

$$\begin{aligned}(2a + 5b)^3 &= (2a)^3 + 3(2a)^2(5b) + 3(2a)(5b)^2 + (5b)^3 \\ &= 8a^3 + 60a^2b + 150ab^2 + 125b^3\end{aligned}$$

EXAMPLE 16

$$\begin{aligned}(4x + y)^3 &= (4x)^3 + 3(4x)^2(y) + 3(4x)(y)^2 + (y)^3 \\ &= 64x^3 + 48x^2y + 12xy^2 + y^3\end{aligned}$$

$$(x - y)^3 = x^3 - 3x^2y + 3xy^2 - y^3$$

EXAMPLE 17

$$(4a - b)^3 = (4a)^3 - 3(4a)^2(b) + 3(4a)(b)^2 - (b)^3$$
$$= 64a^3 - 48a^2b + 12ab^2 - b^3$$

EXAMPLE 18

$$(3x - 2y)^3 = (3x)^3 - 3(3x)^2(2y) + 3(3x)(2y)^2 - (2y)^3$$
$$= 27x^3 - 54x^2y + 36xy^2 - 8y^3$$

Exercises 5.1

Find each product mentally.

1. $-8x^2(5x^3 - 9x^2 + 10x)$
2. $5x^2(3x^5 - 4x^3 + 2x^2 - 6x + 7)$
3. $6x^3y^5(4xy - 7x^3y^2 + 9x^4)$
4. $6a^2b^4(-2a^2b^2 + 5ab^2 - 9b^3)$
5. $(2x + 7)(2x - 7)$
6. $(3x^2 + 8yz^2)(3x^2 - 8yz^2)$
7. $(7x^2 + 2y)^2$
8. $(8a^2b + 5c)^2$
9. $(2a^2 - 3b)^2$
10. $(9x - 4y^3)^2$
11. $(x + 4)(x + 12)$
12. $(x - 9)(x + 8)$
13. $(2x + 5)(-3x - 8)$
14. $(4x - 7)(5x - 6)$
15. $(2a - b + 3c)^2$
16. $(-3a + 2b - c)^2$
17. $(2a + b)^3$
18. $(4x + 5y)^3$
19. $(5 - 2x^2)^3$
20. $(2x - 4y^3)^3$
21. $(6a + 5b)(7a - 3b)$
22. $(5x - y)(4x - 8y)$
23. $(4x + 7y)^2$
24. $(2x - 10y)^2$
25. $(2a + 9b)(6a - 11b)$
26. $(2x^2y + 7z^3)^2$
27. $(3a^3 + 10b^2)^3$
28. $(8 - a^2b^3)^2$
29. $-4a^2b^3(6a^3b^4 - 9a^5b^7 + 20a^5)$
30. $2x^2y^3z(3x^4y - 5x^2yz^3 - 6yz + 4x^2z^2)$
31. $(9a^3 - 12b^2)^2$
32. $(12a - 5b)(15a + 2b)$
33. $(x - 3y - 5z)^2$
34. $(-3x + 7)(7x + 3)$
35. $(9a - 4b)(6a - 10b)$
36. $(80 - a^5)(80 + a^5)$
37. $(6x^3 + 4)(5x^3 - 4)$
38. $(4x^2 - 5)(3x^2 + 7)$
39. $(3xy + 9yz)(5xy - 8yz)$
40. $(-5y + z)^2$
41. $(5abc + 6)(5abc - 6)$
42. $(1 - a^2)^2$
43. $(3x^2 - \frac{1}{2}y)(x^2 + \frac{2}{3}y)$
44. $(2x^2 + \frac{1}{4}y^3)(2x^2 + \frac{1}{2}y^3)$
45. $(x - \frac{2}{3}y)^2$
46. $(4x^2 - \frac{1}{2}y)^2$
47. $x^2(1 + x^2)^2$
48. $x^2(1 - x^2)^3$

5.2 FACTORING ALGEBRAIC EXPRESSIONS

Factoring is the process of writing an algebraic expression as a product of two or more factors*—usually prime factors. The process is the opposite of multiplying algebraic expressions; that is, the reverse of the product forms in the previous section.

The first type of factoring involves finding the largest monomial factor that divides each factor "evenly,"† dividing each term by this factor, and then writing the result as a product.

GREATEST COMMON FACTOR

$$ax + ay + az = a(x + y + z)$$

*Recall that in the product ab, a and b are called **factors.**
†The remainder is zero after dividing.

EXAMPLE 1

Factor $4x + 8y$.

$$4x + 8y = 4(x + 2y)$$

EXAMPLE 2

Factor $15ab - 6ac$.

$$15ab - 6ac = 3a(5b - 2c)$$

EXAMPLE 3

Factor $21x^3 - 14x^2 + 28x$.

$$21x^3 - 14x^2 + 28x = 7x(3x^2 - 2x + 4)$$

EXAMPLE 4

Factor $15xy^2 + 45x^2y^2 - 5xy$.

$$15xy^2 + 45x^2y^2 - 5xy = 5xy(3y + 9xy - 1)$$

EXAMPLE 5

Factor $50xy^2z - 80xyz - 32xz$.

$$50xy^2z - 80xyz - 32xz = 2xz(25y^2 - 40y - 16)$$

Note: You should always try this type of factoring first because many times it simplifies the remaining factor into an expression something that factors more easily.

The difference of two perfect squares is the product of the square root of the first term *plus* the square root of the second term times the square root of the first term *minus* the square root of the second term.

DIFFERENCE OF TWO PERFECT SQUARES

$$x^2 - y^2 = (x + y)(x - y)$$

EXAMPLE 6

Factor $9a^2 - b^2$.

$$9a^2 - b^2 = (3a + b)(3a - b)$$

EXAMPLE 7

Factor $36x^2 - 49y^4$.

$$36x^2 - 49y^4 = (6x + 7y^2)(6x - 7y^2)$$

EXAMPLE 8

Factor $x^2 - 4a^2b^2$.

$$x^2 - 4a^2b^2 = (x + 2ab)(x - 2ab)$$

Note: The *sum* of two perfect squares, $x^2 + y^2$, cannot be factored.

EXAMPLE 9

Factor $81x^4 - y^8$.

$$\begin{aligned} 81x^4 - y^8 &= (9x^2 + y^4)(9x^2 - y^4) \\ &= (9x^2 + y^4)(3x + y^2)(3x - y^2) \end{aligned}$$

Note: To factor this expression completely, we must also factor $9x^2 - y^4$.

The factors of a general trinomial are often binomial factors. To find these binomial factors, you "undo" the FOIL multiplication.

EXAMPLE 10

Factor $x^2 + 7x + 10$.

This type of factoring involves some trial and error. Here we need two numbers whose sum is 7 and whose product is 10. They are +2 and +5.

$$x^2 + 7x + 10 = (x + 2)(x + 5)$$

EXAMPLE 11

Factor $x^2 - 9x + 18$.

We need two numbers whose sum is -9 and whose product is 18: -3 and -6.

$$x^2 - 9x + 18 = (x - 3)(x - 6)$$

EXAMPLE 12

Factor $x^2 - 3x - 28$.

We need two numbers whose sum is -3 and whose product is -28: $+4$ and -7.

$$x^2 - 3x - 28 = (x + 4)(x - 7)$$

A short summary about the signs in trinomials might be helpful. If the trinomial to be factored is one of the following forms, use the corresponding sign patterns.

1. $x^2 + px + q = (x + \quad)(x + \quad)$
2. $x^2 - px + q = (x - \quad)(x - \quad)$
3. $x^2 + px - q = (x + \quad)(x - \quad)$
4. $x^2 - px - q = (x + \quad)(x - \quad)$

EXAMPLE 13

Factor $x^2 + 10x + 24$.

We need two numbers whose sum is 10 and whose product is 24: $+4$ and $+6$.

$$x^2 + 10x + 24 = (x + 4)(x + 6)$$

EXAMPLE 14

Factor $x^2 - 15x + 56$.

We need two numbers whose sum is -15 and whose product is 56: -7 and -8.

$$x^2 - 15x + 56 = (x - 7)(x - 8)$$

EXAMPLE 15

Factor $x^2 + 3x - 40$.

We need two numbers whose sum is 3 and whose product is -40: $+8$ and -5.

$$x^2 + 3x - 40 = (x + 8)(x - 5)$$

EXAMPLE 16

Factor $x^2 - 7x - 18$.

We need two numbers whose sum is -7 and whose product is -18: $+2$ and -9.

$$x^2 - 7x - 18 = (x + 2)(x - 9)$$

EXAMPLE 17

Factor $8x^2 + 14x + 3$.

When the coefficient of the squared term is not 1, the factoring involves more trial and error. Here, we need the following:

1. Pairs of factors whose product is $8x^2$. There are two possible pairs:

$$(8x \quad)(x \quad) \text{ and } (4x \quad)(2x \quad)$$

2. Pairs of factors whose product is 3. There is only one pair: $3 \cdot 1$.
3. To arrange these factors so that the middle term is $14x$. If the coefficient of the squared term is positive, the signs in the factors take the same forms as discussed in the summary after Example 12. The possibilities are as follows.

Factors	*Middle term*	
$(8x + 1)(x + 3)$	$25x$	
$(8x + 3)(x + 1)$	$11x$	
$(4x + 3)(2x + 1)$	$10x$	
$(4x + 1)(2x + 3)$	$14x$	(the correct combination)

Therefore, $8x^2 + 14x + 3 = (4x + 1)(2x + 3)$.

EXAMPLE 18

Factor $12x^2 - 28x + 15$.

Step 1: The pairs of factors of $12x^2$ are

$$(12x \quad)(x \quad), \quad (6x \quad)(2x \quad), \text{ and } (4x \quad)(3x \quad)$$

Step 2: The pairs of factors of 15 are $15 \cdot 1$ and $5 \cdot 3$.

Step 3: Arrange these factors so that the middle term is $-28x$. The possibilities are as follows:

Factors	*Middle term*	
$(12x - 1)(x - 15)$	$-181x$	
$(12x - 15)(x - 1)$	$-27x$	
$(12x - 3)(x - 5)$	$-63x$	
$(12x - 5)(x - 3)$	$-41x$	
$(6x - 15)(2x - 1)$	$-36x$	
$(6x - 1)(2x - 15)$	$-92x$	
$(6x - 5)(2x - 3)$	$-28x$	(the correct combination)
$(6x - 3)(2x - 5)$	$-36x$	
$(4x - 15)(3x - 1)$	$-49x$	
$(4x - 1)(3x - 15)$	$-63x$	
$(4x - 5)(3x - 3)$	$-27x$	
$(4x - 3)(3x - 5)$	$-29x$	

Note: Both signs inside the factors must be negative. Therefore,

$$12x^2 - 28x + 15 = (6x - 5)(2x - 3)$$

Of course, we do not go through this long listing process for every factoring. The work is often done mentally, and we stop when we have found the combination of factors that gives the desired middle term.

EXAMPLE 19

Factor $12x^2 - 7x - 12$.

$$12x^2 - 7x - 12 = (4x + 3)(3x - 4)$$

EXAMPLE 20

Factor $30x^2 - 59x + 9$.

$$30x^2 - 59x + 9 = (5x - 9)(6x - 1)$$

EXAMPLE 21

Factor $10x^2 + 27xy - 9y^2$.

$$10x^2 + 27xy - 9y^2 = (10x - 3y)(x + 3y)$$

EXAMPLE 22

Factor $4x^2 + 28xy^2 + 45y^4$.

$$4x^2 + 28xy^2 + 45y^4 = (2x + 5y^2)(2x + 9y^2)$$

EXAMPLE 23

Factor $-6x^2 - 11x + 10$.

$$-6x^2 - 11x + 10 = (-3x + 2)(2x + 5)$$

An alternate approach to factoring a trinomial whose first term is negative involves dividing each term first by -1 and then proceeding as usual. Using this approach in this example leads to

$$\begin{aligned}-6x^2 - 11x + 10 &= -(6x^2 + 11x - 10)\\ &= -(3x - 2)(2x + 5)\end{aligned}$$

Do you see that this result is equivalent to our previous result?

EXAMPLE 24

Factor $60x^2 + 20x - 105$.

$$\begin{aligned}60x^2 + 20x - 105 &= 5(12x^2 + 4x - 21)\\ &= 5(2x + 3)(6x - 7)\end{aligned}$$

Note: Dividing each term by the common monomial factor, 5, simplified the remaining trinomial factor.

EXAMPLE 25

Factor $180x^2 + 279x + 108$.

$$\begin{aligned}180x^2 + 279x + 108 &= 9(20x^2 + 31x + 12)\\ &= 9(5x + 4)(4x + 3)\end{aligned}$$

Or, by calculator, we have

F2 2

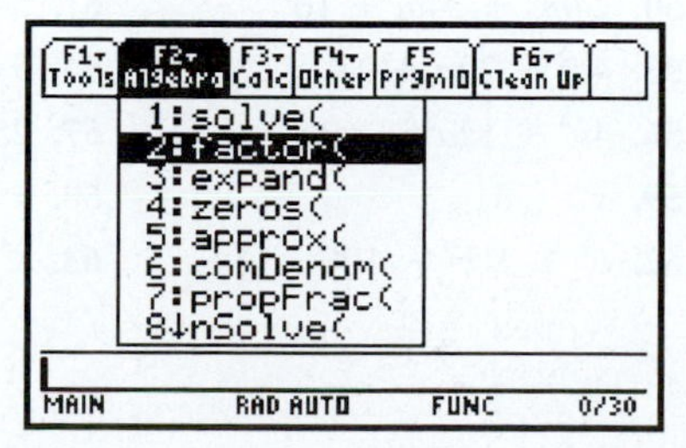

180x^2+279x+108) **ENTER**

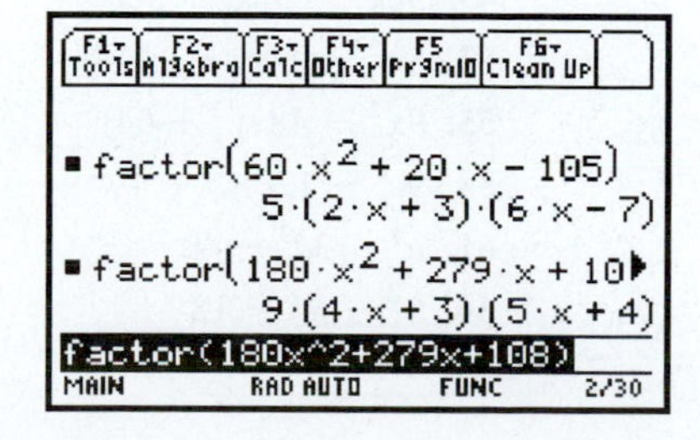

EXAMPLE 26

Factor $9x^2 + 24xy + 16y^2$.

$$9x^2 + 24xy + 16y^2 = (3x + 4y)(3x + 4y) = (3x + 4y)^2$$

EXAMPLE 27

Factor $16x^2 - 8xy + y^2$.

$$16x^2 - 8xy + y^2 = (4x - y)(4x - y) = (4x - y)^2$$

When the factors of a trinomial are the same two binomial factors, the trinomial is called a **perfect square trinomial.** The general forms are as follows:

PERFECT SQUARE TRINOMIALS

$$x^2 + 2xy + y^2 = (x + y)^2$$
$$x^2 - 2xy + y^2 = (x - y)^2$$

Exercises 5.2

Factor each expression completely.

1. $6x + 9y$
2. $15x - 20$
3. $10x + 25y - 45z$
4. $8x^2 - 12xy + 10xz$
5. $12x^2 - 30xy + 6xz$
6. $10x^2 + 25x$
7. $6x^4 - 12x^2 + 3x$
8. $2x^2 - 100x^3 + 4x^4$
9. $27x^3 + 54x$
10. $-16x^3 - 48x^2 - 32x$
11. $8x^3y^2 - 6xy^3 + 12xy^2z$
12. $15x^7y^3 - 6x^6y^2 + 12x^3yz$
13. $x^2 - 16$
14. $x^2 - 64$
15. $9x^2 - 25y^2$
16. $49x^2 - 100y^2$
17. $2e^2 - 72$
18. $20x^2 - 45$
19. $16d^2 - 100$
20. $64x^2 - 4y^2$
21. $4R^2 - 4r^2$
22. $49 - a^4$
23. $x^2 + 6x + 8$
24. $x^2 + 8x + 15$
25. $b^2 + 11b + 24$
26. $m^2 + 11m + 30$
27. $x^2 - 9x + 18$
28. $x^2 - 9x + 14$
29. $a^2 - 18a + 32$
30. $c^2 + 22c + 40$
31. $x^2 - 2x - 35$
32. $x^2 + 4x - 12$
33. $a^2 + 3a - 4$
34. $q^2 - 3q - 28$
35. $2x^2 - x - 6$
36. $5x^2 + 32x + 12$
37. $15x^2 - 31x + 14$
38. $8x^2 + 26x - 45$
39. $45y^2 + 59y + 6$
40. $8m^2 - 14m + 3$
41. $35a^2 + 2a - 1$
42. $16g^2 + 8g + 1$
43. $9c^2 - 24c + 16$
44. $6m^2 - 13m + 5$
45. $25b^2 + 60b + 20$
46. $15t^2 + 69t - 30$
47. $35t^2 - 4ts - 15s^2$
48. $12a^2 + 8ab - 15b^2$
49. $30k^2 - 95kt + 50t^2$
50. $24m^2 + 34m + 10$
51. $54x^2 + 27x - 42$
52. $8x^2 - 24xy - 1440y^2$
53. $4a^2 + 20a + 25$
54. $9x^2 + 42x + 49$
55. $9x^2 - 48xy^2 + 64y^4$
56. $4a^4 + 12a^2b + 9b^2$
57. $25x^2 + 10xy^3 + y^6$
58. $9c^4 - 12c^2y^2 + 4y^4$
59. $x^4 - 81$
60. $x^4 - 16$
61. $a^4 - 3a^2 - 40$
62. $b^4 + 21b^2 - 100$
63. $t^4 - 13t^2 + 36$
64. $x^4 - x^2 - 12$

5.3 OTHER FORMS OF FACTORING

Some algebraic expressions may be factored by grouping their terms so that they are of the types we have already studied. We shall consider those which can be grouped in one of the following ways:

1. With terms having a common binomial factor.
2. As the difference of two squares.
3. In the form of a trinomial.

Finding a common binomial factor is similar to finding a common monomial factor.

EXAMPLE 1

Factor $ax + bx$.

$$ax + bx = x(a + b)$$

EXAMPLE 2

Factor $a(x - y) + b(x - y)$.

$$a(x - y) + b(x - y) = (x - y)(a + b)$$

EXAMPLE 3

Factor $h(x - 1) - 2(x - 1)$.

$$h(x - 1) - 2(x - 1) = (x - 1)(h - 2)$$

In most cases, however, the algebraic expressions are not grouped, so the common binomial factors are not as obvious.

EXAMPLE 4

Factor $3x - 3y + ax - ay$.

Group the first two terms and the last two terms, factor out common monomial factors, and then factor out the resulting binomial factors.

$$\begin{aligned} 3x - 3y + ax - ay &= 3(x - y) + a(x - y) \\ &= (x - y)(3 + a) \end{aligned}$$

EXAMPLE 5

Factor $a^2x - x - 3a^2 + 3$.

$$\begin{aligned} a^2x - x - 3a^2 + 3 &= x(a^2 - 1) - 3(a^2 - 1) \\ &= (a^2 - 1)(x - 3) \\ &= (a + 1)(a - 1)(x - 3) \end{aligned}$$

EXAMPLE 6

Factor $x^3 + x^2 + x + 1$.

$$\begin{aligned} x^3 + x^2 + x + 1 &= x^2(x + 1) + (x + 1) \\ &= (x + 1)(x^2 + 1) \end{aligned}$$

If an algebraic expression can be expressed as the difference of two squares, it can be factored similarly to the form $x^2 - y^2 = (x + y)(x - y)$.

EXAMPLE 7

Factor $(a + b)^2 - 16$.

$$(a + b)^2 - 16 = (a + b + 4)(a + b - 4)$$

EXAMPLE 8

Factor $25a^2 - (a + 2b)^2$.

$$\begin{aligned} 25a^2 - (a + 2b)^2 &= (5a)^2 - (a + 2b)^2 \\ &= [5a + (a + 2b)][5a - (a + 2b)] \\ &= (6a + 2b)(4a - 2b) \\ &= 2(3a + b)(2)(2a - b) \\ &= 4(3a + b)(2a - b) \end{aligned}$$

EXAMPLE 9

Factor $y^2 - 4y + 4 - 25x^2$.

Group the first three terms, which form a perfect square trinomial, factor into a perfect square, and then factor the resulting difference of two squares.

$$\begin{aligned} (y^2 - 4y + 4) - 25x^2 &= (y - 2)^2 - 25x^2 \\ &= (y - 2 + 5x)(y - 2 - 5x) \end{aligned}$$

The last type of group factoring involves algebraic expressions that can be grouped in the form $ax^2 + bx + c$.

EXAMPLE 10

Factor $(a + b)^2 + 8(a + b) + 15$.

Note that $(a + b)^2 + 8(a + b) + 15$ is in the form $x^2 + 8x + 15$, which factors as $(x + 3)(x + 5)$. Therefore,

$$(a + b)^2 + 8(a + b) + 15 = (a + b + 3)(a + b + 5)$$

The last type of factoring we shall study here is factoring the sum and difference of two perfect cubes. The general forms and three examples of each follow.

SUM OF TWO PERFECT CUBES

$$x^3 + y^3 = (x + y)(x^2 - xy + y^2)$$

EXAMPLE 11

Factor $a^3 + 8$.

$$a^3 + 8 = a^3 + 2^3 = (a + 2)(a^2 - 2a + 4)$$

EXAMPLE 12

Factor $y^3 + 27$.

$$y^3 + 27 = y^3 + 3^3 = (y + 3)(y^2 - 3y + 9)$$

EXAMPLE 13

Factor $27x^3 + 64y^6$.

$$27x^3 + 64y^6 = (3x)^3 + (4y^2)^3 = (3x + 4y^2)(9x^2 - 12xy^2 + 16y^4)$$

DIFFERENCE OF TWO PERFECT CUBES

$$x^3 - y^3 = (x - y)(x^2 + xy + y^2)$$

Note the similarities in these two forms.

EXAMPLE 14

Factor $a^3 - 1$.

$$a^3 - 1 = a^3 - 1^3 = (a - 1)(a^2 + a + 1)$$

EXAMPLE 15

Factor $125x^3 - 8y^3$.

$$125x^3 - 8y^3 = (5x)^3 - (2y)^3 = (5x - 2y)(25x^2 + 10xy + 4y^2)$$

EXAMPLE 16

Factor $x^6 - 64$.

$$x^6 - 64 = (x^3 + 8)(x^3 - 8)$$
$$= (x + 2)(x^2 - 2x + 4)(x - 2)(x^2 + 2x + 4)$$

Or, by calculator

F2 2

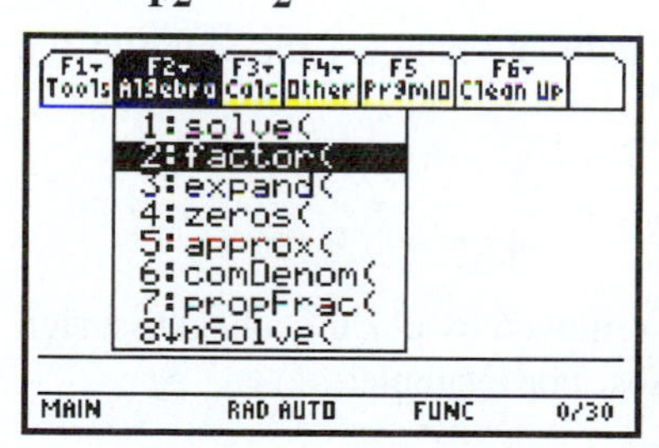

x^6-64) **ENTER**

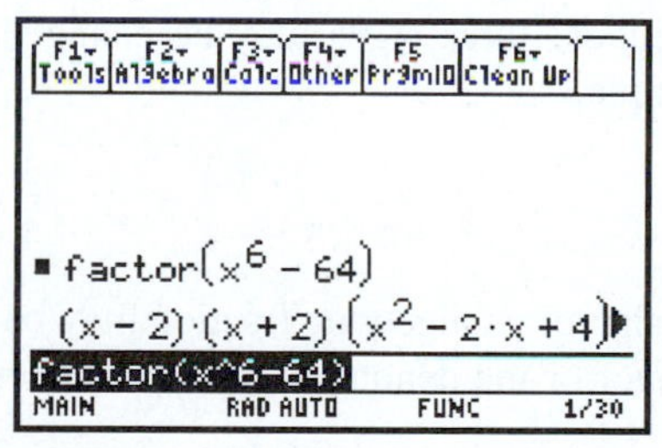

up arrow **2nd** right arrow

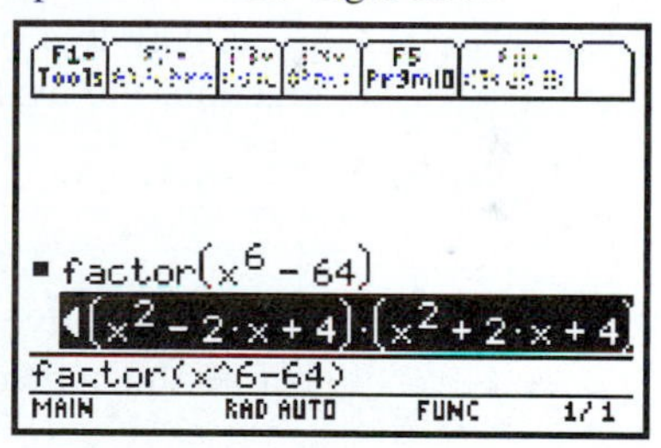

Exercises 5.3

Factor each expression completely.

1. $a(m + n) - b(m + n)$

2. $m(x^2 - 16) - 3(x^2 - 16)$

3. $mx + my + nx + ny$

4. $ab + 3a - bc - 3c$

5. $3x + y - 6x^2 - 2xy$

6. $x^2 + ax + xy + ay$

7. $6x^3 - 4x^2 + 3x - 2$

8. $x^4 - 8x + x^3y - 8y$

9. $(x + y)^2 - 4z^2$

10. $(x - y)^2 - 16a^4$

11. $100 - 49(x - y)^2$

12. $9(a - 4)^2 - 64b^4$

13. $x^2 - 6x + 9 - 4y^2$

14. $x^2 + 10x + 25 - 36y^2$

15. $4x^2 - 4y^2 + 4y - 1$

16. $4 - a^2 - 2ab - b^2$

17. $(x + y)^2 + 13(x + y) + 36$

18. $3(x - y)^2 - 14(x - y) + 8$

19. $24(a + b)^2 - 14(a + b) - 5$

20. $x^2 + 2x(y + z) + (y + z)^2$

21. $a^3 + b^3$

22. $27x^3 + 8y^3$

23. $x^3 - 64$

24. $27x^3 - 8y^3$

25. $a^3b^3 + c^3$

26. $x^3 - y^3z^3$

27. $a^3 - 27b^3$

28. $8a^3 + b^3$

29. $27a^3 + 64b^3$

30. $64x^6 - 125y^3$

31. $27a^6 - 8b^9$

32. $8a^3b^6 + 125c^{12}$

33. $x^6 - y^6$

34. $a^6b^6 - c^6$

5.4 EQUIVALENT FRACTIONS

In order to solve many problems that occur in the technical and applied sciences, it is necessary to use fractions. A **fraction** may be defined as the quotient of two numbers or algebraic expressions. Both the numerator and the denominator of a fraction may be multiplied or divided by the same nonzero number without changing the value of the fraction.

Two fractions are **equivalent** when both the numerator and the denominator of one fraction can be multiplied or divided by the same nonzero number in order to change one fraction to the other. For example, $\frac{2}{3}$ is equivalent to $\frac{6}{9}$, and $\frac{12}{16}$ is equivalent to $\frac{3}{4}$, because

$$\frac{2}{3} = \frac{2 \cdot 3}{3 \cdot 3} = \frac{6}{9} \quad \text{and} \quad \frac{12}{16} = \frac{12 \div 4}{16 \div 4} = \frac{3}{4}$$

A fraction has three signs associated with it:

1. The sign of the fraction.
2. The sign of the numerator.
3. The sign of the denominator.

Any two of these three signs may be changed without changing the value of the fraction. For example,

$$-\frac{-3}{+4} = -\frac{+3}{-4} = +\frac{+3}{+4} = +\frac{-3}{-4}$$

A negative sign of an algebraic fraction may be removed by placing a negative sign before the numerator or the denominator in parentheses. For example,

$$-\frac{a + b - c}{d - 3} = \frac{-(a + b - c)}{d - 3} = \frac{-a - b + c}{d - 3}$$

or

$$-\frac{a + b - c}{d - 3} = \frac{a + b - c}{-(d - 3)} = \frac{a + b - c}{-d + 3} \quad \text{or} \quad \frac{a + b - c}{3 - d}$$

A fraction is in **lowest terms** when its numerator and denominator have no common factors except 1. To reduce a fraction to lowest terms, divide the numerator and the denominator by their common factors.

EXAMPLE 1

Reduce $\dfrac{8xy^3}{12x^4y^2}$ to lowest terms.

The common factors of the numerator and the denominator are $4xy^2$. Dividing both numerator and denominator by $4xy^2$, we have

$$\frac{8xy^3}{12x^4y^2} = \frac{2y}{3x^3}$$

Some prefer to use "cancellation marks" to indicate this division; that is,

$$\frac{\overset{2}{\cancel{8}}\cancel{x}\overset{y}{\cancel{y^3}}}{\underset{3x^3}{\cancel{12}\cancel{x^4}\cancel{y^2}}} = \frac{2y}{3x^3}$$

Remember, *cancellation* is a nontechnical word for the mathematical operation of division. Think *division* whenever these marks are used.

EXAMPLE 2

Reduce $\dfrac{x^2 + 5x + 6}{x^2 + 6x + 9}$ to lowest terms.

First factor the numerator and the denominator.

$$\frac{x^2 + 5x + 6}{x^2 + 6x + 9} = \frac{(x + 2)(x + 3)}{(x + 3)(x + 3)} = \frac{x + 2}{x + 3}$$

EXAMPLE 3

Reduce $\dfrac{4 - 9x^2}{9x^2 + 3x - 2}$ to lowest terms.

$$\frac{4 - 9x^2}{9x^2 + 3x - 2} = \frac{(2 - 3x)(2 + 3x)}{(3x + 2)(3x - 1)} = \frac{2 - 3x}{3x - 1}$$

Note: $2 + 3x = 3x + 2$.

When factors *differ only in sign,* such as $x - y$ and $y - x$, one may be replaced by the opposite of the other, that is,

$$x - y = -(y - x) \quad \text{because} \quad -(y - x) = -y + x = x - y$$

or

$$y - x = -(x - y) \quad \text{because} \quad -(x - y) = -x + y = y - x$$

EXAMPLE 4

Reduce $\dfrac{x^2 - 2x - 8}{16 - x^2}$ to lowest terms.

$$\begin{aligned}\frac{x^2 - 2x - 8}{16 - x^2} &= \frac{(x - 4)(x + 2)}{(4 + x)(4 - x)}\\ &= \frac{-(4 - x)(x + 2)}{(4 + x)(4 - x)} \qquad [\textit{Note: } x - 4 = -(4 - x)]\\ &= -\frac{x + 2}{4 + x} \quad \text{or} \quad \frac{-x - 2}{4 + x}\end{aligned}$$

The TI-89 **factor** command reduces a fraction to lowest terms, writing the simplified answer in factored form. The **comDenom** command also reduces a fraction, but writes the simplified answer in expanded form.

F2 2 (x^2-2x-8)/(16-x^2)) **ENTER** **F2 6** (x^2-2x-8)/(16-x^2)) **ENTER**

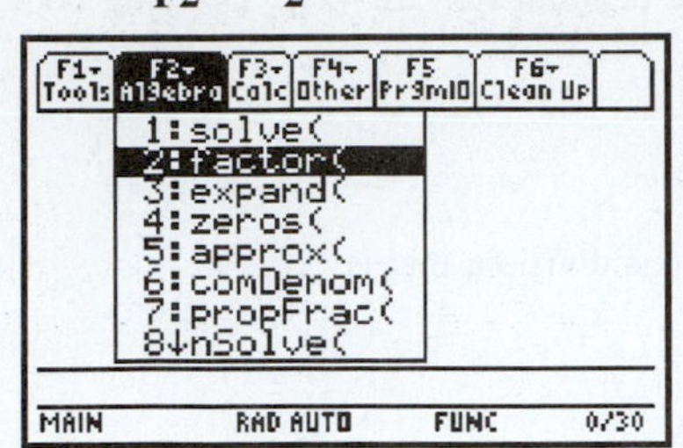

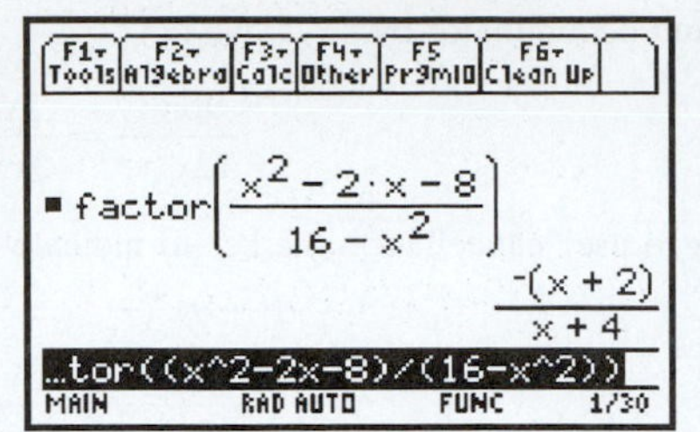

EXAMPLE 5

Reduce $\dfrac{30 - 19x - 5x^2}{5x^2 + 14x - 24}$ to lowest terms.

$$\frac{30 - 19x - 5x^2}{5x^2 + 14x - 24} = \frac{(5 + x)(6 - 5x)}{(5x - 6)(x + 4)}$$

$$= \frac{(5 + x)(6 - 5x)}{-(6 - 5x)(x + 4)} \qquad [\textit{Note: } 5x - 6 = -(6 - 5x)]$$

$$= -\frac{5 + x}{x + 4} \quad \text{or} \quad \frac{-x - 5}{x + 4}$$

Exercises 5.4

Reduce each fraction to lowest terms.

1. $\dfrac{3x}{18x}$
2. $\dfrac{18xy}{24y}$
3. $\dfrac{30x^2y^4}{54x^2y^5}$
4. $\dfrac{27x^2y^3z^4}{57x^5y^7}$
5. $\dfrac{8(x + 4)^3}{2(x + 4)^5}$
6. $\dfrac{6x^2(2x - 1)^6}{2x(2x - 1)^2}$
7. $\dfrac{18(x - 3)^3(1 - 5x)^2}{9(1 - 5x)^2(x - 3)}$
8. $\dfrac{12(3x - 4)^3(2x + 5)^2}{3(2x + 5)^3(3x - 4)^5}$
9. $\dfrac{5m + 15}{6m + 18}$
10. $\dfrac{6x^2 + 3x}{12x + 6}$
11. $\dfrac{6m + 6}{m^2 - 1}$
12. $\dfrac{2t - t^2}{4t - t^2}$
13. $\dfrac{x^2 + x}{x + 1}$
14. $\dfrac{3x^2 + 21x}{6x^2 - 12x}$
15. $\dfrac{x^2 + 5x + 4}{x^2 - 6x - 7}$
16. $\dfrac{x^2 - 2x - 24}{2x^2 + 7x - 4}$
17. $\dfrac{6x^2 + x - 12}{6x^2 + 19x - 36}$
18. $\dfrac{6x^2 + 19x + 10}{6x^2 - 5x - 6}$
19. $\dfrac{18 - 12x}{6x - 9}$
20. $\dfrac{15x - 25}{50 - 30x}$
21. $\dfrac{x^2 - 1}{1 - x}$
22. $\dfrac{6x^3 - 5x^2 - 4x}{16 - 9x^2}$
23. $\dfrac{a^2 - 4}{a^2 + 4a + 4}$
24. $\dfrac{4c^2 + 12c + 9}{4c^2 - 9}$
25. $\dfrac{m^2 - 16m}{16 - m}$
26. $\dfrac{1 - y}{y^2 - y}$
27. $\dfrac{t^2 - t^3}{t^2 - 1}$
28. $\dfrac{1 - 4x + 4x^2}{4x^2 - 1}$
29. $\dfrac{x^2 + xy}{x^2 + xy - 2x - 2y}$
30. $\dfrac{a^2 - ab + 3a - 3b}{a^2 - ab}$
31. $\dfrac{y^3 - 8}{y^2 + 2y + 4}$
32. $\dfrac{a^2b^2 - 16b^2}{a^2b + 9ab + 20b}$
33. $\dfrac{xy - 3x - 2y + 6}{y^3 - 27}$
34. $\dfrac{x^3 - y^3}{x^2 - y^2}$
35. $\dfrac{x^2 + 4x + 16}{x^3 - 64}$
36. $\dfrac{x^4 - 16}{x^4 - 2x^2 - 8}$

37. $\dfrac{3 - 4x - 4x^2}{4x^2 - 8x + 3}$ **38.** $\dfrac{y^2 - x^2}{x^2 - 2xy + y^2}$

39. $\dfrac{3a^3 + 3b^3}{3a^2 + 6ab + 3b^2}$ **40.** $\dfrac{x^2 - 2xy + y^2 - z^2}{z^2 + y^2 - x^2 + 2yz}$

5.5 MULTIPLICATION AND DIVISION OF ALGEBRAIC FRACTIONS

The product of two or more fractions is a fraction whose numerator is the product of the numerators and whose denominator is the product of the denominators, that is,

$$\frac{a}{b} \cdot \frac{c}{d} = \frac{ac}{bd}$$

It is usually helpful to first factor each of the terms of each numerator and denominator; divide by those common factors, if any; and then multiply the numerators and multiply the denominators. Cancellation marks are useful for showing each completed division.

EXAMPLE 1

Multiply $\dfrac{3xy^2}{5ab} \cdot \dfrac{20a^2}{6x^2y}$

$$\frac{\overset{1}{\cancel{3}}\,\overset{1}{\cancel{x}}\,\overset{y}{\cancel{y^2}}}{\underset{1}{\cancel{5}}\,\underset{1}{\cancel{a}}\,b} \cdot \frac{\overset{\overset{2}{\cancel{4}}}{\cancel{20}}\,\overset{a}{\cancel{a^2}}}{\underset{\underset{1}{\cancel{2}}}{\cancel{6}}\,\underset{x}{\cancel{x^2}}\,\underset{1}{\cancel{y}}} = \frac{2ay}{bx}$$

EXAMPLE 2

Multiply $\dfrac{x^2 - 4x + 4}{x^2 + x - 6} \cdot \dfrac{x^2 + 3x}{x^2 - x}$.

Factor each numerator and denominator.

$$\frac{x^2 - 4x + 4}{x^2 + x - 6} \cdot \frac{x^2 + 3x}{x^2 - x} = \frac{\overset{x-2}{\cancel{(x-2)^2}}}{\underset{1}{\cancel{(x+3)}}\underset{1}{\cancel{(x-2)}}} \cdot \frac{\overset{1}{\cancel{x}}\overset{1}{\cancel{(x+3)}}}{\underset{1}{\cancel{x}}(x-1)} = \frac{x-2}{x-1}$$

The **comDenom** command on the TI-89 combines several fractional expressions into a single, simplified fraction in expanded form. If factored form is desired, use **factor(comDenom(***expression***))**. Despite its name, this feature does *not* compute lowest common denominators or least common multiples. Instead, it is intended to add, subtract, multiply, and divide fractions.

F2 6 (x^2-4x+4)/(x^2+x-6)*((x^2+3x)/(x^2-x)) **ENTER** up arrow (*twice*) **2nd** right arrow

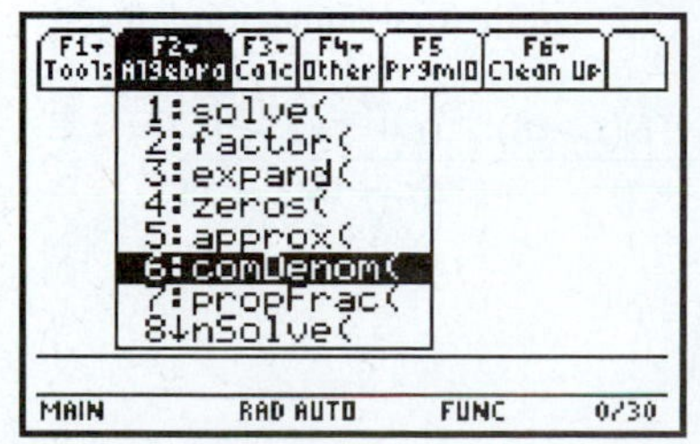

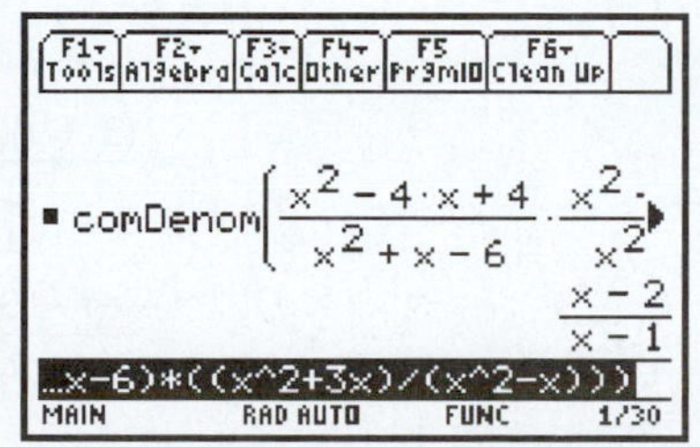

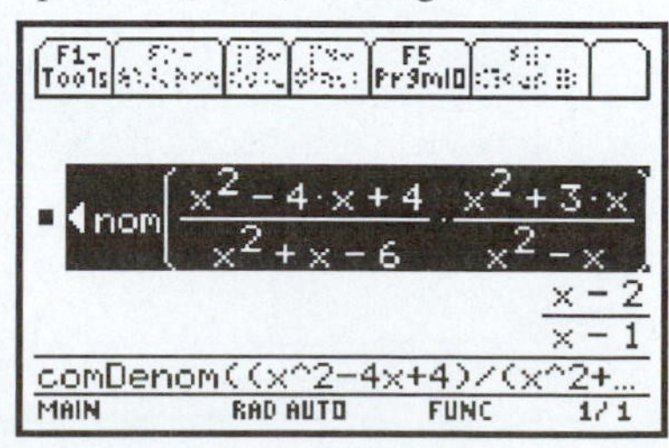

To divide one fraction by another, "invert the divisor" and multiply, or multiply the first fraction by the reciprocal of the second.

$$\frac{a}{b} \div \frac{c}{d} = \frac{a}{b} \cdot \frac{d}{c} = \frac{ad}{bc}$$

EXAMPLE 3

Divide $\frac{xy}{a^2b} \div \frac{x^2y}{ab^3}$.

$$\frac{xy}{a^2b} \div \frac{x^2y}{ab^3} = \frac{xy}{a^2b} \cdot \frac{ab^3}{x^2y} = \frac{\overset{1\,1}{\cancel{x}\cancel{y}}}{\underset{a\,1}{\cancel{a^2}\cancel{b}}} \cdot \frac{\overset{1\,b^2}{\cancel{a}\cancel{b^3}}}{\underset{x\,1}{\cancel{x^2}\cancel{y}}} = \frac{b^2}{ax}$$

EXAMPLE 4

Divide $\frac{x^2 - 9}{4x^2 - 9} \div \frac{2x^2 - x - 15}{2x^2 + x - 6}$.

$$\frac{x^2 - 9}{4x^2 - 9} \div \frac{2x^2 - x - 15}{2x^2 + x - 6} = \frac{x^2 - 9}{4x^2 - 9} \cdot \frac{2x^2 + x - 6}{2x^2 - x - 15}$$

$$= \frac{(x + 3)\overset{1}{\cancel{(x - 3)}}}{(2x + 3)\underset{1}{\cancel{(2x - 3)}}} \cdot \frac{\overset{1}{\cancel{(2x - 3)}}(x + 2)}{\underset{1}{\cancel{(x - 3)}}(2x + 5)}$$

$$= \frac{(x + 3)(x + 2)}{(2x + 3)(2x + 5)}$$

Note: Results left in factored form are usually preferred.

EXAMPLE 5

Divide $\frac{18 + 3t - t^2}{1 + 2t + t^2} \div \frac{2t^2 - 14t + 12}{t^2 - 1}$.

$$\frac{18 + 3t - t^2}{1 + 2t + t^2} \div \frac{2t^2 - 14t + 12}{t^2 - 1} = \frac{18 + 3t - t^2}{1 + 2t + t^2} \cdot \frac{t^2 - 1}{2t^2 - 14t + 12}$$

$$= \frac{(3 + t)(6 - t)}{(1 + t)^2} \cdot \frac{(t + 1)(t - 1)}{2(t - 6)(t - 1)}$$

[Replace $6 - t$ by $(-1)(t - 6)$ in the first numerator.]

$$= \frac{(3 + t)(-1)\overset{1}{\cancel{(t - 6)}}}{\underset{1 + t}{\cancel{(1 + t)^2}}} \cdot \frac{\overset{1}{\cancel{(t + 1)}}\overset{1}{\cancel{(t - 1)}}}{2\underset{1}{\cancel{(t - 6)}}\underset{1}{\cancel{(t - 1)}}}$$

$$= \frac{-t - 3}{2(t + 1)}$$

EXAMPLE 6

Perform the indicated operations.

$$\frac{x^2+xy}{x^2+3xy+2y^2}\cdot\frac{x^2-xy+y^2}{x^2-2xy+y^2}\div\frac{x^4+xy^3}{(x^2-y^2)^2}$$

$$=\frac{x^2+xy}{x^2+3xy+2y^2}\cdot\frac{x^2-xy+y^2}{x^2-2xy+y^2}\cdot\frac{(x^2-y^2)^2}{x^4+xy^3}$$

$$=\frac{\overset{1}{\cancel{x}}\overset{1}{\cancel{(x+y)}}}{\underset{1}{\cancel{(x+y)}}(x+2y)}\cdot\frac{\overset{1}{\cancel{x^2-xy+y^2}}}{\underset{1}{\cancel{(x-y)^2}}}\cdot\frac{\overset{x+y}{\cancel{(x+y)^2}}\overset{1}{\cancel{(x-y)^2}}}{\underset{1}{\cancel{x}}\underset{1}{\cancel{(x+y)}}\underset{1}{\cancel{(x^2-xy+y^2)}}}$$

$$=\frac{x+y}{x+2y}$$

Exercises 5.5

Perform the indicated operations and simplify.

1. $\frac{9}{16}\cdot\frac{4}{3}$
2. $\frac{16}{25}\cdot\frac{5}{32}$
3. $\frac{15}{32}\div\frac{3}{16}$
4. $\frac{5}{9}\div\frac{5}{18}$
5. $\frac{x^4}{3}\cdot\frac{6x}{x^5}$
6. $\frac{4t^4}{6t}\cdot\frac{12t^2}{9t^3}$
7. $\frac{6a^4}{a^2}\div\frac{18a^2}{a^5}$
8. $\frac{8c^2}{4c^3}\div\frac{6c^4}{9c}$
9. $\frac{3ab}{4x^2y}\cdot\frac{6x^2}{5ab^3}$
10. $\frac{15pq^2}{13m^5n^3}\cdot\frac{39mn^4}{5p^4q^3}$
11. $\frac{4ax^2}{9b^2y}\div\frac{5(ax)^2}{18(by^2)^2}$
12. $\frac{(5a)^2}{6x}\div\frac{5a^3}{-6x^2}$
13. $\frac{(2a^2b^3)^2}{(4ab^2)^3}\cdot\frac{(8a^2b)^2}{12a^3b}$
14. $\frac{(3x^2y)^3}{(6x^2y^2)^2}\div\frac{(12x^3y^4)^2}{(9x^2)^3}$
15. $\frac{4x+8}{8xy}\cdot\frac{16x}{32x+64}$
16. $\frac{2x}{8x+4}\cdot\frac{16x+8}{6}$
17. $\frac{5k+5}{12}\div\frac{9k+9}{4}$
18. $(8c-16)\div\frac{3c-6}{12}$
19. $\frac{(y+2)^2}{y}\cdot\frac{y^2}{y^2-4}$
20. $\frac{9a}{(a-1)^2}\cdot\frac{a^2-1}{a^2}$
21. $(m^2-1)\cdot\frac{1+m}{1-m}$
22. $(s^2-9)\div\frac{3-s}{3s}$
23. $\frac{x^2-64}{x+2}\div\frac{8-x}{x}$
24. $\frac{y^2-25}{5y}\cdot\frac{10}{5-y}$
25. $\frac{x-2}{x-y}\cdot\frac{x^2-y^2}{x^2-4x+4}$
26. $\frac{x+1}{x+y}\cdot\frac{x^2-y^2}{x^2-1}$
27. $\frac{x}{(x-3)^2}\cdot\frac{9-x^2}{x^4}$
28. $\frac{x^2-y^2}{z^2-y^2}\cdot\frac{xy-xz}{y-x}$
29. $\frac{x^2+6x+9}{x^2-9}\cdot\frac{x^2-6x+9}{x-3}$
30. $\frac{t^2-10t+25}{t^2-25}\div\frac{t^2+10t+25}{t+5}$
31. $\frac{x^2-x-12}{x^2+2x+1}\cdot\frac{x^2-4x-5}{x^2-9x+20}$
32. $\frac{x^2-x-2}{x^2+7x+6}\cdot\frac{x^2+3x-18}{x^2-4x+4}$
33. $\frac{9x^2-4}{6x^2-5x-6}\div\frac{9x^2-12x+4}{6x^2-13x+6}$
34. $\frac{10x^2+11x+3}{3x^2-19x+28}\div\frac{4x^2-8x-5}{6x^2-29x+35}$
35. $\frac{15a^2+7ab-2b^2}{6a^2-11ab-10b^2}\cdot\frac{2a^2-13ab+20b^2}{25a^2-b^2}$
36. $\frac{18m^2+27mn+10n^2}{6m^2+19mn+10n^2}\div\frac{12m^2-8mn-15n^2}{16m^2+34mn-15n^2}$

37. $\dfrac{x^2 - y^2}{x^2 + 2xy + y^2} \div \dfrac{x^2 - 2xy + y^2}{xy - y^2}$

38. $\dfrac{x^2y - 16y^3}{xy^2 - 4y^3} \div \dfrac{x^2 + 3xy - 4y^2}{y - x}$

39. $\dfrac{x^2 - 16}{x^2 - 9} \div \dfrac{3x^2 + 13x + 4}{2x^2 + x - 21}$

40. $\dfrac{25a^2b^2 - 49}{4ab + 1} \div \dfrac{5ab + 7}{16a^2b^2 + 16ab + 3}$

41. $\dfrac{x^3 - 8}{x^2 - 2x + 4} \cdot \dfrac{x^3 + 8}{x^2 + 2x + 4}$

42. $\dfrac{(x - y)^3}{(x + y)^2} \cdot \dfrac{x^2 + 2xy + y^2}{x^2 - 2xy + y^2}$

43. $\dfrac{xy + y - x^2 - x}{12xy} \div \dfrac{x^2 + x}{4x^2}$

44. $\dfrac{2x^2 + 10x + xy + 5y}{x^2 - 25} \div \dfrac{x^2 + 5x + 25}{x^3 - 125}$

45. $\dfrac{x^3 + 27}{x^2 - 9} \cdot \dfrac{5x^2 + x - 18}{5x^2 - 9x} \div \dfrac{x^2 - 3x + 9}{5x^2 - 6x - 27}$

46. $\dfrac{x^2 - 16}{x^2 - x - 12} \cdot \left(\dfrac{x + 3}{x + 4}\right)^2 \div \dfrac{x^3 + 27}{x^3 + 64}$

47. $\dfrac{18x^2 + 9x - 20}{6x^2 - 5x - 4} \div \left(\dfrac{2x^2 - 9x - 5}{9x^2 - 16} \cdot \dfrac{12x^2 - 10x}{4x^2 + 4x + 1}\right)$

48. $\dfrac{2x^2 - 9x + 7}{x^2 - 1} \div \left(\dfrac{2x^2 + 9x + 4}{x^2 + 4x + 3} \cdot \dfrac{x + 3}{x + 4}\right)^2$

5.6 ADDITION AND SUBTRACTION OF ALGEBRAIC FRACTIONS

Fractions may be added or subtracted if they have a common denominator. If they do not have a common denominator, the lowest common denominator is the most convenient to use because it involves the least amount of computation.

FINDING THE LOWEST COMMON DENOMINATOR (L.C.D.)

1. Factor each denominator into its prime factors, that is, factor each denominator completely.
2. Then the L.C.D. is the product formed by using each of the different factors the greatest number of times that it occurs in any *one* of the given denominators.

EXAMPLE 1

Find the L.C.D. for $\dfrac{3}{18}, \dfrac{15}{60}$, and $\dfrac{19}{27}$.

The prime factors of the denominators are

$$18 = 2 \cdot 3 \cdot 3$$
$$60 = 2 \cdot 2 \cdot 3 \cdot 5$$
$$27 = 3 \cdot 3 \cdot 3$$

Thus,

$$\text{L.C.D.} = 2 \cdot 2 \cdot 3 \cdot 3 \cdot 3 \cdot 5 = 540$$

because 2 occurs at most twice, 3 occurs at most three times, and 5 occurs at most one time in any *one* given denominator.

EXAMPLE 2

Find the L.C.D. for $\dfrac{1}{4x}, \dfrac{5}{6x^2y}$, and $\dfrac{-7}{15y^3}$.

The prime factors of the denominators are

$$4x = 2 \cdot 2 \cdot x$$
$$6x^2y = 2 \cdot 3 \cdot x \cdot x \cdot y$$
$$15y^3 = 3 \cdot 5 \cdot y \cdot y \cdot y$$

So,

$$\text{L.C.D.} = 2 \cdot 2 \cdot 3 \cdot 5 \cdot x \cdot x \cdot y \cdot y \cdot y = 60x^2y^3$$

because 2 occurs at most twice, 3 occurs at most once, 5 occurs at most once, x occurs at most twice, and y occurs at most three times in any *one* given denominator.

EXAMPLE 3

Find the L.C.D. for

$$\frac{3x}{x^2 - 2xy + y^2}, \quad \frac{4y}{x^2 - y^2}, \quad \text{and} \quad \frac{6x + 2y}{x^2 - xy - 2y^2}$$

First, find the prime factors of the denominators.

$$x^2 - 2xy + y^2 = (x - y)(x - y)$$
$$x^2 - y^2 = (x + y)(x - y)$$
$$x^2 - xy - 2y^2 = (x + y)(x - 2y)$$

Then,

$$\text{L.C.D.} = (x - y)(x - y)(x + y)(x - 2y)$$

because in any *one* given denominator the factor $x - y$ occurs at most twice, the factor $x + y$ occurs at most once, and the factor $x - 2y$ occurs at most once.

ADDING OR SUBTRACTING FRACTIONS

1. If the fractions do not have a common denominator, change each fraction to an equivalent fraction having the L.C.D. (For each fraction, divide the L.C.D. by the denominator. Then multiply both numerator and denominator by this result in order to obtain an equivalent fraction having the L.C.D.)
2. Add or subtract the numerators in the order they occur and place this result over the L.C.D.
3. Reduce the resulting fraction to lowest terms, if possible.

EXAMPLE 4

Perform the indicated operations.

$$\frac{5}{6a^2b} - \frac{3}{ab^2} + \frac{1}{4a^3}$$

The L.C.D. is $12a^3b^2$. Divide $12a^3b^2$ by each denominator.

$$\frac{12a^3b^2}{6a^2b} = 2ab \qquad \frac{12a^3b^2}{ab^2} = 12a^2 \qquad \frac{12a^3b^2}{4a^3} = 3b^2$$

Multiply each numerator and denominator by the corresponding result.

$$\frac{5}{6a^2b} - \frac{3}{ab^2} + \frac{1}{4a^3} = \frac{5}{6a^2b} \cdot \frac{2ab}{2ab} - \frac{3}{ab^2} \cdot \frac{12a^2}{12a^2} + \frac{1}{4a^3} \cdot \frac{3b^2}{3b^2}$$
$$= \frac{10ab - 36a^2 + 3b^2}{12a^3b^2}$$

EXAMPLE 5

Perform the indicated operations and simplify.

$$\frac{x-4}{2x}+\frac{x+2}{x+1}-\frac{3x+2}{x^2+x}$$

The L.C.D. is $2x(x+1)$.

$$\frac{x-4}{2x}+\frac{x+2}{x+1}-\frac{3x+2}{\underset{x(x+1)}{x^2+x}}=\frac{x-4}{2x}\cdot\frac{x+1}{x+1}+\frac{x+2}{x+1}\cdot\frac{2x}{2x}-\frac{3x+2}{x(x+1)}\cdot\frac{2}{2}$$

$$=\frac{(x^2-3x-4)+(2x^2+4x)-(6x+4)}{2x(x+1)}$$

$$=\frac{3x^2-5x-8}{2x(x+1)}$$

$$=\frac{(3x-8)(x+1)}{2x(x+1)}$$

$$=\frac{3x-8}{2x}$$

Or, by calculator, we have

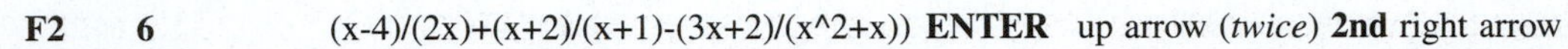
F2 **6** (x-4)/(2x)+(x+2)/(x+1)-(3x+2)/(x^2+x)) **ENTER** up arrow (*twice*) **2nd** right arrow

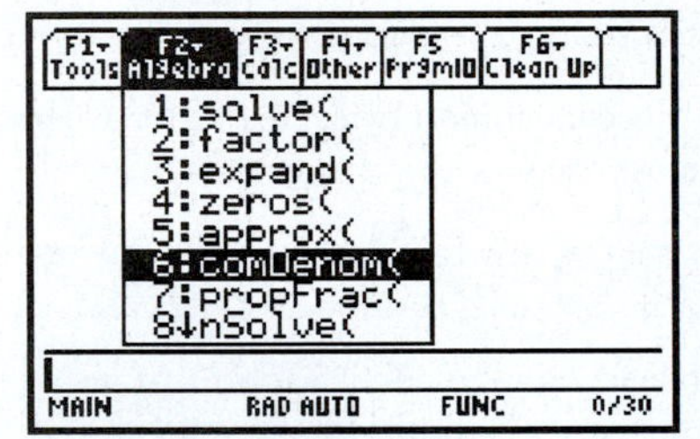

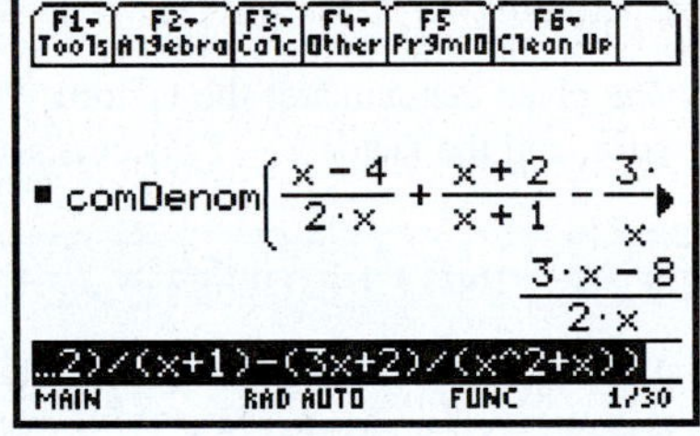

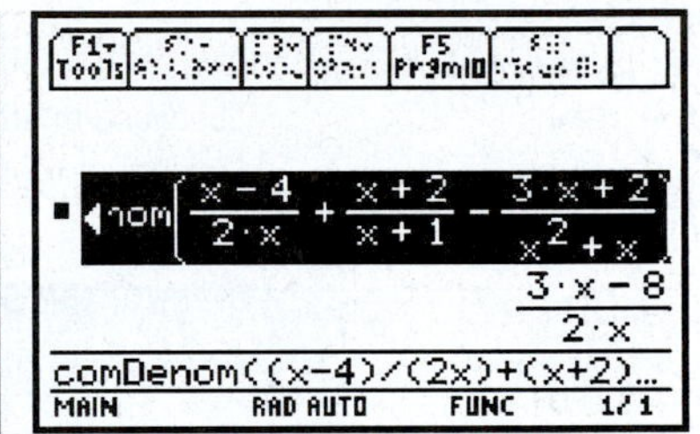

(See the note about the **comDenom** command in Example 2 of Section 5.5.)

EXAMPLE 6

Perform the indicated operations and simplify.

$$\frac{4x}{\underset{(2x+1)(2x-1)}{4x^2-1}}-\frac{3x+1}{\underset{-(2x-1)}{1-2x}}+\frac{2x}{1+2x}$$

Note: If we write $1-2x$ as $-(2x-1)$, the L.C.D. is $(2x+1)(2x-1)$.

$$\frac{4x}{(2x+1)(2x-1)}+\frac{3x+1}{2x-1}\cdot\frac{2x+1}{2x+1}+\frac{2x}{1+2x}\cdot\frac{2x-1}{2x-1}$$

$$=\frac{4x+(6x^2+5x+1)+(4x^2-2x)}{(2x+1)(2x-1)}$$

$$=\frac{10x^2+7x+1}{(2x+1)(2x-1)}$$

$$=\frac{(2x+1)(5x+1)}{(2x+1)(2x-1)}$$

$$=\frac{5x+1}{2x-1}$$

Exercises 5.6

Find the lowest common denominator for each set of denominators.

1. $6x, 8$

2. $15, 20m$

3. $2k, 4k^2$

4. $3t, 12t^3$

5. $6ab^2, 5a^2b$

6. $4ab^3c^4, 12a^2bc, 16ac^2$

7. $5x, x - 1$

8. $x + 3, 3x$

9. $3x + 6, 6x + 12$

10. $4a - 12, 6a - 18$

11. $x^2 - 25, (x - 5)^2$

12. $b^2 - 16, (b + 4)^2$

13. $x^2 + 6x + 8, x^2 - x - 6$

14. $6x^2 + 3x - 1, 6x^2 - 5x - 4$

15. $3x^3 - 12x^2, 6x^2 + 12x, 2x^3 - 4x^2 - 16x$

16. $18a(a + 5)^2, 6(a^2 - 25), 5a - 25$

17. $6(c + 2)(c - 4)^2, 12c(c + 2)(c - 4), 3c(c - 4)^2$

18. $t^2(t + 3)^2, t^2(t - 3)^2, t(t^2 - 9)$

Perform the indicated operations and simplify.

19. $\frac{3}{14} + \frac{5}{14}$

20. $\frac{9}{20} - \frac{3}{20}$

21. $\frac{16}{a} - \frac{9}{a}$

22. $\frac{x}{3a} + \frac{5x}{3a}$

23. $\frac{6}{x + 1} + \frac{9}{x + 1}$

24. $\frac{6a}{2x + 3} - \frac{8a}{2x + 3}$

25. $\frac{7}{36} + \frac{7}{45}$

26. $\frac{19}{24} - \frac{31}{60}$

27. $\frac{2x - 1}{12} + \frac{3x + 5}{20} - \frac{x}{4}$

28. $\frac{3a + b}{6} - \frac{2a - b}{8} + \frac{5a}{12}$

29. $\frac{5}{6y} - \frac{7}{2y}$

30. $\frac{2}{3y} + \frac{1}{y^2}$

31. $\frac{4}{3p} + \frac{2}{p^2}$

32. $\frac{3}{4t} - \frac{5}{3t}$

33. $\frac{5}{4x^2} + \frac{1}{3x^3} - \frac{3}{2x}$

34. $\frac{5}{3a^2b} - \frac{3}{2a} - \frac{1}{4ab^3}$

35. $\frac{1}{a} + \frac{1}{b} + \frac{1}{c}$

36. $\frac{a}{b} + \frac{c}{a} + 1$

37. $3x + \frac{4 - x}{2x}$

38. $9 - \frac{2x - 1}{3y}$

39. $\frac{2}{c - 1} + \frac{1}{c}$

40. $\frac{4}{s - 3} - \frac{1}{s}$

41. $\frac{6}{t + 2} - \frac{3}{t - 2}$

42. $\frac{5}{x - 5} + \frac{2}{x + 5}$

43. $\frac{2}{a} + \frac{1}{a + 1}$

44. $\frac{2}{a + 3} - \frac{4}{a + 2}$

45. $\frac{d}{d + 4} - \frac{2d}{d - 5}$

46. $\frac{3t}{t - 2} + \frac{4t}{t + 4}$

47. $\frac{8}{x - 4} + \frac{2}{4 - x}$

48. $\frac{c}{c - 6} + \frac{2c}{6 - c}$

49. $\frac{4}{3a + 9} - \frac{6}{4a + 12}$

50. $\frac{12}{5r - 25} - \frac{5}{3r - 15}$

51. $\frac{5a + 1}{a^2 - 1} - \frac{2a - 3}{a + 1}$

52. $\frac{2x + 5}{x + 6} + \frac{14x}{x^2 - 36}$

53. $\frac{6}{x - 3} - \frac{5}{3 - x}$

54. $\frac{2x}{x^2 - 1} + \frac{2 - x}{1 - x}$

55. $\frac{2}{r^2 - 4r + 4} + \frac{1}{r^2 + r - 6}$

56. $\frac{4}{y^2 - 5y + 6} - \frac{3}{y^2 - y - 2}$

57. $\frac{1}{x - 2} + \frac{1}{x^2 - 5x + 6}$

58. $\frac{1}{6x} + \frac{1}{3x - 6} - \frac{1}{2x + 4}$

59. $\frac{t + 4}{2t - 4} - \frac{2t + 5}{t^2 - t - 2} + \frac{3}{4}$

60. $\frac{2}{x^2 - y^2} - \frac{1}{x^2 - 3xy + 2y^2} - \frac{3}{x^2 - xy - 2y^2}$

61. $\frac{3x - 2}{x^2 - 3x - 4} + \frac{2x - 3}{x^2 - x - 12}$

62. $\frac{3x + 7}{6x^2 + x - 35} - \frac{3x - 7}{6x^2 + 29x + 35}$

63. $\frac{a^2 + b^2}{a^2 - b^2} - \frac{a}{a + b} + \frac{b}{b - a}$

64. $\frac{2x}{x^2 - 4} - \frac{2 + x}{2 - x} - \frac{x - 2}{x + 2}$

65. $\dfrac{x+3}{x^2-7x+12} + \dfrac{x+3}{16-x^2}$

66. $\dfrac{3x^2-18x+5}{2x^2+x-1} - \dfrac{2x+1}{1-2x} - \dfrac{3x-7}{x+1}$

67. $\dfrac{5x}{8x+2} - \dfrac{3}{4x^2+x} + \dfrac{1}{2x}$

68. $\dfrac{1}{x^2+2x+1} + \dfrac{3}{2x+2} - \dfrac{5}{4x+4}$

69. $\dfrac{1}{t-2} - \dfrac{6t}{t^3-8}$

70. $\dfrac{c}{1-c^3} + \dfrac{1}{c^2-1} - \dfrac{2}{c^2+c+1}$

71. $\dfrac{x^2+x}{x^3+1} + 1$

72. $\dfrac{x^2-1}{x^3-1} + \dfrac{x-1}{x^2+x+1}$

5.7 COMPLEX FRACTIONS

A **complex fraction** is a fraction that contains a fraction in the numerator, the denominator, or both. There are basically two ways to simplify a complex fraction.

Method 1: Multiply both the numerator and the denominator of the complex fraction by the L.C.D. of all the fractions that appear in the numerator and the denominator.

Method 2: Simplify the numerator and the denominator separately. Then divide the simplified numerator by the simplified denominator. Finally, simplify that result, if possible.

EXAMPLE 1

Simplify.

$$\frac{1-\frac{1}{x}}{x-\frac{1}{x^2}}$$

Method 1: The L.C.D. of all the denominators of the fractions in the complex fraction is x^2.

$$\frac{\left(1-\frac{1}{x}\right)x^2}{\left(x-\frac{1}{x^2}\right)x^2} = \frac{x^2-x}{x^3-1} \qquad \text{(Multiply numerator and denominator by } x^2.)$$

$$= \frac{x(x-1)}{(x-1)(x^2+x+1)} = \frac{x}{x^2+x+1}$$

Method 2:

$$\frac{1-\frac{1}{x}}{x-\frac{1}{x^2}} = \frac{\frac{x-1}{x}}{\frac{x^3-1}{x^2}} = \frac{x-1}{x} \div \frac{x^3-1}{x^2}$$

$$= \frac{x-1}{x} \cdot \frac{x^2}{(x-1)(x^2+x+1)} = \frac{x}{x^2+x+1}$$

EXAMPLE 2

Simplify.

$$\frac{x-2+\frac{x-2}{x+2}}{x-\frac{3x+12}{x+2}}$$

Method 1:

$$\frac{\left(x - 2 + \frac{x-2}{x+2}\right)(x+2)}{\left(x - \frac{3x+12}{x+2}\right)(x+2)} = \frac{(x-2)(x+2) + (x-2)}{x(x+2) - (3x+12)}$$

[Multiply numerator and denominator by $(x + 2)$.]

$$= \frac{x^2 - 4 + x - 2}{x^2 + 2x - 3x - 12}$$

$$= \frac{x^2 + x - 6}{x^2 - x - 12}$$

$$= \frac{\cancel{(x+3)}(x-2)}{\cancel{(x+3)}(x-4)}$$

$$= \frac{x-2}{x-4}$$

Method 2:

$$\frac{x - 2 + \frac{x-2}{x+2}}{x - \frac{3x+12}{x+2}} = \frac{\frac{(x-2)(x+2) + (x-2)}{x+2}}{\frac{x(x+2) - (3x+12)}{x+2}}$$

$$= \frac{\frac{x^2 - 4 + x - 2}{x+2}}{\frac{x^2 + 2x - 3x - 12}{x+2}}$$

$$= \frac{\frac{x^2 + x - 6}{x+2}}{\frac{x^2 - x - 12}{x+2}}$$

$$= \frac{(x+3)(x-2)}{x+2} \div \frac{(x-4)(x+3)}{x+2}$$

$$= \frac{\cancel{(x+3)}(x-2)}{\cancel{x+2}} \cdot \frac{\cancel{x+2}}{(x-4)\cancel{(x+3)}}$$

$$= \frac{x-2}{x-4}$$

To simplify an expression containing a complex fraction, first simplify the complex fraction and then perform the remaining indicated operations.

EXAMPLE 3

Simplify.

$$x - \frac{x}{1 - \frac{x}{1-x}} = x - \frac{x}{\frac{1 - x - x}{1-x}}$$

$$= x - \frac{x}{\frac{1-2x}{1-x}} \cdot \frac{(1-x)}{(1-x)}$$

$$= x - \frac{x(1-x)}{1-2x}$$

$$= \frac{x(1-2x)}{1-2x} - \frac{x(1-x)}{1-2x}$$

$$= \frac{(x - 2x^2) - (x - x^2)}{1-2x} = \frac{-x^2}{1-2x}$$

Or, by calculator, we have

F2 **6** x-x/(1-x/(1-x))) **ENTER**

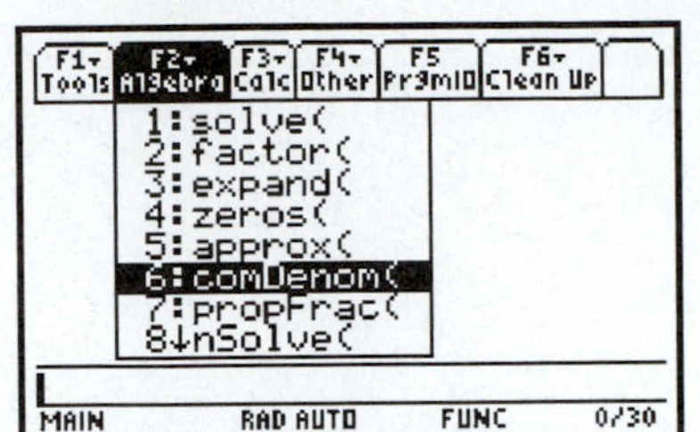

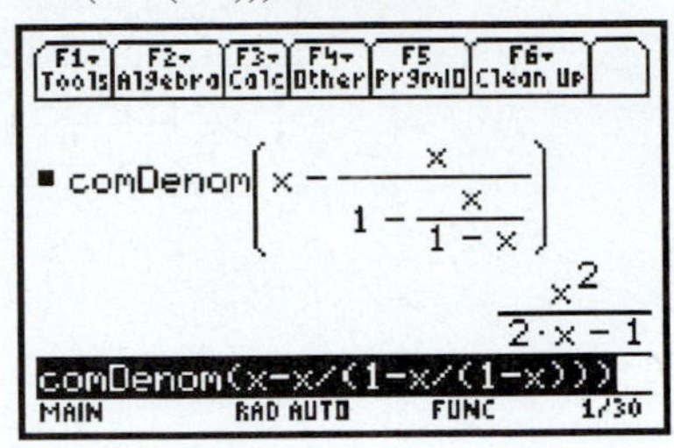

Exercises 5.7

Perform the indicated operations and simplify.

1. $\dfrac{\frac{5}{x}}{\frac{15}{x+2}}$ **2.** $\dfrac{\frac{12}{r-3}}{\frac{3}{2r}}$ **3.** $\dfrac{\frac{t-1}{6t}}{\frac{t+1}{9t}}$ **4.** $\dfrac{\frac{3x^2y}{9z}}{\frac{12xy}{3z}}$

5. $\dfrac{\frac{s+t}{4k}}{\frac{2s+2t}{8k}}$ **6.** $\dfrac{\frac{x-4y}{6x}}{\frac{8x-32y}{12}}$ **7.** $\dfrac{2+\frac{7}{16}}{2-\frac{1}{4}}$ **8.** $\dfrac{\frac{2}{9}-6}{\frac{1}{3}+1}$

9. $\dfrac{a-4}{2-\frac{8}{a}}$ **10.** $\dfrac{a-12}{1-\frac{12}{a}}$ **11.** $\dfrac{a-\frac{25}{a}}{a+5}$ **12.** $\dfrac{a+\frac{27}{a^2}}{a+3}$

13. $\dfrac{\frac{3}{x}+\frac{5}{y}}{\frac{x}{3}+\frac{y}{5}}$ **14.** $\dfrac{\frac{x}{y}-\frac{y}{z}}{\frac{y}{x}-\frac{z}{y}}$ **15.** $\dfrac{1-y}{1-\frac{1}{y}}$ **16.** $\dfrac{\frac{1}{y}-1}{y-\frac{1}{y}}$

17. $\dfrac{1+\frac{1}{a}}{1-\frac{1}{a}}$ **18.** $\dfrac{\frac{2}{c}-1}{\frac{2}{c}+1}$ **19.** $\dfrac{\frac{1}{m}+\frac{1}{n}}{\frac{1}{m}-\frac{1}{n}}$ **20.** $\dfrac{\frac{x}{y}+2}{\frac{x}{y}-\frac{4y}{x}}$

21. $\dfrac{x-3-\frac{28}{x}}{x+10+\frac{24}{x}}$ **22.** $\dfrac{12x+16+\frac{5}{x}}{18x-27-\frac{35}{x}}$ **23.** $\dfrac{\frac{x^2-y^2}{x^3+y^3}}{\frac{x^2+2xy+y^2}{x^2-xy+y^2}}$

24. $\dfrac{\frac{8x^2-2x-21}{6x^2+11x+3}}{\frac{4x^2+20x+25}{12x^3+19x^2+5x}}$ **25.** $\dfrac{\frac{x}{x+y}+\frac{y}{x-y}}{\frac{x^2+y^2}{x^2-y^2}}$ **26.** $\dfrac{\frac{x-y}{x+y}-\frac{x+y}{x-y}}{\frac{x+y}{x-y}-\frac{x-y}{x+y}}$

27. $\dfrac{\frac{6a^2-5ab-4b^2}{a-b}}{\frac{3a^2-7ab+4b^2}{2a+b}}$ **28.** $\dfrac{x+\frac{y^2}{x+y}}{y+\frac{x^2}{x+y}}$ **29.** $\dfrac{1+\frac{1}{x^2-1}}{1+\frac{1}{x-1}}$

30. $\dfrac{3 + \dfrac{5x}{x^2 - 4}}{3 + \dfrac{2}{x - 2}}$

31. $1 - \dfrac{1}{2 - \dfrac{1}{x}}$

32. $3 + \dfrac{1}{4 - \dfrac{2}{y}}$

33. $1 + \dfrac{1}{x + \dfrac{1}{x - 1}}$

34. $1 + \dfrac{x + 1}{x + \dfrac{1}{x + 2}}$

35. $\dfrac{\dfrac{1}{x} - \dfrac{2}{x^2} - \dfrac{3}{x^3}}{x + 6 + \dfrac{5}{x}} - \dfrac{\dfrac{1}{x^2} - \dfrac{9}{x^4}}{1 + \dfrac{2}{x} - \dfrac{15}{x^2}}$

36. $\dfrac{\dfrac{x^2 + 7x + 10}{x + 2}}{x - 6} + \dfrac{\dfrac{x - 2}{x^2 - 2x}}{x - 3}$

5.8 EQUATIONS WITH FRACTIONS

To solve an equation with fractions, multiply each side by the L.C.D. The resulting equation will not contain any fractions and thus may be solved as those equations in Chapter 1.

EXAMPLE 1

Solve.

$$\frac{x}{12} + \frac{1}{6} = \frac{x + 1}{8}$$

The L.C.D. of 12, 6, and 8 is 24; therefore, multiply each side of the preceding equation by 24.

$$24\left(\frac{x}{12} + \frac{1}{6}\right) = 24\left(\frac{x + 1}{8}\right)$$
$$2x + 4 = 3(x + 1)$$
$$2x + 4 = 3x + 3$$
$$1 = x$$

Check: Substitute 1 for x each time it occurs in the *original equation.*

$$\frac{(1)}{12} + \frac{1}{6} = \frac{(1) + 1}{8}$$
$$\frac{1}{12} + \frac{2}{12} = \frac{2}{8}$$
$$\frac{3}{12} = \frac{2}{8}$$
$$\frac{1}{4} = \frac{1}{4}$$

It is common practice in science and technology to solve a given formula for the unknown letter before the substitution of the data is made. To solve a formula involving fractions, follow the same procedure as above.

EXAMPLE 2

Solve for r.

$$\frac{E}{e} = \frac{R + r}{r}$$

The L.C.D. is er; therefore, multiply each side of the preceding equation by er.

$$er\left(\frac{E}{e}\right) = er\left(\frac{R + r}{r}\right)$$

$$Er = eR + er$$

$$Er - er = eR \qquad \text{(Subtract } er \text{ from each side.)}$$

$$(E - e)r = eR \qquad \text{(Factor.)}$$

$$r = \frac{eR}{E - e} \qquad \text{(Divide each side by } E - e.\text{)}$$

The TI-89 is *not* case-sensitive, so there is no difference between an uppercase E and a lowercase e. To write the formula in this example, the capital E will be coded as *ecap* and the capital R will be coded as *rcap* (variable names on the TI-89 can be up to eight letters in length).

ecap/e=(rcap+r)/r,r) **ENTER**

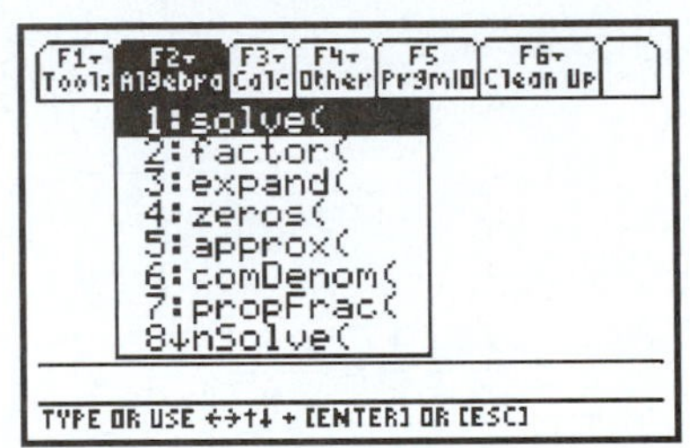

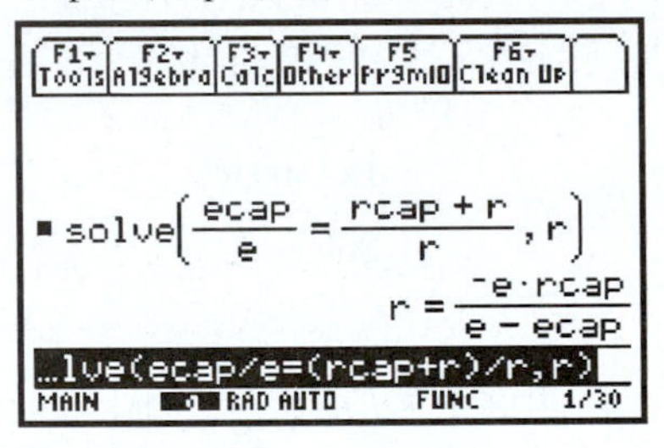

EXAMPLE 3

Solve.

$$\frac{x}{x - 5} = \frac{5}{x - 5} + 2$$

The L.C.D. is $x - 5$; therefore, multiply each side by $x - 5$.

$$(x - 5)\left(\frac{x}{x - 5}\right) = (x - 5)\left(\frac{5}{x - 5} + 2\right)$$

$$x = 5 + 2(x - 5)$$

$$x = 5 + 2x - 10$$

$$x = 2x - 5$$

$$5 = x$$

Check:

$$\frac{5}{5 - 5} = \frac{5}{5 - 5} + 2$$

$$\frac{5}{0} \neq \frac{5}{0} + 2$$

because $\frac{5}{0}$ is meaningless.

The only apparent solution, $x = 5$, does not satisfy the *original* equation. Therefore, the original equation has no solution; that is, there is no real number that can replace x in the original equation and produce a true statement.

The TI-89 indicates that an equation has no solutions by stating that the equation is **false** (not true for any value of the variable).

F2 **1** x/(x-5)=5/(x-5)+2,x) **ENTER**

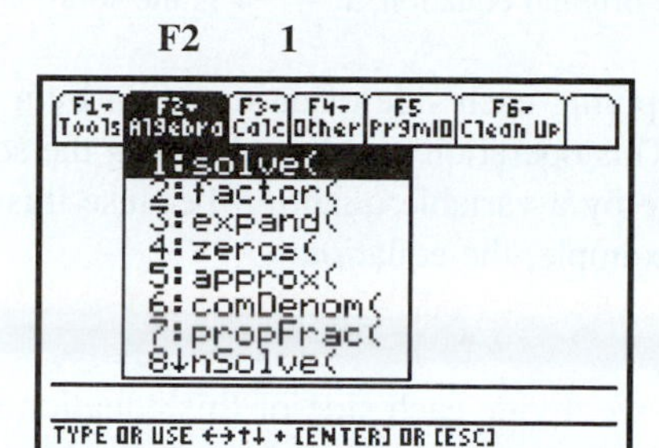

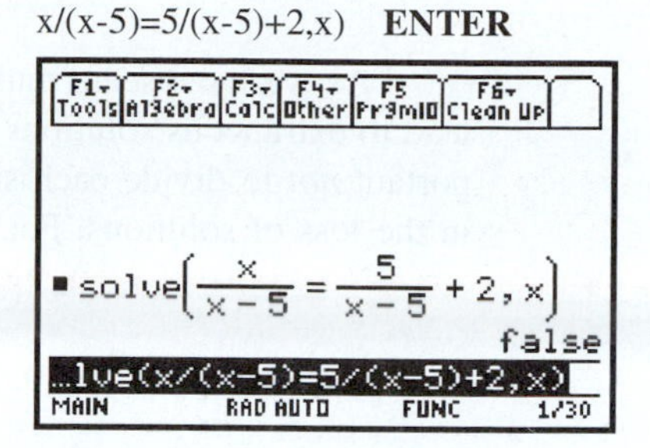

The technique of multiplying each side of an equation by a **variable** quantity gives the possibility of obtaining *extraneous,* or false, solutions.

For example, the simple equation

$$x = 3$$

obviously has only one solution. If we multiply each side by x, we have

$$x^2 = 3x$$

But this equation has two solutions: 3 and 0. Here, 0 is an extraneous solution.

In solving an equation using the technique of multiplying each side by a variable quantity, the only way to determine whether or not an apparent solution is extraneous is to check each apparent solution in the *original* equation.

EXAMPLE 4

Solve.

$$\frac{x}{x+2} + \frac{4}{x-2} = \frac{x^2}{\underset{(x+2)(x-2)}{x^2-4}}$$

The L.C.D. is $(x + 2)(x - 2)$.

$$\begin{aligned} x(x-2) + 4(x+2) &= x^2 \\ x^2 - 2x + 4x + 8 &= x^2 \\ 2x + 8 &= 0 \\ 2x &= -8 \\ x &= -4 \end{aligned}$$

Check: Substitute the apparent solution, $x = -4$, in the original equation.

$$\frac{(-4)}{(-4)+2} + \frac{4}{(-4)-2} = \frac{(-4)^2}{(-4)^2 - 4}$$

$$\frac{-4}{-2} + \frac{4}{-6} = \frac{16}{12}$$

$$2 - \frac{2}{3} = \frac{4}{3}$$

$$\frac{4}{3} = \frac{4}{3}$$

Since $x = -4$ satisfies the original equation, $x = -4$ is the solution, or root.

As we have seen, multiplying each side of an equation by a variable quantity can lead to extraneous solutions. This operation requires checking the solutions. It is also important *not* to divide each side by a variable quantity, because this operation may result in the loss of solutions. For example, the equation

$$x^2 = 5x$$

has two solutions: 5 and 0. If we divide each side of this equation by x, we have

$$x = 5$$

and we have lost a solution.

You should note that there is a great difference between a fractional expression involving addition, subtraction, multiplication, and division of algebraic fractions and an algebraic fractional equation. The only time that you multiply by the L.C.D. to eliminate fractions is when an equal sign is present, which indicates a fractional equation.

Exercises 5.8

Solve and check.

1. $\frac{x}{3} - \frac{x}{6} = 2$

2. $\frac{y}{4} + \frac{y}{12} = 5$

3. $\frac{3m}{4} + \frac{2m}{5} = \frac{23}{2}$

4. $\frac{5t}{4} - \frac{4t}{9} = \frac{29}{4}$

5. $\frac{x}{6} + 2 = \frac{x+3}{4}$

6. $\frac{t-6}{6} + \frac{1}{10} = \frac{t}{15}$

7. $\frac{s+4}{8} - 1\frac{1}{8} = \frac{3s}{4}$

8. $\frac{4m}{5} + 7\frac{2}{3} = \frac{m+5}{15}$

9. $\frac{x+5}{2} = \frac{x+4}{3}$

10. $\frac{x+3}{5} = \frac{x-7}{4}$

11. $\frac{x}{20} + \frac{1}{4} = \frac{x+2}{5}$

12. $\frac{x}{15} + \frac{1}{3} = \frac{x-3}{5}$

13. $\frac{2x+6}{4} + 1 = \frac{3x+1}{8}$

14. $\frac{4x-3}{9} - 2 = \frac{x-4}{3}$

15. $\frac{5-x}{7} - \frac{2x-1}{14} = 1$

16. $\frac{2-x}{24} - \frac{3x+4}{16} = 2$

In Exercises 17 through 28, solve for the indicated letter.

17. $F = \frac{mv^2}{r}$ for r

18. $R = \frac{\rho L}{A}$ for ρ

19. $I_L = \frac{V}{R + R_L}$ for R

20. $\frac{1}{f} = \frac{1}{p} + \frac{1}{q}$ for f

21. $\frac{1}{R} = \frac{1}{R_1} + \frac{1}{R_2}$ for R_1

22. $\frac{1}{R} = \frac{1}{R_1} + \frac{1}{R_2} + \frac{1}{R_3}$ for R_2

23. $y = \frac{x+y}{x}$ for y

24. $y = \frac{x-y}{x}$ for x

25. $\frac{x}{a+b} = \frac{x+a}{b}$ for x

26. $\frac{1}{x+a} - \frac{a}{x-a} = \frac{ax}{x^2 - a^2}$ for x

27. $V = \frac{Q}{R_1} - \frac{Q}{R_2}$ for Q

28. $V = \frac{Q}{R_1} - \frac{Q}{R_2}$ for R_2

Solve and check.

29. $\frac{3}{2x} = \frac{5}{x} - 7$

30. $\frac{7}{2x} - \frac{5}{3} = \frac{1}{x}$

31. $\frac{x-3}{x} = \frac{1}{x} + \frac{2}{3}$

32. $\frac{2x+5}{x} = \frac{2}{x} + \frac{1}{4}$

33. $\frac{1}{x-2} = \frac{3}{x+4}$

34. $\frac{6}{3-2x} = \frac{2}{5-3x}$

35. $\frac{x+2}{x-3} = \frac{x+1}{x-1}$

36. $\frac{6t-3}{2t+1} = \frac{9t}{3t+4}$

37. $\frac{2x^2+8}{x^2-1} - \frac{x-4}{x+1} = \frac{x}{x-1}$

38. $\frac{x-3}{x+3} = \frac{2x+3}{x-3} - \frac{x^2}{x^2-9}$

39. $\frac{x+2}{x+7} = \frac{2}{x-2} + \frac{x^2+3x-28}{x^2+5x-14}$

40. $\frac{1}{x-4} + \frac{1}{x-5} + \frac{1}{x^2-9x+20} = 0$

41. $\frac{x}{x+4} + \frac{x+1}{x-5} = 2$

42. $\frac{-3x^2+5x}{x^2+6x+9} + 4 = \frac{x+5}{x+3}$

43. $\frac{x^2+3x+7}{x^2-x-12} + \frac{x+6}{x+3} = \frac{2x+1}{x-4}$

44. $\frac{7x^2+x-17}{2x^2-5x+3} - \frac{3x-4}{2x-3} = \frac{2x+1}{x-1}$

CHAPTER 5 SUMMARY

1. *Special products:*

(a) *Product of a monomial and a polynomial:*

$$a(x + y + z) = ax + ay + az$$

(b) $(x + y)(x - y) = x^2 - y^2$

(c) *Square of a binomial:*

$$(x + y)^2 = x^2 + 2xy + y^2$$
$$(x - y)^2 = x^2 - 2xy + y^2$$

(d) *Square of a trinomial:*

$$(x + y + z)^2 = x^2 + y^2 + z^2 + 2xy + 2xz + 2yz$$

(e) *Cube of a binomial:*

$$(x + y)^3 = x^3 + 3x^2y + 3xy^2 + y^3$$
$$(x - y)^3 = x^3 - 3x^2y + 3xy^2 - y^3$$

2. *Factoring:*

(a) *Greatest common factor:*

$$ax + ay + az = a(x + y + z)$$

(b) *Difference of two perfect squares:*

$$x^2 - y^2 = (x + y)(x - y)$$

(c) *Trinomials by trial and error:* For a trinomial to be factored in one of the following forms, use the corresponding sign patterns.

$$x^2 + px + q = (x + \quad)(x + \quad)$$
$$x^2 - px + q = (x - \quad)(x - \quad)$$
$$x^2 + px - q = (x + \quad)(x - \quad)$$
$$x^2 - px - q = (x + \quad)(x - \quad)$$

(d) *Perfect square trinomials:*

$$x^2 + 2xy + y^2 = (x + y)^2$$
$$x^2 - 2xy + y^2 = (x - y)^2$$

(e) *Sum and difference of two perfect cubes:*

$$x^3 + y^3 = (x + y)(x^2 - xy + y^2)$$
$$x^3 - y^3 = (x - y)(x^2 + xy + y^2)$$

3. *Equivalent fractions:* Two fractions are equivalent when both numerator and denominator of one fraction can be multiplied or divided by the same nonzero number in order to change one fraction into the other.

4. *Signs of fractions:* A fraction has three signs—the sign of the fraction, the sign of the numerator, and the sign of the denominator. Any two of these three signs may be changed without changing the value of the fraction. For example,

$$-\frac{-a}{+b} = -\frac{+a}{-b} = +\frac{+a}{+b} = +\frac{-a}{-b}$$

5. *Lowest terms:* A fraction is in lowest terms when its numerator and denominator have no common factors except 1.

6. *Factors that differ in sign:* When factors differ in sign, one may be replaced by the opposite of the other, that is,

$$x - y = -(y - x) \quad \text{or} \quad y - x = -(x - y)$$

7. *To multiply fractions:*

$$\frac{a}{b} \cdot \frac{c}{d} = \frac{ac}{bd}$$

8. *To divide fractions:*

$$\frac{a}{b} \div \frac{c}{d} = \frac{a}{b} \cdot \frac{d}{c} = \frac{ad}{bc}$$

9. *To find the lowest common denominator (L.C.D.):*

(a) Factor each denominator completely.

(b) Then the L.C.D. is the product formed by using each of the different factors the greatest number of times that it occurs in any *one* of the given denominators.

10. *Adding or subtracting fractions:*

(a) If the fractions do not have a common denominator, change each fraction to an equivalent fraction having the L.C.D. (For each fraction, divide the L.C.D. by the denominator. Then multiply both numerator and denominator by this result in order to obtain an equivalent fraction having the L.C.D.)

(b) Add or subtract the numerators in the order they occur and place this result over the L.C.D.

(c) Reduce the resulting fraction to lowest terms, if possible.

11. *Complex fraction:* A fraction that contains a fraction in the numerator, the denominator, or both. There are basically two ways to simplify a complex fraction:

Method 1: Multiply both numerator and denominator of the complex fraction by the L.C.D. of all the fractions that appear in the numerator and denominator.

Method 2: Simplify the numerator and the denominator separately. Then divide the simplified numerator by the simplified denominator. Finally, simplify that result, if possible.

12. *To solve an equation with fractions:* Multiply each side by the L.C.D. When you multiply each side by a variable quantity, you *must check* each apparent solution in the original equation.

CHAPTER 5 REVIEW

Find each product mentally.

1. $-6a^2b^4(a^3 - 12ab^2 - 9b^5)$ **2.** $(3x + 13)(5x - 4)$ **3.** $(4x - 7)(4x + 7)$

4. $(5a^2 - 9b^2)^2$ **5.** $(4a - b^2)^3$ **6.** $(6x + 7yz^2)(5x - 2yz^2)$

Factor each expression ***completely.***

7. $18a^3b - 9a^2b^4 + 9a^3b^2$ **8.** $16x^2 - 9y^2$ **9.** $5x^2 + 28x + 32$

10. $14x^2 + 59x - 18$ **11.** $4a^2 - 20a + 25$ **12.** $m^3 - n^3$

13. $3x^2 - 18x - 21$ **14.** $x^4 - 81$ **15.** $15x^2 + 23x - 90$

16. $10x^2 + 25xy - 60y^2$ **17.** $x^2 + 6x + 9 - 25y^2$ **18.** $ax + 2bx - 4ay - 8by$

Reduce each fraction to lowest terms.

19. $\frac{90x^4y^9}{48x^8y^3}$ **20.** $\frac{16x^2 - 40x}{8x^2 - 20x}$ **21.** $\frac{x^2 - 4x - 32}{2x^2 + 11x + 12}$ **22.** $\frac{-2x^2 + 7x + 4}{x^2 - 16}$

Perform the indicated operations and simplify.

23. $\frac{2x^2 + 5x - 25}{4x^2 + 21x + 5} \cdot \frac{4x^2 + x}{4x^2 - 25}$

24. $\frac{6x^2 + 11x - 7}{4x^2 + 4x + 1} \div \frac{3x^2 + 4x - 7}{-2x^2 + x + 1}$

25. $\frac{y^6 + 125}{4y^2 + 20} \div \frac{y^4 - 5y^2 + 25}{12y}$

26. $\frac{2 - x}{x^2 + 10x + 21} \cdot \frac{3x^2 + 21x}{18x + 36} \div \frac{x^3 - 4x}{x + 3}$

27. $\frac{4}{3a} + \frac{9}{4a^2} - \frac{5}{2}$

28. $4 - \frac{3x + 4}{2x + 1}$

29. $\frac{x}{x^2 - 7x + 10} + \frac{x}{2x^2 - 5x + 2}$

30. $\frac{x - 1}{2x^2 + 3x + 1} + \frac{x - 3}{3x^2 - x - 4}$

31. $\dfrac{\frac{1}{x} + \frac{1}{x^2}}{x + \frac{1}{x^2}}$

32. $4 + \dfrac{2}{3 + \frac{1}{y}}$

Solve for x and check.

33. $\frac{x + 2}{5} + \frac{x + 4}{30} = \frac{x}{6}$

34. $\frac{ax + b}{b} = \frac{bx}{a} + \frac{a}{b}$

35. $\frac{x + 1}{x - 3} + 1 = \frac{4}{x - 3}$

36. $\frac{4}{x - 4} - \frac{x^2}{16 - x^2} = 1$

6 Systems of Linear Equations

INTRODUCTION

Joyce owns a field with an irrigation system consisting of two networks of pipes. During the month of July she contracted to have the water turned on the first and third Fridays. On the first Friday, the larger network operated 9 h and the other 8 h, delivering 730 gal of water. On the third Friday, the larger network operated 5 h and the other 7 h, delivering 495 gal. What is the rate of flow (gallons per hour) in each network?

To solve this problem, we need to use two equations with two unknowns to represent all the information. This very powerful problem-solving method involves the use of a *system of equations.* It allows separate conditions or limitations to be expressed in a relatively simple manner while tying all the conditions together in a system.

In this chapter we describe methods of solving systems of linear equations. In the preceding problem, the larger network delivers 50 gal/h, while the smaller delivers 35 gal/h.

Objectives

- Solve systems of equations by substitution.
- Solve systems of equations by the addition-subtraction method.
- Evaluate determinants using determinant properties.
- Use Cramer's rule.
- Use the method of partial fractions to rewrite rational expressions as the sum or the difference of simpler expressions.

6.1 SOLVING A SYSTEM OF TWO LINEAR EQUATIONS

Systems of linear equations in the form

$$a_1x + b_1y = c_1$$
$$a_2x + b_2y = c_2$$

may be solved in a number of ways. In this section we shall study solutions by graphing, by the addition-subtraction method, and by the method of substitution. Any ordered pair (x, y) that satisfies both equations is called a **solution,** or root, of the system.

We saw in Chapter 4 that the graph of a linear equation with two variables is a straight line. The solution of a system of two linear equations with two variables can be discussed in terms of their graphs as follows:

GRAPHS OF LINEAR SYSTEMS OF EQUATIONS WITH TWO VARIABLES

1. The two lines may intersect at a common, single point. This point, in ordered pair form (x, y), is the solution of the system.
2. The two lines may be parallel with no points in common; hence, the system has no solution.
3. The two lines may coincide; the solution of the system is the set of all points on the common line.

These results are illustrated in Fig. 6.1.

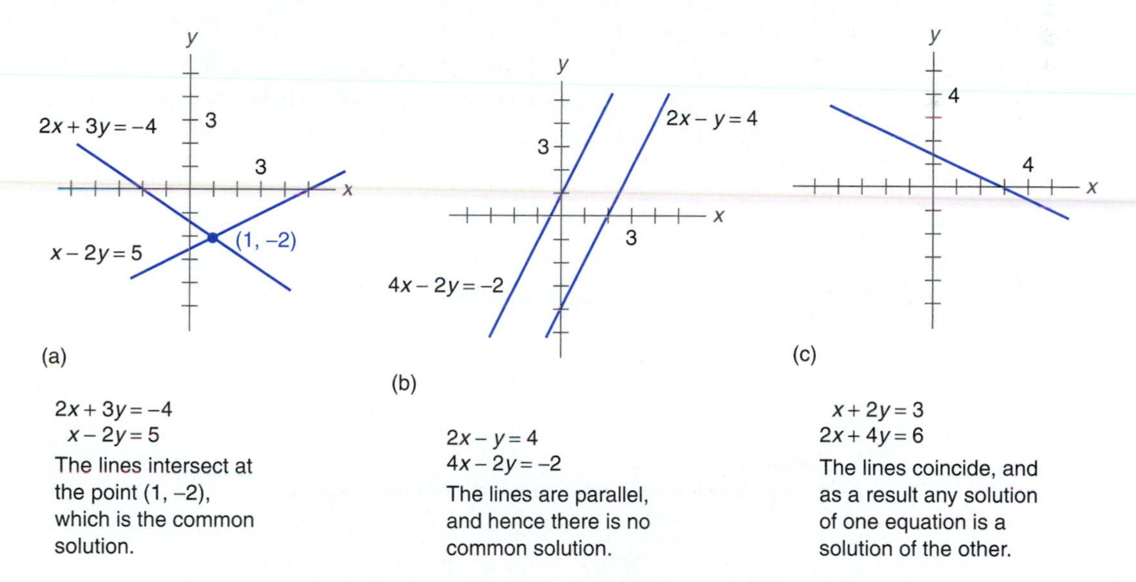

Figure 6.1 General relationships of two straight lines.

When the two lines intersect, the system of equations is called **independent and consistent.** When the two lines are parallel, the system of equations is called **inconsistent.** When the two lines coincide, the system of equations is called **dependent.**

Graphical methods of solving systems of linear equations usually give only approximate solutions. Algebraic methods give exact solutions.* The first algebraic method we shall study is called the **addition-subtraction method** (sometimes called the elimination method).

*Do not use the rules for calculating with measurements in this chapter.

SOLVING A PAIR OF LINEAR EQUATIONS BY THE ADDITION-SUBTRACTION METHOD

1. If necessary, multiply each side of one or both equations by some number so that the numerical coefficients of one of the variables are of equal absolute value.
2. If these coefficients of equal absolute value have like signs, subtract one equation from the other. If they have unlike signs, add the equations. That is, do whatever is necessary to eliminate that variable.
3. Solve the resulting equation for the remaining variable.
4. Substitute the solution for the variable found in Step 3 in either of the original equations, and solve this resulting equation for the second variable.
5. Check by substituting the solution in both original equations.

EXAMPLE 1

Solve

$$\begin{aligned} 2x + 3y &= -4 \\ x - 2y &= 5 \end{aligned}$$

using the addition-subtraction method.

$$\begin{aligned} 2x + 3y &= -4 \\ x - 2y &= 5 \end{aligned}$$

(Multiply each side of this equation by 2, and the numerical coefficients of x will be of equal absolute value.)

$$\begin{aligned} 2x + 3y &= -4 \\ \underline{2x - 4y} &\underline{= 10} \quad \text{(Subtract the equations.)} \\ 7y &= -14 \quad \text{(Solve for } y\text{.)} \\ y &= -2 \end{aligned}$$

Substitute $y = -2$ in either original equation.

$$\begin{aligned} x - 2y &= 5 \\ x - 2(-2) &= 5 \\ x &= 1 \end{aligned}$$

The solution is $(1, -2)$.

Check: Substitute the solution in both original equations.

$$\begin{aligned} 2x + 3y &= -4 & x - 2y &= 5 \\ 2(1) + 3(-2) &= -4 & (1) - 2(-2) &= 5 \\ -4 &= -4 & 5 &= 5 \end{aligned}$$

EXAMPLE 2

Solve

$$\begin{aligned} 3x - 4y &= -36 \\ 10x + 3y &= -22 \end{aligned}$$

using the addition-subtraction method.

$$\begin{aligned} 3x - 4y &= -36 \\ 10x + 3y &= -22 \end{aligned}$$

(Multiply each side of the first equation by 3 and each side of the second equation by 4; the numerical coefficients of y will be of equal absolute value.)

$$
\begin{aligned}
9x - 12y &= -108 \\
\underline{40x + 12y} &\underline{= -88} \qquad \text{(Add the equations.)} \\
49x \qquad &= -196 \qquad \text{(Solve for } x\text{.)} \\
x &= -4
\end{aligned}
$$

Substitute $x = -4$ in either original equation.

$$
\begin{aligned}
3x - 4y &= -36 \\
3(-4) - 4y &= -36 \\
-4y &= -24 \\
y &= 6
\end{aligned}
$$

The solution is $(-4, 6)$.

Check: Substitute the solution in both original equations as before.

The second algebraic method of solving systems of linear equations is called the **method of substitution.**

SOLVING A PAIR OF LINEAR EQUATIONS BY THE METHOD OF SUBSTITUTION

1. From either of the two given equations, solve for one variable in terms of the other.
2. Substitute this result from Step 1 in the *other* equation. Note that this step eliminates one variable.
3. Solve the equation obtained from Step 2 for the remaining variable.
4. From the equation obtained in Step 1, substitute the solution for the variable found in Step 3, and solve this resulting equation for the second variable.
5. Check by substituting the solution in both original equations.

EXAMPLE 3

Solve

$$
\begin{aligned}
3x + 4y &= -15 \\
y &= -2x
\end{aligned}
$$

using the method of substitution.

Since the second equation is already solved for y, substitute $y = -2x$ in the first equation.

$$
\begin{aligned}
3x + 4y &= -15 \\
3x + 4(-2x) &= -15 \qquad \text{(Substitute for } y\text{.)} \\
3x - 8x &= -15 \qquad \text{(Solve for } x\text{.)} \\
-5x &= -15 \\
x &= 3
\end{aligned}
$$

Substitute $x = 3$ in the second equation.

$$
\begin{aligned}
y &= -2x \\
y &= -2(3) \\
y &= -6
\end{aligned}
$$

The solution is $(3, -6)$.

Check: Substitute in both original equations.

EXAMPLE 4

Solve

$$\begin{aligned} 3x + y &= 3 \\ 2x - 4y &= 16 \end{aligned}$$

by the method of substitution.

Solve the first equation for y.

$$\begin{aligned} 3x + y &= 3 \\ y &= -3x + 3 \end{aligned}$$

Substitute $y = -3x + 3$ in the second equation.

$$\begin{aligned} 2x - 4y &= 16 \\ 2x - 4(-3x + 3) &= 16 \qquad \text{(Substitute for } y.) \\ 2x + 12x - 12 &= 16 \qquad \text{(Solve for } x.) \\ 14x &= 28 \\ x &= 2 \end{aligned}$$

Substitute $x = 2$ in the equation $y = -3x + 3$.

$$\begin{aligned} y &= -3x + 3 \\ y &= -3(2) + 3 \\ y &= -3 \end{aligned}$$

The solution is $(2, -3)$.

Check: Substitute the solution in both original equations.

EXAMPLE 5

Solve

$$\begin{aligned} 8x + 5y &= -16 \\ 2x - 3y &= 13 \end{aligned}$$

by the method of substitution.

Solve the second equation for x.

$$\begin{aligned} 2x - 3y &= 13 \\ 2x &= 3y + 13 \\ x &= \frac{3y + 13}{2} \end{aligned}$$

Substitute $x = \dfrac{3y + 13}{2}$ in the first equation.

$$\begin{aligned} 8x + 5y &= -16 \\ 8\left(\frac{3y + 13}{2}\right) + 5y &= -16 \qquad \text{(Substitute for } x.) \\ 4(3y + 13) + 5y &= -16 \qquad \text{(Solve for } y.) \\ 12y + 52 + 5y &= -16 \\ 17y &= -68 \\ y &= -4 \end{aligned}$$

Substitute $y = -4$ in the equation $x = \dfrac{3y + 13}{2}$.

$$x = \frac{3y + 13}{2}$$

$$x = \frac{3(-4) + 13}{2}$$

$$x = \frac{1}{2}$$

The solution is $\left(\frac{1}{2}, -4\right)$.

Check: Substitute the solution in both original equations.

Linear systems of equations (and some nonlinear systems) can be solved directly by the TI-89 using the template for systems found in the **CUSTOM** home screen menus. Just fill in the equations on either side of the word **and** (be sure to leave a blank space before and after **and**).

2nd CUSTOM **F3** **4** (fill in the equations around **and**) **ENTER** up arrow (*twice*) **2nd** right arrow

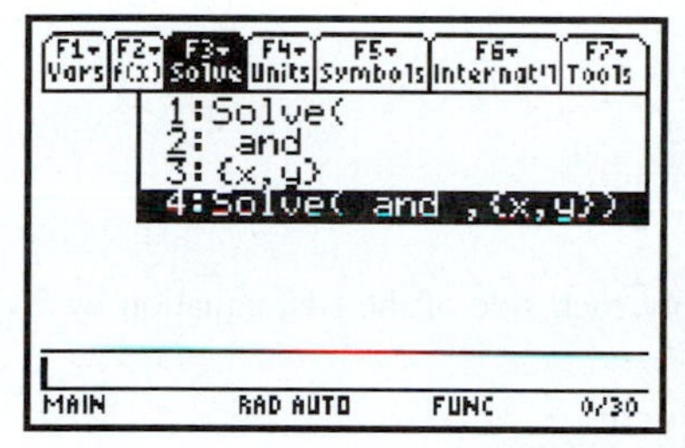

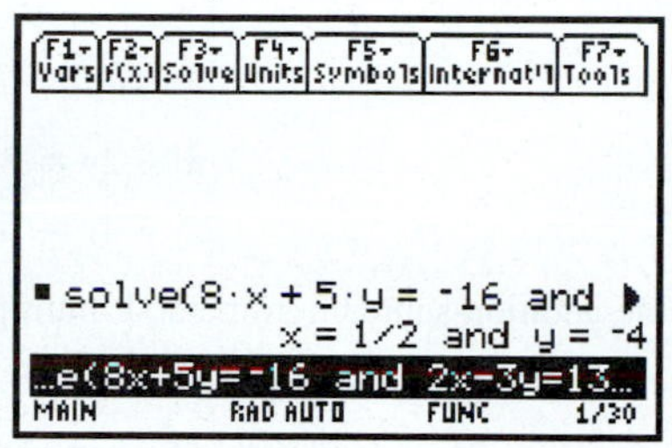

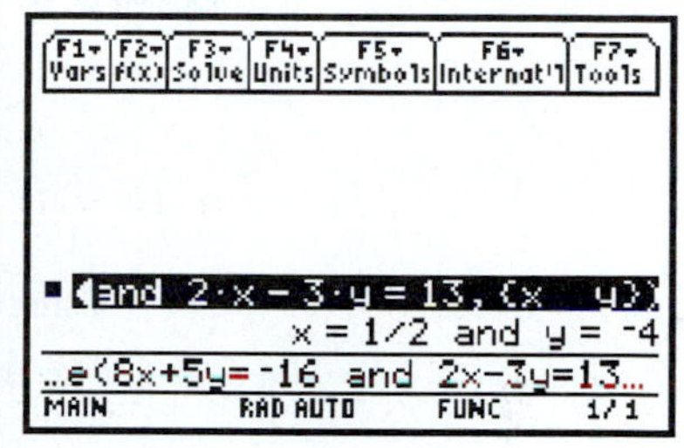

A special case of the substitution method is the **comparison method:**

$$\text{if} \quad a = c \quad \text{and} \quad b = c$$
$$\text{then} \quad a = b$$

EXAMPLE 6

Solve

$$3x - 4 = 5y$$
$$6 - 2x = 5y$$

Since the left side of each equation equals the same quantity, we have

$$3x - 4 = 6 - 2x$$

which eliminates a variable. Then

$$5x = 10$$
$$x = 2$$

Substitute $x = 2$ in the first equation.

$$3x - 4 = 5y$$
$$3(2) - 4 = 5y$$
$$2 = 5y$$
$$\frac{2}{5} = y$$

The solution is $\left(2, \frac{2}{5}\right)$.

Check: Substitute the solution in both original equations.

Note: The basic idea of solving any system of equations is to choose the method that eliminates a variable most easily. This is determined by the form of the given equations.

EXAMPLE 7

Solve

$$\begin{aligned} 2x - y &= 4 \\ 4x - 2y &= -2 \end{aligned}$$

Using the addition-subtraction method, multiply each side of the first equation by 2.

$$\begin{aligned} 4x - 2y &= 8 \\ \underline{4x - 2y} &\underline{= -2} \quad \text{(Subtract the equations.)} \\ 0 &= 10 \end{aligned}$$

When we obtain the result $0 = a$ ($a \neq 0$), this means that there is no solution, the lines when graphed are parallel, and the system of equations is inconsistent.

EXAMPLE 8

Solve

$$\begin{aligned} x + 2y &= 3 \\ 2x + 4y &= 6 \end{aligned}$$

Using the addition-subtraction method, multiply each side of the first equation by 2.

$$\begin{aligned} 2x + 4y &= 6 \\ \underline{2x + 4y} &\underline{= 6} \quad \text{(Subtract the equations.)} \\ 0 &= 0 \end{aligned}$$

When we obtain the result $0 = 0$, this means that the system is dependent and the lines coincide when graphed. Since one equation is a multiple of the other, the equations are equivalent and therefore form the same straight line when graphed.

Many technical applications can be expressed mathematically as a system of linear equations. Examples will be given in this section and in the following sections to demonstrate how to set up an appropriate system of equations for a given problem. Except for Steps 4, 5, and 6, the steps in approaching the solution of a problem stated verbally are the same as for problems involving only one linear equation. You may wish to review Section 1.13 at this time.

STEPS FOR PROBLEM SOLVING

1. Read the problem carefully at least twice.
2. If possible, draw a picture or a diagram.
3. Write what facts are given and what unknown quantities are to be found.
4. Choose a symbol to represent each quantity to be found (there will be more than one). Be sure to label each symbol to indicate what it represents.
5. Write appropriate equations relating these variables from the information given in the problem.* Watch for information that is not stated but which should be assumed. For example, rate $\times$ time $=$ distance.
6. Solve for the unknown variables by the methods presented in this chapter.
7. Check your solution in the original equations.
8. Check your solution in the original verbal problem.

*There should be as many equations as there are unknown variables.

EXAMPLE 9

The sum of two currents is 200 mA. The larger current is three times the smaller. What are the two currents?

Let

$$x = \text{the smaller current}$$
$$y = \text{the larger current}$$

The first sentence gives the first equation that follows; the second sentence gives the second equation.

$$x + y = 200$$
$$y = 3x$$

Substitute the second equation into the first.

$$x + 3x = 200$$
$$4x = 200$$
$$x = 50$$

Substitute $x = 50$ into $y = 3x$.

$$y = 3(50) = 150$$

The solution is (50, 150). That is, the smaller current is 50 mA and the larger current is 150 mA.

The solution checks in both original equations and in the problem as it is originally stated.

EXAMPLE 10

A boat can travel 24 mi upstream in 6 h. It takes 4 h to make the return trip downstream. Find the rate of the current and the speed of the boat (in still water).

Let

$$x = \text{the speed of the boat}$$
$$y = \text{the rate of the current}$$

Then

$x + y =$ the speed of the boat going downstream (the boat is traveling with the current)

$x - y =$ the speed of the boat going upstream (the boat is traveling against the current)

Let's make a chart of the information:

	d	r	t
Traveling downstream	24	$x + y$	4
Traveling upstream	24	$x - y$	6

We now have

$$4(x + y) = 24 \quad \text{(time} \times \text{rate} = \text{distance)}$$
$$6(x - y) = 24$$

This gives us the system

$$4x + 4y = 24$$
$$6x - 6y = 24$$

Use the addition-subtraction method.

$$\begin{aligned} 12x + 12y &= 72 \\ \underline{12x - 12y} &\underline{= 48} \\ 24x \qquad &= 120 \\ x &= 5 \end{aligned}$$

Substitute $x = 5$ in one of the original equations.

$$\begin{aligned} 6x - 6y &= 24 \\ 6(5) - 6y &= 24 \\ 30 - 6y &= 24 \\ -6y &= -6 \\ y &= 1 \end{aligned}$$

The solution is (5, 1). The rate of the current is 1 mi/h and the speed of the boat is 5 mi/h. The solution should be checked in both original equations and in the original problem.

Exercises 6.1

Solve each system of equations graphically by graphing each equation and finding the point of intersection. Check by substituting the solution in both original equations.

1. $2x + 3y = 5$; $x - 2y = 6$

2. $3x - 2y = 0$; $3x + y = -9$

3. $5x + 2y = -5$; $2x + 3y = 9$

4. $-2x + 4y = -6$; $3x - 2y = 9$

5. $9x - 6y = 15$; $-3x + 2y = -5$

6. $-3x + y = 4$; $y = 3x$

Solve each system of equations by the addition-subtraction method. Check by substituting the solution in both original equations.

7. $4x + y = 15$; $3x + y = 13$

8. $-2x + 5y = -30$; $2x - 3y = 22$

9. $-4x + 3y = -22$; $4x - 5y = 34$

10. $7x + 3y = 1$; $2x + 3y = 11$

11. $4x - 5y = -34$; $2x + 3y = 16$

12. $3x - 2y = -3$; $7x - 10y = 1$

13. $3x + 2y = -15$; $x + 5y = -5$

14. $x + 3y = 6$; $5x - 2y = -4$

15. $4x - 5y = 7$; $-2x + 3y = -3$

16. $6x + 7y = -8$; $18x - 4y = 26$

17. $4x + 3y = -3$; $12x + 9y = -12$

18. $2x - y = 6$; $10x - 5y = 30$

19. $3x + 4y = 11$; $-2x + 3y = 21$

20. $5x - 3y = -16$; $4x + 2y = 18$

21. $12x + 5y = -18$; $8x - 7y = -74$

22. $7x - 10y = -6$; $24x - 16y = 16$

23. $-12x + 15y = -43$; $9x - 12y = 34$

24. $24x + 84y = 185$; $42x + 12y = 155$

Solve each system of equations by the method of substitution and check.

25. $2x - 5y = -36$; $y = 4x$

26. $6x - 11y = -3$; $x = 2y$

27. $8x + 9y = -5$; $x = -3y$

28. $12x + 3y = -6$; $y = -5x$

29. $3x - 5y = 64$; $y = 6x - 2$

30. $6x + 8y = 154$; $x = 3y + 4$

31. $3x + 4y = -6$; $2x - y = -15$

32. $2x + 3y = -29$; $x + 4y = -32$

33. $5x - 7y = 23$; $3x + 2y = -11$

34. $6x + 7y = 16$; $-5x - 6y = -13$

Solve each system of equations by any method and check.

35. $\frac{2}{3}x + \frac{1}{2}y = \frac{4}{9}$
$\frac{3}{4}x - \frac{2}{3}y = -\frac{23}{72}$

36. $\frac{5}{4}x - \frac{2}{5}y = -1$
$\frac{1}{4}x + \frac{2}{3}y = 11$

37. $\frac{3}{5}x + 2y = \frac{5}{2}$
$\frac{2}{5}x - 4y = 3$

38. $\frac{7}{3}x + \frac{2}{5}y = -17$
$\frac{2}{5}x + \frac{4}{3}y = 54$

39. $1.4x - 2.7y = 5.66$
$0.5x + 2y = 3.8$

40. $x + y = 40.3$
$0.02x + 0.05y = 1.46$

41. $0.002x + 0.008y = 2.28$
$0.04x + 0.09y = 28.8$

42. $1.57x + 2.04y = 20.262$
$2.16x - 8.42y = -47.342$

43. The sum of two capacitors is 55 microfarads (μF). The difference between them is 25 μF. What is the size of each capacitor?

44. Find the two acute angles in a right triangle if the difference of their measures is 20°.

45. In one concrete mix there is four times as much gravel as cement. If the total volume of the mix is 11.5 m^3, how much of each ingredient is in the mix?

46. The sum of two inductors is 90 millihenries (mH). The larger is 3.5 times the smaller. What is the size of each inductor?

47. A farmer has a 3% solution and an 8% solution of a pesticide. How much of each must he mix to get 2000 L of 4% solution for his sprayer?

48. In testing gasohol mixtures, two mixtures are on hand: mixture A (90% gasoline, 10% alcohol) and mixture B (80% gasoline, 20% alcohol). How much of each mixture must be combined to get 100 gal of an 84% gasoline, 16% alcohol mixture?

49. A boat can travel 20 km upstream in 5 h. It takes 4 h to make the return trip downstream. Find the rate of current and the speed of the boat (in still water).

50. Two planes are 60 km apart. Flying toward each other, they meet in 10 min. When flying in the same direction (at the same speeds as before), the faster plane overtakes the slower plane in 30 min. What is the speed of each plane in kilometres per hour?

51. A rectangular yard has been fenced using 96 m of fence. Find the length and the width of the yard if the length of the yard is 12 m longer than the width.

52. The perimeter of an isosceles triangle is 87 cm. The shortest side is 12 cm less than the longer sides. Find the length of the three sides of the triangle.

53. In a parallel electric circuit, the products of the current and the resistance are equal in all branches. If the total current is 840 mA through any two branches, find the current flowing through branches having resistances of 50 Ω and 300 Ω.

54. The total current in a parallel circuit equals the sum of the currents in the branches. The total current in six parallel branches is 1.90 A. Some of the branches have currents of 0.25 A and others have 0.45 A. How many of each type of branch are in the circuit?

55. The circuit in Fig. 6.2 gives the following system of linear equations:

$$4.3I_1 + 2.3(I_1 + I_2) = 27$$
$$2.8I_2 + 2.3(I_1 + I_2) = 35.2$$

Find I_1 and I_2 in milliamperes.

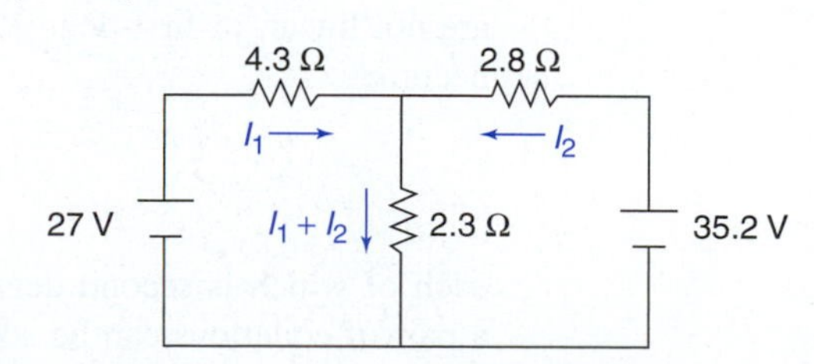

Figure 6.2

56. Harold needs a loan for \$40,000. He receives part at 11% at a credit union. He receives the rest at 14.5% at a bank. If the total annual interest he must pay is \$5520, how much did he receive from each lending institution?

57. Find the distance between the parallel lines $3x + 4y = 25$ and $6x + 8y = 15$.

58. Find the distance between the parallel lines $y = 3x - 4$ and $y = 3x + 5$.

6.2 OTHER SYSTEMS OF EQUATIONS

A **literal equation** is one in which letter coefficients are used in place of numerical coefficients. For example, in the following equations, a and b represent known quantities or coefficients, and x and y are the variables or unknown quantities. To solve such systems of literal equations, use one of the algebraic methods discussed in Section 6.1.

EXAMPLE 1

Solve

$$ax + by = ab$$
$$bx - ay = b^2$$

using the addition-subtraction method.

$$a^2x + aby = a^2b \qquad \text{(Multiply each side of the first equation by } a.)$$
$$b^2x - aby = b^3 \qquad \text{(Multiply each side of the second equation by } b.)$$
$$a^2x + b^2x = a^2b + b^3 \qquad \text{(Add the two equations.)}$$
$$(a^2 + b^2)x = b(a^2 + b^2) \qquad \text{(Factor.)}$$
$$x = \frac{b(a^2 + b^2)}{a^2 + b^2} \qquad \text{(Solve for } x.)$$
$$x = b$$

Substitute $x = b$ in the first equation.

$$ax + by = ab$$
$$a(b) + by = ab$$
$$by = 0$$
$$y = 0$$

The solution is $(b, 0)$.

The equations in the system

$$\frac{6}{x} + \frac{4}{y} = -2$$
$$\frac{9}{x} - \frac{7}{y} = 10$$

are not linear, or first-degree, equations. When each is multiplied by the L.C.D., xy, they become

$$6y + 4x = -2xy$$
$$9y - 7x = 10xy$$

each of which is second degree. However, we must assume that $x \neq 0$ and $y \neq 0$. Such a pair of equations can be solved using the algebraic methods from Section 6.1.

EXAMPLE 2

Solve

$$\frac{6}{x} + \frac{4}{y} = -2$$

$$\frac{9}{x} - \frac{7}{y} = 10$$

using the addition-subtraction method.

$$\frac{18}{x} + \frac{12}{y} = -6 \qquad \text{(Multiply each side of the first equation by 3.)}$$

$$\frac{18}{x} - \frac{14}{y} = 20 \qquad \text{(Multiply each side of the second equation by 2.)}$$

$$\frac{26}{y} = -26 \qquad \text{(Subtract.)}$$

$$26 = -26y \qquad \text{(Solve for } y\text{.)}$$

$$-1 = y$$

Substitute $y = -1$ in the first equation.

$$\frac{6}{x} + \frac{4}{y} = -2$$

$$\frac{6}{x} + \frac{4}{(-1)} = -2$$

$$\frac{6}{x} = 2$$

$$6 = 2x$$

$$3 = x$$

The solution is $(3, -1)$.

Check: Substitute the solution in both original equations.

Note: A system of equations such as the preceding one can also be solved by first letting $A = \frac{1}{x}$ and $B = \frac{1}{y}$. Then solve for A and B, and finally solve for x and y.

Some nonlinear systems are too difficult for direct solution by the TI-89. The system in this example, if entered as originally shown, will take about 1 min of computation time before the calculator (incorrectly) concludes that the system has no solution! The following method makes the substitutions $a = 1/x$ and $b = 1/y$ and solves the resulting linear system for a and b. The calculator then solves those results for x and y.

F4 1

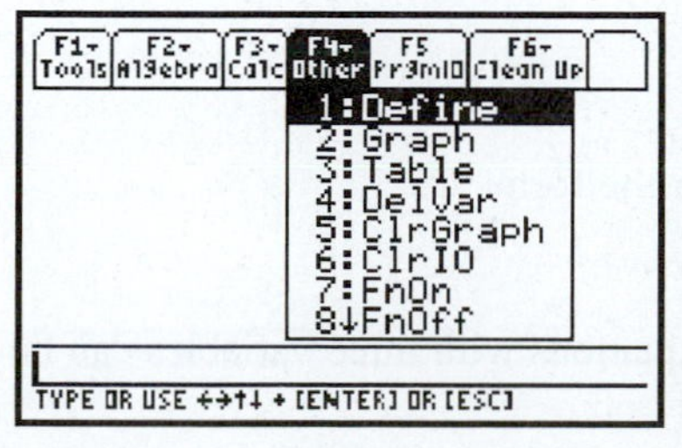

a=1/x **ENTER F4 1** b=1/y **ENTER**

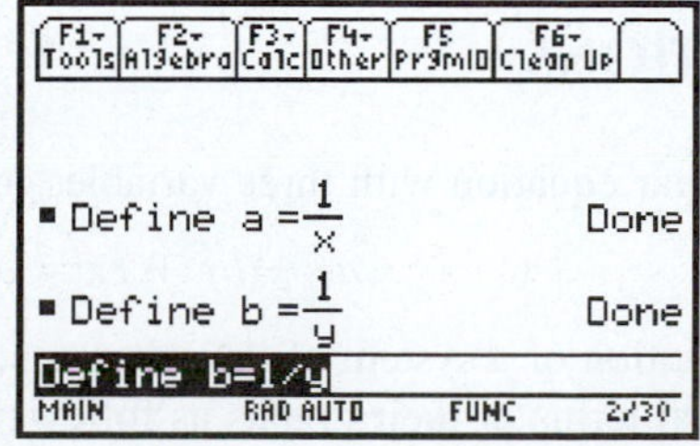

2nd CUSTOM F3 4

F1 Vars | F2 f(x) | F3 Solve | F4 Units | F5 Symbols | F6 Internat'l | F7 Tools
1:Solve(
2: and
3: {x,y}
4:Solve(and ,{x,y})
■ Define b = 1/y Done
Define b=1/y
MAIN RAD AUTO FUNC 2/30

(fill in equations and change to **{a,b}**) **ENTER** **F3** **1** **2nd ANS** **,{x,y})** **ENTER**

```
F1▾ F2▾ F3▾ F4▾ F5▾ F6▾ F7▾
Vars F(x) Solve Units Symbols Internat'l Tools
■ Define a = 1/x                    Done
■ Define b = 1/y                    Done
■ solve(6·a + 4·b = -2 and 9▸
           a = 1/3 and b = -1
…b=-2 and 9a-7b=10, {a,b})
MAIN    RAD AUTO    FUNC    3/30
```

```
F1▾ F2▾ F3▾ F4▾ F5▾ F6▾ F7▾
Vars F(x) Solve Units Symbols Internat'l Tools
■ Def 1:Solve(
      2: and
      3:{x,y}
      4:Solve( and ,{x,y})
■ Define b = 1/y                    Done
■ solve(6·a + 4·b = -2 and 9▸
           a = 1/3 and b = -1
…b=-2 and 9a-7b=10, {a,b})
TYPE OR USE ←→↑↓ + [ENTER] OR [ESC]
```

```
F1▾ F2▾ F3▾ F4▾ F5▾ F6▾ F7▾
Vars F(x) Solve Units Symbols Internat'l Tools
■ Define b = 1/y                    Done
■ solve(6·a + 4·b = -2 and 9▸
           a = 1/3 and b = -1
■ solve(a = 1/3 and b = -1, {▸
           x = 3 and y = -1
Solve(ans(1), {x,y})
MAIN    RAD AUTO    FUNC    4/30
```

Exercises 6.2

Solve each system of equations.

1. $ax + y = b$
$bx - y = a$

2. $x + y = a - b$
$x - y = a + b$

3. $x + y = a^2 + ab$
$ax = by$

4. $2ax + b^2y = a^2b^4 + 2ab^2$
$y = a^2x$

5. $ax - by = 0$
$bx + ay = 1$

6. $ax + by = 1$
$bx + ay = 1$

7. $ax + by = c$
$bx + ay = c$

8. $ax + by = b - a$
$bx - ay = b - a$

9. $(a + 3b)x - by = a$
$(a - b)x - ay = b$

10. $(a + b)x - (a - b)y = 4ab$
$(a - b)x - (a + b)y = 0$

11. $\frac{3}{x} + \frac{4}{y} = 2$
$\frac{6}{x} + \frac{12}{y} = 5$

12. $\frac{12}{x} - \frac{15}{y} = 7$
$\frac{9}{x} + \frac{30}{y} = -3$

13. $\frac{3}{x} + \frac{2}{y} = 0$
$\frac{2}{x} - \frac{5}{y} = 19$

14. $\frac{6}{x} - \frac{4}{y} = 12$
$\frac{5}{x} - \frac{3}{y} = 11$

15. $\frac{1}{x} + \frac{1}{y} = \frac{13}{6}$
$\frac{1}{x} - \frac{1}{y} = \frac{5}{6}$

16. $\frac{3}{x} + \frac{12}{y} = 32$
$\frac{4}{x} + \frac{9}{y} = 21\frac{2}{3}$

17. $\frac{6}{s} + \frac{4}{t} = 16$
$\frac{9}{s} - \frac{5}{t} = 2$

18. $\frac{24}{m} + \frac{18}{n} = 72$
$\frac{36}{m} + \frac{45}{n} = 116$

19. $\frac{1}{x} + \frac{1}{y} = a$
$\frac{1}{x} - \frac{1}{y} = b$

20. $\frac{a}{x} + \frac{b}{y} = 2$
$\frac{b}{x} + \frac{a}{y} = 1$

6.3 SOLVING A SYSTEM OF THREE LINEAR EQUATIONS

The graph of a linear equation with three variables in the form

$$ax + by + cz = d$$

is a **plane.** The solution of a system of three linear equations with three variables can be discussed in general terms of their graphs as follows:

GRAPHS OF LINEAR SYSTEMS OF EQUATIONS WITH THREE VARIABLES

1. The three planes may intersect at a common, single point. This point, in ordered triple form (x, y, z), is then the solution of the system.
2. The three planes may intersect along a common line. The infinite set of points that satisfy the equation of the line is the solution of the system.
3. The three planes may not have any points in common; the system has no solution. For example, the planes may be parallel, or they may intersect triangularly with no points common to all three planes.
4. The three planes may coincide; the solution of the system is the set of all points in the common plane.

Graphical solutions of three linear equations with three unknowns are not used because three-dimensional graphing is required and is not practical by hand. However, three-dimensional graphing is useful to illustrate the general cases (see Fig. 6.3).

A linear equation with four or more variables cannot be represented in three-dimensional space. However, systems of linear equations with four or more variables have many applications, and the methods presented in this chapter will be extended to solve such systems.

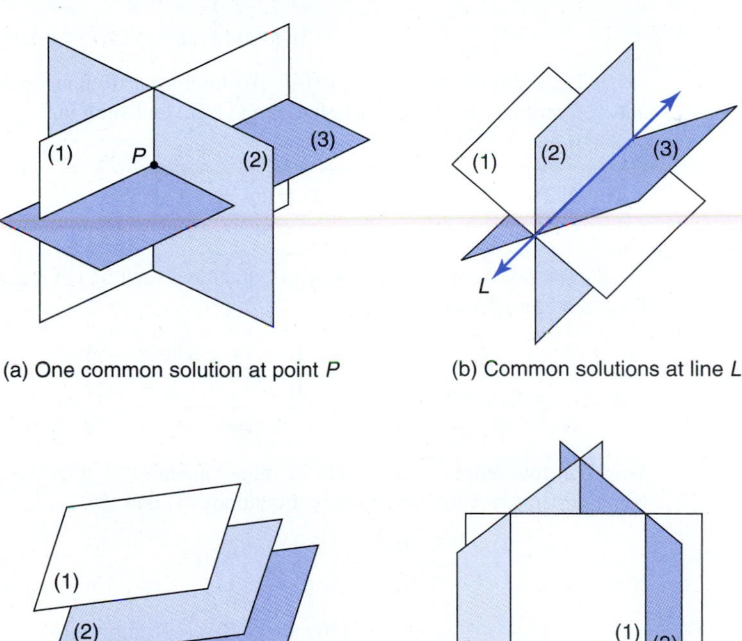

(a) One common solution at point P

(b) Common solutions at line L

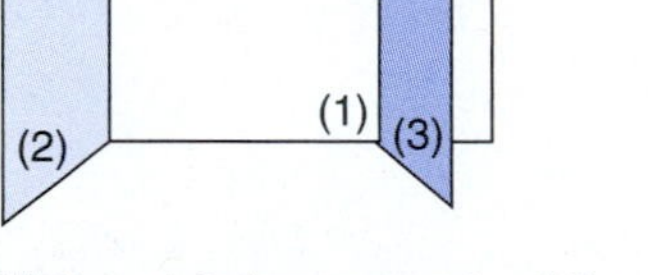

(c) No points in common (parallel planes)

(d) No points in common to all three planes

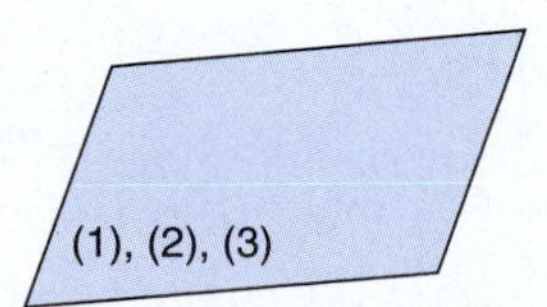

(e) All points in common (planes coincide)

Figure 6.3 General relationships of three planes.

The first method presented here for solving a system of linear equations with three variables is the **addition-subtraction method.**

THE ADDITION-SUBTRACTION METHOD FOR A SYSTEM OF THREE LINEAR EQUATIONS WITH THREE VARIABLES

1. Choose a variable to be eliminated. Eliminate it from any pair of equations by using the techniques of the addition-subtraction method from Section 6.1.
2. Eliminate this same variable from *any other pair* of equations.
3. The result of Steps 1 and 2 is a pair of linear equations in two unknowns. Solve this pair for the two variables.
4. Solve for the third variable by substituting the results from Step 3 in any one of the original equations.
5. Check by substituting the solution in all three original equations.

EXAMPLE 1

Solve

$$x + 2y - 6z = -17 \qquad (1)$$
$$2x - 5y + z = 28 \qquad (2)$$
$$-3x + 4y + 2z = -21 \qquad (3)$$

Let's choose to eliminate x first. To eliminate x from any pair of equations, such as (1) and (2), multiply each side of Equation (1) by 2 and subtract.

$$\begin{aligned} 2x + 4y - 12z &= -34 \\ 2x - 5y + z &= 28 \\ \hline 9y - 13z &= -62 \end{aligned} \qquad (4)$$

To eliminate x from any other pair of equations, such as (1) and (3), multiply each side of Equation (1) by 3 and add.

$$\begin{aligned} 3x + 6y - 18z &= -51 \\ -3x + 4y + 2z &= -21 \\ \hline 10y - 16z &= -72 \end{aligned} \qquad (5)$$

We have now reduced the system of three equations in three variables to a system of two equations in two variables, namely, Equations (4) and (5).

$$9y - 13z = -62 \qquad (4)$$
$$10y - 16z = -72 \qquad (5)$$

$$\begin{aligned} 90y - 130z &= -620 && \text{[Multiply (4) by 10.]} \\ 90y - 144z &= -648 && \text{[Multiply (5) by 9.]} \\ \hline 14z &= 28 && \text{(Subtract.)} \\ z &= 2 \end{aligned}$$

Substitute $z = 2$ in (5).

$$\begin{aligned} 10y - 16z &= -72 \\ 10y - 16(2) &= -72 \\ 10y &= -40 \\ y &= -4 \end{aligned}$$

Substitute $z = 2$ and $y = -4$ in any original equation; we shall use Equation (1).

$$x + 2y - 6z = -17$$
$$x + 2(-4) - 6(2) = -17$$
$$x = 3$$

The solution is $(3, -4, 2)$.

Check: Substitute the solution in all three original equations.

EXAMPLE 2

Solve

$$4x - 2y + z = 12 \quad (1)$$
$$-y + 3z = -11 \quad (2)$$
$$5x + 3y = 45 \quad (3)$$

Let's choose to eliminate z first. To eliminate z from Equations (1) and (2), multiply each side of the first equation by 3 and subtract.

$$\begin{array}{r} 12x - 6y + 3z = 36 \\ -y + 3z = -11 \\ \hline 12x - 5y \quad = 47 \end{array} \quad (4)$$

We now have reduced the system to Equations (3) and (4):

$$5x + 3y = 45 \quad (3)$$
$$12x - 5y = 47 \quad (4)$$

Thus,

$$\begin{array}{rl} 25x + 15y = 225 & \text{[Multiply (3) by 5.]} \\ 36x - 15y = 141 & \text{[Multiply (4) by 3.]} \\ \hline 61x \quad = 366 & \text{(Add.)} \\ x = 6 & \end{array}$$

Substitute $x = 6$ in (3).

$$5x + 3y = 45$$
$$5(6) + 3y = 45$$
$$3y = 15$$
$$y = 5$$

Substitute $x = 6$ and $y = 5$ in any original equation; we shall use Equation (1).

$$4x - 2y + z = 12$$
$$4(6) - 2(5) + z = 12$$
$$z = -2$$

The solution is $(6, 5, -2)$.

Check: Substitute the solution in all three original equations.

Or, to solve linear systems of three variables, modify the **Solve** template to include the word **and** between each of the equations and change the variable list to **{x,y,z}**. The **and** operator is obtained by pressing **F3** then **2** in the **CUSTOM** menu (or press **2nd MATH 8 8**).

2nd CUSTOM F3 4 (fill in equations, an extra **and**, and change to {x,y,z}) **ENTER** up arrows right arrows

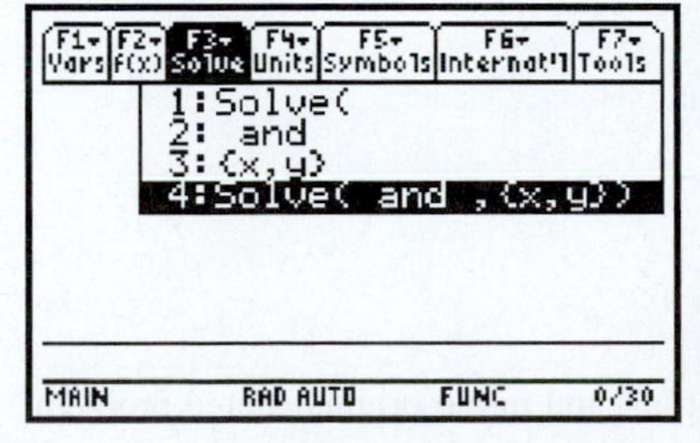

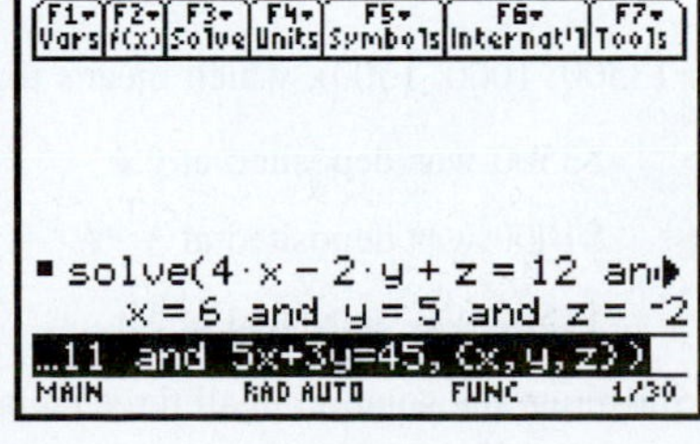

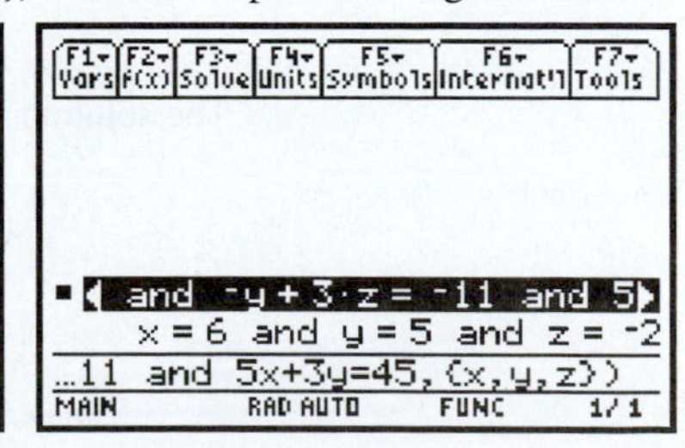

EXAMPLE 3

A person deposits \$6000 into three different savings accounts. One account earns interest at 5%, another at $5\frac{1}{2}\%$, and another at 6%. The total interest earned at the end of 1 year was \$340. How much was deposited into each account if the 6% account earned \$135 more than the 5% account?

Let

$$\begin{aligned} x &= \text{the amount deposited at } 6\% \\ y &= \text{the amount deposited at } 5\tfrac{1}{2}\% \\ z &= \text{the amount deposited at } 5\% \end{aligned}$$

Then

$$x + y + z = 6000 \quad \text{(the total sum deposited)}$$

$$0.06x + 0.055y + 0.05z = 340 \quad \text{(the sum of the interest earned from each account)}$$

$$0.06x - 0.05z = 135 \quad \text{(the difference in earnings between the 6\% and 5\% accounts)}$$

This gives us a system of three linear equations in three unknowns. Multiply each equation by some quantity which will result in removing the decimal points.

$$x + y + z = 6000 \qquad \textbf{(1)}$$

$$60x + 55y + 50z = 340{,}000 \qquad \textbf{(2)}$$

$$6x - 5z = 13{,}500 \qquad \textbf{(3)}$$

Let us choose to eliminate y first. To eliminate y from Equations (1) and (2), multiply each side of (1) by 55 and subtract.

$$\begin{aligned} 55x + 55y + 55z &= 330{,}000 \\ 60x + 55y + 50z &= 340{,}000 \\ \hline -5x + 5z &= -10{,}000 \qquad \textbf{(4)} \end{aligned}$$

We have now reduced the system to Equations (3) and (4).

$$\begin{aligned} 6x - 5z &= 13{,}500 \qquad \textbf{(3)} \\ -5x + 5z &= -10{,}000 \qquad \textbf{(4)} \\ \hline x &= 3500 \qquad \text{(Add.)} \end{aligned}$$

Substitute $x = 3500$ in (3).

$$\begin{aligned} 6x - 5z &= 13{,}500 \\ 6(3500) - 5z &= 13{,}500 \\ 21{,}000 - 5z &= 13{,}500 \\ -5z &= -7500 \\ z &= 1500 \end{aligned}$$

Substitute $x = 3500$ and $z = 1500$ in (1).

$$\begin{aligned} x + y + z &= 6000 \\ (3500) + y + (1500) &= 6000 \\ y &= 1000 \end{aligned}$$

The solution is (3500, 1000, 1500), which means that

\$3500 was deposited at 6%

\$1000 was deposited at $5\frac{1}{2}\%$

\$1500 was deposited at 5%

Check: Substitute the solution in all three equations and in the original stated problem.

Exercises 6.3

Solve each system of equations and check.

1. $$\begin{aligned} x + y + z &= 4 \\ x - y + z &= 0 \\ x - y - z &= 2 \end{aligned}$$

2. $$\begin{aligned} x + y + z &= 9 \\ x - 2y + 3z &= 36 \\ 2x - y + z &= 27 \end{aligned}$$

3. $$\begin{aligned} 3x - 5y - 6z &= -19 \\ 3y + 6z &= 15 \\ 4x - 2y - 5z &= -2 \end{aligned}$$

4. $$\begin{aligned} x + 2z &= -8 \\ y - 3z &= 2 \\ x - y + z &= 2 \end{aligned}$$

5. $$\begin{aligned} 2x + 3y - 5z &= 56 \\ 6x - 4y + 7z &= -42 \\ x - 2y + 3z &= -26 \end{aligned}$$

6. $$\begin{aligned} 5x - y + 3z &= 28 \\ 8x + 2y - 4z &= 24 \\ 9x - 4y + 3z &= 32 \end{aligned}$$

7. $$\begin{aligned} x + y + z &= 16 \\ y + z &= 3 \\ x - z &= 11 \end{aligned}$$

8. $$\begin{aligned} 9x - 7y + 3z &= -48 \\ 2x + 3y + 4z &= -4 \\ -3x + 2y - 5z &= 8 \end{aligned}$$

9. $$\begin{aligned} 2x - 4y + z &= 17 \\ 4x + 5y - z &= -8 \\ x - 3y + 5z &= 16 \end{aligned}$$

10. $$\begin{aligned} x + 3y - 6z &= -1 \\ 2x - y + z &= 10 \\ 5x - 2y + 3z &= 27 \end{aligned}$$

11. $$\begin{aligned} \frac{1}{x} + \frac{1}{y} + \frac{1}{z} &= 3\frac{5}{6} \\ \frac{3}{x} - \frac{5}{y} - \frac{2}{z} &= \frac{10}{3} \\ \frac{2}{x} + \frac{3}{y} + \frac{6}{z} &= 7 \end{aligned}$$

12. $$\begin{aligned} \frac{3}{x} + \frac{2}{y} - \frac{3}{z} &= -\frac{7}{2} \\ \frac{5}{x} + \frac{3}{y} - \frac{4}{z} &= -\frac{53}{12} \\ -\frac{2}{y} + \frac{8}{z} &= 11\frac{1}{6} \end{aligned}$$

13. $$\begin{aligned} 2x + 3y + 4z - w &= 7 \\ -x - 2y + 3z + 2w &= -3 \\ 3x + y - 3w &= -5 \\ 4x + 2z - 5w &= -19 \end{aligned}$$

14. $$\begin{aligned} 5x - 2y - z + 2w &= -44 \\ 3x + 3y - 4z + w &= -27 \\ -2x + 4y + 5z - 2w &= 48 \\ 4x - 5y + 6z + 4w &= -23 \end{aligned}$$

15. The sum of three resistors R_1, R_2, and R_3 connected in series is 1950 ohms (Ω). The sum of R_1 and R_3 is 1800 Ω. If R_1 is eight times R_2, what is the size of each resistor?

16. The following equations are derived from a circuit diagram. Solve for the currents I_1, I_2, and I_3 in amperes (A).

$$\begin{aligned} 5I_1 + 9I_2 \phantom{{}+5I_3} &= 37 \\ 15I_2 + 5I_3 &= 70 \\ I_1 + I_2 - I_3 &= 0 \end{aligned}$$

17. The perimeter of a triangle is 65 cm. The longest side is 5 cm longer than the medium side. The medium side is twice the length of the shortest side. Find the length of each side of the triangle.

18. Seventy-five acres of land were purchased for \$142,500. The land facing the highway cost \$2700 per acre. The land facing the railroad cost \$2200 per acre, and the remainder cost \$1450 per acre. There were 5 acres more facing the railroad than the highway. How much land was sold at each price?

19. Three cylindrical rods are welded together as shown in Fig. 6.4. Find the diameter of each rod.

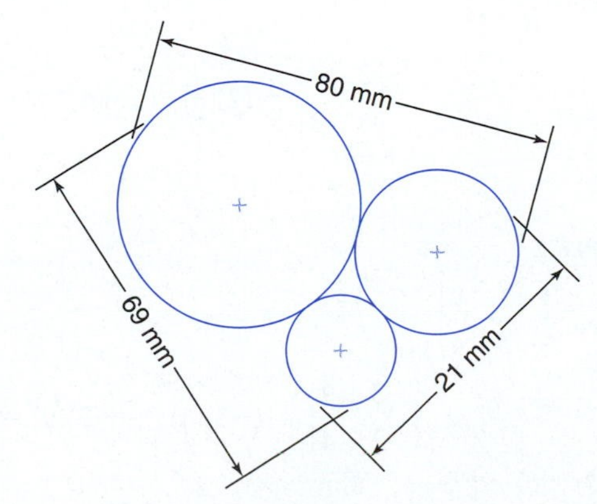

Figure 6.4

20. A chemist has three acid solutions. The first contains 20% acid, the second contains 30% acid, and the third contains 60% acid. She uses all three solutions to obtain a mixture of 80 L containing 40% acid. She uses twice as much of the 60% solution as of the 20% solution. How many litres of each solution does she use?

21. If $f(x) = ax^2 + bx + c$, find a, b, and c such that the graph passes through the points $P_1(1, 4)$, $P_2(3, 10)$, and $P_3(-1, 14)$.

6.4 DETERMINANTS

A **determinant** is a square array of numbers such as

$$\begin{vmatrix} a_1 & b_1 \\ a_2 & b_2 \end{vmatrix}$$

where the array has as many rows as columns. The number of rows or columns determines the order of the determinant. The preceding one is a *determinant of second order.* The numbers a_1, b_1, a_2, and b_2 are called **elements.** The elements a_1 and b_1 form the first row; the elements a_2 and b_2 form the second row; the elements a_1 and a_2 form the first column; and the elements b_1 and b_2 form the second column. The value of a second-order determinant is given by the following:

VALUE OF SECOND-ORDER DETERMINANT

$$\begin{vmatrix} a_1 & b_1 \\ a_2 & b_2 \end{vmatrix} = a_1b_2 - a_2b_1$$

The elements a_1 and b_2 form what is called the **principal diagonal.** The value of a second-order determinant is the product of the elements of the principal diagonal minus the product of the elements of the other diagonal. This can also be shown using the following diagram:

$$\overset{+\qquad\qquad -}{\begin{vmatrix} a_1 & b_1 \\ a_2 & b_2 \end{vmatrix}}$$

EXAMPLE 1

Evaluate $\begin{vmatrix} 2 & 4 \\ 6 & 3 \end{vmatrix}$.

$$\begin{vmatrix} 2 & 4 \\ 6 & 3 \end{vmatrix} = (2)(3) - (6)(4) = 6 - 24 = -18$$

EXAMPLE 2

Evaluate $\begin{vmatrix} -3 & 9 \\ -2 & 4 \end{vmatrix}$.

$$\begin{vmatrix} -3 & 9 \\ -2 & 4 \end{vmatrix} = (-3)(4) - (-2)(9) = -12 + 18 = 6$$

EXAMPLE 3

Evaluate $\begin{vmatrix} 0 & -4 \\ -2 & 7 \end{vmatrix}$.

$$\begin{vmatrix} 0 & -4 \\ -2 & 7 \end{vmatrix} = (0)(7) - (-2)(-4) = 0 - 8 = -8$$

A determinant of third order has three rows and three columns. A **third-order** determinant is in the form

$$\begin{vmatrix} a_1 & b_1 & c_1 \\ a_2 & b_2 & c_2 \\ a_3 & b_3 & c_3 \end{vmatrix}$$

The elements, rows, and columns of a determinant of third order are defined in the same way as those for a determinant of second order. The value of a third-order determinant is defined as

$$\begin{vmatrix} a_1 & b_1 & c_1 \\ a_2 & b_2 & c_2 \\ a_3 & b_3 & c_3 \end{vmatrix} = a_1b_2c_3 + a_3b_1c_2 + a_2b_3c_1 - a_3b_2c_1 - a_1b_3c_2 - a_2b_1c_3$$

Let's rearrange the terms on the right side of the equation as

$$a_1(b_2c_3 - b_3c_2) - a_2(b_1c_3 - b_3c_1) + a_3(b_1c_2 - b_2c_1)$$

These terms can then be expressed as the following 2×2 determinants:

$$a_1\begin{vmatrix} b_2 & c_2 \\ b_3 & c_3 \end{vmatrix} - a_2\begin{vmatrix} b_1 & c_1 \\ b_3 & c_3 \end{vmatrix} + a_3\begin{vmatrix} b_1 & c_1 \\ b_2 & c_2 \end{vmatrix}$$

Each 2×2 determinant is called a **minor** of an element in the 3×3 determinant. In general, the minor of a given element of a determinant is the resulting determinant after the row and the column that contain the element have been deleted, as shown in Table 6.1.

TABLE 6.1

Element	Minor	Original determinant
a_1	$\begin{vmatrix} b_2 & c_2 \\ b_3 & c_3 \end{vmatrix}$	$\begin{vmatrix} a_1 & b_1 & c_1 \\ a_2 & b_2 & c_2 \\ a_3 & b_3 & c_3 \end{vmatrix}$
a_2	$\begin{vmatrix} b_1 & c_1 \\ b_3 & c_3 \end{vmatrix}$	$\begin{vmatrix} a_1 & b_1 & c_1 \\ a_2 & b_2 & c_2 \\ a_3 & b_3 & c_3 \end{vmatrix}$
a_3	$\begin{vmatrix} b_1 & c_1 \\ b_2 & c_2 \end{vmatrix}$	$\begin{vmatrix} a_1 & b_1 & c_1 \\ a_2 & b_2 & c_2 \\ a_3 & b_3 & c_3 \end{vmatrix}$

The value of any determinant of any order may be found by finding the sums and differences of *any* row or column of the products of the elements and the corresponding minors. The following diagram should help you remember how to determine the signs (sum or difference) of the various products of elements and minors:

$$\begin{vmatrix} + & - & + & - & \cdots \\ - & + & - & + & \cdots \\ + & - & + & - & \cdots \\ - & + & - & + & \cdots \\ \cdot & \cdot & \cdot & \cdot & \\ \cdot & \cdot & \cdot & \cdot & \\ \cdot & \cdot & \cdot & \cdot & \end{vmatrix}$$

Note: Begin with a + sign in the upper-left position. Then alternate the signs across each row and down each column.

EXAMPLE 4

Evaluate the determinant

$$\begin{vmatrix} 1 & 3 & -5 \\ 6 & -7 & -2 \\ -4 & 8 & 2 \end{vmatrix}$$

using expansion by minors (a) down the first column and (b) across the second row.

(a) The elements and the corresponding minors down the first column are determined as shown in Table 6.2.

TABLE 6.2

Element	Minor	Original determinant
1	$\begin{vmatrix} -7 & -2 \\ 8 & 2 \end{vmatrix}$	$\begin{vmatrix} 1 & 3 & -5 \\ 6 & -7 & -2 \\ -4 & 8 & 2 \end{vmatrix}$
6	$\begin{vmatrix} 3 & -5 \\ 8 & 2 \end{vmatrix}$	$\begin{vmatrix} 1 & 3 & -5 \\ 6 & -7 & -2 \\ -4 & 8 & 2 \end{vmatrix}$
−4	$\begin{vmatrix} 3 & -5 \\ -7 & -2 \end{vmatrix}$	$\begin{vmatrix} 1 & 3 & -5 \\ 6 & -7 & -2 \\ -4 & 8 & 2 \end{vmatrix}$

The signs down the first column are + − +. Therefore,

$$\begin{aligned}
\begin{vmatrix} 1 & 3 & -5 \\ 6 & -7 & -2 \\ -4 & 8 & 2 \end{vmatrix} &= +1\begin{vmatrix} -7 & -2 \\ 8 & 2 \end{vmatrix} - 6\begin{vmatrix} 3 & -5 \\ 8 & 2 \end{vmatrix} + (-4)\begin{vmatrix} 3 & -5 \\ -7 & -2 \end{vmatrix} \\
&= 1[(-7)(2) - (8)(-2)] - 6[(3)(2) - (8)(-5)] + (-4)[(3)(-2) - (-7)(-5)] \\
&= 1[-14 + 16] - 6[6 + 40] + (-4)[-6 - 35] \\
&= 2 - 276 + 164 \\
&= -110
\end{aligned}$$

(b) The elements and the corresponding minors across the second row are determined as shown in Table 6.3.

TABLE 6.3

Element	Minor	Original determinant
6	$\begin{vmatrix} 3 & -5 \\ 8 & 2 \end{vmatrix}$	$\begin{vmatrix} 1 & 3 & -5 \\ 6 & -7 & -2 \\ -4 & 8 & 2 \end{vmatrix}$
-7	$\begin{vmatrix} 1 & -5 \\ -4 & 2 \end{vmatrix}$	$\begin{vmatrix} 1 & 3 & -5 \\ 6 & -7 & -2 \\ -4 & 8 & 2 \end{vmatrix}$
-2	$\begin{vmatrix} 1 & 3 \\ -4 & 8 \end{vmatrix}$	$\begin{vmatrix} 1 & 3 & -5 \\ 6 & -7 & -2 \\ -4 & 8 & 2 \end{vmatrix}$

The signs across the second row are $- + -$. Therefore,

$$\begin{vmatrix} 1 & 3 & -5 \\ 6 & -7 & -2 \\ -4 & 8 & 2 \end{vmatrix} = -6\begin{vmatrix} 3 & -5 \\ 8 & 2 \end{vmatrix} + (-7)\begin{vmatrix} 1 & -5 \\ -4 & 2 \end{vmatrix} - (-2)\begin{vmatrix} 1 & 3 \\ -4 & 8 \end{vmatrix}$$

$$\begin{aligned}
&= -6[(3)(2) - (8)(-5)] + (-7)[(1)(2) - (-4)(-5)] - (-2)[(1)(8) - (-4)(3)] \\
&= -6[6 + 40] - 7[2 - 20] + 2[8 + 12] \\
&= -276 + 126 + 40 \\
&= -110
\end{aligned}$$

On a TI-89, use a semicolon (press **2nd** then the **9** key) to separate rows in a determinant and a comma to separate the entries within a given row. The entire determinant must be enclosed by brackets [ ] as shown.

2nd MATH 4

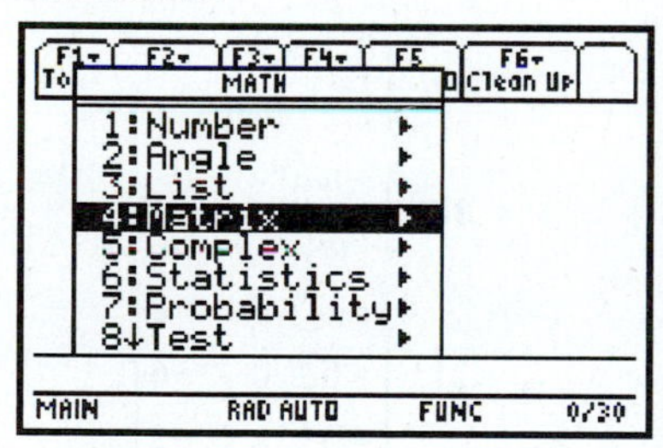

2

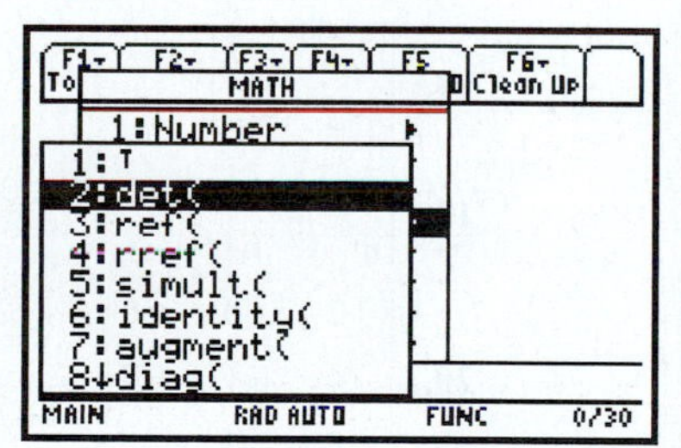

[1,3,-5;6,-7,-2;-4,8,2]) **ENTER**

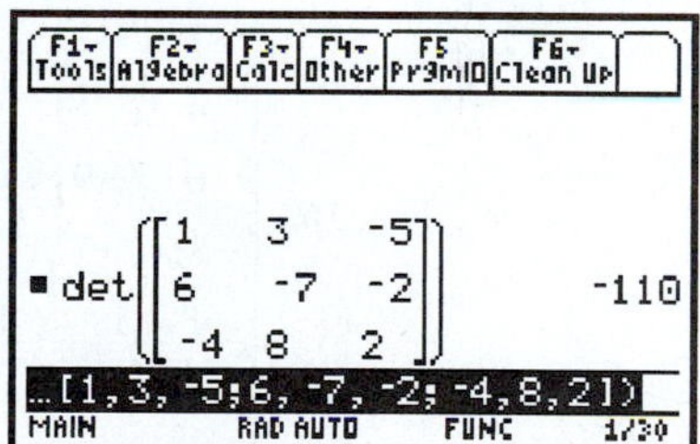

EXAMPLE 5

Evaluate the determinant

$$\begin{vmatrix} 3 & 0 & 5 & 4 \\ 2 & 1 & 3 & 4 \\ 0 & -2 & -1 & -3 \\ 1 & 0 & 2 & 1 \end{vmatrix}$$

Since the value of a determinant may be found by expansion by minors across any row or down any column, let's expand down the second column; it has the most zero elements and therefore involves the least work.

The signs down the second column are $- + - +$. Therefore,

$$\begin{vmatrix} 3 & 0 & 5 & 4 \\ 2 & 1 & 3 & 4 \\ 0 & -2 & -1 & -3 \\ 1 & 0 & 2 & 1 \end{vmatrix} = -0\begin{vmatrix} 2 & 3 & 4 \\ 0 & -1 & -3 \\ 1 & 2 & 1 \end{vmatrix} + 1\begin{vmatrix} 3 & 5 & 4 \\ 0 & -1 & -3 \\ 1 & 2 & 1 \end{vmatrix} - (-2)\begin{vmatrix} 3 & 5 & 4 \\ 2 & 3 & 4 \\ 1 & 2 & 1 \end{vmatrix}$$

$$+ 0\begin{vmatrix} 3 & 5 & 4 \\ 2 & 3 & 4 \\ 0 & -1 & -3 \end{vmatrix}$$

$$= 0 + 1(4) + 2(-1) + 0$$

$$= 2$$

Note: The second and third 3×3 determinants may be evaluated by expansion of three 2×2 minors.

The method for evaluating determinants by expansion by minors works for *any* order determinant.

Exercises 6.4

Evaluate each determinant.

1. $\begin{vmatrix} 4 & 3 \\ 2 & 5 \end{vmatrix}$ **2.** $\begin{vmatrix} 6 & -1 \\ 4 & 5 \end{vmatrix}$ **3.** $\begin{vmatrix} 2 & -3 \\ -4 & 5 \end{vmatrix}$

4. $\begin{vmatrix} 5 & 1 \\ 3 & -4 \end{vmatrix}$ **5.** $\begin{vmatrix} 3 & -7 \\ 6 & -1 \end{vmatrix}$ **6.** $\begin{vmatrix} -5 & -2 \\ 7 & 3 \end{vmatrix}$

7. $\begin{vmatrix} -5 & 8 \\ 9 & -6 \end{vmatrix}$ **8.** $\begin{vmatrix} 4 & 12 \\ 5 & 8 \end{vmatrix}$ **9.** $\begin{vmatrix} 4 & 0 \\ 6 & -3 \end{vmatrix}$

10. $\begin{vmatrix} 2 & -5 \\ -7 & 0 \end{vmatrix}$ **11.** $\begin{vmatrix} -4 & -2 \\ -6 & -7 \end{vmatrix}$ **12.** $\begin{vmatrix} -1 & 5 \\ -4 & 2 \end{vmatrix}$

13. $\begin{vmatrix} -7 & -2 \\ 4 & 1 \end{vmatrix}$ **14.** $\begin{vmatrix} 5 & -7 \\ -2 & -6 \end{vmatrix}$ **15.** $\begin{vmatrix} 8 & -9 \\ 7 & 4 \end{vmatrix}$

16. $\begin{vmatrix} 6 & -9 \\ -8 & 7 \end{vmatrix}$ **17.** $\begin{vmatrix} m & -n \\ n^2 & n \end{vmatrix}$ **18.** $\begin{vmatrix} -mn & -n \\ m & -mn \end{vmatrix}$

19. $\begin{vmatrix} 1 & 1 & -4 \\ -3 & 7 & 11 \\ 2 & 1 & -5 \end{vmatrix}$ **20.** $\begin{vmatrix} 2 & 0 & -6 \\ 4 & -1 & -7 \\ -3 & 3 & 2 \end{vmatrix}$ **21.** $\begin{vmatrix} -1 & 3 & 8 \\ 0 & 0 & -6 \\ -5 & 2 & -3 \end{vmatrix}$

22. $\begin{vmatrix} 2 & -5 & 7 \\ 0 & -8 & 1 \\ 3 & -2 & -1 \end{vmatrix}$ **23.** $\begin{vmatrix} 1 & 5 & -3 \\ 2 & -2 & 2 \\ -5 & -6 & 1 \end{vmatrix}$ **24.** $\begin{vmatrix} 2 & 3 & -6 \\ -5 & 1 & -2 \\ 0 & 3 & 0 \end{vmatrix}$

25. $\begin{vmatrix} 1 & -3 & 4 \\ 4 & 6 & -2 \\ 1 & -3 & 4 \end{vmatrix}$ **26.** $\begin{vmatrix} -2 & -5 & -2 \\ 1 & -7 & 1 \\ 3 & 2 & 3 \end{vmatrix}$ **27.** $\begin{vmatrix} 3 & 6 & -9 \\ 1 & -5 & 2 \\ 0 & 0 & 0 \end{vmatrix}$

28. $\begin{vmatrix} -6 & 0 & 5 \\ 3 & 0 & -2 \\ 7 & 0 & 4 \end{vmatrix}$ **29.** $\begin{vmatrix} 1 & 3 & -7 & 2 \\ -5 & 3 & 0 & -2 \\ 1 & -2 & 5 & 1 \\ 3 & 1 & 0 & 2 \end{vmatrix}$ **30.** $\begin{vmatrix} 2 & -1 & 7 & -5 \\ -3 & -2 & 7 & 4 \\ -1 & 0 & 3 & 2 \\ -2 & 0 & 5 & 1 \end{vmatrix}$

31. $\begin{vmatrix} 3 & -1 & 6 & 2 \\ -5 & 3 & -8 & 7 \\ 1 & 0 & -5 & 0 \\ 2 & -6 & 3 & 1 \end{vmatrix}$ **32.** $\begin{vmatrix} 1 & 1 & 0 & 1 \\ 1 & 0 & 1 & 1 \\ 1 & 1 & 1 & 0 \\ 0 & 1 & 1 & 1 \end{vmatrix}$ **33.** $\begin{vmatrix} 1 & 3 & 6 & -2 \\ 0 & 2 & -5 & 7 \\ 0 & 0 & 3 & 1 \\ 0 & 0 & 0 & 4 \end{vmatrix}$

34. $\begin{vmatrix} 1 & 1 & 1 & 4 \\ 0 & 1 & 1 & 1 \\ 0 & 0 & 1 & 1 \\ 0 & 0 & 0 & 1 \end{vmatrix}$ **35.** $\begin{vmatrix} 3 & -2 & 0 & 2 & -1 \\ 1 & 2 & -3 & 1 & 2 \\ 0 & -1 & 0 & 6 & 1 \\ -3 & 2 & 0 & 6 & -7 \\ 0 & 0 & 0 & 5 & 2 \end{vmatrix}$ **36.** $\begin{vmatrix} -1 & 1 & 5 & 6 & 2 \\ 3 & 2 & 0 & 0 & -1 \\ -2 & 0 & -3 & 1 & 1 \\ 0 & 2 & 0 & 0 & 0 \\ 6 & -1 & 0 & 1 & 2 \end{vmatrix}$

6.5 PROPERTIES OF DETERMINANTS

Evaluation of determinants using expansion by minors allows us to evaluate a determinant of any order. As you saw in Section 6.4, the amount of work is considerable. However, a few basic properties of determinants significantly lessen the effort and time needed to evaluate a determinant.

PROPERTY 1

If every element in a row (or a column) of a determinant is zero, the value of the determinant is zero.

EXAMPLE 1

(a) $\begin{vmatrix} 3 & -2 & 6 \\ 0 & 0 & 0 \\ 1 & 5 & -8 \end{vmatrix} = 0$ (b) $\begin{vmatrix} 5 & 3 & 0 \\ -7 & 2 & 0 \\ 8 & -6 & 0 \end{vmatrix} = 0$

PROPERTY 2

If two rows (or two columns) of a determinant are identical, the value of the determinant is zero.

EXAMPLE 2

(a) $\begin{vmatrix} 1 & -3 & 5 \\ -2 & -6 & 3 \\ 1 & -3 & 5 \end{vmatrix} = 0$ (b) $\begin{vmatrix} 6 & 2 & 2 \\ 0 & 5 & 5 \\ 2 & -3 & -3 \end{vmatrix} = 0$

PROPERTY 3

If any two rows (or two columns) of a determinant are interchanged, only the sign of the value of the determinant is changed.

EXAMPLE 3

(a) $\begin{vmatrix} 3 & -2 & 7 \\ 5 & 8 & 4 \\ 0 & 6 & -9 \end{vmatrix} = -\begin{vmatrix} 5 & 8 & 4 \\ 3 & -2 & 7 \\ 0 & 6 & -9 \end{vmatrix}$ (b) $\begin{vmatrix} 0 & 3 & -5 \\ -6 & 2 & 3 \\ 9 & 7 & 1 \end{vmatrix} = -\begin{vmatrix} -5 & 3 & 0 \\ 3 & 2 & -6 \\ 1 & 7 & 9 \end{vmatrix}$

PROPERTY 4

If every element of a row (or a column) is multiplied by the same real number k the value of the determinant is multiplied by k.

EXAMPLE 4

(a) $\begin{vmatrix} 3 & 5 & -6 \\ 4 & -8 & 12 \\ 3 & 0 & 7 \end{vmatrix} = 4\begin{vmatrix} 3 & 5 & -6 \\ 1 & -2 & 3 \\ 3 & 0 & 7 \end{vmatrix}$ (b) $3\begin{vmatrix} 1 & -4 & 6 \\ 3 & 0 & 8 \\ 2 & 4 & -3 \end{vmatrix} = \begin{vmatrix} 1 & -4 & 18 \\ 3 & 0 & 24 \\ 2 & 4 & -9 \end{vmatrix}$

The next property is probably the most useful. As you saw in the previous section, it was nice to have zero elements. This last property allows us to make more zeros before we expand by minors.

PROPERTY 5

If every element of a row (or a column) is multiplied by the same real number k and if the resulting products are added to another row (or another column), the value of the determinant remains the same.

EXAMPLE 5

Evaluate

$$\begin{vmatrix} 3 & -2 & 6 \\ -6 & 0 & 4 \\ 5 & 1 & -1 \end{vmatrix}$$

Let's multiply the third row by 2 and add the resulting products to the first row to make another zero in the second column.

$$\begin{vmatrix} (2)(5)+3 & (2)(1)+(-2) & (2)(-1)+6 \\ -6 & 0 & 4 \\ 5 & 1 & -1 \end{vmatrix} = \begin{vmatrix} 13 & 0 & 4 \\ -6 & 0 & 4 \\ 5 & 1 & -1 \end{vmatrix}$$

Expanding down the second column, we find that the first two minors are multiplied by zero. Thus,

$$\begin{vmatrix} 13 & 0 & 4 \\ -6 & 0 & 4 \\ 5 & 1 & -1 \end{vmatrix} = -1\begin{vmatrix} 13 & 4 \\ -6 & 4 \end{vmatrix} = -1[52-(-24)] = -76$$

Note that if we evaluate this determinant by expansion by minors, we obtain the same result:

$$\begin{vmatrix} 3 & -2 & 6 \\ -6 & 0 & 4 \\ 5 & 1 & -1 \end{vmatrix}$$

Let's expand by minors down the second column as follows:

$$\begin{aligned} &-(-2)\begin{vmatrix} -6 & 4 \\ 5 & -1 \end{vmatrix} + 0\begin{vmatrix} 3 & 6 \\ 5 & -1 \end{vmatrix} - 1\begin{vmatrix} 3 & 6 \\ -6 & 4 \end{vmatrix} \\ &= 2(6-20) \quad + 0(-3-30) \quad - 1(12+36) \\ &= -28 \quad + 0 \quad - 48 \\ &= -76 \end{aligned}$$

EXAMPLE 6

Evaluate

$$\begin{vmatrix} -7 & -6 & 5 & 1 \\ 5 & 3 & 0 & 7 \\ 0 & -1 & 4 & -2 \\ -8 & -3 & 8 & -10 \end{vmatrix}$$

Let's make two more zeros in the third row. First multiply the second column by 4 and add the resulting products to the third column. Then multiply the second column by -2 and add the resulting products to the fourth column.

$$\begin{vmatrix} -7 & -6 & (4)(-6)+5 & (-2)(-6)+1 \\ 5 & 3 & (4)(3)+0 & (-2)(3)+7 \\ 0 & -1 & (4)(-1)+4 & (-2)(-1)+(-2) \\ -8 & -3 & (4)(-3)+8 & (-2)(-3)+(-10) \end{vmatrix} = \begin{vmatrix} -7 & -6 & -19 & 13 \\ 5 & 3 & 12 & 1 \\ 0 & -1 & 0 & 0 \\ -8 & -3 & -4 & -4 \end{vmatrix}$$

Expanding across the third row, we find that the first, third, and fourth minors result in zero. Therefore,

$$\begin{vmatrix} -7 & -6 & -19 & 13 \\ 5 & 3 & 12 & 1 \\ 0 & -1 & 0 & 0 \\ -8 & -3 & -4 & -4 \end{vmatrix} = -(-1)\begin{vmatrix} -7 & -19 & 13 \\ 5 & 12 & 1 \\ -8 & -4 & -4 \end{vmatrix} = (+1)(-4)\begin{vmatrix} -7 & -19 & 13 \\ 5 & 12 & 1 \\ 2 & 1 & 1 \end{vmatrix}$$

Let's make two zeros in the third row. First multiply the third column by -2 and add the resulting products to the first column. Then multiply the third column by -1 and add the resulting products to the second column.

$$(-4)\begin{vmatrix} (-2)(13)+(-7) & (-1)(13)+(-19) & 13 \\ (-2)(1)+5 & (-1)(1)+12 & 1 \\ (-2)(1)+2 & (-1)(1)+1 & 1 \end{vmatrix} = (-4)\begin{vmatrix} -33 & -32 & 13 \\ 3 & 11 & 1 \\ 0 & 0 & 1 \end{vmatrix}$$

Expanding across the third row, we find that the first two minors result in zero. Thus,

$$(-4)\begin{vmatrix} -33 & -32 & 13 \\ 3 & 11 & 1 \\ 0 & 0 & 1 \end{vmatrix} = (-4)(+1)\begin{vmatrix} -33 & -32 \\ 3 & 11 \end{vmatrix} = (-4)(-363+96) = 1068$$

Note: You could have chosen to make zeros in any row or column. In any case, the result is 1068.

Exercises 6.5

Evaluate each determinant.

1. $\begin{vmatrix} 4 & 3 & -8 \\ 2 & -7 & 9 \\ 0 & 0 & 0 \end{vmatrix}$

2. $\begin{vmatrix} -3 & 6 & -3 \\ 1 & 5 & 1 \\ 7 & 0 & 7 \end{vmatrix}$

3. $\begin{vmatrix} 3 & 6 & 0 & -8 \\ 2 & 7 & 5 & 4 \\ 0 & -5 & 8 & 1 \\ 2 & 7 & 5 & 4 \end{vmatrix}$

4. $\begin{vmatrix} 6 & 5 & 0 & -9 \\ -3 & -4 & 0 & 2 \\ 2 & 3 & 0 & -7 \\ 7 & -1 & 0 & 12 \end{vmatrix}$

5. $\begin{vmatrix} 4 & 6 & -6 \\ 0 & 0 & -9 \\ 1 & -2 & 7 \end{vmatrix}$

6. $\begin{vmatrix} 2 & 6 & -5 \\ -3 & -5 & 0 \\ 7 & 2 & 0 \end{vmatrix}$

7. $\begin{vmatrix} 3 & 0 & 8 & 0 \\ 0 & 7 & 0 & 0 \\ 0 & 0 & -7 & 5 \\ 0 & 0 & 1 & 0 \end{vmatrix}$

8. $\begin{vmatrix} 0 & 0 & 5 & 1 \\ 1 & 0 & 3 & 0 \\ 2 & 6 & 8 & 0 \\ 3 & 0 & 0 & 0 \end{vmatrix}$

9. $\begin{vmatrix} 3 & 0 & 0 & 0 \\ 0 & 1 & 0 & 0 \\ 0 & 0 & -2 & 0 \\ 0 & 0 & 0 & 5 \end{vmatrix}$

10. $\begin{vmatrix} 0 & 0 & 0 & -5 \\ 0 & 0 & -7 & 0 \\ 0 & 2 & 0 & 0 \\ 3 & 0 & 0 & 0 \end{vmatrix}$

11. $\begin{vmatrix} 1 & 0 & 4 \\ 6 & 3 & 2 \\ 5 & 4 & 1 \end{vmatrix}$

12. $\begin{vmatrix} 2 & 7 & 4 \\ 3 & 0 & 1 \\ 6 & 3 & 2 \end{vmatrix}$

13. $\begin{vmatrix} 3 & 1 & -6 \\ -4 & 0 & 7 \\ 5 & 6 & 2 \end{vmatrix}$

14. $\begin{vmatrix} -3 & 2 & 5 \\ 4 & 7 & 0 \\ 6 & -3 & -1 \end{vmatrix}$

15. $\begin{vmatrix} 5 & 3 & 7 \\ 1 & 2 & 3 \\ 6 & 2 & 5 \end{vmatrix}$

16. $\begin{vmatrix} 2 & -4 & 6 \\ 4 & -2 & 1 \\ -8 & 2 & 1 \end{vmatrix}$

17. $\begin{vmatrix} 3 & -7 & 5 \\ 2 & 4 & 6 \\ -9 & 7 & -2 \end{vmatrix}$

18. $\begin{vmatrix} 4 & 3 & -9 \\ 5 & -6 & -4 \\ -7 & 9 & 2 \end{vmatrix}$

19. $\begin{vmatrix} 1 & 0 & 5 & 0 \\ 6 & 2 & -3 & 7 \\ -1 & 2 & 3 & -4 \\ -3 & 2 & -7 & 1 \end{vmatrix}$

20. $\begin{vmatrix} -5 & -6 & 3 & 7 \\ 3 & -1 & 0 & 0 \\ 2 & 1 & 0 & 5 \\ 4 & -4 & -7 & 0 \end{vmatrix}$

21. $\begin{vmatrix} 1 & 3 & -7 & 2 \\ 5 & 6 & 1 & 5 \\ 4 & -8 & 9 & -2 \\ -6 & 7 & -4 & 3 \end{vmatrix}$

22. $\begin{vmatrix} 2 & 5 & 2 & -3 \\ 3 & 4 & -4 & 4 \\ 2 & -8 & 1 & 9 \\ -2 & 6 & 3 & -7 \end{vmatrix}$

23. $\begin{vmatrix} 1 & 0 & 3 & 2 & 5 \\ 3 & 2 & 3 & -2 & 3 \\ -4 & -2 & 4 & 9 & -6 \\ -3 & 7 & 6 & 2 & 3 \\ 0 & 2 & -2 & -4 & 5 \end{vmatrix}$

24. $\begin{vmatrix} 2 & 2 & 3 & 5 & -7 \\ 3 & -6 & 9 & 12 & 15 \\ 4 & -5 & 3 & 7 & 2 \\ 2 & 0 & -4 & 2 & 7 \\ 3 & 3 & 4 & -3 & 3 \end{vmatrix}$

6.6 SOLVING A SYSTEM OF LINEAR EQUATIONS USING DETERMINANTS

Suppose we were to solve the general system of two linear equations in two variables:

$$a_1x + b_1y = c_1 \quad \textbf{(1)}$$
$$a_2x + b_2y = c_2 \quad \textbf{(2)}$$

By the addition-subtraction method we would proceed as follows:

$$\begin{aligned} a_1b_2x + b_1b_2y &= b_2c_1 && \text{[Multiply (1) by } b_2\text{.]} \\ a_2b_1x + b_1b_2y &= b_1c_2 && \text{[Multiply (2) by } b_1\text{.]} \\ \hline a_1b_2x - a_2b_1x &= b_2c_1 - b_1c_2 && \text{(Subtract.)} \\ (a_1b_2 - a_2b_1)x &= b_2c_1 - b_1c_2 && \text{(Factor.)} \\ x &= \frac{b_2c_1 - b_1c_2}{a_1b_2 - a_2b_1} \end{aligned}$$

By similarly eliminating x, we can show that

$$y = \frac{a_1c_2 - a_2c_1}{a_1b_2 - a_2b_1}$$

Note the following:

1. $a_1b_2 - a_2b_1$ is the value of and may be represented by the determinant

$$\begin{vmatrix} a_1 & b_1 \\ a_2 & b_2 \end{vmatrix}$$

2. $b_2c_1 - b_1c_2$ is the value of

$$\begin{vmatrix} c_1 & b_1 \\ c_2 & b_2 \end{vmatrix}$$

3. $a_1c_2 - a_2c_1$ is the value of

$$\begin{vmatrix} a_1 & c_1 \\ a_2 & c_2 \end{vmatrix}$$

Therefore, the solution of

$$a_1x + b_1y = c_1$$
$$a_2x + b_2y = c_2$$

may be written in determinant form as follows:

DETERMINANT SOLUTION OF TWO LINEAR EQUATIONS

$$x = \frac{\begin{vmatrix} c_1 & b_1 \\ c_2 & b_2 \end{vmatrix}}{\begin{vmatrix} a_1 & b_1 \\ a_2 & b_2 \end{vmatrix}} \qquad y = \frac{\begin{vmatrix} a_1 & c_1 \\ a_2 & c_2 \end{vmatrix}}{\begin{vmatrix} a_1 & b_1 \\ a_2 & b_2 \end{vmatrix}}$$

If the determinant of the numerator is not zero and the determinant of the denominator is zero, the system is *inconsistent.* If the determinants of the numerator and the denominator are both zero, the system is *dependent.* If the determinant of the denominator is not zero, there is a unique solution and the system is *independent and consistent.*

Note that the determinant of each denominator is made up of the coefficients of x and y. This determinant is denoted D. To find the determinant of the numerator, use these steps:

1. For x, take the determinant of the denominator and replace the coefficients of x, a's, by the corresponding constants, c's. This determinant is denoted D_x.
2. For y, take the determinant of the denominator and replace the coefficients of y, b's, by the corresponding constants, c's. This determinant is denoted D_y.

This method of solution is called **Cramer's rule** for solving a system of two linear equations in two variables. To use this method, the equations must be in the general form shown in Equations (1) and (2) at the beginning of this section.

EXAMPLE 1

Solve using determinants.

$$2x - 3y = 22$$
$$5x + 4y = -14$$

$$x = \frac{D_x}{D} = \frac{\begin{vmatrix} 22 & -3 \\ -14 & 4 \end{vmatrix}}{\begin{vmatrix} 2 & -3 \\ 5 & 4 \end{vmatrix}} = \frac{(22)(4) - (-14)(-3)}{(2)(4) - 5(-3)} = \frac{46}{23} = 2$$

$$y = \frac{D_y}{D} = \frac{\begin{vmatrix} 2 & 22 \\ 5 & -14 \end{vmatrix}}{\begin{vmatrix} 2 & -3 \\ 5 & 4 \end{vmatrix}} = \frac{(2)(-14) - (5)(22)}{23} = \frac{-138}{23} = -6$$

The solution is $(2, -6)$.

Check: Substitute the solution in both original equations.

EXAMPLE 2

Solve using determinants.

$$3x + 4y = 10$$
$$6x + 8y = -5$$

$$x = \frac{\begin{vmatrix} 10 & 4 \\ -5 & 8 \end{vmatrix}}{\begin{vmatrix} 3 & 4 \\ 6 & 8 \end{vmatrix}} = \frac{(10)(8) - (-5)(4)}{(3)(8) - (6)(4)} = \frac{100}{0}$$

Since the numerator of the determinant is not zero and the denominator is zero, the system is inconsistent (the lines are parallel). There are no solutions.

EXAMPLE 3

Solve using determinants.

$$mx + ny = mn$$
$$nx - my = n^2$$

$$x = \frac{\begin{vmatrix} mn & n \\ n^2 & -m \end{vmatrix}}{\begin{vmatrix} m & n \\ n & -m \end{vmatrix}} = \frac{(mn)(-m) - (n^2)(n)}{m(-m) - (n)(n)}$$

$$= \frac{-m^2n - n^3}{-m^2 - n^2} = \frac{-n(m^2 + n^2)}{-(m^2 + n^2)} = n$$

$$y = \frac{\begin{vmatrix} m & mn \\ n & n^2 \end{vmatrix}}{\begin{vmatrix} m & n \\ n & -m \end{vmatrix}} = \frac{(m)(n^2) - (n)(mn)}{-m^2 - n^2}$$

$$= \frac{mn^2 - mn^2}{-m^2 - n^2} = \frac{0}{-m^2 - n^2} = 0$$

The solution is $(n, 0)$.

The general system of three linear equations in three variables is given as

$$a_1x + b_1y + c_1z = d_1$$
$$a_2x + b_2y + c_2z = d_2$$
$$a_3x + b_3y + c_3z = d_3$$

If we used the addition-subtraction method, we would find the following solutions for x, y, and z:

$$x = \frac{d_1b_2c_3 + d_3b_1c_2 + d_2b_3c_1 - d_3b_2c_1 - d_1b_3c_2 - d_2b_1c_3}{a_1b_2c_3 + a_3b_1c_2 + a_2b_3c_1 - a_3b_2c_1 - a_1b_3c_2 - a_2b_1c_3}$$

$$y = \frac{a_1d_2c_3 + a_3d_1c_2 + a_2d_3c_1 - a_3d_2c_1 - a_1d_3c_2 - a_2d_1c_3}{a_1b_2c_3 + a_3b_1c_2 + a_2b_3c_1 - a_3b_2c_1 - a_1b_3c_2 - a_2b_1c_3}$$

$$z = \frac{a_1b_2d_3 + a_3b_1d_2 + a_2b_3d_1 - a_3b_2d_1 - a_1b_3d_2 - a_2b_1d_3}{a_1b_2c_3 + a_3b_1c_2 + a_2b_3c_1 - a_3b_2c_1 - a_1b_3c_2 - a_2b_1c_3}$$

This general solution may be written in terms of determinants as follows:

DETERMINANT SOLUTION OF THREE LINEAR EQUATIONS

$$x = \frac{\begin{vmatrix} d_1 & b_1 & c_1 \\ d_2 & b_2 & c_2 \\ d_3 & b_3 & c_3 \end{vmatrix}}{\begin{vmatrix} a_1 & b_1 & c_1 \\ a_2 & b_2 & c_2 \\ a_3 & b_3 & c_3 \end{vmatrix}} \qquad y = \frac{\begin{vmatrix} a_1 & d_1 & c_1 \\ a_2 & d_2 & c_2 \\ a_3 & d_3 & c_3 \end{vmatrix}}{\begin{vmatrix} a_1 & b_1 & c_1 \\ a_2 & b_2 & c_2 \\ a_3 & b_3 & c_3 \end{vmatrix}} \qquad z = \frac{\begin{vmatrix} a_1 & b_1 & d_1 \\ a_2 & b_2 & d_2 \\ a_3 & b_3 & d_3 \end{vmatrix}}{\begin{vmatrix} a_1 & b_1 & c_1 \\ a_2 & b_2 & c_2 \\ a_3 & b_3 & c_3 \end{vmatrix}}$$

If the determinant of the numerator is not zero and the determinant of the denominator is zero, the system is *inconsistent*. If the determinants of the numerator and the denominator are both zero, the system is *dependent* or *inconsistent*. If the determinant of the denominator is not zero, there is a unique solution, and the system is *independent and consistent*.

Note that the determinant of each denominator is made up of the coefficients of x, y, and z. This determinant is denoted D. To find the determinant of the numerator, use these steps:

1. For x, take the determinant of the denominator and replace the coefficients for x, a's, by the corresponding constants, d's. This determinant is denoted D_x.
2. For y, take the determinant of the denominator and replace the coefficients of y, b's, by the corresponding constants, d's. This determinant is denoted D_y.
3. For z, take the determinant of the denominator and replace the coefficients of z, c's, by the corresponding constants, d's. This determinant is denoted D_z.

This method is called **Cramer's rule** for solving a system of three linear equations in three variables.

EXAMPLE 4

Solve using determinants.

$$\begin{aligned} 3x - y + 4z &= -15 \\ 2x + 5y &= 29 \\ x - 6y - z &= -24 \end{aligned}$$

Using Cramer's rule, we have

$$x = \frac{D_x}{D} = \frac{\begin{vmatrix} -15 & -1 & 4 \\ 29 & 5 & 0 \\ -24 & -6 & -1 \end{vmatrix}}{\begin{vmatrix} 3 & -1 & 4 \\ 2 & 5 & 0 \\ 1 & -6 & -1 \end{vmatrix}} = \frac{\begin{vmatrix} -111 & -25 & 0 \\ 29 & 5 & 0 \\ -24 & -6 & -1 \end{vmatrix}}{\begin{vmatrix} 7 & -25 & 0 \\ 2 & 5 & 0 \\ 1 & -6 & -1 \end{vmatrix}}$$

(In both numerator and denominator, multiply the third row by 4, and add the result to the first row.)

$$= \frac{(-1)\begin{vmatrix} -111 & -25 \\ 29 & 5 \end{vmatrix}}{(-1)\begin{vmatrix} 7 & -25 \\ 2 & 5 \end{vmatrix}}$$

(Expand down the third column.)

$$= \frac{(-1)(-555 + 725)}{(-1)(35 + 50)}$$

$$= \frac{-170}{-85}$$

$$= 2$$

Next, find y as follows:

$$y = \frac{D_y}{D} = \frac{\begin{vmatrix} 3 & -15 & 4 \\ 2 & 29 & 0 \\ 1 & -24 & -1 \end{vmatrix}}{\begin{vmatrix} 3 & -1 & 4 \\ 2 & 5 & 0 \\ 1 & -6 & -1 \end{vmatrix}} = \frac{\begin{vmatrix} 7 & -111 & 0 \\ 2 & 29 & 0 \\ 1 & -24 & -1 \end{vmatrix}}{-85}$$

(Multiply the third row by 4 and add the result to the first row.)

$$= \frac{(-1)\begin{vmatrix} 7 & -111 \\ 2 & 29 \end{vmatrix}}{-85}$$

(Expand down the third column.)

$$= \frac{(-1)(203 + 222)}{-85}$$

$$= \frac{-425}{-85} = 5$$

Then, find z as follows:

$$z = \frac{D_z}{D} = \frac{\begin{vmatrix} 3 & -1 & -15 \\ 2 & 5 & 29 \\ 1 & -6 & -24 \end{vmatrix}}{\begin{vmatrix} 3 & -1 & 4 \\ 2 & 5 & 0 \\ 1 & -6 & -1 \end{vmatrix}} = \frac{\begin{vmatrix} 0 & 17 & 57 \\ 0 & 17 & 77 \\ 1 & -6 & -24 \end{vmatrix}}{-85}$$

(Multiply the third row by -3 and add the result to the first row. Then multiply the third row by -2 and add the result to the second row.)

$$= \frac{(1)\begin{vmatrix} 17 & 57 \\ 17 & 77 \end{vmatrix}}{-85}$$

(Expand down the first column.)

$$= \frac{(1)(17)\begin{vmatrix} 1 & 57 \\ 1 & 77 \end{vmatrix}}{-85} \qquad \text{(Property 4)}$$

$$= \frac{17(77 - 57)}{-85} = \frac{340}{-85} = -4$$

The solution is $(2, 5, -4)$.

Check: Substitute the solution in all three original equations.

You may also find the value of the third variable by substituting the first two values into any equation and then solving for the third variable. For example, substitute $x = 2$ and $y = 5$ into the third equation and solve for z as follows:

$$\begin{aligned} x - 6y - z &= -24 \\ 2 - 6(5) - z &= -24 \\ 2 - 30 - z &= -24 \\ -z &= 4 \\ z &= -4 \end{aligned}$$

Exercises 6.6

Solve each system of equations using determinants. Use a calculator or methods discussed in this chapter to evaluate the determinants.

1. $3x + 5y = -1$
$2x - 3y = 12$

2. $6x - 2y = 36$
$5x + 4y = 47$

3. $8x - 3y = -43$
$5x - 7y = -73$

4. $4x - 2y = 6$
$-3x + 4y = -17$

5. $6x - 7y = 28$
$-4x + 5y = -20$

6. $3x - 8y = 9$
$-9x - 6y = -27$

7. $3x + 4y = 5$
$6x + 8y = 10$

8. $-x + 4y = 7$
$3x - 12y = -21$

9. $12x - 16y = 24$
$15x - 20y = 36$

10. $-2x + 3y = 6$
$8x - 12y = 24$

11. $15x - 6y = -15$
$9x + 12y = 4$

12. $3x - 5y = 3$
$-5x + 7y = -2$

13. $8x + 7y = 18$
$y = 4x$

14. $20x - 5y = -13$
$x = -3y$

15. $ax + by = 2$
$bx + ay = 4$

16. $ax - 3y = b$
$bx - 2y = a$

17. $5x + ay = b$
$2x - by = a$

18. $4x - by = b$
$6x - ay = a$

19. $ax + by = c$
$y = bx$

20. $ax - by = a$
$x = ay$

21. The sum of two voltages is 210 volts (V). The larger voltage is 15 V less than twice the smaller voltage. Find the voltages.

22. A man has available two different mixtures of solder. One mixture is 25% tin and the other is 65% tin. How much of each must he use to make a 56-kg mixture of 40% tin?

Solve each system of equations using determinants. Use a calculator or methods discussed in this chapter to evaluate the determinants.

23. $3x - 4y + 7z = -26$
$-2x + y - 3z = 9$
$12x + 15z = -36$

24. $5x - 2y - 3z = -3$
$4y + 3z = -2$
$x - y + 9z = 60$

25. $3x + 2y + 5z = -7$
$8x - 3y + 2z = 10$
$7x - 2y + 4z = 1$

26. $3x - 7y - 2z = -38$
$6x + 5y - z = 63$
$-2x - 4y + 5z = -28$

27. $7x - 5y - 7z = 8$
$9x + 3y - 6z = 33$
$4x - 2y - 8z = 28$

28. $x - 6y - 4z = 0$
$2x - 3y + 5z = 46$
$9x + 7y + 8z = 3$

29. $$\begin{aligned} 3x - 4y - 5z &= 0 \\ 9x + 6y + 10z &= 11 \\ 12x + 2y - 20z &= 36 \end{aligned}$$

30. $$\begin{aligned} 4x - 6y - 8z &= 18 \\ 12x + 15y + 16z &= -13 \\ 20x - 12y - 24z &= 60 \end{aligned}$$

31. $$\begin{aligned} 2x \quad + 5z &= 29 \\ 5y - 7z &= -40 \\ 8x + y \quad &= 15 \end{aligned}$$

32. $$\begin{aligned} -4x + 7y \quad &= 37 \\ 6x \quad - 2z &= 24 \\ 3y - 5z &= 36 \end{aligned}$$

33. $$\begin{aligned} 4x + 6y + 8z &= -8 \\ -x \quad + 5z &= -19 \\ 5y + 7z &= -21 \end{aligned}$$

34. $$\begin{aligned} 3x - 5y \quad &= 20 \\ 9x + 8y - 2z &= -25 \\ 5y - 8z &= 8 \end{aligned}$$

35. The perimeter of a triangle is 21 cm. The longest side is 2 cm longer than the shortest side. These two sides together are twice the size of the remaining side. Find the lengths of the sides of the triangle.

36. As a result of Kirchhoff's laws for the current in a circuit, the following system of equations was obtained:

$$\begin{aligned} I_1 - I_2 + I_3 &= 0 \\ 2.2I_1 + 0.5I_2 &= 12.6 \\ 3.4I_1 - 3.8I_2 &= -10.25 \end{aligned}$$

Determine the indicated currents in amperes (A).

Cramer's rule may be extended to solve a system of n linear equations in n unknowns. Use Cramer's rule to solve each system of equations.

37. $$\begin{aligned} 3x - 2y + z - 3w &= -20 \\ 2x \quad + 5z + w &= -3 \\ 5x + y - z + 4w &= 30 \\ 6x - 3y + 4z \quad &= -11 \end{aligned}$$

38. $$\begin{aligned} 8x - 2y + z - w &= 38 \\ 4y - 7z + w &= -28 \\ 3x \quad - 5z \quad &= -3 \\ 3x - 2y \quad + 7w &= 23 \end{aligned}$$

39. $$\begin{aligned} 2x + 2y - z + 3w - 4v &= 13 \\ 3x + 7y - z \quad - 3v &= -8 \\ 3y + 2z + w - v &= 5 \\ -2x \quad + 3z - w + v &= 0 \\ 5x \quad + 7w \quad &= 38 \end{aligned}$$

40. $$\begin{aligned} 3x - 4y + 2z + 2w + v &= -2 \\ x \quad + 3z \quad + 2v &= -1 \\ 3y + 4z - 8w + 3v &= -13 \\ 7z + 2w - v &= 5 \\ 5x \quad + z + 3w \quad &= 11 \end{aligned}$$

6.7 PARTIAL FRACTIONS

We add two fractions such as

$$\begin{aligned} \frac{5}{x+1} + \frac{6}{x-2} &= \frac{5(x-2)}{(x+1)(x-2)} + \frac{6(x+1)}{(x+1)(x-2)} \\ &= \frac{5x - 10 + 6x + 6}{(x+1)(x-2)} \\ &= \frac{11x - 4}{(x+1)(x-2)} \end{aligned}$$

At times, we need to express a fraction as the sum of two or more fractions that are each simpler than the original; that is, we reverse the operation. Such simpler fractions whose numerators are of lower degree than their denominators are called **partial fractions.**

We separate our study of partial fractions into four cases. In each case we assume that the given fraction is expressed in lowest terms and the degree of each numerator is less than the degree of its denominator.

CASE 1: NONREPEATED LINEAR DENOMINATOR FACTORS

For every nonrepeated factor $ax + b$ of the denominator of a given fraction, there corresponds the partial fraction $\frac{A}{ax + b}$, where A is a constant.

EXAMPLE 1

Find the partial fractions of $\frac{11x - 4}{(x + 1)(x + 2)}$.

The possible partial fractions are $\frac{A}{x + 1}$ and $\frac{B}{x - 2}$, so we have

$$\frac{11x - 4}{(x + 1)(x - 2)} = \frac{A}{x + 1} + \frac{B}{x - 2}$$

Multiply each side of this equation by the L.C.D.: $(x + 1)(x - 2)$.

$$11x - 4 = A(x - 2) + B(x + 1)$$

Removing parentheses and rearranging terms, we have

$$11x - 4 = Ax - 2A + Bx + B$$
$$11x - 4 = Ax + Bx - 2A + B$$
$$11x - 4 = (A + B)x - 2A + B$$

Next, the coefficients of x must be equal and the constant terms must be equal. This gives the following system of linear equations:

$$A + B = 11$$
$$-2A + B = -4$$

Subtracting the two equations gives

$$3A = 15$$
$$A = 5$$

Substituting $A = 5$ into either of the preceding equations gives

$$B = 6$$

Then

$$\frac{11x - 4}{(x + 1)(x - 2)} = \frac{5}{x + 1} + \frac{6}{x - 2}$$

EXAMPLE 2

Find the partial fractions of $\frac{3x^2 - 27x - 12}{x(2x + 1)(x - 4)}$.

The possible partial fractions are

$$\frac{A}{x} \quad \frac{B}{2x + 1} \quad \text{and} \quad \frac{C}{x - 4}$$

So we have

$$\frac{3x^2 - 27x - 12}{x(2x + 1)(x - 4)} = \frac{A}{x} + \frac{B}{2x + 1} + \frac{C}{x - 4}$$

Now multiply each side of this equation by the L.C.D.: $x(2x + 1)(x - 4)$.

$$3x^2 - 27x - 12 = A(2x + 1)(x - 4) + Bx(x - 4) + Cx(2x + 1)$$

Removing parentheses and rearranging terms, we have

$$3x^2 - 27x - 12 = 2Ax^2 - 7Ax - 4A + Bx^2 - 4Bx + 2Cx^2 + Cx$$
$$3x^2 - 27x - 12 = (2A + B + 2C)x^2 + (-7A - 4B + C)x - 4A$$

Then the coefficients of x^2 must be equal, the coefficients of x must be equal, and the constant terms must be equal. This gives the following system of linear equations:

$$\begin{aligned} 2A + B + 2C &= 3 \\ -7A - 4B + C &= -27 \\ -4A &= -12 \end{aligned}$$

Note that $A = 3$ from the third equation. Substituting $A = 3$ into the first two equations gives

$$\begin{aligned} 6 + B + 2C &= 3 \\ -21 - 4B + C &= -27 \end{aligned}$$

or

$$\begin{aligned} B + 2C &= -3 \\ -4B + C &= -6 \end{aligned}$$

Multiplying the second equation by 2 gives

$$\begin{aligned} B + 2C &= -3 \\ -8B + 2C &= -12 \end{aligned}$$

Subtracting these two equations gives

$$\begin{aligned} 9B &= 9 \\ B &= 1 \end{aligned}$$

Then $C = -2$ and

$$\frac{3x^2 - 27x - 12}{x(2x + 1)(x - 4)} = \frac{3}{x} + \frac{1}{2x + 1} - \frac{2}{x - 4}$$

CASE 2: REPEATED LINEAR DENOMINATOR FACTORS

For every factor $(ax + b)^k$ of the denominator of a given fraction, there correspond the possible partial fractions

$$\frac{A_1}{ax + b}, \frac{A_2}{(ax + b)^2}, \frac{A_3}{(ax + b)^3}, \ldots, \frac{A_k}{(ax + b)^k}$$

where $A_1, A_2, A_3, \ldots, A_k$ are constants.

EXAMPLE 3

Find the partial fractions of $\dfrac{-x^2 - 8x + 27}{x(x - 3)^2}$.

The possible partial fractions are

$$\frac{A}{x} \quad \frac{B}{x - 3} \quad \text{and} \quad \frac{C}{(x - 3)^2}$$

So we have

$$\frac{-x^2 - 8x + 27}{x(x - 3)^2} = \frac{A}{x} + \frac{B}{x - 3} + \frac{C}{(x - 3)^2}$$

Then multiply each side of this equation by the L.C.D.: $x(x-3)^2$.

$$-x^2 - 8x + 27 = A(x-3)^2 + Bx(x-3) + Cx$$

Removing parentheses and rearranging terms, we have

$$-x^2 - 8x + 27 = Ax^2 - 6Ax + 9A + Bx^2 - 3Bx + Cx$$
$$-x^2 - 8x + 27 = (A + B)x^2 + (-6A - 3B + C)x + 9A$$

Equating coefficients, we have

$$A + B = -1$$
$$-6A - 3B + C = -8$$
$$9A = 27$$

From the third equation, we have $A = 3$. Substituting $A = 3$ into the first equation gives $B = -4$. Then substituting $A = 3$ and $B = -4$ into the second equation gives $C = -2$. Thus,

$$\frac{-x^2 - 8x + 27}{x(x-3)^2} = \frac{3}{x} - \frac{4}{x-3} - \frac{2}{(x-3)^2}$$

The TI-89 uses the **expand** command to produce partial fraction expansions.

F2 **3** (-x^2-8x+27)/(x*(x-3)^2)) **ENTER**

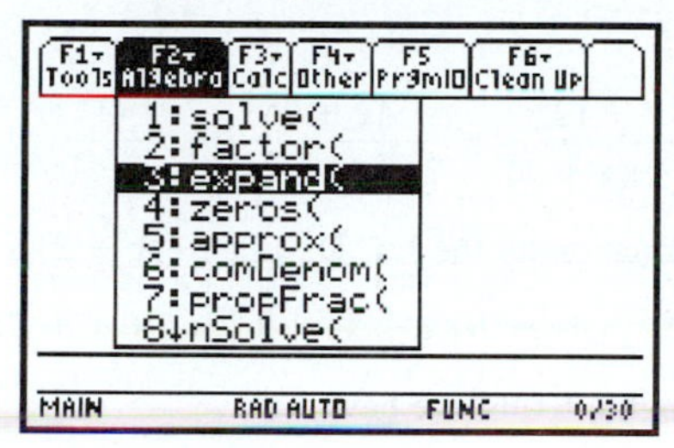

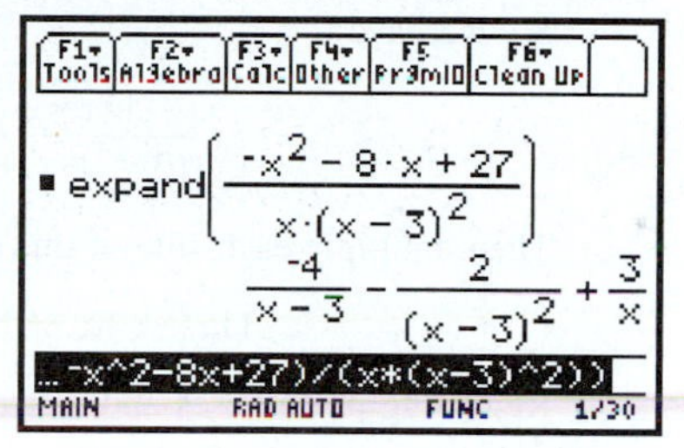

EXAMPLE 4

Find the partial fractions of $\dfrac{3x^2 - 12x + 17}{(x-2)^3}$.

Since $x - 2$ is repeated as a linear factor three times, the possible partial fractions are

$$\frac{A}{x-2} \qquad \frac{B}{(x-2)^2} \quad \text{and} \quad \frac{C}{(x-2)^3}$$

So, we have

$$\frac{3x^2 - 12x + 17}{(x-2)^3} = \frac{A}{x-2} + \frac{B}{(x-2)^2} + \frac{C}{(x-2)^3}$$

Then multiply each side of this equation by the L.C.D.: $(x-2)^3$.

$$3x^2 - 12x + 17 = A(x-2)^2 + B(x-2) + C$$

Removing parentheses and rearranging terms, we have

$$3x^2 - 12x + 17 = Ax^2 - 4Ax + 4A + Bx - 2B + C$$
$$3x^2 - 12x + 17 = Ax^2 + (-4A + B)x + 4A - 2B + C$$

Equating coefficients, we have the system

$$A = 3$$
$$-4A + B = -12$$
$$4A - 2B + C = 17$$

Substituting $A = 3$ into the second equation, we have $B = 0$. Then substituting $A = 3$ and $B = 0$ into the third equation, we have $C = 5$. Thus,

$$\frac{3x^2 - 12x + 17}{(x-2)^3} = \frac{3}{x-2} + \frac{5}{(x-2)^3}$$

CASE 3: NONREPEATED QUADRATIC DENOMINATOR FACTORS

For every nonrepeated factor $ax^2 + bx + c$ of the denominator of a given fraction, there corresponds the partial fraction $\dfrac{Ax + B}{ax^2 + bx + c}$ where A and B are constants.

EXAMPLE 5

Find the partial fractions of $\dfrac{11x^2 + 8x - 12}{(2x^2 + x + 2)(x + 1)}$.

The possible partial fractions are

$$\frac{Ax + B}{2x^2 + x + 2} \quad \text{and} \quad \frac{C}{x + 1}$$

So, we have

$$\frac{11x^2 + 8x - 12}{(2x^2 + x + 2)(x + 1)} = \frac{Ax + B}{2x^2 + x + 2} + \frac{C}{x + 1}$$

Then, multiply each side of this equation by the L.C.D.: $(2x^2 + x + 2)(x + 1)$.

$$11x^2 + 8x - 12 = (Ax + B)(x + 1) + C(2x^2 + x + 2)$$

Removing parentheses and rearranging terms, we have

$$11x^2 + 8x - 12 = Ax^2 + Ax + Bx + B + 2Cx^2 + Cx + 2C$$
$$11x^2 + 8x - 12 = (A + 2C)x^2 + (A + B + C)x + B + 2C$$

Equating coefficients, we have

$$A + 2C = 11$$
$$A + B + C = 8$$
$$B + 2C = -12$$

The solution of this system of linear equations is $A = 17$, $B = -6$, $C = -3$. Then

$$\frac{11x^2 + 8x - 12}{(2x^2 + x + 2)(x + 1)} = \frac{17x - 6}{2x^2 + x + 2} - \frac{3}{x + 1}$$

EXAMPLE 6

Find the partial fractions of $\dfrac{2x^3 - 2x^2 + 8x + 7}{(x^2 + 1)(x^2 + 4)}$.

The possible partial fractions are

$$\frac{Ax + B}{x^2 + 1} \quad \text{and} \quad \frac{Cx + D}{x^2 + 4}$$

So, we have

$$\frac{2x^3 - 2x^2 + 8x + 7}{(x^2 + 1)(x^2 + 4)} = \frac{Ax + B}{x^2 + 1} + \frac{Cx + D}{x^2 + 4}$$

Then, multiply each side of this equation by the L.C.D.: $(x^2 + 1)(x^2 + 4)$.

$$2x^3 - 2x^2 + 8x + 7 = (Ax + B)(x^2 + 4) + (Cx + D)(x^2 + 1)$$

Removing parentheses and rearranging terms, we have

$$2x^3 - 2x^2 + 8x + 7 = Ax^3 + Bx^2 + 4Ax + 4B + Cx^3 + Dx^2 + Cx + D$$
$$2x^3 - 2x^2 + 8x + 7 = (A + C)x^3 + (B + D)x^2 + (4A + C)x + 4B + D$$

Equating coefficients, we have

$$A + C = 2$$
$$B + D = -2$$
$$4A + C = 8$$
$$4B + D = 7$$

The solution of this system of linear equations is $A = 2$, $B = 3$, $C = 0$, $D = -5$. Then,

$$\frac{2x^3 - 2x^2 + 8x + 7}{(x^2 + 1)(x^2 + 4)} = \frac{2x + 3}{x^2 + 1} - \frac{5}{x^2 + 4}$$

CASE 4: REPEATED QUADRATIC DENOMINATOR FACTORS

For every factor $(ax^2 + bx + c)^k$ of the denominator of a given fraction, there correspond the possible partial fractions

$$\frac{A_1x + B_1}{ax^2 + bx + c}, \frac{A_2x + B_2}{(ax^2 + bx + c)^2}, \frac{A_3x + B_3}{(ax^2 + bx + c)^3}, \ldots, \frac{A_kx + B_k}{(ax^2 + bx + c)^k}$$

where $A_1, A_2, A_3, \ldots, A_k, B_1, B_2, B_3, \ldots, B_k$ are constants.

EXAMPLE 7

Find the partial fractions of $\dfrac{5x^4 - x^3 + 44x^2 - 5x + 75}{x(x^2 + 5)^2}$.

The possible partial fractions are

$$\frac{A}{x} \quad \frac{Bx + C}{x^2 + 5} \quad \text{and} \quad \frac{Dx + E}{(x^2 + 5)^2}$$

So, we have

$$\frac{5x^4 - x^3 + 44x^2 - 5x + 75}{x(x^2 + 5)^2} = \frac{A}{x} + \frac{Bx + C}{x^2 + 5} + \frac{Dx + E}{(x^2 + 5)^2}$$

Then, multiply each side of this equation by the L.C.D.: $x(x^2 + 5)^2$.

$$5x^4 - x^3 + 44x^2 - 5x + 75 = A(x^2 + 5)^2 + (Bx + C)(x^2 + 5)(x) + (Dx + E)x$$

Removing parentheses and rearranging terms, we have

$$5x^4 - x^3 + 44x^2 - 5x + 75 = Ax^4 + 10Ax^2 + 25A + Bx^4 + Cx^3 + 5Bx^2 + 5Cx + Dx^2 + Ex$$
$$5x^4 - x^3 + 44x^2 - 5x + 75 = (A + B)x^4 + Cx^3 + (10A + 5B + D)x^2 + (5C + E)x + 25A$$

Equating coefficients, we have

$$A + B = 5$$
$$C = -1$$
$$10A + 5B + D = 44$$

$$5C + E = -5$$
$$25A = 75$$

The solution of this system of linear equations is $A = 3$, $B = 2$, $C = -1$, $D = 4$, $E = 0$. Then,

$$\frac{5x^4 - x^3 + 44x^2 - 5x + 75}{x(x^2 + 5)^2} = \frac{3}{x} + \frac{2x - 1}{x^2 + 5} + \frac{4x}{(x^2 + 5)^2}$$

If the degree of the numerator is greater than or equal to the degree of the denominator of the original fraction, you must first divide the numerator by the denominator using long division. Then find the partial fractions of the resulting remainder.

EXAMPLE 8

Find the partial fractions of $\dfrac{x^3 + 3x^2 + 7x + 4}{x^2 + 2x}$.

Since the degree of the numerator is greater than the degree of the denominator, divide as follows:

$$\begin{array}{r} x + 1 \\ x^2 + 2x \overline{)x^3 + 3x^2 + 7x + 4} \\ \underline{x^3 + 2x^2} \qquad\qquad \\ x^2 + 7x \qquad \\ \underline{x^2 + 2x} \qquad \\ 5x + 4 \end{array}$$

or

$$\frac{x^3 + 3x^2 + 7x + 4}{x^2 + 2x} = x + 1 + \frac{5x + 4}{x^2 + 2x}$$

Now, factor the denominator and find the partial fractions of the remainder.

$$\frac{5x + 4}{x(x + 2)} = \frac{A}{x} + \frac{B}{x + 2}$$
$$5x + 4 = A(x + 2) + Bx$$
$$5x + 4 = Ax + 2A + Bx$$
$$5x + 4 = (A + B)x + 2A$$

Then,

$$A + B = 5$$
$$2A = 4$$

So, $A = 2$ and $B = 3$ and

$$\frac{x^3 + 3x^2 + 7x + 4}{x^2 + 2x} = x + 1 + \frac{2}{x} + \frac{3}{x + 2}$$

By calculator,

F2 **3** (x^3+3x^2+7x+4)/(x^2+2x)) **ENTER**

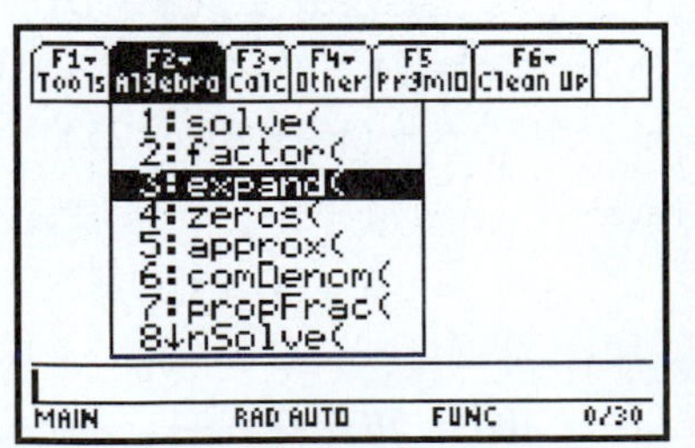

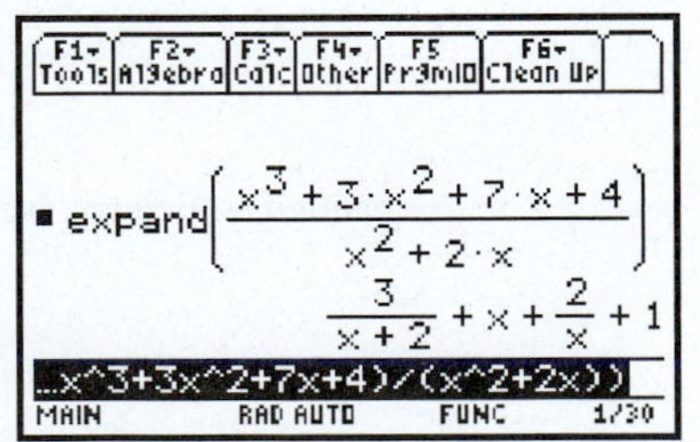

Exercises 6.7

Find the partial fractions of each expression.

1. $\dfrac{8x - 29}{(x + 2)(x - 7)}$
2. $\dfrac{10x - 34}{(x - 4)(x - 2)}$
3. $\dfrac{-x - 18}{2x^2 - 5x - 12}$
4. $\dfrac{17x - 18}{3x^2 + x - 2}$
5. $\dfrac{61x^2 - 53x - 28}{x(3x - 4)(2x + 1)}$
6. $\dfrac{11x^2 - 7x - 42}{(2x + 3)(x^2 - 2x - 3)}$
7. $\dfrac{x^2 + 7x + 10}{(x + 1)(x + 3)^2}$
8. $\dfrac{3x^2 - 18x + 9}{(2x - 1)(x - 1)^2}$
9. $\dfrac{48x^2 - 20x - 5}{(4x - 1)^3}$
10. $\dfrac{x^2 + 8x}{(x + 4)^3}$
11. $\dfrac{11x^2 - 18x + 3}{x(x - 1)^2}$
12. $\dfrac{6x^2 + 4x + 4}{x^3 + 2x^2}$
13. $\dfrac{-x^2 - 4x + 3}{(x^2 + 1)(x^2 - 3)}$
14. $\dfrac{-6x^3 + 2x^2 - 3x + 10}{(2x^2 + 1)(x^2 + 5)}$
15. $\dfrac{4x^3 - 21x - 6}{(x^2 + x + 1)(x^2 - 5)}$
16. $\dfrac{x^3 + 6x^2 + 2x - 2}{(3x^2 - x - 1)(x^2 + 4)}$
17. $\dfrac{4x^3 - 16x^2 - 93x - 9}{(x^2 + 5x + 3)(x^2 - 9)}$
18. $\dfrac{12x^2 + 8x - 72}{(x^2 + x - 1)(x^2 - 16)}$
19. $\dfrac{8x^4 - x^3 + 13x^2 - 6x + 5}{x(x^2 + 1)^2}$
20. $\dfrac{-4x^4 + 6x^3 + 8x^2 - 19x + 17}{(x - 1)(x^2 - 3)^2}$
21. $\dfrac{x^5 - 2x^4 - 8x^2 + 4x - 8}{x^2(x^2 + 2)^2}$
22. $\dfrac{3x^5 + x^4 + 24x^3 + 10x^2 + 48x + 16}{x^2(x^2 + 4)^2}$
23. $\dfrac{6x^2 + 108x + 54}{x^4 - 81}$
24. $\dfrac{x^6 + 2x^4 + 3x^2 + 1}{x^2(x^2 + 1)^3}$
25. $\dfrac{x^3}{x^2 - 1}$
26. $\dfrac{x^4 + x^2}{(x + 1)(x - 2)}$
27. $\dfrac{x^3 - x^2 + 8}{x^2 - 4}$
28. $\dfrac{2x^3 - 2x^2 + 8x - 3}{x(x - 1)}$
29. $\dfrac{3x^4 - 2x^3 - 2x + 5}{x(x^2 + 1)}$
30. $\dfrac{x^5 - x^4 - 3x^3 + 7x^2 + 3x + 20}{(x + 2)(x^2 + 2)}$

CHAPTER 6 SUMMARY

1. *Graphs of linear systems of equations with two variables:*
   (a) The two lines may intersect at a common, single point. This point, in ordered pair form (x, y), is the solution of the system.
   (b) The two lines may be parallel with no points in common; hence, the system has no solution.
   (c) The two lines may coincide; the solution of the system is the set of all points on the common line.
2. *Solving a pair of linear equations by the addition-subtraction method:*
   (a) If necessary, multiply each side of one or both equations by some number so that the numerical coefficients of one of the variables are of equal absolute value.
   (b) If these coefficients of equal absolute value have like signs, subtract one equation from the other. If they have unlike signs, add the equations. That is, do whatever is necessary to eliminate that variable.
   (c) Solve the resulting equation for the remaining variable.
   (d) Substitute the solution for the variable found in Step (c) in either of the original equations, and solve this resulting equation for the second variable.
   (e) Check by substituting the solution in both original equations.

3. *Solving a pair of linear equations by the method of substitution:*
   (a) From either of the two given equations, solve for one variable in terms of the other.
   (b) Substitute this result from Step (a) in the *other* equation. Note that this step eliminates a variable.
   (c) Solve the equation obtained from Step (b) for the remaining variable.
   (d) From the equation obtained in Step (a), substitute the solution for the variable found in Step (c), and solve this resulting equation for the second variable.
   (e) Check by substituting the solution in both original equations.

4. *Steps for problem solving:*
   (a) Read the problem carefully at least twice.
   (b) If possible, draw a picture or a diagram.
   (c) Write what facts are given and what unknown quantities are to be found.
   (d) Choose a symbol to represent each quantity to be found (there will be more than one). Be sure to label each symbol to indicate what it represents.
   (e) Write appropriate equations relating these variables from the information given in the problem. Watch for information that is not stated but which should be assumed.
   (f) Solve for the unknown variables by the methods presented in this chapter.
   (g) Check your solution in the original equations.
   (h) Check your solution in the original verbal problem.

5. *Graphs of linear systems of equations with three variables:*
   (a) The three planes may intersect at a common, single point. This point, in ordered triple form (x, y, z), is then the solution of the system.
   (b) The three planes may intersect along a common line. The infinite set of points that satisfy the equation of the line is the solution of the system.
   (c) The three planes may not have any points in common; the system has no solution. For example, the planes may be parallel, or they may intersect triangularly with no points common to all three planes.
   (d) The three planes may coincide; the solution of the system is the set of all points in the common plane.

6. *The addition-subtraction method for a system of three linear equations with three variables:*
   (a) Choose a variable to be eliminated. Eliminate it from any pair of equations by using the techniques of the addition-subtraction method.
   (b) Eliminate this same variable from *any other pair* of equations.
   (c) The result of Steps (a) and (b) is a pair of linear equations in two unknowns. Solve this pair for the two variables.
   (d) Solve for the third variable by substituting the results from Step (c) in any one of the original equations.
   (e) Check by substituting the solution in all three original equations.

7. *A determinant is a square array of numbers such as*

$$\begin{vmatrix} a_1 & b_1 \\ a_2 & b_2 \end{vmatrix}$$

8. *Value of second-order determinant:*

$$\begin{vmatrix} a_1 & b_1 \\ a_2 & b_2 \end{vmatrix} = a_1b_2 - a_2b_1$$

9. *The value of a third-order determinant is defined as*

$$\begin{vmatrix} a_1 & b_1 & c_1 \\ a_2 & b_2 & c_2 \\ a_3 & b_3 & c_3 \end{vmatrix} = a_1b_2c_3 + a_3b_1c_2 + a_2b_3c_1 - a_3b_2c_1 - a_1b_3c_2 - a_2b_1c_3$$

$$= a_1(b_2c_3 - b_3c_2) - a_2(b_1c_3 - b_3c_1) + a_3(b_1c_2 - b_2c_1)$$

$$= a_1\begin{vmatrix} b_2 & c_2 \\ b_3 & c_3 \end{vmatrix} - a_2\begin{vmatrix} b_1 & c_1 \\ b_3 & c_3 \end{vmatrix} + a_3\begin{vmatrix} b_1 & c_1 \\ b_2 & c_2 \end{vmatrix}$$

Each 2 × 2 determinant is called a *minor* of an element in the 3 × 3 determinant. The minor of a given element is the resulting determinant after the row and the column that contain the element have been deleted, as shown in Table 6.4.

TABLE 6.4

Element	Minor	Original determinant
a_1	$\begin{vmatrix} b_2 & c_2 \\ b_3 & c_3 \end{vmatrix}$	$\begin{vmatrix} a_1 & b_1 & c_1 \\ a_2 & b_2 & c_2 \\ a_3 & b_3 & c_3 \end{vmatrix}$
a_2	$\begin{vmatrix} b_1 & c_1 \\ b_3 & c_3 \end{vmatrix}$	$\begin{vmatrix} a_1 & b_1 & c_1 \\ a_2 & b_2 & c_2 \\ a_3 & b_3 & c_3 \end{vmatrix}$
a_3	$\begin{vmatrix} b_1 & c_1 \\ b_2 & c_2 \end{vmatrix}$	$\begin{vmatrix} a_1 & b_1 & c_1 \\ a_2 & b_2 & c_2 \\ a_3 & b_3 & c_3 \end{vmatrix}$

The value of any determinant of any order may be found by evaluating the sums and differences of *any* row or column of the products of the elements and the corresponding minors. The following diagram shows the signs (sum or difference) of the various products of elements and minors:

$$\begin{vmatrix} + & - & + & - & \cdots \\ - & + & - & + & \cdots \\ + & - & + & - & \cdots \\ - & + & - & + & \cdots \\ \cdot & \cdot & \cdot & \cdot & \\ \cdot & \cdot & \cdot & \cdot & \\ \cdot & \cdot & \cdot & \cdot & \end{vmatrix}$$

10. *Properties of determinants:*

(a) *Property 1:* If every element in a row (or a column) of a determinant is zero, the value of the determinant is zero.

(b) *Property 2:* If two rows (or two columns) of a determinant are identical, the value of the determinant is zero.

(c) *Property 3:* If any two rows (or two columns) of a determinant are interchanged, only the sign of the value of the determinant is changed.

(d) *Property 4:* If every element of a row (or a column) is multiplied by the same real number k, the value of the determinant is multiplied by k.

(e) *Property 5:* If every element of a row (or a column) is multiplied by the same real number k, and if the resulting products are added to another row (or another column), the value of the determinant remains the same.

11. *Cramer's rule:*

(a) The solution of two linear equations in two variables

$$a_1x + b_1y = c_1$$
$$a_2x + b_2y = c_2$$

may be written in determinant form as follows:

$$x = \frac{\begin{vmatrix} c_1 & b_1 \\ c_2 & b_2 \end{vmatrix}}{\begin{vmatrix} a_1 & b_1 \\ a_2 & b_2 \end{vmatrix}} \qquad y = \frac{\begin{vmatrix} a_1 & c_1 \\ a_2 & c_2 \end{vmatrix}}{\begin{vmatrix} a_1 & b_1 \\ a_2 & b_2 \end{vmatrix}}$$

(b) The solution of three linear equations in three variables

$$a_1x + b_1y + c_1z = d_1$$
$$a_2x + b_2y + c_2z = d_2$$
$$a_3x + b_3y + c_3z = d_3$$

may be written in determinant form as follows:

$$x = \frac{\begin{vmatrix} d_1 & b_1 & c_1 \\ d_2 & b_2 & c_2 \\ d_3 & b_3 & c_3 \end{vmatrix}}{\begin{vmatrix} a_1 & b_1 & c_1 \\ a_2 & b_2 & c_2 \\ a_3 & b_3 & c_3 \end{vmatrix}} \qquad y = \frac{\begin{vmatrix} a_1 & d_1 & c_1 \\ a_2 & d_2 & c_2 \\ a_3 & d_3 & c_3 \end{vmatrix}}{\begin{vmatrix} a_1 & b_1 & c_1 \\ a_2 & b_2 & c_2 \\ a_3 & b_3 & c_3 \end{vmatrix}} \qquad z = \frac{\begin{vmatrix} a_1 & b_1 & d_1 \\ a_2 & b_2 & d_2 \\ a_3 & b_3 & d_3 \end{vmatrix}}{\begin{vmatrix} a_1 & b_1 & c_1 \\ a_2 & b_2 & c_2 \\ a_3 & b_3 & c_3 \end{vmatrix}}$$

12. *Partial fractions:*

(a) *Case 1: Nonrepeated linear denominator factors.* For every nonrepeated factor $ax + b$ of the denominator of a given fraction, there corresponds the partial fraction $\dfrac{A}{ax + b}$, where A is a constant.

(b) *Case 2: Repeated linear denominator factors.* For every factor $(ax + b)^k$ of the denominator of the given fraction, there correspond the possible partial fractions

$$\frac{A_1}{ax + b}, \frac{A_2}{(ax + b)^2}, \frac{A_3}{(ax + b)^3}, \ldots, \frac{A_k}{(ax + b)^k}$$

where $A_1, A_2, A_3, \ldots, A_k$ are constants.

(c) *Case 3: Nonrepeated quadratic denominator factors.* For every nonrepeated factor $ax^2 + bx + c$ of the denominator of the given fraction, there corresponds the partial fraction

$$\frac{Ax + B}{ax^2 + bx + c}$$

where A and B are constants.

(d) *Case 4: Repeated quadratic denominator factors.* For every factor $(ax^2 + bx + c)^k$ of the denominator of the given fraction, there correspond the possible partial fractions

$$\frac{A_1x + B_1}{ax^2 + bx + c}, \frac{A_2x + B_2}{(ax^2 + bx + c)^2}, \frac{A_3x + B_3}{(ax^2 + bx + c)^3}, \ldots, \frac{A_kx + B_k}{(ax^2 + bx + c)^k}$$

where $A_1, A_2, A_3, \ldots, A_k, B_1, B_2, B_3, \ldots, B_k$ are constants.

CHAPTER 6 REVIEW

Solve each system of equations graphically and check by substituting the solution ordered pair in both original equations.

1. $x + y = 3$; $x - y = -1$
2. $y - 3x = 2$; $y - x = 2$
3. $2x - y = -10$; $4x + y = 4$
4. $2x - 3y = 6$; $-4x + 6y = 8$

Solve each system of equations by the addition-subtraction method and check.

5. $3x - 4y = 5$; $x + 7y = 10$
6. $2x + 3y = 7$; $4x - 6y = 11$
7. $x - 2y = 6$; $4x + y = 6$
8. $3x - 2y = 5$; $2x + 2y = 15$

Solve each system by the method of substitution and check.

9. $x = -4y$; $3x - 5y = 17$
10. $y = 3x$; $2x - 15y = 86$
11. $5x - y = 7$; $2x + 3y = 13$
12. $3x + 4y = -7$; $-5x + 12y = 14$

Solve each system of equations.

13. $8x + by = 4$; $16x - ay = 12$
14. $\frac{3}{x} - \frac{4}{y} = -5$; $\frac{6}{x} + \frac{5}{y} = 16$
15. $x + y + z = 2$; $x - y - 2z = 3$; $x + 2y - z = \frac{3}{2}$
16. $2x + 2y - z = 4$; $2x - y + 2z = -2$; $x - 5y + z = 8$

Evaluate each determinant.

17. $\begin{vmatrix} 2 & -3 \\ 6 & 2 \end{vmatrix}$
18. $\begin{vmatrix} -1 & 7 \\ 4 & -3 \end{vmatrix}$
19. $\begin{vmatrix} 1 & 3 & 5 \\ 7 & 2 & -1 \\ -4 & 2 & -8 \end{vmatrix}$
20. $\begin{vmatrix} 2 & 0 & -3 \\ 1 & 4 & 2 \\ 5 & 1 & -1 \end{vmatrix}$

Solve each system using determinants and check.

21. $4x + 3y = 46$; $2x - 3y = 14$
22. $3x - 2y = 13$; $-6x + 4y = -26$
23. $2x + 9y = 4$; $5x - 9y = 10$
24. $5x - 2y = 2$; $-3x + 3y = 1$
25. $2x + 2y - z = 11$; $3x + 4y + z = -16$; $4x - 8y + 3z = -113$
26. $x + y + z = 9$; $2x - y + z = 3$; $x + 3y - z = 3$

Solve each system by any method and check.

27. $2x + 7y = 3$
$10x + 3y = -1$

28. $-5x + 2y = 5$
$-3x + 7y = 32$

29. $2x - 3y = 10$
$2x + 5y = 2$

30. $x + y = -25$
$2x - 3y = 70$

31. $x + y + z = 7$
$x + z = 2$
$y - z = 4$

32. $7x - 2y + 5z = 9$
$3x + 4y - 9z = 27$
$10x - 3y + 7z = 7$

33. An alloy contains 10% lead. Another alloy contains 20% lead. How many kilograms of each must be used to make a 30-kg alloy containing 12% lead?

34. A person deposits \$3600 into two different savings accounts. One account earns interest at $5\frac{1}{2}\%$ and the other at $6\frac{1}{2}\%$. The total interest earned from both accounts at the end of one year is \$220. Find the amount deposited at $5\frac{1}{2}\%$ and the amount deposited at $6\frac{1}{2}\%$.

35. It took a plane 1 h 15 min to make a trip of 100 km flying into a head wind. On the return trip with the same wind velocity, it took only 1 h. Find the wind velocity and the speed of the plane (in still air).

36. Jack, John, and Bob work on an assembly line producing hand-cut aluminum parts. Jack and John together can process, on average, three times as many pieces per hour as Bob. Jack can process 2 more pieces per hour than John, and together all three can process 32 pieces per hour. How many pieces per hour does each man process, on average?

Evaluate each determinant.

37. $\begin{vmatrix} 1 & 0 & 3 & 6 \\ 4 & -2 & 1 & -5 \\ -3 & 2 & 5 & 2 \\ 1 & 0 & 7 & -9 \end{vmatrix}$

38. $\begin{vmatrix} 5 & 1 & 6 & -3 \\ -2 & 2 & 5 & 1 \\ 3 & -1 & 7 & 5 \\ 3 & 2 & 3 & -2 \end{vmatrix}$

Find the partial fractions of each expression.

39. $\dfrac{6x + 14}{(x - 3)(x + 5)}$

40. $\dfrac{-x^2 - 6x - 1}{x(x^2 - 1)}$

41. $\dfrac{3x^2 + 5x + 1}{x^2(x + 1)}$

42. $\dfrac{10x + 4}{(x + 1)^2}$

43. $\dfrac{7x^2 - x + 2}{(x^2 + 1)(x - 1)}$

44. $\dfrac{5x^2 + 2x + 21}{(x^2 + 4)^2}$

CHAPTER **6**

APPLICATION
Kirchhoff's Laws for Electric Circuits

Gustav Kirchhoff (1824–1887), a German physicist, formulated two laws that provide a mathematical analysis of the current, voltage, and resistance in an electric circuit and the basis for understanding electric circuit analysis. While *Ohm's law* describes the relationship among voltage V, current I, and resistance R in a circuit, $V = IR$, and simple series and parallel circuits can be solved with Ohm's law, Kirchhoff's laws apply more easily to complex circuits with multiple sources and branches. Note: An electric circuit with only one path for the current to flow is called a *series circuit* and an electric circuit with more than one path for the current to flow is called a *parallel circuit*.

Kirchhoff's two laws are based on two laws of conservation. **Conservation of charge:** In any closed electric circuit, charge cannot be created or destroyed. **Conservation of energy:** In any closed electric circuit, the total energy gain by an electron equals the energy lost. After conducting many experiments with electric circuits, Kirchhoff developed the following two laws, which are named after him:

Kirchhoff's current law (sometimes called the *junction rule*): The sum of currents in a closed electric circuit entering any junction equals the sum of the currents leaving that junction. (This law is a restatement of the law of conservation of charge.) For example, in Fig. 6.5, $I_1 + I_2 = I_3 + I_4 + I_5$.

Kirchhoff's voltage law (sometimes called the *loop rule*): The sum of the voltage drops (potential differences) around any closed loop in an electric circuit equals the sum of the voltage sources of that loop. (This law is a restatement of the law of conservation of energy.) For example, in Fig. 6.6, $E = V_1 + V_2 + V_3$ or $E = I_1R_1 + I_2R_2 + I_3R_3$, where $I_{\text{Total}} = I_1 = I_2 = I_3$.

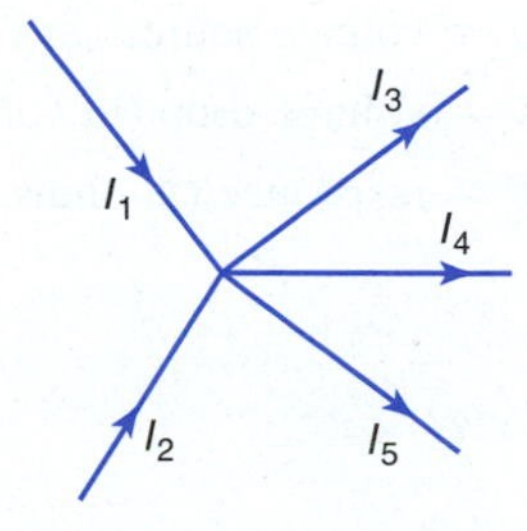

Figure 6.5 The sum of the currents entering the junction equals the sum of the currents leaving the junction; that is, $I_1 + I_2 = I_3 + I_4 + I_5$.

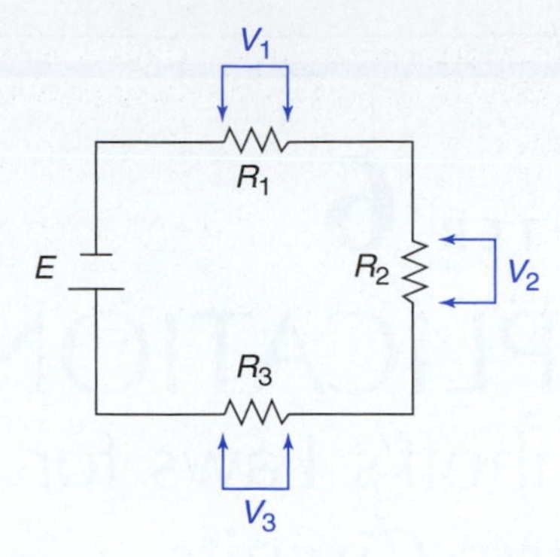

Figure 6.6 The sum of the voltage drops around any closed loop in an electric circuit equals the sum of the voltage sources of that loop; that is, $E = V_1 + V_2 + V_3$ or $E = I_1R_1 + I_2R_2 + I_3R_3$.

To analyze electric circuits using Kirchhoff's laws, you should:

1. Draw the circuit and label the known and unknown quantities. (Assign directions for the currents. If you incorrectly label the direction of a given unknown current, your answer will be negative but will have the correct magnitude.)
2. Apply the current law to as many junctions as possible to obtain several equations involving the current.
3. Apply the voltage law to as many loops as possible to obtain several equations involving voltage, usually in the form of IR.

Your goal is to develop a set of simultaneous equations involving n equations and n unknowns. Then, solve the resulting set of simultaneous equations.

Usually a circuit contains more than n loops. Use an equation from one of these extra loops to check your answer.

The following formulas apply to series and parallel circuits:

Series	*Parallel*
$I_{\text{Total}} = I_1 = I_2 = I_3 = \cdots$	$I_{\text{Total}} = I_1 + I_2 + I_3 + \cdots$
$E_{\text{Total}} = V_1 + V_2 + V_3 + \cdots$	$E_{\text{Total}} = V_1 = V_2 = V_3 = \cdots$
$R_{\text{Total}} = R_1 + R_2 + R_3 + \cdots$	$\dfrac{1}{R_{\text{Total}}} = \dfrac{1}{R_1} + \dfrac{1}{R_2} + \dfrac{1}{R_3} + \cdots$

where I = current (in amperes, A)
E = voltage source (in volts, V)
V = voltage drop (in volts, V)
R = resistance (in ohms, Ω)

EXAMPLE

Find the current through each resistor in the circuit in Fig. 6.7 using Kirchhoff's laws.

Using Kirchhoff's current law at junction A, we have

$$I_1 = I_2 + I_3 \qquad \textbf{(1)}$$

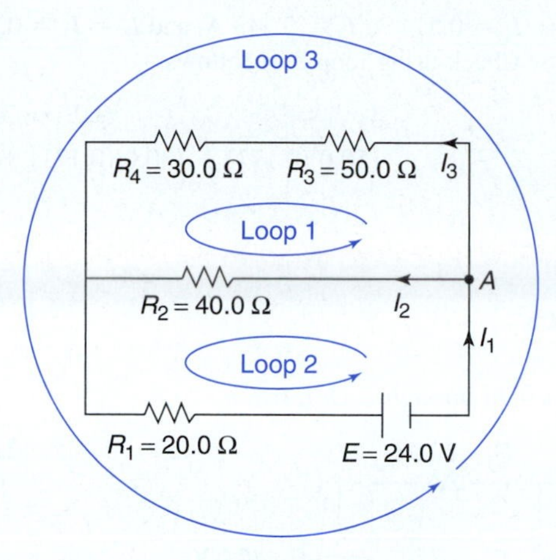

Figure 6.7

Using Kirchhoff's voltage law in loop 1, we have

$$50.0I_3 + 30.0I_4 - 40.0I_2 = 0$$ (*Note:* I_2 is in the opposite direction of I_3 and I_4 in loop 1, and the sum of the voltage drops is 0 because there are no voltage sources in loop 1.)

$$50.0I_3 + 30.0I_3 - 40.0I_2 = 0$$ (*Note:* $I_3 = I_4$ because the current through each resistor in series is the same.)

$$80.0I_3 = 40.0I_2$$

$$2I_3 = I_2 \qquad \textbf{(2)}$$

Using Kirchhoff's voltage law in loop 2, we have

$$40.0I_2 + 20.0I_1 = 24.0$$

$$10I_2 + 5I_1 = 6 \qquad \textbf{(3)}$$

Solve the system of Equations (1), (2), and (3) as follows: Solve Equation (2) for I_3 and substitute into Equation (1).

$$I_1 = I_2 + I_3$$

$$I_1 = I_2 + \tfrac{1}{2}I_2 \qquad [I_3 = \tfrac{1}{2}I_2 \text{ from Equation (2)}]$$

$$I_1 = \tfrac{3}{2}I_2 \qquad \textbf{(4)}$$

Substitute Equation (4) into Equation (3).

$$10I_2 + 5I_1 = 6$$

$$10I_2 + 5(\tfrac{3}{2}I_2) = 6$$

$$17.5I_2 = 6$$

$$I_2 = 0.343 \text{ A}$$

From Equation (4), we then have

$$I_1 = \tfrac{3}{2}I_2$$

$$I_1 = \tfrac{3}{2}(0.343 \text{ A})$$

$$= 0.514 \text{ A}$$

From Equation (1),

$$I_3 = I_1 - I_2$$
$$I_3 = 0.514\text{ A} - 0.343\text{ A}$$
$$= 0.171\text{ A}$$

Thus, $I_1 = 0.514$ A, $I_2 = 0.343$ A, and $I_3 = I_4 = 0.171$ A.

Check using loop 3 as follows:

$$50.0I_3 + 30.0I_4 + 20.0I_1 = 24.0$$
$$(50.0)(0.171) + (30.0)(0.171) + (20.0)(0.514) = 24.0$$
$$24.0 = 24.0$$

Exercises

Find the current through each resistor.

1.

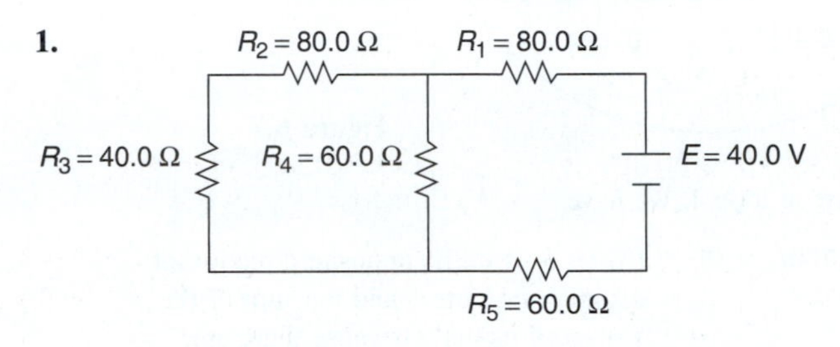

2.

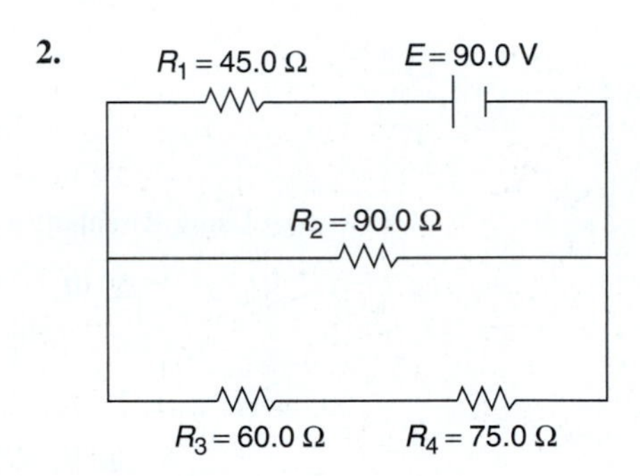

3.

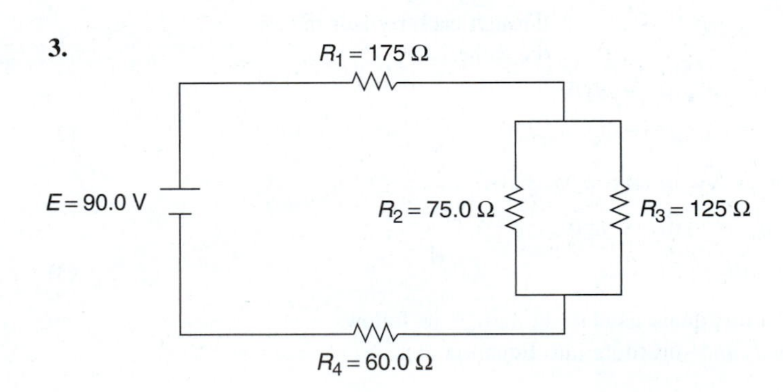

4.

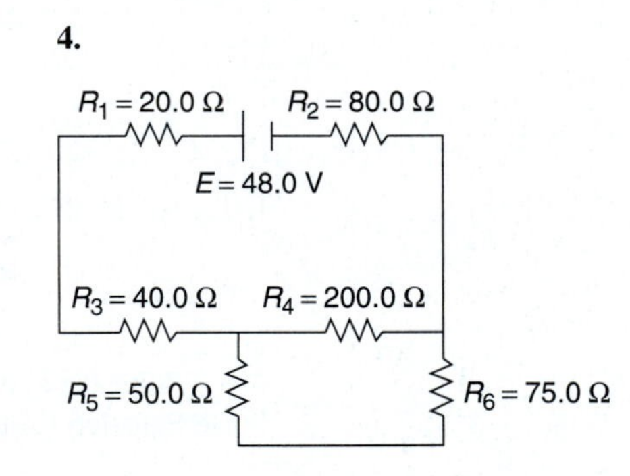

7 Quadratic Equations

INTRODUCTION

A baseball player hits a ball straight up in the air. The ball's height h in feet after t seconds may be expressed by $h = 64t - 16t^2$. Does the ball ever reach a height of 80 feet? How long is the ball in the air?

The formula that gives the height of the ball after t seconds is called a *quadratic equation.* Variable electric current, specification of box sizes, and the path of a projectile shot or thrown in the air can all be represented by a quadratic equation.

In this chapter we learn how to solve quadratic equations and how to interpret some of their characteristics. (The preceding problem is number 25 in the chapter review exercises.)

Objectives

- Solve quadratic equations by factoring.
- Solve quadratic equations by completing the square.
- Solve quadratic equations by using the quadratic formula.

7.1 SOLVING QUADRATIC EQUATIONS BY FACTORING

A **quadratic equation** in one variable is an equation with at least one term of second degree and no other term of higher degree. A quadratic equation is usually expressed in the form $ax^2 + bx + c = 0\ (a \neq 0)$.

In Chapter 4, equations, including quadratic equations, were solved graphically, which was an approximation method. Methods for obtaining exact solutions for quadratic equations are now presented.

One way to solve quadratic equations is by factoring. When using this method, we shall use the following principle:

If $ab = 0$, then either $a = 0$ or $b = 0$, or both.

That is, if the product of two factors is zero, one or both of the factors must also be zero.

EXAMPLE 1

Solve $(x + 4)(x - 3) = 0$.

By the preceding principle,

$$x + 4 = 0 \quad \text{or} \quad x - 3 = 0$$
$$x = -4 \quad \text{or} \quad x = 3$$

SOLVING A QUADRATIC EQUATION BY FACTORING

1. If necessary, write the equation in the form $ax^2 + bx + c = 0$; that is, "set the equation equal to zero."
2. Factor the nonzero side of the equation.
3. Using the preceding principle, set each factor that contains a variable equal to zero.
4. Solve each resulting linear equation.
5. Check.

EXAMPLE 2

Solve $2x^2 = 7x + 15$.

Step 1: $2x^2 - 7x - 15 = 0$

Step 2: $(2x + 3)(x - 5) = 0$

Step 3: $2x + 3 = 0 \quad \text{or} \quad x - 5 = 0$

Step 4: $2x = -3$

$$x = -\frac{3}{2} \quad \text{or} \quad x = 5$$

Check:

Step 5:

$x = -\frac{3}{2}$	$x = 5$
$2\left(-\frac{3}{2}\right)^2 = 7\left(-\frac{3}{2}\right) + 15$	$2(5)^2 = 7(5) + 15$
$\frac{9}{2} = -\frac{21}{2} + 15$	$50 = 35 + 15$
$\frac{9}{2} = \frac{9}{2}$	$50 = 50$

EXAMPLE 3

Solve $6x^2 - 13x - 5 = 0$.

$$(3x + 1)(2x - 5) = 0$$
$$3x + 1 = 0 \quad \text{or} \quad 2x - 5 = 0$$
$$3x = -1 \qquad\qquad 2x = 5$$
$$x = -\frac{1}{3} \quad \text{or} \quad x = \frac{5}{2}$$

Check: Use the same procedure as in Example 2.

EXAMPLE 4

Solve $16x^2 + 4x = 0$.

$$4x(4x + 1) = 0$$

$$4x = 0 \quad \text{or} \quad 4x + 1 = 0$$

$$4x = -1$$

$$x = 0 \quad \text{or} \quad x = -\frac{1}{4}$$

EXAMPLE 5

Solve $4x^2 = 49$.

$$4x^2 - 49 = 0$$

$$(2x + 7)(2x - 7) = 0$$

$$2x + 7 = 0 \quad \text{or} \quad 2x - 7 = 0$$

$$2x = -7 \qquad 2x = 7$$

$$x = -\frac{7}{2} \quad \text{or} \quad x = \frac{7}{2}$$

SOLVING $ax^2 = c$

1. Divide each side by a.
2. Take the square root of each side.
3. Simplify the result, if possible.

EXAMPLE 6

Solve $2x^2 = 32$.

$$2x^2 = 32$$

Step 1: $x^2 = 16$

Step 2: $x = \pm\sqrt{16}$

Step 3: $x = \pm 4$

Note that in Step 2 we introduced the $\pm$ sign because there are **two** numbers whose square is 16, namely, 4 and -4.

EXAMPLE 7

Solve $9y^2 - 13 = 7$.

$$9y^2 - 13 = 7$$

$$9y^2 = 20$$

$$y^2 = \frac{20}{9}$$

$$y = \pm\sqrt{\frac{20}{9}}$$

$$y = \pm\frac{2\sqrt{5}}{3}$$

You may wish to review simplifying radicals in Section 1.7.

Exercises 7.1

Solve each equation.

1. $(x + 4)(x - 7) = 0$
2. $(3x + 1)(x + 5) = 0$
3. $x^2 - 6x + 8 = 0$
4. $x^2 - 9x + 18 = 0$
5. $x^2 + 3x = 10$
6. $x^2 = x + 42$
7. $2x^2 = 3x + 9$
8. $2x^2 + 9x = 5$
9. $14x^2 + 17x + 5 = 0$
10. $36x^2 + 4 = 25x$
11. $18x^2 + 56 = 69x$
12. $14x^2 + 5x = 24$
13. $8x^2 + x = 0$
14. $9x^2 - 6x = 0$
15. $-7x^2 + 21x = 0$
16. $x - 2x^2 = 0$
17. $x^2 - 36 = 0$
18. $16 - x^2 = 0$
19. $16x^2 - 25 = 0$
20. $4x^2 = 9$
21. $5x^2 - 12 = 0$
22. $3x^2 - 8 = 0$
23. $4x^2 = 21$
24. $7x^2 = 4$
25. $5a^2 - 40 = 0$
26. $2c^2 - 36 = 0$
27. $2 = \frac{1}{3}x^2$
28. $\frac{1}{4} = 3x^2$
29. $30x^2 + 16x = 24$
30. $24x^2 + 24 = 50x$
31. $40x^2 + 100x + 40 = 0$
32. $27x^2 + 45x = 18$
33. $\dfrac{x - 1}{2x} = \dfrac{5}{x + 12}$
34. $\dfrac{3x - 1}{\frac{11}{2}} = \dfrac{2}{5 - x}$
35. $\dfrac{2x - 3}{x - 4} = x + 2$
36. $\dfrac{3x + 16}{9x} = x + \dfrac{1}{3}$
37. $\dfrac{x + 1}{x - 2} + \dfrac{x - 1}{x + 1} = \dfrac{9}{2}$
38. $\dfrac{4}{x^2 - 1} + \dfrac{1}{x - 1} = \dfrac{7}{3}$

7.2 SOLVING QUADRATIC EQUATIONS BY COMPLETING THE SQUARE

Factoring is usually the easiest and fastest method of solving a quadratic equation *if* the equation can be factored. If the equation cannot be factored, then one of two methods is normally used. We shall study one of these methods, **completing the square,** in this section. The other method, the *quadratic formula,* is developed using the method of completing the square. This formula and its development are included in the next section.

SOLVING A QUADRATIC EQUATION BY COMPLETING THE SQUARE

1. The coefficient of the second-degree term *must* equal (positive) 1. If not, divide each side of the equation by its coefficient.
2. Write an equivalent equation with the variable terms on the left side and the constant term on the right side; that is, in the form $x^2 + px = q$.
3. Add the square of one-half of the coefficient of the first-degree term to each side, that is, $\left(\frac{1}{2}p\right)^2$.

$$x^2 + px + \frac{p^2}{4} = q + \frac{p^2}{4}$$

4. The left side is now a perfect square trinomial—as the method's name implies. Rewrite the left side as a square.

$$\left(x + \frac{p}{2}\right)^2 = q + \frac{p^2}{4}$$

5. Take the square root of each side.

$$x + \frac{p}{2} = \pm\sqrt{q + \frac{p^2}{4}}$$

6. Solve for the variable x and simplify, if possible.

$$x = -\frac{p}{2} \pm \sqrt{q + \frac{p^2}{4}}$$

7. Check.

EXAMPLE 1

Solve $x^2 + 3x - 10 = 0$ by completing the square.

Step 1: The coefficient of x^2 is 1, so add 10 to each side.

Step 2: $x^2 + 3x = 10$

Add the square of one-half of the coefficient of x, $\left(\frac{3}{2}\right)^2$ or $\frac{9}{4}$, to each side.

Step 3: $x^2 + 3x + \frac{9}{4} = 10 + \frac{9}{4}$

Rewrite the left side as a square and simplify the right side.

$$\left(x + \frac{3}{2}\right)\left(x + \frac{3}{2}\right) = \frac{40}{4} + \frac{9}{4}$$

Step 4: $\left(x + \frac{3}{2}\right)^2 = \frac{49}{4}$

Take the square root of each side.

Step 5: $x + \frac{3}{2} = \pm\frac{7}{2}$

Solve for x and simplify.

Step 6: $x = -\frac{3}{2} \pm \frac{7}{2}$

Note: $x = -\frac{3}{2} \pm \frac{7}{2}$ is a short way of writing

$$x = -\frac{3}{2} + \frac{7}{2} \quad \text{or} \quad x = -\frac{3}{2} - \frac{7}{2}$$

That is, read the equation all the way through a first time by using only the + sign of the ± sign; then read it through a second time using only the − sign. Therefore,

$$x = \frac{4}{2} = 2 \quad \text{or} \quad x = \frac{-10}{2} = -5$$

Check:

Step 7:

$(2)^2 + 3(2) - 10 = 0$	$(-5)^2 + 3(-5) - 10 = 0$
$4 + 6 \quad - 10 = 0$	$25 - 15 \quad - 10 = 0$
$0 = 0$	$0 = 0$

EXAMPLE 2

Solve $4x^2 - 24x + 7 = 0$ by completing the square.

First, divide each side of the equation by the coefficient of x^2, which is 4.

$$x^2 - 6x + \frac{7}{4} = 0$$

$$x^2 - 6x = -\frac{7}{4} \qquad \left(\text{Add } -\frac{7}{4} \text{ to each side.}\right)$$

$$x^2 - 6x + 9 = -\frac{7}{4} + 9 \qquad \left[\text{Add the square of one-half of the coefficient of } x, \left(\frac{-6}{2}\right)^2 \text{ or 9, to each side.}\right]$$

$$(x - 3)^2 = \frac{29}{4} \qquad \text{(Rewrite the left side as a perfect square and simplify the right side.)}$$

$$x - 3 = \pm\frac{\sqrt{29}}{2} \qquad \text{(Take the square root of each side.)}$$

$$x = 3 \pm \frac{\sqrt{29}}{2} \qquad \text{(Solve for } x \text{ and simplify.)}$$

$$x = \frac{6 \pm \sqrt{29}}{2} \qquad \text{(Combine terms using the lowest common denominator.)}$$

EXAMPLE 3

Solve $-2x^2 + 3x + 1 = 0$ by completing the square.

$$x^2 - \frac{3}{2}x - \frac{1}{2} = 0 \qquad \text{(Divide each side by } -2.)$$

$$x^2 - \frac{3}{2}x = \frac{1}{2}$$

$$x^2 - \frac{3}{2}x + \frac{9}{16} = \frac{1}{2} + \frac{9}{16} \qquad \left[\textit{Note: } \left(\frac{1}{2} \cdot -\frac{3}{2}\right)^2 = \frac{9}{16}\right]$$

$$\left(x - \frac{3}{4}\right)^2 = \frac{17}{16}$$

$$x - \frac{3}{4} = \pm\frac{\sqrt{17}}{4}$$

$$x = \frac{3}{4} \pm \frac{\sqrt{17}}{4} \quad \text{or} \quad x = \frac{3 \pm \sqrt{17}}{4}$$

Exercises 7.2

Solve each equation by completing the square.

1. $x^2 + 6x - 16 = 0$
2. $x^2 + 15x + 54 = 0$
3. $x^2 + 35 = 12x$
4. $x^2 = 6x + 27$
5. $2x^2 + x = 1$
6. $2x^2 + 3 = 7x$
7. $25x^2 = 15x + 18$
8. $3x^2 + 17x + 20 = 0$
9. $x^2 + 4x - 7 = 0$
10. $x^2 - 5x + 3 = 0$
11. $2x^2 + 2x - 9 = 0$
12. $5x^2 + 4x = 1$
13. $3x^2 + 9x + 5 = 0$
14. $-3x^2 - 5x + 4 = 0$
15. $-2x^2 + 3x + 9 = 0$
16. $-5x^2 + 8x - 2 = 0$
17. $4x^2 + 11x = 3$
18. $3a^2 + 5a = 2$
19. $6m^2 + 15m + 3 = 0$
20. $8y^2 - 7y - 10 = 0$

7.3 THE QUADRATIC FORMULA

Instead of using the method of completing the square for each quadratic equation, we may use the method to solve the general quadratic equation ($ax^2 + bx + c = 0$) and obtain a formula for solving any quadratic equation in that form. Let's now solve

$$ax^2 + bx + c = 0$$

by the method of completing the square.

First, divide each side by a, the coefficient of x^2.

$$x^2 + \frac{b}{a}x + \frac{c}{a} = 0$$

$$x^2 + \frac{b}{a}x = -\frac{c}{a} \qquad \left(\text{Add } -\frac{c}{a} \text{ to each side.}\right)$$

Add the square of one-half of the coefficient of x, $\left(\frac{1}{2} \cdot \frac{b}{a}\right)^2$ or $\frac{b^2}{4a^2}$, to each side.

$$x^2 + \frac{b}{a}x + \frac{b^2}{4a^2} = \frac{b^2}{4a^2} - \frac{c}{a}$$

$$\left(x + \frac{b}{2a}\right)^2 = \frac{b^2 - 4ac}{4a^2} \qquad \text{(Rewrite the left side as a perfect square and simplify the right side.)}$$

$$x + \frac{b}{2a} = \pm\frac{\sqrt{b^2 - 4ac}}{2a} \qquad \text{(Take the square root of each side.)}$$

Solve for x by adding $-\frac{b}{2a}$ to each side, and simplify.

SOLVING A QUADRATIC EQUATION USING THE QUADRATIC FORMULA

$$x = \frac{-b \pm \sqrt{b^2 - 4ac}}{2a}$$

To solve a quadratic equation by using the quadratic formula, we need only to identify a, b, and c from $ax^2 + bx + c = 0$ and substitute those values in the quadratic formula.

EXAMPLE 1

Solve $x^2 + 2x - 35 = 0$ using the quadratic formula.

First, identify a, b, and c.

$$a = 1$$
$$b = 2$$
$$c = -35$$

Then, substitute these values in the quadratic formula.

$$x = \frac{-b \pm \sqrt{b^2 - 4ac}}{2a}$$

$$x = \frac{-2 \pm \sqrt{(2)^2 - 4(1)(-35)}}{2(1)}$$

$$= \frac{-2 \pm \sqrt{4 + 140}}{2}$$

$$= \frac{-2 \pm \sqrt{144}}{2}$$

$$= \frac{-2 \pm 12}{2}$$

$$x = \frac{-2 + 12}{2} \quad \text{or} \quad x = \frac{-2 - 12}{2}$$

$$x = 5 \quad \text{or} \quad x = -7$$

Check:

$(5)^2 + 2(5) - 35 = 0$	$(-7)^2 + 2(-7) - 35 = 0$
$25 + 10 - 35 = 0$	$49 - 14 - 35 = 0$
$0 = 0$	$0 = 0$

EXAMPLE 2

Solve $3x^2 = 4x + 8$ using the quadratic formula.

First, identify a, b, and c. To find a, b, and c, the equation *must* be in the form

$$ax^2 + bx + c = 0$$

That is,

$$3x^2 - 4x - 8 = 0$$

Therefore, $a = 3$, $b = -4$, and $c = -8$. Second, substitute these values in the quadratic formula,

$$x = \frac{-b \pm \sqrt{b^2 - 4ac}}{2a}$$

$$x = \frac{-(-4) \pm \sqrt{(-4)^2 - 4(3)(-8)}}{2(3)}$$

$$= \frac{4 \pm \sqrt{16 + 96}}{6}$$

$$= \frac{4 \pm \sqrt{112}}{6}$$

$$= \frac{4 \pm 4\sqrt{7}}{6} \qquad (\sqrt{112} = \sqrt{16 \cdot 7} = 4\sqrt{7})$$

$$= \frac{2 \pm 2\sqrt{7}}{3}$$

An examination of the quadratic formula $x = \dfrac{-b \pm \sqrt{b^2 - 4ac}}{2a}$ gives an indication about the type of roots, or solutions, we can expect from the equation $ax^2 + bx + c = 0$, when a, b, and c are integers.

The quantity under the radical sign $(b^2 - 4ac)$ is called the **discriminant** because it discriminates, or determines, whether the roots are real or imaginary, rational or irrational, or one or two in number.

1. If $b^2 - 4ac > 0$, there are two real solutions.
   (a) If $b^2 - 4ac$ is also a perfect square, then the two solutions are rational.
   (b) If $b^2 - 4ac$ is not a perfect square, then both solutions are irrational.
2. If $b^2 - 4ac = 0$, there is only one real, rational solution.
3. If $b^2 - 4ac < 0$, there are two imaginary solutions (see Chapter 14).

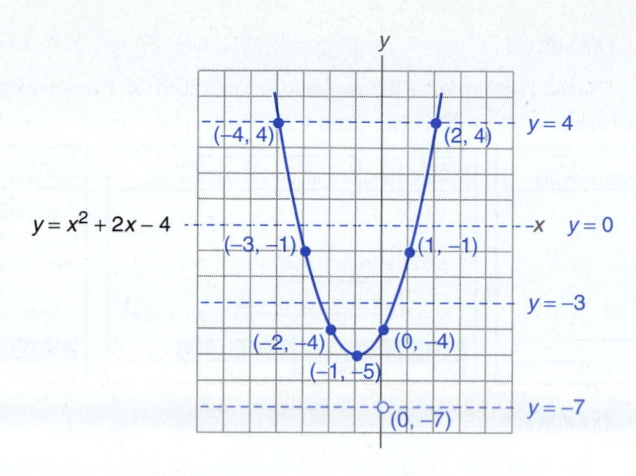

Figure 7.1

In Section 4.2, we solved equations graphically. The equation $y = x^2 + 2x - 4$ was solved graphically for $y = 0$, $y = 4$, $y = -3$, and $y = -7$; the graph is repeated in Fig. 7.1. Let's find each solution either by using the quadratic formula or by factoring.

Quadratic formula	*Factoring*

$$y = x^2 + 2x - 4 \quad \text{for } y = 0$$
$$0 = x^2 + 2x - 4$$
$$x = \frac{-b \pm \sqrt{b^2 - 4ac}}{2a}$$
$$x = \frac{-2 \pm \sqrt{(2)^2 - 4(1)(-4)}}{2(1)}$$
$$= \frac{-2 \pm \sqrt{20}}{2}$$
$$= \frac{-2 \pm 2\sqrt{5}}{2}$$
$$= -1 \pm \sqrt{5} \quad \text{or} \quad 1.24, -3.24$$

$$y = x^2 + 2x - 4 \quad \text{for } y = 4$$
$$4 = x^2 + 2x - 4$$
$$0 = x^2 + 2x - 8$$
$$0 = (x + 4)(x - 2)$$
$$x + 4 = 0 \quad \text{or} \quad x - 2 = 0$$
$$x = -4 \quad \text{or} \quad x = 2$$

$$y = x^2 + 2x - 4 \quad \text{for } y = -3$$
$$-3 = x^2 + 2x - 4$$
$$0 = x^2 + 2x - 1$$
$$x = \frac{-2 \pm \sqrt{(2)^2 - 4(1)(-1)}}{2(1)}$$
$$= \frac{-2 \pm \sqrt{8}}{2}$$
$$= \frac{-2 \pm 2\sqrt{2}}{2}$$
$$= -1 \pm \sqrt{2} \quad \text{or} \quad 0.414, -2.41$$

$$y = x^2 + 2x - 4 \quad \text{for } y = -7$$
$$-7 = x^2 + 2x - 4$$
$$0 = x^2 + 2x + 3$$
$$x = \frac{-2 \pm \sqrt{(2)^2 - 4(1)(3)}}{2(1)}$$
$$= \frac{-2 \pm \sqrt{-8}}{2}$$

The discriminant is negative; therefore, there are no real solutions. *Note:* In Fig. 7.1, the curve does not intersect the line $y = -7$.

The quadratic formula may also be used to solve literal quadratic equations, that is, quadratic equations involving letters. Assume that the letters represent positive real numbers.

EXAMPLE 3

Use the TI-89 to derive the quadratic formula, then use it to solve the equation $2x^2 + 5x - 1 = 0$.

F2 1 a*x^2+b*x+c=0,x) **ENTER** **2nd CUSTOM** right arrow | a=2 **F3 2** b=5 **F3 2** c=-1 **ENTER**

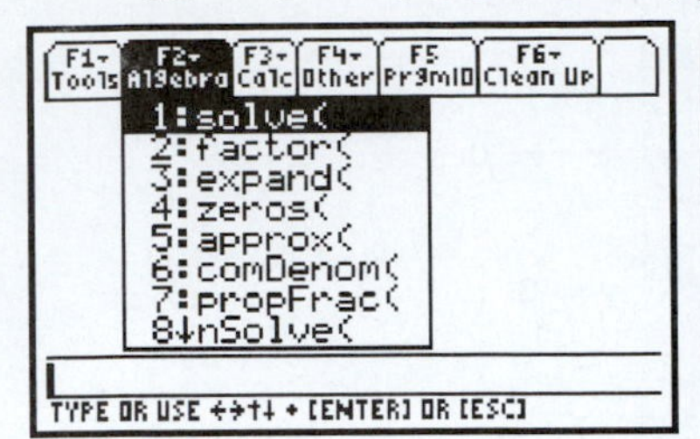

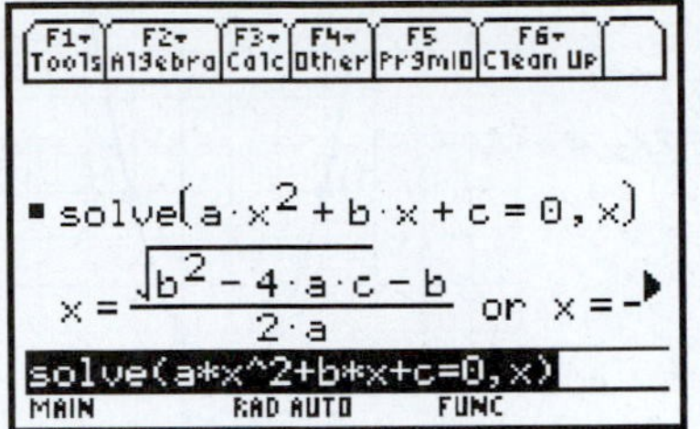

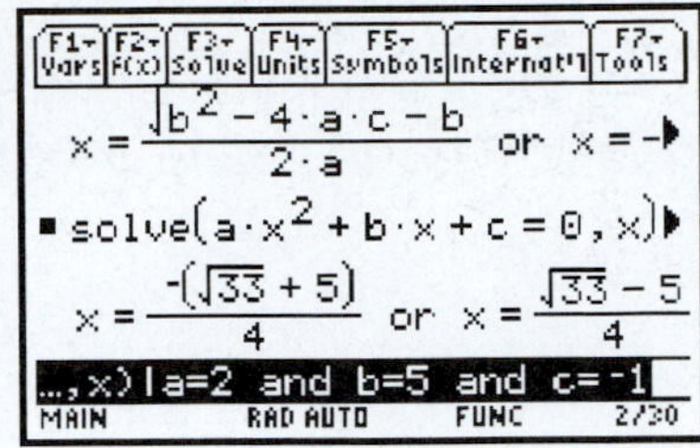

EXAMPLE 4

Solve $10.6x^2 + 11.7x - 28.4 = 0$ using the quadratic formula.

Here $a = 10.6$, $b = 11.7$, and $c = -28.4$.

$$x = \frac{-b \pm \sqrt{b^2 - 4ac}}{2a}$$

$$x = \frac{-11.7 \pm \sqrt{(11.7)^2 - 4(10.6)(-28.4)}}{2(10.6)}$$

$$x = \frac{-11.7 \pm \sqrt{1341.05}}{21.2}$$

$$x = 1.18 \quad \text{or} \quad x = -2.28$$

You may also use the quadratic formula when a quadratic equation has fraction or letter coefficients.

EXAMPLE 5

Solve $m^2x^2 - mnx - n^2 = 0$ for x.

Here $a = m^2$, $b = -mn$, and $c = -n^2$.

$$x = \frac{-b \pm \sqrt{b^2 - 4ac}}{2a}$$

$$x = \frac{-(-mn) \pm \sqrt{(-mn)^2 - 4(m^2)(-n^2)}}{2(m^2)}$$

$$= \frac{mn \pm \sqrt{m^2n^2 + 4m^2n^2}}{2m^2}$$

$$= \frac{mn \pm \sqrt{5m^2n^2}}{2m^2}$$

$$= \frac{mn \pm mn\sqrt{5}}{2m^2}$$

$$= \frac{n \pm n\sqrt{5}}{2m}$$

F2 1 m^2*x^2-m*n*x-n^2=0,x) **ENTER** up arrow **2nd** right arrow

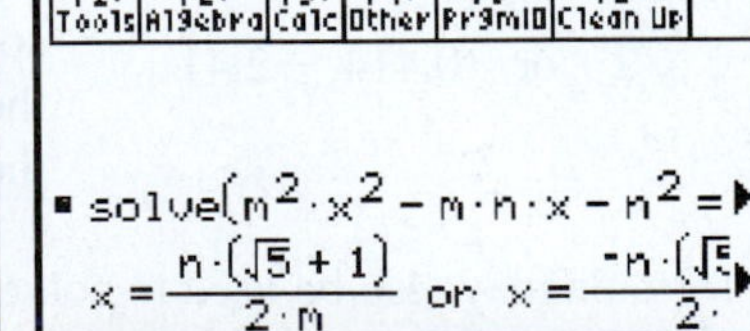

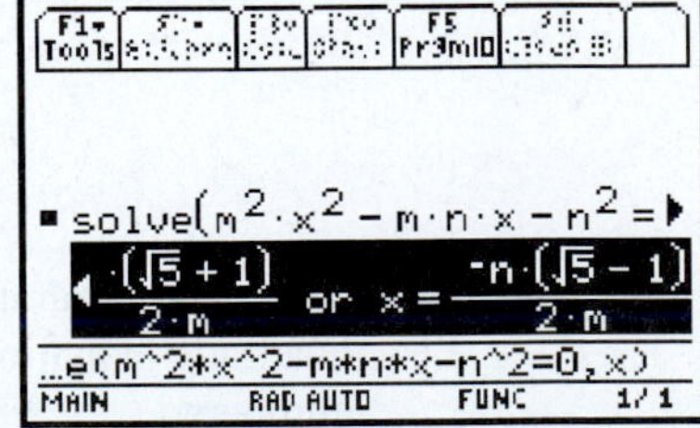

Exercises 7.3

Solve each equation using the quadratic formula.

1. $x^2 - 4x - 32 = 0$

2. $x^2 + 10x + 21 = 0$

3. $2x^2 = 3x + 9$

4. $2x^2 + 9x = 5$

5. $12x^2 + 5x = 28$

6. $6x^2 = 5x + 25$

7. $15x^2 + 75x + 90 = 0$

8. $20x^2 - 30x = 200$

9. $8x^2 + 27 = 30x$

10. $12x^2 - 72x = 135$

11. $6x^2 + 15x = 0$

12. $8x^2 - 24 = 0$

13. $x^2 + 5x = 7$

14. $4x^2 = 4x + 11$

15. $x^2 = 4x + 14$

16. $x^2 = 6x + 19$

17. $4x^2 = 8x + 23$

18. $20x + 3 = 4x^2$

19. $9x^2 + 25 = 42x$

20. $9x^2 = 15x + 59$

21. $x^2 - 4x + 1 = 0$

22. $2x^2 + 7x = -1$

23. $-3x^2 + 7x = 2$

24. $-6x^2 - 4x + 3 = 0$

25. $x^2 + 10x = 24$

26. $x^2 + 7 = 8x$

27. $4x^2 + 12x + 9 = 0$

28. $2x^2 + 5x = 4$

29. $2x^2 - 5x + 1 = 0$

30. $9x^2 + 25 = 30x$

31. $5x^2 + 6x + 4 = 0$

32. $3x^2 - 5x + 8 = 0$

33. $21.4x^2 - 16.4x = 12.4$

34. $3.76x^2 + 8.72x = 6.96$

35. $208.4x^2 + 187.2 = 444.5x$

36. $0.495x^2 + 0.705x = 0.142$

37. $\frac{3}{4}x^2 + \frac{2}{3}x = \frac{1}{2}$

38. $\frac{5}{6}x^2 = \frac{7}{8}x + \frac{1}{12}$

39. $3\frac{5}{6}x^2 + 5\frac{6}{7}x = 2\frac{9}{16}$

40. $24\frac{1}{3}x^2 + 11\frac{1}{4} = 40\frac{3}{4}x$

Solve each equation for the indicated values using any method.

41. $y = -2x^2 + 2x$
for $y = 0, -12$, and 7

42. $y = x^2 + x - 12$
for $y = 0, 8$, and 6

43. $y = 3x^2 + x$
for $y = 0, 4$, and 1

44. $y = 3x^2 + 2x$
for $y = 0, 40, 3$, and -3

Solve for x.

45. $m^2x^2 - 2mx + 1 = 0$

46. $4m^2x^2 + 6mnx - n^2 = 0$

47. $n^2x^2 + 4nx + 4 = 0$

48. $3x^2 + 4x - mn = 0$

49. $x^2 + 5x = 16a^2 - 20a$

50. $x^2 + 6nx + 9n^2 = 0$

Solve for r.

51. $A = \pi r^2$

52. $S = 2\pi r^2 + 2\pi rh$

7.4 APPLICATIONS

Let us now present some applications that involve quadratic equations. For consistency, all final results are rounded to three significant digits in this section.

EXAMPLE 1

Design a rectangular metal plate to meet the following specifications: (a) the length is 4.00 cm less than twice its width and (b) its area is 96.0 cm^2.

First, draw a diagram as in Fig. 7.2 and let

$$\begin{aligned} x &= \text{width} \\ 2x - 4 &= \text{length} \\ A &= lw \\ 96 &= (2x - 4)x \\ 96 &= 2x^2 - 4x \end{aligned}$$

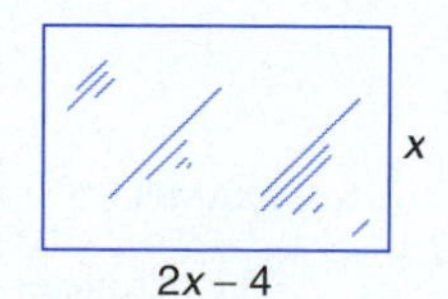

Figure 7.2

$$0 = 2x^2 - 4x - 96$$
$$0 = x^2 - 2x - 48$$
$$0 = (x + 6)(x - 8)$$
$$x + 6 = 0 \quad \text{or} \quad x - 8 = 0$$
$$x = -6 \quad \text{or} \quad x = 8$$

The solution $x = -6$ is not meaningful, as x refers to a length measurement, which must be a positive quantity. Therefore,

$$x = \text{width} = 8.00 \text{ cm}$$
$$2x - 4 = \text{length} = 12.0 \text{ cm}$$

Example 1 represents a situation where we must be very cautious of our interpretation of algebraic solutions. We must be aware that in some cases a numerical solution may not be meaningful to the given physical problem. Such a solution is algebraically correct, but makes no sense when it is applied to the problem.

EXAMPLE 2

A variable voltage in a given electric circuit is given by the formula $V = t^2 - 12t + 40$. At what values of t (in seconds) is the voltage V equal to 8.00 V? To 25.0 V?

If $V = 8.00$,

$$8 = t^2 - 12t + 40$$
$$0 = t^2 - 12t + 32$$
$$0 = (t - 4)(t - 8)$$
$$t - 4 = 0 \quad \text{or} \quad t - 8 = 0$$
$$t = 4.00 \text{ s} \quad \text{or} \quad t = 8.00 \text{ s}$$

If $V = 25.0$,

$$25 = t^2 - 12t + 40$$
$$0 = t^2 - 12t + 15 \qquad \text{(Does not factor.)}$$
$$t = \frac{-b \pm \sqrt{b^2 - 4ac}}{2a}$$
$$t = \frac{-(-12) \pm \sqrt{(-12)^2 - 4(1)(15)}}{2(1)}$$
$$= \frac{12 \pm \sqrt{144 - 60}}{2}$$
$$= \frac{12 \pm \sqrt{84}}{2} \qquad (\sqrt{84} = \sqrt{4 \cdot 21} = 2\sqrt{21})$$
$$= \frac{12 \pm 2\sqrt{21}}{2}$$
$$= 6 \pm \sqrt{21}$$

If decimal solutions are needed ($\sqrt{21}$ is approximately 4.58),

$$t = 6 + \sqrt{21} \quad \text{or} \quad t = 6 - \sqrt{21}$$
$$t = 6 + 4.58 \quad \text{or} \quad t = 6 - 4.58$$
$$t = 10.6 \text{ s} \quad \text{or} \quad t = 1.42 \text{ s}$$

EXAMPLE 3

The perimeter of a rectangle is 24.0 m and its area is 20.0 m^2. Find the dimensions (the length and the width).

First, draw a diagram as in Fig. 7.3 and let

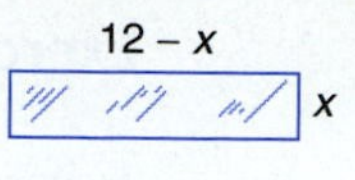

Figure 7.3

$$x = \text{width}$$
$$12 - x = \text{length}$$

(*Note:* The perimeter of a rectangle is the sum of the measurements of all four sides. Therefore, if the perimeter is 24.0 m, one width plus one length is 12.0 m.)

$$A = lw$$
$$20 = (12 - x)x$$
$$20 = 12x - x^2$$
$$x^2 - 12x + 20 = 0$$
$$(x - 2)(x - 10) = 0$$
$$x - 2 = 0 \quad \text{or} \quad x - 10 = 0$$
$$x = 2 \qquad\qquad x = 10$$

If $x = 2$ If $x = 10$

$$x = \text{width} = 2.00 \text{ m} \qquad x = \text{width} = 10.0 \text{ m}$$
$$12 - x = \text{length} = 10.0 \text{ m} \qquad 12 - x = \text{length} = 2.00 \text{ m}$$

Since the length is greater than the width, the width is 2.00 m and the length is 10.0 m.

EXAMPLE 4

A square is cut out of each corner of a rectangular sheet of metal 40.0 cm × 60.0 cm. The sides are folded up to form a rectangular container. What are the dimensions of the square if the area of the bottom of the container is $15\bar{0}0 \text{ cm}^2$? What is the volume of the container?

Draw a diagram as in Fig. 7.4 and let

$$x = \text{side of square cutout}$$
$$40 - 2x = \text{width of rectangular container}$$
$$60 - 2x = \text{length of rectangular container}$$

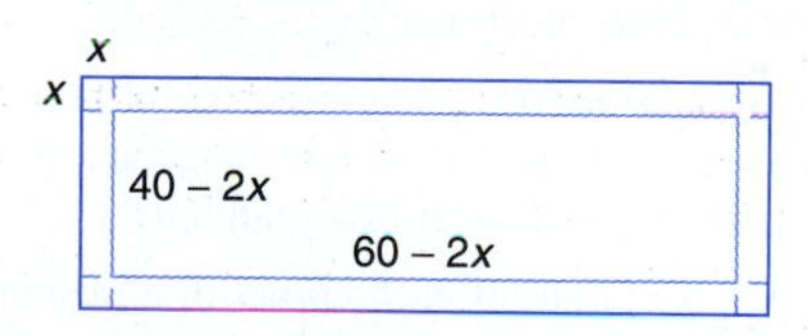

Figure 7.4

Then

$$A = lw$$
$$1500 = (60 - 2x)(40 - 2x)$$
$$1500 = 2400 - 200x + 4x^2$$
$$0 = 4x^2 - 200x + 900$$
$$0 = x^2 - 50x + 225$$
$$0 = (x - 5)(x - 45)$$
$$x - 5 = 0 \quad \text{or} \quad x - 45 = 0$$
$$x = 5 \quad \text{or} \quad x = 45$$

(*Note:* $x = 45$ is physically impossible!)

Therefore, the length of the side of square cut out is 5.00 cm, and

$$V = lwh$$
$$V = (50.0 \text{ cm})(30.0 \text{ cm})(5.00 \text{ cm}) = 75\bar{0}0 \text{ cm}^3$$

Exercises 7.4

Express the final results using three significant digits.

1. Separate 13 into two parts whose product is 40.
2. Separate 10 into two parts whose product is 24.
3. The length of a rectangular metal plate is 3.00 cm greater than twice the width. Find the dimensions if its area is 35.0 cm^2.
4. A rectangular plot of ground contains $54\bar{0}0$ m^2. If the length exceeds the width by 30.0 m, what are the dimensions?
5. The area of a triangle is 66.0 m^2. Find the lengths of the base and the height if the base is one metre longer than the height $\left(A = \frac{1}{2}bh\right)$.
6. A rectangle is 2.00 ft longer than it is wide. Find its dimensions if the area is 48.0 ft^2.
7. A variable electric current is given by the formula $i = t^2 - 7t + 12$. If t is given in seconds, at what times is the current equal to 2.00 amperes (A)? To 0 A? To 4.00 A?
8. A variable voltage is given by the formula $V = t^2 - 14t + 48$. At what times t in seconds is the voltage equal to 3.00 volts (V)? To 35.0 V? To 0 V?
9. A charge in coulombs flows in a given circuit according to the formula $q = 2t^2 - 4t + 4$. Find t in microseconds when $q = 2.00$ and when $q = 3.00$.
10. The work done in a circuit varies with time according to the formula $w = 8t^2 - 12t + 20$. Find t in milliseconds when $w = 16.0$, $w = 18.0$, and $w = 0$.
11. A winding links a magnetic field that varies according to the formula $\phi = 0.4t - 2t^2$. Find t in ms when ϕ is 0.0200 weber.
12. The perimeter of a rectangle is 22.0 cm and the area is 24.0 cm^2. Find the dimensions.
13. The perimeter of a rectangle is 80.0 m and the area is 375 m^2. Find the dimensions.
14. A rectangular yard is made using fencing for three sides of the yard and the house for one of the widths. The area of the yard is 288 m^2. Find its length and its width if the length is twice the width. How much fencing is needed?
15. A square 4.00 in. on a side is cut out of each corner of a square sheet of aluminum. The sides are folded up to form a rectangular container. If the volume is $4\bar{0}0$ in^3, what was the size of the original sheet of aluminum?
16. A square is cut out of each corner of a rectangular sheet of aluminum 20.0 cm $\times$ 30.0 cm. The sides are folded up to form a rectangular container. What are the dimensions of the square if the area of the bottom of the container is 416 cm^2? What is the volume of the container?
17. A rectangular field is fenced in by using a river as one side. If $25\bar{0}0$ m of fencing are used for the $72\bar{0},000$-m^2 field, what are its dimensions?
18. A projectile is shot vertically upward. Its height h in feet after time t in seconds may be expressed by the formula $h = 96t - 16t^2$.
    - **(a)** Find t when $h = 80.0$ ft.
    - **(b)** Find t when $h = 128$ ft.
    - **(c)** Find t when $h = 0$.
    - **(d)** Find the maximum height reached by the projectile.
19. How wide a strip must be mowed around a rectangular grass plot 40.0 m by 60.0 m for one-half of the grass to be mowed?
20. A rectangular sheet of metal 24.0 in. wide is formed into a rectangular trough with an open top and no ends. If the cross-sectional area is 70.0 in^2, find the depth of the trough.
21. A rectangular sheet of metal 48.0 in. wide is formed into a rectangular closed tube. If the cross-sectional area is 108 in^2, find the length and the width of the cross-section of the tube.

22. The cross-sectional area of the L-shaped beam in Fig. 7.5 is 24.0 in^2. What is the thickness x of the metal?

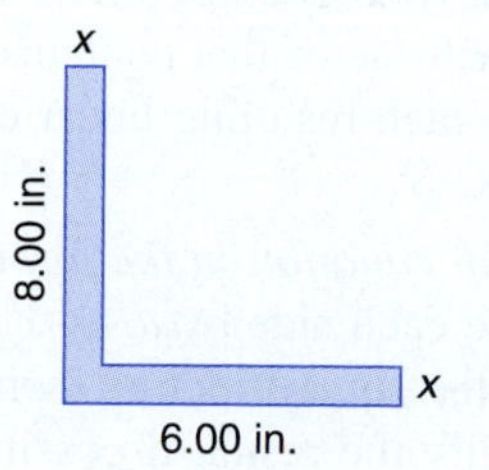

Figure 7.5

23. The area of a rectangular lot, 80.0 m $\times$ $1\bar{0}0$ m, is increased by $40\bar{0}0$ m^2. If the length and the width are increased by the same amount, what are the dimensions of the larger lot?
24. The area of a rectangular lot, $24\bar{0}$ ft $\times$ $40\bar{0}$ ft, is increased by 15,800 ft^2. If the width and the length are increased at a ratio of 2 : 3, what are the dimensions of the larger lot?
25. The equivalent resistance of two resistors connected in parallel is 45.0 ohms (Ω). If the resistance of one resistor is three times that of the other, what is the value of each? The formula for two resistors in parallel is $R = \dfrac{R_1R_2}{R_1 + R_2}$, where R is the equivalent resistance and R_1 and R_2 are the individual resistors.
26. A $1\bar{0}0$-W lamp and a 20.0-Ω resistor are connected in series with a $12\bar{0}$-V power supply. By Kirchhoff's law, we have

$$E = IR + \frac{P}{I}$$

After substituting the known values, we have

$$120 = 20I + \frac{100}{I}$$

Solve for I (in amperes).
27. A parking lot $2\bar{0}0$ m $\times$ $1\bar{0}0$ m is to be tripled in area by adding the region as shown in Fig. 7.6. Find x.
28. A parking lot $2\bar{0}0$ m $\times$ $1\bar{0}0$ m is to be tripled in area by adding the region as shown in Fig. 7.7. Find x.

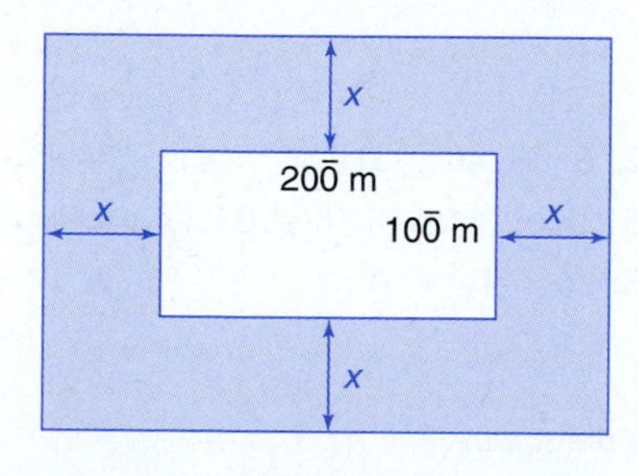

Figure 7.6

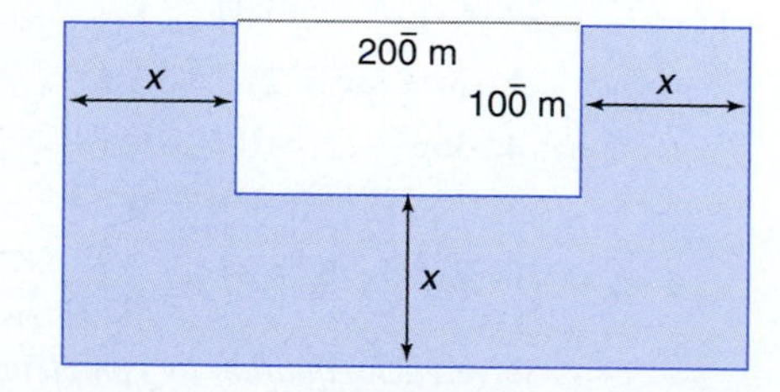

Figure 7.7

CHAPTER 7 SUMMARY

1. *A quadratic equation in one variable is an equation in the form*

$$ax^2 + bx + c = 0 \quad (a \neq 0)$$

2. *To solve a quadratic equation by factoring:*
   (a) If necessary, write the equation in the form $ax^2 + bx + c = 0$.
   (b) Factor. (See Section 5.2 on factoring quadratic trinomials.)
   (c) Set each factor that contains a variable equal to zero.
   (d) Solve each resulting linear equation.
   (e) Check.
3. *To solve an equation in the form $ax^2 = c$:*
   (a) Divide each side by a.
   (b) Take the square root of each side.
   (c) Simplify the result, if possible.
4. *To solve a quadratic equation by completing the square:*
   (a) The coefficient of the second-degree term *must* equal 1. If not, divide each side by its coefficient.
   (b) Write the equation with the variable terms on one side and the constant term on the other.
   (c) Add the square of one-half of the coefficient of the first-degree term to each side.
   (d) Rewrite the variable side, now a perfect square, as the square of a binomial.
   (e) Take the square root of each side.
   (f) Solve for the variable and simplify, if possible.
   (g) Check.
5. *The quadratic formula:* To solve a quadratic equation in the form $ax^2 + bx + c = 0$, substitute the values for a, b, and c into the formula

$$x = \frac{-b \pm \sqrt{b^2 - 4ac}}{2a}$$

6. The *discriminant* $b^2 - 4ac$ is the quantity under the radical in the quadratic formula. Assume a, b, and c are integers.
   (a) If $b^2 - 4ac > 0$, there are two real solutions.
      (i) If $b^2 - 4ac$ is also a perfect square, then the two solutions are rational.
      (ii) If $b^2 - 4ac$ is not a perfect square, then both solutions are irrational.
   (b) If $b^2 - 4ac = 0$, there is only one real, rational solution.
   (c) If $b^2 - 4ac < 0$, there are two imaginary solutions.

CHAPTER 7 REVIEW

Solve each equation.

1. $x^2 + 4x = 21$
2. $6x^2 + 40 = 31x$
3. $20x^2 + 21x + 4 = 0$
4. $36x^2 - 1 = 0$
5. $18x^2 + 45x + 18 = 0$
6. $6x^2 - 36x = 0$
7. $-8x^2 + 6x + 9 = 0$
8. $36 - 4x^2 = 0$
9. $x^2 + 10x + 25 = 0$
10. $9x^2 + 16 = 24x$

Solve each equation by completing the square.

11. $2x^2 + 13x + 20 = 0$
12. $3x^2 = 4x + 7$
13. $x^2 - 8x + 9 = 0$
14. $-5x^2 + 6x + 1 = 0$

Solve each equation for x using the quadratic formula.

15. $3x^2 - 16x + 20 = 0$
16. $4x^2 = 6x + 3$
17. $-2x^2 + 3x + 1 = 0$
18. $6x^2 + 8x - 5 = 0$
19. $mnx^2 - mx + n = 0$
20. $n^4x^2 - 2m^2n^2x + m^4 = 0$

21. Solve $s = vt + \frac{1}{2}at^2$ for t.

22. Divide 15 into two parts whose product is 36.

23. Three thousand feet of fence are used to enclose a rectangular plot of ground. Find the dimensions if the area is $54\bar{0},000$ ft^2.

24. A square is cut out of each corner of a square sheet of aluminum 50.0 cm on a side. The sides are folded up to form a rectangular container. What are the dimensions of the square cutout if the area of the bottom of the container is $9\bar{0}0$ cm^2? What is the volume of the container?

25. A ball is batted vertically upward. Its height h in feet after time t in seconds may be expressed by $h = 64t - 16t^2$.

(a) At what times is the ball 48.0 ft above the ground?

(b) Does the ball ever reach a height of 80.0 ft?

(c) Find the time that the ball is in the air.

(d) What is the maximum height reached?

26. Find the resistance R (in ohms).

$$\frac{1}{R} + \frac{1}{55\bar{0} - R} = \frac{1}{12\bar{0}}$$

8 Exponents and Radicals

INTRODUCTION

Radicals are contained in many formulas, such as trigonometric functions of half angles, for example, $\cos\frac{x}{2} = \pm\sqrt{\frac{1 + \cos x}{2}}$; the area of a triangle when the sides are known, $A = \sqrt{s(s - a)(s - b)(s - c)}$; the magnitude of impedance in an electric circuit, $|Z| = \sqrt{R^2 + X^2}$; and angular velocity, $\omega_n = \sqrt{\frac{kg}{W}}$.

In this chapter we will simplify radicals, rewrite them using fractional exponents, and solve equations involving them.

Objectives

- Simplify expressions with integral and rational exponents.
- Simplify radical expressions.
- Add, subtract, and multiply radical expressions.
- Rationalize denominators of algebraic expressions.
- Solve equations containing radical expressions.
- Solve equations that are quadratic in form.

8.1 INTEGRAL EXPONENTS

In Chapter 1, the laws of exponents were discussed in terms of positive integral exponents. Before proceeding, we must develop additional properties and further uses for exponents. The laws are repeated here for convenient reference.

LAWS OF EXPONENTS

1. $a^m \cdot a^n = a^{m+n}$
2. $\dfrac{a^m}{a^n} = a^{m-n} \quad (a \neq 0)$
3. $(a^m)^n = a^{mn}$
4. $(ab)^n = a^n b^n$
5. $\left(\dfrac{a}{b}\right)^n = \dfrac{a^n}{b^n} \quad (b \neq 0)$
6. $a^0 = 1 \quad (a \neq 0)$

Next, we discuss negative exponents. If Law 1 is to hold for negative exponents, we have

$$a^m \cdot a^{-m} = a^0 = 1 \quad (a \neq 0)$$

If we divide each side by a^m, we have the following equation:

$$a^{-m} = \frac{1}{a^m}$$

If we divide each side by a^{-m}, we have

$$a^m = \frac{1}{a^{-m}}$$

Another way of showing $\dfrac{1}{a^{-m}} = a^m$ is

$$\frac{1}{a^{-m}} = \frac{1}{\dfrac{1}{a^m}} = 1 \div \frac{1}{a^m} = 1 \times \frac{a^m}{1} = a^m$$

EXAMPLE 1

Find each product and write the result using positive exponents.

(a) $a^2 \cdot a^4 = a^{2+4} = a^6$

(b) $3^{-4} \cdot 3^7 = 3^{-4+7} = 3^3$

(c) $5^3 \cdot 5^{-6} = 5^{3+(-6)} = 5^{-3} = \dfrac{1}{5^3}$

(d) $x^{-5} \cdot x^{-4} = x^{(-5)+(-4)} = x^{-9} = \dfrac{1}{x^9}$

EXAMPLE 2

Find each quotient and write the result using positive exponents.

(a) $\dfrac{4^8}{4^6} = 4^{8-6} = 4^2$

(b) $\dfrac{a^{-2}}{a^3} = a^{(-2)-3} = a^{-5} = \dfrac{1}{a^5}$

(c) $\dfrac{c^4}{c^{-3}} = c^{4-(-3)} = c^7$

(d) $\dfrac{8^{-5}}{8^{-2}} = 8^{(-5)-(-2)} = 8^{-3} = \dfrac{1}{8^3}$

EXAMPLE 3

Find each power and write the result using positive exponents.

(a) $(3^2)^4 = 3^{(2)(4)} = 3^8$

(b) $(x^{-3})^2 = x^{(-3)(2)} = x^{-6} = \dfrac{1}{x^6}$

(c) $(c^5)^{-3} = c^{(5)(-3)} = c^{-15} = \dfrac{1}{c^{15}}$

(d) $(m^{-2})^{-3} = m^{(-2)(-3)} = m^6$

EXAMPLE 4

Simplify.

(a) $\left(\dfrac{2}{3}\right)^{-1} = \dfrac{1}{\frac{2}{3}} = 1 \div \dfrac{2}{3} = 1 \cdot \dfrac{3}{2} = \dfrac{3}{2}$

(b) $\left(\dfrac{3}{4}\right)^{-2} = \dfrac{1}{(\frac{3}{4})^2} = \dfrac{1}{\frac{9}{16}} = 1 \div \dfrac{9}{16} = 1 \cdot \dfrac{16}{9} = \dfrac{16}{9}$

(c) $2^{-1} + 4^{-2} = \dfrac{1}{2^1} + \dfrac{1}{4^2} = \dfrac{1}{2} + \dfrac{1}{16} = \dfrac{8}{16} + \dfrac{1}{16} = \dfrac{9}{16}$

EXAMPLE 5

Simplify and write the result using positive exponents.

(a)
$$
\begin{aligned}
(3a^{-2}b^4c^{-5})(6a^6b^{-4}c^{-1}) &= (3 \cdot 6)a^{(-2)+6}b^{4+(-4)}c^{(-5)+(-1)} \\
&= 18a^4b^0c^{-6} \\
&= \frac{18a^4}{c^6} \qquad \text{(Note: } b^0 = 1\text{)}
\end{aligned}
$$

(b)
$$
\begin{aligned}
\frac{2^2x^5y^{-4}z^{-2}}{4^{-2}x^{-2}y^7z^{-6}} &= 2^2 \cdot 4^2x^{5-(-2)}y^{(-4)-7}z^{(-2)-(-6)} \\
&= 4 \cdot 16x^7y^{-11}z^4 \\
&= \frac{64x^7z^4}{y^{11}} \qquad \text{(Write using positive exponents.)}
\end{aligned}
$$

(c)
$$
\begin{aligned}
(2x^{-3}y^2z^0)^{-3} &= 2^{-3}x^{(-3)(-3)}y^{(2)(-3)}z^{(0)(-3)} \\
&= 2^{-3}x^9y^{-6}z^0 \\
&= \frac{x^9}{2^3y^6} \qquad \text{(Write using positive exponents.)} \\
&= \frac{x^9}{8y^6}
\end{aligned}
$$

EXAMPLE 6

Simplify $xy^{-1} + x^{-1}y$ and write the result using positive exponents.

$$
\begin{aligned}
xy^{-1} + x^{-1}y &= \frac{x}{y} + \frac{y}{x} \\
&= \frac{x^2}{xy} + \frac{y^2}{xy} \qquad \text{(The L.C.D. is } xy\text{.)} \\
&= \frac{x^2 + y^2}{xy}
\end{aligned}
$$

Exercises 8.1

Simplify and write the result using positive exponents.

1. $x^2 \cdot x^3$
2. $c^4 \cdot c^6$
3. $\frac{5^6}{5^3}$
4. $\frac{y^2}{y^5}$
5. $(7^2)^3$
6. $(z^3)^3$
7. $m^{-2} \cdot m^5$
8. $c^4 \cdot c^{-6}$
9. $\frac{d^{-3}}{d^4}$
10. $\frac{p^5}{p^{-2}}$
11. $(2^{-3})^2$
12. $(4^{-3})^{-3}$
13. $y^{-3} \cdot y^{-5}$
14. $\frac{r^{-4}}{r^{-2}}$
15. $(s^{-2})^{-5}$
16. $7^{-2} \cdot 7^{-1}$
17. $\frac{t^{-2}}{t^{-5}}$
18. $(2^4)^{-2}$
19. $\left(\frac{4}{7}\right)^2$
20. $\left(\frac{5}{2}\right)^3$
21. $\left(\frac{1}{3}\right)^{-1}$
22. $\left(\frac{1}{4}\right)^{-1}$
23. $\left(\frac{5}{6}\right)^{-1}$
24. $\left(\frac{5}{4}\right)^{-1}$
25. $\left(\frac{1}{2}\right)^{-3}$
26. $\left(\frac{1}{5}\right)^{-2}$
27. $\left(\frac{2}{3}\right)^{-2}$
28. $\left(\frac{3}{5}\right)^{-3}$
29. $\frac{a^5}{a^{-5}}$
30. $\frac{a^{-7}}{a^{-2}}$
31. $a^6 \cdot a^{-4}$
32. $a^{-2} \cdot a^{-4}$
33. $(a^3)^{-4}$
34. $(a^{-2})^{-3}$
35. $(3a^2)^2$
36. $(4a^2b)^2$
37. $(2a^{-2}b)^{-2}$
38. $(4x^2y^{-1})^{-3}$
39. $6k^2(-2k)(4k^{-5})$
40. $(3b^2)(-4b^{-6})(-2b)$
41. $(3a^{-2}b)(5a^3b^{-3})$
42. $(6s^2t^{-4})(3^{-1}st^{-2})$
43. $w^{-4} \cdot w^2 \cdot w^{-3}$
44. $p^5 \cdot p^{-3} \cdot p^4$
45. $\frac{x^3 \cdot x^{-4}}{x^{-2} \cdot x^{-5}}$
46. $\frac{t^{-4} \cdot t^{-3}}{t^3 \cdot t^{-1}}$
47. $\frac{x^{-2}y^3}{x^3y^{-2}}$
48. $\frac{a^{-4}b}{a^{-1}b^{-3}}$
49. $\left(\frac{1}{t^4}\right)^{-2}$
50. $\left(\frac{1}{s^4}\right)^{-3}$
51. $\left(\frac{1}{b^{-2}}\right)^{-1}$
52. $\left(\frac{1}{s^{-1}}\right)^{-1}$
53. $\frac{ab^{-4}c^5}{a^2b^{-2}c}$
54. $\frac{a^4b^2c^{-4}}{a^6b^{-3}c^{-4}}$
55. $\left(\frac{a^2}{a^{-4}}\right)^3$
56. $\left(\frac{x^{-2}}{x^{-4}}\right)^{-2}$
57. $\frac{(3a^{-2})^3}{(2a^{-4})^{-2}}$
58. $\frac{(2a^2)^{-2}}{(4a^{-3})^2}$
59. $(2a^4b^0c^{-2})^3$
60. $(3a^4b^{-5}c^{-2})^4$
61. $\left(\frac{14a^3b^{-8}}{2a^{-2}b^{-4}}\right)^2$
62. $\left(\frac{12a^{-2}b^{-3}}{4a^{-6}b^0}\right)^{-3}$
63. $a^{-1} + b^{-1}$
64. $(a + b)^{-1}$
65. $a^{-2}b + a^{-1}b$
66. $ab^{-2} + a^{-1}b$
67. $3^{-2} + 9^{-1}$
68. $2^{-1} - 4^{-1}$
69. $(2^{-1} + 3^{-1})^{-1}$
70. $(x^{-1} + y^{-1})^{-1}$
71. $\frac{1}{a^0 + b^0}$
72. $(a^{-1} + b^{-1})^0$

8.2 FRACTIONAL EXPONENTS

We have now used the laws of exponents for cases when the exponents were positive integers, zero, and negative integers. Let's next extend their use to rational or fractional exponents.

FRACTIONAL EXPONENTS

Definition 1: $a^{1/n} = \sqrt[n]{a}$

where n is called the index and a is called the radicand.

Definition 2: $a^{m/n} = \sqrt[n]{a^m} = (\sqrt[n]{a})^m$

Note: m is an integer, n is a positive integer, and a is a real number. If n is even, $a \geq 0$.

The examples in Table 8.1 have been worked both in fractional exponential form and in radical form. These examples show that the laws of exponents can be extended to fractional exponents with equivalent results.

TABLE 8.1 Examples

Fractional exponential form	Equivalent expression	Radical form
1. $3^{1/2} \cdot 3^{1/2} = 3^{1/2+1/2}$ $= 3^1 = 3$	$3^{1/2} = \sqrt{3}$	**1.** $\sqrt{3} \cdot \sqrt{3} = \sqrt{9}$ $= 3$
2. $4^{1/3} \cdot 4^{1/3} \cdot 4^{1/3} = 4^{1/3+1/3+1/3}$ $= 4^1 = 4$	$4^{1/3} = \sqrt[3]{4}$	**2.** $\sqrt[3]{4} \cdot \sqrt[3]{4} \cdot \sqrt[3]{4} = \sqrt[3]{64}$ $= 4$
3. $(12^{1/2})^2 = 12^{(1/2)(2)}$ $= 12^1 = 12$	$12^{1/2} = \sqrt{12}$	**3.** $(\sqrt{12})^2 = \sqrt{12} \cdot \sqrt{12}$ $= \sqrt{144} = 12$
4. $(5^{1/3})^3 = 5^{(1/3)(3)}$ $= 5^1 = 5$	$5^{1/3} = \sqrt[3]{5}$	**4.** $(\sqrt[3]{5})^3 = \sqrt[3]{5} \cdot \sqrt[3]{5} \cdot \sqrt[3]{5}$ $= \sqrt[3]{125} = 5$
5. $\dfrac{8^{2/3}}{8^{1/3}} = 8^{2/3-1/3} = 8^{1/3} = (2^3)^{1/3}$ $= 2^{(3)(1/3)} = 2^1 = 2$	$8^{2/3} = \sqrt[3]{8^2}$ $8^{1/3} = \sqrt[3]{8}$	**5.** $\dfrac{\sqrt[3]{8^2}}{\sqrt[3]{8}} = \sqrt[3]{\dfrac{8^2}{8}} = \sqrt[3]{8} = 2$

For the two preceding definitions, note the restrictions on n. First, let us examine the definitions when n is even. For example,

$25^{1/2} = \sqrt{25} = 5$ but $(-25)^{1/2} = \sqrt{-25}$, which is not a real number.
Also, $16^{1/4} = \sqrt[4]{16} = 2$ but $(-16)^{1/4} = \sqrt[4]{-16}$, which is not a real number.
And $4^{3/2} = \sqrt{4^3} = \sqrt{64} = 8$ but $(-4)^{3/2} = \sqrt{(-4)^3} = \sqrt{-64}$, which is not a real number.

However, there is no problem if n is odd. For example,

$$(27)^{1/3} = \sqrt[3]{27} = 3 \quad \text{and} \quad (-27)^{1/3} = \sqrt[3]{-27} = -3$$
$$(32)^{1/5} = \sqrt[5]{32} = 2 \quad \text{and} \quad (-32)^{1/5} = \sqrt[5]{-32} = -2$$
$$8^{2/3} = \sqrt[3]{8^2} = \sqrt[3]{64} = 4 \quad \text{and} \quad (-8)^{2/3} = \sqrt[3]{(-8)^2} = \sqrt[3]{64} = 4$$

EXAMPLE 1

Evaluate $9^{3/2}$.

(a) $9^{3/2} = (9^{1/2})^3$
$= (3)^3 \quad (9^{1/2} = 3)$
$= 27$

(b) $9^{3/2} = (3^2)^{3/2}$
$= 3^3$
$= 27$

EXAMPLE 2

Evaluate $16^{3/4}$.

(a) $16^{3/4} = (16^{1/4})^3$
$= (2)^3 \quad (16^{1/4} = 2)$
$= 8$

(b) $16^{3/4} = (2^4)^{3/4}$
$= 2^3$
$= 8$

EXAMPLE 3

Evaluate $27^{-2/3}$.

$$\begin{aligned} 27^{-2/3} &= (27^{1/3})^{-2} \\ &= (3)^{-2} \qquad (27^{1/3} = 3) \\ &= \frac{1}{3^2} \\ &= \frac{1}{9} \end{aligned}$$

EXAMPLE 4

Simplify (a) $\dfrac{x^{5/6}}{x^{2/3}}$ and (b) $(x^{5/6})(x^{2/3})$.

(a) $$\begin{aligned} \frac{x^{5/6}}{x^{2/3}} &= x^{5/6-2/3} \\ &= x^{5/6-4/6} \\ &= x^{1/6} \end{aligned}$$

(b) $$\begin{aligned} (x^{5/6})(x^{2/3}) &= x^{5/6+2/3} \\ &= x^{5/6+4/6} \\ &= x^{9/6} \\ &= x^{3/2} \end{aligned}$$

EXAMPLE 5

Simplify $(x^{3/4})^{2/3}$.

$$\begin{aligned} (x^{3/4})^{2/3} &= x^{(3/4)(2/3)} \\ &= x^{1/2} \end{aligned}$$

EXAMPLE 6

Simplify $\left(\dfrac{a^3b^{-6}}{c^{12}}\right)^{-2/3}$.

$$\begin{aligned} \left(\frac{a^3b^{-6}}{c^{12}}\right)^{-2/3} &= \frac{a^{(3)(-2/3)}b^{(-6)(-2/3)}}{c^{(12)(-2/3)}} \\ &= \frac{a^{-2}b^4}{c^{-8}} \\ &= \frac{b^4c^8}{a^2} \end{aligned}$$

EXAMPLE 7

Simplify $a^{2/3}(a^{-1/3} + 2a^{1/3})$.

$$\begin{aligned} a^{2/3}(a^{-1/3} + 2a^{1/3}) &= a^{2/3} \cdot a^{-1/3} + 2a^{2/3} \cdot a^{1/3} \\ &= a^{1/3} + 2a \end{aligned}$$

Exercises 8.2

Evaluate each expression.

1. $36^{1/2}$ **2.** $49^{1/2}$ **3.** $64^{1/3}$ **4.** $125^{1/3}$
5. $64^{2/3}$ **6.** $8^{2/3}$ **7.** $8^{5/3}$ **8.** $32^{2/5}$
9. $16^{-1/2}$ **10.** $25^{-1/2}$ **11.** $27^{-1/3}$ **12.** $8^{-1/3}$
13. $27^{2/3}$ **14.** $64^{4/3}$ **15.** $9^{-3/2}$ **16.** $16^{-3/4}$

17. $\left(\frac{16}{25}\right)^{1/2}$ **18.** $\left(\frac{4}{9}\right)^{1/2}$ **19.** $\left(\frac{4}{9}\right)^{3/2}$ **20.** $\left(\frac{16}{81}\right)^{3/4}$

21. $\left(\frac{25}{16}\right)^{-3/2}$ **22.** $\left(\frac{8}{27}\right)^{-2/3}$ **23.** $(-8)^{-1/3}$ **24.** $(-27)^{-2/3}$

25. $3^{3/5} \cdot 3^{7/5}$ **26.** $5^{9/4} \cdot 5^{3/4}$ **27.** $\frac{9^{7/4}}{9^{5/4}}$ **28.** $\frac{8^{2/3}}{8^{1/3}}$

29. $(16^{1/2})^{1/2}$ **30.** $(32^{1/3})^{3/5}$

Perform the indicated operations. Simplify and express with positive exponents.

31. $x^{4/3} \cdot x^{2/3}$ **32.** $m^{-5/4} \cdot m^{1/4}$ **33.** $x^{3/4} \cdot x^{-1/2}$

34. $a^{-2/3} \cdot a^{-1/2}$ **35.** $\frac{x^{2/3}}{x^{1/6}}$ **36.** $\frac{x^{1/2}}{x^{3/4}}$

37. $(x^{2/3})^{-1/2}$ **38.** $(x^{-3/2})^{-2/3}$ **39.** $(x^{2/5} \cdot x^{-4/5})^{5/6}$

40. $(x^{-1/2} \cdot x^{-1/3})^{-1/5}$ **41.** $\left(\frac{x^{3/2}}{x^{3/4}}\right)^{1/3}$ **42.** $\left(\frac{x^{1/4}y^{-3/4}}{x^{1/2}y^{3/2}}\right)^{-1/2}$

43. $\left(\frac{a^4b^{-8}}{c^{16}}\right)^{3/4}$ **44.** $\left(\frac{a^{-6}}{b^9c^{-15}}\right)^{-5/3}$ **45.** $(a^{-12}b^9c^{-6})^{-4/3}$

46. $(a^5b^{10}c^{-20})^{1/5}$ **47.** $x^{2/3}(x^{1/3} + 4x^{4/3})$ **48.** $a^{1/4}(a^{4/3} - a^{1/2})$

49. $4t^{1/2}(2t^{1/2} + \frac{1}{2}t^{-1/2})$ **50.** $2s^{3/4}(s^{1/2} - 3s^{-3/4})$ **51.** $2c^{-1/3}(3c^{2/3} + 4c^{-2/3})$

52. $5p^{-1/4}(3p^{-3/4} - 6p^{1/4})$ **53.** $(x^{1/2} + y^{1/2})(x^{1/2} - y^{1/2})$ **54.** $(x^{1/2} + y^{-1/2})(x^{1/2} - y^{-1/2})$

55. $(x^{1/2} + y^{1/2})^2$ **56.** $(2x^{1/2} + 3y^{1/2})(3x^{1/2} - 4y^{1/2})$

The following formulas involving fractional exponents are taken from various technical fields. Evaluate as indicated.

57. $v = 14t^{2/5}$ Find v when $t = 32$. **58.** $w = 3t^{4/3}$ Find w when $t = 8$.

59. $y = 4^{2t}$ Find y when $t = \frac{5}{4}$. **60.** $\phi = 0.3t^{5/4}$ Find ϕ when $t = 0.0001$.

61. $Q = \frac{bH^{3/2}}{3}$ Find Q when $b = 12$ and $H = 16$.

62. $V = \frac{3R^{2/3}S^{1/2}}{2n}$ Find V when $R = 8$, $S = 9$, and $n = 0.01$.

63. $H = \frac{4M^{3/4}N^{1/3}}{5Q}$ Find H when $M = 16$, $N = 64$, and $Q = 0.04$.

64. $V = \frac{5D^{1/2}G^{2/5}}{2R^{3/4}}$ Find V when $D = 36$, $G = 32$, and $R = 81$.

65. $f = \frac{2}{\pi}\left(\frac{3EIg}{w}\right)^{1/2} l^{-3/2}$ Find f when $E = 4$, $I = 1$, $g = 32$, $w = 6$, and $l = 4$.

66. $J = (2 \times 10^{-6})\frac{E^{3/2}}{d^2}$ Find J when $E = 9$ and $d = 2$.

8.3 SIMPLEST RADICAL FORM

In Section 1.7 we introduced the nth root of a number a, written $\sqrt[n]{a}$, but we limited our discussion mostly to square roots ($n = 2$). In Section 8.2 we saw the relationship between radicals and fractional exponents, $\sqrt[n]{a} = a^{1/n}$, where n is a positive integer greater than one. Now we need to extend and build on our work with radicals.

Sometimes problems involving rational or fractional exponents may be worked more easily in radical form. The rules in this section for radical expressions define the operations with expressions in radical form. These rules are extensions of and consistent with the laws of exponents as discussed in the previous section.

Let's begin our discussion of simplifying radicals with square roots. First, recall that the square root of a nonnegative real number is a nonnegative real number. Whenever variables are used with square roots, we must assume that they represent positive real numbers.

SIMPLIFYING SQUARE ROOT QUANTITIES

A quantity involving a square root is simplified when the following hold:

1. The quantity under the radical contains no perfect square factors.
2. The radicand contains no fractions.
3. The denominator of a fraction contains no radical expression.

The following properties are used to simplify square roots:

1. $\sqrt{ab} = \sqrt{a}\sqrt{b}$
2. $\sqrt{\frac{a}{b}} = \frac{\sqrt{a}}{\sqrt{b}} \quad (b \neq 0)$

Note: a and b are nonnegative real numbers.

EXAMPLE 1

Simplify $\sqrt{160}$.

$$\begin{aligned} \sqrt{160} &= \sqrt{16 \cdot 10} && \text{(Find the largest perfect square factor of 160.)} \\ &= \sqrt{16}\sqrt{10} && \text{(Property 1)} \\ &= 4\sqrt{10} && \text{(Find the square root of the perfect square.)} \end{aligned}$$

EXAMPLE 2

Simplify $\sqrt{9a^3b^4c^7}$.

$$\begin{aligned} \sqrt{9a^3b^4c^7} &= \sqrt{9 \cdot a^2 \cdot a \cdot b^4 \cdot c^6 \cdot c} && \text{(Find each largest perfect square factor.)} \\ &= \sqrt{9}\sqrt{a^2}\sqrt{b^4}\sqrt{c^6}\sqrt{ac} && \text{(Property 1)} \\ &= 3ab^2c^3\sqrt{ac} && \text{(Find the square root of each perfect square factor.)} \end{aligned}$$

When a square root appears in the denominator of a fractional quantity, multiply numerator and denominator by a quantity that makes the denominator a perfect square.

EXAMPLE 3

Simplify $\sqrt{\frac{2}{5}}$.

$$\sqrt{\frac{2}{5}} = \sqrt{\frac{2}{5} \cdot \frac{5}{5}}$$ (Multiply numerator and denominator by 5 to make the denominator a perfect square.)

$$= \sqrt{\frac{10}{25}}$$

$$= \frac{\sqrt{10}}{\sqrt{25}}$$ (Property 2)

$$= \frac{\sqrt{10}}{5}$$ (Find the square root of the perfect square.)

Note: The procedure of changing a fraction with a radical in the denominator to an equivalent one having no radical in the denominator is called **rationalizing the denominator.**

EXAMPLE 4

Simplify $\frac{\sqrt{14a^2c}}{\sqrt{24b^2}}$.

$$\frac{\sqrt{14a^2c}}{\sqrt{24b^2}} = \sqrt{\frac{14a^2c}{24b^2}}$$ (Use Property 2 and then simplify the resulting factors.)

$$= \sqrt{\frac{7a^2c}{12b^2} \cdot \frac{3}{3}}$$ (Multiply numerator and denominator of the radical by 3 to make the denominator a perfect square.)

$$= \sqrt{\frac{21a^2c}{36b^2}}$$

$$= \frac{a\sqrt{21c}}{6b}$$ (Find the square root of each perfect square factor.)

EXAMPLE 5

Simplify $\frac{3}{\sqrt{24}}$.

$$\frac{3}{\sqrt{24}} = \frac{3}{\sqrt{2^3 \cdot 3}}$$ (Find the prime factorization of 24.)

$$= \frac{3}{\sqrt{2^3 \cdot 3}} \cdot \frac{\sqrt{2 \cdot 3}}{\sqrt{2 \cdot 3}}$$ (Multiply numerator and denominator by $\sqrt{2 \cdot 3}$ to make the denominator a perfect square.)

$$= \frac{3\sqrt{6}}{\sqrt{2^4 \cdot 3^2}}$$ (Combine factors.)

$$= \frac{3\sqrt{6}}{2^2 \cdot 3}$$ (Find the square root of each perfect square factor.)

$$= \frac{\sqrt{6}}{4}$$ (Simplify.)

A table of perfect squares and perfect cubes of the first 12 positive integers follows for your reference and convenience:

Number	*Square*	*Cube*
1	1	1
2	4	8
3	9	27
4	16	64
5	25	125
6	36	216
7	49	343
8	64	512
9	81	729
10	100	1000
11	121	1331
12	144	1728

Next, recall that the cube root of a positive real number is positive and the cube root of a negative real number is negative. Whenever variables are used with cube roots, we do not have to assume that they represent positive real numbers as we did with square roots.

SIMPLIFYING CUBE ROOT QUANTITIES

A quantity involving a cube root is simplified when the following hold:

1. The quantity under the radical contains no perfect cube factors.
2. The radicand contains no fractions.
3. The denominator of a fraction contains no radical expression.

The following properties are used to simplify cube roots:

1. $\sqrt[3]{ab} = \sqrt[3]{a}\sqrt[3]{b}$

2. $\sqrt[3]{\dfrac{a}{b}} = \dfrac{\sqrt[3]{a}}{\sqrt[3]{b}} \quad (b \neq 0)$

Note: a and b are any real numbers.

EXAMPLE 6

Simplify $\sqrt[3]{54}$.

$$\begin{aligned} \sqrt[3]{54} &= \sqrt[3]{27 \cdot 2} && \text{(Find the largest perfect cube factor.)} \\ &= \sqrt[3]{27}\sqrt[3]{2} && \text{(Property 1)} \\ &= 3\sqrt[3]{2} && \text{(Find the cube root of the perfect cube.)} \end{aligned}$$

EXAMPLE 7

Simplify $\sqrt[3]{32}$.

$$\begin{aligned}\sqrt[3]{32} &= \sqrt[3]{8 \cdot 4} && \text{(Find the largest perfect cube factor.)}\\ &= \sqrt[3]{8}\,\sqrt[3]{4} && \text{(Property 1)}\\ &= 2\sqrt[3]{4} && \text{(Find the cube root of the perfect cube.)}\end{aligned}$$

EXAMPLE 8

Simplify $\sqrt[3]{128a^5b^7c^9}$.

$$\begin{aligned}\sqrt[3]{128a^5b^7c^9} &= \sqrt[3]{64 \cdot 2 \cdot a^3 \cdot a^2 \cdot b^6 \cdot b \cdot c^9} && \text{(Find each largest perfect cube factor.)}\\ &= \sqrt[3]{64}\,\sqrt[3]{a^3}\,\sqrt[3]{b^6}\,\sqrt[3]{c^9}\,\sqrt[3]{2a^2b} && \text{(Property 1)}\\ &= 4ab^2c^3\sqrt[3]{2a^2b} && \text{(Find the cube root of each perfect cube factor.)}\end{aligned}$$

When a cube root appears in the denominator of a fractional quantity, multiply numerator and denominator by a quantity that makes the denominator a perfect cube.

EXAMPLE 9

Simplify $\dfrac{\sqrt[3]{2}}{\sqrt[3]{3}}$.

$$\begin{aligned}\frac{\sqrt[3]{2}}{\sqrt[3]{3}} &= \sqrt[3]{\frac{2}{3}} && \text{(Property 2)}\\ &= \sqrt[3]{\frac{2}{3} \cdot \frac{9}{9}} && \text{(Multiply numerator and denominator by 9 to make the denominator a perfect cube.)}\\ &= \sqrt[3]{\frac{18}{27}}\\ &= \frac{\sqrt[3]{18}}{\sqrt[3]{27}} && \text{(Property 2)}\\ &= \frac{\sqrt[3]{18}}{3} && \text{(Find the cube root of the perfect cube.)}\end{aligned}$$

EXAMPLE 10

Simplify $\sqrt[3]{\dfrac{5a}{18b}}$.

$$\begin{aligned}\sqrt[3]{\frac{5a}{18b}} &= \sqrt[3]{\frac{5a}{2 \cdot 3^2 \cdot b} \cdot \frac{2^2 \cdot 3 \cdot b^2}{2^2 \cdot 3 \cdot b^2}} && \text{(Multiply numerator and denominator of the radical by } 2^2 \cdot 3 \cdot b^2 \text{ to make the denominator a perfect cube.)}\\ &= \sqrt[3]{\frac{60ab^2}{2^3 \cdot 3^3 \cdot b^3}} && \text{(Combine factors.)}\\ &= \frac{\sqrt[3]{60ab^2}}{6b} && \text{(Find the cube root of each perfect cube factor.)}\end{aligned}$$

EXAMPLE 11

Simplify $\frac{4}{\sqrt[3]{4}}$.

$\frac{4}{\sqrt[3]{4}} = \frac{4}{\sqrt[3]{4}} \cdot \frac{\sqrt[3]{2}}{\sqrt[3]{2}}$ (Multiply numerator and denominator by $\sqrt[3]{2}$ to make the denominator a perfect cube.)

$= \frac{4\sqrt[3]{2}}{\sqrt[3]{8}}$ (Property 1)

$= \frac{4\sqrt[3]{2}}{2}$ (Find the cube root of the perfect cube.)

$= 2\sqrt[3]{2}$ (Simplify.)

Next, let's extend our properties for radical expressions to any root as follows:

OPERATIONS WITH RADICAL EXPRESSIONS

1. $\sqrt[n]{a} \cdot \sqrt[n]{b} = \sqrt[n]{ab}$
2. $\frac{\sqrt[n]{a}}{\sqrt[n]{b}} = \sqrt[n]{\frac{a}{b}} \quad (b \neq 0)$
3. $\sqrt[m]{\sqrt[n]{a}} = \sqrt[mn]{a}$
4. $\sqrt[cn]{a^{cm}} = \sqrt[n]{a^m}$

where n, m, and c are positive integers. It is important to remember that a and b must be positive real numbers if the index is even. If the index is odd, the rules are valid for all real values of a and b.

Properties 3 and 4 may be shown using fractional exponents as follows:

3. $\sqrt[m]{\sqrt[n]{a}} = (a^{1/n})^{1/m} = a^{1/(mn)} = \sqrt[mn]{a}$
4. $\sqrt[cn]{a^{cm}} = a^{cm/cn} = a^{m/n} = \sqrt[n]{a^m}$

Solutions involving radicals are usually expressed in simplest form so that easy comparison of results can be made.

SIMPLEST RADICAL FORM

1. In a radical with index n, the radicand contains no factor with exponent greater than or equal to n.
2. No radical appears in the denominator of any fraction.
3. No fractions are under a radical sign.
4. The index of a radical is as small as possible.

EXAMPLE 12

Simplify $\sqrt[4]{32x^5}$.

$\sqrt[4]{32x^5} = \sqrt[4]{16 \cdot 2 \cdot x^4 \cdot x}$ (Find each largest perfect fourth-power factor.)

$= \sqrt[4]{16 \cdot x^4}\sqrt[4]{2x}$ (Property 1)

$= 2x\sqrt[4]{2x}$ (Find the fourth root of each perfect fourth-power factor.)

When a fourth root appears in the denominator of a fractional quantity, multiply numerator and denominator by a quantity that makes the denominator a perfect fourth power.

EXAMPLE 13

Simplify $\sqrt[4]{\dfrac{3b}{8a^6}}$.

$$\sqrt[4]{\frac{3b}{8a^6}} = \sqrt[4]{\frac{3b}{2^3a^6} \cdot \frac{2a^2}{2a^2}}$$ (Multiply numerator and denominator by $2a^2$ to make the denominator a perfect fourth power.)

$$= \sqrt[4]{\frac{6a^2b}{2^4a^8}}$$ (Combine factors.)

$$= \frac{\sqrt[4]{6a^2b}}{2a^2}$$ (Find the fourth root of each perfect fourth-power factor.)

EXAMPLE 14

Change each expression to simplest radical form.

(a) $\sqrt[6]{8}$ (b) $\sqrt[4]{a^4b^2}$ (c) $\sqrt{\sqrt[3]{5a}}$ (d) $\sqrt[3]{\sqrt[4]{64}}$

Such radicals may be simplified more easily if they are expressed with fractional exponents.

(a) $\sqrt[6]{8} = 8^{1/6} = (2^3)^{1/6} = 2^{3/6} = 2^{1/2} = \sqrt{2}$

(b) $\sqrt[4]{a^4b^2} = (a^4b^2)^{1/4} = a^{4/4} \cdot b^{2/4} = ab^{1/2} = a\sqrt{b}$

(c) $\sqrt{\sqrt[3]{5a}} = [(5a)^{1/3}]^{1/2} = (5a)^{1/6} = \sqrt[6]{5a}$

(d) $\sqrt[3]{\sqrt[4]{64}} = [(64)^{1/4}]^{1/3} = [(2^6)^{1/4}]^{1/3} = 2^{6/12} = 2^{1/2} = \sqrt{2}$

We summarize the relationship between odd and even roots as follows:

> Let a be a real number and n be a positive integer greater than one.
> In particular,
>
> **1.** $\sqrt{a^2} = |a|$ and $\sqrt[3]{a^3} = a$.
>
> In general,
>
> **2.** if n is an **even** positive integer,
> $$\sqrt[n]{a^n} = |a|$$
> and
>
> **3.** if n is an **odd** positive integer,
> $$\sqrt[n]{a^n} = a.$$

Exercises 8.3

Write each expression in simplest radical form. Assume that all variables represent positive real numbers.

1. $\sqrt{49}$ **2.** $\sqrt{81}$ **3.** $\sqrt{75}$ **4.** $\sqrt{72}$

5. $\sqrt{180}$ **6.** $\sqrt{96}$ **7.** $\sqrt{8a^2}$ **8.** $\sqrt{63x^2}$

9. $\sqrt{72b^2}$ **10.** $\sqrt{45x^3}$ **11.** $\sqrt{80a^5b^2}$ **12.** $\sqrt{60a^3b^4}$

13. $\sqrt{32a^2b^4c^9}$ **14.** $\sqrt{108a^5b^{12}c^7}$ **15.** $\sqrt{\dfrac{3}{4}}$ **16.** $\sqrt{\dfrac{5}{16}}$

17. $\sqrt{\frac{5}{8}}$ 18. $\sqrt{\frac{7}{12}}$ 19. $\frac{\sqrt{6}}{\sqrt{10}}$ 20. $\frac{\sqrt{8}}{\sqrt{12}}$

21. $\frac{5}{\sqrt{24}}$ 22. $\frac{8}{\sqrt{50}}$ 23. $\sqrt{\frac{4a}{15b^2}}$ 24. $\sqrt{\frac{12a}{50b}}$

25. $\frac{2}{\sqrt{8b}}$ 26. $\frac{2y}{\sqrt{20y}}$ 27. $\frac{\sqrt{5a^4b}}{\sqrt{20a^2b^3}}$ 28. $\frac{\sqrt{6x^2y^5}}{\sqrt{8x^5}}$

29. $\sqrt[3]{125}$ 30. $\sqrt[3]{343}$ 31. $\sqrt[3]{16a^4}$ 32. $\sqrt[3]{135c^5}$

33. $\sqrt[3]{40a^8}$ 34. $\sqrt[3]{250c^{10}}$ 35. $\sqrt[3]{54x^5}$ 36. $\sqrt[3]{64x^7}$

37. $\sqrt[3]{56x^7y^5z^3}$ 38. $\sqrt[3]{72x^4y^6z^2}$ 39. $\sqrt[3]{\frac{3}{8}}$ 40. $\sqrt[3]{\frac{5}{27}}$

41. $\sqrt[3]{\frac{5}{4}}$ 42. $\sqrt[3]{\frac{5}{9}}$ 43. $\sqrt[3]{\frac{5}{12}}$ 44. $\sqrt[3]{\frac{7}{24}}$

45. $\frac{1}{\sqrt[3]{2}}$ 46. $\frac{4}{\sqrt[3]{18}}$ 47. $\sqrt[3]{\frac{4a}{50b^2}}$ 48. $\sqrt[3]{\frac{6a}{10a}}$

49. $\sqrt[3]{\frac{8a}{63a^2b^3}}$ 50. $\sqrt[3]{\frac{5xy^4}{10x^2z}}$ 51. $\frac{\sqrt[3]{5a^4b}}{\sqrt[3]{20a^2b^3}}$ 52. $\frac{\sqrt[3]{6x^4y^5}}{\sqrt[3]{18x^2y^4}}$

53. $\sqrt[4]{80x^6}$ 54. $\sqrt[4]{162a^8}$ 55. $\sqrt[4]{25a^5b^4}$ 56. $\sqrt[5]{64a^2b^7}$

57. $\sqrt[5]{\frac{32a^8b^{12}}{c^6}}$ 58. $\sqrt[4]{\frac{3bc^5}{4a^3}}$ 59. $\sqrt[4]{a^2}$ 60. $\sqrt[6]{a^3b^6}$

61. $\sqrt[6]{27a^3b^9}$ 62. $\sqrt[5]{32a^5b^{15}}$ 63. $\sqrt{\sqrt{3}}$ 64. $\sqrt{\sqrt[3]{5}}$

65. $\sqrt[3]{\sqrt{64}}$ 66. $\sqrt{\sqrt{32}}$ 67. $\sqrt[4]{64x^6}$ 68. $\sqrt[9]{27a^6b^3c^{12}}$

69. $\sqrt{4a^2 - 4b^2}$ 70. $\sqrt[3]{27a^6 + 27b^9}$ 71. $\sqrt{4(a-b)^2}$ 72. $\sqrt[3]{27(a^2+b^3)^3}$

8.4 ADDITION AND SUBTRACTION OF RADICALS

Radical quantities, like algebraic terms, may be added or subtracted only if they are like terms. Two radical quantities are **like terms** when their radicands and indices are the same (written in simplest form).

EXAMPLE 1

Combine $2\sqrt{5} - 7\sqrt{5} + 3\sqrt{5}$.

$$2\sqrt{5} - 7\sqrt{5} + 3\sqrt{5} = -2\sqrt{5}$$

EXAMPLE 2

Combine $\sqrt{48} - \sqrt{27} + 5\sqrt{3}$.

$$\begin{aligned}\sqrt{48} - \sqrt{27} + 5\sqrt{3} &= 4\sqrt{3} - 3\sqrt{3} + 5\sqrt{3} \\ &= 6\sqrt{3}\end{aligned} \qquad \left(\begin{aligned}\sqrt{48} &= \sqrt{16\cdot 3} = 4\sqrt{3}\\ \sqrt{27} &= \sqrt{9\cdot 3} = 3\sqrt{3}\end{aligned}\right)$$

EXAMPLE 3

Combine $\sqrt{50} + 3\sqrt{\frac{1}{2}}$.

$$\begin{aligned}\sqrt{50} + 3\sqrt{\frac{1}{2}} &= 5\sqrt{2} + \frac{3\sqrt{2}}{2} \\ &= \frac{10\sqrt{2} + 3\sqrt{2}}{2} \\ &= \frac{13\sqrt{2}}{2}\end{aligned} \qquad \left(\begin{aligned}\sqrt{50} &= \sqrt{25\cdot 2} = 5\sqrt{2}\\ \sqrt{\frac{1}{2}} &= \sqrt{\frac{1}{2}}\cdot\sqrt{\frac{2}{2}} = \frac{\sqrt{2}}{2}\end{aligned}\right)$$

EXAMPLE 4

Combine $4\sqrt[3]{24} - \dfrac{3}{\sqrt[3]{9}} - \sqrt[4]{9}$.

$$= 4 \cdot 2\sqrt[3]{3} - \frac{3\sqrt[3]{3}}{3} - \sqrt{3}$$
$$= 8\sqrt[3]{3} - \sqrt[3]{3} - \sqrt{3}$$
$$= 7\sqrt[3]{3} - \sqrt{3}$$

$$\left(\begin{array}{l} \sqrt[3]{24} = \sqrt[3]{8 \cdot 3} = 2\sqrt[3]{3} \\ \dfrac{1}{\sqrt[3]{9}} = \dfrac{1}{\sqrt[3]{9}} \cdot \sqrt[3]{\dfrac{3}{3}} = \dfrac{\sqrt[3]{3}}{\sqrt[3]{27}} = \dfrac{\sqrt[3]{3}}{3} \\ \sqrt[4]{9} = (3^2)^{1/4} = 3^{2/4} = 3^{1/2} = \sqrt{3} \end{array}\right)$$

Note: $\sqrt[3]{3}$ and $\sqrt{3}$ are not like terms and therefore cannot be combined.

Exercises 8.4

Perform the indicated operations and simplify when possible.

1. $3\sqrt{2} - 5\sqrt{2} + \sqrt{2}$

2. $6\sqrt{5} - \sqrt{5} + 7\sqrt{5}$

3. $\sqrt{8} - \sqrt{2}$

4. $\sqrt{3} - \sqrt{27}$

5. $3\sqrt{48} + 4\sqrt{75}$

6. $4\sqrt{32} - 2\sqrt{8}$

7. $5\sqrt{80} - 6\sqrt{45}$

8. $2\sqrt{63} + 4\sqrt{28}$

9. $4\sqrt{12} - \sqrt{27} - 3\sqrt{48}$

10. $2\sqrt{18} - \sqrt{72} + 3\sqrt{75}$

11. $\sqrt{32} + 3\sqrt{50} - 6\sqrt{18}$

12. $4\sqrt{50} - 5\sqrt{72} + 3\sqrt{8}$

13. $3\sqrt{27} - 9\sqrt{48} + \sqrt{75}$

14. $3\sqrt{5} - 6\sqrt{45} - 2\sqrt{20}$

15. $2\sqrt{54} + 3\sqrt{24} - 3\sqrt{96}$

16. $2\sqrt{90} + 3\sqrt{40} - \sqrt{160}$

17. $5\sqrt{2x} - \sqrt{18x} + \sqrt{8x}$

18. $2\sqrt{18c} - \sqrt{72c} + 3\sqrt{50c}$

19. $4\sqrt{32x^2} - \sqrt{72x^2} + 3\sqrt{50x^2}$

20. $6\sqrt{27t^2} - 5\sqrt{108t^2} - 4\sqrt{48t^2}$

21. $3\sqrt[3]{6} - 7\sqrt[3]{6} + \sqrt[3]{6}$

22. $4\sqrt[3]{5} - 6\sqrt[3]{5} + 10\sqrt[3]{5}$

23. $2\sqrt[3]{54} - 2\sqrt[3]{16}$

24. $2\sqrt[3]{81} + 3\sqrt[3]{24}$

25. $\dfrac{2\sqrt[3]{40} + 8\sqrt[3]{5}}{4}$

26. $\dfrac{3\sqrt[3]{32} + 4\sqrt[3]{108}}{6}$

27. $\sqrt[3]{81} + 3\sqrt[3]{3} - \sqrt{3}$

28. $\sqrt[3]{16} + 3\sqrt[3]{2} - \sqrt{2}$

29. $\sqrt[3]{54} - 4\sqrt[3]{16} + \sqrt[3]{128}$

30. $\sqrt[3]{48} - 2\sqrt[3]{162} + \sqrt[3]{750}$

31. $\dfrac{5}{\sqrt{2}} - \dfrac{\sqrt{2}}{2}$

32. $\dfrac{\sqrt{5}}{10} + \dfrac{3}{\sqrt{5}}$

33. $\dfrac{3\sqrt{3}}{2} + \dfrac{2}{\sqrt{3}}$

34. $\dfrac{6\sqrt{2}}{5} + \dfrac{2}{\sqrt{2}}$

35. $\sqrt{\dfrac{2}{3}} + \sqrt{\dfrac{1}{6}} - \sqrt{24}$

36. $\sqrt{\dfrac{5}{8}} + \sqrt{40} - \sqrt{\dfrac{2}{5}}$

37. $\sqrt{3} - \dfrac{1}{\sqrt{3}}$

38. $\dfrac{\sqrt{6}}{2} + \dfrac{1}{\sqrt{6}}$

39. $\dfrac{\sqrt{18}}{6} + \dfrac{\sqrt{2}}{2} + \sqrt{\dfrac{1}{2}}$

40. $\sqrt{\dfrac{1}{3}} + \dfrac{\sqrt{3}}{3} - \dfrac{\sqrt{12}}{6}$

41. $\sqrt{10} - \sqrt{\dfrac{2}{5}}$

42. $\sqrt{\dfrac{7}{3}} + \sqrt{21}$

43. $4\sqrt[3]{3} - \dfrac{6}{\sqrt[3]{9}}$

44. $-2\sqrt[3]{2} + \dfrac{5}{\sqrt[3]{4}}$

45. $\sqrt[3]{\dfrac{3}{4}} - \sqrt[3]{\dfrac{2}{9}} - \dfrac{1}{\sqrt[3]{6}}$

46. $\sqrt[3]{\dfrac{3}{2}} - \sqrt[3]{\dfrac{4}{9}} - \sqrt[6]{144}$

47. $\sqrt{\frac{1}{8}} - \sqrt{50} + \sqrt[4]{\frac{1}{4}}$

48. $\sqrt[4]{36} - \frac{1}{\sqrt[4]{36}} + \sqrt{54}$

49. $\sqrt{50ax^3} + \sqrt{72a^3x^3} - \sqrt{8a^5x}$

50. $\sqrt{27x^3} - \sqrt{48x^3} - \sqrt{3x^5}$

51. $\sqrt[3]{16x^4} + \sqrt[3]{54x^7} + \sqrt[3]{250x}$

52. $\sqrt[3]{a^4b^2} - \sqrt[3]{ab^5} + \sqrt[3]{ab^2}$

53. $\sqrt{\frac{a}{6b}} + \sqrt{\frac{6b}{a}} - \sqrt{\frac{2}{3ab}}$

54. $\sqrt{\frac{a}{2}} + \frac{1}{\sqrt{2a}} - \sqrt{32a^3}$

55. $\sqrt[3]{\frac{a}{b^2}} + \sqrt[3]{\frac{b}{a^2}} - \frac{1}{\sqrt[3]{a^2b^2}}$

56. $\sqrt[3]{\frac{a^2b}{9}} - \sqrt[3]{\frac{3a^2}{b^2}} + \sqrt[3]{\frac{3b}{a}} - \sqrt[3]{24a^5b^7}$

8.5 MULTIPLICATION AND DIVISION OF RADICALS

When two radicals have the same index, the radicands may be multiplied or divided by using the properties of radicals given in Section 8.3.

1. $\sqrt[n]{a} \cdot \sqrt[n]{b} = \sqrt[n]{ab}$
2. $\frac{\sqrt[n]{a}}{\sqrt[n]{b}} = \sqrt[n]{\frac{a}{b}}$

Do you recall the restrictions on a, b, and n?

EXAMPLE 1

Multiply and simplify $\sqrt[4]{4} \cdot \sqrt[4]{8}$.

$$\begin{aligned} \sqrt[4]{4} \cdot \sqrt[4]{8} &= \sqrt[4]{4 \cdot 8} && \text{(Property 1)} \\ &= \sqrt[4]{32} \\ &= \sqrt[4]{16 \cdot 2} && \text{(Find the largest perfect fourth-power factor.)} \\ &= 2\sqrt[4]{2} && \text{(Find the fourth root of the perfect fourth-power factor.)} \end{aligned}$$

EXAMPLE 2

Divide and simplify $\frac{\sqrt[3]{54}}{\sqrt[3]{2}}$.

$$\begin{aligned} \frac{\sqrt[3]{54}}{\sqrt[3]{2}} &= \sqrt[3]{\frac{54}{2}} && \text{(Property 2)} \\ &= \sqrt[3]{27} && \text{(Simplify.)} \\ &= 3 && \text{(Find the cube root of the perfect cube.)} \end{aligned}$$

When two radicals have different indices, it is necessary to change them so that they have a common index before the product or the quotient may be written as a single radical. Expressing the radicals in fractional exponent form is especially useful here. We then can express the exponents with a common denominator.

EXAMPLE 3

Multiply and simplify $\sqrt[3]{4} \cdot \sqrt{6}$.

$\sqrt[3]{4} \cdot \sqrt{6} = 4^{1/3} \cdot 6^{1/2}$	(Express each radical in fractional exponent form.)
$= 4^{2/6} \cdot 6^{3/6}$	(Express the fractional exponents with a common denominator.)
$= \sqrt[6]{4^2} \cdot \sqrt[6]{6^3}$	(Express in radical form.)
$= \sqrt[6]{4^2 \cdot 6^3}$	(Property 1)
$= \sqrt[6]{(2^2)^2(2 \cdot 3)^3}$	(Find the prime factors.)
$= \sqrt[6]{2^4 \cdot 2^3 \cdot 3^3}$	(Remove parentheses.)
$= \sqrt[6]{2^6 \cdot 2 \cdot 3^3}$	(Find the largest perfect sixth-power factor.)
$= 2\sqrt[6]{2 \cdot 3^3}$	(Find the sixth root of the perfect sixth-power factor.)
$= 2\sqrt[6]{54}$	(Simplify.)

EXAMPLE 4

Divide and simplify $\dfrac{\sqrt{12}}{\sqrt[4]{3}}$.

$\dfrac{\sqrt{12}}{\sqrt[4]{3}} = \dfrac{12^{1/2}}{3^{1/4}}$	(Express each radical in fractional exponent form.)
$= \dfrac{12^{2/4}}{3^{1/4}}$	(Express the fractional exponents with a common denominator.)
$= \dfrac{\sqrt[4]{12^2}}{\sqrt[4]{3}}$	(Express in radical form.)
$= \sqrt[4]{\dfrac{144}{3}}$	(Property 2)
$= \sqrt[4]{48}$	(Simplify.)
$= \sqrt[4]{16 \cdot 3}$	(Find the largest perfect fourth-power factor.)
$= 2\sqrt[4]{3}$	(Find the fourth root of the perfect fourth-power factor.)

To multiply expressions containing radicals, use the same procedures that we used to multiply algebraic expressions in Chapter 1.

EXAMPLE 5

Expand and simplify $(5 + 2\sqrt{2})(1 - 3\sqrt{2})$.

$$
\begin{aligned}
(5 + 2\sqrt{2})(1 - 3\sqrt{2}) &= 5 - 13\sqrt{2} - 6(\sqrt{2})^2 \qquad \text{(Multiply as binomials.)} \\
&= 5 - 13\sqrt{2} - 12 \\
&= -7 - 13\sqrt{2}
\end{aligned}
$$

EXAMPLE 6

Expand and simplify $(2 + 3\sqrt{6})^2$.

$$
\begin{aligned}
(2 + 3\sqrt{6})^2 &= 4 + 12\sqrt{6} + (3\sqrt{6})^2 \qquad \text{(Square as a binomial.)} \\
&= 4 + 12\sqrt{6} + 54 \\
&= 58 + 12\sqrt{6}
\end{aligned}
$$

When the denominator is in the form $a + \sqrt{b}$ or $a - \sqrt{b}$, we need to find an expression which, when multiplied by the denominator, will give a product free of radicals.

For example, by what do we multiply numerator and denominator in $\frac{4}{1+\sqrt{3}}$ to rationalize the denominator? How about $\sqrt{3}$?

$$\frac{4}{1+\sqrt{3}} \cdot \frac{\sqrt{3}}{\sqrt{3}} = \frac{4\sqrt{3}}{\sqrt{3}+3} \qquad \text{No!}$$

How about $1 + \sqrt{3}$?

$$\frac{4}{1+\sqrt{3}} \cdot \frac{1+\sqrt{3}}{1+\sqrt{3}} = \frac{4+4\sqrt{3}}{1+2\sqrt{3}+(\sqrt{3})^2} = \frac{4+4\sqrt{3}}{4+2\sqrt{3}} \qquad \text{No!}$$

How about $1 - \sqrt{3}$?

$$\frac{4}{1+\sqrt{3}} \cdot \frac{1-\sqrt{3}}{1-\sqrt{3}} = \frac{4-4\sqrt{3}}{1-(\sqrt{3})^2} = \frac{4-4\sqrt{3}}{-2} = -2+2\sqrt{3} \qquad \text{Yes!}$$

In general, to rationalize a denominator in the form $a + \sqrt{b}$, multiply numerator and denominator by $a - \sqrt{b}$. To rationalize a denominator in the form $a - \sqrt{b}$, multiply numerator and denominator by $a + \sqrt{b}$.

The radical expressions $a + \sqrt{b}$ and $a - \sqrt{b}$ are called **conjugates.** Specific examples of conjugates are as follows:

Radical expression	*Conjugate*
$1 + \sqrt{3}$	$1 - \sqrt{3}$
$2 - 3\sqrt{5}$	$2 + 3\sqrt{5}$
$4 + 3\sqrt{a}$	$4 - 3\sqrt{a}$
$2\sqrt{5} - \sqrt{3}$	$2\sqrt{5} + \sqrt{3}$

EXAMPLE 7

Simplify $\frac{4+\sqrt{3}}{2-\sqrt{3}}$.

$$\frac{4+\sqrt{3}}{2-\sqrt{3}} \cdot \frac{2+\sqrt{3}}{2+\sqrt{3}} = \frac{8+6\sqrt{3}+3}{4-3}$$

(The conjugate of $2 - \sqrt{3}$ is $2 + \sqrt{3}$. Multiply numerator and denominator by $2 + \sqrt{3}$.)

$$= 11 + 6\sqrt{3}$$

By calculator, we have

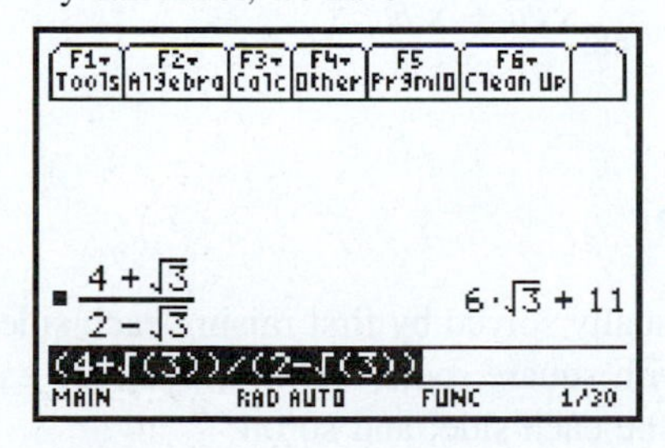

EXAMPLE 8

Simplify $\frac{3\sqrt{5}-\sqrt{2}}{2\sqrt{5}+3\sqrt{2}}$.

$$\frac{3\sqrt{5}-\sqrt{2}}{2\sqrt{5}+3\sqrt{2}}\cdot\frac{2\sqrt{5}-3\sqrt{2}}{2\sqrt{5}-3\sqrt{2}}=\frac{30-11\sqrt{10}+6}{20-18}$$

(The conjugate of $2\sqrt{5}+3\sqrt{2}$ is $2\sqrt{5}-3\sqrt{2}$. Multiply numerator and denominator by $2\sqrt{5}-3\sqrt{2}$.)

$$=\frac{36-11\sqrt{10}}{2}$$

Exercises 8.5

Perform the indicated operations and simplify when possible.

1. $\sqrt{3}\cdot\sqrt{6}$ **2.** $\sqrt{12}\cdot\sqrt{15}$ **3.** $\sqrt[3]{4}\cdot\sqrt[3]{18}$

4. $\sqrt[3]{16}\cdot\sqrt[3]{54}$ **5.** $\dfrac{\sqrt[4]{36}}{\sqrt[4]{9}}$ **6.** $\dfrac{\sqrt[3]{32}}{\sqrt[3]{4}}$

7. $\sqrt[3]{2}\cdot\sqrt{5}$ **8.** $\sqrt{2}\cdot\sqrt[4]{2}$ **9.** $\sqrt{2}\cdot\sqrt[3]{4}$

10. $\sqrt[3]{4}\cdot\sqrt[6]{8}$ **11.** $\sqrt{27}\cdot\sqrt[3]{9}$ **12.** $\sqrt{12}\cdot\sqrt[4]{9}$

13. $\dfrac{\sqrt{3}}{\sqrt[3]{3}}$ **14.** $\dfrac{\sqrt[4]{64}}{\sqrt[3]{4}}$ **15.** $\dfrac{2\sqrt[3]{3}}{3\sqrt{2}}$

16. $\dfrac{2\sqrt{3}}{6\sqrt[4]{2}}$ **17.** $(2+\sqrt{5})(1-\sqrt{5})$ **18.** $(6-\sqrt{3})(5-\sqrt{3})$

19. $(5-\sqrt{3})(5+\sqrt{3})$ **20.** $(6+2\sqrt{5})(6-2\sqrt{5})$ **21.** $(3\sqrt{3}+2)(4\sqrt{3}+3)$

22. $(2\sqrt{3}+2)(3\sqrt{3}-1)$ **23.** $(a+\sqrt{b})(2a-4\sqrt{b})$ **24.** $(3a+\sqrt{ab})(2a+3\sqrt{ab})$

25. $(2\sqrt{7}+2\sqrt{3})(-2\sqrt{7}-\sqrt{3})$ **26.** $(4\sqrt{5}-\sqrt{2})(\sqrt{5}-4\sqrt{2})$

27. $(2\sqrt{a}-3\sqrt{b})(\sqrt{a}+2\sqrt{b})$ **28.** $(3\sqrt{a}-4\sqrt{b})(\sqrt{a}-\sqrt{b})$

29. $(2-\sqrt{3})^2$ **30.** $(4+2\sqrt{5})^2$ **31.** $(2\sqrt{3}-\sqrt{5})^2$

32. $(3\sqrt{5}-2\sqrt{7})^2$ **33.** $(a+\sqrt{b})^2$ **34.** $(3a+2\sqrt{b})^2$

35. $(2\sqrt{a}+3\sqrt{b})^2$ **36.** $(4\sqrt{a}-2\sqrt{b})^2$ **37.** $\dfrac{\sqrt{2}}{3+\sqrt{2}}$

38. $\dfrac{-1}{2-\sqrt{3}}$ **39.** $\dfrac{2+\sqrt{5}}{3-2\sqrt{5}}$ **40.** $\dfrac{4-\sqrt{6}}{2+4\sqrt{6}}$

41. $\dfrac{4\sqrt{6}+2\sqrt{3}}{2\sqrt{6}+5\sqrt{3}}$ **42.** $\dfrac{2\sqrt{3}-\sqrt{2}}{3\sqrt{3}-2\sqrt{2}}$ **43.** $\sqrt{3+\sqrt{5}}\cdot\sqrt{3-\sqrt{5}}$

44. $\sqrt{4-\sqrt{6}}\cdot\sqrt{4+\sqrt{6}}$ **45.** $\dfrac{\sqrt{3+\sqrt{5}}}{\sqrt{3-\sqrt{5}}}$ **46.** $\dfrac{\sqrt{6-\sqrt{11}}}{\sqrt{6+\sqrt{11}}}$

47. $\dfrac{a}{a+\sqrt{b}}$ **48.** $\dfrac{\sqrt{a}+\sqrt{b}}{\sqrt{a}-\sqrt{b}}$

8.6 EQUATIONS WITH RADICALS

Equations with radicals are usually solved by first raising each side of the equation to the same power. For equations with square roots, we would square each side; for equations with cube roots, we would cube each side; and so on.

This process of raising each side to a power has a risk involved: We run the risk of introducing extraneous roots. Recall that an extraneous root is one that is introduced by means of an equation-solving procedure but does not check in the original equation. The only way to discard or cull out the **extraneous** solutions is by checking. For this reason, it is necessary that *all solutions be checked in the original equation.*

The potential for introducing extraneous roots may be shown using the following very simple equation:

$$x = 3$$

Square each side.

$$x^2 = 9$$

Solve the resulting equation.

$$x = \pm 3$$

By checking both solutions in the *original* equation, we see that we have introduced the extraneous root -3 by using the equation-solving procedure of squaring each side.

EXAMPLE 1

Solve $\sqrt{x + 5} = 4$.

$$\begin{aligned} \sqrt{x + 5} &= 4 \\ x + 5 &= 16 \qquad \text{(Square each side.)} \\ x &= 11 \end{aligned}$$

Check:

$$\begin{aligned} \sqrt{(11) + 5} &= 4 \\ \sqrt{16} &= 4 \\ 4 &= 4 \end{aligned}$$

Therefore, 11 is a solution.

EXAMPLE 2

Solve $\sqrt[3]{2x - 5} = 5$.

$$\begin{aligned} \sqrt[3]{2x - 5} &= 5 \\ 2x - 5 &= 125 \qquad \text{(Cube each side.)} \\ 2x &= 130 \\ x &= 65 \end{aligned}$$

Check:

$$\begin{aligned} \sqrt[3]{2(65) - 5} &= 5 \\ \sqrt[3]{130 - 5} &= 5 \\ \sqrt[3]{125} &= 5 \\ 5 &= 5 \end{aligned}$$

Therefore, 65 is a solution.

EXAMPLE 3

Solve $\sqrt{8x + 17} - 2x = 3$.

The radical must be isolated on one side before each side is squared.

$$\begin{aligned} \sqrt{8x + 17} &= 2x + 3 \\ 8x + 17 &= 4x^2 + 12x + 9 \qquad \text{(Square each side.)} \\ 0 &= 4x^2 + 4x - 8 \\ 0 &= x^2 + x - 2 \\ 0 &= (x + 2)(x - 1) \\ x + 2 = 0 \quad &\text{or} \quad x - 1 = 0 \\ x = -2 \quad &\text{or} \quad x = 1 \end{aligned}$$

Check:

$$\sqrt{8(-2)+17}-2(-2)=3$$
$$\sqrt{-16+17}+4=3$$
$$5\neq 3$$

Therefore, the apparent root -2 is extraneous and is not a solution.

$$\sqrt{8(1)+17}-2(1)=3$$
$$\sqrt{25}-2=3$$
$$3=3$$

Therefore, 1 is a solution.

By calculator, we have

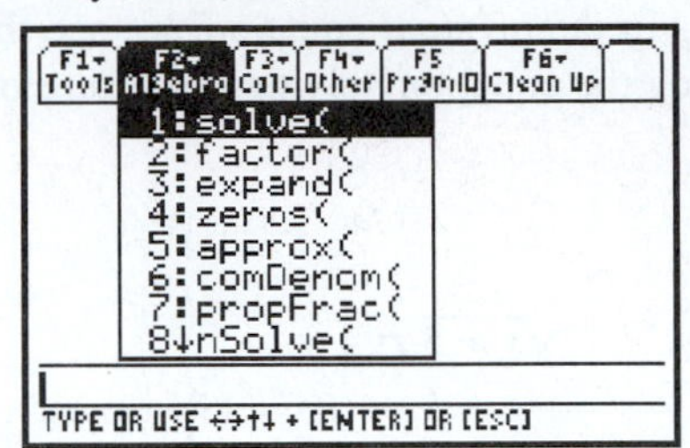

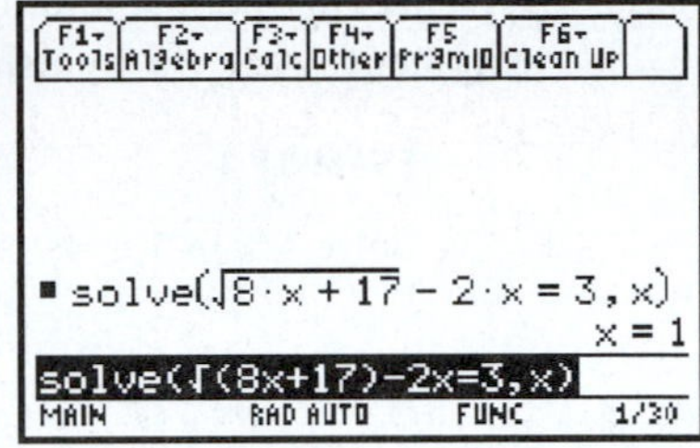

EXAMPLE 4

Solve $\sqrt{x+16}-\sqrt{x-4}=2$.

This type of radical equation can be solved most easily by first placing one radical on each side of the equation. The remaining steps are as follows:

$$\sqrt{x+16}-\sqrt{x-4}=2$$
$$\sqrt{x+16}=2+\sqrt{x-4}$$
$$x+16=4+4\sqrt{x-4}+x-4 \quad \text{(Square each side.)}$$
$$16=4\sqrt{x-4} \quad \text{(Simplify.)}$$
$$4=\sqrt{x-4} \quad \text{(Divide each side by 4.)}$$
$$16=x-4 \quad \text{(Square each side again.)}$$
$$20=x$$

Check:

$$\sqrt{(20)+16}-\sqrt{(20)-4}=2$$
$$\sqrt{36}-\sqrt{16}=2$$
$$6-4=2$$
$$2=2$$

Therefore, 20 is a solution.

By calculator, we have

(Type the equation; press **ENTER**.)

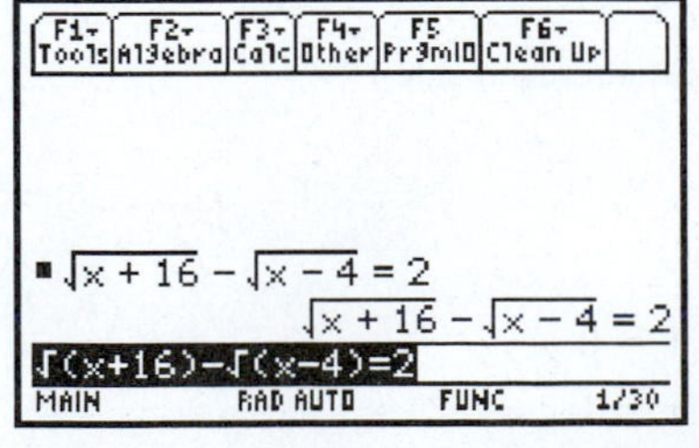

+ **2nd** multiplication sign x-4) **ENTER** ^2 **ENTER**

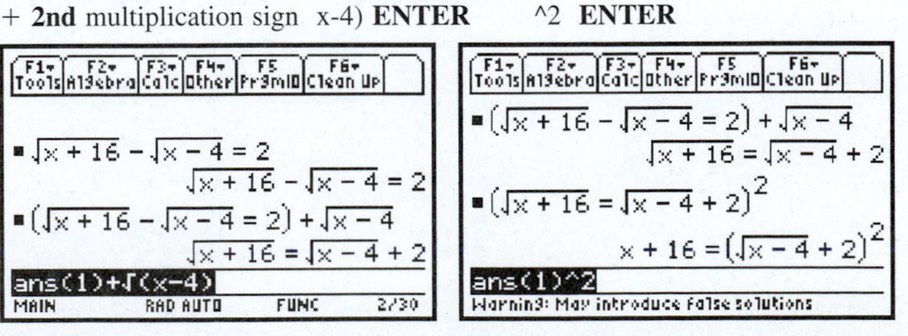

F2 3 2nd ANS) ENTER

-x ENTER

/4 ENTER

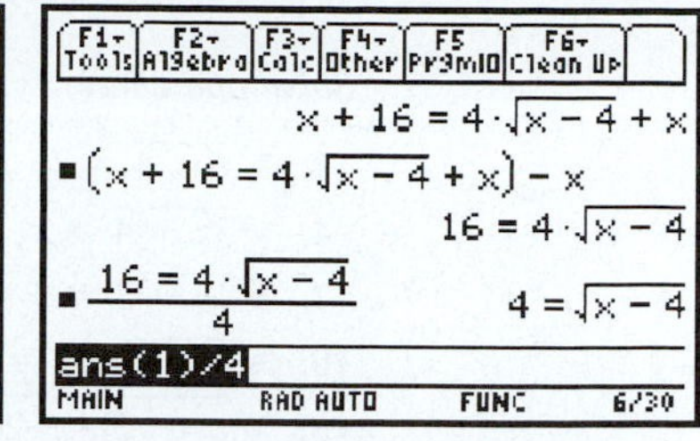

^2 ENTER

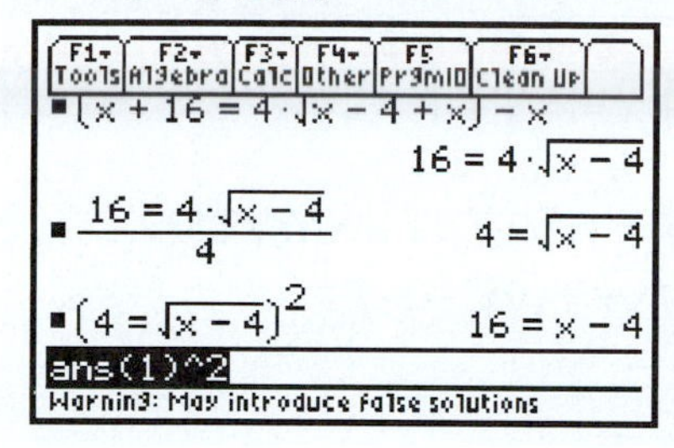

+4 ENTER

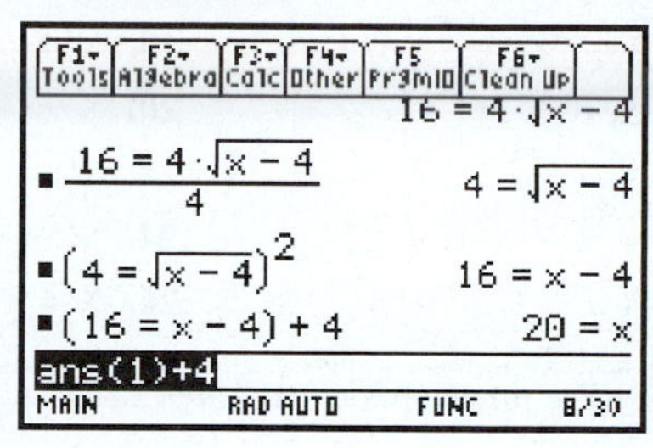

(check the candidate solution)

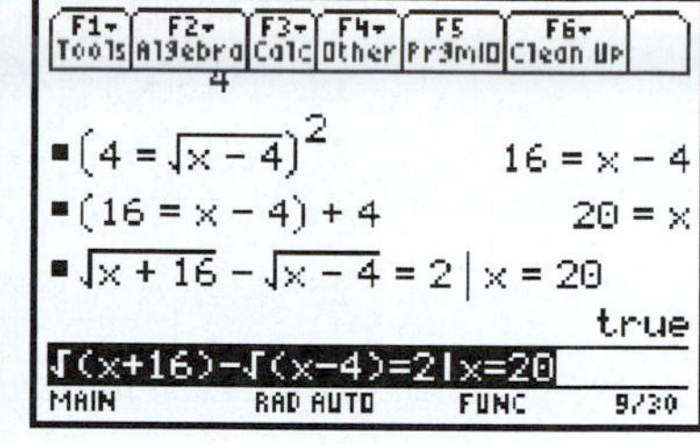

EXAMPLE 5

Solve $f = \dfrac{1}{2\pi\sqrt{LC}}$ for L.

$$f = \frac{1}{2\pi\sqrt{LC}}$$

$$f^2 = \frac{1}{4\pi^2 LC} \qquad \text{(Square each side.)}$$

$$4\pi^2 f^2 LC = 1 \qquad \text{(Multiply each side by } 4\pi^2 LC.\text{)}$$

$$L = \frac{1}{4\pi^2 f^2 C} \qquad \text{(Divide each side by } 4\pi^2 f^2 C.\text{)}$$

EXAMPLE 6

Solve $S = \pi r\sqrt{r^2 + h^2}$ for h.

$$S = \pi r\sqrt{r^2 + h^2}$$

$$S^2 = \pi^2 r^2 (r^2 + h^2) \qquad \text{(Square each side.)}$$

$$S^2 = \pi^2 r^4 + \pi^2 h^2 \qquad \text{(Remove parentheses.)}$$

$$S^2 - \pi^2 r^4 = \pi^2 h^2 \qquad \text{(Subtract } \pi^2 r^4 \text{ from each side.)}$$

$$\frac{S^2 - \pi^2 r^4}{\pi^2} = h^2 \qquad \text{(Divide each side by } \pi^2.\text{)}$$

$$h = \frac{\sqrt{S^2 - \pi^2 r^4}}{\pi} \qquad \text{(Take the square root of each side and simplify.)}$$

EXAMPLE 7

The impedance Z of an ac circuit containing resistance R, capacitance X_C, and inductance X_L is given by the equation $Z = \sqrt{R^2 + (X_L - X_C)^2}$. Solve for X_L.

$$Z = \sqrt{R^2 + (X_L - X_C)^2}$$

$$Z^2 = R^2 + (X_L - X_C)^2 \qquad \text{(Square each side.)}$$

$$Z^2 - R^2 = (X_L - X_C)^2 \qquad \text{(Subtract } R^2 \text{ from each side.)}$$

$$\sqrt{Z^2 - R^2} = X_L - X_C \qquad \text{(Take the square root of each side.)}$$

$$\sqrt{Z^2 - R^2} + X_C = X_L \qquad \text{(Add } X_C \text{ to each side.)}$$

Exercises 8.6

Solve and check.

1. $\sqrt{x-4}=7$
2. $4\sqrt{x-2}=22$
3. $\sqrt{3-x}=2x-3$
4. $\sqrt{3x+4}=x$
5. $\sqrt[3]{3-x}=5$
6. $3\sqrt[3]{x+1}=12$
7. $\sqrt[4]{2x-3}=3$
8. $\sqrt[5]{x-1}=2$
9. $\sqrt{x+6}=x$
10. $\sqrt{x+1}-4=7-x$
11. $\sqrt{x^2+2x+6}=x+2$
12. $\sqrt{x^2-4x+29}=2x+1$
13. $\sqrt{a^2-5a+20}-4a=0$
14. $\sqrt{t^2+10t+12}-3t=0$
15. $\sqrt{m^2+3m}-3m+1=0$
16. $\sqrt{s^2-4s+28}-4s-7=0$
17. $\sqrt{3x^2+4x+2}=\sqrt{x^2+x+11}$
18. $\sqrt{4x^2+2x+9}=\sqrt{3x^2-2x+5}$
19. $\sqrt[3]{3x-4}=\sqrt[3]{5x+8}$
20. $\sqrt[4]{x^2+x}=\sqrt{x+1}$
21. $\sqrt{x+6}-\sqrt{x}=4$
22. $\sqrt{x+7}+\sqrt{x+4}=3$
23. $\sqrt{x+9}-\sqrt{x+2}=\sqrt{4x-27}$
24. $\sqrt{5x+1}+\sqrt{3x+4}=\sqrt{16x+9}$
25. $\sqrt{13+\sqrt{x}}=\sqrt{x}+1$
26. $\sqrt{4+\sqrt{x}}=\sqrt{x}-2$

Given each formula, solve for the indicated letter.

27. $v=\sqrt{v_0^2-2gh}$ for h
28. $v=\sqrt{\dfrac{2GM}{r}}$ for r
29. $P=2\pi\sqrt{\dfrac{l}{g}}$ for l
30. $v=\sqrt{\dfrac{RT}{mN}}$ for m
31. $f=\dfrac{1}{2\pi\sqrt{LC}}$ for C
32. $w_d=\sqrt{w_n^2+\dfrac{c^2g^2}{w^2}}$ for g
33. $Z=\sqrt{R^2+(2\pi fL)^2}$ for L
34. $I=\dfrac{E}{\sqrt{R^2+(X_L-X_C)^2}}$ for X_L
35. $T=2\pi\sqrt{\dfrac{R_2C}{R_1+R_2}}$ for C
36. $T=2\pi\sqrt{\dfrac{R_2C}{R_1+R_2}}$ for R_2
37. $v=\sqrt{Gm\left(\dfrac{1}{D}-\dfrac{1}{r}\right)}$ for m
38. $v=\sqrt{Gm\left(\dfrac{1}{D}-\dfrac{1}{r}\right)}$ for r
39. Find the lengths of the sides of the right triangle in Fig. 8.1.
40. Find the lengths of the base and the height of the triangle in Fig. 8.2 if its area is 10.

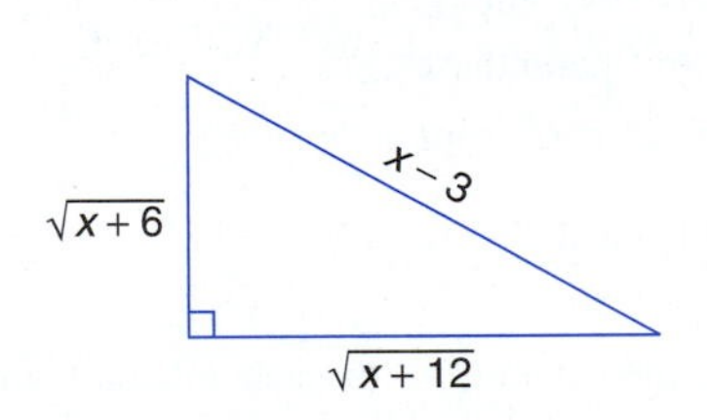

Figure 8.1

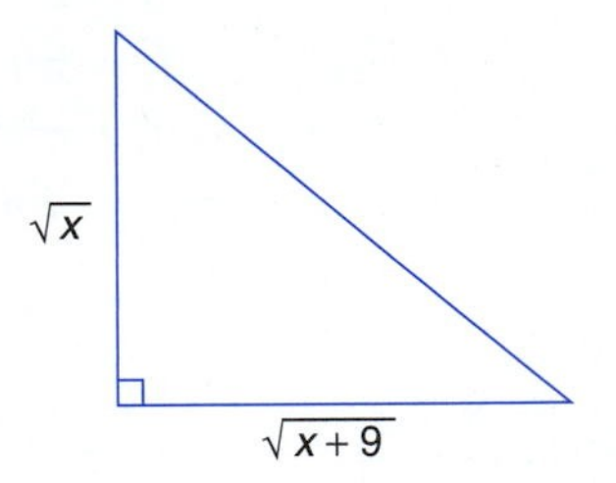

Figure 8.2

8.7 EQUATIONS IN QUADRATIC FORM

Equations in the form $ax^{2n}+bx^n+c=0$ are said to be in **quadratic form.** The following equations are in quadratic form:

$$x^4-9x^2+20=0$$
$$x^{-4}-20x^{-2}+64=0$$
$$(x+1)^{2/3}+3(x+1)^{1/3}-4=0$$
$$x-2\sqrt{x}-3=0$$

Such equations may be solved using any of the quadratic methods studied earlier.

EXAMPLE 1

Solve $x^4 - 9x^2 + 20 = 0$.

$$x^4 - 9x^2 + 20 = 0$$
$$(x^2 - 5)(x^2 - 4) = 0 \quad \text{(Factor.)}$$
$$x^2 - 5 = 0 \quad \text{or} \quad x^2 - 4 = 0$$
$$x^2 = 5 \qquad x^2 = 4$$
$$x = \pm\sqrt{5} \qquad x = \pm 2$$

EXAMPLE 2

Solve $x^{-4} - 20x^{-2} + 64 = 0$.

$$x^{-4} - 20x^{-2} + 64 = 0$$
$$(x^{-2} - 4)(x^{-2} - 16) = 0 \quad \text{(Factor.)}$$
$$x^{-2} - 4 = 0 \quad \text{or} \quad x^{-2} - 16 = 0$$
$$x^{-2} = 4 \qquad x^{-2} = 16$$
$$\frac{1}{x^2} = 4 \qquad \frac{1}{x^2} = 16 \qquad \left(x^{-2} = \frac{1}{x^2}\right)$$
$$x^2 = \frac{1}{4} \qquad x^2 = \frac{1}{16}$$
$$x = \pm\frac{1}{2} \qquad x = \pm\frac{1}{4}$$

Sometimes a substitution may be helpful.

EXAMPLE 3

Solve $(x + 1)^{2/3} + 3(x + 1)^{1/3} - 4 = 0$.

If we let $p = x + 1$, then

$$(x + 1)^{2/3} + 3(x + 1)^{1/3} - 4 = 0$$
$$p^{2/3} + 3p^{1/3} - 4 = 0$$
$$(p^{1/3} + 4)(p^{1/3} - 1) = 0 \quad \text{(Factor.)}$$
$$p^{1/3} + 4 = 0 \quad \text{or} \quad p^{1/3} - 1 = 0$$
$$p^{1/3} = -4 \qquad p^{1/3} = 1$$
$$p = -64 \qquad p = 1$$

Recall that $p = x + 1$. Thus,

$$x + 1 = -64 \qquad x + 1 = 1$$
$$x = -65 \qquad x = 0$$

If the equation does not factor, as before we can use the quadratic formula.

EXAMPLE 4

Solve $3x^4 - 8x^2 + 2 = 0$.

Note: This is a quadratic equation in x^2.

$$x^2 = \frac{-b \pm \sqrt{b^2 - 4ac}}{2a}$$
$$x^2 = \frac{-(-8) \pm \sqrt{(-8)^2 - 4(3)(2)}}{2(3)}$$

$$x^2 = \frac{8 \pm \sqrt{64 - 24}}{6}$$

$$x^2 = \frac{8 \pm 2\sqrt{10}}{6} = \frac{4 \pm \sqrt{10}}{3}$$

Therefore,

$$x = \pm\sqrt{\frac{4 \pm \sqrt{10}}{3}} \quad \text{or} \quad \pm 1.55, \pm 0.528$$

Exercises 8.7

Solve each equation.

1. $x^4 - 11x^2 + 18 = 0$

2. $x^4 - 10x^2 + 24 = 0$

3. $x^{-4} - 17x^{-2} + 16 = 0$

4. $x^{-4} - 13x^{-2} + 36 = 0$

5. $(x + 2)^2 + 3(x + 2) + 2 = 0$

6. $(x + 1)^{-2} + 8(x + 1)^{-1} + 15 = 0$

7. $(x - 1)^4 - 5(x - 1)^2 + 4 = 0$

8. $(2x - 5)^4 - (2x - 5)^2 = 0$

9. $x - 2\sqrt{x} - 3 = 0$

10. $4x - 4\sqrt{x} + 1 = 0$

11. $(3x + 2)^{-4} - 1 = 0$

12. $(2x - 1)^4 - 9 = 0$

13. $x^{2/3} + 2x^{1/3} - 8 = 0$

14. $x^{2/3} - x^{1/3} - 12 = 0$

15. $(3x + 1)^{4/3} - 2(3x + 1)^{2/3} - 8 = 0$

16. $(x - 1)^{2/3} - 5(x - 1)^{1/3} + 6 = 0$

17. $x^4 - 3x^2 + 1 = 0$

18. $3x^4 - 6x^2 + 2 = 0$

19. $\sqrt[3]{x} - 3\sqrt[6]{x} + 2 = 0$

20. $\sqrt[4]{x} + 2\sqrt{x} = 3$

CHAPTER 8 SUMMARY

1. *Laws of exponents:*
(a) $a^m \cdot a^n = a^{m+n}$
(b) $\frac{a^m}{a^n} = a^{m-n} \quad (a \neq 0)$
(c) $(a^m)^n = a^{mn}$
(d) $(ab)^n = a^n b^n$
(e) $\left(\frac{a}{b}\right)^n = \frac{a^n}{b^n} \quad (b \neq 0)$
(f) $a^0 = 1 \quad (a \neq 0)$
(g) $a^{-m} = \frac{1}{a^m} \quad (a \neq 0)$

2. *Fractional exponents:*
(a) $a^{1/n} = \sqrt[n]{a}$
(b) $a^{m/n} = \sqrt[n]{a^m} = (\sqrt[n]{a})^m$
Note: m is an integer, n is a positive integer, and a is a real number. If n is even, $a \geq 0$.

3. *Operations with radical expressions:*
(a) $\sqrt[n]{a} \cdot \sqrt[n]{b} = \sqrt[n]{ab}$
(b) $\frac{\sqrt[n]{a}}{\sqrt[n]{b}} = \sqrt[n]{\frac{a}{b}} \quad (b \neq 0)$
(c) $\sqrt[m]{\sqrt[n]{a}} = \sqrt[mn]{a}$
(d) $\sqrt[cn]{a^{cm}} = \sqrt[n]{a^m}$

4. *Simplest radical form:*
   (a) In a radical with index n, the radicand contains no factor with exponent greater than or equal to n.
   (b) No radical appears in the denominator of any fraction.
   (c) No fractions are under a radical sign.
   (d) The index of a radical is as small as possible.
5. The radical expressions $a + \sqrt{b}$ and $a - \sqrt{b}$ are *conjugates.*
6. An equation in the form $ax^{2n} + bx^n + c = 0$ is in *quadratic form.*

CHAPTER 8 REVIEW

Simplify and rewrite using only positive exponents.

1. $3a^{-2}$ **2.** $(2a)^{-4}$ **3.** $a^{-5} \cdot a^{10}$

4. $(a^{-2})^{-4}$ **5.** $\dfrac{a^3b^0c^{-3}}{a^5b^{-2}c^{-9}}$ **6.** $a^{-2} + a^{-1}$

Evaluate.

7. $49^{1/2}$ **8.** $16^{3/2}$ **9.** $8^{-2/3}$

Perform the indicated operations. Simplify and express with positive exponents.

10. $(x^{2/3}y^{-1/3})^{-3/5}$ **11.** $\dfrac{x^{-2/3}}{x^{-1/4}}$ **12.** $(x^{-3/4})^{-2/3}$

13. Evaluate $w = 2t^{-2/3}$ for $t = 27$.

Simplify.

14. $\sqrt{80}$ **15.** $\sqrt{72a^2b^3}$ **16.** $\sqrt{80x^5y^6}$

17. $\sqrt{48x^3y}$ **18.** $\sqrt{\dfrac{5}{54}}$ **19.** $\sqrt{\dfrac{45a^2}{7b^4}}$

20. $\sqrt[3]{250}$ **21.** $\sqrt[3]{108a^4b^2}$ **22.** $\sqrt[3]{256a^5b^{10}}$

23. $\sqrt[3]{\dfrac{9a}{20b^2}}$ **24.** $\dfrac{8}{\sqrt[3]{40}}$ **25.** $\sqrt[4]{9a^2}$

26. $\dfrac{6}{\sqrt{12a}}$ **27.** $\dfrac{6}{\sqrt[3]{12a}}$ **28.** $\sqrt{\sqrt{5}}$

29. $\sqrt[4]{\sqrt[3]{10}}$

Perform the indicated operations and simplify when possible.

30. $\sqrt{12} + \sqrt{27} - \sqrt[4]{9}$ **31.** $\sqrt{\dfrac{2}{5}} + \dfrac{1}{\sqrt{10}} - \sqrt{40}$ **32.** $\sqrt[3]{54} - \sqrt[3]{250} + \sqrt[3]{16}$

33. $2\sqrt[3]{\dfrac{3}{4}} - \dfrac{3}{\sqrt[3]{36}} + 5\sqrt[3]{\dfrac{2}{9}}$ **34.** $\sqrt{8} \cdot \sqrt{20}$ **35.** $\sqrt{3} \cdot \sqrt[4]{9}$

36. $\sqrt{2} \cdot \sqrt[3]{4}$ **37.** $(2 + \sqrt{3})(-4 - \sqrt{3})$ **38.** $(4 - 3\sqrt{5})^2$

39. $\dfrac{\sqrt{3}}{2 - 3\sqrt{3}}$

Solve each equation.

40. $\sqrt{x + 2} = 8$ **41.** $\sqrt[3]{x} + 2 = -1$ **42.** $\sqrt{x - 5} + \sqrt{x} = 5$

43. $\sqrt{x + 9} = \sqrt[4]{x^2 + 9x}$ **44.** $4x^4 - 41x^2 + 45 = 0$ **45.** $x^{2/3} - 2x^{1/3} - 8 = 0$

9 Exponentials and Logarithms

INTRODUCTION

A construction crew has placed flashing caution lights around its site and has timed them to operate 24 h per day. A battery for each bank of lights diminishes in power by 0.65% for each hour it is operated. The lights will cease to operate when the battery power is reduced to $\frac{1}{10}$ of its original power. How often will the batteries need to be changed?

We use an **exponential function** to calculate that the batteries will last about 354 h. Thus, they need to be changed every 14.7 days, or for practical purposes, every 2 weeks. Exponential functions are also used to express rates of growth and decay for items such as money, bacteria, and populations. The inverses of such functions, called **logarithms,** are used in formulas for decibels and pH values and in determining flow of heat in an insulated pipe.

In this chapter, we learn the rules of operating with exponential and logarithmic functions.

Objectives

- Graph exponential functions.
- Graph logarithmic functions.
- Use a calculator to evaluate exponential and logarithmic functions.
- Use the properties of logarithms.
- Evaluate expressions containing logarithms and exponential functions.
- Solve exponential equations.
- Solve logarithmic equations.
- Use logarithmic and semilogarithmic graph paper.

9.1 THE EXPONENTIAL FUNCTION

We have previously considered equations with a constant exponent in the form

$$y = x^n$$

These are called **power functions;** two examples are $y = x^2$ and $y = x^3$.

EXPONENTIAL FUNCTION

Equations with a variable exponent in the form

$$y = b^x$$

where $b > 0$ and $b \neq 1$ are called **exponential functions.**

Two examples of exponential functions are $y = 2^x$ and $y = (\frac{3}{4})^x$.

EXAMPLE 1

Graph $y = 2^x$ by plotting points.

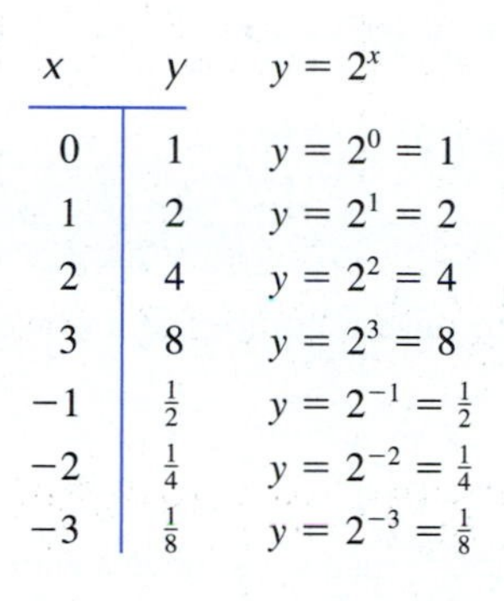

x	y	$y = 2^x$
0	1	$y = 2^0 = 1$
1	2	$y = 2^1 = 2$
2	4	$y = 2^2 = 4$
3	8	$y = 2^3 = 8$
-1	$\frac{1}{2}$	$y = 2^{-1} = \frac{1}{2}$
-2	$\frac{1}{4}$	$y = 2^{-2} = \frac{1}{4}$
-3	$\frac{1}{8}$	$y = 2^{-3} = \frac{1}{8}$

Now plot the points as in Fig. 9.1.

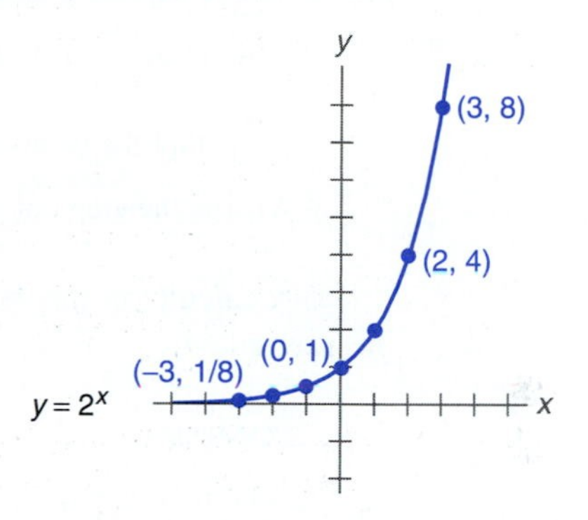

Figure 9.1

In general, for $b > 1$, $y = b^x$ is an **increasing** exponential function. That is, as x increases, y increases.

EXAMPLE 2

Graph $y = (\frac{1}{2})^x$ by plotting points.

x	y	$y = (\frac{1}{2})^x$
0	1	$y = (\frac{1}{2})^0 = 1$
1	$\frac{1}{2}$	$y = (\frac{1}{2})^1 = \frac{1}{2}$
2	$\frac{1}{4}$	$y = (\frac{1}{2})^2 = \frac{1}{4}$
3	$\frac{1}{8}$	$y = (\frac{1}{2})^3 = \frac{1}{8}$
-1	2	$y = (\frac{1}{2})^{-1} = 2$
-2	4	$y = (\frac{1}{2})^{-2} = 4$
-3	8	$y = (\frac{1}{2})^{-3} = 8$

Plot the points as in Fig. 9.2.

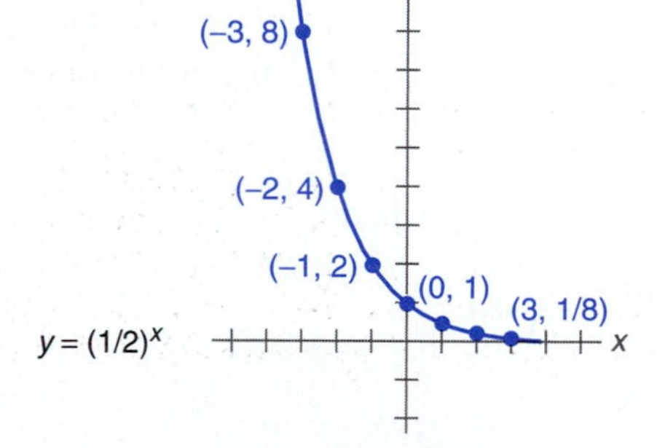

Figure 9.2

In general, for $0 < b < 1$, $y = b^x$ is a **decreasing** exponential function. That is, as x increases, y decreases.

EXAMPLE 3

Graph $y = 3^{-x}$ by plotting points.

x	y	$y = 3^{-x}$
0	1	$y = 3^{-0} = 3^0 = 1$
1	$\frac{1}{3}$	$y = 3^{-1} = \frac{1}{3}$
2	$\frac{1}{9}$	$y = 3^{-2} = \frac{1}{3^2} = \frac{1}{9}$
3	$\frac{1}{27}$	$y = 3^{-3} = \frac{1}{3^3} = \frac{1}{27}$
-1	3	$y = 3^{-(-1)} = 3^1 = 3$
-2	9	$y = 3^{-(-2)} = 3^2 = 9$
-3	27	$y = 3^{-(-3)} = 3^3 = 27$

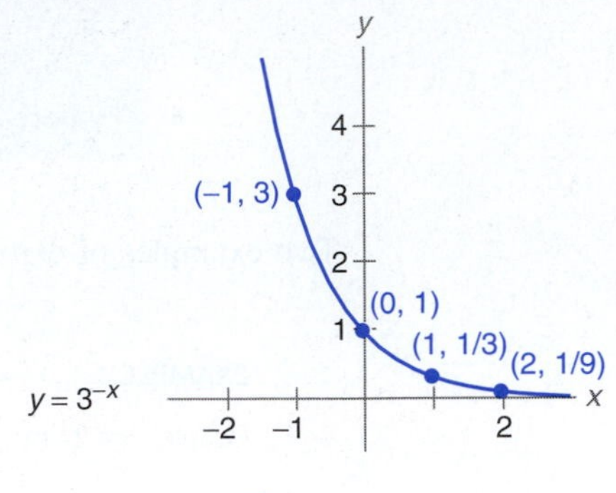

Figure 9.3

Plot the points as in Fig. 9.3.

What is the graph of $y = (\frac{1}{3})^x$?

A calculator may be used to raise a number to a power. Use the ^ button as shown in Example 4.

EXAMPLE 4

Find the values of $2.5^{1.4}$ and $150^{2/3}$ rounded to three significant digits.

2.5^1.4 **ENTER** 150^(2/3) **green diamond ENTER**

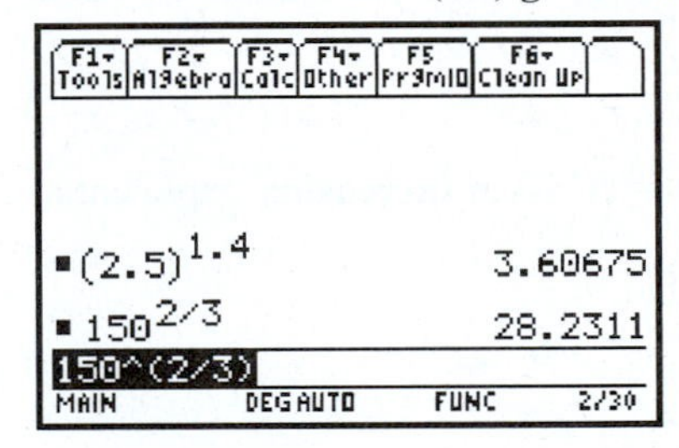

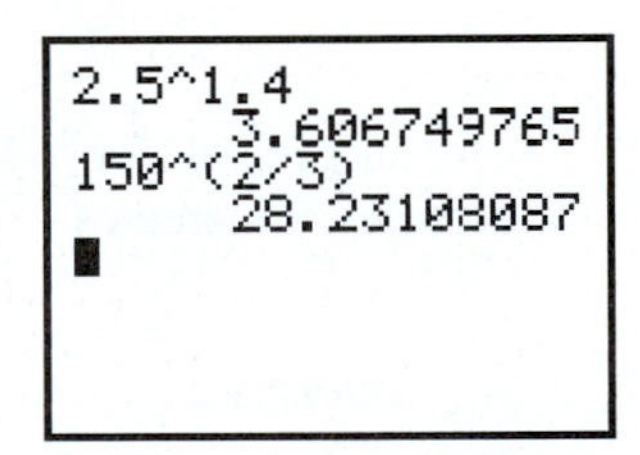

2.5^1.4 **ENTER** 150^(2/3) **ENTER**

That is, $2.5^{1.4} = 3.61$ and $150^{2/3} = 28.2$ rounded to three significant digits.

In growth situations, we could use the exponential function

$$y = A(1 + r)^n$$

where

r = the rate of growth (in decimal form)

n = the time interval

A = the initial amount

y = the new amount

EXAMPLE 5

According to the records of a utility company, the demand for electricity in its area is growing at a constant annual rate of 12%. During this current year, its customers used 750 billion kilowatt-hours (7.5×10^{11} kWh) of electricity. Assuming no conservation efforts on the part of its customers, how much electric power will be needed in 8 years?

In this example,

$$r = 12\% = 0.12$$
$$n = 8 \text{ years}$$
$$A = 7.5 \times 10^{11} \text{ kWh}$$

We are to find y.

$$y = A(1 + r)^n$$
$$y = (7.5 \times 10^{11} \text{ kWh})(1 + 0.12)^8$$
$$= 1.9 \times 10^{12} \text{ kWh} \qquad \text{(rounded to two significant digits)}$$

or 1900 billion kWh.

EXAMPLE 6

Bill and Mary plan to retire this year with a combined annual pension of \$60,000. Because of inflation, their purchasing power is constantly decreasing. Assuming a 7% annual rate of inflation, what will their purchasing power be in 10 years?

In this example,

$$r = -7\% = -0.07 \qquad \text{(negative growth)}$$
$$n = 10 \text{ years}$$
$$A = \$60{,}000$$

We are to find y.

$$y = A(1 + r)^n$$
$$y = \$60{,}000[1 + (-0.07)]^{10}$$
$$= \$29{,}000 \qquad \text{(rounded to two significant digits)}$$

Exercises 9.1

Graph each equation.

1. $y = 4^x$ **2.** $y = 3^x$ **3.** $y = 10^x$ **4.** $y = 5^x$

5. $y = (\frac{1}{3})^x$ **6.** $y = (\frac{1}{4})^x$ **7.** $y = (\frac{1}{10})^x$ **8.** $y = (\frac{1}{5})^x$

9. $y = (\frac{3}{4})^x$ **10.** $y = (\frac{2}{3})^x$ **11.** $y = (\frac{5}{6})^x$ **12.** $y = (\frac{2}{5})^x$

13. $y = 4^{-x}$ **14.** $y = 5^{-x}$ **15.** $y = (\frac{4}{3})^{-x}$ **16.** $y = (\frac{3}{4})^{-x}$

17. $y = (1.2)^x$ **18.** $y = (5.5)^{-x}$ **19.** $y = 3^{-x+2}$ **20.** $y = 4^{2x}$

21. $y = 2^{x-3}$ **22.** $y = 2^{-2x}$ **23.** $y = 2^{x^2}$ **24.** $y = 2^x - 3$

25. $y = 3^x + 2$

Find the value of each power rounded to three significant digits.

26. $12^{0.3}$ **27.** $5^{0.2}$ **28.** $3^{2.7}$ **29.** $10^{5.5}$

30. 4^{π} **31.** $5^{2\pi}$ **32.** $6^{2/3}$ **33.** $8^{\pi/3}$

34. $15^{\pi/4}$ **35.** $9^{3/4}$ **36.** $\sqrt[3]{12}$ **37.** $\sqrt[4]{6}$

38. $\sqrt[5]{9}$ **39.** $\sqrt[6]{140}$ **40.** $\sqrt[3]{46{,}656}$

41. According to the records of a utility company, the demand for electricity in its area is constantly growing at an annual rate of 8%. During this current year, its customers used

$5\bar{0}0$ billion kWh of electricity. Assuming no conservation efforts on the part of its customers, how much electric power will be needed in 6 years?

42. Doris plans to retire this year with an annual pension of \$68,000. Because of inflation, her purchasing power is constantly decreasing. Assuming a 7% annual rate of inflation, find her purchasing power in 10 years.

43. The formula for compound interest is $A = P\left(1 + \frac{r}{x}\right)^{xn}$,

where

A = the amount of money in the account (principal and interest)

P = the original principal (amount invested)

r = the yearly rate of interest (in decimal form)

x = the number of times that interest is compounded per year

n = the number of years that the money is invested

Assume that you have \$1000 to invest at 8% interest. Find to the nearest dollar the amount in the account after 3 years if the interest is compounded **(a)** annually, **(b)** semiannually, **(c)** quarterly, and **(d)** daily.

44. Assume that you owe a credit company \$1000. The interest rate is $1\frac{1}{2}\%$ per month. Assume that you do not charge any more for the next year.
(a) If you pay nothing for 6 months, what is your balance?
(b) If you pay \$100 per month, what do you owe in 6 months?

45. When a gas undergoes an adiabatic (constant heat) process, the final and initial absolute temperatures and pressures are related according to the formula

$$\frac{T_2}{T_1} = \left(\frac{P_1}{P_2}\right)^{(1-\gamma)/\gamma}$$

where γ is the ratio of specific heats at constant pressure and volume. Find T_2 when $T_1 = 575°\text{R}$ absolute, $P_1 = 25.0$ psi (pounds per square inch) absolute, $P_2 = 6\bar{5}0$ psi absolute, and $\gamma = 1.50$.

46. The formula for heat flow by convection is

$$h = 0.0230\frac{k}{D}\left(\frac{DV\rho}{\mu}\right)^{0.8}\left(\frac{\mu c}{k}\right)^{n}$$

Find h when $k = 0.0650$, $D = 2.00$, $V = 1\bar{6}0$, $\rho = 1.45$, $\mu = 0.108$, $c = 0.720$, and $n = 0.3$.

9.2 THE LOGARITHM

When the values of x and y are interchanged in an equation, the resulting equation is called the **inverse** of the given equation. The inverse of the exponential equation $y = b^x$ is the exponential equation $x = b^y$. We define this inverse equation to be the logarithmic equation. The following middle and right equations show how to express this logarithmic equation in either exponential form or logarithmic form:

Exponential equation	*Logarithmic equation in exponential form*	*Logarithmic equation in logarithmic form*
$y = b^x$	$x = b^y$	$y = \log_b x$

That is, $x = b^y$ and $y = \log_b x$ are equivalent equations for $b > 0$ but $b \neq 1$.

The logarithm of a number is the *exponent* indicating the power to which the base must be raised to equal that number. The expression $\log_b x$ is read "the logarithm, base b, of x."

> *Remember:* A logarithm is an exponent.

EXAMPLE 1

Write each equation in logarithmic form.

	Exponential form	*Logarithmic form*
(a)	$2^3 = 8$	$\log_2 8 = 3$
(b)	$5^2 = 25$	$\log_5 25 = 2$
(c)	$4^{-2} = \frac{1}{16}$	$\log_4 (\frac{1}{16}) = -2$
(d)	$36^{1/2} = 6$	$\log_{36} 6 = \frac{1}{2}$
(e)	$p^q = r$	$\log_p r = q$

EXAMPLE 2

Write each equation in exponential form.

	Logarithmic form	*Exponential form*
(a)	$\log_7 49 = 2$	$7^2 = 49$
(b)	$\log_4 64 = 3$	$4^3 = 64$
(c)	$\log_{10} 0.01 = -2$	$10^{-2} = 0.01$
(d)	$\log_{27} 3 = \dfrac{1}{3}$	$27^{1/3} = 3$
(e)	$\log_m p = n$	$m^n = p$

EXAMPLE 3

Graph $y = \log_2 x$ by plotting points.

First, change the equation from logarithmic form to exponential form. That is, $y = \log_2 x$ is equivalent to $x = 2^y$. Then choose values for y and compute values for x.

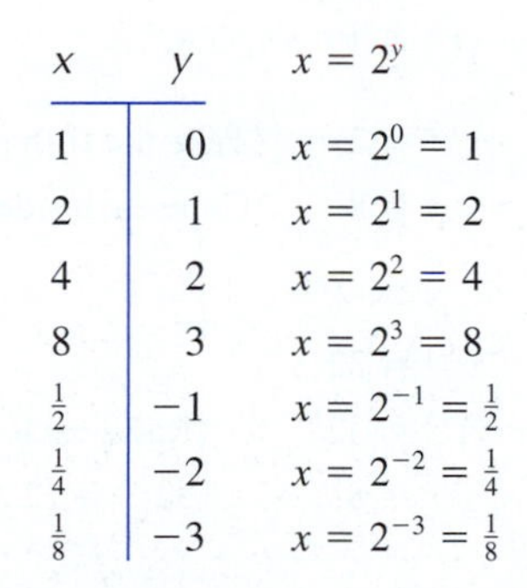

x	y	$x = 2^y$
1	0	$x = 2^0 = 1$
2	1	$x = 2^1 = 2$
4	2	$x = 2^2 = 4$
8	3	$x = 2^3 = 8$
$\frac{1}{2}$	-1	$x = 2^{-1} = \frac{1}{2}$
$\frac{1}{4}$	-2	$x = 2^{-2} = \frac{1}{4}$
$\frac{1}{8}$	-3	$x = 2^{-3} = \frac{1}{8}$

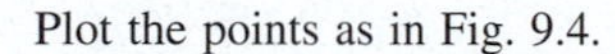

Plot the points as in Fig. 9.4.

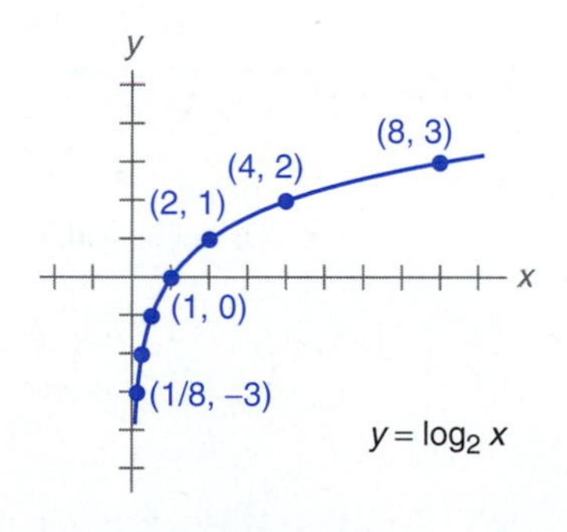

Figure 9.4

EXAMPLE 4

Graph $y = \log_{1/2} x$ by plotting points.

Again, change the equation from logarithmic form to exponential form. That is, $y = \log_{1/2} x$ is equivalent to $x = (\frac{1}{2})^y$.

x	y	$x = (\frac{1}{2})^y$
1	0	$x = (\frac{1}{2})^0 = 1$
$\frac{1}{2}$	1	$x = (\frac{1}{2})^1 = \frac{1}{2}$
$\frac{1}{4}$	2	$x = (\frac{1}{2})^2 = \frac{1}{4}$
$\frac{1}{8}$	3	$x = (\frac{1}{2})^3 = \frac{1}{8}$
2	-1	$x = (\frac{1}{2})^{-1} = 2$
4	-2	$x = (\frac{1}{2})^{-2} = 4$
8	-3	$x = (\frac{1}{2})^{-3} = 8$

Figure 9.5

Plot the points as in Fig. 9.5.

EXAMPLE 5

Given $\log_3 81 = x$, find x.

The exponential form of $\log_3 81 = x$ is

$$3^x = 81$$

We know that

$$3^4 = 81$$

Therefore,

$$x = 4$$

EXAMPLE 6

If $\log_3 x = -2$, find x.

$$\log_3 x = -2 \quad \text{or} \quad 3^{-2} = x$$

$$x = \frac{1}{9}$$

EXAMPLE 7

If $\log_x 32 = \frac{5}{3}$, find x.

$$\log_x 32 = \frac{5}{3} \quad \text{or} \quad x^{5/3} = 32$$

$$x^{1/3} = 2 \quad \text{(Take the fifth root of each side.)}$$

$$x = 8 \quad \text{(Cube each side.)}$$

Or begin with

$$x^{5/3} = 32$$

$$(x^{5/3})^{3/5} = 32^{3/5} \quad \text{(Raise each side to the } \tfrac{3}{5} \text{ power.)}$$

$$x = 8 \quad [32^{3/5} = (2^5)^{3/5} = 2^3 = 8]$$

Exercises 9.2

Write each equation in logarithmic form.

1. $3^2 = 9$ **2.** $7^2 = 49$ **3.** $5^3 = 125$

4. $10^3 = 1000$ **5.** $2^5 = 32$ **6.** $3^4 = 81$

7. $9^{1/2} = 3$ **8.** $16^{1/2} = 4$ **9.** $5^{-2} = \frac{1}{25}$

10. $4^0 = 1$ **11.** $10^{-5} = 0.00001$ **12.** $d^e = f$

Write each equation in exponential form.

13. $\log_5 25 = 2$ **14.** $\log_8 64 = 2$ **15.** $\log_2 16 = 4$

16. $\log_5 125 = 3$ **17.** $\log_{25} 5 = \frac{1}{2}$ **18.** $\log_{81} 9 = \frac{1}{2}$

19. $\log_8 2 = \frac{1}{3}$ **20.** $\log_{27} 3 = \frac{1}{3}$ **21.** $\log_2 (\frac{1}{4}) = -2$

22. $\log_2 (\frac{1}{8}) = -3$ **23.** $\log_{10} 0.01 = -2$ **24.** $\log_g h = k$

Graph each equation.

25. $y = \log_4 x$ **26.** $y = \log_3 x$ **27.** $y = \log_{10} x$

28. $y = \log_5 x$ **29.** $y = \log_{1/4} x$ **30.** $y = \log_{1/3} x$

Solve for x.

31. $\log_4 x = 3$ **32.** $\log_2 x = -1$ **33.** $\log_9 3 = x$

34. $\log_6 36 = x$ **35.** $\log_2 8 = x$ **36.** $\log_3 27 = x$

37. $\log_{25} 5 = x$ **38.** $\log_{27} 3 = x$ **39.** $\log_x 25 = 2$

40. $\log_x (\frac{1}{27}) = -3$ **41.** $\log_{1/2} (\frac{1}{8}) = x$ **42.** $\log_x 3 = \frac{1}{2}$

43. $\log_{12} x = 2$ **44.** $\log_8 (\frac{1}{64}) = x$ **45.** $\log_x 9 = \frac{2}{3}$

46. $\log_x 64 = \frac{3}{2}$ **47.** $\log_x (\frac{1}{8}) = -\frac{3}{2}$ **48.** $\log_x (\frac{1}{27}) = -\frac{3}{4}$

9.3 PROPERTIES OF LOGARITHMS

The most common uses of logarithms today include expressing the exponential and logarithmic relationships in business and between certain natural phenomena in electronics, biology, and radioactivity; solving exponential and logarithmic equations; and rewriting certain algebraic expressions. Before studying these concepts, we need to develop the following three basic logarithmic properties.

1. **Multiplication:** If M and N are positive real numbers,

$$\log_a (M \cdot N) = \log_a M + \log_a N, \qquad \text{where } a > 0 \quad \text{and} \quad a \neq 1$$

To prove this, let $p = \log_a M$ and $q = \log_a N$. Writing each in exponential form, we have $a^p = M$ and $a^q = N$. Forming the product MN, we have $M \cdot N = a^p \cdot a^q = a^{p+q}$.

Now write $M \cdot N = a^{p+q}$ in logarithmic form.

$$\begin{aligned} \log_a (M \cdot N) &= p + q \\ &= \log_a M + \log_a N \end{aligned}$$

That is, the logarithm of a product equals the sum of the logarithms of its factors.

2. **Division:** If M and N are positive real numbers,

$$\log_a \left(\frac{M}{N}\right) = \log_a M - \log_a N, \qquad \text{where } a > 0 \quad \text{and} \quad a \neq 1$$

Again let $p = \log_a M$ and $q = \log_a N$. Writing each in exponential form, we have $a^p = M$ and $a^q = N$. Forming the quotient $\frac{M}{N}$, we have $\frac{M}{N} = \frac{a^p}{a^q} = a^{p-q}$.

Now write $\frac{M}{N} = a^{p-q}$ in logarithmic form.

$$\log_a\left(\frac{M}{N}\right) = p - q$$
$$= \log_a M - \log_a N$$

That is, the logarithm of a quotient equals the difference of the logarithms of its factors.

3. **Powers:** If M is a positive real number and n is any real number,

$$\log_a M^n = n \log_a M, \qquad \text{where } a > 0 \quad \text{and} \quad a \neq 1$$

Let $p = \log_a M$, which in exponential form is

$$a^p = M$$

Taking the nth power of each side, we have

$$(a^p)^n = M^n$$
$$a^{np} = M^n$$

which, in logarithmic form, is

$$\log_a M^n = np$$
$$= n \log_a M$$

That is, the logarithm of a power of a number equals the product of the exponent times the logarithm of the number.

There are three special cases of the power property that are helpful.

(a) Roots: If M is any positive real number and n is any positive integer,

$$\log_a \sqrt[n]{M} = \frac{1}{n} \cdot \log_a M$$

Note that this is a special case where $\sqrt[n]{M} = M^{1/n}$. That is, the logarithm of the root of a number equals the logarithm of the number divided by the index of the root.

(b) For $n = 0$:

$$\log_a M^0 = \log_a 1 \qquad (M^0 = 1)$$
$$\log_a M^0 = 0 \cdot \log_a M \qquad \text{(Property 3)}$$
$$= 0$$

Therefore,

$$\log_a 1 = 0$$

That is, the logarithm of 1 is zero, regardless of the base.

(c) For $n = -1$:

$$\log_a M^{-1} = \log_a \frac{1}{M} \qquad \left(M^{-1} = \frac{1}{M}\right)$$

$$\log_a M^{-1} = (-1)\log_a M \qquad \text{(Property 3)}$$

Therefore,

$$\log_a \frac{1}{M} = -\log_a M$$

That is, the logarithm of the reciprocal of a number is the negative of the logarithm of the number.

EXAMPLE 1

Write $\log_4 2x^5y^2$ as a sum of multiples of single logarithms.

$$\begin{aligned} \log_4 2x^5y^2 &= \log_4 2 + \log_4 x^5 + \log_4 y^2 && \text{(Property 1)} \\ &= \log_4 2 + 5\log_4 x + 2\log_4 y && \text{(Property 3)} \end{aligned}$$

EXAMPLE 2

Write $\log_3 \dfrac{\sqrt{x(x-2)}}{(x+3)^2}$ as a sum or difference of multiples of the logarithms of x, $x - 2$, and $x + 3$.

$$\begin{aligned} \log_3 \frac{\sqrt{x(x-2)}}{(x+3)^2} &= \log_3 \frac{[x(x-2)]^{1/2}}{(x+3)^2} \\ &= \log_3 [x(x-2)]^{1/2} - \log_3 (x+3)^2 && \text{(Property 2)} \\ &= \tfrac{1}{2}\log_3 [x(x-2)] - 2\log_3 (x+3) && \text{(Property 3)} \\ &= \tfrac{1}{2}[\log_3 x + \log_3 (x-2)] - 2\log_3 (x+3) && \text{(Property 1)} \\ &= \tfrac{1}{2}\log_3 x + \tfrac{1}{2}\log_3 (x-2) - 2\log_3 (x+3) \end{aligned}$$

EXAMPLE 3

Write $3\log_2 x + 4\log_2 y - 2\log_2 z$ as a single logarithmic expression.

$$\begin{aligned} 3\log_2 x + 4\log_2 y - 2\log_2 z &= \log_2 x^3 + \log_2 y^4 - \log_2 z^2 && \text{(Property 3)} \\ &= \log_2 (x^3y^4) - \log_2 z^2 && \text{(Property 1)} \\ &= \log_2 \frac{x^3y^4}{z^2} && \text{(Property 2)} \end{aligned}$$

EXAMPLE 4

Write $3\log_{10}(x-1) - \frac{1}{3}\log_{10} x - \log_{10}(2x+3)$ as a single logarithmic expression.

$3\log_{10}(x-1) - \frac{1}{3}\log_{10} x - \log_{10}(2x+3)$

$$\begin{aligned} &= \log_{10}(x-1)^3 - \log_{10} x^{1/3} - \log_{10}(2x+3) && \text{(Property 3)} \\ &= \log_{10}(x-1)^3 - [\log_{10} x^{1/3} + \log_{10}(2x+3)] \\ &= \log_{10}(x-1)^3 - \log_{10}[(x^{1/3})(2x+3)] && \text{(Property 1)} \end{aligned}$$

$$= \log_{10} \frac{(x-1)^3}{x^{1/3}(2x+3)} \qquad \text{(Property 2)}$$

$$= \log_{10} \frac{(x-1)^3}{\sqrt[3]{x}(2x+3)}$$

There are two other logarithmic properties that are useful in simplifying expressions:

$$\log_a a^x = x \quad \text{and} \quad a^{\log_a x} = x$$

To show the first one, we begin with the identity

$$(a^x) = a^x$$

Then, writing this exponential equation in logarithmic form, we have

$$\textbf{4.} \quad \log_a (a^x) = x$$

To show the second one, we begin with the identity

$$\log_a x = (\log_a x)$$

Then, writing this logarithmic equation in exponential form, we have

$$\textbf{5.} \quad a^{(\log_a x)} = x$$

EXAMPLE 5

Find the value of $\log_2 16$.

$$\log_2 16 = \log_2 2^4$$

$$= 4$$ (Property 4. Note that this simplification is possible because 16 is a power of 2.)

EXAMPLE 6

Find the value of $\log_{10} 0.01$.

$$\log_{10} 0.01 = \log_{10} 10^{-2} \qquad \left(\textit{Note: } 0.01 = \frac{1}{100} = 10^{-2}\right)$$

$$= -2 \qquad \text{(Property 4)}$$

EXAMPLE 7

Find the value of $5^{\log_5 8}$.

$$5^{\log_5 8} = 8$$

EXAMPLE 8

Find the value of $9^{\log_3 25}$.

$$9^{\log_3 25} = (3^2)^{\log_3 25}$$

$$= 3^{2 \log_3 25} \qquad [(a^m)^n = a^{mn}]$$

$$= 3^{\log_3 25^2} \qquad \text{(Property 3)}$$

$$= 25^2 \qquad \text{(Property 5)}$$

$$= 625$$

EXAMPLE 9

Find the value of $\log_2 (8^{\log_2 8})$.

$$\begin{aligned}\log_2 (8^{\log_2 8}) &= \log_2 (8^{\log_2 2^3}) \\ &= \log_2 (8^3) \quad (\log_2 2^3 = 3, \text{ Property 4}) \\ &= \log_2 (2^3)^3 \\ &= \log_2 2^9 \\ &= 9 \quad \text{(Property 4)}\end{aligned}$$

Exercises 9.3

Write each expression as a sum or difference of multiples of single logarithms.

1. $\log_2 5x^3y$
2. $\log_3 \dfrac{8x^2y^3}{z^4}$
3. $\log_{10} \dfrac{2x^2}{y^3z}$
4. $\log_4 \dfrac{y^3}{x\sqrt{z}}$
5. $\log_b \dfrac{y^3\sqrt{x}}{z^2}$
6. $\log_b \dfrac{7xy}{\sqrt[3]{z}}$
7. $\log_b \sqrt[3]{\dfrac{x^2}{y}}$
8. $\log_5 \sqrt[4]{xy^2z}$
9. $\log_2 \dfrac{1}{x}\sqrt{\dfrac{y}{z}}$
10. $\log_b \dfrac{1}{z^2}\sqrt[3]{\dfrac{x^2}{y}}$
11. $\log_b \dfrac{z^3\sqrt{x}}{\sqrt[3]{y}}$
12. $\log_b \dfrac{\sqrt{y\sqrt{x}}}{z^2}$
13. $\log_b \dfrac{x^2(x+1)}{\sqrt{x+2}}$
14. $\log_b \dfrac{\sqrt{x(x+4)}}{x^2}$

Write each as a single logarithmic expression.

15. $\log_b x + 2\log_b y$
16. $2\log_b z - 3\log_b x$
17. $\log_b x + 2\log_b y - 3\log_b z$
18. $3\log_7 x - 4\log_7 y - 5\log_7 z$
19. $\log_3 x + \frac{1}{3}\log_3 y - \frac{1}{2}\log_3 z$
20. $\frac{1}{2}\log_2 x - \frac{1}{3}\log_2 y - \log_2 z$
21. $2\log_{10} x - \frac{1}{2}\log_{10}(x-3) - \log_{10}(x+1)$
22. $\log_3 (x+1) + \frac{1}{2}\log_3 (x+2) - 3\log_3 (x-1)$
23. $5\log_b x + \frac{1}{3}\log_b (x-1) - \log_b (x+2)$
24. $\log_b (x+1) + \frac{1}{3}\log_b (x-7) - 2\log_b x$
25. $\log_{10} x + 2\log_{10}(x-1) - \frac{1}{3}[\log_{10}(x+2) + \log_{10}(x-5)]$
26. $\frac{1}{2}\log_b (x+1) - 3[\log_b x + \log_b (x-1) + \log_b (2x-1)]$

Find the value of each expression.

27. $\log_b b^3$
28. $\log_2 2^5$
29. $\log_3 9$
30. $\log_2 16$
31. $\log_5 125$
32. $\log_4 64$
33. $\log_2 \frac{1}{4}$
34. $\log_3 \frac{1}{27}$
35. $\log_{10} 0.001$
36. $\log_{10} 0.1$
37. $\log_3 1$
38. $\log_{10} 1$
39. $6^{\log_6 5}$
40. $3^{\log_3 9}$
41. $25^{\log_5 6}$
42. $27^{\log_3 2}$
43. $4^{\log_2 (1/5)}$
44. $8^{\log_2 (1/3)}$
45. $\log_3 9^{\log_3 27}$
46. $\log_4 16^{\log_4 64}$
47. $\log_2 16^{\log_4 16}$

9.4 COMMON LOGARITHMS

In the previous sections we used a variety of bases of logarithms. Actually, only two bases are in general use: base 10 and base e, where e is an irrational number approximately equal to 2.71828. Logarithms that use 10 as a base are called **common logarithms.**

Logarithms that use e as a base are called **natural logarithms** and are discussed in the next section.

Before the electronic handheld calculator, common logarithms were routinely used for a wide variety of computations, such as multiplication, division, and finding powers and roots of numbers. A variety of technical and scientific measurements using common logarithms remain.

Let's list some powers of 10 and compare each with its equivalent logarithmic form:

Exponential form	*Logarithmic form*
$10^2 = 100$	$\log_{10} 100 = 2$
$10^4 = 10000$	$\log_{10} 10000 = 4$
$10^{-3} = 0.001$	$\log_{10} 0.001 = -3$
$10^{-1} = 0.1$	$\log_{10} 0.1 = -1$
$10^0 = 1$	$\log_{10} 1 = 0$

For nonintegral powers of 10, approximations can be calculated. For example:

Exponential form	*Logarithmic form*
$10^{0.3010} = 2.00$	$\log_{10} 2.00 = 0.3010$
$10^{0.5263} = 3.36$	$\log_{10} 3.36 = 0.5263$
$10^{0.8451} = 7.00$	$\log_{10} 7.00 = 0.8451$
$10^{0.9258} = 8.43$	$\log_{10} 8.43 = 0.9258$

When working exclusively with common logarithms, we shall follow the common practice of not writing the base 10; for example, $\log_{10} 456 = \log 456$. A calculator may be used to find the common logarithm of a number as shown in the following examples.

EXAMPLE 1

Find $\log_{10} 456$ and $\log 0.0596$, rounded to four significant digits.

CATALOG L down arrows **ENTER** 456) **green diamond ENTER CATALOG ENTER** .0596) **ENTER**

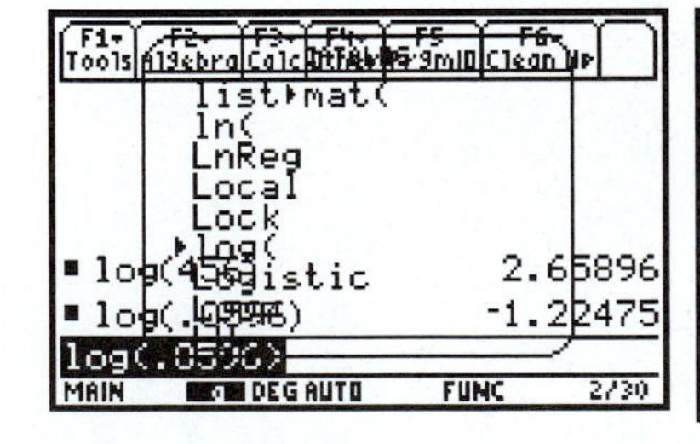

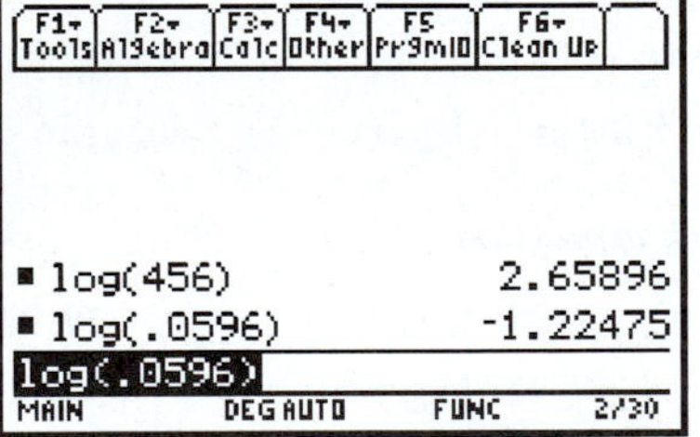

LOG 456) **ENTER LOG** .0596) **ENTER**

That is, $\log_{10} 456 = 2.659$ and $\log 0.0596 = -1.225$.

A calculator may also be used to find the number N when its logarithm is known. Here N is called the **antilogarithm.** Finding the antilogarithm of a number is the reverse process of finding the logarithm of a number.

EXAMPLE 2

Given $\log N = 1.6845$, find N, rounded to three significant digits.

10^1.6845 **ENTER**

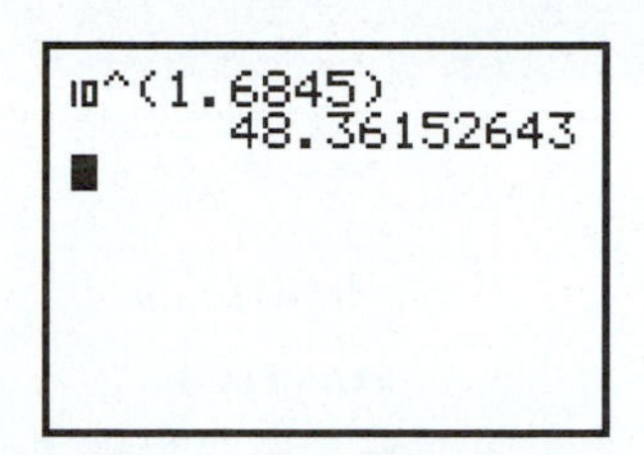

2nd 10^x 1.6845) **ENTER**

That is, $N = 48.4$ rounded to three significant digits.

Next, we find the value of a rational logarithm expression.

EXAMPLE 3

Find the value of $\dfrac{\log 275}{\log 5}$, rounded to three significant digits.

CATALOG L down arrows **ENTER** 275)/ **CATALOG ENTER** 5) **green diamond ENTER**

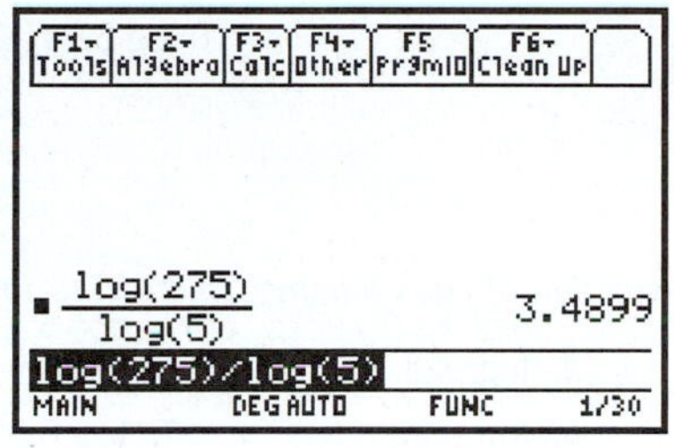

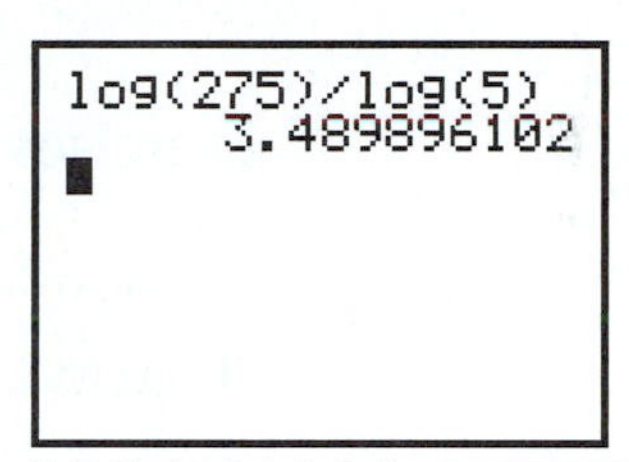

LOG 275)/ **LOG** 5) **ENTER**

The result rounded to three significant digits is 3.49.

EXAMPLE 4

In chemistry, the pH (hydrogen potential) of a solution is a measure of its acidity and is defined as

$$\text{pH} = -\log(\text{H}^+)$$

where H^+ is a numerical value for the concentration of hydrogen ions in moles per litre. Water has a pH of 7. Acids have pH values less than 7, and alkaline solutions (bases) have pH values greater than 7. For beer, if $\text{H}^+ = 6.3 \times 10^{-5}$ moles/litre (M/L), find its pH.

$$\begin{aligned}\text{pH} &= -\log(\text{H}^+)\\ &= -\log(6.3 \times 10^{-5})\\ &= 4.2 \qquad \text{(pH values are usually rounded to the nearest tenth.)}\end{aligned}$$

EXAMPLE 5

The intensity level of sound is given by the formula

$$\beta = 10 \log \frac{I}{I_0}$$

where β is the intensity level in decibels (dB) of a sound of intensity I measured in watts per square centimetre (W/cm^2) and I_0 is the intensity of the threshold of hearing, 10^{-16} W/cm^2. What is the intensity level of a sound with an intensity of 10^{-11} W/cm^2? (Normal hearing ranges between 0 and 120 dB.)

$$\beta = 10 \log \frac{I}{I_0}$$

$$\beta = 10 \log \left(\frac{10^{-11} \text{ W/cm}^2}{10^{-16} \text{ W/cm}^2} \right)$$

$$= 10 \log 10^5$$

$$= 10 \cdot 5 = 50 \text{ dB}$$

EXAMPLE 6

The power gain or loss of an amplifier or other electronic device is given by

$$n = 10 \log \left(\frac{P_o}{P_i} \right)$$

where n is the power gain or loss in decibels (dB), P_o is the power output, and P_i is the power input. Find the power gain when $P_o = 8.0$ W and $P_i = 0.25$ W.

$$n = 10 \log \left(\frac{P_o}{P_i} \right)$$

$$n = 10 \log \left(\frac{8.0 \text{ W}}{0.25 \text{ W}} \right)$$

$$= 10 \log 32 = 15 \text{ dB}$$

Exercises 9.4

Find the common logarithm of each number, rounded to four significant digits.

1. log 68.1
2. log 928
3. log 45
4. log 14,000
5. log 0.142
6. log 0.026
7. log 0.00621
8. log 0.0000497
9. log 805
10. log 0.608
11. log 9.25
12. log 1

Find N, the antilogarithm, rounded to three significant digits.

13. $\log N = 1.4048$
14. $\log N = 2.6191$
15. $\log N = 2.8484$
16. $\log N = 4.7400$
17. $\log N = 0.2782$
18. $\log N = 0.5690$
19. $\log N = -1.6050$
20. $\log N = -2.7376$
21. $\log N = -3.6345$
22. $\log N = -4.805$
23. $\log N = -4.8145$
24. $\log N = -6.8163$

Find the value of each expression, rounded to three significant digits.

25. $\frac{\log 685}{\log 6}$
26. $\frac{\log 984}{\log 4}$
27. $\frac{\log 1675}{\log 12.5}$
28. $\frac{\log 64.5}{\log 207}$
29. $\frac{\log 16.5}{\log 1350}$
30. $\frac{\log 8}{\log 12.5}$
31. $\log \left(\frac{596}{45}\right)$
32. $\log \left(\frac{654}{7}\right)$
33. $\log \left(\frac{4}{15}\right)$
34. $\log \left(\frac{16.5}{2.7}\right)$
35. $\log 14.6 - \log 3.75$
36. $\log 45 + \log 20$
37. $\log 145 + \log 25$
38. $\log 9.46 - \log 0.48$
39. $\frac{\log 486 + \log 680}{\log 14}$
40. $\frac{\log 276 + \log 98.4}{\log 2 - \log 14.5}$

We defined pH in Example 4. Compute the pH for each value.

41. $H^+ = 10^{-6}$
42. $H^+ = 10^{-8}$
43. $H^+ = 3.2 \times 10^{-7}$
44. $H^+ = 2.0 \times 10^{-6}$
45. $H^+ = 5.5 \times 10^{-8}$
46. $H^+ = 6.1 \times 10^{-3}$

The intensity of sound level is defined in Example 5. Compute β for each value.

47. 10^{-6} W/cm^2 **48.** 10^{-14} W/cm^2 **49.** 10^{-10} W/cm^2 **50.** 10^{-9} W/cm^2

We defined electronic power gain or loss in Example 6. Find each power gain or loss.

51. $P_o = 12$ W, $P_i = 0.50$ W

52. $P_o = 150$ W, $P_i = 3.0$ W

53. $P_o = 0.60$ W, $P_i = 0.80$ W

54. $P_o = 1.0$ W, $P_i = 1.2$ W

9.5 NATURAL LOGARITHMS

While common logarithms have a base of 10, **natural logarithms** have a base of e. The number e is irrational and is approximately equal to 2.71828. Although it may seem strange to have a system of logarithms based on such a number, many applications are based on powers of e, especially those involving growth and decay relationships. The form $\log_e x$ is usually written in its special notation $\ln x$.

To raise the number e to a power on a calculator, use the e^x button, as illustrated in Example 1.

EXAMPLE 1

Find the values of e^5 and $e^{-1.75}$, rounded to three significant digits.

green diamond e^x 5) **green diamond ENTER green diamond e^x** -1.75) **ENTER**

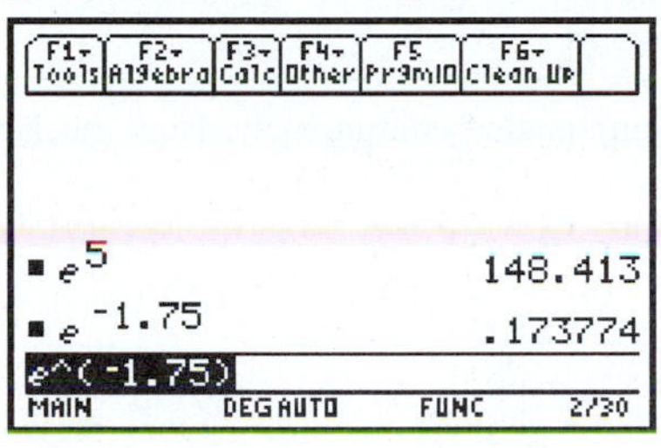

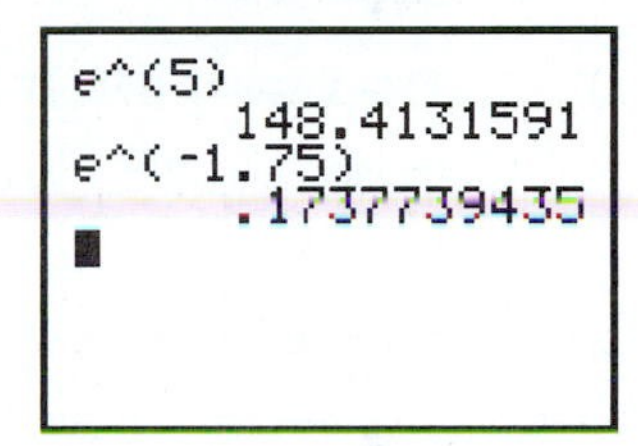

2nd e^x 5) **ENTER 2nd e^x** -1.75) **ENTER**

That is, $e^5 = 148$ and $e^{-1.75} = 0.174$ rounded to three significant digits.

A calculator offers an easy way to find natural logs and antilogs.

EXAMPLE 2

Find ln 4350, rounded to four significant digits.

2nd LN 4350) **ENTER**

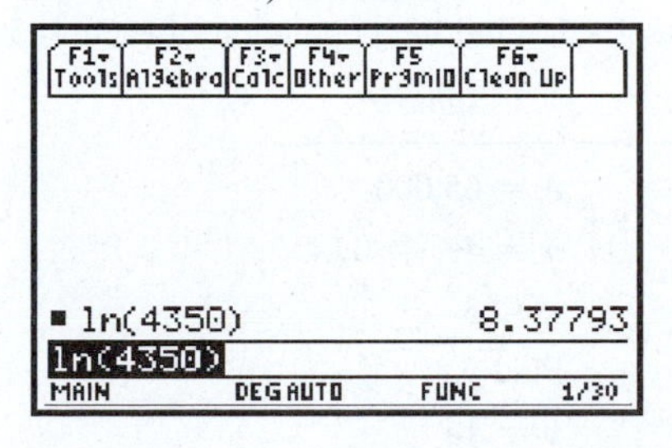

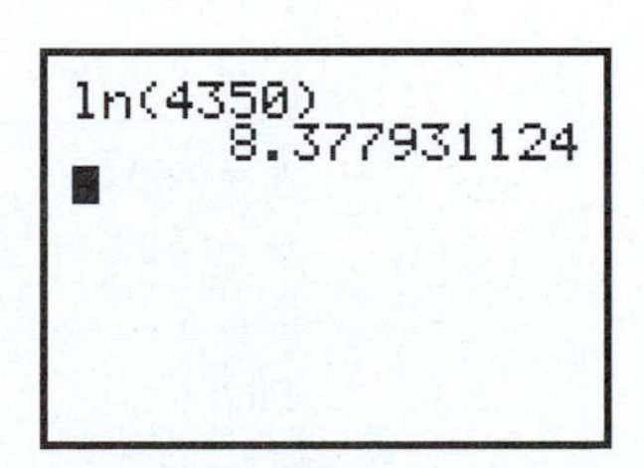

LN 4350) **ENTER**

That is, ln 4350 = 8.378 rounded to four significant digits.

EXAMPLE 3

Given $\ln x = 5.468$, find x, rounded to three significant digits.

green diamond $\mathbf{e^x}$ 5.468) **ENTER**

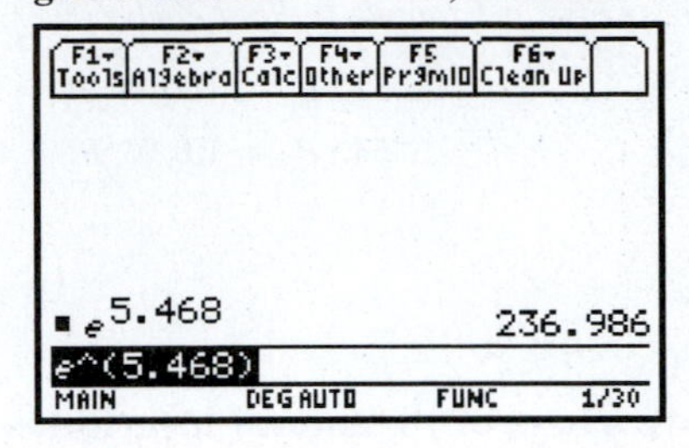

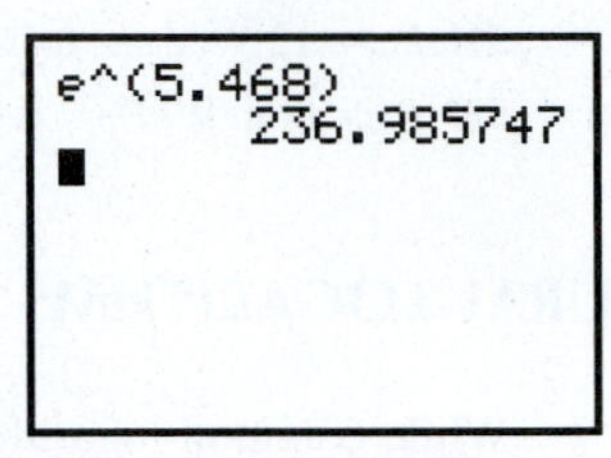

2nd $\mathbf{e^x}$ 5.468) **ENTER**

That is, $x = 237$ rounded to three significant digits.

If money is invested in an account that pays interest compounded continuously, the amount accumulated is given by the exponential function

$$y = Ae^{rn}$$

where A is the initial invested amount, r is the interest rate, n is the number of years that the initial amount is invested, e is the natural base for logarithms, and y is the amount accumulated.

The exponential function $y = Ae^{rn}$ is also an alternate growth function.

EXAMPLE 4

Joan invests \$1000 at $7\frac{3}{4}\%$ compounded continuously. How much money will she have in the account after 5 years?

Here,

$$A = \$1000$$
$$r = 7\tfrac{3}{4}\% = 7.75\% = 0.0775$$
$$n = 5$$

We are to find y.

$$y = Ae^{rn}$$
$$y = \$1000e^{(0.0775)(5)}$$
$$= \$1473 \qquad \text{(rounded to the nearest dollar)}$$

EXAMPLE 5

A city has a population of 850,000 and a nearby suburb has a population of 65,000. Studies have determined that the city will lose population at the rate of 2%, while the suburb will gain population at the rate of 4%. Find the population of each in 5 years.

City	*Suburb*	
$A = 850{,}000$	$A = 65{,}000$	
$r = -2\% = -0.02$	$r = 4\% = 0.04$	
$n = 5$	$n = 5$	
Find y.	Find y.	
$y = Ae^{rn}$	$y = Ae^{rn}$	
$y = 850{,}000e^{(-0.02)(5)}$	$y = 65{,}000e^{(0.04)(5)}$	
$= 770{,}000$	$= 79{,}000$	(each rounded to two significant digits)

EXAMPLE 6

A culture contains $10\bar{0},000$ bacteria. It grows at the rate of 15% per hour. How many bacteria will be present in 12 h?

Using $y = Ae^{rn}$, we have

$$A = 10\bar{0},000$$
$$r = 15\% = 0.15$$
$$n = 12 \text{ h}$$

Find y.

$$y = Ae^{rn}$$
$$y = 10\bar{0},000e^{(0.15)(12)}$$
$$= 6\bar{0}0,000 \text{ bacteria} \quad \text{(rounded to two significant digits)}$$

EXAMPLE 7

The current in an electric circuit is given by

$$i = 2.4e^{-4.0t}$$

where i is the current in amperes (A) and t is the time in seconds. Find the current when $t = 0.35$ s.

$$i = 2.4e^{-4.0t}$$
$$= 2.4e^{(-4.0)(0.35)}$$
$$= 0.59 \text{ A} \quad \text{(rounded to two significant digits)}$$

Logarithms with Other Bases

To find the logarithm with another base, use the following formula:

$$\log_b x = \frac{\log_a x}{\log_a b}$$

where a is base 10 or base e and b is another base. To verify this formula, let

$$u = \log_b x$$

Then

$$b^u = x$$
$$\log_a b^u = \log_a x$$
$$u \log_a b = \log_a x$$
$$u = \frac{\log_a x}{\log_a b}$$

Since $u = \log_b x$,

$$\log_b x = \frac{\log_a x}{\log_a b}$$

EXAMPLE 8

Find $\log_3 370$ to three significant digits.

Let us use base 10.

$$\log_3 370 = \frac{\log_{10} 370}{\log_{10} 3} = \frac{2.5682}{0.4771} = 5.38$$

What is the relationship between the common logarithm and the natural logarithm of a number? In general,

$$\begin{aligned}\ln x = \frac{\log x}{\log e} &= \frac{\log x}{\log 2.718} \\ &= \frac{\log x}{0.4343} \\ &= 2.303 \log x\end{aligned}$$

That is,

$$\ln x = 2.303 \log x$$

Exercises 9.5

Find the value of each power of e, rounded to three significant digits.

1. e^2 **2.** e^3 **3.** e^6 **4.** e^{10}
5. e^{-2} **6.** e^{-3} **7.** e^{-6} **8.** e^{-10}
9. $e^{3.5}$ **10.** $e^{2.1}$ **11.** $e^{0.15}$ **12.** $e^{0.75}$
13. $e^{-2.5}$ **14.** $e^{-1.4}$ **15.** $e^{-0.08}$ **16.** $e^{-0.65}$
17. $e^{2/3}$ **18.** $e^{5/8}$ **19.** $e^{-1/3}$ **20.** $e^{-5/6}$

Find the natural logarithm of each number, rounded to four significant digits.

21. ln 56 **22.** ln 92 **23.** ln 406
24. ln 1845 **25.** ln 4.3 **26.** ln 0.705
27. ln 0.00582 **28.** ln 0.00000114 **29.** ln 1

Find x, rounded to three significant digits.

30. $\ln x = 1.605$ **31.** $\ln x = 0.475$ **32.** $\ln x = -0.1463$
33. $\ln x = -1.445$ **34.** $\ln x = -3.77$ **35.** $\ln x = 14.75$
36. $\ln x = -25$

37. Rework Exercise 41 in Section 9.1 using the growth function $y = Ae^{rn}$. Compare the results.

38. Rework Exercise 42 in Section 9.1 using the growth function $y = Ae^{rn}$. Compare the results.

39. Vera Alice invests \$3500 at $8\frac{1}{4}\%$ compounded continuously. How much money will she have in the account after 6 years?

40. Many savings institutions quote both *interest rates* and *effective interest rates.* The effective interest rate is defined as

$$e^r - 1$$

where r is the rate of interest compounded continuously. Find the effective interest rate in Exercise 39.

41. The population of a city is 95,000. Assuming a growth rate of 3.1%, what will its population be in 10 years?

42. The public school enrollment of a city is 14,300. Assuming a decline in enrollment at the rate of 4.5%, what will the enrollment be in 5 years?

43. The growing amount of bacteria in a culture is given by

$$N = N_0 e^{0.04t}$$

where t is the time in hours, 0.04 is the growth rate per hour, N_0 is the initial amount, and N is the amount after time t. If we begin with a culture of $3\bar{0}00$ bacteria, how many do we have after 5.0 h? Give your answer to two significant digits.

44. A culture contains 25,000 bacteria and grows at the rate of 7.5% per hour. How many bacteria will be present in 24 h?

45. A person invests \$10,000 at 8% for 20 years compounded continuously. Assuming an annual 12% inflation rate, what will the person's purchasing power be after 20 years?

46. The amount of a decaying radioactive element is given by

$$y = y_0 e^{-0.4t}$$

where t is the time in seconds, -0.4 is the decay rate per second, y_0 is the initial amount, and y is the amount remaining. If a given sample has a mass of $15\bar{0}$ g, how much remains after 5.00 min?

47. A given radioactive sample of mass 27.0 g decays at the rate of 2.50% per second. How much remains after 1.00 min?

48. In certain types of dc circuits, current increases exponentially according to the formula

$$i = \frac{E}{R}(1 - e^{-Rt/L})$$

where

i = the instantaneous current in amperes (A)
E = the voltage in volts (V)
R = the resistance in ohms (Ω)
t = the time
L = the inductance in henries (H)

Find i when $E = 12.0$ V, $R = 90.0\ \Omega$, $t = 0.0120$ s, and $L = 8.50$ H.

An $8\bar{0}$-mg sample of radioactive radium decays at the rate of $\ln A - \ln 8\bar{0} = kt$,

where

A = the amount remaining
t = the time in years
k = a constant (-4.10×10^{-4})

Find how long it takes the sample to decay to each amount in Exercises 49 through 53.

49. 79 mg **50.** 75 mg **51.** $6\bar{0}$ mg **52.** $4\bar{0}$ mg **53.** $2\bar{0}$ mg

54. In determination of the flow of heat in an insulated pipe, the heat loss is given by the formula

$$Q = \frac{2\pi k L(\Delta T)}{\ln\left(\frac{D_2}{D_1}\right)}$$

where

Q = the heat loss in British thermal units per hour (Btu/h)
L = length of pipe in feet
ΔT = the difference in temperature between the inner and outer surfaces of insulation in degrees Fahrenheit

D_2 = the outer diameter of insulation in inches

D_1 = the inner diameter of insulation in inches

k = a constant

Find the heat loss in Btu per hour from a pipe with a 12 in. outside diameter. The pipe is $10\bar{0}$ ft long and is covered with 3.0 in. of insulation with a thermal conductivity of $k = 0.045$ Btu/h °F ft. The inner temperature is 790°F and the outer temperature is 140°F.

Use the following equation for Exercises 55 through 58:

$$V_o = E_{AS} - (E_{AS} - E_o)e^{-t/(TC)}$$

55. Given:

$E_{AS} = 10.0$ V
$E_o = 0$ V
$TC = 20.0\ \mu s$
$t = 50.0\ \mu s$

Find V_o.

56. Given:

$E_{AS} = 30.0$ V
$E_o = 10.0$ V
$TC = 50.0\ \mu s$
$t = 75.0\ \mu s$

Find V_o.

57. Given:

$E_{AS} = -40.0$ V
$E_o = 60.0$ V
$TC = 18.0\ \mu s$
$t = 10.0\ \mu s$

Find V_o.

58. Given:

$E_{AS} = -75.0$ V
$E_o = -32.0$ V
$TC = 175\ \mu s$
$t = 225\ \mu s$

Find V_o.

59. Soil permeability as derived from a field pumping test is given by the formula

$$k = \frac{q}{\pi(H_2^2 - H_1^2)} \ln\left(\frac{R_2}{R_1}\right)$$

where k is the coefficient of permeability. Find the coefficient of permeability in ft/min for $H_2 = 48.5$ ft, $H_1 = 44.5$ ft, $R_2 = 24$ ft, $R_1 = 11$ ft, and $q = 51$ gal/min.

60. Using the formula in Exercise 59, find the coefficient of permeability in cm/min for $H_2 = 15.0$ m, $H_1 = 12.5$ m, $R_2 = 7.5$ m, $R_1 = 3.5$ m, and $q = 185$ L/min.

Find each logarithm, rounded to three significant digits.

61. $\log_3 84.1$
62. $\log_2 297$
63. $\log_4 2360$
64. $\log_{12} 5.72$
65. $\log_5 374$
66. $\log_4 4.19$
67. $\log_6 9600$
68. $\log_7 16.5$

9.6 SOLVING EXPONENTIAL EQUATIONS

The solution of the exponential equation $2^x = 8$ may be done by inspection ($x = 3$). However, the solution of $2^x = 6$ is not integral, and trial-and-error attempts at a solution would be complicated, to say the least. Exponential equations are used in such diverse fields as electronics, biology, chemistry, psychology, and economics.

The solution of the general exponential equation $b^x = a$ ($a > 0$, $b > 0$) is based on the fact that if two numbers are equal, their logarithms to the same base (any base) are equal.

If $x = y$
then $\log_b x = \log_b y$

EXAMPLE 1

Solve $2^x = 60$ to three significant digits.

$$2^x = 60$$

$$\log 2^x = \log 60 \qquad \text{(Take the common log of each side.)}$$

$$x \cdot \log 2 = \log 60 \qquad \text{(Property 3)}$$

$$x = \frac{\log 60}{\log 2} = \frac{1.7782}{0.3010} = 5.91$$

Now, let's solve the same equation by using natural logarithms.

$$2^x = 60$$

$$\ln 2^x = \ln 60 \qquad \text{(Take the natural log of each side.)}$$

$$x \cdot \ln 2 = \ln 60 \qquad \text{(Property 3)}$$

$$x = \frac{\ln 60}{\ln 2} = \frac{4.0943}{0.6931} = 5.91$$

As you can see, the solution to this equation can be done quite easily using either common or natural logarithms.

EXAMPLE 2

Solve $4^{x+2} = 36$ using common logarithms. (Give the result to three significant digits.)

$$4^{x+2} = 36$$

$$\log 4^{x+2} = \log 36 \qquad \text{(Take the common log of each side.)}$$

$$(x + 2)\log 4 = \log 36 \qquad \text{(Property 3)}$$

$$x + 2 = \frac{\log 36}{\log 4}$$

$$x = \frac{\log 36}{\log 4} - 2$$

$$= 0.585$$

EXAMPLE 3

Solve $4^{2x} = 12^{x+1}$ using natural logarithms. (Give the result to three significant digits.)

$$4^{2x} = 12^{x+1}$$

$$\ln 4^{2x} = \ln 12^{x+1} \qquad \text{(Take the natural log of each side.)}$$

$$2x \ln 4 = (x + 1)\ln 12 \qquad \text{(Property 3)}$$

$$2x \ln 4 = x \ln 12 + \ln 12$$

$$2x \ln 4 - x \ln 12 = \ln 12$$

$$x(2 \ln 4 - \ln 12) = \ln 12$$

$$x = \frac{\ln 12}{2 \ln 4 - \ln 12}$$

$$= 8.64$$

Since $\log 10^x = x$ and $\ln e^x = x$, using common logs is easier when solving an exponential equation involving a power of 10, and using natural logs is easier when solving an exponential equation involving a power of e.

EXAMPLE 4

Solve $10^{2x} = 1450$. (Give the result to three significant digits.)

$$
\begin{aligned}
10^{2x} &= 1450 \\
\log 10^{2x} &= \log 1450 && \text{(Take the common log of each side.)} \\
2x &= \log 1450 && \text{(Property 4)} \\
x &= \frac{\log 1450}{2} \\
&= 1.58
\end{aligned}
$$

EXAMPLE 5

Solve $e^{-3x} = 0.725$. (Give the result to three significant digits.)

$$
\begin{aligned}
e^{-3x} &= 0.725 \\
\ln e^{-3x} &= \ln 0.725 && \text{(Take the natural log of each side.)} \\
-3x &= \ln 0.725 && \text{(Property 4)} \\
x &= \frac{\ln 0.725}{-3} \\
&= 0.107
\end{aligned}
$$

EXAMPLE 6

Solve $36 = 45(1 - e^{x/2})$. (Give the result to three significant digits.)

$$
\begin{aligned}
36 &= 45(1 - e^{x/2}) \\
\frac{36}{45} &= 1 - e^{x/2} \\
e^{x/2} &= 1 - \frac{36}{45} \\
e^{x/2} &= 1 - 0.8 = 0.2 \\
\ln e^{x/2} &= \ln 0.2 \\
\frac{x}{2} &= \ln 0.2 \\
x &= 2 \ln 0.2 \\
&= -3.22
\end{aligned}
$$

By calculator, we have

F2 **1** 36=45(1- **green diamond** $\mathbf{e^x}$ x/2)),x) **ENTER** **2nd ANS** **green diamond** **ENTER**

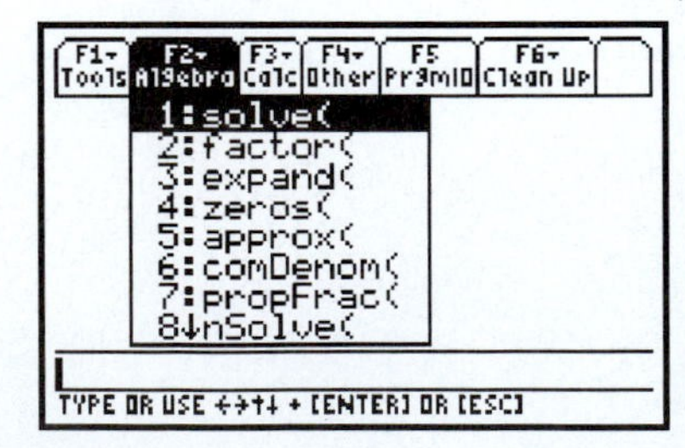

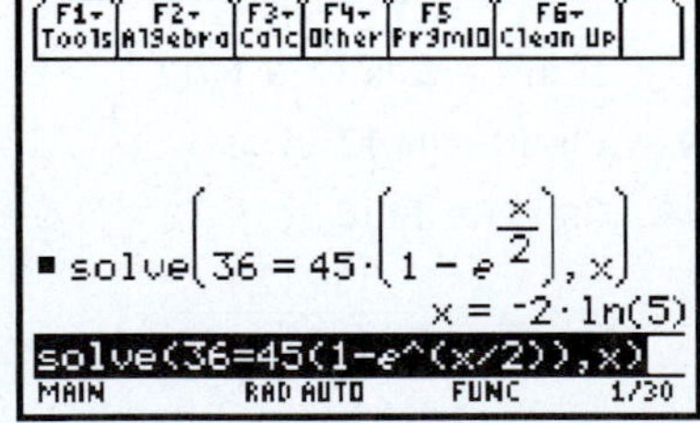

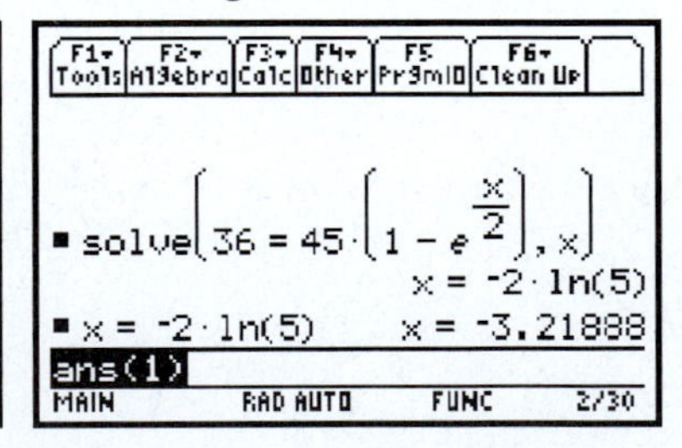

Exercises 9.6

Solve each exponential equation. (Give each result to three significant digits.)

1. $3^x = 12$ **2.** $5^x = 38.1$ **3.** $2^{-x} = 43.7$

4. $4^x = 0.439$
5. $3^{2x} = 0.21$
6. $5^{-3x} = 100$
7. $5^{x+1} = 3^x$
8. $2^{2x+1} = 5^x$

Solve for x. (Give each result to four significant digits.)

9. $4^{-20x} = 50$
10. $(5)2^{x+16} = 382$
11. $3^{2x+1} = 5^{-x}$
12. $8^{-3x} = 3^{2x-1}$
13. $e^x = 23$
14. $10^x = 1.35$
15. $10^{-x} = 0.146$
16. $e^{-x} = 14.7$
17. $e^{2x} = 40.5$
18. $10^{5x} = 2.63$
19. $e^{-3x} = 850$
20. $e^{-2x} = 0.00448$
21. $10^{x/2} = 0.45$
22. $e^{x/3} = 47.2$
23. $4e^x = 94.7$
24. $5.3e^{-4x} = 49.7$
25. $e^x = 5^{x-1}$
26. $e^{-2x} = 10^{x+1}$
27. $e^x = 4^{2x+1}$
28. $4e^{3x} = (10)2^{x+3}$
29. $5e^{-2x} = (12)3^{x-1}$
30. $(40)5^{2x+1} = (250)e^{x+1}$
31. $e^{x^2} = 600$
32. $10^{x^2} = 4650$
33. $3 = 4(1 - e^x)$
34. $8 = 20(1 - e^x)$
35. $35 = 80(1 - e^{-x})$
36. $125 = 175(1 - e^{-x})$
37. $8 = 10(1 - e^{x/2})$
38. $150 = 425(1 - e^{x/5})$
39. $175 = 225(1 - e^{-x/3})$
40. $28 = 56(1 - e^{-x/10})$
41. $48 = 64(1 - e^{-3x})$
42. $120 = 125(1 - e^{3x})$
43. $135 = 145(1 - e^{-4x/3})$
44. $225 = 375(1 - e^{-5x/6})$
45. $50 = 75(1 + e)^x$
46. $16 = 18(1 + e)^{4x}$
47. $9 = 15(1 + e)^{-x}$
48. $1 = 2(1 + e)^{-2x}$
49. $7 = 10(1 + e)^{x/2}$
50. $8 = 12(1 + e)^{-3x/2}$

Use the following equation for Exercises 51 through 56:

$$V_o = E_{AS} - (E_{AS} - E_o)e^{-t/(TC)}$$

51. Given:

$E_{AS} = 20.0$ V
$E_o = -20.0$ V
$TC = 50.0\ \mu$s
$V_o = 0$

Find t.

52. Given:

$E_{AS} = 10.0$ V
$E_o = 20.0$ V
$TC = 15.0\ \mu$s
$V_o = 15.0$ V

Find t.

53. Given:

$E_{AS} = 27.0$ V
$E_o = 30.0$ V
$V_o = 28.5$ V
$t = 145\ \mu$s

Find TC.

54. Given:

$E_{AS} = 72.0$ V
$E_o = 95.0$ V
$V_o = 90.0$ V
$t = 10.5$ ms

Find TC.

55. Given:

$E_o = 15.0$ V
$V_o = 50.0$ V
$TC = 10.0$ ms
$t = 32.0$ ms

Find E_{AS}.

56. Given:

$E_{AS} = 50.0$ V
$V_o = 30.0$ V
$TC = 90.0\ \mu$s
$t = 150\ \mu$s

Find E_o.

9.7 SOLVING LOGARITHMIC EQUATIONS

The following principle is used in solving a logarithmic equation:

If $\log_b x = \log_b y$
then $x = y$

EXAMPLE 1

Solve $\log(x + 7) = \log(3x - 5)$.

$$\log(x + 7) = \log(3x - 5)$$
$$x + 7 = 3x - 5$$
$$12 = 2x$$
$$6 = x$$

EXAMPLE 2

Solve $\ln 2 + \ln(x - 4) = \ln(x + 6)$.

$$\begin{aligned} \ln 2 + \ln(x - 4) &= \ln(x + 6) \\ \ln 2(x - 4) &= \ln(x + 6) \qquad \text{(Property 1)} \\ 2(x - 4) &= x + 6 \\ 2x - 8 &= x + 6 \\ x &= 14 \end{aligned}$$

Another technique used in solving a logarithmic equation is changing the logarithmic equation to exponential form:

If	$y = \log_b x$
then	$x = b^y$

Then solve the resulting equation.

EXAMPLE 3

Solve $\log(x - 3) = 2$.

$$\begin{aligned} \log(x - 3) &= 2 \\ x - 3 &= 10^2 \qquad \text{(Write in exponential form.)} \\ x - 3 &= 100 \\ x &= 103 \end{aligned}$$

EXAMPLE 4

Solve $\log(x + 3) + \log x = 1$.

$$\begin{aligned} \log(x + 3) + \log x &= 1 \\ \log x(x + 3) &= 1 \qquad \text{(Property 1)} \\ x(x + 3) &= 10^1 \qquad \text{(Write in exponential form.)} \\ x^2 + 3x - 10 &= 0 \\ (x + 5)(x - 2) &= 0 \end{aligned}$$

$$x + 5 = 0 \quad \text{or} \quad x - 2 = 0$$

$$x = -5 \qquad\qquad x = 2$$

Since logarithms of negative numbers are not defined, the solution is $x = 2$.

EXAMPLE 5

Solve $\ln(x + 1) - \ln x = 3$.

$$\begin{aligned} \ln(x + 1) - \ln x &= 3 \\ \ln \frac{x + 1}{x} &= 3 \qquad \text{(Property 2)} \\ \frac{x + 1}{x} &= e^3 \qquad \text{(Write in exponential form.)} \\ x + 1 &= xe^3 \\ 1 &= xe^3 - x \\ 1 &= x(e^3 - 1) \\ \frac{1}{e^3 - 1} &= x \\ x &= 0.0524 \end{aligned}$$

Solving by calculator, we have

F2 **1** **2nd LN** x+1)– **2nd LN** x)=3,x) **ENTER** **2nd ANS** **green diamond** **ENTER**

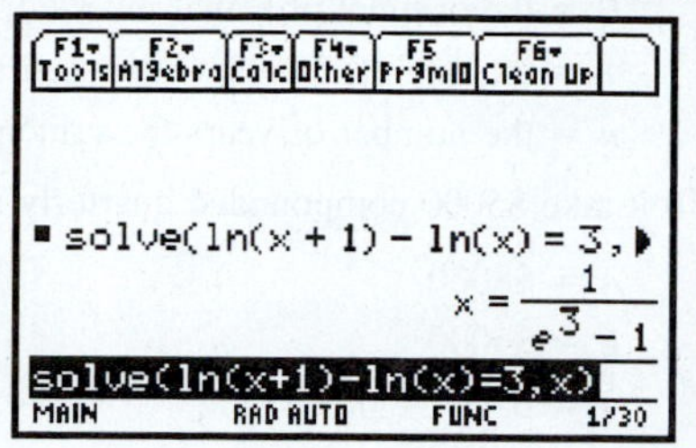

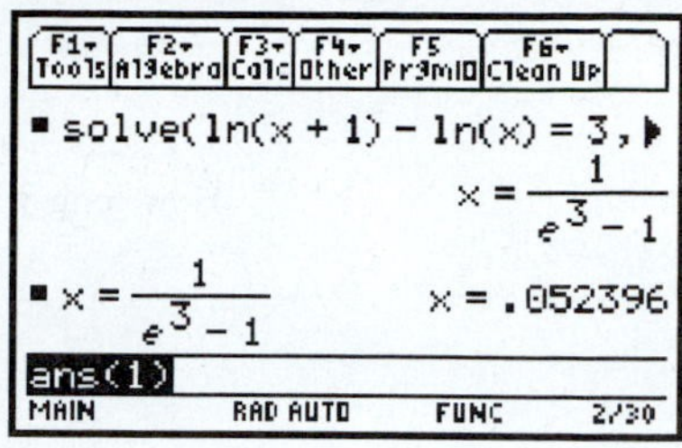

Exercises 9.7

Solve each equation.

1. $\log(3x - 4) = \log(x + 6)$
2. $\log(2x + 1) = \log(4x - 5)$
3. $\ln(x - 4) = \ln(2x + 5)$
4. $\ln(5x - 100) = \ln x$
5. $\log 2 + \log(x - 3) = \log(x + 1)$
6. $\log 3 + \log(x + 1) = \log 5 + \log(x - 3)$
7. $2 \ln x = \ln 49$
8. $3 \ln x = \ln 8$
9. $\log x + \log(2x) = \log 72$
10. $\log(3x) + \log(2x) = \log 150$
11. $\log(x + 4) - \log(x - 2) = \log 2$
12. $\log(2x + 1) - \log(x - 4) = \log 3$
13. $\ln x + \ln(x - 2) = \ln 3$
14. $\ln(2x) + \ln(x + 4) = \ln 24$
15. $\ln x + \ln(x - 4) = \ln(x + 6)$
16. $2 \ln x = \ln(12 - 5x)$
17. $2 \log x = 2$
18. $2 \ln x = 1$
19. $\log(x + 1) = 1$
20. $\log(2x + 4) = 2$
21. $2 \log(x - 1) = 1$
22. $\ln(x - 1) = 2$
23. $\log(x + 1) - 2 \log 3 = 2$
24. $\log(2x - 3) - 3 \log 2 = 1$
25. $\ln(x - 3) - \ln 2 = 1$
26. $\ln(2x - 1) + \ln 2 = 1$
27. $\ln(2x + 1) + \ln 3 = 2$
28. $\ln(2x + 3) - \ln 4 = 2$
29. $\log(x + 3) + 2 \log 4 = 2$
30. $\log(2x + 1) + 3 \log 3 = 3$
31. $\log(2x + 1) + \log(x - 1) = 2$
32. $\log(x + 3) + \log(x + 1) = 1$
33. $\ln x + \ln(x + 2) = 1$
34. $\ln(2x) + \ln(x - 3) = 2$
35. $\ln x + \ln(2x - 3e) - 2 = \ln 2$
36. $\ln x + \ln(2x - e) - 2 = \ln 3$

9.8 APPLICATIONS: SOLVING EXPONENTIAL AND LOGARITHMIC EQUATIONS

Of the many applications involving exponential and logarithmic equations, we present a variety involving several subject fields.

EXAMPLE 1

The value of money compounded quarterly is given by the formula

$$A = P\left(1 + \frac{r}{4}\right)^{4n}$$

where

$$A = \text{the amount of original principal plus accrued interest}$$
$$P = \text{the original principal invested}$$
$$r = \text{the rate of interest (expressed as a decimal)}$$
$$n = \text{the number of years the principal is invested}$$

How long will it take \$5000 compounded quarterly at 6.4% to accrue to \$8000?

$$A = \$8000$$
$$P = \$5000$$
$$r = 6.4\% = 0.064$$

We are to find n.

$$A = P\left(1 + \frac{r}{4}\right)^{4n}$$

$$8000 = 5000\left(1 + \frac{0.064}{4}\right)^{4n}$$

$$\frac{8000}{5000} = (1 + 0.016)^{4n} \qquad \text{(Divide each side by 5000.)}$$

$$1.6 = 1.016^{4n}$$

$$\log 1.6 = \log 1.016^{4n} \qquad \text{(Take the common log of each side.)}$$

$$\log 1.6 = 4n \log 1.016$$

$$n = \frac{\log 1.6}{4 \log 1.016}$$

$$= 7.4 \text{ years} \qquad \text{(rounded to two significant digits)}$$

When a quantity *increases* so that the amount of increase is proportional to the amount present, we have **exponential growth.** One of the most commonly used general equations for exponential or continuous growth is

EXPONENTIAL GROWTH

$$y = Ae^{rt}$$

where

$$y = \text{the amount present at any given time } t$$
$$A = \text{the original amount}$$
$$e = \text{the natural number}$$
$$r = \text{the rate of growth}$$
$$t = \text{the time the growth has occurred}$$

A graph of the general exponential or continual growth equation is shown in Fig. 9.6.

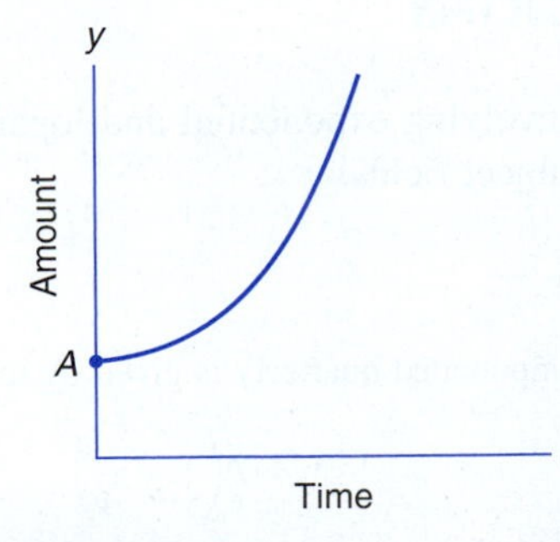

Figure 9.6 Graph of the general exponential growth equation $y = Ae^{rt}$.

EXAMPLE 2

Mary and Sid deposit \$2500 in a savings account that pays $3\frac{1}{4}\%$ compounded continuously. In how many years will the amount in the account double?

Using $y = Ae^{rn}$, we know

$$A = \$2500$$
$$y = \$5000$$
$$r = 3\tfrac{1}{4}\% = 3.25\% = 0.0325$$

We are to find n.

$$
\begin{aligned}
y &= Ae^{rn} \\
5000 &= 2500e^{(0.0325)n} \\
\frac{5000}{2500} &= e^{0.0325n} && \text{(Divide each side by 2500.)} \\
2 &= e^{0.0325n} \\
\ln 2 &= \ln e^{0.0325n} && \text{(Take the natural log of each side.)} \\
\ln 2 &= 0.0325n \\
n &= \frac{\ln 2}{0.0325} \\
&= 21 \text{ years} && \text{(rounded to two significant digits)}
\end{aligned}
$$

When a quantity *decreases* so that the amount of decrease is proportional to the amount present, we have **exponential decay.** One of the most commonly used general equations for exponential or continuous decay is

EXPONENTIAL DECAY

$$y = Ae^{-rt}$$

where

$y =$ the amount remaining at any given time t
$A =$ the original amount
$e =$ the natural number
$r =$ the rate of decay
$t =$ the time the decay has occurred

A graph of the general exponential or continual decay equation is shown in Fig. 9.7.

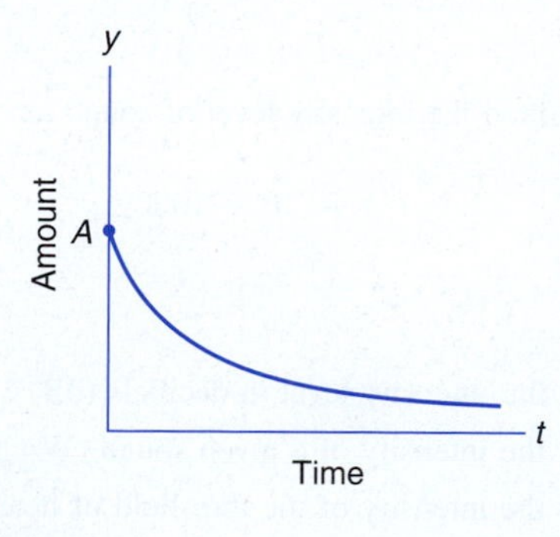

Figure 9.7 Graph of the general exponential decay equation $y = Ae^{-rt}$.

EXAMPLE 3

The amount of a decaying radioactive element is given by

$$y = y_0e^{-0.4t}$$

where

$$t = \text{the time in seconds (s)}$$
$$y_0 = \text{the initial amount}$$
$$y = \text{the amount remaining}$$

How long will it take $5\bar{0}$ g to decay to 25 g?

Here,

$$y_0 = 5\bar{0}\text{ g}$$
$$y = 25\text{ g}$$

We are to find t.

$$y = y_0 e^{-0.4t}$$
$$25 = 50e^{-0.4t}$$
$$\frac{25}{50} = e^{-0.4t} \quad \text{(Divide each side by 50.)}$$
$$0.5 = e^{-0.4t}$$
$$\ln 0.5 = \ln e^{-0.4t} \quad \text{(Take the natural log of each side.)}$$
$$\ln 0.5 = -0.4t$$
$$t = \frac{\ln 0.5}{-0.4}$$
$$= 1.7\text{ s} \quad \text{(rounded to two significant digits)}$$

Note: This time is called the **half-life** of the substance.

EXAMPLE 4

In Section 9.4 we defined pH as

$$\text{pH} = -\log(\text{H}^+)$$

where H^+ is the numerical value for the concentration of hydrogen ions in moles per litre (M/L). Vinegar has a pH of 2.2. Find its H^+.

$$\text{pH} = -\log(\text{H}^+)$$
$$2.2 = -\log(\text{H}^+)$$
$$-2.2 = \log(\text{H}^+) \quad \text{(Multiply each side by } -1.)$$
$$\text{H}^+ = 10^{-2.2} \quad \text{(Write in exponential form.)}$$
$$\text{H}^+ = 6.3 \times 10^{-3}\text{ M/L}$$

EXAMPLE 5

In Section 9.4 we defined the intensity level of sound as

$$\beta = 10 \log \frac{I}{I_0}$$

where

$$\beta = \text{the intensity level in decibels (dB)}$$
$$I = \text{the intensity of a given sound (W/cm}^2)$$
$$I_0 = \text{the intensity of the threshold of hearing, } 10^{-16}\text{ W/cm}^2$$

What is the intensity (in W/cm^2) of street traffic sound that has an intensity level of 75 dB?

$$\beta = 75\text{ dB}$$
$$I_0 = 10^{-16}\text{ W/cm}^2$$

We are to find I.

$$\beta = 10 \log \frac{I}{I_0}$$

$$75 = 10 \log \frac{I}{10^{-16}}$$

$$7.5 = \log \frac{I}{10^{-16}} \qquad \text{(Divide each side by 10.)}$$

$$7.5 = \log I - \log 10^{-16} \qquad \text{(Property 2)}$$

$$7.5 = \log I - (-16) \qquad \text{(Property 4)}$$

$$-8.5 = \log I$$

$$I = 10^{-8.5} \qquad \text{(Write in exponential form.)}$$

$$= 3.2 \times 10^{-9} \text{ W/cm}^2 \qquad \text{(rounded to two significant digits)}$$

There are applications in which exponential growth occurs but the amount of growth reaches an upper limit.

EXPONENTIAL GROWTH TO AN UPPER LIMIT

$$y = A(1 - e^{-rt})$$

where

y = the amount present at any given time t

A = the upper limit

e = the natural number

r = the rate of growth

t = the time the growth has occurred

A graph of the equation for exponential growth to an upper limit is shown in Fig. 9.8.

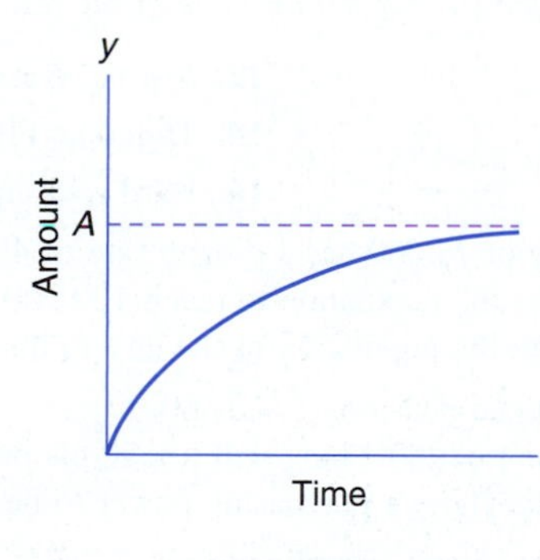

Figure 9.8 Graph of the general equation for exponential growth to an upper limit $y = A(1 - e^{-rt})$.

EXAMPLE 6

The voltage across a capacitor increases exponentially from 0 V to a maximum according to the formula $V_c = V(1 - e^{-t/(RC)})$ where V_c = the capacitor voltage

V = the constant voltage of the source

t = the time

R = the circuit resistance

C = the capacitance

Find the time t when $R = 3.00 \times 10^6\ \Omega$, $C = 6.5 \times 10^{-6}$ F, $V_c = 45.0$ V, and $V = 75.0$ V.

Substituting the data above into the equation, we have

$$\begin{aligned}
V_c &= V(1 - e^{-t/(RC)}) \\
45.0 &= 75.0(1 - e^{-t/[(3.00\times10^6)(6.50\times10^{-6})]}) \\
0.6 &= 1 - e^{-0.0513t} \\
e^{-0.0513t} &= 0.4 \\
\ln(e^{-0.0513t}) &= \ln 0.4 \\
-0.0513t &= \ln 0.4 \\
t &= \frac{\ln 0.4}{-0.0513} = 17.9 \text{ s}
\end{aligned}$$

Exercises 9.8

1. Rework Example 1 to find how long it will take **(a)** \$2500 compounded quarterly at 7.5% to accrue to \$4500 and **(b)** \$2500 to triple.
2. Rework Example 2 to find how long it will take **(a)** \$5000 compounded continuously at $8\frac{1}{4}\%$ to accrue to \$12,500 and **(b)** \$5000 to double.
3. How long will it take 175 g of the element in Example 3 to decay to 25 g?
4. The amount of bacterial growth in a given culture is given by $N = N_0e^{0.04t}$, where t is the time in hours, N_0 is the initial amount, and N is the amount after time t. How long will it take a 6500-bacteria culture to grow to $10,\bar{0}00$?

Approximate the hydrogen-ion concentration in moles per litre (H^+) for each of the following (see Example 4).

5. pH = 3.5
6. pH = 4.6
7. pH = 9.5
8. pH = 8.1
9. pH = 2.7
10. pH = 6.5

Compute the intensity of sound (in W/cm²) for each of the following (see Example 5).

11. Whisper: 15 dB
12. Normal conversation: $6\bar{0}$ dB
13. Pain threshold: $12\bar{0}$ dB
14. Thunder: $11\bar{0}$ dB
15. Soft music: $3\bar{0}$ dB
16. Hard rock music: $10\bar{0}$ dB
17. A city with a population of 75,000 has a growth rate of 4%.
    (a) How long will it take the population to reach $10\bar{0},000$?
    (b) How long will it take the population of the city to double?
18. Harry retires with an annual pension of \$65,000.
    (a) With an 8% inflation rate, how long will it take his purchasing power to fall to \$50,000?
    (b) How long will it take Harry's purchasing power to be halved?
19. A current flows in a given circuit according to $i = 2.7e^{-0.2t}$. Find t in seconds when $i = 0.35$ A.
20. The discharge current of a capacitor is $i = 0.01e^{-75t}$. Find t in seconds when $i = 0.0040$ A.
21. For a given circuit the instantaneous current i at any time t in milliseconds is given by the formula

$$i = \frac{E}{R}e^{-t/(RC)}$$

where E is the voltage in volts (V), R is the resistance in ohms (Ω), and C is the capacitance in farads (F). Find t when $i = 3.91 \times 10^{-5}$ A, $E = 10\bar{0}$ V, $R = 2.00 \times 10^4\ \Omega$, and $C = 2.00 \times 10^{-8}$ F.

22. When an electric capacitor is discharged through a resistor, the voltage across the capacitor decays according to the formula

$$E = E_0 e^{-t/(RC)}$$

where E_0 is the original voltage, t is the time for E_0 to decrease to E, R is the resistance in ohms (Ω), and C is the capacitance in farads (F). Find the time t in milliseconds for the voltage across a given capacitor to decay to 1.0% of its original value when $R = 30,\bar{0}00\ \Omega$ and $C = 5.00\ \mu\text{F}$.

23. A capacitance C in farads (F) is charged through a resistance R by a source having a constant voltage V. The capacitor voltage V_c varies according to the formula

$$V_c = V(1 - e^{-t/(RC)})$$

Find t in microseconds when $R = 2.00 \times 10^6\ \Omega$, $C = 4.00\ \mu\text{F}$, $V_c = 75.0$ V, and $V = 10\bar{0}$ V.

24. In certain types of dc circuits, current increases exponentially to an upper limit according to $i = \dfrac{E}{R}(1 - e^{-Rt/L})$, where $i =$ the instantaneous current in amperes (A), $E =$ the voltage in volts (V), $R =$ the resistance in ohms (Ω), $t =$ the time, and $L =$ the inductance in henries (H). Find L when $i = 0.0175$ A, $E = 18.0$ V, $R = 90.0\ \Omega$, and $t = 0.0150$ s.

Under certain conditions, the pressure and volume of a gas are related as

$$\ln P = C - \gamma \ln V$$

where C and γ are constants.

25. Find P if $C = 2.5$, $\gamma = 1.6$, and $V = 2.5$.

26. Find V if $C = 3.2$, $\gamma = 1.4$, and $P = 8.5$.

A 50.0-mg sample of radioactive radium decays at the rate of

$$\ln A - \ln 50.0 = kt$$

where A is the amount remaining, t is the time in years, and k is a constant (-4.10×10^{-4}). Find the amount remaining after each interval of time in Exercises 27 through 30.

27. 1.00 year
28. 25.0 years
29. $10\bar{0}$ years
30. $50\bar{0}$ years

9.9 DATA ALONG A STRAIGHT LINE

Using Rectangular Graph Paper

Sometimes the graph of data results in a set of points that closely approximates a straight line. We can approximate this straight line by drawing the "best straight line" through the given points. The slope of the line can be found graphically by choosing two convenient points and by computing $m = \dfrac{y_2 - y_1}{x_2 - x_1}$. The y-intercept can be read directly where the line crosses the y-axis. The equation of this straight line can be written using the slope-intercept form, $y = mx + b$.

EXAMPLE 1

Given the following data, graph the best straight line and find its equation:

x	1	2	3	4	5	6
y	2	5	4	7	8	10

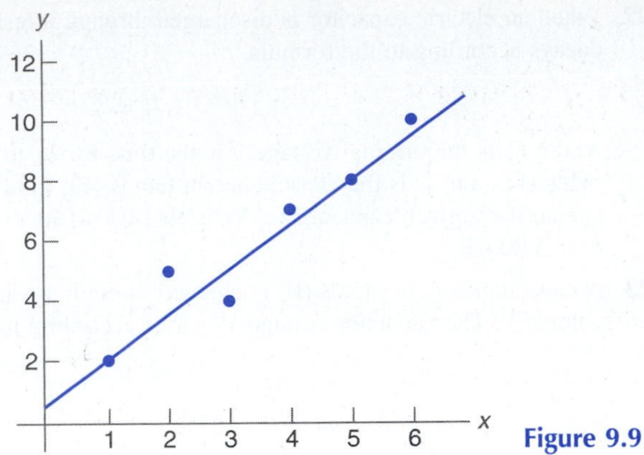

Figure 9.9

Choose two points on the line in Fig. 9.9, such as (1, 2) and (5, 8). The slope of the line is found by

$$m = \frac{y_2 - y_1}{x_2 - x_1} = \frac{8 - 2}{5 - 1} = \frac{6}{4} = 1.5$$

From the graph the y-intercept is approximately 1. Therefore, the equation of the preceding data is

$$y = 1.5x + 1$$

Obviously, the difficulty of this method lies with the inaccuracy involved in drawing the best straight line. Different individuals will probably get slightly different results for the equation of the straight line, but any one of these results is usually accurate enough for most work.

Graphing calculators have built-in routines that are designed to accurately calculate the slope and the y-intercept of this line of best fit (see Fig. 9.10). Statisticians call this process linear regression, which explains why the name of the routine is LinReg(ax + b).

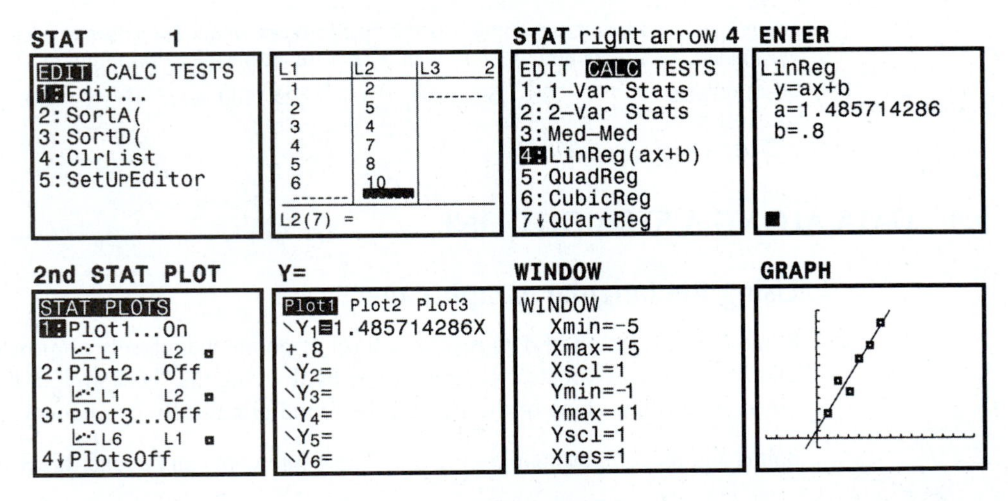

Figure 9.10

For a TI-89 example, see Appendix D, Section D.13, The Line of Best Fit (Linear Regression).

Using Logarithmic Graph Paper

Logarithmic graph paper is ruled or scaled logarithmically both vertically and horizontally rather than linearly, as have been all our previous graphs.

If you plot a set of given data on logarithmic graph paper and it yields a straight line, the equation of the given data may be expressed in the form $y = ax^k$, where $a > 0$ and $x > 0$. To see why $y = ax^k$ is a straight line on logarithmic graph paper, take the logarithm of each side.

$$y = ax^k$$
$$\log y = \log ax^k$$
$$\log y = \log a + \log x^k$$
$$\log y = k \log x + \log a$$

which is a linear equation in slope-intercept form where k, the exponent, is the slope of the line and $\log a$ is the y-intercept.

On logarithmic graph paper we plot $\log y$ versus $\log x$. Therefore, the slope of a straight line on logarithmic graph paper is found by using

$$m = \frac{\log y_2 - \log y_1}{\log x_2 - \log x_1}$$

Also, the value of a can be read directly from the graph where the line crosses the log y-axis.

EXAMPLE 2

Plot the graph of $y = 4x^2$ on logarithmic graph paper.

Generate some ordered pairs that satisfy the equation and then plot them as in Fig. 9.11.

x	2	3	4	5	7	8	10
y	16	36	64	100	196	256	400

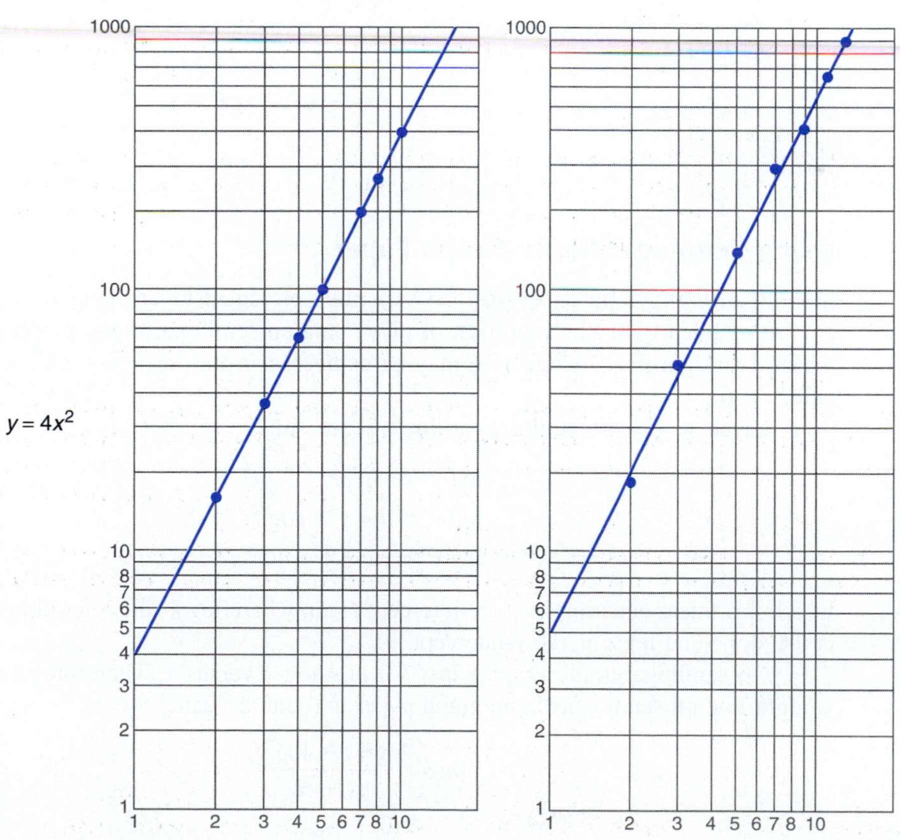

Figure 9.11

Figure 9.12

To verify our graph, note that if we choose two points on the line, such as (4, 64) and (8, 256), we can compute the slope as follows:

$$m = \frac{\log y_2 - \log y_1}{\log x_2 - \log x_1}$$

$$= \frac{\log 256 - \log 64}{\log 8 - \log 4} = 2$$

which is the value of the exponent, k.

Note: The line crosses the log y-axis at 4, which is the value of a.

EXAMPLE 3

Plot the following set of data on logarithmic graph paper, graph the best straight line, and find its equation:

x	2	3	5	7	9	11	13
y	18	51	130	275	408	650	890

Choose two points *on the line* in Fig. 9.12, such as (2, 20) and (5, 130). The value of k, the exponent, equals the slope of the line.

$$m = \frac{\log y_2 - \log y_1}{\log x_2 - \log x_1}$$

$$= \frac{\log 130 - \log 20}{\log 5 - \log 2} = 2$$

which is k. The line crosses the log y-axis at approximately 5, which is a. The equation of the preceding data is

$$y = 5x^2$$

Using Semilogarithmic Graph Paper

Semilogarithmic graph paper has one axis ruled or scaled logarithmically and the other axis ruled linearly. If you plot a set of given data on semilogarithmic graph paper and it yields a straight line, its equation may be expressed in the form $y = ak^x$, where $a > 0$ and $k > 0$.

Given $y = ak^x$, take the logarithm of each side.

$$\log y = \log ak^x$$

$$\log y = \log a + \log k^x$$

$$\log y = x \log k + \log a$$

which is a linear equation in slope-intercept form, where $\log k$ (the logarithm of the base) is the slope and $\log a$ is the y-intercept.

On semilogarithmic graph paper we plot $\log y$ versus x. Therefore, the slope of a straight line on semilogarithmic graph paper is found by using

$$m = \frac{\log y_2 - \log y_1}{x_2 - x_1}$$

EXAMPLE 4

Plot the graph of $y = 3^x$ on semilogarithmic graph paper.

Generate some ordered pairs that satisfy the equation and then plot them as in Fig. 9.13.

x	1	3	4	6
y	3	27	81	729

To verify our graph, note that if we choose two points on the line, such as (3, 27) and (6, 729), we can find the slope as follows:

$$\begin{aligned} m &= \frac{\log y_2 - \log y_1}{x_2 - x_1} \\ &= \frac{\log 729 - \log 27}{6 - 3} \\ &= 0.4771 = \log k \end{aligned}$$

We find the antilogarithm, $k = 3$, as above.

From the graph of Fig. 9.13, note that at $x = 1$, $y = 3$. Substituting the known values into

$$y = ak^x$$
$$3 = a3^1$$

we find $1 = a$ as above.

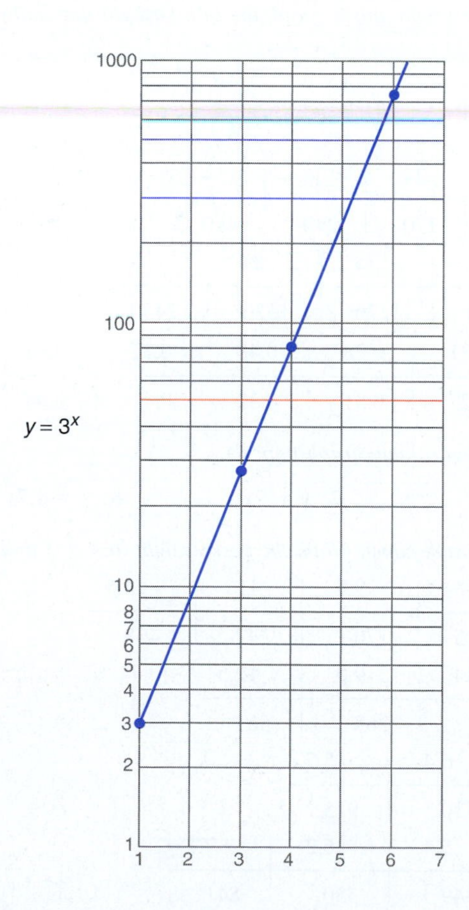

Figure 9.13

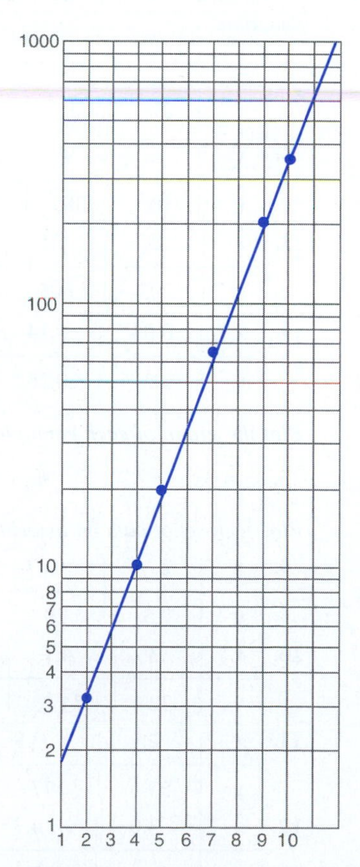

Figure 9.14

EXAMPLE 5

Plot the following set of data on semilogarithmic graph paper, graph the best straight line, and find its equation:

x	2	4	5	7	9	10
y	3.3	10.6	19.8	64.2	205	346

Choose two points on the line in Fig. 9.14, such as (3, 6) and (8, 110). The value of m is

$$\log k = m = \frac{\log y_2 - \log y_1}{x_2 - x_1}$$

$$= \frac{\log 110 - \log 6}{8 - 3}$$

$$= 0.253 = \log k$$

We find the antilogarithm, $k = 1.8$.

From the graph of Fig. 9.14, note that at $x = 3$, $y = 6$. Substituting the known values into

$$y = ak^x$$

$$6 = a(1.8^3)$$

we find $1.0 = a$. The equation of the preceding data is

$$y = 1.8^x$$

Exercises 9.9

Given each set of data on rectangular graph paper, graph the best straight line and find its equation.

1.

x	2	4	5	11	15	24
y	21	30	31	57	74	104

2.

x	6	9	15	24	36	54
y	10	38	88	170	240	410

3.

x	20	35	55	75	85	120
y	295	605	920	1270	1550	2450

4.

x	0.08	0.14	0.21	0.28	0.40	0.52	0.61	0.72
y	0.21	0.26	0.27	0.31	0.32	0.41	0.44	0.49

Plot the graph of each equation on logarithmic graph paper.

5. $y = 3x^2$ **6.** $y = \frac{3}{2}x^2$ **7.** $y = 2x^3$ **8.** $y = 1.7x^4$

Plot each set of data on logarithmic graph paper, graph the best straight line, and find its equation.

9.

x	2	3.7	8	9.5	11	14.6
y	7.5	29	125	170	247	438

10.

x	3	4.7	6.4	9.0	12.5
y	32	74	151	268	561

11.

x	2.1	3.9	6.4	8.7
y	53	177	510	915

12.

x	3.0	4.3	5.4	6.2	7.1
y	61	192	349	550	841

Plot the graph of each equation on semilogarithmic graph paper.

13. $y = 4^x$ **14.** $y = 2^x$ **15.** $y = e^x$

16. $y = (\frac{3}{2})^x$ **17.** $y = 4(3^x)$ **18.** $y = 2.5(4^x)$

Plot each set of data on semilogarithmic graph paper, graph the best straight line, and find its equation.

19.

x	2	4	5	6	7
y	7.4	54	152	410	995

20.

x	2	3	4	5
y	13.1	57.2	174	610

21.

x	2	4	8	10	12	15
y	0.82	0.63	0.45	0.35	0.27	0.21

22.

x	2	4	8	10	12	15	20
y	1.5	2.0	4.4	6.0	8.8	15.5	38.0

23.

x	2	3	4	5
y	19	60	160	515

24.

x	2	3	4	5	6
y	13	33	81	235	612

CHAPTER 9 SUMMARY

1. *Exponential function:*

$$y = b^x \quad \text{where } b > 0 \quad \text{and} \quad b \neq 1$$

(a) For $b > 1$, $y = b^x$ is an increasing exponential function.
(b) For $0 < b < 1$, $y = b^x$ is a decreasing exponential function.

2. The logarithmic function is the inverse of the exponential function. The following middle and right equations show how to express this logarithmic equation in either exponential form or logarithmic form:

Exponential equation	*Logarithmic equation in exponential form*	*Logarithmic equation in logarithmic form*
$y = b^x$	$x = b^y$	$y = \log_b x$

That is, $x = b^y$ and $y = \log_b x$ are equivalent equations for $b > 0$ but $b \neq 1$.
Remember: A logarithm is an exponent.

3. *Properties of logarithms:* M, N, and a are positive real numbers and $a \neq 1$.
(a) *Multiplication:*

$$\log_a (MN) = \log_a M + \log_a N$$

(b) *Division:*

$$\log_a \left(\frac{M}{N}\right) = \log_a M - \log_a N$$

(c) *Powers:* For real numbers n,

$$\log_a M^n = n \log_a M$$

(i) *Roots:*

$$\log_a \sqrt[n]{M} = \frac{1}{n} \cdot \log_a M$$

(ii) For $n = 0$,

$$\log_a 1 = 0$$

(iii) For $n = -1$,

$$\log_a \frac{1}{M} = -\log_a M$$

(d) $\log_a (a^x) = x$
(e) $a^{(\log_a x)} = x$

4. *Common logarithms:* Base 10, written $\log_{10} x$ or $\log x$.

5. *Natural logarithms:* Base e, written $\log_e x$ or $\ln x$, where e is an irrational number approximately equal to 2.71828.

6. The following *exponential functions* have many common applications and uses:
(a) *Exponential growth:* $y = Ae^{rt}$
(b) *Exponential decay:* $y = Ae^{-rt}$
where y = the amount present at any time t, A = the original amount, e = the natural number, and r = the rate of change.

7. To find the logarithm with another base, use the formula

$$\log_b x = \frac{\log_a x}{\log_a b}$$

where a is base 10 or base e, and b is another base.

8. Finding the solution of the general exponential equation $b^x = a$ is usually begun by taking the common logarithm or the natural logarithm of each side of the equation.

9. The solution of a logarithmic equation is usually based on one of the following:
(a) If $\log_b x = \log_b y$, then $x = y$.
(b) If $y = \log_b x$, then $x = b^y$.

10. Review Section 9.9 for using rectangular graph paper, logarithmic graph paper, and semilogarithmic graph paper.

CHAPTER 9 REVIEW

Graph each equation.

1. $y = 3^x$

2. $y = \log_3 x$

Write each equation in logarithmic form.

3. $2^4 = 16$

4. $10^{-3} = 0.001$

Write each equation in exponential form.

5. $\log_{10} 7.389 = 0.8686$

6. $\log_4 (\frac{1}{16}) = -2$

Solve for x.

7. $\log_9 x = 2$

8. $\log_x 8 = 3$

9. $\log_2 32 = x$

Write each expression as a sum or difference of multiples of single logarithms.

10. $\log_4 6x^2y$

11. $\log_3 \dfrac{5x\sqrt{y}}{z^3}$

12. $\log \dfrac{x^2(x+1)^3}{\sqrt{x-4}}$

13. $\ln \dfrac{[x(x-1)]^3}{\sqrt{x+1}}$

Write each expression as a single logarithmic expression.

14. $\log_2 x + 3\log_2 y - 2\log_2 z$

15. $\frac{1}{2}\log(x+1) - 3\log(x-2)$

16. $4\ln x - 5\ln(x+1) - \ln(x+2)$

17. $\frac{1}{2}[\ln x + \ln(x+2)] - 2\ln(x-5)$

Simplify.

18. $\log 1000$

19. $\log 10^{x^2}$

20. $\ln e^2$

21. $\ln e^x$

Find the common logarithm of each number. (Round to four significant digits.)

22. $\log 664.8$

23. $\log 0.04046$

24. $\log 14{,}420$

Find N, the antilogarithm. (Round to three significant digits.)

25. $\log N = 3.0737$

26. $\log N = -2.4289$

27. $\log N = -1.7522$

Find the natural logarithm of each number, rounded to four significant digits.

28. $\ln 72$

29. $\ln 421$

30. $\ln 0.00185$

Find x rounded to three significant digits.

31. $\ln x = 1.315$

32. $\ln x = 3.45$

33. $\ln x = -0.24$

34. Evaluate $\log_4 20$.

Solve for x. (Round to three significant digits.)

35. $6^{-2x} = 48.1$

36. $3^{4x-1} = 14^x$

37. $26.5 = 3.81e^{4x}$

38. $48 = 72(1 - e^{-x/2})$

Solve each logarithmic equation.

39. $\log(x+4) = 2$

40. $\log(2x+3) - 3\log 2 = 2\log 2$

41. $\log(x+1) + \log(x-2) = 1$

42. $\ln x = \ln(3x-2)$

43. $\ln(x+1) - \ln x = \ln 3$

44. $2\ln x = 3$

45. Suppose the population of a certain city is given by

$$y = 125{,}000e^{-0.03t}$$

where t is the time in years. Find its population in 5.0 years.

46. If the annual inflation rate is 8%, how long would it take the average price level to double?

47. The energy of an expanding gas at a constant temperature is given by

$$E = P_oV_o \ln\left(\frac{V_1}{V_o}\right)$$

where E is the energy, P_o is a constant, V_o is the initial volume, and V_1 is the new volume. Find V_1 if $E = 15{,}100$, $P_o = 85.0$, and $V_o = 265$.

For each set of data, graph the best straight line on rectangular graph paper and find its equation.

48.

x	6	11	17	24	28	39	45
y	53	52	44	41	35	32	20

49.

x	20	55	71	102	110	139	150
y	110	210	300	360	420	510	525

Plot the graph of each equation on logarithmic graph paper.

50. $y = 4x^2$

51. $y = 2.5x^3$

Plot each set of data on logarithmic graph paper, graph the best straight line, and find its equation.

52.

x	1.5	2	2.5	3	4	5	6
y	9.8	13	21	35	50	118	112

53.

x	1.5	1.8	2	2.5	3	4	5	7
y	4.9	6.2	11.9	12.2	21.5	31	49.5	111

Plot the graph of each equation on semilogarithmic graph paper.

54. $y = 5^x$

55. $y = 3(4^x)$

Plot each set of data on semilogarithmic graph paper, graph the best straight line, and find its equation.

56.

x	1	1.5	2	3	3.6	4
y	3.7	9	14.8	64	130	230

57.

x	1	1.6	2	3	4
y	4.8	8	9.5	28	38.2

CHAPTER 9

SPECIAL NASA APPLICATION

The Space Shuttle Landing, Part II*

Part I appears at the end of Chapter 4. Reading that Introduction will be helpful to this application.

As the Shuttle comes out of its communication blackout, it travels a circular path around an imaginary cone that will line it up with the center line of the runway. Coming out of its turn, the Space Shuttle should be at an altitude of 13,365 ft, have a speed of 424 mi/h, and be 7.5 mi (horizontal distance) from the runway. It is now 86 s to touchdown. The nose is down so that the Space Shuttle can descend steeply to a point 7500 ft from the runway threshold, where its altitude should be 1750 ft. The vehicle then enters a transitional phase. The Shuttle's nose is raised as it heads for a position where its altitude is 131 ft and its distance from the runway threshold is 2650 ft. The Shuttle is now 17 s to touchdown. From here the Space Shuttle enters the final phase, aiming at a point 2200 ft down the runway. See Fig. 4.33. From point (39600, 13365) to point (7500, 1750), we can use a linear function (model) to describe the path; its glide slope is 19.9º. We can also use a linear function between the last two points (2650, 131) and (−2200, 0); its glide slope is 1.5°. The path from a steep glide slope to a shallow one is not a linear function. See Fig. 9.15.

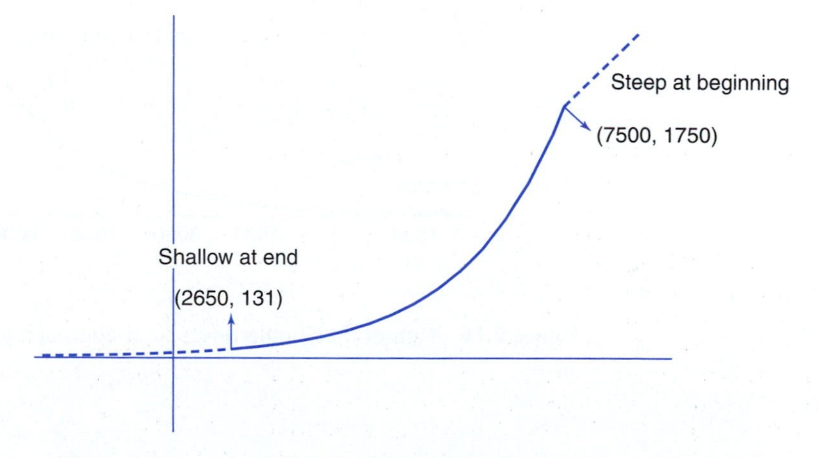

Figure 9.15 Path of the Shuttle on its final approach to the runway.

*From NASA–AMATYC–NSF Project Mathematics Explorations II, grant principals John S. Pazdar, Patricia L. Hirschy, and Peter A. Wursthorn; copyright Capital Community College, 2000.

Because an exponential function has a very flat part and a very steep part, we will try an exponential model. Substituting the points into the form $y = ab^x$, we have

$$1750 = ab^{7500}$$
$$131 = ab^{2650}$$

Then,

$$\frac{1750}{131} = \frac{ab^{7500}}{ab^{2650}} = b^{4850}$$

Thus,

$$b = \left(\frac{1750}{131}\right)^{\frac{1}{4850}} = 1.000534612$$

$$a = \frac{131}{(1.000534612)^{2650}} = 31.78057$$

The exponential function is $y = 31.78057(1.000534612)^x$. The domain of this transitional model is $2650 \leq x \leq 7500$.

This equation can also be found using a graphing calculator. For the TI-83 Plus, press **STAT, ENTER** to enter the two ordered pairs x in L_1 and y in L_2. Press **STAT,** right arrow, and choose **ExpReg** (option **0**), then press **ENTER.**

In Fig. 9.16 we show the two straight lines and the exponential function which connects them. If you examine the graph carefully, you may notice that where the exponential curve meets the line of the first phase there is a fairly sharp "corner." In reality, the Shuttle cannot make instantaneous changes in its direction. As a result, the Shuttle must follow a path at these boundary points that smooths out the curve. A background in calculus is necessary to develop the equation for this path.

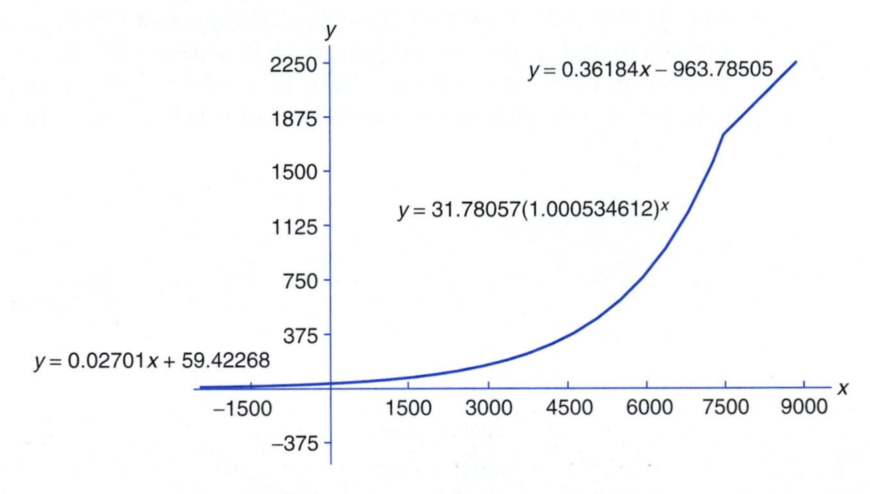

Figure 9.16 Path of the Shuttle on its final approach to the runway.

10 Trigonometric Functions

INTRODUCTION

A technician is trying to determine whether the acceleration of an airplane propeller is related to the density of engine lubricant needed. An instrument reading determines that in 10.0 s, the angular velocity of a propeller increased from $18\bar{0}0$ rpm to $22\bar{0}0$ rpm. She uses her knowledge of the relationship between angular speed and revolutions per minute and a formula to determine the angular acceleration. (This is Exercise 15 in Exercises 10.4.)

Objectives

- Given an angle, find a coterminal angle.
- Find trigonometric functions of an angle in standard position, given a point on its terminal side.
- Find an angle given a trigonometric function of the angle and its quadrant.
- Convert the measure of angles between radians and degrees.
- Work with angular speed.

10.1 THE TRIGONOMETRIC FUNCTIONS

In Chapter 3, we defined six trigonometric ratios in terms of the relationships between an acute angle of a right triangle and the lengths of two of its sides. Before extending these six definitions to all angles, we need to discuss the following.

Using a coordinate plane, an angle is in **standard position** when its vertex is located at the origin and its initial side is lying on the positive x-axis. An angle resulting from a counterclockwise rotation, as indicated by the direction of the arrow, is a **positive angle.** But if the rotation is clockwise, the angle is **negative** (see Fig. 10.1).

EXAMPLE 1

Draw angles of (a) 120° and (b) −240° in standard position.

The angles are shown in Fig. 10.2.

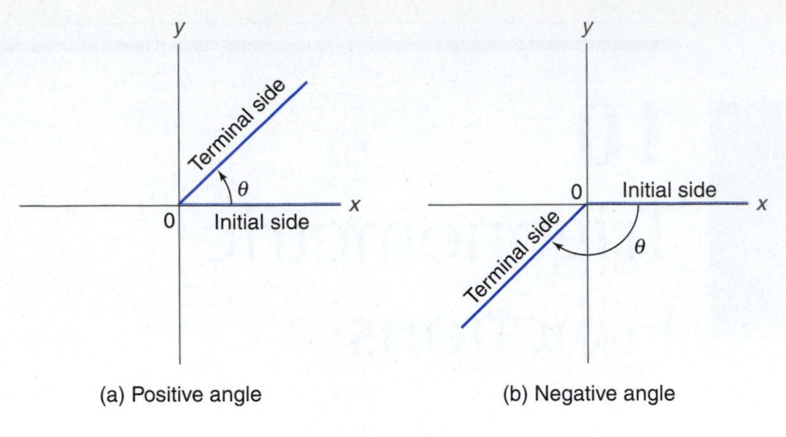

Figure 10.1

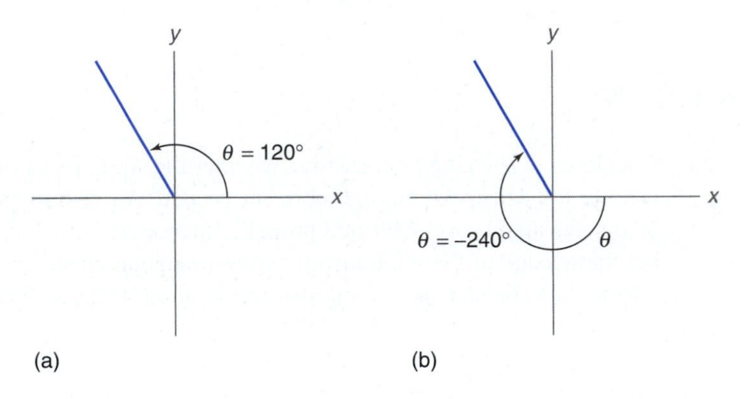

Figure 10.2

Note that the two angles in Example 1 have the same initial side and the same terminal side. However, one was formed by a counterclockwise rotation (120°) and the other by a clockwise rotation (−240°). Angles that share the same initial side and terminal side are **coterminal.** For example, 320° and −40° are coterminal angles, as are 60° and 420° as in Fig. 10.3.

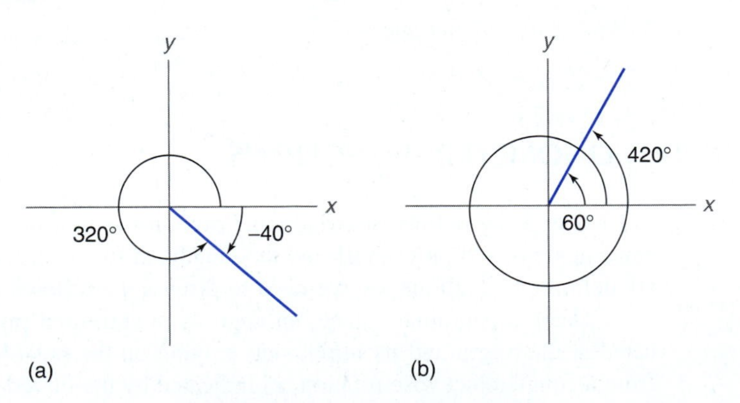

Figure 10.3 Coterminal angles.

An angle in standard position is determined by the position of its terminal side in the xy-plane. In fact, from geometry we know the position of the terminal side once we know a point on the terminal side other than the origin. Thus, knowing the coordinates (a, b) of the point P determines the angle θ (see Fig. 10.4).

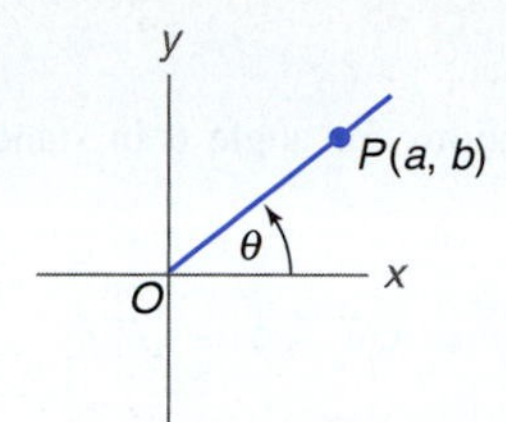

Figure 10.4 Angle θ in standard position is determined by any point $P(a, b)$ on its terminal side.

EXAMPLE 2

Draw the graph of an angle in standard position whose terminal side passes through the point $(-2, 5)$.

The graph is shown in Fig. 10.5.

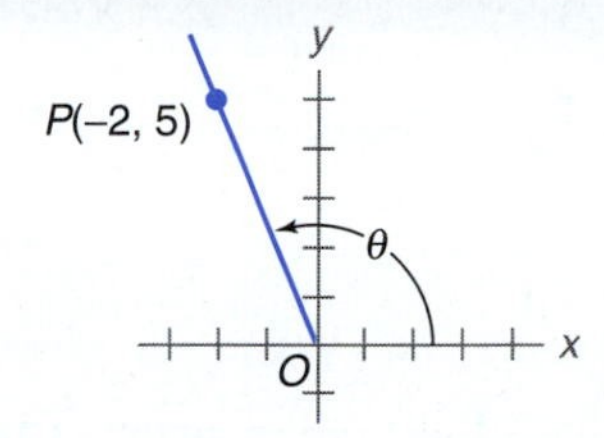

Figure 10.5

Now we define the trigonometric functions as functions of angles. Consider an angle θ in standard position and a point P with coordinates (x, y) on the terminal side of the angle (see Fig. 10.6). Points O, P, and Q form the vertices of a right triangle. Angle Q is a right angle. In a right triangle, the side r opposite the right angle Q is called the **hypotenuse** of the right triangle. We can find r if we know the coordinates of P.

$$r = \sqrt{x^2 + y^2}$$

Note: $r > 0$.

This formula is a direct application of the **Pythagorean theorem,** which states that the square of the hypotenuse of a right triangle is equal to the sum of the squares of the lengths of the other two sides.

There are six trigonometric functions associated with angle θ in standard position. They are expressed in terms of the coordinates of point P, where point P is on the terminal side of angle θ, as follows:

TRIGONOMETRIC FUNCTIONS

$$\text{sine } \theta = \sin\theta = \frac{y}{r} = \frac{\text{ordinate of } P}{r}$$

$$\text{cosine } \theta = \cos\theta = \frac{x}{r} = \frac{\text{abscissa of } P}{r}$$

$$\text{tangent } \theta = \tan\theta = \frac{y}{x} = \frac{\text{ordinate of } P}{\text{abscissa of } P} \quad (x \neq 0)$$

$$\text{cotangent } \theta = \cot\theta = \frac{x}{y} = \frac{\text{abscissa of } P}{\text{ordinate of } P} \quad (y \neq 0)$$

$$\text{secant } \theta = \sec\theta = \frac{r}{x} = \frac{r}{\text{abscissa of } P} \quad (x \neq 0)$$

$$\text{cosecant } \theta = \csc\theta = \frac{r}{y} = \frac{r}{\text{ordinate of } P} \quad (y \neq 0)$$

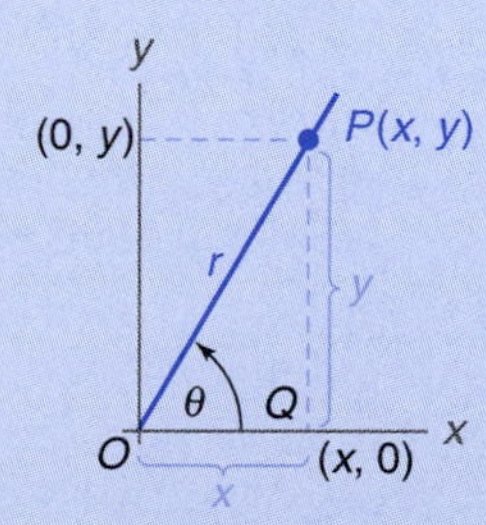

Figure 10.6

EXAMPLE 3

Find the values of the trigonometric functions for angle θ in standard position with point (2, 3) on the terminal side.

First, find r in Fig. 10.7.

$$r = \sqrt{2^2 + 3^2} = \sqrt{4 + 9} = \sqrt{13}$$

Then we have, by the definitions,

$$\sin\theta = \frac{y}{r} = \frac{3}{\sqrt{13}} \qquad \csc\theta = \frac{r}{y} = \frac{\sqrt{13}}{3}$$

$$\cos\theta = \frac{x}{r} = \frac{2}{\sqrt{13}} \qquad \sec\theta = \frac{r}{x} = \frac{\sqrt{13}}{2}$$

$$\tan\theta = \frac{y}{x} = \frac{3}{2} \qquad \cot\theta = \frac{x}{y} = \frac{2}{3}$$

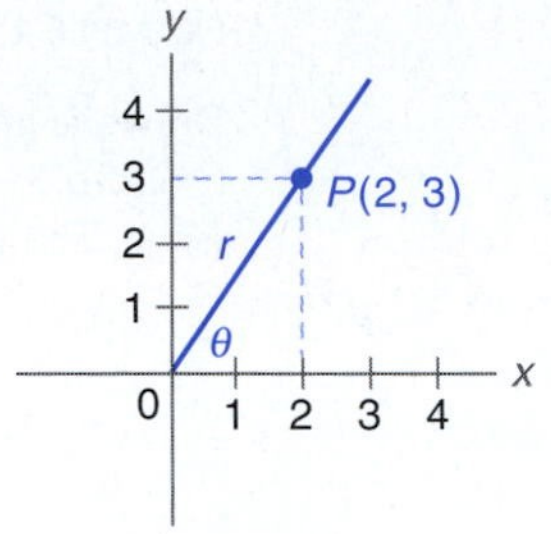

Figure 10.7

We have defined functions of angles. This means, for example, that $\tan\theta$ is determined by θ and does not depend upon the choice of the point P lying on the terminal side. Let (a, b) and (c, d) be two different points on the terminal side of an angle θ as in Fig. 10.8. Triangles OPQ and ORS are similar. From geometry we know that the corresponding sides of similar triangles are proportional. This means that $\frac{a}{b} = \frac{c}{d}$. Thus, no matter which point we use, P or R, the numerical value of the ratio used to define $\tan\theta$ is the same. Similar arguments can be given to show that the other five trigonometric functions of θ do not depend upon the choice of the point used on the terminal side to compute the defining ratios.

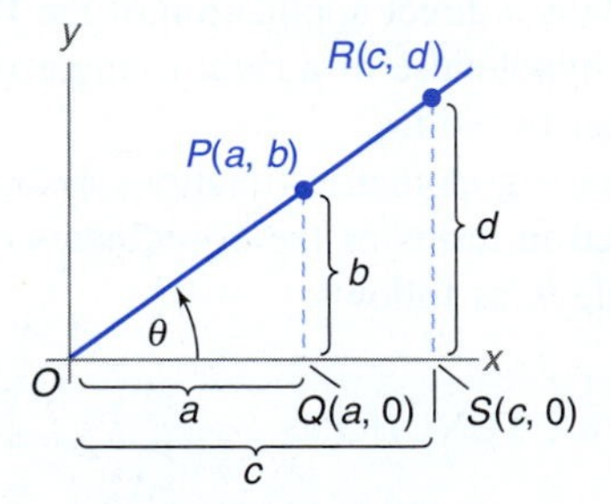

Figure 10.8

Let's demonstrate this fact for two particular points, P with coordinates (3, 4) and R with coordinates (6, 8), which both lie on the terminal side of the same angle θ in standard position (see Fig. 10.9).

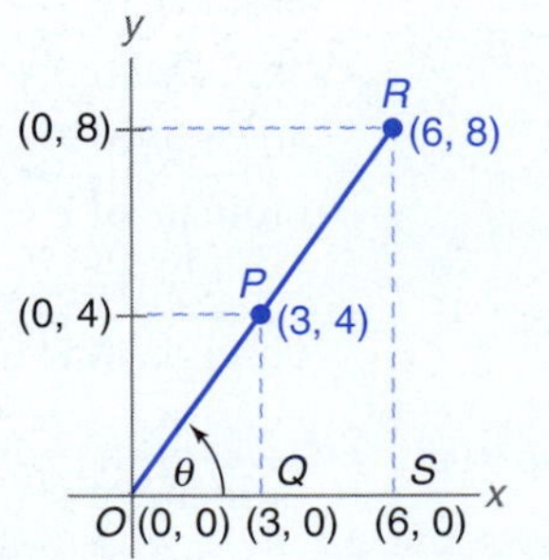

Figure 10.9

We shall first use P to find the trigonometric functions of θ. Find hypotenuse r of right triangle OPQ as follows:

$$r = \sqrt{3^2 + 4^2} = \sqrt{9 + 16} = \sqrt{25} = 5$$

Then

$$\sin\theta = \frac{y}{r} = \frac{4}{5} \qquad \csc\theta = \frac{r}{y} = \frac{5}{4}$$

$$\cos\theta = \frac{x}{r} = \frac{3}{5} \qquad \sec\theta = \frac{r}{x} = \frac{5}{3}$$

$$\tan\theta = \frac{y}{x} = \frac{4}{3} \qquad \cot\theta = \frac{x}{y} = \frac{3}{4}$$

Similarly, using R, we find hypotenuse r of right triangle ORS.

$$r = \sqrt{6^2 + 8^2} = \sqrt{36 + 64} = \sqrt{100} = 10$$

Again,

$$\sin\theta = \frac{y}{r} = \frac{8}{10} = \frac{4}{5} \qquad \csc\theta = \frac{r}{y} = \frac{10}{8} = \frac{5}{4}$$

$$\cos\theta = \frac{x}{r} = \frac{6}{10} = \frac{3}{5} \qquad \sec\theta = \frac{r}{x} = \frac{10}{6} = \frac{5}{3}$$

$$\tan\theta = \frac{y}{x} = \frac{8}{6} = \frac{4}{3} \qquad \cot\theta = \frac{x}{y} = \frac{6}{8} = \frac{3}{4}$$

Since the ratios computed using point R reduce to the values of the ratios computed using point P, we see that the choice between P and R does not affect the ultimate value of the trigonometric functions of θ.

We know from algebra that $\frac{y}{r}$ and $\frac{r}{y}$ are reciprocals of each other; that is,

$$\sin\theta = \frac{y}{r} = \frac{1}{\frac{r}{y}} = \frac{1}{\csc\theta}$$

For this reason, $\sin\theta$ and $\csc\theta$ are called **reciprocal trigonometric functions.** In much the same way, we can complete the following table using the defining ratios:

RECIPROCAL TRIGONOMETRIC FUNCTIONS

$$\sin\theta = \frac{1}{\csc\theta} \qquad \csc\theta = \frac{1}{\sin\theta}$$

$$\cos\theta = \frac{1}{\sec\theta} \qquad \sec\theta = \frac{1}{\cos\theta}$$

$$\tan\theta = \frac{1}{\cot\theta} \qquad \cot\theta = \frac{1}{\tan\theta}$$

If a given angle θ in standard position is in the second quadrant as in Fig. 10.10, then $x < 0$ and $y > 0$. Of course, distance $r > 0$. The sign of each trigonometric function is given as follows:

$$\sin\theta = \frac{y}{r} > 0 \qquad \csc\theta = \frac{r}{y} > 0$$

$$\cos\theta = \frac{x}{r} < 0 \qquad \sec\theta = \frac{r}{x} < 0$$

$$\tan\theta = \frac{y}{x} < 0 \qquad \cot\theta = \frac{x}{y} < 0$$

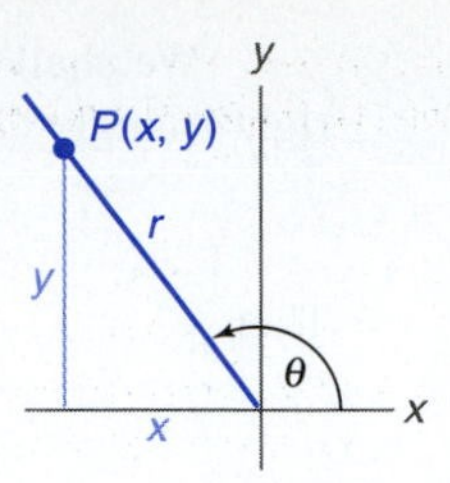

Figure 10.10 Angle θ in the second quadrant.

Note that r, a distance, is always a positive quantity.

In summary, if angle θ is in standard position and its terminal side lies in a given quadrant, the sign of each trigonometric function according to its definition may be tabulated as shown in Table 10.1.

TABLE 10.1 Signs of the Trigonometric Functions of Nonquadrantal Angles

Quadrant	$\sin\theta = \frac{y}{r}$	$\cos\theta = \frac{x}{r}$	$\tan\theta = \frac{y}{x}$	$\cot\theta = \frac{x}{y}$	$\sec\theta = \frac{r}{x}$	$\csc\theta = \frac{r}{y}$
I $x > 0$ $y > 0$	+	+	+	+	+	+
II $x < 0$ $y > 0$	+	−	−	−	−	+
III $x < 0$ $y < 0$	−	−	+	+	−	−
IV $x > 0$ $y < 0$	−	+	−	−	+	−

EXAMPLE 4

Find the values of $\sin\theta$, $\cos\theta$, and $\tan\theta$ if θ is in standard position and its terminal side passes through the point $(-3, 4)$ (see Fig. 10.11).

$$r = \sqrt{x^2 + y^2} = \sqrt{9 + 16} = 5$$

$$\sin\theta = \frac{y}{r} = \frac{4}{5}$$

$$\cos\theta = \frac{x}{r} = \frac{-3}{5} = -\frac{3}{5}$$

$$\tan\theta = \frac{y}{x} = \frac{4}{-3} = -\frac{4}{3}$$

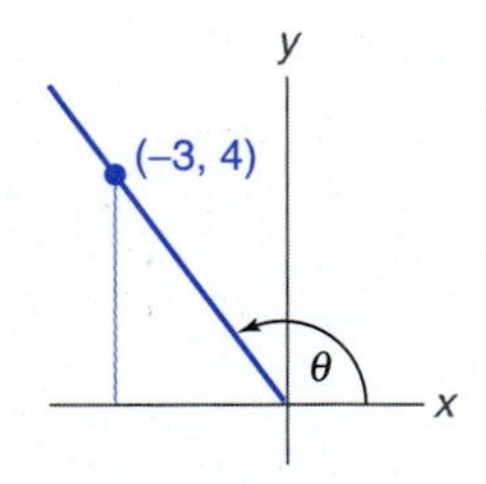

Figure 10.11

A **quadrantal angle** is one which, when in standard position, has its terminal side coinciding with one of the axes. Again using the definitions and Fig. 10.12, we can generate

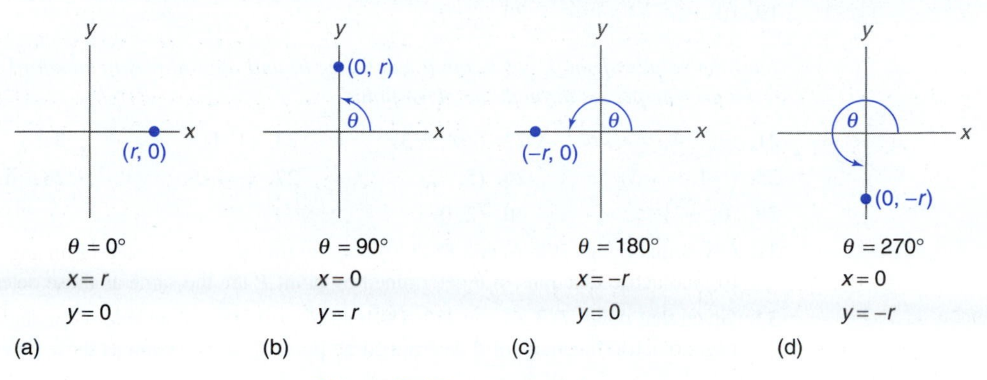

Figure 10.12 Quadrantal angles.

the values of the trigonometric functions of θ when θ is a quadrantal angle. (Remember, $r > 0$.) The values are given in Table 10.2. For example,

$$\sin 180° = \frac{y}{r} = \frac{0}{r} = 0$$

$$\cos 180° = \frac{x}{r} = \frac{-r}{r} = -1$$

$$\tan 90° = \frac{y}{x} = \frac{r}{0} \qquad \text{(undefined)}$$

TABLE 10.2 Values of Quadrantal Angles

θ	0°	90°	180°	270°	360°
$\sin\theta$	0	1	0	−1	0
$\cos\theta$	1	0	−1	0	1
$\tan\theta$	0	Undefined	0	Undefined	0
$\cot\theta$	Undefined	0	Undefined	0	Undefined
$\sec\theta$	1	Undefined	−1	Undefined	1
$\csc\theta$	Undefined	1	Undefined	−1	Undefined

Exercises 10.1

Draw a graph for each angle in standard position.

1. 30°
2. −60°
3. 225°
4. 390°
5. 540°
6. −420°

Find the smallest positive and the largest negative coterminal angle in standard position for each angle.

7. 60°
8. 175°
9. −86°
10. −270°
11. 225°
12. 300°
13. 412°
14. −500°

Draw a graph for the angle in standard position whose terminal side passes through each point.

15. $(2, 3)$ **16.** $(4, -2)$ **17.** $(-1, 3)$ **18.** $(-3, -1)$

19. $(0, -3)$ **20.** $(2, 0)$

Find the values of sin θ, cos θ, tan θ, cot θ, sec θ, and csc θ if θ is in standard position and its terminal side passes through the given point.

21. $(3, -4)$ **22.** $(-4, -3)$ **23.** $(1, 1)$ **24.** $(-3, 3)$

25. $(-1, -\sqrt{3})$ **26.** $(5, 12)$ **27.** $(-4, 5)$ **28.** $(6, -2)$

29. $(0, -3)$ **30.** $(2, 0)$

31. Given that points $P(8, 6)$ and $R(12, 9)$ both lie on the terminal side of an angle θ, show that the trigonometric functions of θ determined by point P are the same as those determined by point R.

32. Given that points $P(3, 6)$ and $R(9, 18)$ both lie on the terminal side of an angle θ, show that the trigonometric functions of θ determined by point P are the same as those determined by point R.

Using the defining ratios, show each equality.

33. $\sec \theta = \dfrac{1}{\cos \theta}$ **34.** $\tan \theta = \dfrac{1}{\cot \theta}$ **35.** $\sin \theta = \dfrac{1}{\csc \theta}$

36. $\csc \theta = \dfrac{1}{\sin \theta}$ **37.** $\tan \theta = \dfrac{\sin \theta}{\cos \theta}$ **38.** $\cot \theta = \dfrac{\cos \theta}{\sin \theta}$

10.2 TRIGONOMETRIC FUNCTIONS OF ANY ANGLE

The **reference angle** α of any nonquadrantal angle θ in standard position is the *acute* angle between the terminal side of θ and the x-axis. Angle α is always considered to be a positive angle less than 90°; that is, $0° < \alpha < 90°$.

EXAMPLE 1

Find the reference angle α for each given angle θ in Fig. 10.13.

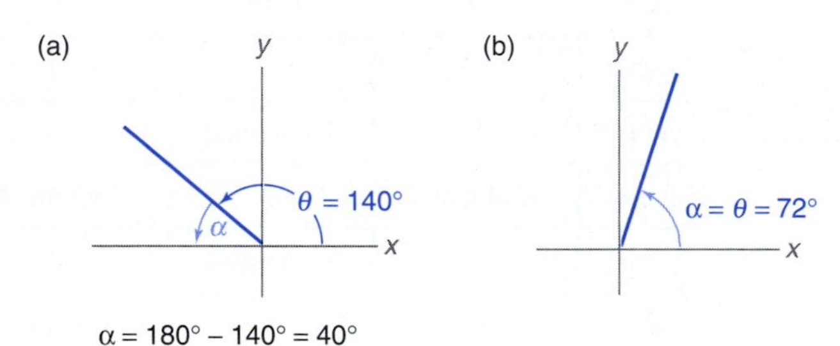

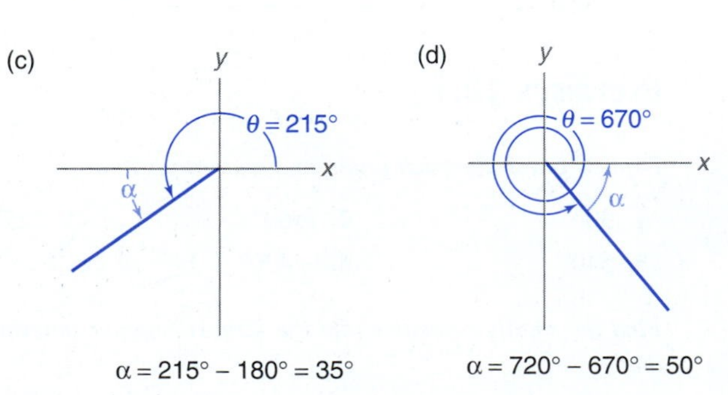

Figure 10.13

Note that if angle θ is in standard position and

1. $0° < \theta < 90°$, then $\alpha = \theta$.
2. $90° < \theta < 180°$, then $\alpha = 180° - \theta$.
3. $180° < \theta < 270°$, then $\alpha = \theta - 180°$.
4. $270° < \theta < 360°$, then $\alpha = 360° - \theta$.

Consider the four angles shown in Figs. 10.14 to 10.17 in Table 10.3. The angles are in standard position, where the terminal sides pass through the points (a, b), $(-a, b)$, $(-a, -b)$, and $(a, -b)$, respectively, with $a > 0$ and $b > 0$. Find the values of each of the trigonometric functions for each of the angles.

Note the following:

1. All four triangles in the four cases shown in Table 10.3 are congruent; that is, all corresponding sides and angles are equal. Thus, the reference angle α is the same in all four quadrants.
2. The absolute value of each corresponding trigonometric function is the same in all four cases.
3. The sign of each trigonometric ratio is determined by the quadrant in which the terminal side of angle θ lies, as in Table 10.1.

Calculators are used to evaluate the trigonometric function of *any* angle in the same way that we evaluated the trigonometric ratios in Section 3.2. Make certain that your calculator is in the degree mode.

EXAMPLE 2

Find sin $(-114°)$ and sec $250°$, rounded to four significant digits.

MODE down arrows right arrow **2 ENTER**

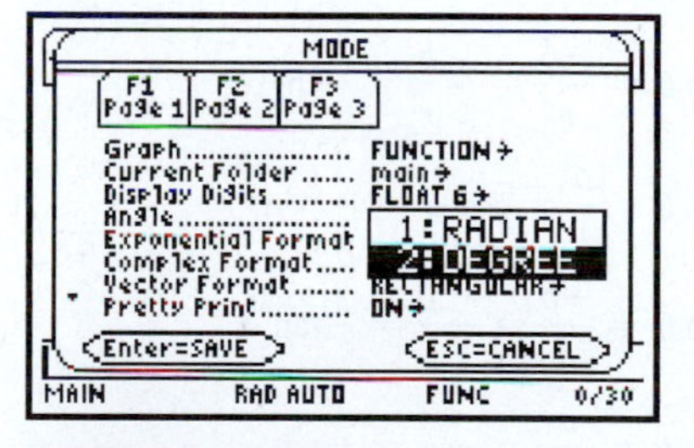

2nd SIN -114) **green diamond ENTER** 1/ **2nd COS** 250.) **ENTER**

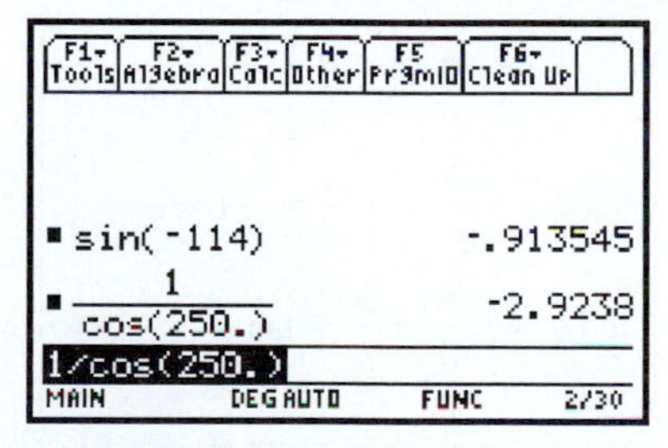

MODE **2nd QUIT**

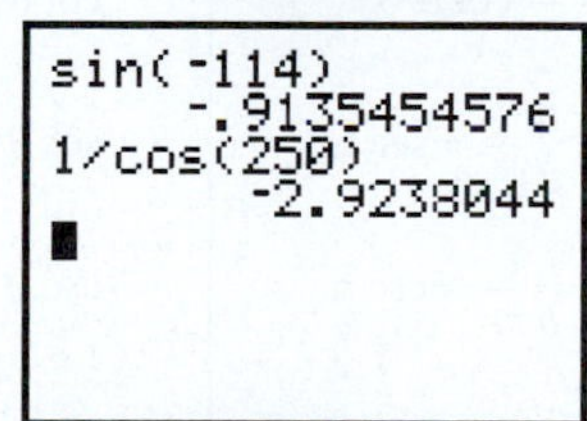

SIN -114) **ENTER** 1/ **COS** 250) **ENTER**

That is, sin $(-114°) = -0.9135$ and sec $250° = -2.924$, rounded to four significant digits.

Note: Since calculators do not have secant buttons, we must use the cosine button and the reciprocal relationship $\sec\theta = \dfrac{1}{\cos\theta}$.

TABLE 10.3

First-quadrant angle	Second-quadrant angle
$\sin\theta = \frac{b}{r}$	$\sin\theta = \frac{b}{r} = \sin\alpha$
$\cos\theta = \frac{a}{r}$	$\cos\theta = \frac{-a}{r} = -\frac{a}{r} = -\cos\alpha$
$\tan\theta = \frac{b}{a}$	$\tan\theta = \frac{b}{-a} = -\frac{b}{a} = -\tan\alpha$
$\cot\theta = \frac{a}{b}$	$\cot\theta = \frac{-a}{b} = -\frac{a}{b} = -\cot\alpha$
$\sec\theta = \frac{r}{a}$	$\sec\theta = \frac{r}{-a} = -\frac{r}{a} = -\sec\alpha$
$\csc\theta = \frac{r}{b}$	$\csc\theta = \frac{r}{b} = \csc\alpha$

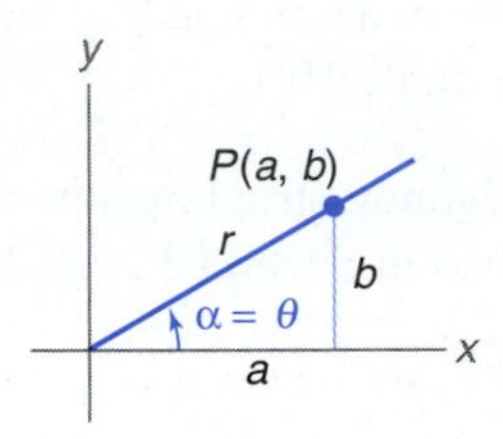

Figure 10.14

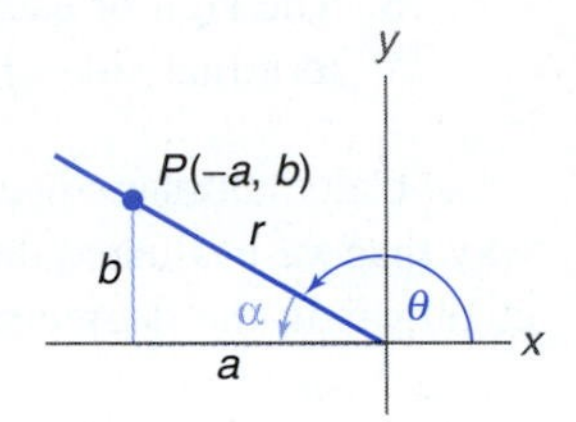

Figure 10.15

Third-quadrant angle	Fourth-quadrant angle
$\sin\theta = \frac{-b}{r} = -\frac{b}{r} = -\sin\alpha$	$\sin\theta = \frac{-b}{r} = -\frac{b}{r} = -\sin\alpha$
$\cos\theta = \frac{-a}{r} = -\frac{a}{r} = -\cos\alpha$	$\cos\theta = \frac{a}{r} = \cos\alpha$
$\tan\theta = \frac{-b}{-a} = \frac{b}{a} = \tan\alpha$	$\tan\theta = \frac{-b}{a} = -\frac{b}{a} = -\tan\alpha$
$\cot\theta = \frac{-a}{-b} = \frac{a}{b} = \cot\alpha$	$\cot\theta = \frac{a}{-b} = -\frac{a}{b} = -\cot\alpha$
$\sec\theta = \frac{r}{-a} = -\frac{r}{a} = -\sec\alpha$	$\sec\theta = \frac{r}{a} = \sec\alpha$
$\csc\theta = \frac{r}{-b} = -\frac{r}{b} = -\csc\alpha$	$\csc\theta = \frac{r}{-b} = -\frac{r}{b} = -\csc\alpha$

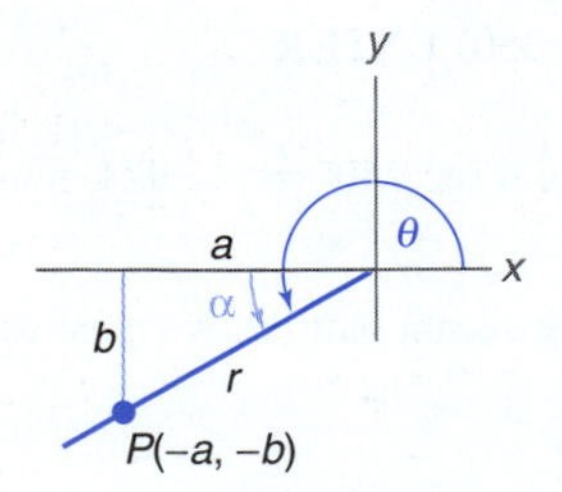

Figure 10.16

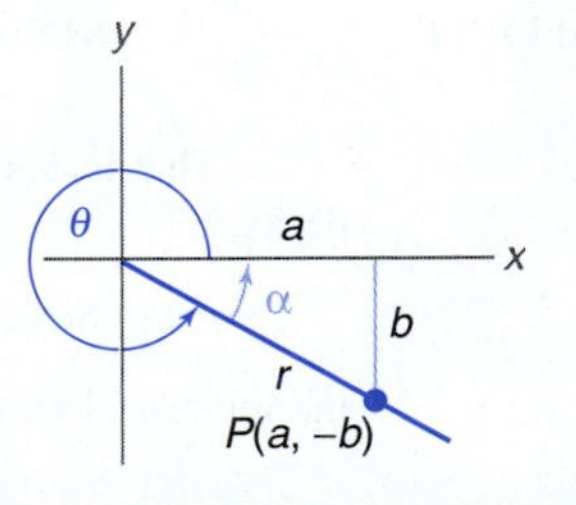

Figure 10.17

To use a calculator to find angles when the value of the trigonometric function is given:

1. Find the reference angle. Enter the inverse trigonometric function followed by the absolute value of the given trigonometric ratio. The angle displayed is the reference angle α.
2. From the sign of the given value, determine the quadrants in which the angles lie.
3. Knowing the reference angle and the quadrants, find the angles.

EXAMPLE 3

Given $\sin\theta = 0.4772$, find θ for $0° \le \theta < 360°$, rounded to the nearest tenth of a degree.

green diamond SIN^{-1} .4772) **ENTER**

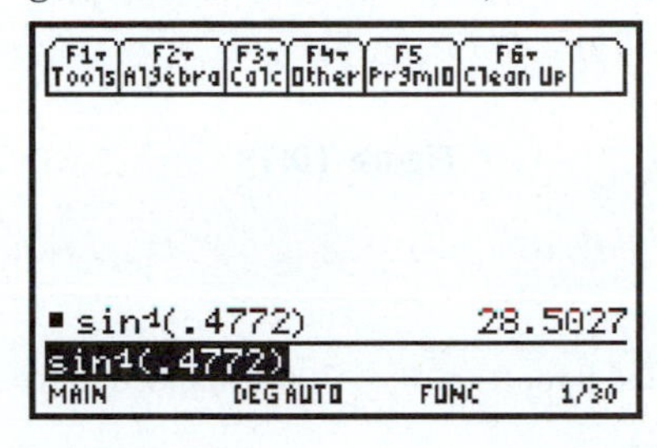

2nd SIN^{-1} .4772) **ENTER**

Thus, the reference angle $\alpha = 28.5°$, rounded to the nearest tenth of a degree.

The sine function is positive in Quadrants I and II.

The first-quadrant angle is 28.5° (see Fig. 10.18).

The second-quadrant angle is $180° - 28.5° = 151.5°$.

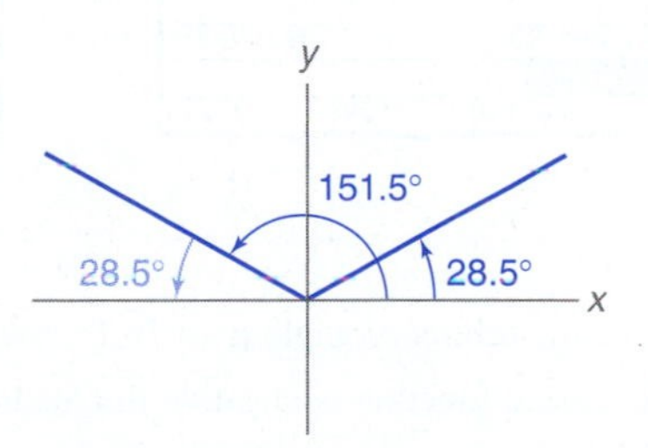

Figure 10.18

EXAMPLE 4

Given $\tan\theta = -0.3172$, find θ for $0° \le \theta < 360°$, rounded to the nearest tenth of a degree.

green diamond TAN^{-1} .3172) **ENTER**

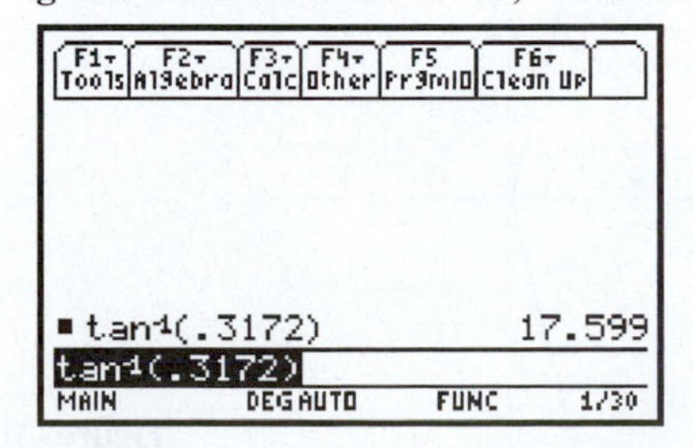

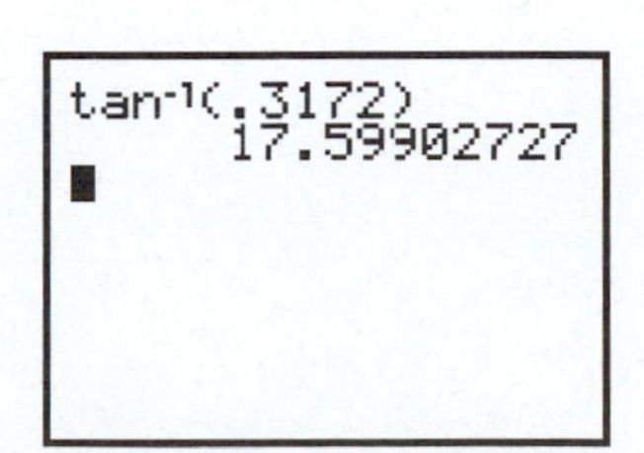

2nd TAN^{-1} .3172) **ENTER**

Thus, the reference angle $\alpha = 17.6°$, rounded to the nearest tenth of a degree.

The tangent function is negative in Quadrants II and IV.

The second-quadrant angle is $180° - 17.6° = 162.4°$ (see Fig. 10.19).

The fourth-quadrant angle is $360° - 17.6° = 342.4°$.

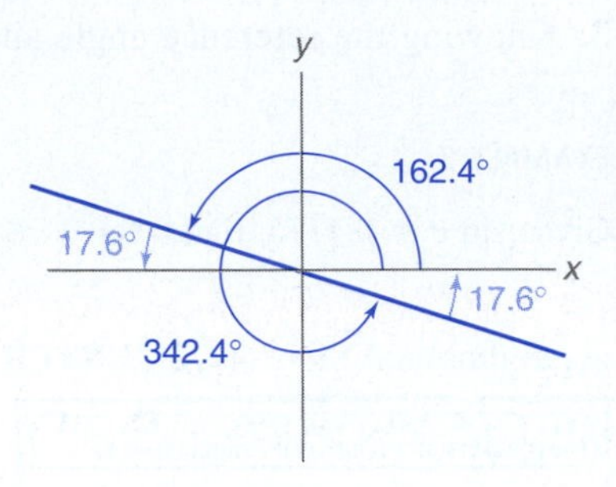

Figure 10.19

EXAMPLE 5

Given $\cos \theta = -0.2405$, find θ for $0° \leq \theta < 360°$, rounded to the nearest tenth of a degree.

green diamond COS^{-1} .2405) **ENTER**

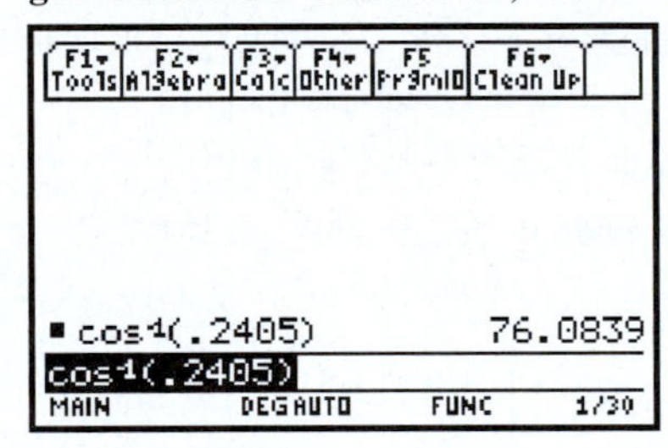

2nd COS^{-1} .2405) **ENTER**

Thus, the reference angle $\alpha = 76.1°$, rounded to the nearest tenth of a degree.

The cosine function is negative in Quadrants II and III.

The second-quadrant angle is $180° - 76.1° = 103.9°$ (see Fig. 10.20).

The third-quadrant angle is $180° + 76.1° = 256.1°$.

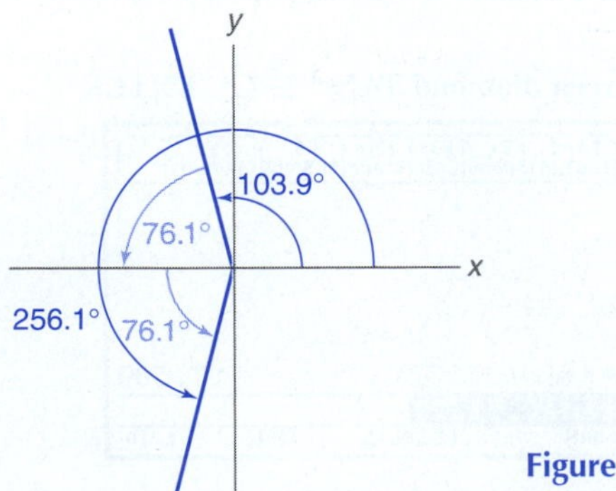

Figure 10.20

EXAMPLE 6

Given $\cot \theta = -1.650$, find θ for $0° \le \theta < 360°$, rounded to the nearest tenth of a degree.

green diamond TAN^{-1} 1/1.650) **ENTER**

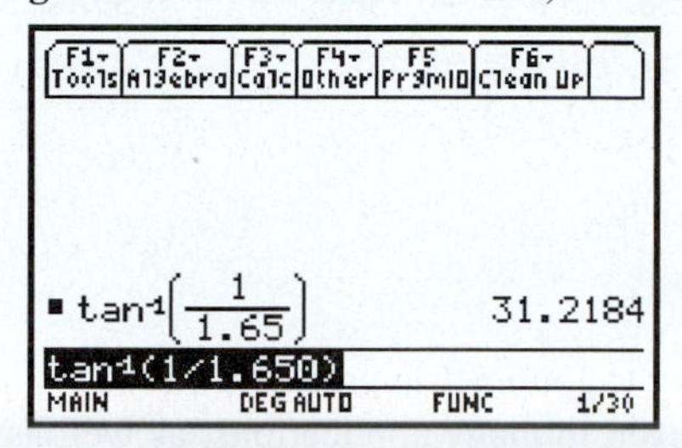

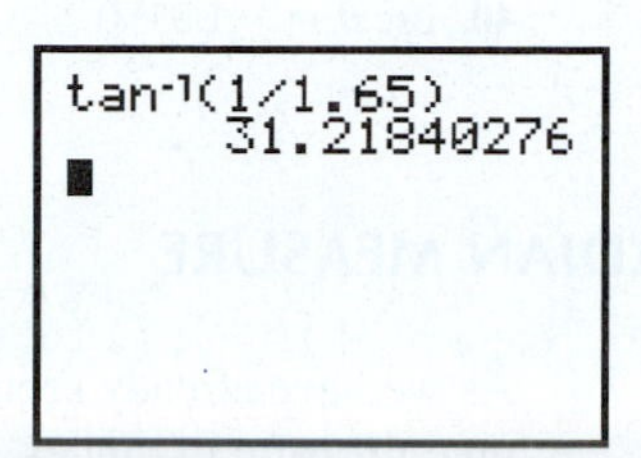

2nd TAN^{-1} 1/1.65) **ENTER**

Thus, the reference angle $\alpha = 31.2°$, rounded to the nearest tenth of a degree.

Note: Since calculators do not have inverse cotangent buttons, we must use the inverse tangent button and the reciprocal relationship $\cot \theta = \dfrac{1}{\tan \theta}$.

The cotangent function is negative in Quadrants II and IV.

The second-quadrant angle is $180° - 31.2° = 148.8°$ (see Fig. 10.21).

The fourth-quadrant angle is $360° - 31.2° = 328.8°$.

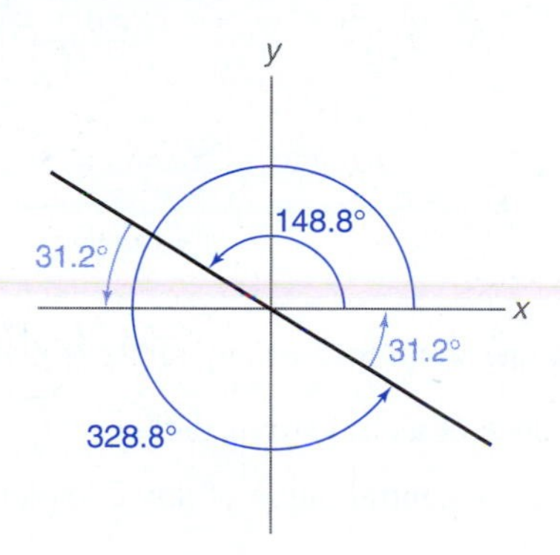

Figure 10.21

Exercises 10.2

For each angle, find the reference angle.

1. 120° **2.** 312° **3.** 253° **4.** 19°

5. 293.4° **6.** 192.5° **7.** −116.7° **8.** −274.8°

9. 462°4′ **10.** 597°13′ **11.** 1920° **12.** 2134°

Find the value of each trigonometric function, rounded to four significant digits.

13. sin 125.7° **14.** cos 217.4° **15.** tan 349.7°

16. tan 98.3° **17.** cos 265.7° **18.** sin 293.9°

19. cos (−143.5°) **20.** sin (−275.6°) **21.** sec 192.0°

22. csc 318.3° **23.** cot (−36.5°) **24.** sec (−105.0°)

Find θ for $0° \le \theta < 360°$, rounded to the nearest tenth of a degree.

25. $\sin \theta = 0.3684$ **26.** $\cos \theta = 0.1849$ **27.** $\tan \theta = 0.7250$

28. $\tan\theta = -1.8605$

29. $\cos\theta = -0.1050$

30. $\sin\theta = -0.8760$

31. $\sin\theta = -0.9111$

32. $\sin\theta = 0.5009$

33. $\sec\theta = -1.7632$

34. $\csc\theta = -2.4105$

35. $\cot\theta = 3.6994$

36. $\sec\theta = 2.8200$

37. $\csc\theta = -1.3250$

38. $\cot\theta = -0.1365$

39. $\sec\theta = 2.3766$

40. $\csc\theta = -1.9130$

41. $\cos\theta = 0.7140$

42. $\tan\theta = 2.3670$

10.3 RADIAN MEASURE

As we have already seen, many trigonometric problems may be solved in terms of the degree measure of angles. However, in many applications, as well as in our development of mathematics, another angular measurement is needed, namely, the radian. A **radian** is the measure of an angle with its vertex at the center of a circle whose intercepted arc is equal in length to the radius of the circle (see Fig. 10.22). That is, the angle θ is defined as the ratio of the length of arc PQ to the length of the radius r. As a result, the unit *radian* has no physical dimensions since it is the ratio of two lengths.

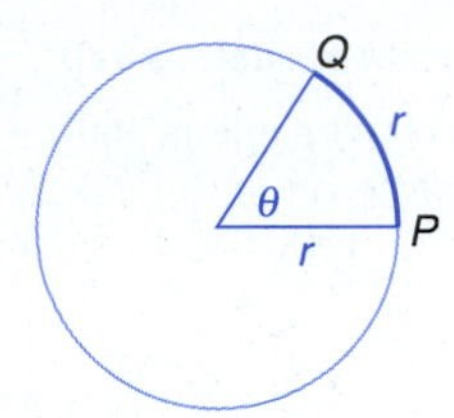

θ = 1 radian

Figure 10.22

The circumference of any circle is given by the formula $C = 2\pi r$. The ratio of the circumference of a circle to its radius is $\frac{2\pi r}{r} = 2\pi$ (about 6.28). That is, 2π radians is the measure of the central angle of any complete circle (one complete revolution).

For comparison purposes, we note the following:

$$2\pi \text{ radians} = 360^\circ$$
$$\pi \text{ rad} = 180^\circ$$

When we divide each side of the above relation by π, we find

$$1 \text{ rad} = \frac{180^\circ}{\pi} = 57.2958^\circ \qquad \text{(approximately)}$$

When we divide each side of the same relation by 180, we find

$$1^\circ = \frac{\pi}{180} \text{ rad} = 0.01745 \text{ rad} \qquad \text{(approximately)}$$

To convert from one unit of measure to another, multiply the first unit by a conversion factor where the numerator given in one unit equals the denominator given in another unit. That is, multiplying by this fraction (conversion factor) equal to one does not change the quantity but changes only the units. (See Appendix B, Section B.2, for more information about conversion factors.) From the relation π rad = 180°, we can form the conversion factors

$$\frac{\pi \text{ rad}}{180°} \quad \text{and} \quad \frac{180°}{\pi \text{ rad}}$$

EXAMPLE 1

Change each angle measure from degrees to radians.

(a) $30° = 30° \times \dfrac{\pi \text{ rad}}{180°} = \dfrac{\pi}{6} \text{ rad} = 0.524 \text{ rad}$

(b) $45° = 45° \times \dfrac{\pi \text{ rad}}{180°} = \dfrac{\pi}{4} \text{ rad} = 0.785 \text{ rad}$

(c) $120° = 120° \times \dfrac{\pi \text{ rad}}{180°} = \dfrac{2\pi}{3} \text{ rad} = 2.09 \text{ rad}$

(d) $36° = 36° \times \dfrac{\pi \text{ rad}}{180°} = \dfrac{\pi}{5} \text{ rad} = 0.628 \text{ rad}$

For some of our work, we find it convenient to express radian measure in terms of π. At other times the decimal expression is more beneficial. *Note: When no unit of angle measure is given, it is understood that the angle is expressed in radians.* Thus, $\theta = \dfrac{\pi}{6}$ is understood to be $\theta = \dfrac{\pi}{6}$ rad.

EXAMPLE 2

Change each angle measure from radians to degrees.

(a) $\dfrac{\pi}{3} \text{ rad} = \dfrac{\pi}{3} \text{ rad} \times \dfrac{180°}{\pi \text{ rad}} = 60°$

(b) $\dfrac{5\pi}{4} \text{ rad} = \dfrac{5\pi}{4} \text{ rad} \times \dfrac{180°}{\pi \text{ rad}} = 225°$

(c) $6\pi \text{ rad} = 6\pi \text{ rad} \times \dfrac{180°}{\pi \text{ rad}} = 1080°$

(d) $1.2 \text{ rad} = 1.2 \text{ rad} \times \dfrac{180°}{\pi \text{ rad}} = 68.8°$

EXAMPLE 3

Given each angle in Fig. 10.23 in radians, find the reference angle α.

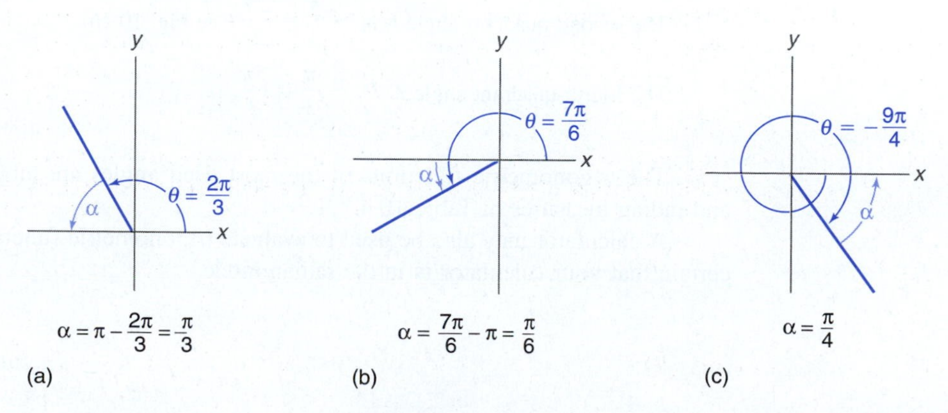

Figure 10.23

EXAMPLE 4

Find $\cos \frac{4\pi}{3}$.

Draw $\theta = \frac{4\pi}{3}$ as in Fig. 10.24.

$$\cos \frac{4\pi}{3} = -\cos \frac{\pi}{3} \quad \text{(The cosine function is negative in Quadrant III.)}$$

$$= -\frac{1}{2}$$

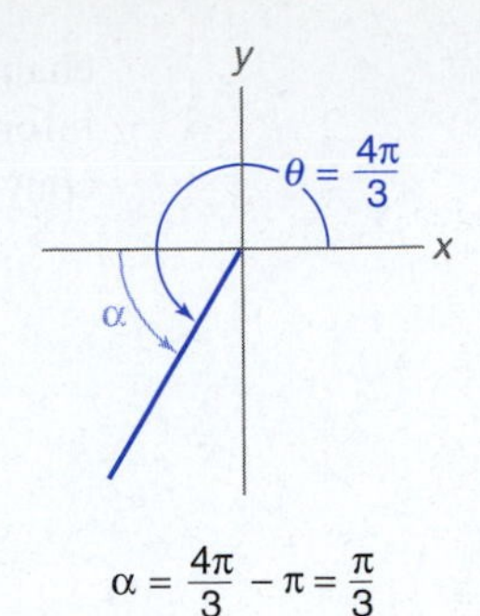

$\alpha = \frac{4\pi}{3} - \pi = \frac{\pi}{3}$

Figure 10.24

EXAMPLE 5

Find $\tan \frac{3\pi}{4}$.

Draw $\theta = \frac{3\pi}{4}$ as in Fig. 10.25.

$$\tan \frac{3\pi}{4} = -\tan \frac{\pi}{4} \quad \text{(The tangent function is negative in Quadrant II.)}$$

$$= -1$$

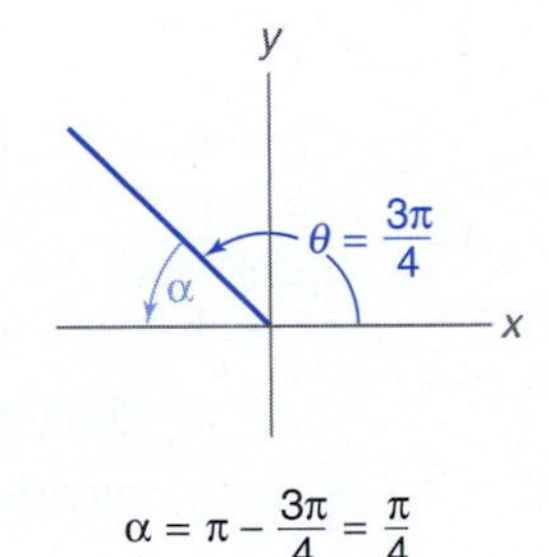

$\alpha = \pi - \frac{3\pi}{4} = \frac{\pi}{4}$

Figure 10.25

EXAMPLE 6

If $\tan \theta = -\sqrt{3}$, find θ for $0 \le \theta < 2\pi$.

$$|-\sqrt{3}| = \sqrt{3} = \tan \frac{\pi}{3}$$

The reference angle is $\alpha = \frac{\pi}{3}$.

The tangent function is negative in Quadrants II and IV.

The second-quadrant angle is $\pi - \frac{\pi}{3} = \frac{2\pi}{3}$ (see Fig. 10.26).

The fourth-quadrant angle is $2\pi - \frac{\pi}{3} = \frac{5\pi}{3}$.

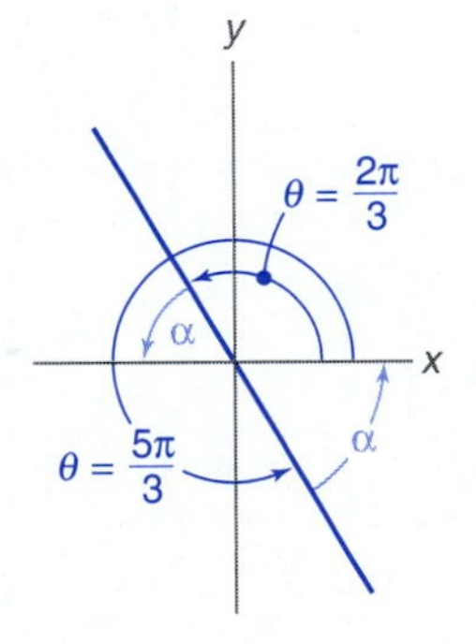

Figure 10.26

The trigonometric functions of the most used angles are tabulated in both degree and radian measures in Table 10.4.

A calculator may also be used to evaluate trigonometric functions in radians. Make certain that your calculator is in the radian mode.

TABLE 10.4

	$0°$ / 0	$30°$ / $\frac{\pi}{6}$	$45°$ / $\frac{\pi}{4}$	$60°$ / $\frac{\pi}{3}$	$90°$ / $\frac{\pi}{2}$	$180°$ / π
$\sin\theta$	0	$\frac{1}{2}$	$\frac{\sqrt{2}}{2}$	$\frac{\sqrt{3}}{2}$	1	0
$\cos\theta$	1	$\frac{\sqrt{3}}{2}$	$\frac{\sqrt{2}}{2}$	$\frac{1}{2}$	0	-1
$\tan\theta$	0	$\frac{\sqrt{3}}{3}$	1	$\sqrt{3}$	Undefined	0
$\cot\theta$	Undefined	$\sqrt{3}$	1	$\frac{\sqrt{3}}{3}$	0	Undefined
$\sec\theta$	1	$\frac{2\sqrt{3}}{3}$	$\sqrt{2}$	2	Undefined	-1
$\csc\theta$	Undefined	2	$\sqrt{2}$	$\frac{2\sqrt{3}}{3}$	1	Undefined

EXAMPLE 7

Find $\sin 0.4$, $\cos 1.684$, and $\tan\frac{\pi}{12}$, rounded to four significant digits.

MODE down arrows right arrow **1 ENTER**

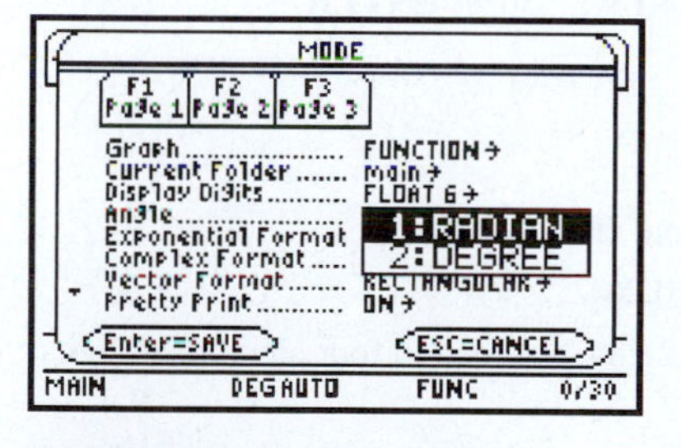

2nd SIN .4) **ENTER 2nd COS** 1.684) **ENTER 2nd TAN 2nd** π/12) **green diamond ENTER**

```
sin(.4)          .389418
cos(1.684)      -.112962
tan(π/12)        .267949
tan(π/12)
MAIN   RAD AUTO   FUNC   3/30
```

MODE 2nd QUIT

```
sin(.4)
      .3894183423
cos(1.684)
     -.1129620426
tan(π/12)
      .2679491924
```

SIN .4) **ENTER COS** 1.684) **ENTER TAN 2nd** π/12) **ENTER**

Thus, $\sin 0.4 = 0.3894$, $\cos 1.684 = -0.1130$, and $\tan\frac{\pi}{12} = 0.2679$, rounded to four significant digits.

To evaluate the cotangent, secant, and cosecant functions, follow the same procedure as in Section 10.2. You may find it helpful to remember the quadrantal angles in radians as shown in Fig. 10.27.

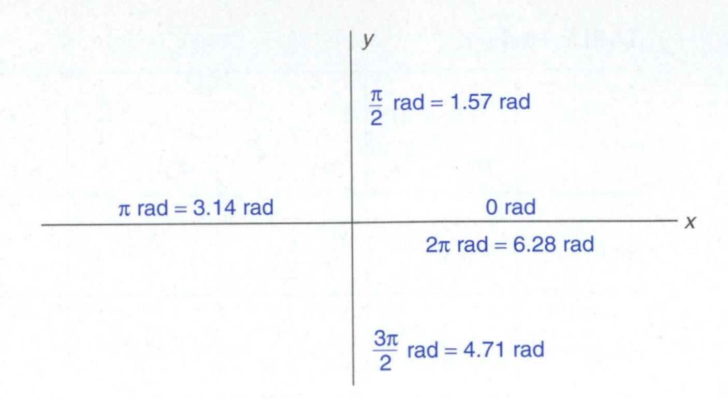

Figure 10.27 Quadrantal angles in radians.

EXAMPLE 8

Given $\sin \theta = 0.7565$, find θ in radians for $0 \le \theta < 2\pi$, rounded to four significant digits.

green diamond SIN^{-1} .7565) **ENTER**

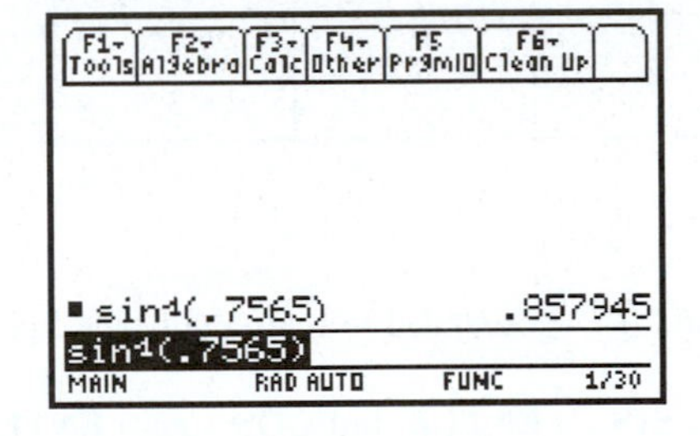

2nd SIN^{-1} .7565) **ENTER**

The reference angle is $\alpha = 0.8579$.

The sine function is positive in Quadrants I and II.

The first-quadrant angle is 0.8579 (see Fig. 10.28).

The second-quadrant angle is $\pi - 0.8579 = 2.284$, rounded to four significant digits.

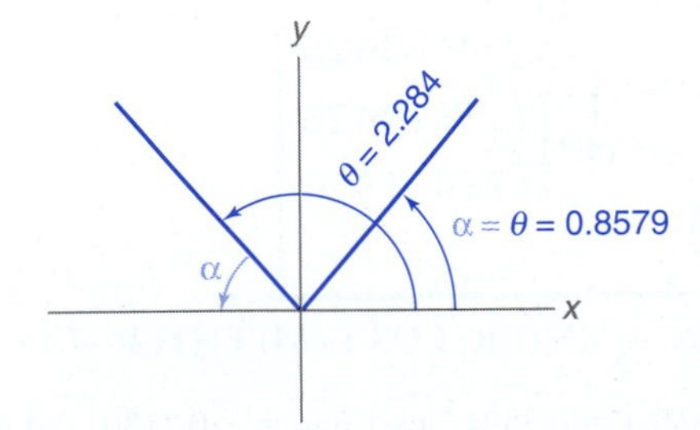

Figure 10.28

EXAMPLE 9

Given $\tan \theta = -1.632$, find θ in radians for $0 \le \theta < 2\pi$, rounded to four significant digits.

green diamond TAN^{-1} 1.632) ENTER

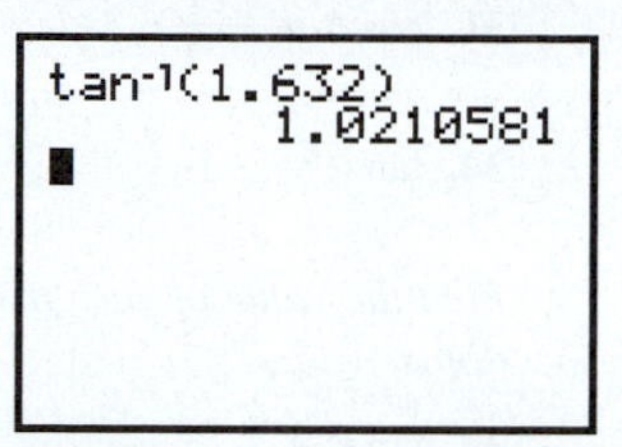

2nd TAN^{-1} 1.632) ENTER

The reference angle is $\alpha = 1.021$.

The tangent function is negative in Quadrants II and IV.

The second-quadrant angle is $\pi - 1.021 = 2.121$ (see Fig. 10.29).

The fourth-quadrant angle is $2\pi - 1.021 = 5.262$.

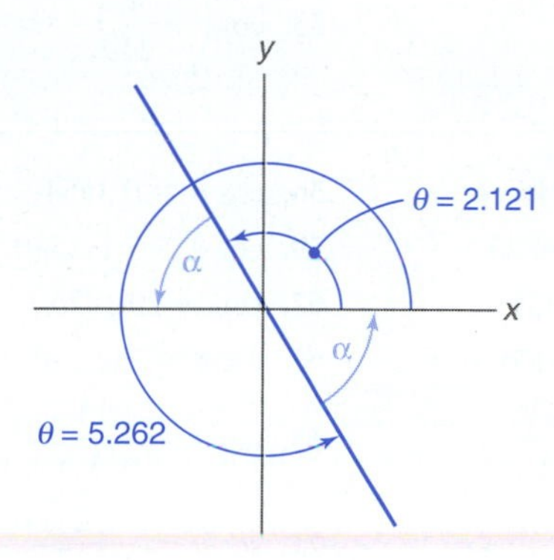

Figure 10.29

Exercises 10.3

Change each angle measure from degrees to radians. Express in terms of π.

1. 135° **2.** 210° **3.** 90° **4.** 270°
5. −75° **6.** −230° **7.** 1260° **8.** 2490°

Change each angle measure from radians to degrees.

9. $\frac{7\pi}{4}$ **10.** $\frac{7\pi}{6}$ **11.** $\frac{19\pi}{4}$ **12.** $\frac{27\pi}{8}$
13. 9π **14.** 12π **15.** 3.7 **16.** 1.82

Given each angle in radians, find the reference angle in radians.

17. $\frac{7\pi}{4}$ **18.** $\frac{5\pi}{6}$ **19.** $\frac{9\pi}{4}$ **20.** $\frac{4\pi}{3}$
21. $\frac{11\pi}{12}$ **22.** $\frac{17\pi}{12}$ **23.** $-\frac{8\pi}{5}$ **24.** $-\frac{15\pi}{7}$

Using Table 10.4, find the value of each trigonometric function.

25. $\sin\frac{3\pi}{4}$ **26.** $\cos\frac{7\pi}{6}$ **27.** $\tan\frac{4\pi}{3}$
28. $\sec\frac{9\pi}{4}$ **29.** $\csc\left(-\frac{3\pi}{4}\right)$ **30.** $\cot\left(-\frac{5\pi}{4}\right)$

Using Table 10.4, find θ for $0 \le \theta < 2\pi$.

31. $\cos\theta = -\dfrac{1}{2}$ **32.** $\sin\theta = \dfrac{\sqrt{3}}{2}$ **33.** $\sec\theta = 2$

34. $\tan\theta = -1$ **35.** $\cot\theta = 1$ **36.** $\csc\theta = -\dfrac{2\sqrt{3}}{3}$

Find the value of each trigonometric function (expressed in radians), rounded to four significant digits.

37. $\sin 0.8$ **38.** $\cos 0.4$ **39.** $\tan 1.2$

40. $\tan 0.15$ **41.** $\cos 1.0$ **42.** $\sin 0.75$

43. $\sin(-1.65)$ **44.** $\cos(-2.11)$ **45.** $\tan 18.7$

46. $\cot 23.6$ **47.** $\sec(-5.6)$ **48.** $\csc(-21.9)$

49. $\sin\left(\dfrac{3\pi}{4}\right)$ **50.** $\cos\left(\dfrac{5\pi}{8}\right)$ **51.** $\tan\left(\dfrac{3\pi}{5}\right)$

52. $\sin 8\pi$ **53.** $\cos\left(-\dfrac{\pi}{24}\right)$ **54.** $\sin\left(-\dfrac{5\pi}{6}\right)$

Find θ for $0 \le \theta < 2\pi$, rounded to four significant digits.

55. $\sin\theta = 0.9845$ **56.** $\cos\theta = 0.3554$ **57.** $\tan\theta = 1.685$

58. $\sin\theta = -0.6825$ **59.** $\cos\theta = -0.7540$ **60.** $\tan\theta = -0.1652$

61. $\cos\theta = 0.6924$ **62.** $\sin\theta = 0.1876$ **63.** $\tan\theta = -4.672$

64. $\sec\theta = -4.006$ **65.** $\csc\theta = -2.140$ **66.** $\cot\theta = -0.1066$

67. $\cos\theta = -\dfrac{\sqrt{3}}{2}$ **68.** $\tan\theta = -\dfrac{1}{\sqrt{3}}$ **69.** $\sin\theta = \dfrac{1}{2}$

70. $\cot\theta = -1$ **71.** $\sec\theta = -\dfrac{2}{\sqrt{3}}$ **72.** $\csc\theta = -2$

10.4 USE OF RADIAN MEASURE

Radian measure has many applications in mathematics and various areas of technology. We illustrate some of these applications by examples and exercises.

The length s of an intercepted arc of a circle equals the product of the radius r and the measure of the central angle θ in radians (see Fig. 10.30). That is,

$$s = r\theta$$

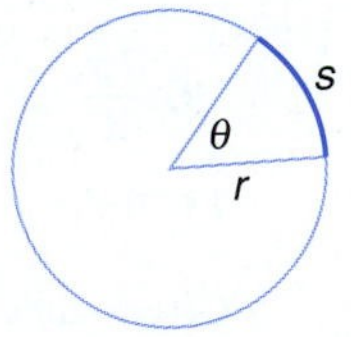

Note that when $\theta = 2\pi$, $s = 2\pi r$, where s is the circumference.

Figure 10.30 Length s of arc of a circle.

EXAMPLE 1

Find the length of the arc of a circle of radius 4.00 ft and with a central angle of $24\bar{0}°$.

$$\theta = 24\bar{0}° = 24\bar{0}° \times \frac{\pi}{180°} = \frac{4\pi}{3}$$

$s = r\theta$ (See Fig. 10.31.)

$$= (4.00\text{ ft})\left(\frac{4\pi}{3}\right)$$

$$= 16.8\text{ ft}$$

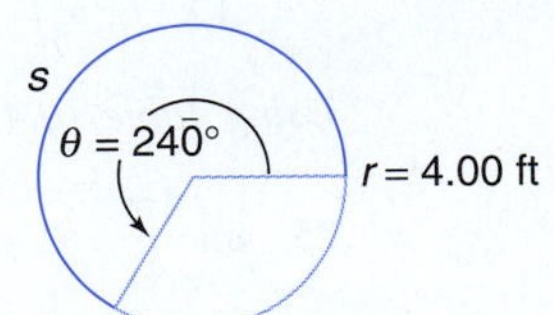

Figure 10.31

The area of a sector of a circle (see Fig. 10.32) is given by the formula

$$A = \frac{1}{2}r^2\theta$$

where r is the radius and θ is the measure of the central angle in radians.

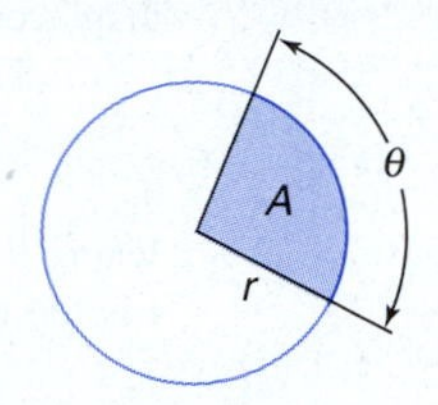

Figure 10.32 Area of a sector of a circle.

EXAMPLE 2

Find the area of the sector of the circle from Example 1.

$$\begin{aligned} A &= \frac{1}{2}r^2\theta \\ &= \frac{1}{2}(4.00 \text{ ft})^2\left(\frac{4\pi}{3}\right) \\ &= 33.5 \text{ ft}^2 \end{aligned}$$

Other illustrations of the use of radian measure from different areas of technology are given in the following examples.

EXAMPLE 3

An automobile engine is rated at $20\bar{0}$ hp at $40\bar{0}0$ revolutions per minute (rpm). Find its torque given by the formula

$$G = \frac{P}{\omega}$$

where G is the torque, P is the power, and ω is the angular speed in radians per unit time, usually radians per second.

The horsepower measure needs to be converted to foot-pounds per second, as follows:

$$\begin{aligned} P = 20\bar{0} \text{ hp} &= 20\bar{0} \cancel{\text{hp}} \times \frac{55\bar{0}\,\frac{\text{ft-lb}}{\text{s}}}{1 \cancel{\text{hp}}} \\ &= 11\bar{0},000 \frac{\text{ft-lb}}{\text{s}} \end{aligned}$$

The angular speed needs to be converted to radians per second, as follows:

$$\omega = 40\bar{0}0 \frac{\cancel{\text{rev}}}{\cancel{\text{min}}} \times \frac{2\pi \text{ rad}}{\cancel{\text{rev}}} \times \frac{1 \cancel{\text{min}}}{60 \text{ s}} = 419 \frac{\text{rad}}{\text{s}}$$

Therefore,

$$G = \frac{P}{\omega} = \frac{11\bar{0},000 \frac{\text{ft-lb}}{\cancel{\text{s}}}}{419 \frac{\text{rad}}{\cancel{\text{s}}}} = 263 \text{ ft-lb}$$

Note: Recall that the *radian* unit has no physical dimensions.

EXAMPLE 4

The flywheel of a gasoline engine rotates at an angular speed of $30\bar{0}0$ rpm. Find its angular displacement in 5.00 s.

The angular displacement of a revolving body is given by the formula

$$\theta = \omega t$$

where θ is the angular displacement in radians, ω is angular speed in radians per second, and t is the time in seconds.

$$\omega = 30\bar{0}0 \frac{\cancel{\text{rev}}}{\cancel{\text{min}}} \times \frac{2\pi \text{ rad}}{\cancel{\text{rev}}} \times \frac{1 \cancel{\text{min}}}{60 \text{ s}} = 314 \frac{\text{rad}}{\text{s}}$$

$$\begin{aligned} \theta &= \omega t \\ &= \left(314 \frac{\text{rad}}{\cancel{\text{s}}}\right)(5.00 \cancel{\text{s}}) \\ &= 1570 \text{ rad} \end{aligned}$$

or

$$1570 \cancel{\text{rad}} \times \frac{1 \text{ rev}}{2\pi \cancel{\text{rad}}} = 25\bar{0} \text{ revolutions}$$

EXAMPLE 5

An airplane propeller is rotating at $18\bar{0}0$ rpm. If the blades of the propeller are 2.00 m long, find the linear speed of each point:

(a) 1.00 m from the axis of rotation.

(b) On the end of a blade.

The linear speed of a revolving body is given by the formula

$$v = \omega r$$

where v is the linear speed, ω is the angular speed, and r is the distance from the axis of rotation.

$$\omega = 18\bar{0}0 \frac{\cancel{\text{rev}}}{\cancel{\text{min}}} \times \frac{2\pi \text{ rad}}{\cancel{\text{rev}}} \times \frac{1 \cancel{\text{min}}}{60 \text{ s}} = 188 \frac{\text{rad}}{\text{s}}$$

(a) $v = \omega r = \left(188 \frac{\text{rad}}{\text{s}}\right)(1.00 \text{ m}) = 188 \text{ m/s}$

(b) $v = \omega r = \left(188 \frac{\text{rad}}{\text{s}}\right)(2.00 \text{ m}) = 376 \text{ m/s}$

To illustrate the difference between linear and angular velocity, consider a common ice-skating routine performed by many traveling shows. A line of skaters moves around in a circle as shown in Fig. 10.33. Who skates the fastest? As you may recall, the person

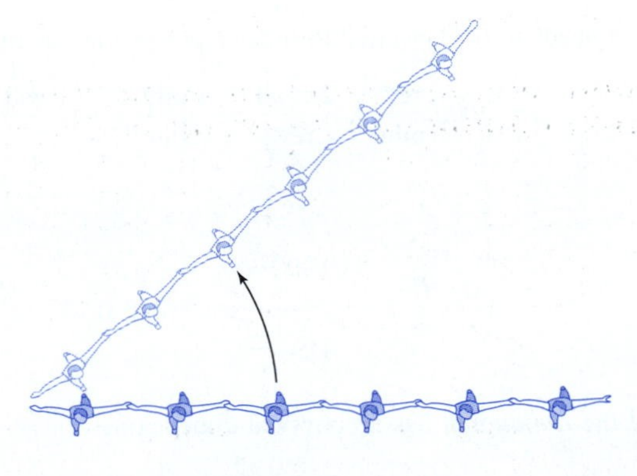

Figure 10.33 A rotating line of ice skaters.

on the end usually has difficulty in keeping up and in maintaining a straight line. Each successive person out from the center skates faster (has a greater linear velocity) than the person nearer the center. Note that the angular velocity of the line of skaters and of each skater remains the same, as they each make the same number of rotations per unit of time.

Exercises 10.4

1. From a circle of radius 12.0 in., find the length of intercepted arc when the central angle is $\frac{2\pi}{3}$ radians.

2. From a circle of radius 8.00 in., find the length of intercepted arc when the central angle is 48°.

3. Find the area of the sector of the circle from Exercise 1.

4. Find the area of the sector of the circle from Exercise 2.

5. Find the central angle, in degrees, of a circle of radius 6.00 in. when the intercepted arc length is 5.00 in.

6. Find the radius of a circle in which a central angle of $\frac{3\pi}{4}$ radians intercepts an arc of 24.0 in.

7. Given the two concentric circles in Fig. 10.34, where $\theta = \frac{\pi}{6}$ rad, $r_1 = 3.00$ m, and $r_2 = 5.00$ m, find the shaded area of the figure.

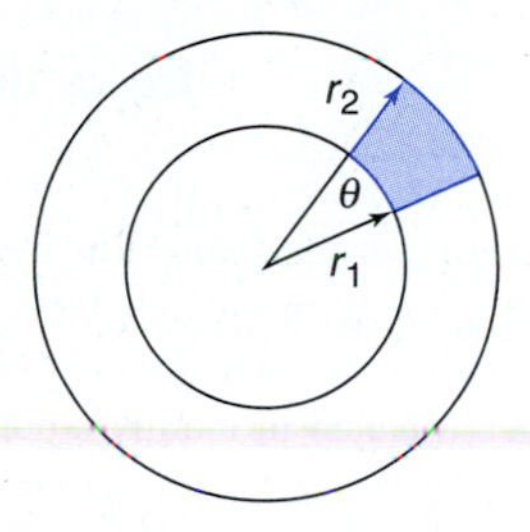

Figure 10.34

8. If a sector of a given circle has an area of 36.0 in^2 and its radius is 6.00 in., find the measure of the central angle.

9. A pendulum of length 5.00 m swings through an angle of 6°. Find the length of arc through which the pendulum swings in one complete swing.

10. An airplane travels in a circular path at $40\bar{0}$ mi/h for 3.00 min. What is the radius of the circle when the central angle is $1\bar{0}$°?

11. An automobile engine is rated at 275 hp at $42\bar{0}0$ rpm. Find the torque developed.

12. A gasoline engine develops a torque of $30\bar{0}$ ft-lb at $36\bar{0}0$ rpm. What is its horsepower rating?

13. The flywheel of a steam engine is rotating at $42\bar{0}$ rpm.

(a) Express this angular speed in radians per second.

(b) Find the angular displacement of the wheel in 10.0 s.

(c) Find the linear speed of a point on the rim of the wheel which has a radius of 1.75 ft.

14. An airplane propeller whose blades are 6.00 ft long is rotating at $22\bar{0}0$ rpm.

(a) Express its angular speed in radians per second.

(b) Find the angular displacement in 3.00 s.

(c) Find the linear speed of a point on the end of the blade.

15. The angular velocity of an airplane propeller is increased from $18\bar{0}0$ rpm to $22\bar{0}0$ rpm in 10.0 s. Find its angular acceleration. The angular acceleration of a rotating body is given by the formula

$$\alpha = \frac{\Delta\omega}{\Delta t}$$

where α is the angular acceleration, $\Delta\omega$ is the change in angular velocity, and Δt is the change in time.

16. Find the angular acceleration of the airplane propeller in Exercise 15 when its angular velocity is increased as given:
(a) From $18\bar{0}0$ rpm to $22\bar{0}0$ rpm in 6.00 s.
(b) From $18\bar{0}0$ rpm to $26\bar{0}0$ rpm in 10.0 s.

17. The earth rotates on its axis at 1 rev per 24 h. Assume that the average radius of the earth is 3960 mi. Find the linear speed in miles per hour of a point on the equator.

18. The earth is revolving about the sun in 1 rev per 365 days. Assume an average circular orbit of radius 1.50×10^8 km. Find the linear speed in kilometres per second of the earth about the sun.

19. In the pulley system in Fig. 10.35, find the following.
(a) The linear velocity of a point on the rim of pulley A in metres per second.
(b) The linear velocity of a point on the rim of pulley B in metres per second.
(c) The angular velocity of pulley B in radians per second.

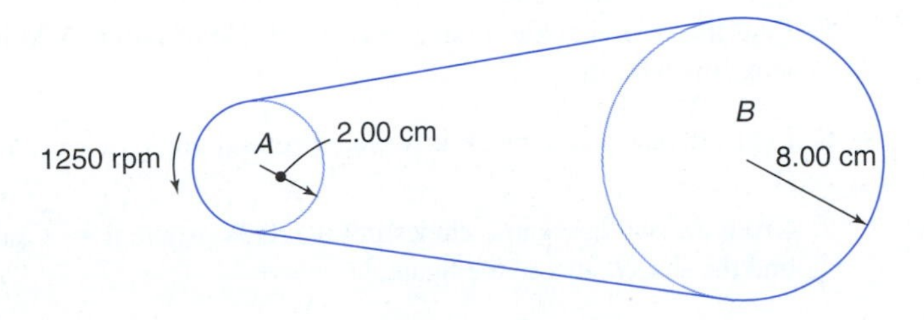

Figure 10.35

20. An automobile is traveling at 55.0 mi/h. Its tires have a radius of 13.0 in. Find the following.
(a) The tires' angular velocity in radians per second.
(b) The number of revolutions each tire completes in 10.0 s.
(c) The linear distance traveled in feet by a point on the tread in 10.0 s.

CHAPTER 10 SUMMARY

1. An angle is in *standard position* when its vertex is located at the origin and its initial side is lying on the positive x-axis. A positive angle is measured counterclockwise; a negative angle is measured clockwise.

2. Angles that share the same initial side and the same terminal side are called *coterminal.*

3. When angle θ is in standard position and $P(x, y)$ is on the terminal side, the six trigonometric functions are defined as follows:

$$\sin\theta = \frac{y}{r} \qquad \cos\theta = \frac{x}{r} \qquad \tan\theta = \frac{y}{x}$$

$$\csc\theta = \frac{r}{y} \qquad \sec\theta = \frac{r}{x} \qquad \cot\theta = \frac{x}{y}$$

4. The *reciprocal trigonometric functions* are as follows:

$$\csc\theta = \frac{1}{\sin\theta} \qquad \sec\theta = \frac{1}{\cos\theta} \qquad \cot\theta = \frac{1}{\tan\theta}$$

5. A *quadrantal angle* is an angle in standard position with its terminal side on one of the coordinate axes.

6. The *reference angle* α of any nonquadrantal angle θ in standard position is the positive acute angle between the terminal side of θ and the x-axis.

7. To use a calculator to find angles when the trigonometric function is given, review the story boards and examples in Section 10.2 for angles in degrees and in Section 10.3 for angles in radians.
8. A *radian* is the measure of an angle with its vertex at the center of a circle whose intercepted arc is equal to the length of the radius of the circle.
9. To form conversion factors to change from degrees to radians or from radians to degrees, use the relation π rad $= 180°$.
10. The *length s of an intercepted arc of a circle* equals the product of the radius r and the measure of the central angle θ in radians:

$$s = r\theta$$

11. The *area of a sector of a circle* is given by the formula

$$A = \frac{1}{2}r^2\theta$$

where θ is in radians.
12. The *angular velocity of a revolving body* is given by

$$\omega = \frac{\theta}{t}$$

where θ is the angular displacement in radians, ω is the angular speed in radians per second, and t is the time.
13. The *linear speed of a revolving body* is given by

$$v = \omega r$$

where v is the linear speed, ω is the angular speed, and r is the distance from the axis of rotation.
14. Another useful conversion is 1 revolution $= 2\pi$ radians.

CHAPTER 10 REVIEW

Find the values of sin θ, cos θ, tan θ, cot θ, sec θ, *and* csc θ *if* θ *is in standard position and its terminal side passes through the given point.*

1. (4, 3) **2.** $(-\sqrt{3}, -1)$ **3.** $(-4, 0)$

For each angle, find the reference angle.

4. 135° **5.** 208°20′ **6.** −125° **7.** 1250°

Find the value of each trigonometric function, rounded to four significant digits.

8. sin 244.3° **9.** tan 337.5° **10.** sec 98.7°
11. cos (−297.4°) **12.** cot 402.1° **13.** csc (−168.0°)

Find θ *for* $0° \le \theta < 360°$ *rounded to the nearest tenth of a degree.*

14. $\sin\theta = 0.3448$ **15.** $\cos\theta = -0.5495$ **16.** $\tan\theta = -1.050$
17. $\sec\theta = 1.956$ **18.** $\cot\theta = -1.855$ **19.** $\csc\theta = 1.353$

Change each angular measurement from degrees to radians. Express in terms of π.

20. 72° **21.** 315°

Change each angular measurement from radians to degrees.

22. $\dfrac{5\pi}{6}$ **23.** $\dfrac{3\pi}{4}$

Given each angle in radians, find the reference angle in radians.

24. $\dfrac{5\pi}{3}$ **25.** $\dfrac{3\pi}{5}$

Using Table 10.4, find the value for each trigonometric function.

26. $\cos \dfrac{5\pi}{6}$ **27.** $\tan \dfrac{2\pi}{3}$ **28.** $\sec \dfrac{7\pi}{4}$

29. $\sin \dfrac{7\pi}{6}$ **30.** $\cot\left(-\dfrac{\pi}{4}\right)$ **31.** $\cos\left(-\dfrac{11\pi}{3}\right)$

Using Table 10.4, find each θ for $0 \le \theta < 2\pi$.

32. $\cos\theta = -\dfrac{\sqrt{2}}{2}$ **33.** $\tan\theta = -1$ **34.** $\csc\theta = \dfrac{2\sqrt{3}}{3}$

35. $\sin\theta = -\dfrac{\sqrt{3}}{2}$ **36.** $\sec\theta = -2$ **37.** $\tan\theta = \dfrac{\sqrt{3}}{3}$

Using a calculator, find the value of each trigonometric function (expressed in radians), rounded to four significant digits.

38. $\sin 1.5$ **39.** $\cos 0.25$ **40.** $\tan \dfrac{\pi}{6}$ **41.** $\sin\left(-\dfrac{2\pi}{3}\right)$

Using a calculator, find each θ for $0 \le \theta < 2\pi$, rounded to four significant digits.

42. $\cos\theta = 0.1981$ **43.** $\sin\theta = -0.6472$

44. $\tan\theta = 1.6182$ **45.** $\sec\theta = -2.8061$

46. Given a sector of a circle of radius 9.00 in. with a central angle of 56.0°, find the length of arc and area of the sector.

47. An automobile engine is rated at 325 hp at $40\bar{0}0$ rpm. Find the torque developed.

48. A flywheel is rotating at $63\bar{0}$ rpm.

(a) Express this angular speed in rad/s.

(b) Find the angular displacement in 5.00 s.

(c) Find the linear speed of a point on the wheel that is 1.50 ft from the axis of rotation.

11 Oblique Triangles and Vectors

INTRODUCTION

A lack of rain for 6 weeks, a wind speed of 19 mph, and an abundance of campers had increased the likelihood of a forest fire in the Smoky Mountains. Rangers in observation towers and helicopter crews were on the alert. The ranger at tower A was the first to spot a fire in a direction 51.5° east of north. Minutes later the ranger in tower B, 5.0 mi due east of A, sighted the fire at 17.2° west of north. The ranger in station A radioed the helicopter crews and determined that there was a helicopter 4.5 mi north of station A. Because of cloud cover, the pilot could not see the fire. Did the ranger have enough information to give the pilot the location of the fire?

Solving the preceding problem requires the use of the law of sines and the law of cosines. Part of the information needed is requested in Problem 7 in Exercises 11.4. The ranger radioed the pilot to fly in a direction of 72.1° east of south for 4.2 miles to find the fire.

Objectives

- Find the sides and angles of a triangle given three parts, one of which is a side.
- Use the law of sines.
- Use the law of cosines.
- Add vectors, using trigonometric methods.
- Express vectors in terms of components.
- Use vectors to solve problems involving displacement, velocity, and force.

11.1 LAW OF SINES

An **oblique,** or **general,** triangle is a triangle that contains no right angles. We shall use the standard notation of labeling the vertices of a triangle by the capital letters A, B, and C and using the lowercase letters a, b, and c as the labels for the sides opposite angles A, B, and C, respectively (see Fig. 11.1).

Solving a triangle means finding all those sides and angles that are not given or known. To solve a triangle, we need three parts (including at least one side). Solving any

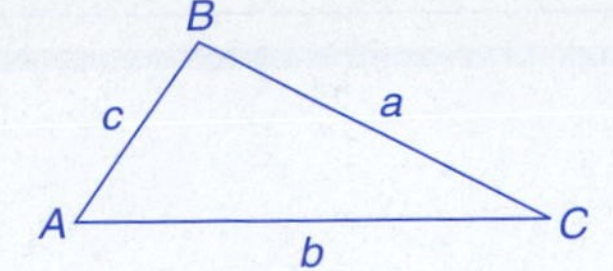

Figure 11.1 Oblique, or general, triangle *ABC*.

oblique triangle falls into one of four cases where the following parts of a triangle are known:

1. Two sides and an angle opposite one of them (SSA).
2. Two angles and a side opposite one of them (AAS).
3. Two sides and the included angle (SAS).
4. Three sides (SSS).

One law that we use to solve triangles is called the **law of sines.** In words, for any triangle the ratio of any side to the sine of the opposite angle is a constant. The formula for the law of sines is as follows:

LAW OF SINES

$$\frac{a}{\sin A} = \frac{b}{\sin B} = \frac{c}{\sin C}$$

To derive the law of sines, draw any general triangle *ABC*. From *C* draw *CD* perpendicular to *AB*. Note that every oblique triangle is either acute (all three angles are between 0° and 90°) or obtuse (one angle is between 90° and 180°). Both cases are shown in Fig. 11.2. Line *CD* is called the altitude *h* and forms two right triangles in each case.

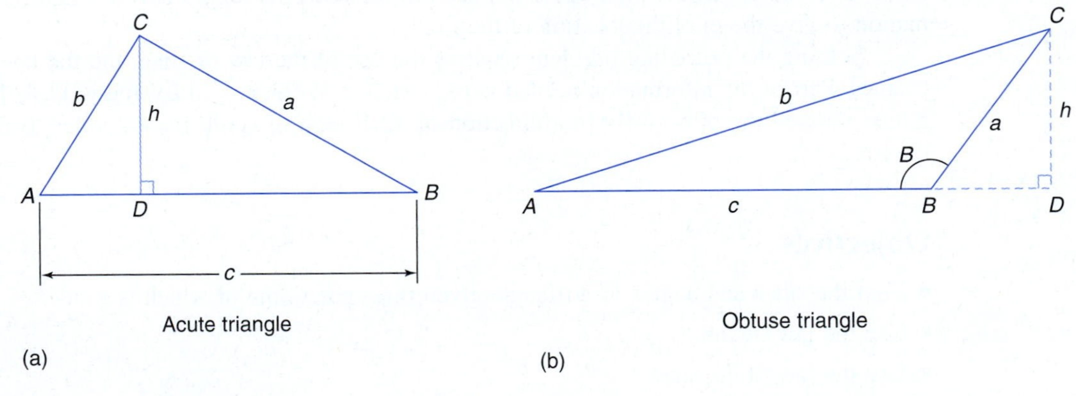

Figure 11.2 Draw altitude *h* from *C* to the opposite side in each triangle.

In right triangle *ADC* in Figs. 11.2(a) and 11.2(b),

$$\sin A = \frac{h}{b} \quad \text{or} \quad h = b \sin A$$

In right triangle *BCD* in Fig. 11.2(a),

$$\sin B = \frac{h}{a} \quad \text{or} \quad h = a \sin B$$

In Fig. 11.2(b), note that angle *DBC* is the reference angle for angle *ABC* or angle *B*. Then in right triangle *BDC*,

$$\sin B = \sin DBC = \frac{h}{a} \quad \text{or} \quad h = a \sin B$$

In either case we have

$$h = a \sin B = b \sin A$$

Dividing each side by $(\sin B)(\sin A)$, we have

$$\frac{a}{\sin A} = \frac{b}{\sin B}$$

By drawing a perpendicular line from A and then from B to the opposite side, we can show in a similar manner that $\frac{c}{\sin C} = \frac{b}{\sin B}$ and $\frac{c}{\sin C} = \frac{a}{\sin A}$. By combining these equations, we have the law of sines:

$$\frac{a}{\sin A} = \frac{b}{\sin B} = \frac{c}{\sin C}$$

In order to use the law of sines, we must know either of the following:

1. Two sides and an angle opposite one of them (SSA).
2. Two angles and a side opposite one of them (AAS). *Note:* Knowing two angles and any side is sufficient because knowing two angles, we can easily find the third.

You must select the proportion that contains three parts that are known and the unknown part.

As in Chapter 3, when calculations with measurements involve a trigonometric function, we shall use the following rule for significant digits:

Angles expressed to the nearest	*Lengths of sides of a triangle will contain*
1°	Two significant digits
0.1° or 1′	Three significant digits
0.01° or 1″	Four significant digits

The following relationship is very helpful as a check when solving a general triangle: The longest side of any triangle is opposite the largest angle, and the shortest side is opposite the smallest angle.

EXAMPLE 1

If $A = 65.0°$, $a = 20.0$ m, and $b = 15.0$ m, solve the triangle.

First, draw a triangle as in Fig. 11.3 and find angle B by using the law of sines.

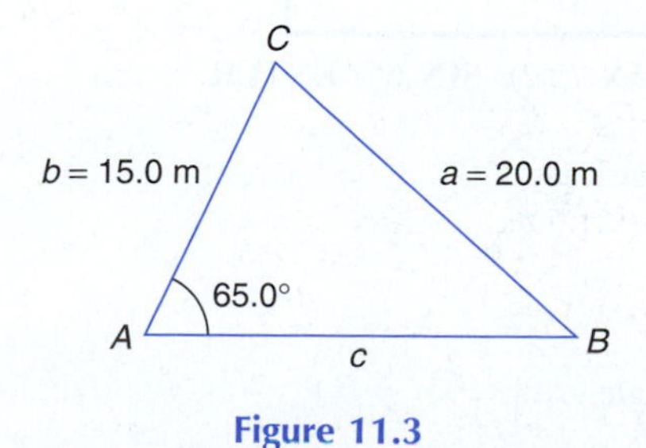

Figure 11.3

$$\frac{a}{\sin A} = \frac{b}{\sin B}$$

$$\frac{20.0 \text{ m}}{\sin 65.0°} = \frac{15.0 \text{ m}}{\sin B}$$

$$\sin B = \frac{(15.0 \text{ m})(\sin 65.0°)}{20.0 \text{ m}} = 0.6797$$

$$B = 42.8°$$

This angle may be found using a calculator as follows:

MODE ...**ENTER ENTER**

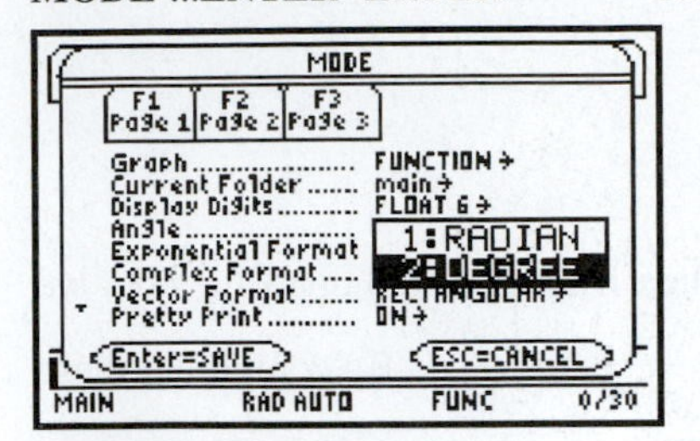

15.0* **2nd SIN** 65.0)/20.0 **ENTER green diamond SIN^{-1} 2nd ANS**) **ENTER**

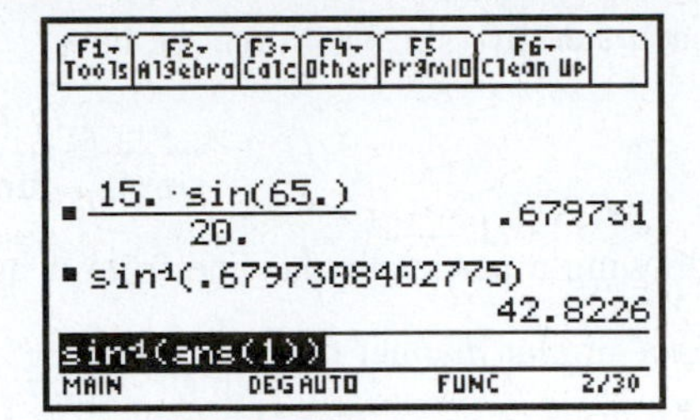

MODE **2nd QUIT**

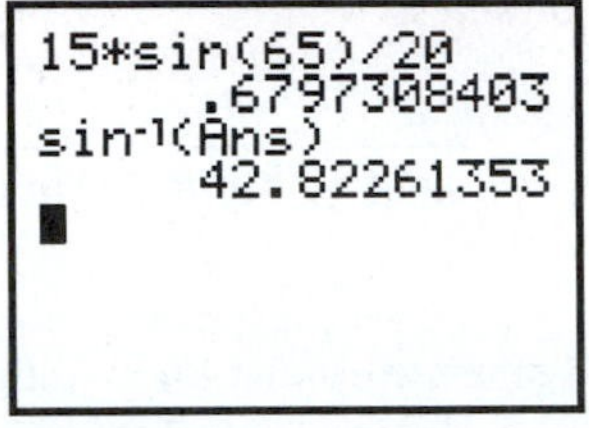

15* **SIN** 65)/20 **ENTER 2nd SIN^{-1} 2nd ANS**) **ENTER**

So $B = 42.8°$ rounded to the nearest tenth of a degree.

To find C, use the fact that the sum of the angles of any triangle is 180°. Therefore,

$$\begin{aligned} C &= 180° - 65.0° - 42.8° \\ &= 72.2° \end{aligned}$$

Finally, find c using the law of sines.

$$\frac{a}{\sin A} = \frac{c}{\sin C}$$

$$\frac{20.0 \text{ m}}{\sin 65.0°} = \frac{c}{\sin 72.2°}$$

$$c = \frac{(20.0 \text{ m})(\sin 72.2°)}{\sin 65.0°} = 21.0 \text{ m} \qquad \text{(rounded to three significant digits)}$$

This side may be found using a calculator as follows:

20.0* **2nd SIN** 72.2)/ **2nd SIN** 65.0) **ENTER**

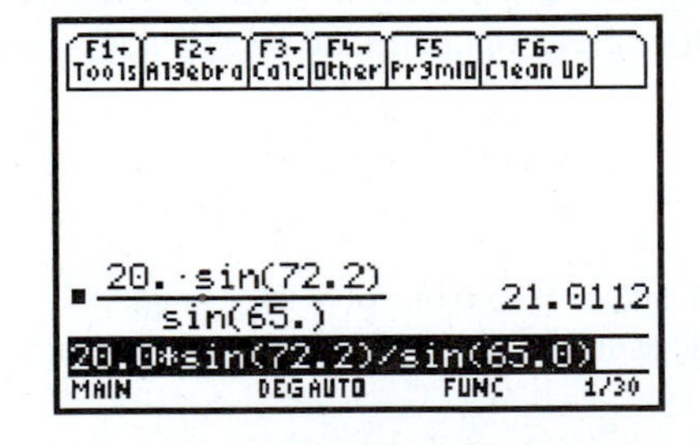

20* **SIN** 72.2)/ **SIN** 65) **ENTER**

That is, side $c = 21.0$ m, rounded to three significant digits.

The solution is $B = 42.8°$, $C = 72.2°$, and $c = 21.0$ m.

EXAMPLE 2

If $C = 25°$, $c = 59$ ft, and $B = 108°$, solve the triangle.

First, draw a triangle as in Fig. 11.4 and find b.

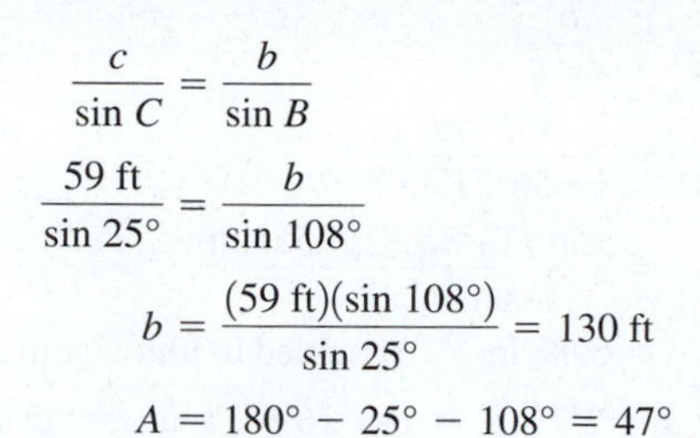

$$\frac{c}{\sin C} = \frac{b}{\sin B}$$

$$\frac{59 \text{ ft}}{\sin 25°} = \frac{b}{\sin 108°}$$

$$b = \frac{(59 \text{ ft})(\sin 108°)}{\sin 25°} = 130 \text{ ft} \qquad \text{(rounded to two significant digits)}$$

$$A = 180° - 25° - 108° = 47°$$

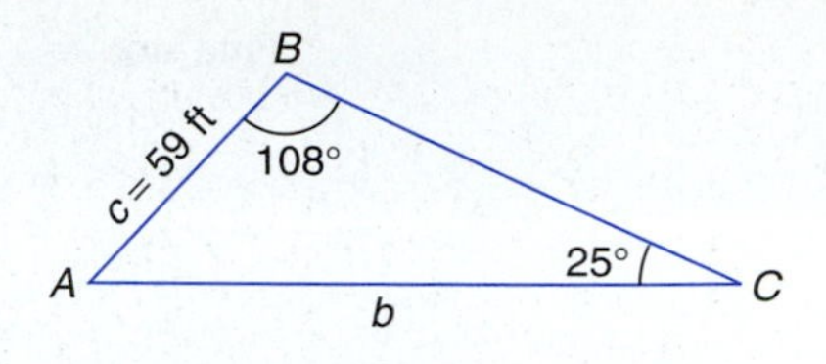

Figure 11.4

Find a.

$$\frac{a}{\sin A} = \frac{c}{\sin C}$$

$$\frac{a}{\sin 47°} = \frac{59 \text{ ft}}{\sin 25°}$$

$$a = \frac{(59 \text{ ft})(\sin 47°)}{\sin 25°} = 1\bar{0}0 \text{ ft}$$

The solution is $A = 47°$, $1\bar{0}0$ ft, and $b = 130$ ft.

EXAMPLE 3

If $B = 36°21'45''$, $a = 3745$ m, and $b = 4551$ m, solve the triangle.

Draw a triangle as in Fig. 11.5 and find angle A.

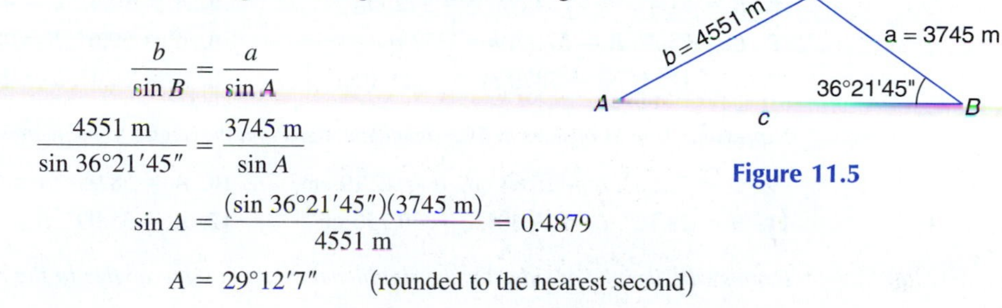

$$\frac{b}{\sin B} = \frac{a}{\sin A}$$

$$\frac{4551 \text{ m}}{\sin 36°21'45''} = \frac{3745 \text{ m}}{\sin A}$$

$$\sin A = \frac{(\sin 36°21'45'')(3745 \text{ m})}{4551 \text{ m}} = 0.4879$$

$$A = 29°12'7'' \qquad \text{(rounded to the nearest second)}$$

Figure 11.5

This angle may be found using a calculator as follows:

2nd SIN 36 **2nd** | 21 **2nd** = 45 **2nd 1**)*3745/4551 **ENTER green diamond SIN**$^{-1}$ **2nd ANS**) **2nd MATH 2 8 ENTER**

SIN 36 **2nd ANGLE 1** 21 **2nd ANGLE 2** 45 **ALPHA +**)*3745/4551 **ENTER**
2nd SIN^{-1} **2nd ANS**) **2nd ANGLE 4 ENTER**

So, angle $A = 29°12'7''$, rounded to the nearest second. Then

$$C = 180° - A - B$$

$$= 180° - 29°12'7'' - 36°21'45'' = 114°26'8''$$

Find side c.

$$\frac{c}{\sin C} = \frac{b}{\sin B}$$

$$\frac{c}{\sin 114°26'8''} = \frac{4551 \text{ m}}{\sin 36°21'45''}$$

$$c = \frac{(\sin 114°26'8'')(4551 \text{ m})}{\sin 36°21'45''}$$

$$= 6988 \text{ m} \quad \text{(rounded to four significant digits)}$$

The solution is $A = 29°12'7''$, $C = 114°26'8''$, and $c = 6988$ m.

Note: Because of differences in rounding, your answers may differ slightly from the answers in the text if you choose to solve for the parts of a triangle in an order different from that chosen by the authors.

Exercises 11.1

Solve each triangle using the labels as shown in Fig. 11.6.

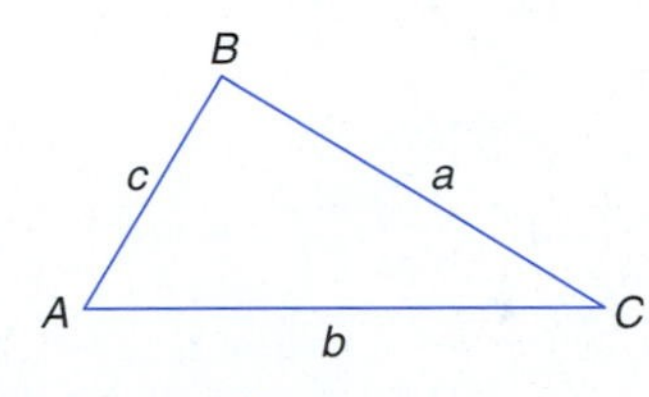

Figure 11.6

Express the lengths of sides to three significant digits and the angles to the nearest tenth of a degree.

1. $A = 69.0°$, $a = 25.0$ m, $b = 16.5$ m

2. $C = 57.5°$, $c = 166$ mi, $b = 151$ mi

3. $B = 61.4°$, $b = 124$ cm, $c = 112$ cm

4. $A = 19.5°$, $a = 487$ km, $c = 365$ km

5. $B = 75.3°$, $A = 57.1°$, $b = 257$ ft

6. $C = 59.6°$, $B = 43.9°$, $b = 4760$ m

7. $A = 115.0°$, $a = 5870$ m, $b = 4850$ m

8. $A = 16.4°$, $a = 205$ ft, $b = 187$ ft

Express the lengths of sides to four significant digits and the angles to the nearest hundredth of a degree.

9. $C = 72.58°$, $b = 28.63$ cm, $c = 42.19$ cm

10. $A = 58.95°$, $a = 3874$ m, $c = 2644$ m

11. $B = 28.76°$, $C = 19.30°$, $c = 39{,}750$ mi

12. $A = 35.09°$, $B = 48.64°$, $a = 8.362$ km

Express the lengths of sides to two significant digits and the angles to the nearest degree.

13. $A = 25°$, $a = 5\bar{0}$ cm, $b = 4\bar{0}$ cm

14. $B = 42°$, $b = 5.3$ km, $c = 4.6$ km

15. $C = 8°$, $c = 16$ m, $a = 12$ m

16. $A = 105°$, $a = 460$ mi, $c = 380$ mi

Express the lengths of sides to four significant digits and the angles to the nearest second.

17. $B = 51°17''$, $b = 1948$ ft, $c = 1525$ ft

18. $A = 49°31'50''$, $a = 37{,}560$ ft, $b = 24{,}350$ ft

19. $A = 31°14'35''$, $B = 85°45'15''$, $c = 4.575$ mi

20. $B = 75°30'6''$, $C = 70°12'18''$, $c = 93.45$ m

Find the indicated lengths of sides and angles consistent with the rules for calculations with measurements.

21. **(a)** side PQ
(b) side QR

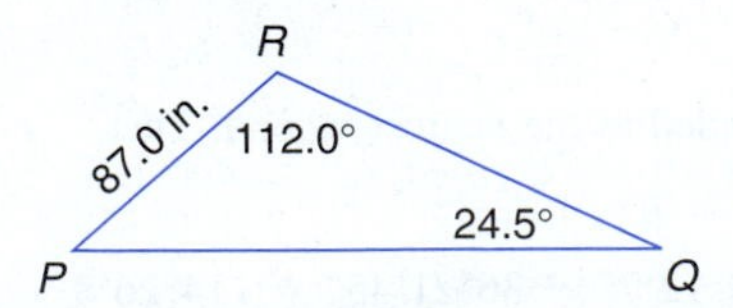

22. **(a)** $\angle R$
(b) side RS

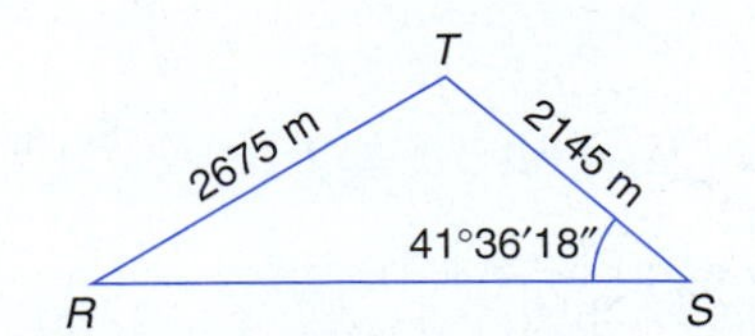

23. **(a)** side EF
(b) side DF

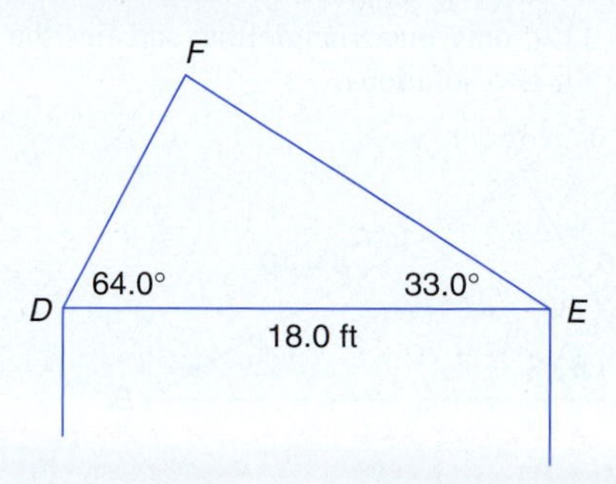

24. **(a)** side RS
(b) $\angle 1$

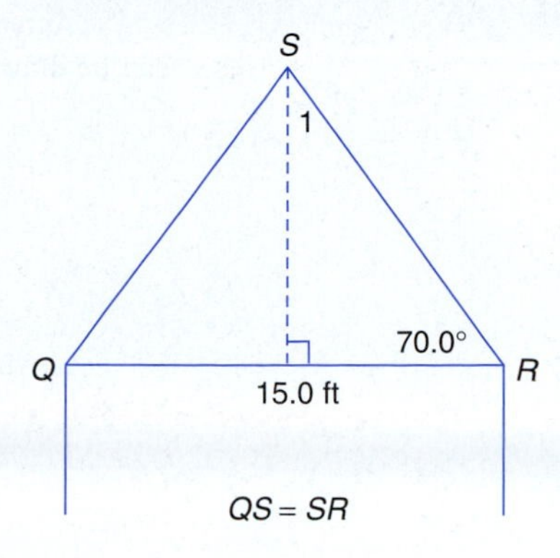

11.2 THE AMBIGUOUS CASE

The solution of a triangle when two sides and an angle opposite one of the sides (SSA) are given requires special care. There may be one, two, or no triangles formed from the given data. By construction and discussion, let's study the possibilities.

EXAMPLE 1

Construct a triangle given that $A = 35°$, $b = 10$, and $a = 7$.

As you can see from Fig. 11.7, two triangles that satisfy the given information can be drawn: triangles ACB and ACB'. Note that in one triangle angle B is acute and in the other triangle angle B is obtuse.

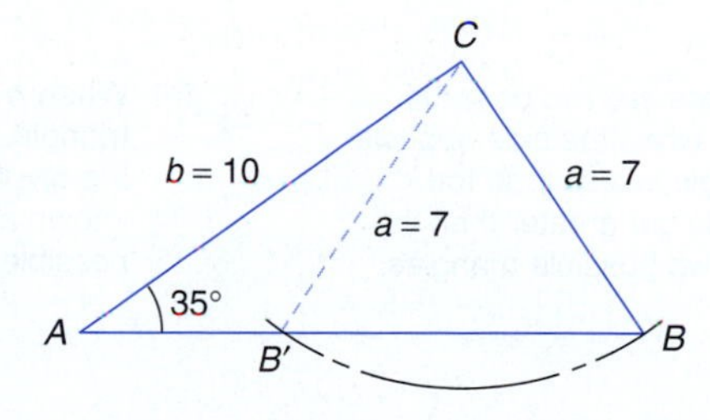

Figure 11.7

EXAMPLE 2

Construct a triangle given that $A = 45°$, $b = 10$, and $a = 5$.

As you can see from Fig. 11.8, no triangle can be drawn that satisfies the given information. Side a is simply not long enough to reach the side opposite angle C.

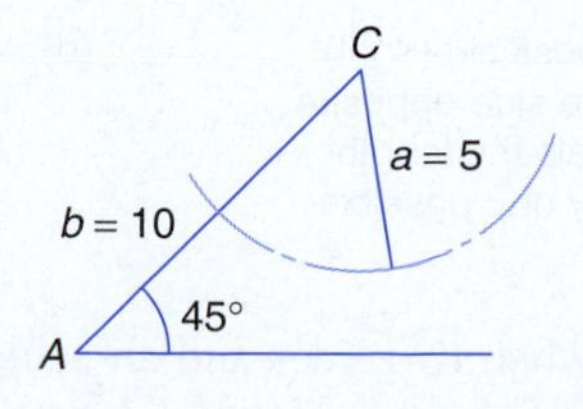

Figure 11.8

EXAMPLE 3

Construct a triangle given that $A = 60°$, $b = 6$, and $a = 10$.

As you can see from Fig. 11.9, only one triangle that satisfies the given information can be drawn. Side a is too long for two solutions.

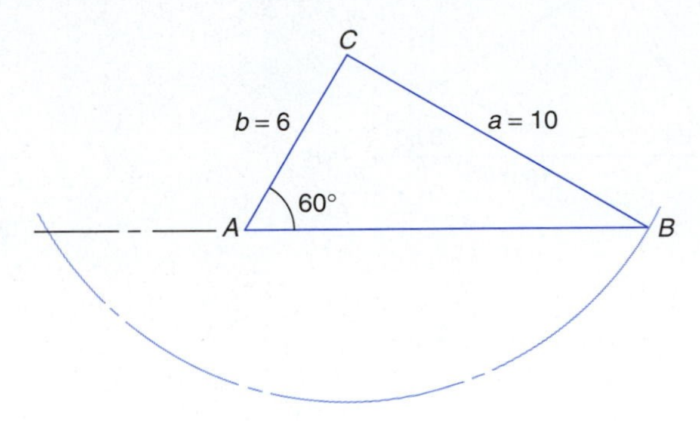

Figure 11.9

In summary, let's list the possible cases when two sides and an angle opposite one of the sides are given. Assume that *acute* angle A and adjacent side b are given. As a result of $h = b \sin A$, h is also determined. Depending on the length of the opposite side, a, we have the four cases shown in Fig. 11.10.

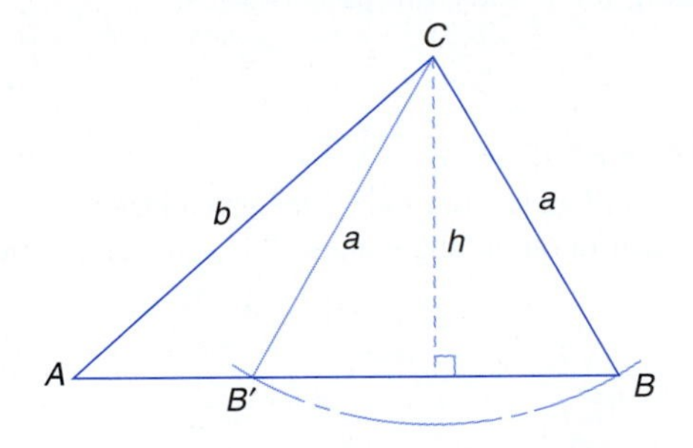

(a) When $h < a < b$, there are two possible triangles. In words, when the side opposite the given *acute* angle is less than the known adjacent side but greater than the altitude, there are two possible triangles.

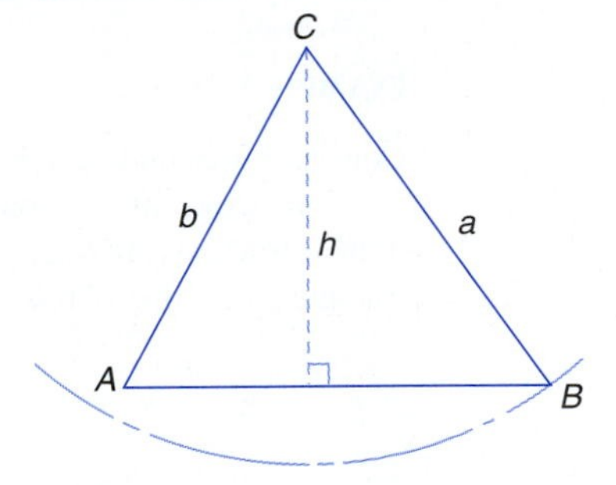

(b) When $h < b < a$, there is only one possible triangle. In words, when the side opposite the given *acute* angle is greater than the known adjacent side, there is only one possible triangle.

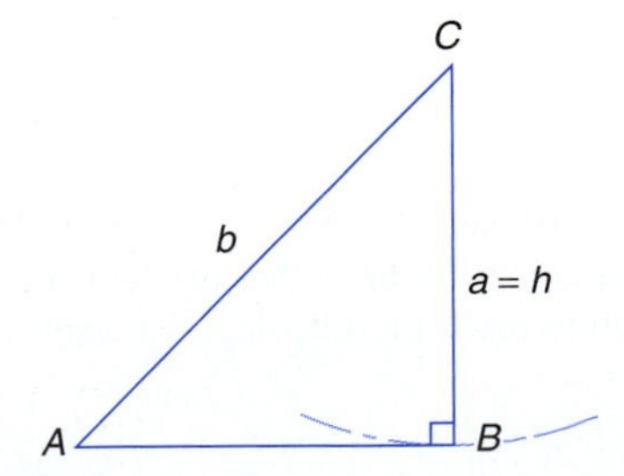

(c) When $a = h$, there is one possible (right) triangle. In words, when the side opposite the given *acute* angle equals the length of the altitude, there is only one possible (right) triangle.

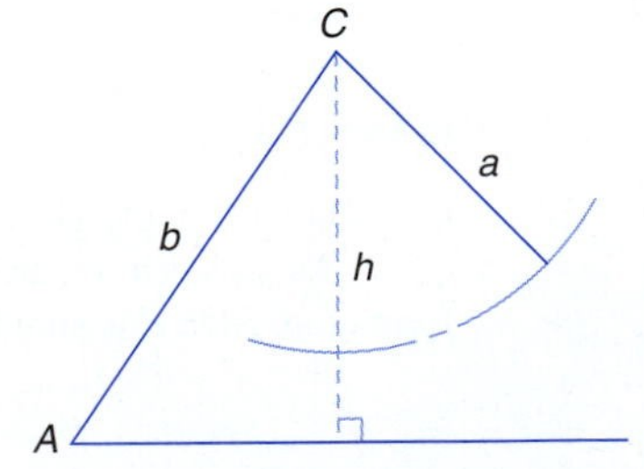

(d) When $a < h$, there is no possible triangle. In words, when the side opposite the given *acute* angle is less than the length of the altitude, there is no possible triangle.

Figure 11.10 Possible triangles when two sides and an acute angle opposite one of the sides are given.

If angle A is *obtuse,* we have two possible cases (Fig. 11.11).

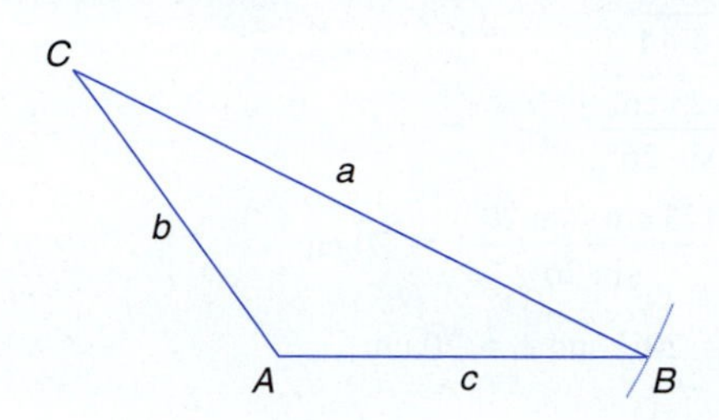

(a) When $a > b$, there is one possible triangle. In words, when the side opposite the given *obtuse* angle is greater than the known adjacent side, there is only one possible triangle.

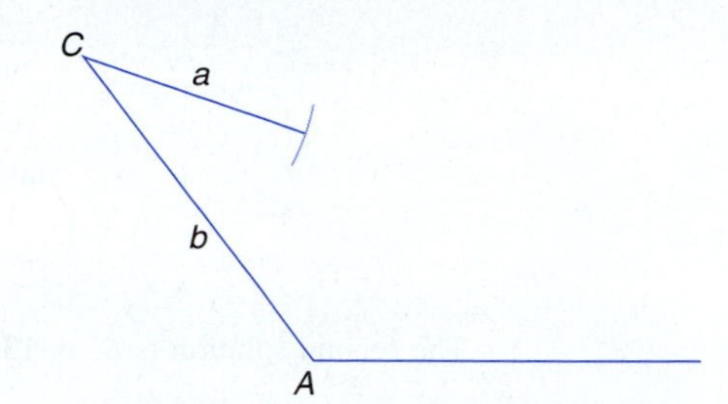

(b) When $a \le b$, there is no possible triangle. In words, when the side opposite the given *obtuse* angle is less than or equal to the known adjacent side, there is no possible triangle.

Figure 11.11 Possible triangles when two sides and an obtuse angle opposite one of the sides are given.

Note: If the given parts are not angle A, side opposite a, and side adjacent b as in our preceding discussions, then you must substitute the given angle and sides accordingly. This is why it is so important to understand the general word description corresponding to each case.

EXAMPLE 4

If $A = 26°$, $a = 25$ cm, and $b = 41$ cm, solve the triangle.

First, find h.

$$h = b \sin A = (41 \text{ cm})(\sin 26°) = 18 \text{ cm}$$

Since $h < a < b$, there are two solutions. First, let's find B in triangle ACB in Fig. 11.12.

$$\frac{a}{\sin A} = \frac{b}{\sin B}$$

$$\frac{25 \text{ cm}}{\sin 26°} = \frac{41 \text{ cm}}{\sin B}$$

$$\sin B = \frac{(41 \text{ cm})(\sin 26°)}{25 \text{ cm}} = 0.7189$$

$$B = 46°$$

$$C = 180° - 26° - 46° = 108°$$

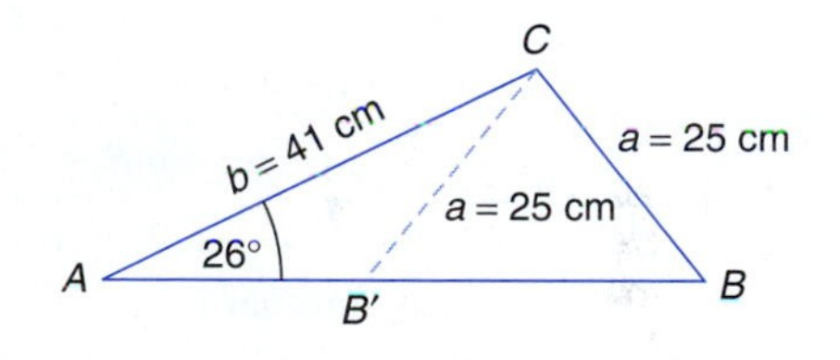

Figure 11.12

Find c.

$$\frac{c}{\sin C} = \frac{a}{\sin A}$$

$$\frac{c}{\sin 108°} = \frac{25 \text{ cm}}{\sin 26°}$$

$$c = \frac{(25 \text{ cm})(\sin 108°)}{\sin 26°} = 54 \text{ cm}$$

Therefore, the first solution is $B = 46°$, $C = 108°$, and $c = 54$ cm.

The second solution occurs when B is obtuse, as in triangle ACB'. That is, find the obtuse angle whose sine is 0.7189.

$$B' = 180° - 46° = 134°$$

Then, $C = 180° - 26° - 134° = 2\bar{0}°$.

For c,

$$\frac{c}{\sin C} = \frac{a}{\sin A}$$

$$\frac{c}{\sin 2\bar{0}^\circ} = \frac{25 \text{ cm}}{\sin 26^\circ}$$

$$c = \frac{(25 \text{ cm})(\sin 2\bar{0}^\circ)}{\sin 26^\circ} = 2\bar{0} \text{ cm}$$

The second solution is $B' = 134^\circ$, $C = 2\bar{0}^\circ$, and $c = 2\bar{0}$ cm.

Note: The triangles in Section 11.1 were carefully chosen so that they had only one solution.

EXAMPLE 5

If $A = 62.0^\circ$, $a = 415$ m, and $b = 855$ m, solve the triangle.

First, find h.

$$h = b \sin A$$

$$h = (855 \text{ m})(\sin 62.0^\circ)$$

$$= 755 \text{ m}$$

Since $a < h$, there is no possible solution. What would happen if you applied the law of sines anyway?

$$\frac{a}{\sin A} = \frac{b}{\sin B}$$

$$\frac{415 \text{ m}}{\sin 62.0^\circ} = \frac{855 \text{ m}}{\sin B}$$

$$\sin B = \frac{(855 \text{ m})(\sin 62.0^\circ)}{415 \text{ m}} = 1.819 \qquad \text{(Tilt!)}$$

Note: $\sin B = 1.819$ is impossible because $-1 \leq \sin B \leq 1$.

In summary,

1. **Given two angles and one side (AAS):** There is only one possible triangle.
2. **Given two sides and an angle opposite one of them (SSA):** There are three possibilities. If the side opposite the given angle is
   (a) greater than the known adjacent side, there is only one possible triangle.
   (b) less than the known adjacent side but greater than the altitude, there are two possible triangles.
   (c) less than the altitude, there is no possible triangle.

Since solving a general triangle requires several operations, errors are often introduced. The following points may be helpful in avoiding some of these errors:

1. Always choose a given value over a calculated value when doing calculations.
2. Always check your results to see that the largest angle is opposite the largest side and the smallest angle is opposite the smallest side.
3. Avoid finding the largest angle by the law of sines whenever possible because it is often not clear whether the resulting angle is acute or obtuse.

Exercises 11.2

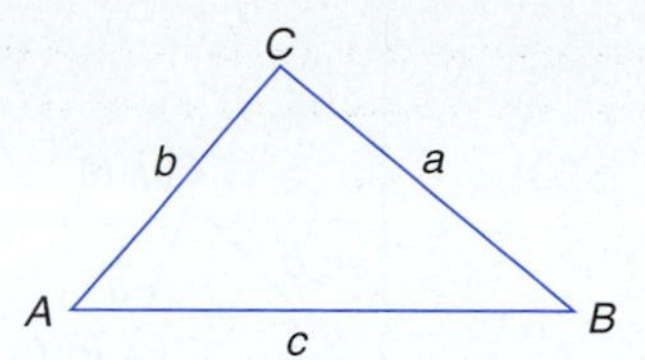

Figure 11.13

For each general triangle as in Fig. 11.13,
(a) determine the number of solutions, and
(b) solve the triangle, if possible.

Express the lengths of sides to three significant digits and the angles to the nearest tenth of a degree.

1. $A = 37.0°$, $a = 21.5$ cm, $b = 16.4$ cm
2. $B = 55.0°$, $b = 182$ m, $c = 203$ m
3. $C = 26.5°$, $c = 42.7$ km, $a = 47.2$ km
4. $B = 40.4°$, $b = 81.4$ m, $c = 144$ m
5. $A = 71.5°$, $a = 3.45$ m, $c = 3.50$ m
6. $C = 17.2°$, $c = 2.20$ m, $b = 2.00$ m
7. $B = 105.0°$, $b = 16.5$ mi, $a = 12.0$ mi
8. $A = 98.8°$, $a = 707$ ft, $b = 585$ ft

Express the lengths of sides to two significant digits and the angles to the nearest degree.

9. $C = 18°$, $c = 24$ mi, $a = 45$ mi
10. $B = 36°$, $b = 75$ cm, $a = 95$ cm
11. $C = \bar{6}0°$, $c = 150$ m, $b = 180$ m
12. $A = 3\bar{0}°$, $a = 4800$ ft, $c = 3600$ ft
13. $B = 8°$, $b = 450$ m, $c = 850$ m
14. $B = 45°$, $c = 2.5$ m, $b = 3.2$ m

Express the lengths of sides to four significant digits and the angles to the nearest hundredth of a degree.

15. $B = 41.50°$, $b = 14.25$ km, $a = 18.50$ km
16. $A = 15.75°$, $a = 642.5$ m, $c = 592.7$ m
17. $C = 63.85°$, $c = 29.50$ cm, $b = 38.75$ cm
18. $B = 50.00°$, $b = 41{,}250$ km, $c = 45{,}650$ km
19. $C = 8.75°$, $c = 89.30$ m, $a = 61.93$ m
20. $A = 31.50°$, $a = 375.0$ mm, $b = 405.5$ mm

Express the lengths of sides to four significant digits and the angles to the nearest second.

21. $B = 29°16'37''$, $b = 215.6$ m, $c = 304.5$ m
22. $A = 61°12'30''$, $a = 3457$ ft, $c = 2535$ ft
23. $C = 25°45''$, $a = 524.5$ ft, $c = 485.6$ ft
24. $A = 21°45'$, $a = 1785$ m, $b = 2025$ m

11.3 LAW OF COSINES

When the law of sines cannot be used, we use the **law of cosines.** In words, the square of any side of a triangle is equal to the sum of the squares of the other two sides minus twice the product of these two sides and the cosine of their included angle (see Fig. 11.14). By formula, the law is stated as follows:

LAW OF COSINES

$$a^2 = b^2 + c^2 - 2bc \cos A$$
$$b^2 = a^2 + c^2 - 2ac \cos B$$
$$c^2 = a^2 + b^2 - 2ab \cos C$$

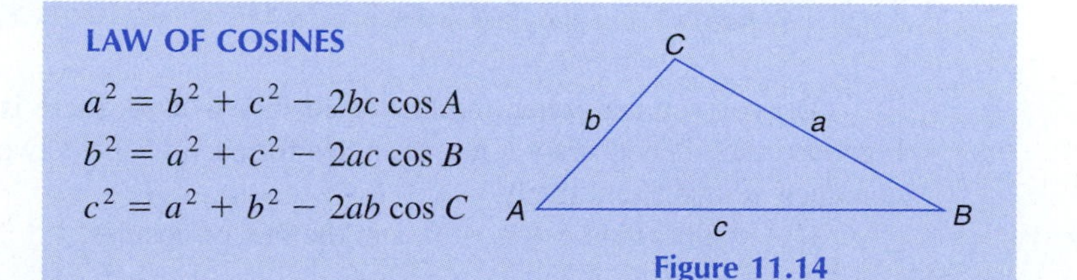

Figure 11.14

To derive the law of cosines, let triangle ABC be any general triangle. Figure 11.15 shows the three possible triangles or cases: point C between A and B, point C to the right of B, and point C to the left of A.

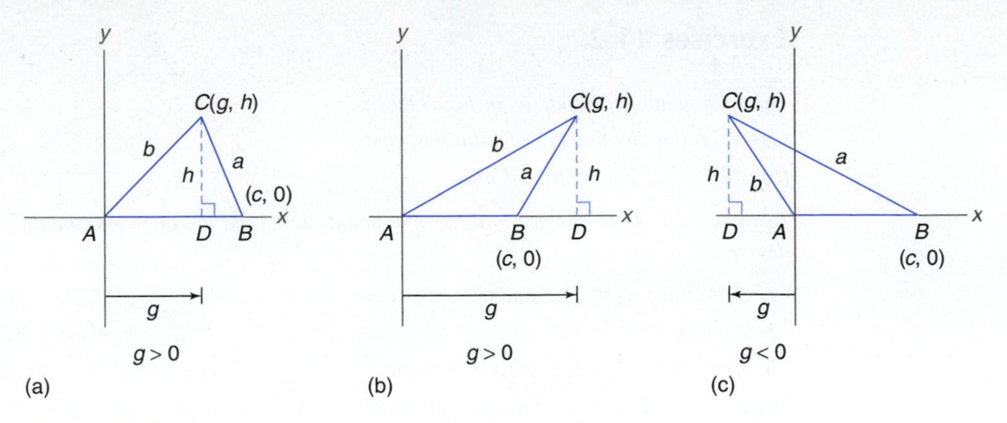

Figure 11.15 Draw altitude h from C to the opposite side on the x-axis in each triangle.

The length of side a in any of the three triangles in Fig. 11.15 is found using the formula for the distance between two points:

$$a = \sqrt{(g - c)^2 + (h - 0)^2}$$

$$a^2 = (g - c)^2 + (h - 0)^2 \qquad \text{(Square each side.)}$$

$$a^2 = g^2 - 2gc + c^2 + h^2 \qquad \textbf{(1)}$$

From Fig. 11.15, note that $b^2 = g^2 + h^2$, and make this substitution in Equation (1):

$$a^2 = b^2 + c^2 - 2gc \qquad \textbf{(2)}$$

Note also in each triangle that $\cos A = \frac{g}{b}$ or $g = b \cos A$. Now, make this substitution in Equation (2):

$$a^2 = b^2 + c^2 - 2bc \cos A$$

Note: If angle A is acute, $\cos A > 0$; if angle A is obtuse, $\cos A < 0$.

The other two forms of the law of cosines may be derived in a similar manner by relabeling the vertices.

There are two cases when the law of sines does not apply and we use the law of cosines to solve triangles:

1. Two sides and the included angle are known (SAS).
2. All three sides are known (SSS).

Do you see that when the law of cosines is used, there is no possibility of an ambiguous case? If not, draw a few triangles for each of these two cases (SAS and SSS) to convince yourself intuitively.

If $A = 90°$, then $\cos A = 0$, and the law of cosines

$$a^2 = b^2 + c^2 - 2bc \cos A$$

reduces to

$$a^2 = b^2 + c^2$$

which is the Pythagorean theorem. The Pythagorean theorem is thus a special case of the law of cosines.

EXAMPLE 1

If $a = 112$ m, $b = 135$ m, and $C = 104.3°$, solve the triangle.

First, draw a triangle as in Fig. 11.16 and find c by using the law of cosines.

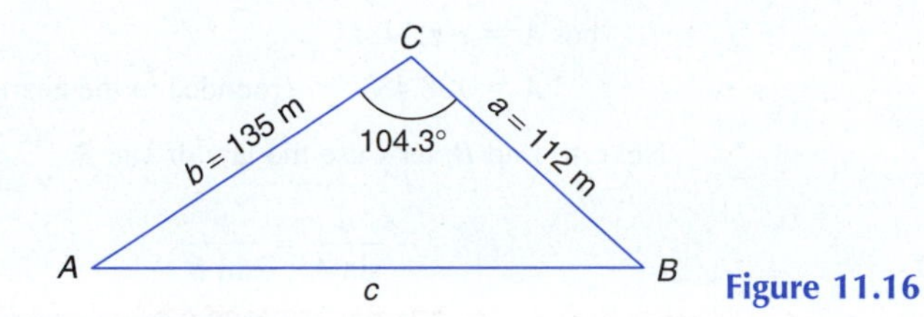

Figure 11.16

$$c^2 = a^2 + b^2 - 2ab \cos C$$
$$c^2 = (112 \text{ m})^2 + (135 \text{ m})^2 - 2(112 \text{ m})(135 \text{ m})(\cos 104.3°)$$
$$c = 196 \text{ m}$$

This side may be found using a calculator as follows:

2nd multiplication sign 112^2+135^2-2*112*135 **2nd COS** 104.3)) **ENTER**

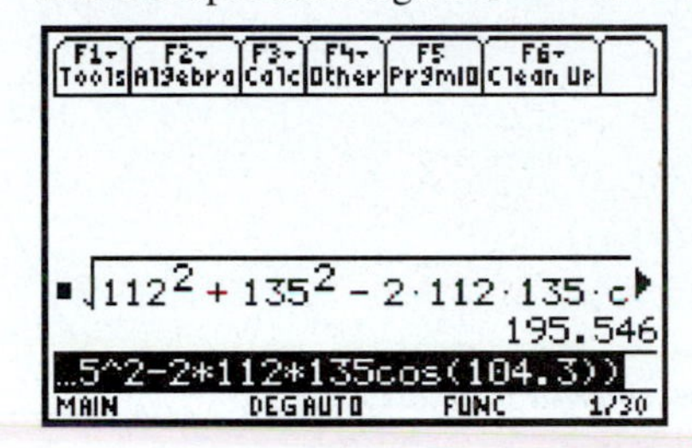

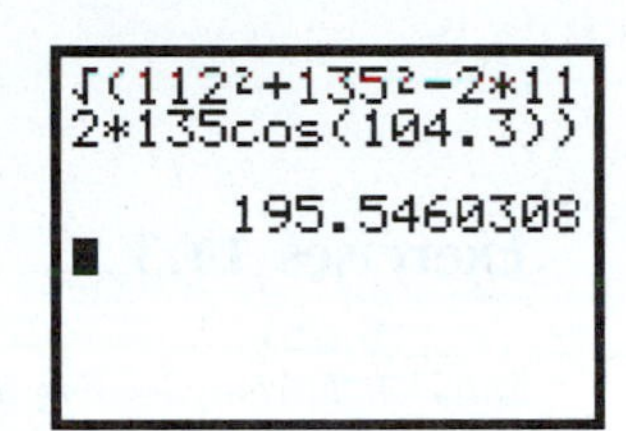

2nd $\mathbf{x^2}$ 112 $\mathbf{x^2}$ +135 $\mathbf{x^2}$ -2*112*135 **COS** 104.3)) **ENTER**

So, $c = 196$ m rounded to three significant digits.

To find A, use the law of sines since it requires less computation.

$$\frac{a}{\sin A} = \frac{c}{\sin C}$$
$$\frac{112 \text{ m}}{\sin A} = \frac{196 \text{ m}}{\sin 104.3°}$$
$$\sin A = \frac{(112 \text{ m})(\sin 104.3°)}{196 \text{ m}} = 0.5537$$
$$A = 33.6°$$
$$B = 180° - 104.3° - 33.6° = 42.1°$$

The solution is $A = 33.6°$, $B = 42.1°$, and $c = 196$ m.

EXAMPLE 2

If $a = 375.0$ ft, $b = 282.0$ ft, and $c = 114.0$ ft, solve the triangle.

First, draw a triangle as in Fig. 11.17 and find A by using the law of cosines.

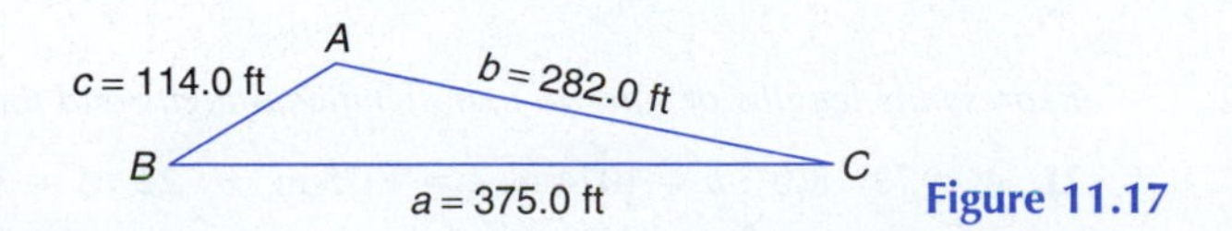

Figure 11.17

$$a^2 = b^2 + c^2 - 2bc \cos A$$

$$(375.0 \text{ ft})^2 = (282.0 \text{ ft})^2 + (114.0 \text{ ft})^2 - 2(282.0 \text{ ft})(114.0 \text{ ft}) \cos A$$

$$\cos A = \frac{(375.0 \text{ ft})^2 - (282.0 \text{ ft})^2 - (114.0 \text{ ft})^2}{-2(282.0 \text{ ft})(114.0 \text{ ft})}$$

$$\cos A = -0.7482$$

$$A = 138.43° \qquad \text{(rounded to the nearest hundredth of a degree)}$$

Next, to find B, let's use the law of sines.

$$\frac{a}{\sin A} = \frac{b}{\sin B}$$

$$\frac{375.0 \text{ ft}}{\sin 138.43°} = \frac{282.0 \text{ ft}}{\sin B}$$

$$\sin B = \frac{(282.0 \text{ ft})(\sin 138.43°)}{375.0 \text{ ft}} = 0.4990$$

$$B = 29.93°$$

$$C = 180° - 138.43° - 29.93° = 11.64°$$

The solution is $A = 138.43°$, $B = 29.93°$, and $C = 11.64°$.

Exercises 11.3

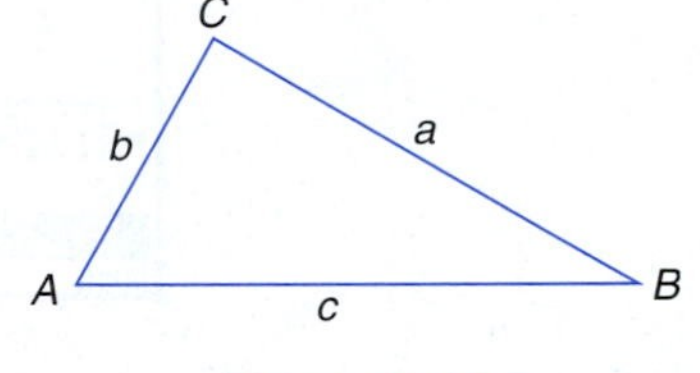

Figure 11.18

Solve each triangle using the labels shown in Fig. 11.18.

Express the lengths of sides to three significant digits and the angles to the nearest tenth of a degree.

1. $A = 60.0°$, $b = 19.5$ m, $c = 25.0$ m

2. $B = 19.5°$, $a = 21.5$ ft, $c = 12.5$ ft

3. $C = 109.0°$, $a = 14\bar{0}$ km, $b = 215$ km

4. $A = 94.7°$, $c = 875$ yd, $b = 185$ yd

5. $a = 19.2$ m, $b = 21.3$ m, $c = 27.2$ m

6. $a = 125$ km, $b = 195$ km, $c = 145$ km

7. $a = 4.25$ ft, $b = 7.75$ ft, $c = 5.50$ ft

8. $a = 3590$ m, $b = 7950$ m, $c = 4650$ m

Express the lengths of sides to two significant digits and the angles to the nearest degree.

9. $A = 45°$, $b = 51$ m, $c = 39$ m

10. $B = 6\bar{0}°$, $a = 160$ cm, $c = 230$ cm

11. $a = 7\bar{0}00$ m, $b = 5600$ m, $c = 4800$ m

12. $a = 5.8$ cm, $b = 5.8$ cm, $c = 9.6$ cm

13. $C = 135°$, $a = 36$ ft, $b = 48$ ft

14. $A = 5°$, $b = 19$ m, $c = 25$ m

Express the lengths of sides to four significant digits and the angles to the nearest hundredth of a degree.

15. $B = 19.25°$, $a = 4815$ m, $c = 1925$ m

16. $C = 75.00°$, $a = 37{,}550$ mi, $b = 45{,}250$ mi

17. $C = 108.75°$, $a = 405.0$ mm, $b = 325.0$ mm

18. $A = 111.05°$, $b = 1976$ ft, $c = 325\bar{0}$ ft

19. $a = 207.5$ km, $b = 105.6$ km, $c = 141.5$ km

20. $a = 19.45$ m, $b = 36.50$ m, $c = 25.60$ m

Express the lengths of sides to four significant digits and the angles to the nearest second.

21. $A = 72°18'0''$, $b = 1074$ m, $c = 1375$ m

22. $C = 101°25'30''$, $a = 685.0$ ft, $b = 515.0$ ft

23. $a = 1.250$ mi, $b = 1.975$ mi, $c = 1.250$ mi

24. $a = 375.1$ m, $b = 286.0$ m, $c = 305.0$ m

Find the indicated lengths of sides and angles consistent with the rules for calculations with measurements.

25. **(a)** $\angle Q$
(b) $\angle S$

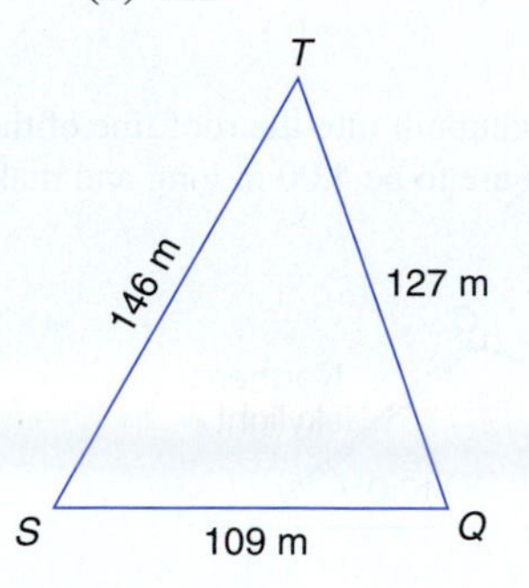

26. **(a)** side PQ
(b) $\angle Q$

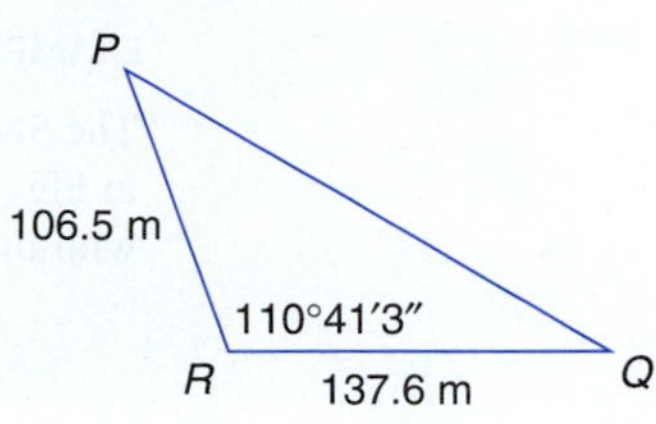

27. **(a)** side SM
(b) $\angle MST$

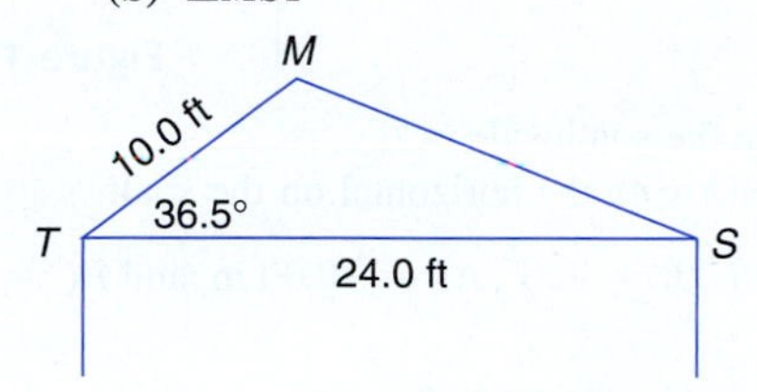

28. **(a)** side QS
(b) $\angle RSQ$

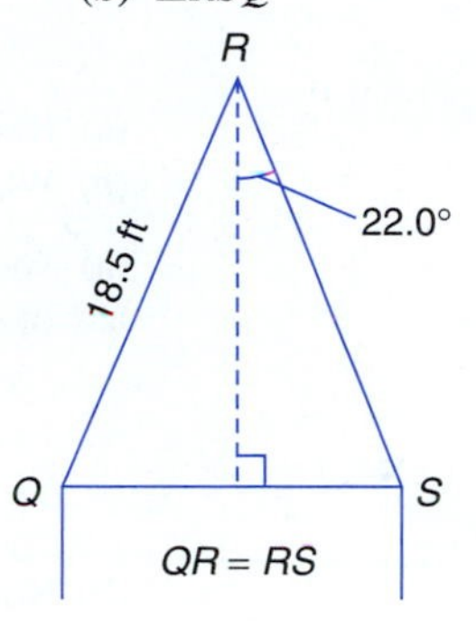

11.4 APPLICATIONS OF OBLIQUE TRIANGLES

In Section 3.4, we saw several applications of right triangles. Similarly, there are many applications of oblique, or general, triangles in the world around us. The following examples and exercises illustrate some of these applications.

EXAMPLE 1

The ground of a subdivision lot slopes upward from the street at an angle of 8°. The builder wants to build a house on a level lot that is $6\bar{0}$ ft deep. To control erosion, the back of the lot must be cut to have a slope of 25°. How far from the street, measured along the present slope, will the excavation extend?

First, draw a diagram as in Fig. 11.19 and find AC. In $\triangle ABC$, $A = 8°$ and

$$\angle ABC = 180° - 25° = 155°$$

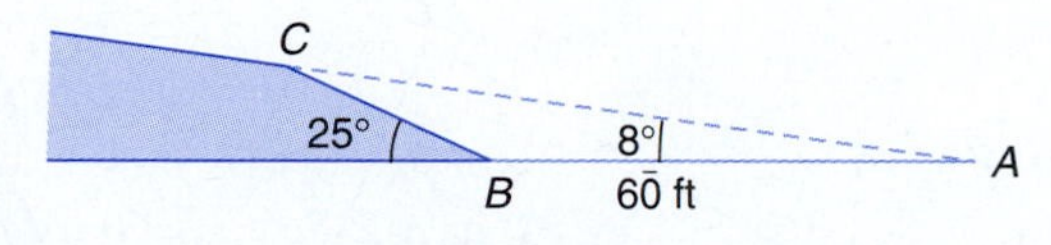

Figure 11.19

Thus, $C = 180° - 8° - 155° = 17°$.

Using the law of sines, we have

$$\frac{AC}{\sin(\angle ABC)} = \frac{AB}{\sin C}$$

$$\frac{AC}{\sin 155°} = \frac{6\bar{0} \text{ ft}}{\sin 17°}$$

$$AC = \frac{(\sin 155°)(6\bar{0} \text{ ft})}{\sin 17°} = 87 \text{ ft}$$

EXAMPLE 2

The Sanchez family wants a northern skylight built into the roof line of their house, as shown in Fig. 11.20. The windows for the skylight are to be 4.00 m long and make an angle of 40.0° with the horizontal.

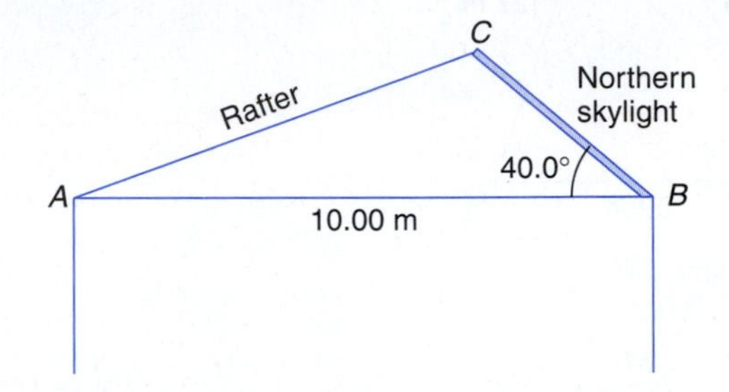

Figure 11.20

(a) How long must the rafters on the south side be?
(b) What angle does the roof make with the horizontal on the south?

(a) We need to find AC. In $\triangle ABC$, $B = 40.0°$, $AB = 10.00$ m, and $BC = 4.00$ m. Using the law of cosines, we have

$$(AC)^2 = (AB)^2 + (BC)^2 - 2(AB)(BC)\cos B$$
$$(AC)^2 = (10.00 \text{ m})^2 + (4.00 \text{ m})^2 - 2(10.00 \text{ m})(4.00 \text{ m})\cos 40.0°$$
$$AC = 7.40 \text{ m}$$

(b) Now find A, using the law of sines.

$$\frac{BC}{\sin A} = \frac{AC}{\sin B}$$

$$\frac{4.00 \text{ m}}{\sin A} = \frac{7.40 \text{ m}}{\sin 40.0°}$$

$$\sin A = \frac{(\sin 40.0°)(4.00 \text{ m})}{7.40 \text{ m}} = 0.3475$$

$$A = 20.3°$$

Exercises 11.4

1. Find the lengths of rafters AC and BC for the roof in Fig. 11.21.
2. The sides of a triangular metal plate measure 17.5 in., 12.3 in., and 21.3 in. Find the measure of the largest angle.
3. Find the distance c across the pond in Fig. 11.22.

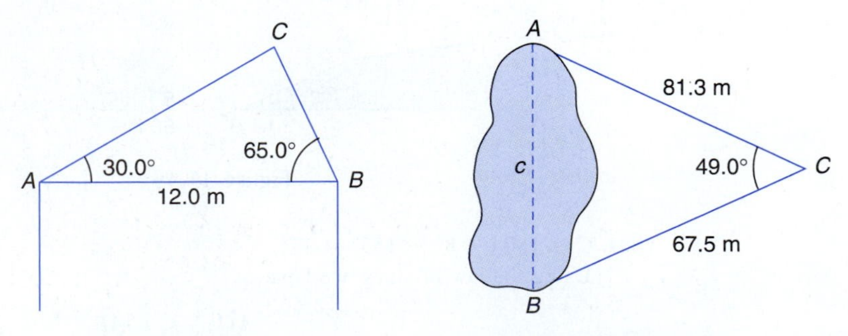

Figure 11.21

Figure 11.22

4. A surveyor on the side of a hill that makes an angle of 25.0° with the horizontal measures the angle of depression to the top of a tree at the bottom of the hill as 10.0°. Find the height of the tree if its base is 90.0 ft from the surveyor.

5. To find the distance between two points A and B lying on opposite banks of a river, a length AC of $3\bar{0}0$ m is measured. Angle BAC measures 58° and $\angle ACB$ measures 49°. Find the distance between A and B.

6. A weather balloon is sighted between points A and B which are 3.6 mi apart on level ground. The angle of elevation of the balloon from A is 28° and its angle of elevation from B is 49°. Find the height (in feet) of the balloon above the ground.

7. A forest ranger in observation tower A sights a fire in a direction 51.5° east of north. Another ranger in tower B 5.00 mi due east of A sights the fire at 17.2° west of north. How far is the fire from each observation tower?

8. The angle at one corner of a triangular plot of ground measures 65.5°. If the sides that meet at this corner measure 225 m and $32\bar{0}$ m, what is the length of the third side?

9. Two automobiles depart at the same time from the intersection of two straight highways, which intersect at 75°. If the automobiles' speeds are $6\bar{0}$ mph and 45 mph, how far apart will they be after 1.5 h?

10. A vertical cable television tower as in Fig. 11.23 is standing on the side of a hill and makes an angle of 25.0° with the horizontal. Guy wires are attached $16\bar{0}$ ft up the tower. What lengths of guy wires are needed to reach points 50.0 ft uphill and 80.0 ft downhill from the base of the tower?

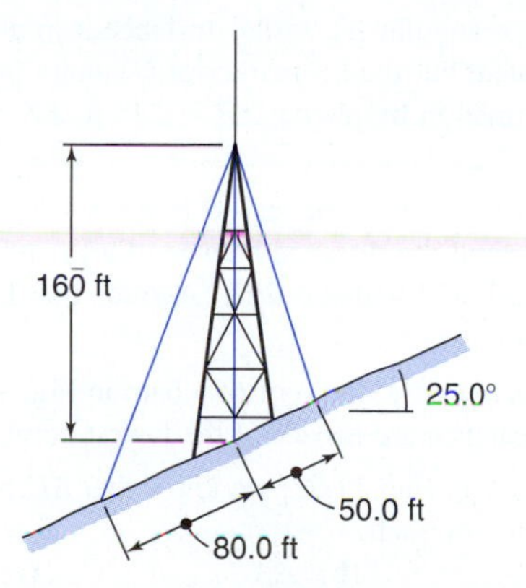

Figure 11.23

11. From point A the angle of elevation to the top of a cliff is 37° as in Fig. 11.24. On level ground the angle of elevation from point B to the top of the cliff is 24°. Point A is 270 ft closer to the cliff than point B. Find the height of the cliff.

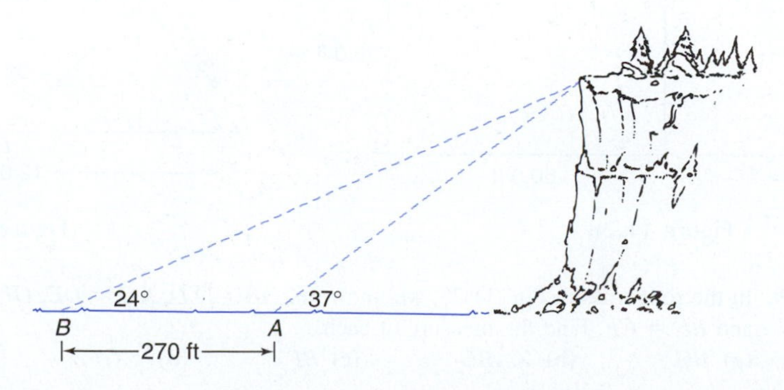

Figure 11.24

12. A lighthouse $15\overline{0}$ ft high is on the edge of a vertical cliff overlooking the ocean. The angle of elevation from a ship to the bottom of the lighthouse measures 15°. The angle of elevation from the ship to the top of the lighthouse measures 24°.
(a) Find the distance from the ship to the cliff.
(b) Find the height of the cliff.

13. A tower 75 m high is on a vertical cliff on the bank of a river. From the top of the tower, the angle of depression to a piece of driftwood on the opposite bank of the river is 28°. From the bottom of the tower, the angle of depression to the same piece of driftwood is 19°.
(a) Find the width of the river.
(b) Find the height of the cliff.

14. Find the lengths L_1 and L_2 of the rafters in Fig. 11.25.

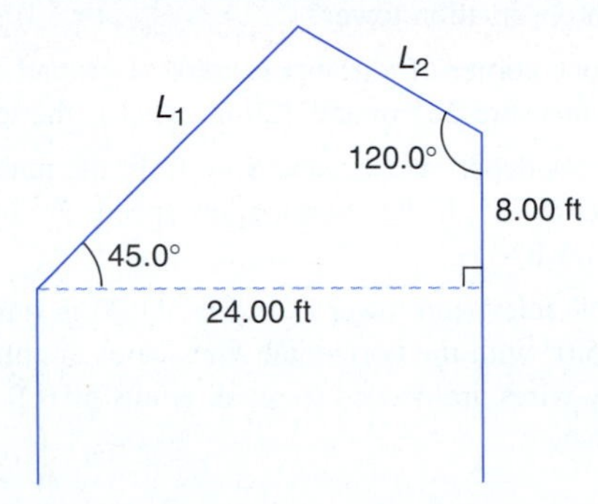

Figure 11.25

15. An owner of a triangular lot wishes to fence it in along the lot lines. Lot markers at A and B have been located, but the lot marker at C cannot be located. The owner's attorney gives the following information by phone: $AB = 245$ ft, $BC = 185$ ft, and $A = 35.0°$. What is the length of AC?

16. The average distance from the sun to Earth is 1.5×10^8 km and from the sun to Venus is 1.1×10^8 km. Find the distance between Earth and Venus when the angle between Earth and the sun and Earth and Venus is 25°. (Assume that Earth and Venus have circular orbits around the sun.)

17. A farmer wants to extend the roof of a barn in Fig. 11.26 to add a storage area for machinery. If 20.0 ft of clearance are needed at the lowest point, how wide can the addition be?

18. In the framework in Fig. 11.27, we know that $AE = CD$, $AB = BC$, $BD = BE$, and $AC \parallel ED$. Find the measure of each.
(a) $\angle AEB$ **(b)** $\angle A$ **(c)** BE **(d)** DE

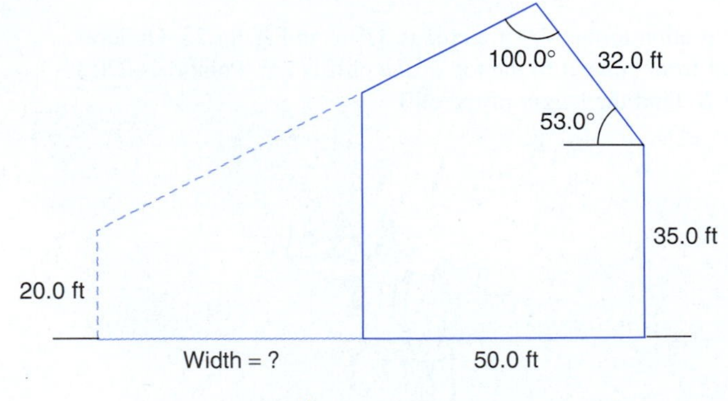

Figure 11.26

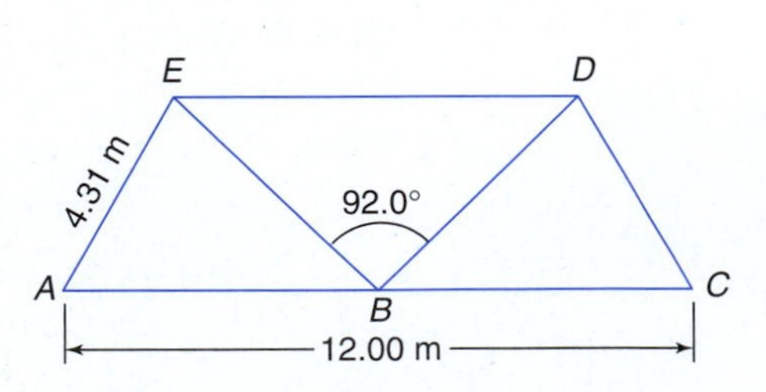

Figure 11.27

19. In the roof truss in Fig. 11.28, we know that $AB = CD$, $AG = DE$, $GF = FE$, $BG = CE$, and $BF = CF$. Find the measure of each.
(a) BG **(b)** $\angle ABG$ **(c)** BF **(d)** $\angle GFB$ **(e)** $\angle FBC$

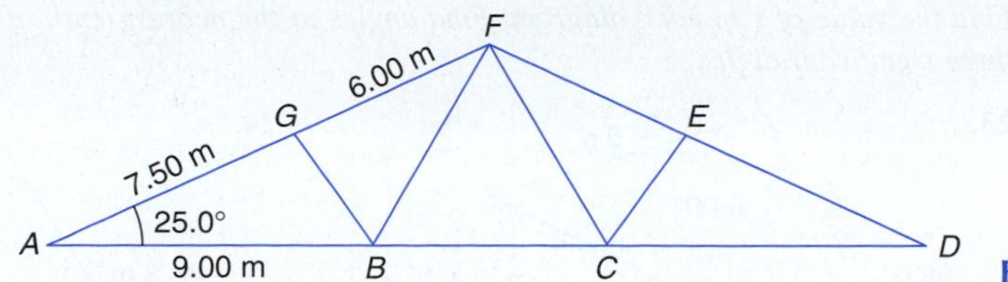

Figure 11.28

20. Find the distance between the peaks of two hills across the gorge in Fig. 11.29. Points A and B are trees on the peaks of the two hills. Point C is where you stand. Measure a length of $10\bar{0}$ m from point C to point D. Then $\angle BCD$ measures 115.0°, $\angle CDA$ measures 120.5°, $\angle BCA$ measures 86.5°, and $\angle ADB$ measures 98.1°.

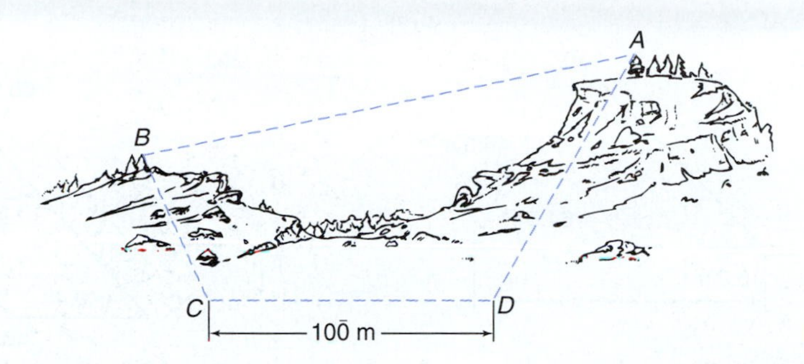

Figure 11.29

21. A deck is to be built on the house in Fig. 11.30. Find the length l of the stairway.

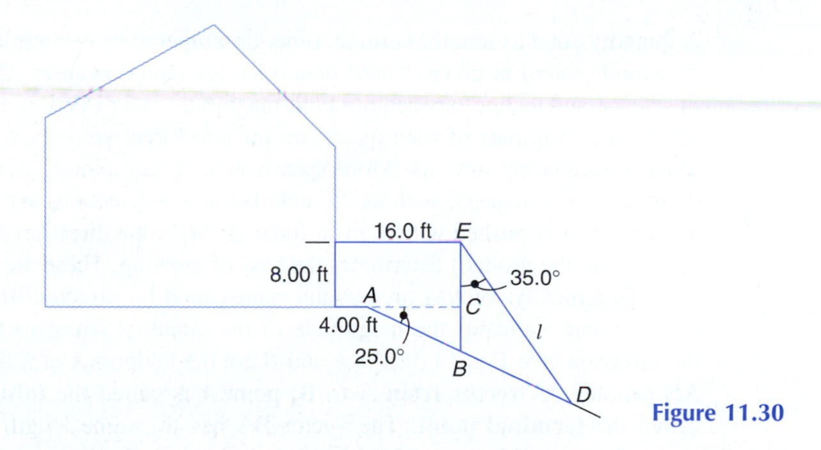

Figure 11.30

22. A dental office building in the shape of a regular pentagon is planned as in Fig. 11.31. The central records department is located at point A. The various offices receive the dental records by a conveyor system.

(a) Find the length of the conveyor from A to D.

(b) Find the length of the conveyor from A to N, the midpoint of side BC.

(c) Find the length of the conveyor from A to M, the midpoint of side CD.

Note: Each interior angle of a regular pentagon is 108°.

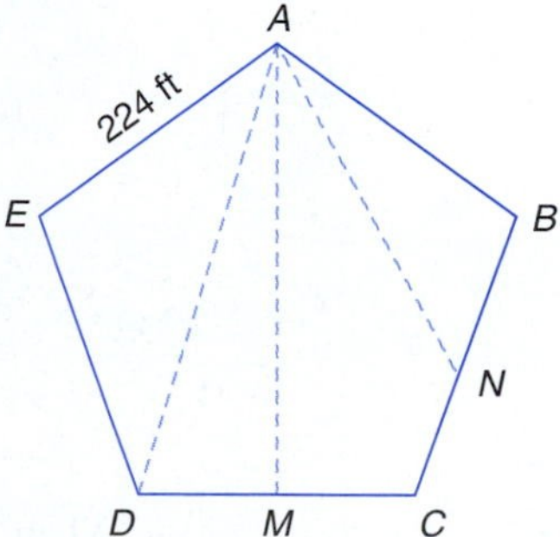

Figure 11.31

Find the value of x in each diagram. Find angles to the nearest tenth of a degree and sides to three significant digits.

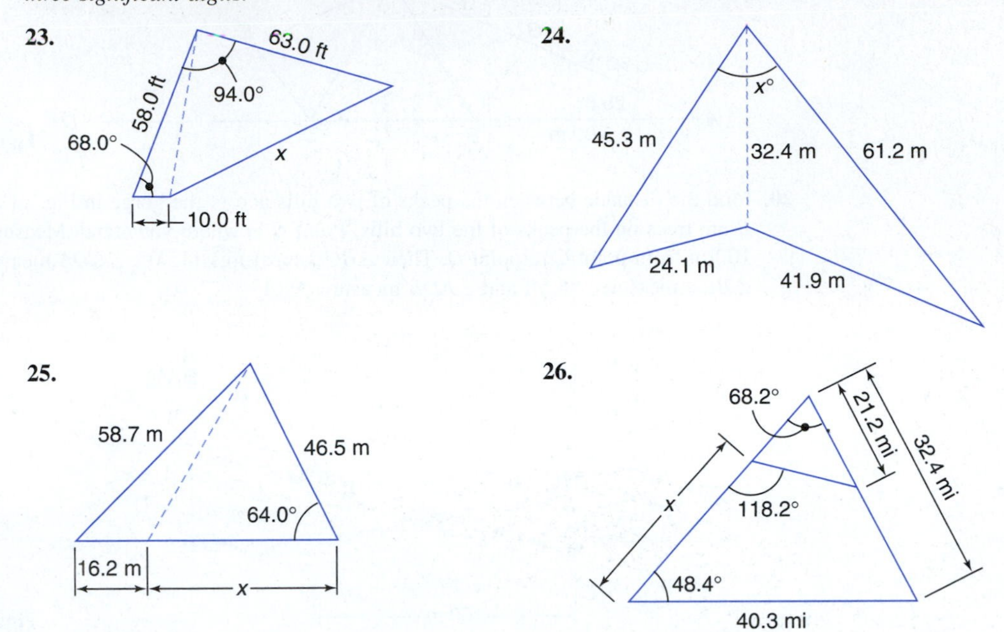

11.5 ADDITION OF VECTORS: GRAPHICAL METHODS

A quantity such as length, volume, time, or temperature is completely described when its magnitude (size) is given. These quantities are called **scalars.** Other physical quantities in science and technology require both magnitude and direction in order to be completely described. Examples of such quantities include force, velocity, torque, and certain quantities from electric circuits. More specifically, to completely describe wind velocity requires not only a speed, such as 25 mph, but also a direction, say 20° east of north. When a lawnmower is pushed with a given force of 20 lb, the direction of the force (angle of the handle with the ground) determines the ease of mowing. These are called **vector** quantities.

Graphically, vectors are usually represented by directed line segments. The length of a segment indicates the magnitude of the quantity. An arrowhead is used to indicate the direction (see Fig. 11.32). If *A* and *B* are the endpoints of a line segment, the symbol **AB** denotes the **vector from A to B;** point *A* is called the **initial point** and point *B* is called the **terminal point.** The vector **BA** has the same length as **AB** but has the opposite direction. Vectors may also be denoted by a single lowercase letter, such as **v, u,** or **w.** The magnitude or length of a vector **v** is denoted as $|\mathbf{v}|$.

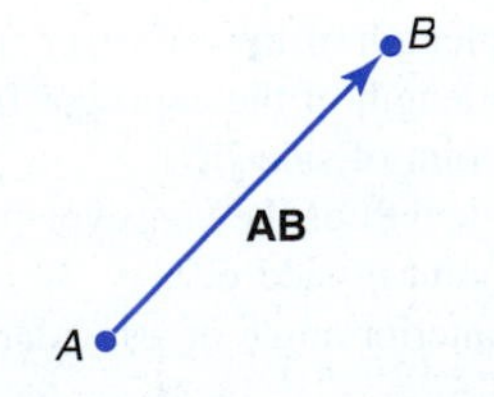

Figure 11.32 Vector **AB**.

When writing vectors on paper or a chalkboard, we use a small arrow above the vector quantity, such as $\vec{v}, \vec{R}$, or $\overrightarrow{AB}$ as in Fig. 11.33.

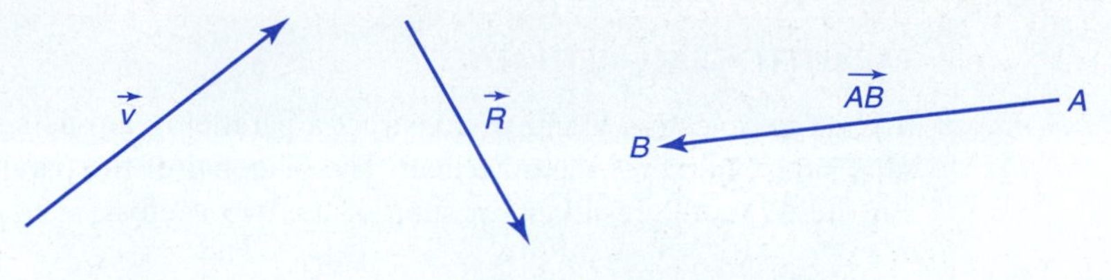

Figure 11.33

Two vectors are equal when they have the same magnitude and the same direction. Two vectors are negatives of each other when they have the same magnitude but opposite directions (see Fig. 11.34).

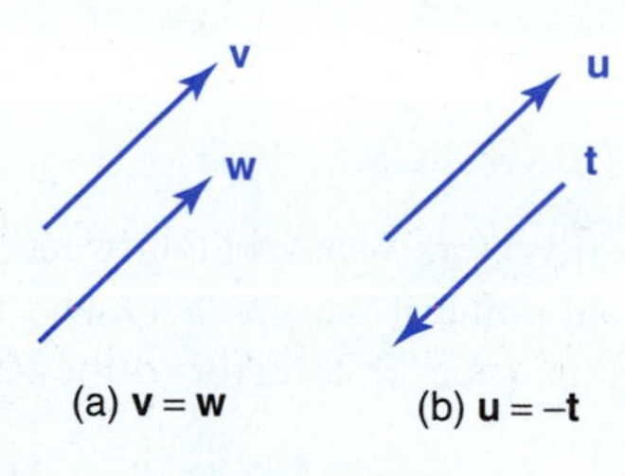

Figure 11.34

Note that a given vector **v** may be placed in any position as long as its magnitude and direction are not changed. Such vectors are called *free vectors* (see Fig. 11.35).

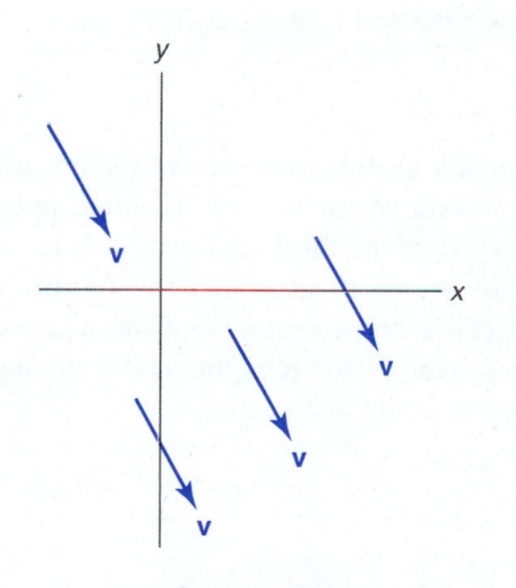

Figure 11.35

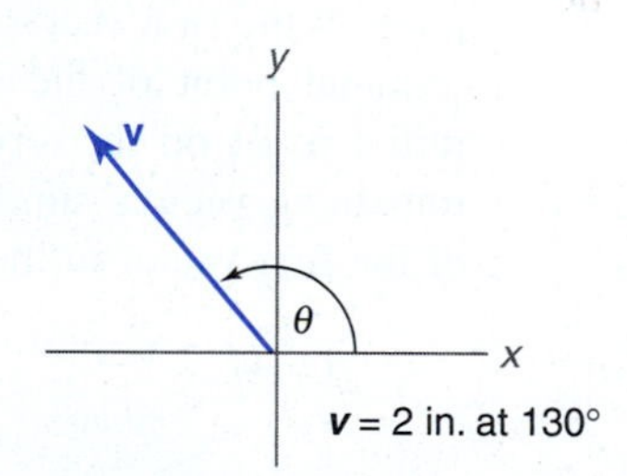

Figure 11.36 A vector in standard position.

A vector is in **standard position** when its initial point is at the origin of the xy-coordinate system. A vector in standard position is expressed in terms of its length and angle θ, where θ is measured counterclockwise from the positive x-axis to the vector. The vector in Fig. 11.36 is expressed in standard position.

The sum of two or more vectors is called the **resultant.** This sum may be obtained graphically using one of two methods: the parallelogram method or the vector triangle method, defined as follows.

PARALLELOGRAM METHOD

To add two vectors **v** and **w**, construct a parallelogram using **v** as one pair of parallel sides and **w** for the other pair. The diagonal of the parallelogram, as shown in Fig. 11.37, is the resultant, or sum, of the two vectors.

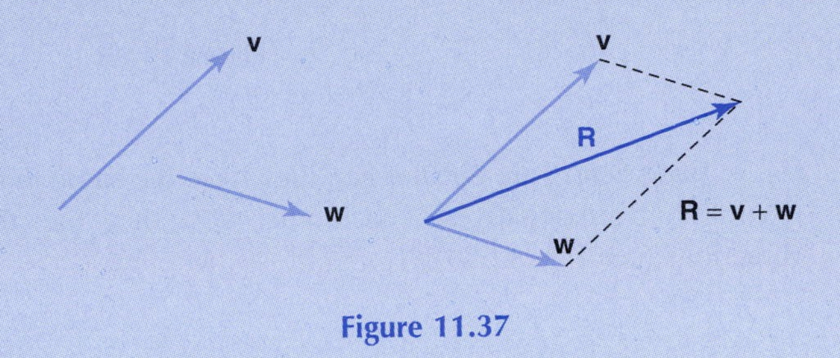

Figure 11.37

VECTOR TRIANGLE METHOD

To add two vectors **v** and **w**, construct the second vector **w** with its initial point on the terminal point of the first vector **v**. The resultant vector is the vector joining the initial point of the first vector to the terminal point of the second vector (see Fig. 11.38).

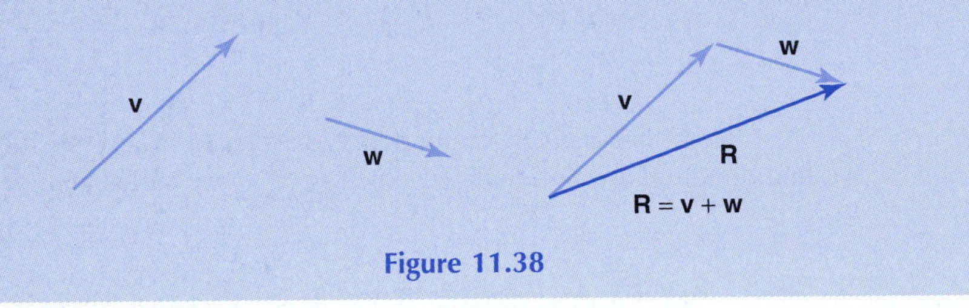

Figure 11.38

The triangle method is particularly useful when several vectors are to be added (see Fig. 11.39). That is, construct the second vector $\mathbf{v}_2$ with its initial point on the terminal point of the first vector $\mathbf{v}_1$. Then, construct the third vector $\mathbf{v}_3$ with its initial point on the terminal point of the second vector $\mathbf{v}_2$. Next, construct the fourth vector $\mathbf{v}_4$ with its initial point on the terminal point of the third vector $\mathbf{v}_3$. Continue constructing all the remaining vectors similarly. The resultant vector **R** is the vector joining the initial point of the first vector to the terminal point of the last vector.

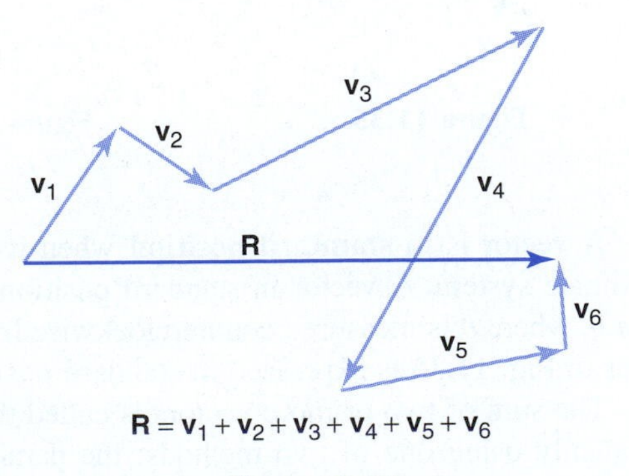

Figure 11.39

A vector may be subtracted by adding its negative. That is, $\mathbf{v} - \mathbf{w} = \mathbf{v} + (-\mathbf{w})$. Construct **v** as usual, construct the negative of **w,** and find the resultant as in Fig. 11.40.

Suppose a friend offers to fly you from Parkland to Kampsville and asks you how to get there. You reply, "It is 250 km." Have you given the friend enough information to find Kampsville? Obviously not! You must also tell him or her in what direction to go. If you reply, "Go 250 km due west," the friend can find Kampsville. This change in position is represented by the vector **PK** in Fig. 11.41.

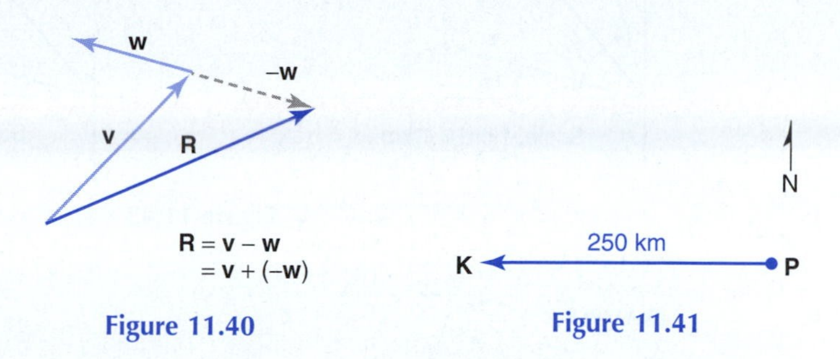

Figure 11.40

Figure 11.41

Perhaps the simplest vector is **displacement,** which is the net change in position. Displacement is a vector because it requires both a magnitude and a direction for its complete description.

Suppose your friend needs to stop at Hillsfield on the way to Kampsville, as shown in the flight plan in Fig. 11.42. Which is the displacement vector now? Since displacement is the net change in position, the displacement is the shortest distance between the beginning point and the ending point, or vector **PK.** The displacement vector is the same no matter which route is taken. This second situation may be expressed by vectors as follows:

$$\mathbf{PH} + \mathbf{HK} = \mathbf{PK}$$

This does not mean the distances are the same. In fact, we know that

$$|\mathbf{PH}| + |\mathbf{HK}| > |\mathbf{PK}|$$

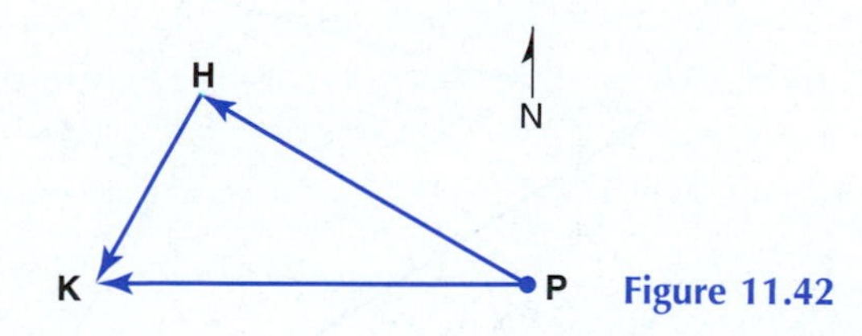

Figure 11.42

EXAMPLE 1

Find the vector sum of the following two vectors, each given in standard position, by using each graphic method:

$$\mathbf{v} = 2.4 \text{ km at } 15^\circ$$
$$\mathbf{w} = 3.3 \text{ km at } 75^\circ$$

Choose a suitable scale as in Fig. 11.43. Using ruler and protractor for measuring, we find that

$$\mathbf{v} + \mathbf{w} = \mathbf{R} = 5.0 \text{ km at } 5\overline{0}^\circ$$

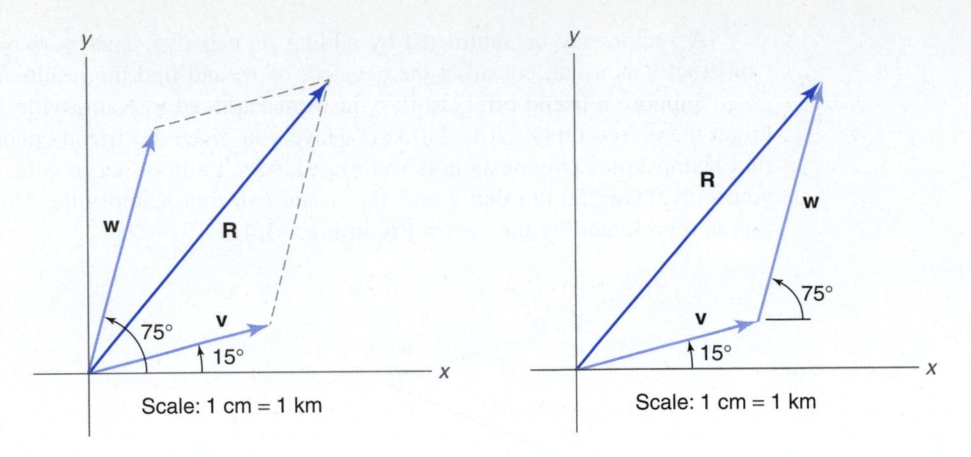

Figure 11.43

EXAMPLE 2

Given the following three vectors, each in standard position, find the vector sum graphically:

$$\mathbf{v}_1 = 38 \text{ mi at } 12\bar{0}°$$
$$\mathbf{v}_2 = 48 \text{ mi at } 35°$$
$$\mathbf{v}_3 = 82 \text{ mi at } 195°$$

Choose a suitable scale as in Fig. 11.44. Using ruler and protractor for measuring, we find that

$$\mathbf{v}_1 + \mathbf{v}_2 + \mathbf{v}_3 = \mathbf{R} = 71 \text{ mi at } 146°$$

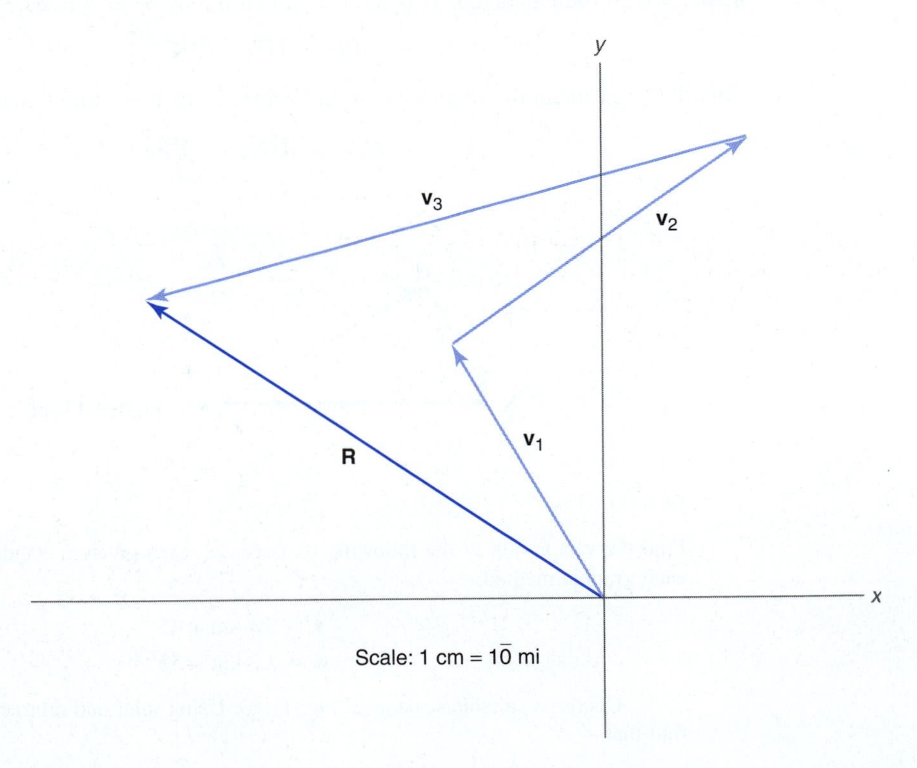

Figure 11.44

EXAMPLE 3

An airplane is flying at 250 mi/h (air speed) on a compass heading of $4\bar{0}°$ east of north. A wind of $5\bar{0}$ mi/h is blowing from the north. What is the true course (the heading with respect to the ground) of the airplane and what is its ground speed?

Let's use the scale 1 cm = $5\bar{0}$ mi/h as in Fig. 11.45. Let

$$\left.\begin{aligned}\mathbf{v} &= 2\bar{5}0 \text{ mi/h at } 5\bar{0}° \\ \mathbf{w} &= 5\bar{0} \text{ mi/h at } 27\bar{0}°\end{aligned}\right\} \quad \text{(standard position)}$$

By measuring the length of **R** with a ruler, we find

$$4.2 \text{ cm} \times \frac{5\bar{0} \text{ mi/h}}{\text{cm}} = 210 \text{ mi/h}$$

Now use a protractor to measure the angle that **R** makes with the north line, 51°. Therefore,

$$\begin{aligned}\mathbf{v} + \mathbf{w} = \mathbf{R} &= 210 \text{ mi/h at } 51° \text{ east of north} \\ &= 210 \text{ mi/h at } 39° \text{ (in standard position)}\end{aligned}$$

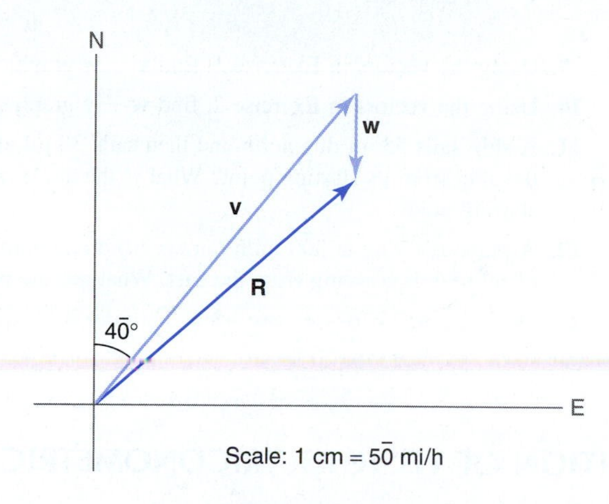

Figure 11.45

Velocity may be defined as the time rate of change of displacement. Thus, velocity is a vector, and both its magnitude (speed) and its direction are required for its complete description.

Bearing is another system of measuring angles as in Fig. 11.46. Angles are measured *clockwise* from north. An angle with bearing 210° is shown in Fig. 11.46(b). To avoid confusion, we shall consistently refer to angles in standard position.

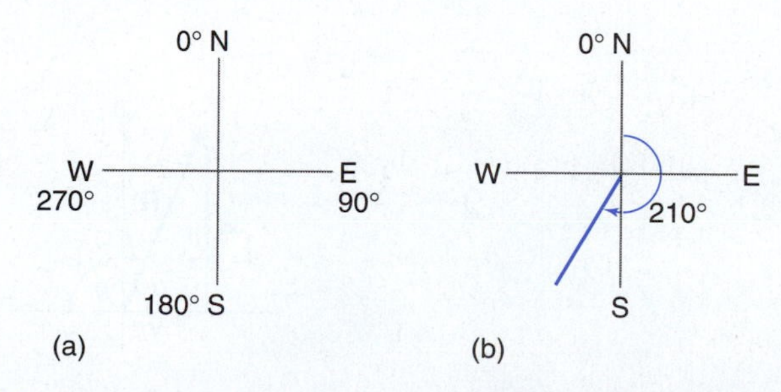

Figure 11.46 Bearing is a system for measuring angles relative to the compass.

Exercises 11.5

(Angles are given in standard position unless stated otherwise.)

Find the sum of each set of vectors by using the parallelogram method. Choose a suitable scale.

1. $\mathbf{v}$ = 4.5 km at 67°
$\mathbf{w}$ = 6.5 km at 105°

2. $\mathbf{v}$ = 150 mi at 175°
$\mathbf{w}$ = 270 mi at 215°

3. $\mathbf{v}$ = 75 mi/h at 34$\bar{5}$°
$\mathbf{w}$ = 25 mi/h at 27$\bar{0}$°

4. $\mathbf{v}$ = 85 km/h at 25°
$\mathbf{w}$ = 25 km/h at 165°

Find the sum of each set of vectors by using the vector triangle method. Choose a suitable scale.

5. $\mathbf{v}$ = 75 mi at 30$\bar{0}$°
$\mathbf{w}$ = 25 mi at 195°

6. $\mathbf{v}$ = 250 km at 25°
$\mathbf{w}$ = 450 km at 345°

7. $\mathbf{v}$ = 18 mi at 45°
$\mathbf{w}$ = 32 mi at 11$\bar{0}$°
$\mathbf{t}$ = 28 mi at 20$\bar{0}$°

8. $\mathbf{v}$ = 95 km at 35$\bar{0}$°
$\mathbf{w}$ = 65 km at 30$\bar{0}$°
$\mathbf{t}$ = 25 km at 255°
$\mathbf{u}$ = 75 km at 18$\bar{0}$°

9. Using the vectors in Exercise 1, find $\mathbf{v} - \mathbf{w}$ graphically.

10. Using the vectors in Exercise 2, find $\mathbf{w} - \mathbf{v}$ graphically.

11. A ship sails 55 mi due north and then sails 3$\bar{0}$ mi at an angle of 45° east of north. How far is the ship from its starting point? What is the angle of the resultant path with respect to the starting point?

12. A plane is flying at 3$\bar{0}$0 mi/h (air speed) on a compass heading of 4$\bar{0}$° west of south. A wind of 6$\bar{0}$ mi/h is blowing from the east. What are the plane's true course and ground speed?

13. Do Exercise 12 for the case when the wind is blowing from the west.

11.6 ADDITION OF VECTORS: TRIGONOMETRIC METHODS

Vector sums may be found more accurately using trigonometry.

EXAMPLE 1

Using trigonometry, find the sum of the given vectors.

$\mathbf{v}$ = 19.5 km due west
$\mathbf{w}$ = 45.0 km due north

The vector triangle method gives the sketch in Fig. 11.47.

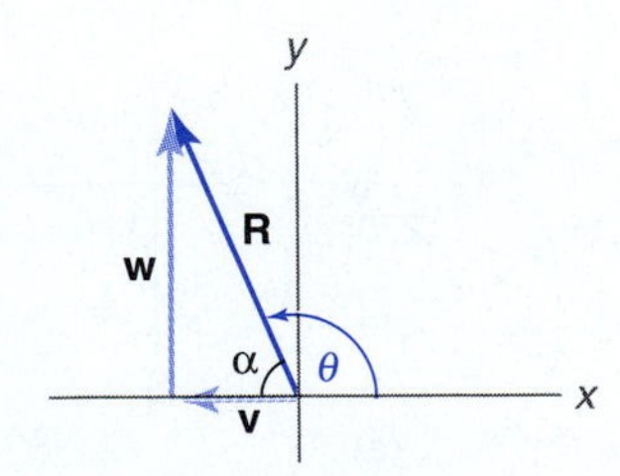

Figure 11.47

Using the Pythagorean theorem to find $|\mathbf{R}|$, we have

$$|\mathbf{R}| = \sqrt{|\mathbf{v}|^2 + |\mathbf{w}|^2}$$

$$|\mathbf{R}| = \sqrt{(19.5 \text{ km})^2 + (45.0 \text{ km})^2} = 49.0 \text{ km}$$

Next, find α.

$$\tan \alpha = \frac{\text{side opposite } \alpha}{\text{side adjacent to } \alpha} = \frac{|\mathbf{w}|}{|\mathbf{v}|}$$

$$\tan \alpha = \frac{45.0 \text{ km}}{19.5 \text{ km}} = 2.308$$

$$\alpha = 66.6°$$

$$\theta = 180° - \alpha = 180° - 66.6° = 113.4°$$

Therefore,

$$\mathbf{R} = 49.0 \text{ km at } 113.4°$$

Vector sums may also be found by using the law of sines and the law of cosines.

EXAMPLE 2

Using trigonometry, find the sum of the given vectors.

$$\mathbf{v} = 2.40 \text{ km at } 15.0°$$

$$\mathbf{w} = 3.30 \text{ km at } 75.0°$$

First, draw the vectors as in Fig. 11.48. Use the law of cosines to find the length of $\mathbf{R}$.

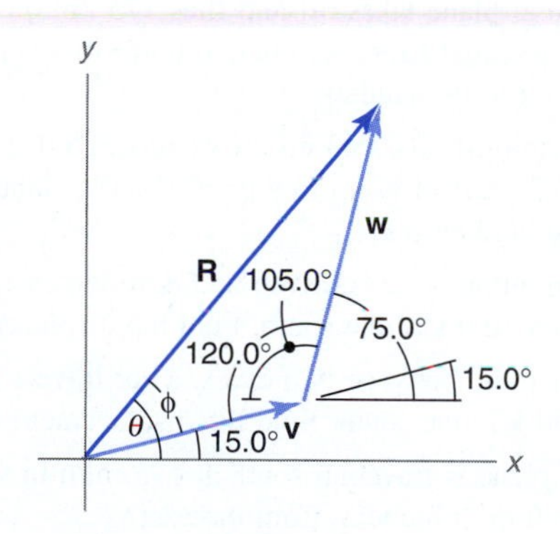

Figure 11.48

$$|\mathbf{R}|^2 = |\mathbf{v}|^2 + |\mathbf{w}|^2 - 2|\mathbf{v}||\mathbf{w}| \cos 120.0°$$

$$|\mathbf{R}|^2 = (2.40 \text{ km})^2 + (3.30 \text{ km})^2 - 2(2.40 \text{ km})(3.30 \text{ km}) \cos 120.0°$$

$$|\mathbf{R}| = 4.96 \text{ km}$$

Using the law of sines to find ϕ, we have

$$\frac{|\mathbf{R}|}{\sin 120.0°} = \frac{|\mathbf{w}|}{\sin \phi}$$

$$\frac{4.96 \text{ km}}{\sin 120.0°} = \frac{3.30 \text{ km}}{\sin \phi}$$

$$\sin\phi = \frac{(\sin 120.0°)(3.30\text{ km})}{4.96\text{ km}} = 0.5762$$

$$\phi = 35.2°$$

$$\theta = \phi + 15.0° = 35.2° + 15.0° = 50.2°$$

Therefore,

$$\mathbf{R} = 4.96\text{ km at }50.2°$$

Exercises 11.6

(Angles are given in standard position unless stated otherwise.)

Using trigonometry, find the sum of each set of vectors.

1. $\mathbf{v} = 65.3$ km/h at 270.0°
$\mathbf{w} = 40.5$ km/h at 180.0°

2. $\mathbf{v} = 6150$ m due south
$\mathbf{w} = 1780$ m due east

3. $\mathbf{v} = 4.50$ km at 67.0°
$\mathbf{w} = 6.50$ km at 105.0°

4. $\mathbf{v} = 1\bar{5}0$ mi at 175.0°
$\mathbf{w} = 2\bar{7}0$ mi at 215.0°

5. $\mathbf{v} = 87.1$ mi/h at 130.5°
$\mathbf{w} = 46.7$ mi/h at 207.0°

6. $\mathbf{v} = 60.0$ km/h at 286.0°
$\mathbf{w} = 60.0$ km/h at 254.0°

7. $\mathbf{v} = 605$ m at 60.0°
$\mathbf{w} = 415$ m at 120.0°
$\mathbf{t} = 295$ m at 90.0°

8. $\mathbf{v} = 15.3$ mi at 135.0°
$\mathbf{w} = 24.5$ mi at 75.0°
$\mathbf{t} = 19.7$ mi at 180.0°

9. Using the vectors in Exercise 3, find $\mathbf{v} - \mathbf{w}$.

10. Using the vectors in Exercise 8, find $\mathbf{w} - \mathbf{v}$.

11. An airplane takes off and flies 175 km on a course of 15.0° west of north and then changes course and flies 105 km due north to where it lands. Find the displacement from the starting point to the landing point.

12. A ship travels 75.0 mi on a course 25.0° north of east; then it travels 45.0 mi on a course 15.0° east of north to a point where it lands. Find the displacement from the starting point to the landing point.

13. An automobile is driven 25.0 km due east, then 10.0 km due north, then 15.0 km due east, then 10.0 km due south. Find the displacement from the starting point to the ending point.

14. En route between two cities, a car travels $10\bar{0}$ km due west, then 125 km southwest, then $1\bar{5}0$ km due south. Find the displacement from the starting point to the ending point.

15. A plane is traveling south at 175 mi/h in still air. What is its velocity with a wind of 65.0 mi/h blowing from the east?

16. A plane is traveling due north at 175 mi/h in still air. What is its velocity with a wind of 65.0 mi/h blowing from 45.0° east of south?

11.7 VECTOR COMPONENTS

We find it desirable to express a given vector as the sum of two vectors, especially two vectors along the coordinate axes. In Fig. 11.49,

$$\mathbf{v} = \mathbf{v}_1 + \mathbf{v}_2$$

$$\mathbf{v} = \mathbf{u}_1 + \mathbf{u}_2$$

$$\mathbf{v} = \mathbf{v}_x + \mathbf{v}_y$$

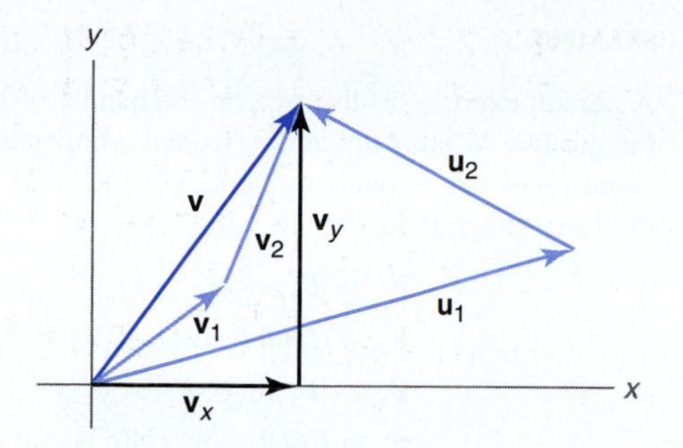

Figure 11.49 Vectors $\mathbf{v}_1$ and $\mathbf{v}_2$ as well as $\mathbf{u}_1$ and $\mathbf{u}_2$ are components of vector **v**. Vectors $\mathbf{v}_x$ and $\mathbf{v}_y$ are the horizontal and vertical components, respectively, of vector **v**.

The vectors $\mathbf{v}_1$, $\mathbf{v}_2$, $\mathbf{u}_1$, $\mathbf{u}_2$, $\mathbf{v}_x$, and $\mathbf{v}_y$ are called **components** of vector **v**. That is, if two or more vectors are added and their sum is the resultant vector **v**, then each of these vectors is called a component of **v**. We call $\mathbf{v}_x$ the **horizontal component** and $\mathbf{v}_y$ the **vertical component.**

In general, the horizontal and vertical components may be found using the definitions for sine and cosine,

$$\mathbf{v}_x = |\mathbf{v}|\cos\theta$$
$$\mathbf{v}_y = |\mathbf{v}|\sin\theta$$

where θ is in standard position (see Fig. 11.50).

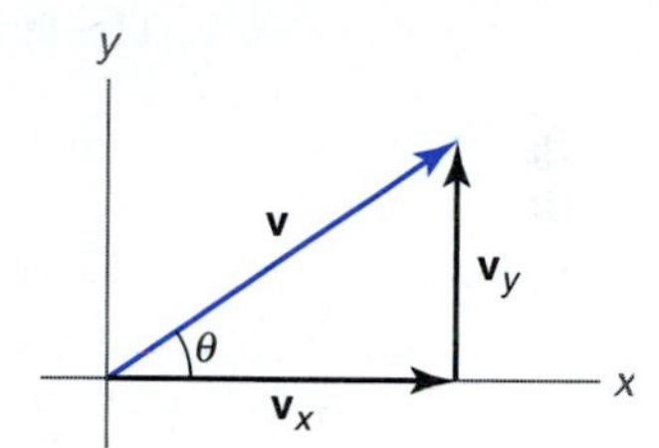

Figure 11.50 Horizontal and vertical components of vector **v** in standard position.

EXAMPLE 1

Find the horizontal and vertical components of the following vector in standard position: 50.0 mi/h at 60.0°.

First, draw the vector as in Fig. 11.51. Thus,

$$\mathbf{v}_x = |\mathbf{v}|\cos\theta$$
$$\mathbf{v}_x = (50.0 \text{ mi/h})(\cos 60.0°)$$
$$= 25.0 \text{ mi/h} \quad \text{(At 0° is understood.)}$$
$$\mathbf{v}_y = |\mathbf{v}|\sin\theta$$
$$\mathbf{v}_y = (50.0 \text{ mi/h})(\sin 60.0°)$$
$$= 43.3 \text{ mi/h} \quad \text{(At 90° is understood.)}$$

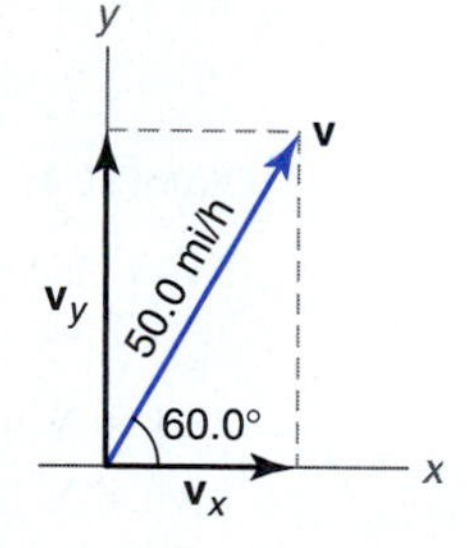

Figure 11.51

EXAMPLE 2

Find the horizontal and vertical components of the following vector in standard position: $1\bar{0}$ m at 153°.

First, draw the vector as in Fig. 11.52. Then

$$\mathbf{v}_x = |\mathbf{v}|\cos\theta$$
$$\mathbf{v}_x = (1\bar{0} \text{ m})(\cos 153°) = -8.9 \text{ m}$$
$$\mathbf{v}_y = |\mathbf{v}|\sin\theta$$
$$\mathbf{v}_y = (1\bar{0} \text{ m})(\sin 153°) = 4.5 \text{ m}$$

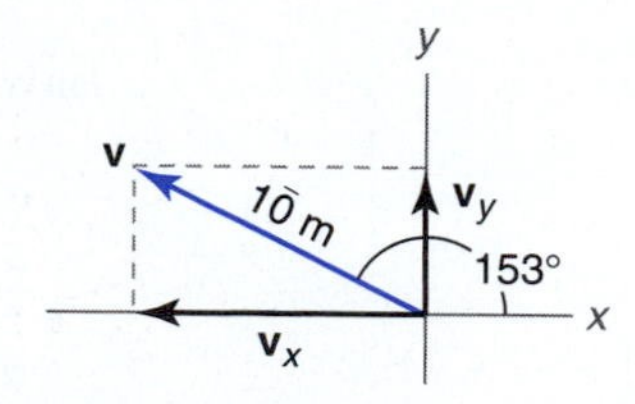

Figure 11.52

EXAMPLE 3

A person exerts a $5\bar{0}$-lb force on the handle of a lawnmower that is at an angle of $4\bar{0}°$ with the ground. What is the net horizontal component of the force that pushes the mower ahead? What is the net vertical component of the force that pushes the mower into the ground? See the force diagram in Fig. 11.53.

$$\mathbf{F}_x = |\mathbf{F}| \cos \theta$$
$$\mathbf{F}_x = (5\bar{0} \text{ lb})[\cos(-40°)] = 38 \text{ lb}$$
$$\mathbf{F}_y = |\mathbf{F}| \sin \theta$$
$$\mathbf{F}_y = (5\bar{0} \text{ lb})[\sin(-40°)] = -32 \text{ lb}$$

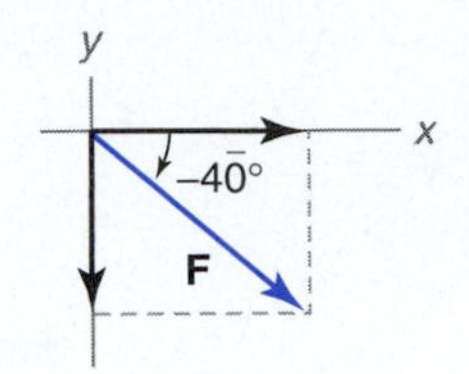

Force diagram

Figure 11.53

That is, a $5\bar{0}$-lb force at $-4\bar{0}°$ and a 38-lb force at 0° produce the same force in keeping the lawnmower moving forward. This same given force also produces an equivalent 32-lb force vertically into the ground.

We must also be able to find a vector **v** when its horizontal and vertical components $\mathbf{v}_x$ and $\mathbf{v}_y$ are given. In general, to express **v** in standard position, use these steps:

1. The length of **v** may be found by using the Pythagorean theorem.

$$|\mathbf{v}| = \sqrt{|\mathbf{v}_x|^2 + |\mathbf{v}_y|^2}$$

2. To find angle θ, first find α, the reference angle, by using the definition of tangent.

$$\tan \alpha = \frac{|\mathbf{v}_y|}{|\mathbf{v}_x|}$$

EXAMPLE 4

If $\mathbf{v}_x = 36$ km/h and $\mathbf{v}_y = 52$ km/h, find **v**.

First, draw the components and **v** as in Fig. 11.54. Then

$$|\mathbf{v}| = \sqrt{|\mathbf{v}_x|^2 + |\mathbf{v}_y|^2}$$
$$|\mathbf{v}| = \sqrt{|36 \text{ km/h}|^2 + |52 \text{ km/h}|^2} = 63 \text{ km/h}$$
$$\tan \alpha = \frac{|\mathbf{v}_y|}{|\mathbf{v}_x|}$$
$$\tan \alpha = \frac{52 \text{ km/h}}{36 \text{ km/h}} = 1.444$$
$$\alpha = 55°$$

Since θ is in the first quadrant,

$$\alpha = \theta = 55°$$

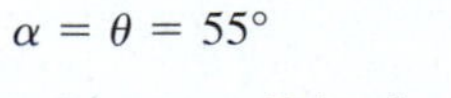

Therefore, $\mathbf{v} = 63$ km/h at 55°

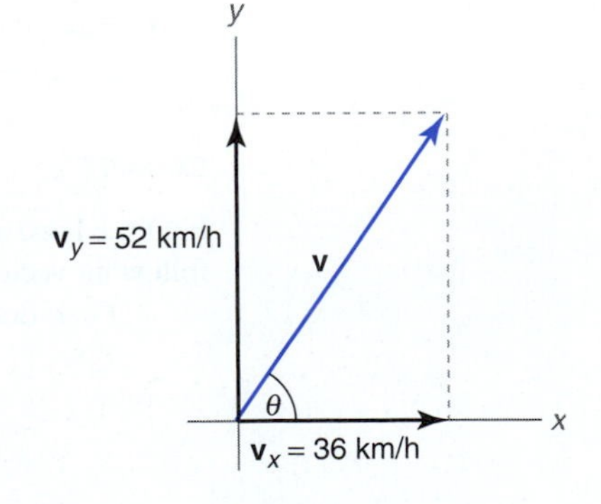

Figure 11.54

EXAMPLE 5

If $\mathbf{v}_x = -26.4$ N and $\mathbf{v}_y = -15.3$ N, find $\mathbf{v}$.

First, draw the components and $\mathbf{v}$ as in Fig. 11.55.

$$|\mathbf{v}| = \sqrt{|\mathbf{v}_x|^2 + |\mathbf{v}_y|^2}$$

$$|\mathbf{v}| = \sqrt{|-26.4\text{ N}|^2 + |-15.3\text{ N}|^2} = 30.5\text{ N}$$

$$\tan\alpha = \frac{|\mathbf{v}_y|}{|\mathbf{v}_x|}$$

$$\tan\alpha = \frac{15.3\text{ N}}{26.4\text{ N}} = 0.5795$$

$$\alpha = 30.1°$$

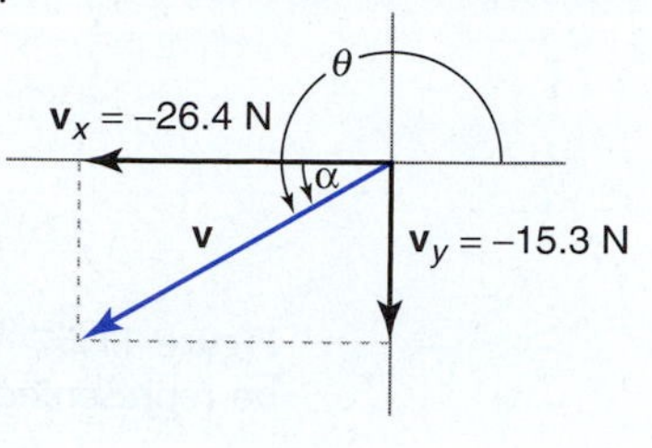

Figure 11.55

Since θ is in the third quadrant,

$$\theta = 180° + \alpha = 180° + 30.1° = 210.1°$$

Therefore,

$$\mathbf{v} = 30.5\text{ N at }210.1°$$

The **impedance** of a series circuit containing a **resistance** and an **inductance** can be represented by the vector diagram in Fig. 11.56, where ϕ is the phase angle, which equals the amount by which the *current lags behind the voltage.* The resistance is always drawn as a vector pointing in the positive x-direction and the inductive reactance is always drawn as a vector pointing in the positive y-direction.

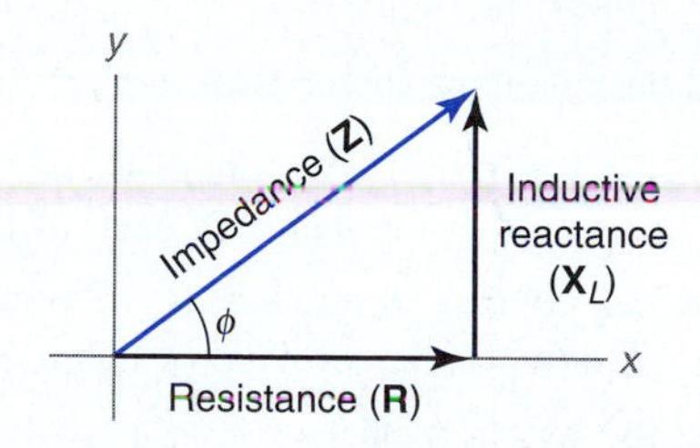

Figure 11.56 Impedance of a series circuit containing a resistance and an inductance may be represented by a vector diagram.

EXAMPLE 6

If the resistance is $6\bar{0}$ ohms (Ω) and the inductive reactance is 36 Ω, find the impedance.

Using Fig. 11.56, we have

$$|\mathbf{Z}| = \sqrt{(\mathbf{R})^2 + (\mathbf{X}_L)^2} = \sqrt{(6\bar{0}\ \Omega)^2 + (36\ \Omega)^2} = 7\bar{0}\ \Omega$$

$$\tan\phi = \frac{36\ \Omega}{6\bar{0}\ \Omega} = 0.6000$$

$$\phi = 31°$$

Therefore,

$$\mathbf{Z} = 7\bar{0}\ \Omega\text{ at }31°$$

The **impedance** of a series circuit containing a **resistance** and a **capacitance** can be represented by the vector diagram in Fig. 11.57, where ϕ is the phase angle, which equals the amount by which the *voltage lags behind the current.* The resistance is always drawn as a vector pointing in the positive x-direction and the capacitive reactance is drawn as a vector pointing in the negative y-direction.

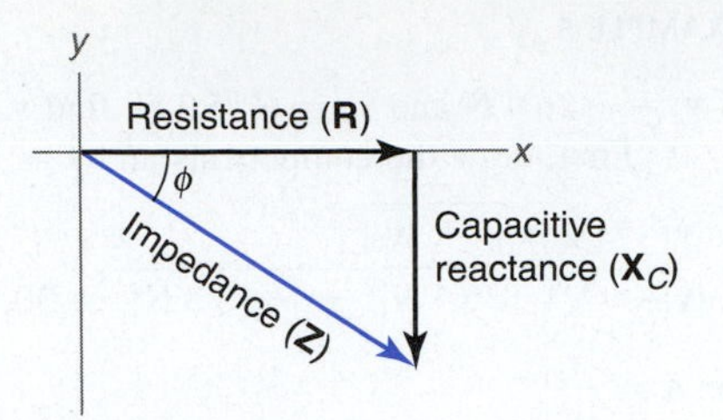

Figure 11.57 Impedance of a series circuit containing a resistance and a capacitance may be represented by a vector diagram.

EXAMPLE 7

If the impedance is 120 Ω and $\phi = 4\bar{0}°$, find the resistance and the capacitive reactance.

Using Fig. 11.57, we have

$$|\mathbf{R}| = |\mathbf{Z}| \cos \phi = (120\ \Omega)(\cos 4\bar{0}°) = 92\ \Omega$$
$$|\mathbf{X}_C| = |\mathbf{Z}| \sin \phi = (120\ \Omega)(\sin 4\bar{0}°) = 77\ \Omega$$

Addition of vectors using graphic methods has a limited degree of accuracy. Also, addition of vectors using the law of sines and the law of cosines, while accurate, is sometimes cumbersome. The following method of adding vectors is accurate and rather efficient.

COMPONENT METHOD

To find the resultant vector **R** of two or more vectors using the component method:

1. Find the horizontal component $\mathbf{R}_x$ of vector **R** by finding the algebraic sum of the horizontal components of each of the vectors being added.
2. Find the vertical component $\mathbf{R}_y$ of vector **R** by finding the algebraic sum of the vertical components of each of the vectors being added.
3. Find the length of **R**:
$$|\mathbf{R}| = \sqrt{|\mathbf{R}_x|^2 + |\mathbf{R}_y|^2}$$
4. To find angle θ, first find α, the reference angle.
$$\tan \alpha = \frac{|\mathbf{R}_y|}{|\mathbf{R}_x|}$$

EXAMPLE 8

Using the component method, find the sum of the given vectors.

$$\mathbf{v} = 25.0 \text{ km at } 121.0°$$
$$\mathbf{w} = 66.0 \text{ km at } 245.0°$$

First, draw the components and **R** as in Fig. 11.58.

$$\mathbf{v}_x = |\mathbf{v}| \cos \theta \qquad \mathbf{v}_y = |\mathbf{v}| \sin \theta$$
$$\mathbf{v}_x = (25.0 \text{ km})(\cos 121.0°) = -12.9 \text{ km} \qquad \mathbf{v}_y = (25.0 \text{ km})(\sin 121.0°) = 21.4 \text{ km}$$
$$\mathbf{w}_x = |\mathbf{w}| \cos \theta \qquad \mathbf{w}_y = |\mathbf{w}| \sin \theta$$
$$\mathbf{w}_x = (66.0 \text{ km})(\cos 245.0°) = \underline{-27.9 \text{ km}} \qquad \mathbf{w}_y = (66.0 \text{ km})(\sin 245.0°) = \underline{-59.8 \text{ km}}$$
$$\mathbf{R}_x\text{: Sum of } x\text{-components} = -40.8 \text{ km} \qquad \mathbf{R}_y\text{: Sum of } y\text{-components} = -38.4 \text{ km}$$

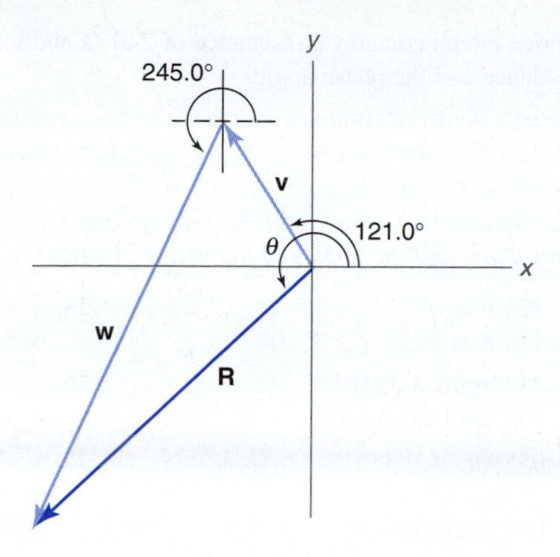

Figure 11.58

Then

$$|\mathbf{R}| = \sqrt{|\mathbf{R}_x|^2 + |\mathbf{R}_y|^2}$$
$$|\mathbf{R}| = \sqrt{|-40.8 \text{ km}|^2 + |-38.4 \text{ km}|^2} = 56.0 \text{ km}$$
$$\tan \alpha = \frac{|\mathbf{R}_y|}{|\mathbf{R}_x|}$$
$$\tan \alpha = \frac{38.4 \text{ km}}{40.8 \text{ km}} = 0.9412$$
$$\alpha = 43.3°$$

Since $\mathbf{R}_x < 0$ and $\mathbf{R}_y < 0$, θ is in the third quadrant. So

$$\theta = 180° + \alpha = 180° + 43.3° = 223.3°$$

Therefore,

$$\mathbf{R} = 56.0 \text{ km at } 223.3°$$

Exercises 11.7

(Angles are given in standard position unless stated otherwise.)

Find the horizontal and vertical components of each vector given in standard position.

1. $\mathbf{v} = 18.2$ km at 85.0°
2. $\mathbf{v} = 27.9$ mi at 138.0°
3. $\mathbf{v} = 135$ mi/h at 270.0°
4. $\mathbf{v} = 448$ m at 319.0°
5. $\mathbf{v} = 2680$ ft at 152.5°
6. $\mathbf{v} = 3620$ ft at 187.3°

For each pair of horizontal and vertical components, find the vector.

7. $\mathbf{v}_x = 8.70$ m, $\mathbf{v}_y = 6.40$ m
8. $\mathbf{v}_x = -2.10$ ft, $\mathbf{v}_y = 3.20$ ft
9. $\mathbf{v}_x = 4.70$ m/s, $\mathbf{v}_y = -6.60$ m/s
10. $\mathbf{v}_x = -925$ m, $\mathbf{v}_y = 125$ m
11. $\mathbf{v}_x = -14.7$ km, $\mathbf{v}_y = 0$
12. $\mathbf{v}_x = 427$ mi/h, $\mathbf{v}_y = 381$ mi/h

13. A series circuit containing a resistance and an inductance has an impedance of $9\overline{0}$ Ω, and the phase angle is $2\overline{0}°$. Find the resistance and the inductive reactance.

14. A series circuit contains a resistance of 85 Ω and has an inductive reactance of 45 Ω. Find the impedance and the phase angle.

15. A series circuit contains a resistance of 240 Ω and a capacitive reactance of 140 Ω. Find the impedance and the phase angle.

16. A series circuit containing a resistance and a capacitance has an impedance of 110 Ω, and $\phi = 49°$. Find the resistance and the capacitive reactance.

Using the component method, find the sum of each set of vectors. Find the magnitudes to three significant digits and the angles to the nearest tenth of a degree.

17. $\mathbf{v} = 324$ ft at 0°
    $\mathbf{w} = 576$ ft at 90.0°

18. $\mathbf{v} = 91.2$ km/h at 180.0°
    $\mathbf{w} = 84.7$ km/h at 270.0°

19. $\mathbf{v} = 28.9$ mi/h at 52.0°
    $\mathbf{w} = 16.2$ mi/h at 310.0°

20. $\mathbf{v} = 59.7$ km at 125.0°
    $\mathbf{w} = 86.4$ km at 298.0°

21. $\mathbf{v} = 655$ km at 108.0°
    $\mathbf{w} = 655$ km at 27.0°
    $\mathbf{u} = 655$ km at 270.0°

22. $\mathbf{v} = 29.7$ mi at 237.0°
    $\mathbf{w} = 16.4$ mi at 180.0°
    $\mathbf{u} = 18.5$ mi at 15.0°

23. $\mathbf{v} = 5020$ m at 0°
    $\mathbf{w} = 3130$ m at 148.0°
    $\mathbf{u} = 6250$ m at 65.0°
    $\mathbf{t} = 4620$ m at 335.0°

24. $\mathbf{v} = 5760$ ft at 90.0°
    $\mathbf{w} = 3940$ ft at 205.0°
    $\mathbf{u} = 6140$ ft at 150.0°
    $\mathbf{t} = 1230$ ft at 330.0°

25. Using the vectors in Exercise 19, find $\mathbf{v} - \mathbf{w}$.

26. Using the vectors in Exercise 20, find $\mathbf{w} - \mathbf{v}$.

27. An airplane flies 165 mi on a course of 25.0° west of south; then it changes course and flies 125 mi on a course of 15.0° north of west to where it lands. Find the displacement from the starting point to the landing point.

28. A ship travels 75.0 mi on a course of 35.0° west of north; then it travels 45.0 mi on a course of 60.0° south of west to where it docks. Find the displacement from the starting point to the docking point.

11.8 VECTOR APPLICATIONS

Many applications involve vector quantities. Displacement, velocity, and force have been chosen here to illustrate some basic applications.

Displacement

Recall that **displacement** is a change of position, that is, the difference between the initial position of a body and any later position.

EXAMPLE 1

A ship travels 175 mi at 52.0°, then 295 mi at 141.0°, and then 225 mi due west. Find the displacement; that is, find the net distance between the starting point and the endpoint and the angle.

Using the component method, let

$$\mathbf{u} = 175 \text{ mi at } 52.0°$$
$$\mathbf{v} = 295 \text{ mi at } 141.0°$$
$$\mathbf{w} = 225 \text{ mi at } 180.0°$$

Draw the components and **R** as in Fig. 11.59. Then we have the following:

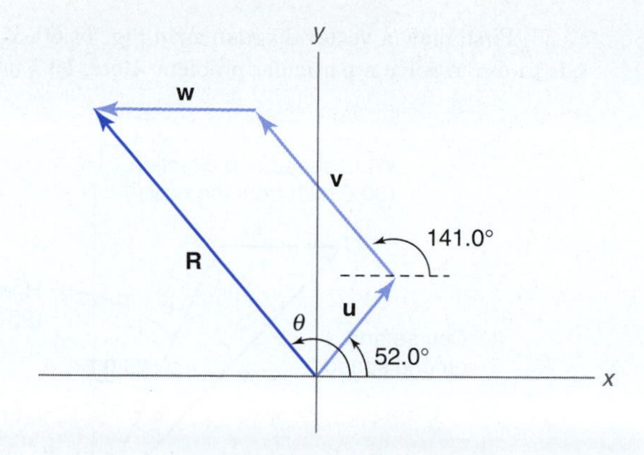

Figure 11.59 Displacement.

$\mathbf{u}_x = \lvert\mathbf{u}\rvert \cos\theta$	$\mathbf{u}_y = \lvert\mathbf{u}\rvert \sin\theta$
$\mathbf{u}_x = (175 \text{ mi})(\cos 52.0°) = 108 \text{ mi}$	$\mathbf{u}_y = (175 \text{ mi})(\sin 52.0°) = 138 \text{ mi}$
$\mathbf{v}_x = \lvert\mathbf{v}\rvert \cos\theta$	$\mathbf{v}_y = \lvert\mathbf{v}\rvert \sin\theta$
$\mathbf{v}_x = (295 \text{ mi})(\cos 141.0°) = -229 \text{ mi}$	$\mathbf{v}_y = (295 \text{ mi})(\sin 141.0°) = 186 \text{ mi}$
$\mathbf{w}_x = \lvert\mathbf{w}\rvert \cos\theta$	$\mathbf{w}_y = \lvert\mathbf{w}\rvert \sin\theta$
$\mathbf{w}_x = (225 \text{ mi})(\cos 180.0°) = -225 \text{ mi}$	$\mathbf{w}_y = (225 \text{ mi})(\sin 180.0°) = 0 \text{ mi}$
$\mathbf{R}_x$: Sum of x-components $= -346$ mi	$\mathbf{R}_y$: Sum of y-components $= 324$ mi

Then

$$|\mathbf{R}| = \sqrt{|\mathbf{R}_x|^2 + |\mathbf{R}_y|^2}$$

$$|\mathbf{R}| = \sqrt{|-346 \text{ mi}|^2 + |324 \text{ mi}|^2} = 474 \text{ mi}$$

$$\tan\alpha = \frac{|\mathbf{R}_y|}{|\mathbf{R}_x|}$$

$$\tan\alpha = \frac{324 \text{ mi}}{346 \text{ mi}} = 0.9364$$

$$\alpha = 43.1°$$

Since $\mathbf{R}_x < 0$ and $\mathbf{R}_y > 0$, θ is in the second quadrant. So,

$$\theta = 180° - \alpha = 180° - 43.1° = 136.9°$$

Therefore, the displacement is

$$\mathbf{R} = 474 \text{ mi at } 136.9°$$

Velocity

Velocity is a vector quantity that is described in terms of both speed and direction. For example, wind typically has an effect on both the speed and the direction of a plane in flight. The *heading* of an airplane is the direction the plane is pointed with respect to the compass. The *course* of the airplane is the direction the plane is actually flying with respect to the ground. The *air speed* is the speed of the plane with respect to the air; the *ground speed* is the speed of the plane with respect to the ground.

EXAMPLE 2

An airplane is flying on a heading of 25.0° west of north at an air speed of 95.0 mi/h. The wind is from the east at 30.0 mi/h. Find the plane's course and ground speed.

First, draw a vector diagram as in Fig. 11.60. You often have a choice of vector methods to use to solve a particular problem. Here, let's use trigonometry.

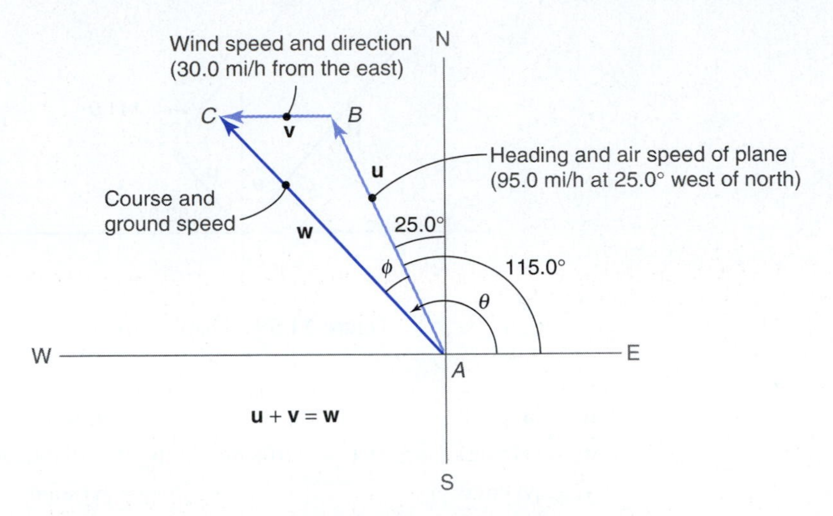

Figure 11.60 Velocity.

Since **BC** is parallel to the x-axis, angle $B = 115.0°$. Then use the law of cosines in triangle ABC.

$$|\mathbf{w}|^2 = |\mathbf{u}|^2 + |\mathbf{v}|^2 - 2|\mathbf{u}||\mathbf{v}|\cos B$$

$$|\mathbf{w}| = \sqrt{(95.0 \text{ mi/h})^2 + (30.0 \text{ mi/h})^2 - 2(95.0 \text{ mi/h})(30.0 \text{ mi/h})(\cos 115.0°)}$$

$$= 111 \text{ mi/h}$$

Next, find angle ϕ using the law of sines.

$$\frac{|\mathbf{w}|}{\sin B} = \frac{|\mathbf{v}|}{\sin \phi}$$

$$\frac{111 \text{ mi/h}}{\sin 115.0°} = \frac{30.0 \text{ mi/h}}{\sin \phi}$$

$$\sin \phi = \frac{(\sin 115.0°)(30.0 \text{ mi/h})}{111 \text{ mi/h}} = 0.2449$$

$$\phi = 14.2°$$

Then

$$\theta = 115.0° + 14.2° = 129.2°$$

The ground speed is 111 mi/h, and the course is 129.2° in standard position or 39.2° west of north.

Force

In physics a **force** is defined as a push or a pull that tends to cause or prevent motion. When a force is applied, it must be applied in some direction. Thus, force is a vector quantity.

EXAMPLE 3

Two forces act on a point at an angle of 70.0°. One force is 1850 newtons (N). The resultant force is 2250 N. Find the second force and the angle it makes with the resultant.

First, draw a force diagram as in Fig. 11.61.

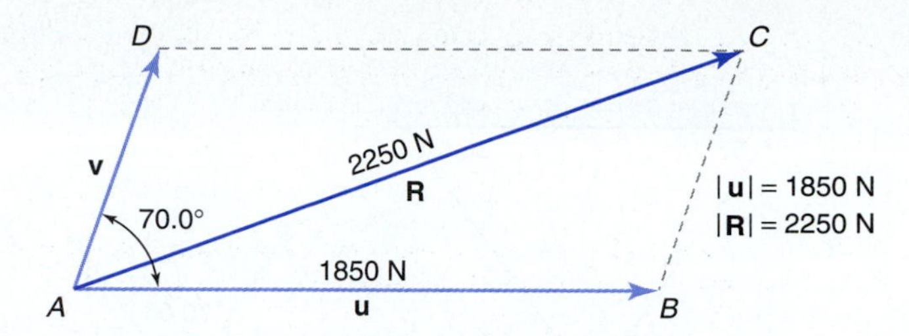

Figure 11.61 Force.

The consecutive angles of a parallelogram are supplementary; $B = 110.0°$. Let's first find angle ACB using the law of sines.

$$\frac{AC}{\sin B} = \frac{|\mathbf{u}|}{\sin ACB}$$

$$\frac{2250 \text{ N}}{\sin 110.0°} = \frac{1850 \text{ N}}{\sin ACB}$$

$$\sin ACB = \frac{(\sin 110.0°)(1850 \text{ N})}{2250 \text{ N}} = 0.7726$$

$$\text{angle } ACB = 50.6°$$

Then angle $CAB = 180° - 110.0° - 50.6° = 19.4°$.

Now find $|\mathbf{v}|$ using the law of sines and noting that $BC = |\mathbf{v}|$ in triangle ABC.

$$\frac{BC}{\sin CAB} = \frac{AC}{\sin B}$$

$$\frac{|\mathbf{v}|}{\sin 19.4°} = \frac{2250 \text{ N}}{\sin 110.0°} \qquad (\textit{Note}: BC = |\mathbf{v}|.)$$

$$|\mathbf{v}| = \frac{(\sin 19.4°)(2250 \text{ N})}{\sin 110.0°} = 795 \text{ N}$$

Thus, $\mathbf{v} = 795$ N at 50.6° from **R**.

When two or more forces act at a point, the force applied at the same point that produces equilibrium is called the *equilibrant force.* Thus, the equilibrant force is *equal* in magnitude but *opposite* in direction to the resultant force.

EXAMPLE 4

A 1250-lb weight hangs from two ropes of equal lengths suspended from a support as shown in Fig. 11.62. Find the tension in each rope.

First, draw the force diagram as in Fig. 11.62(b). Note that the equilibrant force is 1250 lb; the resultant force is shown in Fig. 11.62(c). Find $|\mathbf{T}_1|$ using the law of sines in triangle ABC as follows:

$$\frac{|\mathbf{T}_1|}{\sin ACB} = \frac{|\mathbf{R}|}{\sin B}$$

$$\frac{|\mathbf{T}_1|}{\sin 35.0°} = \frac{1250 \text{ lb}}{\sin 110.0°} \qquad (\textit{Note}: \text{Angle } B = 110.0°.)$$

$$|\mathbf{T}_1| = \frac{(\sin 35.0°)(1250 \text{ lb})}{\sin 110.0°} = 763 \text{ lb} = |\mathbf{T}_2|$$

The tension in each rope is 763 lb.

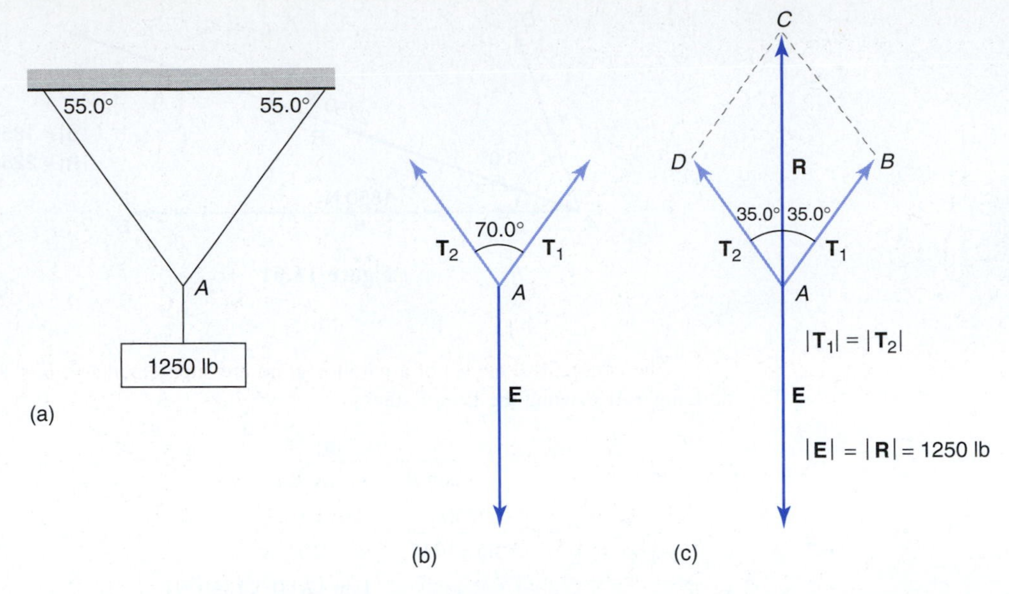

Figure 11.62

A system is in equilibrium when the sum of the x-components is zero and the sum of the y-components is zero. Let's rework Example 4 using this approach. First, redraw the force diagram on the xy-axes as shown in Fig. 11.63. Then, find the x- and y-components of each vector.

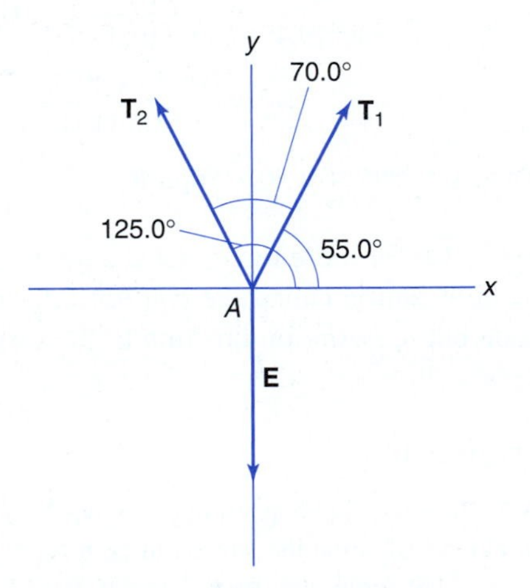

Figure 11.63

$$T_{1x} = |T_1| \cos \theta \qquad T_{1y} = |T_1| \sin \theta$$
$$T_{1x} = |T_1| \cos 55.0° = 0.574\ |T_1| \qquad T_{1y} = |T_1| \sin 55.0° = 0.819\ |T_1|$$

$$T_{2x} = |T_2| \cos \theta \qquad T_{2y} = |T_2| \sin \theta$$
$$T_{2x} = |T_2| \cos 125.0° = -0.574\ |T_2| \qquad T_{2y} = |T_2| \sin 125.0° = 0.819\ |T_2|$$

$$E_x = |E| \cos \theta \qquad E_y = |E| \sin \theta$$
$$E_x = (1250 \text{ lb}) \cos 270.0° = 0 \qquad E_y = (1250 \text{ lb}) \sin 270.0° = -1250 \text{ lb}$$

Thus,

$$\text{Sum of } x\text{-components: } 0.574\,|T_1| - 0.574\,|T_2| = 0 \quad \textbf{(1)}$$
$$\text{Sum of } y\text{-components: } 0.819|T_1| + 0.819\,|T_2| + (-1250 \text{ lb}) = 0 \quad \textbf{(2)}$$

From Equation (1), we see that $|T_1| = |T_2|$, as expected. Then, substituting this into Equation (2), we have

$$0.819\,|T_1| + 0.819\,|T_1| = 1250 \text{ lb}$$
$$1.638\,|T_1| = 1250 \text{ lb}$$
$$|T_1| = 763 \text{ lb} = |T_2|$$

Exercises 11.8

1. An airplane takes off and flies 125 mi on a course of 61.5° south of west; then it changes course and flies 185 mi due south, where it lands. Find the displacement from the starting point to the landing point.
2. A ship travels 18.5 km on a course of 31.2° south of east, then 12.7 km due south, then 21.5 km on a course of 61.3° west of south, where it lands. Find the displacement from the starting point to the landing point.
3. En route between two cities, a person travels by automobile 115 km due south, then 195 km at 45.0° west of south by plane, and then 45.0 km due west by boat. Find the displacement from the starting point to the ending point.
4. Natasha jogs 2.00 km due west, then 1.00 km due north, then 3.00 km due east, and then 2.00 km due south. Find the displacement from her starting point to her ending point.

In Exercises 5 through 8, given displacements with magnitudes of 10.0 mi and 15.0 mi:

5. Find the magnitude of the maximum resultant displacement.
6. Find the magnitude of the minimum resultant displacement.
7. Find the angle between the two original displacements if the magnitude of the resultant displacement is 12.0 mi.
8. Find the angle between the two original displacements if the magnitude of the resultant displacement is 7.50 mi.

In Exercises 9 and 10, given three displacements of magnitudes 10.0 km, 15.0 km, and 20.0 km:

9. Find the magnitude of the maximum resultant displacement.
10. Find the magnitude of the minimum resultant displacement.
11. A boat travels 12 mi/h in still water. The flow of the current is 3 mi/h.
    (a) What is the boat's speed going downstream?
    (b) What is its speed going upstream?
12. A plane's air speed is 125 mi/h. The wind speed is 21 mi/h.
    (a) What is the plane's ground speed when it is flying into the wind?
    (b) What is the plane's ground speed when it is flying with the wind?
13. A plane is flying due west at 175 mi/h. Suddenly a wind of 25.0 mi/h from the south develops. Find the plane's new course and ground speed.
14. A plane is flying north at 145 mi/h in still air. Find the velocity of the plane when the wind is blowing 55.0 mi/h from the east.
15. A plane is flying on a heading of 35.0° west of north at an air speed of 315 km/h. The wind is blowing from the west at 50.0 km/h. Find the plane's course and ground speed.
16. A plane is flying on a heading of 60.0° east of south at an air speed of 165 mi/h. The wind is blowing at 45.0 mi/h from 25.0° south of west. Find the plane's course and ground speed.

17. A pilot is flying on a course of 25.5° north of west at 215 mi/h. The wind is blowing from the east at 45.6 mi/h. Find the plane's heading and air speed.

18. During a model boating competition, the following circumstances are presented to the contestants: width of stream, 125 ft; speed of current, 4.00 ft/s; speed of boat in still water, 12.0 ft/s. At what angle must the boat be steered to reach a point on the opposite side of the stream **(a)** that is directly across the stream, **(b)** that is 45.0 ft upstream, and **(c)** that is 45.0 ft downstream?

19. DeCarlos applies a 25.0-lb force to the handle of a lawnmower, which makes a 35.0° angle with the ground. Find the horizontal and vertical components of this force.

20. Cindy pulls a sled with a rope with a force of 50.0 lb. The rope makes an angle of 30.0° with the ground. Find the horizontal and vertical components of this force.

In Exercises 21 through 24, find the vector sum of each set of forces.

21.

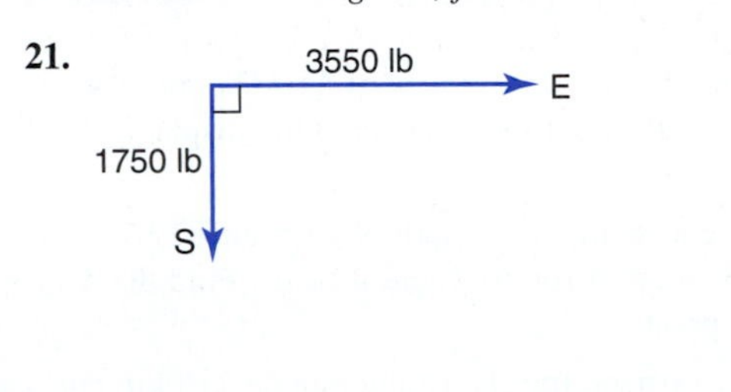

22.

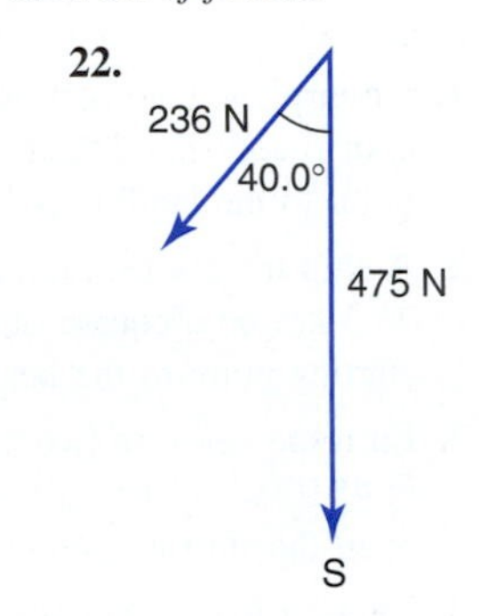

23.

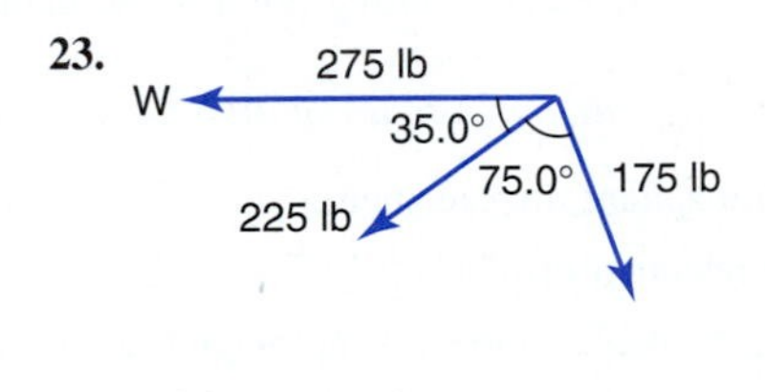

24.

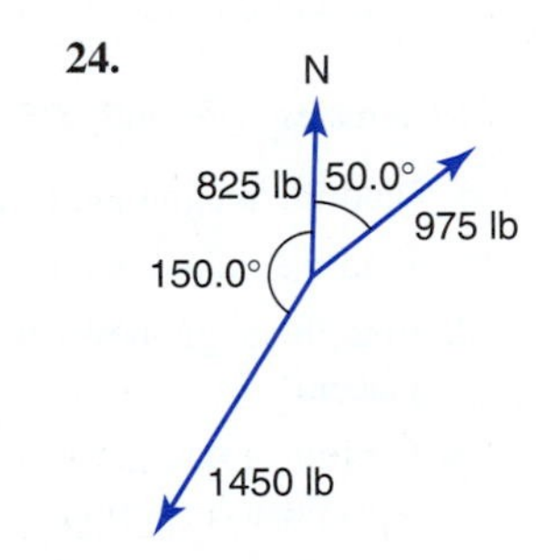

25. Two forces act on a point at an angle of 125.0°. One force is 225 lb. The resultant force is 195 lb. Find the second force and the angle it makes with the resultant.

26. Two forces act on a point at an angle of 65.0°. One force is 175 lb. The resultant force is 215 lb. Find the second force and the angle it makes with the resultant.

In Exercises 27 through 30, find each equilibrant force in standard position.

27.

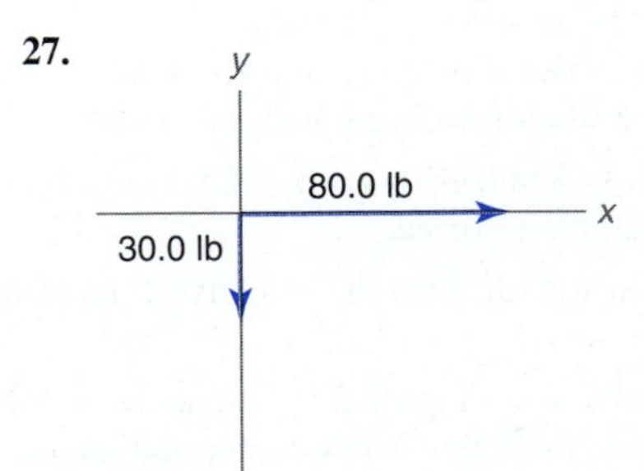

28.

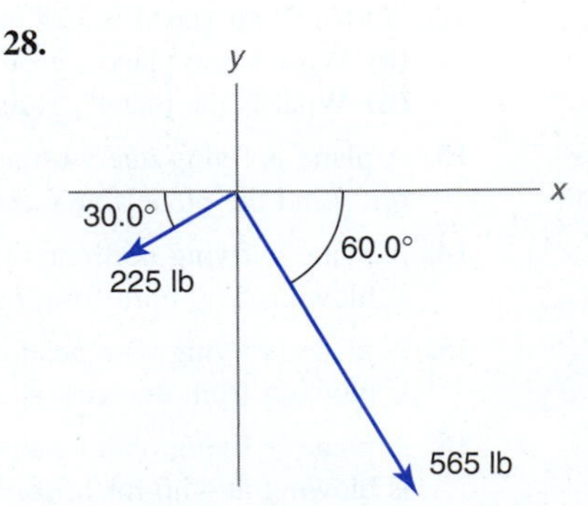

29. **30.**

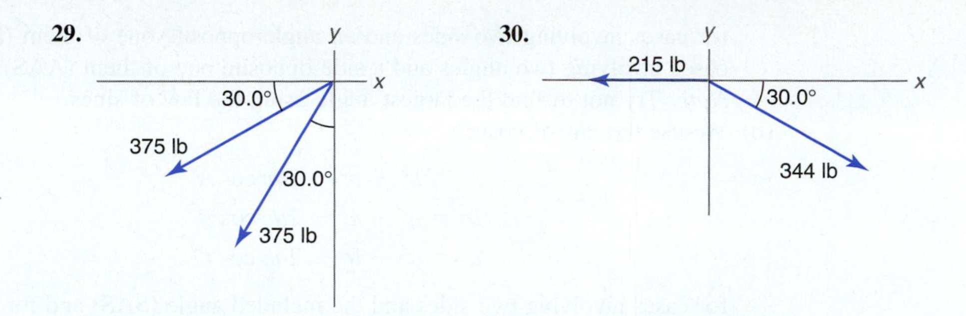

31. A sign weighing 615 lb is supported as shown in Fig. 11.64. Find the tension in the cable.

32. A sign weighing 275 lb is suspended by two cables attached between two buildings at equal heights above the ground. Each cable makes an angle of 65.0° with its building. Find the tension in each cable.

33. A 475-lb sign hangs from two cables as shown in Fig. 11.65. Find the tension in each cable.

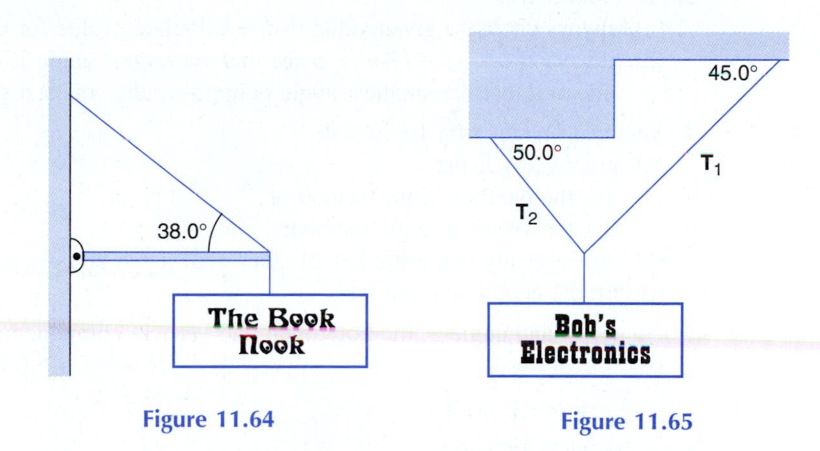

Figure 11.64 Figure 11.65

34. A 1250-lb weight hangs from two cables of equal length as shown in Fig. 11.66. If each cable can withstand a maximum tension of 825 lb, at what angle A are the cables unsafe?

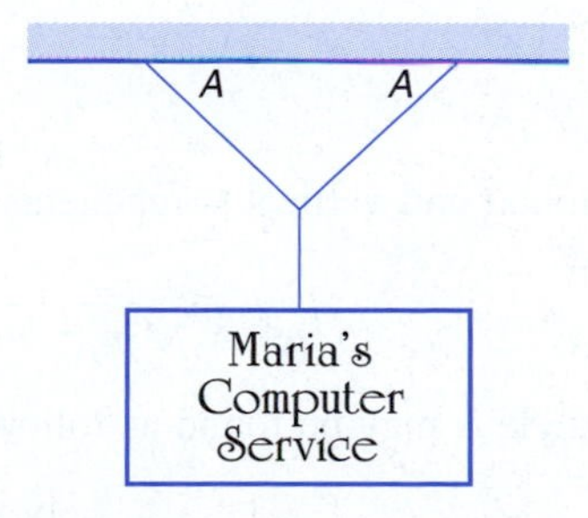

Figure 11.66

CHAPTER 11 SUMMARY

1. *For any general triangle as in Fig. 11.67:*

(a) We use the law of sines

$$\frac{a}{\sin A} = \frac{b}{\sin B} = \frac{c}{\sin C}$$

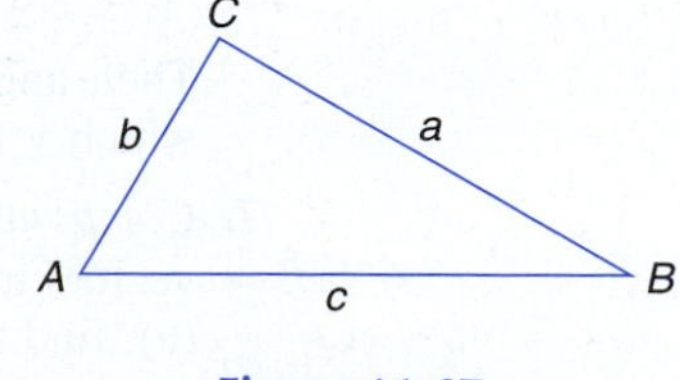

Figure 11.67

for cases involving two sides and an angle opposite one of them (SSA) and for cases involving two angles and a side opposite one of them (AAS).
Note: Try not to find the largest angle using the law of sines.
   (b) We use the law of cosines

$$a^2 = b^2 + c^2 - 2bc \cos A$$
$$b^2 = a^2 + c^2 - 2ac \cos B$$
$$c^2 = a^2 + b^2 - 2ab \cos C$$

for cases involving two sides and the included angle (SAS) and for cases involving three sides (SSS).

2. The SSA case requires special consideration. If the side opposite the given angle is
   (a) greater than the known adjacent side, there is only one possible triangle.
   (b) less than the known adjacent side but greater than the altitude, there are two possible triangles.
   (c) less than the altitude, there is no possible triangle.

3. *As a final check,*
   (a) always choose a given value over a calculated value for doing calculations.
   (b) always check your results to see that the largest angle is opposite the largest side and that the smallest angle is opposite the smallest side.

4. Vector problems may be solved
   (a) graphically using
      (i) the parallelogram method or
      (ii) the vector triangle method;
   (b) algebraically using the law of sines and/or the law of cosines; or
   (c) by the component method.

5. Given $|\mathbf{v}|$ and angle θ, the horizontal and vertical components are found as follows (see Fig. 11.68):

$$\mathbf{v}_x = |\mathbf{v}| \cos \theta$$
$$\mathbf{v}_y = |\mathbf{v}| \sin \theta$$

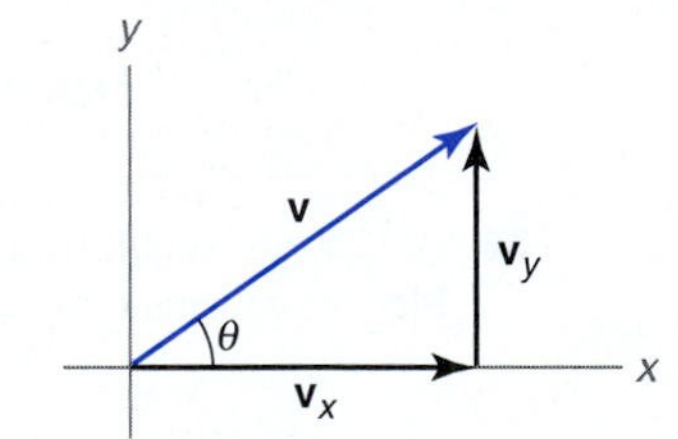

Figure 11.68

6. Given the horizontal and vertical components $\mathbf{v}_x$ and $\mathbf{v}_y$, the magnitude of $\mathbf{v}$ may be found as follows:

$$|\mathbf{v}| = \sqrt{|\mathbf{v}_x|^2 + |\mathbf{v}_y|^2}$$

The reference angle α may be found as follows:

$$\tan \alpha = \frac{|\mathbf{v}_y|}{|\mathbf{v}_x|}$$

Then angle θ in standard position is determined from angle α and the quadrant in which $\mathbf{v}$ lies.

7. *Component method of adding vectors:* To find the resultant vector $\mathbf{R}$ of two or more vectors using the component method,
   (a) find the horizontal component $\mathbf{R}_x$ of vector $\mathbf{R}$ by finding the algebraic sum of the horizontal components of each of the vectors being added.

(b) find the vertical component $\mathbf{R}_y$ of vector $\mathbf{R}$ by finding the algebraic sum of the vertical components of each of the vectors being added.

(c) find the length of $\mathbf{R}$: $|\mathbf{R}| = \sqrt{|\mathbf{R}_x|^2 + |\mathbf{R}_y|^2}$.

(d) To find angle θ, first find α, the reference angle.

$$\tan \alpha = \frac{|\mathbf{R}_y|}{|\mathbf{R}_x|}$$

CHAPTER 11 REVIEW

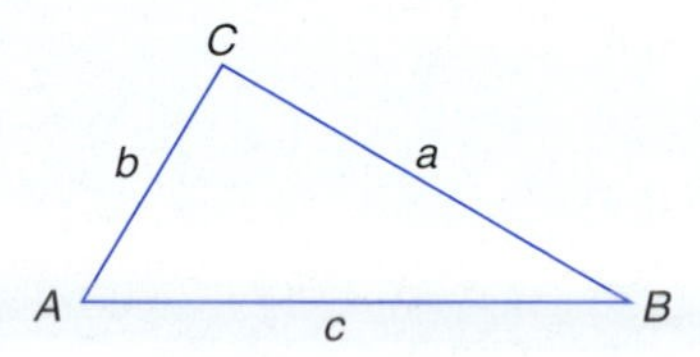

Figure 11.69

Solve each triangle using the labels as in Fig. 11.69.

Express the lengths of sides to three significant digits and the angles to the nearest tenth of a degree.

1. $B = 71.4°$, $b = 409$ ft, $c = 327$ ft

2. $A = 25.1°$, $C = 37.7°$, $a = 15.7$ m

3. $A = 15.5°$, $b = 236$ cm, $c = 209$ cm

4. $a = 25.6$ m, $b = 42.2$ m, $c = 35.2$ m

Express the lengths of sides to two significant digits and the angles to the nearest degree.

5. $B = 44°$, $b = 150$ mi, $c = 240$ mi

6. $A = 29°$, $a = 41$ cm, $b = 49$ cm

7. $C = 36°$, $a = 2100$ ft, $b = 3600$ ft

8. $C = 58°$, $a = 450$ m, $c = 410$ m

Express the lengths of sides to four significant digits and the angles to the nearest hundredth of a degree.

9. $B = 105.15°$, $a = 231.1$ m, $c = 190.7$ m

10. $A = 74.75°$, $a = 22.19$ cm, $c = 15.28$ cm

11. $B = 18.25°$, $a = 1675$ ft, $b = 1525$ ft

12. $C = 40.16°$, $b = 25{,}870$ ft, $c = 10{,}250$ ft

Express the lengths of sides to four significant digits and the angles to the nearest second.

13. $A = 48°15'35''$, $B = 68°7'18''$, $a = 2755$ ft

14. $C = 29°25'16''$, $a = 13{,}560$ ft, $b = 24{,}140$ ft

15. Find the total length of fence needed to enclose the triangular field shown in Fig. 11.70.

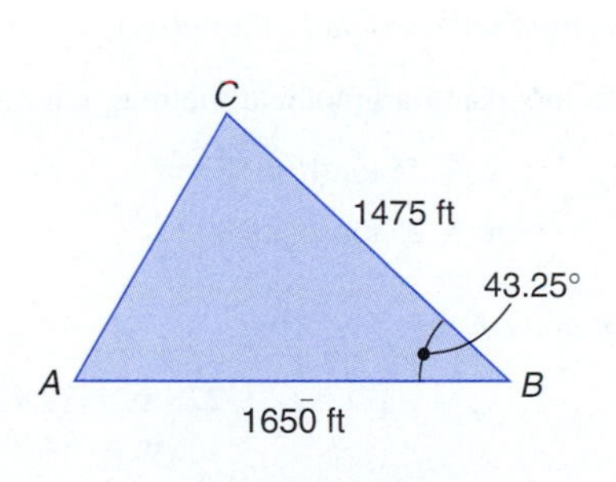

Figure 11.70

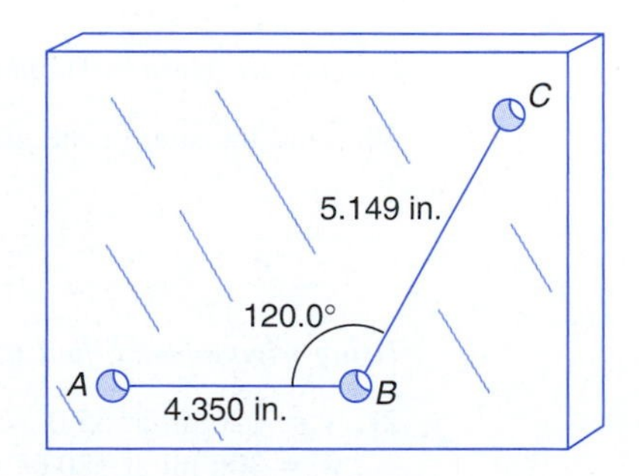

Figure 11.71

16. In surveying a tract of timber, it is necessary to find the distance to an inaccessible, but visible, point C. A distance between two points A and B of 300.0 ft is measured. Then, $\angle ABC$ is measured as 85.00° and $\angle BAC$ is measured as 79.55°.

(a) Find AC.

(b) Find the perpendicular distance from C to line AB.

17. A toolmaker needs to lay out three holes in a plate, as shown in Fig. 11.71. Find the distance between holes A and C.

18. The centers of five holes are equally spaced around a 5.000-in.-diameter circle as shown in Fig. 11.72.
   (a) Find the distance between the centers of holes A and B.
   (b) Find the distance between the centers of holes A and C.

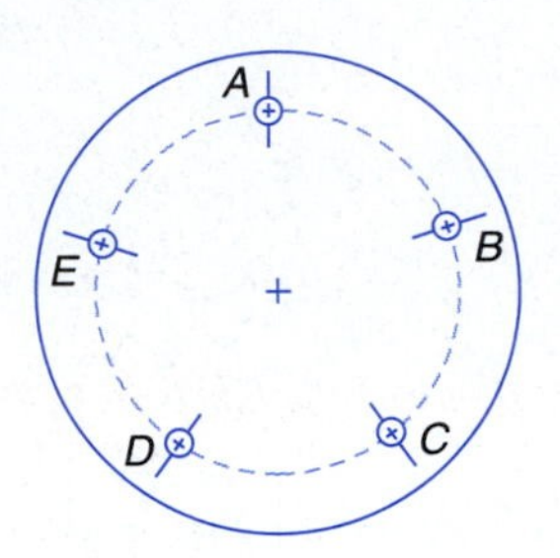

Figure 11.72

19. In the roof truss in Fig. 11.73, we know that $AB = DE$, $BC = CD$, and $AE = 20.0$ m. Find each length.
   (a) AF **(b)** BF **(c)** CF **(d)** BC

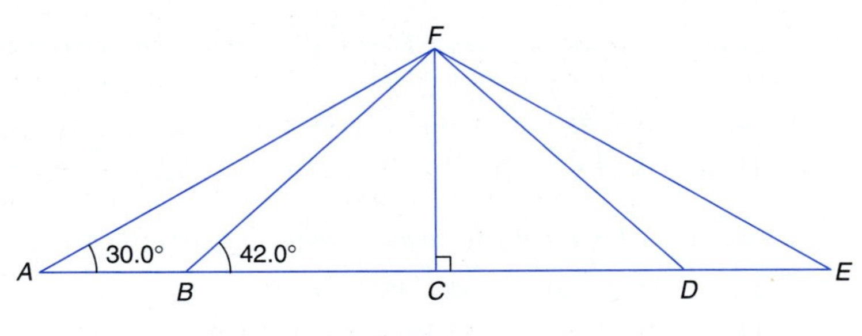

Figure 11.73

(Angles are given in standard position unless stated otherwise.)

20. Find the sum of the given vectors using a graphical method. Choose a suitable scale.

$$\mathbf{v} = 75 \text{ km/h at } 87°$$
$$\mathbf{w} = 25 \text{ km/h at } 142°$$

Using trigonometry, find the sum of each set of vectors.

21. $\mathbf{v} = 126$ mi at 35.0°
    $\mathbf{w} = 306$ mi at 180.0°

22. $\mathbf{v} = 89.4$ mi/h at 142.0°
    $\mathbf{w} = 44.7$ mi/h at 322.0°

Find the horizontal and vertical components of each vector.

23. 258 km at 135.0°

24. 42.2 mi/h at 303.0°

25. A man pushes with 160 N of force on the handle of a pushcart. If the angle between the handle and the ground is 45°, find the horizontal and vertical components of this force.

26. A woman pulls a loaded cart with a pulling force of 110 lb when the handle makes an angle of 23° with the ground. Find the horizontal and vertical components of this force.

27. If $\mathbf{v}_x = 18.5$ N and $\mathbf{v}_y = -31.0$ N, find $\mathbf{v}$.

Using the component method, find the sum of each set of vectors. Find magnitudes to three significant digits and angles to the nearest tenth of a degree.

28. $\mathbf{v}$ = 87.1 mi/h at 120.0°
$\mathbf{w}$ = 25.6 mi/h at 247.0°

29. $\mathbf{v}$ = 2560 N at 237.1°
$\mathbf{w}$ = 3890 N at 346.7°

30. $\mathbf{u}$ = 325 N at 90.0°
$\mathbf{v}$ = 325 N at 162.0°
$\mathbf{w}$ = 325 N at 270.0°

31. $\mathbf{v}$ = 19.7 km at 144.5°
$\mathbf{w}$ = 28.5 km at 180.0°
$\mathbf{u}$ = 10.3 km at 225.5°
$\mathbf{t}$ = 31.7 km at 90.0°

32. An airplane is to maintain a velocity of 550 km/h on a true course of $6\overline{0}°$ east of north. A wind of $9\overline{0}$ km/h is blowing from the east. What should the air speed and compass reading be to offset the wind?

33. A ship travels 16.5 mi at 13.5° east of north, then 24.7 mi at 34.5° west of north, then 30.5 mi due north, where it lands. Find the displacement from the starting point to the landing point.

34. A weight of 850 lb is suspended by two ropes attached to opposite walls at equal heights above the floor. The first rope makes an angle of 25° with the first wall, and the second rope makes an angle of 45° with the second wall. Find the tension in each rope.

CHAPTER 11
SPECIAL NASA APPLICATION
Vector Analysis of Wind Changes Affecting Shuttle Launch*

BACKGROUND

Space vehicles are preprogrammed to fly a certain trajectory in order to reach the correct orbit. This programming takes into account the forces of the winds expected to be encountered by the vehicle on ascent. If the winds actually encountered differ from those anticipated, the forces on the vehicle (called "loads") will differ. This could result either in the inability of the guidance system to fly the required trajectory or excessive loads leading to breakup of the vehicle. Either consequence is unacceptable. NASA/Kennedy Space Center will not launch a spacecraft if there is a perceived danger to the vehicle as a result of wind conditions.

DESCRIPTION OF THE DRWP

The Kennedy Space Center (KSC) Doppler Radar Wind Profiler (DRWP) is an instrument used to measure wind speed and direction as a function of height. It provides a measurement every 150 m of altitude above the 2-acre antenna. You can think of several parallel zones of air, one above the other, each 150 m wide. Each zone is called a "gate," from radar terminology, in which distance along the radar beam is measured in "range gates." The lowest gate, referred to as Gate 1, is at an altitude of 2011 m; there are 112 gates. A "wind profile" is a complete set of data from all 112 gates. That is, a wind profile provides the data needed to determine the velocity (speed and direction) of the wind in all gates at a certain instant of time. These data may be plotted as graphs of wind speed and direction versus height.

The wind velocity has two perpendicular components that are especially important: One component lies in the vertical plane determined by the Shuttle's launch trajectory; the other is perpendicular to the plane of the launch trajectory. These are referred to, respectively, as the in-plane and out-of-plane wind velocity. For a Shuttle launch, the profile

*From NASA–AMATYC–NSF Project Mathematics Explorations I, grant principals John S. Pazdar, Peter A. Wursthorn, and Patricia L. Hirschy; copyright Capital Community College, 1999.

collected at $t - 4$ h (4 h before scheduled launch) is used to create the guidance program. The profiles collected at $t - 2$ h and $t - 1$ h are then compared with the $t - 4$ h profile. If the absolute value of the change in either the in-plane or out-of-plane wind velocity relative to the $t - 4$ h reading is greater than a predetermined amount (typically in the order of 20 m/s), the launch may be delayed or scrubbed. In deciding whether to delay or to scrub a flight, it is important to realize that for each mission there is a window of time within which the Shuttle can be launched. Also, it takes several hours to recalculate the guidance program on the basis of the data taken at $t - 1$ h. The launch will be delayed if it is possible to alter the guidance program within the launch window; otherwise it will be scrubbed.

Suppose, for example, a scheduled launch time for a Shuttle is 10:00 A.M. on Monday, with a launch window that ends at 11:30 A.M. on the same day. The in-plane wind velocity at $t - 4$ h was 37 m/s and at $t - 1$ h it was 15 m/s. The absolute value of the difference of these in-plane velocities is 22 m/s. Because this exceeds 20 m/s, the flight must be scrubbed since the launch window ends 2.5 h from $t - 1$ h and it takes 4 h to recalculate the guidance program.

BUILDING THE MODEL

The DRWP provides the meteorologist with two components, u and v, of the wind velocity. The letter $\boldsymbol{u}$ represents the wind component in the east–west direction, with east taken as the positive direction, and $\boldsymbol{v}$ represents the wind component in the north–south direction, with north being positive. A coordinate system is used to establish a frame of reference with the positive x-axis pointing in the east direction and the positive y-axis pointing in the north direction. A vector can be represented using the rectangular coordinates (x, y) of the terminal point of the vector. If the east component is 5 m/s and the north component is -3 m/s (the negative sign means the wind direction is south), the wind velocity can be represented by the vector $(5, -3)$.

Vectors represented by ordered pairs in rectangular form can be added or subtracted by adding or subtracting x-components and adding or subtracting y-components. To illustrate,

$$(3, -4) + (2, 6) = (3 + 2, -4 + 6) = (5, 2)$$
$$(-7, 5) - (6, -3) = (-7 - 6, 5 - (-3)) = (-13, 8)$$

In order to decide whether the launch should be aborted, the operator must look at wind change in the plane of launch and wind change perpendicular to this plane. These components of the velocity vector are called the in-plane and out-of-plane components. The in-plane component is the vertical plane in which the Shuttle trajectory lies. When the Shuttle is launched, the direction in which it is launched depends on the mission. Common launch directions are directly east (0° north of east), 35° north of east, and 56° north of east. If the launch direction is 0°, the east component u is the in-plane component and the north component v is the out-of-plane component. This is the easiest case, since we do not have to rotate the axes to find the desired components. (See Fig. 11.74.)

EXAMPLE

Suppose the DRWP readings at Gate 40 at time $t - 4$ h are $u = 4$ m/s and $v = -2$ m/s. Therefore, the horizontal velocity vector $\mathbf{A} = (u, v)$ at $t - 4$ h is $(4, -2)$. This vector, along with the readings at the other 111 gates, is used to create a guidance program. The reason for using these data is that it takes almost 4 h to run the program to determine the loads, create a guidance program to ensure a safe launch under these conditions, and load the program. If the readings at Gate 40 at time $t - 1$ h are $u = 20$ m/s and $v = 16$ m/s, is this wind change drastic enough to delay or scrub the launch?

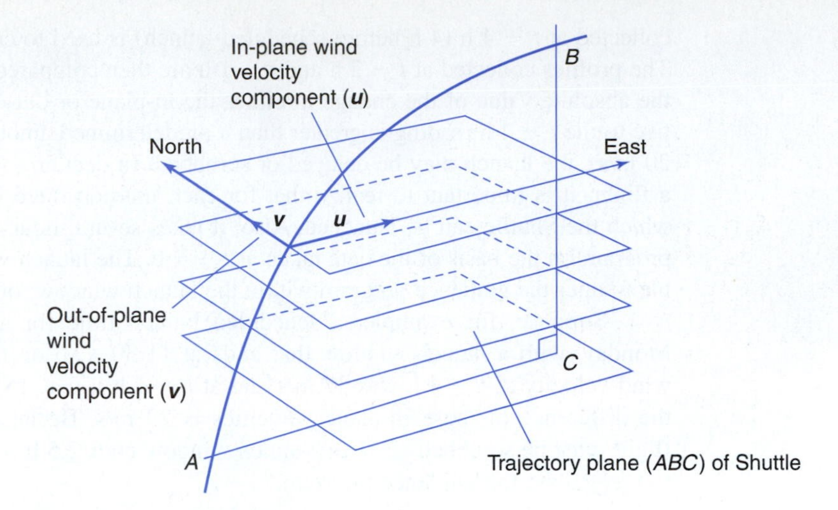

Figure 11.74 Gates and a due east Shuttle trajectory.

The wind velocity vector **A** at time $t - 1$ h is (20, 16) and at $t - 4$ h is (4, −2). The change in wind velocity is the difference between these values of **A.**

$$\text{change in } \mathbf{A} = (20, 16) - (4, -2) = (16, 18)$$

The threshold value for changes in wind-speed in-plane and out-of-plane components is approximately 20 m/s. Checking the vector for the change in wind velocity, we see that the absolute values are 16 m/s for the in-plane and 18 m/s for the out-of-plane component. We are still within the safety envelope and do not need to delay the launch to recalculate the guidance program (if the launch window is more than 3 h) or scrub the launch (if the launch window is less than 3 h).

Question: Suppose at Gate 62 the readings at $t - 4$ h are $u = -5$ m/s and $v = 13$ m/s and those at $t - 1$ h are $u = -5$ m/s and $v = -8$ m/s. Should the launch be delayed or scrubbed based on these readings? Justify your answer.

12 Graphing the Trigonometric Functions

INTRODUCTION

Harmonic motion, voltage, or current can be described using the general equation $y = a \sin(\omega t + \theta)$. In this chapter we will graph trigonometric functions and identify their amplitude, period, and phase shift.

Objectives

- Sketch the graphs of the trigonometric functions.
- Identify the amplitude, period, and phase shift for functions of the form

$$y = a \sin(bx + c) \quad \text{and} \quad y = a \cos(bx + c)$$

- Graph composite curves.
- Solve simple harmonic motion problems.

12.1 GRAPHING THE SINE AND COSINE FUNCTIONS

First, consider the graphs of the trigonometric equations $y = \sin x$ and $y = \cos x$. We find it convenient to express x in radian measure. (Be careful not to confuse the variables x and y in the equation $y = \sin x$ with the coordinates of a point on the terminal side of an angle. That is, in the equation of each trigonometric function, x is an angle in radian measure and y is the trigonometric value for that angle.)

To graph $y = \sin x$, find a large number of values of x and y that satisfy the equation and plot them in the xy-plane. It is convenient to scale the x-axis in multiples of π radians. A table of ordered pairs is as follows:

x	0	$\frac{\pi}{6}$	$\frac{\pi}{4}$	$\frac{\pi}{3}$	$\frac{\pi}{2}$	$\frac{2\pi}{3}$	$\frac{3\pi}{4}$	$\frac{5\pi}{6}$	π	$\frac{7\pi}{6}$	$\frac{5\pi}{4}$	$\frac{4\pi}{3}$	$\frac{3\pi}{2}$	$\frac{5\pi}{3}$	$\frac{7\pi}{4}$	$\frac{11\pi}{6}$	2π
y	0	0.5	0.71	0.87	1	0.87	0.71	0.5	0	−0.5	−0.71	−0.87	−1	−0.87	−0.71	−0.5	0

The graph is shown in Fig. 12.1. Note that the graph is a smooth, continuous curve. The domain (set of x-replacements or inputs) is the set of all real numbers (angles measured in radians). The range (set of y-replacements or outputs) is $-1 \le y \le 1$.

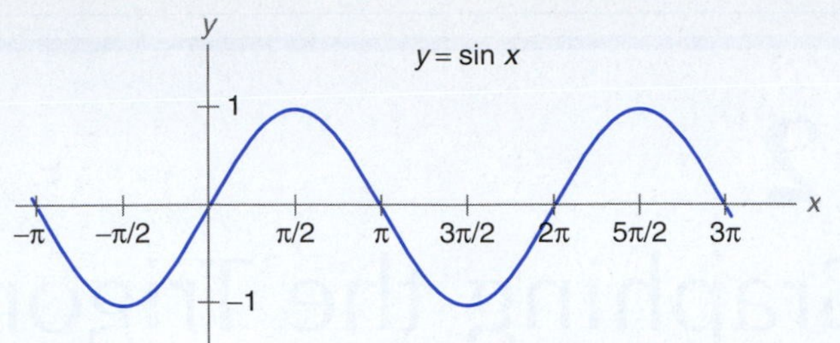

Figure 12.1

The graph of $y = \cos x$ is found in the same manner and is shown in Fig. 12.2.

x	0	$\frac{\pi}{6}$	$\frac{\pi}{4}$	$\frac{\pi}{3}$	$\frac{\pi}{2}$	$\frac{2\pi}{3}$	$\frac{3\pi}{4}$	$\frac{5\pi}{6}$	π	$\frac{7\pi}{6}$	$\frac{5\pi}{4}$	$\frac{4\pi}{3}$	$\frac{3\pi}{2}$	$\frac{5\pi}{3}$	$\frac{7\pi}{4}$	$\frac{11\pi}{6}$	2π
y	1	0.87	0.71	0.5	0	−0.5	−0.71	−0.87	−1	−0.87	−0.71	−0.5	0	0.5	0.71	0.87	1

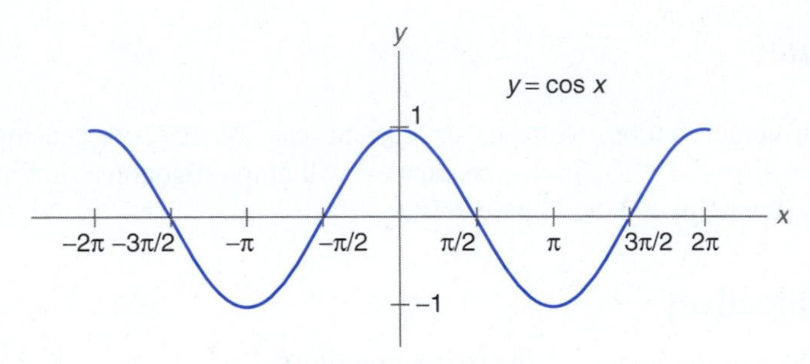

Figure 12.2

The domain and the range for $y = \cos x$ are the same as for $y = \sin x$. Note how the two graphs differ. Could we shift the cosine curve so that it coincides with the sine curve? Show $\sin\left(x + \frac{\pi}{2}\right) = \cos x$ for all values of x by choosing various values of x and checking them in each side of the equation.

It is important to note the points at which these two curves cross the axes. We shall "sketch" similar curves based on the fact that the basic shape of the sine and cosine curves always remains the same. This will save time and work in that it will be unnecessary to plot a large number of points each time we need to sketch a curve.

The basic sine curve is an equation of the form $y = a \sin x$, where a is a real number. We have already graphed this equation for $a = 1$. For $y = 2 \sin x$, each ordered pair satisfying this relationship has a y-value that is two times the corresponding y value for $y = \sin x$. Likewise, for $y = \frac{1}{2} \sin x$, each ordered pair satisfying this relationship has a y-value that is one-half the corresponding y-value for $y = \sin x$.

EXAMPLE 1

Graph $y = \sin x$, $y = 2 \sin x$, $y = \frac{1}{2} \sin x$, and $y = 4 \sin x$ on the same set of coordinate axes.

The graph is drawn in Fig. 12.3.

The coefficient a of $a \sin x$ or $a \cos x$ in the equations $y = a \sin x$ and $y = a \cos x$ determines the amplitude of the curve. The **amplitude** is the maximum y-value of the curve. The amplitudes for the curves in Example 1 are 1, 2, $\frac{1}{2}$, and 4, respectively. The

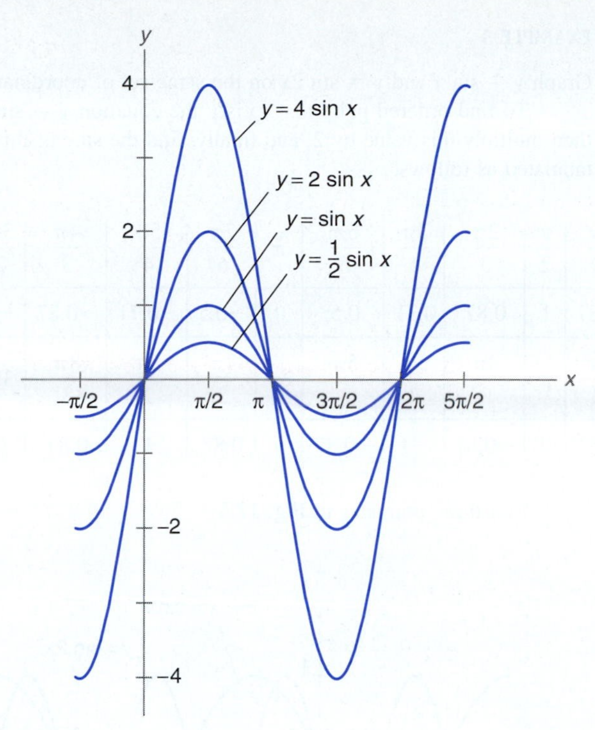

Figure 12.3

effect of a negative value of a in $y = a \sin x$ or $y = a \cos x$ is to "flip," or invert, the curve about the x-axis. That is, for $y = -\cos x$, each ordered pair satisfying this relationship has a y-value that is opposite the corresponding y-value for $y = \cos x$.

EXAMPLE 2

Graph $y = \cos x$, $y = -\cos x$, and $y = -3 \cos x$ on the same set of coordinate axes.

The graph is shown in Fig. 12.4. The amplitudes for the curves in this example are 1, 1, and 3, respectively.

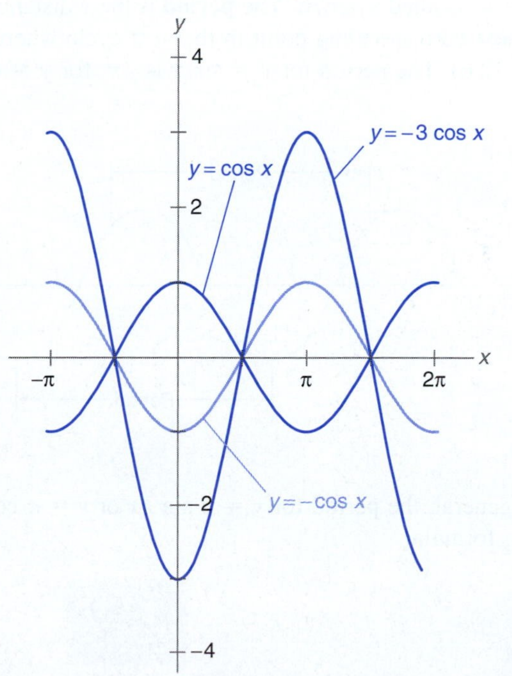

Figure 12.4

EXAMPLE 3

Graph $y = \sin x$ and $y = \sin 2x$ on the same set of coordinate axes.

To find ordered pairs that satisfy the equation $y = \sin 2x$, first choose a value for x, then multiply this value by 2, and finally, find the sine of this result. Some of the values are tabulated as follows:

x	0	$\frac{\pi}{6}$	$\frac{\pi}{4}$	$\frac{\pi}{3}$	$\frac{\pi}{2}$	$\frac{2\pi}{3}$	$\frac{3\pi}{4}$	$\frac{5\pi}{6}$	π	$\frac{7\pi}{6}$	$\frac{5\pi}{4}$	$\frac{4\pi}{3}$	$\frac{3\pi}{2}$	$\frac{5\pi}{3}$	$\frac{7\pi}{4}$	$\frac{11\pi}{6}$	2π
$\sin x$	0	0.5	0.71	0.87	1	0.87	0.71	0.5	0	−0.5	−0.71	−0.87	−1	−0.87	−0.71	−0.5	0
$2x$	0	$\frac{\pi}{3}$	$\frac{\pi}{2}$	$\frac{2\pi}{3}$	π	$\frac{4\pi}{3}$	$\frac{3\pi}{2}$	$\frac{5\pi}{3}$	2π	$\frac{7\pi}{3}$	$\frac{5\pi}{2}$	$\frac{8\pi}{3}$	3π	$\frac{10\pi}{3}$	$\frac{7\pi}{2}$	$\frac{11\pi}{3}$	4π
$\sin 2x$	0	0.87	1	0.87	0	−0.87	−1	−0.87	0	0.87	1	0.87	0	−0.87	−1	−0.87	0

Plot these points as in Fig. 12.5.

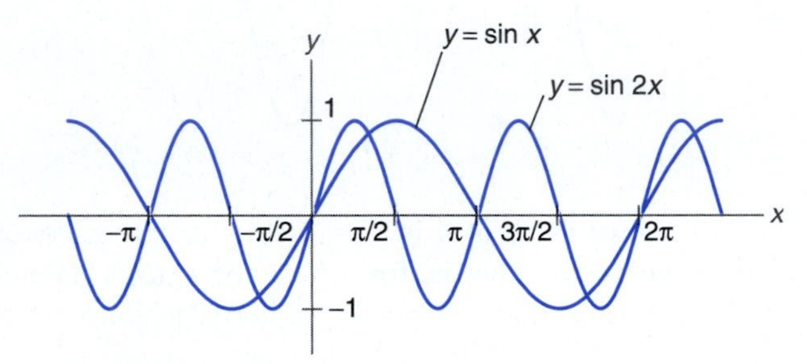

Figure 12.5

Note that the two curves in Fig. 12.5 have the same basic shape, but $y = \sin 2x$ goes through a complete cycle twice as $y = \sin x$ completes one cycle. The length of each of these cycles is called a *period*. The **period** is the x-distance between any point on the curve and the next corresponding point in the next cycle where the graph starts repeating itself (see Fig. 12.6). The period for $y = \sin x$ is 2π; for $y = \cos x$, 2π; and for $y = \sin 2x$, π.

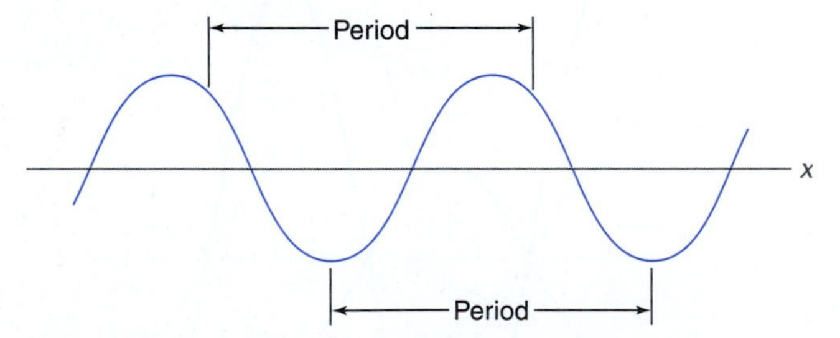

Figure 12.6

In general, the period for $y = a \sin bx$ or $y = a \cos bx$ may be found by using the following formula:

$$P = \frac{2\pi}{b}$$

EXAMPLE 4

Graph $y = 2 \cos 3x$.

The amplitude is 2 and the period is

$$P = \frac{2\pi}{b} = \frac{2\pi}{3}$$

Sketch a cosine graph with amplitude 2 that completes one complete cycle each $\frac{2\pi}{3}$ radians as in Fig. 12.7.

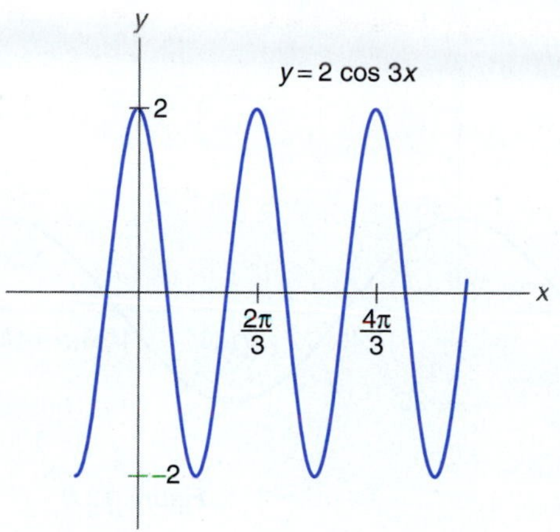

Figure 12.7

EXAMPLE 5

Graph $y = -3 \sin \frac{1}{2}x$.

The amplitude is 3; the period is

$$P = \frac{2\pi}{b} = \frac{2\pi}{\frac{1}{2}} = 4\pi$$

The effect of the negative sign is to flip, or invert, the curve $y = 3 \sin \frac{1}{2}x$ about the x-axis. Sketch a sine graph with amplitude 3 that completes one complete cycle each 4π radians as shown by the dashed graph in Fig. 12.8. Then flip the dashed curve about the x-axis to obtain the solid-line graph of the given function.

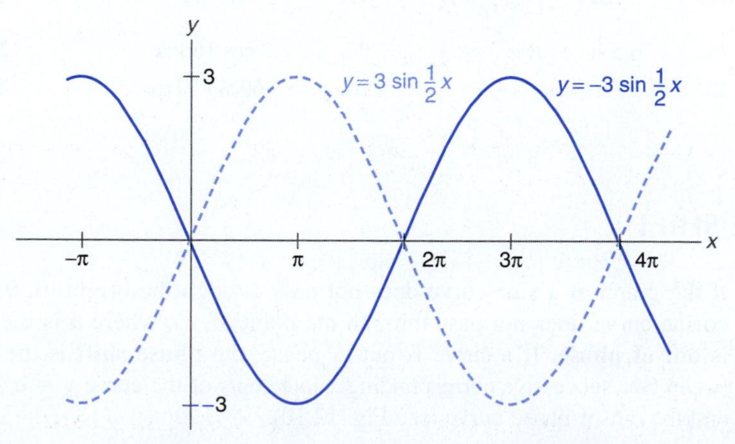

Figure 12.8

EXAMPLE 6

Graph $y = 25 \cos 120\pi x$.

The amplitude is 25; the period is

$$P = \frac{2\pi}{b} = \frac{2\pi}{120\pi} = \frac{1}{60}$$

Sketch a cosine graph with amplitude 25 that completes one complete cycle each $\frac{1}{60}$ radian as shown in Fig. 12.9.

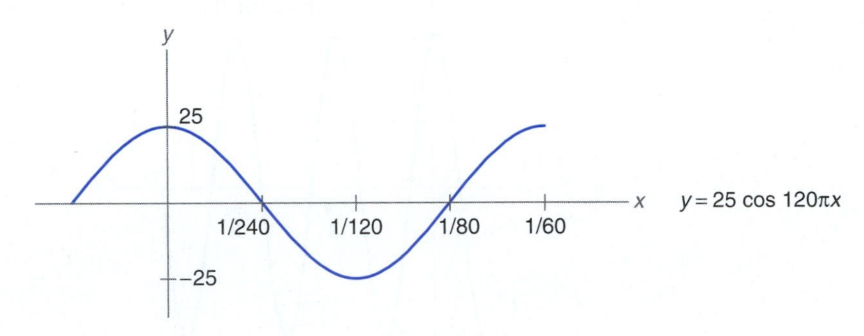

Figure 12.9

Exercises 12.1

Find the amplitude and the period of each function, and sketch its graph through at least two cycles.

1. $y = 2 \cos x$

2. $y = 5 \sin x$

3. $y = -3 \sin x$

4. $y = -4 \cos x$

5. $y = \sin 3x$

6. $y = \sin 4x$

7. $y = \cos 2x$

8. $y = \cos 3x$

9. $y = 2 \sin 4x$

10. $y = 3 \cos 6x$

11. $y = \frac{5}{2} \cos \frac{1}{2}x$

12. $y = \frac{3}{2} \sin \frac{3}{2}x$

13. $y = -\frac{1}{2} \sin \frac{2}{3}x$

14. $y = -\frac{5}{2} \cos \frac{3}{4}x$

15. $y = 2 \sin 3\pi x$

16. $y = 6 \cos \frac{4\pi x}{3}$

17. $y = -3 \cos \pi x$

18. $y = -\sin \frac{\pi x}{2}$

19. $y = 6.5 \sin 120\pi x$

20. $y = 12 \cos 160\pi x$

21. $y = 40 \cos 60x$

22. $y = 60 \sin 40x$

23. $y = -60 \sin 80\pi x$

24. $y = -240 \cos 120x$

12.2 PHASE SHIFT

If the graph of a sine curve does not pass through the origin (0, 0), or if the graph of a cosine curve does not pass through the point $(0, a)$, where a is the amplitude, the curve is **out of phase.** If a curve is out of phase, the **phase shift** is the directed distance between two successive corresponding x-intercepts of the curve $y = a \sin bx$ or $y = a \cos bx$ and the out-of-phase curve (see Fig. 12.10).

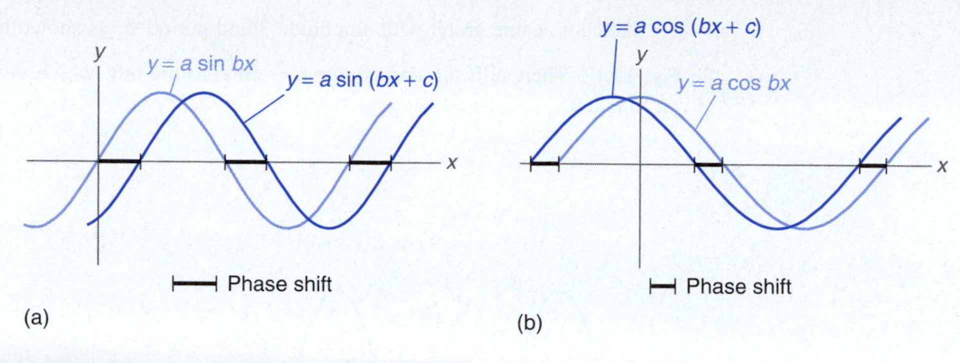

Figure 12.10 Phase shift.

EXAMPLE 1

Graph $y = \sin x$ and $y = \sin\left(x - \frac{\pi}{4}\right)$ on the same set of coordinate axes.

For $y = \sin\left(x - \frac{\pi}{4}\right)$, complete the following table:

x	$-\frac{\pi}{4}$	0	$\frac{\pi}{4}$	$\frac{\pi}{2}$	$\frac{3\pi}{4}$	π	$\frac{5\pi}{4}$	$\frac{3\pi}{2}$	$\frac{7\pi}{4}$	2π	$\frac{9\pi}{4}$	$\frac{5\pi}{2}$	$\frac{11\pi}{4}$
y	-1	-0.71	0	0.71	1	0.71	0	-0.71	-1	-0.71	0	0.71	1

Plot the points as in Fig. 12.11.

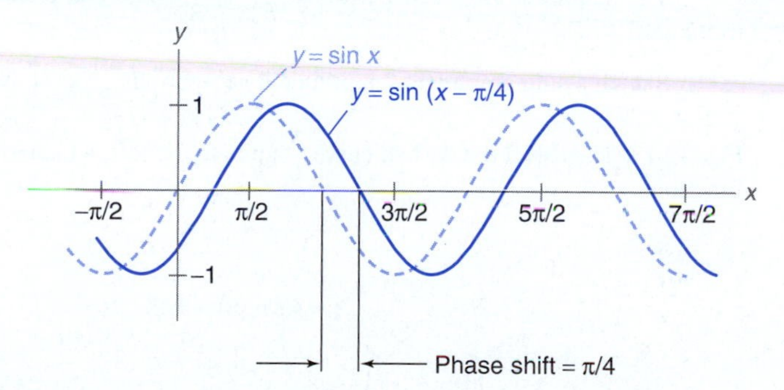

Figure 12.11

A consideration of phase shifts is helpful when graphing equations in the form $y = a \sin(bx + c)$ or $y = a \cos(bx + c)$. The effect of c in these equations is to shift the curve $y = a \sin bx$ or $y = a \cos bx$ as follows:

1. To the *left* $\frac{c}{b}$ units if $\frac{c}{b} > 0$.
2. To the *right* $\frac{c}{b}$ units if $\frac{c}{b} < 0$.

EXAMPLE 2

Graph $y = 3 \cos\left(2x + \frac{\pi}{4}\right)$.

The amplitude is 3. The period is $\frac{2\pi}{b} = \frac{2\pi}{2} = \pi$. The phase shift is $\frac{c}{b} = \frac{\pi/4}{2} = \frac{\pi}{8}$, or $\frac{\pi}{8}$ to the left.

Sketch a cosine graph with amplitude 3 and period π, as shown by the dashed graph in Fig. 12.12. Then shift the dashed curve $\frac{\pi}{8}$ units to the left, which gives the graph of the given function.

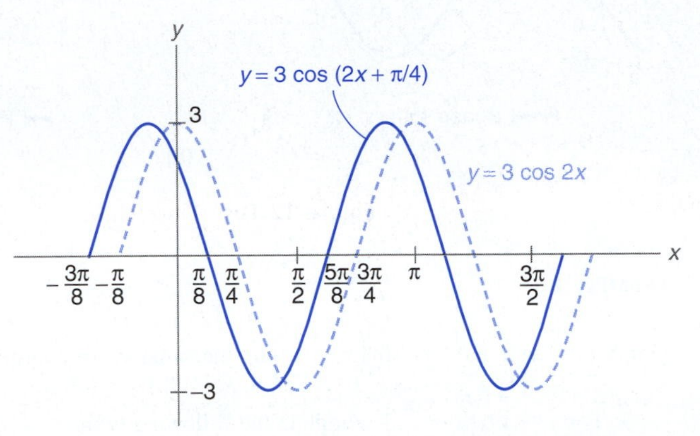

Figure 12.12

EXAMPLE 3

Graph $y = 4\sin\left(3x + \frac{\pi}{2}\right)$.

The amplitude is 4. The period is $\frac{2\pi}{b} = \frac{2\pi}{3}$. The phase shift is $\frac{c}{b} = \frac{\pi/2}{3} = \frac{\pi}{6}$, or $\frac{\pi}{6}$ to the left.

Sketch a sine graph with amplitude 4 and period $\frac{2\pi}{3}$ as shown by the dashed graph in Fig. 12.13. Then shift the dashed curve $\frac{\pi}{6}$ units to the left, which gives the graph of the given function.

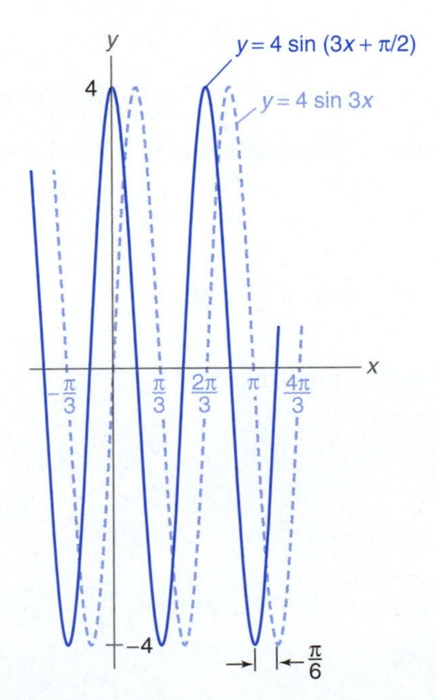

Figure 12.13

EXAMPLE 4

Graph $y = -3 \sin\left(\frac{1}{2}x - \pi\right)$.

The amplitude is 3. The period is $\frac{2\pi}{b} = \frac{2\pi}{\frac{1}{2}} = 4\pi$. The phase shift is $\frac{c}{b} = \frac{-\pi}{\frac{1}{2}} = -2\pi$, or 2π to the right.

First, graph $y = -3 \sin \frac{1}{2}x$, the dashed curve in Fig. 12.14. Then shift the dashed curve 2π units to the right, which gives the graph of the given function.

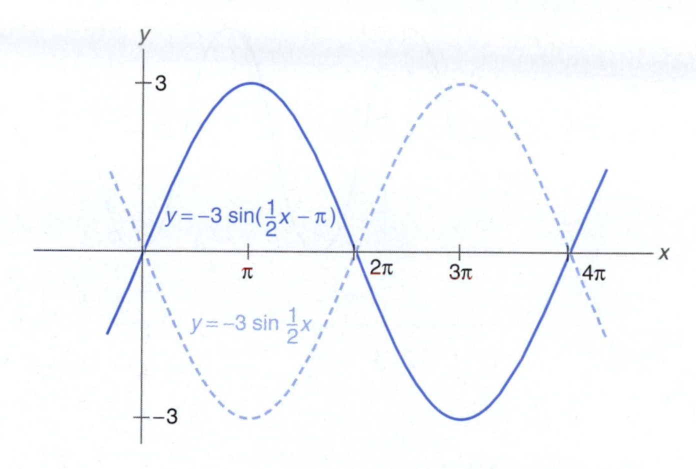

Figure 12.14

AMPLITUDE, PERIOD, AND PHASE SHIFT

In summary, the values of a, b, and c in the following equations determine amplitude, period, and phase shift as follows:

$$y = a \sin(bx + c)$$
$$y = a \cos(bx + c)$$

Amplitude $= |a|$ Phase shift $= \frac{c}{b}$ $\left(\text{To the } left \text{ if } \frac{c}{b} > 0\right)$

Period $= \frac{2\pi}{b}$ $\left(\text{To the } right \text{ if } \frac{c}{b} < 0\right)$

Vertical Shift

The difference between the graph of any trigonometric equation in the form in the left column below and its counterpart in the right column is that the graph of the equation on the left is shifted vertically *up* d units when $d > 0$ or shifted vertically *down* d units when $d < 0$:

$y = a \sin(bx + c)$	$y = a \sin(bx + c) + d$
$y = a \cos(bx + c)$	$y = a \cos(bx + c) + d$
$y = a \tan(bx + c)$	$y = a \tan(bx + c) + d$

$$y = a \cot(bx + c) \qquad y = a \cot(bx + c) + d$$
$$y = a \sec(bx + c) \qquad y = a \sec(bx + c) + d$$
$$y = a \csc(bx + c) \qquad y = a \csc(bx + c) + d$$

For example, the graph of $y = 3 \cos(2x + \pi/4) + 5$ in Fig. 12.15 is shifted vertically up 5 units in comparison with the graph of $y = 3 \cos(2x + \pi/4)$ in Fig. 12.12.

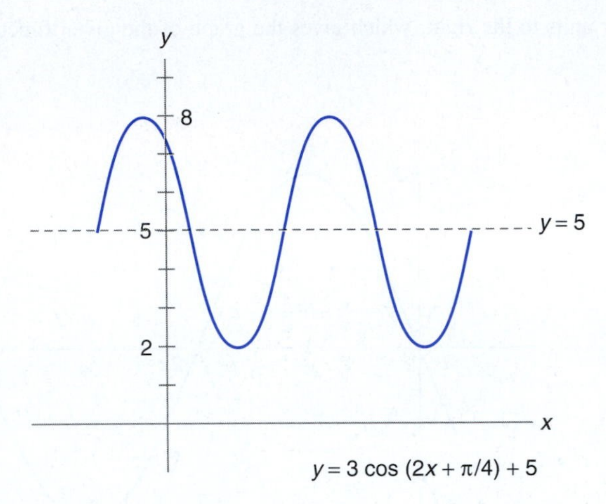

Figure 12.15

Exercises 12.2

Find the amplitude, period, and phase shift of each function and sketch its graph.

1. $y = \sin\left(x + \dfrac{\pi}{3}\right)$
2. $y = \cos\left(x + \dfrac{\pi}{4}\right)$
3. $y = 2\cos\left(x - \dfrac{\pi}{6}\right)$
4. $y = 3\sin\left(x - \dfrac{\pi}{3}\right)$
5. $y = \sin(3x - \pi)$
6. $y = \cos\left(2x + \dfrac{\pi}{3}\right)$
7. $y = -\cos(4x + \pi)$
8. $y = -\sin\left(4x - \dfrac{2\pi}{3}\right)$
9. $y = 3\sin\left(\dfrac{1}{2}x - \dfrac{\pi}{4}\right)$
10. $y = 4\sin\left(\dfrac{1}{2}x - \dfrac{4\pi}{3}\right)$
11. $y = 2\sin\left(\dfrac{4}{3}x + \dfrac{\pi}{3}\right)$
12. $y = 5\cos\left(\dfrac{2}{3}x + \pi\right)$
13. $y = 3\sin(\pi x + \pi)$
14. $y = 3\cos\left(\dfrac{\pi x}{4} + \dfrac{\pi}{8}\right)$
15. $y = 2\cos\left(\dfrac{\pi x}{2} - \dfrac{\pi}{2}\right)$
16. $y = 4\sin\left(\dfrac{\pi x}{6} - \dfrac{\pi}{3}\right)$
17. $y = 40\cos\left(60x - \dfrac{\pi}{3}\right)$
18. $y = 80\sin\left(40x - \dfrac{\pi}{4}\right)$
19. $y = 120\cos\left(40\pi x - \dfrac{\pi}{2}\right)$
20. $y = 40\sin\left(20\pi x + \dfrac{\pi}{3}\right)$
21. $y = 7.5\sin(220x + \pi)$
22. $y = 4.5\cos\left(180x - \dfrac{\pi}{3}\right)$
23. $y = 20\sin(120\pi x + 4\pi)$
24. $y = 30\cos(160\pi x + 8\pi)$

12.3 GRAPHING THE OTHER TRIGONOMETRIC FUNCTIONS

The remaining trigonometric functions have interesting graphs, but they do not have as many technical applications. We shall graph these remaining functions and discuss them briefly.

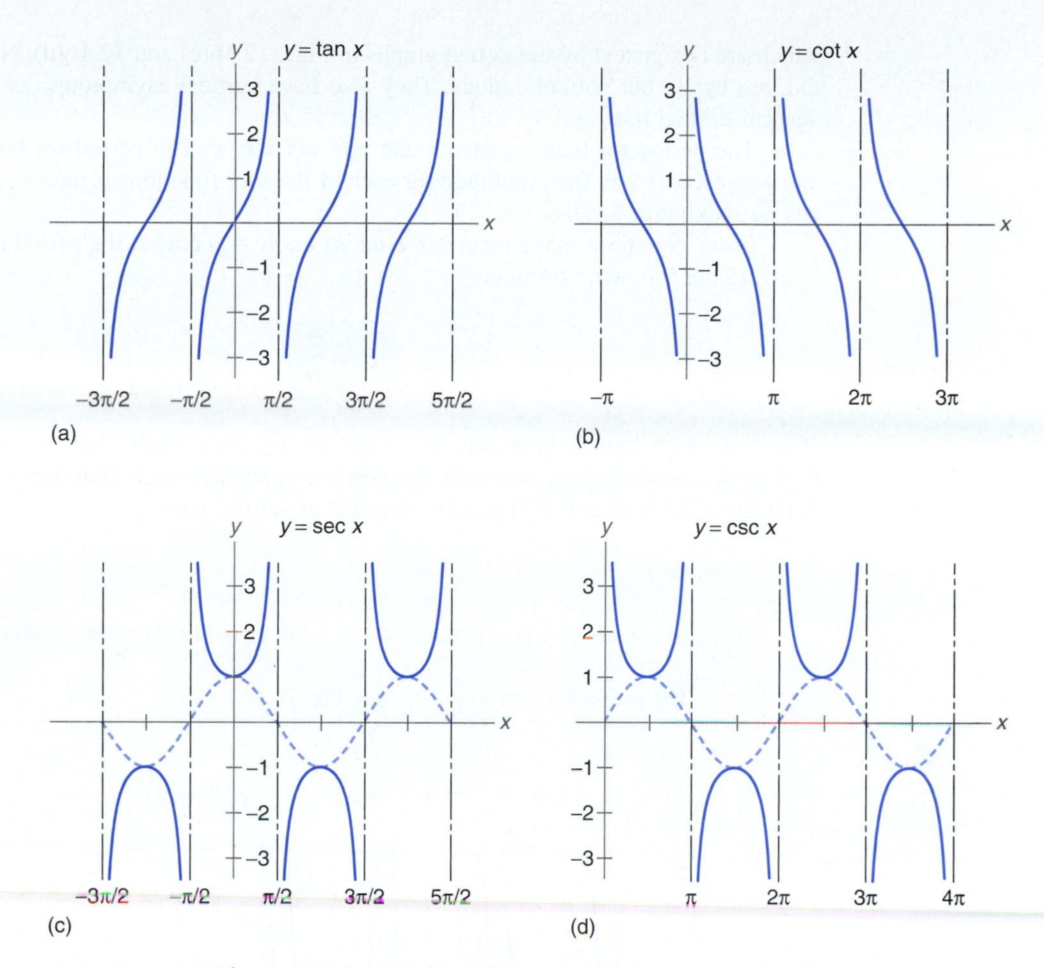

Figure 12.16 Graphs of the other trigonometric functions.

To graph $y = \tan x$, we find a large number of values of x and y that satisfy the equation and plot them in the xy-plane as in Fig. 12.16(a).

x	$-\frac{\pi}{2}$	$-\frac{\pi}{3}$	$-\frac{\pi}{4}$	$-\frac{\pi}{6}$	0	$\frac{\pi}{6}$	$\frac{\pi}{4}$	$\frac{\pi}{3}$	$\frac{\pi}{2}$	$\frac{2\pi}{3}$	$\frac{3\pi}{4}$	$\frac{5\pi}{6}$	π	$\frac{7\pi}{6}$	$\frac{5\pi}{4}$	$\frac{4\pi}{3}$	$\frac{3\pi}{2}$
y	—	−1.7	−1	−0.58	0	0.58	1	1.7	—	−1.7	−1	−0.58	0	0.58	1	1.7	—

Note that $y = \tan x$ is a cyclic curve but is not continuous. At the points

$$x = \cdots, -\frac{\pi}{2}, \frac{\pi}{2}, \frac{3\pi}{2}, \frac{5\pi}{2}, \ldots$$

there are vertical lines called **asymptotes,** which form guidelines to the curve. That is, as x gets larger and closer to $\frac{\pi}{2}$, y becomes larger and larger and the graph approaches the asymptote. The graph never crosses the asymptote because at $x = \frac{\pi}{2}$, the tangent function is undefined.

Likewise, the three other trigonometric functions may be graphed. These graphs are shown in Figs. 12.16(b) through 12.16(d).

Since the secant and cosecant functions are reciprocals of the cosine and sine, respectively, each may be sketched by plotting reciprocals of the cosine and sine functions, which are designated by the dotted graphs in Figs. 12.16(c) and 12.16(d). Note that they, too, are cyclic but not continuous. They also have vertical asymptotes, as noted by the vertical dashed lines.

The period for both $y = \tan x$ and $y = \cot x$ is π. The period for both $y = \sec x$ and $y = \csc x$ is 2π. The amplitude for each of the four functions is undefined since each has no maximum y-value.

Given equations in the form $y = a \tan bx$ and $y = a \cot bx$, the period may be found by using the following formula:

$$P = \frac{\pi}{b}$$

If $a > 1$, each branch intersects the x-axis at a greater angle than for $y = \tan x$. If $0 < a < 1$, each branch intersects the x-axis at a smaller angle than for $y = \tan x$. If a is negative, each branch is flipped or inverted about the x-axis.

EXAMPLE 1

Graph $y = 2 \tan 2x$, $y = \frac{1}{2} \tan 2x$, and $y = -\tan 2x$ on the same set of coordinate axes.

The period for each is $\frac{\pi}{b} = \frac{\pi}{2}$ (see Fig. 12.17).

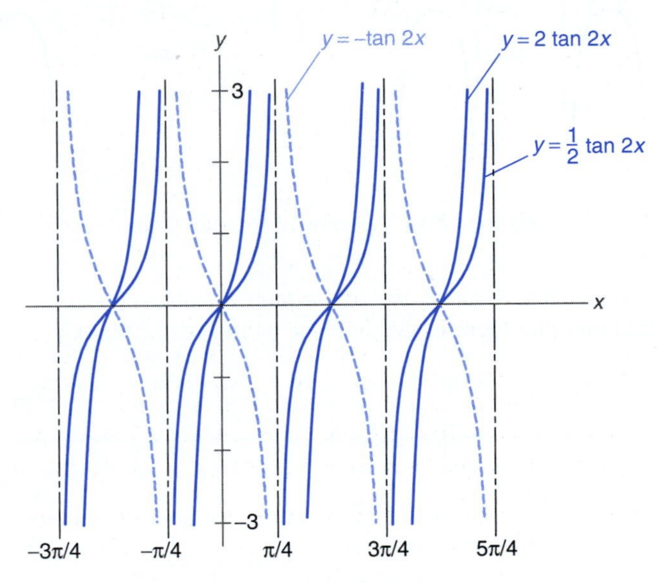

Figure 12.17

Given equations in the form $y = a \sec bx$ and $y = a \csc bx$, the period may be found by using the following formula:

$$P = \frac{2\pi}{b}$$

The quantity $|a|$ is the vertical distance between the x-axis and the low point of each of the branches above the x-axis. If a is negative, each branch is flipped or inverted about the x-axis.

EXAMPLE 2

Graph $y = 2 \sec 4x$.

The period is $\frac{2\pi}{4} = \frac{\pi}{2}$ and $a = 2$. As a graphical aid, you may first graph $y = 2 \cos 4x$ as in Fig. 12.18.

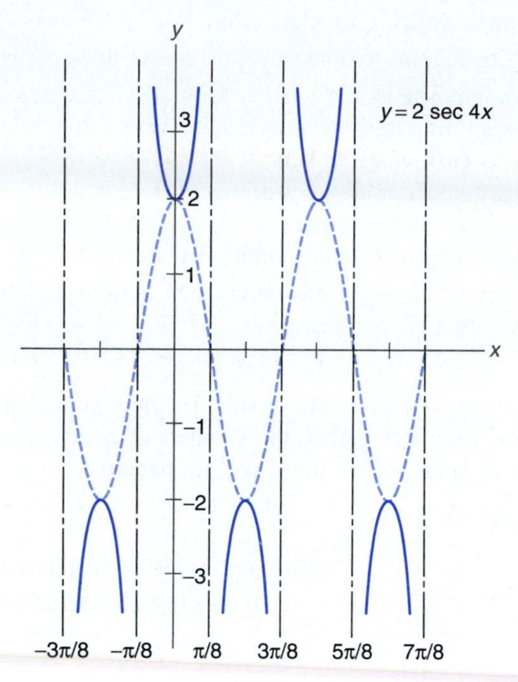

Figure 12.18

Exercises 12.3

Find the period of each function and sketch its graph.

1. $y = \tan 3x$

2. $y = \tan 4x$

3. $y = -2 \tan \frac{1}{2}x$

4. $y = -\tan \frac{3}{2}x$

5. $y = \cot 6x$

6. $y = 2 \cot \frac{1}{3}x$

7. $y = 3 \sec x$

8. $y = 4 \csc x$

9. $y = -2 \csc x$

10. $y = -5 \sec x$

11. $y = 2 \sec 6x$

12. $y = 4 \sec 2x$

13. $y = -3 \sec \frac{1}{2}x$

14. $y = -5 \sec \frac{3}{2}x$

15. $y = \csc 2x$

16. $y = 2 \csc 3x$

Sketch the graph of each function.

17. $y = \tan\left(x + \frac{\pi}{2}\right)$

18. $y = \cot\left(2x - \frac{\pi}{2}\right)$

19. $y = 3 \sec\left(2x - \frac{\pi}{2}\right)$

20. $y = 2 \csc\left(x + \frac{\pi}{3}\right)$

12.4 GRAPHING COMPOSITE CURVES

In applications it is common to find functions that are composites of sums or differences of expressions, some of which are trigonometric, such as $y = \cos x + \sin 2x$ or

$y = x - \sin x$. To graph such functions, a graphical technique called **addition of ordinates** is useful.

ADDITION OF ORDINATES

1. Graph each of the functions that make up the composite on the same set of coordinate axes.
2. It may be helpful to draw several vertical lines perpendicular to the x-axis.
3. If the composite is
   (a) a sum, find the algebraic sum of the y-values where each vertical line intersects each of the graphs.
   (b) a difference, find the algebraic difference of the y-values where each vertical line intersects each of the graphs, or graph the negative (or "flipped") curve and find the algebraic sum as in (a).
4. Plot each resultant y-value from Step 3 on each vertical line and connect the points with a smooth curve.

Or, instead of Step 3, add or subtract the y-values on the graph itself by means of a compass or ruler. Remember, the y-values of points above the x-axis are positive and the y-values of points below the x-axis are negative.

EXAMPLE 1

Graph $y = \cos x + \sin 2x$ using the method of addition of ordinates.

The graph is shown in Fig. 12.19. *Note:* One period of $y = \sin 2x$ is not sufficient since $\cos x$ has a longer period.

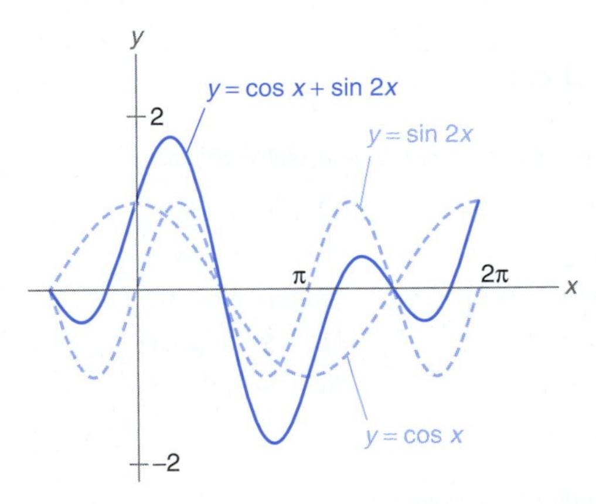

Figure 12.19

By calculator, we have

green diamond Y=

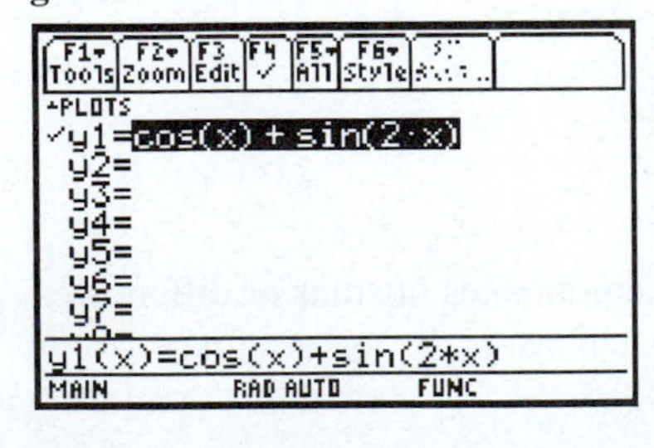

F2 7

(x scale is $\pi/2$; y scale is 0.5)

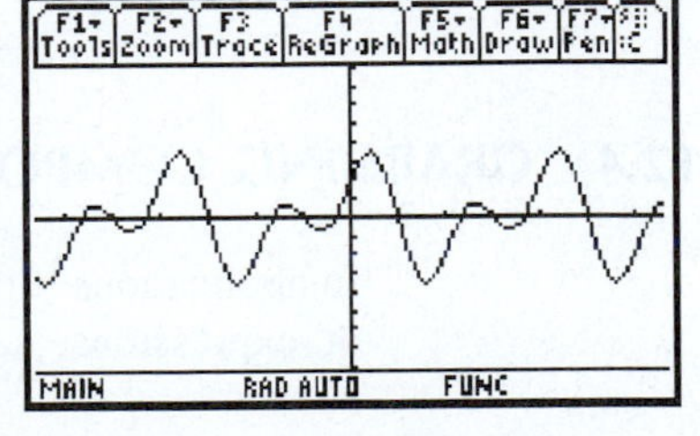

EXAMPLE 2

Graph $y = x - \sin x$ using the method of addition of ordinates.

The graph is shown in Fig. 12.20.

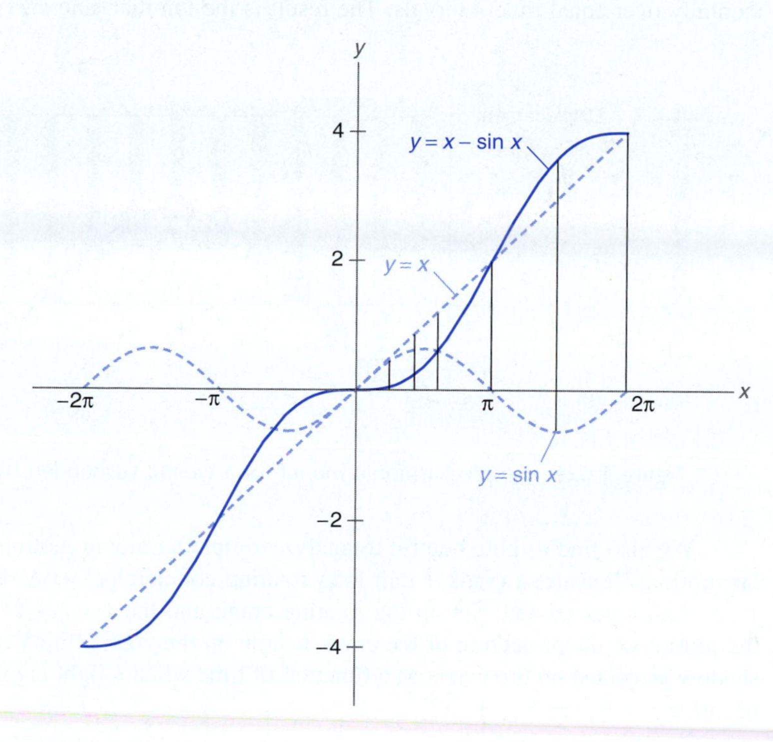

Figure 12.20

Exercises 12.4

Use the method of addition of ordinates or a graphing calculator to graph one complete period of each composite curve.

1. $y = \sin x + \cos x$
2. $y = 2 \cos x + \sin x$
3. $y = 2 \cos x + 2 \sin x$
4. $y = \sin x + 3 \cos x$
5. $y = 2 \sin x + \sin \dfrac{x}{2}$
6. $y = 2 \cos \dfrac{x}{2} + \sin x$
7. $y = 3 \sin x + 2 \cos x$
8. $y = 2 \sin 2x + 2 \cos x$
9. $y = 2 \sin x - \cos x$
10. $y = \cos x - 2 \sin x$
11. $y = 2 \sin 2x - \cos x$
12. $y = \cos 2x - \cos x$
13. $y = x + \sin x$
14. $y = x - 2 \cos x$
15. $y = \sin 4x + \cos \left(2x + \dfrac{\pi}{3}\right)$
16. $y = \sin 2x + \cos \left(2x - \dfrac{\pi}{4}\right)$

12.5 SIMPLE HARMONIC MOTION

Simple harmonic motion is a type of linear periodic motion of a particle between two definite endpoints that is symmetric about the midpoint or point of equilibrium. For example, consider a weight suspended on a spring. First, pull down on the weight and then let go; the weight moves up and down in simple harmonic motion as shown in Fig. 12.21(a). The displacement is the distance between the point of equilibrium and either endpoint. Next, graph the vertical displacement of the weight over equal units of time as in Fig. 12.21(b), where successive vertical positions of the weight are shown displaced horizontally over equal time intervals. The result is the familiar sine curve.

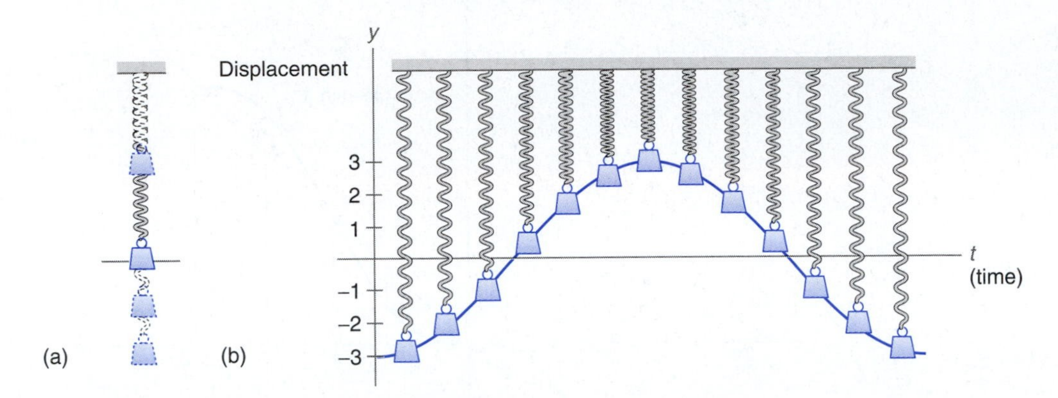

Figure 12.21 Simple harmonic motion of a weight suspended by a spring.

We also find it quite helpful to analyze simple harmonic motion in terms of circular motion. Consider a crank 1 unit long rotating counterclockwise at a constant rate of 1 revolution per second. Set up the rotating crank and the *xy*-axes as follows: Consider the motion of the projection of the crank handle on the *y*-axis (think of the motion of its shadow projected on the *y*-axis as a function of time when a light is placed far out on the negative *x*-axis; see Fig. 12.22).

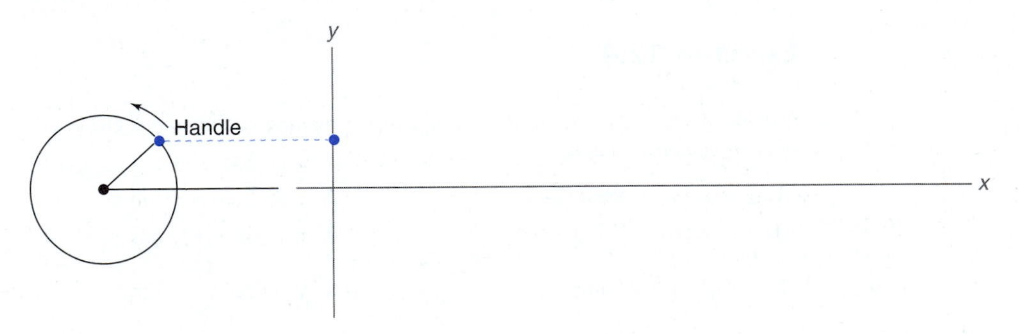

Figure 12.22 Simple harmonic motion of the projection of a rotating crank handle on the *y*-axis.

Let's assume that the crank starts from 0°. At the end of $\frac{1}{12}$ s, it will have rotated 30°, or $\frac{\pi}{6}$ rad, to position P_1 as in Fig. 12.23; at the end of $\frac{2}{12}$ s, it will have rotated 60°, or $\frac{\pi}{3}$ rad, to position P_2; and so forth; until after 1 s, or 1 rev, it will have rotated back to its original position P_0.

The projection of the crank handle on the *y*-axis (that is, its *y*-value per unit time) can be plotted as a curve. We let the *x*-axis be the *time* axis and divide it into as many intervals as there are angles to be plotted—in this case 12, every $\frac{\pi}{6}$ rad from 0 to 2π rad. Through

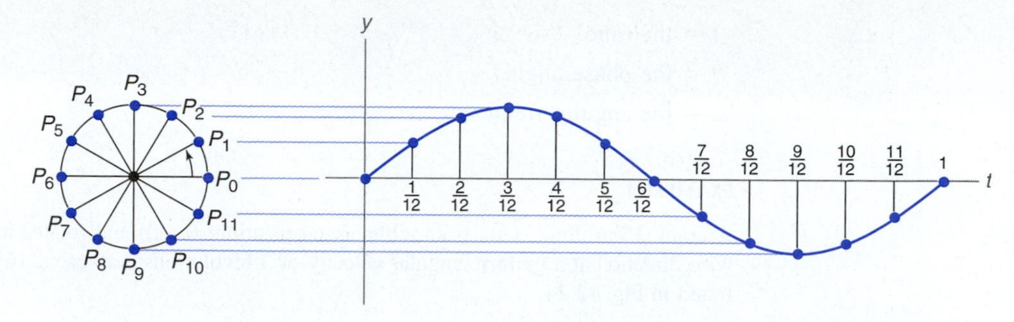

Figure 12.23 Simple harmonic motion of the projection of a rotating crank handle on the y-axis with the crank starting at 0°.

the points of division on the time axis (x-axis), construct vertical lines. Through the position points $P_0, P_1, P_2, \ldots$, construct horizontal lines. Draw a small closed circle around the intersection of the corresponding vertical and horizontal lines, that is, P_1 and $\frac{1}{12}$, P_2 and $\frac{2}{12}$, and so on. Draw a smooth curve through these points of intersection, which results in a sine curve. If we were to continue rotating the crank, successive rotations would generate successive periods of the sine curve. The equation is $y = \sin x$. Since the crank rotates through 2π rad in 1 s, $x = 2\pi$ in 1 s, $x = 4\pi$ in 2 s, and $x = 2\pi t$ in t s. Therefore, the y-value at any time t is given by $y = \sin 2\pi t$.

More generally, we let a be the length of a vector **B** rotating at a uniform angular velocity of ω rad per unit time. If we further assume a horizontal starting position at $t = 0$, the projections of vector **B** onto the y-axis, which are the vertical components of **B**, will be in simple harmonic motion. Their motion may be represented by the equation

$$y = a \sin \omega t$$

This rotating position vector is sometimes called a *phasor.*

This equation is of fundamental importance in that it describes the motion of any object or quantity that is in simple harmonic motion. Familiar examples of motion that is simple harmonic motion (or very nearly so) are (1) the motion of a satellite in a circular orbit; (2) the motion of a pendulum bob if the displacement is small; (3) the motion of the prongs of a vibrating tuning fork (sound waves); (4) the motion of a rotating wire through a magnetic field—alternating current; and (5) the motion of a mass vibrating up and down on a spring. If the rotating vector is not in position P_0 at $t = 0$, the simple harmonic motion equation becomes

$$y = a \sin (\omega t + \theta)$$

where θ is the initial position of the vector at $t = 0$. The graph results in a phase shift of $\frac{\theta}{\omega}$.

The general equation of a particle in simple harmonic motion is

$$y = a \sin (2\pi f t + \theta) = a \sin (\omega t + \theta)$$

where

$y =$ the displacement of motion at any time t

$a =$ the amplitude or maximum displacement of motion

$f =$ the frequency $= \frac{1}{\text{period}} =$ the number of oscillations per unit of time

t = the time
θ = the phase angle
ω = the angular frequency

EXAMPLE 1

A crank 15 in. long starts from a horizontal position (0 rad) and rotates in a counterclockwise direction at a uniform angular velocity of 3 revolutions per second (6π rad/s), as illustrated in Fig. 12.24.

(a) Plot the curve (as shown previously in Fig. 12.23) that shows the projection of the crank handle on the y-axis per unit time.
(b) Find the equation of this curve.
(c) Through how many radians does the crank turn in 0.15 s?
(d) What is the distance of the crank handle from the horizontal axis (t-axis) at $\frac{1}{9}$ s? That is, what is the length of the projection of the crank on the y-axis at $\frac{1}{9}$ s?

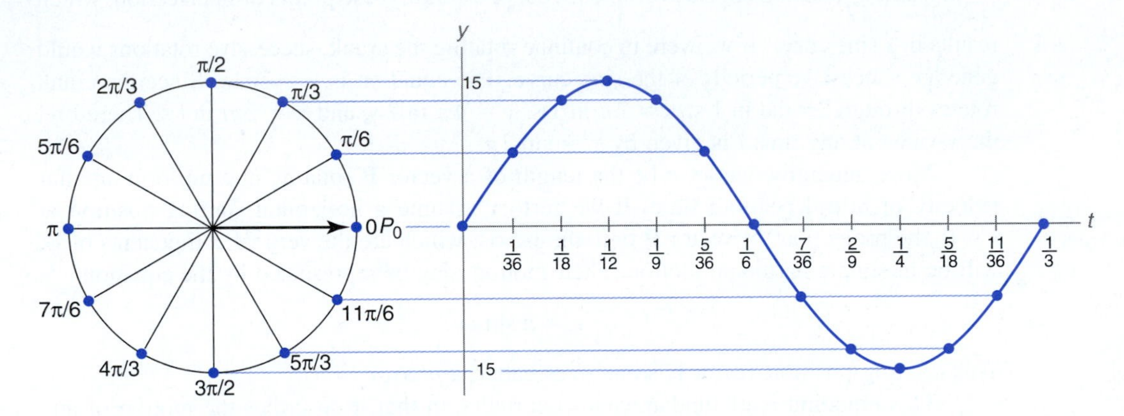

Figure 12.24

(a) The amplitude a is 15. The period is $\frac{2\pi}{6\pi} = \frac{1}{3}$. There is no phase shift, and $\omega = 6\pi$ rad/s.

(b) $y = a \sin \omega t$
$y = 15 \sin 6\pi t$

(c) $\theta = \omega t$
$\theta = (6\,\pi \text{ rad/s})(0.15 \text{ s}) = 2.8 \text{ rad}$

(d) Length of projection $= (15 \text{ in.}) \cdot \sin \frac{2\pi}{3} = 13$ in.

EXAMPLE 2

A wire rotating through a magnetic field generates an alternating current (ac). The instantaneous current i is given by the equation

$$i = I \sin(\omega t + \theta)$$

where I is the maximum instantaneous current, ω is the angular velocity of the rotating wire, t is the time, and θ is the phase angle. Given $I = 12$ A, $\omega = 120\pi$ rad/s (60 cycles/s), and $\theta = \frac{\pi}{3}$ (see Fig. 12.25):

(a) Graph the resulting equation $i = 12 \sin\left(120\pi t + \frac{\pi}{3}\right)$.

(b) From the graph find the time at which the current is +10 A, −6 A, and 0 A.

(c) From the graph find the current at $t = \frac{1}{240}$ s and $\frac{1}{120}$ s.

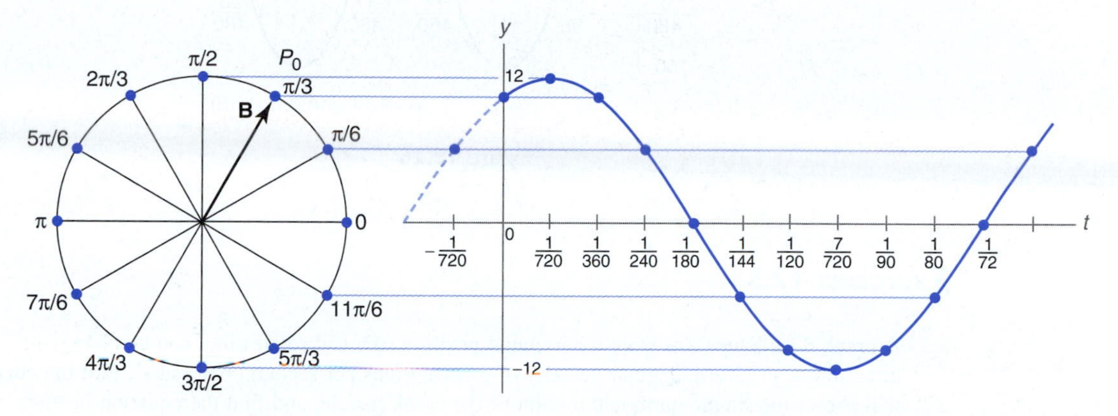

Figure 12.25

(a) The amplitude is 12. The period is $\frac{2\pi}{120\pi} = \frac{1}{60}$. The phase shift is $\frac{\pi/3}{120\pi} = \frac{1}{360}$ to the left.

(b) At $i = +10$ A, $t = 0, \frac{1}{360}, \frac{1}{60}, \frac{7}{360}, \ldots$.

At $i = -6$ A, $t = \frac{1}{144}, \frac{1}{80}, \frac{17}{720}, \ldots$.

At $i = 0$, $t = \frac{1}{180}, \frac{1}{72}, \frac{1}{45}, \ldots$.

(c) At $t = \frac{1}{240}$ s, $i = +6$ A.

At $t = \frac{1}{120}$ s, $i = -10$ A.

EXAMPLE 3

An ac generator produces a voltage that varies according to the equation

$$e = E \sin(\omega t + \theta)$$

where e is the instantaneous voltage at time t, the peak or maximum voltage E is 150 V, the frequency f is 40 Hz, and the phase angle θ is $\pi/6$. (a) Write the equation for the variable voltage. (b) Graph two periods of the resulting equation.

(a) $E = 150$ V, $\omega = 40$ Hz $= 40$ cycles/s $= 80\pi$ rad/s, and $\theta = \pi/6$. Then, $e = 150 \sin(80\pi t + \pi/6)$.

(b) The amplitude is 150. The period is $\frac{2\pi}{80\pi} = \frac{1}{40}$. The phase shift is $\frac{\pi/6}{80\pi} = \frac{1}{480}$ to the left. Two periods of the graph are shown in Fig. 12.26.

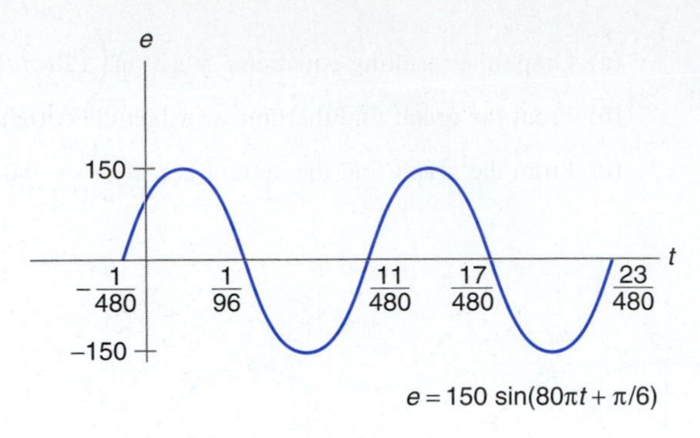

Figure 12.26

Exercises 12.5

1. A crank 8 in. long starts from a horizontal position (0°) and rotates in a counterclockwise direction at a uniform angular velocity of 5 revolutions per second (10π rad/s). Plot the curve that shows the simple harmonic motion of the crank handle, and find the equation of this curve.
2. Complete Exercise 1 for when the crank starts from a vertical position (90°).
3. If the rotating wire from Example 2 produces a maximum current of 15 A at 60 cycles/s, graph the resulting equation when $\theta = \frac{\pi}{4}$.
4. Complete Exercise 3 for $\theta = 0$.
5. A rotating wire through a magnetic field generates an alternating current whose instantaneous voltage e is given by the equation

$$e = E \cos(\omega t + \theta)$$

where E is the maximum instantaneous voltage, ω is the angular velocity of the rotating wire, t is the time, and θ is the phase angle. Graph the resulting equation when $E = 110$ V, $\omega = 60$ cycles/s, and $\theta = 0°$.

6. Complete Exercise 5 for $\theta = -\frac{\pi}{3}$.
7. A mass hanging on a spring is pulled down 10 cm from its equilibrium position and then released $\left(\theta = -\frac{\pi}{2}\right)$. The resulting oscillation of the mass is simple harmonic motion. The period of one complete cycle is observed to be 0.80 s. Find and graph the resulting simple harmonic motion equation.
8. A weight is attached to a spring and vibrates at a frequency of 4 vibrations per second and an observed amplitude of 3 in. Find and graph the resulting equation beginning with $t = 0$ at $\theta = 0$.
9. A thin reed fixed at one end vibrates at 200 vibrations per second with an amplitude of 0.25 cm. Find and graph the resulting equation beginning with $t = 0$ at $\theta = 0$.
10. The equation of motion for a particle in simple harmonic motion is

$$y = 5.0 \cos 8.0t$$

where y is in metres and t is in seconds. Graph the equation. Find the distance from the horizontal at $t = \frac{\pi}{4}$.

CHAPTER 12 SUMMARY

1. The values of a, b, and c determine *amplitude, period,* and *phase shift* for the sine and cosine functions as follows:

$$y = a \sin (bx + c)$$
$$y = a \cos (bx + c)$$

amplitude $= |a|$ phase shift $= \dfrac{c}{b}$ $\left(\text{to the } \textit{left} \text{ if } \dfrac{c}{b} > 0\right)$

period $= \dfrac{2\pi}{b}$ $\left(\text{to the } \textit{right} \text{ if } \dfrac{c}{b} < 0\right)$

If a is negative, each branch is flipped or inverted about the x-axis.

2. For $y = a \tan bx$ and $y = a \cot bx$, the *period* may be found by using

$$P = \frac{\pi}{b}$$

If $a > 1$, each branch intersects the x-axis at a greater angle than for $y = \tan x$. If $0 < a < 1$, each branch intersects the x-axis at a smaller angle than for $y = \tan x$. If a is negative, each branch is flipped or inverted about the x-axis.

3. For $y = a \sec bx$ and $y = a \csc bx$, the *period* may be found from

$$P = \frac{2\pi}{b}$$

The quantity $|a|$ is the vertical distance between the x-axis and the low point of each of the branches above the x-axis. If a is negative, each branch is flipped or inverted about the x-axis.

4. To *add or subtract two functions graphically,* use the method of addition of ordinates as follows:

(a) Graph each of the functions that make up the composite on the same set of coordinate axes.

(b) It may be helpful to draw several vertical lines perpendicular to the x-axis.

(c) If the composite is

(i) a sum, find the algebraic sum of the y-values where each vertical line intersects each of the graphs.

(ii) a difference, find the algebraic difference of the y-values where each vertical line intersects each of the graphs, or graph the negative (or "flipped") curve and find the algebraic sum as in (i).

(d) Plot each resultant y-value from Step (c) on each vertical line and connect with a smooth curve.

5. *Simple harmonic motion* is described by the equations

$$y = a \sin (2\pi ft + \theta) = a \sin (\omega t + \theta)$$

CHAPTER 12 REVIEW

Find the amplitude and period of each function and sketch its graph.

1. $y = 4 \cos 6x$ **2.** $y = -2 \sin \frac{1}{3}x$ **3.** $y = 3 \cos 2\pi x$

Find the amplitude, period, and phase shift of each function and sketch its graph.

4. $y = 3 \sin\left(x - \frac{\pi}{4}\right)$

5. $y = \cos\left(2x + \frac{2\pi}{3}\right)$

6. $y = 4 \sin\left(\pi x + \frac{\pi}{2}\right)$

Find the period of each function and sketch its graph.

7. $y = \tan 5x$

8. $y = -\cot 3x$

9. $y = 2 \sec 4x$

Use the method of addition of ordinates or a graphing calculator to graph one complete period of each composite curve.

10. $y = 3 \sin x + 2 \cos 2x$

11. $y = 2 \sin x + \cos \frac{x}{2}$

12. $y = \sin 2x - 2 \cos \frac{x}{2}$

13. A weight is hanging on a spring and oscillating vertically in simple harmonic motion. The spring's displacement from its equilibrium position as a function of time is given by $y = 6 \cos 12t$. Sketch the curve of this described motion as a function of time.

14. An alternating voltage has a peak voltage of 220 V and a frequency of 60 Hz at $\theta = \pi/4$. **(a)** Write the equation for the variable voltage. **(b)** Graph two periods of the resulting equation.

CHAPTER **12**

SPECIAL NASA APPLICATION

Fitting a Sine Curve to Energy Use Data*

Monthly energy-related data may fit a sine curve. The following exercises refer to the record of monthly electric energy use for the Engineering and Operations Building, an administrative office building at the Cape Canaveral Air Station in Florida. Before undertaking this application you should read the Background Information in the Chapter 2 Special Application.

In the following table, the data for fiscal year 1997 (FY97) and October to March FY98 are actual data from NASA. Other data are estimated from NASA data.

Engineering and Operations Building Electricity Use

Month	FY96 (kW)	FY97 (kW)	FY98 (kW)
October	79,130	70,320	87,120
November	67,910	61,920	60,720
December	58,010	56,280	60,650
January	51,870	80,720	56,280
February	50,990	52,560	47,880
March	55,570	69,000	47,040
April	64,500	58,560	59,020
May	75,600	71,760	69,210
June	75,600	77,520	80,400
July	86,140	90,240	89,870
August	93,550	93,240	95,290
September	96,010	97,920	95,330

To study the data, use a graphing calculator (here a TI-83 Plus) and complete the following steps to fit a sine curve to the NASA electric usage data for 18 months (FY97 and the first 6 months of FY98) for the Engineering and Operations Building:

*From NASA–AMATYC–NSF Project Mathematics Explorations II, grant principals John S. Pazdar, Patricia L. Hirschy, and Peter A. Wursthorn; copyright Capital Community College, 2000.

(a) Number the first list L_1 in your graphing calculator with 1 to 18 (**STAT**, **ENTER**).

(b) Enter the energy use data for all 12 months of FY97 and the first 6 months of FY98 in the second list L_2.

(c) Calculate one-variable statistics on the second list (**2nd QUIT**, **STAT**, right arrow, **ENTER**, **2nd 2**, **ENTER**).

(d) Record the mean and range. The mean is $\bar{x} = 68{,}874$. The range is the difference between the highest and lowest values in L_2; that is, $97{,}920 - 47{,}040 = 50{,}880$.

(e) Set a viewing window with *X*min = 0, *X*max = number of data points (18), *X*scl = 2, *Y*min = lowest number in second list (47,040), *Y*max = highest number in second list (97,920), and *Y*scl = range / 10 = 5088. Use Statistical Plot to graph the data.

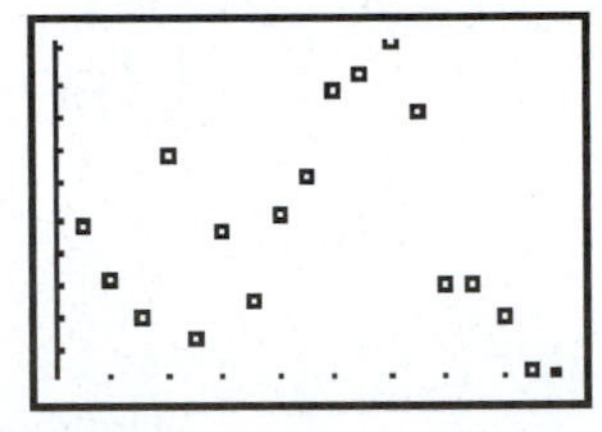

(f) Enter the mean as one equation, Y = mean, that is, $Y = 68874$, in [Y =]. Graph the line. Notice the mean passes horizontally through the middle of the data.

(g) In [MODE], select [Radian] option. Set up a first guess for the equation $Y = a \sin (bx + c) + d$ using $a = \text{range}/2 = 25{,}440$. We let $b = 0.5$ because we expect there to be a 12-month seasonal cycle or period. The period for $y = \sin x$ is 2π or approximately 6.28 radians. The value of b, the coefficient of x, determines the period according to the following formula: period $= 2\pi/b$. We want a period close to 12 rad, so we let $b = 0.5$. The coefficient $b = 0.523599$ is more precise but 0.5 works adequately. Let d = mean = 68,874.

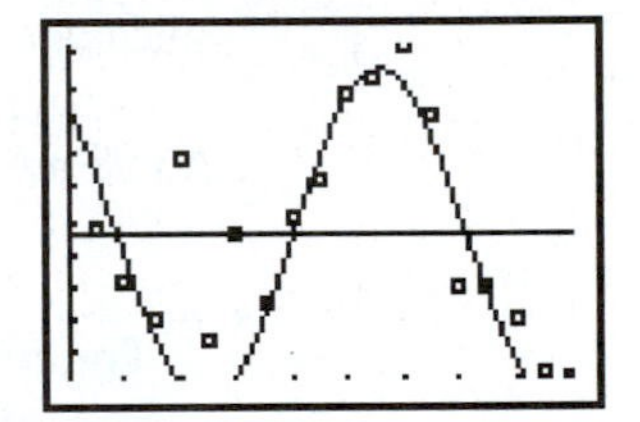

(h) Enter the sine function into a second equation in [Y =].

(i) Use guess and check to set the parameter c (the source of the shift left or right) so as to match any cyclic behavior in the graph of the data. Our graph reflects the choice of $c = -4$. Our function is $Y = 25{,}440 \sin (0.5x - 4) + 68{,}874$.

13 Trigonometric Formulas and Identities

INTRODUCTION

Thus far, we have solved algebraic, exponential, and logarithmic equations. Trigonometric equations are more complicated than any we have studied. They may involve more than one trigonometric function or contain functions with different (but related) arguments. In this chapter we develop trigonometric identities that will be used to simplify expressions and convert equations into a form that we can readily solve.

Objectives

- Develop basic trigonometric identities.
- Simplify trigonometric expressions.
- Develop a strategy for proving trigonometric identities.
- Use trigonometric identities to solve equations.
- Work with inverse trigonometric functions.
- Identify the domain of an inverse trigonometric function.

13.1 BASIC TRIGONOMETRIC IDENTITIES

An equation that is true for all values of the variables for which both sides are defined is called an **identity.** The following are examples of algebraic identities:

1. $2(x - 3) = 2x - 6$ (true for all x)

2. $4(x + y) = 4x + 4y$ (true for all x and all y)

3. $\frac{4x + y}{x} = 4 + \frac{y}{x}$ (true for all y and all x except 0)

The following are examples of trigonometric identities (some have been shown previously; the rest will be shown later):

4. $\sin\theta = \dfrac{1}{\csc\theta}$ (true for all θ except where $\csc\theta = 0$)

5. $\sin^2\theta + \cos^2\theta = 1$ (true for all θ)

6. $\sin(\theta + \phi) = \sin\theta\cos\phi + \cos\theta\sin\phi$ (true for all θ and all ϕ)

Identities are used to prove other identities or to simplify a given trigonometric expression. They are also used to change a given trigonometric expression into a different but equivalent expression which is more useful for a particular problem, either technical or mathematical.

A rather long list of trigonometric identities exists. We shall first prove (show the validity of) the basic identities. Before you use an identity, you must be sure that it comes from a list of valid identities or be able to show that it is valid. At times, just a change in the form of an identity will greatly simplify the solution of a given problem.

A given identity may be proven by any of a number of ways, as long as all the steps are valid. We shall illustrate five basic ways to prove trigonometric identities. When a number of valid ways may be used to prove a given identity, each is *correct;* the one most preferred is the one that is shortest and most efficient. Generally, a helpful suggestion is to try to simplify the more complicated side to the less complicated side of the identity.

Method 1: Use of Trigonometric Definitions

The trigonometric definitions from Chapter 10 are shown in the following box with Fig. 13.1; they may be used to prove identities.

$\sin\theta = \dfrac{y}{r}$

$\cos\theta = \dfrac{x}{r}$

$\tan\theta = \dfrac{y}{x} \quad (x \neq 0)$

$\cot\theta = \dfrac{x}{y} \quad (y \neq 0)$

$\sec\theta = \dfrac{r}{x} \quad (x \neq 0)$

$\csc\theta = \dfrac{r}{y} \quad (y \neq 0)$

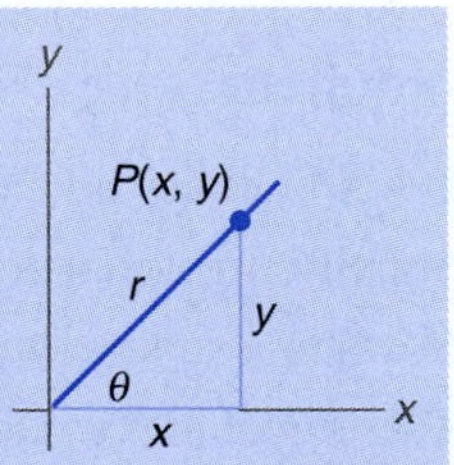

Figure 13.1

EXAMPLE 1

Prove $\cos\theta = \dfrac{1}{\sec\theta}$.

$$\frac{1}{\sec\theta} = \frac{1}{\frac{r}{x}} = \frac{x}{r} = \cos\theta$$

All of the reciprocal identities may be proven in a similar way; they are listed in the following box. All basic identities are listed and numbered for easy reference.

$$\sin\theta = \frac{1}{\csc\theta} \quad (1)$$

$$\cos\theta = \frac{1}{\sec\theta} \quad (2)$$

$$\tan\theta = \frac{1}{\cot\theta} \quad (3)$$

$$\cot\theta = \frac{1}{\tan\theta} \quad (4)$$

$$\sec\theta = \frac{1}{\cos\theta} \quad (5)$$

$$\csc\theta = \frac{1}{\sin\theta} \quad (6)$$

EXAMPLE 2

Prove $\tan\theta = \dfrac{\sin\theta}{\cos\theta}$.

$$\frac{\sin\theta}{\cos\theta} = \frac{\frac{y}{r}}{\frac{x}{r}} = \frac{y}{x} = \tan\theta$$

$$\tan\theta = \frac{\sin\theta}{\cos\theta} \quad (7)$$

$$\cot\theta = \frac{\cos\theta}{\sin\theta} \quad (8)$$

The next three identities are called the **Pythagorean identities** because they use the Pythagorean relationship, $x^2 + y^2 = r^2$.

EXAMPLE 3

Prove $\sin^2\theta + \cos^2\theta = 1$.

Given $x^2 + y^2 = r^2$,

$$\left(\frac{x}{r}\right)^2 + \left(\frac{y}{r}\right)^2 = 1 \qquad \text{(Divide each side by } r^2.\text{)}$$

$$\cos^2\theta + \sin^2\theta = 1 \qquad \left(\text{From the trigonometric definitions, } \frac{x}{r} = \cos\theta \text{ and } \frac{y}{r} = \sin\theta.\right)$$

Note: $\cos^2\theta$ may also be written $(\cos\theta)^2$.

EXAMPLE 4

Prove $1 + \tan^2\theta = \sec^2\theta$.

Given $x^2 + y^2 = r^2$,

$$1 + \left(\frac{y}{x}\right)^2 = \left(\frac{r}{x}\right)^2 \qquad \text{(Divide each side by } x^2.\text{)}$$

$$1 + \tan^2\theta = \sec^2\theta$$

The three Pythagorean identities are as follows:

$$\sin^2\theta + \cos^2\theta = 1 \qquad (9)$$

$$1 + \tan^2\theta = \sec^2\theta \qquad (10)$$

$$\cot^2\theta + 1 = \csc^2\theta \qquad (11)$$

Method 1 is typically used only to develop or prove the first 11 identities. Other methods are then needed to prove the more complex identities because Method 1 commonly results in cumbersome fractional expressions. We suggest that you use Method 1 only in Exercises 1 through 4 at the end of this section.

Method 2: Substitution of Known Identities

We may also show that an identity is valid by changing one of the sides of the given expression or any of its parts by substitution of a known identity or any of its forms one or more times.

EXAMPLE 5

Prove $\tan x = \sin x \sec x$.

Let's start with the right-hand side and show that it is equal to the left-hand side.

$$\begin{aligned} \sin x \sec x &= \sin x\left(\frac{1}{\cos x}\right) && \text{[Identity (5)]} \\ &= \frac{\sin x}{\cos x} \\ &= \tan x && \text{[Identity (7)]} \end{aligned}$$

Therefore, $\tan x = \sin x \sec x$.

EXAMPLE 6

Prove $(1 - \sin^2\theta)\tan^2\theta = \sin^2\theta$.

Let's start with the left-hand side and show that it is equal to the right-hand side.

$$\begin{aligned} (1 - \sin^2\theta)\tan^2\theta &= \cos^2\theta\tan^2\theta && \text{[Form of Identity (9)]} \\ &= \cos^2\theta\left(\frac{\sin^2\theta}{\cos^2\theta}\right) && \text{[Identity (7)]} \\ &= \sin^2\theta \end{aligned}$$

Therefore, $(1 - \sin^2\theta)\tan^2\theta = \sin^2\theta$.

EXAMPLE 7

Prove $\cos^2 x\,(1 + \tan^2 x) = 1$.

Let's start with the left-hand side and show that it is equal to the right-hand side.

$$\begin{aligned} \cos^2 x\,(1 + \tan^2 x) &= \cos^2 x \sec^2 x && \text{[Identity (10)]} \\ &= \cos^2 x\left(\frac{1}{\cos^2 x}\right) && \text{[Identity (5)]} \\ &= 1 \end{aligned}$$

Therefore, $\cos^2 x\,(1 + \tan^2 x) = 1$.

EXAMPLE 8

Prove $\sin\theta + \cot\theta\cos\theta = \csc\theta$.

Again, let's start with the left-hand side and show that it is equal to the right-hand side.

$$\begin{aligned}
\sin\theta + \cot\theta\cos\theta &= \sin\theta + \frac{\cos\theta}{\sin\theta}\cdot\cos\theta && \text{[Identity (8)]}\\
&= \sin\theta + \frac{\cos^2\theta}{\sin\theta}\\
&= \frac{\sin^2\theta + \cos^2\theta}{\sin\theta}\\
&= \frac{1}{\sin\theta} && \text{[Identity (9)]}\\
&= \csc\theta && \text{[Identity (6)]}
\end{aligned}$$

Therefore, $\sin\theta + \cot\theta\cos\theta = \csc\theta$.

Method 3: Factoring One Side of the Identity

EXAMPLE 9

Prove $\sin^2\theta + \sin^2\theta\tan^2\theta = \tan^2\theta$.

$$\begin{aligned}
\sin^2\theta + \sin^2\theta\tan^2\theta &= \sin^2\theta\,(1 + \tan^2\theta) && \text{(Factor.)}\\
&= \sin^2\theta\sec^2\theta && \text{[Identity (10)]}\\
&= \sin^2\theta\left(\frac{1}{\cos^2\theta}\right) && \text{[Identity (5)]}\\
&= \tan^2\theta && \text{[Identity (7)]}
\end{aligned}$$

Therefore, $\sin^2\theta + \sin^2\theta\tan^2\theta = \tan^2\theta$.

EXAMPLE 10

Prove $\sin^4\theta - \cos^4\theta = 2\sin^2\theta - 1$.

$$\begin{aligned}
\sin^4\theta - \cos^4\theta &= (\sin^2\theta + \cos^2\theta)(\sin^2\theta - \cos^2\theta) && \text{(Factor.)}\\
&= (1)(\sin^2\theta - \cos^2\theta) && \text{[Identity (9)]}\\
&= \sin^2\theta - (1 - \sin^2\theta) && \text{[Identity (9)]}\\
&= 2\sin^2\theta - 1
\end{aligned}$$

Therefore, $\sin^4\theta - \cos^4\theta = 2\sin^2\theta - 1$.

Method 4: Multiplication of the Numerator and Denominator of a Fractional Expression

Multiplication of the numerator and the denominator by some trigonometric quantity may change the form of one side of the identity so that you may then use one of the other methods to complete the proof.

EXAMPLE 11

Prove $\dfrac{\cos x}{1 + \sin x} = \dfrac{1 - \sin x}{\cos x}$.

Multiply the numerator and the denominator of the right-hand side by $1 + \sin x$.

$$\frac{1 - \sin x}{\cos x} \cdot \frac{1 + \sin x}{1 + \sin x} = \frac{1 - \sin^2 x}{\cos x\,(1 + \sin x)}$$

$$= \frac{\cos^2 x}{\cos x\,(1 + \sin x)} \qquad \text{[Identity (9)]}$$

$$= \frac{\cos x}{1 + \sin x}$$

Therefore, $\dfrac{\cos x}{1 + \sin x} = \dfrac{1 - \sin x}{\cos x}$.

Method 5: Expressing All Functions on One Side in Terms of Sines and Cosines and Using Any of Methods 1 Through 4

EXAMPLE 12

Prove $\tan x + \cot x = \sec x \csc x$.

$$\tan x + \cot x = \frac{\sin x}{\cos x} + \frac{\cos x}{\sin x} \qquad \text{[Identities (7) and (8)]}$$

$$= \frac{\sin^2 x + \cos^2 x}{\cos x \sin x}$$

$$= \frac{1}{\cos x \sin x} \qquad \text{[Identity (9)]}$$

$$= \sec x \csc x \qquad \text{[Identities (5) and (6)]}$$

Therefore, $\tan x + \cot x = \sec x \csc x$.

Note: In any method, introduce radicals only as a last resort!

Proving identities and using identities to simplify a trigonometric expression are quite different from solving an equation. You usually work from only one side of the identity to show that it is equivalent to the other. When you use identities to simplify a trigonometric expression, you begin with the expression and use identities to change from one equivalent expression to another until the expression is simplified or in the form that is most useful to you. That is, you may *not* use equation-solving principles such as adding the same quantity to each side or multiplying each side by the same quantity.

Exercises 13.1

Use the trigonometric definitions in the box with Fig. 13.1 to prove each identity; that is, use Method 1.

1. $\dfrac{1}{\csc \theta} = \sin \theta$

2. $\dfrac{1}{\tan \theta} = \cot \theta$

3. $\dfrac{\cos \theta}{\sin \theta} = \cot \theta$

4. $\cot^2 \theta + 1 = \csc^2 \theta$

Prove each identity without using the trigonometric definitions.

5. $\cos \theta \sec \theta = 1$

6. $\tan \theta \cot \theta = 1$

7. $\cos x \tan x = \sin x$

8. $\csc x \tan x = \sec x$

9. $\dfrac{\csc\theta}{\sec\theta} = \cot\theta$

10. $\dfrac{\tan\theta}{\cot\theta} = \tan^2\theta$

11. $\dfrac{\tan\theta}{\sin\theta} = \sec\theta$

12. $\dfrac{\cot\theta}{\cos\theta} = \csc\theta$

13. $\sec\theta\cot\theta = \csc\theta$

14. $\sin\theta\cot\theta = \cos\theta$

15. $(1 - \cos^2 x)\csc^2 x = 1$

16. $(\cot^2 x + 1)\tan^2 x = \sec^2 x$

17. $(1 - \sin^2\theta)\cos^2\theta = \cos^4\theta$

18. $(1 + \tan^2 x)\sin^2 x = \tan^2 x$

19. $\dfrac{\cos^2 x - 1}{\sin x} = -\sin x$

20. $\dfrac{\sec^2 x - 1}{\sin^2 x} = \sec^2 x$

21. $\cos\theta(\csc\theta - \sec\theta) = \cot\theta - 1$

22. $\cot\theta - \cos\theta = \cot\theta(1 - \sin\theta)$

23. $\tan^2\theta - \tan^2\theta\sin^2\theta = \sin^2\theta$

24. $\tan\theta\sin\theta + \tan\theta\cot\theta\cos\theta = \sec\theta$

25. $\dfrac{\sec x - \cos x}{\sin x} = \tan x$

26. $\dfrac{\csc x - \sin x}{\cot x} = \cos x$

27. $\dfrac{\sec^2\theta - 1}{\sec^2\theta} = \sin^2\theta$

28. $\sin^2\theta - \cos^2\theta = 1 - 2\cos^2\theta$

29. $\dfrac{\csc^2 x}{1 + \tan^2 x} = \cot^2 x$

30. $\dfrac{1 + \tan^2 x}{\tan^2 x} = \csc^2 x$

31. $\dfrac{(1 + \sin\theta)(1 - \sin\theta)}{\sin^2\theta} = \cot^2\theta$

32. $\dfrac{\sin x\cos x}{(1 + \cos x)(1 - \cos x)} = \cot x$

33. $\dfrac{\sin^2 x}{1 + \cos x} = 1 - \cos x$

34. $\dfrac{\cos^2 x}{1 + \sin x} = 1 - \sin x$

35. $\cos^4\theta - \sin^4\theta = 2\cos^2\theta - 1$

36. $\sec^4 x - \tan^4 x = 2\tan^2 x + 1$

37. $\dfrac{1 + \sec x}{\csc x} = \sin x + \tan x$

38. $\dfrac{\cos x\tan x + \sin x}{\tan x} = 2\cos x$

39. $(\sec x - \tan x)(\csc x + 1) = \cot x$

40. $\sec x + \tan x + \sin x = \dfrac{1 + \sin x + \sin x\cos x}{\cos x}$

41. $\dfrac{1 - \sin^2 x}{1 - \cos^2 x} = \cot^2 x$

42. $\dfrac{1 - \tan x}{1 + \tan x} = \dfrac{\cot x - 1}{\cot x + 1}$

43. $\dfrac{\sin x}{1 + \cos x} = \dfrac{1 - \cos x}{\sin x}$

44. $\dfrac{\sin\theta}{1 + \cos\theta} = \dfrac{1 - \cos\theta}{\cos\theta\tan\theta}$

45. $\dfrac{\cos^2 x}{1 - \sin x} = 1 + \sin x$

46. $\dfrac{\sin\theta + \cos\theta}{\cos\theta - \sin\theta} = \dfrac{\cot\theta + 1}{\cot\theta - 1}$

47. $\dfrac{1}{\sec x - 1} - \dfrac{1}{\sec x + 1} = 2\cot^2 x$

48. $\dfrac{1}{\sin x + 1} - \dfrac{1}{\sin x - 1} = 2\sec^2 x$

49. $\cos^4 x - \sin^4 x = 1 - 2\sin^2 x$

50. $\sec^4 x - 1 = \tan^2 x(\sec^2 x + 1)$

51. $\dfrac{\tan^2 x - 1}{1 - \cot^2 x} = \tan^2 x$

52. $\sec\theta - \tan\theta = \dfrac{1}{\sec\theta + \tan\theta}$

53. $\dfrac{\tan x + \tan y}{\cot x + \cot y} = \tan x\tan y$

54. $\dfrac{\tan\theta}{\sec\theta - \cos\theta} = \csc\theta$

13.2 FORMULAS FOR THE SUM AND THE DIFFERENCE OF TWO ANGLES

The sine, the cosine, and the tangent of the sum and the difference of two angles have practical applications. These formulas are also used to develop the double- and half-angle formulas in the next section.

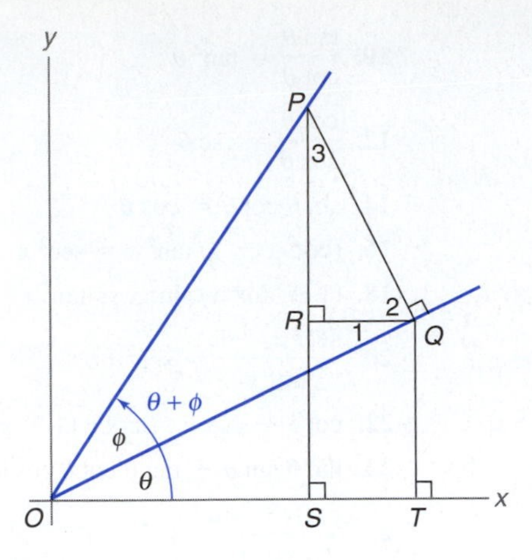

Figure 13.2 Angle $\theta + \phi$ in standard position.

First, we want to show

$$\sin(\theta + \phi) = \sin\theta\cos\phi + \cos\theta\sin\phi$$

In Fig. 13.2, angles θ and $\theta + \phi$ are constructed so that they are in standard position. The initial side of ϕ coincides with the terminal side of θ; its terminal side coincides with the terminal side of $\theta + \phi$. Choose any point P on the terminal side of $\theta + \phi$. Then, drop perpendicular lines to the terminal side of θ at Q and to the initial side of θ at S. Next, draw perpendiculars QR and QT. Because RQ and OT are parallel, $\theta = \angle 1$ (they are alternate interior angles).

$$\angle 1 + \angle 2 = 90° \quad (PQ \perp OQ)$$

$$\angle 3 + \angle 2 = 90° \quad \text{(The sum of the two acute angles of a right triangle equals 90°.)}$$

$$\angle 1 - \angle 3 = 0 \quad \text{(Subtract.)}$$

or

$$\angle 1 = \angle 3$$

We have already shown that

$$\angle 1 = \theta$$

Therefore, $\theta = \angle 3$. That is, $\angle SPQ$ and θ are equal. In right triangle OSP of Fig. 13.2,

$$\begin{aligned}\sin(\theta + \phi) = \frac{SP}{OP} &= \frac{SR + RP}{OP} \\ &= \frac{SR}{OP} + \frac{RP}{OP} \\ &= \frac{TQ}{OP} + \frac{RP}{OP} \quad \text{(In rectangle } SRQT,\ SR = TQ.)\end{aligned}$$

Next, multiply numerator and denominator of the first term by OQ and rearrange the terms; then, multiply numerator and denominator of the second term by PQ and rearrange the terms.

$$\begin{aligned}\frac{TQ}{OP} + \frac{RP}{OP} &= \frac{TQ}{OP} \cdot \frac{OQ}{OQ} + \frac{RP}{OP} \cdot \frac{PQ}{PQ} \\ &= \frac{TQ}{OQ} \cdot \frac{OQ}{OP} + \frac{RP}{PQ} \cdot \frac{PQ}{OP}\end{aligned}$$

From Fig. 13.2 and each of the following right triangles, note that

$$\text{Right } \triangle OTQ: \quad \sin\theta = \frac{TQ}{OQ}$$

$$\text{Right } \triangle OPQ: \quad \cos\phi = \frac{OQ}{OP}$$

$$\text{Right } \triangle PRQ: \quad \cos\angle 3 = \frac{RP}{PQ}$$

$$\text{Right } \triangle OPQ: \quad \sin\phi = \frac{PQ}{OP}$$

Then,

$$\begin{aligned} \sin(\theta + \phi) &= \frac{TQ}{OQ} \cdot \frac{OQ}{OP} + \frac{RP}{PQ} \cdot \frac{PQ}{OP} \\ &= \sin\theta\cos\phi + \cos\angle 3 \sin\phi \\ &= \sin\theta\cos\phi + \cos\theta\sin\phi \qquad (\angle 3 = \theta) \end{aligned}$$

Therefore, we have the following:

$$\boldsymbol{\sin(\theta + \phi) = \sin\theta\cos\phi + \cos\theta\sin\phi} \qquad \textbf{(12)}$$

Using Fig. 13.2 and a similar procedure, we can show the following.

$$\boldsymbol{\cos(\theta + \phi) = \cos\theta\cos\phi - \sin\theta\sin\phi} \qquad \textbf{(13)}$$

To illustrate these formulas let's use the following examples.

EXAMPLE 1

Simplify $\cos(\theta + 90°)$ using Formula (13).

$$\begin{aligned} \cos(\theta + 90°) &= \cos\theta\cos 90° - \sin\theta\sin 90° \\ &= \cos\theta \quad (0) \quad - \sin\theta \quad (1) \\ &= -\sin\theta \end{aligned}$$

EXAMPLE 2

Simplify $\sin(\theta + 90°)$ using Formula (12).

$$\begin{aligned} \sin(\theta + 90°) &= \sin\theta\cos 90° + \cos\theta\sin 90° \\ &= \sin\theta \quad (0) \quad + \cos\theta \quad (1) \\ &= \cos\theta \end{aligned}$$

Before proceeding, we need to show the following identities:

$$\boldsymbol{\sin(-\theta) = -\sin\theta} \qquad \textbf{(14)}$$

$$\boldsymbol{\cos(-\theta) = \cos\theta} \qquad \textbf{(15)}$$

First, construct any angle θ and its negative, $-\theta$, as in Fig. 13.3. Label any point P_1 on the terminal side of θ with coordinates (x_1, y_1). Let r be the distance from P_1 to the origin. On

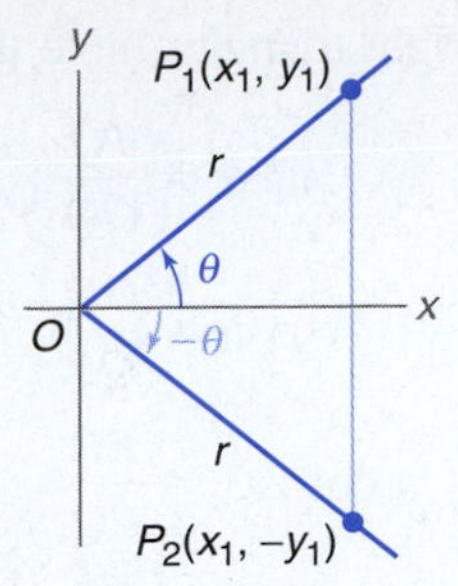

Figure 13.3 Angles θ and $-\theta$ in standard position.

the terminal side of $-\theta$, label a point at a distance of r from the origin as P_2. Note that its coordinates are $(x_1, -y_1)$.

Then,

$$\sin\theta = \frac{y_1}{r} \quad \text{and} \quad \sin(-\theta) = \frac{-y_1}{r} \qquad \text{Thus, } \sin(-\theta) = -\sin\theta.$$

$$\cos\theta = \frac{x_1}{r} \quad \text{and} \quad \cos(-\theta) = \frac{x_1}{r} \qquad \text{Thus, } \cos(-\theta) = \cos\theta.$$

Similarly, the following may also be shown:

$$\tan(-\theta) = -\tan\theta \tag{16}$$
$$\cot(-\theta) = -\cot\theta \tag{17}$$
$$\sec(-\theta) = \sec\theta \tag{18}$$
$$\csc(-\theta) = -\csc\theta \tag{19}$$

To find the formula for $\sin(\theta - \phi)$, use Formula (12), and write $\theta - \phi$ as $\theta + (-\phi)$.

$$\begin{aligned}\sin[\theta + (-\phi)] &= \sin\theta\cos(-\phi) + \cos\theta\sin(-\phi)\\ &= \sin\theta\cos\phi - \cos\theta\sin\phi\end{aligned}$$

Therefore,

$$\sin(\theta - \phi) = \sin\theta\cos\phi - \cos\theta\sin\phi \tag{20}$$

Likewise, to find the formula for $\cos(\theta - \phi)$, use Formula (13), and again write $\theta - \phi$ as $\theta + (-\phi)$.

$$\begin{aligned}\cos[\theta + (-\phi)] &= \cos\theta\cos(-\phi) - \sin\theta\sin(-\phi)\\ &= \cos\theta\cos\phi + \sin\theta\sin\phi\end{aligned}$$

Therefore,

$$\cos(\theta - \phi) = \cos\theta\cos\phi + \sin\theta\sin\phi \tag{21}$$

To find a formula for $\tan(\theta + \phi)$, observe the following:

$$\begin{aligned}\tan(\theta + \phi) &= \frac{\sin(\theta + \phi)}{\cos(\theta + \phi)} && \text{[Identity (7)]}\\ &= \frac{\sin\theta\cos\phi + \cos\theta\sin\phi}{\cos\theta\cos\phi - \sin\theta\sin\phi} && \text{[Identities (12) and (13)]}\end{aligned}$$

$$= \frac{\dfrac{\sin\theta\cos\phi}{\cos\theta\cos\phi} + \dfrac{\cos\theta\sin\phi}{\cos\theta\cos\phi}}{\dfrac{\cos\theta\cos\phi}{\cos\theta\cos\phi} - \dfrac{\sin\theta\sin\phi}{\cos\theta\cos\phi}} \quad \text{(Divide numerator and denominator by } \cos\theta\cos\phi.)$$

$$= \frac{\tan\theta + \tan\phi}{1 - \tan\theta\tan\phi} \quad \text{[Identity (7)]}$$

Therefore,

$$\boldsymbol{\tan(\theta + \phi) = \frac{\tan\theta + \tan\phi}{1 - \tan\theta\tan\phi}} \qquad \textbf{(22)}$$

Similarly, we can show the following:

$$\boldsymbol{\tan(\theta - \phi) = \frac{\tan\theta - \tan\phi}{1 + \tan\theta\tan\phi}} \qquad \textbf{(23)}$$

The formulas for cotangent, secant, and cosecant of the sum and the difference of two angles are not as frequently used but may be derived similarly.

EXAMPLE 3

Prove $\sin(x - 180°) = -\sin x$.

$$\begin{aligned} \sin(x - 180°) &= \sin x \cos 180° - \cos x \sin 180° \quad \text{[Identity (20)]} \\ &= (\sin x)(-1) - (\cos x)(0) \\ &= -\sin x \end{aligned}$$

EXAMPLE 4

Simplify $\cos 2\theta \cos 3\theta - \sin 2\theta \sin 3\theta$.

By Formula (13),

$$\cos 2\theta \cos 3\theta - \sin 2\theta \sin 3\theta = \cos(2\theta + 3\theta) = \cos 5\theta$$

Exercises 13.2

Prove each identity.

1. $\sin(x + \pi) = -\sin x$

2. $\cos(x + 180°) = -\cos x$

3. $\sin(x + 2\pi) = \sin x$

4. $\cos(x + 2\pi) = \cos x$

5. $\tan(x + \pi) = \tan x$

6. $\tan(\pi - x) = -\tan x$

7. $\sin(90° - \theta) = \cos\theta$

8. $\cos\left(\frac{\pi}{2} - \theta\right) = \sin\theta$

9. $\cos\left(\frac{\pi}{2} + \theta\right) = -\sin\theta$

10. $\sin(90° + \theta) = \cos\theta$

11. $\tan(180° - \theta) = -\tan\theta$

12. $\cos(2\pi - \theta) = \cos\theta$

13. $\cos\left(\frac{\pi}{4} + \theta\right) = \frac{\cos\theta - \sin\theta}{\sqrt{2}}$

14. $\sin\left(\frac{\pi}{3} + \theta\right) = \frac{\sqrt{3}\cos\theta + \sin\theta}{2}$

15. $\tan(90° - x) = \cot x$

16. $\sec\left(\frac{\pi}{2} - x\right) = \csc x$

17. $\tan(x + 45°) = \frac{1 + \tan x}{1 - \tan x}$

18. $\tan(\theta + 270°) = -\cot \theta$

19. $\cos(x + y)\cos(x - y) = \cos^2 x - \sin^2 y$

20. $\sin(x + y)\sin(x - y) = \sin^2 x - \sin^2 y$

Simplify each expression.

21. $\cos M \cos N - \sin M \sin N$

22. $\sin 4C \cos C - \cos 4C \sin C$

23. $\sin \theta \cos 3\theta + \cos \theta \sin 3\theta$

24. $\cos 2\theta \cos \theta - \sin 2\theta \sin \theta$

25. $\cos 4\theta \cos 3\theta + \sin 4\theta \sin 3\theta$

26. $\sin 2\theta \cos 3\theta - \cos 2\theta \sin 3\theta$

27. $\frac{\tan 3\theta + \tan 2\theta}{1 - \tan 3\theta \tan 2\theta}$

28. $\frac{\tan \theta - \tan 2\theta}{1 + \tan \theta \tan 2\theta}$

29. $\sin(\theta + \phi) + \sin(\theta - \phi)$

30. $\cos(\theta + \phi) + \cos(\theta - \phi)$

31. $\sin(A + B)\cos B + \cos(A + B)\sin B$

32. $\cos(A + B)\cos B + \sin(A + B)\sin B$

13.3 DOUBLE- AND HALF-ANGLE FORMULAS

To find a formula for $\sin 2\theta$, we use Formula (12) and let $\phi = \theta$, then substitute as follows:

$$\begin{aligned} \sin(\theta + \phi) &= \sin \theta \cos \phi + \cos \theta \sin \phi \\ \sin 2\theta = \sin(\theta + \theta) &= \sin \theta \cos \theta + \cos \theta \sin \theta \\ &= 2 \sin \theta \cos \theta \end{aligned}$$

Therefore, we have the following formula:

$$\mathbf{\sin 2\theta = 2 \sin \theta \cos \theta} \qquad \textbf{(24)}$$

Likewise,

$$\begin{aligned} \cos 2\theta = \cos(\theta + \theta) &= \cos \theta \cos \theta - \sin \theta \sin \theta \qquad \text{[Identity (13)]} \\ &= \cos^2 \theta - \sin^2 \theta \end{aligned}$$

This formula has two other forms:

$$\begin{aligned} \cos^2 \theta - \sin^2 \theta &= \cos^2 \theta - (1 - \cos^2 \theta) \qquad \text{[Identity (9)]} \\ &= 2\cos^2 \theta - 1 \end{aligned}$$

and

$$\begin{aligned} \cos^2 \theta - \sin^2 \theta &= (1 - \sin^2 \theta) - \sin^2 \theta \qquad \text{[Identity (9)]} \\ &= 1 - 2\sin^2 \theta \end{aligned}$$

Therefore,

$$\begin{aligned} \cos 2\theta &= \cos^2 \theta - \sin^2 \theta \qquad \textbf{(25a)} \\ &= 2\cos^2 \theta - 1 \qquad \textbf{(25b)} \\ &= 1 - 2\sin^2 \theta \qquad \textbf{(25c)} \end{aligned}$$

The geometric notion of an identity is that the individual formulas would have identical graphs. This is illustrated in the following frames for the four different ways to write $\cos 2x$:

green diamond Y= **F2 7** (All four have the same graph.)

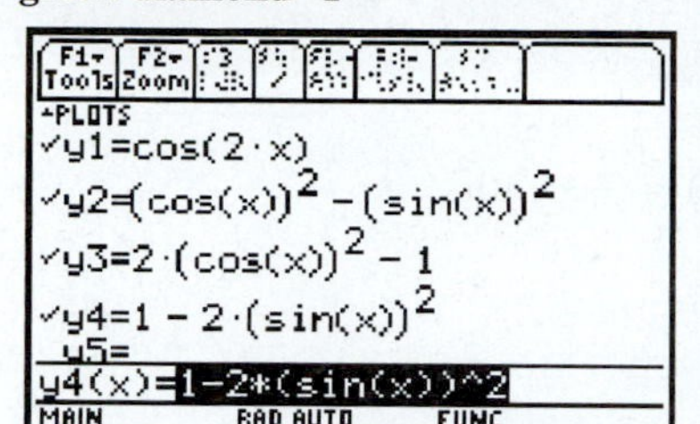

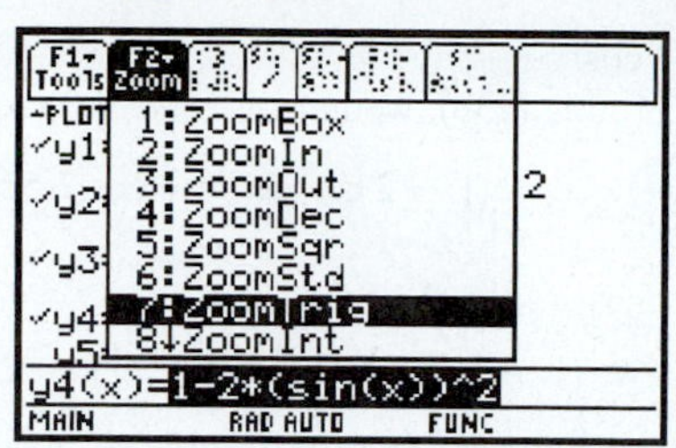

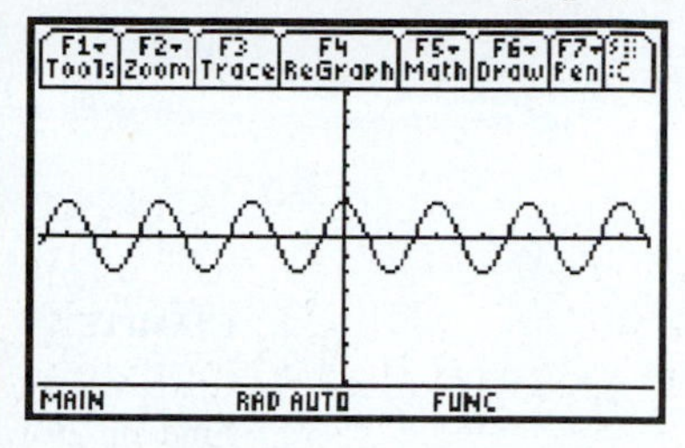

Similarly,

$$\tan 2\theta = \tan(\theta + \theta) = \frac{\tan\theta + \tan\theta}{1 - \tan\theta\tan\theta} \qquad \text{[Identity (22)]}$$

$$= \frac{2\tan\theta}{1 - \tan^2\theta}$$

Therefore,

$$\boldsymbol{\tan 2\theta = \frac{2\tan\theta}{1 - \tan^2\theta}} \tag{26}$$

EXAMPLE 1

Find sin 120° using Formula (24).

$$\sin 120° = \sin 2(60°) = 2\sin 60°\cos 60°$$
$$= 2\left(\frac{\sqrt{3}}{2}\right)\left(\frac{1}{2}\right)$$
$$= 0.8660$$

EXAMPLE 2

Find cos 120° using Formula (25a).

$$\cos 120° = \cos 2(60°) = \cos^2 60° - \sin^2 60°$$
$$= \left(\frac{1}{2}\right)^2 - \left(\frac{\sqrt{3}}{2}\right)^2$$
$$= \frac{1}{4} - \frac{3}{4} = -\frac{1}{2}$$

EXAMPLE 3

Simplify $2\sin 3x\cos 3x$.

Using Formula (24), we obtain

$$2\sin 3x\cos 3x = \sin 2(3x) = \sin 6x$$

Using a calculator, we have

F2 9 2 2 **2nd SIN** 3x) **2nd COS** 3x)) **ENTER**

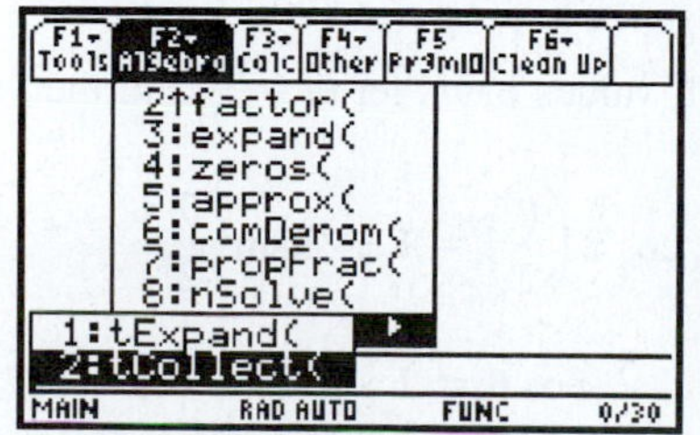

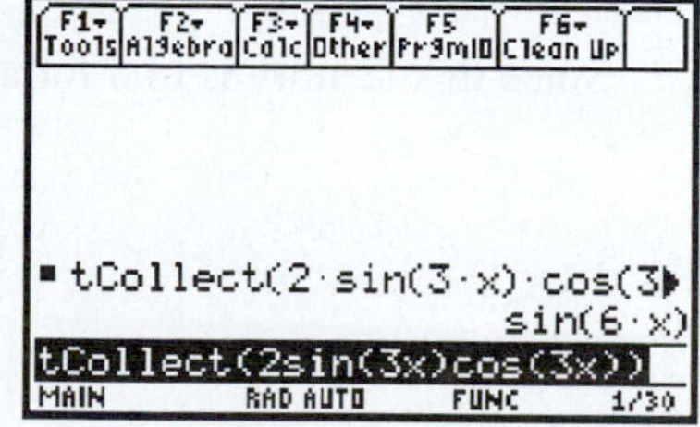

EXAMPLE 4

Simplify $1 - 2\cos^2 5x$.

Using Formula (25b), we have

$$1 - 2\cos^2 5x = -\cos 2(5x) = -\cos 10x$$

EXAMPLE 5

Find $\sin 2\theta$ when $\cos\theta = \frac{4}{5}$ (θ in the fourth quadrant).

If $\cos\theta = \frac{4}{5}$ in the fourth quadrant (see Fig. 13.4), then $\sin\theta = -\frac{3}{5}$.

$$\begin{aligned}\sin 2\theta &= 2\sin\theta\cos\theta \qquad \text{[Identity (24)]}\\ &= 2\left(-\frac{3}{5}\right)\left(\frac{4}{5}\right)\\ &= -\frac{24}{25}\end{aligned}$$

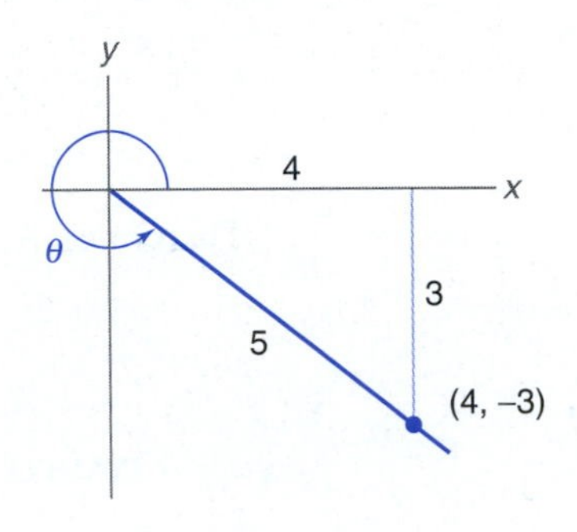

Figure 13.4

EXAMPLE 6

Find $\cos 2\theta$ when $\sin\theta = \frac{1}{2}$.

Using Formula (25c), we have

$$\begin{aligned}\cos 2\theta &= 1 - 2\sin^2\theta\\ &= 1 - 2\left(\frac{1}{2}\right)^2 = 1 - \frac{1}{2} = \frac{1}{2}\end{aligned}$$

In an identity the choice of specific variables does not matter. For example, the identity

$$\sin^2\theta + \cos^2\theta = 1$$

may also be written in the form

$$\sin^2 m + \cos^2 m = 1$$

To find a formula for $\sin\frac{\theta}{2}$, let's begin with Formula (25c) in the form

$$\cos 2m = 1 - 2\sin^2 m$$

Since this identity is true for all values of m, let $m = \frac{\theta}{2}$ and then solve for $\sin\frac{\theta}{2}$.

$$\begin{aligned}\cos 2\left(\frac{\theta}{2}\right) &= 1 - 2\sin^2\frac{\theta}{2}\\ \cos\theta &= 1 - 2\sin^2\frac{\theta}{2}\end{aligned}$$

$$2\sin^2\frac{\theta}{2} = 1 - \cos\theta$$

$$\sin^2\frac{\theta}{2} = \frac{1-\cos\theta}{2}$$

$$\boldsymbol{\sin\frac{\theta}{2} = \pm\sqrt{\frac{1-\cos\theta}{2}}} \qquad \textbf{(27)}$$

To find a formula for $\cos\frac{\theta}{2}$, let's begin with Formula (25b) in the form

$$\cos 2m = 2\cos^2 m - 1$$

Let $m = \frac{\theta}{2}$ and then solve for $\cos\frac{\theta}{2}$.

$$\cos 2\left(\frac{\theta}{2}\right) = 2\cos^2\frac{\theta}{2} - 1$$

$$\cos\theta = 2\cos^2\frac{\theta}{2} - 1$$

$$\frac{1+\cos\theta}{2} = \cos^2\frac{\theta}{2}$$

$$\boldsymbol{\cos\frac{\theta}{2} = \pm\sqrt{\frac{1+\cos\theta}{2}}} \qquad \textbf{(28)}$$

Note: In both Formulas (27) and (28), the sign used depends on the quadrant in which $\frac{\theta}{2}$ lies.

Now find $\tan\frac{\theta}{2}$.

$$\tan\frac{\theta}{2} = \frac{\sin\frac{\theta}{2}}{\cos\frac{\theta}{2}} \qquad \text{[Identity (7)]}$$

$$^*\tan\frac{\theta}{2} = \frac{2\sin^2\frac{\theta}{2}}{2\sin\frac{\theta}{2}\cos\frac{\theta}{2}} \qquad \text{(Multiply numerator and denominator by } 2\sin\tfrac{\theta}{2}.)$$

Note that

$$\cos 2\left(\frac{\theta}{2}\right) = 1 - 2\sin^2\frac{\theta}{2}$$

That is,

$$2\sin^2\frac{\theta}{2} = 1 - \cos\theta$$

Then, note that

$$\sin 2\left(\frac{\theta}{2}\right) = 2 \sin \frac{\theta}{2} \cos \frac{\theta}{2}$$

That is,

$$\sin \theta = 2 \sin \frac{\theta}{2} \cos \frac{\theta}{2}$$

Substituting both into the equation marked by an *, we have

$$\tan \frac{\theta}{2} = \frac{1 - \cos \theta}{\sin \theta} \qquad \textbf{(29)}$$

EXAMPLE 7

Find $\cos \frac{\theta}{2}$ when $\sin \theta = -\frac{3}{5}$ (θ in the third quadrant).

If $\sin \theta = -\frac{3}{5}$ in the third quadrant as in Fig. 13.5, then $\cos \theta = -\frac{4}{5}$.

$$\begin{aligned}\cos \frac{\theta}{2} &= -\sqrt{\frac{1 + \cos \theta}{2}} \\ &= -\sqrt{\frac{1 + \left(-\frac{4}{5}\right)}{2}} \\ &= -\sqrt{\frac{1}{10}} \quad \text{or} \quad -\frac{\sqrt{10}}{10}\end{aligned}$$

The sign is negative because if $180° < \theta < 270°$, then $90° < \frac{\theta}{2} < 135°$ where $\cos \frac{\theta}{2}$ is negative.

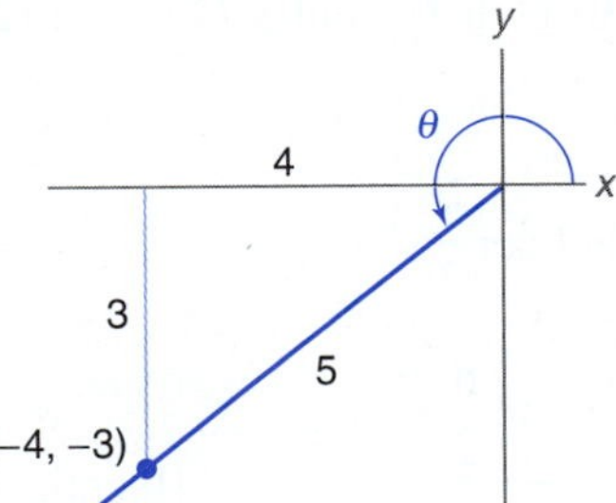

Figure 13.5

EXAMPLE 8

Prove $\frac{1 - \cos 2x}{\sin 2x} = \tan x$.

$$\begin{aligned}\frac{1 - \cos 2x}{\sin 2x} &= \frac{1 - (1 - 2\sin^2 x)}{2 \sin x \cos x} && \text{[Identities (25c) and (24)]} \\ &= \frac{2 \sin^2 x}{2 \sin x \cos x} \\ &= \frac{\sin x}{\cos x} \\ &= \tan x && \text{[Identity (7)]}\end{aligned}$$

Therefore, $\frac{1 - \cos 2x}{\sin 2x} = \tan x$.

The following trigonometric formulas are given without proof for reference:

$$\sin\theta + \sin\phi = 2\sin\left(\frac{\theta+\phi}{2}\right)\cos\left(\frac{\theta-\phi}{2}\right) \tag{30}$$

$$\sin\theta - \sin\phi = 2\cos\left(\frac{\theta+\phi}{2}\right)\sin\left(\frac{\theta-\phi}{2}\right) \tag{31}$$

$$\cos\theta + \cos\phi = 2\cos\left(\frac{\theta+\phi}{2}\right)\cos\left(\frac{\theta-\phi}{2}\right) \tag{32}$$

$$\cos\theta - \cos\phi = -2\sin\left(\frac{\theta+\phi}{2}\right)\sin\left(\frac{\theta-\phi}{2}\right) \tag{33}$$

$$\sin(\theta+\phi) + \sin(\theta-\phi) = 2\sin\theta\cos\phi \tag{34}$$

$$\sin(\theta+\phi) - \sin(\theta-\phi) = 2\cos\theta\sin\phi \tag{35}$$

$$\cos(\theta+\phi) + \cos(\theta-\phi) = 2\cos\theta\cos\phi \tag{36}$$

$$\cos(\theta+\phi) - \cos(\theta-\phi) = -2\sin\theta\sin\phi \tag{37}$$

Exercises 13.3

Simplify each expression.

1. $2\sin\frac{x}{4}\cos\frac{x}{4}$

2. $20\sin^2 x\cos^2 x$

3. $1 - 2\sin^2 3x$

4. $\sqrt{\frac{1-\cos 6\theta}{2}}$

5. $\sqrt{\frac{1+\cos\frac{\theta}{4}}{2}}$

6. $\cos 2x + 2\sin^2 x$

7. $\cos^2\frac{x}{6} - \sin^2\frac{x}{6}$

8. $10\cos^2 44° - 10\sin^2 44°$

9. $20\sin 4\theta\cos 4\theta$

10. $2\sin 3x\cos 3x$

11. $-\sqrt{\frac{1+\cos 250°}{2}}$

12. $\sqrt{\frac{1-\cos 16\theta}{2}}$

13. $4 - 8\sin^2\theta$

14. $1 - 2\sin^2 7t$

15. $100\sin 30t\cos 30t$

16. $15\sin\frac{x}{6}\cos\frac{x}{6}$

17. Find $\sin 2\theta$ when $\cos\theta = \frac{3}{5}$ (θ in the first quadrant).

18. Find $\cos 2\theta$ when $\sin\theta = \frac{3}{5}$ (θ in the second quadrant).

19. Find $\tan 2\theta$ when $\cos\theta = -\frac{12}{13}$ (θ in the third quadrant).

20. Find $\sin\frac{\theta}{2}$ when $\sin\theta = -\frac{5}{13}$ (θ in the fourth quadrant).

21. Find $\cos\frac{\theta}{2}$ when $\sin\theta = -\frac{2}{3}$ (θ in the third quadrant).

22. Find $\tan \frac{\theta}{2}$ when $\tan \theta = -\frac{3}{4}$ (θ in the fourth quadrant).

Prove each identity.

23. $(\sin x + \cos x)^2 = 1 + \sin 2x$

24. $\frac{2 \tan x}{\sin 2x} = 1 + \tan^2 x$

25. $\cos^4 x - \sin^4 x = \cos 2x$

26. $\sin 2x = \frac{2 \tan x}{1 + \tan^2 x}$

27. $\frac{1 - \tan^2 x}{1 + \tan^2 x} = \cos 2x$

28. $2 \tan x \csc 2x = \sec^2 x$

29. $\cot 2x = \frac{\cot^2 x - 1}{2 \cot x}$

30. $\sec 2x = \frac{\sec^2 x}{2 - \sec^2 x}$

31. $\tan x + \cot 2x = \csc 2x$

32. $\tan (x + 45°) + \tan (x - 45°) = 2 \tan 2x$

33. $\sin^2 \frac{x}{2} = \frac{\sec x - 1}{2 \sec x}$

34. $2 \cos^2 \frac{\theta}{2} = \frac{1 + \sec \theta}{\sec \theta}$

35. $\sec^2 \frac{x}{2} = \frac{2}{1 + \cos x}$

36. $\csc^2 \frac{x}{2} = \frac{2 \sec x}{\sec x - 1}$

37. $\tan \frac{x}{2} = \frac{\sin x}{1 + \cos x}$

38. $\tan \left(\frac{\theta}{2} + \frac{\pi}{4}\right) = \frac{1 + \sin \theta}{\cos \theta}$

39. $2 \cos \frac{x}{2} = (1 + \cos x) \sec \frac{x}{2}$

40. $\frac{\sin 3x}{\sin x} - \frac{\cos 3x}{\cos x} = 2$

41. $\tan \frac{x}{2} + \cot \frac{x}{2} = 2 \csc x$

42. $\frac{\sin^3 x - \cos^3 x}{\sin x - \cos x} = 1 + \frac{1}{2} \sin 2x$

43. $\left(\sin \frac{\theta}{2} - \cos \frac{\theta}{2}\right)^2 = 1 - \sin \theta$

44. $8 \sin^2 \frac{x}{2} \cos^2 \frac{x}{2} = 1 - \cos 2x$

45. $\sin 3x = 3 \sin x - 4 \sin^3 x$

46. $\cos 3x = 4 \cos^3 x - 3 \cos x$

13.4 TRIGONOMETRIC EQUATIONS

A trigonometric equation that is not an identity may have a number of solutions. The solutions may be given in terms of degrees or radians. There are no general procedures to follow when solving a trigonometric equation. You may try algebraic methods or use identities to write the equation in terms of a single trigonometric function.

EXAMPLE 1

Solve $2 \sin \theta + 1 = 0$ for $0° \le \theta < 360°$.

$$2 \sin \theta + 1 = 0$$

$$\sin \theta = -\frac{1}{2}$$

$$\theta = 210°, 330° \quad \text{(Sine is negative in the third and fourth quadrants.)}$$

Note: If we did not have the restriction on θ, $0° \le \theta < 360°$, there would be an infinite number of solutions, that is, all angles that are coterminal with 210° and 330°.

EXAMPLE 2

Solve $\sin 2\theta - \sin \theta = 0$ for $0 \le \theta < 2\pi$.

Replace $\sin 2\theta$ by $2 \sin \theta \cos \theta$ [Formula (24)] and factor.

$$2 \sin \theta \cos \theta - \sin \theta = 0$$
$$\sin \theta (2 \cos \theta - 1) = 0$$
$$\sin \theta = 0 \quad \text{or} \quad \cos \theta = \frac{1}{2}$$
$$\theta = 0, \pi \quad \text{or} \quad \theta = \frac{\pi}{3}, \frac{5\pi}{3}$$

EXAMPLE 3

Solve $2 \cos^2 x + 5 \sin x = 4$ for $0° \leq x < 360°$.

Replace $\cos^2 x$ by $1 - \sin^2 x$ [from Formula (9)].

$$2(1 - \sin^2 x) + 5 \sin x = 4$$
$$0 = 2 \sin^2 x - 5 \sin x + 2$$
$$0 = (2 \sin x - 1)(\sin x - 2) \qquad \text{(Factor.)}$$
$$\sin x = \frac{1}{2} \quad \text{or} \quad \sin x = 2$$
$$x = 30°, 150° \qquad \text{No solution because } -1 \leq \sin x \leq 1$$

EXAMPLE 4

Solve $2 \cos^2 x + 6 \cos x + 3 = 0$ for $0° \leq x < 360°$.

Since the left-hand side does not factor, we must use the quadratic formula with $a = 2$, $b = 6$, and $c = 3$.

$$\cos x = \frac{-b \pm \sqrt{b^2 - 4ac}}{2a}$$
$$\cos x = \frac{-6 \pm \sqrt{6^2 - 4(2)(3)}}{2(2)}$$
$$= \frac{-6 \pm \sqrt{12}}{4}$$
$$= \frac{-6 \pm 2\sqrt{3}}{4}$$
$$= \frac{-3 \pm \sqrt{3}}{2}$$

Changing each solution to a decimal value, we have

$$\cos x = \frac{-3 + \sqrt{3}}{2} = -0.6340 \quad \text{or} \quad \cos x = \frac{-3 - \sqrt{3}}{2} = -2.366$$
$$x = 129.3°, 230.7° \qquad \text{No solution because } -1 \leq \cos x \leq 1$$

EXAMPLE 5

Solve $\tan 2\theta + \cot 2\theta + 2 = 0$ for $0 \leq \theta < 2\pi$.

Replace $\cot 2\theta$ by $\dfrac{1}{\tan 2\theta}$ [from Formula (4)].

$$\tan 2\theta + \frac{1}{\tan 2\theta} + 2 = 0$$
$$\tan^2 2\theta + 1 + 2 \tan 2\theta = 0 \qquad \text{(Multiply each side by } \tan 2\theta.\text{)}$$

$$(\tan 2\theta + 1)^2 = 0$$
$$\tan 2\theta + 1 = 0$$
$$\tan 2\theta = -1$$

Hence,

$$2\theta = \frac{3\pi}{4}, \frac{7\pi}{4}, \frac{11\pi}{4}, \frac{15\pi}{4} \quad (0 \le 2\theta < 4\pi)$$
$$\theta = \frac{3\pi}{8}, \frac{7\pi}{8}, \frac{11\pi}{8}, \frac{15\pi}{8} \quad (0 \le \theta < 2\pi)$$

Note: The restriction $0 \le \theta < 2\pi$ is on θ, not 2θ.

EXAMPLE 6

Solve $\cos x = \sin \frac{x}{2}$ for $0 \le x < 2\pi$.

Replace $\sin \frac{x}{2}$ by $\pm\sqrt{\frac{1 - \cos x}{2}}$ [from Formula (27)].

$$\cos x = \pm\sqrt{\frac{1 - \cos x}{2}}$$
$$\cos^2 x = \frac{1 - \cos x}{2} \qquad \text{(Square each side.)}$$
$$2\cos^2 x + \cos x - 1 = 0$$
$$(2\cos x - 1)(\cos x + 1) = 0 \qquad \text{(Factor.)}$$
$$\cos x = \frac{1}{2} \quad \text{or} \quad \cos x = -1$$
$$x = \frac{\pi}{3}, \frac{5\pi}{3} \quad \text{or} \quad x = \pi$$

Since we squared each side, we *must* check for possible extraneous roots.

(a) $\cos \frac{\pi}{3} = \sin \frac{\frac{\pi}{3}}{2}$

$\cos \frac{\pi}{3} = \sin \frac{\pi}{6}$

$\frac{1}{2} = \frac{1}{2}$

$\frac{\pi}{3}$ is a solution.

(b) $\cos \frac{5\pi}{3} = \sin \frac{\frac{5\pi}{3}}{2}$

$\cos \frac{5\pi}{3} = \sin \frac{5\pi}{6}$

$\frac{1}{2} = \frac{1}{2}$

$\frac{5\pi}{3}$ is a solution.

(c) $\cos \pi = \sin \frac{\pi}{2}$

$-1 \ne 1$

π is not a solution.

Therefore, the solutions are $\frac{\pi}{3}$ and $\frac{5\pi}{3}$.

Using a calculator, we have

F2 1 2nd COS x)= **2nd SIN** x/2),x)|x **green diamond >** 0 **2nd MATH 8 8** x **2nd <** 2π **ENTER**

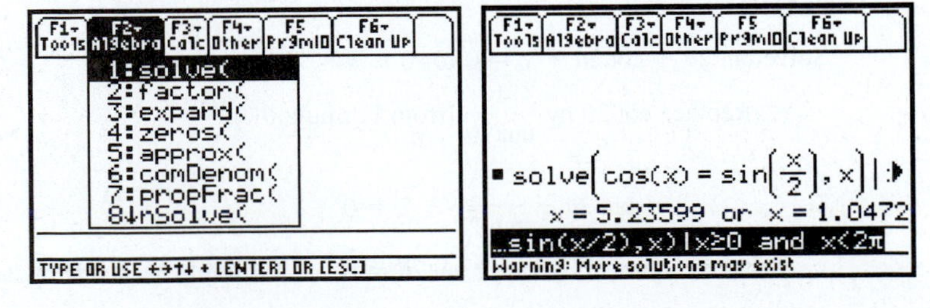

Exercises 13.4

Solve each trigonometric equation for $0° \leq x < 360°$.

1. $\cos x - 1 = 0$

2. $2 \sin x + \sqrt{3} = 0$

3. $\tan x - 1 = 0$

4. $2 \sin^2 x - 1 = 0$

5. $4 \cos^2 x - 3 = 0$

6. $\tan^2 x - 3 = 0$

7. $\sin 2x = 1$

8. $2 \cos 3x + 1 = 0$

9. $3 \tan^2 3x - 1 = 0$

10. $4 \sin^2 2x = 3$

11. $\sin 2x + \cos x = 0$

12. $\cos 2x - \sin x = 0$

13. $\sin^2 x + \sin x = 0$

14. $\cos x \tan x + \cos x = 0$

15. $2 \sin^2 x = 1 - 2 \sin x$

16. $\cos^2 x - 3 \cos x + 1 = 0$

Solve each trigonometric equation for $0 \leq x < 2\pi$.

17. $2 \cos^2 x + \sin x = 1$

18. $\tan x + 2 \cos x = \sec x$

19. $\cos^2 x - \cos x \sec x = 0$

20. $\sin^2 x + \sin x \cos x = 0$

21. $2 \sin^2 x + 2 \cos 2x = 1$

22. $6 \sin^2 x + \cos 2x = 4$

23. $2 \sin^2 2x - \sin 2x - 1 = 0$

24. $\cos 2x - 2 \sin^2 2x + 1 = 0$

25. $4 \tan^2 x = 3 \sec^2 x$

26. $2 \sin x - \tan x = 0$

27. $4 \sin 2x \cos 2x = 1$

28. $\cos^2 3x - \sin^2 3x = 1$

29. $\cos x = \cos \frac{x}{2}$

30. $\sin x = \cos \frac{x}{2}$

31. $\cos \frac{x}{2} = 1 + \cos x$

32. $\sin \frac{x}{2} = 1 - \cos x$

33. $1 + \cos^2 \frac{x}{2} = 2 \cos x$

34. $1 + \sin^2 \frac{x}{2} = \cos x$

13.5 INVERSE TRIGONOMETRIC FUNCTIONS

In Section 10.2, we defined the inverse of a given equation to be the resulting equation when the variables x and y are interchanged. Recall that

$$\text{The inverse of } y = b^x \text{ is } x = b^y.$$
$$\text{The inverse of } y = \sqrt[3]{x} \text{ is } x = \sqrt[3]{y}.$$

Likewise, each basic trigonometric equation has an inverse:

The inverse of	*Is*
$y = \sin x$	$x = \sin y$
$y = \cos x$	$x = \cos y$
$y = \tan x$	$x = \tan y$
$y = \cot x$	$x = \cot y$
$y = \sec x$	$x = \sec y$
$y = \csc x$	$x = \csc y$

Inverse functions are found by solving the inverse equation for y. As previously mentioned, $y = \sqrt[3]{x}$ has an inverse equation, $x = \sqrt[3]{y}$. Solving it for y (by cubing both sides) gives the inverse function $y = x^3$. This inverse function is often written as $f^{-1}(x) = x^3$

when the original function is written as $f(x) = \sqrt[3]{x}$. The notation "$f^{-1}(x)$" is pronounced "f inverse of x" and must not be confused with the reciprocal function $\frac{1}{f(x)}$. That is, the -1 in "$f^{-1}(x)$" should *not* be interpreted as an exponent.

On your calculator, the inverse trig functions are marked using this raised -1, since it gives a very short, convenient notation that fits comfortably on the calculator's keyboard. For example, the inverse of $f(x) = \sin x$ is shown as $f^{-1}(x) = \sin^{-1} x$. However, this is awkward for computer programming languages and other software, which require commands to be typewritten on a computer keyboard. When working with computers, expect to use the notation "arcsin x" (or an abbreviation based on this idea). Learning to read and write both notations will provide the best preparation for modern technical work. Capitalized versions of these notations, $\text{Sin}^{-1} x$ and Arcsin x, are also commonly used, but are no longer considered standard.

There are two common forms of the inverse trigonometric equations solved for y:

		Solved for *y*,	
The inverse of	*Is*	*Is*	*Is*
$y = \sin x$	$x = \sin y$	$y = \arcsin x$*	$y = \sin^{-1} x$
$y = \cos x$	$x = \cos y$	$y = \arccos x$	$y = \cos^{-1} x$
$y = \tan x$	$x = \tan y$	$y = \arctan x$	$y = \tan^{-1} x$
$y = \cot x$	$x = \cot y$	$y = \text{arccot}\, x$	$y = \cot^{-1} x$
$y = \sec x$	$x = \sec y$	$y = \text{arcsec}\, x$	$y = \sec^{-1} x$
$y = \csc x$	$x = \csc y$	$y = \text{arccsc}\, x$	$y = \csc^{-1} x$

EXAMPLE 1

What is the meaning of each equation?

(a) $y = \arctan x$
(b) $y = \arccos 3x$
(c) $y = 4\, \text{arccsc}\, 5x$

(a) y is the angle whose tangent is x.
(b) y is the angle whose cosine is $3x$.
(c) y is four times the angle whose cosecant is $5x$.

EXAMPLE 2

Solve the equation $y = \cos 2x$ for x.

The equation $y = \cos 2x$ is equivalent to

$$\arccos y = 2x$$

So,

$$x = \frac{1}{2} \arccos y$$

*This is read "y equals the arcsine of x" and means that y is the angle whose sine is x.

EXAMPLE 3

Solve the equation $2y = \arctan 3x$ for x.

The equation $2y = \arctan 3x$ is equivalent to

$$\tan 2y = 3x$$

So,

$$x = \frac{1}{3}\tan 2y$$

EXAMPLE 4

Solve the equation $y = \frac{1}{3}\text{ arcsec } 2x$ for x.

First, multiply both sides by 3.

$$3y = \text{arcsec } 2x$$

The equation $3y = \text{arcsec } 2x$ is equivalent to

$$\sec 3y = 2x$$

Thus,

$$x = \frac{1}{2}\sec 3y$$

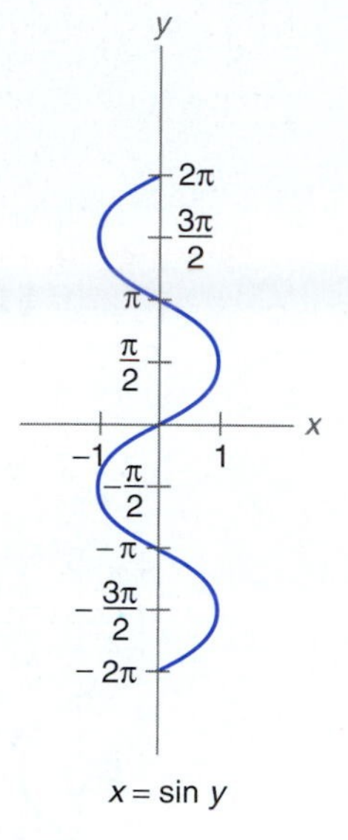

Figure 13.6

To understand an inverse trigonometric function like

$$y = \arcsin x$$

first graph the corresponding inverse trigonometric equation solved for x, that is,

$$x = \sin y$$

as shown in Fig. 13.6. The graphs of all six inverse trigonometric equations are shown in Fig. 13.7. To easily produce these graphs yourself, use a transparency (or a thin piece of paper with markers that will bleed through enough to be visible on the back side) and draw the graph of the original trigonometric function, like $y = \sin x$, carefully labeling the (positive) x- and y-axes. Rotate the transparency (or paper) 90° counterclockwise and notice that the positive x-axis is where the positive y-axis used to be; unfortunately, the positive y-axis is where the *negative* x-axis was. Now turn the transparency over (looking at it from the back) and you will finally see the perfect reversal of the roles of x and y, and thus you will be viewing a graph of $x = \sin y$ instead of your drawing of $y = \sin x$. This transformation can also be performed with a graphing calculator and a mirror. Graph $y = \sin x$ (using **ZoomTrig**), then tilt the calculator sideways so that the keyboard is on your right as you are viewing the screen. Now look at the back of the calculator, viewing the image in the mirror. This will show you the graph with the positive x-axis replacing the positive y-axis and vice versa, so you will be viewing $x = \sin y$ instead of $y = \sin x$. To go directly to this correct mirror image position, you should be looking at the back of the calculator in a sideways orientation, with its tapered (battery chamber) end on the left-hand side.

As you can see from the graphs in Fig. 13.7, each x-value in each domain corresponds to (infinitely) many values of y. Thus, none of the inverse trigonometric equations describes a function unless we restrict the y-values in an appropriate way. The customary restrictions are as follows:

INVERSE TRIGONOMETRIC FUNCTIONS

$y = \arcsin x \qquad -\dfrac{\pi}{2} \le y \le \dfrac{\pi}{2}$

$y = \arccos x \qquad 0 \le y \le \pi$

$y = \arctan x \qquad -\dfrac{\pi}{2} < y < \dfrac{\pi}{2}$

$y = \operatorname{arccot} x \qquad 0 < y < \pi$

$y = \operatorname{arcsec} x \qquad 0 \le y \le \pi \quad y \ne \dfrac{\pi}{2}$

$y = \operatorname{arccsc} x \qquad -\dfrac{\pi}{2} \le y \le \dfrac{\pi}{2} \quad y \ne 0$

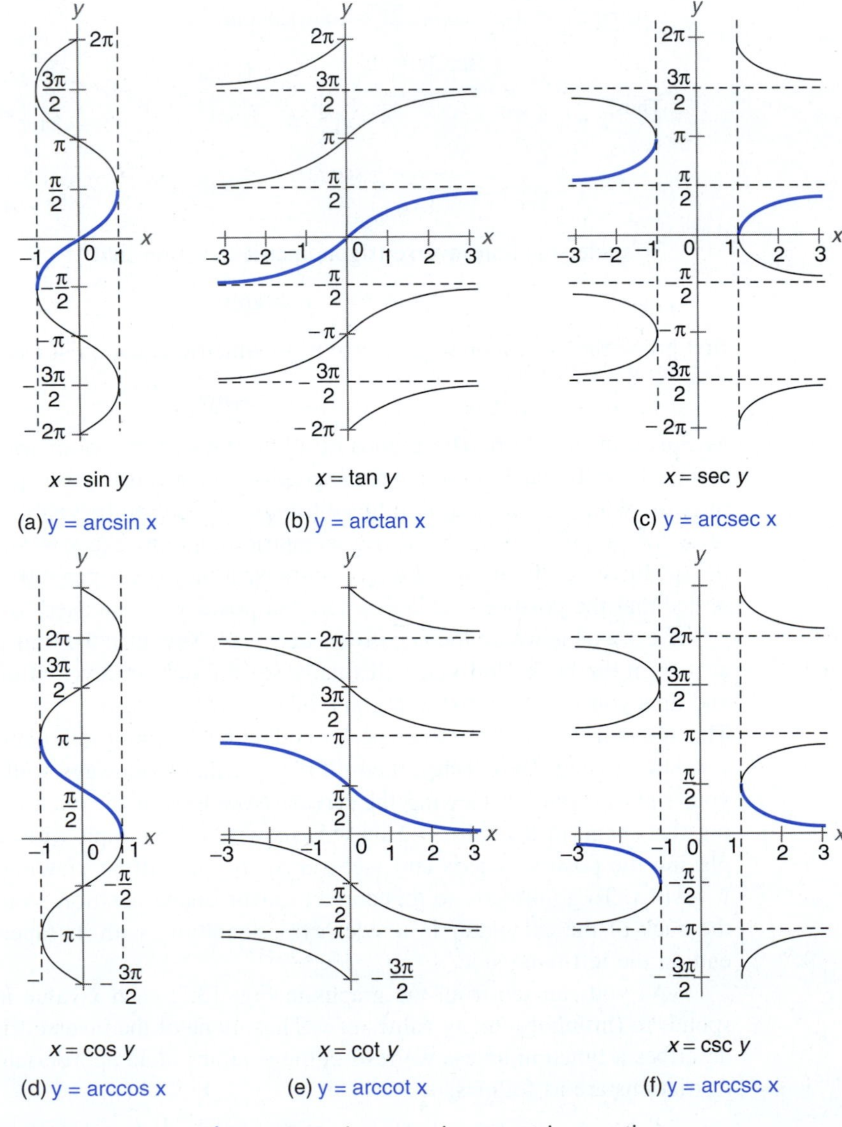

Figure 13.7 Inverse trigonometric equations.

Look at the graphs of the inverse trigonometric equations in Fig. 13.7. The colored lines indicate the portions of the graphs that correspond to the inverse trigonometric *functions.*

Note: The three inverse trigonometric functions on calculators are programmed to these same restricted ranges. When using a calculator to find the value of arccot x, arcsec x, or arccsc x, use the following:

1. $$\text{arccot}\, x = \begin{cases} \arctan \dfrac{1}{x} & \text{if } x > 0 \\ \pi + \arctan \dfrac{1}{x} & \text{if } x < 0 \end{cases}$$

2. $$\text{arcsec}\, x = \arccos \frac{1}{x} \quad \text{where } x \geq 1 \quad \text{or} \quad x \leq -1$$

3. $$\text{arccsc}\, x = \arcsin \frac{1}{x} \quad \text{where } x \geq 1 \quad \text{or} \quad x \leq -1$$

EXAMPLE 5

Find $\arcsin\left(\frac{1}{2}\right)$.

$$\arcsin\left(\frac{1}{2}\right) = \frac{\pi}{6}$$

This is the only value in the defined range of $-\frac{\pi}{2} \leq y \leq \frac{\pi}{2}$.

EXAMPLE 6

Find $\arctan(-1)$.

$$\arctan(-1) = -\frac{\pi}{4}$$

This is the only value in the defined range of $-\frac{\pi}{2} < y < \frac{\pi}{2}$.

EXAMPLE 7

Find $\cos^{-1}\left(-\frac{1}{2}\right)$.

$$\cos^{-1}\left(-\frac{1}{2}\right) = \frac{2\pi}{3}$$

This is the only value in the defined range of $0 \leq y \leq \pi$.

EXAMPLE 8

Find $\tan\, [\arccos(-1)]$.

$$\tan\, [\arccos(-1)] = \tan \pi = 0$$

EXAMPLE 9

Find $\cos\, (\sec^{-1} 2)$.

$$\cos\, (\sec^{-1} 2) = \cos \frac{\pi}{3} = \frac{1}{2}$$

EXAMPLE 10

Find $\sin\left[\arctan\left(-\frac{1}{\sqrt{3}}\right)\right]$.

$$\sin\left[\arctan\left(-\frac{1}{\sqrt{3}}\right)\right] = \sin\left(-\frac{\pi}{6}\right) = -\frac{1}{2}$$

EXAMPLE 11

Find an algebraic expression for sin (arccos x).

Let $\theta = \arccos x$. Then

$$\cos\theta = x = \frac{x}{1}$$

Draw a right triangle with θ as an acute angle, x as the adjacent side, and 1 as the hypotenuse as in Fig. 13.8(a). Using the Pythagorean theorem, we have

$$c^2 = a^2 + b^2$$
$$1^2 = x^2 + (\text{side opposite } \theta)^2$$

and

$$\text{side opposite } \theta = \sqrt{1 - x^2} \qquad \text{[Fig. 13.8(b)]}$$

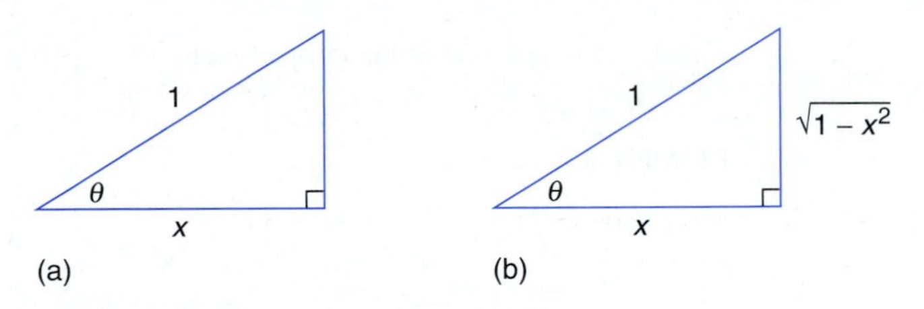

Figure 13.8

Now we see that

$$\begin{aligned}\sin(\arccos x) &= \sin\theta \\ &= \frac{\text{side opposite } \theta}{\text{hypotenuse}} \\ &= \frac{\sqrt{1 - x^2}}{1} \\ &= \sqrt{1 - x^2}\end{aligned}$$

Using a calculator, we have

2nd SIN green diamond COS^{-1} x)) **ENTER**

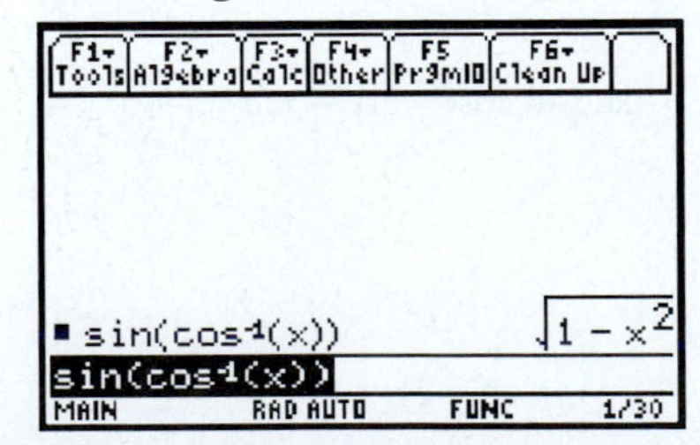

EXAMPLE 12

Find an algebraic expression for sec (arctan x).

Let $\theta = \arctan x$. Then

$$\tan\theta = x = \frac{x}{1}$$

Draw a right triangle with θ as an acute angle, x as the opposite side, and 1 as the adjacent side as in Fig. 13.9. Using the Pythagorean theorem, we find that the hypotenuse is $\sqrt{x^2+1}$.

$$\begin{aligned}\sec(\arctan x) &= \sec\theta \\ &= \frac{\text{hypotenuse}}{\text{side adjacent to }\theta} \\ &= \frac{\sqrt{x^2+1}}{1} \\ &= \sqrt{x^2+1}\end{aligned}$$

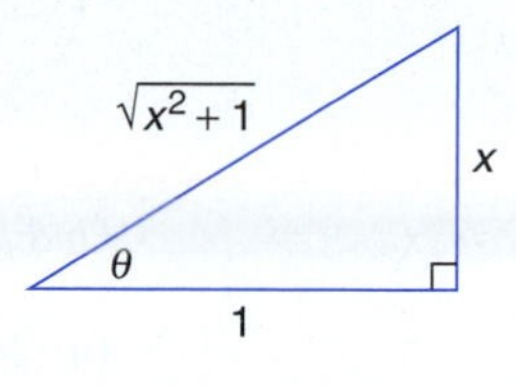

Figure 13.9

To use a calculator, you must write secant in terms of cosine.

1/ 2nd COS green diamond TAN $^{-1}$ x)) ENTER

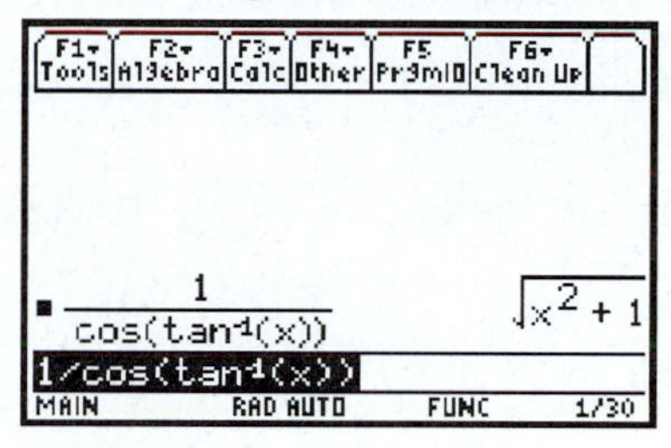

EXAMPLE 13

Find an algebraic expression for cos ($2\sin^{-1} x$).

Let $\theta = \sin^{-1} x$. Then

$$\sin\theta = \frac{x}{1}$$

Draw a right triangle with θ as an acute angle, x as the opposite side, and 1 as the hypotenuse as in Fig. 13.10. Using the Pythagorean theorem, we find that the side adjacent to θ is $\sqrt{1-x^2}$.

$$\begin{aligned}\cos(2\sin^{-1} x) &= \cos 2\theta \\ &= 1 - 2\sin^2\theta \qquad \text{[Formula (25c)]} \\ &= 1 - 2x^2\end{aligned}$$

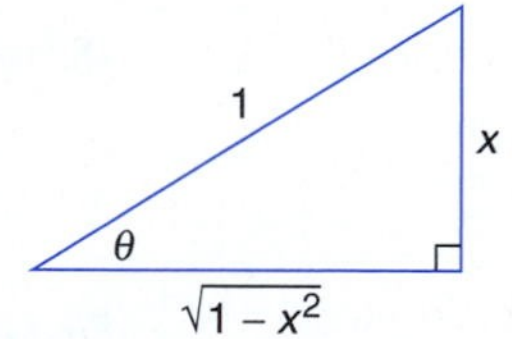

Figure 13.10

Exercises 13.5

Write the meaning of each equation.

1. $y = \arcsin x$

2. $y = \operatorname{arcsec} x$

3. $y = \cot^{-1} 4x$

4. $y = \cos^{-1} 2x$

5. $y = 3 \operatorname{arccsc} \frac{1}{2}x$

6. $y = \frac{1}{2}\arctan 3x$

Solve each equation for x.

7. $y = \sin 3x$ **8.** $y = \tan 4x$ **9.** $y = 4 \cos x$

10. $y = 3 \sec x$ **11.** $y = 5 \tan \frac{x}{2}$ **12.** $y = \frac{1}{2} \cos 3x$

13. $y = \frac{3}{2} \cot \frac{x}{4}$ **14.** $y = \frac{5}{2} \sin \frac{2x}{3}$ **15.** $y = 3 \sin (x - 1)$

16. $y = 4 \tan (2x + 1)$ **17.** $y = \frac{1}{2} \cos (3x + 1)$ **18.** $y = \frac{1}{3} \sec (1 - 4x)$

Find the value of each expression in radians.

19. $\arcsin\left(\frac{\sqrt{3}}{2}\right)$ **20.** $\arccos\left(\frac{1}{2}\right)$ **21.** $\tan^{-1}\left(-\frac{1}{\sqrt{3}}\right)$

22. $\sin^{-1}\left(-\frac{1}{2}\right)$ **23.** $\arccos\left(-\frac{\sqrt{3}}{2}\right)$ **24.** $\arctan\left(\frac{1}{\sqrt{3}}\right)$

25. $\text{arccsc}\ \sqrt{2}$ **26.** $\arcsin (-1)$ **27.** $\arctan \sqrt{3}$

28. $\arccos 0$ **29.** $\cos^{-1}\left(\frac{1}{\sqrt{2}}\right)$ **30.** $\sin^{-1} 1$

31. $\sin^{-1}\left(-\frac{\sqrt{3}}{2}\right)$ **32.** $\tan^{-1}(-\sqrt{3})$ **33.** $\text{arccot}\ (-1.5)$

34. $\text{arccsc}\ 2.5$ **35.** $\text{arcsec}\ (-3.2)$ **36.** $\arcsin 0.75$

37. $\csc^{-1}(-1.15)$ **38.** $\cos^{-1}(-0.55)$

Find the value of each expression.

39. $\cos (\arctan \sqrt{3})$ **40.** $\tan\left[\arcsin\left(\frac{1}{\sqrt{2}}\right)\right]$ **41.** $\sin\left[\arccos\left(-\frac{1}{\sqrt{2}}\right)\right]$

42. $\sin [\arctan (-1)]$ **43.** $\tan [\cos^{-1}(-1)]$ **44.** $\sec\left[\cos^{-1}\left(-\frac{1}{2}\right)\right]$

45. $\sin\left[\arcsin\left(\frac{\sqrt{3}}{2}\right)\right]$ **46.** $\tan [\arctan (-\sqrt{3})]$ **47.** $\cos\left[\sin^{-1}\left(\frac{3}{5}\right)\right]$

48. $\tan\left[\sin^{-1}\left(\frac{12}{13}\right)\right]$ **49.** $\tan [\arcsin (-0.1560)]$ **50.** $\sin [\text{arccot}\ (1.635)]$

Find an algebraic expression for each.

51. $\cos (\arcsin x)$ **52.** $\tan (\arccos x)$ **53.** $\sin (\text{arcsec}\ x)$

54. $\cot (\text{arcsec}\ x)$ **55.** $\sec (\cos^{-1} x)$ **56.** $\sin (\tan^{-1} x)$

57. $\tan (\arctan x)$ **58.** $\sin (\arcsin x)$ **59.** $\cos (\arcsin 2x)$

60. $\tan (\cos^{-1} 3x)$ **61.** $\sin (2 \sin^{-1} x)$ **62.** $\cos (2 \arctan x)$

Graph each equation.

63. $y = \arcsin 2x$ **64.** $y = 3 \arccos x$ **65.** $y = 2 \arctan 3x$

CHAPTER 13 SUMMARY

1. The *basic identities* developed in this chapter are as follows:

$$\sin\theta = \frac{1}{\csc\theta} \qquad (1)$$

$$\cos\theta = \frac{1}{\sec\theta} \qquad (2)$$

$$\tan\theta = \frac{1}{\cot\theta} \qquad (3)$$

$$\cot\theta = \frac{1}{\tan\theta} \qquad (4)$$

$$\sec\theta = \frac{1}{\cos\theta} \qquad (5)$$

$$\csc\theta = \frac{1}{\sin\theta} \qquad (6)$$

$$\tan\theta = \frac{\sin\theta}{\cos\theta} \qquad (7)$$

$$\cot\theta = \frac{\cos\theta}{\sin\theta} \qquad (8)$$

$$\sin^2\theta + \cos^2\theta = 1 \qquad (9)$$

$$1 + \tan^2\theta = \sec^2\theta \qquad (10)$$

$$\cot^2\theta + 1 = \csc^2\theta \qquad (11)$$

$$\sin(\theta + \phi) = \sin\theta\cos\phi + \cos\theta\sin\phi \qquad (12)$$

$$\cos(\theta + \phi) = \cos\theta\cos\phi - \sin\theta\sin\phi \qquad (13)$$

$$\sin(-\theta) = -\sin\theta \qquad (14)$$

$$\cos(-\theta) = \cos\theta \qquad (15)$$

$$\tan(-\theta) = -\tan\theta \qquad (16)$$

$$\cot(-\theta) = -\cot\theta \qquad (17)$$

$$\sec(-\theta) = \sec\theta \qquad (18)$$

$$\csc(-\theta) = -\csc\theta \qquad (19)$$

$$\sin(\theta - \phi) = \sin\theta\cos\phi - \cos\theta\sin\phi \qquad (20)$$

$$\cos(\theta - \phi) = \cos\theta\cos\phi + \sin\theta\sin\phi \qquad (21)$$

$$\tan(\theta + \phi) = \frac{\tan\theta + \tan\phi}{1 - \tan\theta\tan\phi} \qquad (22)$$

$$\tan(\theta - \phi) = \frac{\tan\theta - \tan\phi}{1 + \tan\theta\tan\phi} \qquad (23)$$

$$\sin 2\theta = 2\sin\theta\cos\theta \qquad (24)$$

$$\cos 2\theta = \cos^2\theta - \sin^2\theta \qquad (25a)$$

$$= 2\cos^2\theta - 1 \qquad (25b)$$

$$= 1 - 2\sin^2\theta \qquad (25c)$$

$$\tan 2\theta = \frac{2\tan\theta}{1 - \tan^2\theta} \qquad (26)$$

$$\sin\frac{\theta}{2} = \pm\sqrt{\frac{1-\cos\theta}{2}} \tag{27}$$

$$\cos\frac{\theta}{2} = \pm\sqrt{\frac{1+\cos\theta}{2}} \tag{28}$$

$$\tan\frac{\theta}{2} = \frac{1-\cos\theta}{\sin\theta} \tag{29}$$

2. The general methods for *proving a trigonometric identity* are as follows:
   (a) Substitute one expression for another using a known identity.
   (b) Factor one side of the identity.
   (c) Multiply numerator and denominator of a fractional expression by some trigonometric expression.
   (d) Express all functions on one side in terms of sines and cosines, and use one of the preceding methods.
3. The *inverse trigonometric functions* are defined as follows:

$$y = \arcsin x \qquad -\frac{\pi}{2} \le y \le \frac{\pi}{2}$$

$$y = \arccos x \qquad 0 \le y \le \pi$$

$$y = \arctan x \qquad -\frac{\pi}{2} < y < \frac{\pi}{2}$$

$$y = \operatorname{arccot} x \qquad 0 < y < \pi$$

$$y = \operatorname{arcsec} x \qquad 0 \le y \le \pi \quad y \ne \frac{\pi}{2}$$

$$y = \operatorname{arccsc} x \qquad -\frac{\pi}{2} \le y \le \frac{\pi}{2} \quad y \ne 0$$

See Fig. 13.7 for the colored graphs of the inverse trigonometric functions.

CHAPTER 13 REVIEW

Prove each identity.

1. $\sec x \cot x = \csc x$
2. $\sec^2\theta + \tan^2\theta + 1 = \dfrac{2}{\cos^2\theta}$
3. $\dfrac{\cos\theta}{\cos\theta + \sin\theta} = \dfrac{\cot\theta}{1+\cot\theta}$
4. $\cos\left(\theta - \dfrac{3\pi}{2}\right) = -\sin\theta$
5. $\left(\sin\dfrac{1}{2}x + \cos\dfrac{1}{2}x\right)^2 = 1 + \sin x$
6. $2\cos^2\dfrac{\theta}{2} = \dfrac{1+\sec\theta}{\sec\theta}$
7. $\dfrac{2\cot\theta}{1+\cot^2\theta} = \sin 2\theta$
8. $\csc x - \cot x = \tan\dfrac{1}{2}x$
9. $\tan 2x = \dfrac{2\cos x}{\csc x - 2\sin x}$
10. $\tan^2\dfrac{x}{2} + 1 = 2\tan\dfrac{x}{2}\csc x$

Simplify each expression.

11. $\sin\theta\cos\theta$
12. $\cos^2 3\theta - \sin^2 3\theta$
13. $\dfrac{1+\cos 4\theta}{2}$
14. $1 - 2\sin^2\dfrac{\theta}{3}$
15. $\cos 2x\cos 3x - \sin 2x\sin 3x$
16. $\sin 2x\cos x - \cos 2x\sin x$

17. Find $\sin 2\theta$ when $\cos\theta = -\frac{5}{13}$ (θ in the second quadrant).

18. Find $\cos\frac{\theta}{2}$ when $\tan\theta = \frac{4}{3}$ (θ in the third quadrant).

Solve each trigonometric equation for $0 \le x < 2\pi$.

19. $2\cos^2 x = \cos x$

20. $4\sin^2 x - 1 = 0$

21. $2\cos^2 3x + \sin 3x - 1 = 0$

22. $\tan^2 x = \sin^2 x$

23. $\sin\frac{x}{2} + \cos\frac{x}{2} = 0$ (*Hint:* Square each side.)

24. $\sin 2x = \cos^2 x - \sin^2 x$

25. Solve for x: $y = \frac{1}{2}\sin\frac{3x}{4}$.

Find the value of each expression in radians.

26. $\arcsin\left(\frac{1}{\sqrt{2}}\right)$

27. $\arctan\left(-\frac{1}{\sqrt{3}}\right)$

28. $\sec^{-1}(-1)$

29. $\cos^{-1}\left(-\frac{1}{2}\right)$

Find the value of each expression.

30. $\sin\left[\arccos\left(-\frac{1}{2}\right)\right]$

31. $\tan(\tan^{-1}\sqrt{3})$

32. Find an algebraic expression for $\sin(\text{arccot}\, x)$.

33. Graph $y = 1.5 \arccos 2x$

14 Complex Numbers

INTRODUCTION

Mathematicians apply the laws of logic to a few assumptions and definitions to produce theorems and truths about mathematical systems. Often these systems have no basis in our physical world when they are first studied, but as our knowledge of the physical world increases, scientists often find real phenomena that match the mathematics already discovered! Such is the case with complex numbers, which are used extensively in the study of electricity and electronics.

Objectives

- Know when two complex numbers are equal.
- Write complex numbers in trigonometric form.
- Write complex numbers in exponential form.
- Add, subtract, multiply, and divide complex numbers in all forms.

14.1 COMPLEX NUMBERS IN RECTANGULAR FORM

Up to this point we have considered only problems with real-number solutions. When we studied roots, we restricted the radicand of even roots to be nonnegative. For equations whose solutions are not real, such as $x^2 + 1 = 0$, or are even roots with negative radicands, we must extend our number system to include the imaginary and complex numbers.

The square root of a negative number is an imaginary number. (Historically, the term *imaginary* was, indeed, a poor choice of terms. It was meant to distinguish such numbers from the "real" numbers—also a poor choice.) The imaginary unit is defined as $\sqrt{-1}$ and in many mathematics texts is denoted by the symbol i. However, in technical work i is used to denote current. To avoid confusion, many technical books use j to denote $\sqrt{-1}$, which is what we shall do.

IMAGINARY UNIT

$$j = \sqrt{-1}$$

Thus, we may write $\sqrt{-16}$ as $4j$ [since $\sqrt{-16} = \sqrt{(-1)(16)} = \sqrt{-1}\sqrt{16} = 4j$].

We define an **imaginary number** as any number in the form bj where b is a real number. We define a **complex number** as any number in the form $a + bj$ where a and b are real numbers. Note that when $a = 0$, we have an imaginary number; when $b = 0$, we have a real number.

RECTANGULAR FORM OF A COMPLEX NUMBER

$$a + bj$$

is called the *rectangular form* of a complex number, where a is the **real** part and bj is the **imaginary** part.

Two complex numbers $a + bj$ and $c + dj$ are **equal** only when $a = c$ and $b = d$.

EXAMPLE 1

Express each number in terms of j and simplify.

(a) $\sqrt{-36} = \sqrt{(-1)(36)}$
$= \sqrt{(-1)}\sqrt{36}$
$= 6j$

(b) $\sqrt{-45} = \sqrt{(-1)(9)(5)}$
$= \sqrt{-1}\sqrt{9}\sqrt{5}$
$= 3\sqrt{5}j$

Or, by calculator, we have

MODE *five* down arrows　　right arrow **2 ENTER**　　(The TI-89 uses **i** instead of ***j***.)

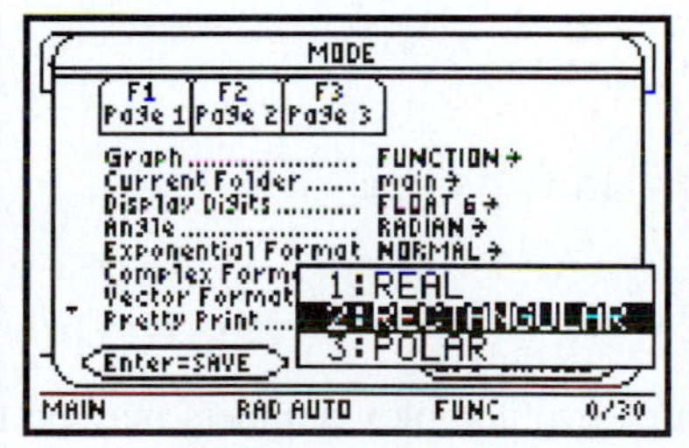

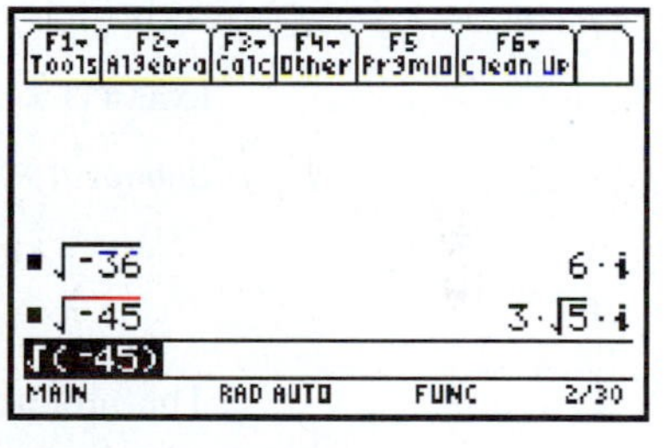

Next, we need to consider powers of j, or $\sqrt{-1}$. Using the properties of exponents and the definition of j, we find some powers of j:

$$\begin{aligned}
j &= j \\
j^2 &= (\sqrt{-1})^2 = -1 \\
j^3 &= j^2 \cdot j = (-1)j = -j \\
j^4 &= j^2 \cdot j^2 = (-1)(-1) = 1 \\
j^5 &= j^4 \cdot j = (1)j = j \\
j^6 &= j^4 \cdot j^2 = (1)(-1) = -1 \\
j^7 &= j^4 \cdot j^3 = (1)(-j) = -j \\
j^8 &= (j^4)^2 = 1^2 = 1
\end{aligned}$$

The integral powers of j are cyclic in the order of $j, -1, -j, 1, j, -1, -j, 1, \ldots$. It is helpful to note that any power of j evenly divisible by four is equal to one.

EXAMPLE 2

Simplify.

(a) $j^{14} = j^{12} \cdot j^2$
$= (1)(-1)$
$= -1$

(b) $j^{37} = j^{36} \cdot j$
$= (1)j$
$= j$

(c) $j^{323} = j^{320} \cdot j^3$
$= (1)(-j)$
$= -j$

The imaginary numbers and the complex numbers, unlike the real numbers, are not ordered. That is, given two unequal complex numbers, one is not larger or smaller than the other.

Complex numbers may be added by finding the sum of the real parts and the sum of the imaginary parts.

$$(a + bj) + (c + dj) = (a + c) + (b + d)j$$

EXAMPLE 3

Add $(2 + 3j) + (4 - 2j)$.

$$\begin{aligned}(2 + 3j) + (4 - 2j) &= (2 + 4) + [3 + (-2)]j \\ &= 6 + j\end{aligned}$$

Complex numbers may be subtracted by finding the difference of the real parts and the difference of the imaginary parts.

$$(a + bj) - (c + dj) = (a - c) + (b - d)j$$

EXAMPLE 4

Subtract $(-3 + 4j) - (7 - 2j)$.

$$\begin{aligned}(-3 + 4j) - (7 - 2j) &= (-3 - 7) + [4 - (-2)]j \\ &= -10 + 6j\end{aligned}$$

The product of two complex numbers may be found as if they were two ordinary binomials. Simplifying, we have

$$\begin{aligned}(a + bj)(c + dj) &= ac + (ad + bc)j + bdj^2 \qquad (j^2 = -1) \\ &= (ac - bd) + (ad + bc)j\end{aligned}$$

Or, using the calculator, we have

(a+b **2nd CATALOG**)(c+d **2nd CATALOG**) **ENTER**

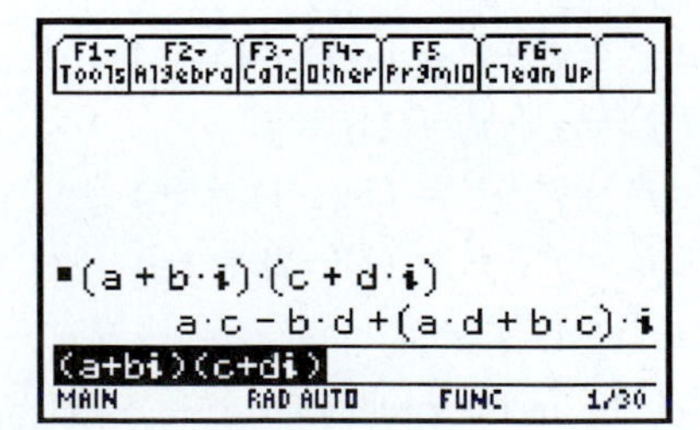

EXAMPLE 5

Multiply $(5 - 3j)(-2 + 7j)$.

$$\begin{aligned}(5 - 3j)(-2 + 7j) &= -10 + 41j - 21j^2 \qquad (j^2 = -1)\\ &= -10 + 41j + 21\\ &= 11 + 41j\end{aligned}$$

Before we consider the division of complex numbers, we need to define and discuss the conjugate of a complex number. The **conjugate** of the complex number $a + bj$ is $a - bj$ and the conjugate of $a - bj$ is $a + bj$. That is, the conjugate is formed by changing the sign of the imaginary part.

The product of two conjugates is always a real number.

$$(a + bj)(a - bj) = a^2 - b^2j^2 = a^2 + b^2$$

Complex numbers may be divided by multiplying numerator and denominator by the conjugate of the denominator.

$$\begin{aligned}\frac{a + bj}{c + dj} &= \frac{a + bj}{c + dj} \cdot \frac{c - dj}{c - dj}\\ &= \frac{(ac + bd) + (bc - ad)j}{c^2 + d^2} \quad \text{or} \quad \frac{ac + bd}{c^2 + d^2} + \left(\frac{bc - ad}{c^2 + d^2}\right)j\end{aligned}$$

EXAMPLE 6

Divide $\frac{4 + j}{2 - 3j}$.

$$\begin{aligned}\frac{4 + j}{2 - 3j} = \frac{4 + j}{2 - 3j} \cdot \frac{2 + 3j}{2 + 3j} &= \frac{8 + 14j + 3j^2}{4 - 9j^2}\\ &= \frac{5 + 14j}{13} \quad \text{or} \quad \frac{5}{13} + \frac{14}{13}j\end{aligned}$$

Note: This is the same technique used in rationalizing binomial denominators in Section 8.5.

EXAMPLE 7

Divide $\frac{5 - 5j}{3 + 7j}$.

$$\begin{aligned}\frac{5 - 5j}{3 + 7j} = \frac{5 - 5j}{3 + 7j} \cdot \frac{3 - 7j}{3 - 7j} &= \frac{15 - 50j + 35j^2}{9 - 49j^2}\\ &= \frac{-20 - 50j}{58}\\ &= -\frac{20}{58} - \frac{50j}{58}\\ &= -\frac{10}{29} - \frac{25}{29}j\end{aligned}$$

When we solved the quadratic equation $ax^2 + bx + c = 0$, its solutions were $x = \frac{-b \pm \sqrt{b^2 - 4ac}}{2a}$ with the restriction that the *discriminant* $b^2 - 4ac \geq 0$. We now remove this restriction. If $b^2 - 4ac < 0$, the solutions are complex.

EXAMPLE 8

Solve $x^2 + 1 = 0$.

$$x^2 + 1 = 0$$
$$x^2 = -1$$
$$x = \pm\sqrt{-1} = \pm j$$

EXAMPLE 9

Solve $3x^2 + 4x + 2 = 0$.

Using the quadratic formula, we have

$$x = \frac{-b \pm \sqrt{b^2 - 4ac}}{2a}$$

$$x = \frac{-4 \pm \sqrt{4^2 - 4(3)(2)}}{2(3)}$$

$$= \frac{-4 \pm \sqrt{-8}}{6}$$

$$= \frac{-4 \pm 2j\sqrt{2}}{6}$$

$$= \frac{-2 \pm j\sqrt{2}}{3} \quad \text{or} \quad -\frac{2}{3} \pm \frac{\sqrt{2}}{3}j$$

Using a calculator, we have

F2 alpha A 1 | 3x^2+4x+2=0,x) **ENTER** | up arrow **2nd** right arrow

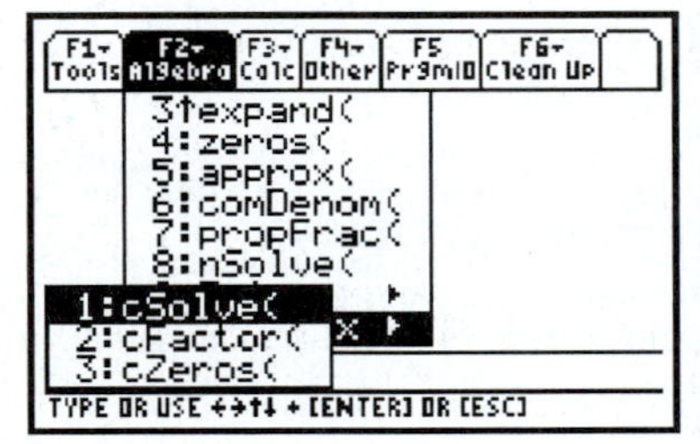

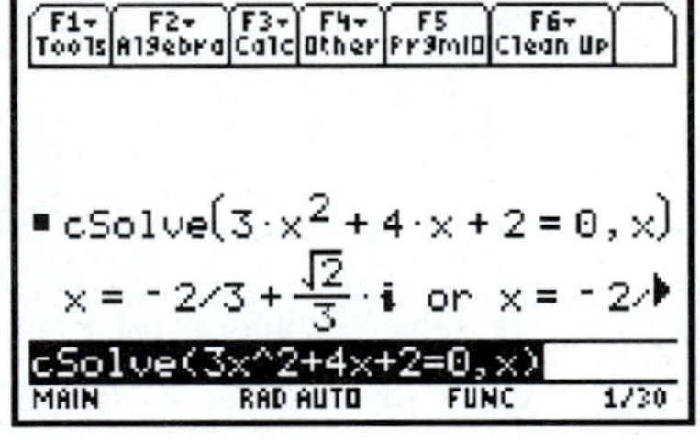

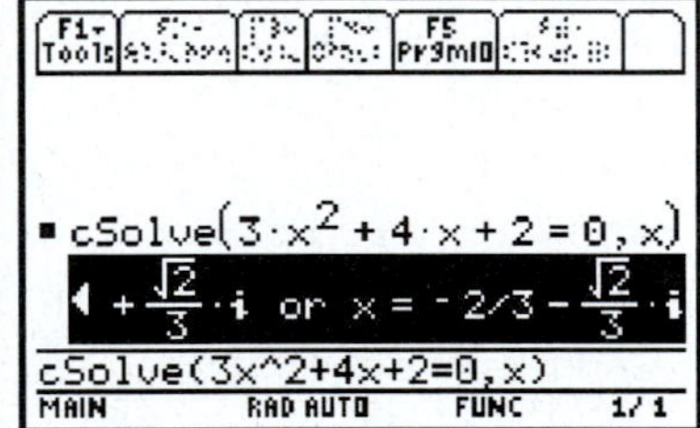

The complex number $a + bj$ may also be written as an ordered pair of real numbers (a, b). This allows us to associate points in the coordinate plane with the complex numbers. To do so, we modify the coordinate axes by letting the horizontal axis be the real axis and the vertical axis be the imaginary axis. This plane is called the **complex plane.**

To find the point that corresponds to a given complex number $a + bj$, plot the ordered pair (a, b). For example, the complex numbers $4 + 2j$ and $-3 - j$ are plotted in Fig. 14.1.

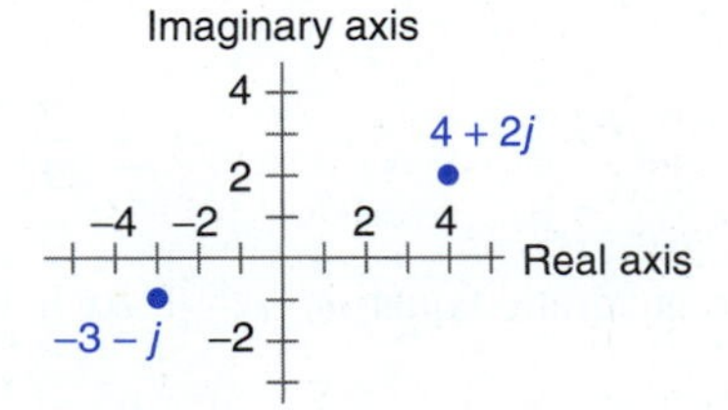

Figure 14.1 Complex plane.

The sum of two complex numbers may also be found graphically, which is shown by the following example.

EXAMPLE 10

Add $5 + 4j$ and $3 - j$ graphically.

First plot $5 + 4j$ and $3 - j$ in the complex plane, as in Fig. 14.2. Each may be drawn as a vector from the origin to the points (5, 4) and (3, −1), respectively. The graphical sum is the diagonal of the parallelogram which is the resultant of the two vectors. The endpoint of the resultant is (8, 3), which corresponds to the complex number $8 + 3j$, which is also the algebraic sum.

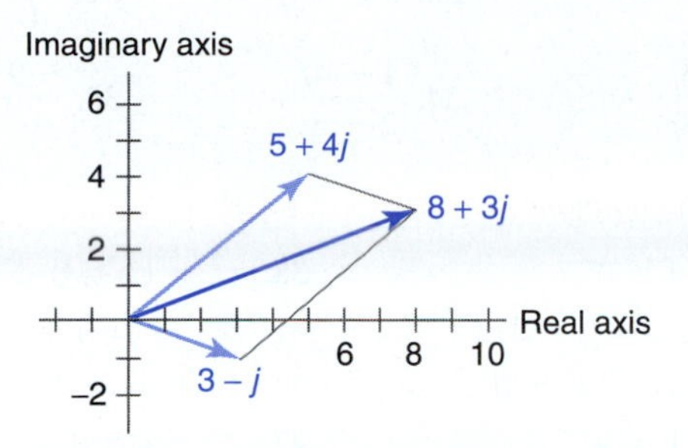

Figure 14.2

To subtract one complex number from another graphically, add the negative of the number being subtracted (the second number or subtrahend).

EXAMPLE 11

Find the difference $(-2 - 4j) - (-3 + 5j)$ graphically.

First, plot $-2 - 4j$ and $-3 + 5j$ in the complex plane as in Fig. 14.3. Then, plot the negative of $-3 + 5j$, which is $3 - 5j$. The endpoint of the resultant is (1, −9), which corresponds to the complex number $1 - 9j$, which is also the algebraic difference.

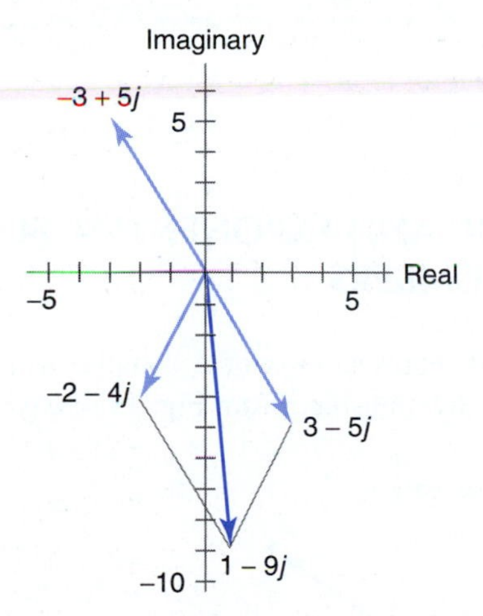

Figure 14.3

Exercises 14.1

Express each number in terms of j and simplify.

1. $\sqrt{-49}$
2. $\sqrt{-100}$
3. $\sqrt{-64}$
4. $\sqrt{-121}$
5. $\sqrt{-12}$
6. $\sqrt{-72}$
7. $\sqrt{-54}$
8. $\sqrt{-500}$

Simplify.

9. j^{19}
10. j^{34}
11. j^{22}
12. j^{39}
13. j^{81}
14. j^{97}
15. j^{246}
16. j^{308}

Perform the indicated operations and simplify.

17. $(3 + 4j) + (9 + 2j)$ **18.** $(-2 + 5j) + (6 - 7j)$ **19.** $(4 - 9j) - (2 - j)$

20. $(-6 + 3j) - (9 - 2j)$ **21.** $(4 + 2j) + (-4 - 3j)$ **22.** $(6 - j) - (1 - j)$

23. $(2 + j)(8 - 3j)$ **24.** $(4 - 6j)(9 - 2j)$ **25.** $(-4 + 5j)(3 + 2j)$

26. $(-3 - j)(8 + j)$ **27.** $(2 + 5j)(2 - 5j)$ **28.** $(-6 - 2j)(-6 + 2j)$

29. $(-3 + 4j)^2$ **30.** $(5 - 3j)^2$ **31.** $\dfrac{3 + 7j}{4 - j}$ **32.** $\dfrac{-3 + j}{2 + j}$

33. $\dfrac{6 - 3j}{4 + 8j}$ **34.** $\dfrac{1 - 4j}{2 - 3j}$ **35.** $\dfrac{-9 + 8j}{6 - 2j}$ **36.** $\dfrac{4 + 3j}{4 - 3j}$

Solve each equation.

37. $x^2 + 4 = 0$ **38.** $x^2 + 12 = 0$ **39.** $3x^2 + 4x + 9 = 0$

40. $2x^2 - x + 6 = 0$ **41.** $5x^2 - 2x + 5 = 0$ **42.** $9x^2 + 9x + 9 = 0$

43. $1 - 2x + 3x^2 = 0$ **44.** $10 + 4x + 2x^2 = 0$ **45.** $x^3 + 1 = 0$

46. $x^3 - 1 = 0$ **47.** $x^4 - 1 = 0$ **48.** $x^3 + 9x = 0$

49. $x^4 + 80x^2 = 0$ **50.** $x^4 + 25x^2 = 0$ **51.** $2x^4 + 54x = 0$

52. $4x^4 - 32x = 0$ **53.** $x^5 = x^2$ **54.** $2x^5 + 128x^2 = 0$

Plot each complex number in the complex plane.

55. $4 + 2j$ **56.** $-1 - 5j$ **57.** $-2 + 3j$

58. $3 - 3j$ **59.** $-4j$ or $0 - 4j$ **60.** 5 or $5 + 0j$

Add or subtract the complex numbers graphically.

61. $(3 + 2j) + (-2 + j)$ **62.** $(-4 - 3j) + (-3 + 2j)$ **63.** $(5 - 3j) - (1 + 5j)$

64. $(-6 - 3j) - (5 + 3j)$ **65.** $(-3 - 4j) - (-6 + 2j)$ **66.** $(8 + j) + (0 - 4j)$

67. $(2 + 4j) + (-2 + 3j)$ **68.** $(6 - 4j) - (2 - 4j)$

14.2 TRIGONOMETRIC AND EXPONENTIAL FORMS OF COMPLEX NUMBERS

The graphical representation of a complex number $a + bj$ leads us into a very useful trigonometric relationship. From Fig. 14.4, we can see that

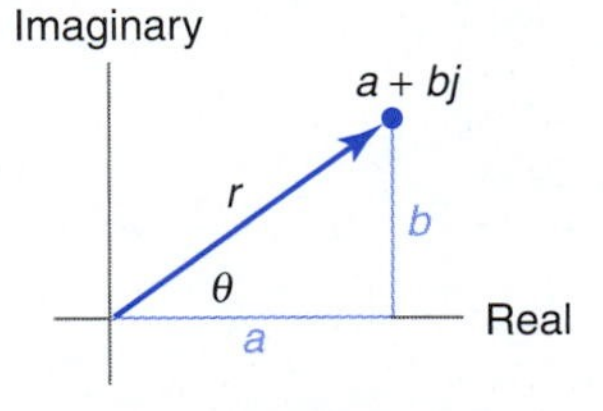

Figure 14.4 Graphical representation of a complex number.

$$\cos \theta = \frac{a}{r} \quad \text{(from the definition of cosine)}$$

$$a = r \cos \theta \tag{1}$$

and

$$\sin \theta = \frac{b}{r} \quad \text{(from the definition of sine)}$$

$$b = r \sin \theta \tag{2}$$

Substituting these values into $a + bj$, we have the following:

$$\begin{aligned} a + bj &= r\cos\theta + (r\sin\theta)j \\ &= r(\cos\theta + j\sin\theta) \end{aligned} \tag{3}$$

TRIGONOMETRIC FORM OF A COMPLEX NUMBER

$$r(\cos\theta + j\sin\theta)$$

is called the *trigonometric form* of a complex number. Here, r is the *absolute value* or the *modulus* and corresponds to the length of the vector when the complex number is expressed as a vector. Angle θ is called the *argument* of the complex number and is given in standard position.

Sometimes this form is called the *polar form* of a complex number. Other notations for the trigonometric form include $r \text{ cis } \theta$ and $r \underline{/\theta}$. (Recall that $a + bj$ was called the rectangular form of a complex number.)

If we know r and θ, we can find a and b using Equations (1) and (2). If we know a and b, we can find r by using the Pythagorean theorem.

$$r = \sqrt{a^2 + b^2} \tag{4}$$

Also, we can find θ using the definition of tangent.

$$\tan\theta = \frac{b}{a} \tag{5}$$

Using Equations (1) through (5), we can change from rectangular to trigonometric form, and vice versa.

EXAMPLE 1

Write $2 - 2j$ in trigonometric form.

First, graph $2 - 2j$ as in Fig. 14.5.

$$r = \sqrt{a^2 + b^2} = \sqrt{2^2 + (-2)^2} = \sqrt{8} = 2\sqrt{2}$$

$$\tan\theta = \frac{b}{a} = \frac{-2}{2} = -1$$

$$\theta = 315^\circ \quad \text{(Normally we choose } \theta \text{ so that } 0 \le \theta < 360^\circ.)$$

Therefore,

$$2 - 2j = 2\sqrt{2}(\cos 315^\circ + j\sin 315^\circ)$$

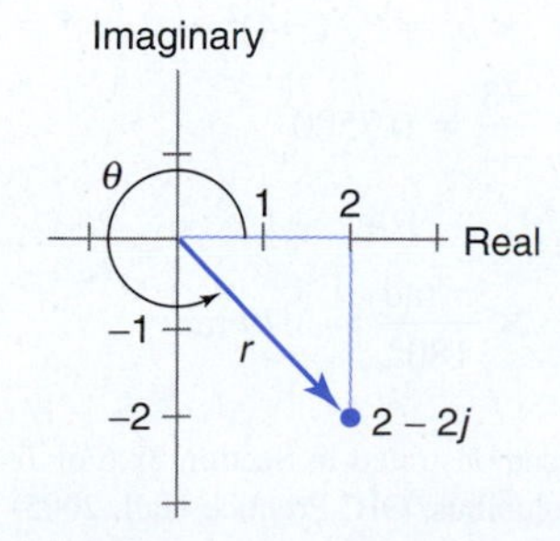

Figure 14.5

EXAMPLE 2

Write $4(\cos 120° + j \sin 120°)$ in rectangular form.

First, graph $4(\cos 120° + j \sin 120°)$ as in Fig. 14.6.

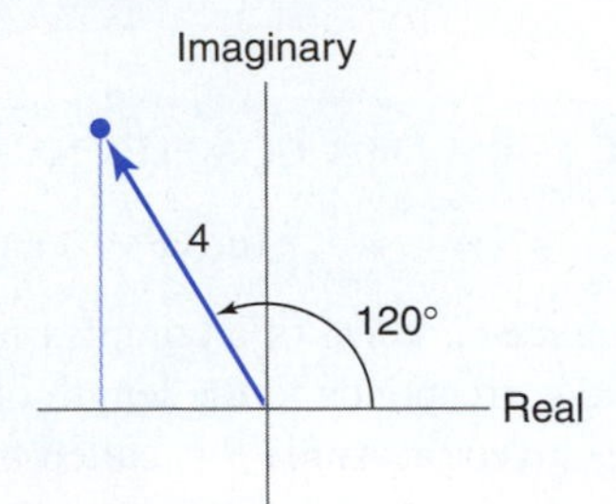

Figure 14.6

$$a = r \cos \theta = 4 \cos 120° = 4\left(-\frac{1}{2}\right) = -2$$

$$b = r \sin \theta = 4 \sin 120° = 4\left(\frac{\sqrt{3}}{2}\right) = 2\sqrt{3}$$

Therefore,

$$4(\cos 120° + j \sin 120°) = -2 + 2\sqrt{3}\,j$$

Another very useful form of a complex number was discovered by the Swiss mathematician Leonhard Euler. In equation form,*

$$e^{j\theta} = \cos \theta + j \sin \theta \qquad \textbf{(6)}$$

where e is an irrational number whose approximation is 2.71828. If we multiply each side of Equation (6) by r, we have the following:

$$re^{j\theta} = r(\cos \theta + j \sin \theta)$$

EXPONENTIAL FORM OF A COMPLEX NUMBER

$$re^{j\theta}$$

is called the *exponential form* of a complex number, where r and θ are defined in the same way as for the trigonometric form. When θ is expressed in radians, $j\theta$ is the actual exponent of the complex number and follows all the laws of exponents. Here, as in most applications, θ is expressed in radians when using exponential form.

EXAMPLE 3

Write $-4 - 3j$ in exponential form.

$$r = \sqrt{a^2 + b^2} = \sqrt{(-4)^2 + (-3)^2} = \sqrt{25} = 5$$

$$\tan \theta = \frac{b}{a} = \frac{-3}{-4} = 0.7500 \qquad (\textit{Note: } \theta \text{ is in the third quadrant.})$$

$$\theta = 217° \qquad \text{(to the nearest degree)}$$

$$\theta = 217° \times \frac{\pi \text{ rad}}{180°} = 3.79 \text{ rad}$$

*This relationship is demonstrated in Section 28.6 of *Technical Mathematics with Calculus,* Second Edition, by D. Ewen et al. (Columbus, OH: Prentice Hall, 2005).

Therefore, the exponential form is

$$-4 - 3j = 5e^{3.79j}$$

EXAMPLE 4

Write $3e^{2.86j}$ in trigonometric and rectangular forms.

Note that $r = 3$ and $\theta = 2.86 \text{ rad} \times \dfrac{180°}{\pi \text{ rad}} = 164°$ (to the nearest degree).
The trigonometric form is $3(\cos 164° + j \sin 164°)$.
The rectangular form is $3(\cos 164° + j \sin 164°) = -2.88 + 0.827j$.

Using a calculator, we have

MODE

3 **green diamond** $\mathbf{e^x}$ 2.86 **2nd CATALOG)** **ENTER**

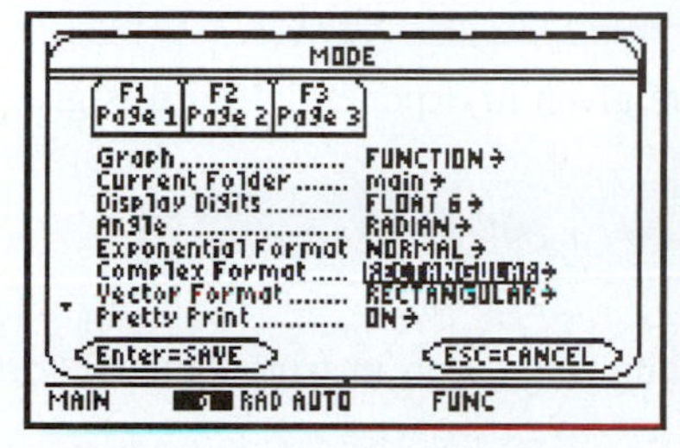

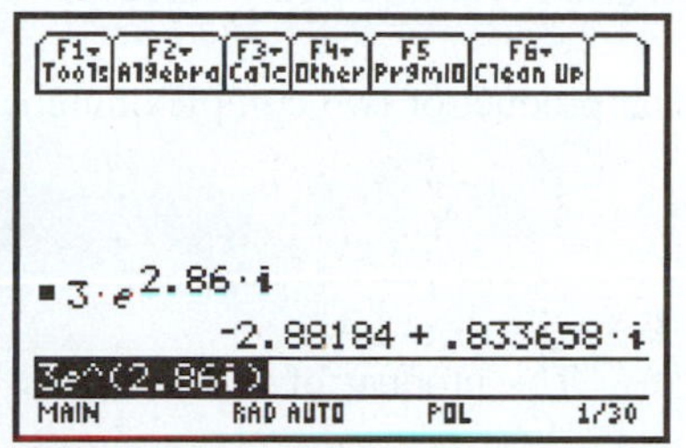

3 **2nd** $\mathbf{e^x}$ 2.86 **2nd decimal point)** **ENTER** right arrows

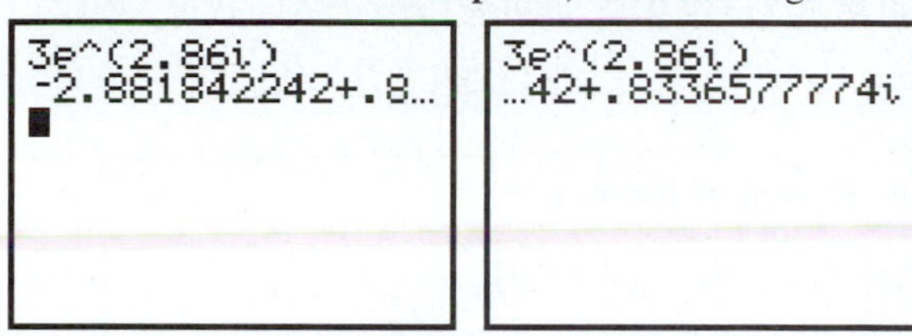

Explain the difference in results.

Exercises 14.2

Write each complex number in trigonometric form. Give angles to the nearest degree.

1. $2 + 2j$ **2.** $4 - 4j$ **3.** $-1 - \sqrt{3}j$ **4.** $-4\sqrt{3} + 4j$
5. $4j$ **6.** -3 **7.** $-6 - 6j$ **8.** $5 + 12j$
9. $-2 + 3j$ **10.** $5 - 3j$

Write each complex number in rectangular form. Use simplest radical form, where possible; otherwise use decimal values rounded to three significant digits.

11. $4(\cos 60° + j \sin 60°)$
12. $6(\cos 150° + j \sin 150°)$
13. $2(\cos 330° + j \sin 330°)$
14. $3(\cos 240° + j \sin 240°)$
15. $3\sqrt{2}(\cos 135° + j \sin 135°)$
16. $\sqrt{3}(\cos 225° + j \sin 225°)$
17. $3\underline{/270°}$
18. $4\underline{/65°}$
19. $\sqrt{53}\ \underline{/344°}$
20. $2\sqrt{11}\ \underline{/243.5°}$

Write each complex number in exponential form.

21. $\sqrt{3} - j$ **22.** $-3 + 3j$ **23.** $-\sqrt{2} - \sqrt{2}j$

24. $2 + 2\sqrt{3}j$ **25.** $4 + 6j$ **26.** $5 - 7j$

Write each complex number in trigonometric and rectangular forms.

27. $3e^{1.35j}$ **28.** $5e^{3.02j}$ **29.** $4e^{5.76j}$

30. $e^{6.91j}$ **31.** $2e^{j}$ **32.** $6e^{\pi j}$

14.3 MULTIPLICATION AND DIVISION OF COMPLEX NUMBERS IN EXPONENTIAL AND TRIGONOMETRIC FORMS

The product of two complex numbers given in exponential form is given as follows:

$$r_1 e^{j\theta_1} \cdot r_2 e^{j\theta_2} = r_1 r_2 e^{j\theta_1 + j\theta_2} = r_1 r_2 e^{j(\theta_1 + \theta_2)}$$

The product of two complex numbers given in trigonometric form is as follows:

$$r_1(\cos\theta_1 + j\sin\theta_1) \cdot r_2(\cos\theta_2 + j\sin\theta_2) = r_1 r_2[\cos(\theta_1 + \theta_2) + j\sin(\theta_1 + \theta_2)]$$

$$\text{or} \quad (r_1\underline{/\theta_1})(r_2\underline{/\theta_2}) = r_1 r_2\underline{/\theta_1 + \theta_2}$$

This formula may be proven as follows:

$$\begin{aligned}
&[r_1(\cos\theta_1 + j\sin\theta_1)][r_2(\cos\theta_2 + j\sin\theta_2)] \\
&\quad = r_1 r_2[\cos\theta_1\cos\theta_2 + j\sin\theta_1\cos\theta_2 + j\cos\theta_1\sin\theta_2 + j^2\sin\theta_1\sin\theta_2] \\
&\quad = r_1 r_2[\cos\theta_1\cos\theta_2 - \sin\theta_1\sin\theta_2 + j(\sin\theta_1\cos\theta_2 + \cos\theta_1\sin\theta_2)] \\
&\quad = r_1 r_2[\cos(\theta_1 + \theta_2) + j\sin(\theta_1 + \theta_2)] \qquad \text{[Formulas (13) and (12) in Section 13.2]}
\end{aligned}$$

EXAMPLE 1

Find the product $(3e^{2.5j})(4e^{1.7j})$ and write the result in exponential form.

$$\begin{aligned}
(3e^{2.5j})(4e^{1.7j}) &= (3)(4)e^{2.5j + 1.7j} \\
&= 12e^{4.2j}
\end{aligned}$$

EXAMPLE 2

Find the product $(5e^{\pi j})(7e^{\pi j/2})$ and write the result in rectangular form.

$$\begin{aligned}
(5e^{\pi j})(7e^{\pi j/2}) &= (5)(7)e^{\pi j + \pi j/2} \\
&= 35e^{3\pi j/2}
\end{aligned}$$

Note that $r = 35$ and $\theta = \dfrac{3\pi}{2} = 270°$. Thus, the trigonometric form is

$$\begin{aligned}
35(\cos 270° + j\sin 270°) &= 35[0 + j(-1)] \\
&= -35j
\end{aligned}$$

EXAMPLE 3

Find the product

$$[3(\cos 235° + j \sin 235°)][6(\cos 175° + j \sin 175°)]$$

and write the result in trigonometric form.

$$\begin{aligned}
&[3(\cos 235° + j \sin 235°)][6(\cos 175° + j \sin 175°)] \\
&\qquad = (3)(6)[\cos (235° + 175°) + j \sin (235° + 175°)] \\
&\qquad = 18(\cos 410° + j \sin 410°) \\
&\qquad = 18(\cos 50° + j \sin 50°)
\end{aligned}$$

Note that $18(\cos 410° + j \sin 410°)$ and $18(\cos 50° + j \sin 50°)$ both have the same coordinates in the complex plane.

In general,

$$r(\cos \theta + j \sin \theta) = r[\cos (\theta \pm 360°) + j \sin (\theta \pm 360°)]$$

since they have the same coordinates in the complex plane. Unless stated otherwise, we shall write $0° \leq \theta < 360°$.

EXAMPLE 4

Find the product

$$[5(\cos 60° + j \sin 60°)][4(\cos 120° + j \sin 120°)]$$

and write the result in rectangular form.

$$\begin{aligned}
&[5(\cos 60° + j \sin 60°)][4(\cos 120° + j \sin 120°)] \\
&\qquad = (5)(4)[\cos (60° + 120°) + j \sin (60° + 120°)] \\
&\qquad = 20(\cos 180° + j \sin 180°) \\
&\qquad = 20[-1 + j(0)] \\
&\qquad = -20
\end{aligned}$$

The quotient of two complex numbers in exponential form is as follows:

$$\frac{r_1 e^{j\theta_1}}{r_2 e^{j\theta_2}} = \frac{r_1}{r_2} e^{j(\theta_1 - \theta_2)}$$

The quotient of two complex numbers given in trigonometric form is as follows:

$$\frac{r_1(\cos \theta_1 + j \sin \theta_1)}{r_2(\cos \theta_2 + j \sin \theta_2)} = \frac{r_1}{r_2}[\cos (\theta_1 - \theta_2) + j \sin (\theta_1 - \theta_2)]$$

$$\text{or} \quad \frac{r_1 \angle \theta_1}{r_2 \angle \theta_2} = \frac{r_1}{r_2} \angle \theta_1 - \theta_2$$

This formula may be proven as follows:

$$\begin{aligned}
\frac{r_1(\cos \theta_1 + j \sin \theta_1)}{r_2(\cos \theta_2 + j \sin \theta_2)} &= \frac{r_1(\cos \theta_1 + j \sin \theta_1)}{r_2(\cos \theta_2 + j \sin \theta_2)} \cdot \frac{(\cos \theta_2 - j \sin \theta_2)}{(\cos \theta_2 - j \sin \theta_2)} \\
&= \frac{r_1}{r_2} \cdot \frac{\cos \theta_1 \cos \theta_2 + j \sin \theta_1 \cos \theta_2 - j \cos \theta_1 \sin \theta_2 - j^2 \sin \theta_1 \sin \theta_2}{\cos^2 \theta_2 - j^2 \sin^2 \theta_2} \\
&= \frac{r_1}{r_2} \cdot \frac{\cos \theta_1 \cos \theta_2 + \sin \theta_1 \sin \theta_2 + j(\sin \theta_1 \cos \theta_2 - \cos \theta_1 \sin \theta_2)}{\cos^2 \theta_2 + \sin^2 \theta_2} \\
&= \frac{r_1}{r_2}[\cos (\theta_1 - \theta_2) + j \sin (\theta_1 - \theta_2)] \quad \text{[Formulas (21) and (20) in Section 13.2]}
\end{aligned}$$

EXAMPLE 5

Find the quotient $\dfrac{18e^{3.10j}}{6e^{5.50j}}$ and write the result in exponential form.

$$\begin{aligned}\frac{18e^{3.10j}}{6e^{5.50j}} &= \frac{18}{6}e^{3.10j-5.50j}\\ &= 3e^{-2.40j}\\ &= 3e^{3.88j} \qquad (\textit{Note: } -2.40 + 2\pi = 3.88)\end{aligned}$$

Note that $3e^{-2.40j}$ and $3e^{3.88j}$ have the same coordinates when plotted in the complex plane.

In general,

$$re^{j\theta} = re^{(\theta \pm 2\pi)j}$$

since they have the same coordinates when plotted in the complex plane. Unless stated otherwise, we shall write $0 \le \theta < 2\pi$.

EXAMPLE 6

Find the quotient $\dfrac{18e^{3.94j}}{24e^{-1.72j}}$ and write the result in rectangular form.

$$\begin{aligned}\frac{18e^{3.94j}}{24e^{-1.72j}} &= \frac{18}{24}e^{3.94j-(-1.72j)}\\ &= \frac{3}{4}e^{5.66j}\end{aligned}$$

Note that $r = \dfrac{3}{4}$ and $\theta = 5.66 \text{ rad} \times \dfrac{180°}{\pi \text{ rad}} = 324°$ (rounded to the nearest degree). Thus, the trigonometric form is

$$\frac{3}{4}(\cos 324° + j \sin 324°) = 0.607 - 0.441j$$

EXAMPLE 7

Find the quotient $\dfrac{28(\cos 59° + j \sin 59°)}{16(\cos 135° + j \sin 135°)}$ and write the result in trigonometric form.

$$\begin{aligned}\frac{28(\cos 59° + j \sin 59°)}{16(\cos 135° + j \sin 135°)} &= \frac{28}{16}[\cos (59° - 135°) + j \sin (59° - 135°)]\\ &= 1.75[\cos (-76°) + j \sin (-76°)]\\ &= 1.75(\cos 284° + j \sin 284°)\\ &\quad (\textit{Note: } -76° + 360° = 284°)\end{aligned}$$

EXAMPLE 8

Find the quotient $\dfrac{18(\cos 300° + j \sin 300°)}{9(\cos 60° + j \sin 60°)}$ and write the result in rectangular form.

$$\begin{aligned}\frac{18(\cos 300° + j \sin 300°)}{9(\cos 60° + j \sin 60°)} &= \frac{18}{9}[\cos (300° - 60°) + j \sin (300° - 60°)]\\ &= 2(\cos 240° + j \sin 240°)\\ &= 2\left[-\frac{1}{2} + j\left(-\frac{\sqrt{3}}{2}\right)\right]\\ &= -1 - j\sqrt{3}\end{aligned}$$

Using a calculator, we have

18(**2nd** COS 300 **2nd** |)+ **2nd i 2nd SIN** 300 **2nd** |))/(9(**2nd** COS 60 **2nd** |)+ **2nd i 2nd SIN** 60 **2nd** |)))
ENTER

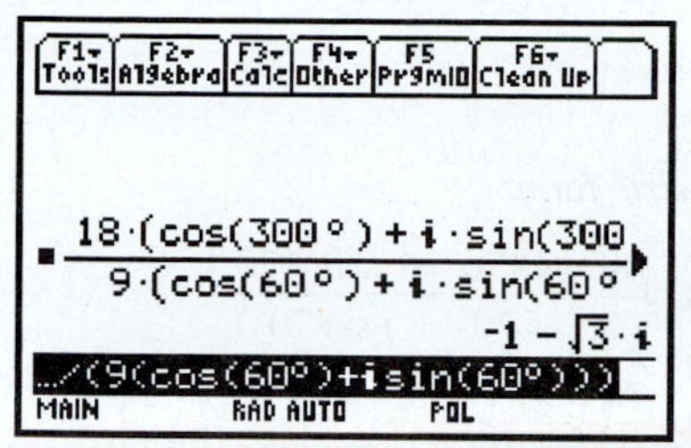

18(cos(300°)+isin(300°))/(9(cos(60°)+isin(60°)))
-1-1.732050808i

18(**COS** 300 **2nd ANGLE 1**)+ **2nd i 2nd SIN** 300 **2nd ANGLE 1**))/
(9(**COS** 60 **2nd ANGLE 1**)+ **2nd i 2nd SIN** 60 **2nd ANGLE 1**))) **ENTER**

Exercises 14.3

Find each product and write the result in exponential form.

1. $(4e^{j})(7e^{3j})$ **2.** $(6e^{2j})(8e^{3j})$ **3.** $(9e^{5j})(3e^{-3j})$
4. $(2e^{4j})(12e^{2j})$ **5.** $(6e^{5.6j})(4e^{3.7j})$ **6.** $(3e^{-9.4j})(7e^{6.7j})$

Find each product and write the result in rectangular form.

7. $(3e^{\pi j})(8e^{\pi j/3})$ **8.** $(3e^{\pi j/2})(5e^{2\pi j/3})$ **9.** $(4e^{3.7j})(20e^{6.1j})$
10. $(5e^{1.2j})(9e^{3.6j})$ **11.** $(6e^{-1.4j})(9e^{-2.5j})$ **12.** $(3e^{-j})(1e^{4.3j})$

Find each product and write the result in trigonometric form.

13. $[3(\cos 75° + j \sin 75°)][4(\cos 38° + j \sin 38°)]$
14. $[5(\cos 145° + j \sin 145°)][2(\cos 153° + j \sin 153°)]$
15. $[3(\cos 150° + j \sin 150°)][3(\cos 150° + j \sin 150°)]$
16. $[6(\cos 240° + j \sin 240°)][3(\cos 300° + j \sin 300°)]$
17. $(1\underline{/180°})(7\underline{/315°})$ **18.** $(8\underline{/168°})(9\underline{/-215°})$

Find each product and write the result in rectangular form.

19. $[5(\cos 50° + j \sin 50°)][5(\cos 10° + j \sin 10°)]$
20. $[4(\cos 105° + j \sin 105°)][1(\cos 120° + j \sin 120°)]$
21. $[2(\cos 120° + j \sin 120°)][6(\cos 60° + j \sin 60°)]$
22. $[3(\cos 145° + j \sin 145°)][9(\cos 125° + j \sin 125°)]$
23. $(7\underline{/162°})(8\underline{/213°})$ **24.** $(3\underline{/305°})(7\underline{/215°})$

Find each quotient and write the result in exponential form.

25. $\dfrac{3e^{6j}}{9e^{2j}}$ **26.** $\dfrac{6e^{3j}}{4e^{j}}$ **27.** $\dfrac{20e^{-4j}}{5e^{3j}}$

28. $\dfrac{24e^{-2j}}{2e^{-4j}}$ **29.** $\dfrac{8e^{1.6j}}{24e^{3.8j}}$ **30.** $\dfrac{12e^{-2.1j}}{3e^{4.8j}}$

Find each quotient and write the result in rectangular form.

31. $\dfrac{10e^{\pi j/6}}{5e^{\pi j}}$ **32.** $\dfrac{6e^{\pi j}}{9e^{3\pi j}}$ **33.** $\dfrac{14e^{4.6j}}{2e^{1.3j}}$

34. $\dfrac{36e^{5.2j}}{4e^{-1.8j}}$ **35.** $\dfrac{35e^{6.7j}}{7e^{-5.2j}}$ **36.** $\dfrac{15e^{-3.6j}}{12e^{-5.8j}}$

Find each quotient and write the result in trigonometric form.

37. $\dfrac{25(\cos 120° + j \sin 120°)}{5(\cos 50° + j \sin 50°)}$ **38.** $\dfrac{18(\cos 170° + j \sin 170°)}{2(\cos 70° + j \sin 70°)}$

39. $\dfrac{42(\cos 275° + j \sin 275°)}{7(\cos 156° + j \sin 156°)}$ **40.** $\dfrac{49(\cos 318° + j \sin 318°)}{14(\cos 251° + j \sin 251°)}$

41. $\dfrac{40\underline{/86°}}{5\underline{/215°}}$ **42.** $\dfrac{54\underline{/140°}}{9\underline{/350°}}$

Find each quotient and write the result in rectangular form.

43. $\dfrac{72(\cos 240° + j \sin 240°)}{8(\cos 120° + j \sin 120°)}$ **44.** $\dfrac{6(\cos 295° + j \sin 295°)}{24(\cos 160° + j \sin 160°)}$

45. $\dfrac{8(\cos 185° + j \sin 185°)}{40(\cos 35° + j \sin 35°)}$ **46.** $\dfrac{60(\cos 240° + j \sin 240°)}{12(\cos 330° + j \sin 330°)}$

47. $\dfrac{96\underline{/85°}}{16\underline{/145°}}$ **48.** $\dfrac{80\underline{/30°}}{25\underline{/300°}}$

Find each power and write the result in rectangular form.

49. $[2(\cos 60° + j \sin 60°)]^3$ **50.** $[3(\cos 150° + j \sin 150°)]^3$

51. $[3(\cos 157.5° + j \sin 157.5°)]^4$ **52.** $[2(\cos 315° + j \sin 315°)]^4$

14.4 POWERS AND ROOTS

The nth power of a complex number in exponential form is given by the following equation:

$$(re^{j\theta})^n = r^n e^{jn\theta}$$

The trigonometric form of $(re^{j\theta})^n$ is

$$[r(\cos \theta + j \sin \theta)]^n$$

The trigonometric form of $r^n e^{jn\theta}$ is

$$r^n(\cos n\theta + j \sin n\theta)$$

Therefore, we have the following result:

DEMOIVRE'S THEOREM

$$[r(\cos \theta + j \sin \theta)]^n = r^n(\cos n\theta + j \sin n\theta)$$
$$\text{or} \quad (r\underline{/\theta})^n = r^n\underline{/n\theta}$$

which is the nth power of a complex number in trigonometric form. This theorem is valid for all real values of n.

EXAMPLE 1

Find $(3e^{2j})^4$ and write the result in exponential form.

$$\begin{aligned}(3e^{2j})^4 &= 3^4(e^{2j})^4 \\ &= 81e^{8j} \\ &= 81e^{1.72j} \qquad (\textit{Note: } 8 - 2\pi = 1.72)\end{aligned}$$

EXAMPLE 2

Find $[2(\cos 125° + j \sin 125°)]^5$ and write the result in trigonometric form.

$$\begin{aligned}&[2(\cos 125° + j \sin 125°)]^5 \\ &\quad = 2^5(\cos 5 \cdot 125° + j \sin 5 \cdot 125°) \\ &\quad = 32(\cos 625° + j \sin 625°) \\ &\quad = 32(\cos 265° + j \sin 265°) \qquad (\textit{Note: } 625° - 360° = 265°)\end{aligned}$$

EXAMPLE 3

Find $(-2 + 2j)^6$ and write the result in rectangular form.

First,

$$r = \sqrt{a^2 + b^2} = \sqrt{(-2)^2 + 2^2} = \sqrt{8} = 2\sqrt{2}$$

$$\tan\theta = \frac{b}{a} = \frac{2}{-2} = -1$$

Since θ is in the second quadrant, $\theta = 135°$. Thus,

$$-2 + 2j = 2\sqrt{2}(\cos 135° + j \sin 135°)$$

and

$$\begin{aligned}(-2 + 2j)^6 &= [2\sqrt{2}(\cos 135° + j \sin 135°)]^6 \\ &= (2\sqrt{2})^6(\cos 6 \cdot 135° + j \sin 6 \cdot 135°) \\ &= 512(\cos 810° + j \sin 810°) \\ &= 512(\cos 90° + j \sin 90°) \qquad (\textit{Note: } 810° - 2 \cdot 360° = 90°) \\ &= 512(0 + j \cdot 1) \\ &= 512j\end{aligned}$$

DeMoivre's theorem is also valid for negative integers; that is,

$$(a + bj)^{-n} = [r(\cos\theta + j \sin\theta)]^{-n} = r^{-n}[\cos(-n\theta) + j \sin(-n\theta)]$$

where n is a positive integer.

EXAMPLE 4

Find $(-1 - j\sqrt{3})^{-4}$ and write the result in rectangular form.

First,

$$r = \sqrt{a^2 + b^2} = \sqrt{(-1)^2 + (-\sqrt{3})^2} = \sqrt{4} = 2$$

$$\tan\theta = \frac{b}{a} = \frac{-\sqrt{3}}{-1} = \sqrt{3}$$

Since θ is in the third quadrant, $\theta = 240°$. Thus,

$$-1 - j\sqrt{3} = 2(\cos 240° + j \sin 240°)$$

and

$$\begin{aligned}(-1 - j\sqrt{3})^{-4} &= [2(\cos 240° + j \sin 240°)]^{-4} \\ &= 2^{-4}[\cos(-4 \cdot 240°) + j \sin(-4 \cdot 240°)] \\ &= \frac{1}{16}[\cos(-960°) + j \sin(-960°)] \\ &= \frac{1}{16}(\cos 120° + j \sin 120°) \qquad (\textit{Note: } -960° + 3 \cdot 360° = 120°) \\ &= \frac{1}{16}\left(-\frac{1}{2} + j\frac{\sqrt{3}}{2}\right) \\ &= -\frac{1}{32} + j\frac{\sqrt{3}}{32}\end{aligned}$$

By calculator, we have

(-1– **2nd CATALOG 2nd** multiplication sign 3))^-4 **ENTER**

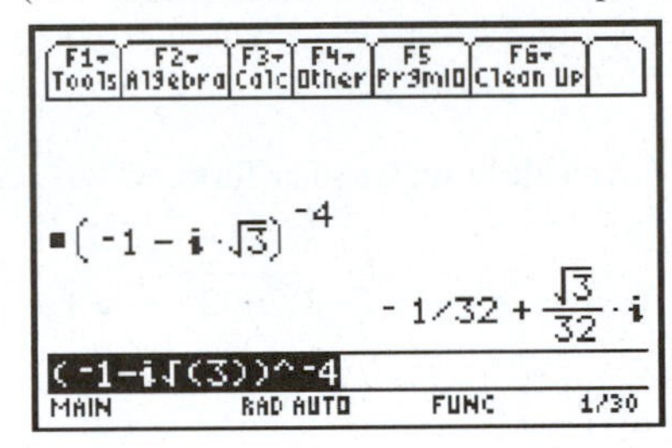

To find an nth root of the complex number z means to find a number $a + bj$ such that

$$(a + bj)^n = [r(\cos \theta + j \sin \theta)]^n = z$$

For example, consider the complex number

$$r^{1/n}\left(\cos\frac{\theta}{n} + j \sin\frac{\theta}{n}\right)$$

where $r > 0$ and n is a positive integer. By DeMoivre's theorem, we have

$$\begin{aligned}\left[r^{1/n}\left(\cos\frac{\theta}{n} + j \sin\frac{\theta}{n}\right)\right]^n &= (r^{1/n})^n\left[\cos\left(n \cdot \frac{\theta}{n}\right) + j \sin\left(n \cdot \frac{\theta}{n}\right)\right] \\ &= r(\cos \theta + j \sin \theta)\end{aligned}$$

Thus, $r^{1/n}\left(\cos\frac{\theta}{n} + j \sin\frac{\theta}{n}\right)$ is an nth root of z.

In general, there are n distinct nth roots of a given complex number. The following formula may be used to find the n distinct roots of a complex number written in trigonometric form:

DEMOIVRE'S THEOREM FOR ROOTS

$$[r(\cos \theta + j \sin \theta)]^{1/n} = r^{1/n}\left[\cos\left(\frac{\theta + k \cdot 360°}{n}\right) + j \sin\left(\frac{\theta + k \cdot 360°}{n}\right)\right]$$

where $k = 0, 1, 2, \ldots, n - 1$.

EXAMPLE 5

Find the four fourth roots of 1. That is, solve the equation $x^4 = 1$.

In trigonometric form, $1 = 1(\cos 0° + j \sin 0°)$.

For the first root ($k = 0$),

$$1^{1/4} = 1^{1/4}\left[\cos \frac{1}{4}(0°) + j \sin \frac{1}{4}(0°)\right]$$
$$= \cos 0° + j \sin 0° = 1$$

For the second root ($k = 1$), $\theta = 0° + (1)360° = 360°$, and

$$1^{1/4} = 1^{1/4}\left[\cos \frac{1}{4}(360°) + j \sin \frac{1}{4}(360°)\right]$$
$$= \cos 90° + j \sin 90°$$
$$= 0 + j = j$$

For the third root ($k = 2$), $\theta = 0° + 2(360°) = 720°$, and

$$1^{1/4} = 1^{1/4}\left[\cos \frac{1}{4}(720°) + j \sin \frac{1}{4}(720°)\right]$$
$$= \cos 180° + j \sin 180°$$
$$= -1 + 0j = -1$$

For the fourth root ($k = 3$), $\theta = 0° + 3(360°) = 1080°$, and

$$1^{1/4} = 1^{1/4}\left[\cos \frac{1}{4}(1080°) + j \sin \frac{1}{4}(1080°)\right]$$
$$= \cos 270° + j \sin 270°$$
$$= 0 - j = -j$$

As you can see in Fig. 14.7, when the four roots of Example 5 are shown graphically in the complex plane, they differ by $\frac{360°}{4} = 90°$. In general, the roots differ by $\frac{360°}{n}$, where n is the number of roots.

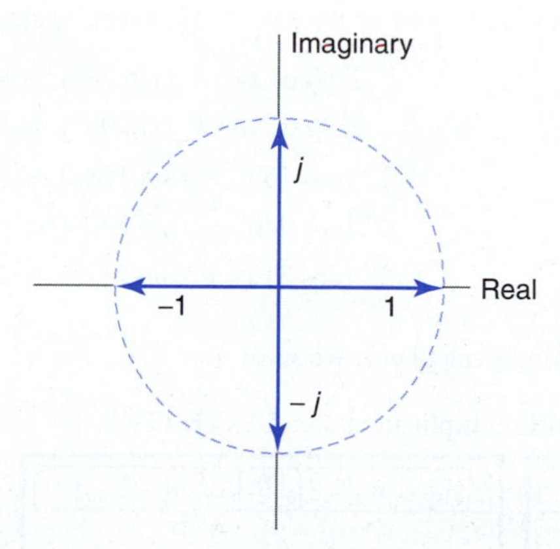

Figure 14.7

EXAMPLE 6

Find all complex solutions of $x^5 = -1 + j\sqrt{3}$.

In trigonometric form,

$$-1 + j\sqrt{3} = 2(\cos 120° + j \sin 120°)$$

For $k = 0$,

$$2^{1/5}\left[\cos\left(\frac{120° + 0 \cdot 360°}{5}\right) + j \sin\left(\frac{120° + 0 \cdot 360°}{5}\right)\right] = 2^{1/5}(\cos 24° + j \sin 24°)$$

By letting $k = 1, 2, 3,$ and 4, we obtain the other roots as follows:

$$k = 1:\quad 2^{1/5}(\cos 96° + j \sin 96°)$$
$$k = 2:\quad 2^{1/5}(\cos 168° + j \sin 168°)$$
$$k = 3:\quad 2^{1/5}(\cos 240° + j \sin 240°)$$
$$k = 4:\quad 2^{1/5}(\cos 312° + j \sin 312°)$$

Note: These last four roots could also have been found by adding $\frac{360°}{n} = \frac{360°}{5} = 72°$ to the root for $k = 0$ and then to each successive root through $k = 4$. These five roots are shown graphically in Fig. 14.8.

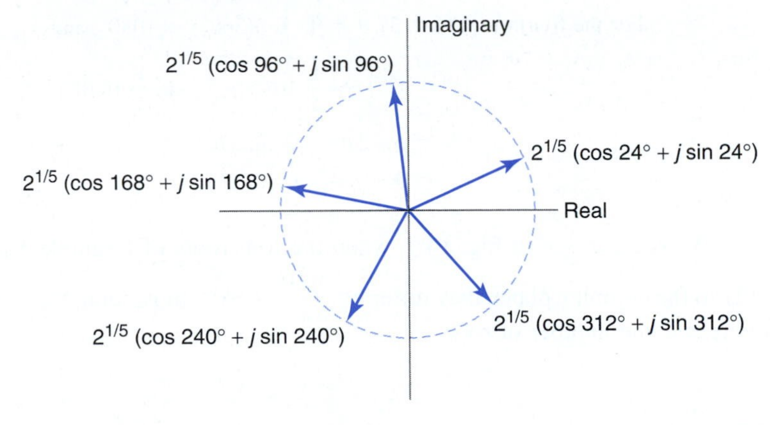

Figure 14.8

Approximations of the five roots in rectangular form are

$$2^{1/5}(\cos 24° + j \sin 24°) = 1.05 + 0.467j$$
$$2^{1/5}(\cos 96° + j \sin 96°) = -0.120 + 1.14j$$
$$2^{1/5}(\cos 168° + j \sin 168°) = -1.12 + 0.239j$$
$$2^{1/5}(\cos 240° + j \sin 240°) = -0.574 - 0.995j$$
$$2^{1/5}(\cos 312° + j \sin 312°) = 0.769 - 0.854j$$

Using a calculator, we have

F2 alpha A 1 x^5=-1+ **2nd i 2nd** multiplication sign 3),x) **ENTER** up arrow right arrows... etc.

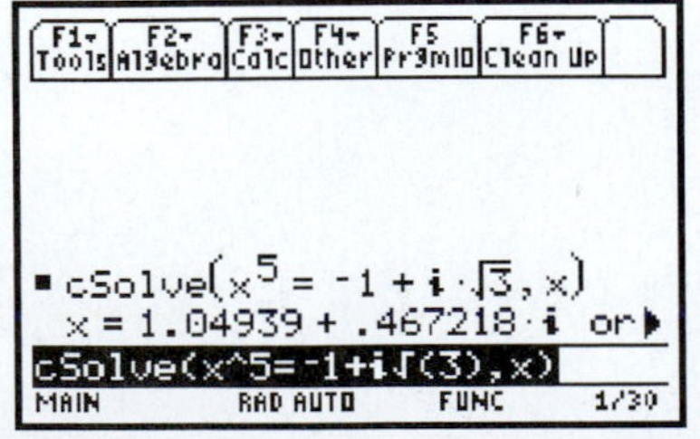

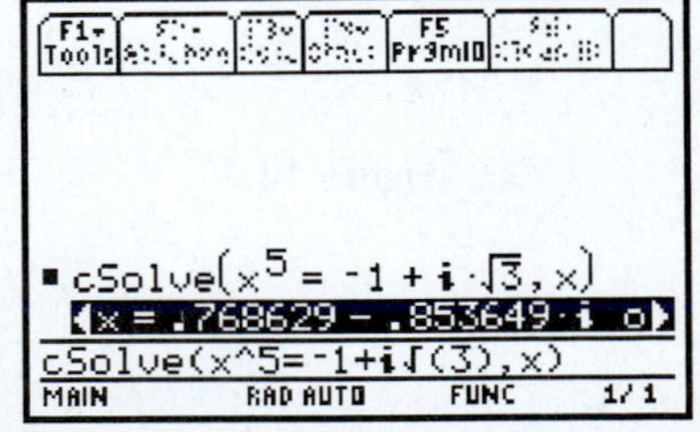

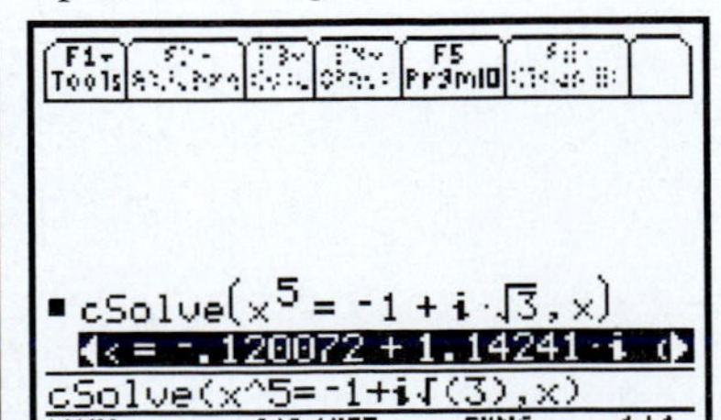

Exercises 14.4

Find each power and write the result in exponential form.

1. $(3e^{1.4j})^5$ 2. $(2e^{2.75j})^3$ 3. $(5e^{4.6j})^2$ 4. $(2e^{1.7j})^6$

Find each power and write the result in trigonometric form.

5. $[3(\cos 20° + j \sin 20°)]^4$
6. $[1(\cos 25° + j \sin 25°)]^7$
7. $(2\underline{/150°})^5$
8. $(5\underline{/275°})^3$
9. $[2(\cos 240° + j \sin 240°)]^{-3}$
10. $[1(\cos 220° + j \sin 220°)]^{-4}$

Find each power and write the result in rectangular form.

11. $(1 - j)^8$ 12. $(\sqrt{3} + j)^4$ 13. $(-2\sqrt{3} - 2j)^3$ 14. $(-2 + 2j)^6$
15. $(1 + j\sqrt{3})^5$ 16. $(-4 - 3j)^5$ 17. $(-3 + 3j)^{-4}$ 18. $(-1 - j\sqrt{3})^{-3}$

In Exercises 19 through 22, write each result in rectangular form.

19. Find the three cube roots of 1.
20. Find the four fourth roots of j.
21. Find the five fifth roots of j.
22. Find the six sixth roots of -1.

Find all complex solutions of each equation. Write each solution in both trigonometric and rectangular forms.

23. $x^3 = 27(\cos 405° + j \sin 405°)$
24. $x^4 = 81(\cos 180° + j \sin 180°)$
25. $x^2 = j$ 26. $x^3 = -1$ 27. $x^5 = 1$ 28. $x^3 = -j$
29. $x^5 = -1$ 30. $x^6 = j$ 31. $x^4 = -16$ 32. $x^3 = 8$
33. $x^3 = -2 + 11j$ 34. $x^2 = 1 + j$

CHAPTER 14 SUMMARY

1. A *complex number* is any number in the form $a + bj$, where a and b are real numbers and $j = \sqrt{-1}$. If $a = 0$, $a + bj$ is an imaginary number; if $b = 0$, $a + bj$ is a real number.
2. The *rectangular form* of a complex number is $a + bj$, where a and b are real numbers; a is called the *real part* and bj is called the *imaginary part*.
3. Two complex numbers $a + bj$ and $c + dj$ are equal only when $a = c$ and $b = d$.
4. *Operations with complex numbers in rectangular form:*
   (a) *Addition:* $(a + bj) + (c + dj) = (a + c) + (b + d)j$.
   (b) *Subtraction:* $(a + bj) - (c + dj) = (a - c) + (b - d)j$.
   (c) *Multiplication:* $(a + bj)(c + dj) = (ac - bd) + (ad + bc)j$; or multiply the two complex numbers as if they were binomials, and simplify.
   (d) *Division:* To divide two complex numbers, multiply both numerator and denominator by the conjugate of the denominator and simplify.
   Note: The complex numbers $a + bj$ and $a - bj$ are **conjugates** of each other.
5. The *complex plane* is shown in Fig. 14.9. Complex numbers may be added or subtracted graphically as vectors.

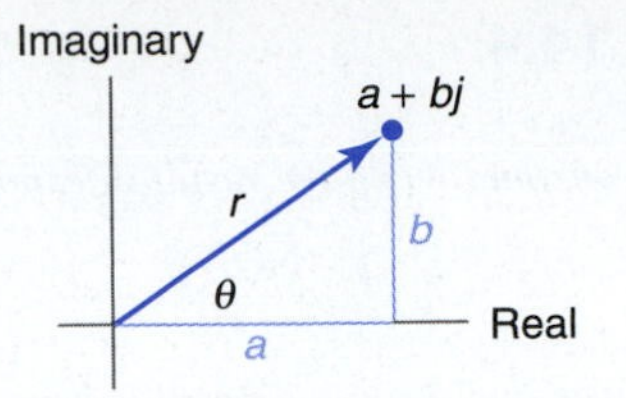

Figure 14.9

6. The *trigonometric form of a complex number:*

$$a + bj = r(\cos\theta + j\sin\theta) \quad \text{or} \quad r\underline{/\theta}$$

7. The *exponential form of a complex number:*

$$r(\cos\theta + j\sin\theta) = re^{j\theta} \quad \text{where } \theta \text{ is in radians}$$

8. *The product of two complex numbers:*

(a) *In exponential form:*

$$r_1e^{j\theta_1} \cdot r_2e^{j\theta_2} = r_1r_2e^{j(\theta_1+\theta_2)}$$

(b) *In trigonometric form:*

$$r_1(\cos\theta_1 + j\sin\theta_1) \cdot r_2(\cos\theta_2 + j\sin\theta_2) = r_1r_2[\cos(\theta_1 + \theta_2) + j\sin(\theta_1 + \theta_2)]$$

$$\text{or} \quad (r_1\underline{/\theta_1})(r_2\underline{/\theta_2}) = r_1r_2\underline{/\theta_1 + \theta_2}$$

9. *The quotient of two complex numbers:*

(a) *In exponential form:*

$$\frac{r_1e^{j\theta_1}}{r_2e^{j\theta_2}} = \frac{r_1}{r_2}e^{j(\theta_1-\theta_2)}$$

(b) *In trigonometric form:*

$$\frac{r_1(\cos\theta_1 + j\sin\theta_1)}{r_2(\cos\theta_2 + j\sin\theta_2)} = \frac{r_1}{r_2}[\cos(\theta_1 - \theta_2) + j\sin(\theta_1 - \theta_2)]$$

$$\text{or} \quad \frac{r_1\underline{/\theta_1}}{r_2\underline{/\theta_2}} = \frac{r_1}{r_2}\underline{/\theta_1 - \theta_2}$$

10. *The nth power of a complex number:*

(a) *In exponential form:*

$$(re^{j\theta})^n = r^ne^{jn\theta}$$

(b) *In trigonometric form:*

$$[r(\cos\theta + j\sin\theta)]^n = r^n(\cos n\theta + j\sin n\theta) \qquad \text{(DeMoivre's theorem)}$$

$$\text{or} \quad (r\underline{/\theta})^n = r^n\,\underline{/n\theta}$$

11. *DeMoivre's theorem for roots:*

$$[r(\cos\theta + j\sin\theta)]^{1/n} = r^{1/n}\left[\cos\left(\frac{\theta + k\cdot 360°}{n}\right) + j\sin\left(\frac{\theta + k\cdot 360°}{n}\right)\right]$$

where $k = 0, 1, 2, \ldots, n - 1$.

CHAPTER 14 REVIEW

Express in terms of j and simplify.

1. $\sqrt{-81}$
2. $\sqrt{-18}$

Simplify.

3. j^{18}
4. j^{23}
5. j^{48}
6. j^{145}

Perform the indicated operations in rectangular form and simplify.

7. $(9 + 3j) + (-4 + 7j)$
8. $(-1 + j) - (-4 + 5j)$
9. $(5 + 2j)(6 - 7j)$
10. $(3 - 2j)^2$
11. $\dfrac{1 - 2j}{4 + j}$
12. $\dfrac{5 + j}{3 - 7j}$
13. $\dfrac{4 + 3j}{1 - 2j}$
14. $\dfrac{7 - 2j}{3 + j}$

Solve.

15. $x^2 + 36 = 0$
16. $2x^2 + 3x + 2 = 0$
17. $3x^2 - 5x + 4 = 0$
18. $2x^2 + 50 = 0$

Write in trigonometric and exponential forms.

19. $-1 + j$
20. $1 - \sqrt{3}j$
21. $3 - 5j$
22. $-2 - 8j$

Write in rectangular and exponential forms.

23. $6(\cos 315° + j \sin 315°)$
24. $4(\cos 210° + j \sin 210°)$
25. $5.52(\cos 105° + j \sin 105°)$
26. $6.50(\cos 253° + j \sin 253°)$

Write in trigonometric and rectangular forms.

27. $2e^{0.489j}$
28. $3e^{8.75j}$
29. $7e^{1.46j}$
30. $8e^{8.59j}$

Do as indicated and write the result in the given form.

31. $(5e^{2j})(3e^{3j})$
32. $[2(\cos 150° + j \sin 150°)][4(\cos 300° + j \sin 300°)]$
33. $\dfrac{12e^{3j}}{3e^{-2j}}$
34. $\dfrac{24\angle 150°}{8\angle 275°}$
35. $(4e^{2j})^3$
36. $[2(\cos 60° + j \sin 60°)]^7$

Find each power and write the result in rectangular form.

37. $(-2 + 2j)^4$
38. $(1 + j\sqrt{3})^6$
39. $(1 + j)^{-4}$

Find all complex solutions of each equation. Write each solution in both trigonometric form and rectangular form.

40. $x^3 = j$
41. $x^4 = -1$
42. $x^4 = 16$
43. $x^5 = 4 - 4j$

15 Matrices

INTRODUCTION

Matrices are a convenient way of denoting and working with equations, systems of equations, and vectors. Areas in which they are used include air traffic control, petroleum engineering, computer system simulation, circuit physics, fluid flow, and curve fitting (as demonstrated in Section 19.5). In this chapter we will study operations and properties of matrices.

Objectives

- Learn matrix terminology.
- Add, subtract, and multiply compatible matrices.
- Find the inverse of a matrix.
- Apply properties of matrices.
- Solve systems of equations using matrices.

15.1 BASIC OPERATIONS

A **matrix** is a rectangular array of numbers, usually shown between brackets. Some examples are

$$\begin{bmatrix} 3 & 2 & 1 \\ 5 & -6 & 0 \end{bmatrix} \quad \begin{bmatrix} -1 & 0 \\ 5 & 0 \end{bmatrix} \quad \begin{bmatrix} 3 & 6 \\ 4 & 1 \\ 8 & -7 \end{bmatrix} \quad \begin{bmatrix} 3 \\ 5 \end{bmatrix} \quad \begin{bmatrix} 3 & 1 & -7 \end{bmatrix}$$

The individual numbers that make up a matrix are called *entries,* or *elements,* of the matrix. Such entries are restricted to real numbers in this chapter. The **order,** or **dimension,** of a matrix is given by stating the number of rows and then the number of columns in the matrix. A matrix with m rows and n columns has order $m \times n$, read "order m by n." *Capital letters* are often used to name a matrix. For example,

$$A = \begin{bmatrix} 3 & 2 & 1 \\ 5 & -6 & 0 \end{bmatrix}; A \text{ is a } 2 \times 3 \text{ matrix (2 rows and 3 columns).}$$

$$B = \begin{bmatrix} -1 & 0 \\ 5 & 0 \end{bmatrix}; B \text{ is a } 2 \times 2 \text{ matrix (2 rows and 2 columns).}$$

$$C = \begin{bmatrix} 3 & 6 \\ 4 & 1 \\ 8 & -7 \end{bmatrix}$$; C is a 3×2 matrix (3 rows and 2 columns).

$$D = \begin{bmatrix} 3 \\ 5 \end{bmatrix}$$; D is a 2×1 matrix (2 rows and 1 column).

$E = [3 \quad 1 \quad -7]$; E is a 1×3 matrix (1 row and 3 columns).

A **square matrix** has the same number of rows and columns. The **principal diagonal** of a square matrix is the diagonal containing the numbers from the upper left to the lower right. For example,

$$\begin{bmatrix} 3 & 6 & 5 \\ 2 & -9 & 0 \\ -5 & 3 & 1 \end{bmatrix} \quad \text{principal diagonal}$$

A *double-subscript notation* is often used to name specific elements in a matrix as follows:

$$A = \begin{bmatrix} a_{11} & a_{12} & a_{13} \\ a_{21} & a_{22} & a_{23} \\ a_{31} & a_{32} & a_{33} \end{bmatrix}$$

Note: The *first subscript* refers to the *row* where the element is located, and the *second subscript* refers to the *column* where the element is located.

A **zero matrix** is a matrix with all entries of zero.

Two matrices are *equal* if they have the same order and the same corresponding entries. That is, they must have the same number of rows and the same number of columns, and the corresponding entries must be identical.

EXAMPLE 1

Find the value of each variable that makes the matrices equal.

$$\begin{bmatrix} x & -y & 2z \\ a - 4 & \frac{b}{2} & c + 2 \end{bmatrix} = \begin{bmatrix} 3 & 5 & -8 \\ 6 & 4 & 7 \end{bmatrix}$$

Since the matrices are equal, the corresponding entries are equal. That is,

$$x = 3 \qquad -y = 5 \qquad 2z = -8 \qquad a - 4 = 6 \qquad \frac{b}{2} = 4 \qquad c + 2 = 7$$

$$y = -5 \qquad z = -4 \qquad a = 10 \qquad b = 8 \qquad c = 5$$

EXAMPLE 2

Find the value of each variable that makes the matrices equal.

$$[x - y \quad x - 2z \quad x + y + z] = [3 \quad 0 \quad 7]$$

Since the matrices are equal, the corresponding entries are equal. This gives the following system of equations.

$$\begin{aligned} x - y \qquad &= 3 \qquad (1) \\ x \qquad - 2z &= 0 \qquad (2) \\ x + y + \ z &= 7 \qquad (3) \end{aligned}$$

Adding Equations (1) and (3) gives $2x + z = 10$. Multiplying Equation (2) by 2 gives the following system:

$$\begin{array}{rl} 2x + z = 10 & \\ \underline{2x - 4z = 0} & \\ 5z = 10 & \text{(Subtract.)} \\ z = 2 & \end{array}$$

Substituting $z = 2$ into $x - 2z = 0$ gives $x = 4$. Then, substituting $x = 4$ into Equation (1) gives $y = 1$.

The *sum* of two matrices A and B of the *same order* is the matrix with entries that are the sums of the corresponding entries of A and B.

Note: The sum of two matrices of different orders is not defined.

EXAMPLE 3

Add the matrices, if possible.

$$\begin{bmatrix} 3 & 6 & -1 \\ 2 & -4 & 11 \end{bmatrix} + \begin{bmatrix} 0 & -5 & -7 \\ 8 & 10 & 9 \end{bmatrix} = \begin{bmatrix} 3+0 & 6+(-5) & (-1)+(-7) \\ 2+8 & (-4)+10 & 11+9 \end{bmatrix}$$

$$= \begin{bmatrix} 3 & 1 & -8 \\ 10 & 6 & 20 \end{bmatrix}$$

EXAMPLE 4

Add the matrices, if possible.

$$\begin{bmatrix} 2 & 4 & 6 \\ 1 & 3 & 5 \end{bmatrix} + \begin{bmatrix} 8 & 4 \\ 5 & 7 \end{bmatrix}$$

These two matrices cannot be added because they are of different orders.

The *negative* or *additive inverse* of matrix A is defined as the matrix, $-A$, in which each entry is the negative of each corresponding entry in A. For example,

$$\text{if} \quad A = \begin{bmatrix} 1 & -4 \\ -3 & 0 \end{bmatrix}, \quad \text{then} \quad -A = \begin{bmatrix} -1 & 4 \\ 3 & 0 \end{bmatrix}$$

The *difference* of two matrices $A - B$ *of the same order* is defined as follows:

$$A - B = A + (-B)$$

EXAMPLE 5

If $A = \begin{bmatrix} 3 & -2 \\ -7 & -8 \end{bmatrix}$ and $B = \begin{bmatrix} -4 & 5 \\ 6 & -2 \end{bmatrix}$, find $A - B$.

$$\begin{bmatrix} 3 & -2 \\ -7 & -8 \end{bmatrix} - \begin{bmatrix} -4 & 5 \\ 6 & -2 \end{bmatrix} = \begin{bmatrix} 3 & -2 \\ -7 & -8 \end{bmatrix} + \begin{bmatrix} 4 & -5 \\ -6 & 2 \end{bmatrix} = \begin{bmatrix} 7 & -7 \\ -13 & -6 \end{bmatrix}$$

In matrix algebra, as in vector algebra, a real number is called a **scalar.** The product of a scalar k and a matrix A is defined as the matrix kA, where each entry in kA is k times the corresponding entry in A.

EXAMPLE 6

If $A = \begin{bmatrix} -2 & 4 \\ 7 & 5 \\ -1 & 9 \end{bmatrix}$, find $3A$.

$$3\begin{bmatrix} -2 & 4 \\ 7 & 5 \\ -1 & 9 \end{bmatrix} = \begin{bmatrix} 3(-2) & 3(4) \\ 3(7) & 3(5) \\ 3(-1) & 3(9) \end{bmatrix} = \begin{bmatrix} -6 & 12 \\ 21 & 15 \\ -3 & 27 \end{bmatrix}$$

EXAMPLE 7

If $A = \begin{bmatrix} 2 & 4 & 6 & 8 \\ 1 & 3 & 5 & 7 \end{bmatrix}$ and $B = \begin{bmatrix} -3 & 4 & 7 & -2 \\ 0 & -1 & 5 & 8 \end{bmatrix}$, find $3A - 2B$.

$$3A = \begin{bmatrix} 6 & 12 & 18 & 24 \\ 3 & 9 & 15 & 21 \end{bmatrix} \quad \text{and} \quad -2B = \begin{bmatrix} 6 & -8 & -14 & 4 \\ 0 & 2 & -10 & -16 \end{bmatrix}$$

$$3A + (-2B) = \begin{bmatrix} 12 & 4 & 4 & 28 \\ 3 & 11 & 5 & 5 \end{bmatrix}$$

Using a calculator, we have

3[2,4,6,8;1,3,5,7]-2[-3,4,7,-2;0,-1,5,8] **ENTER**

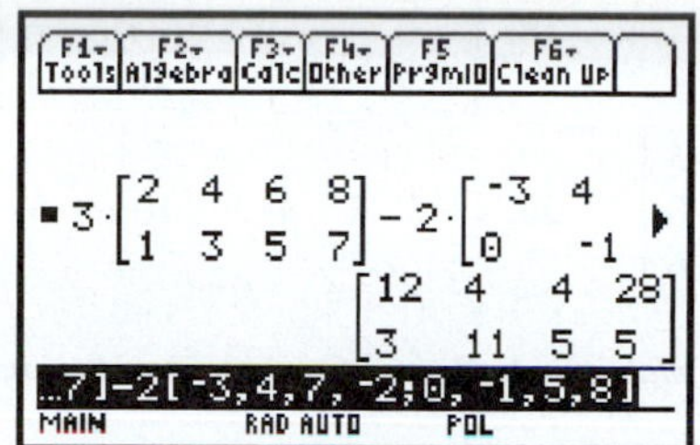

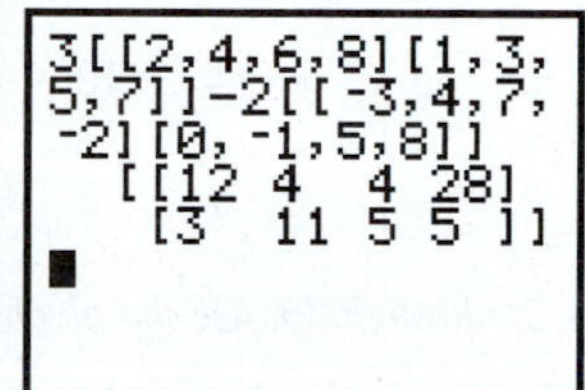

3[[2,4,6,8][1,3,5,7]]-2[[-3,4,7,-2][0,-1,5,8]] **ENTER**

In general, we have the following matrix properties, where A, B, and C are matrices, k is a real number, and O is the zero matrix.

$A + B = B + A$ (commutative property of addition)

$(A + B) + C = A + (B + C)$ (associative property of addition)

$k(A + B) = kA + kB$

$A + O = A$ (identity element of addition)

$A + (-A) = O$ (additive inverse property)

Note: O is the zero matrix.

Exercises 15.1

Give the order of each matrix.

1. $\begin{bmatrix} 4 & -5 \\ 6 & 3 \end{bmatrix}$

2. $\begin{bmatrix} -5 & 2 \\ 9 & 6 \\ 5 & -1 \end{bmatrix}$

3. $\begin{bmatrix} 3 & -1 & 4 & 6 \\ 9 & 1 & 2 & -3 \end{bmatrix}$

4. $[1 \quad -2 \quad -5 \quad 6 \quad 7]$

5. $\begin{bmatrix} 3 \\ 4 \\ -1 \\ 12 \end{bmatrix}$

6. $\begin{bmatrix} 1 & 4 & 6 & -9 \\ 2 & -7 & 4 & -1 \\ -6 & 3 & 0 & 0 \end{bmatrix}$

7. $[9 \quad -16]$

8. $\begin{bmatrix} 0 & -5 & 2 \\ -7 & 1 & 1 \end{bmatrix}$

Find the value of each variable that makes the matrices equal.

9. $\begin{bmatrix} x & 4 \\ 3 & -5 \end{bmatrix} = \begin{bmatrix} 2 & y \\ 3 & z \end{bmatrix}$

10. $\begin{bmatrix} x & 3 & 4 \\ 0 & -6 & 7 \end{bmatrix} = \begin{bmatrix} 9 & y & 4 \\ 0 & -6 & z \end{bmatrix}$

11. $\begin{bmatrix} x & 2y & 3z \\ a+1 & b-3 & \frac{c}{4} \end{bmatrix} = \begin{bmatrix} 1 & 6 & -12 \\ 5 & 9 & -3 \end{bmatrix}$

12. $\begin{bmatrix} 4x & 3y \\ 2a+1 & 1-3b \end{bmatrix} = \begin{bmatrix} 16 & 24 \\ -15 & 19 \end{bmatrix}$

13. $\begin{bmatrix} 2x+y \\ x-y \end{bmatrix} = \begin{bmatrix} 2 \\ 10 \end{bmatrix}$

14. $[2x \quad x+2y \quad x+y+z] = [12 \quad 22 \quad 16]$

15. $\begin{bmatrix} x & x-y \\ x+z & y-2z \\ x-2w & y+w \end{bmatrix} = \begin{bmatrix} 3 & -2 \\ 13 & -15 \\ -9 & 11 \end{bmatrix}$

16. $\begin{bmatrix} x & x+y & 3x-4z \\ 2a & 3a-2b & a+5c \end{bmatrix} = \begin{bmatrix} 9 & 14 & 3 \\ 16 & 18 & 43 \end{bmatrix}$

Add the matrices, if possible.

17. $\begin{bmatrix} 4 & 7 & -2 \\ -9 & 0 & 6 \end{bmatrix} + \begin{bmatrix} 4 & -3 & -1 \\ 5 & 8 & 10 \end{bmatrix}$

18. $\begin{bmatrix} 2 & -8 \\ 0 & -1 \\ -7 & 5 \end{bmatrix} + \begin{bmatrix} 3 & -9 \\ 4 & 6 \\ 10 & 4 \end{bmatrix}$

19. $\begin{bmatrix} 1 & 0 & 5 \\ 11 & -9 & 1 \\ 4 & -2 & -8 \end{bmatrix} + \begin{bmatrix} 4 & -10 & 3 \\ -7 & 2 & 0 \\ 6 & 3 & -6 \end{bmatrix}$

20. $[-3 \quad 1 \quad 4] + [10 \quad -5 \quad 6]$

In Exercises 21 through 36, use the given matrices A, B, and C to perform the indicated operations:

$$A = \begin{bmatrix} 1 & 10 & 6 \\ 5 & -2 & 0 \end{bmatrix} \quad B = \begin{bmatrix} 0 & -2 & -8 \\ 4 & 7 & 5 \end{bmatrix} \quad C = \begin{bmatrix} 1 & -9 \\ 4 & 3 \end{bmatrix}$$

21. $-A$
22. $-C$
23. $3A$
24. $2C$
25. $-2C$
26. $-5B$
27. $A + B$
28. $A - B$
29. $B - A$
30. $A + C$
31. $B + C$
32. $2A + 3B$
33. $3A - 4B$
34. $5B - 2A$
35. $2A + 3C$
36. $4C - 2B$

In Exercises 37 through 41, use the given matrices D, E, and F to verify the indicated matrix properties:

$$D = \begin{bmatrix} 3 & -4 & 5 \\ -2 & 0 & 10 \end{bmatrix} \quad E = \begin{bmatrix} 4 & -6 & 2 \\ 1 & 5 & -8 \end{bmatrix} \quad F = \begin{bmatrix} 3 & -7 & 4 \\ 0 & 10 & 6 \end{bmatrix}$$

37. $D + E = E + D$
38. $(D + E) + F = D + (E + F)$
39. $E + (-E) = O$, where O is the zero matrix.
40. $4(D + E) = 4D + 4E$
41. $D + O = D$, where O is the zero matrix.
42. A small company makes ceiling fans and keeps its inventory in a matrix format on a spreadsheet as follows:

	Finish		
No. of blades	Polished brass	Antique brass	Painted
4	*a*	*b*	*c*
5	*d*	*e*	*f*
6	*g*	*h*	*i*

Matrix A is the inventory on March 1. B is the production matrix for March. C is the sales matrix for March. Find the inventory matrix for April 1.

$$A = \begin{bmatrix} 40 & 30 & 24 \\ 15 & 21 & 5 \\ 10 & 18 & 0 \end{bmatrix} \quad B = \begin{bmatrix} 85 & 65 & 48 \\ 70 & 60 & 25 \\ 34 & 25 & 15 \end{bmatrix} \quad C = \begin{bmatrix} 90 & 60 & 54 \\ 60 & 60 & 20 \\ 29 & 36 & 10 \end{bmatrix}$$

43. In anticipation of a special promotion for May, the manager of the company in Exercise 42 decides to double the March production during April. Find the production matrix for April.

44. A small mattress factory keeps its inventory in a matrix format on a spreadsheet as follows:

	Firmness			
Mattress size	Regular	Firm	Extra firm	Super firm
Twin	a	b	c	d
Full	e	f	g	h
Queen	i	j	k	l
King	m	n	o	p

Matrix A is the inventory on June 1. B is the production matrix for June. C is the sales matrix for June. Find the matrix that shows the difference between sales and production for the month of June.

$$A = \begin{bmatrix} 6 & 3 & 4 & 5 \\ 7 & 4 & 3 & 2 \\ 1 & 3 & 0 & 2 \\ 2 & 3 & 4 & 0 \end{bmatrix} \qquad B = \begin{bmatrix} 24 & 30 & 14 & 10 \\ 20 & 23 & 12 & 8 \\ 15 & 19 & 11 & 4 \\ 15 & 17 & 16 & 7 \end{bmatrix} \qquad C = \begin{bmatrix} 27 & 25 & 12 & 9 \\ 25 & 22 & 14 & 0 \\ 12 & 6 & 3 & 4 \\ 10 & 15 & 18 & 6 \end{bmatrix}$$

45. Find the inventory matrix for July 1 for the mattress company in Exercise 44.

15.2 MULTIPLICATION OF MATRICES

Two matrices may be multiplied only when the number of columns in the first matrix equals the number of rows in the second matrix. If A is an $m \times n$ matrix and B is an $n \times p$ matrix, then the product $AB = C$, where C is an $m \times p$ matrix.

$$\underset{m \times n}{A} \cdot \underset{n \times p}{B} = \underset{m \times p}{C}$$

must be the same

To find the product matrix C:

1. Multiply the elements of the first row of A by the corresponding elements of the first column of B and add these products. This sum gives the first element of C.
2. Multiply the elements of the first row of A by the corresponding elements of the second column of B and add these products. This sum gives the second element of the first row of C.
3. Continue multiplying the corresponding elements of each row and column and adding these products until all the entries of C are determined.

We now give a formal definition of matrix multiplication:

MATRIX MULTIPLICATION

The product of an $m \times n$ matrix A times an $n \times p$ matrix B is the $m \times p$ matrix C, whose entry in the ath row and bth column is the sum of the products of the corresponding entries in the ath row of A and the bth column of B.

EXAMPLE 1

Multiply: $\begin{bmatrix} 1 & 2 & 4 \end{bmatrix} \begin{bmatrix} 0 & 5 \\ 7 & -4 \\ -3 & 6 \end{bmatrix}$.

First, determine if the product is defined.

	first matrix	second matrix	product
order	1×3	3×2	1×2

(the 3 of the first matrix and the 3 of the second matrix are marked: same)

Step 1: Multiply the elements of the first row of the first matrix times the corresponding elements of the first column of the second matrix and add these products. This result is the first element of the product matrix.

$$\begin{bmatrix} 1 & 2 & 4 \end{bmatrix} \begin{bmatrix} 0 & 5 \\ 7 & -4 \\ -3 & 6 \end{bmatrix} = \begin{bmatrix} 2 & \ \end{bmatrix} \quad \text{(row 1} \times \text{column 1)}$$

$$1 \cdot 0 + 2 \cdot 7 + 4(-3) = 2$$

Step 2: Multiply the elements of the first row of the first matrix times the corresponding elements of the second column of the second matrix and add these products. This result is the second element of the product matrix.

$$\begin{bmatrix} 1 & 2 & 4 \end{bmatrix} \begin{bmatrix} 0 & 5 \\ 7 & -4 \\ -3 & 6 \end{bmatrix} = \begin{bmatrix} 2 & 21 \end{bmatrix} \quad \text{(row 1} \times \text{column 2)}$$

$$1 \cdot 5 + 2(-4) + 4 \cdot 6 = 21$$

EXAMPLE 2

Multiply: $\begin{bmatrix} 1 & 3 & -5 \\ 6 & 0 & 8 \end{bmatrix} \begin{bmatrix} 4 & 2 & 5 & 1 \\ 8 & 4 & 3 & -2 \end{bmatrix}$.

First, determine if the product is defined.

	first matrix	second matrix
order	2×3	2×4

(the 3 of the first matrix and the 2 of the second matrix are marked: not same)

Since the number of columns of the first matrix does not equal the number of rows of the second matrix, the product is not defined.

EXAMPLE 3

Multiply: $\begin{bmatrix} 3 & 2 & 1 \\ 4 & 0 & -3 \end{bmatrix} \begin{bmatrix} 1 & 2 & -6 \\ -7 & 3 & 8 \\ -4 & 0 & -5 \end{bmatrix}$.

First, determine if the product is defined.

	first matrix	second matrix	product
order	2×3	3×3	2×3

(the 3 of the first matrix and the 3 of the second matrix are marked: same)

Step 1: Multiply the elements of the first row of the first matrix by the corresponding elements of the first column of the second matrix and add these products. This result is the first element of the product matrix.

$$\begin{bmatrix} 3 & 2 & 1 \\ 4 & 0 & -3 \end{bmatrix}\begin{bmatrix} 1 & 2 & -6 \\ -7 & 3 & 8 \\ -4 & 0 & -5 \end{bmatrix} = \begin{bmatrix} -15 & & \\ & & \end{bmatrix} \quad \text{(row 1} \times \text{column 1)}$$

$$3 \cdot 1 + 2(-7) + 1(-4) = -15$$

Step 2: Multiply the elements of the first row of the first matrix by the corresponding elements of the second column of the second matrix and add these products. This result is the second element of the first row of the product matrix.

$$\begin{bmatrix} 3 & 2 & 1 \\ 4 & 0 & -3 \end{bmatrix}\begin{bmatrix} 1 & 2 & -6 \\ -7 & 3 & 8 \\ -4 & 0 & -5 \end{bmatrix} = \begin{bmatrix} -15 & 12 & \\ & & \end{bmatrix} \quad \text{(row 1} \times \text{column 2)}$$

$$3 \cdot 2 + 2 \cdot 3 + 1 \cdot 0 = 12$$

Step 3: Multiply the elements of the first row of the first matrix by the corresponding elements of the third column of the second matrix and add these products. This result is the third element of the first row of the product matrix.

$$\begin{bmatrix} 3 & 2 & 1 \\ 4 & 0 & -3 \end{bmatrix}\begin{bmatrix} 1 & 2 & -6 \\ -7 & 3 & 8 \\ -4 & 0 & -5 \end{bmatrix} = \begin{bmatrix} -15 & 12 & -7 \\ & & \end{bmatrix} \quad \text{(row 1} \times \text{column 3)}$$

$$3(-6) + 2 \cdot 8 + 1(-5) = -7$$

Step 4: Multiply the elements of the second row of the first matrix times the corresponding elements of the first column of the second matrix and add these products. This result is the first element of the second row of the product matrix.

$$\begin{bmatrix} 3 & 2 & 1 \\ 4 & 0 & -3 \end{bmatrix}\begin{bmatrix} 1 & 2 & -6 \\ -7 & 3 & 8 \\ -4 & 0 & -5 \end{bmatrix} = \begin{bmatrix} -15 & 12 & -7 \\ 16 & & \end{bmatrix} \quad \text{(row 2} \times \text{column 1)}$$

$$4 \cdot 1 + 0(-7) + (-3)(-4) = 16$$

Step 5: Multiply the elements of the second row of the first matrix times the corresponding elements of the second column of the second matrix and add these products. This result is the second element of the second row of the product matrix.

$$\begin{bmatrix} 3 & 2 & 1 \\ 4 & 0 & -3 \end{bmatrix}\begin{bmatrix} 1 & 2 & -6 \\ -7 & 3 & 8 \\ -4 & 0 & -5 \end{bmatrix} = \begin{bmatrix} -15 & 12 & -7 \\ 16 & 8 & \end{bmatrix} \quad \text{(row 2} \times \text{column 2)}$$

$$4 \cdot 2 + 0 \cdot 3 + (-3)(0) = 8$$

Step 6: Multiply the elements of the second row of the first matrix times the corresponding elements of the third column of the second matrix and add these products. This result is the third element in the second row of the product matrix.

$$\begin{bmatrix} 3 & 2 & 1 \\ 4 & 0 & -3 \end{bmatrix}\begin{bmatrix} 1 & 2 & -6 \\ -7 & 3 & 8 \\ -4 & 0 & -5 \end{bmatrix} = \begin{bmatrix} -15 & 12 & -7 \\ 16 & 8 & -9 \end{bmatrix} \quad \text{(row 2} \times \text{column 3)}$$

$$4(-6) + 0 \cdot 8 + (-3)(-5) = -9$$

For square matrices A and B, both products AB and BA are defined but AB is generally **not** equal to BA. For any square matrix A, the powers $A^2, A^3, A^4, \ldots$ may be found by successive multiplication by the matrix A.

EXAMPLE 4

If $A = \begin{bmatrix} 1 & 0 & -1 \\ -2 & 2 & 5 \\ 6 & 4 & 3 \end{bmatrix}$ and $B = \begin{bmatrix} 4 & -2 & 1 \\ 2 & 0 & 8 \\ 0 & 3 & 5 \end{bmatrix}$, find (a) AB, (b) BA, (c) A^2.

$$\text{(a) } AB = \begin{bmatrix} 1 & 0 & -1 \\ -2 & 2 & 5 \\ 6 & 4 & 3 \end{bmatrix} \begin{bmatrix} 4 & -2 & 1 \\ 2 & 0 & 8 \\ 0 & 3 & 5 \end{bmatrix} = \begin{bmatrix} 4 & -5 & -4 \\ -4 & 19 & 39 \\ 32 & -3 & 53 \end{bmatrix}$$

$$\text{(b) } BA = \begin{bmatrix} 4 & -2 & 1 \\ 2 & 0 & 8 \\ 0 & 3 & 5 \end{bmatrix} \begin{bmatrix} 1 & 0 & -1 \\ -2 & 2 & 5 \\ 6 & 4 & 3 \end{bmatrix} = \begin{bmatrix} 14 & 0 & -11 \\ 50 & 32 & 22 \\ 24 & 26 & 30 \end{bmatrix}$$

$$\text{(c) } A^2 = A \cdot A = \begin{bmatrix} 1 & 0 & -1 \\ -2 & 2 & 5 \\ 6 & 4 & 3 \end{bmatrix} \begin{bmatrix} 1 & 0 & -1 \\ -2 & 2 & 5 \\ 6 & 4 & 3 \end{bmatrix} = \begin{bmatrix} -5 & -4 & -4 \\ 24 & 24 & 27 \\ 16 & 20 & 23 \end{bmatrix}$$

Using a calculator, we have

[1,0,-1;-2,2,5;6,4,3]*[4,-2,1;2,0,8;0,3,5] **ENTER,** etc.

[1,0,-1;-2,2,5;6,4,3]^2 **ENTER**

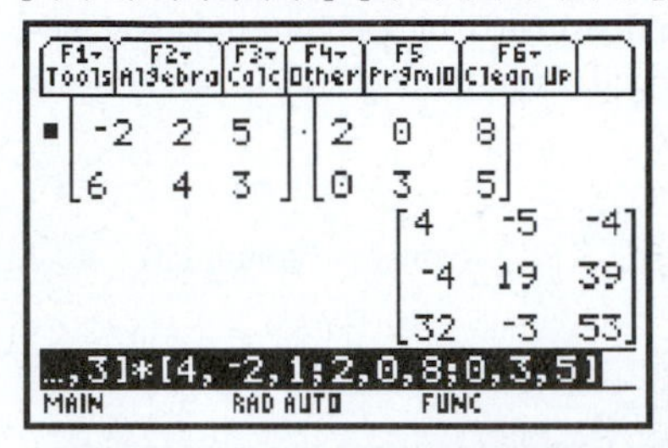

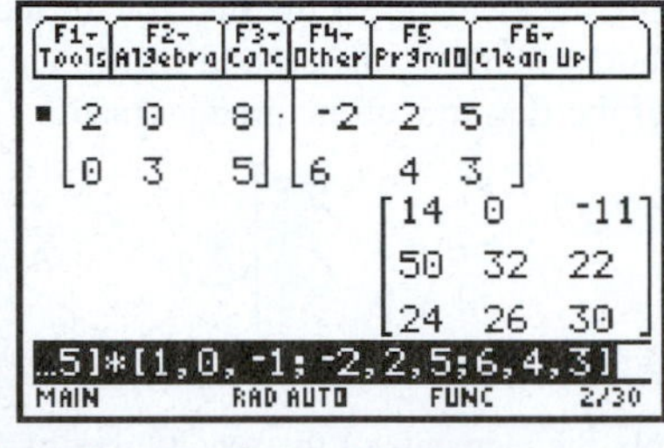

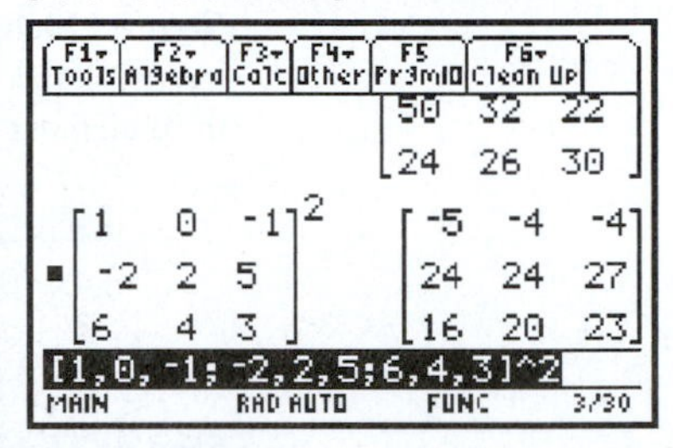

MATRX† left arrow **1** 3 **ENTER** 3 **ENTER** 1 **ENTER** 0 **ENTER**, etc. **2nd QUIT** **MATRX** left arrow **2** 3 **ENTER** 3 **ENTER** 4 **ENTER** -2 **ENTER**, etc. (to leave the matrix editor, press **2nd QUIT**)

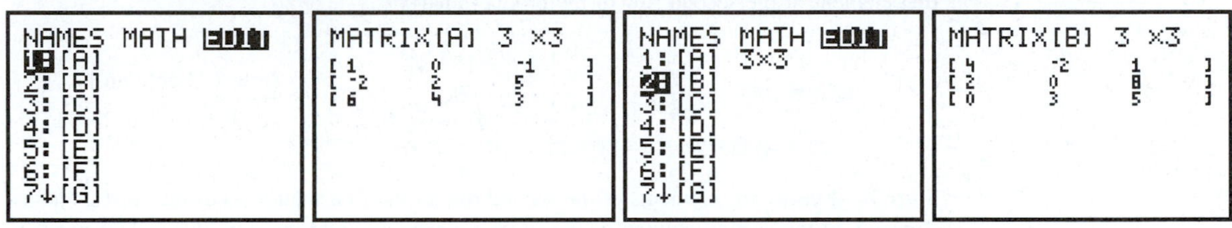

MATRX 1 MATRX 2 ENTER MATRX 2 MATRX 1 ENTER MATRX 1 x² ENTER

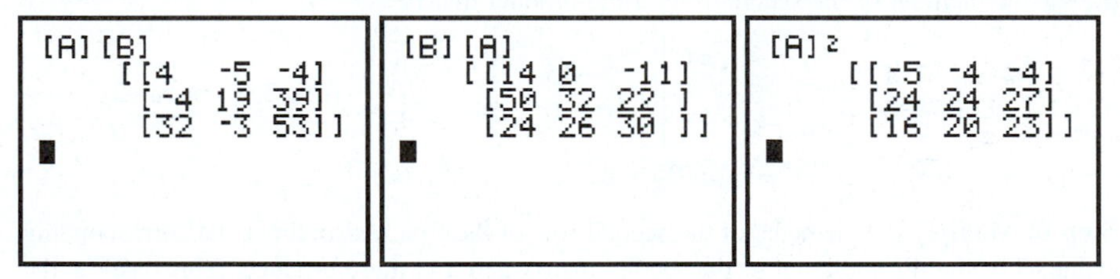

The **identity matrix,** I, is a square matrix whose principal diagonal contains 1 (one) as each of its elements and all other elements are zero. The following matrices are identity matrices:

$$\begin{bmatrix} \mathbf{1} & 0 \\ 0 & \mathbf{1} \end{bmatrix} \quad \text{and} \quad \begin{bmatrix} \mathbf{1} & 0 & 0 \\ 0 & \mathbf{1} & 0 \\ 0 & 0 & \mathbf{1} \end{bmatrix}$$

EXAMPLE 5

Given $A = \begin{bmatrix} -3 & 2 \\ 4 & 5 \end{bmatrix}$, show that $AI = IA = A$.

†If you are using a TI-83, **MATRX** is the key directly above **SIN.** However, on a TI-83 Plus, **MATRX** is the **2nd** shifted version of the $\mathbf{x^{-1}}$ key. So always read **MATRX** as **2nd MATRX** if you are using a TI-83 Plus.

$$AI = \begin{bmatrix} -3 & 2 \\ 4 & 5 \end{bmatrix}\begin{bmatrix} 1 & 0 \\ 0 & 1 \end{bmatrix} = \begin{bmatrix} (-3)(1) + 2 \cdot 0 & (-3)(0) + 2 \cdot 1 \\ 4 \cdot 1 + 5 \cdot 0 & 4 \cdot 0 + 5 \cdot 1 \end{bmatrix}$$

$$= \begin{bmatrix} -3 & 2 \\ 4 & 5 \end{bmatrix} = A$$

$$IA = \begin{bmatrix} 1 & 0 \\ 0 & 1 \end{bmatrix}\begin{bmatrix} -3 & 2 \\ 4 & 5 \end{bmatrix} = \begin{bmatrix} (1)(-3) + 0 \cdot 4 & 1 \cdot 2 + 0 \cdot 5 \\ (0)(-3) + 1 \cdot 4 & 0 \cdot 2 + 1 \cdot 5 \end{bmatrix}$$

$$= \begin{bmatrix} -3 & 2 \\ 4 & 5 \end{bmatrix} = A$$

Exercises 15.2

Find each product, if possible.

1. $\begin{bmatrix} 5 & 2 & 4 \end{bmatrix}\begin{bmatrix} 1 \\ -1 \\ 3 \end{bmatrix}$

2. $\begin{bmatrix} -1 & 4 & 2 \end{bmatrix}\begin{bmatrix} 0 \\ 2 \\ 5 \end{bmatrix}$

3. $\begin{bmatrix} 3 & 1 & -1 \end{bmatrix}\begin{bmatrix} -2 & 4 \\ 6 & 5 \\ 0 & 2 \end{bmatrix}$

4. $\begin{bmatrix} 1 & -3 \end{bmatrix}\begin{bmatrix} 1 & -2 \\ 6 & 3 \end{bmatrix}$

5. $\begin{bmatrix} 3 & 0 \\ 2 & 1 \end{bmatrix}\begin{bmatrix} -1 & 2 \\ 0 & 5 \end{bmatrix}$

6. $\begin{bmatrix} 1 & -3 \\ -1 & 6 \end{bmatrix}\begin{bmatrix} 2 & -1 \\ 0 & 3 \end{bmatrix}$

7. $\begin{bmatrix} 3 & -1 \\ 0 & 4 \end{bmatrix}\begin{bmatrix} 4 & 3 & -1 \\ 5 & 0 & 2 \end{bmatrix}$

8. $\begin{bmatrix} 2 & 0 \\ 1 & -1 \end{bmatrix}\begin{bmatrix} 2 & 3 & 4 & 5 \\ 1 & 0 & -1 & -3 \end{bmatrix}$

9. $\begin{bmatrix} 4 & 1 & 2 \\ 6 & 0 & -3 \end{bmatrix}\begin{bmatrix} 8 \\ -1 \\ -3 \end{bmatrix}$

10. $\begin{bmatrix} 3 & 0 & 2 \\ 5 & 1 & -2 \end{bmatrix}\begin{bmatrix} 1 & 0 \\ -4 & 2 \\ 3 & 5 \end{bmatrix}$

11. $\begin{bmatrix} 1 & 3 & 5 \\ 6 & 8 & -4 \end{bmatrix}\begin{bmatrix} 2 & 8 \\ 6 & -7 \end{bmatrix}$

12. $\begin{bmatrix} 1 & 2 \\ 3 & -6 \\ 5 & 0 \end{bmatrix}\begin{bmatrix} 1 & -5 & 0 \\ 0 & 2 & 5 \\ 3 & -1 & 4 \end{bmatrix}$

13. $\begin{bmatrix} 1 & 2 \\ 4 & -3 \\ 0 & 5 \end{bmatrix}\begin{bmatrix} 2 & -5 & 0 \\ 2 & -3 & 4 \end{bmatrix}$

14. $\begin{bmatrix} 3 & 0 \\ 1 & -4 \\ 5 & 6 \end{bmatrix}\begin{bmatrix} 5 & 3 \\ -1 & 2 \end{bmatrix}$

15. $\begin{bmatrix} 1 & 7 \\ 6 & -3 \\ 0 & 4 \\ -2 & 5 \end{bmatrix}\begin{bmatrix} 4 & 6 & -4 & 0 \\ -2 & 5 & 1 & 3 \end{bmatrix}$

16. $\begin{bmatrix} 6 & 5 \\ 1 & -9 \\ 8 & 2 \\ -3 & 0 \end{bmatrix}\begin{bmatrix} 2 & 4 & 6 \\ 0 & 7 & -1 \end{bmatrix}$

17. $\begin{bmatrix} 5 & 6 & 1 & 3 \\ -5 & 2 & 0 & 4 \end{bmatrix}\begin{bmatrix} 0 & -1 & 4 & 10 \\ 3 & 8 & -2 & -7 \end{bmatrix}$

18. $\begin{bmatrix} -3 & 2 & 5 \end{bmatrix}\begin{bmatrix} -6 & 1 & 0 \end{bmatrix}$

19. $\begin{bmatrix} 1 & 2 & -2 \\ -5 & 3 & 0 \\ 6 & 4 & -1 \end{bmatrix}\begin{bmatrix} -3 & 0 & 2 \\ 4 & 1 & -4 \\ 7 & 5 & 6 \end{bmatrix}$

20. $\begin{bmatrix} -3 & 6 & -2 \\ 1 & -9 & 4 \\ 10 & 5 & 0 \end{bmatrix}\begin{bmatrix} 0 \\ -4 \\ 5 \end{bmatrix}$

21. $\begin{bmatrix} 6 & 2 & 1 & 10 & -4 \\ -5 & 0 & 3 & -7 & 6 \end{bmatrix}\begin{bmatrix} 3 \\ -2 \\ 5 \\ -1 \\ 4 \end{bmatrix}$

22. $\begin{bmatrix} 1 & 5 & 0 & 6 \end{bmatrix}\begin{bmatrix} 1 & 3 \\ 4 & -5 \\ -6 & 2 \\ 3 & 3 \end{bmatrix}$

23. Given $A = \begin{bmatrix} 3 & 2 \\ -4 & 1 \end{bmatrix}$ and $B = \begin{bmatrix} 0 & 1 \\ 5 & 2 \end{bmatrix}$, find **(a)** AB, **(b)** BA, and **(c)** A^2.

24. Use matrices A and B in Exercise 23 to find **(a)** B^2, **(b)** A^3, and **(c)** $(-A)^2$.

25. Given $A = \begin{bmatrix} 1 & 2 & 3 \\ 0 & 4 & 1 \\ 5 & -2 & 6 \end{bmatrix}$ and $B = \begin{bmatrix} 2 & 0 & 1 \\ -1 & 3 & 4 \\ 0 & -5 & 8 \end{bmatrix}$, find **(a)** AB, **(b)** BA, and **(c)** A^2.

26. Use matrices A and B in Exercise 25 to find **(a)** B^2, **(b)** A^3, and **(c)** $(-B)^2$.

27. Given $A = \begin{bmatrix} 1 & 0 & 5 \\ -3 & 2 & 6 \end{bmatrix}$ and $B = \begin{bmatrix} 3 & 2 \\ -5 & 0 \\ 1 & 1 \end{bmatrix}$, find **(a)** AB, **(b)** BA, and **(c)** A^2.

28. Given $A = \begin{bmatrix} 4 & -1 \\ -6 & 3 \\ 2 & 5 \end{bmatrix}$ and $B = \begin{bmatrix} 2 & 6 \\ 5 & 0 \end{bmatrix}$, find **(a)** AB, **(b)** BA, and **(c)** B^2.

For each matrix, show that $AI = IA = A$, where I is the identity matrix.

29. $\begin{bmatrix} 5 & 3 \\ 2 & -6 \end{bmatrix}$ **30.** $\begin{bmatrix} -4 & -3 \\ 1 & 0 \end{bmatrix}$ **31.** $\begin{bmatrix} 1 & -4 & 1 \\ 5 & 0 & 3 \\ 3 & 1 & 2 \end{bmatrix}$ **32.** $\begin{bmatrix} 4 & 1 & 0 & 3 \\ -2 & 3 & 6 & 3 \\ 1 & 3 & 1 & 5 \\ 1 & 0 & -1 & 2 \end{bmatrix}$

15.3 FINDING THE INVERSE OF A MATRIX

Some, but not all, square matrices A have an **inverse,** written A^{-1}**,** with the property that

$$A \cdot A^{-1} = A^{-1} \cdot A = I$$

where I is the identity matrix with the same order as A. A matrix cannot have more than one inverse, but the inverse may not exist.

EXAMPLE 1

Show that $\begin{bmatrix} 7 & 5 \\ 4 & 3 \end{bmatrix}$ and $\begin{bmatrix} 3 & -5 \\ -4 & 7 \end{bmatrix}$ are inverses.

$$\begin{bmatrix} 7 & 5 \\ 4 & 3 \end{bmatrix}\begin{bmatrix} 3 & -5 \\ -4 & 7 \end{bmatrix} = \begin{bmatrix} 7 \cdot 3 + 5(-4) & 7(-5) + 5 \cdot 7 \\ 4 \cdot 3 + 3(-4) & 4(-5) + 3 \cdot 7 \end{bmatrix} = \begin{bmatrix} 1 & 0 \\ 0 & 1 \end{bmatrix}$$

$$\begin{bmatrix} 3 & -5 \\ -4 & 7 \end{bmatrix}\begin{bmatrix} 7 & 5 \\ 4 & 3 \end{bmatrix} = \begin{bmatrix} 3 \cdot 7 + (-5)4 & 3 \cdot 5 + (-5)3 \\ (-4)7 + 7 \cdot 4 & (-4)5 + 7 \cdot 3 \end{bmatrix} = \begin{bmatrix} 1 & 0 \\ 0 & 1 \end{bmatrix}$$

Unfortunately, it is much more difficult to find the inverse of a matrix than it is to verify that $A \cdot A^{-1} = I$. The first method of finding the inverse of a matrix applies **only** to 2×2 matrices and has the following steps:

1. Interchange the elements of the principal diagonal.
2. Change the signs of the other two elements.
3. Divide each resulting element of this new matrix by the determinant of the original matrix.

That is,

$$\text{if } A = \begin{bmatrix} a & b \\ c & d \end{bmatrix}, \quad \text{then } A^{-1} = \frac{1}{D}\begin{bmatrix} d & -b \\ -c & a \end{bmatrix}$$

2. Change the signs of the other two elements.

1. Interchange the elements of the principal diagonal.

3. Divide by D,

where $D = \begin{vmatrix} a & b \\ c & d \end{vmatrix}$.

This product can be shown as follows:

$$\begin{aligned} A \cdot A^{-1} &= \begin{bmatrix} a & b \\ c & d \end{bmatrix} \cdot \frac{1}{D}\begin{bmatrix} d & -b \\ -c & a \end{bmatrix} \\ &= \frac{1}{D}\begin{bmatrix} ad + b(-c) & a(-b) + ba \\ cd + d(-c) & c(-b) + da \end{bmatrix} \end{aligned}$$

$$= \frac{1}{D}\begin{bmatrix} ad - bc & 0 \\ 0 & -bc + ad \end{bmatrix}$$

$$= \frac{1}{D}(ad - bc)\begin{bmatrix} 1 & 0 \\ 0 & 1 \end{bmatrix}$$

$$= \frac{ad - bc}{ad - bc}\begin{bmatrix} 1 & 0 \\ 0 & 1 \end{bmatrix} \qquad (D = ad - bc)$$

$$= \begin{bmatrix} 1 & 0 \\ 0 & 1 \end{bmatrix}$$

$$= I$$

EXAMPLE 2

Find the inverse of $A = \begin{bmatrix} 5 & 6 \\ 3 & 4 \end{bmatrix}$.

First, find D.

$$D = \begin{vmatrix} 5 & 6 \\ 3 & 4 \end{vmatrix} = 5 \cdot 4 - 3 \cdot 6 = 2$$

Then, $A^{-1} = \frac{1}{2}\begin{bmatrix} 4 & -6 \\ -3 & 5 \end{bmatrix} = \begin{bmatrix} 2 & -3 \\ -\frac{3}{2} & \frac{5}{2} \end{bmatrix}$.

Check by multiplication as follows:

$$\begin{bmatrix} 5 & 6 \\ 3 & 4 \end{bmatrix}\begin{bmatrix} 2 & -3 \\ -\frac{3}{2} & \frac{5}{2} \end{bmatrix} = \begin{bmatrix} 5 \cdot 2 + 6(-\frac{3}{2}) & 5(-3) + 6(\frac{5}{2}) \\ 3 \cdot 2 + 4(-\frac{3}{2}) & 3(-3) + 4(\frac{5}{2}) \end{bmatrix}$$

$$= \begin{bmatrix} 1 & 0 \\ 0 & 1 \end{bmatrix}$$

In general, for any square matrix A, its inverse A^{-1} exists and is unique when its determinant $|A|$ does not equal zero. A^{-1} does not exist if $|A| = 0$ or if A is not a square matrix.

The second method for finding the inverse of a matrix is more general, can be applied to any square matrix, and is based on the following **row operations:**

ELEMENTARY ROW OPERATIONS

1. Interchange any two rows.
2. Multiply all the elements of a row by the same nonzero real number.
3. Change a given row by multiplying all the elements of another row by the same nonzero real number and add each product to the corresponding element of the given row.

To find the inverse of the square matrix A, begin by forming the *augmented matrix,* which consists of the combination of A on the left and its corresponding identity matrix I on the right in the form $[A \mid I]$. Then apply a series of elementary row operations until the matrix on the left is in the form of I, the identity matrix. The matrix on the right is then A^{-1}, the inverse of A.

To make this method more efficient, the following outline may be used:

1. Obtain a 1 in the first position of the principal diagonal by interchanging rows or by multiplying the first row by the reciprocal of the first entry. (If any column contains all zeros, the matrix does not have an inverse.)
2. Use the new first row with 1 in the first position and the third elementary row operation above to obtain zeros for all column entries in the first column under the 1.

3. Obtain a 1 in the second position of the principal diagonal (second row, second column) using the procedures in Step 1.
4. Use the new second row and the 1 entry to obtain zeros for all column entries in the second column under this 1.
5. Obtain a 1 in the third position of the principal diagonal, and then obtain zeros for all column entries in the third column under this 1. Continue this process until the principal diagonal of the left half of the matrix contains all 1's and all entries below the 1's are zeros.
6. Use the 1 in the last row of the principal diagonal to obtain zeros in all column entries above the 1 in the last row of the left half of the matrix.
7. Use the 1 in the next-to-last row of the principal diagonal to obtain zeros in all column entries above this 1. Continue this process until all entries above the 1's in the principal diagonal are zero (the left half of the matrix has been changed into the identity matrix).
8. The matrix on the right is the inverse A^{-1}.

EXAMPLE 3

Find the inverse of the matrix $A = \begin{bmatrix} 3 & -5 \\ 1 & -2 \end{bmatrix}$.

First, set up the augmented matrix with matrix A on the left and the 2×2 identity matrix I on the right as follows:

$$\left[\begin{array}{cc|cc} 3 & -5 & 1 & 0 \\ 1 & -2 & 0 & 1 \end{array}\right]$$

$$\left[\begin{array}{cc|cc} 1 & -2 & 0 & 1 \\ 3 & -5 & 1 & 0 \end{array}\right] \quad \text{(Interchange the rows to obtain a 1 in the upper left position.)}$$

Next, multiply the first row by -3 and add each product to the corresponding element of the second row. We shall use the shorthand notation in the margin as follows:

$$\left[\begin{array}{cc|cc} 1 & -2 & 0 & 1 \\ 0 & 1 & 1 & -3 \end{array}\right] \quad (-3R_1 + R_2)$$

$$\left[\begin{array}{cc|cc} 1 & 0 & 2 & -5 \\ 0 & 1 & 1 & -3 \end{array}\right] \quad (2R_2 + R_1)$$

Thus, $A^{-1} = \begin{bmatrix} 2 & -5 \\ 1 & -3 \end{bmatrix}$.

EXAMPLE 4

Find the inverse of the matrix $A = \begin{bmatrix} 3 & 6 \\ 5 & 9 \end{bmatrix}$.

First, set up the augmented matrix with matrix A on the left and the 2×2 identity matrix I on the right as follows:

$$\left[\begin{array}{cc|cc} 3 & 6 & 1 & 0 \\ 5 & 9 & 0 & 1 \end{array}\right]$$

Multiply the first row by $\frac{1}{3}$ to obtain a 1 in the upper left position.

$$\left[\begin{array}{cc|cc} 1 & 2 & \frac{1}{3} & 0 \\ 5 & 9 & 0 & 1 \end{array}\right] \quad (\tfrac{1}{3}R_1)$$

$$\left[\begin{array}{cc|cc} 1 & 2 & \frac{1}{3} & 0 \\ 0 & -1 & -\frac{5}{3} & 1 \end{array}\right] \quad (-5R_1 + R_2)$$

Multiply the second row by -1 to obtain a 1 in the principal diagonal.

$$\left[\begin{array}{cc|cc} 1 & 2 & \frac{1}{3} & 0 \\ 0 & 1 & \frac{5}{3} & -1 \end{array}\right] \quad (-1R_2)$$

$$\left[\begin{array}{cc|cc} 1 & 0 & -3 & 2 \\ 0 & 1 & \frac{5}{3} & -1 \end{array}\right] \quad (-2R_2 + R_1)$$

Thus, $A^{-1} = \begin{bmatrix} -3 & 2 \\ \frac{5}{3} & -1 \end{bmatrix}$.

EXAMPLE 5

Find the inverse of the matrix $A = \begin{bmatrix} 0 & 1 & 2 \\ 2 & 3 & 1 \\ -1 & -1 & 0 \end{bmatrix}$.

First, set up the augmented matrix with matrix A on the left and the 3×3 identity matrix on the right as follows:

$$\left[\begin{array}{ccc|ccc} 0 & 1 & 2 & 1 & 0 & 0 \\ 2 & 3 & 1 & 0 & 1 & 0 \\ -1 & -1 & 0 & 0 & 0 & 1 \end{array}\right]$$

$$\left[\begin{array}{ccc|ccc} -1 & -1 & 0 & 0 & 0 & 1 \\ 2 & 3 & 1 & 0 & 1 & 0 \\ 0 & 1 & 2 & 1 & 0 & 0 \end{array}\right] \quad \text{(Interchange } R_1 \text{ and } R_3.)$$

$$\left[\begin{array}{ccc|ccc} 1 & 1 & 0 & 0 & 0 & -1 \\ 2 & 3 & 1 & 0 & 1 & 0 \\ 0 & 1 & 2 & 1 & 0 & 0 \end{array}\right] \quad (-1R_1)$$

$$\left[\begin{array}{ccc|ccc} 1 & 1 & 0 & 0 & 0 & -1 \\ 0 & 1 & 1 & 0 & 1 & 2 \\ 0 & 1 & 2 & 1 & 0 & 0 \end{array}\right] \quad (-2R_1 + R_2)$$

$$\left[\begin{array}{ccc|ccc} 1 & 1 & 0 & 0 & 0 & -1 \\ 0 & 1 & 1 & 0 & 1 & 2 \\ 0 & 0 & 1 & 1 & -1 & -2 \end{array}\right] \quad (-1R_2 + R_3)$$

$$\left[\begin{array}{ccc|ccc} 1 & 1 & 0 & 0 & 0 & -1 \\ 0 & 1 & 0 & -1 & 2 & 4 \\ 0 & 0 & 1 & 1 & -1 & -2 \end{array}\right] \quad (-1R_3 + R_2)$$

$$\left[\begin{array}{ccc|ccc} 1 & 0 & 0 & 1 & -2 & -5 \\ 0 & 1 & 0 & -1 & 2 & 4 \\ 0 & 0 & 1 & 1 & -1 & -2 \end{array}\right] \quad (-1R_2 + R_1)$$

Thus, $A^{-1} = \begin{bmatrix} 1 & -2 & -5 \\ -1 & 2 & 4 \\ 1 & -1 & -2 \end{bmatrix}$.

Using a calculator, we have

[0,1,2;2,3,1;-1,-1,0]^-1 **ENTER**

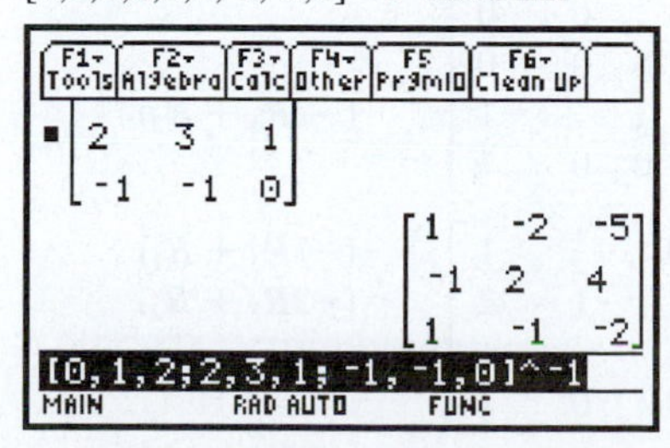

MATRX left arrow **1** 3 **ENTER** 3 **ENTER** 0 **ENTER** 1 **ENTER**, etc. **2nd QUIT MATRX 1** $\mathbf{x^{-1}}$ **ENTER**

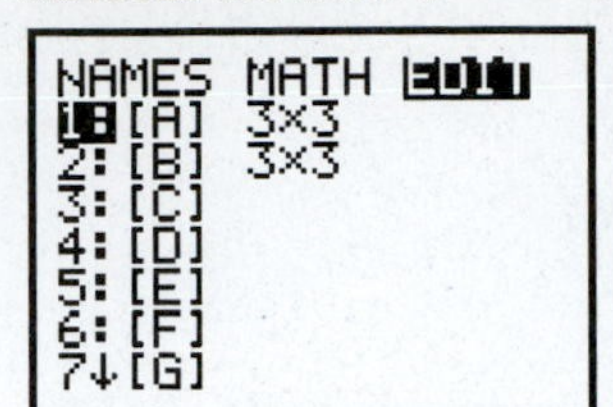

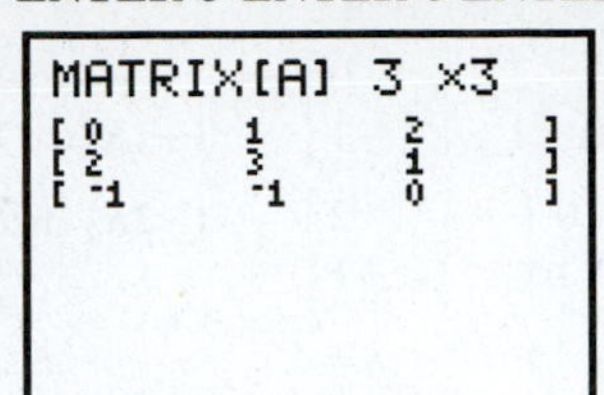


```
[A]-1
  [[1  -2 -5]
   [-1 2  4 ]
   [1  -1 -2]]
```

EXAMPLE 6

Find the inverse of the matrix $A = \begin{bmatrix} 0 & 1 & 2 & 0 \\ 5 & 0 & 5 & 0 \\ -2 & 0 & 0 & 4 \\ 0 & 2 & 4 & 2 \end{bmatrix}$.

First, set up the augmented matrix with matrix A on the left and the 4×4 identity matrix I on the right as follows:

$$\left[\begin{array}{cccc|cccc} 0 & 1 & 2 & 0 & 1 & 0 & 0 & 0 \\ 5 & 0 & 5 & 0 & 0 & 1 & 0 & 0 \\ -2 & 0 & 0 & 4 & 0 & 0 & 1 & 0 \\ 0 & 2 & 4 & 2 & 0 & 0 & 0 & 1 \end{array}\right]$$

$$\left[\begin{array}{cccc|cccc} 0 & 1 & 2 & 0 & 1 & 0 & 0 & 0 \\ 1 & 0 & 1 & 0 & 0 & \frac{1}{5} & 0 & 0 \\ -2 & 0 & 0 & 4 & 0 & 0 & 1 & 0 \\ 0 & 2 & 4 & 2 & 0 & 0 & 0 & 1 \end{array}\right] \quad (\tfrac{1}{5}R_2)$$

$$\left[\begin{array}{cccc|cccc} 1 & 0 & 1 & 0 & 0 & \frac{1}{5} & 0 & 0 \\ 0 & 1 & 2 & 0 & 1 & 0 & 0 & 0 \\ -2 & 0 & 0 & 4 & 0 & 0 & 1 & 0 \\ 0 & 2 & 4 & 2 & 0 & 0 & 0 & 1 \end{array}\right] \quad (\text{Interchange } R_1 \text{ and } R_2.)$$

$$\left[\begin{array}{cccc|cccc} 1 & 0 & 1 & 0 & 0 & \frac{1}{5} & 0 & 0 \\ 0 & 1 & 2 & 0 & 1 & 0 & 0 & 0 \\ 0 & 0 & 2 & 4 & 0 & \frac{2}{5} & 1 & 0 \\ 0 & 2 & 4 & 2 & 0 & 0 & 0 & 1 \end{array}\right] \quad (2R_1 + R_3)$$

$$\left[\begin{array}{cccc|cccc} 1 & 0 & 1 & 0 & 0 & \frac{1}{5} & 0 & 0 \\ 0 & 1 & 2 & 0 & 1 & 0 & 0 & 0 \\ 0 & 0 & 2 & 4 & 0 & \frac{2}{5} & 1 & 0 \\ 0 & 0 & 0 & 2 & -2 & 0 & 0 & 1 \end{array}\right] \quad (-R_2 + R_4)$$

$$\left[\begin{array}{cccc|cccc} 1 & 0 & 1 & 0 & 0 & \frac{1}{5} & 0 & 0 \\ 0 & 1 & 2 & 0 & 1 & 0 & 0 & 0 \\ 0 & 0 & 1 & 2 & 0 & \frac{1}{5} & \frac{1}{2} & 0 \\ 0 & 0 & 0 & 1 & -1 & 0 & 0 & \frac{1}{2} \end{array}\right] \quad \begin{array}{l} (\tfrac{1}{2}R_3) \\ (\tfrac{1}{2}R_4) \end{array}$$

$$\left[\begin{array}{cccc|cccc} 1 & 0 & 1 & 0 & 0 & \frac{1}{5} & 0 & 0 \\ 0 & 1 & 2 & 0 & 1 & 0 & 0 & 0 \\ 0 & 0 & 1 & 0 & 2 & \frac{1}{5} & \frac{1}{2} & -1 \\ 0 & 0 & 0 & 1 & -1 & 0 & 0 & \frac{1}{2} \end{array}\right] \quad (-2R_4 + R_3)$$

$$\left[\begin{array}{cccc|cccc} 1 & 0 & 0 & 0 & -2 & 0 & -\frac{1}{2} & 1 \\ 0 & 1 & 0 & 0 & -3 & -\frac{2}{5} & -1 & 2 \\ 0 & 0 & 1 & 0 & 2 & \frac{1}{5} & \frac{1}{2} & -1 \\ 0 & 0 & 0 & 1 & -1 & 0 & 0 & \frac{1}{2} \end{array}\right] \quad \begin{array}{l} (-1R_3 + R_1) \\ (-2R_3 + R_2) \end{array}$$

$$\text{Thus, } A^{-1} = \begin{bmatrix} -2 & 0 & -\frac{1}{2} & 1 \\ -3 & -\frac{2}{5} & -1 & 2 \\ 2 & \frac{1}{5} & \frac{1}{2} & -1 \\ -1 & 0 & 0 & \frac{1}{2} \end{bmatrix}$$

Each result may be checked using the property $A \cdot A^{-1} = I$.

Matrix Representation of Linear Systems of Equations

Given the system of linear equations

$$\begin{aligned} y + 2z &= 9 \\ 2x + 3y + z &= 1 \\ -x - y &= 1 \end{aligned}$$

If we let $A = \begin{bmatrix} 0 & 1 & 2 \\ 2 & 3 & 1 \\ -1 & -1 & 0 \end{bmatrix}$, $X = \begin{bmatrix} x \\ y \\ z \end{bmatrix}$, and $B = \begin{bmatrix} 9 \\ 1 \\ 1 \end{bmatrix}$, then $AX = B$ is a matrix equation representing the given system.

To solve any matrix equation representing a system of n linear equations with n unknowns whose square matrix of coefficients A has an inverse A^{-1}, X is an $n \times 1$ matrix of n unknowns, and B is an $n \times 1$ matrix of the constants, we have the following general solution:

$$\begin{aligned} AX &= B \\ (A^{-1})(AX) &= (A^{-1})B && \text{(Multiply each side by } A^{-1}.) \\ (A^{-1}A)X &= A^{-1}B && \text{(associative property of multiplication)} \\ IX &= A^{-1}B && (A^{-1}A = I) \\ X &= A^{-1}B && \text{(identity property)} \end{aligned}$$

That is, to solve a system of linear equations, multiply A^{-1} and B to find X.

EXAMPLE 7

Solve the system of linear equations

$$\begin{aligned} y + 2z &= 9 \\ 2x + 3y + z &= 1 \\ -x - y &= 1 \end{aligned}$$

Write this system in matrix form as follows:

$$A = \begin{bmatrix} 0 & 1 & 2 \\ 2 & 3 & 1 \\ -1 & -1 & 0 \end{bmatrix} \qquad X = \begin{bmatrix} x \\ y \\ z \end{bmatrix} \qquad B = \begin{bmatrix} 9 \\ 1 \\ 1 \end{bmatrix}$$

From Example 5, $A^{-1} = \begin{bmatrix} 1 & -2 & -5 \\ -1 & 2 & 4 \\ 1 & -1 & -2 \end{bmatrix}$.

Then, $X = A^{-1}B = \begin{bmatrix} 1 & -2 & -5 \\ -1 & 2 & 4 \\ 1 & -1 & -2 \end{bmatrix} \begin{bmatrix} 9 \\ 1 \\ 1 \end{bmatrix} = \begin{bmatrix} 9 - 2 - 5 \\ -9 + 2 + 4 \\ 9 - 1 - 2 \end{bmatrix} = \begin{bmatrix} 2 \\ -3 \\ 6 \end{bmatrix}$.

The solution is $x = 2$, $y = -3$, $z = 6$, or $(2, -3, 6)$.

Using a calculator, we have

[0,1,2;2,3,1;-1,-1,0]^-1*[9;1;1] **ENTER**

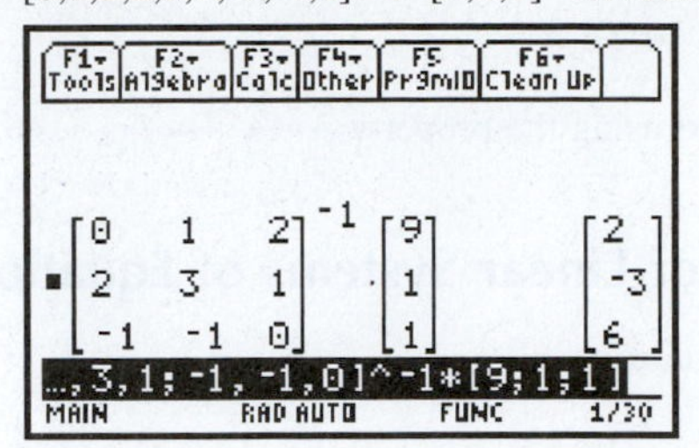

MATRX left arrow **1,** etc. **2nd QUIT** **MATRX** left arrow **2** etc.
2nd QUIT MATRX 1 x^{-1} MATRX 2 ENTER

It is also possible to skip the matrix editor on the TI-83 or TI-83 Plus, specifying each matrix on the home screen instead. Note that *rows* are enclosed in brackets and that each matrix is also enclosed within brackets.

```
[[0,1,2][2,3,1][
-1,-1,0]]⁻¹[[9][1
][1]]
                [[2 ]
                 [-3]
                 [6 ]]
```

EXAMPLE 8

Solve the system of linear equations

$$\begin{aligned} y + 2z \quad &= 3 \\ 5x + \quad 5z \quad &= 25 \\ -2x + \quad\quad 4w &= 14 \\ 2y + 4z + 2w &= 16 \end{aligned}$$

Write this system of equations in matrix form as follows:

$$A = \begin{bmatrix} 0 & 1 & 2 & 0 \\ 5 & 0 & 5 & 0 \\ -2 & 0 & 0 & 4 \\ 0 & 2 & 4 & 2 \end{bmatrix} \qquad X = \begin{bmatrix} x \\ y \\ z \\ w \end{bmatrix} \qquad B = \begin{bmatrix} 3 \\ 25 \\ 14 \\ 16 \end{bmatrix}$$

From Example 6, $A^{-1} = \begin{bmatrix} -2 & 0 & -\frac{1}{2} & 1 \\ -3 & -\frac{2}{5} & -1 & 2 \\ 2 & \frac{1}{5} & \frac{1}{2} & -1 \\ -1 & 0 & 0 & \frac{1}{2} \end{bmatrix}$

Then

$$X = A^{-1}B = \begin{bmatrix} -2 & 0 & -\frac{1}{2} & 1 \\ -3 & -\frac{2}{5} & -1 & 2 \\ 2 & \frac{1}{5} & \frac{1}{2} & -1 \\ -1 & 0 & 0 & \frac{1}{2} \end{bmatrix} \begin{bmatrix} 3 \\ 25 \\ 14 \\ 16 \end{bmatrix} = \begin{bmatrix} -6+0-7+16 \\ -9-10-14+32 \\ 6+5+7-16 \\ -3+0+0+8 \end{bmatrix} = \begin{bmatrix} 3 \\ -1 \\ 2 \\ 5 \end{bmatrix}$$

The solution is $x = 3$, $y = -1$, $z = 2$, $w = 5$ or $(3, -1, 2, 5)$.

Exercises 15.3

Find the inverse of each matrix using the method in Example 2.

1. $\begin{bmatrix} 2 & 6 \\ 1 & 4 \end{bmatrix}$ **2.** $\begin{bmatrix} 5 & 2 \\ -4 & 3 \end{bmatrix}$ **3.** $\begin{bmatrix} 5 & -2 \\ 11 & -4 \end{bmatrix}$ **4.** $\begin{bmatrix} -6 & -2 \\ 7 & 2 \end{bmatrix}$

5. $\begin{bmatrix} -6 & -4 \\ 3 & 5 \end{bmatrix}$ **6.** $\begin{bmatrix} -3 & 1 \\ 2 & 1 \end{bmatrix}$ **7.** $\begin{bmatrix} -4 & 2 \\ 5 & -2 \end{bmatrix}$ **8.** $\begin{bmatrix} 10 & -6 \\ -1 & 2 \end{bmatrix}$

Find the inverse of each matrix as directed by your instructor.

9. $\begin{bmatrix} 1 & 4 \\ 1 & 3 \end{bmatrix}$ **10.** $\begin{bmatrix} 1 & 5 \\ 2 & 4 \end{bmatrix}$ **11.** $\begin{bmatrix} 1 & 3 \\ -2 & 6 \end{bmatrix}$ **12.** $\begin{bmatrix} 1 & -2 \\ -4 & 6 \end{bmatrix}$

13. $\begin{bmatrix} -2 & 4 \\ 3 & -5 \end{bmatrix}$ **14.** $\begin{bmatrix} -5 & -6 \\ 4 & 4 \end{bmatrix}$ **15.** $\begin{bmatrix} 3 & 6 \\ 4 & 7 \end{bmatrix}$ **16.** $\begin{bmatrix} -2 & 1 \\ 3 & 3 \end{bmatrix}$

17. $\begin{bmatrix} 1 & 0 & 4 \\ 0 & 2 & 4 \\ 2 & 3 & 6 \end{bmatrix}$ **18.** $\begin{bmatrix} 1 & 2 & 3 \\ 2 & 5 & 7 \\ 3 & 7 & 8 \end{bmatrix}$ **19.** $\begin{bmatrix} -1 & -2 & 3 \\ 0 & -1 & 2 \\ 3 & 1 & 0 \end{bmatrix}$

20. $\begin{bmatrix} 2 & 0 & 1 \\ 0 & 4 & 2 \\ -1 & 2 & 2 \end{bmatrix}$ **21.** $\begin{bmatrix} 1 & 1 & 2 \\ 4 & 0 & 1 \\ 2 & 2 & 5 \end{bmatrix}$ **22.** $\begin{bmatrix} 2 & -1 & 0 \\ 2 & 2 & 0 \\ 0 & 0 & 1 \end{bmatrix}$

23. $\begin{bmatrix} 9 & -5 & 1 \\ -5 & 1 & 1 \\ 1 & 1 & -1 \end{bmatrix}$ **24.** $\begin{bmatrix} 1 & 2 & 3 \\ -1 & 2 & 0 \\ 2 & -1 & 1 \end{bmatrix}$ **25.** $\begin{bmatrix} 1 & 0 & 2 & 0 \\ 2 & 0 & 0 & 2 \\ -1 & 0 & -1 & 1 \\ 0 & 1 & 4 & 1 \end{bmatrix}$

26. $\begin{bmatrix} 1 & 1 & 0 & 2 \\ 0 & 1 & -4 & 2 \\ -1 & 2 & 4 & -2 \\ -2 & 1 & 2 & 1 \end{bmatrix}$ **27.** $\begin{bmatrix} 2 & 1 & -2 & 2 \\ 1 & -2 & 4 & 1 \\ 0 & 1 & 2 & 2 \\ 2 & 2 & -4 & 4 \end{bmatrix}$ **28.** $\begin{bmatrix} 1 & 0 & 2 & 1 \\ 1 & 0 & 4 & 1 \\ 2 & 4 & 0 & -2 \\ 4 & 0 & -1 & 2 \end{bmatrix}$

Solve each system of linear equations. Note: Each exercise is keyed to an exercise containing the matrix of the coefficients of the variables.

29. $\begin{aligned} 2x + 6y &= 34 \\ x + 4y &= 22 \end{aligned}$ (1)

30. $\begin{aligned} -6x - 2y &= 42 \\ 7x + 2y &= -48 \end{aligned}$ (4)

31. $\begin{aligned} 5x - 2y &= -18 \\ 11x - 4y &= -40 \end{aligned}$ (3)

32. $\begin{aligned} -3x + y &= 21 \\ 2x + y &= -4 \end{aligned}$ (6)

33. $\begin{aligned} x + 4y &= 33 \\ x + 3y &= 25 \end{aligned}$ (9)

34. $\begin{aligned} x - 2y &= 6 \\ -4x + 6y &= -20 \end{aligned}$ (12)

35. $\begin{aligned} -2x + 4y &= 26 \\ 3x - 5y &= -34 \end{aligned}$ (13)

36. $\begin{aligned} -5x - 6y &= -39 \\ 4x + 4y &= 28 \end{aligned}$ (14)

37. $\begin{aligned} x \qquad + 4z &= -27 \\ 2y + 4z &= -16 \\ 2x + 3y + 6z &= -30 \end{aligned}$ (17)

38. $\begin{aligned} x + 2y + 3z &= 30 \\ 2x + 5y + 7z &= 71 \\ 3x + 7y + 8z &= 87 \end{aligned}$ (18)

39. $$\begin{aligned} -x - 2y + 3z &= 37 \\ -y + 2z &= 26 \\ 3x + y &= 7 \end{aligned} \quad (19)$$

40. $$\begin{aligned} x + 2y + 3z &= -1 \\ -x + 2y &= -17 \\ 2x - y + z &= 18 \end{aligned} \quad (24)$$

41. $$\begin{aligned} 2x + y - 2z + 2w &= -3 \\ x - 2y + 4z + w &= 61 \\ y + 2z + 2w &= 21 \\ 2x + 2y - 4z + 4w &= -22 \end{aligned} \quad (27)$$

15.4 SOLVING A SYSTEM OF EQUATIONS BY A MATRIX METHOD

The last method that we shall study for solving systems of linear equations is similar to the method in Section 15.3. Instead of using the augmented matrix $[A \,\vdots\, I]$, we use the augmented matrix $[A \,\vdots\, B]$, where A is the $n \times n$ square matrix of coefficients of the variables and B is the $n \times 1$ matrix of the constants. Again, apply a series of elementary row operations until the matrix on the left is in the form I, the identity matrix. The solution is the resulting matrix on the right.

EXAMPLE 1

Solve the system of linear equations

$$\begin{aligned} 3x - 2y + 4z &= 32 \\ -4x - y + 6z &= 25 \\ x + 2y - 5z &= -29 \end{aligned}$$

First, set up the augmented matrix with the coefficient matrix A on the left and the constant matrix on the right as follows:

$$\left[\begin{array}{rrr|r} 3 & -2 & 4 & 32 \\ -4 & -1 & 6 & 25 \\ 1 & 2 & -5 & -29 \end{array}\right]$$

$$\left[\begin{array}{rrr|r} 1 & 2 & -5 & -29 \\ -4 & -1 & 6 & 25 \\ 3 & -2 & 4 & 32 \end{array}\right] \quad \text{(Interchange } R_1 \text{ and } R_3.\text{)}$$

$$\left[\begin{array}{rrr|r} 1 & 2 & -5 & -29 \\ 0 & 7 & -14 & -91 \\ 3 & -2 & 4 & 32 \end{array}\right] \quad \mathbf{(4R_1 + R_2)}$$

$$\left[\begin{array}{rrr|r} 1 & 2 & -5 & -29 \\ 0 & 7 & -14 & -91 \\ 0 & -8 & 19 & 119 \end{array}\right] \quad \mathbf{(-3R_1 + R_3)}$$

$$\left[\begin{array}{rrr|r} 1 & 2 & -5 & -29 \\ 0 & 1 & -2 & -13 \\ 0 & -8 & 19 & 119 \end{array}\right] \quad \mathbf{(\tfrac{1}{7}R_2)}$$

$$\left[\begin{array}{rrr|r} 1 & 2 & -5 & -29 \\ 0 & 1 & -2 & -13 \\ 0 & 0 & 3 & 15 \end{array}\right] \quad \mathbf{(8R_2 + R_3)}$$

$$\left[\begin{array}{rrr|r} 1 & 2 & -5 & -29 \\ 0 & 1 & -2 & -13 \\ 0 & 0 & 1 & 5 \end{array}\right] \quad \mathbf{(\tfrac{1}{3}R_3)}$$

$$\left[\begin{array}{rrr|r} 1 & 2 & -5 & -29 \\ 0 & 1 & 0 & -3 \\ 0 & 0 & 1 & 5 \end{array}\right] \quad \mathbf{(2R_3 + R_2)}$$

$$\left[\begin{array}{ccc|c} 1 & 2 & 0 & -4 \\ 0 & 1 & 0 & -3 \\ 0 & 0 & 1 & 5 \end{array}\right] \quad (5R_3 + R_1)$$

$$\left[\begin{array}{ccc|c} 1 & 0 & 0 & 2 \\ 0 & 1 & 0 & -3 \\ 0 & 0 & 1 & 5 \end{array}\right] \quad (-2R_2 + R_1)$$

The solution is found in the matrix on the right: $x = 2$, $y = -3$, $z = 5$, or $(2, -3, 5)$.

This final form of the matrix is known as its "row-reduced echelon form," which is why the calculator feature that finds it directly is called **rref.**

2nd MATH 4 4

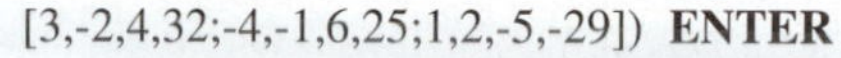

[3,-2,4,32;-4,-1,6,25;1,2,-5,-29]) **ENTER**

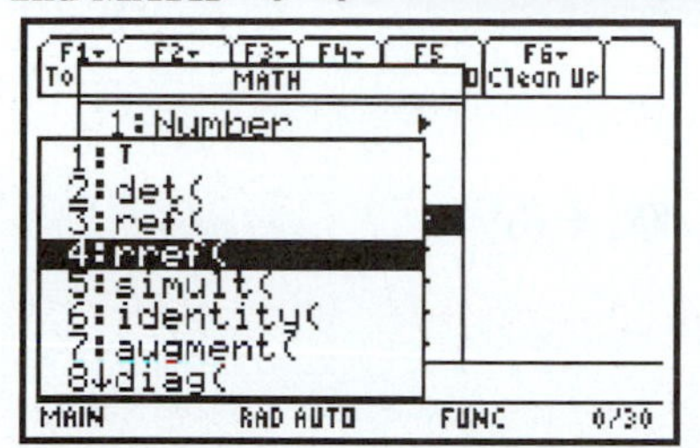

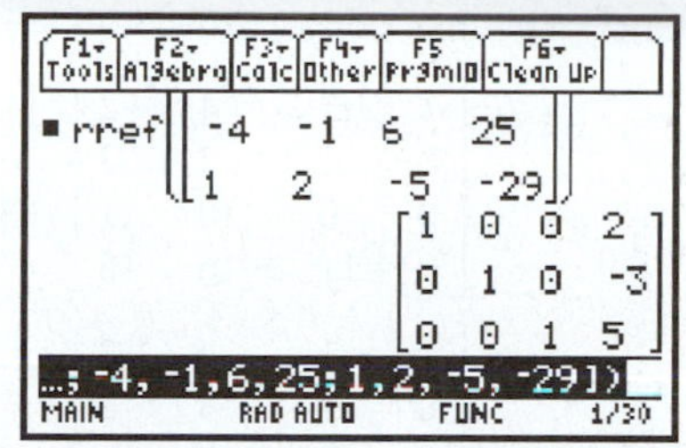

MATRX right arrow **ALPHA B** [[3,-2,4,32][-4,-1,6,25][1,2,-5,-29]])

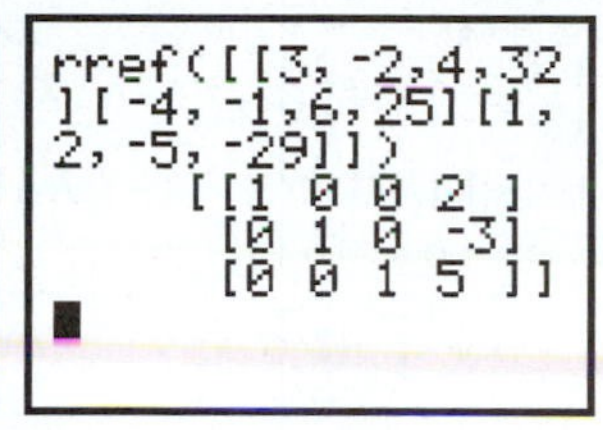

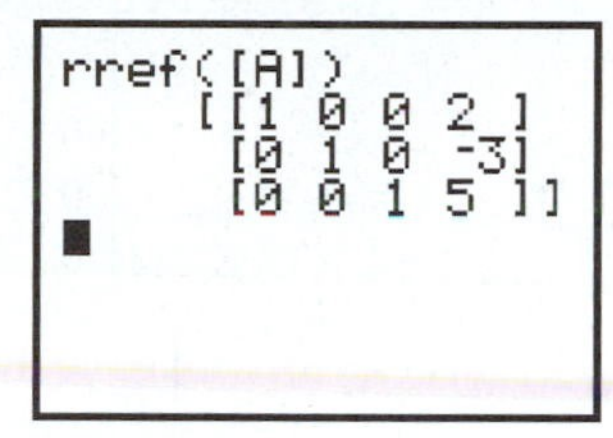

if the editor is used to store the matrix in [A]

EXAMPLE 2

Solve the system of linear equations

$$\begin{aligned} 4x + 2y - 2z + 4w &= 8 \\ x - y + 4z - 2w &= 2 \\ -3x - 5y + 2z + w &= -40 \\ 5x + 6y + 4z + 6w &= 17 \end{aligned}$$

First, set up the augmented matrix with the coefficient matrix A on the left and the constant matrix on the right as follows:

$$\left[\begin{array}{cccc|c} 4 & 2 & -2 & 4 & 8 \\ 1 & -1 & 4 & -2 & 2 \\ -3 & -5 & 2 & 1 & -40 \\ 5 & 6 & 4 & 6 & 17 \end{array}\right]$$

$$\left[\begin{array}{cccc|c} 1 & -1 & 4 & -2 & 2 \\ 4 & 2 & -2 & 4 & 8 \\ -3 & -5 & 2 & 1 & -40 \\ 5 & 6 & 4 & 6 & 17 \end{array}\right] \quad \text{(Interchange } R_1 \text{ and } R_2.)$$

$$\left[\begin{array}{cccc|c} 1 & -1 & 4 & -2 & 2 \\ 0 & 6 & -18 & 12 & 0 \\ -3 & -5 & 2 & 1 & -40 \\ 5 & 6 & 4 & 6 & 17 \end{array}\right] \quad (-4R_1 + R_2)$$

$$\left[\begin{array}{rrrr|r} 1 & -1 & 4 & -2 & 2 \\ 0 & 6 & -18 & 12 & 0 \\ 0 & -8 & 14 & -5 & -34 \\ 5 & 6 & 4 & 6 & 17 \end{array}\right] \quad (\mathbf{3R_1 + R_3})$$

$$\left[\begin{array}{rrrr|r} 1 & -1 & 4 & -2 & 2 \\ 0 & 6 & -18 & 12 & 0 \\ 0 & -8 & 14 & -5 & -34 \\ 0 & 11 & -16 & 16 & 7 \end{array}\right] \quad (\mathbf{-5R_1 + R_4})$$

$$\left[\begin{array}{rrrr|r} 1 & -1 & 4 & -2 & 2 \\ 0 & 1 & -3 & 2 & 0 \\ 0 & -8 & 14 & -5 & -34 \\ 0 & 11 & -16 & 16 & 7 \end{array}\right] \quad (\tfrac{1}{6}\mathbf{R_2})$$

$$\left[\begin{array}{rrrr|r} 1 & -1 & 4 & -2 & 2 \\ 0 & 1 & -3 & 2 & 0 \\ 0 & 0 & -10 & 11 & -34 \\ 0 & 11 & -16 & 16 & 7 \end{array}\right] \quad (\mathbf{8R_2 + R_3})$$

$$\left[\begin{array}{rrrr|r} 1 & -1 & 4 & -2 & 2 \\ 0 & 1 & -3 & 2 & 0 \\ 0 & 0 & -10 & 11 & -34 \\ 0 & 0 & 17 & -6 & 7 \end{array}\right] \quad (\mathbf{-11R_2 + R_4})$$

$$\left[\begin{array}{rrrr|r} 1 & -1 & 4 & -2 & 2 \\ 0 & 1 & -3 & 2 & 0 \\ 0 & 0 & 1 & -1.1 & 3.4 \\ 0 & 0 & 17 & -6 & 7 \end{array}\right] \quad (-\tfrac{1}{10}\mathbf{R_3})$$

$$\left[\begin{array}{rrrr|r} 1 & -1 & 4 & -2 & 2 \\ 0 & 1 & -3 & 2 & 0 \\ 0 & 0 & 1 & -1.1 & 3.4 \\ 0 & 0 & 0 & 12.7 & -50.8 \end{array}\right] \quad (\mathbf{-17R_3 + R_4})$$

$$\left[\begin{array}{rrrr|r} 1 & -1 & 4 & -2 & 2 \\ 0 & 1 & -3 & 2 & 0 \\ 0 & 0 & 1 & -1.1 & 3.4 \\ 0 & 0 & 0 & 1 & -4 \end{array}\right] \quad (\tfrac{1}{12.7}\mathbf{R_4})$$

$$\left[\begin{array}{rrrr|r} 1 & -1 & 4 & -2 & 2 \\ 0 & 1 & -3 & 2 & 0 \\ 0 & 0 & 1 & 0 & -1 \\ 0 & 0 & 0 & 1 & -4 \end{array}\right] \quad (\mathbf{1.1R_4 + R_3})$$

$$\left[\begin{array}{rrrr|r} 1 & -1 & 4 & -2 & 2 \\ 0 & 1 & -3 & 0 & 8 \\ 0 & 0 & 1 & 0 & -1 \\ 0 & 0 & 0 & 1 & -4 \end{array}\right] \quad (\mathbf{-2R_4 + R_2})$$

$$\left[\begin{array}{rrrr|r} 1 & -1 & 4 & 0 & -6 \\ 0 & 1 & -3 & 0 & 8 \\ 0 & 0 & 1 & 0 & -1 \\ 0 & 0 & 0 & 1 & -4 \end{array}\right] \quad (\mathbf{2R_4 + R_1})$$

$$\left[\begin{array}{rrrr|r} 1 & -1 & 4 & 0 & -6 \\ 0 & 1 & 0 & 0 & 5 \\ 0 & 0 & 1 & 0 & -1 \\ 0 & 0 & 0 & 1 & -4 \end{array}\right] \quad (\mathbf{3R_3 + R_2})$$

$$\left[\begin{array}{cccc|c} 1 & -1 & 0 & 0 & -2 \\ 0 & 1 & 0 & 0 & 5 \\ 0 & 0 & 1 & 0 & -1 \\ 0 & 0 & 0 & 1 & -4 \end{array}\right] \quad (-4R_3 + R_1)$$

$$\left[\begin{array}{cccc|c} 1 & 0 & 0 & 0 & 3 \\ 0 & 1 & 0 & 0 & 5 \\ 0 & 0 & 1 & 0 & -1 \\ 0 & 0 & 0 & 1 & -4 \end{array}\right] \quad (1R_2 + R_1)$$

The solution is found in the matrix on the right: $x = 3$, $y = 5$, $z = -1$, and $w = -4$, or $(3, 5, -1, -4)$.

EXAMPLE 3

Solve the system of linear equations

$$\begin{aligned} 2x + y + 3z &= 26 \\ 4x + 2y + 5z &= 46 \\ -2x - y + z &= -2 \end{aligned}$$

First, set up the augmented matrix with the coefficient matrix A on the left and the constant matrix on the right as follows:

$$\left[\begin{array}{ccc|c} 2 & 1 & 3 & 26 \\ 4 & 2 & 5 & 46 \\ -2 & -1 & 1 & -2 \end{array}\right]$$

$$\left[\begin{array}{ccc|c} 1 & \frac{1}{2} & \frac{3}{2} & 13 \\ 4 & 2 & 5 & 46 \\ -2 & -1 & 1 & -2 \end{array}\right] \quad (\tfrac{1}{2}R_1)$$

$$\left[\begin{array}{ccc|c} 1 & \frac{1}{2} & \frac{3}{2} & 13 \\ 0 & 0 & -1 & -6 \\ 0 & 0 & 4 & 24 \end{array}\right] \quad \begin{array}{l} \\ (-4R_1 + R_2) \\ (2R_1 + R_3) \end{array}$$

$$\left[\begin{array}{ccc|c} 1 & \frac{1}{2} & 0 & 4 \\ 0 & 0 & 1 & 6 \\ 0 & 0 & 1 & 6 \end{array}\right] \quad \begin{array}{l} (\frac{3}{2}R_2 + R_1) \\ (-1R_2) \\ (\frac{1}{4}R_3) \end{array}$$

The last two rows are identical. Operations which result in two identical rows indicate that the system is dependent. We could multiply the second row by -1 and add the result to the third row, replacing the third row with the answer. This would result in a row of all zeros, which is another indication of a dependent system. There are infinitely many ordered triples that satisfy this system. Two are $(2, 4, 6)$ and $(6, -4, 6)$.

An inconsistent system is indicated by a row with zeros in the coefficient positions and a nonzero value in the constant position. Such a system has no solutions.

Exercises 15.4

Each augmented matrix corresponds to a system of linear equations. Find each solution.

1. $\left[\begin{array}{cc|c} 1 & 0 & 3 \\ 0 & 1 & -4 \end{array}\right]$

2. $\left[\begin{array}{cc|c} 1 & 0 & 0 \\ 0 & 1 & -6 \end{array}\right]$

3. $\left[\begin{array}{ccc|c} 1 & 0 & 0 & 17 \\ 0 & 1 & 0 & -2 \\ 0 & 0 & 1 & 5 \end{array}\right]$

4. $\left[\begin{array}{ccc|c} 1 & 0 & 0 & -6 \\ 0 & 1 & 0 & 12 \\ 0 & 0 & 1 & 40 \end{array}\right]$

5. $\left[\begin{array}{cccc|c} 1 & 0 & 0 & 0 & 3 \\ 0 & 1 & 0 & 0 & 15 \\ 0 & 0 & 1 & 0 & -6 \\ 0 & 0 & 0 & 1 & 10 \end{array}\right]$

6. $\left[\begin{array}{cccc|c} 1 & 0 & 0 & 0 & -4 \\ 0 & 1 & 0 & 0 & 6 \\ 0 & 0 & 1 & 0 & 20 \\ 0 & 0 & 0 & 1 & -12 \end{array}\right]$

Find the solution of each system of linear equations.

7. $\begin{aligned} 3x - y &= 3 \\ -4x + 2y &= -2 \end{aligned}$

8. $\begin{aligned} 5x + 4y &= 53 \\ -x + 3y &= 16 \end{aligned}$

9. $\begin{aligned} 4x - 6y &= -64 \\ x + 5y &= 36 \end{aligned}$

10. $\begin{aligned} -3x + 5y &= 51 \\ 4x + y &= 1 \end{aligned}$

11. $\begin{aligned} 2x - 8y &= 46 \\ 3x + 2y &= -1 \end{aligned}$

12. $\begin{aligned} 4x + 12y &= -52 \\ 5x + 4y &= 1 \end{aligned}$

13. $\begin{aligned} 6x + 8y &= 98 \\ 2x + 5y &= 49 \end{aligned}$

14. $\begin{aligned} 5x - 3y &= 1 \\ 7x - 12y &= 27 \end{aligned}$

15. $\begin{aligned} x + 2y - z &= 9 \\ 3x + y - 2z &= 13 \\ y + z &= -2 \end{aligned}$

16. $\begin{aligned} x + y - 3z &= 10 \\ 4x - y + z &= 12 \\ 3x + 2y - 6z &= 24 \end{aligned}$

17. $\begin{aligned} x + y - z &= -4 \\ x + 2y + 3z &= 14 \\ 4x + 3y - 2z &= -4 \end{aligned}$

18. $\begin{aligned} x + y - z &= -7 \\ 2x - y + 5z &= 24 \\ x - 2y - 3z &= 13 \end{aligned}$

19. $\begin{aligned} 2x - 3y + z &= 33 \\ x - 4y - 2z &= 29 \\ x - 6y - 4z &= 39 \end{aligned}$

20. $\begin{aligned} 2x + 3y - 4z &= 5 \\ x - 2y + z &= 9 \\ 3x + y + z &= 18 \end{aligned}$

21. $\begin{aligned} x + 2y - 2z + 3w &= 8 \\ 2x - 3y - 4z + w &= -7 \\ -x + 4y - z + 2w &= 11 \\ 4x + 5y - 2z + 4w &= 28 \end{aligned}$

22. $\begin{aligned} 3x - 4y + z - w &= -11 \\ 2x + y - z + w &= 16 \\ -x + 2y - 3z + 2w &= 14 \\ x - 3y + 2z - 4w &= -19 \end{aligned}$

23. $\begin{aligned} 2x + 4y + 6z + 2w &= 32 \\ x - y + 2z - 3w &= -2 \\ -3x - 2y + 4z + w &= 16 \\ 3x + 5y + z - 2w &= -7 \end{aligned}$

24. $\begin{aligned} 6x + 4y + 3z + 2w &= 19 \\ 2x - 3y + 4z - w &= 19 \\ x + y - z + 5w &= 12 \\ 3x - 2y - 5z + w &= -2 \end{aligned}$

25. The perimeter of a triangle is 34 in. One side is twice as long as the shortest side. The third side is 6 in. longer than the shortest side. Find the lengths of the three sides.

26. A factory makes three calculator models at a cost of \$4670 for 835 calculators. The manufacturing costs for the three models are \$6, \$4, and \$7, respectively. They sell for \$20, \$18, and \$24, respectively. How many of each model are made if the profit is \$12,560? (Profit = income − cost)

CHAPTER 15 SUMMARY

1. A *matrix* is a rectangular array of numbers. Its order, or dimension, is given by stating the number of rows and then the number of columns.
2. *Two matrices are equal* if they have the same order and the same corresponding entries.
3. The *sum of two matrices* A and B of the same order is the matrix with entries that are the sums of the corresponding entries of A and B.
4. The *negative or additive inverse* of matrix A is $-A$, the matrix with each entry the negative of each corresponding entry in A.
5. The *difference of two matrices* A and B of the same order is defined as $A - B = A + (-B)$.
6. For each real number or scalar k and matrix A, the matrix kA is defined as the matrix where each entry is k times each corresponding entry in A.

7. The *product* of an $m \times n$ matrix A times an $n \times p$ matrix B is the $m \times p$ matrix C whose entry in the ath row and bth column is the sum of the products of the corresponding entries in the ath row of A and the bth column of B.
8. The *identity matrix* is a square matrix whose principal diagonal contains 1 (one) as each of its entries and all of whose other entries are zero.
9. *Matrix properties* (A, B, and C are matrices, k is a real number, O is the zero matrix, and I is the identity matrix):

(a) $A + B = B + A$	commutative property of addition
(b) $(A + B) + C = A + (B + C)$	associative property of addition
(c) $k(A + B) = kA + kB$	
(d) $A + O = A$	identity element of addition
(e) $A + (-A) = O$	additive inverse property
(f) $AB \neq BA$	Multiplication is **not** commutative.
(g) $A \cdot A^{-1} = A^{-1} \cdot A = I$	multiplicative inverse property (for square matrices only)
(h) $AI = IA = A$	identity element of multiplication (for square matrices only)

10. *To find the inverse of a 2×2 matrix:*
    (a) Interchange the elements of the principal diagonal.
    (b) Change the signs of the other two elements.
    (c) Divide each resulting element of this new matrix by the determinant of the original matrix.
11. *Elementary row operations:*
    (a) Interchange any two rows.
    (b) Multiply all the elements of a row by the same nonzero real number.
    (c) Change a given row by multiplying all the elements of another row by the same nonzero real number and add each product to the corresponding element of the given row.
12. The *inverse of a square matrix* may be found using the procedure outlined on pages 509–510.
13. *To solve any matrix equation* representing a system of n linear equations with n unknowns whose square matrix of coefficients A has an inverse A^{-1}, X is an $n \times 1$ matrix of n unknowns, and B is an $n \times 1$ matrix of the constants, find the product $A^{-1}B$; that is, $X = A^{-1}B$.
14. *To solve a system of n linear equations with n unknowns:*
    (a) Form the augmented matrix $[A \mid B]$, where A is the $n \times n$ matrix of coefficients of the variables and B is the $n \times 1$ matrix of the constants.
    (b) Apply a series of elementary row operations until the matrix on the left is in the form I, the identity matrix. The solution is the resulting matrix on the right.

CHAPTER 15 REVIEW

Give the order of each matrix.

1. $\begin{bmatrix} 2 & 5 & 0 \\ -1 & 6 & 11 \end{bmatrix}$

2. $\begin{bmatrix} 9 \\ -1 \\ 8 \end{bmatrix}$

3. $\begin{bmatrix} 0 & 3 & 4 & -1 \\ 1 & -2 & 0 & -5 \\ 2 & 6 & 6 & 10 \end{bmatrix}$

Find the value of each variable that makes the matrices equal.

4. $\begin{bmatrix} x & 9 & 10 \\ -1 & y & 12 \end{bmatrix} = \begin{bmatrix} 12 & 9 & z \\ -1 & -3 & 12 \end{bmatrix}$

5. $\begin{bmatrix} x + y - z \\ x - y + z \\ x + \quad z \end{bmatrix} = \begin{bmatrix} 0 \\ 2 \\ 5 \end{bmatrix}$

Add the matrices, if possible.

6. $\begin{bmatrix} 5 & 0 & -3 \\ 2 & 6 & 4 \end{bmatrix} + \begin{bmatrix} 2 & 7 & 10 \\ -2 & 9 & -5 \end{bmatrix}$

7. $\begin{bmatrix} 2 & 4 \\ -1 & -5 \\ -3 & 7 \end{bmatrix} + \begin{bmatrix} -2 & 5 \\ 0 & -8 \\ 5 & 10 \end{bmatrix}$

In Exercises 8 through 15, use the given matrices A, B, and C to perform the indicated operations.

$$A = \begin{bmatrix} 2 & -3 & 5 \\ 7 & 0 & -1 \end{bmatrix} \quad B = \begin{bmatrix} 6 & -2 & -5 \\ 10 & 4 & 8 \end{bmatrix} \quad C = \begin{bmatrix} 2 & -4 \\ 8 & 11 \end{bmatrix}$$

8. $-C$ **9.** $3A$ **10.** $A + B$ **11.** $A + C$

12. $B - A$ **13.** $2A + 3B$ **14.** $5A - 2B$ **15.** $2C - B$

Find each product, if possible.

16. $\begin{bmatrix} 1 & 2 & 3 \\ -2 & 4 & 0 \end{bmatrix} \begin{bmatrix} 4 & -5 \\ -6 & 2 \\ 1 & 1 \end{bmatrix}$

17. $\begin{bmatrix} 1 & 2 & -3 \\ -4 & 5 & 2 \\ 0 & 1 & -1 \end{bmatrix} \begin{bmatrix} 0 & 1 & 2 \\ -5 & 2 & 0 \\ 1 & 2 & -2 \end{bmatrix}$

18. $\begin{bmatrix} 3 & -4 \\ -5 & 6 \\ 7 & 2 \end{bmatrix} \begin{bmatrix} 5 & 1 \\ -2 & 3 \\ 1 & 1 \end{bmatrix}$

19. $\begin{bmatrix} 2 & 0 & 1 & -1 \\ 3 & 1 & 2 & 5 \end{bmatrix} \begin{bmatrix} 1 & 2 & -3 \\ 4 & 1 & 6 \\ 1 & 0 & 1 \\ -2 & 1 & 2 \end{bmatrix}$

20. Given $A = \begin{bmatrix} 0 & -2 \\ 1 & -3 \end{bmatrix}$, find A^2.

Find the inverse of each matrix.

21. $\begin{bmatrix} -3 & -2 \\ 8 & 5 \end{bmatrix}$

22. $\begin{bmatrix} 8 & -5 \\ 4 & -3 \end{bmatrix}$

23. $\begin{bmatrix} 2 & 4 & -2 \\ 1 & 2 & 0 \\ 4 & 2 & 2 \end{bmatrix}$

24. $\begin{bmatrix} 2 & 3 & -1 \\ 1 & 2 & 0 \\ 4 & 5 & 0 \end{bmatrix}$

25. $\begin{bmatrix} 1 & 0 & -1 & 0 \\ 2 & 1 & 4 & 0 \\ -1 & 1 & 0 & -2 \\ 2 & 0 & 1 & 1 \end{bmatrix}$

Solve each system of linear equations using the inverse of a matrix. Note: Each exercise is keyed to an earlier exercise containing the matrix of the coefficients of the variables.

26. $\begin{aligned} -3x - 2y &= -17 \\ 8x + 5y &= 44 \end{aligned}$ (21)

27. $\begin{aligned} 8x - 5y &= 50 \\ 4x - 3y &= 26 \end{aligned}$ (22)

28. $\begin{aligned} 2x + 4y - 2z &= -14 \\ x + 2y &= -3 \\ 4x + 2y + 2z &= 8 \end{aligned}$ (23)

29. $\begin{aligned} 2x + 3y - z &= 10 \\ x + 2y &= 7 \\ 4x + 5y &= 22 \end{aligned}$ (24)

30. $\begin{aligned} x - z &= 5 \\ 2x + y + 4z &= -1 \\ -x + y - 2w &= -10 \\ 2x + z + w &= 8 \end{aligned}$ (25)

Each augmented matrix corresponds to a system of linear equations. Find each solution.

31. $\left[\begin{array}{cc|c} 1 & 0 & -6 \\ 0 & 1 & 10 \end{array}\right]$

32. $\left[\begin{array}{ccc|c} 1 & 0 & 0 & 4 \\ 0 & 1 & 0 & -3 \\ 0 & 0 & 1 & 15 \end{array}\right]$

33. $\left[\begin{array}{cccc|c} 1 & 0 & 0 & 0 & 4 \\ 0 & 1 & 0 & 0 & -11 \\ 0 & 0 & 1 & 0 & 8 \\ 0 & 0 & 0 & 1 & 0 \end{array}\right]$

Find the solution of each system of linear equations using one of the methods in Section 15.4.

34.
$$\begin{aligned} x + 4y &= 15 \\ -2x + 3y &= 3 \end{aligned}$$

35.
$$\begin{aligned} 3x + 8y &= -4 \\ 6x + y &= 22 \end{aligned}$$

36.
$$\begin{aligned} x + 3y - z &= 8 \\ 3x - 4y + 2z &= -9 \\ -2x + y - 4z &= 15 \end{aligned}$$

37.
$$\begin{aligned} 3x - 4y + 2z &= 49 \\ -x + y + 5z &= 31 \\ 2x - 3y + 2z &= 40 \end{aligned}$$

38.
$$\begin{aligned} 3x + 2y - z + w &= 8 \\ 3x + y - 2z - 2w &= 1 \\ x + 3y + 4z + 8w &= 17 \\ -x + y + z &= -3 \end{aligned}$$

CHAPTER 15

APPLICATION

Matrices Used to Encode Information

Cryptography is the science of writing a message in code so that only the intended receiver will understand it. Initially, keeping information secure concerned only governments and the military. Now the widespread use of the Internet makes privacy an issue for the general public. Individuals are interested in keeping personal messages and electronic payments (credit card numbers) private. Corporations want to guard company information that would help their competitors, such as unpatented ideas and corporate takeover plans.

One of the methods used to encipher and decipher messages involves matrices. Initially, a numerical value is assigned to each letter, such as 1 for A, 2 for B, and so on:

A	B	C	D	...	Y	Z	Blank
1	2	3	4	...	25	26	27

The message "Gauss Wins" would be represented by the sequence 7, 1, 21, 19, 19, 27, 23, 9, 14, 19. This is too obvious to be considered secure, but we need the numeric values for the encryption procedure, which uses a nonsingular matrix (i.e., one that has an inverse). For ease of computation use a matrix whose inverse consists of integers.

Let $A = \begin{bmatrix} 1 & 1 \\ 4 & 5 \end{bmatrix}$. Using pairs of numbers (in order), we multiply by matrix A. The product of A and the first pair becomes $\begin{bmatrix} 1 & 1 \\ 4 & 5 \end{bmatrix}\begin{bmatrix} 7 \\ 1 \end{bmatrix} = \begin{bmatrix} 8 \\ 33 \end{bmatrix}$. The numbers 8 and 33 have replaced 7 and 1 as representatives of the first two letters of our message. Continuing the pattern with each pair of numbers yields

$$\begin{bmatrix} 1 & 1 \\ 4 & 5 \end{bmatrix}\begin{bmatrix} 21 \\ 19 \end{bmatrix} = \begin{bmatrix} 40 \\ 179 \end{bmatrix}$$

$$\begin{bmatrix} 1 & 1 \\ 4 & 5 \end{bmatrix}\begin{bmatrix} 19 \\ 27 \end{bmatrix} = \begin{bmatrix} 46 \\ 211 \end{bmatrix}$$

$$\begin{bmatrix} 1 & 1 \\ 4 & 5 \end{bmatrix}\begin{bmatrix} 23 \\ 9 \end{bmatrix} = \begin{bmatrix} 32 \\ 137 \end{bmatrix}$$

$$\begin{bmatrix} 1 & 1 \\ 4 & 5 \end{bmatrix}\begin{bmatrix} 14 \\ 19 \end{bmatrix} = \begin{bmatrix} 33 \\ 151 \end{bmatrix}$$

Writing the results of the matrix multiplication in order yields 8, 33, 40, 179, 46, 211, 32, 137, 33, 151, the encryption for "Gauss Wins." If the original code has an odd number of characters, use the blank, 27, for the last entry.

The multiplication could have been written more simply using

$$\begin{bmatrix} 1 & 1 \\ 4 & 5 \end{bmatrix}\begin{bmatrix} 7 & 21 & 19 & 23 & 14 \\ 1 & 19 & 27 & 9 & 19 \end{bmatrix}$$

where the second matrix is formed by the pairs of numbers in order as the columns. Name the 2×5 matrix B and the product C. Then

$$AB = \begin{bmatrix} 1 & 1 \\ 4 & 5 \end{bmatrix}\begin{bmatrix} 7 & 21 & 19 & 23 & 14 \\ 1 & 19 & 27 & 9 & 19 \end{bmatrix} = \begin{bmatrix} 8 & 40 & 46 & 32 & 33 \\ 33 & 179 & 211 & 137 & 151 \end{bmatrix} = C$$

Matrix A is the encoding matrix, while A^{-1} is the decoding matrix. Since

$$A^{-1}AB = A^{-1}C$$
$$B = A^{-1}C$$

the receiver of message C needs to know A^{-1} to decode the message.

$$A^{-1} = \begin{bmatrix} 5 & -1 \\ -4 & 1 \end{bmatrix}$$

Thus, $B = A^{-1}C = \begin{bmatrix} 5 & -1 \\ -4 & 1 \end{bmatrix}\begin{bmatrix} 8 & 40 & 46 & 32 & 33 \\ 33 & 179 & 211 & 137 & 151 \end{bmatrix} = \begin{bmatrix} 7 & 21 & 19 & 23 & 14 \\ 1 & 19 & 27 & 9 & 19 \end{bmatrix}$.

Now the receiver can write the sequence appearing in the columns 7, 1, 21, 19, 19, 27, 23, 9, 14, 19, which translates to "Gauss Wins."

For longer messages, use larger encoding matrices. The only restriction is that the matrix must have an inverse.

16 Polynomials of Higher Degree

INTRODUCTION

Polynomial functions occur in almost every aspect of applied mathematics. They are involved in formulas for electric resistance and energy flow and uniform motion problems. In addition, they are used to approximate values of transcendental functions.

Objectives

- Determine the degree of a polynomial.
- Find the x- and y-intercepts of a polynomial function.
- Interpret graphs of polynomial functions.
- Solve polynomial equations for their real and complex solutions.

16.1 POLYNOMIAL FUNCTIONS

Any function in the form

$$f(x) = a_n x^n + a_{n-1} x^{n-1} + \cdots + a_1 x + a_0 \tag{1}$$

is called a *polynomial function of degree n*, where n is a nonnegative integer and $a_n \neq 0$. The numbers $a_n, a_{n-1}, \ldots, a_1, a_0$ are called the *coefficients* of the polynomial. The coefficient a_n in Equation (1) is called the *leading coefficient*.

In Section 4.2, we graphed *linear functions* in the form

$$f(x) = a_1 x + a_0 \quad \text{or} \quad f(x) = ax + b$$

and *quadratic functions* in the form

$$f(x) = a_2 x^2 + a_1 x + a_0 \quad \text{or} \quad f(x) = ax^2 + bx + c$$

which are special cases of polynomial functions. Each expression $a_k x^k$ is called a *term* of the polynomial.

Intercepts

The *intercepts* are those points where the graph of a function intersects either the x-axis or the y-axis. Any point on the x-axis has a y-coordinate of zero, and any point on the y-axis has an x-coordinate of zero. Thus,

1. To find the x-intercept(s), substitute $y = f(x) = 0$ into the function's equation and solve for x. Each resulting value of x corresponds to an x-intercept of the form $(x, 0)$. *Note:* These values of x that satisfy the equation $f(x) = 0$ are called *zeros* of the function $f(x)$. In general, if the polynomial can be factored in the form

$$f(x) = (x - a)(x - b)g(x)$$

then $x = a$ and $x = b$ are zeros of $f(x)$ and the points $(a, 0)$ and $(b, 0)$ are x-intercepts of its graph. This strong connection between factors, zeros, and x-intercepts of a polynomial will be one of the central themes of this chapter.

2. To find the y-intercept, substitute $x = 0$ into the function's equation and evaluate the corresponding y-value, namely $f(0)$. The resulting point $(0, f(0))$ is the y-intercept.

EXAMPLE 1

Find the intercepts for the graph of $f(x) = (x + 1)(x - 4)^2$.

(a) To find the x-intercepts, substitute 0 for $f(x)$ and solve for x.

$$f(x) = (x + 1)(x - 4)^2$$
$$0 = (x + 1)(x - 4)^2$$
$$x + 1 = 0 \quad \text{or} \quad (x - 4)^2 = 0$$
$$x = -1 \quad \text{or} \quad x = 4$$

So the x-intercepts are the points $(-1, 0)$ and $(4, 0)$.

We may also say that -1 and 4 are the *zeros* of the polynomial. Actually, 4 is a zero of multiplicity 2 and -1 is a zero of multiplicity 1. These *multiplicities* correspond to the exponents of each factor.

(b) To find the y-intercept, substitute $x = 0$ and evaluate.

$$f(x) = (x + 1)(x - 4)^2$$
$$f(0) = (0 + 1)(0 - 4)^2$$
$$f(0) = 16$$

So the point $(0, 16)$ is the y-intercept.

(c) Use a graphing calculator to see how these intercepts relate to the overall graph of the polynomial:

green diamond Y=

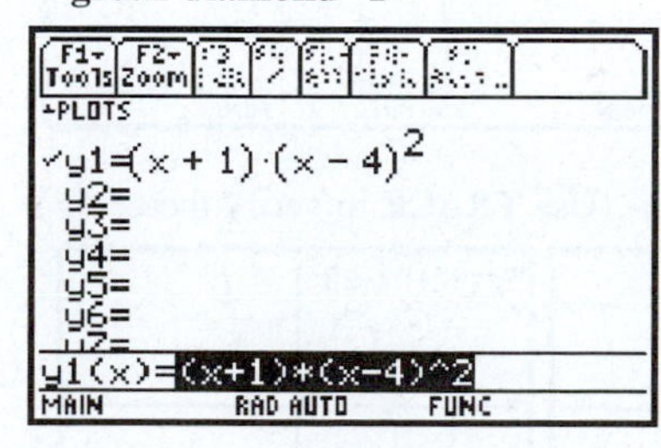

green diamond WINDOW

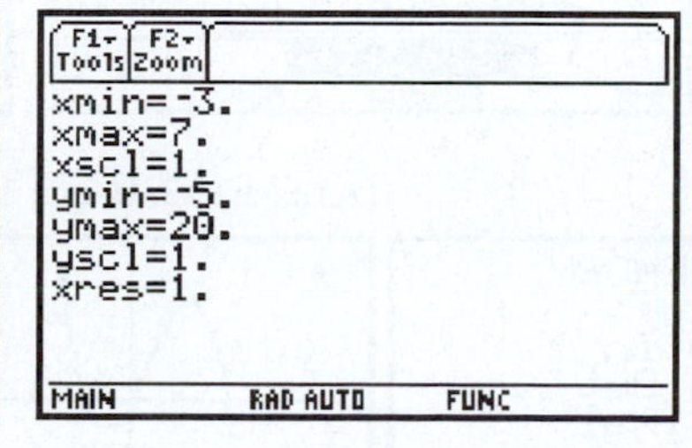

green diamond GRAPH

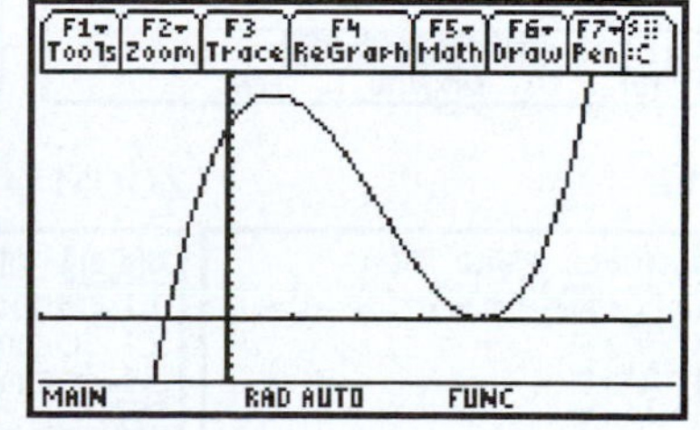

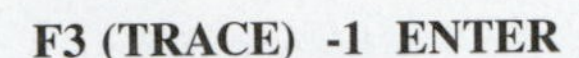

0 ENTER

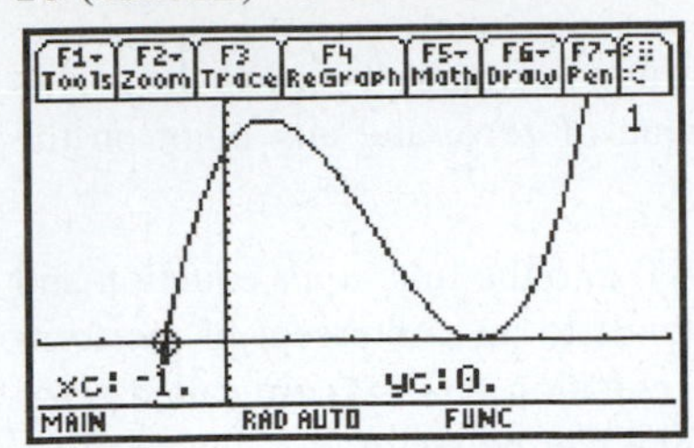

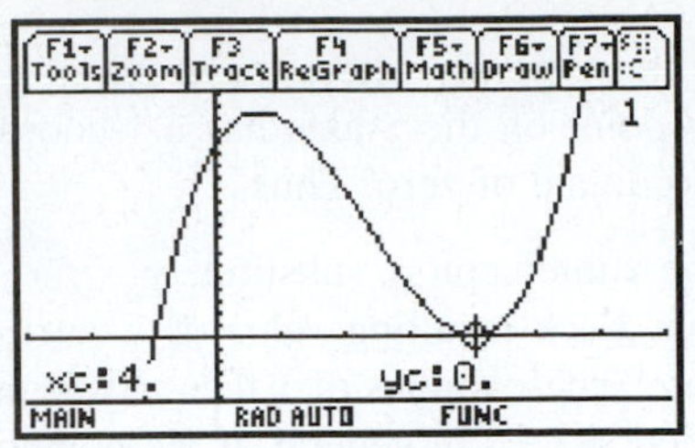

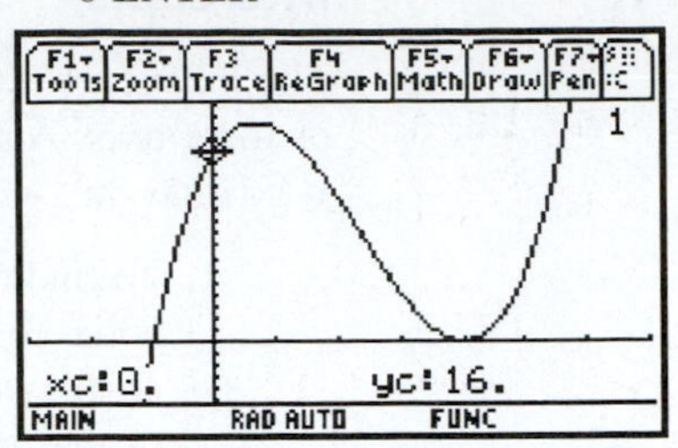

Y= **WINDOW** **GRAPH**

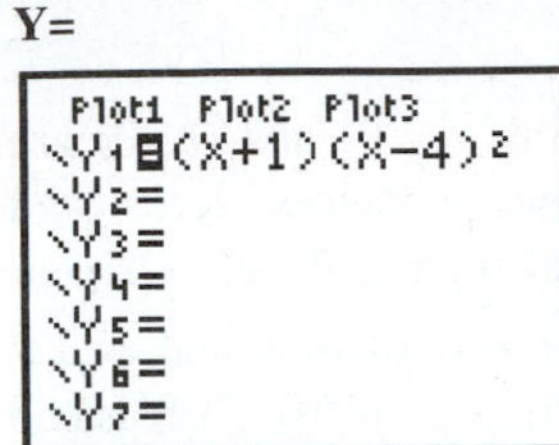

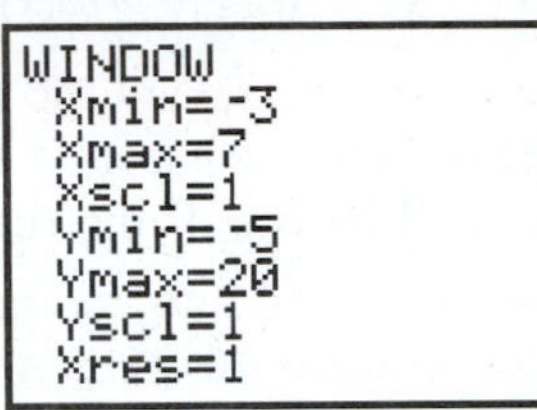

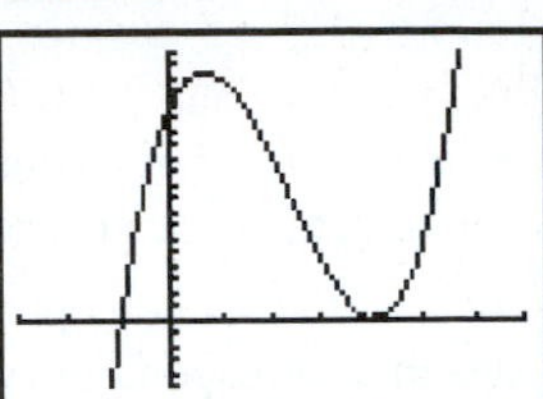

TRACE -1 ENTER **4 ENTER** **0 ENTER**

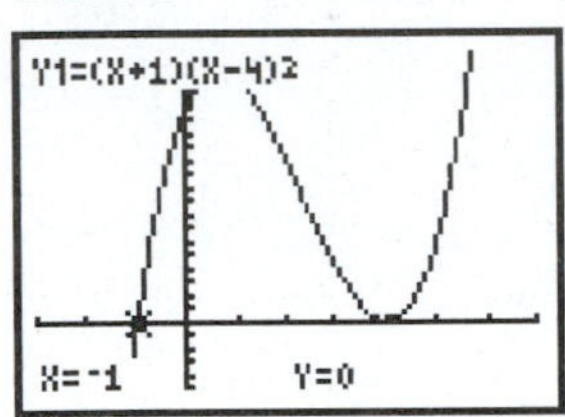

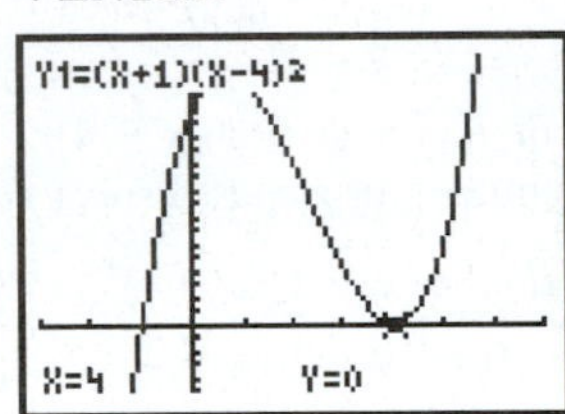

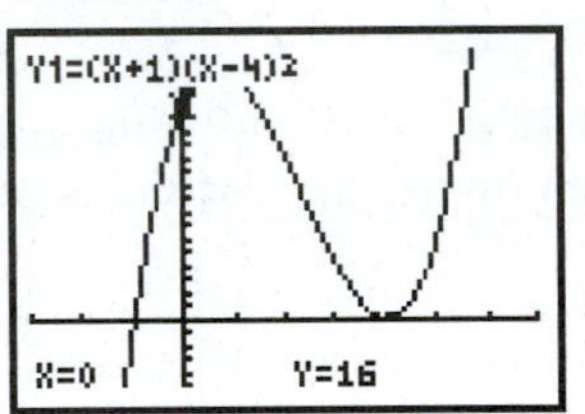

Note that the graph *crosses* the x-axis at a zero of *odd* multiplicity (-1 is a zero of multiplicity *one*), but the graph just *touches* the x-axis at a zero of *even* multiplicity (4 is a zero of multiplicity *two*).

EXAMPLE 2

(a) Graph the function $f(x) = x^3 + x^2 - 2x$.

green diamond Y= **F2 4** (apparently, the zeros are -2, 0, and 1)

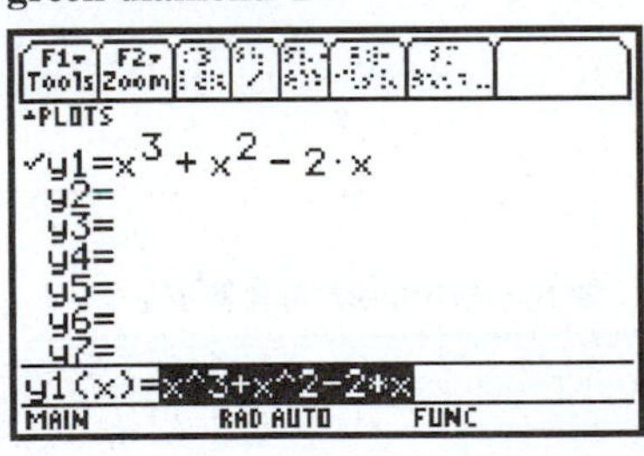

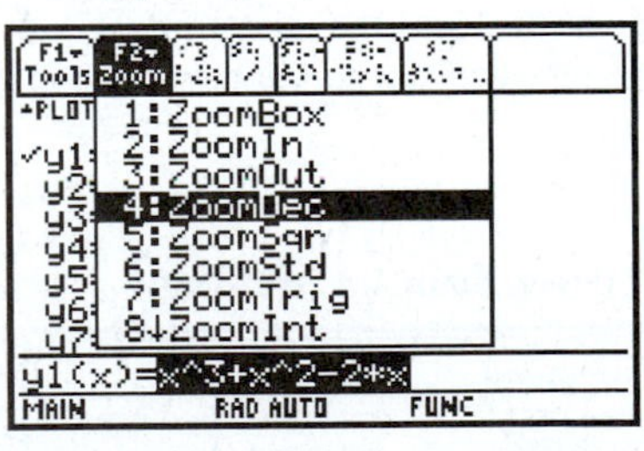

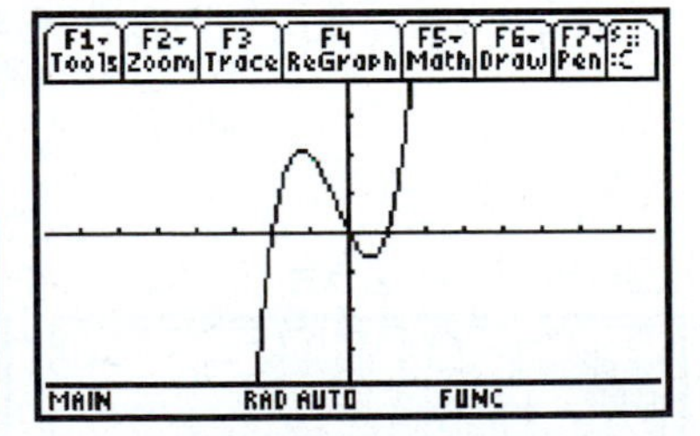

Y=

(zeros are -2, 0, and 1) (Use **TRACE** to verify the zeros.)

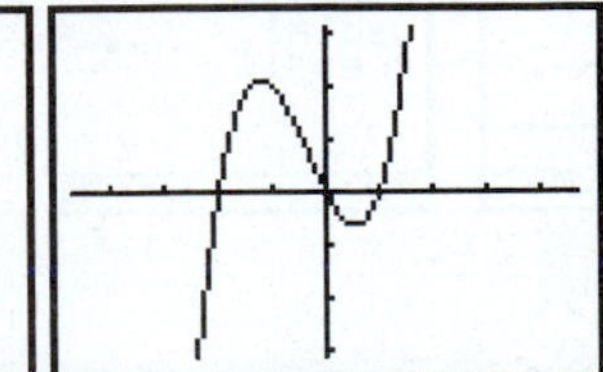

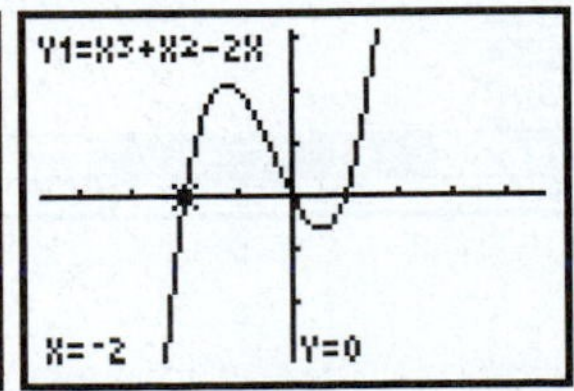

(b) Use the graph to identify the zeros and the corresponding factors of the polynomial.

From the graph, the x-intercepts appear to be the points $(-2, 0)$, $(0, 0)$, and $(1, 0)$. These can be verified using your calculator's **TRACE** feature (see Example 1). So the zeros of the polynomial are -2, 0, and 1.

Working backwards, we can speculate that a factorization of the polynomial might be

$$f(x) = (x - (-2))\,(x - 0)\,(x - 1) = (x + 2)\,x\,(x - 1) = x\,(x + 2)\,(x - 1)$$

Factoring in a more traditional way verifies this connection:

$$f(x) = x^3 + x^2 - 2x = x\,(x^2 + x - 2) = x\,(x + 2)\,(x - 1)$$

We already know that if $x - a$ is a factor of $f(x)$, then $(a, 0)$ is an x-intercept (a is a zero) of $f(x)$. Now it appears that the converse is also true.

FACTOR THEOREM

If the point $(a, 0)$ is an x-intercept of the polynomial function $f(x)$, then $x - a$ is a factor of $f(x)$. In other words, if a is a zero of the polynomial function $f(x)$, then $x - a$ must be a factor of $f(x)$.

Proof: Divide the function $f(x)$ by $x - a$, calling the quotient $Q(x)$ and the remainder R. Using the "check" for division, we can write

$$f(x) = Q(x)\,(x - a) + R \qquad \textbf{(2)}$$

(If this seems unfamiliar, try dividing 17 by 3, getting a quotient of 5 and a remainder of 2, which is checked by showing that $17 = 5 \cdot 3 + 2$.) If the point $(a, 0)$ is an x-intercept of $f(x)$, then we know that $f(a) = 0$. Replacing x with a in Equation (2) and combining it with our knowledge that $f(a) = 0$ gives

$$0 = f(a) = Q(a)\,(a - a) + R = Q(a)\,(0) + R = 0 + R = R \qquad \textbf{(3)}$$

So R must be zero; thus, from Equation (2),

$$f(x) = Q(x)\,(x - a) + 0$$
$$f(x) = Q(x)\,(x - a)$$

which shows that $x - a$ is one of the factors of $f(x)$.

Ignoring the "0 =" at the beginning of Equation (3) in the preceding proof also shows that if you divide a polynomial $f(x)$ by $x - a$, the remainder will be $f(a)$. This result is known as the *Remainder Theorem*:

REMAINDER THEOREM

If a polynomial $f(x)$ is divided by $x - a$ until the result is a constant remainder R, then $f(a) = R$. Thus, the remainder equals the value of the polynomial at $x = a$.

Thus, if $f(a) = 0$, then the remainder is zero when $f(x)$ is divided by $x - a$, which implies that $x - a$ is a factor of $f(x)$.

Before graphing calculators were in common use, polynomial division and the Remainder Theorem were among the primary tools used to investigate a polynomial, its factors, and its graph. For example, if you divided a polynomial by $x - 7$ and the remainder was zero, you knew that $x - 7$ was a factor of the polynomial and that $(7, 0)$ was a point on its graph. If the remainder had been 19, you would have known that (7, 19) was a point on the graph of the polynomial (and that $x - 7$ was *not* one of its factors). Unfortunately, the use of polynomial division to test potential factors and evaluate points easily becomes very tedious, especially in practical work. Instead, consider the method illustrated in Example 3.

1. Use a graphing calculator to locate the x-intercepts (or zeros) of the polynomial.
2. Use the Factor Theorem to infer a corresponding factor from each x-intercept (or zero).
3. Use the calculator's **TRACE** feature to evaluate other points that you may wish to investigate.

EXAMPLE 3

Graph the polynomial $f(x) = 2x^4 + x^3 - 6x^2 + x + 2$ to locate its x-intercepts (its zeros).

green diamond Y=

F2 6

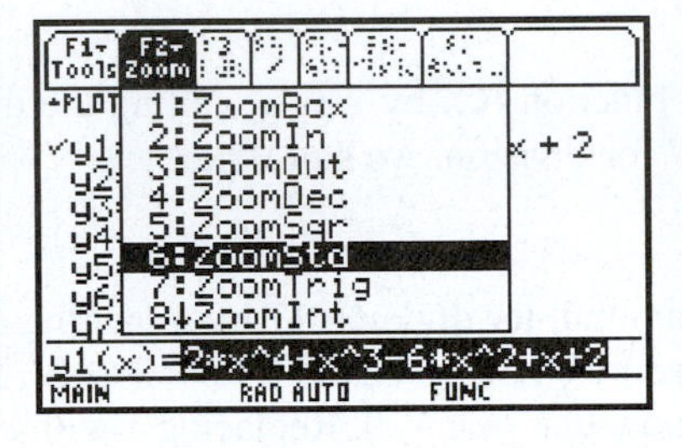

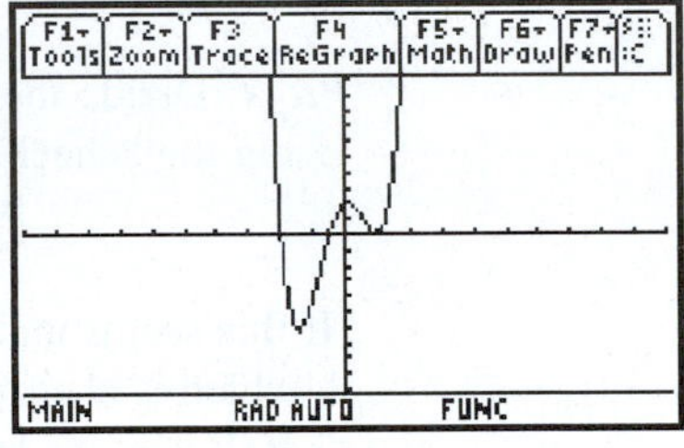

Y=

ZOOM 6

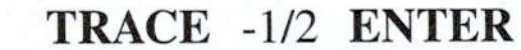

TRACE -1/2 ENTER

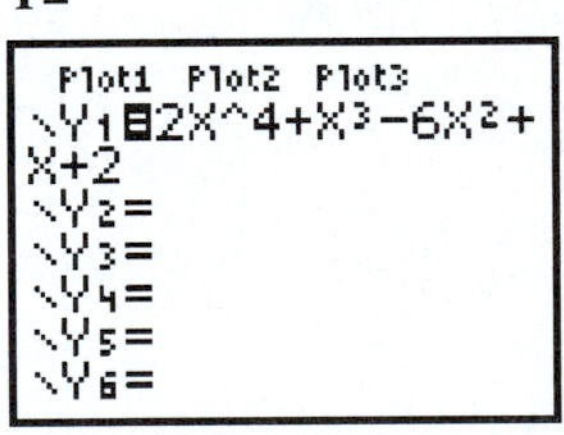

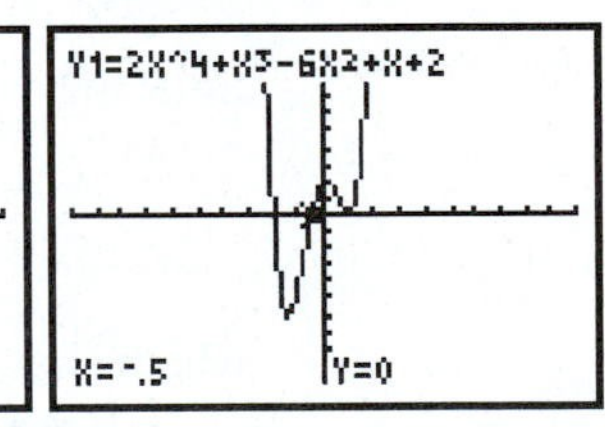

etc.

The x-intercepts are $(-2, 0)$, $\left(-\frac{1}{2}, 0\right)$, and $(1, 0)$, so $x - (-2)$, $x - \left(-\frac{1}{2}\right)$, and $x - 1$ are factors of the polynomial $2x^4 + x^3 - 6x^2 + x + 2$. At (1, 0) the graph just touches the x-axis rather than crossing it, so $x - 1$ should have an even exponent (probably 2). This leads us to infer that

$$2x^4 + x^3 - 6x^2 + x + 2 = 2\,(x + 2)\,(x + \tfrac{1}{2})\,(x - 1)^2$$

Note that the leading coefficient (in this case, 2) must be included as a constant factor to reproduce the original polynomial in factored form. Some people prefer to multiply $2\left(x + \frac{1}{2}\right) = 2x + 1$ giving

$$2x^4 + x^3 - 6x^2 + x + 2 = (x + 2)\,(2x + 1)\,(x - 1)^2$$

Exercises 16.1

Find the x-intercepts and the y-intercept of each polynomial function.

1. $f(x) = 3x(x-5)(x-2)$

2. $f(x) = (x+3)(x-2)(x+4)$

3. $f(x) = (x-4)(2x-5)(x-1)$

4. $f(x) = (3+x)(5-4x)(x-1)$

5. $f(x) = x^3 - 9x$

6. $f(x) = (x+1)(x^2-4)$

Graph each of the following functions using a graphing calculator. Use the graph's x-intercepts (or zeros) and the Factor Theorem to write the polynomial in factored form.

7. $f(x) = x^3 + 5x^2 - 2x - 24$

8. $f(x) = x^3 + x^2 - 4x - 4$

9. $f(x) = x^3 - x^2 - 9x + 9$

10. $f(x) = x^3 + x^2 - 14x - 24$

11. $f(x) = x^3 + 3x^2 - 4x$

12. $f(x) = x^3 - 2x^2 - 8x$

13. $f(x) = x^3 - 7x^2 + 15x - 9$

14. $f(x) = x^3 + x^2 - 5x + 3$

15. $f(x) = x^3 - 6x^2 + 32$

16. $f(x) = x^3 - 6x^2 + 9x$

17. $f(x) = x^4 - 2x^3 - 13x^2 + 14x + 24$

18. $f(x) = x^4 - 15x^2 + 10x + 24$

19. $f(x) = x^4 - 5x^3 + x^2 + 21x - 18$

20. $f(x) = x^4 + 5x^3 - x^2 - 17x + 12$

21. $f(x) = x^4 - 2x^3 - 8x^2 + 18x - 9$

22. $f(x) = x^4 - x^3 - 6x^2$

23. $f(x) = x^4 + 4x^3 - 2x^2 - 12x + 9$

24. $f(x) = x^4 - 6x^3 + x^2 + 24x + 16$

16.2 REAL SOLUTIONS OF POLYNOMIAL EQUATIONS

In the preceding section, polynomial equations of the form $f(x) = 0$ were solved by examining the graph of $f(x)$ and locating its x-intercepts. This method works reasonably well if the solutions are "nice" numbers such as integers and easy fractions. To find an irrational solution accurate to 6 or 10 significant digits, our graphical method can be used to get an initial estimate, but more powerful methods are needed to improve that estimate. Graphing calculators have built-in numerical estimation features designed specifically for this purpose. Near the end of this section, Newton's method will be explained and illustrated for those who would like to understand how the calculator is able to improve a reasonably good estimate of a solution.

EXAMPLE 1

Use a graphing calculator to solve the equation

$$x^3 + 1 = 3x$$

Step 1: Set one side of the equation equal to zero:

$$x^3 - 3x + 1 = 0$$

Step 2: Graph the left-hand side and estimate its zeros (locate its x-intercepts):

green diamond Y=

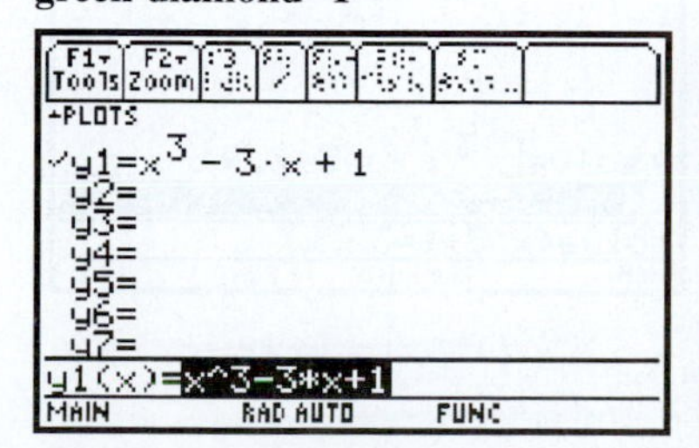

F2 4

(The zeros are near -2, 0.3, and 1.5.)

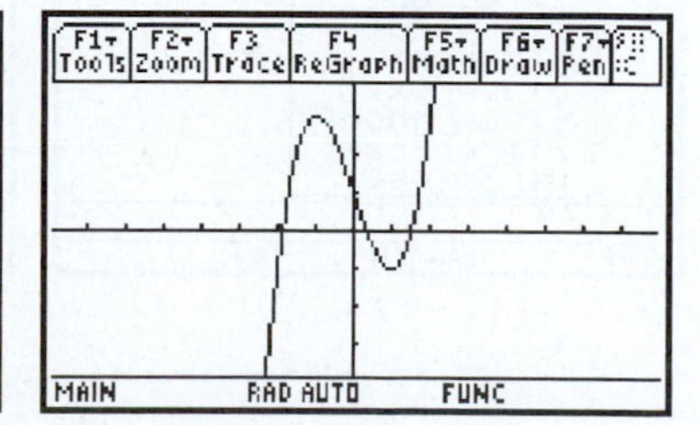

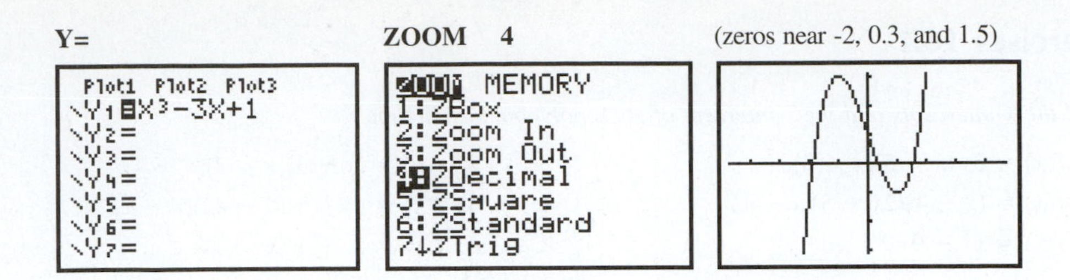

Step 3: Use the **Solve** feature to improve the estimates.

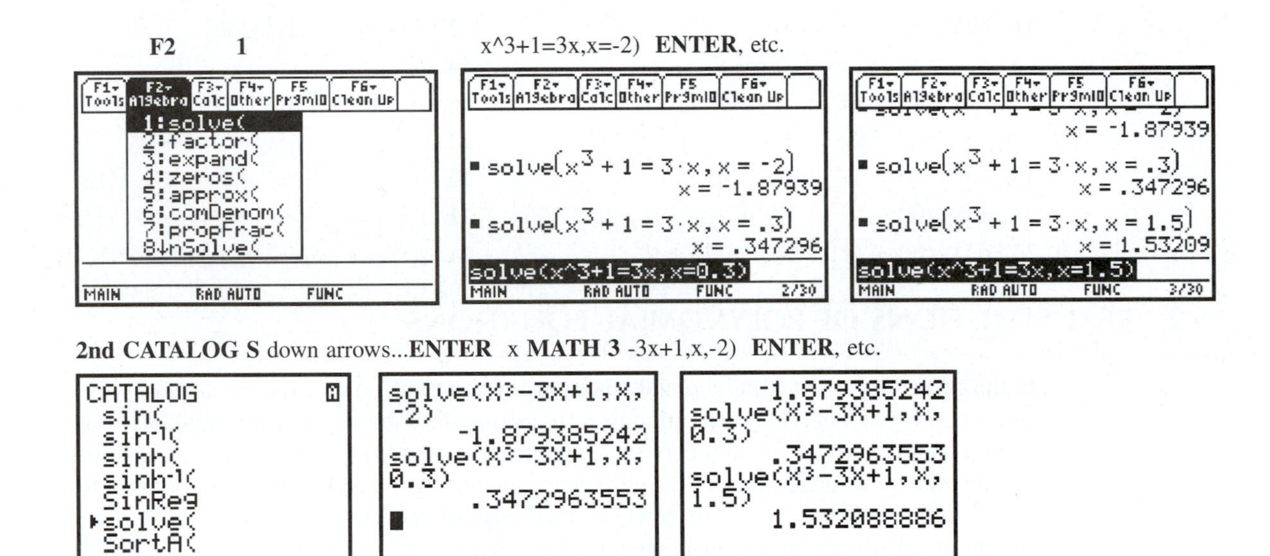

Note that the TI-89 requires that the entire *equation* be written, while the TI-83 Plus wants only the *polynomial* which has been set equal to zero. In each case, notice how the initial estimate is very significant in determining which solution will be found (wild guesses are not recommended, so don't skip Step 2).

The TI-89's **Solve** feature can also be used without a specific estimate. In that case, it will try to find all real solutions of the equation.

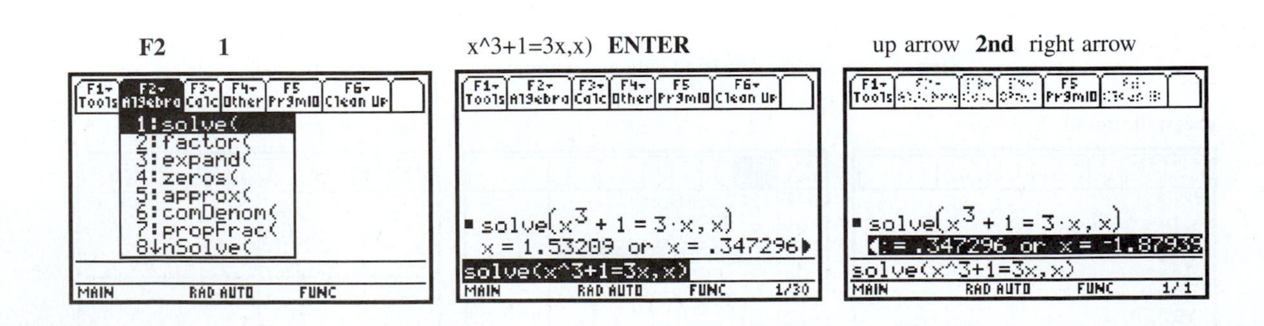

EXAMPLE 2

Use a graphing calculator to solve the equation $x^4 - 7x^3 + 3x - 2 = 0$.

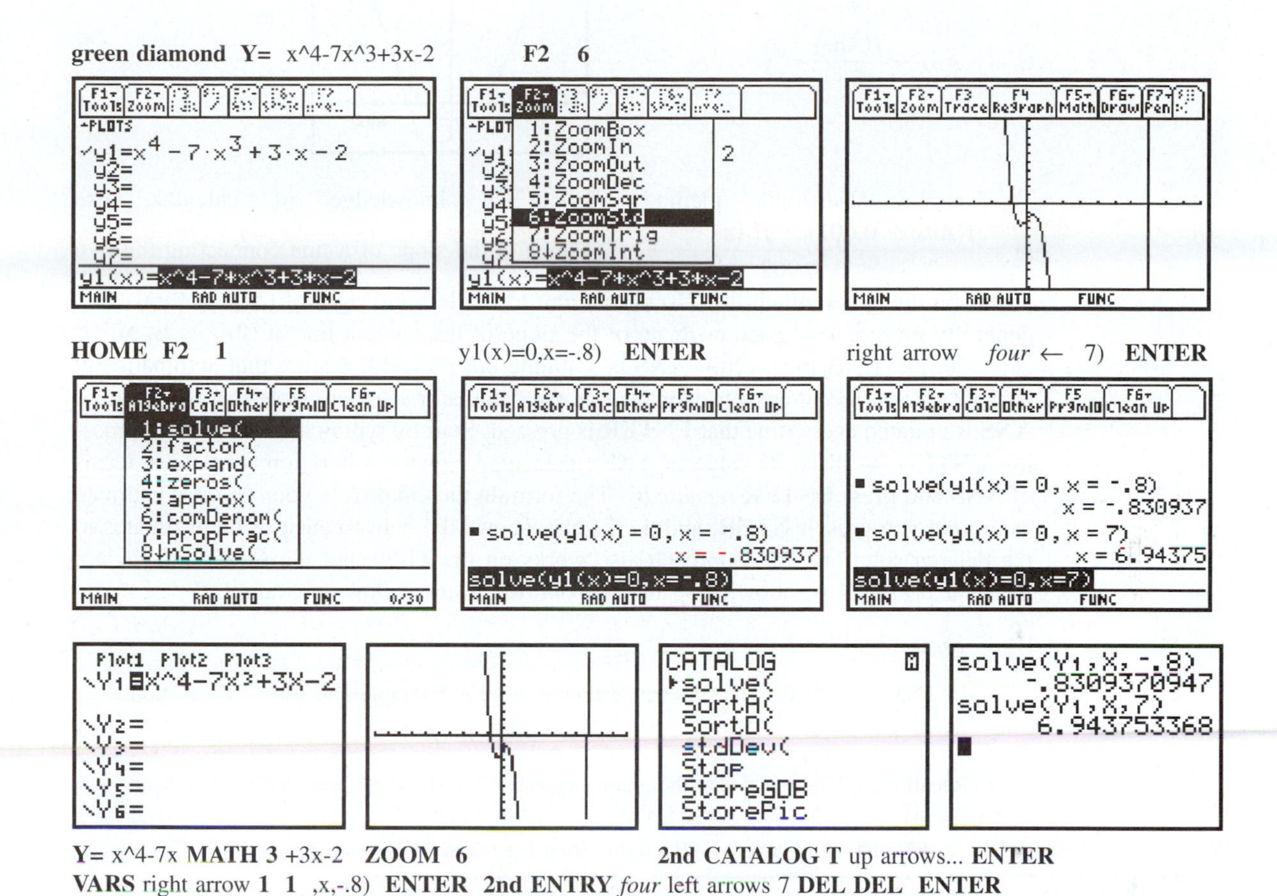

For further examples showing how graphing calculators can be used to solve equations numerically and graphically, see Appendix C, Sections C.6 and C.7 (TI-83 Plus), or Appendix D, Sections D.8 and D.10 (TI-89).

Newton's Method

This *optional* subsection is designed to give you an idea of what your calculator is doing with the initial estimates that you provide when you use the **Solve** feature. Sir Isaac Newton invented this famous method for improving estimated solutions. Though Newton's method is usually categorized as a numerical procedure, it is based on a powerful geometric insight. Let *Ans* represent an estimated solution of the equation $f(x) = 0$. Newton's method constructs the point $(Ans, f(Ans))$ and uses the x-intercept of its tangent line as the next estimate. Using the point-slope equation of the tangent line at $(Ans, f(Ans))$, we obtain

$$y - f(Ans) = m(x - Ans) \quad (m \text{ is the slope of the tangent line.})$$

$$0 - f(Ans) = m(x - Ans) \quad (\text{Set } y = 0 \text{ to find the } x\text{-intercept.})$$

$$-\frac{f(Ans)}{m} = x - Ans$$

$$Ans - \frac{f(Ans)}{m} = x$$

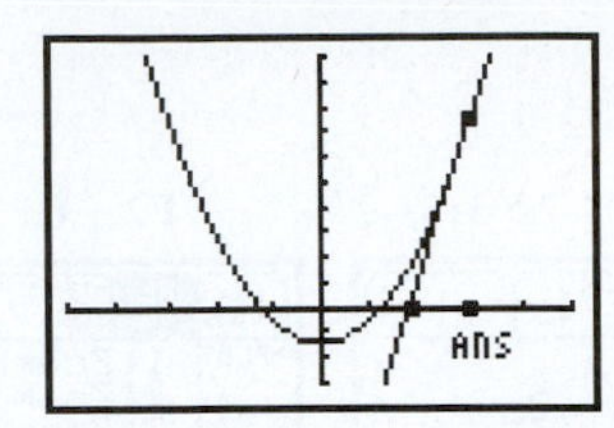

To employ Newton's method without a knowledge of calculus, use $m = \dfrac{f(Ans + 0.001) - f(Ans - 0.001)}{0.002}$, which is the slope of a line connecting the two points on the curve which lie 1/1000 of a unit to the left and right of $(Ans, f(Ans))$. This generally gives a very good estimate of the slope of the tangent line at $(Ans, f(Ans))$.

On the TI-83 Plus, (**2nd**) **ANS** is a simple but powerful feature that automatically stores the most recent answer displayed on the calculator's screen. Note that the value of **ANS** is updated every time that **ENTER** is pressed. Start by typing an estimate and pressing **ENTER** (to place its value in **ANS**), then type Newton's iteration formula in terms of **ANS** and press **ENTER** *repeatedly*. The formula thus improves your *updated* estimate each time you press **ENTER**, and the display shows the convergence of the estimates to the desired solution. This approach also works on the TI-89, but the screen images are messy and harder to follow (they are thus omitted in the following example).

EXAMPLE 3

Use Newton's method to refine the estimates used in Example 1 to solve the equation

$$x^3 - 3x + 1 = 0$$

Recall from Example 1 that the graph suggested that the zeros were near −2, 0.3, and 1.5.

Y= **2nd QUIT** -2 **ENTER** **2nd ANS** - **VARS** right arrow **1 1** (**2nd ANS**), etc. **ENTER**, **ENTER**, etc.

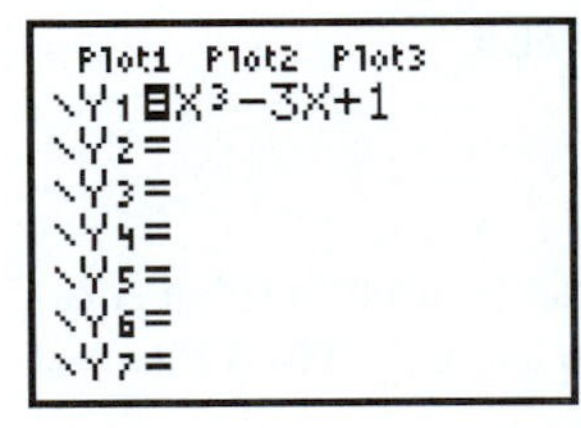

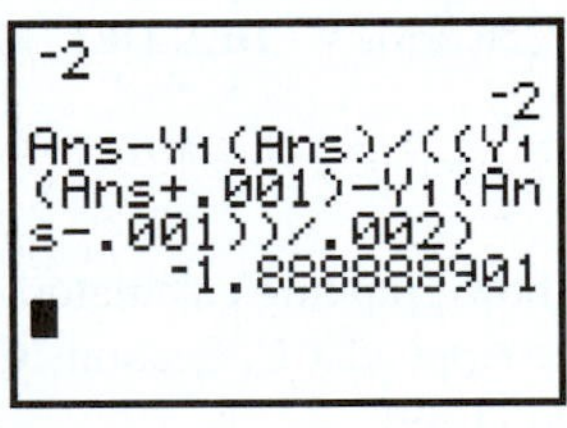

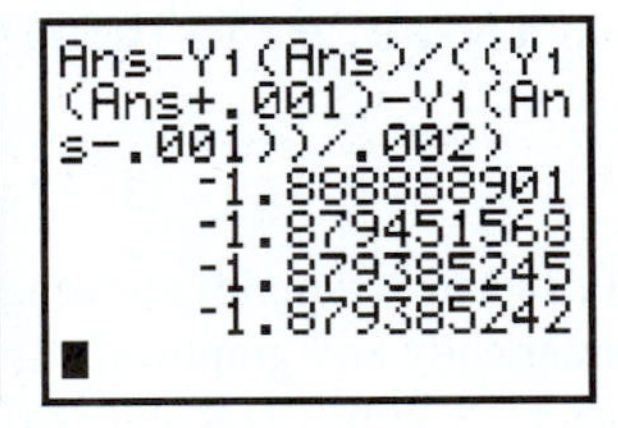

(for the estimate −2)

(for the estimate 0.3)

(for the estimate 1.5)

After entering a new estimate, you can bring back the iteration formula by pressing **2nd ENTRY** *twice*.

Comparing our results with Example 1 shows that the estimated solutions were improved to 10 significant digits by using only a few iterations (repetitions). This is typical of Newton's method when the solutions are of multiplicity 1. Zeros of higher multiplicities are more difficult to estimate accurately because the tangent lines flatten out to a nearly horizontal position as the solution is approached. Unfortunately, this yields a nearly zero slope, which happens to occupy the denominator of the iteration formula. Just as a zero denominator would make the iteration formula undefined, a denominator very close to zero can cause difficulties with numeric accuracy.

Exercises 16.2

Use a graph to find estimates for the real solutions of each equation, then use the **Solve** *feature of your graphing calculator (or Newton's method) to improve your estimates to at least six significant digits.*

1. $x^3 + x^2 - 5x + 2 = 0$

2. $x^3 - 5x^2 + 2x + 7 = 0$

3. $2x^3 - 5x^2 + x + 1 = 0$

4. $3x^3 - 7x^2 + 2x + 1 = 0$

5. $2x^3 - x^2 + x - 1 = 0$

6. $x^3 - 7x^2 + 15x - 8 = 0$

7. $x^4 + x^3 - 11x^2 - 9x + 21 = 0$

8. $x^4 - 2x^3 - x^2 + 3x - 1 = 0$

9. $x^4 - x^3 - 4x^2 - 1 = 0$

10. $x^4 - 5x^3 + x^2 + 25x - 28 = 0$

16.3 COMPLEX SOLUTIONS OF POLYNOMIAL EQUATIONS

Next, we extend the definition of a polynomial function of degree n,

$$f(x) = a_nx^n + a_{n-1}x^{n-1} + \cdots + a_1x + a_0, \quad a_n \neq 0$$

to include complex number coefficients $a_n, a_{n-1}, \ldots, a_1, a_0$.

The next theorem is one of the most important in the development of the theory of equations.

FUNDAMENTAL THEOREM OF ALGEBRA

A polynomial function $f(x)$ of degree $n \geq 1$ has at least one complex zero (this zero may be a real number).

That is, every polynomial equation has at least one complex solution. The graph of the polynomial will touch or cross the x-axis at real solutions; it will not touch or cross the x-axis at imaginary solutions. *Note:* The Fundamental Theorem of Algebra addresses only the existence of such solutions and does not address how to find them.

THEOREM

Every polynomial function $f(x)$ of degree $n \geq 1$, where x is a complex number, can be expressed as a product of its leading coefficient (a constant) and n linear factors. That is,

$$\text{if} \quad f(x) = a_nx^n + a_{n-1}x^{n-1} + \cdots + a_1x + a_0$$
$$\text{then} \quad f(x) = a_n(x - c_1)(x - c_2) \cdots (x - c_n)$$

If a factor $x - c$ appears k times in such a linear factorization, then c is called a zero of *multiplicity k*.

THEOREM

A polynomial function $f(x)$ of degree $n \geq 1$ with complex number coefficients has precisely n zeros if a zero of multiplicity k is counted k times.

EXAMPLE 1

Find the number of complex solutions of each polynomial equation (counting multiplicities).

(a) The equation $6x^5 + 8x^4 - 3x^2 + 7x - 11 = 0$ has precisely five complex solutions.
(b) The equation $8(x - 5)(x - 5)(x + 3j)(x - 3j) = 0$ has precisely four complex solutions.

Recall that each complex number $a + bj$ has a conjugate $a - bj$. The next theorem states that complex solutions of a polynomial equation with real coefficients occur in conjugate pairs.

THEOREM

If a polynomial equation $f(x) = 0$ has real coefficients and $a + bj$ is a solution, then its conjugate $a - bj$ is also a solution (complex zeros occur in conjugate pairs).

EXAMPLE 2

Given that $2 - j$ is a solution of the equation $x^4 - 5x^3 + 7x^2 + 3x - 10 = 0$, find all of its complex solutions.

Since $2 - j$ is a solution, so is $2 + j$. Thus, by the Factor Theorem, we know that $x - (2 - j)$ and $x - (2 + j)$ are both factors. Multiplying, we find

$$\begin{aligned}[x - (2 - j)]\,[x - (2 + j)] &= x^2 - (2 + j)\,x - (2 - j)\,x + (2 - j)(2 + j) \\ &= x^2 - 2x - jx - 2x + jx + 4 + 2j - 2j - j^2 \\ &= x^2 - 4x + 4 - (-1) \qquad \text{(Recall that } j^2 = -1.) \\ &= x^2 - 4x + 5\end{aligned}$$

Dividing the original polynomial by $x^2 - 4x + 5$ gives

(To review polynomial division, see Section 1.9.)

$$\begin{array}{r} x^2 - x \quad - 2 \\ x^2 - 4x + 5 \overline{)\, x^4 - 5x^3 + 7x^2 + 3x - 10} \\ \underline{x^4 - 4x^3 + 5x^2} \qquad\quad \\ -x^3 + 2x^2 + 3x \qquad \\ \underline{-x^3 + 4x^2 - 5x} \qquad \\ -2x^2 + 8x - 10 \\ \underline{-2x^2 + 8x - 10} \end{array}$$

So, the equation can be written as

$$\begin{aligned} x^4 - 5x^3 + 7x^2 + 3x - 10 &= 0 \\ (x^2 - x - 2)\,(x^2 - 4x + 5) &= 0 \\ (x - 2)(x + 1)\,[x - (2 - j)]\,[x - (2 + j)] &= 0 \end{aligned}$$

$$\text{so,} \quad x = 2 \quad \text{or} \quad x = -1 \quad \text{or} \quad x = 2 - j \quad \text{or} \quad x = 2 + j$$

Using a calculator, we have

F2 alpha A 1 x^4–5x^3+7x^2+3x–10=0,x) **ENTER** up arrow **2nd** right arrow

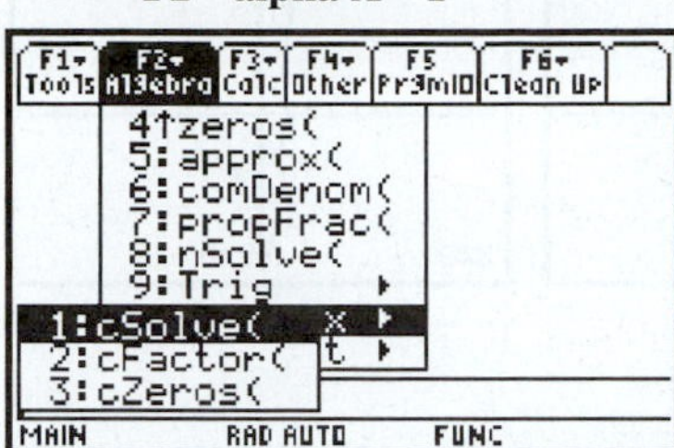

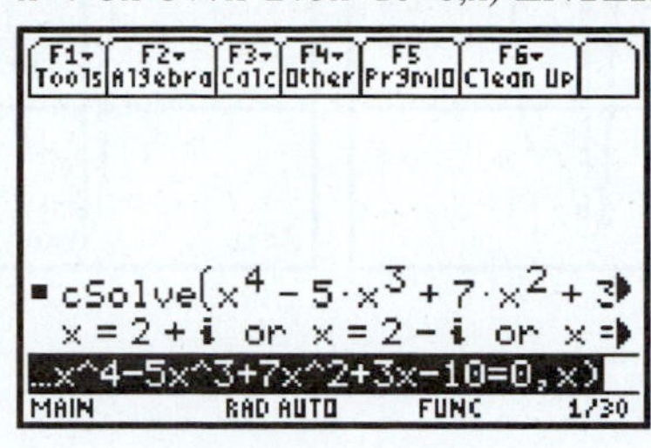

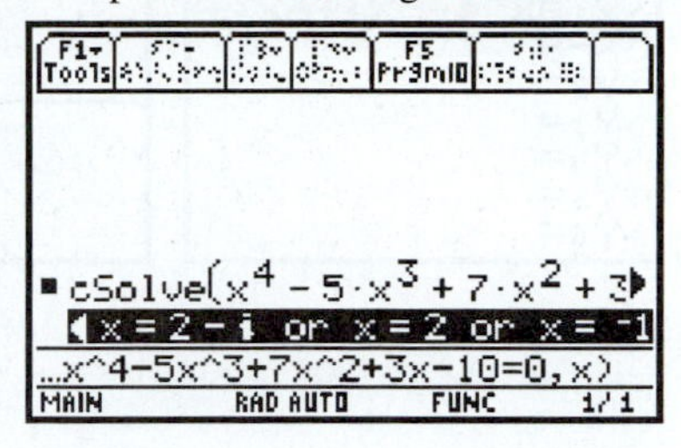

Note that the TI-89 uses **i** for the imaginary unit (instead of j).

The graph can be used to confirm that -1 and 2 are real zeros (x-intercepts) of the polynomial.

green diamond Y= x^4-5x^3+7x^2+3x-10 **F2 6**

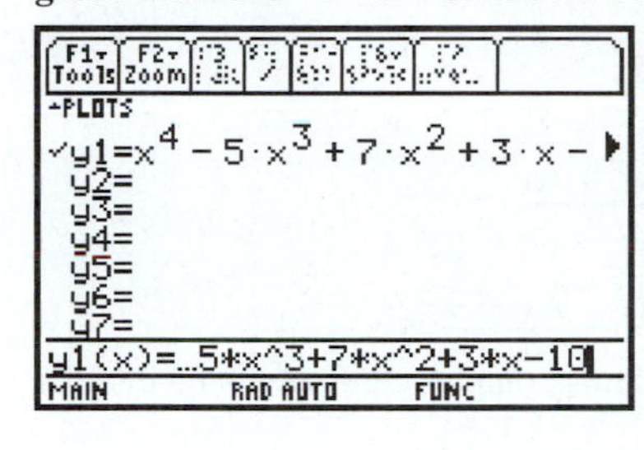

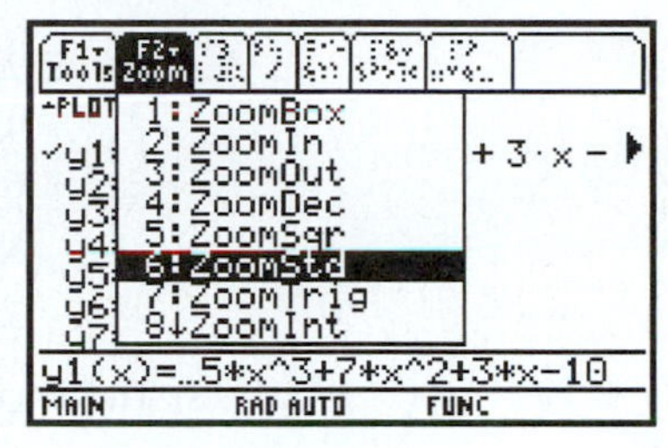

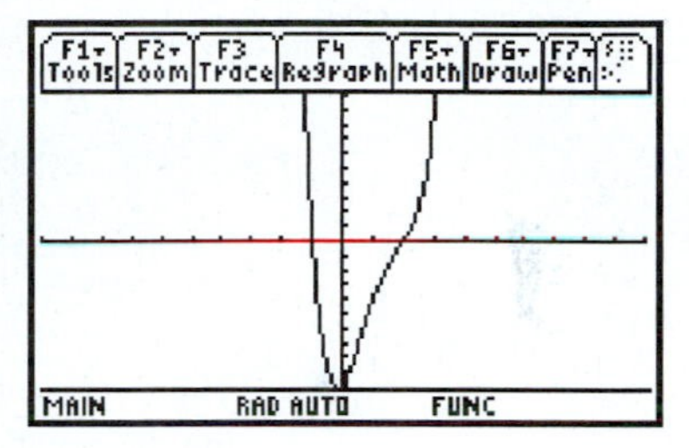

EXAMPLE 3

From the graph of $f(x) = x^4 - 5x^3 + 6x^2 + 3x - 9$, we can see that -1 and 3 are real zeros of the function. From this information, find the remaining complex roots.

green diamond Y= x^4-5x^3+6x^2+3x-9 **F2 6**

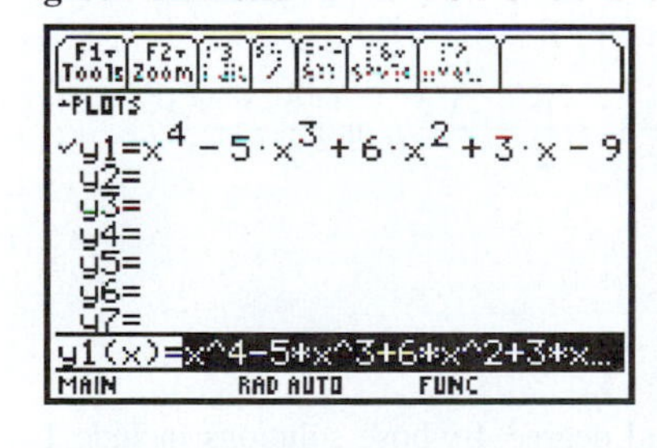

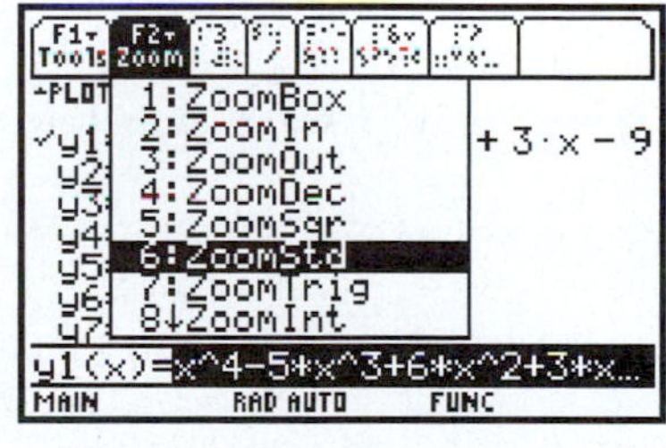

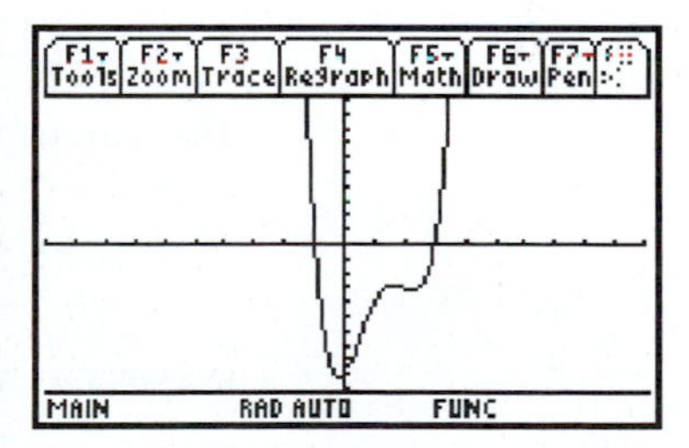

F3 (Trace) -1 **ENTER** 3 **ENTER**

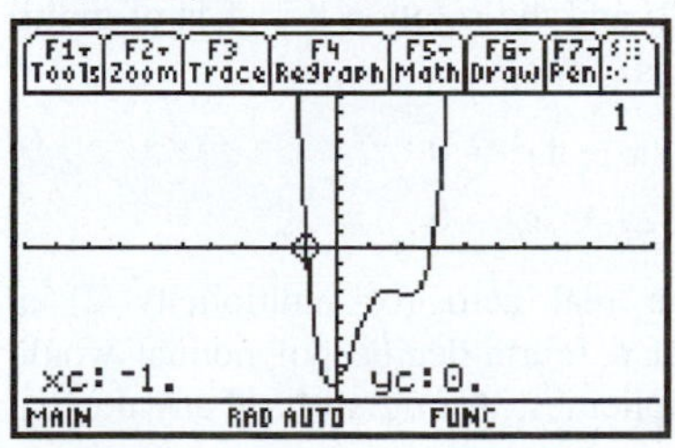

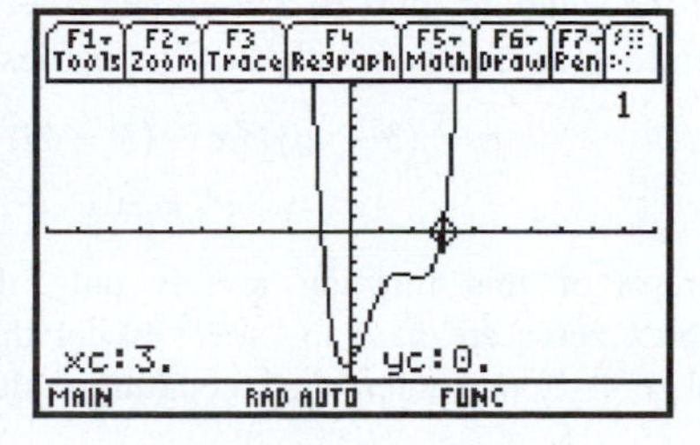

Y= x^4-5x **MATH 3** +6x $\mathbf{x^2}$ +3x-9 **ZOOM 6** **TRACE** -1 **ENTER** 3 **ENTER**

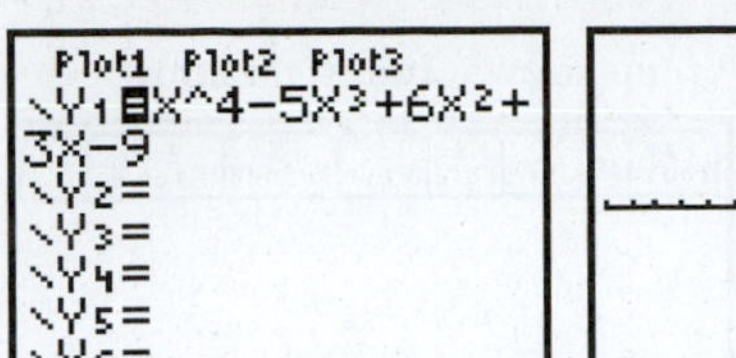

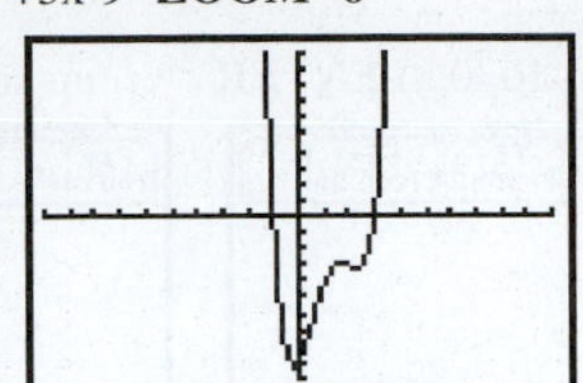

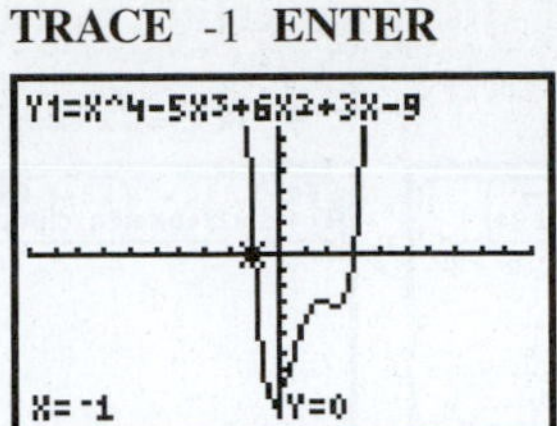

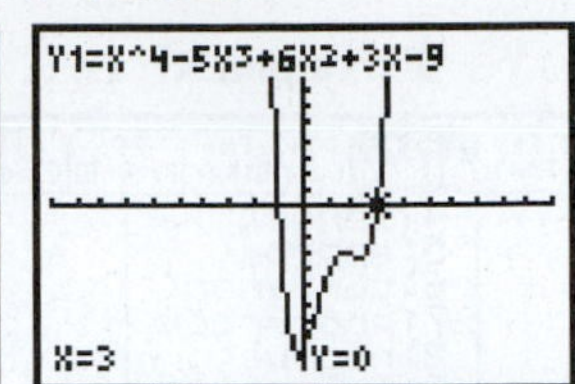

By the Factor Theorem, we know that $x - (-1) = x + 1$ and $x - 3$ are factors of $f(x)$. Multiplying them gives

$$(x + 1)(x - 3) = x^2 - 3x + x - 3 = x^2 - 2x - 3$$

Now, divide by $x^2 - 2x - 3$ to reveal the other factor of $f(x)$.

$$\begin{array}{r}
x^2 - 3x + 3 \\
x^2 - 2x - 3 \overline{\smash{\big)}\, x^4 - 5x^3 + 6x^2 + 3x - 9} \\
\underline{x^4 - 2x^3 - 3x^2} \\
-3x^3 + 9x^2 + 3x \\
\underline{-3x^3 + 6x^2 + 9x} \\
3x^2 - 6x - 9 \\
\underline{3x^2 - 6x - 9}
\end{array}$$

Thus, $f(x) = (x + 1)(x - 3)(x^2 - 3x + 3)$. The remaining complex zeros are found by solving the equation

$$x^2 - 3x + 3 = 0$$

$$x = \frac{-(-3) \pm \sqrt{(-3)^2 - 4(1)(3)}}{2(1)}$$

$$x = \frac{3 \pm \sqrt{-3}}{2}$$

$$x = \frac{3}{2} \pm \frac{\sqrt{3}}{2}j$$

The zeros of $f(x) = x^4 - 5x^3 + 6x^2 + 3x - 9$ are -1, 3, $\frac{3}{2} + \frac{\sqrt{3}}{2}j$, and $\frac{3}{2} - \frac{\sqrt{3}}{2}j$.

EXAMPLE 4

Find a polynomial equation with real coefficients and degree 4 whose solutions include 1 (multiplicity 2) and $3 + 2j$.

Two of the solutions must be $3 + 2j$ and $3 - 2j$ and the solution $x = 1$ is of multiplicity 2. Therefore, an equation with these properties is

$$[x - (3 + 2j)]\,[x - (3 - 2j)]\,(x - 1)^2 = 0$$
$$x^4 - 8x^3 + 26x^2 - 32x + 13 = 0$$

The graph of this function reveals only the real zero (of multiplicity 2) at $x = 1$. Complex zeros are hard to "see," except that a fourth-degree polynomial would have four real zeros (four x-intercepts), counting multiplicities, if it didn't have any nonreal zeros.

green diamond Y= x^4-8x^3+26x^2-32x+13 **F2 6**

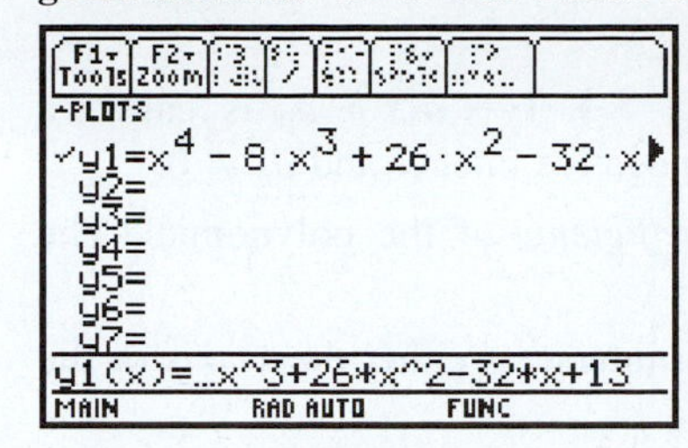

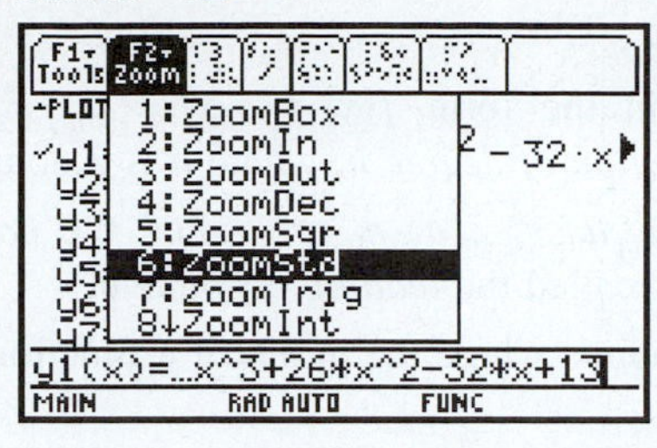

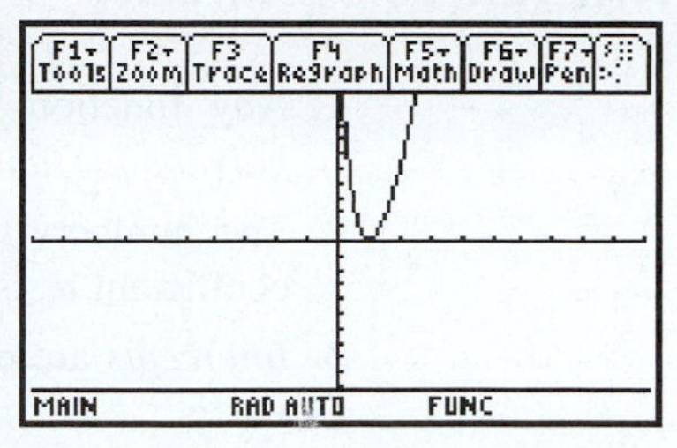

Exercises 16.3

State the number of complex zeros (counting multiplicities) of each polynomial equation.

1. $x^5 - 3x^4 + 7x - 2 = 0$

2. $x^3 - 5x^2 + 8x + 19 = 0$

3. $9x^4 - 16 = 0$

4. $8x^3 + 125 = 0$

5. $2x^{10} - x^5 + 7x - 5 = 0$

6. $3x^8 - 4x^7 + 2x - 11 = 0$

Using the given information, find all complex solutions of each equation.

7. $x^3 - 6x^2 + x + 34 = 0$; $4 + j$ is a solution.

8. $x^3 + 6x^2 - 14x - 104 = 0$; $-5 - j$ is a solution.

9. $x^4 - 2x^3 + 3x^2 - 2x + 2 = 0$; j and $1 - j$ are solutions.

10. $x^4 - 2x^3 + 26x^2 - 32x + 160 = 0$; $-4j$ and $1 - 3j$ are solutions.

11. $x^5 - 3x^4 + 2x^3 - 6x^2 + x - 3 = 0$; j (multiplicity 2) and 3 are solutions.

12. $x^4 - 4x^3 + 10x^2 - 12x + 5 = 0$; 1 (multiplicity 2) and $1 + 2j$ are solutions.

13. $x^4 - 4x^3 + 10x^2 + 68x - 75 = 0$; $3 - 4j$ is a solution.

14. $x^4 - x^3 + 6x^2 + 14x - 20 = 0$; $1 + 3j$ is a solution.

Find a polynomial equation $f(x) = 0$ with real coefficients that satisfies the given conditions.

15. Degree 3, solutions include 4 and $2j$.

16. Degree 3, solutions include 1 and $2 - 3j$.

17. Degree 4, solutions include 2, -1, and $1 + j$.

18. Degree 4, solutions include $\frac{3}{2}$ (multiplicity 2), 1, and $-\frac{1}{2}$.

19. Degree 4, solutions include $1 + j$ and $1 - 2j$.

20. Degree 4, solutions include 6 (multiplicity 2) and j.

21. Degree 5, solutions include 0, $1 - 3j$, and $2j$.

22. Degree 6, solutions include 1 (multiplicity 3), $\frac{2}{3}$, and $-j$.

23. Why is it impossible to have a polynomial equation with real coefficients of degree 4 with solutions 6, 2, $-\frac{2}{3}$, and $2j$?

24. Why is it impossible to have a polynomial equation with real coefficients of degree 5 with solutions 4 (multiplicity 3), -6, and $1 + 2j$?

25. Why is it impossible to have a polynomial equation with real coefficients of degree 3 with solutions $6j$ and $3 - 4j$?

CHAPTER 16 SUMMARY

1. Any function in the form $f(x) = a_n x^n + a_{n-1}x^{n-1} + \cdots + a_1 x + a_0$ is called a *polynomial function of degree n*, where n is a nonnegative integer and $a_n \neq 0$.
2. The numbers $a_n, a_{n-1}, \ldots, a_1, a_0$ are called the *coefficients* of the polynomial. The coefficient a_n is called the *leading coefficient*.
3. *Intercepts* are points where the graph of a function intersects either the x-axis or the y-axis.
   (a) To find the x-intercept(s), substitute $y = f(x) = 0$ into the function's equation and solve for x. Each resulting value of x gives an x-intercept of the form $(x, 0)$.
   (b) To find the y-intercept, substitute $x = 0$ into the function's equation and evaluate the corresponding y-value, namely $f(0)$. The resulting point $(0, f(0))$ is the y-intercept.
4. Values of x that satisfy the equation $f(x) = 0$ are called *zeros* of the function $f(x)$.
5. *Factor Theorem:* If the point $(a, 0)$ is an x-intercept (if a is a zero) of the polynomial function $f(x)$, then $x - a$ is a factor of $f(x)$.
6. *Remainder Theorem*: If a polynomial $f(x)$ is divided by $x - a$ until the result is a constant remainder R, then $f(a) = R$.
7. The *multiplicity* of a zero is the exponent of its corresponding factor in the polynomial's factored form.
8. The graph of a polynomial *crosses* the x-axis at a zero of odd multiplicity, but it just *touches* the x-axis at a zero of even multiplicity.
9. Method for investigating a polynomial, its graph, and its (real) factors:
   (a) Use a graphing calculator to locate the x-intercepts (or zeros) of the polynomial.
   (b) Use the Factor Theorem to infer a corresponding factor from each x-intercept (or zero).
   (c) Use the calculator's **TRACE** feature to evaluate other points that you may wish to investigate.
10. Method for using a graphing calculator to accurately estimate real solutions of a polynomial equation:
    (a) Set one side of the equation equal to zero.
    (b) Use your calculator to graph the polynomial function on the other side of the equation.
    (c) Use the graph to get reasonable estimates of the zeros of the polynomial function.
    (d) Use the calculator's **Solve** feature (or Newton's method) to improve your estimates.
11. *Fundamental Theorem of Algebra*: A polynomial of degree $n \geq 1$ has at least one complex zero (this zero may be a real number).
12. Every polynomial function $f(x)$ of degree $n \geq 1$, where x is a complex number, can be expressed as a product of its leading coefficient (a constant) and n linear factors. That is,

$$\text{if} \quad f(x) = a_n x^n + a_{n-1}x^{n-1} + \cdots + a_1 x + a_0$$
$$\text{then} \quad f(x) = a_n(x - c_1)(x - c_2)\cdots(x - c_n)$$

13. A polynomial function $f(x)$ of degree $n \geq 1$ with complex number coefficients has precisely n zeros if a zero of multiplicity k is counted k times.
14. *Complex conjugate solutions*: If a polynomial equation $f(x) = 0$ has real coefficients and $a + bj$ is a solution, then its conjugate $a - bj$ is also a solution (complex zeros occur in conjugate pairs).

CHAPTER 16 REVIEW

Find the degree of each polynomial.

1. $3x^2 - 8x^4 + x^7$

2. $-9x^5 + 2x^3 + 84$

3. $4x^8 - x^{12} + 3x^7 - 8$

Find the x- and y-intercepts of each function.

4. $f(x) = 4x(x + 7)(2x - 1)$

5. $f(x) = (x - 2)(x^2 + 6x + 9)$

6. $f(x) = x^3 - 6x^2$

7. $f(x) = 2x^3 + 3x^2 - 44x - 105$

8. $f(x) = x^3 - 3x^2 - 22x + 24$

Graph each function using a graphing calculator. Use the graph's x-intercepts (or zeros) and the Factor Theorem to write the polynomial in factored form.

9. $f(x) = x^3 - 7x^2 + 14x - 8$

10. $f(x) = x^3 + 4x^2 + x - 6$

11. $f(x) = x^4 + 9x^3 + 9x^2 - 49x + 30$

12. $f(x) = x^4 - 4x^3 - 7x^2 + 34x - 24$

13. $f(x) = x^4 + 3x^3 - 13x^2 - 51x - 36$

14. $f(x) = 2x^4 - 35x^3 + 166x^2 + 77x - 1470$

Find the zeros and their multiplicity for each function.

15. $f(x) = (2x + 9)^2(x - 4)^3$

16. $f(x) = (x - 6)(x + 10)^3$

17. $f(x) = x^4(x - 1)^2$

18. $f(x) = (x - 3)(x + 5)^2$

Use a graph to find estimates for the real solutions of each equation, then use the solve feature of your graphing calculator (or Newton's method) to improve your estimates to at least six significant digits.

19. $x^2 + 3x - 1 = 0$

20. $x^3 - 5x^2 + 6x - 3 = 0$

21. $2x^3 - 8x^2 + 3x + 5 = 0$

22. $4x^3 - x^2 + 2x - 8 = 0$

23. $x^3 + 2x^2 - 5x + 1 = 0$

24. $5x^3 - x^2 - 8x + 3 = 0$

State the number of complex zeros (counting multiplicities) of each polynomial equation.

25. $x^8 - 3x^5 + 2x - 5 = 0$

26. $7x^5 - 8x^3 + 2x - 10 = 0$

Using the given information, find all complex solutions of each equation.

27. $x^3 - 11x^2 + 40x - 50 = 0$; $3 + j$ is a solution.

28. $x^3 - 7x^2 + 25x - 39 = 0$; $2 - 3j$ is a solution.

29. $x^4 + 8x^3 + 25x^2 + 72x + 144 = 0$; -4 (multiplicity 2) and $-3j$ are solutions.

30. $x^4 - 2x^3 - 38x^2 + 150x - 175 = 0$; $2 + j$ is a solution.

Find a polynomial equation $f(x) = 0$ with real coefficients that satisfies the given conditions.

31. Degree 3, solutions include 8 and $7j$

32. Degree 3, solutions include -4 and $3 - 2j$

33. Degree 4, solutions include $2j$ and $1 + j$

34. Degree 5, solutions include $\frac{3}{5}$ (multiplicity 2), $\frac{1}{2}$, and j

17 Inequalities and Absolute Value

INTRODUCTION

An inequality is a way of expressing a relationship between two quantities when one is larger than the other. Usually an infinite number of values satisfy an inequality. Businesses use systems of inequalities to solve problems such as maximizing profits or minimizing costs.

The concept of absolute value helps us to express the distance between two objects or the size of an object. Absolute value equations and inequalities occur often in the development of theories that are used to solve applied mathematical problems.

Objectives

- Use the trichotomy property of real numbers.
- Graph linear inequalities in one variable.
- Solve simple and compound linear inequalities in one variable.
- Solve absolute value equations and inequalities.
- Solve nonlinear inequalities in one variable.
- Solve inequalities of rational expressions in one variable.
- Solve inequalities in two variables.

17.1 INEQUALITIES

Earlier, we used such statements as $x \geq 0$, $y < 4$, and $0° \leq \theta < 360°$. These are examples of simple inequalities. We now need to study more about inequalities and in some depth.

An **inequality** is a statement in one of the following forms:

1. $a < b$: One quantity is *less than* another.
2. $a > b$: One quantity is *greater than* another.
3. $a \leq b$: One quantity is *less than or equal to* another.
4. $a \geq b$: One quantity is *greater than or equal to* another.

It is also true that $a < b$ if $b - a$ is positive and $a > b$ if $a - b$ is positive.

First, we have two basic properties of inequalities.

1. **Trichotomy property:** For any two real numbers a and b, **exactly one** of the following is true
   (a) $a < b$
   (b) $a = b$
   (c) $a > b$
2. **Transitive property:** If a, b, and c are real numbers such that $a < b$ and $b < c$, then $a < c$.

Such properties are called **order** properties.

The **sense** of an inequality refers to the direction (greater than or less than) in which the inequality sign points. The inequalities $a < b$ and $c < d$ or $e > f$ and $g > h$ have the *same sense,* while $a < b$ and $e > f$ have *opposite* senses.

Graphing Inequalities

All real numbers greater than 4 are solutions of the inequality $x > 4$. Its graph on the number line is shown in Fig. 17.1.

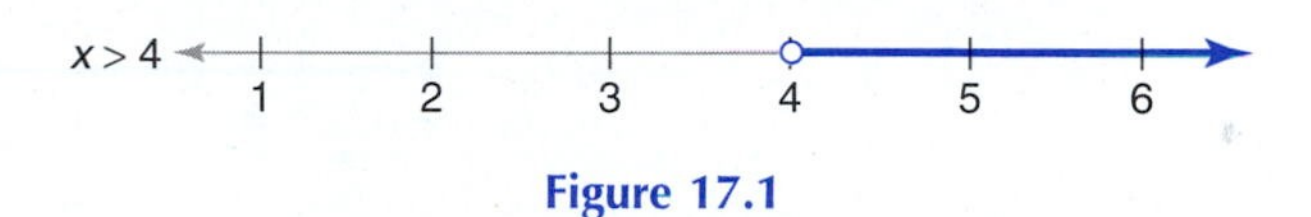

Figure 17.1

Note: An open circle shows that the endpoint is *not* included.

All real numbers greater than or equal to 11 are solutions of the inequality $x \geq 11$. Its graph is shown in Fig. 17.2.

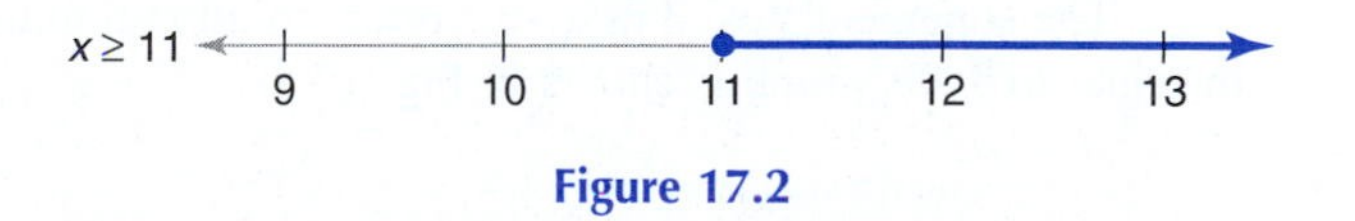

Figure 17.2

Note: A closed circle shows that the endpoint *is* included.

All real numbers less than -5 are solutions of the inequality $x < -5$. Its graph is shown in Fig. 17.3.

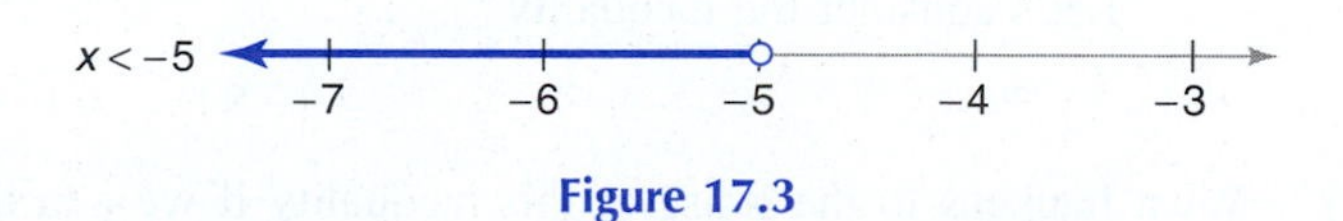

Figure 17.3

Similarly, the graph of $x \leq 7$ is given in Fig. 17.4.

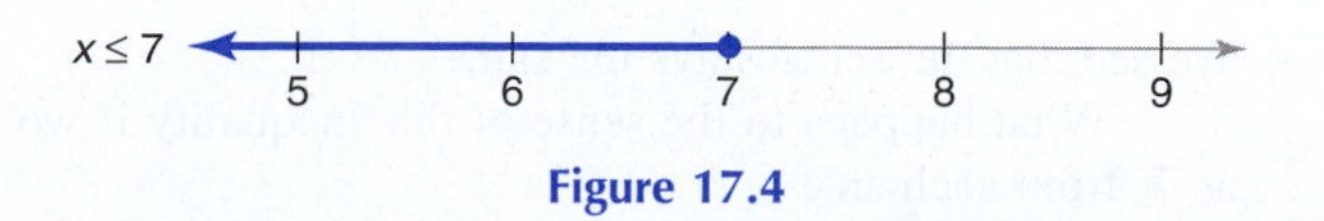

Figure 17.4

All real numbers greater than 6 *and* less than 10 are solutions of the inequality $6 < x < 10$. Its graph is shown in Fig. 17.5.

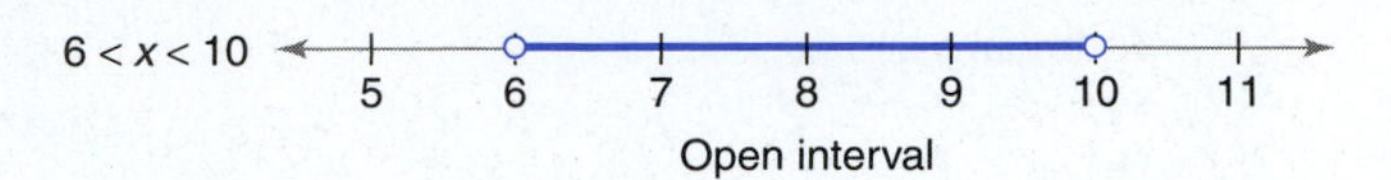

Figure 17.5

This is called an **open interval.** The endpoints are *not* included.

Similarly, the graph of $-4 \le x \le 2$ is given in Fig. 17.6.

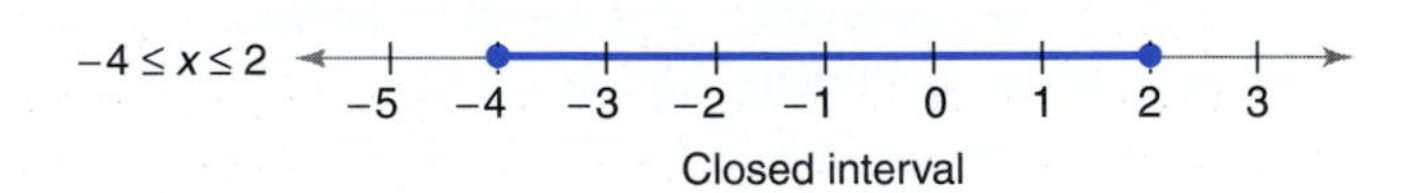

Figure 17.6

This is called a **closed interval.** The endpoints *are* included.

Examples of **half-open intervals** are given in Fig. 17.7.

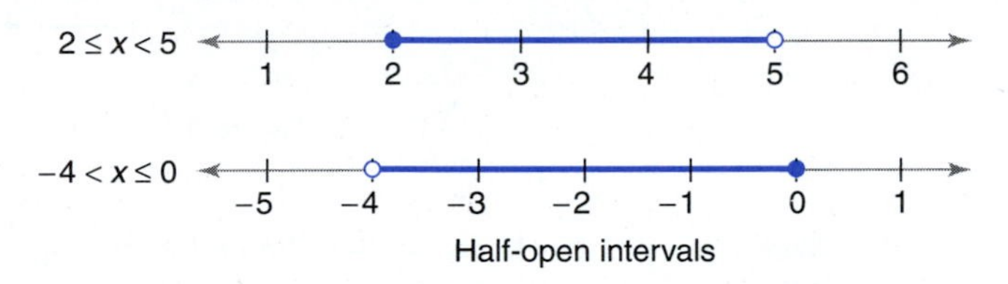

Figure 17.7

The statement $x < 3$ or $x \ge 7$ refers to all real numbers less than 3 *or* greater than or equal to 7. Its graph is shown in Fig. 17.8.

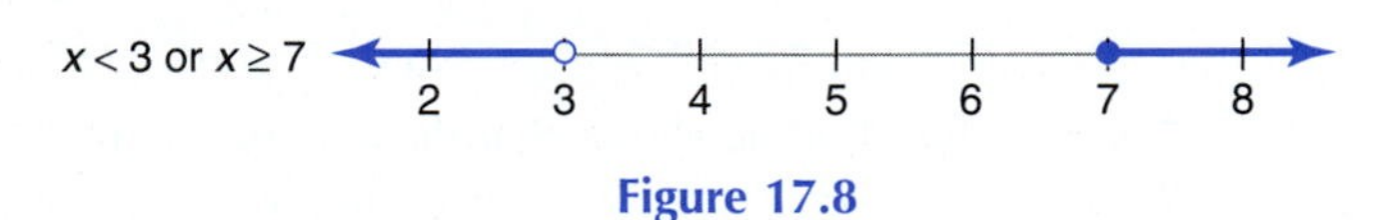

Figure 17.8

Let's consider the inequality

$$3 < 8$$

What happens to the sense of this inequality if we add the same number, such as 4, to each side?

$$3 + 4 < 8 + 4$$
$$7 < 12$$

We see that the sense stays the same.

What happens to the sense of this inequality if we subtract the same number, such as 7, from each side?

$$3 < 8$$
$$3 - 7 < 8 - 7$$
$$-4 < 1$$

Again, the sense stays the same.

If we consider subtracting a number to be the same as adding its opposite, we may summarize this discussion as follows:

ADDITION PROPERTY

If a, b, and c are real numbers and

$$a < b$$

then

$$a + c < b + c$$

In words, *the same number may be added to each side of an inequality and the sense remains the same.*

EXAMPLE 1

Solve $3x - 4 > 2x + 1$.

$$3x - 4 > 2x + 1$$
$$3x - 4 - 2x > 2x + 1 - 2x \quad \text{(Subtract } 2x \text{ from each side.)}$$
$$x - 4 > 1$$
$$x - 4 + 4 > 1 + 4 \quad \text{(Add 4 to each side.)}$$
$$x > 5$$

The graph of this solution on the number line is shown in Fig. 17.9.

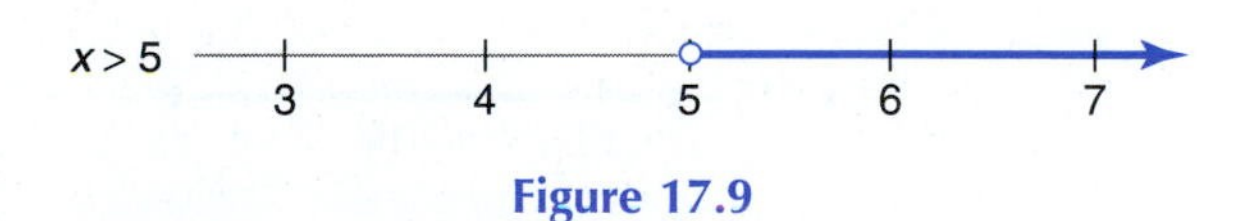

Figure 17.9

Now, let's consider the inequality

$$2 < 5$$

What happens if we multiply each side by 3?

$$(2)(3) < (5)(3)$$
$$6 < 15$$

We see that the sense stays the same.

What happens to the sense of this inequality if we multiply each side by -4?

$$2 < 5$$
$$(2)(-4) < (5)(-4)$$
$$-8 > -20$$

To obtain a true statement here, we must *reverse* the sense of the inequality.

If we consider dividing by a nonzero number to be the same as multiplying by its reciprocal, we may summarize as follows:

MULTIPLICATION PROPERTY

Let a, b, and c be real numbers.

1. If $a < b$ and $c > 0$, then $ac < bc$.

2. If $a < b$ and $c < 0$, then $ac > bc$.

In words, *each side of an inequality may be multiplied by the same*

1. *positive number and the sense remains the same, or*

2. *negative number and the sense is reversed.*

The Addition and Multiplication properties may be summarized as follows:

1. If the *same* number is added to or subtracted from each side of an inequality or if each side of an inequality is multiplied or divided by the *same positive* number, the sense remains the same.

2. If each side of an inequality is multiplied or divided by the *same negative* number, the sense is reversed.

EXAMPLE 2

Solve $\frac{x}{3} \le 5$.

$$\frac{x}{3} \le 5$$

$$x \le 15 \qquad \text{(Multiply each side by 3.)}$$

The graph of this solution is shown in Fig. 17.10.

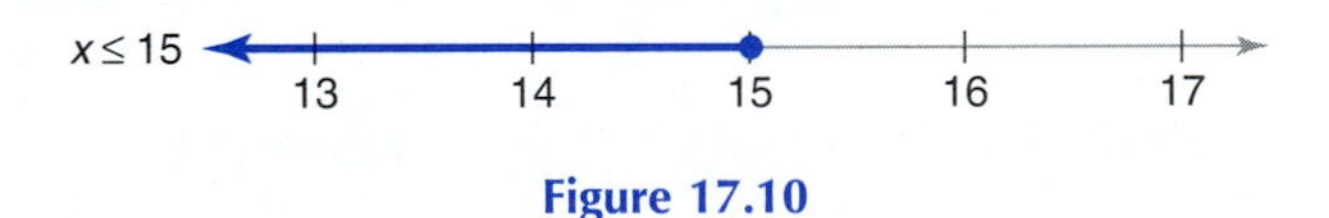

Figure 17.10

EXAMPLE 3

Solve $-4x \ge 28$.

$$-4x \ge 28$$

$$x \le -7 \qquad \text{(Divide each side by } -4.\text{)}$$

The graph of this solution is shown in Fig. 17.11.

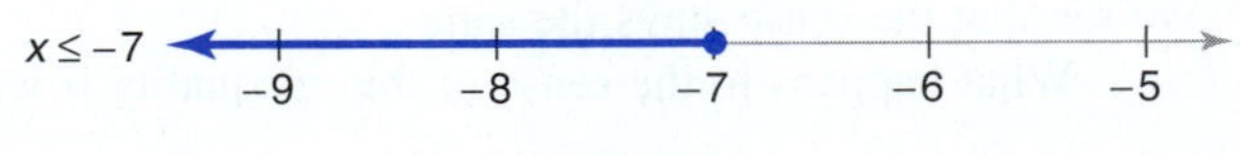

Figure 17.11

EXAMPLE 4

Solve $3x + 4 < 5x - 8$.

$$\begin{aligned} 3x + 4 &< 5x - 8 & \\ 4 &< 2x - 8 & \text{(Subtract } 3x \text{ from each side.)} \\ 12 &< 2x & \text{(Add 8 to each side.)} \\ 6 &< x & \text{(Divide each side by 2.)} \end{aligned}$$

The graph of the solution is shown in Fig. 17.12.

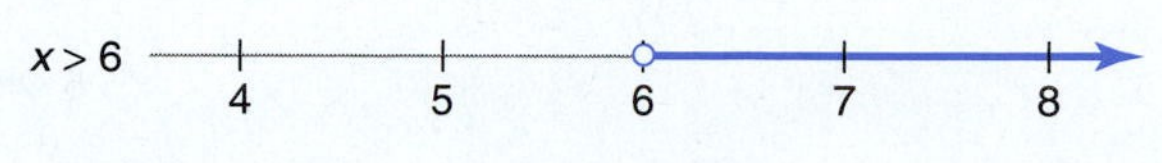

Figure 17.12

EXAMPLE 5

Solve $\frac{x}{6} - \frac{1}{3} \geq x - \frac{1}{4}$.

$$\begin{aligned} \frac{x}{6} - \frac{1}{3} &\geq x - \frac{1}{4} & \\ 12\left(\frac{x}{6} - \frac{1}{3}\right) &\geq 12\left(x - \frac{1}{4}\right) & \text{(Multiply each side by the L.C.D., 12.)} \\ 2x - 4 &\geq 12x - 3 & \\ -10x - 4 &\geq -3 & \text{(Subtract } 12x \text{ from each side.)} \\ -10x &\geq 1 & \text{(Add 4 to each side.)} \\ x &\leq -\frac{1}{10} & \text{(Divide each side by } -10.\text{)} \end{aligned}$$

The graph of the solution is shown in Fig. 17.13.

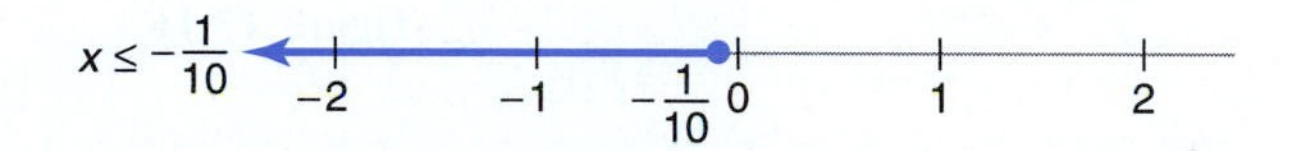

Figure 17.13

Compound Statements

EXAMPLE 6

Consider the statement $x > 3$ and $x > 5$. What is its solution?

Let's first look at the graph of each part on the number line as in Fig. 17.14. The solution is all real numbers that are *both* greater than 3 *and* greater than 5, that is, those points shaded twice. This solution can be written

$$x > 5$$

x > 5

x > 3

x > 5 and x > 3

2 3 4 5 6 7

Figure 17.14

EXAMPLE 7

Consider the statement $x > 5$ or $x > 3$. What is its solution?

Let's look at the graph of each part on the number line as shown in Fig. 17.15. The solution is all real numbers that are *either* greater than 5, *or* greater than 3, that is, those points shaded once or twice. This solution is shown in Fig. 17.15 and can be written

$$x > 3$$

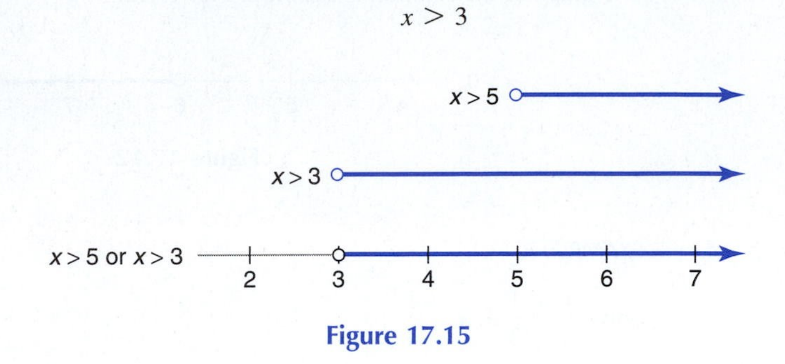

Figure 17.15

EXAMPLE 8

Consider the statement $x > -1$ and $x \leq 2$. What is its solution?

The graph of each part is shown in Fig. 17.16.

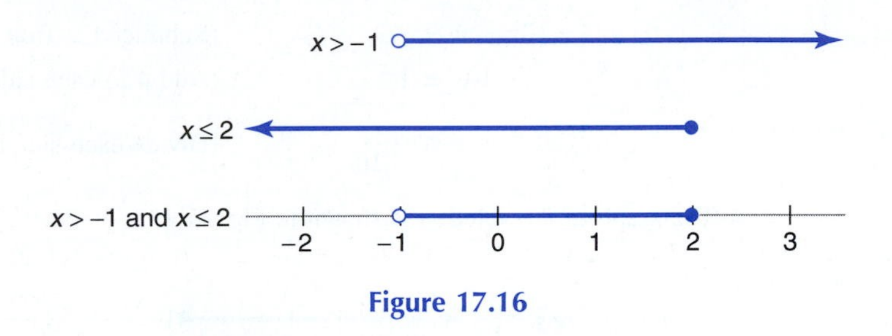

Figure 17.16

The solution is all real numbers that are *both* greater than -1 *and* less than or equal to 2, that is, those points shaded twice. This solution can be written

$$-1 < x \leq 2$$

Note: Do you see that the statement $x > -1$ *or* $x \leq 2$ includes all points on the number line? Its solution is all real numbers.

In summary:

a and b means *both a and b.*

a or b means *either a or b or both.*

EXAMPLE 9

Consider the statement $x \leq 1$ or $x \geq 5$. What is its solution?

The graph of each part is shown in Fig. 17.17.

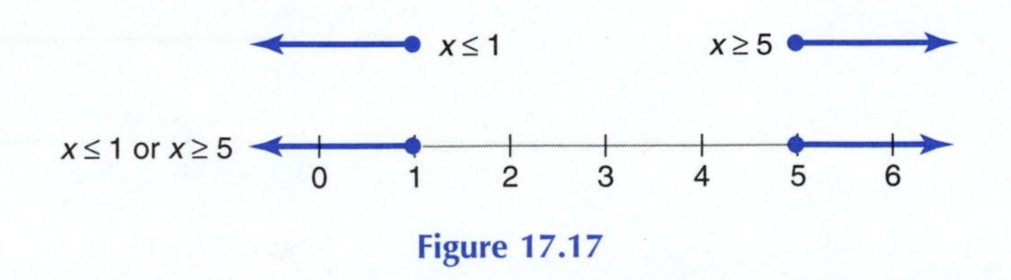

Figure 17.17

The solution is all real numbers that are *either* less than or equal to 1 *or* greater than or equal to 5, that is, those points shaded once or twice. This solution can only be written

$$x \leq 1 \quad \text{or} \quad x \geq 5$$

which is the same as the given statement.

Note: Do you see that the statement $x \leq 1$ *and* $x \geq 5$ includes no points on the number line? There are no real numbers that are *both* less than or equal to 1 *and* greater than or equal to 5. There is no solution, so its solution is the empty set, a set with no members. The empty set is usually written using the symbol $\varnothing$.

Inequalities that have three parts in the form $a < x < b$ may also be solved using these same properties.

EXAMPLE 10

Solve $-8 \leq 2x + 4 < 11$.

$$-8 \leq 2x + 4 < 11$$

$$-12 \leq 2x < 7 \qquad \text{(Subtract 4 from all three parts.)}$$

$$-6 \leq x < \tfrac{7}{2} \qquad \text{(Divide all three parts by 2.)}$$

The graph of this solution is shown in Fig. 17.18.

Figure 17.18

Exercises 17.1

Graph each inequality on the number line.

1. $x > 2$ **2.** $x \leq -4$ **3.** $x \leq -1$ **4.** $x > 15$

5. $4 < x < 8$ **6.** $-3 \leq x \leq 2$ **7.** $0 \leq x < 5$ **8.** $15 < x \leq 20$

9. $x < -1$ or $x \geq 3$ **10.** $x \leq 0$ or $x \geq 5$

Use an inequality to express those real numbers shaded on each number line.

11. **12.**

13. **14.**

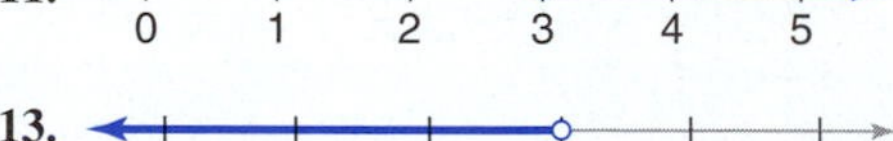

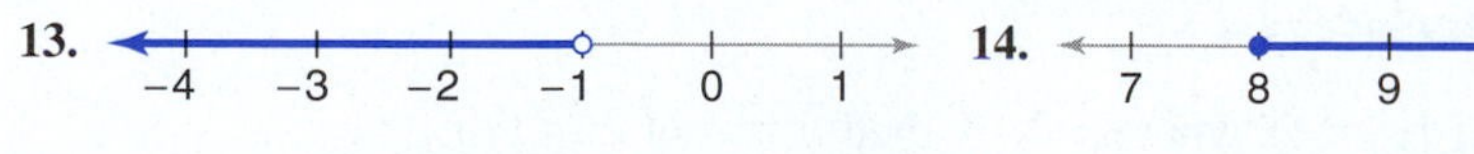

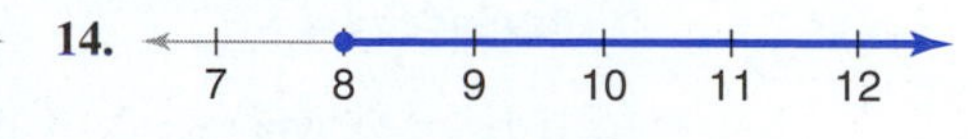

15. **16.**

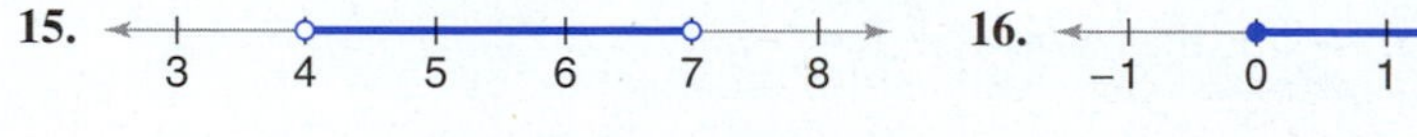

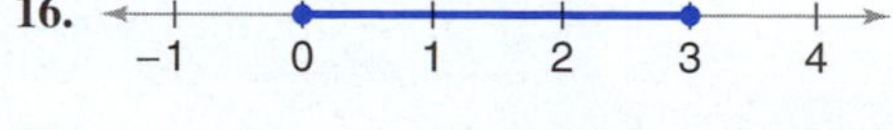

17. **18.**

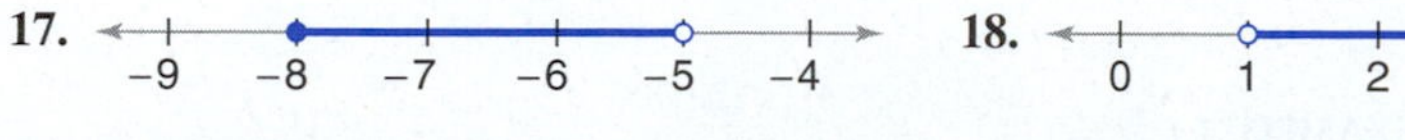

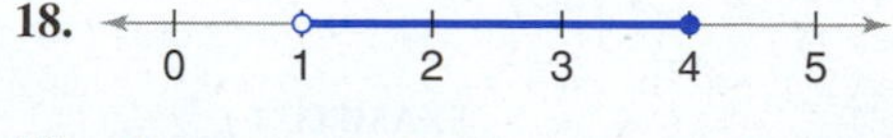

19. **20.**

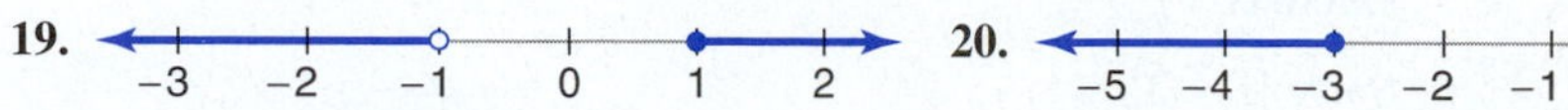

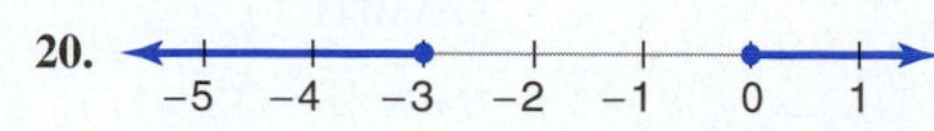

Solve each inequality. Graph the solution on the number line.

21. $4x < 12$ **22.** $5x \geq 30$ **23.** $-3x \geq 9$

24. $-6x < -24$ **25.** $3x + 4 < 22$ **26.** $5x - 7 > -12$

27. $5 - 2x \geq 17$ **28.** $3 - 4x \leq 27$ **29.** $6x - 5 \leq 2x + 7$

30. $8x + 1 > 6x - 11$

31. $3(x - 2) > 5x - 12$

32. $5(2x + 1) < 2(3x - 5)$

33. $\frac{x}{4} \geq 2$

34. $\frac{x}{3} \leq -5$

35. $\frac{x}{6} + \frac{3}{8} > \frac{3x}{4} + \frac{7}{12}$

36. $\frac{2}{3}(2x - 1) \geq 4 + \frac{2x}{5}$

37. $\frac{2x + 3}{4} < 5x$

38. $\frac{4}{3}(x - 7) + x < \frac{x}{9}$

39. $\frac{2}{3}(3x + 4) \leq \frac{4}{5}(x - 2)$

40. $-\frac{5}{6}(2 - 4x) > \frac{8}{3}(2x - 1)$

Graph each compound statement. Simplify the form of the inequality, if possible.

41. $x \geq 5$ and $x \geq 10$

42. $x < 4$ and $x < 7$

43. $x < 4$ or $x \leq 3$

44. $x \geq 2$ or $x \geq 0$

45. $x > 7$ and $x \leq 10$

46. $x \geq 4$ or $x < -4$

47. $x < 4$ or $x > 1$

48. $x \leq -1$ and $x \geq 5$

49. $3x + 4 < 16$ and $5x + 1 \geq -14$

50. $1 - 2x \geq 9$ or $5 - 3x \geq 35$

Solve each inequality. Graph the solution on the number line.

51. $-9 \leq 2x + 3 \leq 7$

52. $8 < 3x - 4 < 20$

53. $-2 < 4x - 2 < 18$

54. $-10 \leq 6x + 2 < 26$

55. $-9 \leq 1 - 2x < -1$

56. $1 < 3 - 4x \leq 15$

57. $15 < 5 - 3x \leq 26$

58. $0 \leq 4 - x \leq 10$

17.2 EQUATIONS AND INEQUALITIES INVOLVING ABSOLUTE VALUE

ABSOLUTE VALUE

The absolute value of a real number x, written $|x|$, is defined as

$$|x| = \begin{cases} x & \text{if } x \geq 0 \\ -x & \text{if } x < 0 \end{cases}$$

For example, $|+6| = 6$, $|-4| = -(-4) = 4$, and $|0| = 0$. We shall study three important properties of equations and inequalities involving absolute value.

PROPERTY 1

If $|x| = a$, where $a > 0$, then $x = a$ or $x = -a$.
If $|x| = |a|$, then $x = a$ or $x = -a$.

EXAMPLE 1

Solve $|x - 3| = 8$.

$$|x - 3| = 8 \text{ is equivalent to}$$
$$x - 3 = 8 \quad \text{or} \quad x - 3 = -8$$
$$x = 11 \quad \text{or} \quad x = -5$$

The solutions are 11 and -5.

EXAMPLE 2

Solve $|x + 4| = |x - 10|$.

$$|x + 4| = |x - 10| \text{ is equivalent to}$$

$$\begin{array}{lcl} x + 4 = x - 10 & \text{or} & x + 4 = -(x - 10) \\ 4 = -10 & & x + 4 = -x + 10 \\ \text{No solution} & & 2x = 6 \\ & & x = 3 \end{array}$$

The only solution is 3.

PROPERTY 2

If $|x| < a$, where $a > 0$, then $-a < x < a$.

EXAMPLE 3

Solve $|2x + 1| < 7$.

$$|2x + 1| < 7 \text{ is equivalent to}$$

$$\begin{aligned} -7 &< 2x + 1 < 7 \\ -8 &< \quad 2x \quad < 6 \\ -4 &< \quad x \quad < 3 \end{aligned}$$

The graph of the solution is shown in Fig. 17.19.

Figure 17.19

EXAMPLE 4

Solve $|3x - 4| \leq 11$.

Property 2 is also valid when the symbol $<$ is replaced by $\leq$, so $|3x - 4| \leq 11$ is equivalent to

$$\begin{aligned} -11 &\leq 3x - 4 \leq 11 \\ -7 &\leq \quad 3x \quad \leq 15 \\ -\tfrac{7}{3} &\leq \quad x \quad \leq 5 \end{aligned}$$

The graph of the solution is shown in Fig. 17.20.

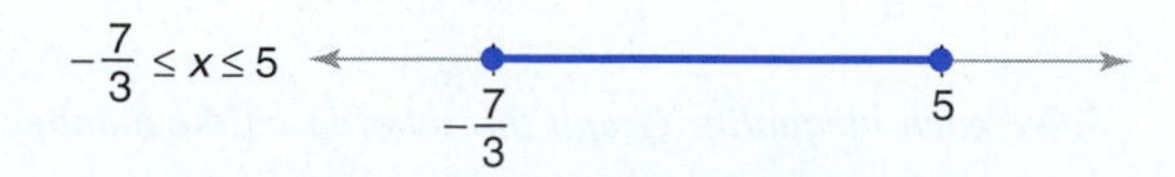

Figure 17.20

PROPERTY 3

If $|x| > a$, where $a > 0$, then $x < -a$ or $x > a$.

EXAMPLE 5

Solve $|2x - 1| > 5$.

$$|2x - 1| > 5 \text{ is equivalent to}$$
$$\begin{array}{rcr} 2x - 1 < -5 & \text{or} & 2x - 1 > 5 \\ 2x < -4 & & 2x > 6 \\ x < -2 & \text{or} & x > 3 \end{array}$$

The graph of this solution is shown in Fig. 17.21.

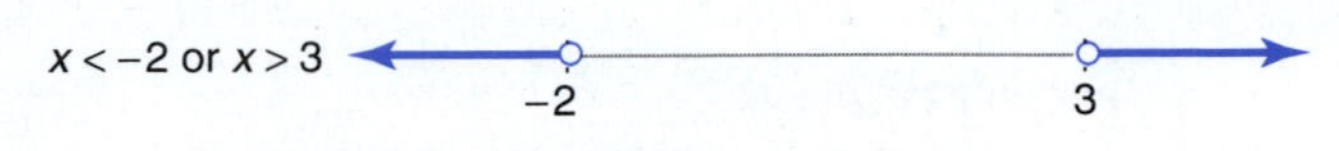

Figure 17.21

EXAMPLE 6

Solve $|5 - 3x| \geq 8$.

Property 3 is also valid when the symbol $>$ is replaced by $\geq$, so $|5 - 3x| \geq 8$ is equivalent to

$$\begin{array}{rcr} 5 - 3x \leq -8 & \text{or} & 5 - 3x \geq 8 \\ -3x \leq -13 & & -3x \geq 3 \\ x \geq \frac{13}{3} & \text{or} & x \leq -1 \end{array}$$

The graph of this solution is shown in Fig. 17.22.

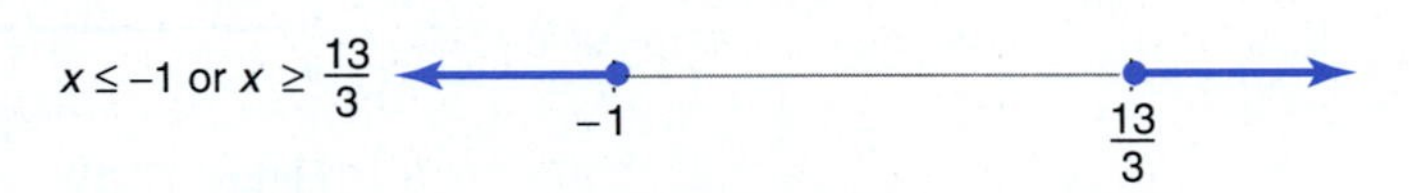

Figure 17.22

Exercises 17.2

Solve each equation.

1. $|x + 4| = 10$

2. $|x - 5| = 7$

3. $|6 - 5x| = 4$

4. $|3 + 4x| = 2$

5. $|x - 2| = |2x + 4|$

6. $|2x - 1| = |4x + 3|$

7. $|3x - 2| = |3x + 4|$

8. $|1 - x| = |1 + x|$

9. $\left|\dfrac{x + 3}{x - 4}\right| = 8$

10. $\left|\dfrac{1 - 2x}{1 + x}\right| = 5$

Solve each inequality. Graph the solution on the number line.

11. $|x| < 3$

12. $|x| > 4$

13. $|x - 2| < 5$

14. $|x + 7| < 9$

15. $|2x + 3| > 1$

16. $|3x - 2| > 4$

17. $|3 - 4x| \leq 19$

18. $|1 - 2x| \leq 13$

19. $|3x + 1| \geq 16$

20. $|2x + 7| \geq 23$

21. $|4x - 5| < 31$

22. $|6x - 2| > 12$

23. $|1 - 6x| \geq 7$

24. $|4x + 9| \leq 49$

25. $|2x + 6| > 18$

26. $|4x + 12| < 8$

27. $|4x - 2| < -3$

28. $|3x + 1| > 0$

17.3 OTHER TYPES OF INEQUALITIES

In Section 17.1, we limited our discussion to linear, first-degree, inequalities. Next, let's consider quadratic, second-degree, inequalities. Examples are $x^2 + 3x > 3$ and $3x^2 + 10x + 8 \leq 0$.

SOLVING A FACTORABLE INEQUALITY

1. Rearrange the inequality so that 0 is on one side.
2. Factor the resulting expression or the numerator and the demoninator of any resulting rational expression.
3. Set each factor from Step 2 equal to zero. Graph each solution on the number line. Each solution is called a **point of division.**

Note: In Step 3, we have found all points on the number line where the product is zero or where the quotient is undefined in the case of a rational expression. In each interval formed by these points of division, substitute a value to determine whether the expression is positive or negative.

4. Choose any number from each interval as a test point and substitute it into the factored form of the inequality. It is enough to test *any one point* in each interval because *all values* in the interval behave the same with regard to sign; that is, all values in the interval give either positive or negative results. Keep a record of the sign of each factor in each interval above the number line. Using the rules for multiplication and division of signed numbers, record the sign of the product or quotient below the number line.
5. Intervals that make the inequality true are included in the solution set. Points of division from the numerator are included if the inequality contains an "or equal to" symbol. Points of division from the denominator are never included.

EXAMPLE 1

Solve $x^2 + 2x > 3$.

Step 1: $x^2 + 2x - 3 > 0$

Step 2: $(x + 3)(x - 1) > 0$

Step 3: $x + 3 = 0$ or $x - 1 = 0$

$x = -3$ $\quad$ $x = 1$

Step 4: *Signs of factors:* $(x + 3)(x - 1)$ $\quad$ $- -$ $\quad$ $+ -$ $\quad$ $+ +$

Signs of intervals: $\quad$ $+$ $\quad$ -3 $\quad$ $-$ $\quad$ 1 $\quad$ $+$

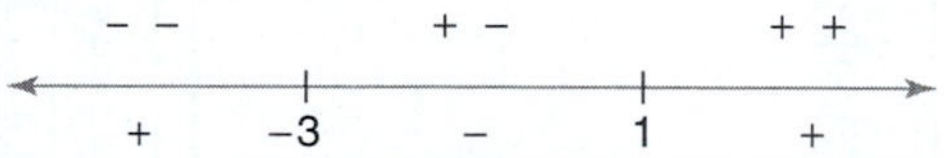

Figure 17.23

From Fig. 17.23, we note that the product is positive in the intervals

$$x < -3 \quad \text{or} \quad x > 1$$

which is the solution of the given inequality.

EXAMPLE 2

Solve $3x^2 + 10x + 8 \leq 0$.

Step 1: Not needed.

Step 2: $(x + 2)(3x + 4) \le 0$

Step 3: $x + 2 = 0 \quad \text{or} \quad 3x + 4 = 0$

$x = -2 \quad \text{or} \quad x = -\frac{4}{3}$

Step 4: *Signs of factors:* $(x + 2)(3x + 4)$ $\quad - - \quad + - \quad + +$

Signs of intervals: $\quad + \quad -2 \quad - \quad -\frac{4}{3} \quad +$

Figure 17.24

From Fig. 17.24, we note that the product is negative in the interval $-2 < x < -\frac{4}{3}$. Since a product equal to zero is also a solution, the solution of the given inequality is

$$-2 \le x \le -\frac{4}{3}$$

EXAMPLE 3

Solve $x^3 < x$.

Step 1: $x^3 - x < 0$

Step 2: $x(x + 1)(x - 1) < 0$

Step 3: $x = 0 \quad \text{or} \quad x + 1 = 0 \quad \text{or} \quad x - 1 = 0$

$x = -1 \qquad x = 1$

Step 4: *Signs of factors:* $x(x + 1)(x - 1)$ $\quad - - - \quad - + - \quad + + - \quad + + +$

Signs of intervals: $\quad - \quad -1 \quad + \quad 0 \quad - \quad 1 \quad +$

Figure 17.25

From Fig. 17.25, note that the product is negative in the intervals

$$x < -1 \quad \text{or} \quad 0 < x < 1$$

which is the solution of the given inequality.

Using a calculator, we have

green diamond **Y=** x^3-x **ENTER**

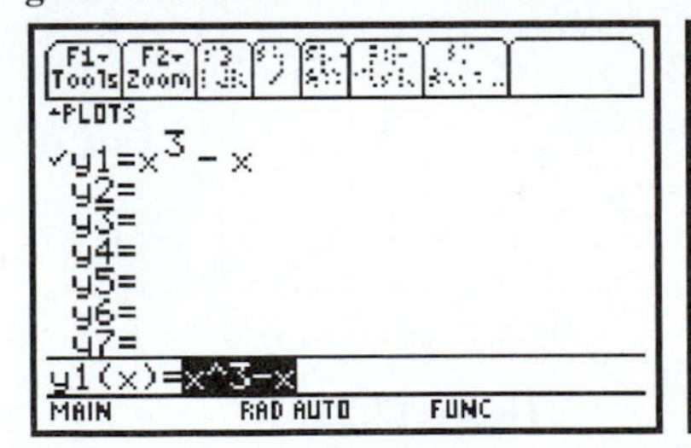

F2 **4**

(Locate where the function is *negative.*)

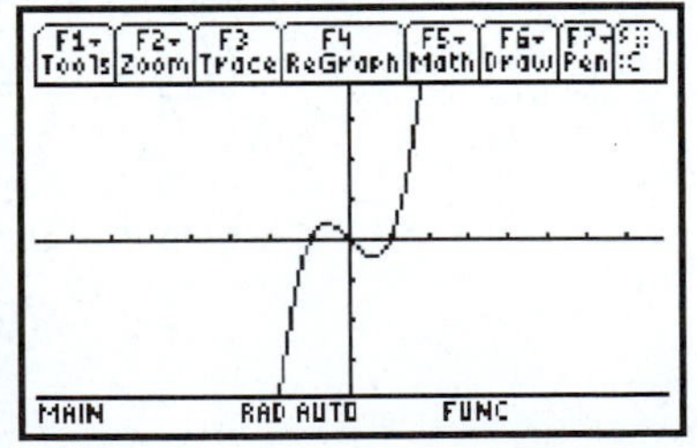

Y= x **MATH** **3** -x

ZOOM **4**

(Locate where the function is *negative.*)

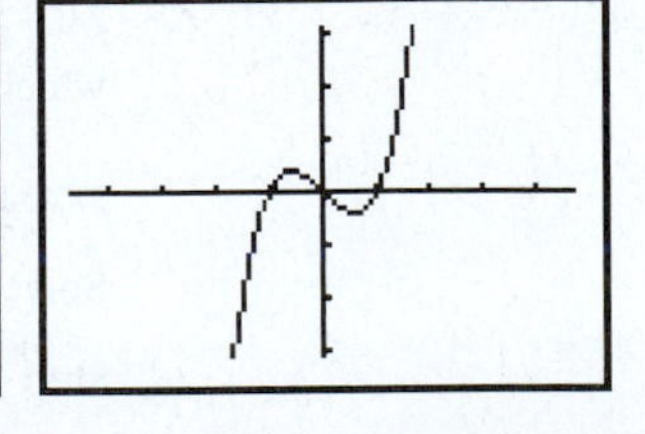

The inequality $x^3 - x < 0$ can be interpreted as finding where the polynomial $x^3 - x$ is *negative* ("less than zero"). This is equivalent to locating the portions of the graph that lie below the x-axis and describing their corresponding x-values (we're solving for x). In this example, the graph lies below the x-axis for x-values to the left of -1 as well as for x-values between 0 and 1; thus, the solution is $x < -1$ or $0 < x < 1$.

Since the rules for division of signed numbers are the same as for multiplication, this method may be extended to rational inequalities, but be careful of division by zero!

EXAMPLE 4

Solve $\dfrac{x - 5}{x + 2} > 0$.

Step 1: Not needed.

Step 2: Already factored.

Step 3: $x - 5 = 0$ or $x + 2 = 0$

$x = 5 \qquad x = -2$

Note: While $x = -2$ makes the given inequality undefined, that point is still a point of division.

Step 4: *Signs of factors:* $\dfrac{x - 5}{x + 2}$

Signs of intervals:

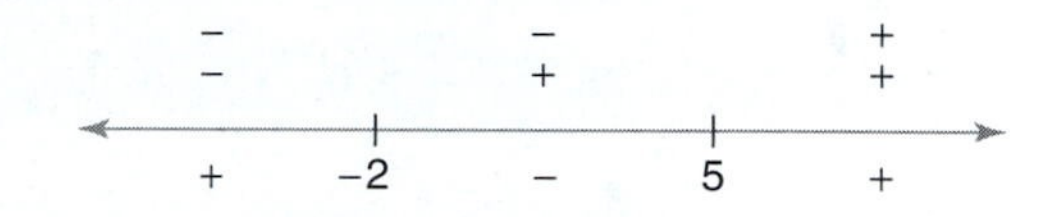

Figure 17.26

From Fig. 17.26, we see that the quotient is positive in the intervals

$$x < -2 \quad \text{or} \quad x > 5$$

which is the solution of the given inequality.

EXAMPLE 5

Solve $\dfrac{1 - x - 2x^2}{x - 3} \geq 0$.

Step 1: Not needed.

Step 2: $\dfrac{(1 + x)(1 - 2x)}{x - 3} \geq 0$

Step 3: $1 + x = 0$ or $1 - 2x = 0$ or $x - 3 = 0$

$x = -1 \qquad x = \frac{1}{2} \qquad x = 3$

Step 4: *Signs of factors:* $\dfrac{(1 + x)(1 - 2x)}{x - 3}$

Signs of intervals:

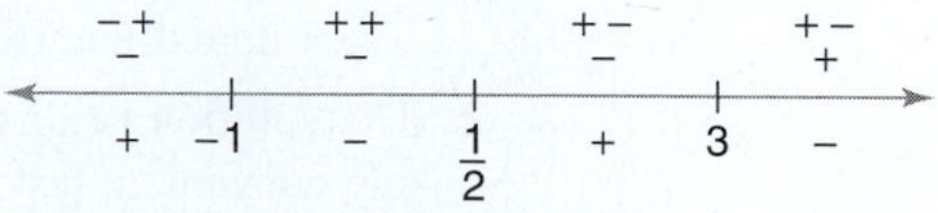

$$\begin{array}{ccccccc} \frac{-\,+}{-} & & \frac{+\,+}{-} & & \frac{+\,-}{-} & & \frac{+\,-}{+} \\ + & -1 & - & \frac{1}{2} & + & 3 & - \end{array}$$

Figure 17.27

From Fig. 17.27, note that the quotient is positive in the intervals

$$x < -1 \quad \text{or} \quad \frac{1}{2} < x < 3$$

Since a quotient equal to zero is also a solution, the solution of the given inequality is then

$$x \le -1 \quad \text{or} \quad \frac{1}{2} \le x < 3$$

Exercises 17.3

Solve each inequality.

1. $(x + 3)(x - 5) > 0$
2. $(x + 7)(x - 2) < 0$
3. $(3x - 7)(4x + 1) \le 0$
4. $(6x - 5)(3x - 1) \ge 0$
5. $x^2 + 9x + 14 > 0$
6. $x^2 + 4x < 5$
7. $2x^2 + 4 < 9x$
8. $3x^2 + 4 \le 8x$
9. $15x^2 \ge 2 - x$
10. $4x^2 > 10x + 6$
11. $4x^2 \le 11x + 3$
12. $3x^2 + 4 \ge 13x$
13. $4x^2 + 12x + 9 > 0$
14. $9x^2 + 25 \le 30x$
15. $x^2 > 4$
16. $x^2 \le 9$
17. $x^2 \le 5$
18. $x^2 > 12$
19. $x^2 > 3x$
20. $x^2 \le 4x$
21. $x^3 + x^2 \ge 12x$
22. $x^3 + 9x^2 + 20x < 0$
23. $x^3 > 6x^2$
24. $x^3 \le x^2$
25. $x^4 - 5x^2 + 4 < 0$
26. $x^4 - 10x^2 + 9 \ge 0$
27. $\frac{x + 4}{x - 2} < 0$
28. $\frac{x + 7}{x + 3} \ge 0$
29. $\frac{x^2 + 3x + 2}{x + 3} \ge 0$
30. $\frac{x - 1}{x^2 + 2x - 8} \le 0$
31. $\frac{9 - x^2}{16 - x^2} \le 0$
32. $\frac{x^2 - 5x + 6}{x^2 + 5x + 6} \le 0$
33. $\frac{9 - 12x + 4x^2}{x^2 - 3x - 10} > 0$
34. $\frac{12 + x - x^2}{18 - 9x + x^2} < 0$
35. $\frac{x^3(x + 1)(x - 2)}{(3 - x)^2(x - 1)} \le 0$
36. $\frac{x^4(x + 1)^2(x - 1)^3}{(x - 4)^3(x - 5)} \ge 0$
37. $\frac{x + 1}{x - 2} \ge 4$
38. $\frac{x - 3}{x - 4} \le 2$
39. $\frac{1 - x}{x} < 1$
40. $\frac{1}{x + 3} \ge 5$

17.4 INEQUALITIES IN TWO VARIABLES

Next, we consider the graphical solution of inequalities in two variables.

To find the graphical solution of an inequality in two variables, such as $y > f(x)$:

1. Graph the equation $y = f(x)$, which is the boundary of the solution of the inequality. Use a solid line if the given inequality is in the form $y \ge f(x)$ or $y \le f(x)$. Use a dotted line if the given inequality is in the form $y > f(x)$ or $y < f(x)$.
2. The solution is all points on one side or the other of the boundary line. Choose any convenient test point (x, y) which is *not* on the boundary line (or close to the boundary line). Substitute this ordered pair into the given inequality.
   - **(a)** If this ordered pair *satisfies* the given inequality, shade all points *on the same side* of the boundary line as the test point.
   - **(b)** If this ordered pair *does not satisfy* the given inequality, shade all points *on the opposite side* of the boundary line as the test point.

EXAMPLE 1

Graph $y \leq 3x - 2$.

Step 1: Graph $y = 3x - 2$ as a solid boundary line as in Fig. 17.28.

Step 2: Choose a test point, such as (4, 0). Substitute this ordered pair into

$$y \leq 3x - 2$$
$$0 \leq 3(4) - 2$$
$$0 \leq 10$$

Since this ordered pair satisfies the given inequality, shade all points *on the same side* of the boundary line as shown in Fig. 17.28.

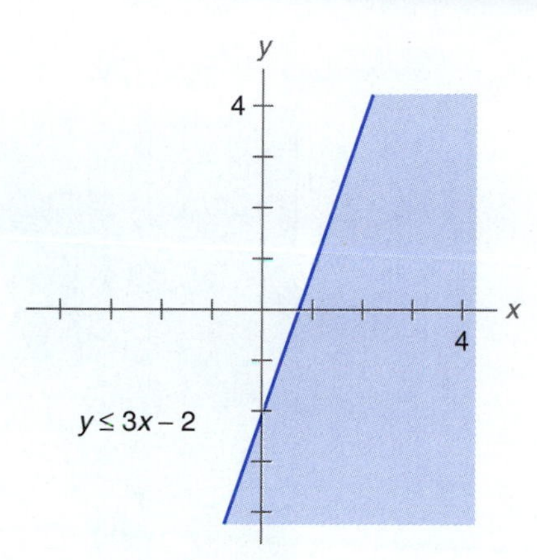

Figure 17.28

EXAMPLE 2

Graph $y > x^2$.

Step 1: Graph $y = x^2$ as a dotted boundary line as in Fig. 17.29.

Step 2: Choose a test point, such as (3, 0). Substitute this ordered pair into

$$y > x^2$$
$$0 > 3^2$$
$$0 > 9$$

Since this ordered pair does not satisfy the given inequality, shade all points *on the opposite side* (inside) of the boundary line as shown in Fig. 17.29.

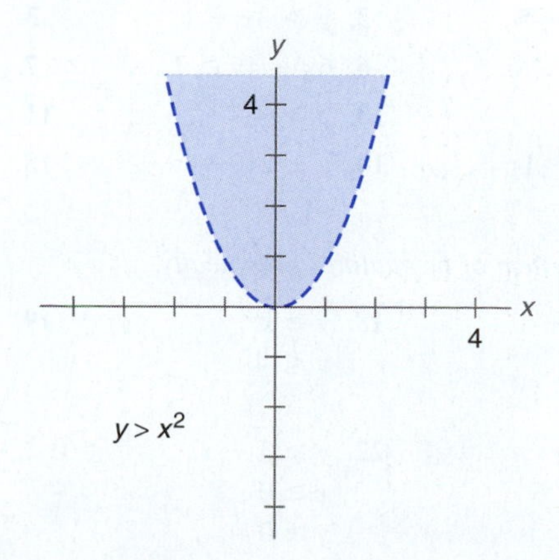

Figure 17.29

Systems of inequalities in two variables may be solved graphically using this method of finding the areas or regions common to all of the given inequalities.

EXAMPLE 3

Solve the system of inequalities graphically.

$$x + 2y < 1$$
$$3x - y \le 6$$
$$x \ge -1$$

The solution of this system is all points common to all three regions, which is all points in the triangle in Fig. 17.30.

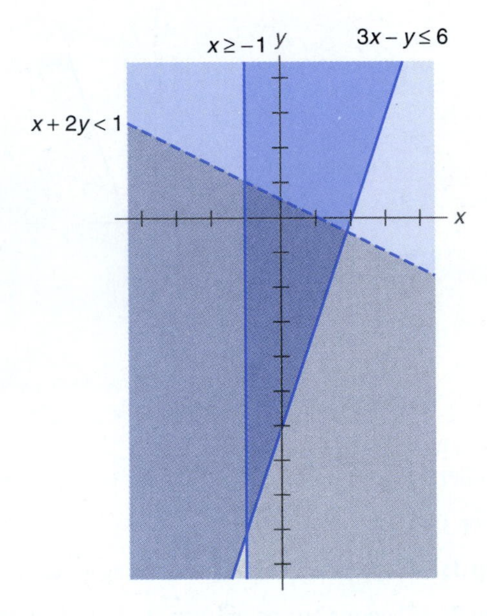

Figure 17.30

Exercises 17.4

Graph each inequality.

1. $y < 2x - 5$
2. $y > 3x + 1$
3. $y \ge 8 - 4x$
4. $y \le 1 - 2x$
5. $2x + 3y \le 6$
6. $6x - 4y \ge 7$
7. $4x - 2y > 8$
8. $3x + 5y < 4$
9. $x \ge 2y$
10. $y > 3x$
11. $x \ge 3$
12. $y < -2$
13. $y \le x^2 + 1$
14. $y \ge 2x - x^2$
15. $y \ge x^3$
16. $y < 1 - x^2$

Solve each system of inequalities graphically.

17. $y > 2x + 1$
    $y < x$
18. $y \ge x^2$
    $y < 4$
19. $x \le 2y$
    $y \le 3x$
    $2x + y \le 5$
20. $y \ge x - 1$
    $x \ge 1$
    $y \le 2$
21. $y > x^2$
    $y < x + 6$
    $y < 5$
22. $y \ge 3 - x$
    $x \ge 0$
    $y \ge 0$

CHAPTER 17 SUMMARY

1. *Inequality properties:*

(a) *Trichotomy property:* For any two real numbers, *exactly one* of the following is true:

$$a < b \quad \text{or} \quad a = b \quad \text{or} \quad a > b$$

(b) *Transitive property:* If a, b, and c are real numbers such that $a < b$ and $b < c$, then $a < c$.

2. *Inequalities* are statements involving *less than* or *greater than* and may be used to describe various intervals on the number line.

3. *Addition property:* If a, b, and c are real numbers and $a < b$, then $a + c < b + c$.

4. *Multiplication property:* Let a, b, and c be real numbers.

(a) If $a < b$ and $c > 0$, then $ac < bc$.

(b) If $a < b$ and $c < 0$, then $ac > bc$.

5. *Compound statements:*

(a) $x > a$ **and** $x > b$ means *both* $x > a$ *and* $x > b$.

(b) $x > a$ **or** $x > b$ means *either* $x > a$ *or* $x > b$ *or both.*

6. The *absolute value* of a real number x, written $|x|$, is defined as

$$|x| = \begin{cases} x & \text{if } x \geq 0 \\ -x & \text{if } x < 0 \end{cases}$$

7. *Absolute value and inequality properties:*

(a) If $|x| = a$, where $a > 0$, then $x = a$ or $x = -a$.

(b) If $|x| = |a|$, then $x = a$ or $x = -a$.

(c) If $|x| < a$, where $a > 0$, then $-a < x < a$.

(d) If $|x| > a$, where $a > 0$, then $x < -a$ or $x > a$.

8. To solve factorable polynomial and rational inequalities, see Section 17.3.

9. To find graphical solutions of an equality in two variables, see Section 17.4.

CHAPTER 17 REVIEW

Graph each inequality on the number line.

1. $x \leq 2$ **2.** $x < -1$ or $x \geq 5$ **3.** $1 < x \leq 4$ **4.** $x > 6$

Use an inequality to express those real numbers shaded on the number line.

5.
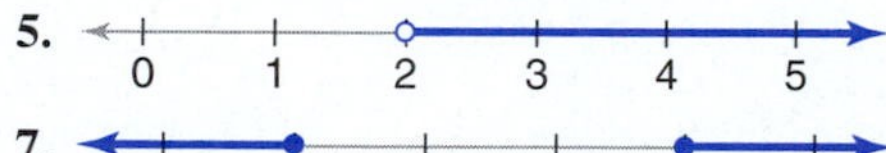

6.
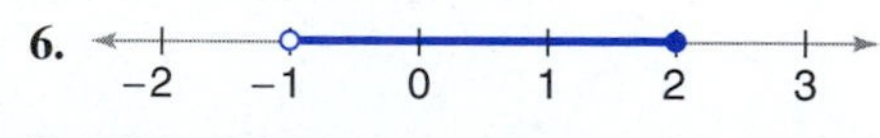

7.
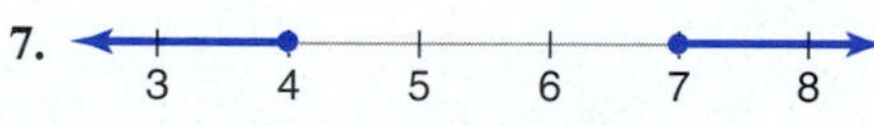

8.
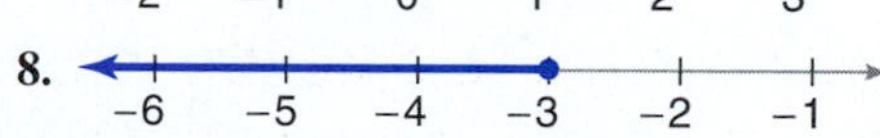

Solve each inequality. Graph the solution on the number line.

9. $7x \leq -28$ **10.** $\dfrac{x}{2} > 7$ **11.** $-2x \geq 10$

12. $2x + 3 < 9$ **13.** $4 - 3x < 34$ **14.** $6x + 15 \geq 4x - 7$

15. $4(x - 3) \geq 3(x + 4)$ **16.** $\dfrac{x}{2} + \dfrac{3}{4} \leq 2x + \dfrac{1}{6}$

Graph each compound statement. Simplify the inequality, if possible.

17. $x \le 4$ or $x \le 6$

18. $x \ge 5$ and $x \le 0$

19. $x \le 0$ or $x > 3$

20. $x \le -1$ and $x \le 2$

Solve each inequality. Graph the solution on the number line.

21. $-7 \le 2x + 5 \le 9$

22. $-2 < 1 - 3x \le 8$

Solve each equation.

23. $|2x + 1| = 9$

24. $|2 - 3x| = 11$

25. $|3x + 2| = |4x - 7|$

26. $\left|\dfrac{x - 2}{x + 3}\right| = 6$

Solve each inequality. Graph the solution on the number line.

27. $|x + 4| < 8$

28. $|x - 3| > 9$

29. $|2x + 5| \ge 17$

30. $|3 - 2x| \le 15$

31. $|1 - 3x| > 16$

32. $|4x - 7| \le 17$

Solve each inequality.

33. $(2x + 1)(x - 4) \le 0$

34. $3x^2 + 12 > 20x$

35. $x^2 \le 49$

36. $x^3 \ge 36x$

37. $x^3 + 8x^2 + 16x > 0$

38. $\dfrac{x + 4}{x - 9} < 0$

39. $\dfrac{3x^2 + 8x - 3}{x - 3} \le 0$

40. $\dfrac{(2x - 3)^2}{16 - 8x - 3x^2} \ge 0$

41. $\dfrac{1}{x + 5} \le 10$

42. $\dfrac{2x + 1}{x - 4} \le 3$

Graph each inequality.

43. $y \ge 2x$

44. $y < 3 - 4x$

45. $2x + 5y \le 5$

46. $x - 4y > 6$

47. $y \le 3x - x^2$

48. $y > x^2 - 4$

Solve each system of inequalities graphically.

49. $y \ge x$
$x \le 2$
$y \le 6$

50. $y < 5x - x^2$
$y > 2x$

18 Progressions and the Binomial Theorem

INTRODUCTION

The fact that the sum of an infinite number of numbers could be finite is somewhat startling, yet we use this concept when we consider a rational number written as a repeating decimal. For example,

$$\frac{2}{3} = 0.666\ldots$$

$$= \frac{6}{10} + \frac{6}{100} + \frac{6}{1000} + \cdots$$

The numbers $\frac{6}{10}, \frac{6}{100}, \frac{6}{1000}, \ldots$ form a progression or sequence, and their sum is called a series. In this chapter we study progressions and series as well as the binomial theorem.

Objectives

- Find the nth term of an arithmetic progression.
- Find the sum of the first n terms of an arithmetic series.
- Find the nth term of a geometric progression.
- Find the sum of the first n terms of a geometric series.
- Use the binomial theorem to expand a binomial raised to the nth power.
- Find a specified term of a binomial expansion.

18.1 ARITHMETIC PROGRESSIONS

Many technical applications make use of infinite series. One in particular, the Fourier series, is very useful in the field of electronics. An *infinite series* is the summation of an infinite sequence of numbers. We will first look, however, at sums of finite sequences.

A *finite sequence* consists of a succession of quantities $a_1, a_2, a_3, \ldots, a_n$, where the three dots represent the quantities or terms between a_3 and a_n. The term a_n is called the

nth term or last term. An *infinite sequence* is a succession of quantities which continues indefinitely and may be written

$$a_1, a_2, a_3, \ldots, a_n, \ldots \quad \text{or} \quad a_1, a_2, a_3, \ldots$$

A basic example of an infinite sequence is an arithmetic progression. An *arithmetic progression* is a sequence of terms where each term differs from the immediately preceding term by a fixed number d, which is called the *common difference* of the progression. For example, the sequence 5, 10, 15, 20, 25, 30, . . . is an arithmetic progression with a common difference of 5. The sequence $1, \frac{1}{3}, -\frac{1}{3}, -1, -\frac{5}{3}, \ldots$ is an arithmetic progression with a common difference of $-\frac{2}{3}$.

In any arithmetic progression if a is the first term and d is the common difference, then the second term is $a + d$. Likewise, the third term is $a + 2d$ and the fourth term is $a + 3d$. Continuing in this manner, we can express the arithmetic progression as follows:

$$a, a + d, a + 2d, a + 3d, \ldots$$

Note that the first n terms of an arithmetic progression, an example of a finite sequence, can be written as

$$a, a + d, a + 2d, a + 3d, \ldots, a + (n - 1)d$$

The nth or last term l of such a finite arithmetic progression is then given by

$$l = a + (n - 1)\,d$$

EXAMPLE 1

Find the sixth term of the arithmetic progression 5, 9, 13, ...

The common difference is the difference between *any* term and the preceding term, that is,

$$d = 9 - 5 = 13 - 9 = 4, \qquad a = 5 \quad \text{and} \quad n = 6$$

We then have

$$\begin{aligned} l &= a + (n - 1)\,d \\ &= 5 + (6 - 1)(4) \\ &= 25 \end{aligned}$$

EXAMPLE 2

Find the 22nd term of the arithmetic progression $1, \frac{1}{3}, -\frac{1}{3}, \ldots$.

Since $d = \frac{1}{3} - 1 = -\frac{2}{3}$, $a = 1$, and $n = 22$, we have

$$\begin{aligned} l &= a + (n - 1)d \\ &= 1 + (22 - 1)(-\tfrac{2}{3}) \\ &= -13 \end{aligned}$$

To find the sum S_n of the first n terms of an arithmetic progression, note that the first n terms can be written as an expression involving l instead of a:

$$a, \ldots, l - 2d, l - d, l$$

In fact, we can indicate the sum S_n of these terms by

$$S_n = a + (a + d) + (a + 2d) + \cdots + (l - 2d) + (l - d) + l$$

or

$$S_n = l + (l - d) + (l - 2d) + \cdots + (a + 2d) + (a + d) + a$$

where the terms of the last equation are written in reverse order.

If we add these two equations for S_n, we have

$$2S_n = (a + l) + (a + l) + (a + l) + \cdots + (a + l) + (a + l) + (a + l)$$

since, term by term, the multiples of d add to zero. Note that we obtain a sum of n terms of the form $(a + l)$. That is,

$$2S_n = n(a + l)$$

or

$$S_n = \frac{n}{2}(a + l)$$

EXAMPLE 3

Find the sum of the first 12 terms of the arithmetic progression 6, 11, 16, ...

Since $d = 11 - 6 = 5$, $a = 6$, and $n = 12$, we have

$$\begin{aligned} l &= a + (n - 1)d \\ l &= 6 + (12 - 1)(5) \\ &= 61 \\ S_n &= \frac{n}{2}(a + l) \\ &= \frac{12}{2}(6 + 61) \\ &= 402 \end{aligned}$$

EXAMPLE 4

Find the sum of the first 500 positive integers.

Since $a = 1$, $d = 1$, $n = 500$, and $l = 500$, we have

$$\begin{aligned} S_n &= \frac{n}{2}(a + l) \\ &= \frac{500}{2}(1 + 500) \\ &= 125{,}250 \end{aligned}$$

Exercises 18.1

Find the nth term of each arithmetic progression.

1. $2, 5, 8, \ldots, n = 6$ **2.** $-3, -7, -11, \ldots, n = 7$ **3.** $3, 4\frac{1}{2}, 6, \ldots, n = 15$

4. $-2, \frac{1}{5}, 2\frac{2}{5}, \ldots, n = 8$ **5.** $4, -5, -14, \ldots, n = 12$ **6.** $10, 50, 90, \ldots, n = 9$

7–12. Find the sum of the first n terms of the progressions in Exercises 1 through 6.

Write the first five terms of each arithmetic progression whose first term is a and whose common difference is d.

13. $a = 2, d = -3$ **14.** $a = -4, d = 2$ **15.** $a = 5, d = \frac{2}{3}$ **16.** $a = 3, d = -\frac{1}{2}$

17. Find the first term of an arithmetic progression whose tenth term is 12 and whose sum of the first 10 terms is 80.

18. Find the common difference of an arithmetic progression whose first term is 7 and whose eighth term is 16.

19. Find the sum of the first 1000 odd positive integers.
20. Find the sum of the first 500 even positive integers.
21. A man is employed at an initial salary of \$36,000. If he receives an annual raise of \$1400, what is his salary for the tenth year?
22. Equipment purchased at an original value of \$13,600 is depreciated \$1200 per year for 10 years. Find the depreciated value after 4 years. Find the scrap value (depreciated value after 10 years).

18.2 GEOMETRIC PROGRESSIONS

A geometric progression is another example of a sequence. A *geometric progression* is a sequence of terms where each term can be obtained by multiplying the preceding term by a fixed number r, which is called the *common ratio*. For example, $1, \frac{1}{2}, \frac{1}{4}, \frac{1}{8}, \ldots$ is a geometric progression with a common ratio of $\frac{1}{2}$, and $-6, -18, -54, -162, \ldots$ is a geometric progression with a common ratio of 3.

For any geometric progression if a is the first term and r is the common ratio, then ar is the second term. Likewise, $(ar)r = ar^2$ is the third term and $(ar^2)r = ar^3$ is the fourth term. Continuing in this manner, we can express a geometric progression as follows:

$$a, ar, ar^2, ar^3, \ldots, ar^{n-1}$$

where ar^{n-1} is the nth or last term l of the progression. That is,

$$l = ar^{n-1}$$

EXAMPLE 1

Find the eighth term of the geometric progression $1, \frac{1}{2}, \frac{1}{4}, \frac{1}{8}, \ldots$.

The common ratio is found by dividing *any* term by the preceding term. So

$$r = \frac{\frac{1}{2}}{1} = \frac{\frac{1}{4}}{\frac{1}{2}} = \frac{\frac{1}{8}}{\frac{1}{4}} = \frac{1}{2}$$

Since $a = 1$ and $n = 8$, we have

$$\begin{aligned} l &= ar^{n-1} \\ &= (1)(\tfrac{1}{2})^{8-1} \\ &= \frac{1}{128} \end{aligned}$$

EXAMPLE 2

Find the 10th term of the geometric progression $3, -6, 12, -24, \ldots$.

In this example $a = 3$, $r = -\frac{6}{3} = -2$, and $n = 10$.

$$\begin{aligned} l &= ar^{n-1} \\ &= (3)(-2)^{10-1} \\ &= -1536 \end{aligned}$$

To find the sum S_n of the first n terms of the geometric progression

$$S_n = a + ar + ar^2 + ar^3 + \cdots + ar^{n-2} + ar^{n-1}$$

multiply each side of this equation by r and subtract as follows:

$$\begin{array}{rl} S_n = & a + ar + ar^2 + ar^3 + \cdots + ar^{n-2} + ar^{n-1} \\ rS_n = & \quad\; ar + ar^2 + ar^3 + ar^4 + \cdots \quad + ar^{n-1} + ar^n \\ \hline S_n - rS_n = & a \qquad\qquad\qquad\qquad\qquad\qquad\qquad - ar^n \end{array}$$

Solving for S_n, we have

$$S_n(1 - r) = a(1 - r^n)$$

or

$$S_n = \frac{a(1 - r^n)}{1 - r}$$

EXAMPLE 3

Find the sum of the first eight terms of the geometric progression $1, \frac{1}{2}, \frac{1}{4}, \frac{1}{8}, \ldots$.

Since $r = \frac{1}{2}$, $a = 1$, and $n = 8$, we have

$$\begin{aligned} S_n &= \frac{a(1 - r^n)}{1 - r} \\ &= \frac{(1)[1 - (\frac{1}{2})^8]}{1 - \frac{1}{2}} \\ &= \frac{1 - \frac{1}{256}}{\frac{1}{2}} = \frac{255}{128} \end{aligned}$$

EXAMPLE 4

Find the sum of the first five terms of the geometric progression $2, -\frac{2}{3}, \frac{2}{9}, -\frac{2}{27}, \ldots$.

In this example $a = 2$, $r = -\frac{1}{3}$, and $n = 5$.

$$\begin{aligned} S_n &= \frac{a(1 - r^n)}{1 - r} \\ &= \frac{(2)[1 - (-\frac{1}{3})^5]}{1 - (-\frac{1}{3})} \\ &= \frac{(2)(1 + \frac{1}{243})}{\frac{4}{3}} = \frac{122}{81} \end{aligned}$$

EXAMPLE 5

If \$3000 is deposited annually in a savings account at 8% interest compounded annually, find the total amount in this account after 4 years.

The total amount in the account is the sum of a geometric progression

$$(3000)(1.08) + (3000)(1.08)^2 + (3000)(1.08)^3 + (3000)(1.08)^4$$

since the value of each dollar in the account increases by 8% each year. Note that the first term $(3000)(1.08) = \$3240$ represents the amount in the account after 1 year. Thus $a = 3240$, $r = 1.08$, and $n = 4$.

$$\begin{aligned} S_n &= \frac{a(1 - r^n)}{1 - r} \\ &= \frac{(3240)(1 - 1.08^4)}{1 - 1.08} \\ &= \$14{,}600 \end{aligned}$$

The term *series* is used to denote the sum of a sequence of terms. Each S_n is thus a finite series. The methods of computing the sums S_n of finite arithmetic and geometric series have already been shown.

An infinite series is the indicated sum of an infinite sequence of terms. For example, $1 + \frac{1}{2} + \frac{1}{4} + \frac{1}{8} + \cdots$ is an infinite series. Since it is the infinite summation of the terms of a geometric sequence, it is called an *infinite geometric series.*

In Example 3 we found that the sum of the first eight terms of the geometric progression $1, \frac{1}{2}, \frac{1}{4}, \frac{1}{8}, \ldots$ is $\frac{255}{128}$. The sum of the first nine terms can be shown to be $\frac{511}{256}$ and the sum of the first ten terms is $\frac{1023}{512}$. This last sum is close to the value 2. In fact, the sum of the first 50 terms is given by

$$S_n = \frac{(1)[1 - (\frac{1}{2})^{50}]}{1 - \frac{1}{2}} = \frac{1 - (\frac{1}{2})^{50}}{\frac{1}{2}} = 2[1 - (\tfrac{1}{2})^{50}]$$

But since $\left(\frac{1}{2}\right)^{50} = \frac{1}{1{,}125{,}899{,}906{,}842{,}624}$, which is practically zero, we conclude that the sum S_n is very close to the value 2. The sum S_n gets closer and closer to 2 as n is given a larger and larger value.

If we denote the sum of the first n terms of a geometric progression by S_n and S as the sum of the terms of an infinite geometric progression, then

$$S = 1 + \tfrac{1}{2} + \tfrac{1}{4} + \tfrac{1}{8} + \cdots = 2$$

Not every infinite series has a finite sum. For example, $3 + 6 + 12 + 24 + \cdots$ has no finite sum. When the sum exists, the series is said to *converge.* When the limit or sum does not exist, the series is said to *diverge.*

In general, if $|r| < 1$, the infinite geometric series

$$a + ar + ar^2 + \cdots + ar^{n-1} + \cdots$$

has the sum

$$S = \frac{a}{1 - r}$$

If $|r| \geq 1$, the infinite geometric series has no sum (it diverges).

EXAMPLE 6

Find the sum of the infinite geometric series $3 + \frac{3}{5} + \frac{3}{25} + \cdots + 3(\frac{1}{5})^{n-1} + \cdots$

Since $r = \frac{1}{5}$ and $a = 3$, we have

$$S = \frac{a}{1 - r}$$

$$= \frac{3}{1 - \frac{1}{5}} = \frac{15}{4}$$

EXAMPLE 7

Find, if possible, the sum of the infinite geometric series

$$1 + 2 + (2)^2 + (2)^3 + \cdots + (2)^{n-1} + \cdots$$

Since $r = 2 \geq 1$, this infinite geometric series diverges.

EXAMPLE 8

Find a fraction that is equivalent to the decimal 0.232323 . . . .

We can write this decimal as the infinite series

$$0.23 + 0.0023 + 0.000023 + \cdots$$

Then $a = 0.23$ and $r = 0.01$. Thus,

$$S = \frac{a}{1 - r} = \frac{0.23}{1 - 0.01} = \frac{0.23}{0.99} = \frac{23}{99}$$

Note that any repeating decimal can be expressed as a fraction. The fractional form can be found by using the method shown in Example 8.

EXAMPLE 9

A tank contains 1000 gal of alcohol. Then 100 gal are drained out and the tank refilled with water. Then 100 gal of the mixture are drained out and the tank refilled with water. Assuming this process continues, how much alcohol remains in the tank after eight 100-gal units are drained out?

Draining the first 100 gal of alcohol leaves 900 gal or 9/10 of the alcohol in the tank. The amount of alcohol left after eight drainings is the ninth term in a geometric progression where $a = 1000$ gal and $r = 9/10$.

Term	*Gallons of alcohol*	*Drain number*
1	1000	0
2	1000 (9/10)	1
3	1000 (9/10)(9/10)	2
.	.	.
.	.	.
.	.	.
9	$1000\ (9/10)^8$	8

The formula $l = ar^{n-1}$ gives

$$l = 1000 \text{ gal} \cdot (9/10)^8 = 430 \text{ gal}$$

Exercises 18.2

Find the nth term of each geometric progression.

1. $20, \frac{20}{3}, \frac{20}{9}, \ldots, n = 8$ **2.** $\frac{1}{8}, -\frac{1}{4}, \frac{1}{2}, \ldots, n = 7$

3. $\sqrt{2}, 2, 2\sqrt{2}, \ldots, n = 6$ **4.** $6, 3, \frac{3}{2}, \ldots, n = 8$

5. $8, -4, 2, \ldots, n = 10$ **6.** $3, 12, 48, \ldots, n = 5$

7–12. Find the sum of the first n terms of the progressions in Exercises 1 through 6.

Write the first five terms of each geometric progression whose first term is a and whose common ratio is r.

13. $a = 3, r = \frac{1}{2}$ **14.** $a = -6, r = \frac{1}{3}$ **15.** $a = 5, r = -\frac{1}{4}$

16. $a = 2, r = -\frac{3}{2}$ **17.** $a = -4, r = 3$ **18.** $a = -5, r = -2$

19. Find the common ratio of a geometric progression whose first term is 6 and whose fourth term is $\frac{3}{4}$.

20. Find the first term of a geometric progression with common ratio $\frac{1}{3}$ if the sum of the first three terms is 13.

21. If $1000 is deposited annually in an account at 5.75% interest compounded annually, find the total amount in the account after 10 years.

22. If $200 is deposited quarterly in an account at 4% annual interest compounded quarterly, find the total amount in the account after 10 years.

23. A ball is dropped from a height of 12 ft. After each bounce, it rebounds to $\frac{1}{2}$ the height of the previous height from which it fell. Find the distance the ball rises after the fifth bounce.

24. The half-life of a chlorine isotope, ^{38}Cl, used in radioisotope therapy is 37 min. This means that half of a given amount will disintegrate in 37 min. This means also that three-fourths will have disintegrated after 74 min. Find how much will have disintegrated in 148 min.

25. A salt solution is being cooled so that the temperature decreases 20% each minute. Find the temperature of the solution after 8 min if the original temperature was 90°C.

26. A tank contains 400 gal of acid. Then 100 gal is drained out and refilled with water. Then 100 gal of the mixture is drained out and refilled with water. Assuming that this process continues, how much acid remains in the tank after five 100-gal units are drained out?

Find the sum, when possible, for each infinite geometric series.

27. $4 + \frac{4}{7} + \frac{4}{49} + \cdots + 4(\frac{1}{7})^{n-1} + \cdots$

28. $6 + \frac{6}{11} + \frac{6}{121} + \cdots + 6(\frac{1}{11})^{n-1} + \cdots$

29. $3 - \frac{3}{8} + \frac{3}{64} - \cdots + 3(-\frac{1}{8})^{n-1} + \cdots$

30. $1 - \frac{1}{9} + \frac{1}{81} - \cdots + (-\frac{1}{9})^{n-1} + \cdots$

31. $4 + 12 + 36 + \cdots + 4(3)^{n-1} + \cdots$

32. $-5 - \frac{5}{2} - \frac{5}{4} - \cdots - 5(\frac{1}{2})^{n-1} - \cdots$

33. $5 + 1 + 0.2 + 0.04 + \cdots + 5(0.2)^{n-1} + \cdots$

34. $2 + 2 + 2 + \cdots + 2(1)^{n-1} + \cdots$

Find the fraction that is equivalent to each decimal.

35. 0.3333 ...

36. 0.135135135 ...

37. 0.0121212 ...

38. 0.6252525 ...

39. 0.86666 ...

40. 0.365365365 ...

18.3 THE BINOMIAL THEOREM

The *binomial theorem* provides us with a convenient means of expressing any power of a binomial as a sum of terms.

For small nonnegative integers n, we can find $(a + b)^n$ by actual multiplication.

$$\begin{aligned}
n = 0:&\quad (a + b)^0 = 1\\
n = 1:&\quad (a + b)^1 = a + b\\
n = 2:&\quad (a + b)^2 = a^2 + 2ab + b^2\\
n = 3:&\quad (a + b)^3 = a^3 + 3a^2b + 3ab^2 + b^3\\
n = 4:&\quad (a + b)^4 = a^4 + 4a^3b + 6a^2b^2 + 4ab^3 + b^4\\
n = 5:&\quad (a + b)^5 = a^5 + 5a^4b + 10a^3b^2 + 10a^2b^3 + 5ab^4 + b^5
\end{aligned}$$

We could continue this process, but the multiplications become more complicated for larger values of n.

No matter what positive integral value of n is chosen, we obtain the following results:

1. $(a + b)^n$ has $n + 1$ terms.
2. The first term is a^n.
3. The second term is $na^{n-1}b$.
4. The exponent of a decreases by 1 and the exponent of b increases by 1 for each successive term.
5. In each term, the sum of the exponents of a and b is n.
6. The last term is b^n.

The kth term is given by the formula

$$\frac{n(n-1)(n-2)\cdots(n-k+2)}{(k-1)!}a^{n-k+1}b^{k-1}$$

where $k!$ (k factorial) indicates the product of the first k positive integers. (For example, $3! = 3 \cdot 2 \cdot 1 = 6$; $5! = 5 \cdot 4 \cdot 3 \cdot 2 \cdot 1 = 120$; and $6! = 6 \cdot 5 \cdot 4 \cdot 3 \cdot 2 \cdot 1 = 720$.) The three dots in the numerator indicate that the multiplication of decreasing numbers is to continue until the number $n - k + 2$ is reached. For example, if $n = 8$ and $k = 4$, then the formula gives

$$\frac{8 \cdot 7 \cdot 6}{3 \cdot 2 \cdot 1} = 56$$

BINOMIAL THEOREM

$$(a+b)^n = a^n + na^{n-1}b + \frac{n(n-1)}{2!}a^{n-2}b^2 + \frac{n(n-1)(n-2)}{3!}a^{n-3}b^3 + \cdots + \frac{n(n-1)(n-2)\cdots(n-k+2)}{(k-1)!}a^{n-k+1}b^{k-1} + \cdots + b^n$$

where the three dots indicate that you are to complete the process of calculating the terms. The expression for the kth term is also given.

EXAMPLE 1

Expand $(x + 4y)^5$ by using the binomial theorem.

Let $a = x$, $b = 4y$, and $n = 5$.

$$\begin{aligned}(x+4y)^5 &= x^5 + 5x^{5-1}(4y) + \frac{5(5-1)}{2!}x^{5-2}(4y)^2 + \frac{5(5-1)(5-2)}{3!}x^{5-3}(4y)^3 \\ &\quad + \frac{5(5-1)(5-2)(5-3)}{4!}x^{5-4}(4y)^4 + (4y)^5 \\ &= x^5 + 20x^4y + 160x^3y^2 + 640x^2y^3 + 1280xy^4 + 1024y^5\end{aligned}$$

For small values of n, it is possible to determine the coefficients of each term of the expansion by the use of *Pascal's triangle* as shown below:

$n = 0$:						1					
$n = 1$:					1		1				
$n = 2$:				1		2		1			
$n = 3$:			1		3		3		1		
$n = 4$:		1		4		6		4		1	
$n = 5$:	1		5		10		10		5		1

Observe the similarity of this triangle to the triangular format shown earlier when expanding $(a + b)^n$ for $n = 0, 1, 2, 3, 4$, and 5. Each row gives the coefficients for all terms of the binomial expansion for a given integer n. Each successive row provides the coefficients for the next integer n. Each row is read from left to right. The first and last coefficients are always 1 as observed in the triangle. Beginning with the third row ($n = 2$), coefficients of terms other than the first and last are found by adding together the two nearest coefficients found in the row above. For example, the coefficient of the fourth term for the expansion with $n = 5$ is 10, which is the sum of 6 and 4. The numbers 6 and 4 appear just above 10 in Pascal's triangle.

We can enlarge Pascal's triangle to obtain a row for any desired integer n. However, this is not practical for large values of n.

EXAMPLE 2

Using Pascal's triangle, find the coefficients of the terms of the binomial expansion for $n = 7$.

We need two more rows of the triangle:

$$n = 6: \quad 1 \quad 6 \quad 15 \quad 20 \quad 15 \quad 6 \quad 1$$

$$n = 7: \quad 1 \quad 7 \quad 21 \quad 35 \quad 35 \quad 21 \quad 7 \quad 1$$

The last row provides the desired coefficients.

Using a calculator, we can show that the entries in the row of Pascal's triangle for $n = 7$ match the decimal digits of $(1.01)^7$. To understand this, think of $(1.01)^7$ as $(1 + 0.01)^7$ and apply the binomial theorem. Note that we can also produce the first several rows of Pascal's triangle by entering the number 1 and multiplying it by 1.01 *repeatedly.* Unfortunately, three-digit numbers appear in the rows of Pascal's triangle for $n \geq 9$, so powers of 1.01 cannot be used beyond the row for $n = 8$.

1.01^7 **ENTER** up arrow **ENTER F1 8** 1 **ENTER** × 1.01 **ENTER ENTER ENTER,** *etc.*

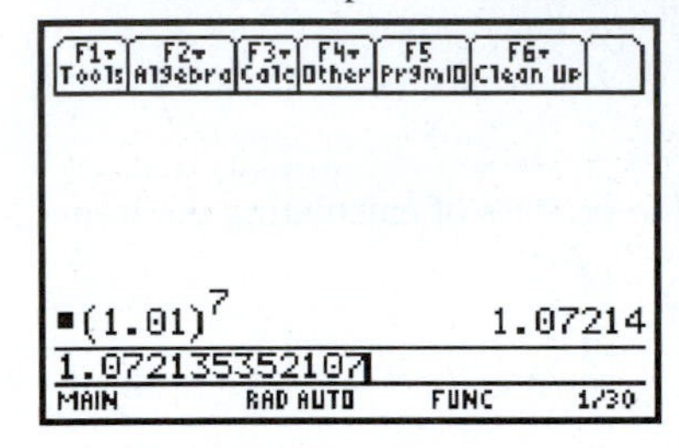

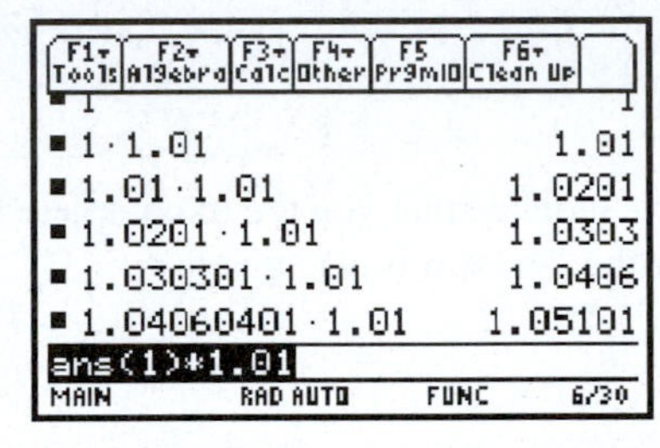

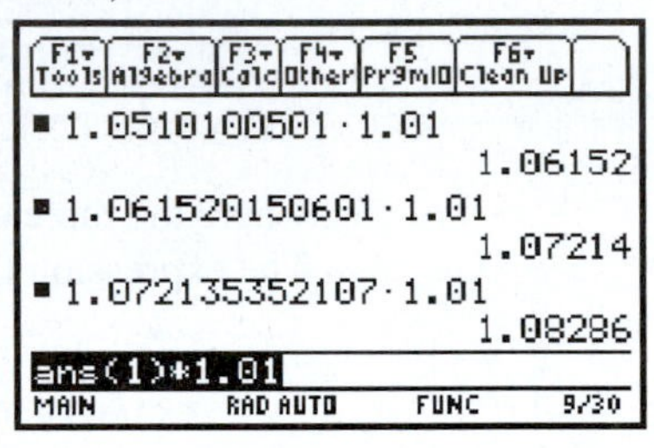

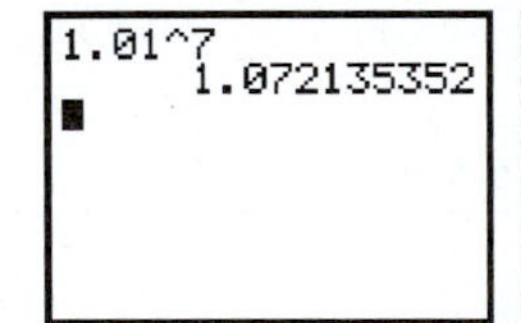

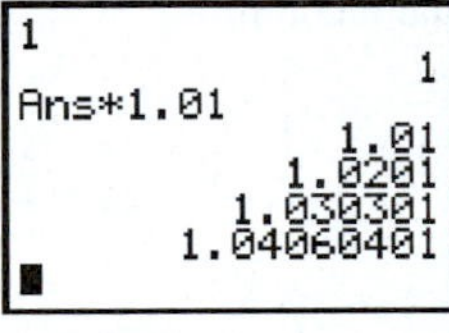

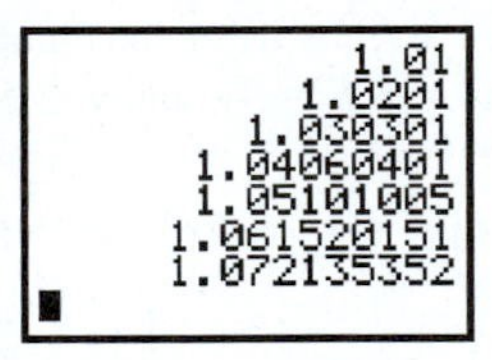

1.01^7 **ENTER**

1 **ENTER** × 1.01 **ENTER ENTER ENTER,** *etc.*

EXAMPLE 3

Expand $(2m + k)^7$ using the binomial theorem.

Let $a = 2m$, $b = k$, and $n = 7$.

$$\begin{aligned}(2m + k)^7 &= 1(2m)^7 + 7(2m)^{7-1}k^1 + 21(2m)^{7-2}k^2 + 35(2m)^{7-3}k^3 \\ &\quad + 35(2m)^{7-4}k^4 + 21(2m)^{7-5}k^5 + 7(2m)^{7-6}k^6 + (1)k^7 \\ &= 128m^7 + 448m^6k + 672m^5k^2 + 560m^4k^3 \\ &\quad + 280m^3k^4 + 84m^2k^5 + 14mk^6 + k^7\end{aligned}$$

EXAMPLE 4

Find the seventh term of $(x^3 - 2y)^{10}$.

First, note that $k = 7$, $n = 10$, $a = x^2$, and $b = -2y$.

$$k\text{th term} = \frac{n(n-1)(n-2)\cdots(n-k+2)}{(k-1)!}a^{n-k+1}b^{k-1}$$

$$\text{seventh term} = \frac{10 \cdot 9 \cdot 8 \cdot 7 \cdot 6 \cdot 5}{6!}(x^3)^4(-2y)^6$$
$$= 210(x^{12})(64y^6)$$
$$= 13{,}440x^{12}y^6$$

Graphing calculators are programmed to calculate the kth binomial coefficient. The notation is ${}_nC_{r-1}$, where r is used instead of k. For $(a + b)^n$ the rth term is ${}_nC_{r-1}(a)^{n-(r-1)}\,(b)^{r-1}$. Doing Example 4 with a calculator, we have $r = 7$, $n = 10$, $a = x^2$, and $b = -2y$.

$$\text{seventh term} = {}_{10}C_{7-1}\,(x^2)^{10-(7-1)}(-2y)^{7-1}$$
$$\text{seventh term} = {}_{10}C_6(x^2)^4(-2y)^6$$

Use your calculator to find ${}_{10}C_6$ as follows:

2nd MATH 7 3

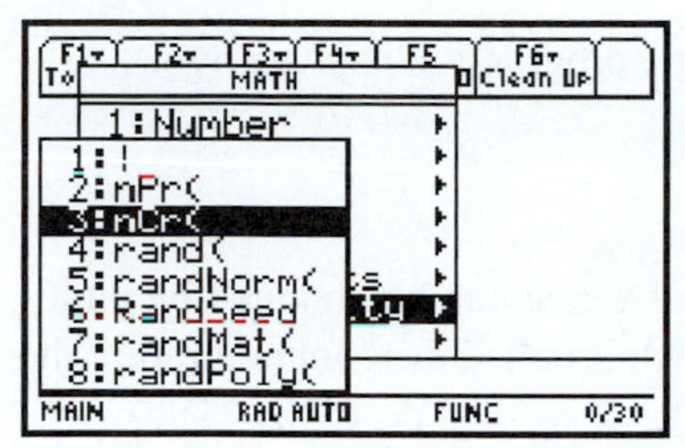

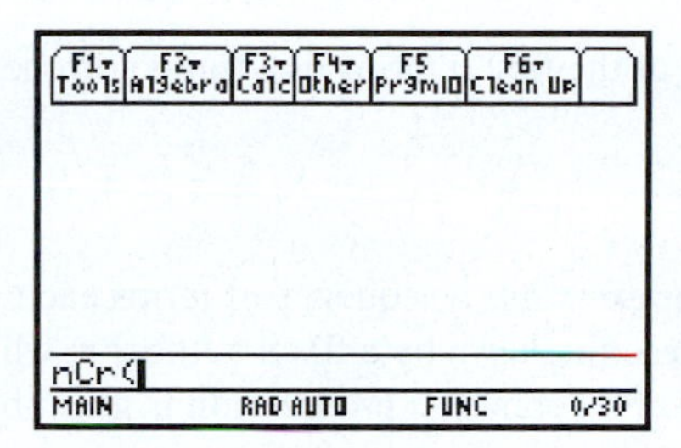

10,6) **ENTER**

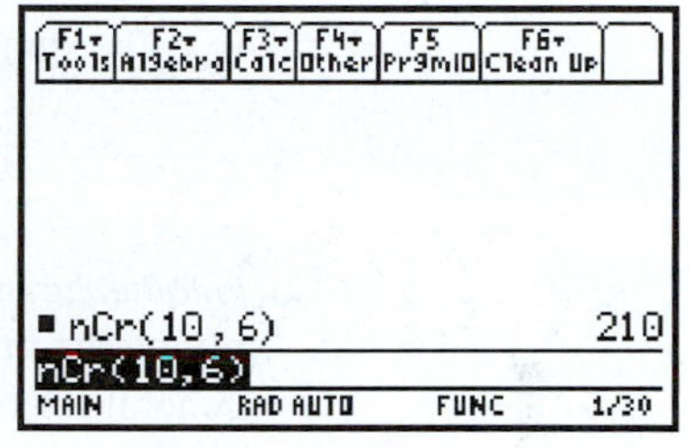

10

MATH left arrow **3**

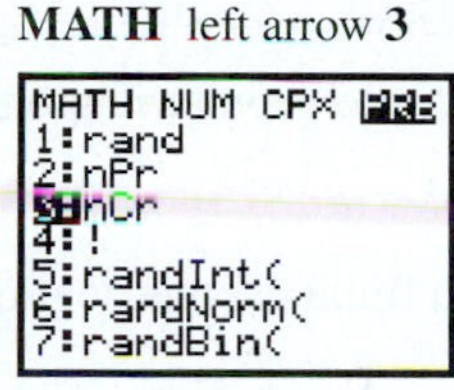

6 ENTER

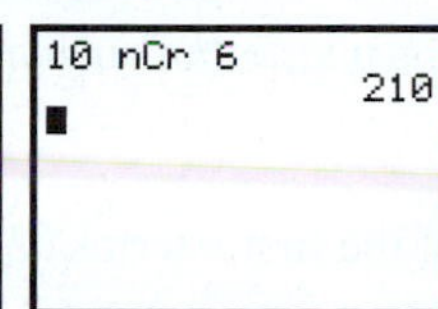

$$\text{seventh term} = 210x^8\,(-2)^6\,y^6$$
$$\text{seventh term} = 210x^8(64)y^6 = 13{,}400x^8y^6$$

Exercises 18.3

Expand each binomial using the binomial theorem.

1. $(3x + y)^3$ **2.** $(2x - 3y)^4$ **3.** $(a - 2)^5$ **4.** $(5x + 1)^4$

5. $(2x - 1)^4$ **6.** $(1 + x)^7$ **7.** $(2a + 3b)^6$ **8.** $(a - 2b)^8$

9. $(\frac{2}{3}x - 2)^5$ **10.** $(\frac{3}{4}m + \frac{2}{3}k)^5$ **11.** $(a^{1/2} + 3b^2)^4$ **12.** $(x^{1/2} - y^{1/2})^4$

13. $\left(\frac{x}{y} - \frac{2}{z}\right)^4$ **14.** $\left(\frac{2x}{y} + \frac{3}{z}\right)^6$

Find the indicated term of each binomial expansion.

15. $(x - y)^9$; 6th term

16. $(4x + 2y)^5$; 3rd term

17. $(2x - y)^{13}$; 9th term

18. $(x^{1/2} + 2)^{10}$; 8th term

19. $(2x + y^2)^7$; 5th term

20. $(x^2 - y^3)^8$; middle term

21. $(3x + 2y)^6$; middle term

22. $(x - 2y)^{10}$; middle term

23. $(2x - 1)^{10}$; term containing x^5

24. $(x^2 + 1)^8$; term containing x^6

CHAPTER 18 SUMMARY

1. *Arithmetic progression:* a sequence of terms where each term differs from the immediately preceding term by a fixed number d, which is called the common difference. The general form of an arithmetic progression is written

$$a, a + d, a + 2d, a + 3d, \ldots, a + (n - 1)d$$

(a) The nth or last term l of such a finite arithmetic progression is given by

$$l = a + (n - 1)d$$

(b) The sum of the first n terms of a finite arithmetic progression is given by

$$S_n = \frac{n}{2}(a + l)$$

2. *Geometric progression:* a sequence of terms each of which can be obtained by multiplying the preceding term by a fixed number r, which is called the common ratio. The general form of a geometric progression is given by

$$a, ar, ar^2, ar^3, \ldots, ar^{n-1}$$

(a) The nth or last term l of such a finite geometric progression is given by

$$l = ar^{n-1}$$

(b) The sum of the first n terms of a finite geometric progression is given by

$$S_n = \frac{a(1 - r^n)}{1 - r}$$

(c) The sum of the infinite geometric series, where $|r| < 1$, is

$$S = \frac{a}{1 - r}$$

3. $k!$ (k factorial) $= k(k - 1)(k - 2) \cdots 4 \cdot 3 \cdot 2 \cdot 1$. For example,

$$5! = 5 \cdot 4 \cdot 3 \cdot 2 \cdot 1$$
$$10! = 10 \cdot 9 \cdot 8 \cdot 7 \cdot 6 \cdot 5 \cdot 4 \cdot 3 \cdot 2 \cdot 1$$

4. *Binomial theorem:*

$$(a + b)^n = a^n + na^{n-1}b + \frac{n(n - 1)}{2!}a^{n-2}b^2 + \frac{n(n - 1)(n - 2)}{3!}a^{n-3}b^3 + \cdots + \frac{n(n - 1)(n - 2) \cdots (n - k + 2)}{(k - 1)!}a^{n-k+1}b^{k-1} + \cdots + b^n$$

5. The *kth term of a binomial expansion* is given by the formula

$$\frac{n(n - 1)(n - 2) \cdots (n - k + 2)}{(k - 1)!}a^{n-k+1}b^{k-1}$$

or $\quad {}_nC_{k-1}\, a^{n-k+1} b^{k-1}$

CHAPTER 18 REVIEW

Find the nth term of each progression.

1. $3, 7, 11, 15, \ldots, n = 12$

2. $4, 2, 1, \frac{1}{2}, \ldots, n = 7$

3. $\sqrt{3}, -3, 3\sqrt{3}, -9, \ldots, n = 8$

4. $4, -2, -8, -14, \ldots, n = 12$

5. $6, 2, \frac{2}{3}, \frac{2}{9}, \ldots, n = 6$

6. $5, 15, 25, 35, \ldots, n = 10$

7–12. Find the sum of the first n terms of the progressions in Exercises 1 through 6.

13. Find the sum of the first 1000 even positive integers.

14. If $500 is deposited annually in a savings account at 6% interest compounded annually, find the total amount in the account after 5 years.

Find the sum, when possible, for each infinite geometric series.

15. $3 + 6 + 12 + \cdots$

16. $5 + \frac{5}{7} + \frac{5}{49} + \cdots$

17. $2 - \frac{2}{3} + \frac{2}{9} - \cdots$

18. $3 + \frac{9}{2} + \frac{27}{4} + \cdots$

19. Find the fraction equivalent to 0.454545 . . . .

20. Find the fraction equivalent to 0.9212121 . . . .

Expand each binomial using the binomial theorem.

21. $(a - b)^6$

22. $(2x^2 - 1)^5$

23. $(2x + 3y)^4$

24. $(1 + x)^8$

Find the indicated term of each binomial expansion.

25. $(1 - 3x)^5$; 3rd term

26. $(a + 4b)^6$; 4th term

27. $(x + 2b^2)^{10}$; middle term

28. $(3x^2 - 1)^{12}$; term containing x^{16}

19 Basic Statistics

INTRODUCTION

Statistical processes affect almost every aspect of our lives, including gauging the popularity of a political candidate, the reliability and effectiveness of a medicine, and the quality of food and other products that we use daily.

To conduct a quality check on every item that is produced would require substantial resources. Also, some items would be ruined in the process. That is why we check a sample of items (i.e., some, rather than all). Statistical methods allow us to draw conclusions about all items based on what we find in the samples. In this chapter we will focus on some of the statistical methods used for quality control.

Objectives

- Use frequency tables, histograms, and frequency polygons to display data.
- Calculate and use the mean, the median, and the mode of a data set.
- Calculate and use the standard deviation of a data set.
- Fit a curve to a set of points.
- Use a normal distribution to find the percent of values within a given interval.

19.1 GRAPHICAL PRESENTATION OF DATA

Gathering data and interpreting it to draw conclusions or predict the future is the basis for studying statistics. A pictorial representation is very helpful in understanding how data is distributed.

EXAMPLE 1

In the month of April an automobile manufacturer collected daily data on the number of cars coming off the line that would not start on the first try (see Table 19.1). Because the data is not grouped or arranged by size, it is difficult to get a sense of its distribution. We can improve the display of data by grouping it and counting the number of occurrences in each group. This is called a **frequency table** (see Table 19.2). Usually data is grouped so that there are between 5 and 15 groups or intervals. Each interval should have the same width. In Table 19.2, there are 5 groups or intervals and each interval has a width of 8.

TABLE 19.1 Cars Not Starting on the First Try for Each Day in April

42	25	12	44	17	18
27	18	48	30	36	9
33	46	29	14	26	28
31	39	35	31	32	21
29	20	16	45	23	35

TABLE 19.2 Cars Not Starting on the First Try in April

First-try nonstarts	Tallies	Frequency
9–16	||||	4
17–24	~~||||~~ |	6
25–32	~~||||~~ ~~||||~~	10
33–40	~~||||~~	5
41–48	~~||||~~	5

The data from Table 19.2 could also be presented in a **histogram**—a graph composed of rectangles whose equal widths represent a group of data and whose heights represent the frequency with which the data occurs. Figure 19.1 is a histogram of the data from Table 19.2.

A **frequency polygon** is another way to graphically represent data. To construct a frequency polygon, we plot points corresponding to the center of an interval and the frequency of the interval and connect the points with straight lines. For the data from Table 19.2, we plot the points (12.5, 4), (20.5, 6), (28.5, 10), (36.5, 5), and (44.5, 5). Figure 19.2 is a frequency polygon.

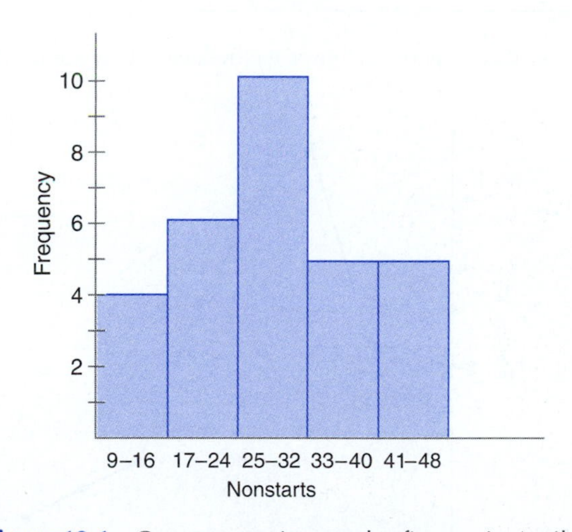

Figure 19.1 Cars not starting on the first try in April: histogram.

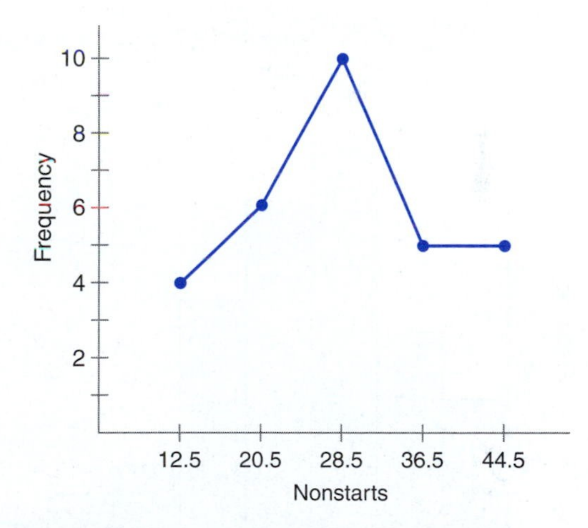

Figure 19.2 Cars not starting on the first try in April: frequency polygon.

EXAMPLE 2

After receiving many complaints about Catchum's 10-lb fishing line breaking, a national consumer advocate group decided to test the line. They suspended 10-lb weights from 5000 different lines. Table 19.3 gives the number of breaks in a line for each 100 lines. Prepare a frequency table, a histogram, and a frequency polygon for the data given in Table 19.3.

TABLE 19.3 **Numbers of Breaks for Each 100 Lines**

15	5	8	3	9	4	10	8	6	4
8	2	11	7	6	7	6	1	4	5
3	5	0	1	4	8	12	3	2	0
7	6	4	5	3	5	5	4	9	5
0	7	2	8	7	10	13	7	6	9

In preparing the frequency table, we need to decide whether to group the data. Since not grouping would give us 16 rectangles for our histogram, we decide to group in two's for all three data representations.

Note: We could make the width of the intervals 5; then the groups would be 0 to 4, 5 to 9, 10 to 14, and 15 to 18. Even though 17 and 18 are not data values, we include them in the last interval so that the interval widths will be the same. Table 19.4 is the frequency table for the data grouped in two's.

TABLE 19.4 **Number of Breaks for Each 100 Lines**

Line breaks per 100	Tallies	Frequency
0–1	~~IIII~~	5
2–3	~~IIII~~ II	7
4–5	~~IIII~~ ~~IIII~~ III	13
6–7	~~IIII~~ ~~IIII~~ I	11
8–9	~~IIII~~ III	8
10–11	III	3
12–13	II	2
14–15	I	1

Figure 19.3 gives the histogram for the data. Figure 19.4 gives the frequency polygon for the data.

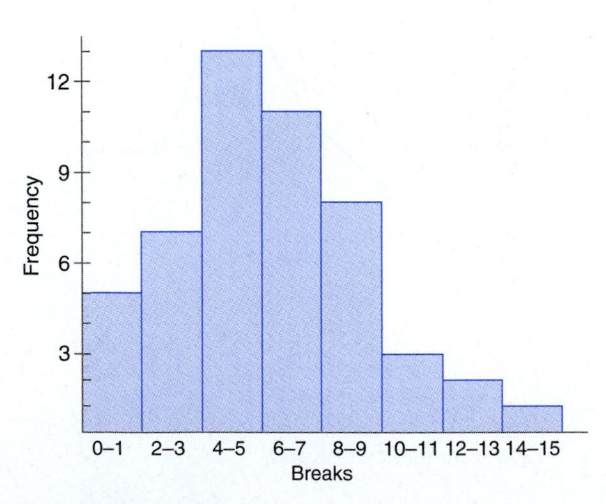

Figure 19.3 Number of breaks for each 100 lines: histogram.

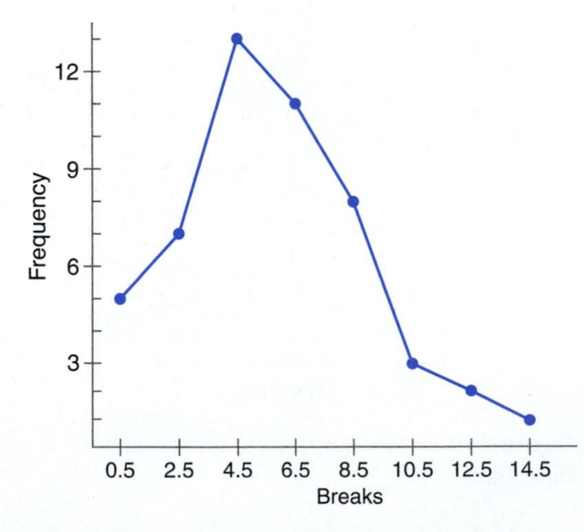

Figure 19.4 Number of breaks for each 100 lines: frequency polygon.

Quality control equipment will often display histograms and frequency polygons to illustrate the distribution of data. This is helpful to technicians who need to interpret the data.

Exercises 19.1

*For Exercises 1 through 7, construct **(a)** a frequency table, **(b)** a histogram, and **(c)** a frequency polygon.*

1. Twenty-five light bulbs were tested for the number of hours they would give light. The data is contained in Table 19.5. Use five intervals each containing 39 numbers: 205 to 243, 244 to 282, and so on.

TABLE 19.5 Number of Hours Light Bulbs Give Light

372	297	320	344	336
258	384	316	351	305
205	375	335	349	382
326	338	291	358	399
362	294	345	360	395

2. A manufacturer of hardware produces ten thousand 18-mm bolts every week. The specifications indicate that a 1% error is acceptable; that is, the bolts can really be between 17.82 mm and 18.18 mm in length. For quality control 500 bolts are checked every week for 35 weeks. The number of bolts that do not meet specifications each week is given in Table 19.6. Use six intervals each containing five numbers.

TABLE 19.6 Number of Bolts Not Acceptable

7	26	17	9	27	3	28
14	17	21	3	16	20	12
8	3	17	18	27	2	27
0	21	5	27	23	28	9
13	16	18	29	9	25	3

3. A candy maker produces 3-oz chocolate almond candy bars. Each bar is supposed to contain between 0.75 oz and 1.00 oz of almonds. Two hundred bars are checked daily for 30 consecutive days to see if they contain the correct amount of almonds. Table 19.7 gives the daily count of the bars that do not have the correct amount of almonds. Use eight intervals each containing three numbers.

TABLE 19.7 Number of Candy Bars That Do Not Have the Correct Amount of Almonds

1	15	24	15	17	21
3	12	13	2	18	19
23	12	23	17	6	10
5	21	18	21	1	9
6	16	3	13	8	11

4. A cheese producer cuts cheddar cheese into 10-oz pieces to be wrapped for retail sale. Five hundred pieces are checked for the correct weight each week for 25 weeks. Table 19.8 gives the weekly number of pieces that did not weigh the correct amount. Use eight intervals each containing six numbers.

TABLE 19.8 Number of Pieces That Did Not Weigh the Correct Amount

6	12	29	12	11
19	11	31	34	26
11	24	20	19	9
48	35	28	44	21
19	18	7	1	22

5. A seed company harvests and packages grass seed for 20 weeks every year. The 60-oz bags of seed should contain no more than 4 oz of weed seed. One hundred fifty samples are tested each week. Table 19.9 gives the number of samples that failed each week. Use five intervals each containing four numbers.

TABLE 19.9 Number of Samples That Had Too Much Weed Seed

5	10	16	7
16	11	3	10
13	0	16	5
19	11	0	18
8	15	2	9

6. A local restaurant counted the number of hamburgers served on 25 consecutive weekends. Table 19.10 gives the count for each weekend. Use 10 intervals each containing six numbers.

TABLE 19.10 Number of Hamburgers Served Each Weekend

268	252	222	279	234
260	261	220	246	268
253	250	228	273	243
268	245	272	254	225
230	240	250	279	231

7. A bottling company fills eight thousand 12-oz juice bottles each week. The actual volume needs to be between 12.0 and 12.6 oz. One hundred bottles were tested for volume for 25 consecutive weeks. Table 19.11 gives the number of bottles per week that did not contain an acceptable volume. Use six intervals each containing three numbers.

TABLE 19.11 Number of Bottles with Unacceptable Volume

11	3	13	3	15
6	17	10	9	6
14	16	3	4	12
7	13	7	14	7
0	15	8	7	5

For Exercises 8 through 14, construct ***(a)*** *a frequency table,* ***(b)*** *a histogram, and* ***(c)*** *a frequency polygon.*

8. Use the data in Table 19.5 with 13 intervals each containing 15 numbers.

9. Use the data in Table 19.6 with 10 intervals each containing 3 numbers.

10. Use the data in Table 19.7 with 6 intervals each containing 4 numbers.

11. Use the data in Table 19.8 with 12 intervals each containing 4 numbers.

12. Use the data in Table 19.9 with 10 intervals each containing 2 numbers.

13. Use the data in Table 19.10 with 4 intervals each containing 15 numbers.

14. Use the data in Table 19.11 with 9 intervals each containing 2 numbers.

19.2 MEASURES OF CENTRAL TENDENCY

Finding one number which describes the center of a collection of numbers is often desirable. A teacher averaging test scores is looking for a single number that describes the collection of scores. Actually, there are three numerical values that are used to describe the central tendency of a set of numbers. They are the mean, the median, and the mode.

The **mean** is commonly referred to as the average of a set of numbers. It is the sum of the numbers divided by the number of numbers.

THE MEAN

The mean, denoted $\bar{x}$, of n numbers $a_1, a_2, \ldots, a_n$ is given by

$$\bar{x} = \frac{a_1 + a_2 + \cdots + a_n}{n}$$

EXAMPLE 1

Nine farmers reported their soybean yield in bushels per acre to their co-op group. The reports were 61, 50, 51, 48, 53, 43, 57, 39, and 57. Find the mean of this set.

$$\bar{x} = \frac{61 + 50 + 51 + 48 + 53 + 43 + 57 + 39 + 57}{9}$$

$$= \frac{459}{9} = 51 \text{ bu/acre}$$

EXAMPLE 2

A production line manager for a bolt manufacturer wrote a daily report for bolts produced in the first week of June. The numbers of bolts produced were 987, 1042, 1005, 946, and 73. Find the mean of this set.

$$\bar{x} = \frac{987 + 1042 + 1005 + 946 + 73}{5}$$

$$\bar{x} = 810.6 \text{ bolts}$$

We round the answer to the nearest whole number of bolts, 811.

The data in Example 2 illustrates that another measure of central tendency may be more meaningful in this situation. Four of the five numbers cluster around 1000. The fifth number represents a breakdown on the line, a very unusual event. If we put the numbers in order, the middle number of the set is the **median.**

MEDIAN

For an ordered set of n numbers $a_1, a_2, \ldots, a_n$, the median is the number in the middle if n is odd and the mean of the two numbers in the middle if n is even.

Putting the numbers from Example 2 in increasing order, we have 73, 946, 987, 1005, and 1042. The median is 987, the number in the middle. The average number of bolts is more accurately described by the median than by the mean.

EXAMPLE 3

The salary committee for a small company is investigating the annual salaries of the employees. The executive informs them that the average (mean) salary of all employees is \$34,149.50. This doesn't sound correct to the committee since none of them earns that much. The employees' salaries in dollars are 30,000; 30,400; 30,900; 31,000; 31,100; 31,250; 31,875; 31,995; 32,125; and 60,850. The median would more accurately describe the distribution of the salaries. The median is the mean of the two middle entries.

$$\text{Median} = \frac{\$31{,}100 + \$31{,}250}{2} = \$31{,}175$$

The third central tendency value we discuss is the **mode.**

MODE

The mode is the element of a data set which occurs most often. A set of data may contain more than one mode.

EXAMPLE 4

A company produces bolts of the following lengths: 0.75 cm, 1.00 cm, 1.25 cm, 1.50 cm, 1.75 cm, and 2.00 cm. The production manager wishes to determine which, if any, is the most in demand. The company sells in lots of 10,000, so he records the bolt size for each lot ordered in July. The data set is given in Table 19.12.

TABLE 19.12

1.25	1.50	1.75	1.75	1.00
0.75	1.25	2.00	1.25	1.75
1.50	1.75	0.75	1.75	2.00
1.75	1.00	1.75	2.00	1.75
1.75	1.75	1.50	1.50	1.50

The mode of the element is 1.75. Ten lots of bolts measuring 1.75 cm were ordered. The next in demand was the 1.50-cm size, with five lots ordered. This information will be used when deciding how many bolts of each size to produce. Creating a frequency table for the information in Table 19.12 would help us to easily identify the mode.

EXAMPLE 5

A technical mathematics class earned the grades listed in Table 19.13 on the second test. Determine the mean, median, and mode for these grades.

TABLE 19.13 Technical Mathematics Grades for Test 2

65	82	65	60	68
83	92	92	97	82
92	82	83	98	60
68	65	60	92	65
26	97	92	65	97

Because the data set is large, we will organize it into a frequency table (see Table 19.14).

TABLE 19.14 Technical Mathematics Grades for Test 2, by Frequency

Grade	26	60	65	68	82	83	92	97	98
Frequency	1	3	5	2	3	2	5	3	1

There are two modes, 65 and 92; each occurs five times. The class appears to be divided into two groups, one that understands the material and one that doesn't.

We can use the frequency table to determine the mean. For example, 60 times 3 is the same as adding three 60's. Thus, we will enter the data into our calculator by using a combination of multiplying and adding. To find the sum of the 25 grades we will use $1(26) + 3(60) + 5(65) + 2(68) + 3(82) + 2(83) + 5(92) + 3(97) + 1(98)$. The sum is 1928. The mean is $\frac{1928}{25} = 77.12$ or approximately 77. Notice that the mean is not as informative as the two modes. No one in the class had a grade in the 70's.

Finally, we wish to find the median. This data set contains 25 entries or numbers. The middle one will be in the 13th position. We find this by dividing 25 by 2 and rounding to the next whole number. We can use the frequency table to find that 82 is in the 13th position, so the median is 82. This indicates that 82 is in the middle of the scores when they are arranged in order.

In general, if the number of data entries is n and n is odd, then the *position* of the median will be $n \div 2$ rounded to the next whole number. If n is an even number, then there will be two middle positions. They will be $n \div 2$ and the next whole number. Remember, when the number of entries is even, the median is the mean of the two middle scores.

EXAMPLE 6

If a data set has 100 numbers, then the two middle entries will be at the positions $100 \div 2$ or 50th and the next position (i.e., 51st).

We have seen examples where each of the mean, the median, and the mode has been the best description of the central tendency of data. When gathering information about a collection of data, we need to determine all three.

Exercises 19.2

In Exercises 1 through 7 find the mean, the median, and the mode.

1. Ten batteries were randomly chosen to be tested for the number of hours they would supply power. The numbers of hours were 25, 17, 28, 21, 19, 17, 26, 23, 17, and 21.
2. Twelve extension cords were measured for the depth of their rubber coating. The measurements in millimetres were 2.3, 2.8, 3.4, 2.9, 4.8, 3.7, 3.1, 2.6, 2.8, 3.5, 2.5, and 2.8.

3. A manufacturer of computer disks checks the thickness of five disks while giving a demonstration on quality control. The measurements in micrometres are 2468, 2594, 2432, 2639, and 2497.
4. The outputs of ten 9-V batteries were measured. The results in volts were 8.9, 9.0, 9.2, 9.1, 9.0, 8.8, 9.1, 9.0, 8.7, and 9.0.
5. A bus company tests the reaction time of its drivers with a machine that records the number of seconds for a driver to step on the brake after seeing a video image which indicates she should. The results in seconds were 2.5, 1.5, 2.1, 3.2, 3.1, 2.0, 1.8, 2.5, and 1.9.
6. A machine is calibrated to cut hoses 15.75 in. long. Each day for a 2-week period 100 hoses are measured. The number of hoses each day which did not meet the specifications were 6, 9, 5, 6, 9, 2, 5, 7, 8, 3, 9, 4, 1, and 8.
7. A manufacturer of spring scales checks them by weighing an item whose weight has been determined by a balance scale. Five hundred scales are checked each week for 9 weeks. The following list gives the number of scales each week that did not meet specifications: 34, 27, 23, 14, 19, 23, 31, 9, and 27.

In Exercises 8 through 14 find the median and the mode for the data in Exercises 19.1.

8. Use the data in Table 19.5 from Exercise 1.
9. Use the data in Table 19.6 from Exercise 2.
10. Use the data in Table 19.7 from Exercise 3.
11. Use the data in Table 19.8 from Exercise 4.
12. Use the data in Table 19.9 from Exercise 5.
13. Use the data in Table 19.10 from Exercise 6.
14. Use the data in Table 19.11 from Exercise 7.
15. Find the mean for the data in Table 19.6 from Exercise 2 in Section 19.1.
16. Find the mean for the data in Table 19.8 from Exercise 4 in Section 19.1.
17. Find the mean for the data in Table 19.9 from Exercise 5 in Section 19.1.
18. Find the mean for the data in Table 19.11 from Exercise 7 in Section 19.1.

For Exercises 19 and 20 find the mean, the median, and the mode.

19. The daily high temperatures in degrees Fahrenheit for Kansas City during the month of March are given in the following frequency table:

Temperature	28	35	42	56	67	75
Frequency	1	2	8	12	5	3

20. The daily reports of lost baggage for a major airline during December are given in the following frequency table:

Reports	206	218	257	268	291	325	346
Frequency	3	7	4	6	8	2	1

19.3 MEASURES OF DISPERSION

In addition to discussing how data clusters (i.e., its central tendencies), we also need to discuss its dispersion, which is to say how it is distributed. One easy measure of dispersion is the **range** of the data.

RANGE

The range of data is the difference between the largest and smallest elements.

EXAMPLE 1

Find the range for the data set $A = \{1, 2, 3, 4, 5, 6, 7, 8, 9, 10\}$.

The range is $10 - 1$ or 9.

EXAMPLE 2

Find the range for the data set $B = \{1, 2, 2, 3, 3, 8, 8, 9, 10, 10\}$.

The range is 9.

Although sets A and B have the same range, the data is distributed quite differently. This is an indication that we need another measure of dispersion. We want one that describes how the data points deviate from the mean. The measure we will use is the **sample standard deviation.**

Quality control is based on choosing *some* items and checking them for whatever criteria we want to test. The collection of all items being considered is called a **population;** the items that are selected are called a **sample.** The manner in which the items are selected is extremely important. When an item is picked at **random,** it means that each item in the population had an equal chance of being chosen. Most quality control procedures involve selecting a random sample.

SAMPLE STANDARD DEVIATION

The sample standard deviation s_x of a set of data $x_1, x_2, \ldots, x_n$ is given by the formula

$$s_x = \sqrt{\frac{(x_1 - \bar{x})^2 + (x_2 - \bar{x})^2 + \cdots + (x_n - \bar{x})^2}{n - 1}}$$

Note: The formula for the standard deviation σ_x of an entire population is similar to the one for a sample. In the formula for σ_x, divide by the population size n, while in calculating s_x, divide by the slightly smaller $n - 1$.

In quality control procedures we expect to make valid generalizations about a population by using random samples. Mathematicians have established that when using a sample, dividing by $n - 1$ gives an unbiased *estimation* of the population standard deviation.

EXAMPLE 3

A company is cutting wire $30\bar{0}$ cm in length. Acceptable lengths are between 295 cm and 305 cm. A random sample of five wires had lengths in centimetres of 296, 299, $30\bar{0}$, 302, and 303. Find the sample standard deviation.

We use the following steps to find the sample standard deviation.

Step 1: Find the mean of the sample. For this example $\bar{x} = 30\bar{0}$.

Step 2: Calculate the deviations from the mean, $x - \bar{x}$, and square each result. See Table 19.15.

TABLE 19.15

x	$x - \bar{x}$	$(x - \bar{x})^2$
296	$296 - 30\bar{0}$	$(-4)^2 = 16$
299	$299 - 30\bar{0}$	$(-1)^2 = 1$
$30\bar{0}$	$30\bar{0} - 30\bar{0}$	$(0)^2 = 0$
302	$302 - 30\bar{0}$	$(2)^2 = 4$
303	$303 - 30\bar{0}$	$(3)^2 = 9$

Step 3: Add the squares and divide the sum by $n - 1$.

$$\frac{16 + 1 + 0 + 4 + 9}{5 - 1} = \frac{30}{4} = 7.5$$

Step 4: Find the square root of the answer to Step 3. $s_x = \sqrt{7.5} = 2.74$. Thus, 2.74 is in some sense the "average" deviation of a data point from the mean.

Data that is closely grouped will have a relatively small sample standard deviation, as in Example 3. Data that is more spread out will have a larger sample standard deviation, as in our next example.

EXAMPLE 4

The situation is the same as in Example 3 except that the random sample had lengths in centimetres of 291, 294, 301, 304, and $31\bar{0}$. To calculate the sample standard deviation, we find the mean, $\bar{x} = 30\bar{0}$. Next we calculate the deviations from the mean and square each result. See Table 19.16.

TABLE 19.16

x	$x - \bar{x}$	$(x - \bar{x})^2$
291	$291 - 30\bar{0}$	$(-9)^2 = 81$
294	$294 - 30\bar{0}$	$(-6)^2 = 36$
301	$301 - 30\bar{0}$	$1^2 = 1$
304	$304 - 30\bar{0}$	$4^2 = 16$
$31\bar{0}$	$31\bar{0} - 30\bar{0}$	$10^2 = 100$

The sum of the squares divided by $n - 1$ is $\frac{234}{4} = 58.5$. Finally, the sample standard deviation is $\sqrt{58.5} = 7.65$, which indicates the variation of the lengths of the cuts with respect to the mean. Thus, there was more variation for the data in Example 4 than for the data in Example 3.

Graphing calculators contain built-in programs that find the mean, the median, and the sample standard deviation. We illustrate with Example 5.

EXAMPLE 5

Use a calculator to find the sample standard deviation for the scores 285, 286, 289, 301, 305, 307, and 311.

APPS 6 3 down arrow *twice* **2nd a-lock S C O R E S ENTER ENTER alpha**

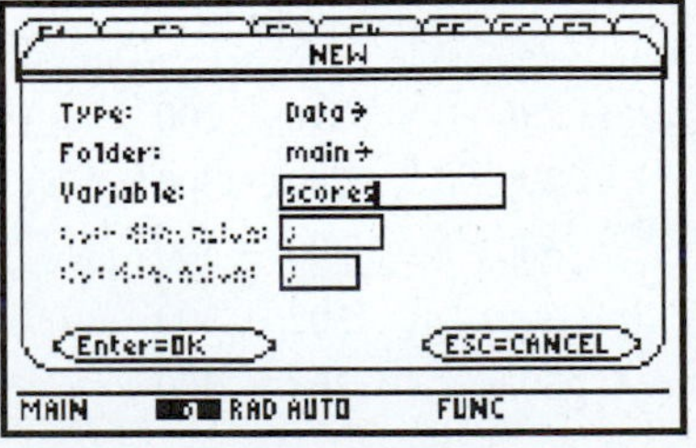

285 **ENTER** 286 **ENTER** 289 **ENTER** 301 **ENTER** 305 **ENTER**, etc.

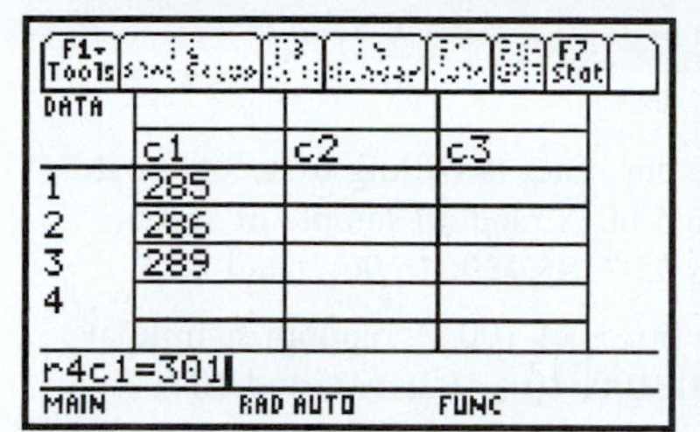

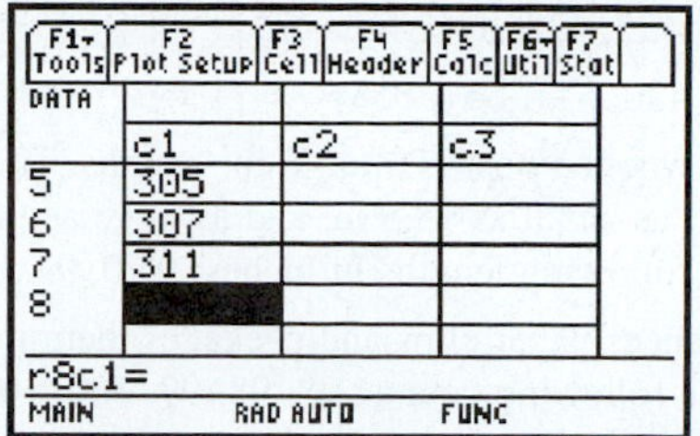

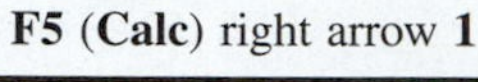

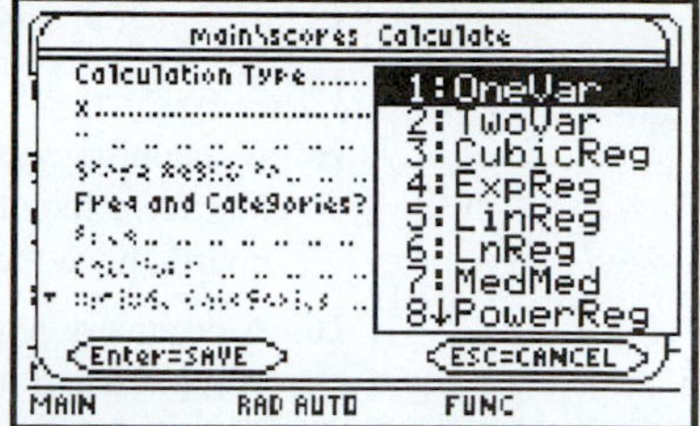

down arrow **alpha C 1 ENTER**

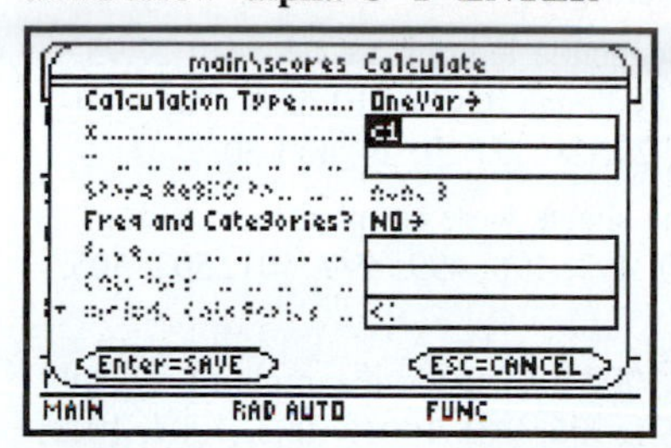

ENTER

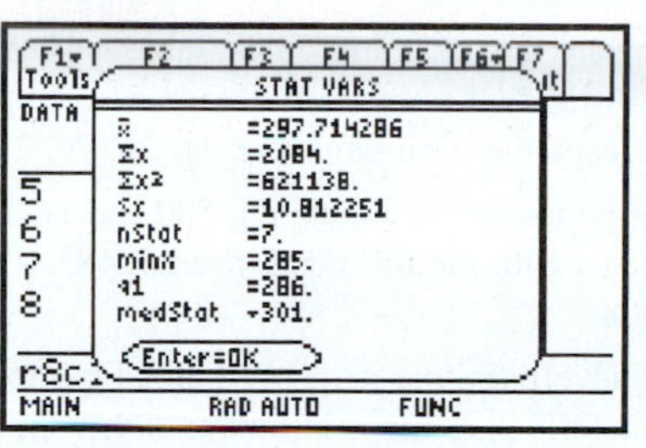

down arrows

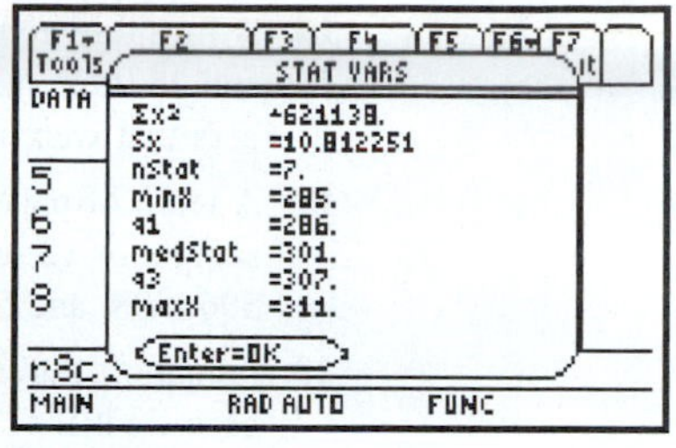

STAT 1

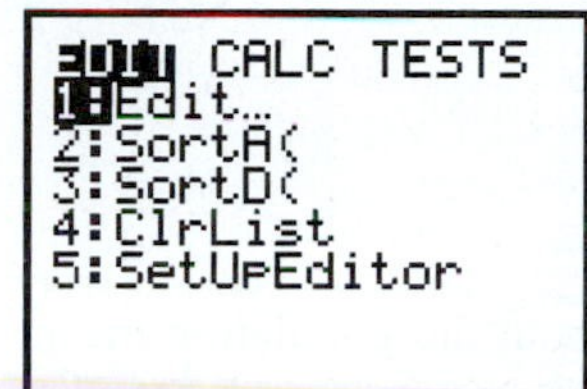

285 **ENTER** 286 **ENTER**, etc.

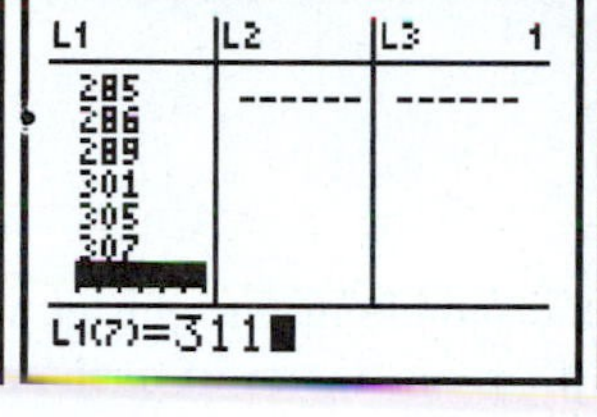

STAT right arrow **1**

2nd 1 ENTER

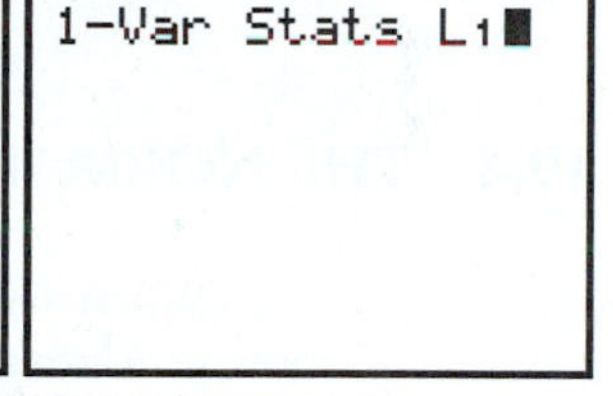

down arrows

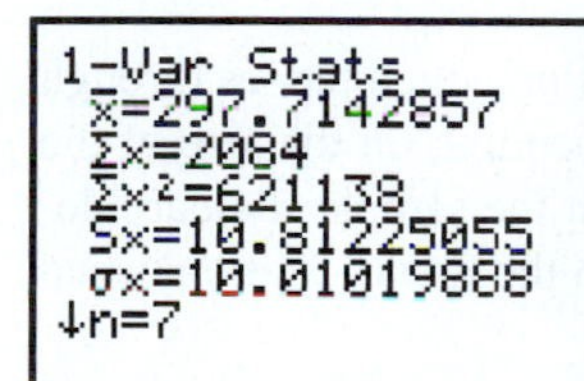

The sample standard deviation is $s_x = 10.8$.

Exercises 19.3

In Exercises 1 through 10 find the range and the sample standard deviation s_x of the given set of numbers. Round your answer for s_x to the nearest hundredth.

1. $A = \{3, 3, 4, 5, 6, 6, 7, 8, 12\}$

2. $B = \{1, 1, 4, 5, 6, 7, 8, 9, 9, 10\}$

3. $C = \{29, 30, 19, 20, 28, 29, 26, 35\}$

4. $D = \{26, 32, 41, 35, 39, 43, 29\}$

5. $E = \{58, 60, 64, 64, 65, 69, 71, 77\}$

6. $F = \{63, 69, 70, 73, 77, 78, 81, 81\}$

7. $G = \{78, 79, 80, 82, 83, 92, 94\}$

8. $H = \{49, 51, 55, 57, 59, 61, 66, 66\}$

9. $I = \{858, 860, 861, 863, 864, 874, 875\}$

10. $J = \{514, 516, 519, 523, 527, 530, 532\}$

In Exercises 11 through 20 find the range and the sample standard deviation of the set of numbers. Round your answers for s_x to three significant digits.

11. $K = \{47, 49, 55, 61, 62, 72, 81, 84, 84, 87, 91\}$

12. $L = \{46, 50, 59, 69, 72, 74, 79, 87, 94\}$

13. $M = \{389, 394, 425, 428, 441, 463, 472\}$

14. $N = \{482, 487, 490, 511, 551, 558\}$

15. A machine cuts wooden boards into 96-in. lengths. The customer is willing to accept boards that have lengths as small as 95.5 in. and as large as 96.5 in. A random sample of seven boards have the following lengths in inches: 94.0, 94.3, 95.7, 95.9, 96.0, 96.2, and 96.5.

16. A company produces paper clips and packages them in boxes of 100. A random sample of 10 boxes had the following counts: 98, 98, 99, 100, 100, 100, 101, 102, 103, and 104.

17. Nails are packaged in boxes that contain 200. A random sample of nine boxes had the following counts: 194, 195, 196, 197, 198, 200, 202, 202, and 203.

18. A furniture company manufactures 28-in. table legs. Acceptable lengths are between 27.9375 and 28.0625 in. A random sample of 30 legs was measured each day for 10 days. The numbers that were acceptable each day were 24, 26, 26, 27, 27, 28, 28, 29, 29, and 30.

19. A ream of paper is labeled as containing 500 sheets. The sheets were counted in a random sample of 12 reams with the following results: 483, 490, 493, 496, 499, 499, 501, 502, 505, 506, 508, and 508.

20. A manufacturer advertises that its 60-W bulbs light for 1000 h. A random sample of 10 bulbs gave light for the following times in hours: 989, 992, 994, 996, 999, 1001, 1003, 1004, 1011, and 1015.

19.4 THE NORMAL DISTRIBUTION

When sample sizes are very large, the sample means cluster about the population mean in a **normal distribution.** If we draw a histogram of such sample means and connect the center points of the tops of the rectangles, we have a curve like the one shown in Fig. 19.5.

Normal distribution curves are symmetric about a vertical line which passes through the mean $\bar{x}$. They may have a high peak or be relatively flat depending on the size of the sample standard deviation. In Fig. 19.6 the mean is the same, but the sample standard deviations differ. Notice that for a smaller s_x, more data is close to the mean. These normal distribution curves are often described as **bell shaped.**

A very important feature of a normal distribution curve is that approximately 68.2% of the data is located within one sample standard deviation of the mean, approximately 95.4% of the data is within two sample standard deviations of the mean, and 99.7% of the data is within three sample standard deviations of the mean (see Fig. 19.7).

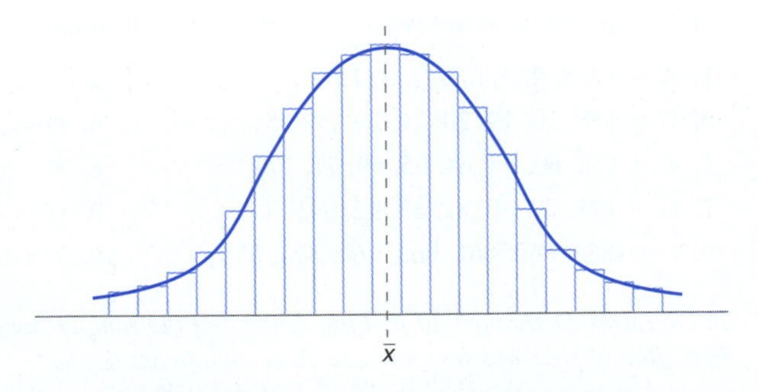

Figure 19.5 Normal distribution curve.

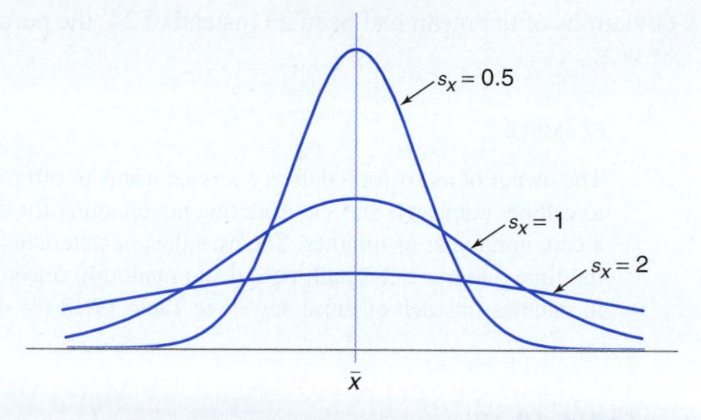

Figure 19.6 Normal distribution curves.

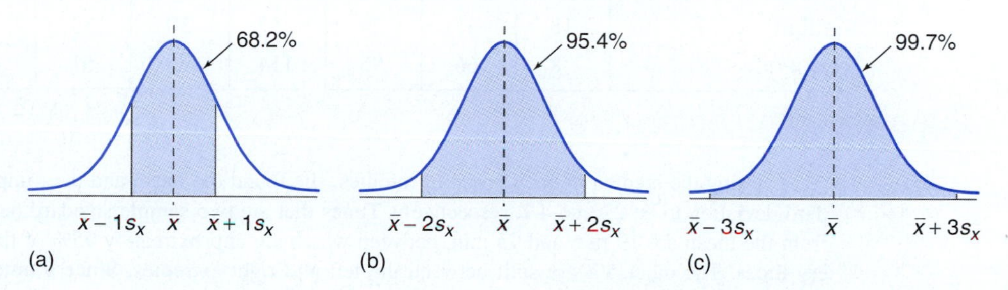

Figure 19.7 Percent of data within one, two, and three sample standard deviations of the mean.

EXAMPLE 1

A logging company measures the length of 25 tree trunks as they are loaded onto trucks. The measurements in feet are given in Table 19.17.

TABLE 19.17 Log Lengths in Feet

26	27	29	29	31
32	32	32	33	33
33	34	34	34	34
35	35	35	35	36
36	39	39	41	43

The sample standard deviation and the mean rounded to two significant digits are 4.0 and 34, respectively.

$$\bar{x} - 1s_x = 30 \quad \text{and} \quad \bar{x} + 1s_x = 38$$

From Table 19.17 the number of logs measuring between 30 and 38 feet is 17. This is 68% of the logs.

$$\bar{x} - 2s_x = 26 \quad \text{and} \quad \bar{x} + 2s_x = 42$$

Again from Table 19.17 the number of logs measuring between 26 and 42 feet is 24, which is 96% of the logs.

For sample data, percents will be approximately equal to the expected values in a normal distribution. In Example 1 if the number of logs within two sample standard deviations of the mean had been 23 instead of 24, the percent would have been 92% instead of 96%.

EXAMPLE 2

The owner of a fast food delivery service wants to offer something that will induce customers to call her company. She's considering not charging for the food if the delivery time is beyond a certain number of minutes. She uses the characteristics of a normal distribution to establish the time. During a 3-month period she randomly selected days and noted the delivery times in minutes for each of those days. See Table 19.18 for the times of 444 deliveries.

TABLE 19.18

Minutes for delivery	17–19	20–22	23–25	26–28	29–31	32–34	35–37
Median	18	21	24	27	30	33	36
Frequency	8	66	95	114	96	51	14

Using the median of each group of minutes, she found the mean and the sample standard deviation to be 27 and 4.2, respectively. Times that are two sample standard deviations from the mean are 19 min and 35 min, between which are approximately 95% of the delivery times. The other 5% are split between the left and right extremes. Since a time that is less than the mean does not cause her to lose her fee, she establishes 35 min as the goal for delivery time. If it takes longer than that she will not charge for the food, which she expects to happen for about 2.5% of the deliveries.

EXAMPLE 3

For the data set {49, 50, 52, 61, 64, 66, 75, 78, 87}, where $\bar{x} = 64.7$ and $s_x = 13.3$, find the percent of data points that are within one sample standard deviation of the mean.

First, we calculate $\bar{x} - 1s_x = 51.4$ and $\bar{x} + 1s_x = 78$. The data points are 52, 61, 64, 66, 75, and 78. Since six of the nine data points are between 51.4 and 78, the percent is 66.7%. This is within 2% of the $\bar{x} \pm 1s_x$ (68%) criterion for a normal distribution.

EXAMPLE 4

The results of a uniform technical mathematics test taken by 1000 students are a mean of 72 with a standard deviation of 8.5. Mark's score is 90. Assuming the scores are normally distributed, what percent of the students does he estimate scored below him?

First, we calculate the mean plus two standard deviations to be $72 + 2(8.5) = 89$. Since we expect 95% of the students to score within two standard deviations of the mean, we expect half of 95% to be between the mean and 89 (i.e., 47.5%). Adding the 50% below the mean to this value, we estimate that 97.5% of the students scored below Mark.

Exercises 19.4

For Exercises 1 through 6 calculate the mean and the sample standard deviation rounded to three significant digits. Find the percent of data points within one sample standard deviation of the mean and determine whether this is within 2% of the $\bar{x} \pm 1s_x$ criterion for a normal distribution.

1. $A = \{27, 27, 38, 43, 44, 57, 76, 82\}$

2. $B = \{36, 39, 40, 51, 55, 61, 66, 67\}$

3. $C = \{35, 36, 38, 43, 47, 48, 48, 55, 67\}$

4. $D = \{48, 50, 51, 57, 62, 69, 70, 73, 73\}$

5. $E = \{18, 24, 32, 37, 38, 45, 49\}$

6. $F = \{43, 47, 52, 56, 57, 69\}$

*For Exercises 7 through 12 the mean and the sample standard deviation are given. Find the percent of data points within **(a)** one sample standard deviation of the mean and **(b)** two sample standard deviations of the mean. Round all values to three significant digits. Are these percents within 2% of the $\bar{x} \pm 1s_x$ (68%) and $\bar{x} \pm 2s_x$ (95%) criteria for a normal distribution?*

7.

Metres	9	10	11	12	13	14	15
Frequency	5	16	31	38	27	19	4

$\bar{x} = 12.0$, $s_x = 1.42$

8.

Feet	35	36	37	38	39	40	41	42	43
Frequency	2	6	21	43	45	46	31	6	1

$\bar{x} = 39.1$, $s_x = 1.50$

9.

Kilometres	65	66	67	68	69	70	71
Frequency	6	21	27	47	29	18	3

$\bar{x} = 67.9$, $s_x = 1.39$

10.

Inches	38	40	42	44	46	48	50
Frequency	8	28	36	62	38	24	4

$\bar{x} = 43.8$, $s_x = 2.79$

11.

Millimetres	23	24	25	26	27	28	29
Frequency	10	35	45	79	48	37	5

$\bar{x} = 26.0$, $s_x = 1.41$

12.

Centimetres	52	54	56	58	60	62	64
Frequency	3	28	37	71	41	29	1

$\bar{x} = 58.0$, $s_x = 2.56$

13. The heights of college basketball players (men and women) are normally distributed, with a mean of 72.5 in. and a standard deviation of 3.95 in. At a convention of 3500 college basketball players, the opening speaker commented that 68% of the players attending were at least 6′4″ (76 in.) tall. Assuming the heights of the attending players are normally distributed, is this a reasonable statement?

14. A small bakery shop has a mean weekly gross income of $2300, with a standard deviation of $180. Approximately what percent of the time does the baker expect to gross at least $2660 per week?

15. A machine cuts lengths of hose with a mean length of 20.0 in. and standard deviation of 0.125 in. A batch of 50 is carefully measured. Assuming normal distribution, 95% of the hoses should measure between what two lengths?

19.5 FITTING CURVES TO DATA SETS

Data we collect often involves two variables, such as fuel consumption compared with travel distance, the pressure of a gas compared with its volume, and the number of bacteria in a culture charted with a time line. We need to find an equation which relates the two variables as closely as possible so that we can make predictions. This is called **regression analysis.**

The **line or curve of best fit** for a given set of ordered pairs is a formula for a line or curve which comes closest to approximating the values of the data set. First we will use matrices to find a line of best fit and a quadratic equation of best fit, then we will use the calculator routine.

In Section 9.9 we used the linear regression procedure on a calculator to find the line of best fit. We will use the same data points to demonstrate the matrix procedure.

EXAMPLE 1

Find the equation for the line of best fit $ax + b = y$ for the data

x	1	2	3	4	5	6
y	2	5	4	7	8	10

Substituting the data points in the equation, we have

$$\begin{aligned} a(1) + b &= 2 \\ a(2) + b &= 5 \\ a(3) + b &= 4 \\ a(4) + b &= 7 \\ a(5) + b &= 8 \\ a(6) + b &= 10 \end{aligned} \quad \text{which in matrix form is} \quad \begin{bmatrix} 1 & 1 \\ 2 & 1 \\ 3 & 1 \\ 4 & 1 \\ 5 & 1 \\ 6 & 1 \end{bmatrix} \begin{bmatrix} a \\ b \end{bmatrix} = \begin{bmatrix} 2 \\ 5 \\ 4 \\ 7 \\ 8 \\ 10 \end{bmatrix}$$

This overconstrained system of six equations in two unknowns has no solutions since there isn't a straight line that passes through all the points. However, we are looking for the best *approximation.* If we multiply both sides of the matrix equation by the transpose of the coefficient matrix, the result will be a 2×2 system that can be solved for a and b. The **transpose** of a matrix is found by interchanging its rows and columns; that is, the first row becomes the first column, the second row becomes the second column, and so on.

$$\begin{bmatrix} 1 & 2 & 3 & 4 & 5 & 6 \\ 1 & 1 & 1 & 1 & 1 & 1 \end{bmatrix} \begin{bmatrix} 1 & 1 \\ 2 & 1 \\ 3 & 1 \\ 4 & 1 \\ 5 & 1 \\ 6 & 1 \end{bmatrix} \begin{bmatrix} a \\ b \end{bmatrix} = \begin{bmatrix} 1 & 2 & 3 & 4 & 5 & 6 \\ 1 & 1 & 1 & 1 & 1 & 1 \end{bmatrix} \begin{bmatrix} 2 \\ 5 \\ 4 \\ 7 \\ 8 \\ 10 \end{bmatrix}$$

$$\begin{bmatrix} 91 & 21 \\ 21 & 6 \end{bmatrix} \begin{bmatrix} a \\ b \end{bmatrix} = \begin{bmatrix} 152 \\ 36 \end{bmatrix}$$

Notice that 91 is the sum of the squares of the x's, 21 is the sum of the x's, 6 is the number of points, 152 is the sum of the corresponding xy products, and 36 is the sum of the y's. These are the same values that are calculated and substituted into formulas to find a and b when the method of least squares is used. Solving our 2×2 system, we find $\begin{bmatrix} a \\ b \end{bmatrix} = \begin{bmatrix} 1.4857 \\ 0.8 \end{bmatrix}$. Thus, the line of best fit is $y = 1.5x + 0.8$.

EXAMPLE 2

The following chart shows data comparing the number of hours spent studying for a math test and the percent grade received on the test. We will let x represent the hours and y represent the grades. Find the equation for the line of best fit for the data.

Hours x	0.5	2.3	3.7	4.5	5.6	6.0	7.9	8.5
Grades y	51	59	68	70	76	80	88	95

Using a calculator, we have

STAT 1 | (enter the data) | **STAT** right arrow 4 | **ENTER**

EDIT CALC TESTS
1:Edit...
2:SortA(
3:SortD(
4:ClrList
5:SetUpEditor

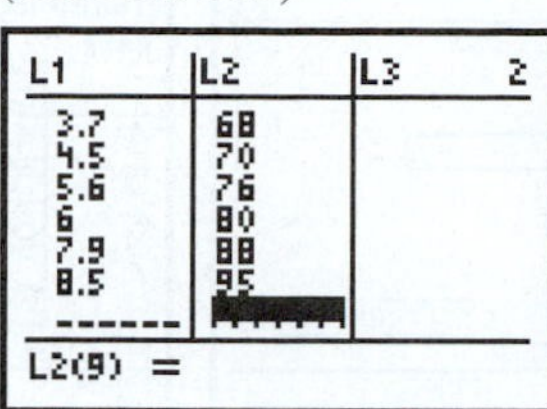

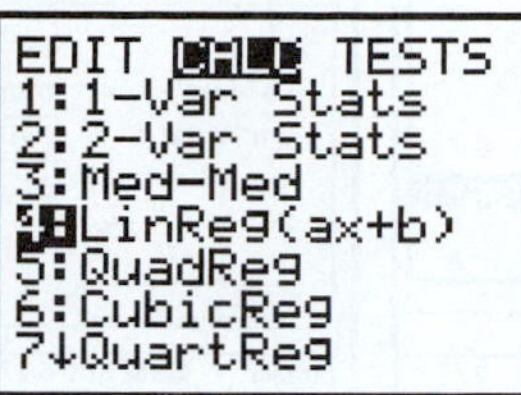

The straight line of best fit is

$$y = 5.3x + 47$$

For each hour of studying, a grade increased an average of 5.3 points.

EXAMPLE 3

Use the matrix method to find the least squares quadratic curve, $ax^2 + bx + c = y$, to fit the data in Example 1.

Substitute the data points in the equation and then write the corresponding matrix equation.

$$\begin{aligned} a(1)^2 + b(1) + c &= 2 \\ a(2)^2 + b(2) + c &= 5 \\ a(3)^2 + b(3) + c &= 4 \\ a(4)^2 + b(4) + c &= 7 \\ a(5)^2 + b(5) + c &= 8 \\ a(6)^2 + b(6) + c &= 10 \end{aligned} \qquad \begin{bmatrix} 1^2 & 1 & 1 \\ 2^2 & 2 & 1 \\ 3^2 & 3 & 1 \\ 4^2 & 4 & 1 \\ 5^2 & 5 & 1 \\ 6^2 & 6 & 1 \end{bmatrix} \begin{bmatrix} a \\ b \\ c \end{bmatrix} = \begin{bmatrix} 2 \\ 5 \\ 4 \\ 7 \\ 8 \\ 10 \end{bmatrix}$$

Next, multiply both sides of the matrix equation by the transpose of the coefficient matrix. The resulting 3×3 system can be solved for a, b, and c.

$$\begin{bmatrix} 1 & 4 & 9 & 16 & 25 & 36 \\ 1 & 2 & 3 & 4 & 5 & 6 \\ 1 & 1 & 1 & 1 & 1 & 1 \end{bmatrix} \begin{bmatrix} 1 & 1 & 1 \\ 4 & 2 & 1 \\ 9 & 3 & 1 \\ 16 & 4 & 1 \\ 25 & 5 & 1 \\ 36 & 6 & 1 \end{bmatrix} \begin{bmatrix} a \\ b \\ c \end{bmatrix} = \begin{bmatrix} 1 & 4 & 9 & 16 & 25 & 36 \\ 1 & 2 & 3 & 4 & 5 & 6 \\ 1 & 1 & 1 & 1 & 1 & 1 \end{bmatrix} \begin{bmatrix} 2 \\ 5 \\ 4 \\ 7 \\ 8 \\ 10 \end{bmatrix}$$

$$\begin{bmatrix} 2275 & 441 & 91 \\ 441 & 91 & 21 \\ 91 & 21 & 6 \end{bmatrix} \begin{bmatrix} a \\ b \\ c \end{bmatrix} = \begin{bmatrix} 730 \\ 152 \\ 36 \end{bmatrix}$$

Notice that the matrix equation for the linear regression from Example 1 is embedded in the quadratic one. The solution to this system is $\begin{bmatrix} a \\ b \\ c \end{bmatrix} = \begin{bmatrix} 0.05357 \\ 1.1107 \\ 1.3 \end{bmatrix}$. Thus, the quadratic curve of best fit is $y = 0.054x^2 + 1.1x + 1.3$.

Using a calculator, we have

APPS **6** **3** down arrow *twice* **2nd a-lock** **P O I N T S** **ENTER ENTER** **alpha** (enter the data)

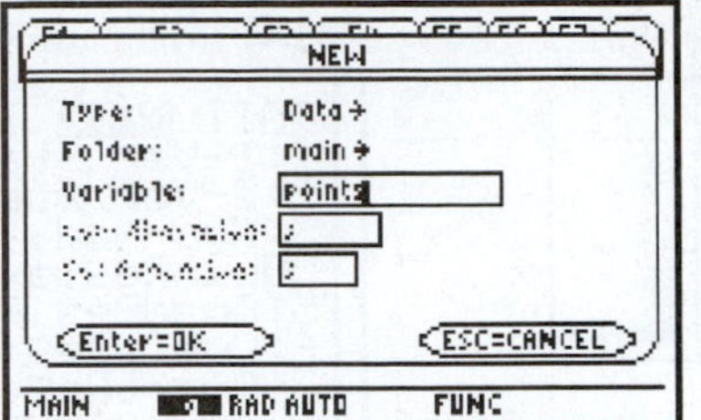

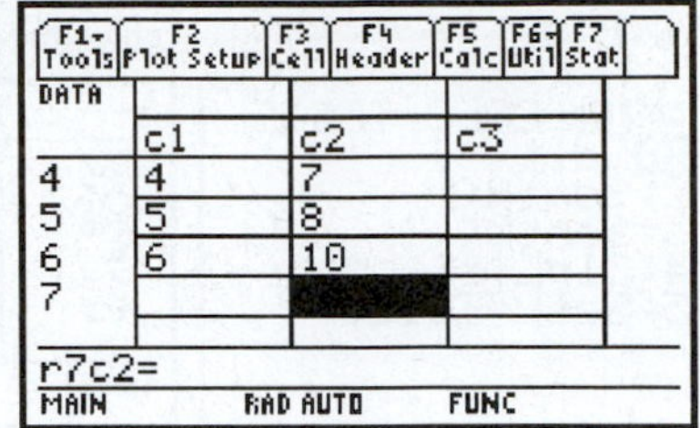

F5 (**Calc**) right arrow **9** down arrow **alpha C 1** down arrow **alpha C 2** **ENTER ENTER**

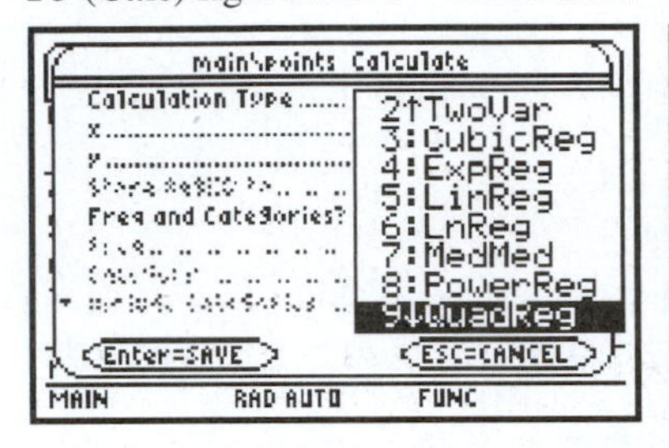

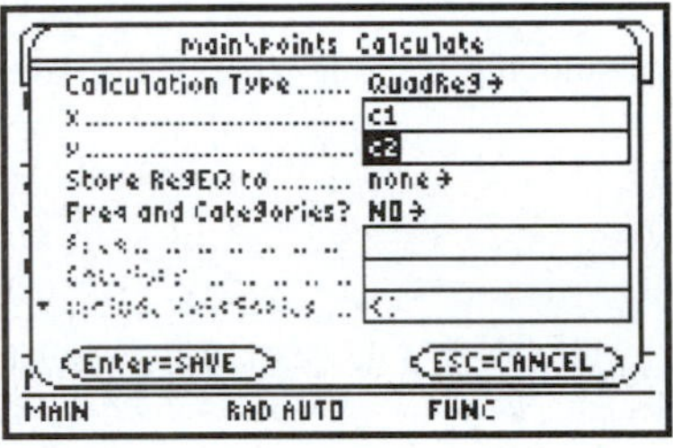

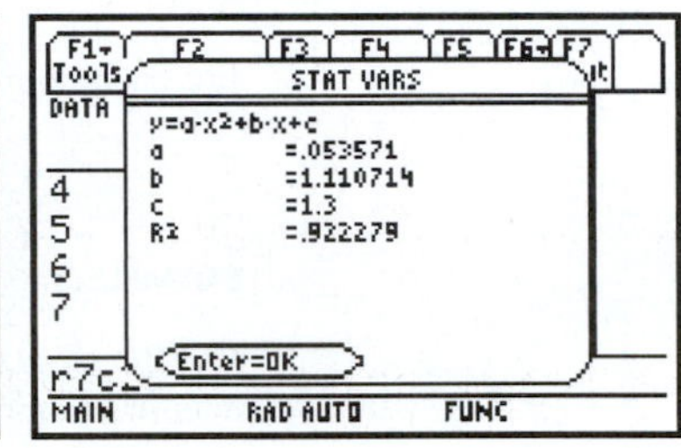

STAT **1** (enter the data) **STAT** right arrow **5** **ENTER**

EDIT CALC TESTS
1:Edit...
2:SortA(
3:SortD(
4:ClrList
5:SetUpEditor

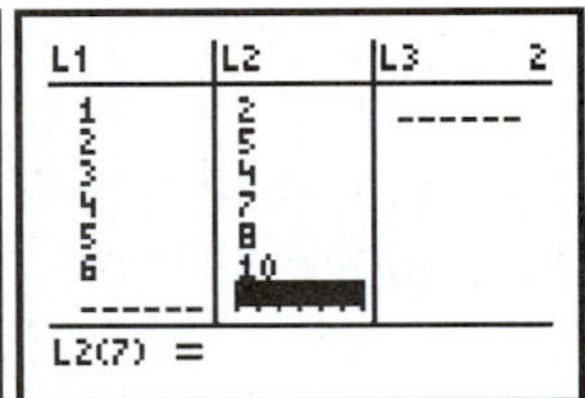

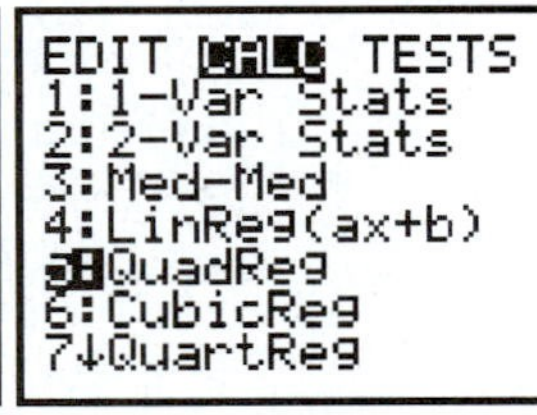

QuadReg
y=ax²+bx+c
a=.0535714286
b=1.110714286
c=1.3

EXAMPLE 4

Using the same data points from Example 1 to find a cubic equation of best fit, $ax^3 + bx^2 + cx + d = y$, again write the matrix equation, multiply both sides by the transpose of the coefficient matrix, and solve the resulting 4×4 system.

$$\begin{bmatrix} 1^3 & 1^2 & 1 & 1 \\ 2^3 & 2^2 & 2 & 1 \\ 3^3 & 3^2 & 3 & 1 \\ 4^3 & 4^2 & 4 & 1 \\ 5^3 & 5^2 & 5 & 1 \\ 6^3 & 6^2 & 6 & 1 \end{bmatrix} \begin{bmatrix} a \\ b \\ c \\ d \end{bmatrix} = \begin{bmatrix} 2 \\ 5 \\ 4 \\ 7 \\ 8 \\ 10 \end{bmatrix}$$

$$\begin{bmatrix} 1 & 8 & 27 & 64 & 125 & 216 \\ 1 & 4 & 9 & 16 & 25 & 36 \\ 1 & 2 & 3 & 4 & 5 & 6 \\ 1 & 1 & 1 & 1 & 1 & 1 \end{bmatrix} \begin{bmatrix} 1 & 1 & 1 & 1 \\ 8 & 4 & 2 & 1 \\ 27 & 9 & 3 & 1 \\ 64 & 16 & 4 & 1 \\ 125 & 25 & 5 & 1 \\ 216 & 36 & 6 & 1 \end{bmatrix} \begin{bmatrix} a \\ b \\ c \\ d \end{bmatrix} = \begin{bmatrix} 1 & 8 & 27 & 64 & 125 & 216 \\ 1 & 4 & 9 & 16 & 25 & 36 \\ 1 & 2 & 3 & 4 & 5 & 6 \\ 1 & 1 & 1 & 1 & 1 & 1 \end{bmatrix} \begin{bmatrix} 2 \\ 5 \\ 4 \\ 7 \\ 8 \\ 10 \end{bmatrix}$$

$$\begin{bmatrix} 67171 & 12201 & 2275 & 441 \\ 12201 & 2275 & 441 & 91 \\ 2275 & 441 & 91 & 21 \\ 441 & 91 & 21 & 6 \end{bmatrix} \begin{bmatrix} a \\ b \\ c \\ d \end{bmatrix} = \begin{bmatrix} 3758 \\ 730 \\ 152 \\ 36 \end{bmatrix}$$

$$\begin{bmatrix} a \\ b \\ c \\ d \end{bmatrix} = \begin{bmatrix} 0.0648 \\ -0.6270 \\ 3.165 \\ -0.333 \end{bmatrix}$$

The equation is $y = 0.065x^3 - 0.63x^2 + 3.2x - 0.33$

We can find many curves of best fit, including polynomials, logarithms, and exponentials. How do we know which curve will be the best choice? Unfortunately, just looking at the data will not suffice, since in a given interval of x, many curves look similar. We need to know something about the behavior of the data we have collected. For example, bacteria grow exponentially, so we would fit points corresponding to time and the size of a culture to an exponential curve.

To find curves of best fit on the calculator, first enter the data, then choose the appropriate curve. After entering **F5 (calc)** right arrow, we have the following choices for a regression equation. The calculator uses the method of least squares to find the given curve.

CubicReg $y = ax^3 + bx^2 + cx + d$. (At least four data points are required.)
ExpReg $y = ab^x$
LinReg $y = ax + b$
LnReg $y = a + b \ln x$
PowerReg $y = ax^b$
QuadReg $y = ax^2 + bx + c$. (At least three data points are required.)
QuartReg $y = ax^4 + bx^3 + cx^2 + dx + e$. (At least five data points are required.)
SinReg $y = a \sin (bx + c) + d$

Note: The line of best fit is demonstrated in Section 9.9 and in Appendix C, Section C.10 (TI-83 Plus), and Appendix D, Section D.13 (TI-89).

Exercises 19.5

In Exercises 1 through 4 use the matrix method to find the line of best fit, $y = ax + b$. Round to two significant digits.

1.

x	2	4	6	7	9
y	1	6	8	12	14

2.

x	1	3	6	7	10
y	5	8	12	13	18

3.

x	1	4	5	8	10	13
y	8	17	19	26	33	41

4.

x	1	2	5	6	8	11
y	1	1.6	2.5	2.8	3.6	4.5

In Exercises 5 through 8 use the matrix method to find the quadratic equation of best fit, $y = ax^2 + bx + c$. Round to two significant digits.

5.

x	2	3	4	6	10
y	3	9	20	41	117

6.

x	0	1	3	5	6
y	2	6	18	43	55

7.

x	1	2	4	6	7	9
y	3	5	20	54	76	132

8.

x	0	2	3	5	6	7
y	8	17	20	18	15	9

9. A rock is thrown upward and its distance from the ground after x seconds is approximated in feet. Using the following data, find the quadratic equation of best fit, $y = ax^2 + bx + c$, where y represents the distance:

x	1	2	3	5	7
y	17	19	21	18	5

10. A missile is launched upward and its distance from the ground after t seconds is approximated in kilometres. Using the following data, find the quadratic equation of best fit, $y = at^2 + bt + c$, where y represents the distance:

t	1	2	3	4	5	6
y	18	25	27	29	30	28

11. At a constant temperature, the volume of nitrogen is compared to its pressure in the following chart. Find the equation relating volume and pressure using the PowerReg formula, $y = ax^b$. *Note:* We expect b to be negative because of Boyle's law. x represents pressure in lb/in^2 and y represents volume in ft^3. Round to three significant digits.

x	14	15	17	20	26
y	3750	3510	3110	2620	2050

12. At a constant temperature, the volume of oxygen is compared to its pressure in the following chart. Find the equation relating volume and pressure using the PowerReg formula, $y = ax^b$, where x represents pressure in lb/in^2 and y represents volume in ft^3. Round to three significant digits.

x	13	15	16	20	24
y	3730	3230	3030	2430	2020

13. The number of bacteria in a culture was estimated every hour for a 5-h period with the results shown in the following chart. Find the equation relating the number of bacteria present after x hours using the ExpReg formula, $y = ab^x$, where y represents the number of bacteria. Round to three significant digits.

x	0	1	2	3	4	5
y	325	1630	8100	40,600	203,000	1,020,000

14. The number of bacteria in a culture was estimated every hour for a 5-h period after an antibiotic was added. Find the equation relating the number of bacteria present after x hours using the ExpReg formula, $y = ab^x$, where y represents the number of bacteria. Round to three significant digits.

x	0	1	2	3	4	5
y	215	48.0	10.8	2.39	0.533	0.119

15. Kathy sells logs for firewood and is interested in finding a relationship between the length of a log and its weight. Find the line of best fit, $y = ax + b$, for the following data, where x is the length in inches and y is the weight in pounds. Round to three significant digits.

x	15.0	16.5	18.2	21.0	28.5	33.4
y	9.75	11.2	11.9	13.8	18.3	21.9

16. Find the line of best fit, $y = ax + b$, for the relationship of area of carpet to its weight. x represents the area in ft^2 and y represents the weight in pounds. Round to three significant digits.

x	80	108	130	180	245	320
y	18.5	25.4	36.1	45.2	80.0	112

19.6 STATISTICAL PROCESS CONTROL

Earlier in this chapter we studied statistical concepts; here we concentrate on how those concepts help to achieve **quality control** in the production process. Quality control can be described in a number of ways:

1. *Specifications:* Does the item meet or exceed given or published specifications?
2. *Durability:* Does the item last (perform) at least as long as expected?
3. *Reliability:* How often does the item need to be repaired?
4. *Service:* Is the item easy to maintain or repair? Are errors in shipping and/or billing rare? Are such errors quickly corrected?
5. *Customer needs:* Does the item meet the needs and expectations of the customer?

Statistical process control (SPC) is used to monitor the production of quality products. It involves managerial methods and philosophy as well as **control charts.** Control charts are the histograms, frequency polygons, and displays of the mean and standard deviation that we have studied in previous sections. They help identify **special-cause problems** which can be detected and controlled. Examples of special-cause problems include tool wear, dust or other foreign objects entering the production process, and technician error.

In the manufacturing process, variation will always occur in the product, which is why specifications have tolerance limits. This is called common-cause variation or **common cause,** which may occur because of small voltage surges, occasional slight vibrations, or other small changes in conditions. A manufacturing process is **in control** when it consistently produces items within its common-cause tolerance limits and the item's measurements fit a normal (bell-shaped) curve. The process limits are called the upper control limit (UCL) and the lower control limit (LCL). The control limits are *approximately* three standard deviations from the mean.

A process in control has a chart such as the one in Figure 19.8 where the data points are within the control limits. Also notice that no group of points seems to be moving toward a trend line.

Once a process is in control, it will be **capable** of doing a specific job if the control limits are within the job specification limits. The specification limits are called upper tolerance limit (UTL) and lower tolerance limit (LTL).

Figure 19.8 Data points of a process in control.

EXAMPLE 1

The JDJ company produces bolts that have a diameter of 7.000 cm with a tolerance of ± 0.010 cm. Thus, the LTL is 6.990 cm and the UTL is 7.010 cm. A capable process for this job will have a UCL less than 7.010 cm and a LCL greater than 6.990 cm, as shown in Figure 19.9.

Figure 19.9 Capable process.

EXAMPLE 2

A process has control limits of 1.997 mm and 2.003 mm for producing 2.000-mm bolts. An order is placed for 2.000-mm bolts with a tolerance of ± 0.005 mm. Is this process capable of filling the order?

To answer the question we calculate the upper and lower tolerance levels at 2.005 mm and 1.995 mm. Since

$$1.995 < \underbrace{1.997 < 2.003}_{\text{control limits}} < 2.005$$

we see that the control limits are within the tolerance limits, so the process is capable of doing this job.

EXAMPLE 3

One section of the Taylor Flour Mill bags whole wheat flour for retail sales. The label on each bag claims the bag contains 5 lb of flour. The machines are set to load each bag with 5.000 lb. The tolerance specifications are $\pm$ 0.040 lb. One set of computer-generated data downloaded at a specified time shows the control chart in Fig. 19.10. (a) What are the control limits of the process? (b) Is the process in control?

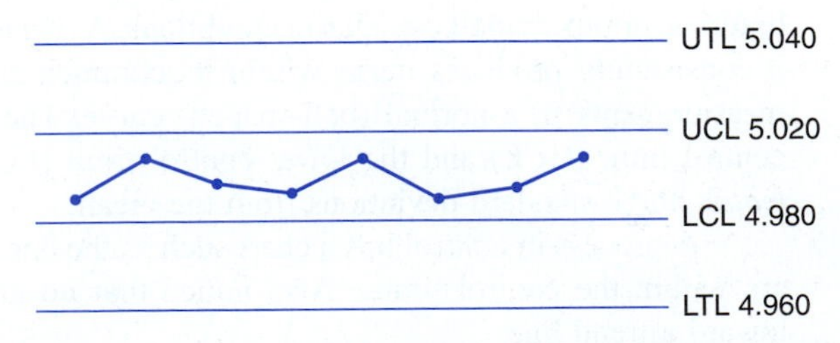

Figure 19.10 Control chart for loading bags of flour.

(a) The control limits are $\pm(5.020 - 4.980)/2$, that is, ± 0.020. (b) The process is in control.

Usually charts like the one in Fig. 19.9 are computer generated, with the data coming directly from machines in the production line (see pages 606–607). For example, bottles may be weighed and the weight of each represented by a data point in the chart. The horizontal line represents time. Periodically, a technician will download the weight data and look for variations. When a data point is outside the control limits, as in Fig. 19.11, it indicates (with about 99.7% accuracy) that the process is not in statistical control, and a search will begin for a special-cause problem.

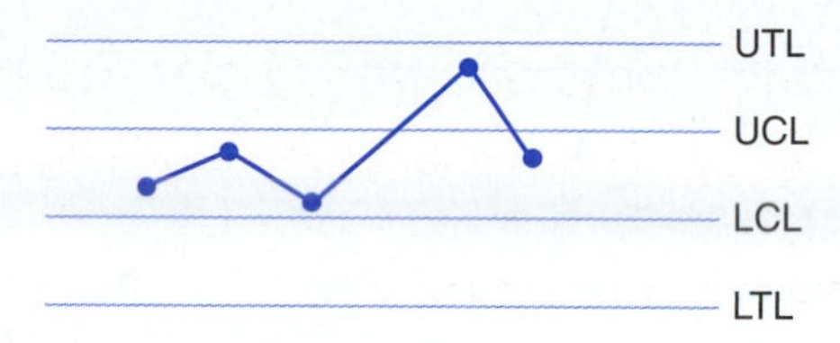

Figure 19.11 Process not in control.

There are other indications that a process may be out of statistical control. Since a stable process is expected to produce a random scatter of points on a control chart, a pattern of points that is not random may indicate special-cause problems. The Shewhart Control Charts and notes on tests for special causes were presented by L. S. Nelson in 1984 in the *Journal of Quality Technology* 16 (no. 4): 237–239. See Figs. 19.12 and 19.13 for these charts and notes.

The desire to produce quality products must be a part of the management philosophy, and its importance needs to be communicated to everyone involved. One way that management can do this is to work for registration with ISO 9001, ISO 9002, or ISO 9003. Companies which have such registration are recognized as operating excellent quality system programs. The International Organization for Standardization (ISO) is an agency composed of the national standards bodies of over 90 countries. The American National Standards Institute (ANSI) is the U.S. representative. The ISO 9000 series is a collection of five separate, but related, guidelines and standards concerning quality process, quality assurance, and quality management. The standards can be applied to any product since they deal with management systems. They guide the user in documenting the elements necessary to maintain an efficient quality operation.

ISO 9000 explains the overall system and gives guidelines for deciding which of the three models, ISO 9001, ISO 9002, or ISO 9003, is most applicable to the user's business. ISO 9001 is the most comprehensive, covering design/development, manufacturing, installation, and servicing. ISO 9002 concerns manufacturing and installation, while ISO 9003 deals only with the final product inspection and test. Finally, ISO 9004 gives guidance for operating a business so that other standards will be met. Neither ISO 9000 nor ISO 9004 is used to audit quality systems. The expression "ISO 9000" is often used to indicate one of ISO 9001, ISO 9002, or ISO 9003. The United States has adopted the ANSI/ASQ Q9000 series, which is identical to the ISO 9000 series.

Foreign customers are beginning to expect U.S. companies to be registered with ISO 9001, ISO 9002, or ISO 9003 or their equivalent. This means having an acceptable independent third party (selected by the company) audit a company's procedures and policies with respect to the requirements of the chosen standard. A company that passes the on-site audit is then registered; a reaudit will occur approximately every 3 to 4 years.

Books, seminars, software packages, and consultants are all available to help a company qualify for ISO 9000 registration. Even if a company is not interested in becoming ISO 9000 compliant, it will be addressing many of the same issues as those that are. Statistical process control will be one of those issues.

Test 1. One point beyond Zone *A*

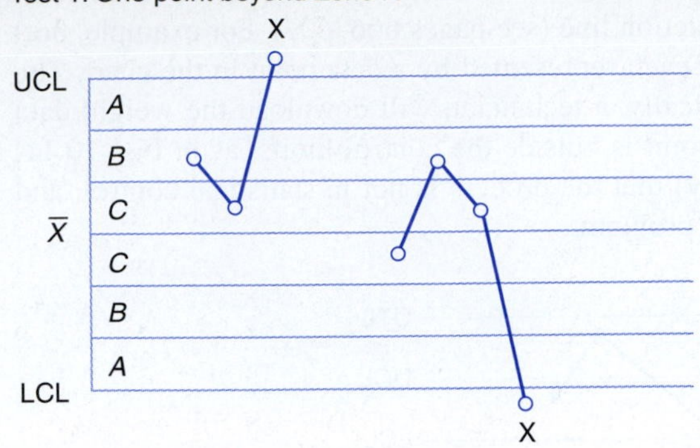

Test 2. Nine points in a row in Zone *C* or beyond

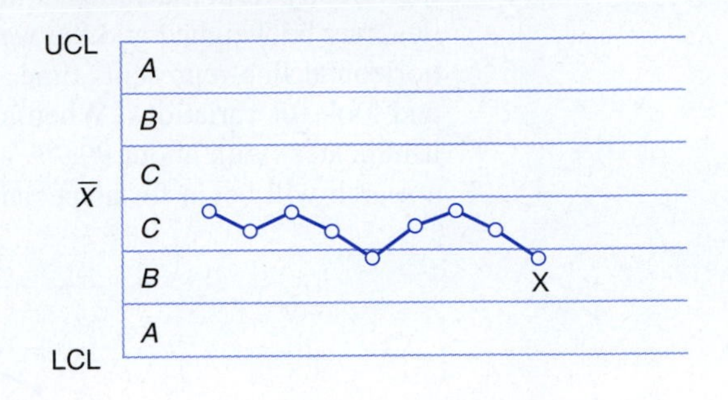

Test 3. Six points in a row steadily increasing or decreasing

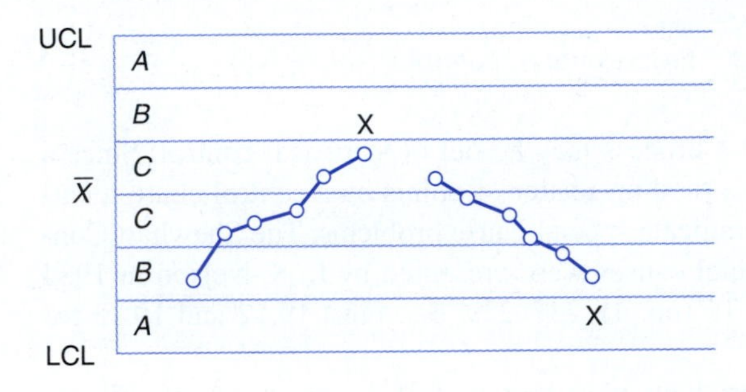

Test 4. Fourteen points in a row alternating up and down

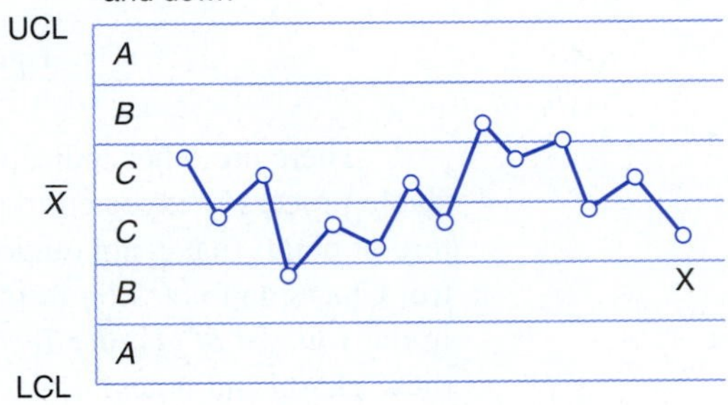

Test 5. Two out of three points in a row in Zone *A* or beyond

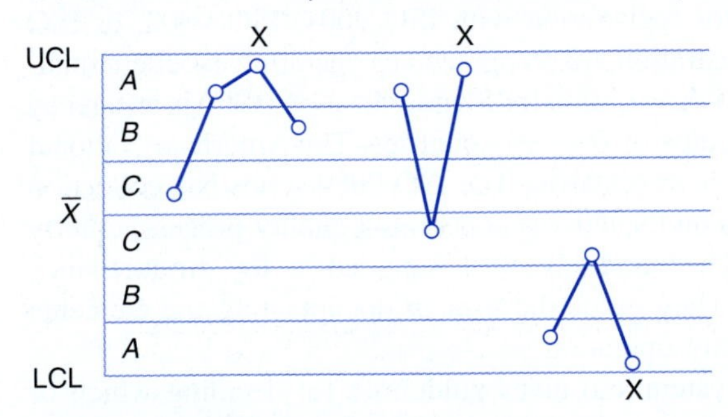

Test 6. Four out of five points in a row in zone *B* or beyond

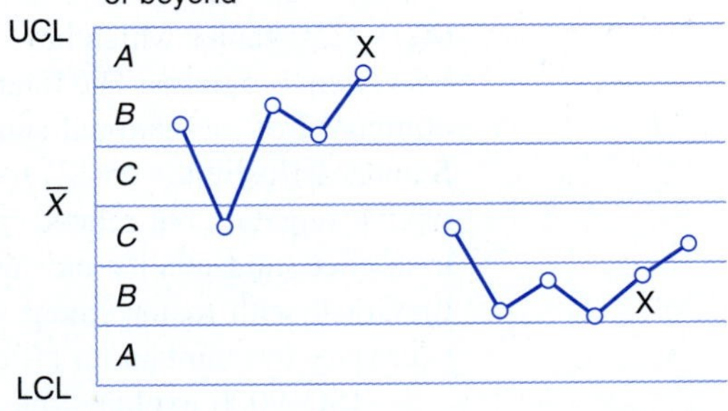

Test 7. Fifteen points in a row in Zone *C* (above and below centerline)

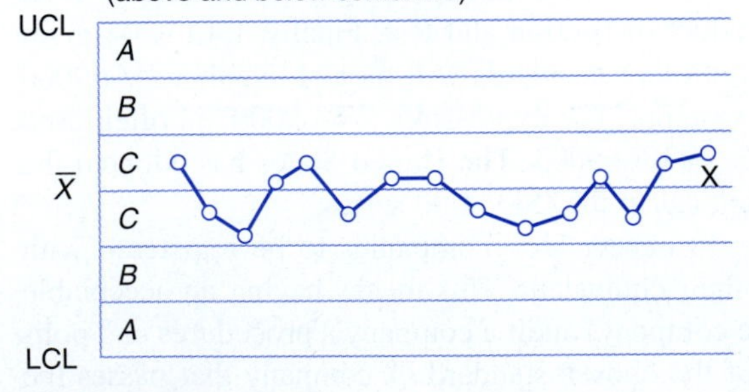

Test 8. Eight points in a row on both sides of centerline with none in Zone *C*

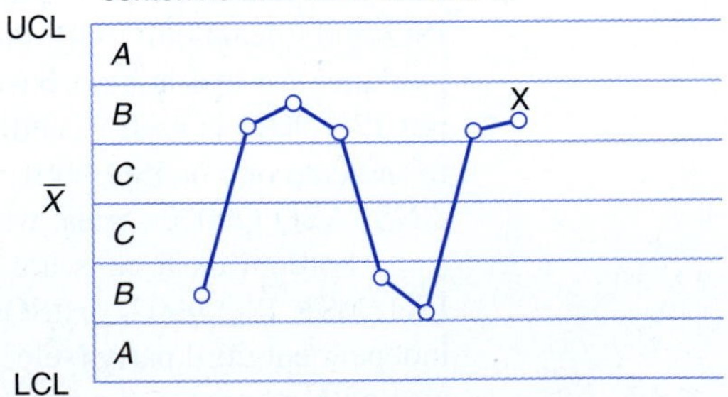

Figure 19.12 Illustrations of tests for special causes applied to Shewhart Control Charts.

SHEWHART CONTROL CHARTS

Notes on Tests for Special Causes

1. These tests are applicable to $\overline{X}$ charts and to individual (X) charts. A normal distribution is assumed. Tests 1, 2, 5, and 6 are to be applied to the upper and lower halves of the chart separately. Tests 3, 4, 7, and 8 are to be applied to the whole chart.
2. The upper control limit and the lower control limit are set at three sigma (standard deviations) above the centerline and three sigma below the centerline. For the purpose of applying the tests, the control chart is equally divided into six zones, each zone being one sigma wide. The upper half of the chart is referred to as A (outer third), B (middle third), and C (inner third). The lower half is taken as the mirror image.
3. When a process is in a state of statistical control, the chance of (incorrectly) getting a signal for the presence of a special cause is less than 5 in 1000 for each of these tests.
4. It is suggested that Tests 1, 2, 3, and 4 be applied routinely by the person plotting the chart. The overall probability of getting a false signal from one or more of these is about 1 in 100.
5. It is suggested that the first four tests be augmented by Tests 5 and 6 when it becomes economically desirable to have earlier warning. This will raise the probability of a false signal to about 2 in 100.
6. Tests 7 and 8 are diagnostic tests for stratification. They are very useful in setting up a control chart. These tests show when the observations in a subgroup have been taken from two (or more) sources with different means. Test 7 reacts when the observations in the subgroup always come from both sources. Test 8 reacts when the subgroups are taken from one source at a time.
7. Whenever the existence of a special cause is signaled by a test, this should be indicated by placing a cross just above the last point if that point lies above the centerline, or just below it if it lies below the centerline.
8. Points can contribute to more than one test. However, no point is ever marked with more than one cross.
9. The presence of a cross indicates that the process is not in statistical control. It means that the point is the last one of a sequence of points (a single point in Test 1) that is very unlikely to occur if the process is in statistical control.
10. Although this can be taken as a basic set of tests, analysts should be alert to any patterns of points that might indicate the influences of special causes in their process.

Figure 19.13 Comments on tests for special causes.

Exercises 19.6

For Exercises 1 through 7 determine if the process which is in control with the given control limits is capable of filling the given order. All measurements are in centimetres.

	Control limits	Order specifications
1.	37.998 to 38.002	38 ± 0.0004
2.	44.9992 to 45.0008	45 ± 0.001

	Control limits	Order specifications
3.	16.997 to 17.003	17 ± 0.002
4.	3.9995 to 4.0005	4 ± 0.0001
5.	27.9994 to 28.0006	28 ± 0.002
6.	64.997 to 65.003	65 ± 0.004
7.	51.994 to 52.006	52 ± 0.005

8. Problems in a production line which can be detected and controlled are called ________.

9. Product variations which are a result of the normal production process are called ________.

10. A manufacturing process is ____________ when it consistently produces items within its tolerance limits and is normally distributed.

11. A process in control which can do a specific job is called ____________.

12. A well-known international standard that is identified with companies that exhibit excellence in products and management is often referred to as ____________.

CHAPTER 19 SUMMARY

1. A *frequency table* displays data in groups.
2. A *histogram* is a graph composed of rectangles whose equal widths represent a group of data and whose heights represent the frequency with which the data occurs.
3. A *frequency polygon* uses line segments to connect plotted data points corresponding to the center of an interval and the frequency of the interval.
4. The *mean,* denoted $\bar{x}$, of n numbers $a_1, a_2, \ldots, a_n$ equals $\dfrac{a_1 + a_2 + \cdots + a_n}{n}$.
5. The *median* of an ordered set of n numbers is the middle number if n is odd and the mean of the two middle numbers if n is even.
6. The *mode* is the element of a data set which occurs most often. A set of data may contain more than one mode.
7. The *range* of data is the difference between the largest and smallest elements.
8. The collection of all items being considered is called a *population.*
9. A proper subset of a population that is selected to be studied is called a *sample.*
10. Picking an item at *random* means each item had an equal chance of being chosen.
11. The *sample standard deviation* s_x of a set of data $x_1, x_2, \ldots, x_n$ is given by the formula

$$s_x = \sqrt{\frac{(x_1 - \bar{x})^2 + (x_2 - \bar{x})^2 + \cdots + (x_n - \bar{x})^2}{n - 1}}$$

12. A *normal distribution* curve is bell shaped, is symmetric with respect to a vertical line through the mean, and contains 68% of the data within one sample standard deviation of the mean, 95% of the data within two sample standard deviations of the mean, and 99.7% of the data within three sample standard deviations of the mean.
13. The *transpose* of an $m \times n$ matrix is an $n \times m$ matrix whose columns $c_1, c_2, \ldots, c_m$ are the rows $r_1, r_2, \ldots, r_m$ from the $m \times n$ matrix.

14. The *line or curve of best fit* for a given set of ordered pairs is a formula for a line or curve which comes closest to approximating the values of the data set.
15. *Special-cause problems* are those that can be detected and controlled.
16. *Common-cause problems* occur as a natural part of the manufacturing process.
17. A manufacturing process is *in control* when it consistently produces items within its common-cause tolerance limits and the item's measurements fit a normal curve.
18. A process in control is *capable* of doing a specific job if the control limits are within the tolerance limits.
19. ISO 9000 is an international standard that is identified with companies that exhibit excellence in products and management.

CHAPTER 19 REVIEW

*For Exercises 1 through 3 construct (**a**) a frequency table, (**b**) a histogram, and (**c**) a frequency polygon.*

1. Thirty batteries were tested for the number of hours they gave usable power. Use five intervals each containing three numbers.

Number of Hours Batteries Gave Power

8	13	16	14	20	13
14	7	12	17	18	11
16	20	8	9	16	19
21	9	7	14	15	12
15	14	10	8	16	15

2. Twenty-five bottles were checked to see how much pressure, in psi (lb/in^2), they could withstand before bursting. Use six intervals each containing five numbers.

Bursting Pressure in psi

123	130	116	111	101
118	126	124	127	112
129	118	101	117	116
107	105	114	104	114
112	113	109	123	119

3. Twenty thousand plastic bottles were checked to see if they leaked. The chart gives the number that leaked in each group of 1000. Use five intervals each containing seven numbers.

Number Leaking for Each Thousand

68	53	79	70	58
82	65	87	56	82
55	59	64	85	66
74	82	75	63	81

For Exercises 4 through 8 find the mean, the median, and the mode for the data set. Round to three significant digits.

4. {27, 18, 31, 24, 21, 16, 23, 25, 18}

5. {8, 15, 9, 14, 19, 15, 10, 13, 15}

6. {35, 38, 33, 37, 38, 34, 36, 40}

7. {18, 21, 25, 21, 23, 19, 29, 20}

8.

Measurement	18	19	20	21	22
Frequency	9	15	16	14	3

For Exercises 9 through 12 find the range and the sample standard deviation. Round to three significant digits.

9. {8, 8, 9, 10, 11, 14}

10. {18, 19, 20, 20, 24, 27, 28}

11. {25, 28, 32, 33, 36, 39, 39}

12. {6, 8, 9, 9, 12, 13, 16, 17}

*For Exercises 13 and 14 the mean and the sample standard deviation are given. Find the percent of data points (**a**) within one sample standard deviation of the mean and (**b**) within two sample standard deviations of the mean. Are these percents within 2% of the $\bar{x} \pm 1s_x$ and $\bar{x} \pm 2s_x$ criteria for normal distribution?*

13.

Centimetres	54	55	56	57	58	59	60
Frequency	5	22	28	47	27	17	3

$\bar{x} = 56.9$, $s_x = 1.38$

14.

Inches	27	28	29	30	31	32	33
Frequency	7	29	35	62	37	23	5

$\bar{x} = 29.9$, $s_x = 1.40$

In Exercises 15 through 18 use the matrix method to find the line of best fit, $y = ax + b$. Round to two significant digits.

15.

x	1	2	3	5	6	7
y	2	8	13	20	26	38

16.

x	−2	1	2	3	4	5
y	−5	7	10	15	19	21

17.

x	1	2	3	4	5	7
y	3	5	7	12	15	21

18.

x	1	2	3	4	5	7
y	2	3	6	8	10	14

In Exercises 19 and 20 use the matrix method to find the quadratic equation of best fit, $y = ax^2 + bx + c$. Round to two significant digits.

19.

x	1	2	3	4	5	6
y	7	13	21	34	45	58

20.

x	2	3	4	5	6	7
y	4	11	22	34	43	58

21. Use a calculator routine to find the exponential curve of best fit, $y = ab^x$, for the following chart of bacteria growth:

x	0	1	2	3	4	5	6
y	2	7	16	55	160	485	1450

22. Use a calculator routine to find the exponential curve of best fit, $y = ab^x$, for the following chart of decay of radioactive material (expect $0 < b < 1$):

x	1	2	3	4	5	6	7
y	1.3	0.80	0.52	0.33	0.20	0.11	0.09

23. At a constant temperature, the volume of oxygen is compared to its pressure in the following chart. Find the equation relating volume and pressure using the PowerReg formula, $y = ax^b$, where x represents pressure in psi and y represents volume in ft^3. Round to three significant digits.

x	20	21	22	23	24	25
y	2400	2300	2150	2050	2000	1915

For Exercises 24 and 25, determine if the process, which is in control, with the given control limits is capable of filling the given order. All measurements are in inches.

	Control limits	Order specifications
24.	13.997 to 14.003	14 ± 0.005
25.	20.995 to 21.005	21 ± 0.001

CHAPTER 19

APPLICATION

Statistics in the Manufacturing Process

To gain some firsthand knowledge about the use of statistics in quality control practices, the authors toured Plastipak Packaging, Inc., in Champaign, Illinois, and talked with Zack Stork, Plant Quality Manager and ISO Management Representative. This plant has over 50 production lines; the fastest can produce 400 to 500 plastic bottles per *minute.* These containers are used for a variety of products including soft drinks, salad dressings, edible oils, and cleaning solutions.

The quality control process begins with inspection of the incoming materials: labels, corrugation (cardboard), and resins (plastics), which are high-density polyethylene (HDPE) and polyethylene terephthalate (PET). The supply vendors are rated as approved, conditional, or nonapproved, and are rerated every 6 months. Goods from nonapproved vendors are 100% inspected; others are checked by sampling.

The plastic is colored (or not) and then either extruded as a flowing parison (HDPE) or injection molded into a test-tube-like shape with a threaded top called a *preform* (PET). Injection presses, which make the preforms, have up to 96 individual cavities. The preforms are heated using infrared lamps prior to being transferred to blow-molding cavities, where they are blown into bottles. A network-driven, computer-aided, statistical process control (SPC) system allows real-time downloading and statistical charting of critical bottle measurements and attributes from the entire mold set as well as from the individual mold cavities. For example, the range of weights and the mean (arithmetic average) of weights from a given mold set are displayed using a control chart. The weights of bottles from a designated cavity are also displayed using a control chart. This cavity-specific capability helps to identify local problems. Charts in the plant are color coded, with a green bar indicating measurements within 1.5 standard deviations of the mean, a yellow bar indicating that a data point is between 1.5 and 3.0 standard deviations of the mean, and a red bar indicating that a data point is within specification but outside the control limits (see Section 19.6). This prompts an investigation into the root cause. A data point in the white space indicates a bottle is out of specification. When this happens, the product is placed "on hold" and the cause is immediately sought. A two-color reproduction of this chart is shown on page 607. The upper control limit (UCL) and lower control limit (LCL) indicated on the charts mark the boundary of two standard deviations above and below the mean. The upper and lower tolerance (specification) limits are indicated by USL and LSL and mark the boundary of three standard deviations above and below the mean. Tgt (target) marks the mean; points in the shaded area above and below it are within one standard deviation of the mean. Histograms, as on page 606, are also used to help identify problems—when a histogram is skewed or bimodal it indicates a *special cause* problem such as tool wear, power surges, or a technician error. Note that the machines are programmed to adjust for humidity and temperature variations.

In addition to the automatically loaded computer data, hourly visual inspections are conducted on bottleneck threads, label position, and in some cases, the dimensions of the cardboard containers for the bottles. One manufacturer of a cleaning solution containing enzymes buys the bottles upright in boxes and fills them in the box. If the boxes are too large, the position of the bottles will shift, resulting in cleaning solution missing the bottle opening. Such a spill would result in a costly cleanup operation, so having the box exactly right is important. Sometimes bottles are stacked ten high on a large pallet, then bound in plastic wrap. For this situation, bottles must be able to withstand a predetermined amount of weight/pressure. Topload (vertical compression) testing provides the amount of weight that a bottle can resist prior to failing. Topload is a variable which is control charted.

Finally, *random* sampling techniques with respect to time are used to obtain information about the entire collection of bottles (the population). The importance of the random selection of bottles to be tested cannot be overemphasized, for it is mathematically necessary in order to draw valid conclusions about the population. Employees are trained to choose samples in a random manner, and often different workers will do the selecting.

Samples are sent to the Quality Laboratory, which contains a spectrophotometer for detailed color analysis, wall thickness testers, top loaders, burst testers, a drop tester, a torque tester, a side impact tester, height gauges, and other test equipment. Technicians operate the equipment while the data is automatically entered into a computer. Histograms, frequency polygons, means, ranges, and standard deviation charts are then available for quality control technicians and statisticians to monitor for possible shifts, trends, and uncommon occurrences. Control charts are also computer produced at numerous quality stations throughout each production area.

Although most measurements and attributes are monitored through random sampling, there is one characteristic that is checked in every single bottle—that is, whether it will leak. This is done by injecting air to a fixed pressure into a bottle, then measuring the air pressure. If the air pressure decreases, then the bottle leaks and is ejected into a recycling bin.

We have seen that histograms, frequency polygons, mean, range, standard deviation, and control charts are the tools used in quality control. In Chapter 19 we examined these as well as some other important features of statistical process control.

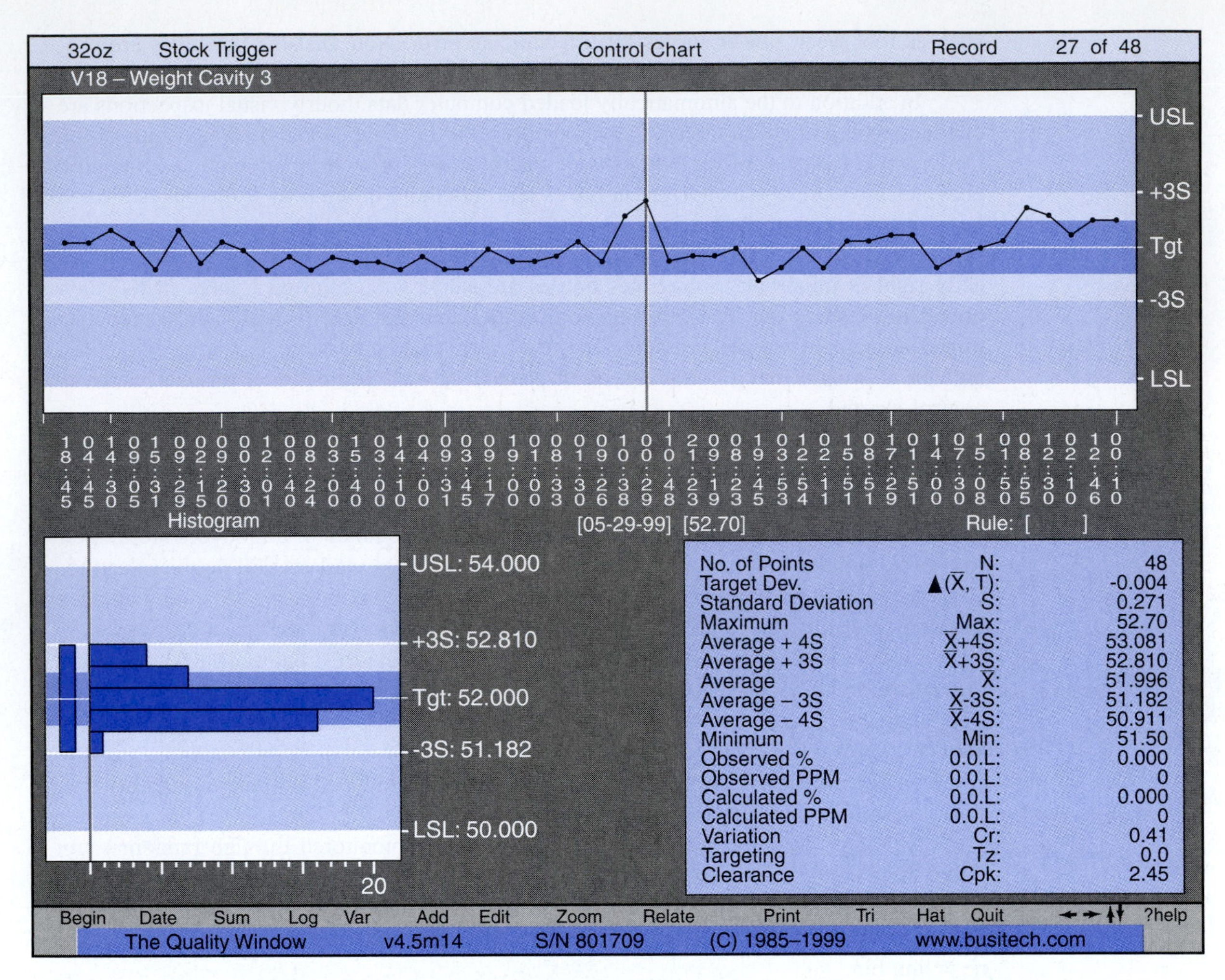

Cavity-specific control charts allow extremely subtle shifts and trends to be observed. Such features may be hidden in standard X-bar charts. (Laboratory data courtesy of Plastipak Packaging, Inc.)

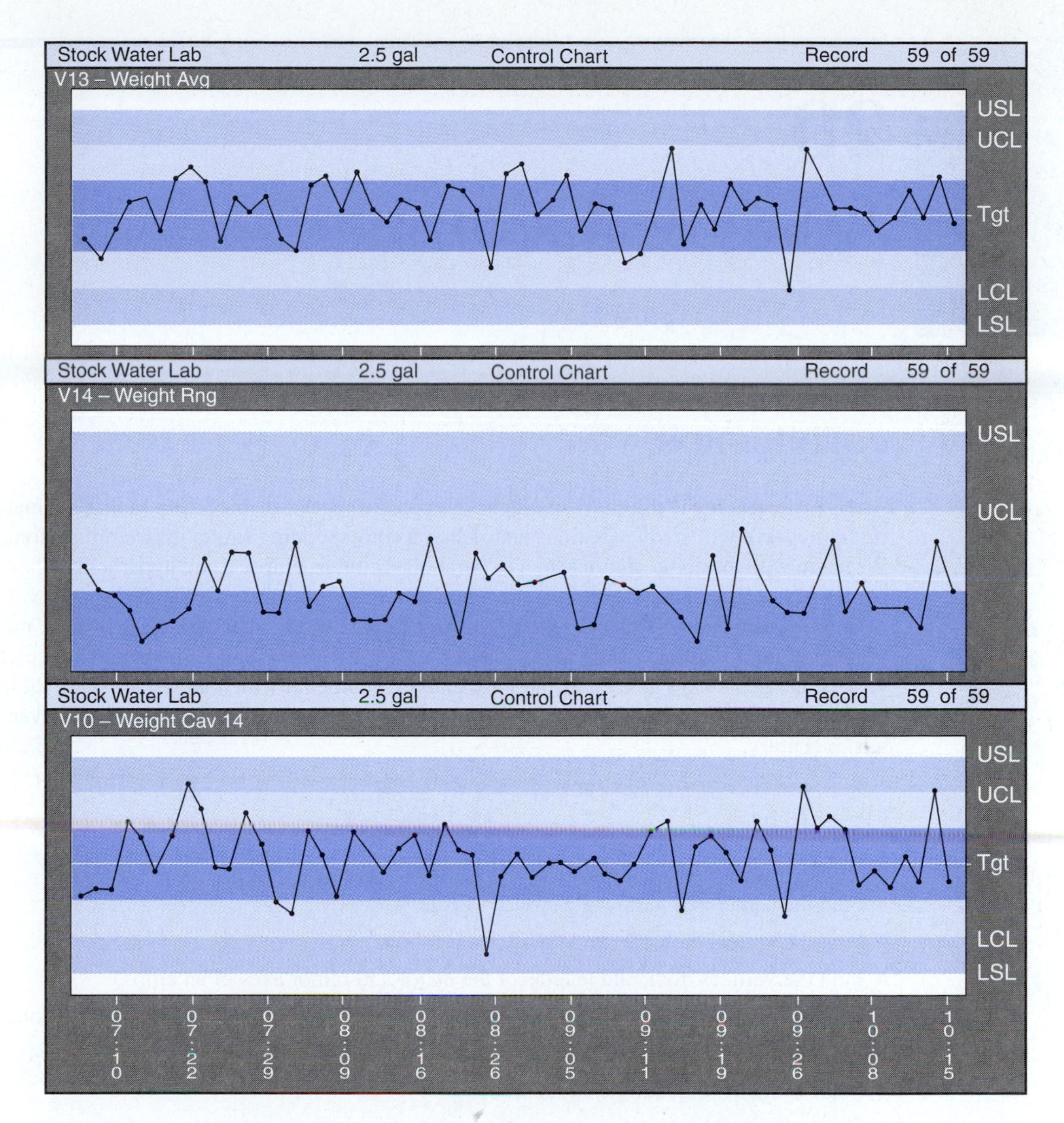

Top: X-bar (average) chart. Each point represents the average for a complete mold round.
Middle: Range chart. Each point represents the range (max-min) for a complete mold round.
Bottom: Cavity-specific chart. Each point represents a single bottle from a specific mold cavity. In this specific case, each point represents the weight of bottle #14 as a function of time.

20 Analytic Geometry*

INTRODUCTION

Analytic geometry is the study of algebraic expressions and their corresponding geometric figures. We will study equations and sketch a corresponding figure in two dimensions. We begin with the conic sections; that is, parabolas, circles, ellipses, and hyperbolas. These curves occur often in nature and play an important role in applied mathematics.

For example, the existence of the focus of a parabola is what makes flashlights, microphones, and satellite dishes work. Planets orbit the sun in elliptical paths, and circular wheels and gears help to keep us mobile. Analytic geometry, which we use to study these curves, evolved from the work of René Descartes, a French mathematician in the seventeenth century.

Objectives

- Graph circles, ellipses, parabolas, and hyperbolas.
- Find the center and radius of a circle.
- Find the vertex, directrix, and focus of a parabola.
- Find the vertices, foci, and lengths of the major and minor axes of an ellipse.
- Find the vertices, foci, and lengths of the transverse and conjugate axes of a hyperbola.
- Use translation of axes in sketching graphs and identifying key features.
- Solve systems of quadratic equations.
- Graph using polar coordinates.
- Convert between polar and rectangular coordinates.

20.1 THE CIRCLE

Equations in two variables of second degree in the form

$$Ax^2 + Bxy + Cy^2 + Dx + Ey + F = 0$$

are called **conics.** We begin a systematic study of conics with the circle.

The **circle** consists of the set of points located the same distance from a given point, called the **center.** The distance at which all points are located from the center is called the **radius.** A circle may thus be graphed in the plane given its center and radius.

*Note: The previous rules for calculations of measurements with significant digits are normally not used in analytic geometry and calculus. We will not use them from this point.

EXAMPLE 1

Graph the circle with center at $(1, -2)$ and radius $r = 3$.

Plot all points in the plane located 3 units away from the point $(1, -2)$ as in Fig. 20.1. (You may wish to use a compass.)

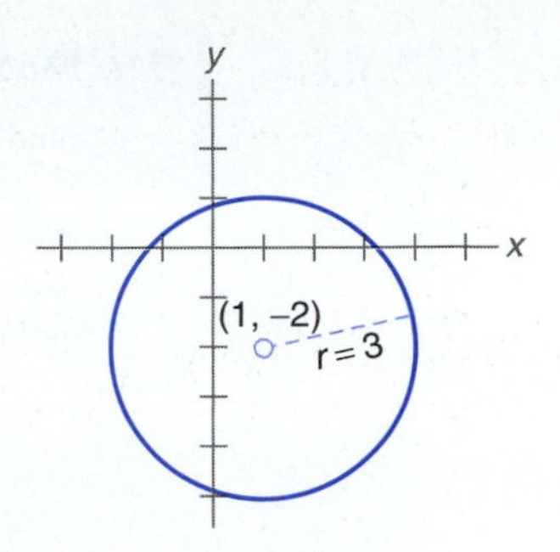

Figure 20.1

From the definition of a circle we can determine its equation. Let (h, k) be the coordinates of the center and let r represent the radius. If any point (x, y) is located on the circle, it must be a distance r from the center (h, k) as in Fig. 20.2.

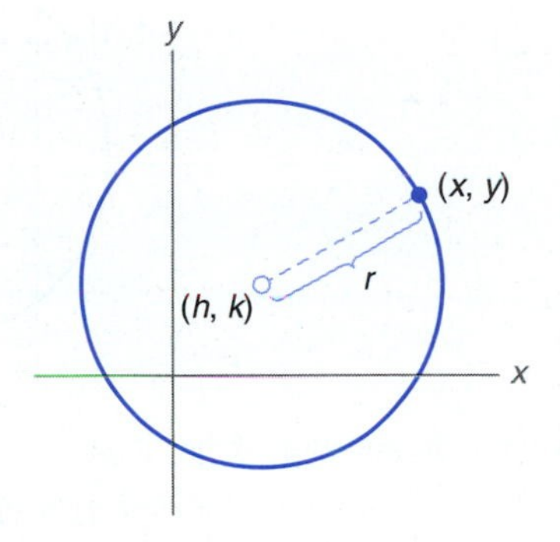

Figure 20.2 The set of points (x, y) located at the same distance r from the given point (h, k).

Using the distance formula, we have

$$\sqrt{(x_2 - x_1)^2 + (y_2 - y_1)^2} = d$$
$$\sqrt{(x - h)^2 + (y - k)^2} = r$$

Squaring each side, we have the following result:

STANDARD FORM OF A CIRCLE

$$(x - h)^2 + (y - k)^2 = r^2$$

where r is the radius and (h, k) is the center.

Any point (x, y) satisfying this equation lies on the circle.

EXAMPLE 2

Find the equation of the circle with radius 3 and center $(1, -2)$ (see Example 1).

Using the standard form of the equation of a circle, we have

$$(x - h)^2 + (y - k)^2 = r^2$$
$$(x - 1)^2 + [y - (-2)]^2 = (3)^2$$
$$(x - 1)^2 + (y + 2)^2 = 9$$

EXAMPLE 3

Find the equation of the circle with center at $(3, -2)$ and passing through $(-1, 1)$.

To write the equation, we need to know the radius r of the circle. Although r has not been stated, we do know that every point on the circle is a distance r from the center, $(3, -2)$. In particular, the point $(-1, 1)$ is a distance r from $(3, -2)$ (see Fig. 20.3). Using the distance formula, we obtain

$$d = \sqrt{(x_2 - x_1)^2 + (y_2 - y_1)^2}$$
$$d = r = \sqrt{(x - h)^2 + (y - k)^2}$$
$$r = \sqrt{(-1 - 3)^2 + [1 - (-2)]^2}$$
$$= \sqrt{(-4)^2 + 3^2} = \sqrt{16 + 9} = \sqrt{25} = 5$$

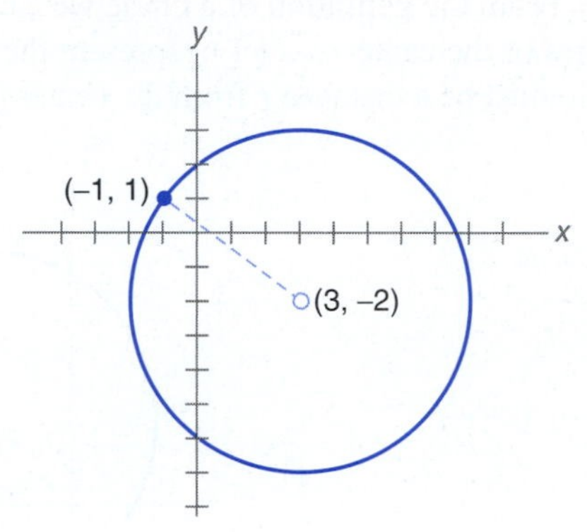

Figure 20.3

Now write the equation of the circle.

$$(x - h)^2 + (y - k)^2 = r^2$$
$$(x - 3)^2 + [y - (-2)]^2 = 5^2$$
$$(x - 3)^2 + (y + 2)^2 = 25$$

If we remove parentheses in the equation

$$(x - h)^2 + (y - k)^2 = r^2$$

we have

$$x^2 - 2xh + h^2 + y^2 - 2yk + k^2 = r^2$$

Rearranging terms, we have

$$x^2 + y^2 - 2hx - 2ky + h^2 + k^2 - r^2 = 0$$

If we let $D = -2h$, $E = -2k$, and $F = h^2 + k^2 - r^2$, we obtain the following equation:

GENERAL FORM OF A CIRCLE

$$x^2 + y^2 + Dx + Ey + F = 0$$

Any equation in this form represents a circle.

EXAMPLE 4

Write the equation $(x - 3)^2 + (y + 2)^2 = 25$ obtained in Example 3 in general form.

$$(x - 3)^2 + (y + 2)^2 = 25$$
$$x^2 - 6x + 9 + y^2 + 4y + 4 = 25$$
$$x^2 + y^2 - 6x + 4y - 12 = 0$$

EXAMPLE 5

Find the center and radius of the circle given by the equation

$$x^2 + y^2 - 4x + 2y - 11 = 0$$

Looking back at how we arrived at the general equation of a circle, we see that if we rearrange the terms of the equation as

$$(x^2 - 4x \qquad) + (y^2 + 2y \qquad) = 11$$

then $(x^2 - 4x \qquad)$ represents the first two terms of

$$(x - h)^2 = x^2 - 2hx + h^2$$

and $(y^2 + 2y \qquad)$ represents the first two terms of

$$(y - k)^2 = y^2 - 2ky + k^2$$

This means that

$$\begin{aligned} -4 &= -2h \quad \text{and} \quad 2 = -2k \\ h &= 2 \qquad\qquad\quad k = -1 \end{aligned}$$

To complete the squares $(x - h)^2$ and $(y - k)^2$, we must add $h^2 = 2^2 = 4$ and $k^2 = (-1)^2 = 1$ to each side of the equation.

$$\begin{aligned} (x^2 - 4x \qquad) + (y^2 + 2y \qquad) &= 11 \\ (x^2 - 4x + 4) + (y^2 + 2y + 1) &= 11 + 4 + 1 \\ (x - 2)^2 + (y + 1)^2 &= 16 \\ (x - 2)^2 + [y - (-1)]^2 &= 16 = 4^2 \end{aligned}$$

From this we see that we have the standard form of the equation of a circle with radius 4 and center at the point $(2, -1)$.

This process is called **completing the square** of the x- and y-terms. In general, if the coefficients of x^2 and y^2 are both equal to 1, then these values can be found as follows: Add h^2 and k^2 to each side of the equation, where

$$\begin{aligned} h^2 &= (\tfrac{1}{2} \text{ the coefficient of } x)^2 = (\tfrac{1}{2}D)^2 \\ k^2 &= (\tfrac{1}{2} \text{ the coefficient of } y)^2 = (\tfrac{1}{2}E)^2 \end{aligned}$$

EXAMPLE 6

Find the center and radius of the circle given by the equation

$$x^2 + y^2 + 6x - 4y - 12 = 0$$

Sketch the graph of the circle. Write

$$\begin{aligned} h^2 &= [(\tfrac{1}{2})(6)]^2 = 3^2 = 9 \\ k^2 &= [\tfrac{1}{2}(-4)]^2 = (-2)^2 = 4 \end{aligned}$$

Rewrite the equation and add 9 and 4 to each side.

$$\begin{aligned} (x^2 + 6x \qquad) + (y^2 - 4y \qquad) &= 12 \\ (x^2 + 6x + 9) + (y^2 - 4y + 4) &= 12 + 9 + 4 \\ (x + 3)^2 + (y - 2)^2 &= 25 = 5^2 \end{aligned}$$

The center is at $(-3, 2)$ and the radius is 5. Plot all points that are at a distance of 5 from the point $(-3, 2)$ as in Fig. 20.4.

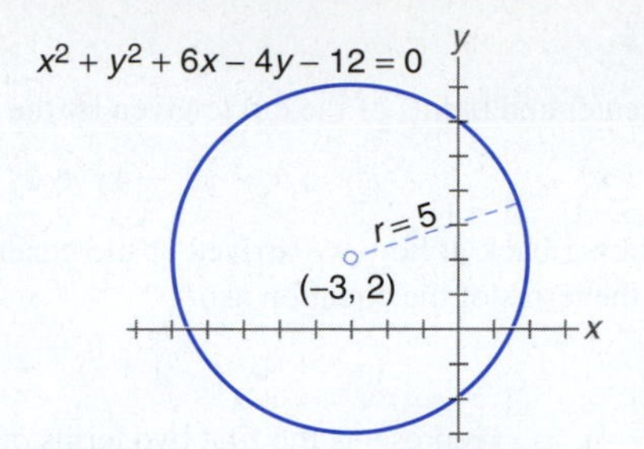

Figure 20.4

If the center of a circle is at the origin, then $h = 0$ and $k = 0$, and its standard equation becomes

$$x^2 + y^2 = r^2$$

where r is the radius and the center is at the origin.

EXAMPLE 7

Find the equation of the circle with radius 3 and center at the origin. Also, graph the circle.

$$x^2 + y^2 = r^2$$
$$x^2 + y^2 = (3)^2$$
$$x^2 + y^2 = 9 \qquad \text{(See Fig. 20.5.)}$$

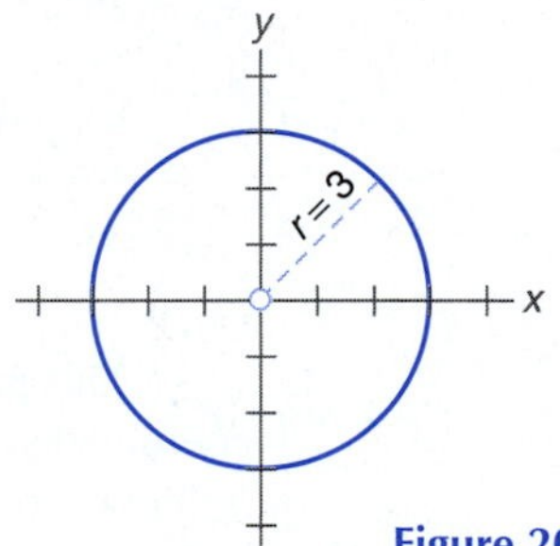

Figure 20.5

Exercises 20.1

Graph the circle with the given center and radius.

1. Center at $(2, -1)$, $r = 3$.

2. Center at $(3, 3)$, $r = 2$.

3. Center at $(0, 2)$, $r = 4$.

4. Center at $(-4, -5)$, $r = 3$.

Find the equation of the circle (in standard form) with the given properties.

5. Center at $(1, -1)$, radius 4.

6. Center at $(-2, 3)$, radius $\sqrt{5}$.

7. Center at $(-2, -4)$, passing through $(1, -9)$.

8. Center at $(5, 2)$, passing through $(-2, -6)$.

9. Center at $(0, 0)$, radius 6.

10. Center at $(0, 0)$, passing through $(3, -4)$.

Find the center and radius of the given circle.

11. $x^2 + y^2 = 16$

12. $x^2 + y^2 - 4x - 5 = 0$

13. $x^2 + y^2 + 6x - 8y - 39 = 0$

14. $x^2 + y^2 - 6x + 14y + 42 = 0$

15. $x^2 + y^2 - 8x + 12y - 8 = 0$

16. $x^2 + y^2 + 10x + 2y - 14 = 0$

17. $x^2 + y^2 - 12x - 2y - 12 = 0$

18. $x^2 + y^2 + 4x - 9y + 4 = 0$

19. $x^2 + y^2 + 7x + 3y - 9 = 0$

20. $x^2 + y^2 - 5x - 8y = 0$

21. Find the equation of the circle or circles whose center is on the y-axis and that contain the points $(1, 4)$ and $(-3, 2)$. Give the center and radius.

22. Find the equation of the circle with center in the first quadrant on the line $y = 2x$, tangent to the x-axis, and radius 6. Give its center.

23. Find the equation of the circle containing the points $(3, 1)$, $(0, 0)$, and $(8, 4)$. Give its center and radius.

24. Find the equation of the circle containing the points $(1, -4)$, $(-3, 4)$, and $(4, 5)$. Give its center and radius.

20.2 THE PARABOLA

While the parabola may not be as familiar a geometric curve as the circle, examples of the parabola are found in many technical applications. A **parabola** consists of all points that are the same distance from a given fixed point and a given fixed line. The fixed point is called the **focus.** The fixed line is called the **directrix.** This relationship is shown in Fig. 20.6 for the points P, Q, and V, which lie on a parabola with focus F and directrix D.

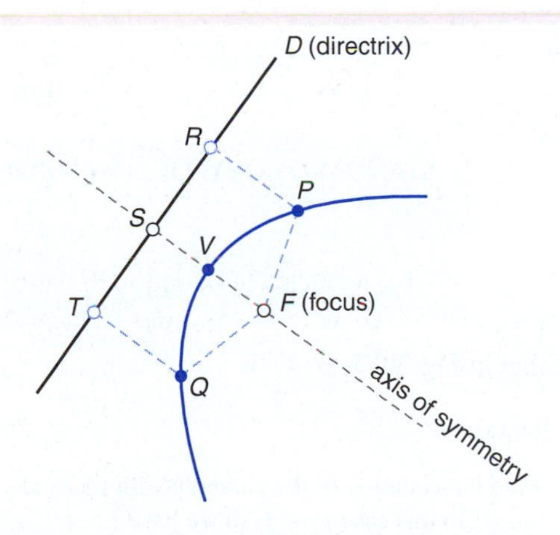

Figure 20.6 A parabola is a set of points that are the same distance from a given fixed point (focus) and a given fixed line (directrix).

Note:

$$RP = PF$$
$$SV = VF$$
$$TQ = QF$$

The point V midway between the directrix and the focus is called the **vertex.** The vertex and the focus lie on a line perpendicular to the directrix, which is called the **axis of symmetry.**

There are two standard forms for the equation of a parabola. The form depends on the position of the parabola in the plane. We first discuss the parabola with focus on the x-axis at $(p, 0)$ and directrix the line $x = -p$ as in Fig. 20.7. Let $P(x, y)$ represent any point on this parabola. Vertex V is then at the origin, and the axis of symmetry is the x-axis.

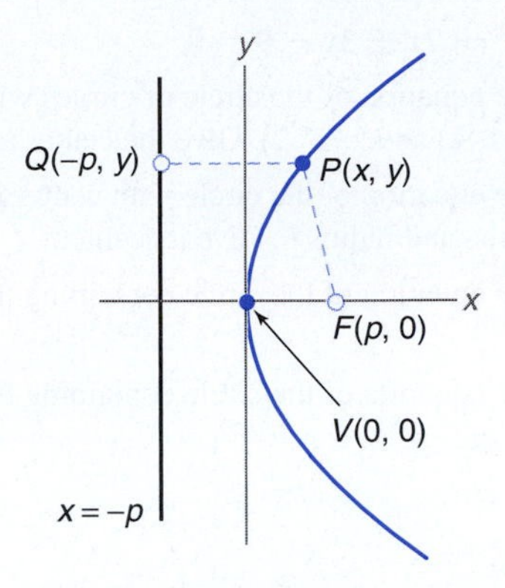

Figure 20.7

By the way we have described the parabola, the distance between P and F must equal the distance between P and Q. Using the distance formula, we have

$$PF = PQ$$

$$\sqrt{(x - p)^2 + (y - 0)^2} = \sqrt{[x - (-p)]^2 + (y - y)^2}$$

$$(x - p)^2 + y^2 = (x + p)^2 \qquad \text{(Square each side.)}$$

$$x^2 - 2xp + p^2 + y^2 = x^2 + 2xp + p^2$$

$$y^2 = 4px$$

STANDARD FORM OF A PARABOLA

$$y^2 = 4px$$

with focus at $(p, 0)$ and with the line $x = -p$ as the directrix.

Note that in Fig. 20.7, $p > 0$.

EXAMPLE 1

Find the equation of the parabola with focus at $(3, 0)$ and directrix $x = -3$.

In this case $p = 3$, so we have

$$y^2 = 4(3)x$$

$$y^2 = 12x$$

EXAMPLE 2

Find the focus and equation of the directrix of the parabola $y^2 = 24x$.

$$y^2 = 24x$$

$$y^2 = 4(6)x$$

$$y^2 = 4px$$

Since p must be 6, the focus is $(6, 0)$ and the directrix is the line $x = -6$.

EXAMPLE 3

Find the equation of the parabola with focus at $(-2, 0)$ and with directrix $x = 2$. Sketch the graph.

Here $p = -2$. The equation becomes

$$y^2 = 4(-2)x$$
$$y^2 = -8x \qquad \text{(See Fig. 20.8.)}$$

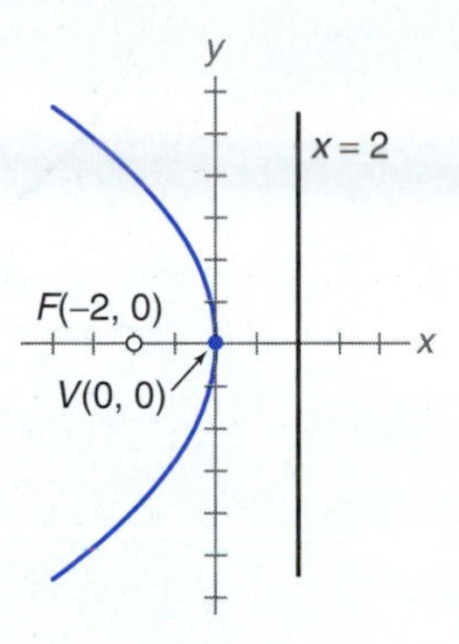

Figure 20.8

Observe the following:

1. If $p > 0$, the coefficient of x in the equation $y^2 = 4px$ is *positive* and the parabola opens to the *right* (see Fig. 20.9a).
2. If $p < 0$, the coefficient of x in the equation $y^2 = 4px$ is *negative* and the parabola opens to the *left* (see Fig. 20.9b).

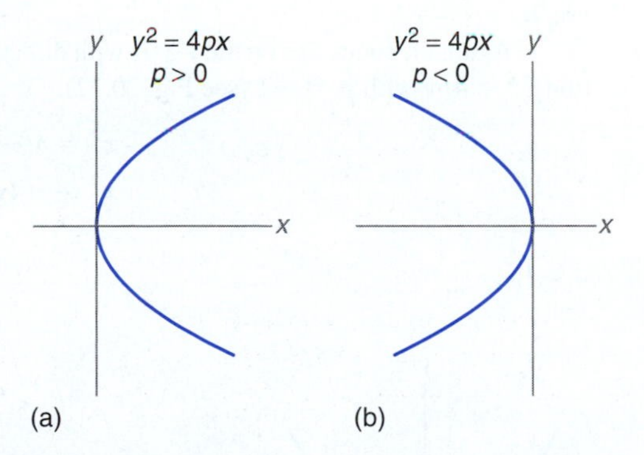

Figure 20.9 Standard form of the parabola $y^2 = 4px$.

We obtain the other standard form of the parabola when the focus lies on the y-axis and the directrix is parallel to the x-axis. Let $(0, p)$ be the focus F and $y = -p$ be the directrix. The vertex is still at the origin, but the axis of symmetry is now the y-axis (see Fig. 20.10).

STANDARD FORM OF A PARABOLA

$$x^2 = 4py$$

with focus at $(0, p)$ and with the line $y = -p$ as the directrix. (See Fig. 20.10.)

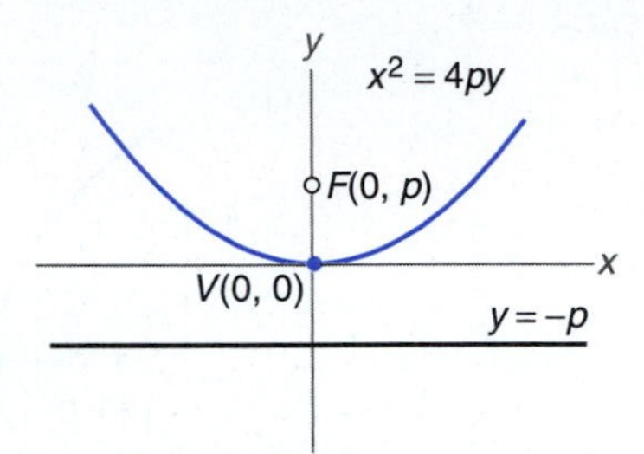

Figure 20.10

EXAMPLE 4

Find the equation of the parabola with focus at $(0, 3)$ and with directrix $y = -3$. Sketch the graph.

Since the focus lies on the y-axis and the directrix is parallel to the x-axis, we use the equation $x^2 = 4py$ with $p = 3$ (see Fig. 20.11).

$$x^2 = 4(3)y$$
$$x^2 = 12y$$

EXAMPLE 5

Find the equation of the parabola with focus at $(0, -1)$ and with directrix $y = 1$. Sketch the graph.

Again the focus lies on the y-axis with directrix parallel to the x-axis, so we use the equation $x^2 = 4py$ with $p = -1$ (see Fig. 20.12).

$$x^2 = 4(-1)y$$
$$x^2 = -4y$$

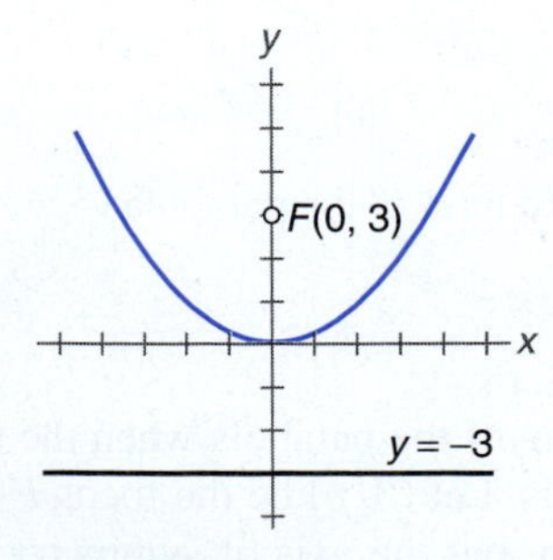

Figure 20.11

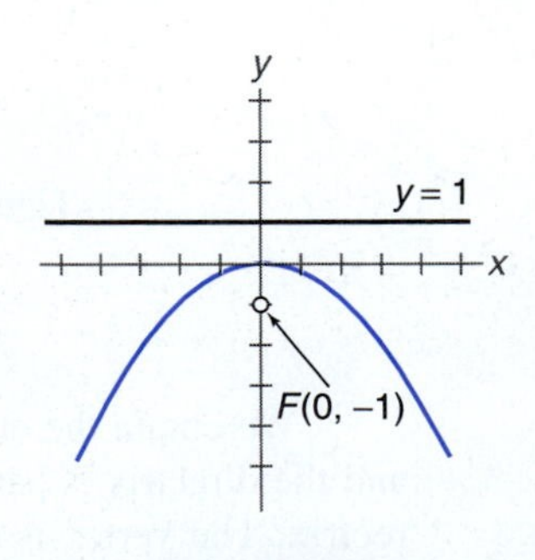

Figure 20.12

Observe the following:

1. **1.** If $p > 0$, the coefficient of y in the equation $x^2 = 4py$ is *positive* and the parabola opens *upward* (see Fig. 20.13a).
2. **2.** If $p < 0$, the coefficient of y in the equation $x^2 = 4py$ is *negative* and the parabola opens *downward* (see Fig. 20.13b).

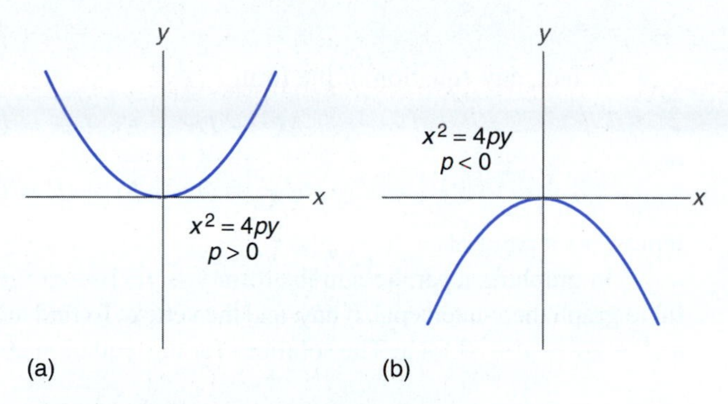

Figure 20.13 Standard form of the parabola $x^2 = 4py$.

We are now able to describe the graph of a parabola by inspection of its equation in standard form. We can also find the focus and directrix.

EXAMPLE 6

Describe the graph of the equation $y^2 = 20x$.

This is an equation of a parabola in the form $y^2 = 4px$. Since $p = 5$, this parabola has its focus at (5, 0) and its directrix is the line $x = -5$. The parabola opens to the right (since $p > 0$).

EXAMPLE 7

Describe the graph of the equation $x^2 = -2y$.

This is an equation of a parabola in the form $x^2 = 4py$, where $p = -\frac{1}{2}$ [as $4(-\frac{1}{2}) = -2$]. The focus is at $(0, -\frac{1}{2})$ and the directrix is the line $y = \frac{1}{2}$. The parabola opens downward (since $p < 0$).

Of course, not all parabolas are given in standard position.

EXAMPLE 8

Find the equation of the parabola with focus at (1, 3) and with the line $y = -1$ as directrix. We must use the definition of the parabola (see Fig. 20.14).

$$PF = PQ$$
$$\sqrt{(x-1)^2 + (y-3)^2} = \sqrt{(x-x)^2 + [y-(-1)]^2}$$
$$x^2 - 2x + 1 + y^2 - 6y + 9 = y^2 + 2y + 1 \qquad \text{(Square each side and}$$
$$x^2 - 2x - 8y + 9 = 0 \qquad \text{remove parentheses.)}$$

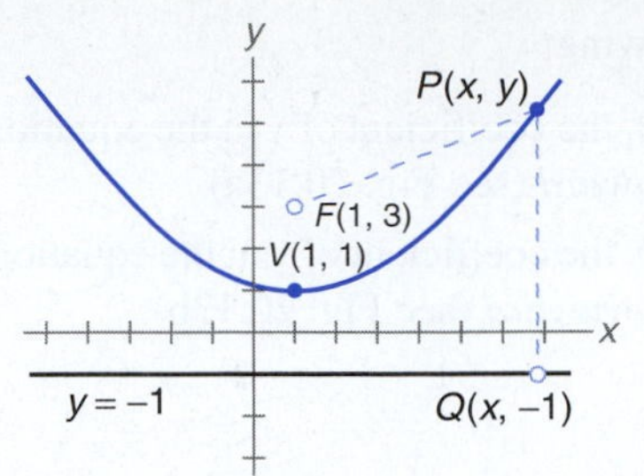

Figure 20.14

In fact, any equation of the form

$$Ax^2 + Dx + Ey + F = 0$$

or

$$Cy^2 + Dx + Ey + F = 0$$

represents a parabola.

In graphing a parabola in the form $y = f(x) = ax^2 + bx + c$, $a \neq 0$, it is most helpful to graph the x-intercepts, if any, and the vertex. To find the x-intercepts, let $y = 0$ and solve $ax^2 + bx + c = 0$ for x. The solutions for this equation are given by the quadratic formula:

$$x = \frac{-b \pm \sqrt{b^2 - 4ac}}{2a}$$

Recall that the solutions are real numbers only if the *discriminant,* $b^2 - 4ac$, is nonnegative and that the solutions are imaginary if $b^2 - 4ac < 0$. Thus, the graph of the parabola $y = f(x) = ax^2 + bx + c$, $a \neq 0$, has

1. two different x-intercepts if $b^2 - 4ac > 0$,
2. only one x-intercept if $b^2 - 4ac = 0$ (the graph is tangent to the x-axis),
3. no x-intercepts if $b^2 - 4ac < 0$.

The **axis of symmetry** of the parabola in the form $y = f(x) = ax^2 + bx + c$ is a vertical line halfway between the x-intercepts. The equation of the axis is the vertical line passing through the midpoint of the line segment joining the two x-intercepts (see Fig. 20.15). This midpoint is

$$\frac{x_1 + x_2}{2} = \frac{\dfrac{-b - \sqrt{b^2 - 4ac}}{2a} + \dfrac{-b + \sqrt{b^2 - 4ac}}{2a}}{2} = \frac{\dfrac{-2b}{2a}}{2} = -\frac{b}{2a}$$

Thus, the equation of the axis is $x = -\dfrac{b}{2a}$.

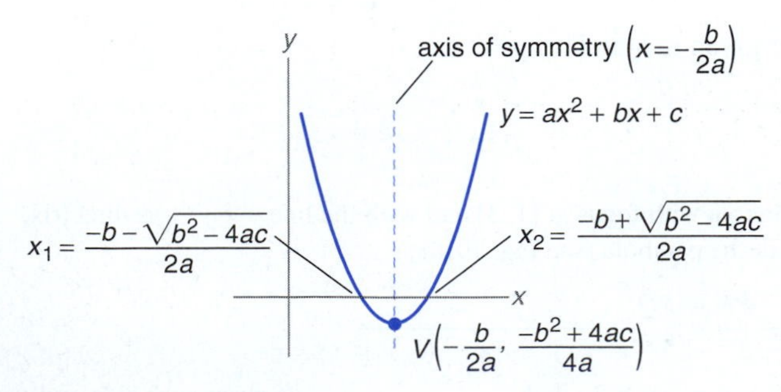

Figure 20.15 Axis of symmetry of the parabola $y = ax^2 + bx + c$.

Since the parabola contains the vertex, its x-coordinate is $-\dfrac{b}{2a}$. Its y-coordinate is then $f\left(-\frac{b}{2a}\right)$. To find the y-coordinate, evaluate

$$\begin{aligned} f(x) &= ax^2 + bx + c \\ f\left(-\frac{b}{2a}\right) &= a\left(-\frac{b}{2a}\right)^2 + b\left(-\frac{b}{2a}\right) + c \\ &= \frac{b^2}{4a} - \frac{b^2}{2a} + c \\ &= \frac{b^2}{4a} - \frac{2b^2}{4a} + \frac{4ac}{4a} \\ &= \frac{-b^2 + 4ac}{4a} \end{aligned}$$

AXIS AND VERTEX OF A PARABOLA

Given the parabola $y = f(x) = ax^2 + bx + c$, its axis is the vertical line $x = -\frac{b}{2a}$ and its vertex is the point

$$\left(-\frac{b}{2a}, f\left(-\frac{b}{2a}\right)\right) = \left(-\frac{b}{2a}, \frac{-b^2 + 4ac}{4a}\right)$$

The vertex is a maximum point if $a < 0$ and a minimum point if $a > 0$.

EXAMPLE 9

Graph $y = f(x) = 2x^2 - 8x + 11$. Find its vertex and the equation of the axis.

First, note that $b^2 - 4ac = (-8)^2 - 4(2)(11) = -24 < 0$, which means that the graph has no x-intercepts. The equation of the axis is

$$x = -\frac{b}{2a} = -\frac{-8}{2(2)} = 2$$

The vertex is the point $(2, f(2))$ or $(2, 2 \cdot 2^2 - 8 \cdot 2 + 11) = (2, 3)$.

This y-coordinate may also be found using the formula

$$\frac{-b^2 + 4ac}{4a} = \frac{-(-8)^2 + 4 \cdot 2 \cdot 11}{4 \cdot 2} = \frac{24}{8} = 3$$

Since $a > 0$, the vertex $(2, 3)$ is a minimum point and the graph opens upward. You may also find some additional ordered pairs to graph this equation depending on whether you need only a rough sketch or a fairly accurate graph. The graph is shown in Fig. 20.16.

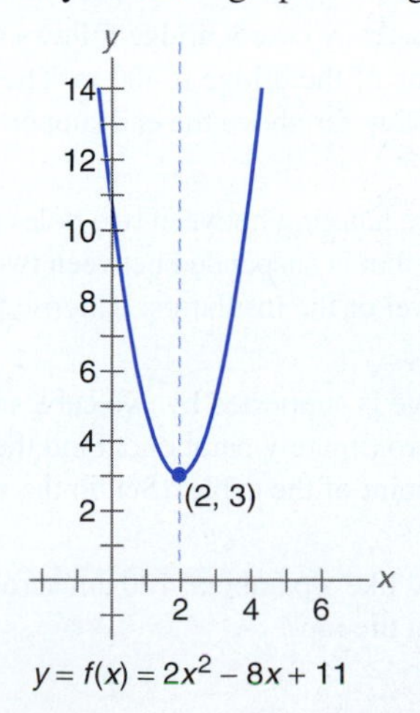

Figure 20.16

Since the vertex of a parabola in the form $y = f(x) = ax^2 + bx + c$ is the highest point or the lowest point on the graph, we can use this fact to find a maximum or a minimum value of a quadratic function.

EXAMPLE 10

An object is thrown upward with an initial velocity of 48 ft/s. Its height after t seconds is given by $h = f(t) = 48t - 16t^2$. Find its maximum height and the time it takes the object to hit the ground.

First, find the vertex.

$$x = -\frac{b}{2a} = -\frac{48}{2(-16)} = \frac{3}{2}$$

Then the vertex is

$$\left(\frac{3}{2}, f\left(\frac{3}{2}\right)\right) = \left(\frac{3}{2}, 48\left(\frac{3}{2}\right) - 16\left(\frac{3}{2}\right)^2\right) = \left(\frac{3}{2}, 36\right)$$

Since $a = -16 < 0$, the vertex is a maximum point, and the maximum height is 36 ft.

The first coordinate of the vertex gives the amount of time it takes for the object to reach its maximum height. The time it takes for such a projectile to reach its maximum height is the same as the time it takes to drop back to the ground. Thus, the object hits the ground $2 \cdot \frac{3}{2}$ or 3 s after it is thrown.

Exercises 20.2

Find the focus and the directrix of each parabola. Sketch each graph.

1. $x^2 = 4y$
2. $x^2 = -8y$
3. $y^2 = -16x$
4. $x^2 = -6y$
5. $y^2 = x$
6. $y^2 = -4x$
7. $x^2 = 16y$
8. $y^2 = -12x$
9. $y^2 = 8x$
10. $x^2 = -y$

Find the equation of the parabola with given focus and directrix.

11. $(2, 0), x = -2$
12. $(0, -3), y = 3$
13. $(-8, 0), x = 8$
14. $(5, 0), x = -5$
15. $(0, 6), y = -6$
16. $(0, -1), y = 1$

17. Find the equation of the parabola with focus at $(-4, 0)$ and vertex at $(0, 0)$.

18. Find the equation of the parabola with vertex at $(0, 0)$ and directrix $y = -2$.

19. Find the equation of the parabola with focus $(-1, 3)$ and directrix $x = 3$.

20. Find the equation of the parabola with focus $(2, -5)$ and directrix $y = -1$.

21. The surface of a roadway over a bridge follows a parabolic curve with vertex at the middle of the bridge. The span of the bridge is 400 m. The roadway is 16 m higher in the middle than at the end supports. How far above the end supports is a point 50 m from the middle? 150 m from the middle?

22. The shape of a wire hanging between two poles closely approximates a parabola. Find the equation of a wire that is suspended between two poles 40 m apart and whose lowest point is 10 m below the level of the insulators. (Choose the lowest point as the origin of your coordinate system.)

23. A suspension bridge is supported by two cables that hang between two supports. The curve of these cables is approximately parabolic. Find the equation of this curve if the focus lies 8 m above the lowest point of the cable. (Set up the xy-coordinate system so that the vertex is at the origin.)

24. A culvert is shaped like a parabola, 120 cm across the top and 80 cm deep. How wide is the culvert 50 cm from the top?

Graph each parabola. Find its vertex and the equation of the axis.

25. $y = 2x^2 + 7x - 15$ **26.** $y = -x^2 - 6x - 8$

27. $f(x) = -2x^2 + 4x + 16$ **28.** $f(x) = 3x^2 + 6x + 10$

29. Starting at (0, 0), a projectile travels along the path $y = f(x) = -\frac{1}{256}x^2 + 4x$, where x is in metres. Find **(a)** the maximum height and **(b)** the range of the projectile.

30. The height of a bullet fired vertically upward is given by $h = f(t) = 1200t - 16t^2$ (initial velocity is 1200 ft/s). Find **(a)** its maximum height and **(b)** the time it takes to hit the ground.

31. Enclose a rectangular area with 240 m of fencing. Find the largest possible area that can be enclosed.

32. A 36-in.-wide sheet of metal is bent into a rectangular trough with a cross section as shown in Fig. 20.17. What dimensions will maximize the flow of water? That is, what dimensions will maximize the cross-sectional area?

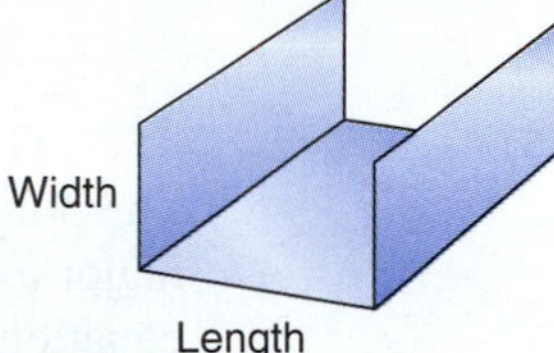

Figure 20.17

20.3 THE ELLIPSE

An **ellipse** consists of the set of points in a plane, the *sum* of whose distances from two fixed points is a positive constant. These two fixed points are called **foci.** As in Fig. 20.18, let the foci lie on the x-axis at $(-c, 0)$ and $(c, 0)$. Then any point $P(x, y)$ lies on the ellipse if its distance d_1 from P to the point $(-c, 0)$ plus its distance d_2 from P to the point $(c, 0)$ is equal to a given constant k. Let the constant be written as $k = 2a$; then

$$d_1 + d_2 = 2a$$

Again using the formula for computing the distance between two points, we have

$$\sqrt{[x - (-c)]^2 + (y - 0)^2} + \sqrt{(x - c)^2 + (y - 0)^2} = 2a$$

Rewrite the previous equation as follows:

$$\sqrt{(x + c)^2 + y^2} = 2a - \sqrt{(x - c)^2 + y^2}$$

$$(x + c)^2 + y^2 = 4a^2 - 4a\sqrt{(x - c)^2 + y^2} + (x - c)^2 + y^2 \quad \text{(Square each side.)}$$

$$x^2 + 2cx + c^2 + y^2 = 4a^2 - 4a\sqrt{(x - c)^2 + y^2} + x^2 - 2cx + c^2 + y^2$$

$$4cx - 4a^2 = -4a\sqrt{(x - c)^2 + y^2}$$

$$a^2 - cx = a\sqrt{(x - c)^2 + y^2} \quad \text{(Divide each side by } -4.)$$

$$(a^2 - cx)^2 = a^2[(x - c)^2 + y^2] \quad \text{(Square each side.)}$$

$$a^4 - 2a^2cx + c^2x^2 = a^2[x^2 - 2cx + c^2 + y^2]$$

$$a^4 - 2a^2cx + c^2x^2 = a^2x^2 - 2a^2cx + a^2c^2 + a^2y^2$$

$$a^4 - a^2c^2 = a^2x^2 - c^2x^2 + a^2y^2$$

$$a^2(a^2 - c^2) = (a^2 - c^2)x^2 + a^2y^2 \quad \text{(Factor.)}$$

$$1 = \frac{x^2}{a^2} + \frac{y^2}{a^2 - c^2} \quad [\text{Divide each side by } a^2(a^2 - c^2).]$$

If we now let $y = 0$ in this equation, we find that $x^2 = a^2$. The points $(-a, 0)$ and $(a, 0)$, which lie on the graph, are called **vertices** of the ellipse. Observe that $a > c$.

If we let $b^2 = a^2 - c^2$, the preceding equation becomes

$$\frac{x^2}{a^2} + \frac{y^2}{b^2} = 1$$

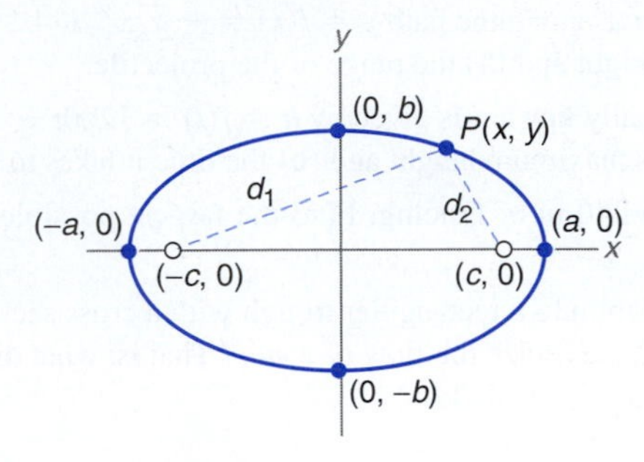

Figure 20.18 An ellipse is a set of points in a plane, the sum of whose distances from two fixed points (called foci) is a positive constant.

The line segment connecting the vertices $(a, 0)$ and $(-a, 0)$ is called the **major axis.** The point midway between the vertices is called the **center** of the ellipse. In this case the major axis lies on the x-axis and the center is at the origin. If we let $x = 0$ in the above equation, we find $y^2 = b^2$. The line connecting $(0, b)$ and $(0, -b)$ is perpendicular to the major axis and passes through the center (see Fig. 20.18). This line is called the **minor axis** of the ellipse. In this case the minor axis lies on the y-axis. Note that $2a$ is the length of the major axis and $2b$ is the length of the minor axis.

STANDARD FORM OF AN ELLIPSE

$$\frac{x^2}{a^2} + \frac{y^2}{b^2} = 1$$

with center at the origin and with the major axis lying on the x-axis.
Note: $a > b$.

One easy way to approximate the curve of an ellipse is to fix a string at two points (foci) on a piece of paper as in Fig. 20.19(a). Then, using a pencil to keep the string taut, trace out the curve as illustrated. Note that $d_1 + d_2$ is always constant—namely, the length of the string. Detach the string and compare the length of the string with the length of the major axis; note that the lengths are the same, $2a$.

The relationship $b^2 = a^2 - c^2$ or $a^2 = b^2 + c^2$ can also be seen from this string demonstration, as in Fig. 20.19(b). Put a pencil inside the taut string and on an end of the minor axis; this bisects the length of string and sets up a right triangle with a as its hypotenuse and b and c as the legs.

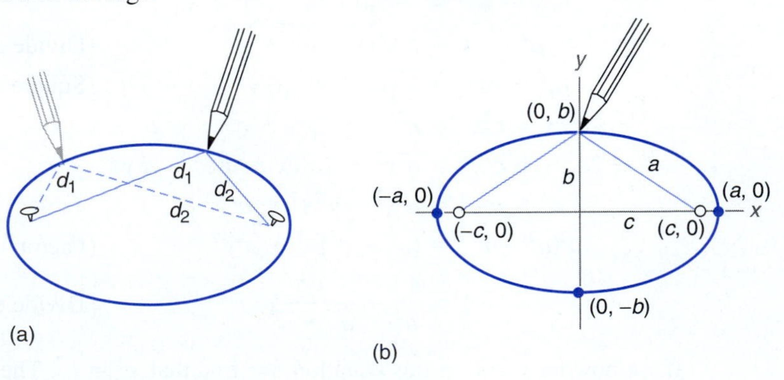

Figure 20.19 Drawing an ellipse with the use of a string.

EXAMPLE 1

Find the vertices, the foci, and the lengths of the major and minor axes of the ellipse

$$\frac{x^2}{25} + \frac{y^2}{9} = 1$$

Sketch the graph.

Since $a^2 = 25$, the vertices are at $(5, 0)$ and $(-5, 0)$. The length of the major axis is $2a = 2(5) = 10$. Since $b^2 = 9$, then $b = 3$, and the length of the minor axis is $2b = 2(3) = 6$. We need the value of c to determine the foci. Since $b^2 = a^2 - c^2$, we can write

$$c^2 = a^2 - b^2 = 25 - 9 = 16$$

$$c = 4$$

The foci are thus $(4, 0)$ and $(-4, 0)$. The graph of the ellipse is shown in Fig. 20.20.

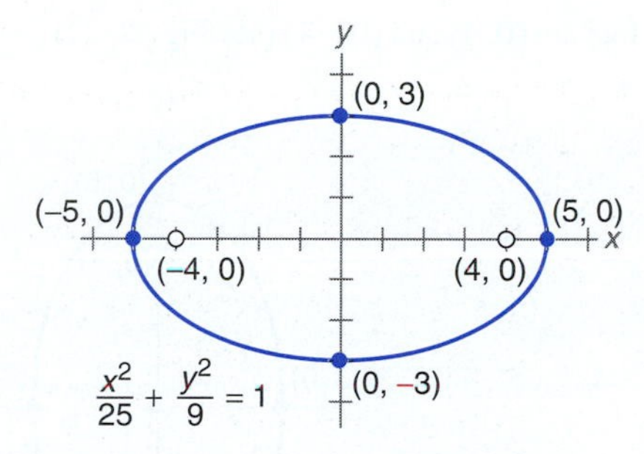

Figure 20.20

You will want to remember the equation relating a, b, and c for the ellipse:

$$c^2 = a^2 - b^2$$

When the major axis lies on the y-axis with center at the origin as in Fig. 20.21, the **standard form** of the equation of the ellipse becomes as follows:

STANDARD FORM OF AN ELLIPSE

$$\frac{y^2}{a^2} + \frac{x^2}{b^2} = 1$$

with center at the origin and with the major axis lying on the y-axis.
Note: $a > b$

This result may be shown similarly as the derivation of the first standard form. Notice that the larger denominator now lies below y^2 instead of below x^2 as in the first case. The vertices are now $(0, a)$ and $(0, -a)$.

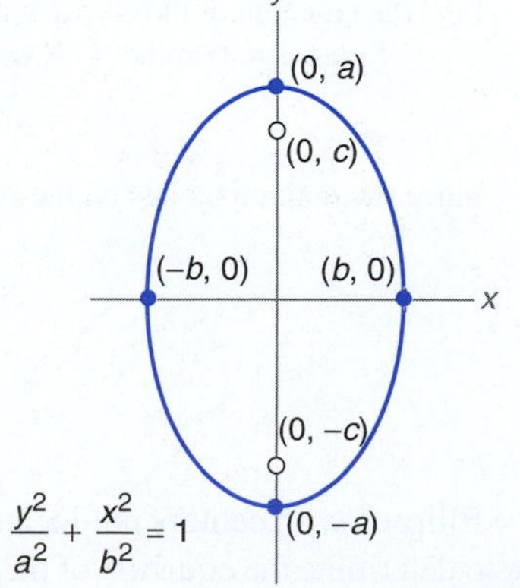

Figure 20.21

EXAMPLE 2

Given the ellipse $25x^2 + 9y^2 = 225$, find the foci, vertices, and lengths of the major and minor axes. Sketch the graph.

First divide each side of the equation by 225 to put the equation in standard form.

$$\frac{x^2}{9} + \frac{y^2}{25} = 1$$

Since the larger denominator belongs to the y^2 term, this ellipse has its major axis on the y-axis and a^2 must then be 25. So $a = 5$ and $b = 3$. The vertices are $(0, 5)$ and $(0, -5)$. The length of the major axis is $2a = 10$ and the length of the minor axis is $2b = 6$.

$$c^2 = a^2 - b^2 = 25 - 9 = 16$$
$$c = 4$$

Thus the foci are $(0, 4)$ and $(0, -4)$ (see Fig. 20.22).

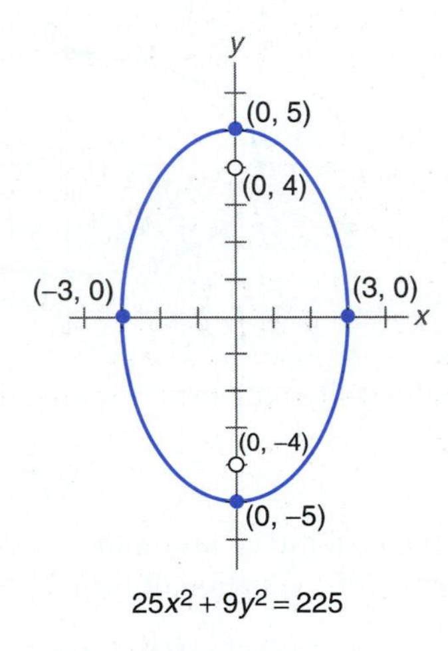

Figure 20.22

In general, a is always greater than b for an ellipse. The following are also true:

1. If the larger denominator belongs to the x^2-term, then its denominator is a^2, the major axis lies on the x-axis, and the vertices are $(a, 0)$ and $(-a, 0)$.
2. If the larger denominator belongs to the y^2-term, then its denominator is a^2, the major axis lies on the y-axis, and the vertices are $(0, a)$ and $(0, -a)$.

EXAMPLE 3

Find the equation of the ellipse with vertices at $(6, 0)$ and $(-6, 0)$ and foci at $(4, 0)$ and $(-4, 0)$.

Since $a = 6$ and $c = 4$, we have $a^2 = 36$ and $c^2 = 16$. Thus

$$b^2 = a^2 - c^2 = 36 - 16 = 20$$

Since the major axis lies on the x-axis, the equation in standard form is

$$\frac{x^2}{a^2} + \frac{y^2}{b^2} = 1$$
$$\frac{x^2}{36} + \frac{y^2}{20} = 1$$

Ellipses with centers not located at the origin will be presented in Section 20.5. If we were to determine the equation of the ellipse where the sum of the distances of all points from the foci $(-2, 3)$ and $(6, 3)$ is always 10, we would have

$$9x^2 + 25y^2 - 36x - 150y + 36 = 0$$

In general, an equation of the form

$$Ax^2 + Cy^2 + Dx + Ey + F = 0$$

represents an ellipse with axes parallel to the coordinate axes, where A and C are both positive (or both negative) and, unlike a circle, $A \neq C$.

Exercises 20.3

Find the vertices, foci, and lengths of the major and minor axes of each ellipse. Sketch each graph.

1. $\frac{x^2}{25} + \frac{y^2}{16} = 1$ **2.** $\frac{x^2}{36} + \frac{y^2}{64} = 1$ **3.** $9x^2 + 16y^2 = 144$

4. $25x^2 + 16y^2 = 400$ **5.** $36x^2 + y^2 = 36$ **6.** $4x^2 + 3y^2 = 12$

7. $16x^2 + 9y^2 = 144$ **8.** $x^2 + 4y^2 = 16$

Find the equation of each ellipse satisfying the given conditions.

9. Vertices at $(4, 0)$ and $(-4, 0)$; foci at $(2, 0)$ and $(-2, 0)$.

10. Vertices at $(0, 7)$ and $(0, -7)$; foci at $(0, 5)$ and $(0, -5)$.

11. Vertices at $(0, 9)$ and $(0, -9)$; foci at $(0, 6)$ and $(0, -6)$.

12. Vertices at $(12, 0)$ and $(-12, 0)$; foci at $(10, 0)$ and $(-10, 0)$.

13. Vertices at $(6, 0)$ and $(-6, 0)$; length of minor axis is 10.

14. Vertices at $(0, 10)$ and $(0, -10)$; length of minor axis is 18.

15. Foci at $(0, 5)$ and $(0, -5)$; length of major axis is 16.

16. Foci at $(3, 0)$ and $(-3, 0)$; length of major axis is 8.

17. A weather satellite with an orbit about the earth reaches a minimum altitude of 1000 mi and a maximum altitude of 1600 mi. The path of its orbit is approximately an ellipse with the center of the earth at one focus. Find the equation of this curve. Assume the radius of the earth is 4000 mi and the x-axis is the major axis.

18. An arch is in the shape of the upper half of an ellipse with a horizontal major axis supporting a foot bridge 40 m long over a stream in a park. The center of the arch is 8 m above the bridge supports. Find an equation of the ellipse. (Choose the point midway between the bridge supports as the origin.)

20.4 THE HYPERBOLA

A **hyperbola** consists of the set of points in a plane, the *difference* of whose distances from two fixed points is a positive constant. The two fixed points are called the **foci.**

Assume now as in Fig. 20.23 that the foci lie on the x-axis at $(-c, 0)$ and $(c, 0)$. Then a point $P(x, y)$ lies on the hyperbola if the difference between its distances to the foci is

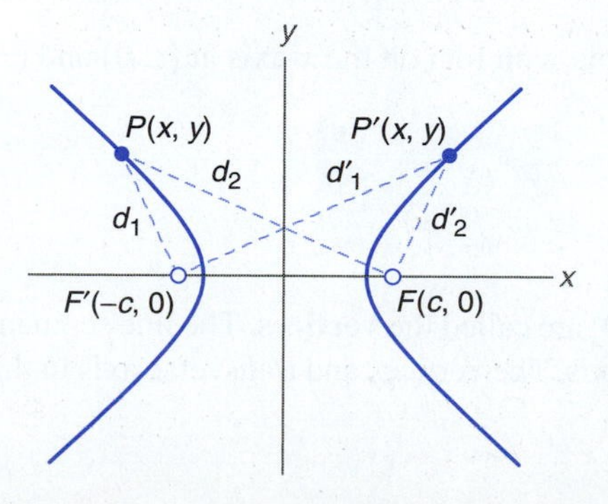

Figure 20.23 A hyperbola is the set of points in a plane, the difference of whose distances from two fixed points (called foci) is a positive constant.

equal to a given constant k. That is, $d_1 - d_2 = k$ or $d_2 - d_1 = k$. Again, this constant k equals $2a$; that is,

$$d_2 - d_1 = 2a$$

To obtain the equation of the hyperbola, use the distance formula.

$$d_2 - d_1 = 2a$$
$$\sqrt{(x-c)^2 + (y-0)^2} - \sqrt{[x-(-c)]^2 + (y-0)^2} = 2a$$

Rewrite the equation above as follows:

$$\sqrt{(x-c)^2 + y^2} = 2a + \sqrt{(x+c)^2 + y^2}$$
$$(x-c)^2 + y^2 = 4a^2 + 4a\sqrt{(x+c)^2 + y^2} + (x+c)^2 + y^2 \quad \text{(Square each side.)}$$
$$x^2 - 2cx + c^2 + y^2 = 4a^2 + 4a\sqrt{(x+c)^2 + y^2} + x^2 + 2cx + c^2 + y^2$$
$$-4a^2 - 4cx = 4a\sqrt{(x+c)^2 + y^2}$$
$$-a^2 - cx = a\sqrt{(x+c)^2 + y^2} \quad \text{(Divide each side by 4.)}$$
$$a^4 + 2a^2cx + c^2x^2 = a^2[(x+c)^2 + y^2] \quad \text{(Square each side.)}$$
$$a^4 + 2a^2cx + c^2x^2 = a^2[x^2 + 2cx + c^2 + y^2]$$
$$a^4 + 2a^2cx + c^2x^2 = a^2x^2 + 2a^2cx + a^2c^2 + a^2y^2$$
$$a^4 - a^2c^2 = a^2x^2 - c^2x^2 + a^2y^2$$
$$a^2(a^2 - c^2) = (a^2 - c^2)x^2 + a^2y^2 \quad \text{(Factor.)}$$
$$1 = \frac{x^2}{a^2} + \frac{y^2}{a^2 - c^2} \quad \text{[Divide each side by } a^2(a^2 - c^2).]$$
$$1 = \frac{x^2}{a^2} - \frac{y^2}{c^2 - a^2}$$

In triangle $F'PF$

$PF' < PF + FF'$ (The sum of any two sides of a triangle is greater than the third side.)

$PF' - PF < FF'$

$2a < 2c$ ($PF' - PF = 2a$ by the definition of a hyperbola and $FF' = 2c$.)

$a < c$

$a^2 < c^2$ (Since $a > 0$ and $c > 0$.)

$0 < c^2 - a^2$

Since $c^2 - a^2$ is positive, we may replace it by the positive number, b^2, as follows:

$$1 = \frac{x^2}{a^2} - \frac{y^2}{b^2}$$

where $b^2 = c^2 - a^2$.

The equation of the hyperbola with foci on the x-axis at $(c, 0)$ and $(-c, 0)$ is

$$\frac{x^2}{a^2} - \frac{y^2}{b^2} = 1$$

The points $(a, 0)$ and $(-a, 0)$ are called the **vertices.** The line segment connecting the vertices is called the **transverse axis.** The vertices and transverse axis in this case lie on the

x-axis. The length of the transverse axis is $2a$. The line segment connecting the points $(0, b)$ and $(0, -b)$ is called the **conjugate axis** and in this case lies on the y-axis. The length of the conjugate axis is $2b$. The **center** lies at the intersection of the conjugate and transverse axes.

STANDARD FORM OF A HYPERBOLA

$$\frac{x^2}{a^2} - \frac{y^2}{b^2} = 1$$

with center at the origin and with the transverse axis lying on the x-axis.

If we draw the central rectangle as in Fig. 20.24 and draw lines passing through opposite vertices of the rectangle, we obtain lines called the **asymptotes** of the hyperbola. In this case the equations of these lines are

$$y = \frac{b}{a}x$$

$$y = -\frac{b}{a}x$$

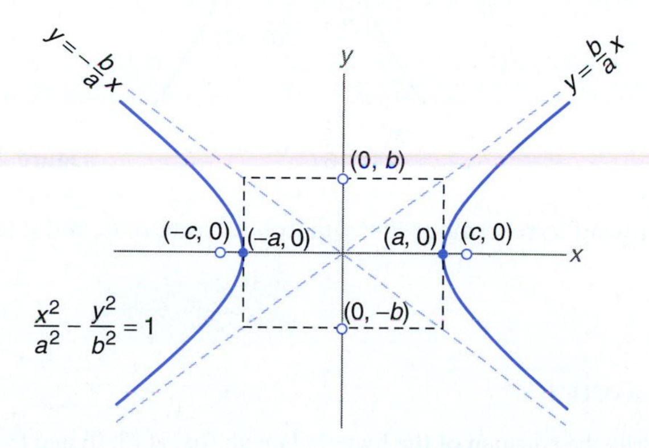

Figure 20.24

Asymptotes serve as guidelines to the branches of the hyperbola. That is, as the distance from the center of the hyperbola increases, the points on the branches get closer and closer to the asymptotes but never cross or touch them.

To sketch the graph of a hyperbola:

1. Locate the vertices $(a, 0)$ and $(-a, 0)$.
2. Locate the points $(0, b)$ and $(0, -b)$.
3. Sketch the central rectangle as in Fig. 20.24. [The coordinates of the vertices are (a, b), $(a, -b)$, $(-a, b)$, and $(-a, -b)$.]
4. Sketch the two asymptotes (the lines passing through the pairs of opposite vertices of the rectangle).
5. Sketch the branches of the hyperbola.

EXAMPLE 1

Find the vertices, foci, and lengths of the transverse and conjugate axes of the hyperbola

$$\frac{x^2}{9} - \frac{y^2}{16} = 1$$

Sketch the graph. Find the equations of the asymptotes.

Since 9 is the denominator of the x^2-term, $a^2 = 9$ and $a = 3$. The vertices are therefore $(3, 0)$ and $(-3, 0)$, and the length of the transverse axis is $2a = 2(3) = 6$. Since 16 is the denominator of the y^2-term, $b^2 = 16$ and $b = 4$. So the length of the conjugate axis is $2b = 2(4) = 8$.

To find the foci we need to know c^2. Since $b^2 = c^2 - a^2$, we have

$$c^2 = a^2 + b^2 = (3)^2 + (4)^2 = 25$$
$$c = 5$$

The foci are $(5, 0)$ and $(-5, 0)$. The asymptotes are $y = \frac{4}{3}x$ and $y = -\frac{4}{3}x$ (see Fig. 20.25).

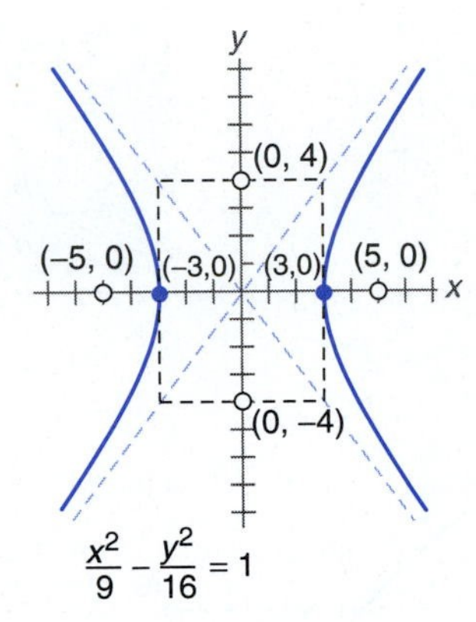

Figure 20.25

You will want to remember the equation relating a, b, and c for the hyperbola.

$$c^2 = a^2 + b^2$$

EXAMPLE 2

Write the equation of the hyperbola with foci at $(5, 0)$ and $(-5, 0)$ and whose transverse axis is 8 units in length.

Here we have $c = 5$. Since $2a = 8$, $a = 4$,

$$c^2 = a^2 + b^2$$
$$25 = 16 + b^2$$
$$b^2 = 9$$

The equation is then

$$\frac{x^2}{16} - \frac{y^2}{9} = 1$$

STANDARD FORM OF A HYPERBOLA

$$\frac{y^2}{a^2} - \frac{x^2}{b^2} = 1$$

with center at the origin and with the transverse axis lying on the y-axis.

We obtain a graph as shown in Fig. 20.26.

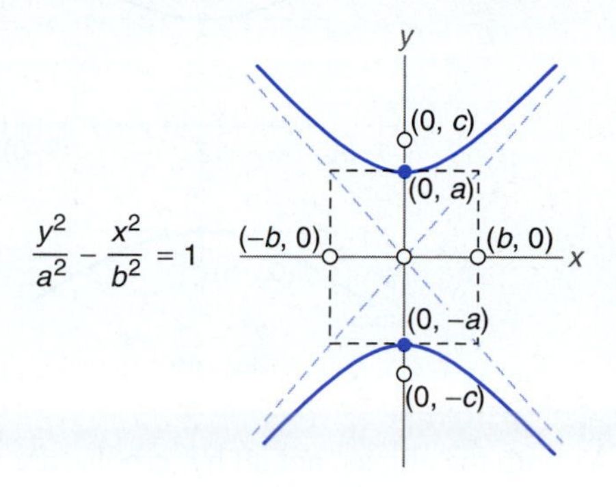

Figure 20.26

Note that the difference between this equation and the first equation is that a^2 is now the denominator of the y^2-term, which is the positive term. This means that the vertices (and transverse axis) now lie on the y-axis.

The equations of the asymptotes are

$$y = \frac{a}{b}x$$

$$y = -\frac{a}{b}x$$

In general, the positive term indicates on which axis the vertices, foci, and transverse axis lie.

1. If the x^2 term is positive, then the denominator of x^2 is a^2 and the denominator of y^2 is b^2. The transverse axis lies along the x-axis and the vertices are $(a, 0)$ and $(-a, 0)$.
2. If the y^2 term is positive, then the denominator of y^2 is a^2 and the denominator of x^2 is b^2. The transverse axis lies along the y-axis and the vertices are $(0, a)$ and $(0, -a)$.

EXAMPLE 3

Sketch the graph of the hyperbola

$$\frac{y^2}{36} - \frac{x^2}{49} = 1$$

Since the y^2-term is positive, the vertices lie on the y-axis and $a^2 = 36$. Then $b^2 = 49$, $a = 6$, and $b = 7$. The graph is sketched in Fig. 20.27.

EXAMPLE 4

Write the equation of the hyperbola with foci at $(0, 8)$ and $(0, -8)$ and vertices at $(0, 6)$ and $(0, -6)$.

In this case $a = 6$ and $c = 8$, so $b^2 = c^2 - a^2 = 64 - 36 = 28$. Since the vertices and foci lie on the y-axis, the y^2-term is positive with denominator a^2. The equation is

$$\frac{y^2}{36} - \frac{x^2}{28} = 1$$

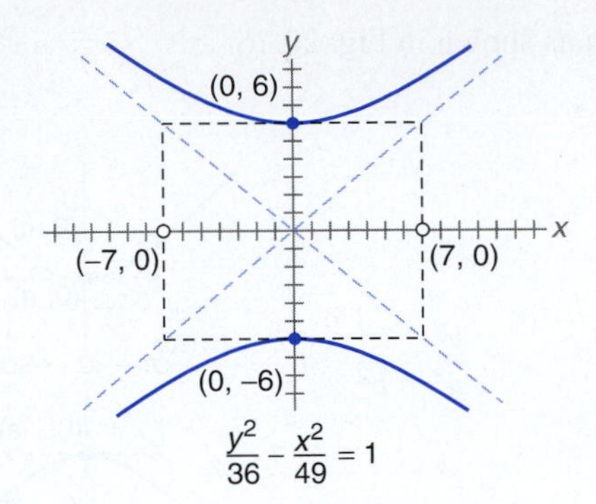

Figure 20.27

As with the ellipse, not all hyperbolas are located with their centers at the origin. We have seen the standard forms of the equation of the hyperbola with center at the origin and whose transverse and conjugate axes lie on the x-axis and y-axis. In general, however, the equation of a hyperbola is of the form

$$Ax^2 + Bxy + Cy^2 + Dx + Ey + F = 0$$

where either (1) $B = 0$ and A and C differ in sign or (2) $A = 0$, $C = 0$, and $B \neq 0$.

A simple example of this last case is the equation $xy = k$. The foci and vertices lie on the line $y = x$ if $k > 0$ or on the line $y = -x$ if $k < 0$.

EXAMPLE 5

Sketch the graph of the hyperbola $xy = -6$.

Since there are no easy clues for sketching this equation (unlike hyperbolas in standard position), set up a table of values for x and y. Then plot the corresponding points in the plane as in Fig. 20.28.

x	y
6	−1
3	−2
2	−3
1	−6
−1	6
−2	3
−3	2
−6	1

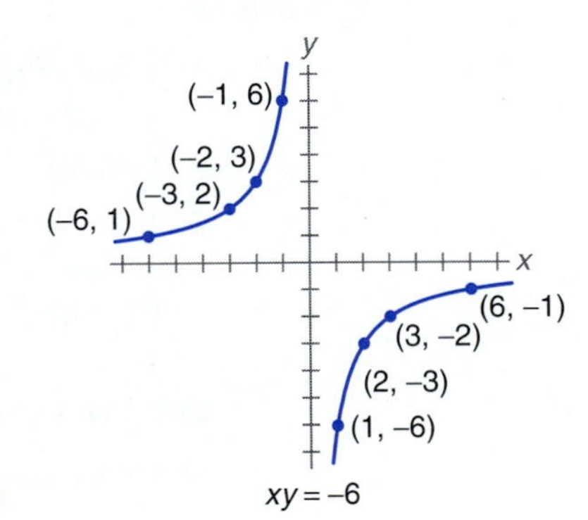

Figure 20.28

Exercises 20.4

Find the vertices, foci, and lengths of the transverse and conjugate axes of each hyperbola. Find the equations of the asymptotes and sketch each graph.

1. $\dfrac{x^2}{25} - \dfrac{y^2}{144} = 1$ **2.** $\dfrac{x^2}{144} - \dfrac{y^2}{25} = 1$ **3.** $\dfrac{y^2}{9} - \dfrac{x^2}{16} = 1$

4. $\dfrac{y^2}{16} - \dfrac{x^2}{9} = 1$ **5.** $5x^2 - 2y^2 = 10$ **6.** $3y^2 - 2x^2 = 6$

7. $4y^2 - x^2 = 4$ **8.** $4x^2 - y^2 = 4$

Find the equation of the hyperbola satisfying each of the given conditions.

9. Vertices at $(4, 0)$ and $(-4, 0)$; foci at $(6, 0)$ and $(-6, 0)$.
10. Vertices at $(0, 5)$ and $(0, -5)$; foci at $(0, 7)$ and $(0, -7)$.
11. Vertices at $(0, 6)$ and $(0, -6)$; foci at $(0, 8)$ and $(0, -8)$.
12. Vertices at $(2, 0)$ and $(-2, 0)$; foci at $(5, 0)$ and $(-5, 0)$.
13. Vertices at $(3, 0)$ and $(-3, 0)$; length of conjugate axis is 10.
14. Vertices at $(0, 6)$ and $(0, -6)$; length of conjugate axis is 8.
15. Foci at $(6, 0)$ and $(-6, 0)$; length of transverse axis is 10.
16. Foci at $(0, 8)$ and $(0, -8)$; length of transverse axis is 12.
17. Sketch the graph of the hyperbola given by $xy = 8$.
18. Sketch the graph of the hyperbola given by $xy = -4$.

20.5 TRANSLATION OF AXES

We have seen the difficulty in determining the equations of the parabola, ellipse, and hyperbola when these are not in standard position in the plane. It is still possible to find the equation of these curves fairly easily if the axes of these curves lie on lines parallel to the coordinate axes. This is accomplished by the translation of axes. We shall demonstrate this method with four examples.

EXAMPLE 1

Find the equation of the ellipse with foci at $(-2, 3)$ and $(6, 3)$ and vertices at $(-3, 3)$ and $(7, 3)$.

The center of the ellipse is at $(2, 3)$, which is midway between the foci or the vertices. The distance between the foci $(-2, 3)$ and $(6, 3)$ is 8. So $c = 4$. The distance between $(-3, 3)$ and $(7, 3)$ is 10. So $a = 5$.

$$c^2 = a^2 - b^2$$
$$16 = 25 - b^2$$
$$b^2 = 9$$
$$b = 3$$

Sketch the graph as in Fig. 20.29(a). Next, plot the same ellipse in another coordinate system with center at the origin as in Fig. 20.29(b). Label the coordinate axes of this new system x' and y'. We know that in this $x'y'$-coordinate system, the equation for this ellipse is

$$\frac{(x')^2}{a^2} + \frac{(y')^2}{b^2} = 1$$

Since $a = 5$ and $b = 3$, we have

$$\frac{(x')^2}{25} + \frac{(y')^2}{9} = 1$$

Each point on the ellipse can now be seen as having coordinates (x, y) in the xy-plane and coordinates (x', y') in the $x'y'$-plane. If we compare coordinates in the two coordinate systems, we see, for example, that the right-hand vertex has coordinates $(7, 3)$ in the xy-plane, but the same point has coordinates $(5, 0)$ in the $x'y'$-plane. Likewise, the point at the upper end of the minor axis has coordinates $(2, 6)$ in the xy-plane, but the same point has coordinates $(0, 3)$ in the $x'y'$-plane.

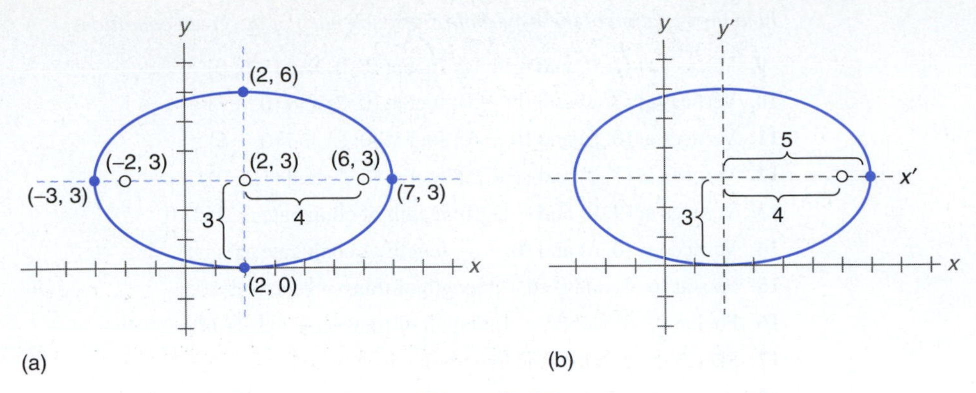

Figure 20.29

In general, the x- and x'-coordinates are related as follows:

$$x = x' + 2$$

That is, the original x-coordinates are 2 larger than the new x'-coordinates. Note that this is the distance that the new origin was moved along the x-axis: the x-coordinate of the center of the ellipse. (See Fig. 20.29b.)

Similarly, the y- and y'-coordinates are related as follows:

$$y = y' + 3$$

Note that 3 is the distance that the new origin was moved along the y-axis: the y-coordinate of the center of the ellipse. (See Fig. 20.29b.) We now rearrange terms and have

$$x' = x - 2$$
$$y' = y - 3$$

Now replace x' by $x - 2$ and y' by $y - 3$ in the equation

$$\frac{(x')^2}{25} + \frac{(y')^2}{9} = 1$$
$$\frac{(x-2)^2}{25} + \frac{(y-3)^2}{9} = 1$$

This is the equation of the ellipse with center at (2, 3) in the xy-plane.

To write an equation for a parabola, ellipse, or hyperbola whose axes are parallel to the x-axis and y-axis:

1. For a parabola, identify (h, k) as the vertex; for an ellipse or hyperbola identify (h, k) as the center.
2. Translate xy-coordinates to a new $x'y'$-coordinate system by using the translation equations
$$x' = x - h$$
$$y' = y - k$$
where (h, k) has been identified as in Step 1.
3. Write the equation of the conic, which is now in standard position in the $x'y'$-coordinate system.

4. Translate the equation derived in Step 3 back into the original coordinate system by making the following substitutions for x and y into the derived equation:

$$x' = x - h$$
$$y' = y - k$$

The resulting equation is an equation for the conic in the xy-coordinate system.

EXAMPLE 2

Find the equation of the parabola with focus $(-1, 2)$ and directrix $x = -7$.

Step 1: The vertex of this parabola is halfway along the line $y = 2$ between the focus $(-1, 2)$ and the directrix $x = -7$ (see Fig. 20.30). Thus the vertex has coordinates $(-4, 2)$. This becomes the origin of the new coordinate system, so $h = -4$ and $k = 2$.

Step 2:

$$x' = x - h = x - (-4) = x + 4$$
$$y' = y - k = y - 2$$

The new $x'y'$-coordinates of the focus become

$$x' = x - h = -1 - (-4) = 3$$
$$y' = y - k = 2 - 2 = 0$$

and the equation of the directrix $x = -7$ becomes

$$x' = x - h = -7 - (-4)$$
$$x' = -3$$

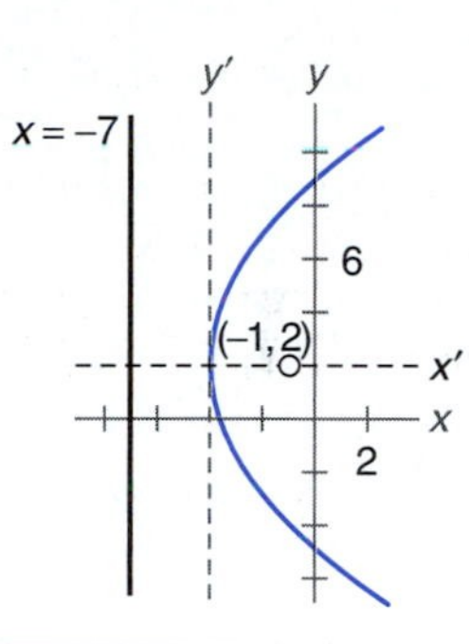

Figure 20.30

Step 3: Since the parabola is now in standard position in the new coordinate system with focus $(3, 0)$, we have $p = 3$. The equation in this system becomes

$$(y')^2 = 4px'$$
$$(y')^2 = 4(3)x'$$
$$(y')^2 = 12x'$$

Step 4: Replace x' with $x + 4$ and y' with $y - 2$.

$$(y')^2 = 12x'$$
$$(y - 2)^2 = 12(x + 4)$$
$$y^2 - 4y - 12x - 44 = 0$$

We sometimes know the equation of a curve and need to identify the curve and sketch its graph, as in the following example.

EXAMPLE 3

Describe and sketch the graph of the equation

$$\frac{(y - 4)^2}{9} - \frac{(x + 2)^2}{16} = 1$$

If we let

$$x' = x - h = x + 2 = x - (-2)$$
$$y' = y - k = y - 4$$

we have

$$\frac{(y')^2}{9} - \frac{(x')^2}{16} = 1$$

This is the equation of a hyperbola with center at $(-2, 4)$. Since $a^2 = 9$ and $b^2 = 16$, we have

$$c^2 = a^2 + b^2 = 9 + 16 = 25$$

so

$$a = 3 \qquad b = 4 \qquad \text{and} \qquad c = 5$$

In terms of the $x'y'$-coordinates, the foci are at $(0, 5)$ and $(0, -5)$, the vertices are at $(0, 3)$ and $(0, -3)$, the length of the transverse axis is 6, and the length of the conjugate axis is 8.

To translate the $x'y'$-coordinates to xy-coordinates, we use the equations

$$x = x' + h \qquad y = y' + k$$

In this case

$$x = x' + (-2) \qquad y = y' + 4$$

So, in the xy-plane the foci are at $(-2, 9)$ and $(-2, -1)$, the vertices are at $(-2, 7)$ and $(-2, 1)$, the length of the transverse axis is 6, and the length of the conjugate axis is 8 (see Fig. 20.31).

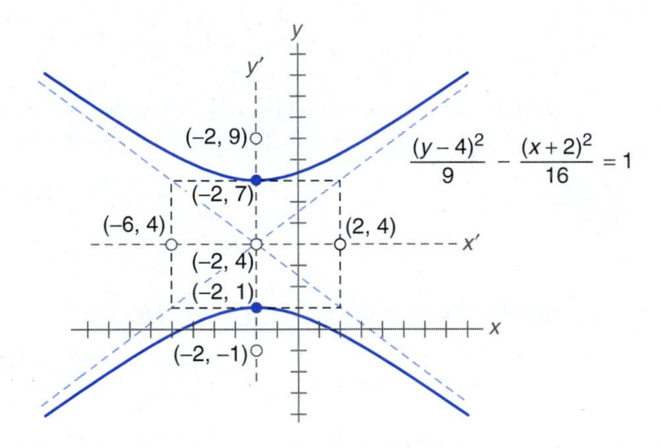

Figure 20.31

EXAMPLE 4

Name the equation $16x^2 + 9y^2 + 64x + 54y + 1 = 0$. Locate the vertex if it is a parabola or the center if it is an ellipse or a hyperbola.

First complete the square for x and y. (See Section 20.1.)

$$(16x^2 + 64x \quad) + (9y^2 + 54y \quad) = -1$$
$$16(x^2 + 4x \quad) + 9(y^2 + 6y \quad) = -1$$

(Factor out the coefficients of x^2 and y^2 before completing the square. The coefficients of x^2 and y^2 must be **one.**)

$$16(x^2 + 4x + 4) + 9(y^2 + 6y + 9) = -1 + 16(4) + 9(9)$$
$$16(x + 2)^2 + 9(y + 3)^2 = 144$$
$$\frac{(x + 2)^2}{9} + \frac{(y + 3)^2}{16} = 1 \qquad \text{(Divide each side by 144.)}$$

This is an equation of an ellipse. Noting that

$$x' = x - h = x + 2 = x - (-2)$$
$$y' = y - k = y + 3 = y - (-3)$$

we see that the center is at $(-2, -3)$.

GENERAL FORMS OF CONICS WITH AXES PARALLEL TO THE COORDINATE AXES

1. $(y - k)^2 = 4p(x - h)$
is a parabola with vertex at (h, k) and axis parallel to the x-axis.

2. $(x - h)^2 = 4p(y - k)$
is a parabola with vertex at (h, k) and axis parallel to the y-axis.

3. $\dfrac{(x - h)^2}{a^2} + \dfrac{(y - k)^2}{b^2} = 1 \qquad (a > b)$
is an ellipse with center at (h, k) and major axis parallel to the x-axis.

4. $\dfrac{(y - k)^2}{a^2} + \dfrac{(x - h)^2}{b^2} = 1 \qquad (a > b)$
is an ellipse with center at (h, k) and major axis parallel to the y-axis.

5. $\dfrac{(x - h)^2}{a^2} - \dfrac{(y - k)^2}{b^2} = 1$
is a hyperbola with center at (h, k) and transverse axis parallel to the x-axis.

6. $\dfrac{(y - k)^2}{a^2} - \dfrac{(x - h)^2}{b^2} = 1$
is a hyperbola with center at (h, k) and transverse axis parallel to the y-axis.

Exercises 20.5

Find the equation of each curve from the given information.

1. Ellipse with center at $(1, -1)$, vertices at $(5, -1)$ and $(-3, -1)$, and foci at $(3, -1)$ and $(-1, -1)$.
2. Parabola with vertex at $(-1, 3)$, focus at $(-1, 4)$, and directrix $y = 2$.
3. Hyperbola with center at $(1, 1)$, vertices at $(1, 7)$ and $(1, -5)$, and foci at $(1, 9)$ and $(1, -7)$.
4. Ellipse with center at $(-2, -3)$, vertices at $(4, -3)$ and $(-8, -3)$, and minor axis length 10.
5. Parabola with vertex at $(3, -1)$, focus at $(5, -1)$, and directrix $x = 1$.
6. Hyperbola with center at $(-2, -2)$, vertices at $(1, -2)$ and $(-5, -2)$, and conjugate axis length 10.

Name and graph each equation.

7. $(x - 2)^2 = 4(y + 3)$

8. $\frac{(x + 1)^2}{36} + \frac{(y - 2)^2}{64} = 1$

9. $\frac{y^2}{9} - \frac{(x + 2)^2}{16} = 1$

10. $y^2 = 8(x + 1)$

11. $9(x - 2)^2 + 16y^2 = 144$

12. $\frac{(x + 1)^2}{9} - \frac{(y + 3)^2}{16} = 1$

13. $\frac{(x - 3)^2}{36} + \frac{(y - 1)^2}{16} = 1$

14. $\frac{(x - 3)^2}{36} - \frac{(y - 1)^2}{16} = 1$

15. $(y + 3)^2 = 8(x - 1)$

16. $(x - 5)^2 = 12(y + 2)$

17. $\frac{(y + 1)^2}{9} - \frac{(x + 1)^2}{9} = 1$

18. $\frac{(y + 4)^2}{4} + \frac{(x - 2)^2}{9} = 1$

Name and sketch the graph of each equation. Locate the vertex if it is a parabola or the center if it is an ellipse or a hyperbola.

19. $x^2 - 4x + 2y + 6 = 0$

20. $9x^2 + 4y^2 - 18x + 24y + 9 = 0$

21. $x^2 + 4y^2 + 4x - 8y - 8 = 0$

22. $-2x^2 + 3y^2 + 8x - 14 = 0$

23. $4x^2 - y^2 - 8x + 2y + 3 = 0$

24. $y^2 + 6y - x + 12 = 0$

25. $25y^2 - 4x^2 - 24x - 150y + 89 = 0$

26. $25x^2 + 9y^2 - 100x - 54y - 44 = 0$

27. $x^2 + 16x - 12y + 40 = 0$

28. $9x^2 - 4y^2 + 54x + 40y - 55 = 0$

29. $4x^2 + y^2 + 48x + 4y + 84 = 0$

30. $y^2 - 10x - 6y + 39 = 0$

20.6 THE GENERAL SECOND-DEGREE EQUATION

The circle, parabola, ellipse, and hyperbola are all special cases of the second-degree equation

$$Ax^2 + Bxy + Cy^2 + Dx + Ey + F = 0$$

When $B = 0$ and at least one of the coefficients A or C is not zero, the following summarizes the conditions for each curve.

1. If $A = C$, we have a *circle*.
   In special cases, the graph of the equation may be a point or there may be no graph. (The equation may have only one or no solution.)
2. If $A = 0$ and $C \neq 0$, or $C = 0$ and $A \neq 0$, then we have a *parabola*.
3. If $A \neq C$, and A and C are either both positive or both negative, then we have an *ellipse*.
   In special cases, the graph of the equation may be a point or there may be no graph. (The equation may have only one or no solution.)
4. If A is positive and C is negative, or A is negative and C is positive, then we have a *hyperbola*.
   In some special cases the graph may be a pair of intersecting lines.

If $D \neq 0$ or $E \neq 0$ or both are not equal to zero, the curve does not have its center (or vertex in the case of the parabola) at the origin (see Section 20.5). If $B \neq 0$, then the axis of the curve does not lie along the x-axis or y-axis. The hyperbola $xy = k$ is the only such example we have studied (see Section 20.4).

EXAMPLE

Identify the curve

$$x^2 + 3y^2 - 2x + 4y - 7 = 0$$

Since $A \neq C, A$ and C are both positive, and $B = 0$, the curve is an ellipse. (The center is not the origin since $D \neq 0$ and $E \neq 0$.)

The curves represented by the second-degree equation

$$Ax^2 + Bxy + Cy^2 + Dx + Ey + F = 0$$

are called **conic sections** because they can be obtained by cutting the cones with a plane, as in Fig. 20.32.

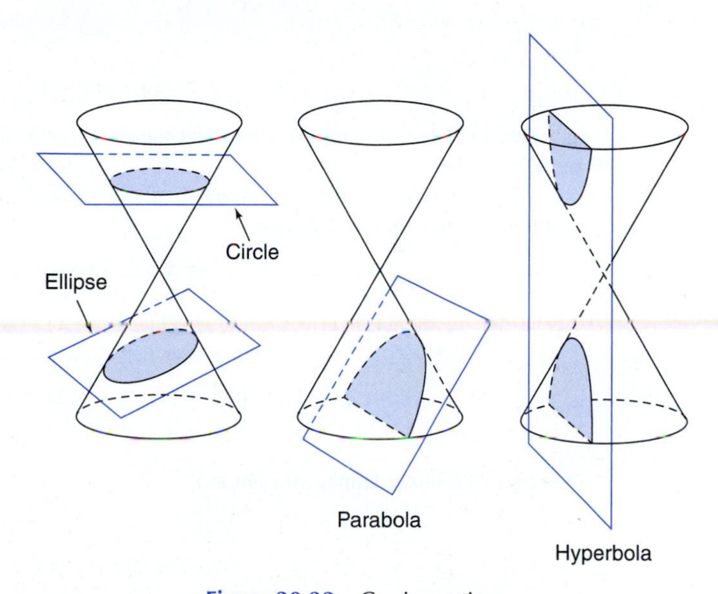

Figure 20.32 Conic sections.

Exercises 20.6

Determine whether each equation represents a circle, a parabola, an ellipse, or a hyperbola.

1. $x^2 + 3y^2 + 4x - 5y - 40 = 0$

2. $x^2 + y^2 + 4x - 6y - 12 = 0$

3. $4y^2 - 8y + 3x - 2 = 0$

4. $9x^2 + 4y^2 + 36x - 8y + 4 = 0$

5. $4x^2 - 5y^2 - 16x + 10y + 20 = 0$

6. $x^2 + y^2 + 3x - 2y - 14 = 0$

7. $3x^2 + 3y^2 + x - y - 6 = 0$

8. $x^2 + 4x - 3y - 52 = 0$

9. $x^2 + y^2 + 2x - 3y - 21 = 0$

10. $x^2 - y^2 - 6x + 3y - 100 = 0$

11. $9x^2 + 4y^2 - 18x + 8y + 4 = 0$

12. $3x^2 - 2y^2 + 6x - 8y - 17 = 0$

13. $3x^2 - 3y^2 - 2x - 4y - 13 = 0$

14. $4x^2 + 4y^2 - 16x - 4y - 5 = 0$

15. $x^2 - 6x - 6y + 3 = 0$

16. $4x^2 - 4x - 4y - 5 = 0$

20.7 SYSTEMS OF QUADRATIC EQUATIONS

To solve systems of equations involving conics algebraically, try to eliminate a variable.

Substitution Method

EXAMPLE 1

Solve the system of equations using the substitution method.

$$y^2 = x$$
$$y = x - 2$$

Since $y^2 = x$, we can substitute y^2 for x in the second equation.

$$y = x - 2$$
$$y = (y^2) - 2$$
$$y^2 - y - 2 = 0$$
$$(y - 2)(y + 1) = 0$$

So

$$y = 2 \quad \text{or} \quad y = -1$$

Substituting these values for y in the second equation, $y = x - 2$, we have $x = 1$ when $y = -1$ and $x = 4$ when $y = 2$.

The solutions of the system are then $(1, -1)$ and $(4, 2)$. Check by substituting the solutions in each original equation.

	$y^2 = x$	$y = x - 2$
$(1, -1)$	$(-1)^2 = (1)$	$(-1) = (1) - 2$
	$1 = 1$	$-1 = -1$
$(4, 2)$	$(2)^2 = (4)$	$(2) = (4) - 2$
	$4 = 4$	$2 = 2$

(See Fig. 20.33 for a graphical solution.)

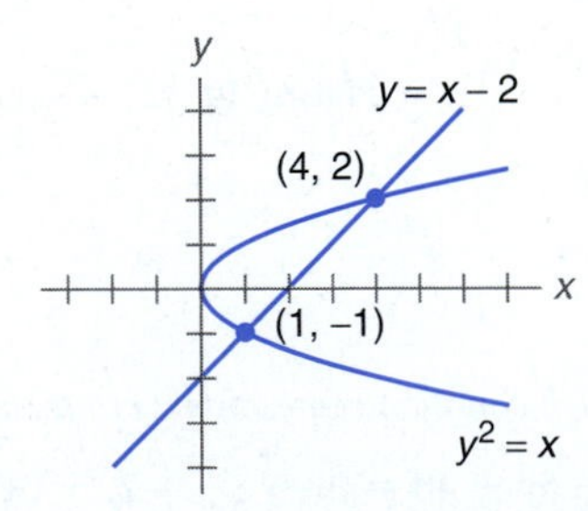

Figure 20.33

EXAMPLE 2

Solve the system of equations using the substitution method.

$$4x^2 - 3y^2 = 4$$
$$x^2 - 4x + y^2 = 0$$

If we solve the second equation for y^2, we have $y^2 = 4x - x^2$. We can now substitute $4x - x^2$ for y^2 in the first equation.

$$4x^2 - 3(4x - x^2) = 4$$
$$4x^2 - 12x + 3x^2 = 4$$
$$7x^2 - 12x - 4 = 0$$
$$(x - 2)(7x + 2) = 0$$
$$x = 2 \quad \text{or} \quad x = -\frac{2}{7}$$

To find y we substitute each of these values for x in one of the original equations. We use the second equation.

For $x = 2$:

$$(2)^2 - 4(2) + y^2 = 0$$
$$4 - 8 + y^2 = 0$$
$$y^2 = 4$$

Thus $y = 2$ or -2 when $x = 2$.

For $x = -\frac{2}{7}$:

$$\left(-\frac{2}{7}\right)^2 - 4\left(-\frac{2}{7}\right) + y^2 = 0$$
$$\frac{4}{49} + \frac{8}{7} + y^2 = 0$$
$$y^2 = -\frac{60}{49}$$

Since y^2 can never be negative, we conclude that there are no real solutions when $x = -\frac{2}{7}$. (This is what we call an *extraneous root.*) The solutions of the system are (2, 2) and (2, −2). The solutions should be checked in each original equation (see Fig. 20.34 for a graphical solution).

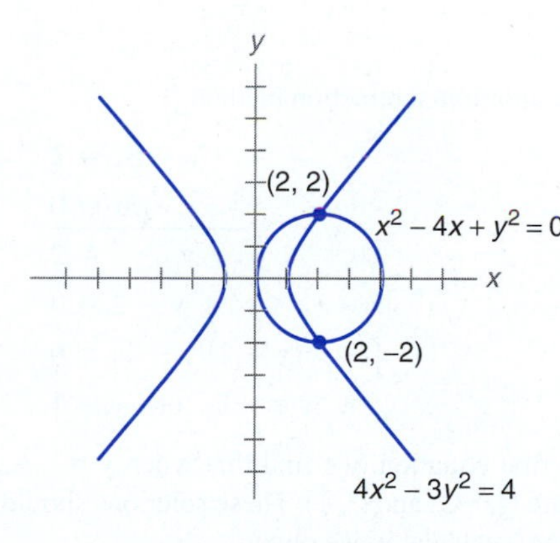

Figure 20.34

Addition-Subtraction Method

EXAMPLE 3

Solve the system of equations using the addition-subtraction method.

$$3x^2 + y^2 = 14$$
$$x^2 - y^2 = 2$$

By adding the two equations we can eliminate the y^2-term.

$$\begin{aligned} 3x^2 + y^2 &= 14 \\ x^2 - y^2 &= 2 \\ \hline 4x^2 \quad &= 16 \end{aligned}$$

Solving for x,

$$x^2 = 4$$

$$x = 2 \quad \text{or} \quad x = -2$$

Using the second equation, $x^2 - y^2 = 2$, we find the corresponding values for y. For $x = 2$:

$$\begin{aligned} (2)^2 - y^2 &= 2 \\ 4 - y^2 &= 2 \\ y^2 &= 2 \\ y = \sqrt{2} \quad \text{or} \quad y &= -\sqrt{2} \end{aligned}$$

For $x = -2$:

$$\begin{aligned} (-2)^2 - y^2 &= 2 \\ 4 - y^2 &= 2 \\ y^2 &= 2 \\ y = \sqrt{2} \quad \text{or} \quad y &= -\sqrt{2} \end{aligned}$$

The solutions are $(2, \sqrt{2})$, $(2, -\sqrt{2})$, $(-2, \sqrt{2})$, and $(-2, -\sqrt{2})$. These should be checked in each original equation.

When solving a system of two equations where one represents a conic and the other a line, the substitution method is usually preferred. Example 4 shows how the addition-subtraction method may be helpful in some cases.

EXAMPLE 4

Solve the system of equations.

$$\begin{aligned} y + 6x &= 2 \\ y^2 &= 6x \end{aligned}$$

We use the addition-subtraction method.

$$\begin{aligned} y + 6x &= 2 \\ y^2 \quad - 6x &= 0 \\ \hline y^2 + y \quad &= 2 \qquad \text{(Add.)} \\ y^2 + y - 2 &= 0 \\ (y + 2)(y - 1) &= 0 \\ y = -2 \quad \text{or} \quad y &= 1 \end{aligned}$$

Using the first equation, we find that when $y = -2$, $x = \frac{2}{3}$, and when $y = 1$, $x = \frac{1}{6}$, so the solutions are $(\frac{2}{3}, -2)$ and $(\frac{1}{6}, 1)$. These solutions should be checked in each original equation.

Using a calculator, we obtain

2nd CUSTOM F3 4 *twelve* left arrows y+6x=2 *five* right arrows y^2=6x **ENTER** up arrow **2nd** right arrow

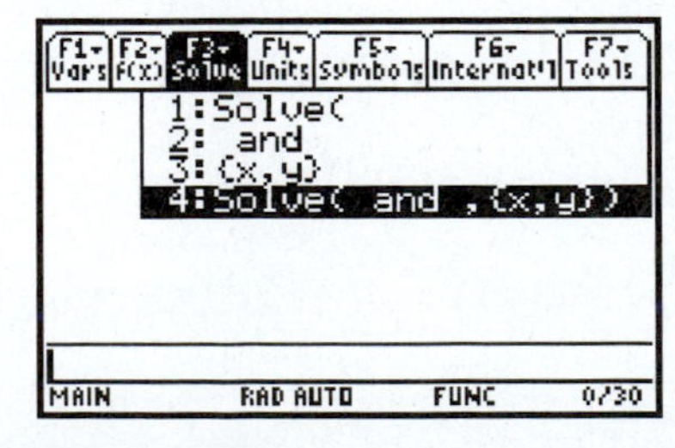

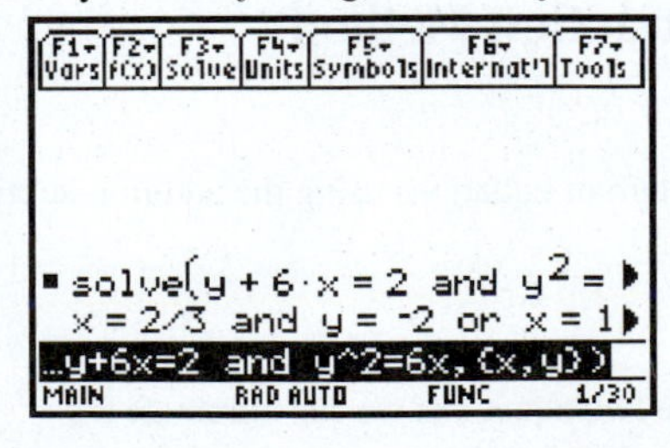

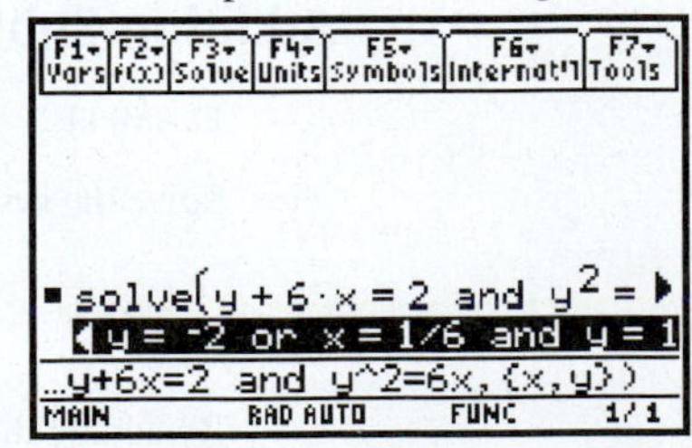

Exercises 20.7

Solve each system of equations.

1. $x^2 = 3y$
   $y = 2x - 3$
2. $x^2 - 2y^2 = 1$
   $3x^2 + 2y^2 = 3$
3. $x^2 + 4x + y^2 - 8 = 0$
   $x^2 + y^2 = 4$
4. $x^2 + 2y^2 = 12$
   $y = -x$
5. $y^2 - x^2 = 12$
   $x^2 = 4y$
6. $x^2 + y^2 = 9$
   $y = 4$
7. $x^2 + y^2 = 4$
   $x^2 - y^2 = 4$
8. $x^2 + y^2 - 6y = 0$
   $y = x$
9. $x^2 = 6y$
   $y = 6$
10. $\dfrac{y^2}{16} + \dfrac{x^2}{9} = 1$
    $4x + 3y = 12$
11. $y^2 = 4x + 12$
    $y^2 = -4x - 4$
12. $x^2 - y^2 = 2$
    $y^2 = x$
13. $x^2 + y^2 = 36$
    $y = x^2$
14. $y = x^2 - 3x - 10$
    $2x + y + 4 = 0$
15. $x^2 - y^2 = 9$
    $x^2 + 9y^2 = 169$
16. $x^2 + 4y^2 = 36$
    $x^2 + y^2 = 16$
17. $x^2 + y^2 = 17$
    $xy = 4$
18. $3x^2 + 4y^2 = 48$
    $xy = 6$

20.8 POLAR COORDINATES

Each point in the number plane has been associated with an ordered pair of real numbers (x, y), which are called rectangular or Cartesian coordinates. Point $P(x, y)$ is shown in Fig. 20.35. Point P can also be located by specifying an angle θ from the positive x-axis and a directed distance r from the origin, and described by the ordered pair (r, θ) called *polar coordinates.* The polar coordinate system has a fixed point in the number plane called the *pole* or *origin*. From the pole draw a horizontal ray directed to the right, which is called the *polar axis* (see Fig. 20.36).

Angle θ is a directed angle: $\theta > 0$ is measured counterclockwise; $\theta < 0$ is measured clockwise. Angle θ is commonly expressed in either degrees or radians. Distance r is a directed distance: $r > 0$ is measured in the direction of the ray (terminal side of θ); $r < 0$ is measured in the direction opposite the direction of the ray.

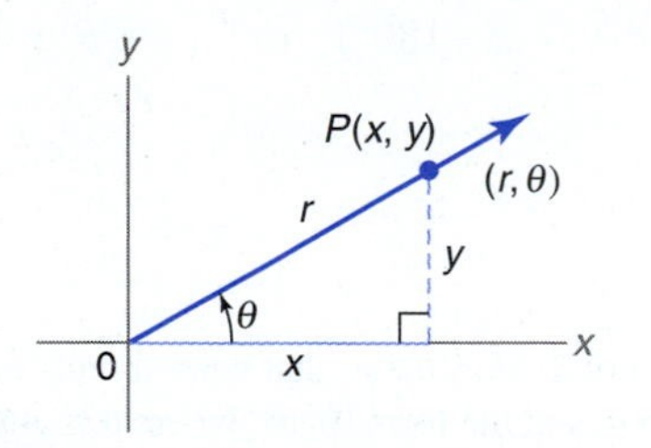

Figure 20.35 Polar coordinates.

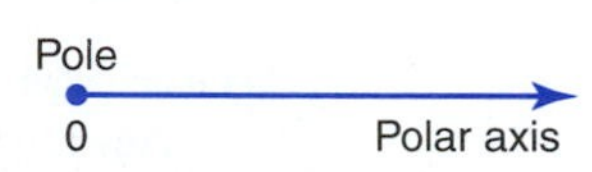

Figure 20.36 Polar axis.

EXAMPLE 1

Graph each point whose polar coordinates are given: (a) $(2, 120°)$, (b) $(4, 4\pi/3)$, (c) $(4, -2\pi/3)$, (d) $(-5, 135°)$, (e) $(-2, -60°)$, (f) $(-3, 570°)$ (see Fig. 20.37).

From the results of Example 1, you can see that there is a major difference between the rectangular coordinate system and the polar coordinate system. In the rectangular system there is a one-to-one correspondence between points in the plane and ordered pairs

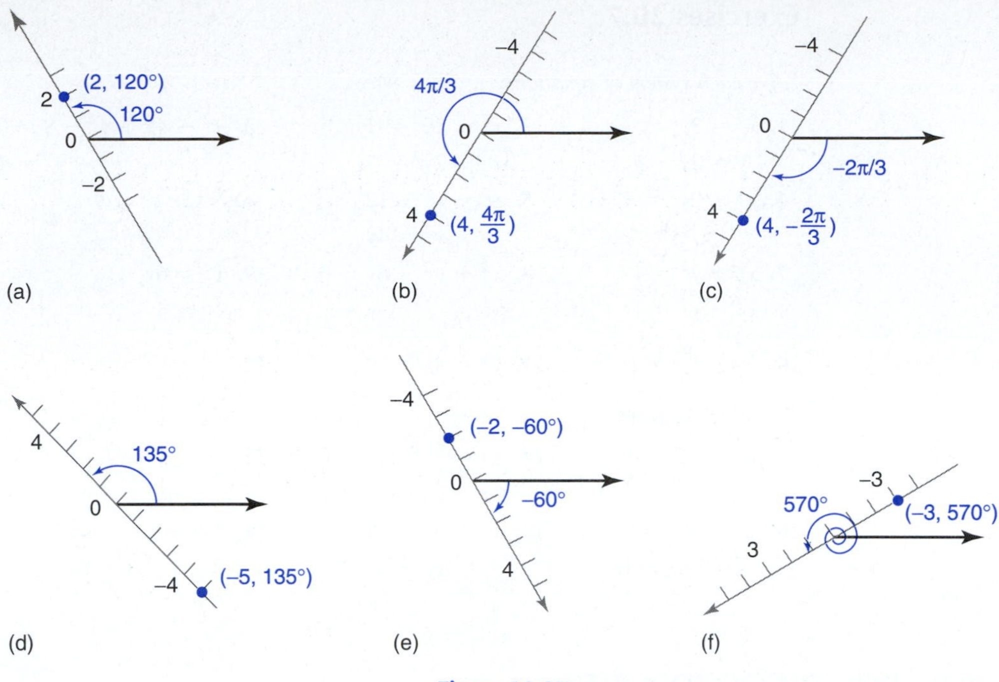

Figure 20.37

of real numbers. That is, each point is named by exactly one ordered pair, and each ordered pair corresponds to exactly one point. This one-to-one correspondence is not a property of the polar coordinate system. In Example 1, the point described by parts (a) and (e) is the same, and the point described by parts (b) and (c) is the same. In fact, each point may be named by infinitely many polar coordinates. In general, the point $P(r, \theta)$ may be represented by

$$(r, \theta + k \cdot 360°) \quad \text{or} \quad (r, \theta + k \cdot 2\pi)$$

where k is any integer. $P(r, \theta)$ may also be represented by

$$(-r, \theta + k \cdot 180°) \quad \text{or} \quad (-r, \theta + k\pi)$$

where k is any odd integer.

EXAMPLE 2

Name an ordered pair of polar coordinates that corresponds to the pole or origin.

Any set of coordinates in the form $(0, \theta)$, where θ is any angle, corresponds to the pole. For example, $(0, 64°)$, $(0, 2\pi/3)$, and $(0, -\pi/6)$ name the pole.

Polar graph paper is available for working with polar coordinates. Figure 20.38 shows graph paper in both degrees and radians.

EXAMPLE 3

Plot each point whose polar coordinates are given. Use polar graph paper in degrees (see Fig. 20.39).

$$A(6, 60°), \quad B(4, 270°), \quad C(3, -210°), \quad D(-6, 45°), \quad E(-2, -150°), \quad F(8, 480°)$$

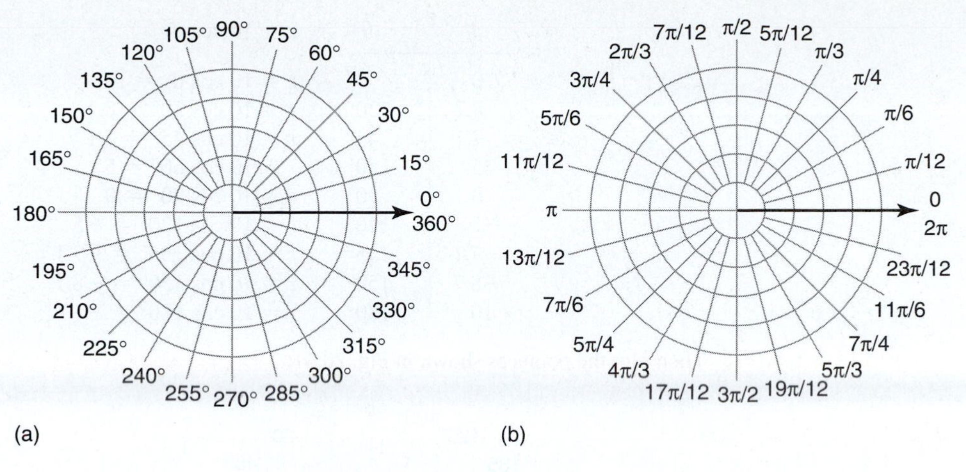

Figure 20.38 Polar graph paper.

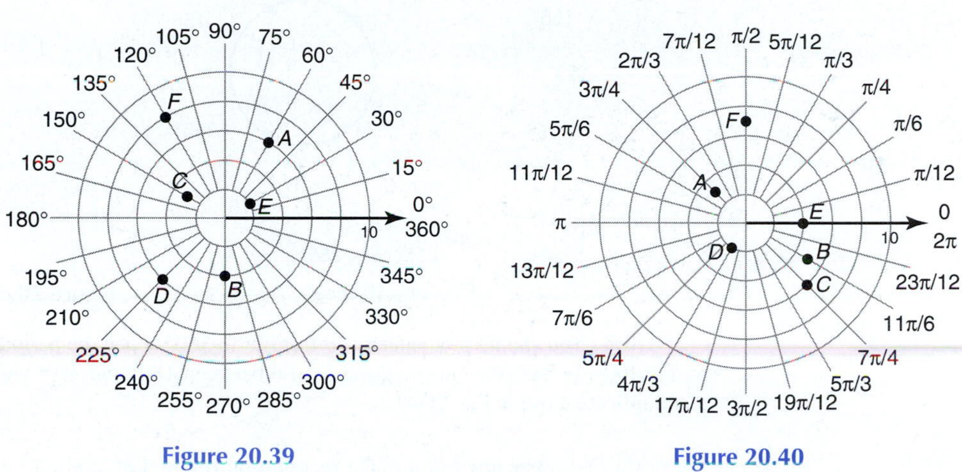

Figure 20.39 **Figure 20.40**

EXAMPLE 4

Plot each point whose polar coordinates are given. Use polar graph paper in radians (see Fig. 20.40).

$$A\left(3, \frac{3\pi}{4}\right),\quad B\left(5, \frac{11\pi}{6}\right),\quad C\left(6, -\frac{\pi}{4}\right),\quad D\left(-2, \frac{\pi}{3}\right),\quad E(-4, -\pi),\quad F\left(7, \frac{13\pi}{2}\right)$$

EXAMPLE 5

Given the point $P(4, 150°)$, name three other sets of polar coordinates for P such that $-360° \leq \theta \leq 360°$.

For $r > 0$ and $\theta < 0$: $(4, -210°)$
For $r < 0$ and $\theta > 0$: $(-4, 330°)$
For $r < 0$ and $\theta < 0$: $(-4, -30°)$

EXAMPLE 6

Graph $r = 10 \cos \theta$ by plotting points. Assign θ values of 0°, 30°, 45°, 60°, and so on until you have a smooth curve.

Make a table for the ordered pairs as follows. *Note:* Although r and θ are given in the same order as the ordered pair (r, θ), θ is actually the independent variable.

r	θ	$r = 10 \cos \theta$
10	0°	$r = 10 \cos 0° = 10$
8.7	30°	$r = 10 \cos 30° = 8.7$
7.1	45°	$r = 10 \cos 45° = 7.1$
5	60°	$r = 10 \cos 60° = 5$
0	90°	$r = 10 \cos 90° = 0$
−5	120°	$r = 10 \cos 120° = -5$
−7.1	135°	$r = 10 \cos 135° = -7.1$
−8.7	150°	$r = 10 \cos 150° = -8.7$
−10	180°	$r = 10 \cos 180° = -10$

Then plot the points as shown in Fig. 20.41.

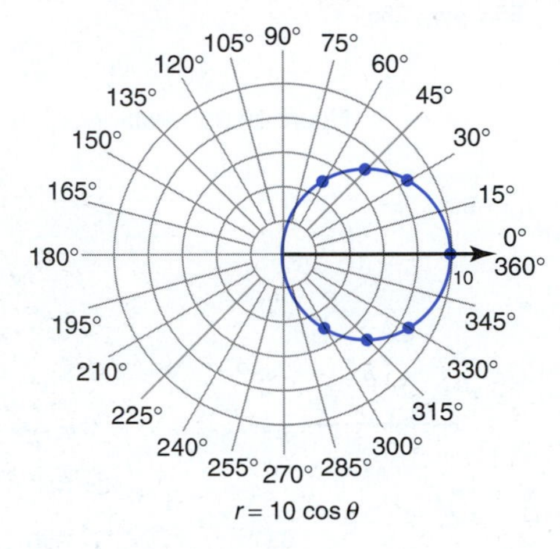

Figure 20.41

Note: You should plot values of θ from 0° to 360°, since the period of the cosine function is 360°. In this case choosing values of θ between 180° and 360° will give ordered pairs that duplicate those in Fig. 20.41.

Let point $P(x, y)$ be any point in the rectangular plane. Let the polar plane coincide with the rectangular plane so that $P(x, y)$ and $P(r, \theta)$ represent the same point, as shown in Fig. 20.42. Note the following relationships:

POLAR–RECTANGULAR RELATIONSHIPS

1. $\cos \theta = \dfrac{x}{r}$ or $x = r \cos \theta$

2. $\sin \theta = \dfrac{y}{r}$ or $y = r \sin \theta$

3. $\tan \theta = \dfrac{y}{x}$ or $\theta = \arctan \dfrac{y}{x}$

4. $x^2 + y^2 = r^2$

5. $\cos \theta = \dfrac{x}{\sqrt{x^2 + y^2}}$

6. $\sin \theta = \dfrac{y}{\sqrt{x^2 + y^2}}$

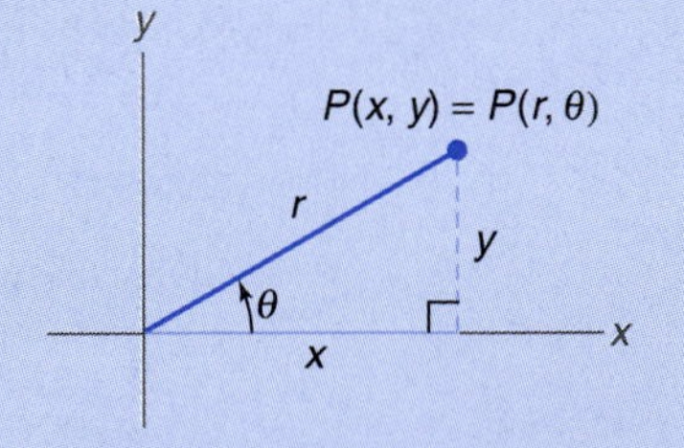

Figure 20.42

Suppose that we wish to change coordinates from one system to the other.

EXAMPLE 7

Change $A(4, 60°)$ and $B(8, 7\pi/6)$ to rectangular coordinates.

For point $A(4, 60°)$:

$$\begin{aligned} x &= r\cos\theta & y &= r\sin\theta \\ x &= 4\cos 60° & y &= 4\sin 60° \\ &= 4\left(\frac{1}{2}\right) & &= 4\left(\frac{\sqrt{3}}{2}\right) \\ &= 2 & &= 2\sqrt{3} \end{aligned}$$

Thus $A(4, 60°) = (2, 2\sqrt{3})$.

For point $B(8, 7\pi/6)$:

$$\begin{aligned} x &= r\cos\theta & y &= r\sin\theta \\ x &= 8\cos\frac{7\pi}{6} & y &= 8\sin\frac{7\pi}{6} \\ &= 8\left(-\frac{\sqrt{3}}{2}\right) & &= 8\left(-\frac{1}{2}\right) \\ &= -4\sqrt{3} & &= -4 \end{aligned}$$

Thus $B(8, 7\pi/6) = (-4\sqrt{3}, -4)$.

EXAMPLE 8

Find polar coordinates for each point: $C(2\sqrt{3}, 2)$ in degrees, $0° \le \theta < 360°$, and $D(6, -6)$ in radians, $0 \le \theta < 2\pi$.

Note: The signs of x and y determine the quadrant for θ. That is, the signs of x and y determine in which quadrant the point lies and hence the quadrant in which θ must lie.

For point $C(2\sqrt{3}, 2)$:

$$\begin{aligned} r^2 &= x^2 + y^2 & \theta &= \arctan\frac{y}{x} \\ r^2 &= (2\sqrt{3})^2 + 2^2 = 16 & \theta &= \arctan\frac{2}{2\sqrt{3}} \\ r &= 4 & \theta &= 30° \end{aligned}$$

Thus $C(2\sqrt{3}, 2) = (4, 30°)$.

For point $D(6, -6)$:

$$\begin{aligned} r^2 &= x^2 + y^2 & \theta &= \arctan\frac{y}{x} \\ r^2 &= 6^2 + (-6)^2 = 72 & \theta &= \arctan\left(\frac{-6}{6}\right) = \arctan(-1) \\ r &= 6\sqrt{2} & \theta &= \frac{7\pi}{4} \quad \left(\textit{Note:}\text{ the reference angle } \alpha = \frac{\pi}{4}.\right) \end{aligned}$$

Thus $D(6, -6) = (6\sqrt{2}, 7\pi/4)$.

Some curves are most simply expressed and easiest to work with in rectangular coordinates; others are most simply expressed and easiest to work with in polar coordinates. As a result, you must be able to change a polar equation to a rectangular equation and to change a rectangular equation to a polar equation.

EXAMPLE 9

Change $x^2 + y^2 - 4x = 0$ to polar form.

Substituting $x^2 + y^2 = r^2$ and $x = r\cos\theta$, we have

$$r^2 - 4r\cos\theta = 0$$
$$r(r - 4\cos\theta) = 0 \qquad \text{(Factor.)}$$

So

$$r = 0 \quad \text{or} \quad r - 4\cos\theta = 0$$

But $r = 0$ (the pole) is a point that is included in the graph of the equation $r - 4\cos\theta = 0$. Note that $(0, \pi/2)$ is an ordered pair that satisfies the second equation and names the pole. Thus the simplest polar equation is

$$r = 4\cos\theta$$

EXAMPLE 10

Change $r = 4\sin\theta$ to rectangular form.

Multiply both sides of the equation by r:

$$r^2 = 4r\sin\theta$$

Substituting $r^2 = x^2 + y^2$ and $r\sin\theta = y$, we have

$$x^2 + y^2 = 4y$$

Note that by multiplying both sides of the given equation by r, we added the root $r = 0$. But the point represented by that root is already included in the original equation. So no new points are added to those represented by the original equation.

EXAMPLE 11

Change $r\cos^2\theta = 6\sin\theta$ to rectangular form.

First multiply both sides by r:

$$r^2\cos^2\theta = 6r\sin\theta$$
$$(r\cos\theta)^2 = 6r\sin\theta$$

Substituting $r\cos\theta = x$ and $r\sin\theta = y$, we have

$$x^2 = 6y$$

EXAMPLE 12

Change $r = \dfrac{2}{1 - \cos\theta}$ to rectangular form.

$$r = \frac{2}{1 - \cos\theta} \tag{1}$$

First multiply both sides by $1 - \cos\theta$:

$$r(1 - \cos\theta) = 2$$
$$r - r\cos\theta = 2$$
$$r = 2 + r\cos\theta \tag{2}$$

Substituting $r = \pm\sqrt{x^2 + y^2}$ and $r\cos\theta = x$, we have

$$\pm\sqrt{x^2 + y^2} = 2 + x$$

Squaring both sides, we have

$$x^2 + y^2 = 4 + 4x + x^2$$
$$y^2 = 4x + 4$$

Note that squaring both sides was a risky operation because we introduced the possible extraneous solutions

$$r = -(2 + r\cos\theta) \qquad \textbf{(3)}$$

However, in this case both Equations (2) and (3) have the same graph. To show this, solve Equation (3) for r:

$$r = \frac{-2}{1 + \cos\theta} \qquad \textbf{(4)}$$

Recall that the ordered pairs (r, θ) and $(-r, \theta + \pi)$ represent the same point. Let us replace (r, θ) by $(-r, \theta + \pi)$ in Equation (4):

$$-r = \frac{-2}{1 + \cos(\theta + \pi)}$$

$$r = \frac{2}{1 - \cos\theta} \qquad [\text{Recall } \cos(\theta + \pi) = -\cos\theta.]$$

Equations (2) and (3) and thus Equations (1) and (4) have the same graph, and no extraneous solutions were introduced when we squared both sides. So our result $y^2 = 4x + 4$ is correct.

Exercises 20.8

Plot each point whose polar coordinates are given.

1. $A(3, 150°)$, $B(7, -45°)$, $C(2, -120°)$, $D(-4, 225°)$

2. $A(5, -90°)$, $B(2, -210°)$, $C(6, -270°)$, $D(-5, 30°)$

3. $A\left(4, \frac{\pi}{3}\right)$, $B\left(5, -\frac{\pi}{4}\right)$, $C\left(3, -\frac{7\pi}{6}\right)$, $D\left(-6, \frac{11\pi}{6}\right)$

4. $A\left(4, \frac{5\pi}{3}\right)$, $B\left(5, -\frac{3\pi}{2}\right)$, $C\left(3, -\frac{19\pi}{12}\right)$, $D\left(-6, -\frac{2\pi}{3}\right)$

For each point, name three other sets of polar coordinates such that $-360° \le \theta \le 360°$.

5. $(3, 60°)$

6. $(2, 240°)$

7. $(-5, 315°)$

8. $(-6, 90°)$

9. $(4, -135°)$

10. $(-1, -180°)$

For each point, name three other sets of polar coordinates such that $-2\pi \le \theta \le 2\pi$.

11. $\left(3, \frac{\pi}{6}\right)$

12. $\left(-7, \frac{\pi}{2}\right)$

13. $\left(-9, \frac{2\pi}{3}\right)$

14. $\left(-2, -\frac{5\pi}{6}\right)$

15. $\left(-4, -\frac{7\pi}{4}\right)$

16. $\left(5, -\frac{5\pi}{3}\right)$

Graph each equation by plotting points. Assign θ values of $0°$, $30°$, $45°$, $60°$, and so on, until you have a smooth curve.

17. $r = 10\sin\theta$

18. $r = -10\sin\theta$

19. $r = 4 + 4\cos\theta$

20. $r = 4 + 4\sin\theta$

21. $r\cos\theta = 4$

22. $r\sin\theta = -4$

Graph each equation by plotting points. Assign θ values of 0, $\pi/6$, $\pi/4$, $\pi/3$, and so on, until you have a smooth curve.

23. $r = -10 \cos \theta$ **24.** $r = 6 \sin \theta$ **25.** $r = 4 - 4 \sin \theta$

26. $r = 4 - 4 \cos \theta$ **27.** $r = \theta, 0 \le \theta \le 4\pi$ **28.** $r = 2\theta, 0 \le \theta \le 2\pi$

Change each set of polar coordinates to rectangular coordinates.

29. $(3, 30°)$ **30.** $(2, 180°)$ **31.** $\left(2, \frac{\pi}{3}\right)$

32. $\left(7, \frac{5\pi}{6}\right)$ **33.** $(-4, 150°)$ **34.** $(1, 420°)$

35. $\left(-6, \frac{3\pi}{2}\right)$ **36.** $(3, -\pi)$ **37.** $(-5, -240°)$

38. $(2, -120°)$ **39.** $\left(2, -\frac{7\pi}{4}\right)$ **40.** $\left(-1, -\frac{5\pi}{3}\right)$

Change each set of rectangular coordinates to polar coordinates in degrees, $0° \le \theta \le 360°$.

41. $(5, 5)$ **42.** $(-\sqrt{3}, 1)$ **43.** $(0, 4)$

44. $(-3, 0)$ **45.** $(-2, -2\sqrt{3})$ **46.** $(-1, 1)$

Change each set of rectangular coordinates to polar coordinates in radians, $0 \le \theta < 2\pi$.

47. $(-4, 4)$ **48.** $(-1, -\sqrt{3})$ **49.** $(-\sqrt{6}, \sqrt{2})$

50. $(5\sqrt{2}, -5\sqrt{2})$ **51.** $(0, -4)$ **52.** $(0, 0)$

Change each equation to polar form.

53. $x = 3$ **54.** $y = 5$ **55.** $x^2 + y^2 = 36$

56. $y^2 = 5x$ **57.** $x^2 + y^2 + 2x + 5y = 0$ **58.** $2x + 3y = 6$

59. $4x - 3y = 12$ **60.** $ax + by = c$ **61.** $9x^2 + 4y^2 = 36$

62. $4x^2 - 9y^2 = 36$ **63.** $x^3 = 4y^2$ **64.** $x^4 - 2x^2y^2 + y^4 = 0$

Change each equation to rectangular form.

65. $r \sin \theta = -3$ **66.** $r \cos \theta = 7$ **67.** $r = 5$

68. $r = 3 \sec \theta$ **69.** $\theta = \frac{\pi}{4}$ **70.** $\theta = -\frac{2\pi}{3}$

71. $r = 5 \cos \theta$ **72.** $r = 6 \sin \theta$ **73.** $r = 6 \cos\left(\theta + \frac{\pi}{3}\right)$

74. $r = 4 \sin\left(\theta - \frac{\pi}{4}\right)$ **75.** $r \sin^2 \theta = 3 \cos \theta$ **76.** $r^2 = \tan^2 \theta$

77. $r^2 \sin 2\theta = 2$ **78.** $r^2 \cos 2\theta = 6$ **79.** $r^2 = \sin 2\theta$

80. $r^2 = \cos 2\theta$ **81.** $r = \tan \theta$ **82.** $r = 4 \tan \theta \sec \theta$

83. $r = \frac{3}{1 + \sin \theta}$ **84.** $r = \frac{-4}{1 + \cos \theta}$ **85.** $r = 4 \sin 3\theta$

86. $r = 4 \cos 2\theta$ **87.** $r = 2 + 4 \sin \theta$ **88.** $r = 1 - \cos \theta$

89. Find the distance between the points whose polar coordinates are $(3, 60°)$ and $(2, 330°)$.

90. Find the distance between the points whose polar coordinates are $(5, \pi/2)$ and $(1, 7\pi/6)$.

91. Find a formula for the distance between two points whose polar coordinates are $P_1(r_1, \theta_1)$ and $P_2(r_2, \theta_2)$.

20.9 GRAPHS IN POLAR COORDINATES

As you undoubtedly know, a graph of any equation may be made by finding and plotting "enough" ordered pairs that satisfy the equation and connecting them with a curve. As you also undoubtedly know, this is often tedious and time-consuming at best. We need a method for sketching the graph of a polar equation that minimizes the number of ordered pairs that must be found and plotted. One such method involves symmetry. We shall present tests for three kinds of symmetry:

SYMMETRY WITH RESPECT TO THE

1. *Horizontal axis:* Replace θ by $-\theta$ in the original equation. If the resulting equation is equivalent to the original equation, then the graph of the original equation is symmetric with respect to the *horizontal* axis.
2. *Vertical axis:* Replace θ by $\pi - \theta$ in the original equation. If the resulting equation is equivalent to the original equation, then the graph of the original equation is symmetric with respect to the *vertical* axis.
3. *Pole:*
   (a) Replace r by $-r$ in the original equation. If the resulting equation is equivalent to the original equation, then the graph of the original equation is symmetric with respect to the *pole*.
   (b) Replace θ by $\pi + \theta$ in the original equation. If the resulting equation is equivalent to the original equation, then the graph of the original equation is symmetric with respect to the *pole*.

You should note that these tests for symmetry are sufficient conditions for symmetry; that is, they are sufficient to assure symmetry. You should also note that these are not necessary conditions for symmetry; that is, symmetry may exist even though the test fails.

If either Test 3(a) or 3(b) is satisfied, then the graph is symmetric with respect to the pole. It is also true that if any two of the three kinds of symmetry hold, then the remaining third symmetry automatically holds. Can you explain why?

To help you quickly test for symmetry, the following identities are listed for your convenience.

POLAR COORDINATE IDENTITIES FOR TESTING SYMMETRY

$$\sin(-\theta) = -\sin\theta$$
$$\cos(-\theta) = \cos\theta$$
$$\tan(-\theta) = -\tan\theta$$
$$\sin(\pi - \theta) = \sin\theta$$
$$\cos(\pi - \theta) = -\cos\theta$$
$$\tan(\pi - \theta) = -\tan\theta$$
$$\sin(\pi + \theta) = -\sin\theta$$
$$\cos(\pi + \theta) = -\cos\theta$$
$$\tan(\pi + \theta) = \tan\theta$$

EXAMPLE 1

Graph $r = 4 + 2 \cos \theta$.

Replacing θ by $-\theta$, we see that the graph is symmetric with respect to the horizontal axis. The other tests fail. Thus we need to make a table as follows (note that because of symmetry with respect to the horizontal axis, we need only generate ordered pairs for $0° \leq \theta \leq 180°$):

r	θ	$r = 4 + 2 \cos \theta$
6	0°	$r = 4 + 2 \cos 0° = 6$
5.7	30°	$r = 4 + 2 \cos 30° = 5.7$
5	60°	$r = 4 + 2 \cos 60° = 5$
4	90°	$r = 4 + 2 \cos 90° = 4$
3	120°	$r = 4 + 2 \cos 120° = 3$
2.3	150°	$r = 4 + 2 \cos 150° = 2.3$
2	180°	$r = 4 + 2 \cos 180° = 2$

Plot the points as shown in Fig. 20.43(a). Because of the symmetry with respect to the horizontal axis, plot the corresponding mirror-image points below the horizontal axis (see Fig. 20.43b).

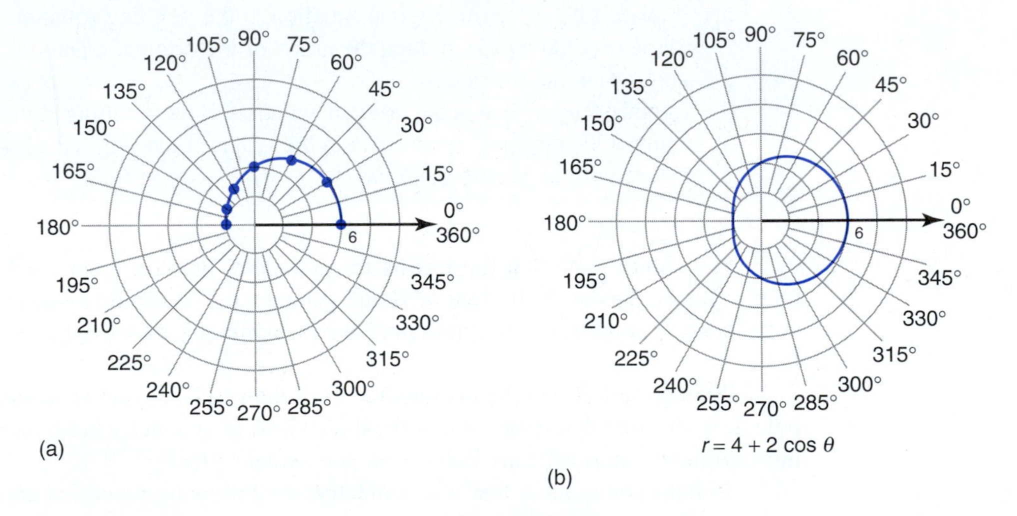

Figure 20.43

EXAMPLE 2

Graph $r = 4 + 4 \sin \theta$.

Replacing θ by $\pi - \theta$, we see that the graph is symmetric with respect to the vertical axis. The other tests fail. Thus, make a table as follows (note that because of symmetry with respect to the vertical axis, we need only generate ordered pairs for $-\pi/2 \leq \theta \leq \pi/2$):

r	θ	$r = 4 + 4 \sin \theta$
4	0	$r = 4 + 4 \sin 0 = 4$
6	$\pi/6$	$r = 4 + 4 \sin \pi/6 = 6$
7.5	$\pi/3$	$r = 4 + 4 \sin \pi/3 = 7.5$
8	$\pi/2$	$r = 4 + 4 \sin \pi/2 = 8$
2	$-\pi/6$	$r = 4 + 4 \sin (-\pi/6) = 2$
0.54	$-\pi/3$	$r = 4 + 4 \sin (-\pi/3) = 0.54$
0	$-\pi/2$	$r = 4 + 4 \sin (-\pi/2) = 0$

Plot the points as shown in Fig. 20.44(a). Because of the symmetry with respect to the vertical axis, plot the corresponding mirror-image points to the left of the vertical axis (see Fig. 20.44(b)).

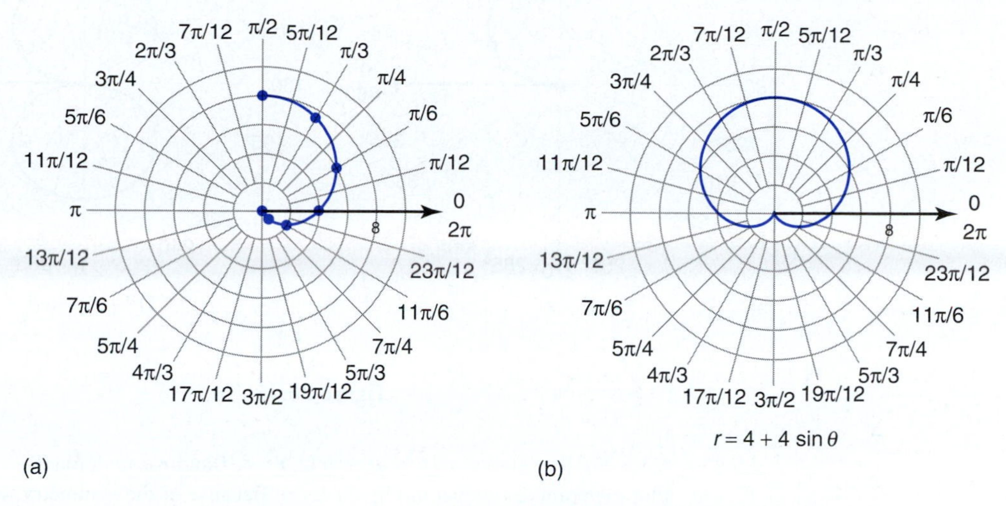

Figure 20.44

EXAMPLE 3

Graph $r^2 = 16 \sin \theta$.

Replacing r by $-r$, we see that the graph is symmetric with respect to the pole. Replacing θ by $\pi - \theta$, we see that the graph is also symmetric with respect to the vertical axis. Since two of the three kinds of symmetry hold, the graph is also symmetric with respect to the horizontal axis. (*Note:* Replacing θ by $-\theta$ gives the resulting equation $r^2 = -16 \sin \theta$, which is different from the original equation. However, its solutions when graphed give the same curve.)

r	θ	$r^2 = 16 \sin \theta$
0	0°	$r^2 = 16 \sin 0° = 0; r = 0$
2.8	30°	$r^2 = 16 \sin 30° = 8; r = 2.8$
3.7	60°	$r^2 = 16 \sin 60° = 13.9; r = 3.7$
4	90°	$r^2 = 16 \sin 90° = 16; r = 4$

Plot the points as shown in Fig. 20.45(a). Because of the symmetry with respect to the horizontal and vertical axes, plot the corresponding mirror-image points below the horizontal axis. Then plot the mirror image points of all resulting points to the left of the vertical axis (see Fig. 20.45(b)).

EXAMPLE 4

Graph $r^2 = 25 \cos 2\theta$.

By replacing θ by $-\theta$ and r by $-r$, we have symmetry with respect to the horizontal axis and the pole, respectively. Thus we also have symmetry with respect to the vertical axis. Working in the first quadrant, we have

r	θ	$r^2 = 25 \cos 2\,\theta$
5	0	$r^2 = 25 \cos 2(0) = 25; r = 5$
4.7	$\pi/12$	$r^2 = 25 \cos 2(\pi/12) = 21.7; r = 4.7$
3.5	$\pi/6$	$r^2 = 25 \cos 2(\pi/6) = 12.5; r = 3.5$
0	$\pi/4$	$r^2 = 25 \cos 2(\pi/4) = 0; r = 0$

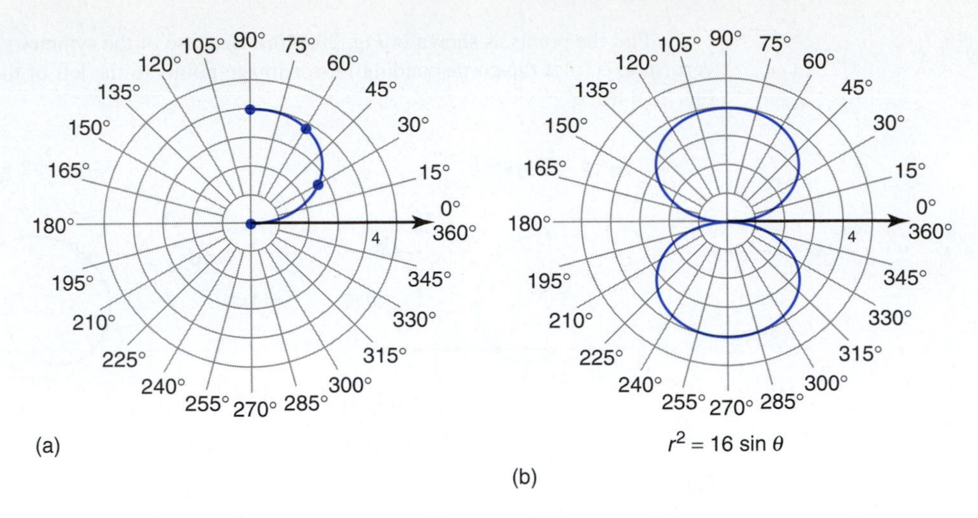

Figure 20.45

Note: For the interval $\pi/4 < \theta \leq \pi/2$, $r^2 < 0$ and r is undefined.

Plot the points as shown in Fig. 20.46(a). Because of the symmetry with respect to the horizontal and vertical axes, plot the corresponding mirror-image points below the horizontal axis and to the left of the vertical axis (see Fig. 20.46b).

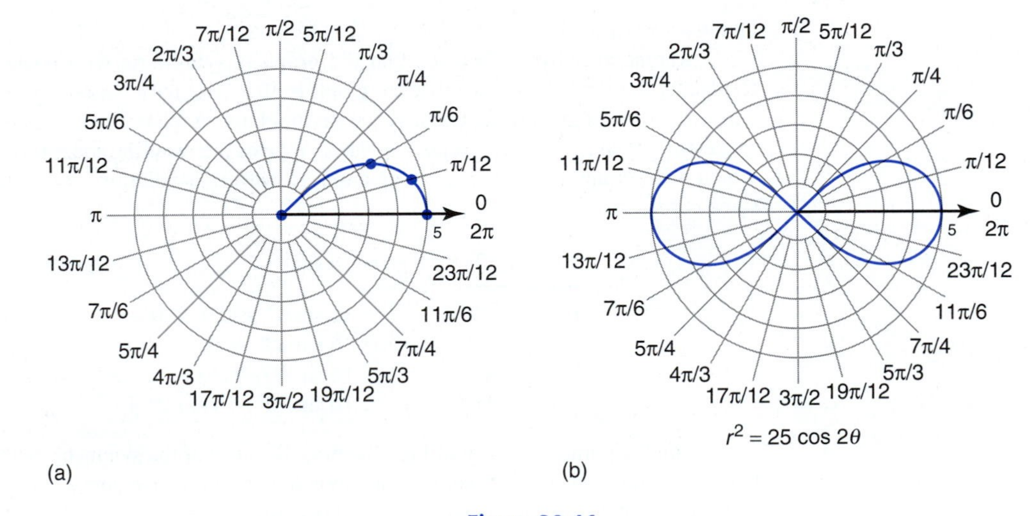

Figure 20.46

There are various general polar equations whose graphs may be classified as shown in Fig. 20.47. What are the graphs of the various forms of $r = a + b \sin \theta$ like?

Equations in the form

$$r = a \sin n\theta \quad \text{or} \quad r = a \cos n\theta$$

where n is a positive integer, are called *petal* or *rose curves*. The number of petals is equal to n if n is an *odd* integer, and is equal to $2n$ if n is an *even* integer. This is because the graph "retraces" itself as θ goes from 0° to 360° when n is odd, so there are only half as many distinct petals. (For $n = 1$ there is one circular petal. See Example 6, Section 20.8). The value of a corresponds to the length of each petal.

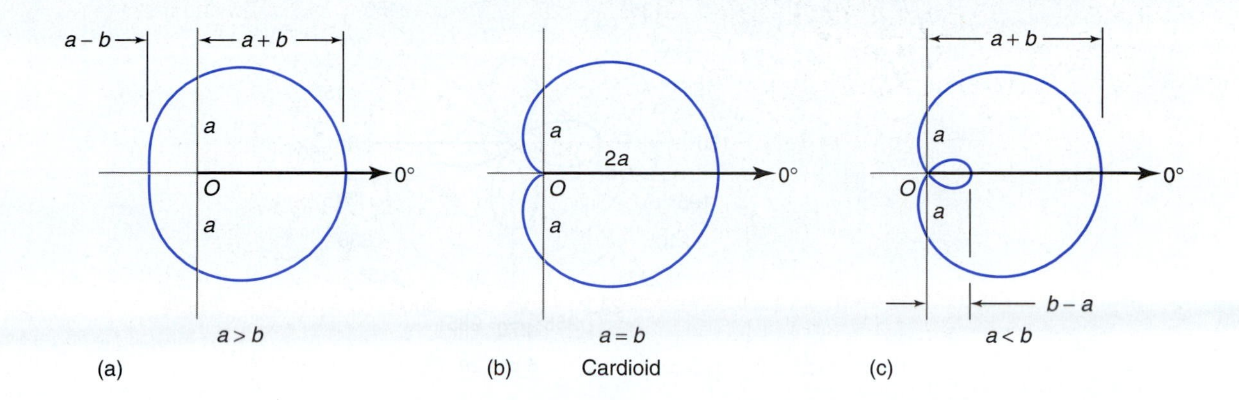

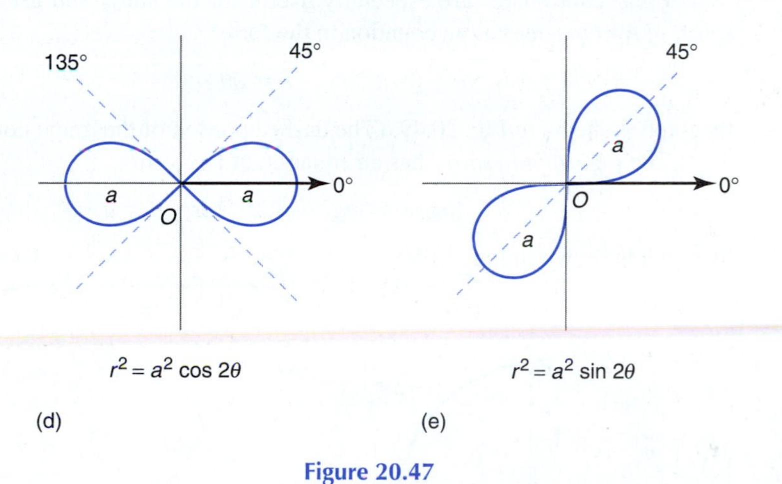

Figure 20.47

The tests for symmetry may be used to graph petal curves. However, we shall illustrate a method that is somewhat different, as well as easier and quicker, for graphing petal curves.

EXAMPLE 5

Graph $r = 6 \cos 2\theta$.

First, note that $n = 2$, which is even. Therefore, we have four petals. The petals are always uniform; each petal occupies $360°/4$, or $90°$, of the polar coordinate system. Next, find the tip of a petal; this occurs when r is maximum or when

$$\cos 2\theta = 1$$
$$2\theta = 0°$$
$$\theta = 0°$$

That is, $r = 6$ when $\theta = 0°$.

Finally, sketch four petals, each having a maximum length of six and occupying 90° (see Fig. 20.48). For more accuracy, you may graph the ordered pairs corresponding to a "half petal" ($0° \leq \theta \leq 45°$ in this case).

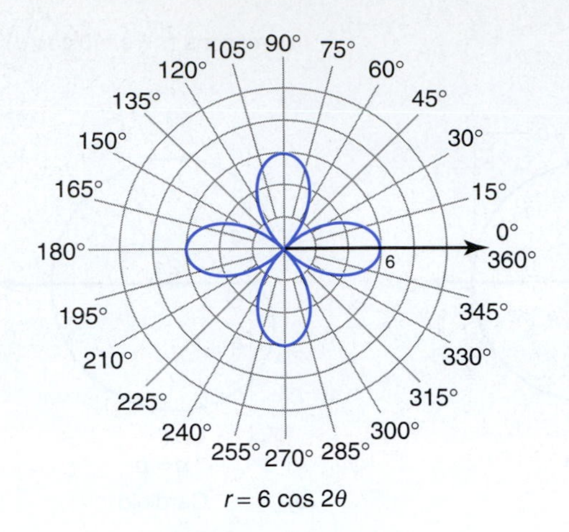

Figure 20.48

Polar coordinates are especially useful for the study and graphing of *spirals*. The *spiral of Archimedes* has an equation in the form

$$r = a\theta$$

Its graph is shown in Fig. 20.49. (The dashed portion of the graph corresponds to $\theta < 0$.)

The *logarithmic spiral* has an equation of the form

$$\log_b r = \log_b a + k\theta \quad \text{or} \quad r = a \cdot b^{k\theta}$$

Its graph is shown in Fig. 20.50.

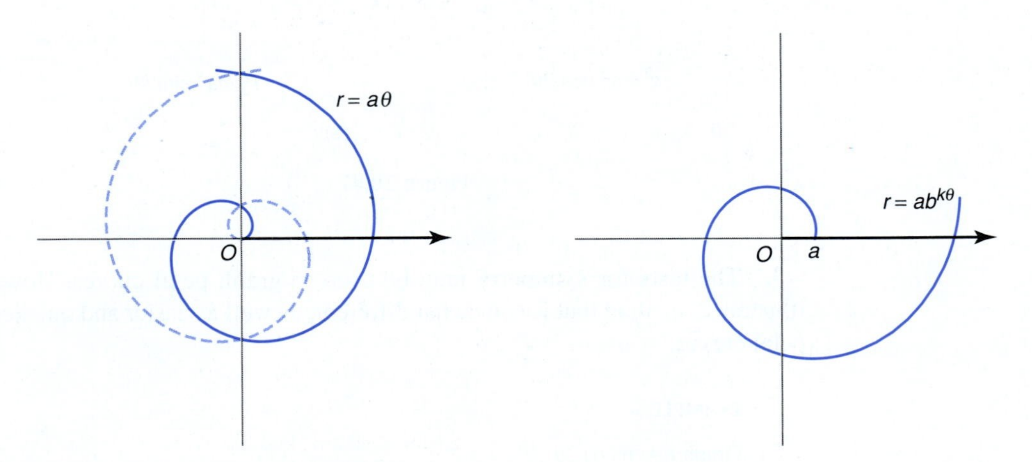

Figure 20.49 Spiral of Archimedes. **Figure 20.50** Logarithmic spiral.

For graphing calculator examples in polar coordinates, see Appendix C, Section C.5 (TI-83 Plus), or Appendix D, Section D.7 (TI-89).

Exercises 20.9

Graph each equation.

1. $r = 6$ **2.** $r = 3$ **3.** $r = -2$

4. $r = -4$ **5.** $\theta = 30°$ **6.** $\theta = -120°$

7. $\theta = -\frac{\pi}{3}$

8. $\theta = \frac{7\pi}{6}$

9. $r = 5 \sin \theta$

10. $r = 8 \cos \theta$

11. $r = 6 \cos\left(\theta + \frac{\pi}{3}\right)$

12. $r = 4 \sin\left(\theta - \frac{\pi}{4}\right)$

13. $r = 4 + 2 \sin \theta$

14. $r = 8 + 2 \cos \theta$

15. $r = 4 - 2 \cos \theta$

16. $r = 4 - 2 \sin \theta$

17. $r = 3 + 3 \cos \theta$

18. $r = 5 + 5 \sin \theta$

19. $r = 2 + 4 \sin \theta$

20. $r = 2 + 8 \cos \theta$

21. $r = 2 - 4 \cos \theta$

22. $r = 2 - 4 \sin \theta$

23. $r = 3 - 3 \cos \theta$

24. $r = 5 - 5 \sin \theta$

25. $r \cos \theta = 6$

26. $r \sin \theta = -4$

27. $r^2 = 25 \cos \theta$

28. $r^2 = -9 \sin \theta$

29. $r^2 = 9 \sin 2\theta$

30. $r^2 = 16 \cos 2\theta$

31. $r^2 = -36 \cos 2\theta$

32. $r = -36 \sin 2\theta$

33. $r = 5 \sin 3\theta$

34. $r = 4 \cos 5\theta$

35. $r = 3 \cos 2\theta$

36. $r = 6 \sin 4\theta$

37. $r = 9 \sin^2 \theta$

38. $r = 16 \cos^2 \theta$

39. $r = 4 \cos \frac{\theta}{2}$

40. $r = 5 \sin^2 \frac{\theta}{2}$

41. $r = \tan \theta$

42. $r = 2 \csc \theta$

43. $r = 3\theta, \theta > 0$

44. $r = \frac{3}{\theta}, \theta > 0$

45. $r = 2^{3\theta}$

46. $r = 2 \cdot 3^{2\theta}$

47. $r = \frac{4}{\sin \theta + \cos \theta}$

48. $r = \frac{-2}{\sin \theta + \cos \theta}$

49. $r(1 + \cos \theta) = 4$

50. $r(1 + 2 \sin \theta) = -4$

CHAPTER 20 SUMMARY

1. *Conics:* Equations in the form $Ax^2 + Bxy + Cy^2 + Dx + Ey + F = 0$ are called conics.
2. *Circle:*
   (a) *Standard form:* $(x - h)^2 + (y - k)^2 = r^2$, where r is the radius and (h, k) is the center.
   (b) *General form:* $x^2 + y^2 + Dx + Ey + F = 0$.
   (c) *Center at the origin:* $x^2 + y^2 = r^2$, where r is the radius.
3. *Parabola with vertex at the origin:*
   (a) $y^2 = 4px$ with focus at $(p, 0)$ and $x = -p$ as the directrix.
      (i) When $p > 0$, the parabola opens to the right.
      (ii) When $p < 0$, the parabola opens to the left.
   (b) $x^2 = 4py$ with focus at $(0, p)$ and $y = -p$ as the directrix.
      (i) When $p > 0$, the parabola opens upward.
      (ii) When $p < 0$, the parabola opens downward.
4. *Ellipse with center at the origin:*
   (a) $\frac{x^2}{a^2} + \frac{y^2}{b^2} = 1$ with the major axis on the x-axis and $a > b$.
   (b) $\frac{y^2}{a^2} + \frac{x^2}{b^2} = 1$ with the major axis on the y-axis and $a > b$.
5. *Hyperbola with center at the origin:*
   (a) $\frac{x^2}{a^2} - \frac{y^2}{b^2} = 1$ with the transverse axis on the x-axis.
   (b) $\frac{y^2}{a^2} - \frac{x^2}{b^2} = 1$ with the transverse axis on the y-axis.

6. *Translation equations:* $x' = x - h$ and $y' = y - k$.

7. *General forms of conics with axes parallel to the coordinate axes:*

   (a) $(y - k)^2 = 4p(x - h)$
   is a parabola with vertex at (h, k) and axis parallel to the x-axis.

   (b) $(x - h)^2 = 4p(y - k)$
   is a parabola with vertex at (h, k) and axis parallel to the y-axis.

   (c) $\dfrac{(x - h)^2}{a^2} + \dfrac{(y - k)^2}{b^2} = 1, \quad (a > b)$
   is an ellipse with center at (h, k) and major axis parallel to the x-axis.

   (d) $\dfrac{(y - k)^2}{a^2} + \dfrac{(x - h)^2}{b^2} = 1, \quad (a > b)$
   is an ellipse with center at (h, k) and major axis parallel to the y-axis.

   (e) $\dfrac{(x - h)^2}{a^2} - \dfrac{(y - k)^2}{b^2} = 1$
   is a hyperbola with center at (h, k) and transverse axis parallel to the x-axis.

   (f) $\dfrac{(y - k)^2}{a^2} - \dfrac{(x - h)^2}{b^2} = 1$
   is a hyperbola with center at (h, k) and transverse axis parallel to the y-axis.

8. *The general second-degree equation:* The circle, parabola, ellipse, and hyperbola are all special cases of the second-degree equation

$$Ax^2 + Bxy + Cy^2 + Dx + Ey + F = 0$$

   When $B = 0$ and at least one of the coefficients A or C is not zero, the following summarizes the conditions for each curve:

   (a) If $A = C$, we have a *circle.*
   In special cases, the graph of the equation may be a point or there may be no graph. (The equation may have only one or no solution.)

   (b) If $A = 0$ and $C \neq 0$ or $C = 0$ and $A \neq 0$, then we have a *parabola.*

   (c) If $A \neq C$, and A and C are either both positive or both negative, then we have an *ellipse.*
   In special cases, the graph of the equation may be a point or there may be no graph. (The equation may have only one or no solution.)

   (d) If A and C differ in sign, then we have a *hyperbola*. In some special cases the graph may be a pair of intersecting lines.
   If $D \neq 0$ or $E \neq 0$ or both are not equal to zero, the curve does not have its center (or vertex in the case of the parabola) at the origin. If $B \neq 0$, then the axis of the curve does not lie along the x-axis or the y-axis.

9. Each point $P(x, y)$ in the rectangular coordinate system may be described by an ordered pair $P(r, \theta)$ in the polar coordinate system.

10. In the rectangular coordinate system, there is a one-to-one correspondence between points in the plane and ordered pairs of real numbers. This one-to-one correspondence is not a property of the polar coordinate system. In general the point $P(r, \theta)$ may be represented by

$$(r, \theta + k \cdot 360°) \quad \text{or} \quad (r, \theta + k \cdot 2\pi)$$

   where k is any integer. $P(r, \theta)$ may also be represented by

$$(-r, \theta + k \cdot 180°) \quad \text{or} \quad (-r, \theta + k\pi)$$

   where k is any odd integer.

11. *Relationships between the rectangular and polar coordinate systems* (see Fig. 20.51):

(a) $x = r\cos\theta$ (b) $y = r\sin\theta$

(c) $\tan\theta = \dfrac{y}{x}$ or $\theta = \arctan\dfrac{y}{x}$ (d) $x^2 + y^2 = r^2$

(e) $\cos\theta = \dfrac{x}{\sqrt{x^2 + y^2}}$ (f) $\sin\theta = \dfrac{y}{\sqrt{x^2 + y^2}}$

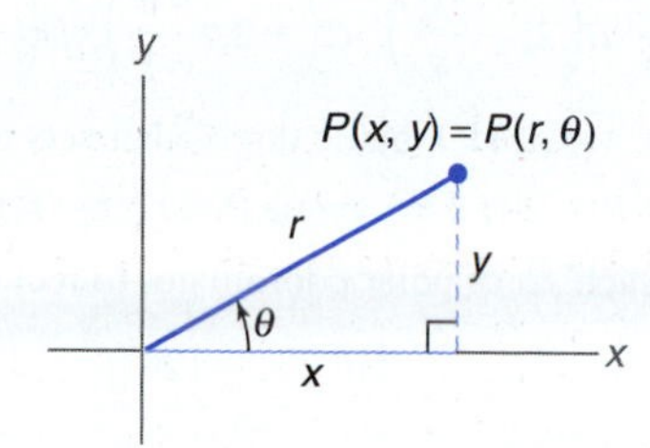

Figure 20.51

12. *Symmetry tests for graphing polar equations:*
    (a) *Horizontal axis:* Replace θ by $-\theta$ in the original equation. If the resulting equation is equivalent to the original equation, then the graph of the original equation is symmetric with respect to the *horizontal* axis.
    (b) *Vertical axis:* Replace θ by $\pi - \theta$ in the original equation. If the resulting equation is equivalent to the original equation, then the graph of the original equation is symmetric with respect to the *vertical* axis.
    (c) *Pole*
        (i) Replace r by $-r$ in the original equation. If the resulting equation is equivalent to the original equation, then the graph of the original equation is symmetric with respect to the *pole*.
        (ii) Replace θ by $\pi + \theta$ in the original equation. If the resulting equation is equivalent to the original equation, then the graph of the original equation is symmetric with respect to the *pole*.

CHAPTER 20 REVIEW

1. Write the equation of the circle with center at $(5, -7)$ and with radius 6.
2. Find the center and radius of the circle $x^2 + y^2 - 8x + 6y - 24 = 0$.
3. Find the focus and directrix of the parabola $x^2 = 6y$ and sketch its graph.
4. Write the equation of the parabola with focus at $(-4, 0)$ and directrix $x = 4$.
5. Write the equation of the parabola with focus at $(4, 3)$ and directrix $x = 0$.
6. Find the vertices and foci of the ellipse $4x^2 + 49y^2 = 196$ and sketch its graph.
7. Find the equation of the ellipse with vertices $(0, 4)$ and $(0, -4)$ and with foci at $(0, 2\sqrt{3})$ and $(0, -2\sqrt{3})$.
8. Find the vertices and foci of the hyperbola $4x^2 - 9y^2 = 144$ and sketch its graph.
9. Write the equation of the hyperbola with vertices at $(0, 5)$ and $(0, -5)$ and with foci at $(0, \sqrt{41})$ and $(0, -\sqrt{41})$.
10. Write the equation of the ellipse with center at $(3, -4)$, vertices at $(3, 1)$ and $(3, -9)$, and foci at $(3, 0)$ and $(3, -8)$.
11. Write the equation of the hyperbola with center at $(-7, 4)$ and vertices at $(2, 4)$ and $(-16, 4)$; the length of the conjugate axis is 6.
12. Name and sketch the graph of $16x^2 - 4y^2 - 64x - 24y + 12 = 0$.

Solve each system of equations.

13. $y^2 + 4y + x = 0$
$x = 2y$

14. $3x^2 - 4y^2 = 36$
$5x^2 - 8y^2 = 56$

Plot each point whose polar coordinates are given.

15. $A(6, 60°)$, $B(3, -210°)$, $C(-2, -270°)$, $D(-4, 750°)$

16. $A\left(5, \frac{\pi}{6}\right)$, $B\left(2, -\frac{5\pi}{4}\right)$, $C\left(-3, -\frac{\pi}{2}\right)$, $D\left(-5, \frac{19\pi}{2}\right)$

17. For point $A(5, 135°)$, name three other sets of polar coordinates for $-360° \le \theta < 360°$.

18. For point $B(-2, 7\pi/6)$, name three other sets of polar coordinates for $-2\pi \le \theta \le 2\pi$.

19. Change each set of polar coordinates to rectangular coordinates.
(a) $(3, 210°)$ (b) $(2, -120°)$ (c) $\left(-5, \frac{11\pi}{6}\right)$ (d) $\left(-6, -\frac{\pi}{2}\right)$

20. Change each set of rectangular coordinates to polar coordinates in degrees for $0° \le \theta < 360°$.
(a) $(-3, 3)$ (b) $(0, -6)$ (c) $(-1, \sqrt{3})$

21. Change each set of rectangular coordinates to polar coordinates in radians for $0 \le \theta < 2\pi$.
(a) $(-5, 0)$ (b) $(-6\sqrt{3}, 6)$ (c) $(1, -1)$

Change each equation to polar form.

22. $x^2 + y^2 = 49$
23. $y^2 = 9x$
24. $5x + 2y = 8$
25. $x^2 - 4y^2 = 12$
26. $y^3 = 6x^2$
27. $y(x^2 + y^2) = x^2$

Change each equation to rectangular form.

28. $r \cos \theta = 12$
29. $r = 9$
30. $\theta = \frac{2\pi}{3}$
31. $r = 8 \cos \theta$
32. $r \sin^2 \theta = 5 \cos \theta$
33. $r^2 \sin 2\theta = 8$
34. $r^2 = 4 \cos 2\theta$
35. $r = \csc \theta$
36. $r = 1 + \sin \theta$
37. $r = \frac{2}{1 - \sin \theta}$

Graph each equation.

38. $r = 7$
39. $\theta = -\frac{\pi}{4}$
40. $r = 5 \cos \theta$
41. $r = 6 + 3 \sin \theta$
42. $r = 6 - 3 \sin \theta$
43. $r = 4 + 4 \cos \theta$
44. $r = 3 - 6 \cos \theta$
45. $r \sin \theta = 5$
46. $r^2 = 36 \cos \theta$
47. $r = 6 \sin 5\theta$
48. $r^2 = 25 \sin 2\theta$
49. $r(1 - \sin \theta) = 6$

APPENDIX A
Tables

TABLE 1 U.S. Weights and Measures

Units of length	Units of weight
Standard unit—inch (in. or ″)	Standard unit—pound (lb)
12 inches = 1 foot (ft or ′)	16 ounces (oz) = 1 pound
3 feet = 1 yard (yd)	2000 pounds = 1 ton (T)
$5\frac{1}{2}$ yards or $16\frac{1}{2}$ feet = 1 rod (rd)	
5280 feet = 1 mile (mi)	

Volume measure
Liquid
16 ounces (fl oz) = 1 pint (pt)
2 pints = 1 quart (qt)
4 quarts = 1 gallon (gal)
Dry
2 pints (pt) = 1 quart (qt)
8 quarts = 1 peck (pk)
4 pecks = 1 bushel (bu)

TABLE 2 Conversion Tables

Length

	cm	*m*	*km*	*in.*	*ft*	*mi*
1 centimetre	1	10^{-2}	10^{-5}	0.394	3.28×10^{-2}	6.21×10^{-6}
1 metre	100	1	10^{-3}	39.4	3.28	6.21×10^{-4}
1 kilometre	10^5	1000	1	3.94×10^4	3280	0.621
1 inch	2.54	2.54×10^{-2}	2.54×10^{-5}	1	8.33×10^{-2}	1.58×10^{-5}
1 foot	30.5	0.305	3.05×10^{-4}	12	1	1.89×10^{-4}
1 mile	1.61×10^5	1610	1.61	6.34×10^4	5280	1

Area

Metric	*U.S.*
$1\ m^2 = 10{,}000\ cm^2$	$1\ ft^2 = 144\ in^2$
$= 1{,}000{,}000\ mm^2$	$1\ yd^2 = 9\ ft^2$
$1\ cm^2 = 100\ mm^2$	$1\ rd^2 = 30.25\ yd^2$
$= 0.0001\ m^2$	$1\ acre = 160\ rd^2$
$1\ km^2 = 1{,}000{,}000\ m^2$	$= 4840\ yd^2$
$1\ ha = 10{,}000\ m^2$	$= 43{,}560\ ft^2$
	$1\ mi^2 = 640\ acres$

	m^2	*cm^2*	*ft^2*	*in^2*
$1\ m^2$	1	10^4	10.8	1550
$1\ cm^2$	10^{-4}	1	1.08×10^{-3}	0.155
$1\ ft^2$	9.29×10^{-2}	929	1	144
$1\ in^2$	6.45×10^{-4}	6.45	6.94×10^{-3}	1

$1\ mi^2 = 2.79 \times 10^7\ ft^2 = 640\ acres$

$1\ circular\ mil = 5.07 \times 10^{-6}\ cm^2 = 7.85 \times 10^{-7}\ in^2$

1 hectare = 2.47 acres

Volume

Metric	U.S.
$1\ m^3 = 10^6\ cm^3$	$1\ ft^3 = 1728\ in^3$
$1\ cm^3 = 10^{-6}\ m^3$	$1\ yd^3 = 27\ ft^3$
$= 10^3\ mm^3$	

	m^3	cm^3	L	ft^3	in^3
$1\ m^3$	1	10^6	1000	35.3	6.10×10^4
$1\ cm^3$	10^{-6}	1	1.00×10^{-3}	3.53×10^{-5}	6.10×10^{-2}
1 L	1.00×10^{-3}	1000	1	3.53×10^{-2}	61.0
$1\ ft^3$	2.83×10^{-2}	2.83×10^4	28.3	1	1728
$1\ in^3$	1.64×10^{-5}	16.4	1.64×10^{-2}	5.79×10^{-4}	1

1 U.S. fluid gallon = 4 U.S. fluid quarts = 8 U.S. pints = 128 U.S. fluid ounces = 231 in^3 = 0.134 ft^3 = 3.79 litres
1 L = 1000 cm^3 = 1.06 qt

Other useful conversion factors

1 newton (N) = 0.225 lb
1 pound (lb) = 4.45 N
1 slug = 14.6 kg
1 joule (J) = 0.738 ft-lb
= 2.39×10^{-4} kcal
1 calorie (cal) = 4.185 J
1 kilocalorie (kcal) = 4185 J
1 foot-pound (ft-lb) = 1.36 J
1 watt (W) = 1 J/s = 0.738 ft-lb/s
1 kilowatt (kW) = 1000 W
= 1.34 hp
1 hp = 550 ft-lb/s = 746 W

1 atm = 101.32 kPa
= 14.7 lb/in^2
1 Btu = 0.252 kcal
1 kcal = 3.97 Btu
$F = \frac{9}{5} C + 32°$
$C = \frac{5}{9}(F - 32°)$
1 kg = 2.20 lb (on the earth's surface)
1 lb = 454 g
= 16 oz
1 metric ton = 1000 kg
= 2200 lb

TABLE 3 Physical Quantities and Their Units

Quantity	Symbol	Unit	
		Metric	*U.S.*
Distance	s	metre (m)	foot (ft)
Time	t	second (s)	second (s)
Mass	m	kilogram (kg)	slug
Force, weight	F, w	newton (N)	pound (lb)
Area	A	m^2	ft^2
Volume	V	m^3 or L	ft^3
Velocity	v	m/s	ft/s
Acceleration	a	m/s^2	ft/s^2
Energy, work	E, W	Nm or joule (J)	ft-lb
Power	P	joule/s or watt (W)	ft-lb/s or hp
Heat	Q	joule (J)	British thermal unit (Btu)
Pressure	p	N/m^2 or pascal (Pa)	lb/in^2
Electric charge	q	coulomb (C)	coulomb (C)
Electric current	I	ampere (A)	ampere (A)
Electric potential	V, E	volt (V)	volt (V)
Capacitance	C	farad (F)	farad (F)
Inductance	L	henry (H)	henry (H)
Resistance	R	ohm (Ω)	ohm (Ω)
Frequency	f	1/s or hertz (Hz)	1/s or Hz

APPENDIX B
The Metric System

B.1 INTRODUCTION

The International System of Units, often referred to as the metric system or the SI metric system, is used throughout the world. Because so many businesses and consumer products (cars, entertainment systems, machinery, etc.) are international, we need to know how to work with the SI metric system. This Appendix explains the units that are used to measure electricity, temperature, and other things that technicians work with, as well as conversion from one unit of measure to another.

The SI metric system has seven **basic units,** which are listed in Table B.1.

TABLE B.1

Basic unit	SI abbreviation	Used for measuring
metre*	m	length
kilogram	kg	mass
second	s	time
ampere	A	electric current
kelvin	K	temperature
candela	cd	light intensity
mole	mol	molecular substance

*See note in Table B.2.

All other SI units are called **derived units;** that is, they can be defined in terms of these seven basic units (see Fig. B.1). For example, the newton (N) is defined as 1 kg m/s^2 (kilogram metre per second per second). Other commonly used derived SI units are listed in Table B.2.

Since the SI metric system is a decimal, or base 10, system, it is very similar to our decimal number system and our decimal money system. It is an easy system to use because calculations are based on the number 10 and its multiples. Special prefixes are used to name these multiples and submultiples, which may be used with almost all SI units. Since the same prefixes are repeatedly used, the task of memorization of many conversions has been significantly reduced. Table B.3 shows these prefixes and their corresponding symbols.

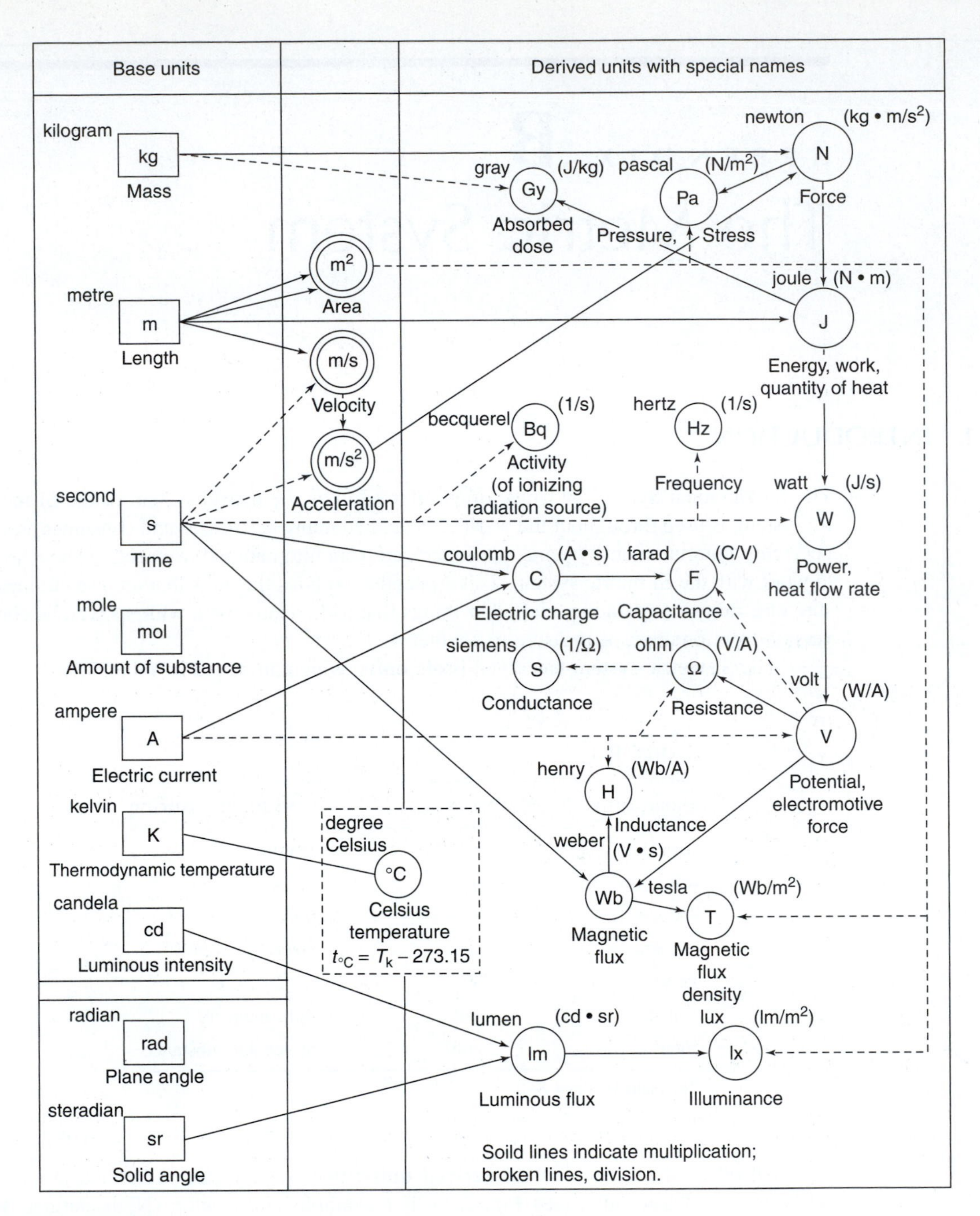

Figure B.1 Relationships of SI units with names.

EXAMPLE 1

Write the SI abbreviation for 23 centimetres.

The symbol for the prefix *centi* is c.

The symbol for the unit *metre* is m.

The SI abbreviation for 23 centimetres is 23 cm.

TABLE B.2

Derived unit	SI abbreviation	Used for measuring
litre*	L	capacity
cubic metre	m^3	volume
square metre	m^2	area
newton	N	force
metre per second	m/s	speed
joule	J	energy
watt	W	power
ohm	Ω	resistance
volt	V	voltage
farad	F	capacitance
henry	H	inductance

*There is some difference of opinion in the United States on the spelling of *metre* and *litre.* We have chosen the *re* spellings for two reasons. First, this is the internationally accepted spelling for all English-speaking countries. Second, the word *meter* already has many different meanings—parking meter, electric meter, odometer, and so on. Many people feel that the metric unit of length should be distinctive and readily recognizable, which the *re* spelling is.

TABLE B.3 Prefixes for SI Units

Multiple or submultiple[a] decimal form	Power of 10	Prefix	Prefix symbol	Pronunciation	Meaning
1,000,000,000,000	10^{12}	tera	T	tĕr′ă	one trillion times
1,000,000,000	10^{9}	giga	G	jĭg′ă	one billion times
1,000,000	10^{6}	mega	M	mĕg′ă	one million times
1,000	10^{3}	kilo[b]	k	kĭl′ō	one thousand times
100	10^{2}	hecto	h	hĕk′tō	one hundred times
10	10^{1}	deka	da	dĕk′ă	ten times
0.1	10^{-1}	deci	d	dĕs′ĭ	one-tenth of
0.01	10^{-2}	centi[b]	c	sĕnt′ĭ	one-hundredth of
0.001	10^{-3}	milli[b]	m	mĭl′ĭ	one-thousandth of
0.000001	10^{-6}	micro	μ	mī′krō	one-millionth of
0.000000001	10^{-9}	nano	n	năn′ō	one-billionth of
0.000000000001	10^{-12}	pico	p	pē′kō	one-trillionth of

[a]Factor by which the unit is multiplied.
[b]Most commonly used prefixes.

EXAMPLE 2

Write the SI metric unit for the abbreviation 65 kg.

The prefix for k is *kilo.*

The unit for g is *gram.*

The SI metric unit for 65 kg is 65 kilograms.

Exercises B.1

Give the metric prefix for each value.

1. 1000 **2.** 0.01 **3.** 100 **4.** 0.1

5. 0.001 **6.** 10 **7.** 1,000,000 **8.** 0.000001

Give the metric symbol, or abbreviation, for each prefix.

9. hecto **10.** kilo **11.** milli **12.** deci

13. mega **14.** deka **15.** centi **16.** micro

Write the abbreviation for each quantity.

17. 133 millimetres **18.** 63 dekagrams **19.** 18 kilolitres

20. 42 centimetres **21.** 19 centigrams **22.** 25 milligrams

23. 72 hectometres **24.** 17 decilitres

Write the SI unit for each abbreviation.

25. 14 m **26.** 182 L **27.** 19 g **28.** 147 kg

29. 17 mm **30.** 23 dL **31.** 25 dam **32.** 17 mg

33. 16 Mm **34.** 250 μg

35. The basic metric unit of length is ____.

36. The basic metric unit of mass is ____.

37. Two common metric units of volume are ____ and ____.

38. The basic metric unit for electric current is ____.

39. The basic metric unit for time is ____.

40. The common metric unit for power is ____.

B.2 LENGTH

The basic SI unit of length is the metre (m) (see Fig. B.2). Long distances are measured in kilometres (km); 1 km = 1000 m (see Fig. B.3). We use the centimetre (cm) to measure short distances, such as the length of a book or the width of a board (see Fig. B.4). The millimetre (mm) is used to measure very small lengths, such as the thickness of a book or the depth of a tire tread (see Fig. B.5).

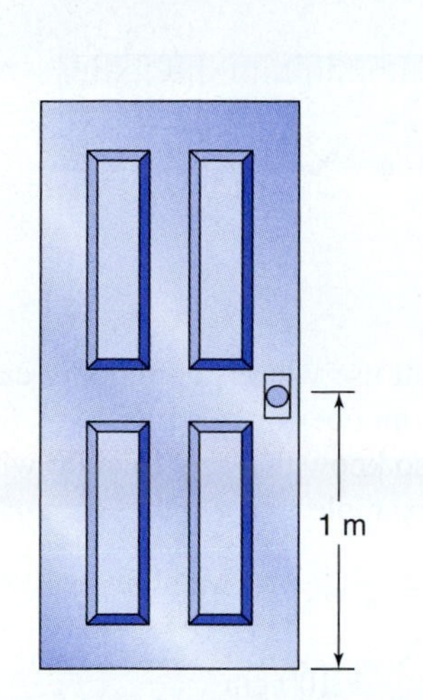

(a) The height of a doorknob is about 1 m.

(b) The length of a person's long pace is also about 1 m.

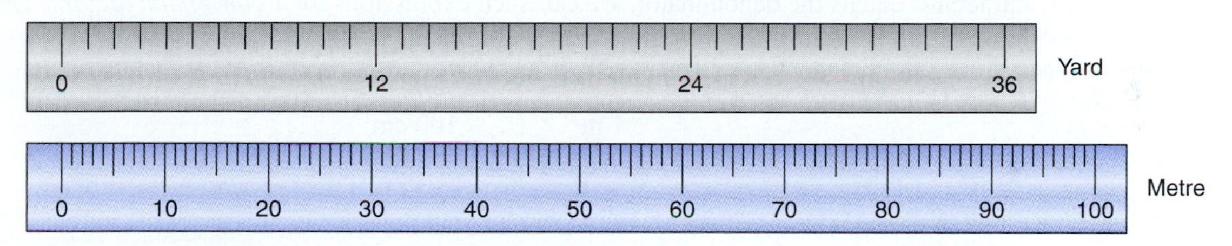

(c) One metre is a little more than 1 yd.

Figure B.2 Length of one metre.

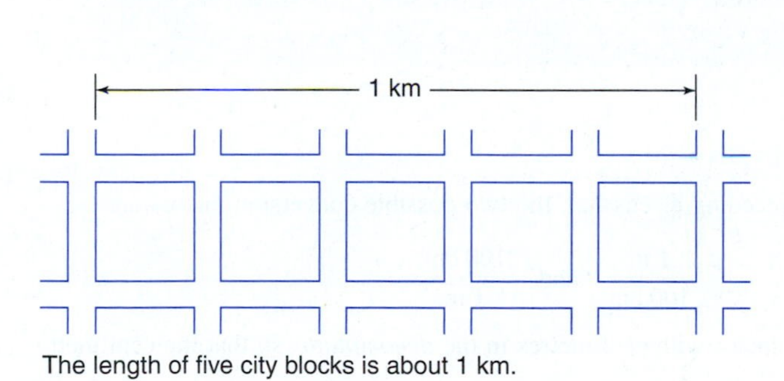

The length of five city blocks is about 1 km.

Figure B.3

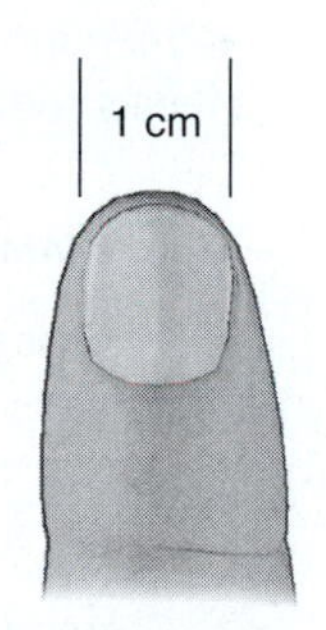

The width of your small fingernail is about 1 cm.

Figure B.4

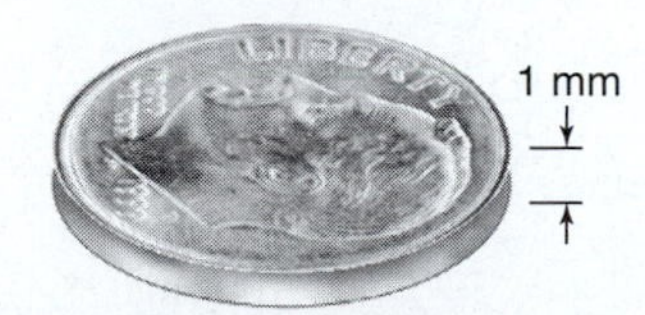

The thickness of a dime is about 1 mm. **Figure B.5**

A metric ruler is shown in Fig. B.6. The large numbered divisions are centimetres. They are divided into 10 equal parts, called millimetres.

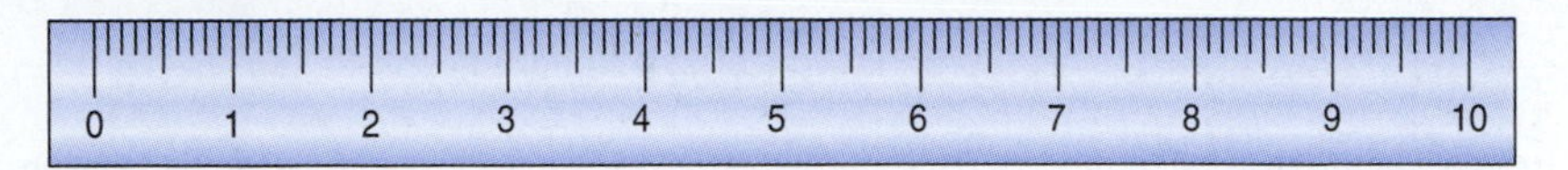

Figure B.6 Metric ruler.

Conversion Factors

To change from one unit or set of units to another, we shall use what is commonly called a **conversion factor.** We know that we can multiply any number or quantity by 1 (one) without changing the value of the original quantity. We also know that any fraction whose numerator and denominator are equal is equal to 1. For example,

$$\frac{3}{3} = 1 \qquad \frac{15 \text{ m}}{15 \text{ m}} = 1 \qquad \frac{4.5 \text{ kg}}{4.5 \text{ kg}} = 1$$

In addition, since 1 m = 100 cm, $\frac{1 \text{ m}}{100 \text{ cm}} = 1$. Similarly, $\frac{100 \text{ cm}}{1 \text{ m}} = 1$, because the numerator equals the denominator. We call such expressions for 1 *conversion factors.* The information necessary for forming a conversion factor is usually found in tables.

As in the case 1 m = 100 cm, there are two conversion factors for each set of data:

$$\frac{1 \text{ m}}{100 \text{ cm}} \quad \text{and} \quad \frac{100 \text{ cm}}{1 \text{ m}}$$

CHOOSING CONVERSION FACTORS

The correct choice for a particular conversion factor is the one in which the old units are in the numerator of the original expression and in the denominator of the conversion factor, or in the denominator of the original expression and in the numerator of the conversion factor. That is, we want the old units to divide (cancel) each other.

EXAMPLE 1

Change 245 cm to metres.

As we saw in the preceding discussion, the two possible conversion factors are

$$\frac{1 \text{ m}}{100 \text{ cm}} \quad \text{and} \quad \frac{100 \text{ cm}}{1 \text{ m}}$$

We choose the conversion factor with centimetres in the *denominator* so that the centimetre units cancel each other.

$$245 \cancel{\text{cm}} \times \frac{1 \text{ m}}{100 \cancel{\text{cm}}} = 2.45 \text{ m}$$

Note: Conversions *within* the metric system involve only moving the decimal point.

EXAMPLE 2

Change 5 m to centimetres.

$$5 \cancel{\text{m}} \times \frac{100 \text{ cm}}{1 \cancel{\text{m}}} = 500 \text{ cm}$$

EXAMPLE 3

Change 29.5 mm to centimetres.

We choose the conversion factor with millimetres in the denominator so that the millimetre units cancel each other.

$$29.5 \cancel{\text{mm}} \times \frac{1 \text{ cm}}{10 \cancel{\text{mm}}} = 2.95 \text{ cm}$$

EXAMPLE 4

Change 0.08 km to centimetres.

First, change to metres and then to centimetres.

$$0.08 \cancel{\text{km}} \times \frac{1000 \text{ m}}{1 \cancel{\text{km}}} = 80 \text{ m}$$

$$80 \cancel{\text{m}} \times \frac{100 \text{ cm}}{1 \cancel{\text{m}}} = 8000 \text{ cm}$$

Exercises B.2

Which unit is longer?

1. 1 metre or 1 centimetre.

2. 1 metre or 1 millimetre.

3. 1 metre or 1 kilometre.

4. 1 centimetre or 1 millimetre.

5. 1 centimetre or 1 kilometre.

6. 1 millimetre or 1 kilometre.

Which metric unit (km, m, cm, or mm) should you use to measure the following?

7. Length of a pipe wrench.

8. Thickness of a saw blade.

9. Height of a house.

10. Distance around an automobile racing track.

11. Diameter of a hypodermic needle.

12. Width of a table.

13. Distance between Boston and New York.

14. Length of a hurdle race.

15. Thread size on a spark plug.

16. Width of a house lot.

Fill in each blank with the most reasonable metric unit (km, m, cm, or mm).

17. Your car is about 6 ____ long.

18. Your pencil is about 20 ____ long.

19. The distance between New York and San Francisco is about 4000 ____.

20. Your pencil is about 7 ____ thick.

21. The ceiling in my bedroom is about 240 ____ high.

22. The length of a football field is about 90 ____.

23. A jet plane usually flies about 9 ____ high.

24. A standard size film for cameras is 35 ____.

25. The diameter of my car tire is about 60 ____.

26. The zipper on my jacket is about 70 ____ long.

27. Maria drives 8 ____ to college each day.

28. Bill, our basketball center, is 203 ____ tall.

29. The width of your hand is about 80 ____.

30. A hand saw is about 70 ____ long.

31. A newborn baby is usually about 45 ____ long.

32. The standard metric piece of plywood is 1200 ____ wide and 2400 ____ long.

Fill in each blank.

33. 1 km = ____ m
34. 1 mm = ____ m
35. 1 m = ____ cm
36. 1 m = ____ hm
37. 1 dm = ____ m
38. 1 dam = ____ m
39. 1 m = ____ mm
40. 1 m = ____ dm
41. 1 hm = ____ m
42. 1 cm = ____ m
43. 1 m = ____ km
44. 1 m = ____ dam
45. 1 cm = ____ mm
46. Change 230 m to cm.
47. Change 230 m to km.
48. Change 576 mm to cm.
49. Change 198 km to m.
50. Change 25 dm to dam.
51. Change 840 cm to m.
52. Change 85 hm to km.
53. Change 475 cm to mm.
54. Change 8.5 mm to μm.
55. Change 4 m to μm.
56. What is your height in centimetres and metres?

B.3 MASS

The mass of an object is the quantity of material making up the object. One unit of mass in the metric system is the *gram* (g). The gram is defined as the mass of one cubic centimetre (cm^3) of water at its maximum density (see Fig. B.7).

(a) A common paper clip has a mass of about 1 g.

(b) Three aspirin have a mass of about 1 g.

Figure B.7

Since the gram is so small, the **kilogram** (kg) is the basic unit of mass in the metric system. One kilogram is defined as the mass of one cubic decimetre (dm^3) of water at its maximum density.

For very, very small masses, such as medicine dosages, we use the *milligram* (mg). One grain of salt has a mass of about 1 milligram.

The metric ton (1000 kg) is used to measure the mass of very large quantities, such as the load of coal on a barge, a trainload of grain, or a shipload of ore.

EXAMPLE 1

Change 84 kg to grams.

We choose the conversion factor with kilograms in the denominator so that the kilogram units cancel each other.

$$84 \cancel{\text{kg}} \times \frac{1000 \text{ g}}{1 \cancel{\text{kg}}} = 84{,}000 \text{ g}$$

EXAMPLE 2

Change 500 mg to grams.

$$500 \cancel{\text{mg}} \times \frac{1 \text{ g}}{1000 \cancel{\text{mg}}} = 0.5 \text{ g}$$

Exercises B.3

Which unit is larger?

1. 1 gram or 1 centigram
2. 1 gram or 1 milligram
3. 1 gram or 1 kilogram
4. 1 centigram or 1 milligram
5. 1 centigram or 1 kilogram
6. 1 milligram or 1 kilogram

Which metric unit (kg, g, mg, or metric ton) should you use to measure the following?

7. Your mass.
8. An aspirin.
9. A bag of fertilizer.
10. A bar of soap.
11. A trainload of grain.
12. A sewing needle.
13. A small can of corn.
14. A channel catfish.
15. A vitamin capsule.
16. A car.

Fill in each blank with the most reasonable metric unit (kg, g, mg, or metric ton).

17. A newborn baby's mass is about 3 ____.
18. The elevator in a department store has a load limit of 2000 ____.
19. Millie's diet calls for 250 ____ of meat.
20. A 200-car train can carry 11,000 ____ of soybeans.
21. A truckload shipment of copper pipe has a mass of about 900 ____.
22. A carrot has a mass of about 75 ____.
23. A candy recipe calls for 150 ____ of chocolate.
24. By just looking, you can tell your best friend has a mass of 70 ____.
25. A pencil weighs about 10 ____.
26. Postage rates for letters would be based on the ____.
27. A heavyweight boxer weighed in at 93 ____.
28. A nickel has a mass of 5 ____.
29. A spaghetti recipe calls for 1 ____ of ground beef.
30. A spaghetti recipe calls for 150 ____ of tomato paste.
31. A grain elevator shipped 10,000 ____ of wheat last year.
32. A slice of bread has a mass of about 25 ____.
33. I bought a 5-____ bag of potatoes at the store today.
34. My grandmother takes 250-____ capsules for her arthritis.

Fill in each blank.

35. 1 kg = ____ g
36. 1 mg = ____ g
37. 1 g = ____ cg
38. 1 g = ____ hg
39. 1 dg = ____ g
40. 1 dag = ____ g
41. 1 g = ____ mg
42. 1 g = ____ dg
43. 1 hg = ____ g
44. 1 cg = ____ g
45. 1 g = ____ kg
46. 1 g = ____ dag
47. 1 g = ____ μg
48. 1 mg = ____ μg
49. Change 565 g to mg.
50. Change 565 g to kg.
51. Change 850 mg to g.
52. Change 275 kg to g.
53. Change 50 dg to g.
54. Change 485 dag to dg.
55. Change 80 kg to mg.
56. Change 5 metric tons to kg.
57. Change 15 hg to kg.
58. Change 57 μg to g.
59. Change 500 μg to mg.
60. Change 40,000 kg to metric tons.

B.4 VOLUME AND AREA

Volume

A common unit of volume in the metric system is the *litre* (L). The litre is commonly used for liquid volumes (see Fig. B.8).

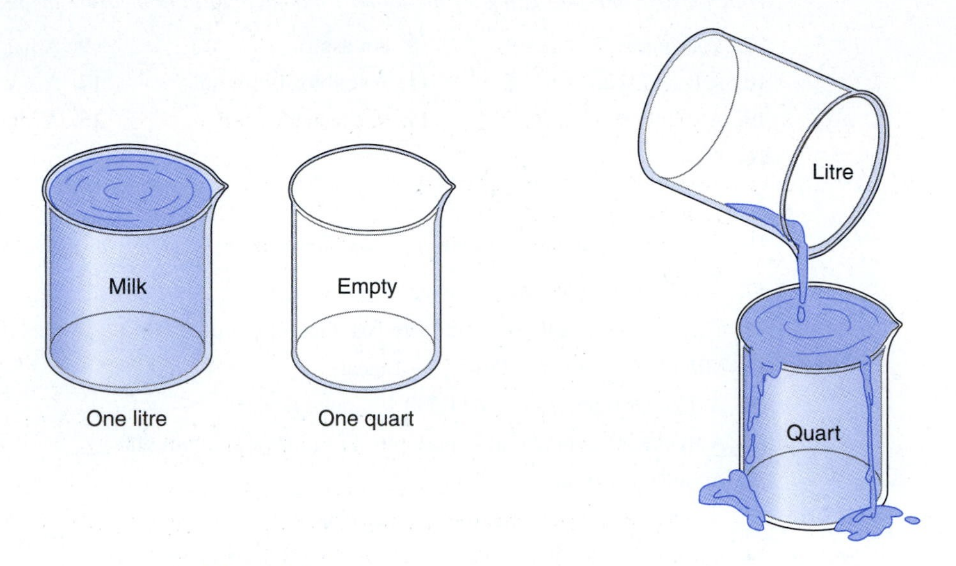

Figure B.8 One litre is a little more than one quart.

The cubic metre (m^3) and the cubic centimetre (cm^3) (see Fig. B.9) are also used to measure volume. The cubic metre is the volume in a cube 1 m on an edge. It is used to measure large volumes. The cubic centimetre is the volume in a cube 1 cm on an edge. For comparison purposes, one teacher's desk and materials could be boxed into 2 m^3 side by side, and the eraser on a pencil could fit into 1 cm^3.

The volume of any solid is the number of cubic units in that solid.

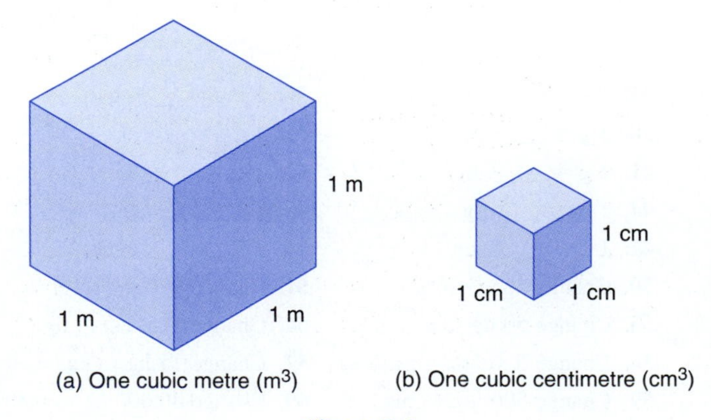

(a) One cubic metre (m^3)   (b) One cubic centimetre (cm^3)

Figure B.9

EXAMPLE 1

Find the volume of a box 4 cm long, 3 cm wide, and 2 cm high.

Each cube in Fig. B.10 shows 1 cm^3. By simply counting the number of cubes (cubic centimetres), we find that the volume of the box is 24 cm^3.

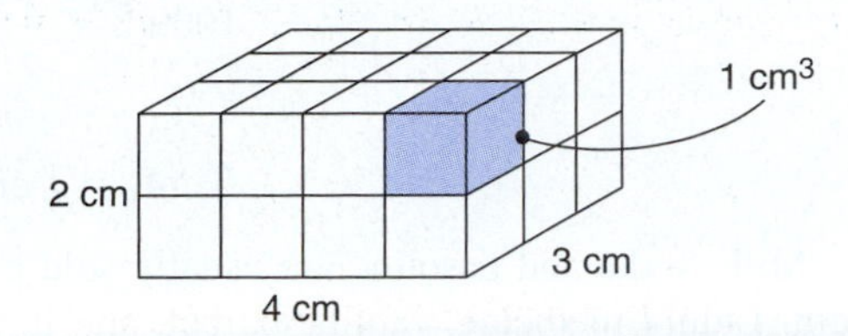

Figure B.10

We can also find the volume of the box by using the formula

$$\begin{aligned} V &= lwh \\ &= (4 \text{ cm})(3 \text{ cm})(2 \text{ cm}) \\ &= 24 \text{ cm}^3 \qquad (\textit{Note: } \text{cm} \times \text{cm} \times \text{cm} = \text{cm}^3) \end{aligned}$$

The relationship between the litre and the cubic centimetre deserves some special mention. The litre is defined as the capacity in a volume of 1 cubic decimetre (dm^3). That is, 1 L of liquid fills a cube 1 dm (10 cm) on an edge (see Fig. B.11).

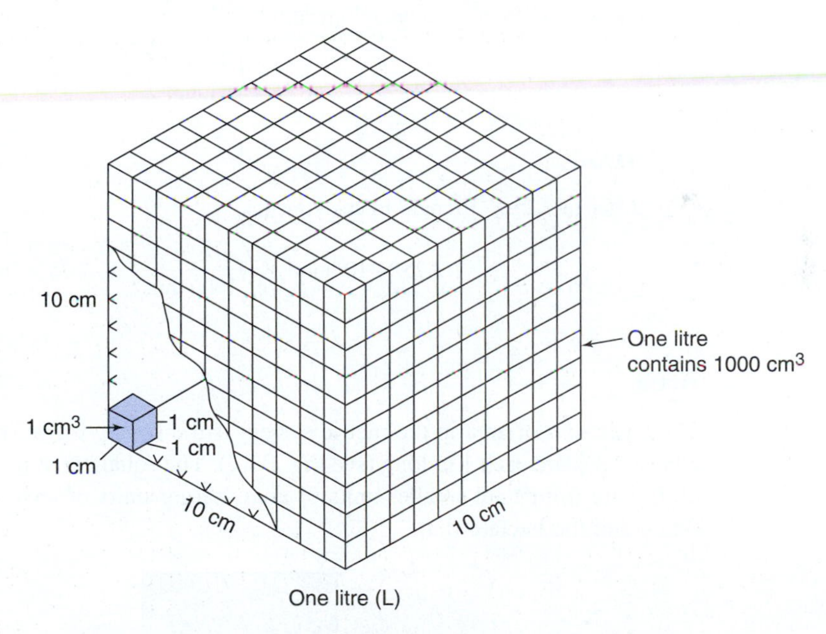

Figure B.11

The volume of this cube can also be found by using the formula

$$\begin{aligned} V &= lwh \\ &= (10 \text{ cm})(10 \text{ cm})(10 \text{ cm}) \\ &= 1000 \text{ cm}^3 \end{aligned}$$

That is,

$$1 \text{ L} = 1000 \text{ cm}^3$$

Then

$$\frac{1}{1000} \text{ L} = 1 \text{ cm}^3$$

or

$$1 \text{ mL} = 1 \text{ cm}^3$$

Milk, soda, and gasoline are usually sold by the litre in countries using the metric system. Liquid medicine, vanilla extract, and lighter fluid are usually sold by the millilitre. Many metric cooking recipes are given in millilitres. Very large quantities of gasoline are sold by the kilolitre (1000 litres).

Recall that the kilogram is defined as the mass of 1 dm^3 of water. Since 1 dm^3 = 1 L, 1 L of water has a mass of 1 kg.

EXAMPLE 2

Change 0.75 L to millilitres.

$$0.75 \cancel{\text{L}} \times \frac{1000 \text{ mL}}{1 \cancel{\text{L}}} = 750 \text{ mL}$$

EXAMPLE 3

Change 0.75 cm^3 to cubic millimetres.

$$0.75 \text{ cm}^3 \times \left(\frac{10 \text{ mm}}{1 \text{ cm}}\right)^3 = 750 \text{ mm}^3$$

EXAMPLE 4

Change 3.25×10^6 mm^3 to cubic metres.

$$3.25 \times 10^6 \text{ mm}^3 \times \left(\frac{1 \text{ m}}{1000 \text{ mm}}\right)^3 = 3.25 \times 10^{-3} \text{ m}^3$$

Area

The basic unit of area in the metric system is the square metre (m^2), the area in a square whose sides are each 1 m long (see Fig. B.12). The square centimetre (cm^2) and the square millimetre (mm^2) are smaller units of area. Larger units of area are the square kilometre (km^2) and the hectare (ha).

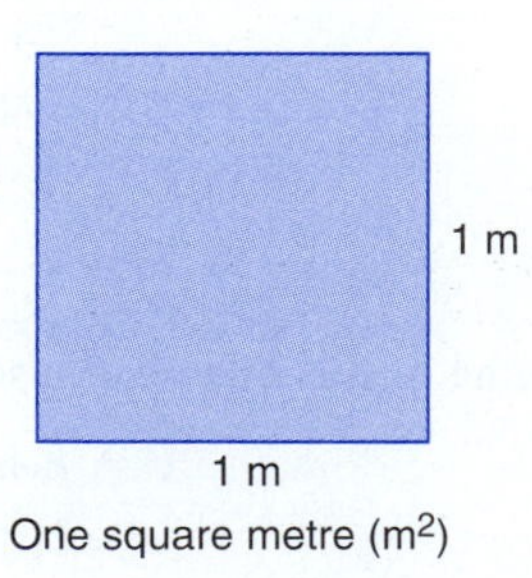

Figure B.12

EXAMPLE 5

Find the area of a rectangle 4 m long and 3 m wide.

Each square in Fig. B.13 represents 1 m^2. By simply counting the number of squares (square metres), we find that the area of the rectangle is 12 m^2.

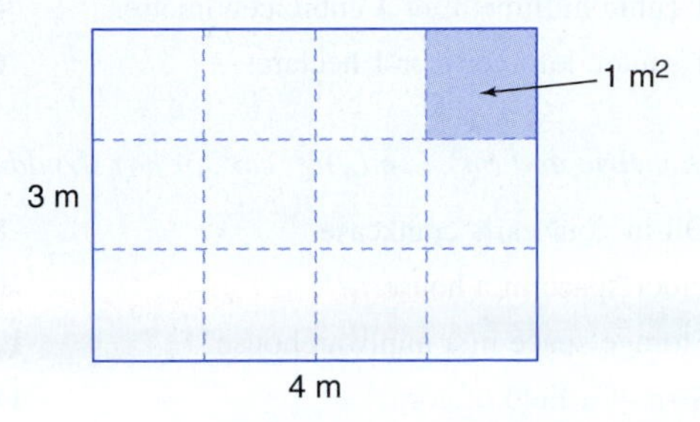

Figure B.13

We can also find the area of the rectangle by using the formula $A = lw$.

$$\begin{aligned} A &= lw \\ &= (4 \text{ m})(3 \text{ m}) \\ &= 12 \text{ m}^2 \qquad (\textit{Note: } \text{m} \times \text{m} = \text{m}^2) \end{aligned}$$

EXAMPLE 6

Change 1500 cm^2 to square metres.

$$1500 \text{ cm}^2 \times \left(\frac{1 \text{ m}}{100 \text{ cm}}\right)^2 = 0.15 \text{ m}^2$$

EXAMPLE 7

Change 0.6 km^2 to square metres.

$$0.6 \text{ km}^2 \times \left(\frac{1000 \text{ m}}{1 \text{ km}}\right)^2 = 600{,}000 \text{ m}^2$$

The hectare (ha) is the fundamental SI unit for land area. An area of 1 ha equals the area of a square 100 m on a side (see Fig. B.14). The hectare is used because it is more convenient to say and use than square hectometre. The metric prefixes are *not* used with the hectare. That is, instead of saying "2 kilohectares," we say "2000 hectares."

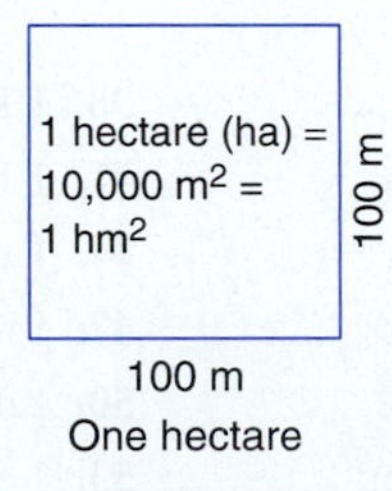

Figure B.14

When converting between metric and U.S. land area units, use the relationship

$$1 \text{ hectare} = 2.47 \text{ acres}$$

Exercises B.4

Which unit is larger?

1. 1 litre or 1 centilitre.
2. 1 millilitre or 1 kilolitre.
3. 1 cubic millimetre or 1 cubic centimetre.
4. 1 cubic centimetre or 1 cubic metre.
5. 1 square kilometre or 1 hectare.
6. 1 square millimetre or 1 square decimetre.

Which metric unit (m^3, L, mL, m^2, cm^2, or ha) should you use to measure the following?

7. Oil in your car's crankcase.
8. Water in a bathtub.
9. Floor space in a house.
10. Cross section of a piston.
11. Storage space in a miniwarehouse.
12. Coffee in an office coffeepot.
13. Size of a field of corn.
14. Page size of a newspaper.
15. A dose of cough syrup.
16. Size of a ranch.
17. Cargo space in a truck.
18. Gasoline in your car's gas tank.
19. Piston displacement of an engine.
20. Paint needed to paint a house.
21. Drops to put in your eyes.
22. Size of a plot of timber.

Fill in the blank with the most reasonable metric unit (m^3, L, mL, m^2, cm^2, or ha).

23. Go to the store and get 4 ____ of root beer for the party.
24. I drank 200 ____ of orange juice for breakfast.
25. Harold bought a 30-____ tarpaulin for his truck.
26. The cross section of a log is 3200 ____.
27. A farmer buys gasoline for a storage tank in bulk. When filled, the storage tank holds 4000 ____.
28. Our city water tower holds 500 ____ of water.
29. Bill planted 60 ____ of soybeans this year.
30. I need some copper tubing with a cross section of 3 ____.
31. A building contractor ordered 15 ____ of concrete for a driveway.
32. I must heat 420 ____ of living space in my house.
33. Our house has 210 ____ of floor space.
34. My little brother mows 5 ____ of lawn each week.
35. Amy is told by her doctor to drink 2 ____ of water each day.
36. My coffee cup holds about 200 ____ of coffee.

Fill in each blank.

37. 1 L = ____ mL
38. 1 kL = ____ L
39. 1 L = ____ daL
40. 1 L = ____ kL
41. 1 L = ____ hL
42. 1 L = ____ dL
43. 1 daL = ____ L
44. 1 mL = ____ L
45. 1 mL = ____ cm^3
46. 1 L = ____ cm^3
47. 1 m^3 = ____ cm^3
48. 1 cm^3 = ____ mL
49. 1 cm^3 = ____ L
50. 1 dm^3 = ____ L
51. 1 m^2 = ____ cm^2
52. 1 km^2 = ____ m^2
53. 1 cm^2 = ____ mm^2
54. 1 mm^2 = ____ m^2
55. 1 dm^2 = ____ m^2
56. 1 ha = ____ m^2
57. 1 km^2 = ____ ha
58. 1 ha = ____ km^2
59. Change 6500 mL to L.
60. Change 0.80 L to mL.
61. Change 1.4 L to mL.
62. Change 8 mL to L.
63. Change 225 cm^3 to mm^3.
64. Change 6 m^3 to cm^3.
65. Change 2 m^3 to mm^3.
66. Change 530 mm^3 to cm^3.

67. Change 175 cm^3 to mL. **68.** Change 145 cm^3 to L. **69.** Change 1 m^3 to L.

70. Change 160 mm^3 to L. **71.** Change 7.5 L to cm^3. **72.** Change 350 L to m^3.

73. Change 5000 mm^2 to cm^2. **74.** Change 1.25 km^2 to m^2.

75. Change 5 m^2 to cm^2. **76.** Change 150 cm^2 to mm^2.

77. Change 4×10^8 m^2 to km^2. **78.** Change 5×10^7 cm^2 to m^2.

79. What is the mass of 750 mL of water? **80.** What is the mass of 1 m^3 of water?

B.5 TEMPERATURE, TIME, CURRENT, AND POWER

Temperature

The basic SI unit for temperature is the *kelvin* (K), which is used mostly in scientific and engineering work. Everyday temperatures are measured in degrees *Celsius* (°C). In the United States, temperatures are also measured in degrees *Fahrenheit* (°F).

The theoretical lowest temperature is called **absolute zero.** At this temperature, there is no heat in an object, and its molecules have stopped moving. The Kelvin scale begins with absolute zero as its zero point, so it has no negative temperature readings. The units on the Kelvin and Celsius scales are the same size. On the Celsius scale, water freezes at 0° and boils at 100° (see Fig. B.15). Each degree Celsius is $\frac{1}{100}$ of the difference between the boiling temperature and the freezing temperature of water.

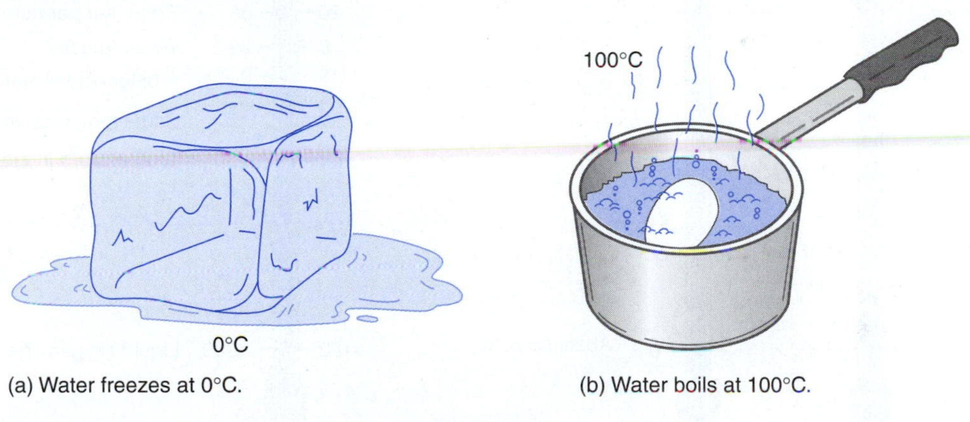

(a) Water freezes at 0°C. (b) Water boils at 100°C.

Figure B.15

Absolute zero on the Celsius scale is about −273°C (see Fig. B.16). Water freezes at 273 K and boils at 373 K. Note that Kelvin temperatures do *not* use the degree symbol (°). In formula form, we have

$$K = C + 273$$

where K = degrees kelvin and C = degrees Celsius.

Figure B.17 shows some approximate temperature readings in degrees Celsius and Fahrenheit with some related activity.

The formulas for converting between degrees Celsius (C) and degrees Fahrenheit (F) are as follows:

$$C = \frac{5}{9}(F - 32)$$

$$F = \frac{9}{5}C + 32$$

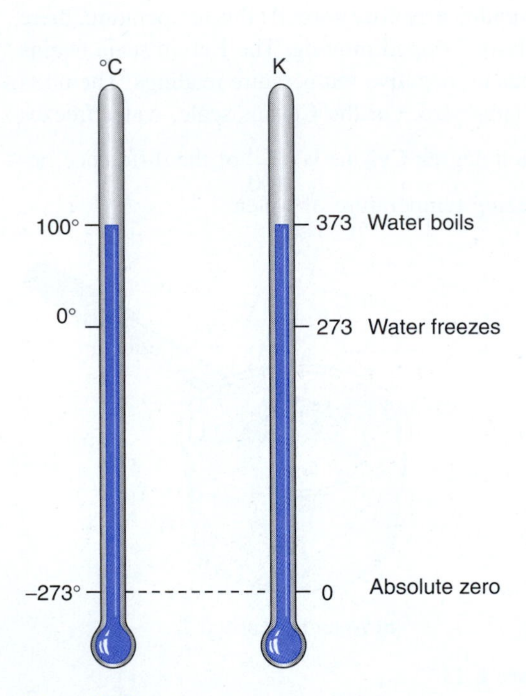

Figure B.16 Comparison of Celsius and Kelvin scales.

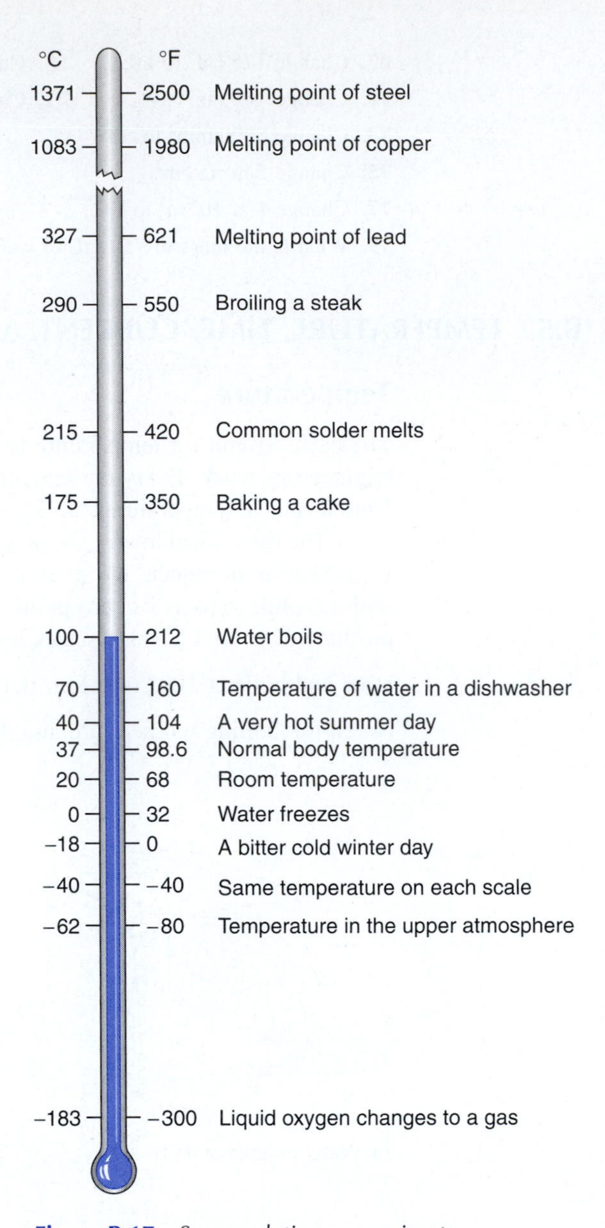

Figure B.17 Some relative approximate temperature readings.

EXAMPLE 1

Change 50°F to degrees Celsius.

$$C = \frac{5}{9}(F - 32)$$

$$C = \frac{5}{9}(50^\circ - 32^\circ)$$

$$= \frac{5}{9}(18^\circ)$$

$$= 10^\circ$$

That is, 50°F = 10°C.

EXAMPLE 2

Change 30°C to degrees Fahrenheit.

$$F = \frac{9}{5}C + 32$$

$$\begin{aligned} F &= \frac{9}{5}(30°) + 32° \\ &= 54° + 32° \\ &= 86° \end{aligned}$$

That is, 30°C = 86°F.

EXAMPLE 3

Change 14°F to degrees Celsius.

$$C = \frac{5}{9}(F - 32)$$

$$\begin{aligned} C &= \frac{5}{9}(14° - 32°) \\ &= \frac{5}{9}(-18°) \\ &= -10° \end{aligned}$$

That is, 14°F = −10°C.

EXAMPLE 4

Change −52°C to degrees Fahrenheit.

$$F = \frac{9}{5}C + 32$$

$$\begin{aligned} F &= \frac{9}{5}(-52°) + 32° \\ &= -93.6° + 32° \\ &= -61.6° \end{aligned}$$

That is, −52°C = −61.6°F.

Time

The basic metric unit of time is the second (s), which is the same in all systems of measurement. We measure longer periods of time in minutes (min), hours (h), and days.

$$\begin{aligned} 1 \text{ day} &= 24 \text{ h} \\ 1 \text{ h} &= 60 \text{ min} \\ 1 \text{ min} &= 60 \text{ s} \end{aligned}$$

EXAMPLE 5

Change 3 h 20 min to seconds.

First,

$$3 \cancel{\text{h}} \times \frac{60 \text{ min}}{1 \cancel{\text{h}}} = 180 \text{ min}$$

So,

$$3 \text{ h } 20 \text{ min} = 180 \text{ min} + 20 \text{ min}$$
$$= 200 \text{ min}$$

Then,

$$200 \cancel{\text{min}} \times \frac{60 \text{ s}}{1 \cancel{\text{min}}} = 12{,}000 \text{ s}$$

Very short periods of time are measured in parts of a second, given with the appropriate metric prefix. These units are commonly used in electronics.

EXAMPLE 6

What is the meaning of each unit?
(a) 1 ms = 1 millisecond = 10^{-3} s and means one-thousandth of a second.
(b) 1 μs = 1 microsecond = 10^{-6} s and means one-millionth of a second.
(c) 1 ns = 1 nanosecond = 10^{-9} s and means one-billionth of a second.
(d) 1 ps = 1 picosecond = 10^{-12} s and means one-trillionth of a second.

EXAMPLE 7

Change 15 ms to seconds.
Since 1 ms = 10^{-3} s,

$$15 \cancel{\text{ms}} \times \frac{10^{-3} \text{ s}}{1 \cancel{\text{ms}}} = 15 \times 10^{-3} \text{ s} = 0.015 \text{ s}$$

EXAMPLE 8

Change 0.000000075 s to nanoseconds.
Since 1 ns = 10^{-9} s,

$$0.000000075 \cancel{\text{s}} \times \frac{1 \text{ ns}}{10^{-9} \cancel{\text{s}}} = 75 \text{ ns}$$

Current

The basic metric unit of electric current is the ampere, A, which is the same as in the U.S. system. The ampere is a fairly large unit of current, so smaller currents are measured in parts of an ampere and are shown by the appropriate metric prefix.

EXAMPLE 9

What is the meaning of each unit?
(a) 1 mA = 1 milliampere = 10^{-3} A and means one-thousandth of an ampere.
(b) 1 μA = 1 microampere = 10^{-6} A and means one-millionth of an ampere.

EXAMPLE 10

Change 150 μA to amperes.
Since 1 μA = 10^{-6} A,

$$150 \cancel{\mu\text{A}} \times \frac{10^{-6} \text{ A}}{1 \cancel{\mu\text{A}}} = 0.00015 \text{ A}$$

EXAMPLE 11

Change 0.0075 A to milliamperes.

Since $1\text{ mA} = 10^{-3}\text{ A}$,

$$0.0075\ \cancel{\text{A}} \times \frac{1\text{ mA}}{10^{-3}\ \cancel{\text{A}}} = 7.5\text{ mA}$$

Power

The common metric unit for both mechanical and electrical power is the watt (W).

EXAMPLE 12

What is the meaning of each unit?

(a) $1\text{ MW} = 1\text{ megawatt} = 10^6\text{ W}$ and means one million watts.
(b) $1\text{ kW} = 1\text{ kilowatt} = 10^3\text{ W}$ and means one thousand watts.
(c) $1\text{ mW} = 1\text{ milliwatt} = 10^{-3}\text{ W}$ and means one-thousandth of a watt.

EXAMPLE 13

Change 5 MW to watts.

Since $1\text{ MW} = 10^6\text{ W}$,

$$5\ \cancel{\text{MW}} \times \frac{10^6\text{ W}}{1\ \cancel{\text{MW}}} = 5 \times 10^6\text{ W} = 5{,}000{,}000\text{ W}$$

EXAMPLE 14

Change 0.275 W to milliwatts.

Since $1\text{ mW} = 10^{-3}\text{ W}$,

$$0.275\ \cancel{\text{W}} \times \frac{1\text{ mW}}{10^{-3}\ \cancel{\text{W}}} = 275\text{ mW}$$

Exercises B.5

Use Fig. B.17 to choose the most reasonable answer for each statement.

1. The freezing temperature of water.
(a) 32°C **(b)** 100°C **(c)** 0°C **(d)** −32°C

2. The boiling temperature of water.
(a) 212°C **(b)** 100°C **(c)** 50°C **(d)** 1000°C

3. Normal body temperature.
(a) 100°C **(b)** 50°C **(c)** 98.6°C **(d)** 37°C

4. The body temperature of a person who is chilled and has a fever.
(a) 100°C **(b)** 101°C **(c)** 39°C **(d)** 37°C

5. The temperature on a cold winter day in Chicago.
(a) 15°C **(b)** 5°C **(c)** 30°C **(d)** −15°C

6. The temperature on a hot summer day in the California desert.
(a) 110°C **(b)** 42°C **(c)** 60°C **(d)** 85°C

7. The temperature on the beach in Florida on a warm winter day.
(a) 85°C **(b)** 30°C **(c)** 20°C **(d)** 70°C

8. The baking temperature of a cherry pie.
(a) 85°C **(b)** 350°C **(c)** 500°C **(d)** 215°C

9. The serving temperature of hot soup.
(a) 20°C **(b)** 115°C **(c)** 50°C **(d)** 150°C

10. The temperature at which freezing rain is most likely to occur.
(a) 32°C **(b)** 25°C **(c)** 0°C **(d)** −15°C

11. The weather forecast calls for a high temperature of 15°C. What should you plan to wear?
(a) A heavy coat. **(b)** A sweater. **(c)** A short-sleeve shirt.

12. The weather forecast calls for a low temperature of 3°C. What should you plan to do?
(a) Sleep with the air conditioner on. **(b)** Sleep with an extra blanket.
(c) Protect your plants from frost. **(d)** Sleep with the windows open.

13. The oven temperature for making pizza.
(a) 100°C **(b)** 80°C **(c)** 450°C **(d)** 225°C

14. The temperature of a lake's water is 28°C. What could you do?
(a) Swim. **(b)** Ice-fish.
(c) Dress warmly and walk along the shore.

15. The setting for the thermostat in your home.
(a) 68°C **(b)** 19°C **(c)** 30°C **(d)** 50°C

Fill in each blank.

16. 59°F = ____ °C **17.** 25°C = ____ °F **18.** 375°C = ____ °F

19. 140°F = ____ °C **20.** 5°F = ____ °C **21.** −12°F = ____ °C

22. −12°C = ____ °F **23.** 550°C = ____ °F **24.** 50°C = ____ K

25. 50 K = ____ °C

26. The temperature in a crowded room is 80°F. What is the Celsius reading?

27. The temperature of an iced-tea drink is 5°C. What is the Fahrenheit reading?

28. The boiling point (temperature at which a liquid changes to a gas) of liquid nitrogen is −196°C. What is the Fahrenheit reading?

29. The melting point (temperature at which a solid changes to a liquid) of mercury is −38°F. What is the Celsius reading?

During the forging and heat-treating of steel, it is important to use the color of heated steel as an indicator of its temperature. Complete the following chart, which shows the color of heat-treated steel and the corresponding approximate temperatures in degrees Celsius and Fahrenheit.

Color	°C	°F
30. White		2200
31. Yellow	1100	
32. Orange		1725
33. Cherry red	718	
34. Dark red	635	
35. Faint red		900
36. Pale blue	310	

Fill in each blank.

37. The basic metric unit of time is the ____. Its abbreviation is ____.

38. The basic metric unit of electric current is the ____. Its abbreviation is ____.

39. The common metric unit of power is the ____. Its abbreviation is ____.

Which is larger?

40. 1 watt or 1 milliwatt

41. 1 milliampere or 1 microampere

42. 1 millisecond or 1 nanosecond

43. 1 megawatt or 1 milliwatt

44. 1 ps or 1 μs

45. 1 mA or 1 A

46. 1 mW or 1 μW

Write the abbreviation for each unit.

47. 2.7 microamperes

48. 3.6 microseconds

49. 9.5 kilowatts

50. 150 milliamperes

51. 15 nanoseconds

52. 12 amperes

53. 135 megawatts

54. 75 picoseconds

Fill in each blank.

55. 1 kW = ____ W

56. 1 A = ____ μA

57. 1 s = ____ ns

58. 1 mA = ____ A

59. 1 MW = ____ W

60. 1 μs = ____ s

61. 1 mA = ____ μA

62. 1 ns = ____ ps

63. 1 MW = ____ μW

64. 1 μs = ____ ns

65. Change 6 A to mA.

66. Change 2500 W to kW.

67. Change 42 mW to μW.

68. Change 245 μs to s.

69. Change 7800 mA to A.

70. Change 1 h 25 min to min.

71. Change 4 h 25 min 15 s to s.

72. Change 7×10^6 s to h.

73. Change 3 s to ns.

74. Change 6×10^{10} μW to W.

75. Change 4×10^{10} μW to MW.

76. Change 1 h to ps.

B.6 OTHER CONVERSIONS

One of the most basic and most useful concepts that you can learn is often called **unit analysis.** Unit analysis can be divided into three areas: (1) converting from one set of units to another, as we saw in the previous section; (2) determining and simplifying the units of a physical quantity obtained from the substitution of data into a formula, as in Section 1.12; and (3) analyzing the derived units in terms of the basic units or other derived units, as introduced in the SI metric system and basic in physics and most technical courses.

Conversion factors can also be used to change units within the U.S. system. The information for forming the conversion factors can be found in the tables in Appendix A.

EXAMPLE 1

(a) Change 8 ft to inches.

$$8 \cancel{\text{ft}} \times \frac{12 \text{ in.}}{1 \cancel{\text{ft}}} = 96 \text{ in.} \qquad \left(\textit{Note: } \frac{12 \text{ in.}}{1 \text{ ft}} = 1 \right)$$

(b) Change 36 ft to yards.

$$36 \cancel{\text{ft}} \times \frac{1 \text{ yd}}{3 \cancel{\text{ft}}} = 12 \text{ yd}$$

(c) Change 2880 in^2 to square feet.

$$2880 \cancel{\text{in}^2} \times \frac{1 \text{ ft}^2}{144 \cancel{\text{in}^2}} = 20 \text{ ft}^2$$

Converting from U.S. to metric or from metric to U.S. units also involves a conversion factor in which the numerator equals the denominator. Three-significant-digit accuracy is sufficient.

EXAMPLE 2

(a) Change 18 lb to kilograms.

$$18\ \cancel{\text{lb}} \times \frac{1\ \text{kg}}{2.20\ \cancel{\text{lb}}} = 8.18\ \text{kg} \qquad \left(\textit{Note}: \frac{1\ \text{kg}}{2.20\ \text{lb}} = 1\right)$$

(b) Change 315 km to miles.

$$315\ \cancel{\text{km}} \times \frac{0.621\ \text{mi}}{1\ \cancel{\text{km}}} = 196\ \text{mi}$$

(c) Change 45 gal to litres.

$$45\ \cancel{\text{gal}} \times \frac{3.79\ \text{L}}{1\ \cancel{\text{gal}}} = 171\ \text{L}$$

EXAMPLE 3

Change 90 km/h to metres per second.

This example involves two conversions: kilometres to metres and hours to seconds. To convert km to m (1 km = 1000 m), we have two possible conversion factors:

$$\frac{1\ \text{km}}{1000\ \text{m}} \quad \text{and} \quad \frac{1000\ \text{m}}{1\ \text{km}}$$

We choose the conversion factor with km in the *denominator* so that the km units cancel each other.

To convert h to s (1 h = 3600 s), we again have two possible conversion factors:

$$\frac{1\ \text{h}}{3600\ \text{s}} \quad \text{and} \quad \frac{3600\ \text{s}}{1\ \text{h}}$$

We choose the conversion factor with h in the *numerator* so that the h units cancel each other.

$$90\ \frac{\cancel{\text{km}}}{\cancel{\text{h}}} \times \frac{1000\ \text{m}}{1\ \cancel{\text{km}}} \times \frac{1\ \cancel{\text{h}}}{3600\ \text{s}} = 25\ \text{m/s}$$

EXAMPLE 4

Change 60 mi/h to ft/s.

$$60\ \frac{\cancel{\text{mi}}}{\cancel{\text{h}}} \times \frac{1\ \cancel{\text{h}}}{60\ \cancel{\text{min}}} \times \frac{1\ \cancel{\text{min}}}{60\ \text{s}} \times \frac{5280\ \text{ft}}{1\ \cancel{\text{mi}}} = 88\ \text{ft/s}$$

or

$$60\ \frac{\cancel{\text{mi}}}{\cancel{\text{h}}} \times \frac{1\ \cancel{\text{h}}}{3600\ \text{s}} \times \frac{5280\ \text{ft}}{1\ \cancel{\text{mi}}} = 88\ \text{ft/s}$$

EXAMPLE 5

The weight density of copper is $555\ \text{lb/ft}^3$. Find its density in newtons per cubic metre.

$$555\ \frac{\cancel{\text{lb}}}{\cancel{\text{ft}^3}} \times \frac{4.45\ \text{N}}{1\ \cancel{\text{lb}}} \times \frac{35.3\ \cancel{\text{ft}^3}}{1\ \text{m}^3} = 87{,}200\ \frac{\text{N}}{\text{m}^3}$$

Exercises B.6

Round each result to three significant digits when necessary.

1. Change 38,000 ft to **(a)** mi, **(b)** km, and **(c)** yd.

2. Change 1290 lb to **(a)** oz, **(b)** kg, and **(c)** g.

3. Change 6.7 km to **(a)** m, **(b)** mi, and **(c)** yd.

4. Change 250 kg to **(a)** g, **(b)** lb, and **(c)** mg.

5. Change 168 in. to **(a)** ft, **(b)** cm, **(c)** m, and **(d)** yd.

6. Change 130 gal to **(a)** qt, **(b)** L, and **(c)** kL.

7. Change 250,000 ft^2 to **(a)** yd^2, **(b)** m^2, **(c)** acres, and **(d)** ha.

8. Change 86,000 m^2 to **(a)** cm^2, **(b)** km^2, and **(c)** yd^2.

9. Change 15 ft^3 to **(a)** in^3, **(b)** m^3, and **(c)** L.

10. Change 25 kL to **(a)** L, **(b)** ft^3, **(c)** in^3, and **(d)** m^3.

11. Change 750 mL to **(a)** L, **(b)** pints, and **(c)** qt.

12. Change 150 ha to **(a)** acres, **(b)** m^2, **(c)** km^2, and **(d)** ft^2.

13. Change 20 ft/s to cm/s.

14. Change 55 mi/h to km/h.

15. Change 6×10^{-4} °C/min to °C/h.

16. Change 0.03°C/h to °C/min.

17. Change 0.006 in./h to in./min.

18. Change 0.45 in./min to in./h.

19. Change 30 m/s^2 to ft/s^2.

20. Change 50 ft/s^2 to km/h^2.

21. Change 2500 N/m^2 to lb/in^2.

22. Change 358 lb/ft^3 to g/cm^3.

23. Change 75 g/cm^3 to oz/in^3.

24. Change 25 mg/cm^3 to g/mm^3.

25. Change 560 kcal to Btu.

26. Change 2.5 kW to hp.

27. Change 1600 ft-lb to **(a)** J (joules) and **(b)** kcal.

28. Change 200 kg to slugs.

29. A road sign states that it is 125 km to Chicago. How far is it in miles?

30. A camera uses film that is 35 mm wide. What is its width in inches?

31. The diameter of a bolt is 0.625 in. What is its diameter in millimetres?

32. Change $4\frac{15}{32}$ in. to cm.

33. A tank has 75 gal of fuel. How many litres does it contain?

34. A satellite weighs 275 kg. How many pounds does it weigh?

35. A football field is 100 yd long. **(a)** How long is it in feet? **(b)** How long is it in metres?

36. A microwheel weighs 0.065 oz. What is its mass in milligrams?

37. A hole 0.325 in. in diameter is drilled in a metal plate. What is its diameter in **(a)** centimetres and **(b)** millimetres?

38. What is your weight in kilograms and grams?

39. A mechanic finds that the fuel consumption of an automobile is 30 mi/gal. How many fluid ounces of fuel does it take to drive the automobile 1 mi?

40. How many cubic centimetres of fuel does it take to drive the automobile 1 km in Exercise 39?

APPENDIX B REVIEW

Give the metric prefix for each decimal.

1. 0.01

2. 1000

Give the metric abbreviation for each unit.

3. millilitre

4. microgram

Write the SI unit for each abbreviation.

5. 16 km

6. 250 mA

7. 1.1 hL

8. 18 MW

Which is larger?

9. 1 litre or 1 millilitre

10. 1 kilometre or 1 millimetre

11. 1 kilogram or 1 gram

12. 1 m^3 or 1 L

13. 1 km^2 or 1 ha

14. 1 ps or 1 ns

Fill in each blank.

15. 180 m = ____ km

16. 250 mg = ____ g

17. 5.7 kL = ____ L

18. 1.5 km = ____ m

19. 650 cm^3 = ____ mL

20. 15 μs = ____ ns

21. 15 MW = ____ W

22. 750 cm^2 = ____ mm^2

23. 0.75 m^3 = ____ cm^3

24. 18,000 m^2 = ____ ha

25. 70°F = ____ °C

26. −5°C = ____ °F

27. Water boils at ____ °C.

28. Water freezes at ____ °C.

Choose the most reasonable quantity.

29. A young couple took a short stroll of 85 cm; 1200 m; 35 km; 1600 mm into a park for a picnic.

30. They ate 1 kg; 50 g; 5 g; 75 kg; 150 mg of chicken.

31. They drank 16 L; 7 mL; 70 mL; 1.5 L; 18 kL of lemonade.

32. The man, being of average height, is 67 cm; 5 m; 170 cm; 3.5 m; 0.5 km; 175 mm tall.

33. Linda weighs 50 kg; 150 kg; 175 μg.

34. Bob's new car averages 350 km/L; 12 km/L; 0.25 km/L; 40 km/L.

35. Bob plans to drive no faster than 50 km/h; 80 km/h; 250 km/h; 650 km/h until the car is "broken in."

Fill in each blank.

36. 8850 μA = ____ mA

37. 0.0775 ns = ____ ps

Round each result in Exercises 38 through 42 to three significant digits when necessary.

38. Change 3600 ft to **(a)** yards and **(b)** metres.

39. Change 53.5 kg to **(a)** grams and **(b)** pounds.

40. Change 3600 yd^2 to **(a)** square feet and **(b)** square centimetres.

41. Change 50 km/h to miles per hour.

42. Change 250 kg/m^3 to pounds per cubic foot.

APPENDIX C

Using a Graphing Calculator

This appendix is included to provide faculty with the flexibility of integrating graphing calculators in their classes. Each section explains and illustrates important features of the Texas Instruments TI-83 and TI-84 Plus. Though this appendix was specifically designed to supplement the graphing calculator examples found throughout the text, the material is organized so that an interested student could also study it as a separate chapter.

C.1 INTRODUCTION TO THE TI-83 KEYBOARD

This section provides a guided tour of the keyboard of the TI-83 Plus and TI-84 Plus graphing calculators (including their Silver Editions). In this and the following sections, please have your calculator in front of you and be sure to try out the features as they are discussed.

First, notice that the keys forming the bottom six rows of the keyboard perform the standard functions of a scientific calculator. The thin blue keys that form the very top row allow functions to be defined and their graphs to be drawn (see Section C.3 for details). The second, third, and fourth rows of keys provide access to menus full of advanced features and perform special tasks such as **INS**ert, **DEL**ete, **CLEAR**, and **QUIT** (to leave a menu, an editor, or a graph, and return to the home screen). Also found in these rows are the **2nd** and **ALPHA** shift keys, which give additional, color-coded meanings to almost every key on the calculator.

The **ON** key is in the lower left-hand corner. Note that pressing the (golden yellow) **2nd** key followed by the **ON** key will turn the calculator **OFF**. If the calculator is left unattended (or no buttons are pressed for a couple of minutes), the calculator will shut itself off. No work is lost when the unit is turned off. Just turn the calculator back **ON** and the display will be exactly as you left it. Due to different lighting conditions and battery strengths, the screen contrast needs adjustment from time to time. Press the **2nd** key, then *press and hold* the up (or down) arrow key to darken (or lighten) the screen contrast.

The **ENTER** key in the lower right-hand corner is like the = key on many scientific calculators; it signals the calculator to perform the calculation that you've been typing. Its (shifted) **2nd** meaning, **ENTRY**, gives you access to previously entered formulas, starting with the most recent one. If you continue to press **2nd ENTRY** you can access previous entries up to an overall memory limit of 128 characters. Depending on the length of your formulas, this means that about 10 to 15 of your most recent entries can be retrieved from the calculator's memory to be reused or modified.

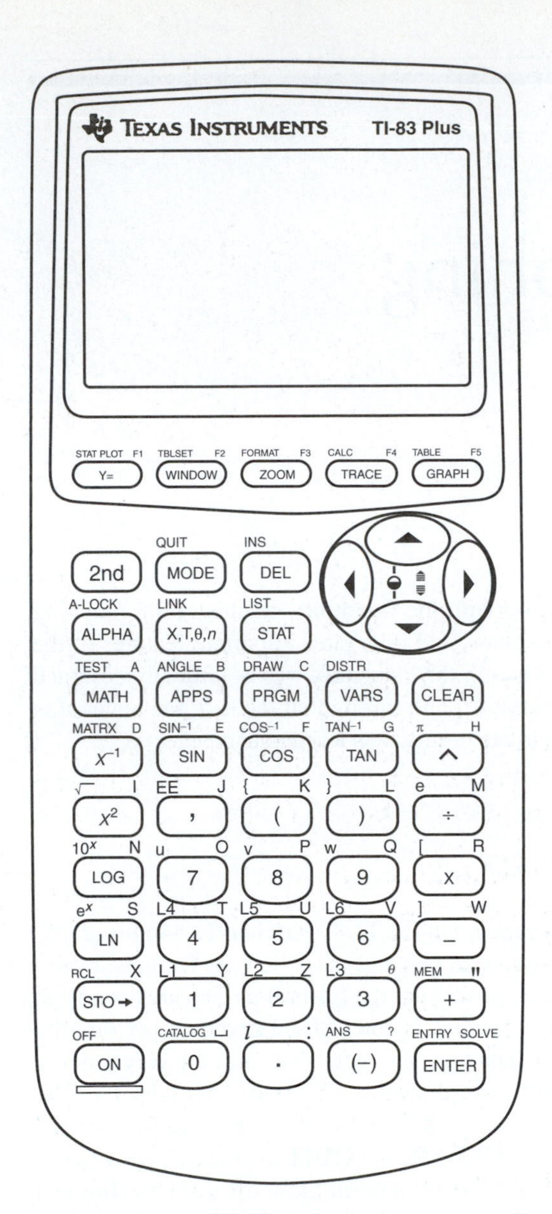

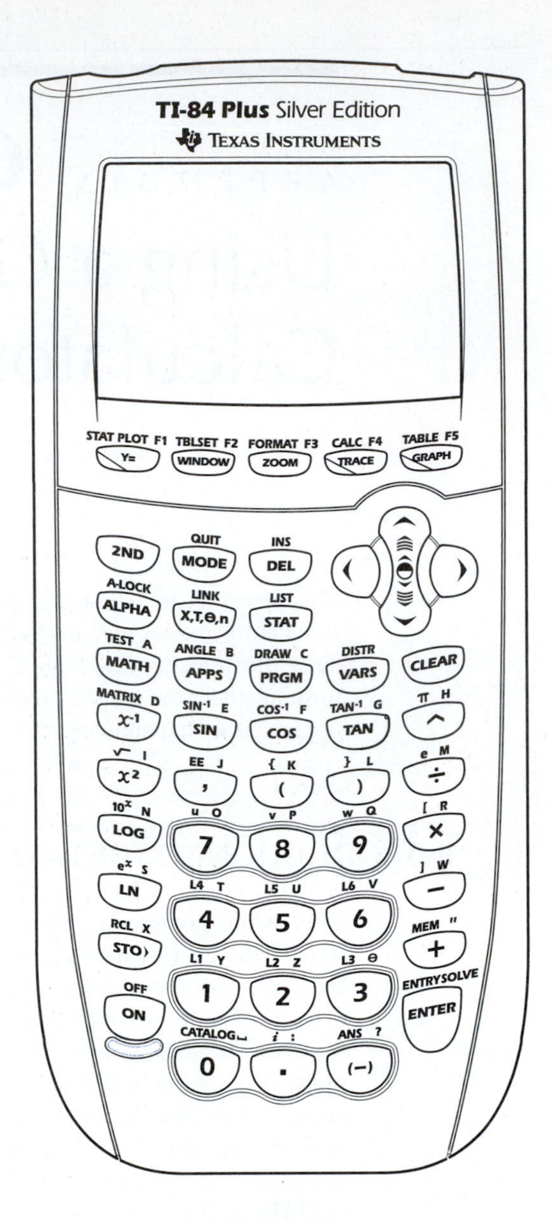

Courtesy of Texas Instruments

Just above **ENTER** is a column of four other blue keys that perform the standard operations of arithmetic. Note, though, that the multiplication key, indicated by an ×, prints an asterisk on the screen and the division key prints a slash on the screen. Just above these four is the ^ key, which indicates that you're raising something to the power that follows; for example, 2^5 would mean 2^5. Moving to the left across that row, you will see the keys for the trigonometric functions: **SIN**, **COS**, and **TAN** (note that their standard setting is radians, but you can specify degrees by using the degree symbol, which is option 1 in the **ANGLE** menu, or the calculator can be set to always think in degrees by specifying that option in the **MODE** menu). Always press the trig key before typing the angle, as in $\cos(\pi)$ or $\sin(30°)$. Notice that the left-hand parenthesis is automatically included when you press any of the trig keys. To the left of these three is a key labeled $\mathbf{x^{-1}}$, which acts as a reciprocal key for ordinary arithmetic. It will also invert a matrix, as in $[A]^{-1}$, which explains why the key isn't labeled **1/x**, as it would be on many scientific calculators. Beneath that key is $\mathbf{x^2}$

(the squaring key), whose shifted **2nd** meaning is square root. Below in that column are keys for logs, whose shifted **2nd** versions give exponential functions. Like the trig keys, the square root, **LOG**, **LN**, and exponential keys also precede their arguments. For example, log(2) will find the common logarithm of 2.

Between **LN** and **ON** is the **STO>** key, which is used to store a number (possibly the result of a calculation) into any of the 27 memory locations whose names are A, B, C, . . . , Z, and θ. First indicate the number or calculation, then press **STO>** (which just prints an arrow on the screen) followed by the (green) **ALPHA** key, then the (green) letter name you want the stored result to have, and finally press **ENTER**. The computation will be performed and the result will be stored in the desired memory location as well as being displayed on the screen. If you have just performed a calculation and now wish that you had stored it, don't worry. Just press **STO>** on the next line followed by **ALPHA** and the letter name you want to give this quantity, then press **ENTER**.

Here are some examples:

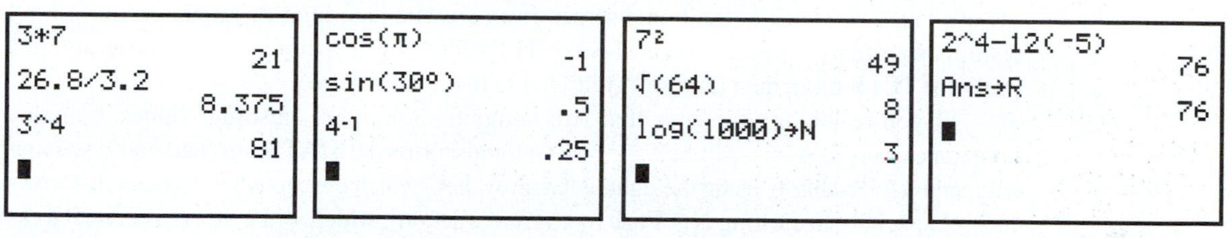

If you watched the last **STO>** example closely, you may have noticed that the calculator prints **Ans** (which stands for "the previous answer") on the screen whenever you don't indicate the first operand on a given line. For example, if you begin a formula with a plus sign, the calculator assumes that you want to add something to the previous result, so it displays "Ans+" instead of just "+." At times, you'll want to refer to the previous result somewhere other than at the beginning of your formula. In that case, press **2nd ANS** (the shifted version of the key to the left of **ENTER**) wherever you want the previous answer to appear in the computation.

The shifted **2nd** meaning of the **STO>** key is **RCL** (recall), as in **RCL Z**, which would display the *contents* of memory location Z at the current cursor position in your formula. It is usually easier to write the letter Z itself (press **ALPHA** followed by **Z**) in formulas instead of the current value that's stored there, so this recall feature isn't the best choice in most computations. However, the **RCL** feature is very useful in creating instant copies of functions (Rcl Y_1) and programs (Rcl prgmSIMPSON) so that newly modified versions don't have to destroy the old ones.

The key that changes the sign of a number is labeled **(-)** and is located just to the left of the **ENTER** key. Don't confuse this white (or gray) key with the dark blue subtraction key! Note also that the calculator consistently views the lack of an indicated operation between two quantities as an intended multiplication.

The parentheses keys are just above the 8 and 9 keys. These are used for all levels of parentheses. Do not be confused by symbols such as { } and [ ], which are the shifted **2nd** versions of these and other nearby keys. Braces { } are used *only* to indicate lists, and brackets [ ] are used *only* for matrices. Once again, these special symbols *cannot* be used to indicate higher levels of parentheses; just nest ordinary parentheses to show several levels of quantification. Also note that the comma key is used only with matrices, lists, multiple-argument functions, and certain commands in the calculator's programming language. Never use commas to separate digits within a number. The number three thousand should always be typed 3000 (not 3,000). The shifted **2nd** meaning of the comma key is **EE** (enter exponent), which is used to enter data in scientific notation; for example, **1.3** followed by **2nd EE (-)8** would be the keystrokes needed to enter 1.3×10^{-8} in a formula. It would be displayed on the screen as 1.3E-8.

The shifted **2nd** versions of the numbers 1 through 9 provide keyboard access to lists and sequences. The shifted **ALPHA** version of the zero key prints a blank space on the display. The shifted **2nd** version of the zero key is **CATALOG,** which provides alphabetical access to every feature of the calculator. Just press the first letter of the desired feature (without pressing **ALPHA**), then scroll from there using the down arrow key. Press **ENTER** when the desired feature is marked by the small arrow. The shifted **2nd** version of the decimal point is ***i***, the imaginary unit (which is often called *j* in electronics applications). This symbol can be used in computations involving imaginary and complex numbers even when **MODE Real** has been selected.

The shifted **2nd** version of the plus sign is **MEM** (the memory management menu), which gives you a chance to erase programs, lists, and anything else stored in memory. Use this menu sparingly (remember, your calculator has a fairly large memory, so you don't usually need to be in a hurry to dispose of things which might prove useful later). If you get into **MEM** by accident, just press **2nd QUIT** to get back to the home screen. **2nd QUIT** always takes you back to the home screen from any menu, editor, or graph, but it will not terminate a running program on a TI-83 or TI-84 Plus. To interrupt a running program, just press the **ON** button, then choose "Quit" in the menu you'll see.

If you're looking for keys that will compute cube roots, absolute values, complex conjugates, permutations, combinations, or factorials, press the **MATH** key, and you'll see four submenus (selectable by using the right or left arrow key) which give you these options and many more. Especially interesting is >**Frac** (convert to fraction), which will convert a decimal to its simplified fractional form, provided that the denominator would be less than 10,000 (otherwise, it just writes the decimal form of the number). Other examples are also included below to give you a better idea of just how many options are available in the **MATH** menu.

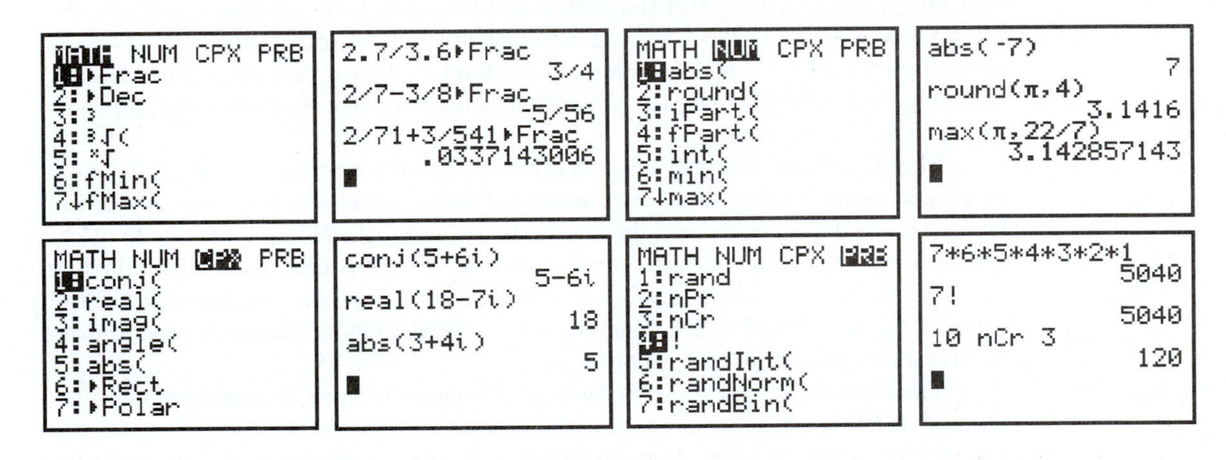

C.2 COMPUTATIONAL EXAMPLES

EXAMPLE 1

Compute the following:

(a) 7×6

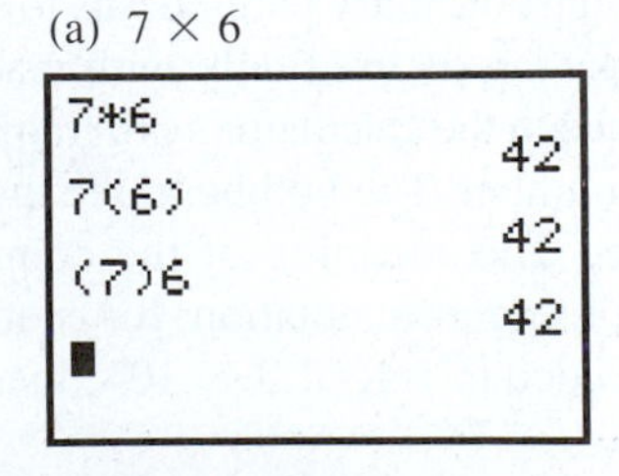

(b) $3 \times 7 + 6\,(3 - 5)$

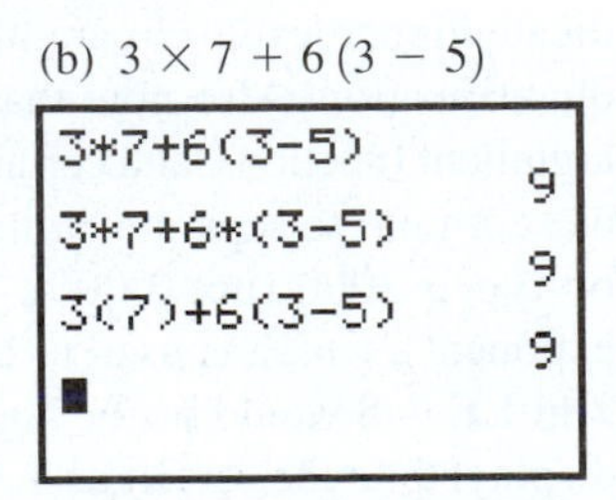

EXAMPLE 2

Compute $8\{3 + 5[2 - 7(8 - 9)]\}$.

```
8(3+5(2-7(8-9)))
                 384
■
```

Note: The calculator uses only ordinary parentheses.

EXAMPLE 3

Express the following as a decimal and as a simplified fraction:

(a) $\frac{105}{100}$ (b) $\frac{3}{8} + \frac{21}{10} - \frac{17}{25}$

```
105/100
                1.05
105/100▶Frac
               21/20
■
```

```
3/8+21/10-17/25
               1.795
3/8+21/10-17/25▶
Frac
             359/200
■
```

Note that a fraction is an indicated division operation and that the division key always prints a diagonal fraction bar line on the screen. The convert to fraction feature is the first item in the **MATH** menu and is accessed by pressing **MATH** then **1** (or **MATH** then **ENTER**) at the end of a formula. Note also that simplified improper fractions are the intended result. *Mixed numbers are not supported.* A decimal result would mean that the answer cannot be written as a simplified fraction with a denominator less than 10,000.

EXAMPLE 4

Compute the following, expressing the answer as a simplified fraction:

(a) $\frac{2^5}{6^2}$ (b) $\frac{5 - (-7)}{-2 - 12}$

```
2^5/6²▶Frac
                 8/9
2^5/6^2▶Frac
                 8/9
■
```

```
(5-(⁻7))/(⁻2-12)
▶Frac
                ⁻6/7
■
```

Squares can be computed by pressing the $\mathbf{x^2}$ key; similarly, a third power can be indicated by pressing **MATH** then **3**. Most other exponents require the use of the ^ key (found between **CLEAR** and the division key). In Part (b), notice the calculator's need for additional parentheses which enclose the numerator and denominator of the fraction. Also notice the difference between the calculator's negative sign (the key below **3**) and its subtraction symbol (the key to the right of **6**).

EXAMPLE 5

Compute the following complex numbers:

(a) $(3 + 4i)(-2 - 5i)$ (b) $\dfrac{7 + 29i}{30 + 10i}$

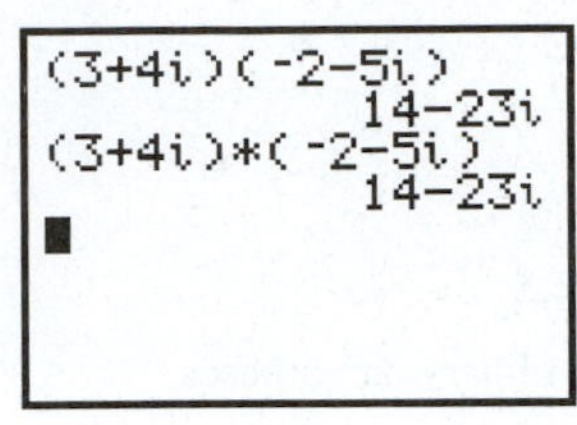


```
(7+29i)/(30+10i)
                .5+.8i
(7+29i)/(30+10i)
▸Frac
              1/2+4/5i
```

Note that the calculator key for the imaginary unit i is the shifted **2nd** version of the decimal point. This imaginary number is often called j in electronics applications.

EXAMPLE 6

Evaluate these expressions:

(a) $x^2 + 5x - 8$ when $x = 4$ (b) $x^2y^3 + 4x - y$ when $x = 2$ and $y = 5$

```
4→X
                     4
X²+5X-8
                    28
4→X:X²+5X-8
                    28
```

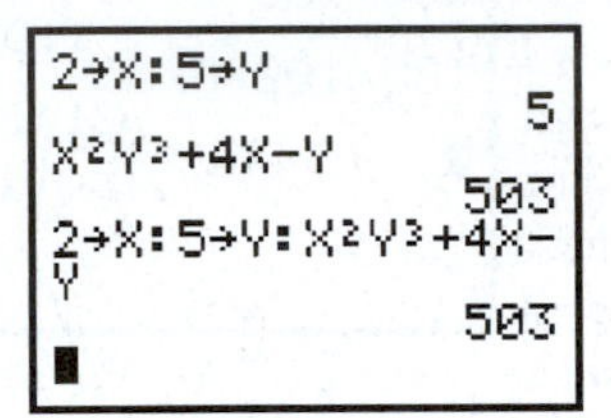

The **STO>** key (just above **ON**) is used to print the arrow symbol on the screen. The letter x can be typed on the screen by pressing the key next to **ALPHA**, labeled **X,T,*θ*,n,** or by pressing **ALPHA**, then **X**. Note that several steps can be performed on one line if the steps are separated by a colon (the shifted **ALPHA** version of the decimal point key). In such cases, all steps are performed in sequence, but only the result of the very last step is displayed on the screen.

EXAMPLE 7

Given that $f(x) = x^4 - 7x + 11$, find $f(3), f(5)$, and $f(-1)$.

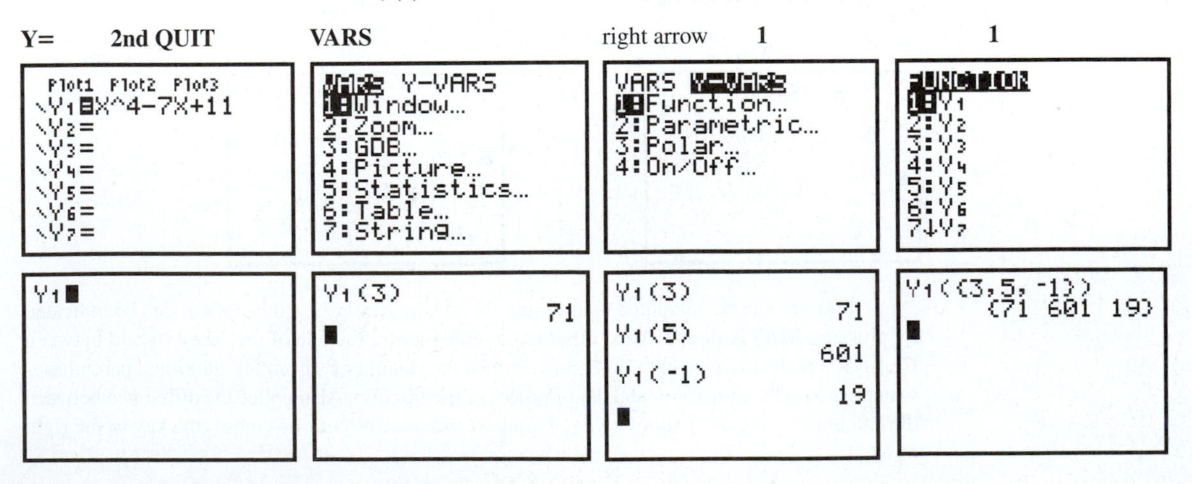

Note that using a stored function requires entering the function in the **Y=** menu, pressing **2nd QUIT** to return to the home screen, then finding the name of that function in the **FUNCTION** submenu under **Y-VARS**. Function evaluation requires the use of parentheses; without parentheses around the argument, multiplication would be assumed. To type the second and third uses of Y_1, press **2nd ENTRY**, then left arrow twice, modifying only the argument from the previous formula. The last screen shows how a list of arguments can be used to calculate a list of function values. Your list entries must be enclosed by braces { } and separated by commas (list entries output by the calculator are separated by spaces instead of commas).

C.3 GRAPHING FEATURES

The thin blue buttons along the top row of the calculator do most of its graphical work. The **Y=** key provides access to the calculator's list of 10 functions (assuming that the calculator is set in **MODE Func**). Pressing the **Y=** key will reveal functions Y_1 through Y_7; the other three functions, Y_8, Y_9, and Y_0, can be seen by pressing the (blue) down arrow key nine times (or just press and hold the down arrow key). These functions are part of the calculator's memory, but the information stored on this screen can be easily edited, overwritten, or **CLEAR**ed. Functions are selected for graphing (turned "on") by highlighting their equal sign. This is done automatically when you type in a new function or modify an old one. In other cases, to change the status of a function (from "off" to "on" or vice versa), you will need to use the arrow keys to position the cursor over the equal sign (making it blink); then press **ENTER**. Functions marked with an ordinary equal sign are stored in memory but will *not* be graphed. To the left of each function name is a symbol indicating how it will be graphed. The normal setting looks like a backslash, \, and simply indicates that the graph will be drawn with a thin line. Other settings include a thicker line, shading above the graph, shading below the graph, two animated settings (one marks the path of motion on the screen; the other just shows the motion without marking its path), and finally an option that graphs with a dotted line. To switch from one option to another, just press the left-arrow key until the cursor is over the option marking (at the far left of that function's name), then press **ENTER** repeatedly until the desired option appears. One warning about the **Y=** menu is that the names Plot1, Plot2, and Plot3 at the top of the function list refer only to the calculator's **STAT**istical **PLOT**s. They have nothing to do with ordinary graphing and should *not* be highlighted if you are just trying to graph some functions.

The **WINDOW** key allows you to *manually* specify the extents of the x- and y-values that will be visible on the calculator's graphing screen (see **ZOOM** for *automatic* ways of doing this). The Xscl and Yscl options specify the meaning of a mark on the x- or y-axis. For example Xscl=5 means that each mark shown on the x-axis will mean an increment of 5 units (Xscl=1 is a common setting for algebraic functions; Xscl=$\pi/2$ is commonly used when graphing trigonometric functions). The last option, Xres, allows you to control how many points will actually be calculated when a graph is drawn. Xres=1 means that an accurate point will be calculated for each pixel on the x-axis (somewhat slow, but very accurate). Xres=2 will calculate only at every other pixel, and so on; Xres=8 only calculates a point for every eighth pixel (this is the fastest setting, but also the least accurate). In the examples that follow, all graphs are shown with Xres=1.

Pressing **2nd FORMAT** (the shifted version of the **ZOOM** key), reveals additional graphing options that allow you to change the way coordinates are displayed (polar instead of rectangular), turn coordinates off completely (inhibiting some **TRACE** features), provide a coordinate grid, hide the axes, label the axes, or inhibit printing expressions which describe the graphs. If you find your graphs looking cluttered or notice that axes, coordinates, or algebraic expressions are missing, the "standard" settings are all in the left-hand column. Like other menus where the options aren't numbered (**MODE** is similar), use your arrow keys to make a new option blink, then select it by pressing **ENTER**.

ZOOM accesses a menu full of *automatic* ways to set the graphical viewing window. **ZStandard** (option 6) is usually a good place to start, but you should consider option 7, **ZTrig**, if you're graphing trigonometric functions. **ZStandard** shows the origin in the exact center of the screen with x- and y-values both ranging from -10 to 10. From here you can **Zoom In** or **Zoom Out** (options 2 and 3), or draw a box around a portion of the graph that you would like magnified to fit the entire screen (option 1, **ZBox**). There is also an option to "square up" your graph so that units along the x-axis are equal in length to units along the y-axis (option 5, **ZSquare**); the *smaller* unit length from the axes of the previous graph will now be used on both axes. This option makes the graph look more like it would on regular graph paper; for example, circles really look like circles. **ZDecimal** and **ZInteger** (options 4 and 8) prepare the screen for **TRACE**s, which will utilize x-coordinates at exact tenths or integer values, respectively. Option 9, **ZoomStat**, makes sure that all of the data in a statistical plot will fit in the viewing window. Option 0, **ZoomFit**, calculates a viewing window using the present x-axis, but adjusts the y-axis so that the function fits neatly within the viewing window. All of these options work by making automatic changes to the **WINDOW** settings. Want to go back to the view you had before? The **MEMORY** submenu (press the right arrow key after pressing **ZOOM**) contains options to go back to your immediately previous view (option 1, **ZPrevious**) or to a window setting you saved a while ago (option 3, **ZoomRcl**). **ZoomSto** (option 2) is the way to save the current window setting for later (note that it can only retain one window setting, so the new information replaces whatever setting you had saved before). Option 4, **SetFactors...**, gives you the chance to control how dramatically your calculator will **Zoom In** or **Zoom Out**. These zoom factors are set by Texas Instruments for a magnification ratio of 4 on each axis. Many people prefer smaller factors, such as 2 on each axis. It is possible to set either factor to any number greater than or equal to 1; they don't need to be whole numbers, and they don't necessarily have to be equal.

The **TRACE** key takes you from any screen or menu to the current graph, displaying the x- and y-coordinates of specific points as you trace along a curve using the left and right arrow keys. Note that in **TRACE**, the up and down arrow keys are used to jump from one curve to another when several curves have been drawn on the same screen. The expression (formula) for the function you are presently tracing is shown in the upper left-hand corner of the screen (or its subscript number is shown in the upper right-hand corner if you have selected the **ExprOff** option from the **FORMAT** menu). If you press **ENTER** while in **TRACE**, the graph will be redrawn with the currently selected point in the exact center of the screen, even if that point is presently outside the current viewing window. This feature is a convenient way to pan up or down to see higher or lower portions of the graph. It is also the easiest way to locate a "lost" graph that doesn't appear anywhere in the current viewing window (just press **TRACE**, then **ENTER**). To pan left or right, just press and hold the left or right arrow key until new portions of the graph come into view. These useful features change the way the graph is centered on the screen without changing its magnification. Note also that these recentering features work *only* in **TRACE**. To exit **TRACE** without disturbing your view of the graph, just press **GRAPH** (or **CLEAR**). To return to the home screen, abandoning both **TRACE** and the graph, just press **2nd QUIT** (or press **CLEAR** twice). Note that using any **ZOOM** feature also causes an exit from **TRACE** (to resume tracing on the new zoomed version, you must press **TRACE** again).

The **GRAPH** key takes the calculator from any screen or menu to the current graph. Note that the calculator is smart enough that it will redraw the curves only if changes have been made to the function list (**Y=**). As previously mentioned, the **GRAPH** key can be used to turn off **TRACE**. You can also hide an unwanted free cursor by pressing **GRAPH**. When you're finished with viewing a graph, press **CLEAR** or **2nd QUIT** to return to the home screen.

C.4 EXAMPLES OF GRAPHING

This appendix is designed to explain graphing calculator features rather than mathematics itself. Accordingly, the example format in this and subsequent sections is different from that found in the body of the text.

EXAMPLE 1

To graph $y = x^2 - 5x$, first press **Y=**, then press **CLEAR** to erase the current formula in Y1 (or use the down arrow key to find a blank function), then press the **X,T,θ,n** key (**ALPHA** then **X** will also work), followed by the **x²** button; now press the (blue) minus sign key, then **5**, followed immediately by the **X,T,θ,n** key (a multiplication sign is not needed). Your screen should look very much like the first one shown below. There is no need to press **ENTER** when you have finished typing a function's formula. To set up a good graphing window, press **ZOOM** and then **6** to choose **ZStandard**. This causes the graph to be immediately drawn on axes that range from −10 to 10. Notice that you did not have to press the **GRAPH** key; the **ZOOM** menu items and the **TRACE** key also activate the graphing screen. *Note:* If you have one or more unwanted graphs drawn on top of this one, go back to your function list (**Y=**) and turn "off" the unwanted functions by placing the cursor over their highlighted equal signs and pressing **ENTER**. After you have turned off the unwanted functions, just press **GRAPH** and you will finally see the last screen below.

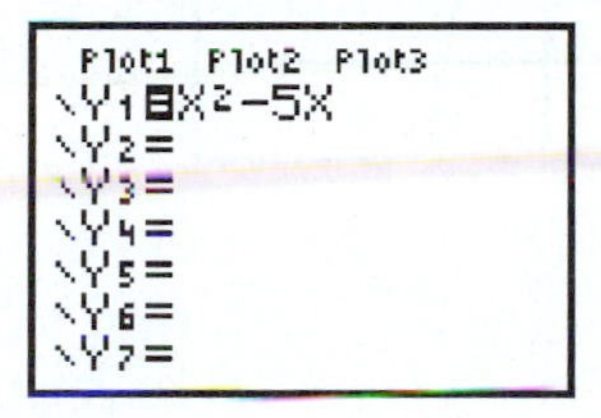

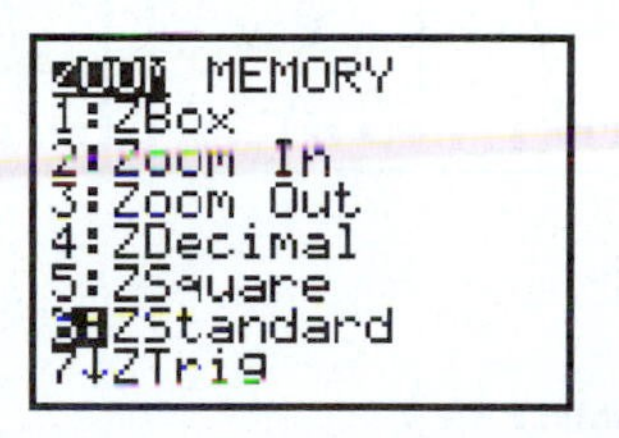

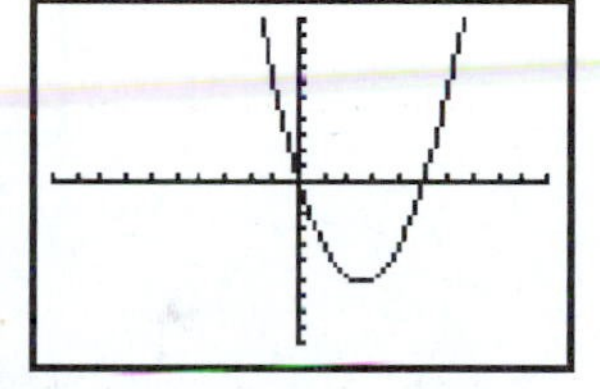

EXAMPLE 2

To modify this function to be $y = -x^2 + 4$, press **Y=**, then insert the negative sign by pressing **2nd INS** followed by the white (or gray) sign change key **(-)**; now press the right arrow key twice to skip over the parts of the formula that are to be preserved. Note that the arrow keys also take you out of insert mode. Now type the plus sign and the 4 (replacing the −5), and finally press **DEL** to delete the extra X at the end of the formula. Press **TRACE** to plot this function. **TRACE** gives the added bonus of a highlighted point, with its coordinates shown at the bottom of the screen. Press the right or left arrow keys to highlight other points on the curve.

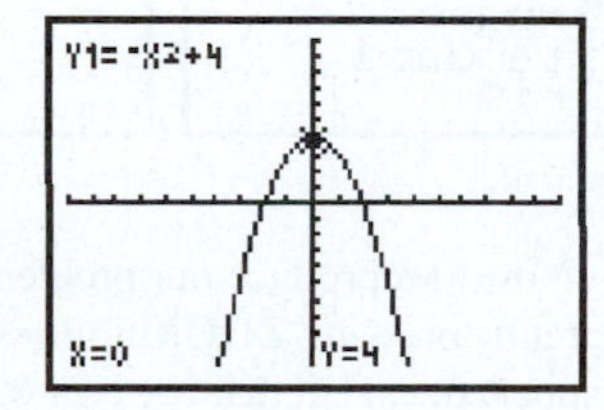

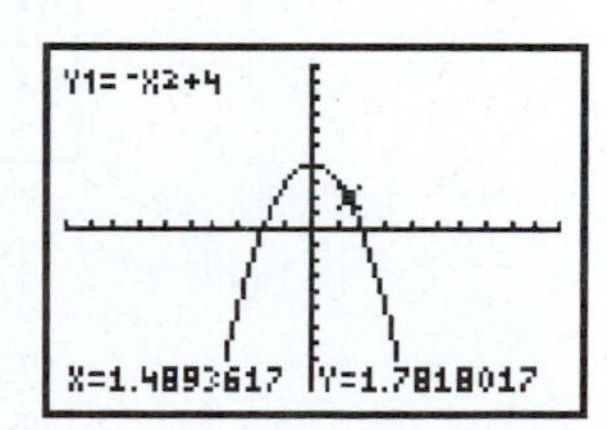

Perhaps $x = 1.4893617$, $y = 1.7818017$ was not a coordinate pair you had expected to investigate. Two special **ZOOM** features (options 4 and 8) can be used to make the **TRACE** option more predictable. Press **ZOOM**, then **4** to select **ZDecimal**; now press **TRACE**.

ZOOM **4** **TRACE** right arrows

ZOOM MEMORY
1:ZBox
2:Zoom In
3:Zoom Out
4:ZDecimal
5:ZSquare
6:ZStandard
7↓ZTrig

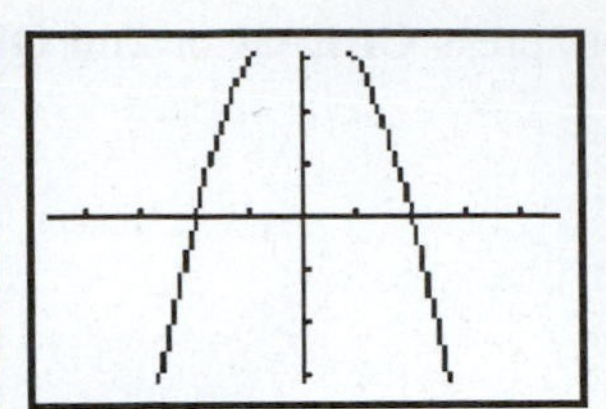

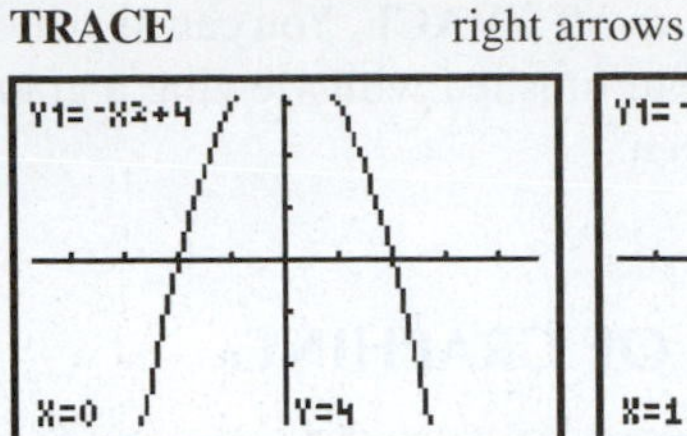

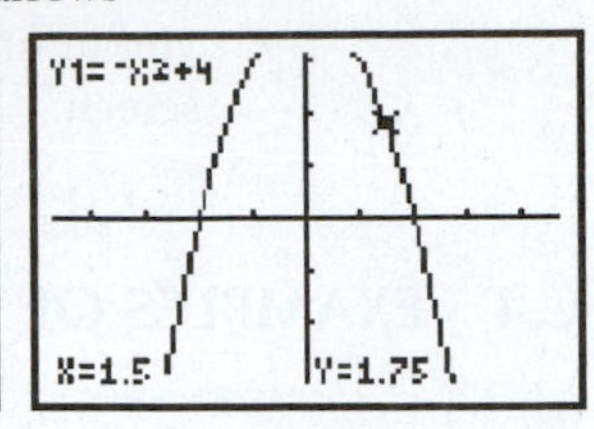

Try pressing the right or left arrow key about 15 times while watching the values at the bottom of the screen. You'll quickly notice that the x-values are now all *exact tenths* (**ZOOM** option 8, **ZInteger**, produces **TRACE**able x-values which are all integers). Another nice thing about **ZDecimal** is that the graph is "square" in the sense that units on the x- and y-axes have the same length. The main disadvantage to **ZDecimal** is that the graphing window is "small," displaying only points with x-values between -4.7 and 4.7 and y-values between -3.1 and 3.1. This disadvantage is apparent on the current graph, which runs off the top of the screen. To demonstrate how this problem can be overcome, **TRACE** the graph to the point $x = 1.5$, $y = 1.75$, and then press **ENTER**. This special feature of **TRACE** causes the graph to be redrawn with the highlighted point in the exact center of the screen (with no change in the magnification of the graph). This is the way to pan up or down from the current viewing window (to pan left or right, see Example 6).

ENTER

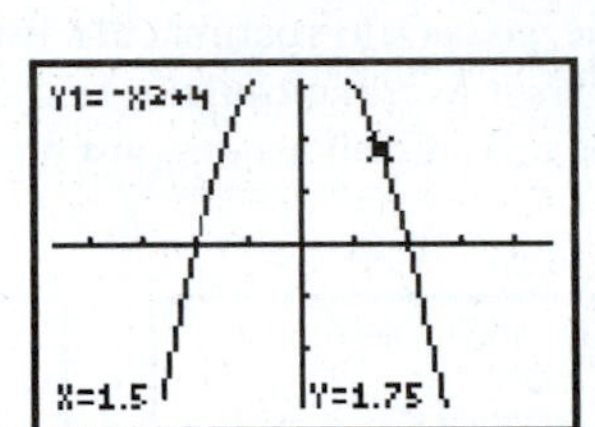

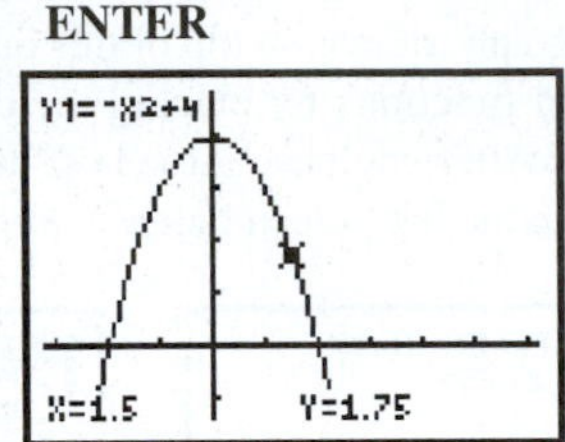

EXAMPLE 3

Another way to deal with the preceding problem is to **Zoom Out**, but first you'll want to set your ZOOM FACTORS to 2 (the factory setting is 4). Press **ZOOM**, then the right arrow key (**MEMORY**), then press **4** (**SetFactors**). To change the factors to 2, just type **2**, press **ENTER**, and then type another **2**.

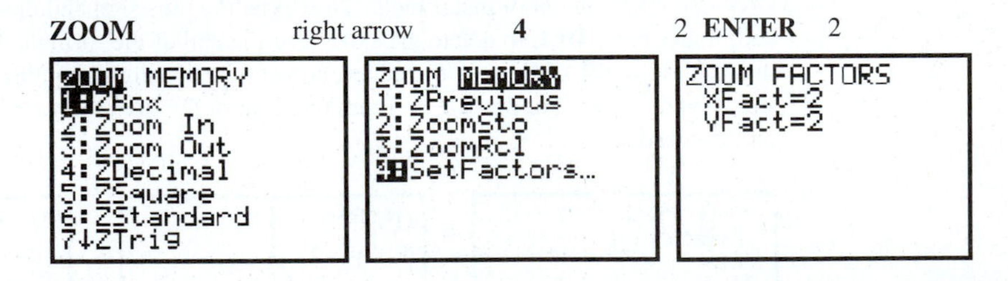

Now to reproduce our problem, press **ZOOM**, then **4** (**ZDecimal**). However, this time correct it by pressing **ZOOM** followed by **3** (**Zoom Out**). At first glance, it looks like nothing has happened, except that X=0 Y=0 is displayed at the bottom of the screen. The calculator is waiting for you to use your arrow keys to locate the point in the current window where you would like the exact center of the new graph to be (then press **ENTER**). Of course, if you like the way the graph is already centered, you will still have to press **ENTER** (you'll just skip pressing the arrow keys).

ZOOM 4 **ZOOM** 3

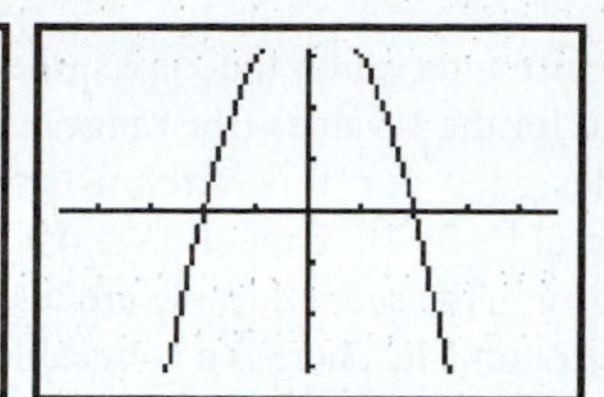

ZOOM MEMORY
1:ZBox
2:Zoom In
3:Zoom Out
4:ZDecimal
5:ZSquare
6:ZStandard
7↓ZTrig

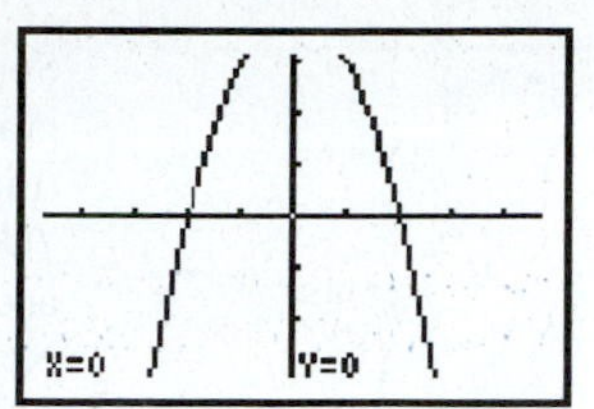

ENTER **TRACE** right arrows

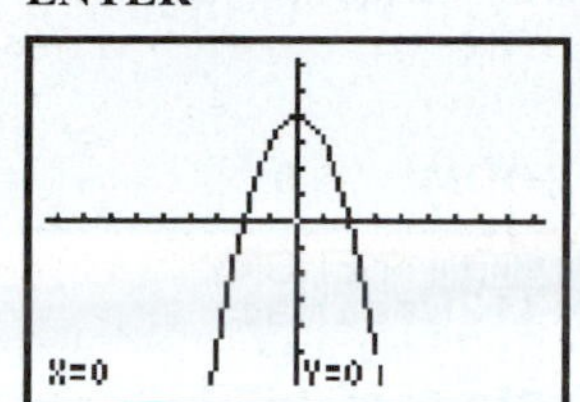

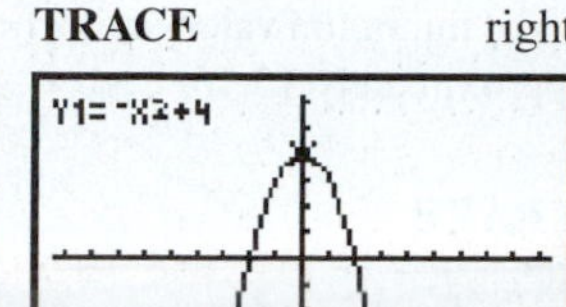

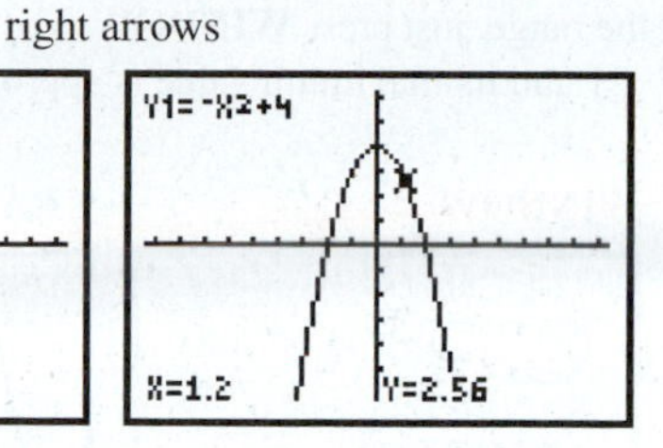

This extra keystroke has proven to be a bit confusing to beginners who think that **Zoom In**, **Zoom Out**, and **ZInteger** should work like the six zoom options (4, 5, 6, 7, 9, and 0) that do their job without pressing **ENTER**. Perhaps more interesting is the fact that you can continue to **Zoom Out** just by pressing **ENTER** again and again (of course, you can also press some arrow keys to recenter between zooms if you wish). Before you experiment with that feature, press **TRACE** and notice, by pressing the left or right arrow key a few times, that the x-values are now changing by .2 (instead of .1) and the graph is still "square." Other popular square window settings can be obtained by repeating this example with both zoom factors set to 2.5 or 5. These give "larger" windows where the x-values change by .25 or .5, respectively, during a **TRACE**.

EXAMPLE 4

The only other zoom option that needs extra keystrokes is **ZBox** (option 1). This is a very powerful option that lets you draw a box around a part of the graph which you would like enlarged to fit the entire screen. After selecting this option, use the arrow keys to locate the position of one corner of the box and press **ENTER**. Now use the arrow keys to locate the *opposite* corner, and press **ENTER** again.

ZOOM 1 left and up arrows **ENTER** right and down arrows

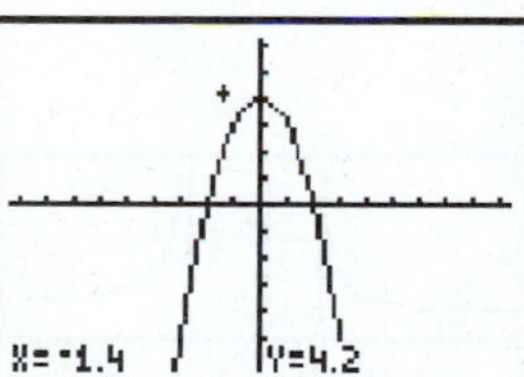

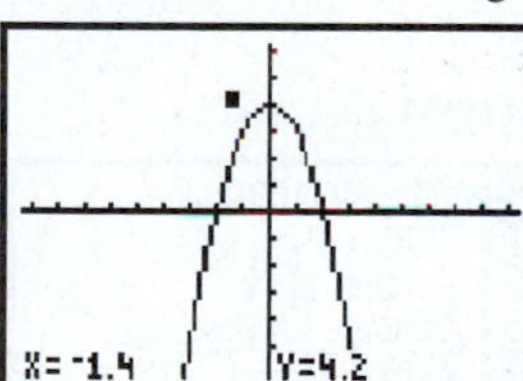

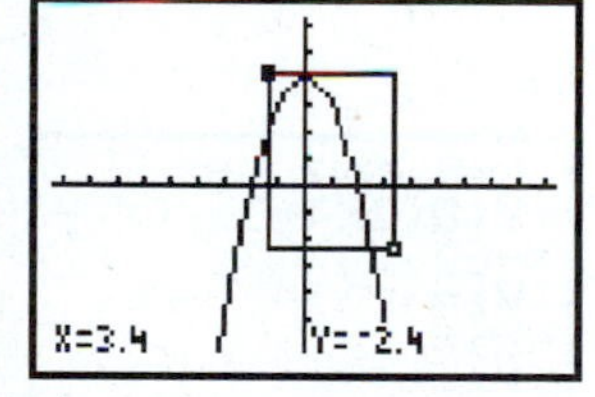

ENTER **GRAPH**

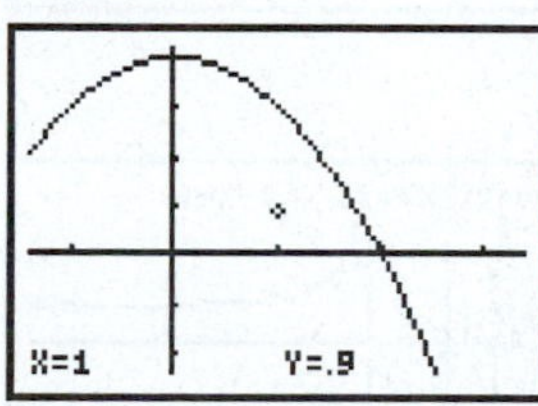

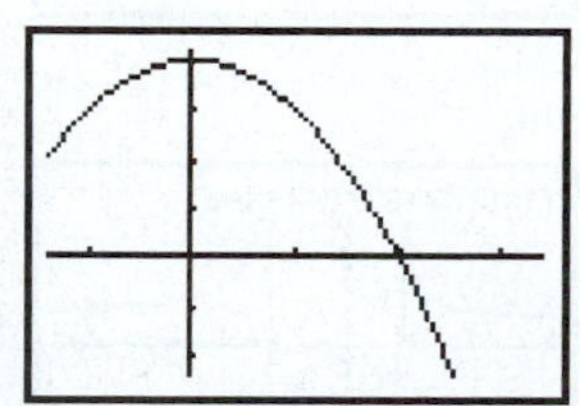

Note that the resulting graph has a free cursor identifying the point in the exact center of the screen. What may not be apparent is that your calculator is ready for you to draw another box if you wish to zoom in closer. To get rid of this free cursor, just press **GRAPH** (or **CLEAR**).

EXAMPLE 5

Sometimes, you will know the precise interval on the x-axis (the domain) that you want for a graph, but a corresponding interval for the y-values (the range) may not be obvious. **ZoomFit** (the 10th **ZOOM** option) is designed for this circumstance. For example, to graph $f(x) = 2x^3 - 8x + 9$ on the interval $[-3, 2]$, manually set the **WINDOW** so that Xmin=–3 and Xmax=2 (the other values shown in the second frame are just leftovers from **ZStandard**). Now press **ZOOM**, then **0** to select **ZoomFit**. There is a noticeable pause while appropriate values of Ymin and Ymax are calculated, then the graph is drawn. To view the values calculated for the range, just press **WINDOW**. The minimum value of this function on the interval $[-3, 2]$ is -21 and its maximum value is approximately 15.16.

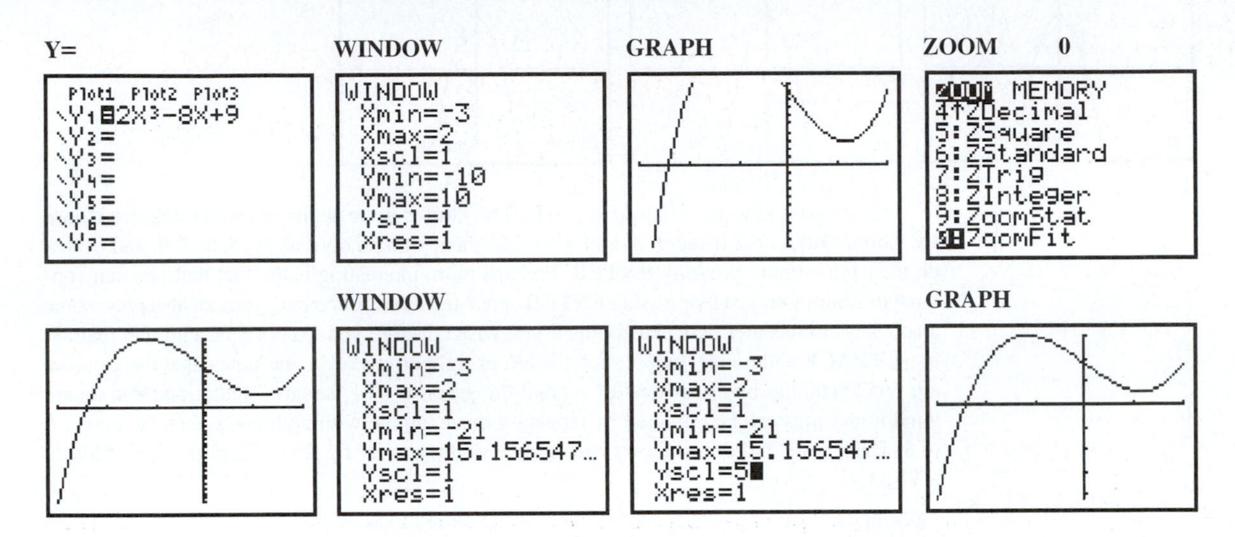

Note that **ZoomFit** changes only Ymin and Ymax. You may also wish to change Yscl.

EXAMPLE 6

Panning to the right is done by tracing a curve off the right-hand edge of the screen:

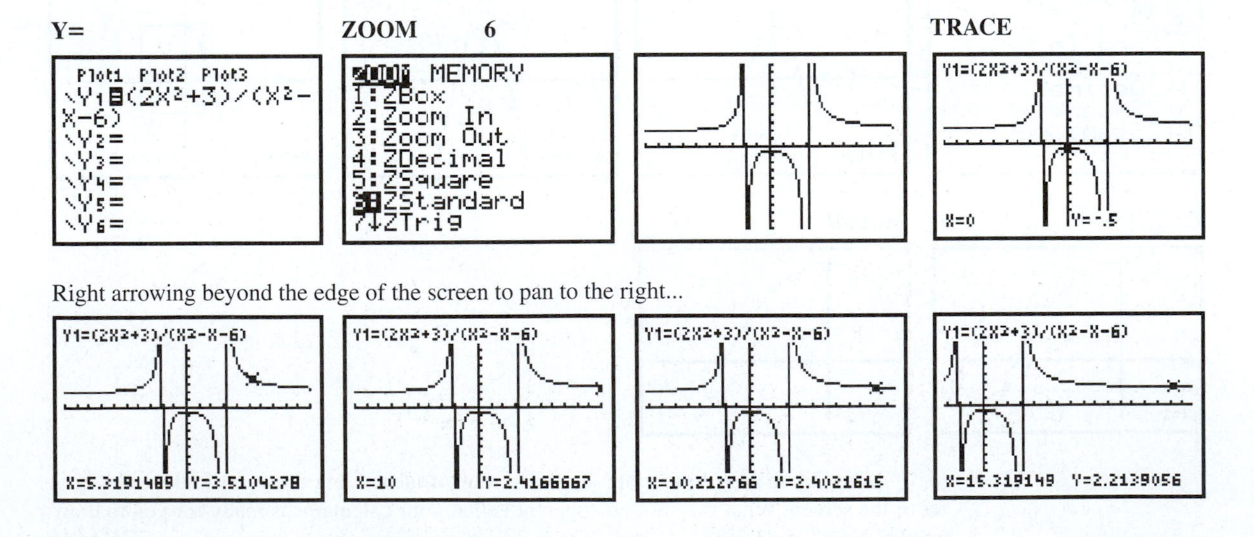

Panning to the left is done similarly. To pan *up or down*, see the last part of Example 2.

EXAMPLE 7

Creating, storing, and retrieving viewing windows:

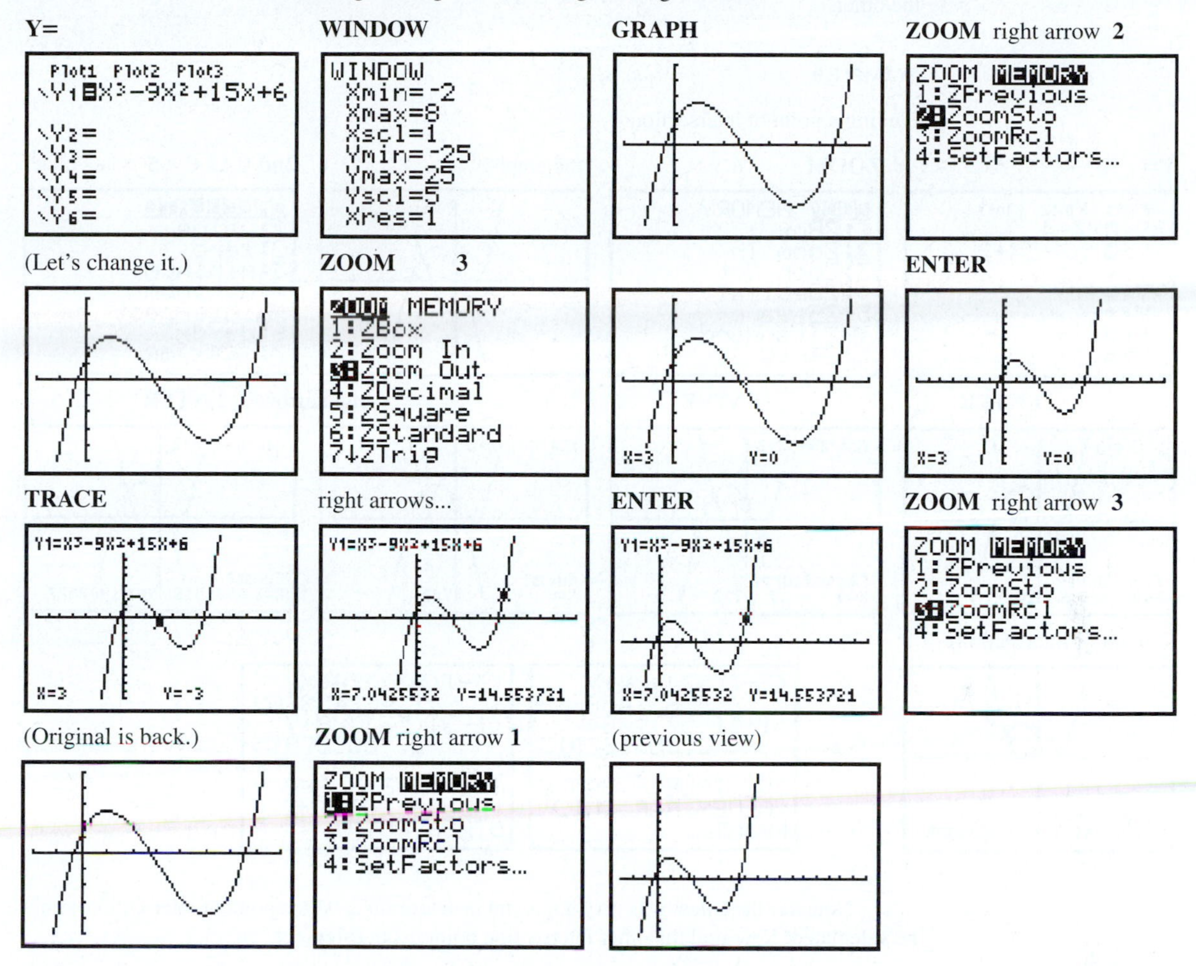

EXAMPLE 8

Graphing and tracing more than one function:

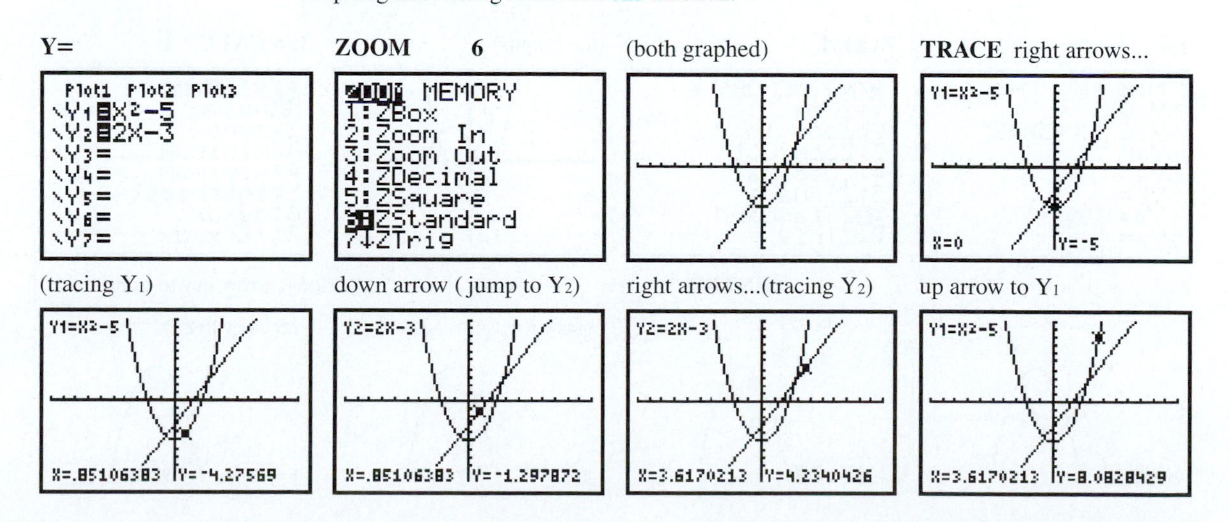

Note that the down arrow increases the subscript of the function being traced and the up arrow decreases the subscript. The result has nothing to do with which graph is above or below the other.

EXAMPLE 9

Finding a point of intersection:

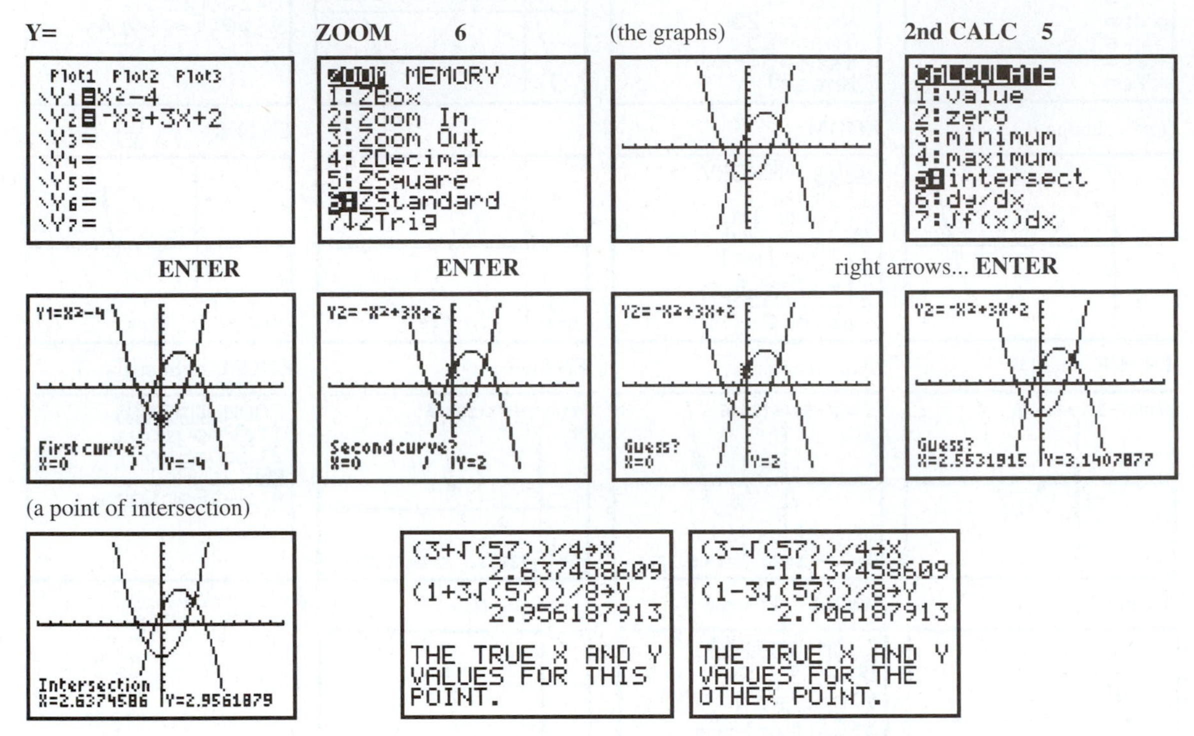

Note that the *Guess* was very important in determining which point of intersection would be calculated. Now find the other intersection point using **intersect**.

EXAMPLE 10

Calculating y-values and locating the resulting points on your graphs:
If you have just completed Example 9, please skip to the fourth frame.

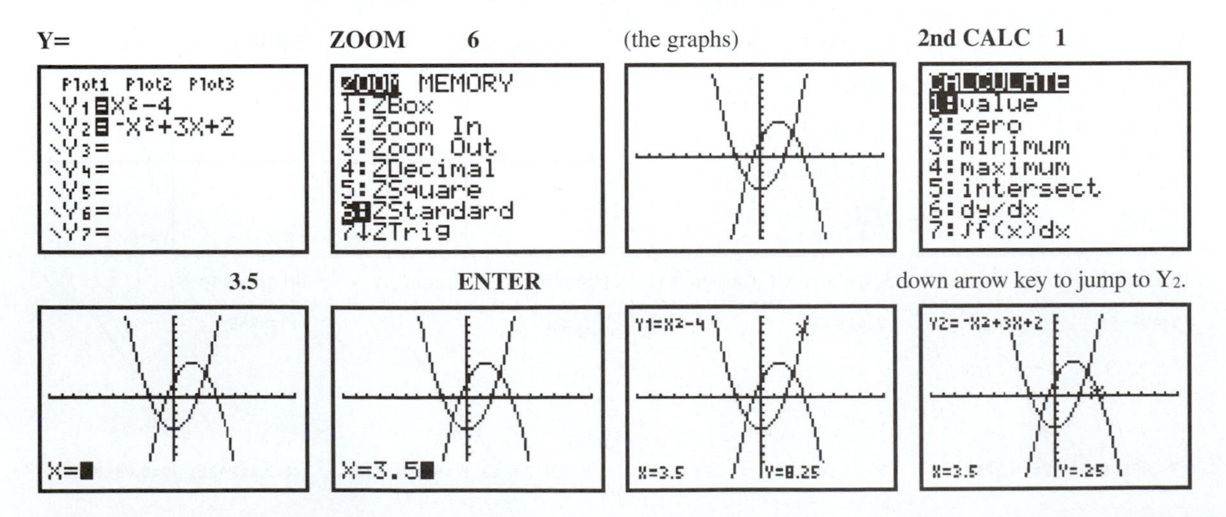

The x-value entered must be in the current viewing **WINDOW** (but the resulting y-value need not be). This x-value can be investigated for any of the functions that are presently graphed by pressing the up or down arrow keys. Note again that the subscript of a function increases as you jump from one curve to another by pressing the down arrow key. The up arrow key decreases this number, and either key can be used to wrap around and start over.

EXAMPLE 11

Another way to calculate a specific value of a function is to press **TRACE**, then type the x-value and press **ENTER**. The main difference between the **CALCULATE value** feature (see Example 10) and this special **TRACE** feature is that **TRACE** does *not* preserve the entered x-value if you jump from one curve to another (you would usually need to retype that x-value). The following frames assume **ZStandard (ZOOM 6) WINDOW** settings.

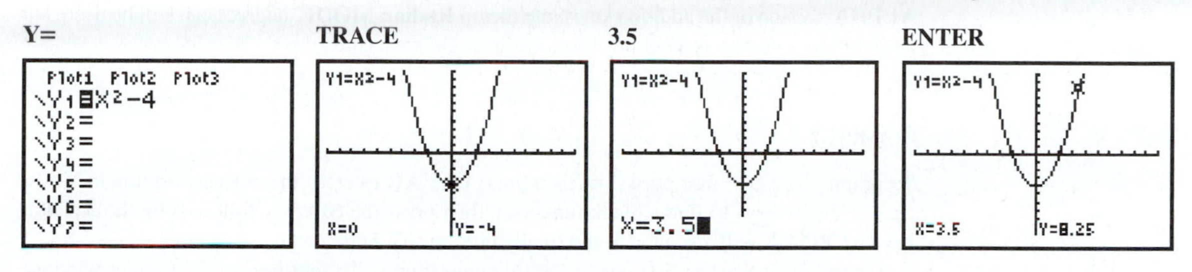

EXAMPLE 12

Using **TRACE** to quickly locate a "lost" graph that doesn't appear anywhere in the current viewing window:

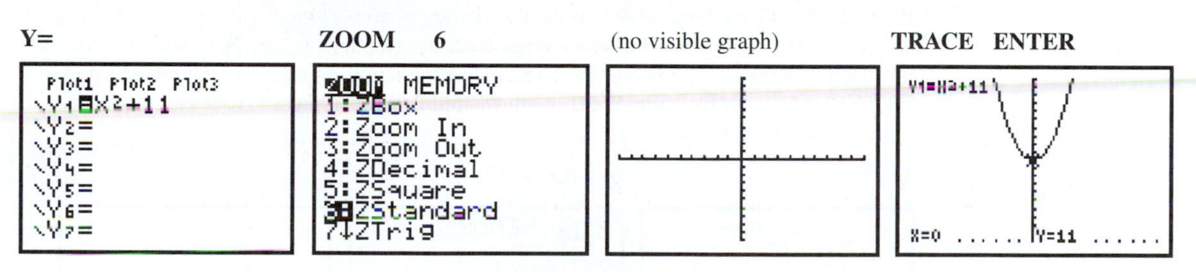

C.5 TRIGONOMETRIC FUNCTIONS AND POLAR COORDINATES

For calculating and graphing trigonometric functions on the TI-83 and TI-84 Plus, the standard default setting is **Radian** mode. To set the calculator to **Degree** mode, press the **MODE** key, arrow down and over to **Degree**, then press **ENTER**.

EXAMPLE 1

Evaluating trigonometric functions using degrees, minutes, and seconds:

```
sin(30)
                .5
sin(29°60')
                .5
sin(29°59'60")
                .5
```

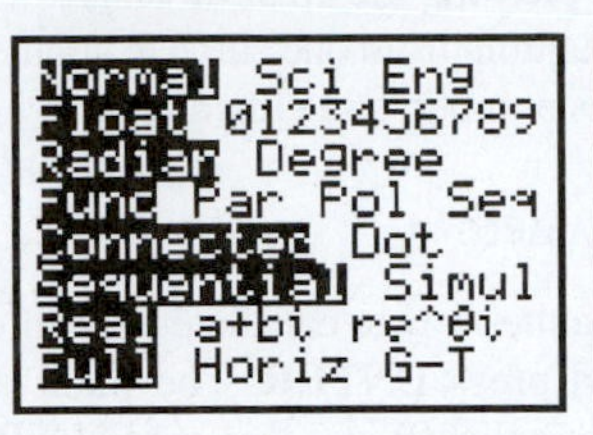

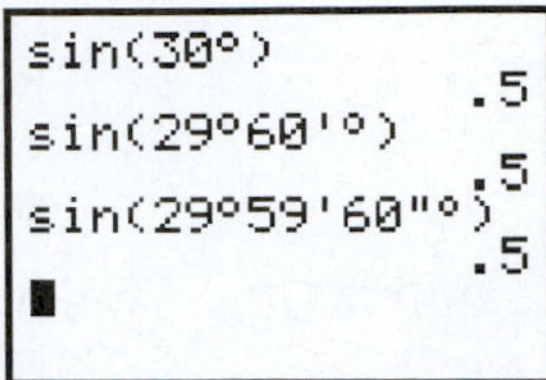

The symbols for degrees and minutes are the first two options in the menu found by pressing **2nd ANGLE** (just to the right of the **MATH** key). The symbol for seconds is the shifted **ALPHA** version of the addition key. Note that in **Radian MODE**, degrees can still be used, but an *additional* degree symbol must follow the angle's measure.

EXAMPLE 2

To graph $y = \sin x$, first press **Y=**, then press **CLEAR** to erase the current formula in Y_1 (or use the arrow keys to find a blank function), then press the **SIN** key followed by the **X,T,θ,n** key (**ALPHA X** will also work), and finally press the right parenthesis key. Your screen should look very much like the first one in the following figure. To set up a good graphing window, press **ZOOM** and then press **7** to choose **ZTrig**. This causes the graph to be immediately drawn on axes that range from roughly -2π to 2π (actually from $-352.5°$ to $352.5°$) in the x-direction and from -4 to 4 in the y-direction. Each mark along the x-axis represents a multiple of $\pi/2$ radians (90°). Notice also that you did not need to press the **GRAPH** key; the **ZOOM** menu items and the **TRACE** key also activate the graphing screen. *Note*: If you have one or more unwanted graphs drawn on top of this one, go back to your function list (**Y=**) and turn "off" the unwanted functions by placing the cursor over their highlighted equal signs and pressing **ENTER**. After you have turned off the unwanted functions, press **GRAPH** and you will finally see the last frame below.

Y=

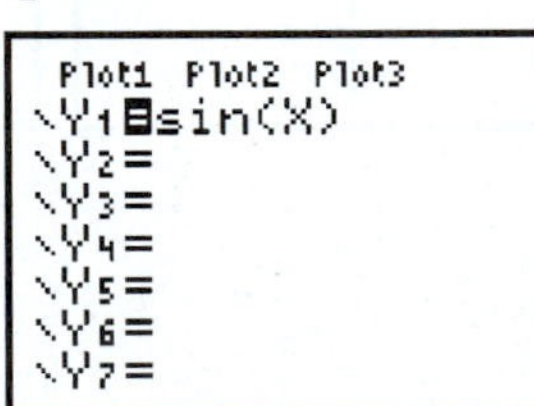

ZOOM 7

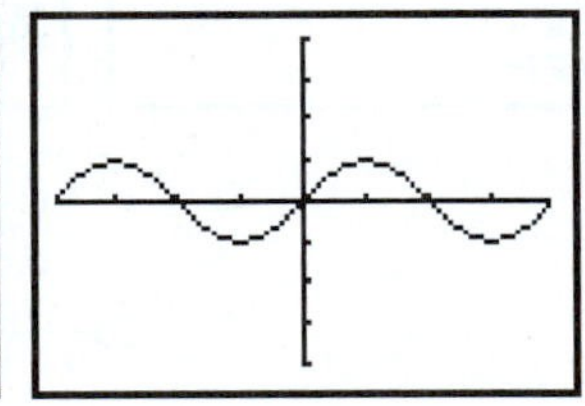

EXAMPLE 3

To modify this function to be $y = -3 \sin 2x$, press **Y=**, then insert the -3 by pressing **2nd INS** followed by the white (or gray) sign change key **(-)** then the number **3**. Now press the right arrow key to skip over the part of the formula that's OK. Note that the arrow keys take you out of insert mode. Now type **2nd INS** then the **2**. Press **ZOOM 7**, then **TRACE** to plot this function. **TRACE** gives the added bonus of a highlighted point with its coordinates shown at the bottom of the screen. Press the right or left arrow keys to highlight other points on the curve. The highlighted coordinate pair in the 4th frame that follows is $x = 5\pi/12$, $y = -1.5$. **ZTrig** allows **TRACE** to display all points whose x-values are multiples of $\pi/24$ (of course, this includes such special values as 0, $\pi/6$, $\pi/4$, $\pi/3$, $\pi/2$, etc.), written in their decimal forms. In degrees, follow the same directions. The only difference is that the traced x-values are now multiples of 7.5° (which is the equivalent of $\pi/24$ radians). As this example illustrates, **ZTrig** has been carefully designed to produce the same graph for both radians and degrees. Other automatic ways of establishing a viewing window, such as **ZStandard** and **ZDecimal**, ignore the **MODE** setting and are not recommended for graphing trigonometric functions in degrees.

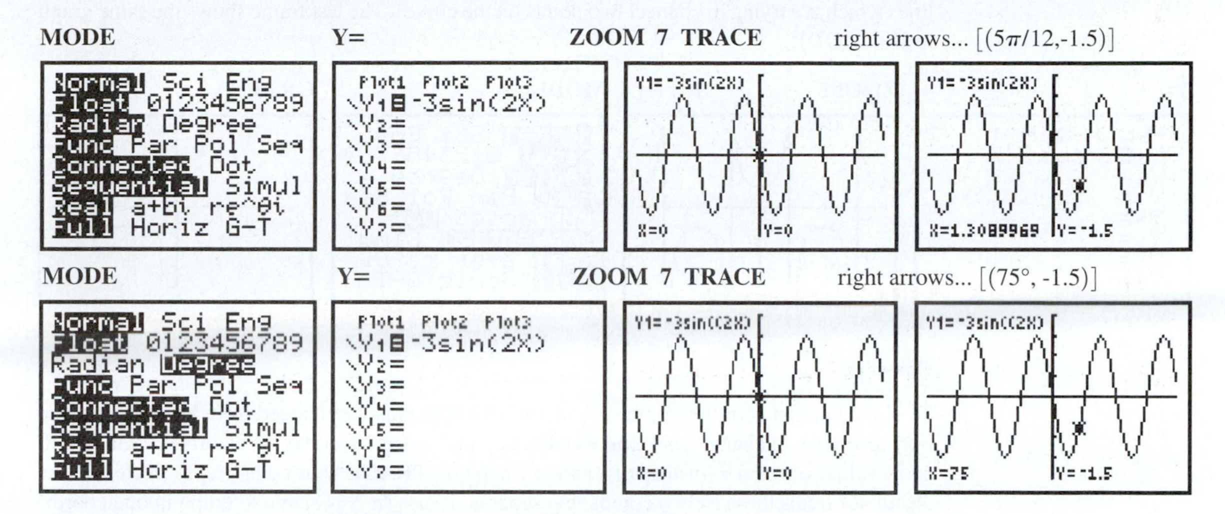

EXAMPLE 4

Several related trig functions can be drawn on the same screen, either by typing them separately in the function list **Y=** or by using a list of coefficients as shown in the following figures. A list consists of numbers separated by commas which are enclosed by braces { }. The braces are the shifted **2nd** versions of the parentheses keys. The first two frames indicate how to efficiently graph $y = \sin(x)$, $y = 2\sin(x)$, and $y = 4\sin(x)$ on the same screen. The last two frames graph $y = 2\sin(x)$ and $y = 2\sin(3x)$.

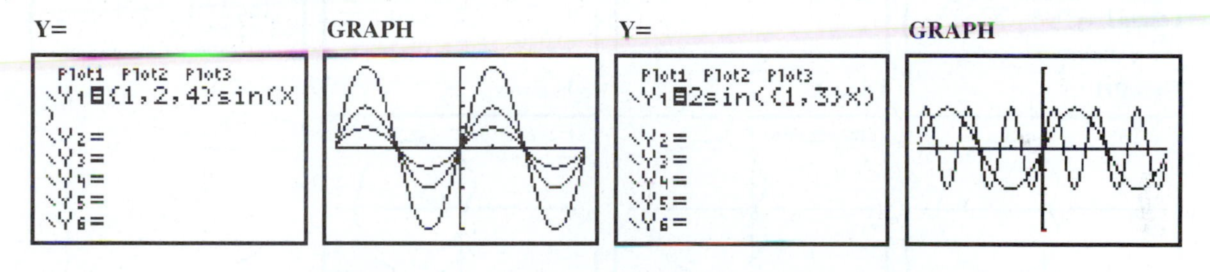

EXAMPLE 5

Multiple lists are allowed, but are not highly recommended. For example, to graph the two functions $y = 2\sin 3x$ and $y = 4\sin x$, you could do what's shown in the first frame or type them separately as shown in the third frame.

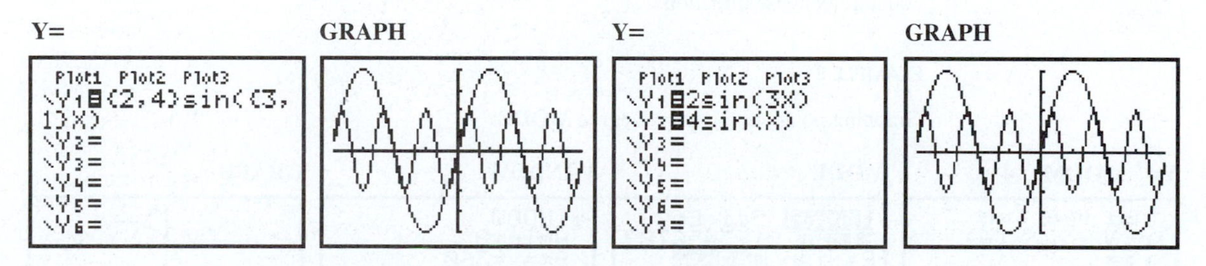

EXAMPLE 6

When graphing trig functions that have vertical asymptotes, remember that your calculator just evaluates individual points and arbitrarily assumes that it should connect those points if it is set in **Connected MODE**. The effect is shown in the following graph of $y = \sec x$. Some people like these "vertical asymptotes" being shown on the graph (they are really just nearly vertical lines which are trying to connect two points on the curve). The last frame shows the same graph in **Dot MODE**.

Y= **ZOOM 7** **MODE** **GRAPH**

EXAMPLE 7

To graph in polar coordinates, the calculator's **MODE** must be changed from **Func** to **Pol**. You may also wish to change your **2nd FORMAT** options from **RectGC** to **PolarGC**, which will show values of r and θ (instead of x and y) when you **TRACE** your polar graphs. Note that the calculator treats these as two completely separate issues (it is possible to graph in one coordinate system and trace in the other). Pressing **Y=** will reveal the calculator's six polar functions $r_1, r_2, \ldots, r_6$. The **X,T,θ,n** key now prints θ on the screen.

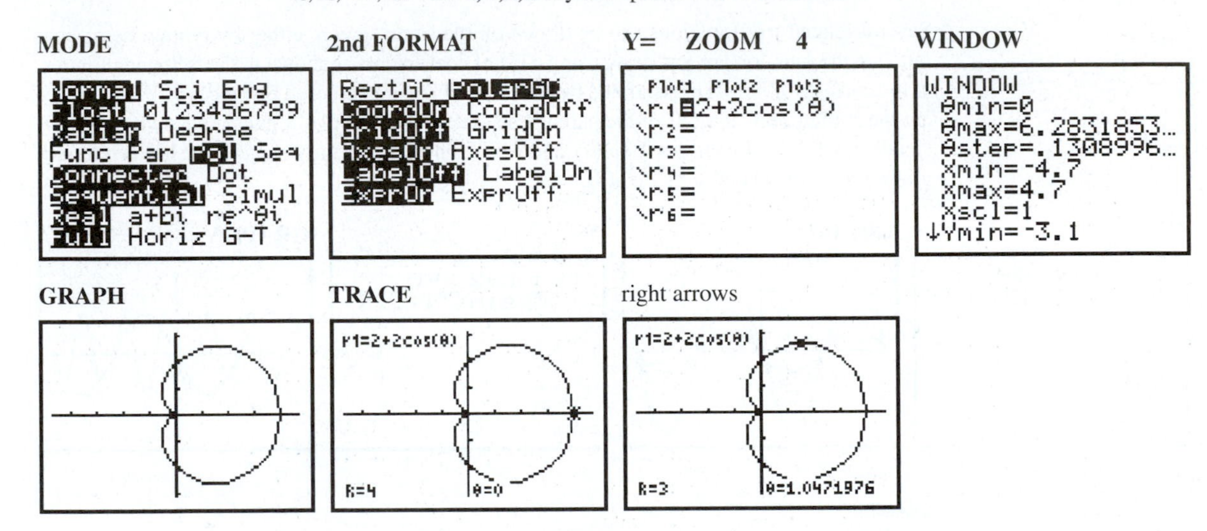

Note in the preceding frames that the standard radian values of θmin, θmax, and θstep are 0, 2π, and $\pi/24$, respectively. A more accurate, smoother graph can be obtained by using θstep$=\pi/48$ or $\pi/96$. Also note that the *right* arrow key is used to **TRACE** in the standard, counterclockwise direction.

EXAMPLE 8

Graphing polar equations in **Degree MODE**:

WINDOW (change θstep)

```
WINDOW
 θmin=0
 θmax=360
 θstep=2
 Xmin=-4.7
 Xmax=4.7
 Xscl=1
↓Ymin=-3.1
```

GRAPH

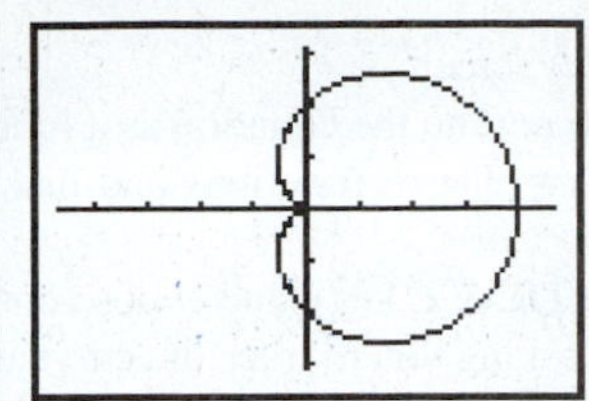

2nd FORMAT

TRACE right arrows

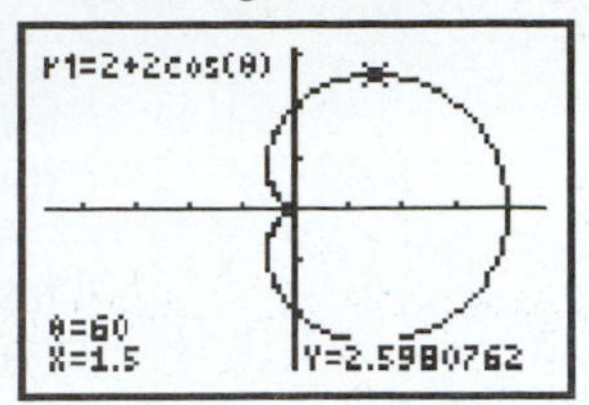

The standard degree values of θmin, θmax, and θstep are 0, 360, and 7.5, respectively. Smoother (but slower) graphs can be obtained by using smaller values of θstep. The last two frames show a polar graph traced in **RectGC FORMAT**.

C.6 EQUATION-SOLVING AND TABLE FEATURES

EXAMPLE 1

Solving an equation on the home screen:

(a) Rewrite the equation on paper in the form $f(x) = 0$; for example, rewrite

$$x^3 + 15x = 9x^2 - 6$$

as

$$x^3 - 9x^2 + 15x + 6 = 0$$

(b) From the home screen, press **2nd CATALOG**, then press the letter **T** (the **4** key), next press the up arrow repeatedly until **solve(** comes into view; then press **ENTER**. You should now see **solve(** on the home screen.

(c) Finish the statement so that it looks like one of the following: **solve(X³ − 9X² + 15X + 6, X, 3)** or **solve (Y1, X, 3)**, presuming you've entered the function in Y1 (to type the symbol Y1 in a formula, press **VARS** then the right arrow key, then **1**, then **1** again).

Y=

```
 Plot1 Plot2 Plot3
\Y1=X³-9X²+15X+6

\Y2=
\Y3=
\Y4=
\Y5=
\Y6=
```

WINDOW

```
WINDOW
 Xmin=-2
 Xmax=8
 Xscl=1
 Ymin=-25
 Ymax=25
 Yscl=5
 Xres=1
```

GRAPH

2nd CATALOG T up arrows

```
CATALOG         A
▸solve(
 SortA(
 SortD(
 stdDev(
 Stop
 StoreGDB
 StorePic
```

ENTER (home screen)

```
solve(
```

VARS right arrow **1**

```
VARS Y-VARS
1:Function...
2:Parametric...
3:Polar...
4:On/Off...
```

1

```
FUNCTION
1:Y1
2:Y2
3:Y3
4:Y4
5:Y5
6:Y6
7↓Y7
```

etc.

```
solve(Y1,X,3)
        2.748677137
solve(Y1,X,7)
         6.58291867
```

(d) Press **ENTER** and the calculator will try to find a zero of this function near 3 (answer: 2.748677137).

(e) Press **2nd ENTRY** to bring back your formula, then arrow left and change the 3 to a 7.

(f) Press **ENTER** and it will now find the zero near 7 (answer: 6.58291867).

(g) See if you can use the **solve** feature to find the other zero (answer: −.3315958073).

Notice that the **solve** feature finds solutions of an equation *one at a time*, with each new solution requiring its own estimate. Graphing the function and noticing where it crosses the x-axis is usually the easiest way to discover good estimates. If you prefer the **TABLE** feature (see Example 3), look for sign changes in the list of y-values; the corresponding x-values should be good estimates. Random guesses, although not recommended, can be effective when the equation has very few solutions.

EXAMPLE 2

Solving an equation on the graphics screen:

(a) As in Example 1(a), be sure to rewrite the equation as a function set equal to zero.

(b) Enter this function in your (**Y=**) list of functions and make sure that it is the only one selected for graphing.

(c) Press **2nd CALC** (the shifted **TRACE** key), and choose option 2, **zero**.

(d) The prompt "Left Bound?" is asking you to trace the curve using the arrow keys until you are just to the *left* of the desired zero (then press **ENTER**). Again, "Left Bound" just refers to an x-value that's too small to be the solution; do not consider whether the curve is above the axis or below the axis at that point. Similarly, the prompt "Right Bound?" is asking you to trace the curve until you are just to the *right* of the desired zero (then press **ENTER**). You'll notice in each case that a bracketing arrow is displayed near the top of the screen to graphically document the interval which will be searched for a solution.

(e) The prompt "Guess?" is asking you to trace the curve to a point as close as possible to where it crosses the axis (then press **ENTER**). This Guess is just an approximate solution like the **solve** feature uses (see Example 1).

(f) The solution (the "Zero") is displayed at the bottom of the screen (using 7 or 8 significant digits rather than the 10 digits you get on the home screen). An added bonus is that the y-value is also included (it should be exactly zero or extremely close to zero like "1E-12," which means 10^{-12}). Two of the solutions are found below. Try the third one on your own. The **WINDOW** from the previous example is assumed.

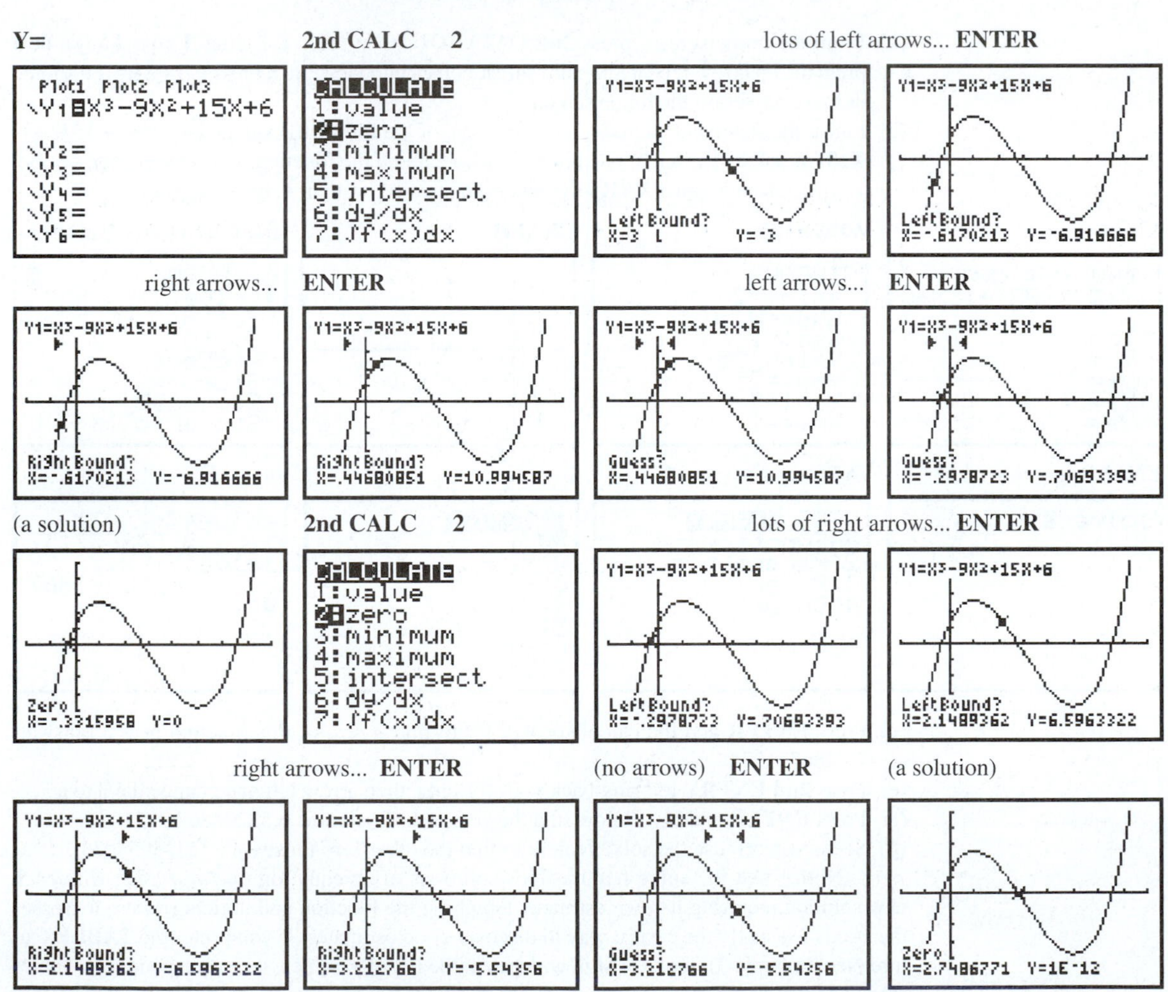

Note that a Right Bound can also be used as the Guess (see the last three frames).

EXAMPLE 3

Basic **TABLE** features:

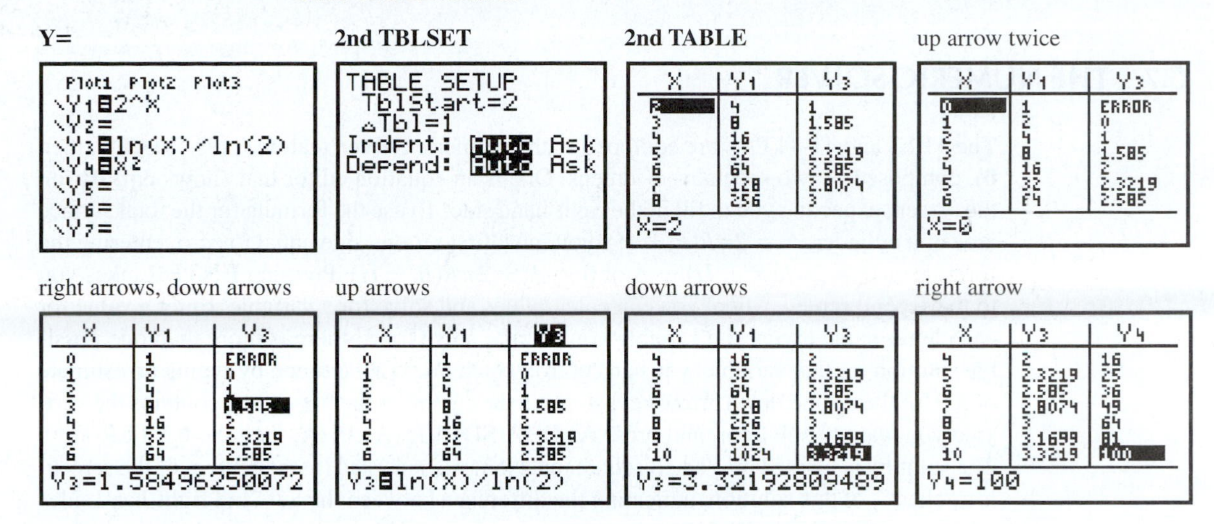

Note that functions to be investigated using a **TABLE** need to be entered and turned on in the same sense as those you want to graph. To get to the TABLE SETUP screen, press **2nd TBLSET** (the shifted version of the **WINDOW** key). TblStart is just a beginning x-value for the table; you can scroll up or down using the arrow keys. ΔTbl is the incremental change in x. You can use ΔTbl=1 (as in the preceding example) to calculate the function at consecutive integers; in calculus, you could use 0.001 to investigate what is happening to a function as it approaches a limit; or you could use a larger number like 10 or 100 to study the function's numerical behavior as x goes to ∞.

EXAMPLE 4

Split-screen graphing with a table (**MODE G-T**):

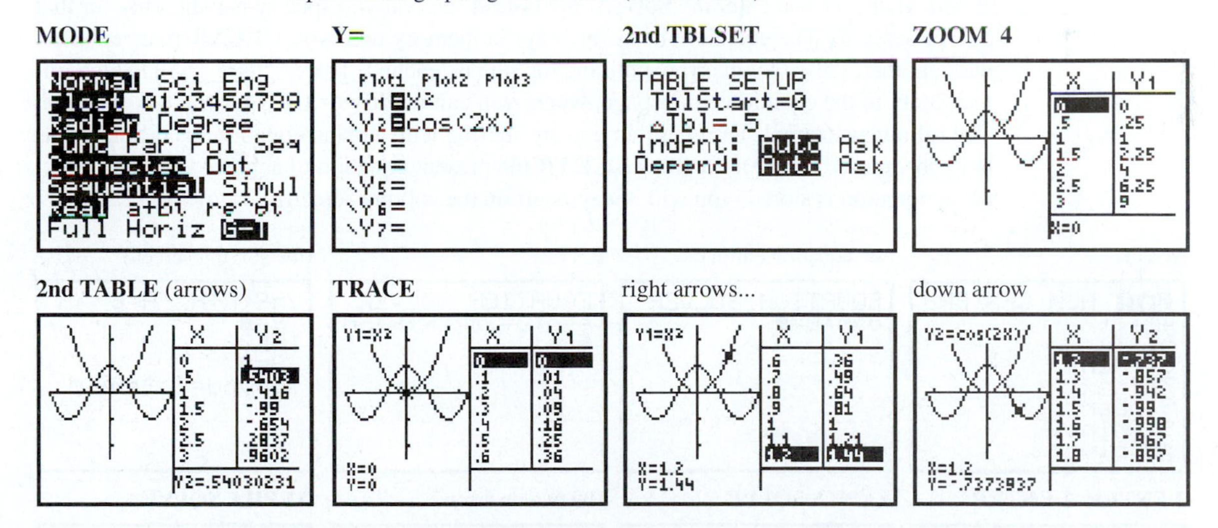

In **MODE G-T**, the graph and a corresponding table share the screen, but only one of them is "active" at any given moment. The **ZOOM** commands and the **GRAPH** key give control to the graphical side of the screen. This just means that the arrow keys refer to the graph rather than the table. Pressing **2nd TABLE** enables the arrow keys to be used to scroll through its values. The **TRACE** key links the table to the graph, with the graph in control (all previous TABLE SETUP specifications are replaced with values related to the **TRACE**).

Note in the last frame that jumping to a different function will display a different column of the table (the same value of x is highlighted).

C.7 THE NUMERIC SOLVER

The TI-83 and TI-84 Plus are equipped with a numeric **Solver** feature (press **MATH**, then **0**), composed of two specialized screens. One is an equation editor that shows **eqn:0=** on the screen, expecting you to fill in the right-hand side. To use the formula for the total surface area of a cylinder, $A = 2\pi R(R + H)$, you must first set one side equal to zero, entering the formula as $0 = 2\pi R(R + H) - A$ or $0 = A - 2\pi R(R + H)$. Pressing **ENTER** takes you to the second screen, where you can enter values and solve for a variable. Enter a value for each letter name except the variable you want to solve for. If there is more than one possible solution for that variable, you can control which one will be found by typing an estimate of it. Use the up and down arrow keys to place the cursor on the line which contains the variable you want to solve for, and press **ALPHA SOLVE** (**ALPHA**, then the **ENTER** key). The solution is marked by a small square to its left. Also marked (at the bottom of the screen) is a "check" of this solution, indicating the difference between the left- and right-hand sides of the equation using all of the values shown (this should be zero or something very close to zero, such as 1E−12). Fourteen significant digits are calculated and displayed for the solution variable. Press **2nd**, then the right arrow key to view the last several digits or to perform arithmetic on the solution; for example, you could divide a solution by π to see *what multiple* of π it is.

The **bound=** option near the bottom of the screen allows you to specify an interval to be searched for the solution. The default interval {-1E99,1E99} essentially considers any number that the calculator is capable of representing. When solving trigonometric equations, a limited **bound** like the interval $\{0,2\pi\}$ might be better, but in most cases, the default interval works well. If you accidentally erase this line and can't remember the syntax, just exit the **Solver** by pressing **2nd QUIT** and reenter it by pressing **MATH 0**. The default **bound** is restored whenever you enter the **Solver** (any **bound** interval you specify is valid only for that **Solver** session). The equation, however, stays in memory until you **CLEAR** it or replace it with another. Press the up arrow until the top line is reached. The calculator will switch immediately to the equation editor page, where you can modify or **CLEAR** the old equation. The following example assumes that you are starting with a blank equation (as if **Solver** has never been used before); you may **CLEAR** the present equation to achieve the same effect (if an equation is stored, you will always start on the solving screen).

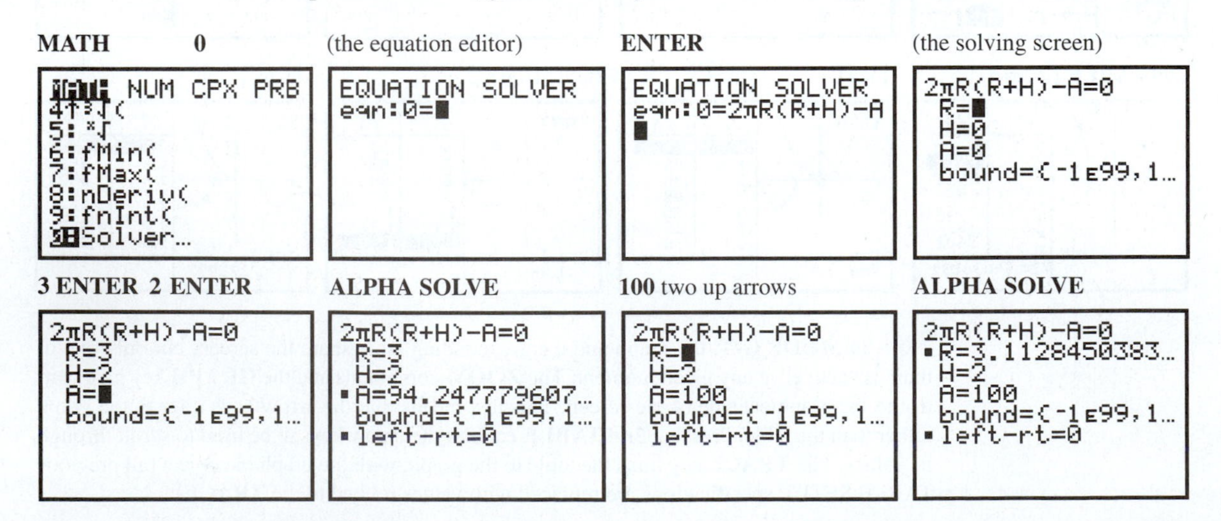

The first two frames in the bottom row of the previous example found the total surface area of a cylinder whose radius is 3 and whose height is 2. The last two frames found the radius of a cylinder whose height is 2 and whose total surface area is 100 square units. Use the **Solver** to find the height of a cylinder whose radius is 5 and whose total surface area is 440 square units (the answer is about 9).

The **Solver** can also be used on equations which contain only one variable, but think about the previous example (which had several variables) to understand why the **Solver** works as it does. In particular, it would seem natural to press **ENTER** after typing an estimate for the solution variable; instead, you must press **ALPHA SOLVE** (press **ALPHA**, *then* the **ENTER** key). If you accidentally press **ENTER** first, you will have to press the up arrow key to get back to the solution variable's line (if the cursor is on **bound=** rather than a variable's line, pressing **ALPHA SOLVE** does *nothing*). For direct comparison, the next example solves the same equation that was solved on both the home screen and the graphics screen in Section C.6. All three methods need one side to be zero and require each solution to be estimated. To solve

$$x^3 + 15x = 9x^2 - 6$$

rewrite it as

$$x^3 - 9x^2 + 15x + 6 = 0$$

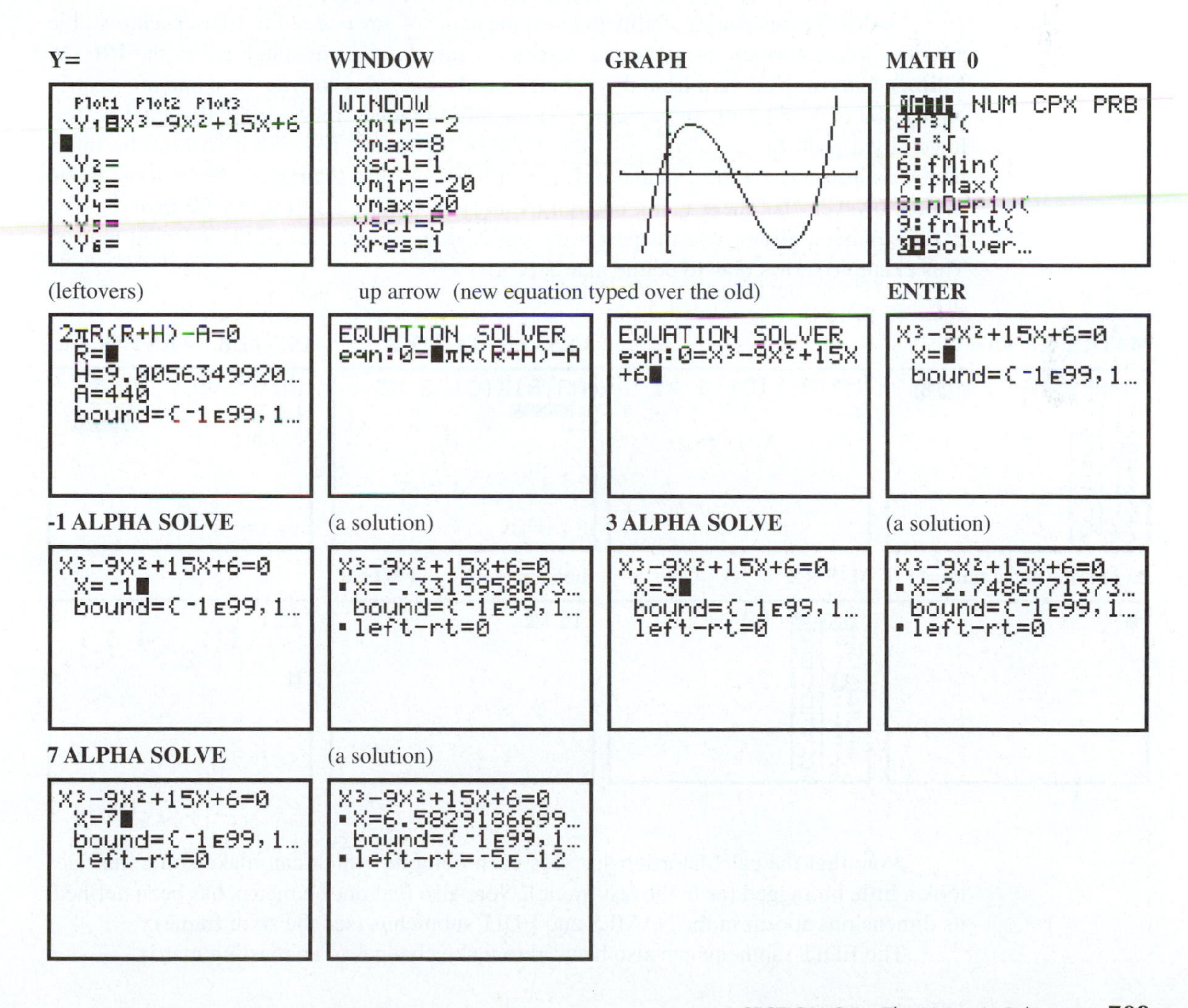

C.8 MATRIX FEATURES

Start by pressing the **MATRX** key on the TI-83 or **2nd MATRX** on the TI-83 Plus. This is the main keyboard difference between these two models. In the remainder of this section, directions will just refer to **MATRX**, which should be interpreted as **2nd MATRX** if you are using a TI-83 Plus (**2nd MATRIX** on Silver Edition). **MATRX** gives you access to three submenus: NAMES, MATH, and EDIT (which can be selected by pressing the right or left arrow keys). The MATH submenu has 16 options, but only 7 of them will fit on the screen at any one time (the arrows next to the 7 in the second frame and the 8 and D in the third frame indicate that there are additional options in those directions).

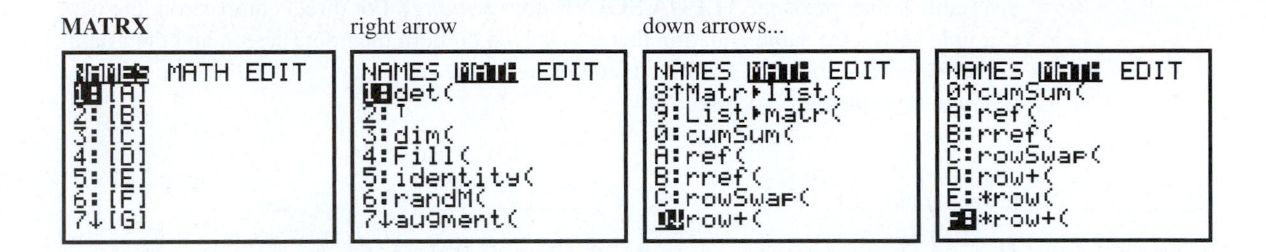

NAMES gives you the ability to insert the name of any one of the 10 user-addressable matrices into a formula on the home screen (or into a program statement in the **PRGM Edit**or). Choosing an item from this submenu is the *only* way to type the name of a matrix on a TI-83 or TI-83 Plus. In particular, typing a left bracket, followed by the letter A, followed by a right bracket, *looks* like the name of the matrix [A], but it will *not* be interpreted as a matrix by the calculator. Either dimension of a matrix can be as large as 99 (note, however, that there is not enough memory to handle a full 99 by 99 matrix). The EDIT submenu allows you to specify the dimensions and entries of the selected matrix. This example shows how to define matrix [C]:

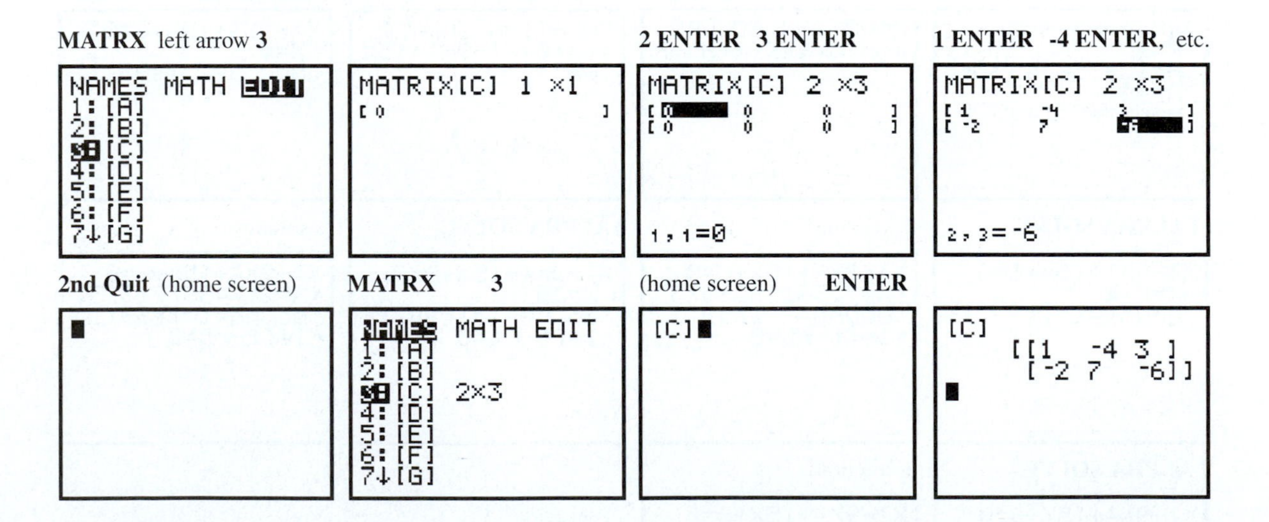

Note that the calculator left-justifies each column, which can make some matrices look a little bit ragged (as in the last frame). Note also that once a matrix has been defined, its dimensions appear in the NAMES and EDIT submenus (see the sixth frame).

The EDIT submenu can also be used to make changes to an existing matrix.

MATRX left arrow **3** arrow to the incorrect entry, type the new value...**ENTER**

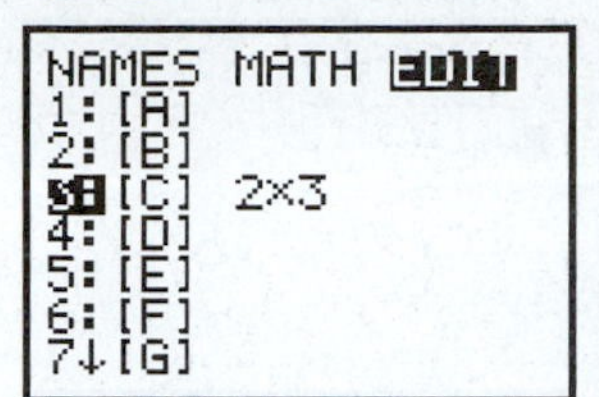

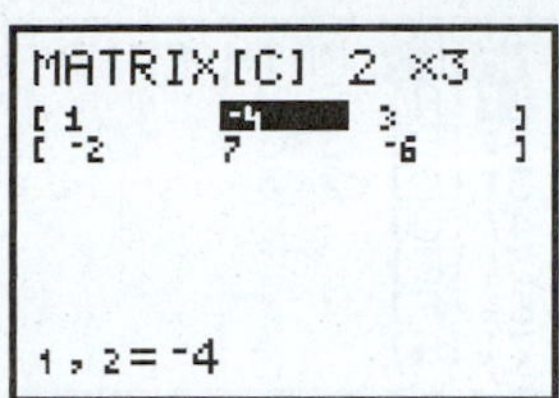

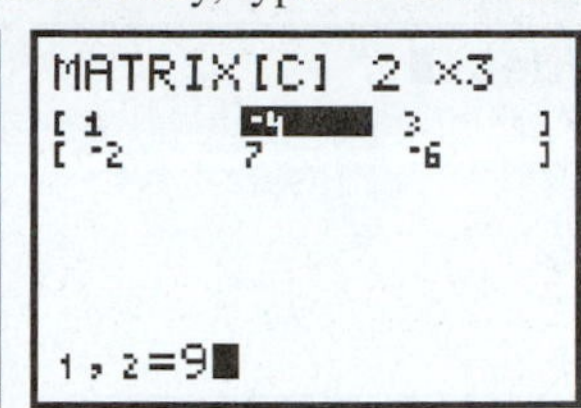

Press **2nd QUIT** to return to the home screen after you finish editing a matrix.

You can also create a new matrix or overwrite an existing one on the home screen (for most purposes, the matrix editor is much more convenient; see the previous example). Type each *row* within brackets, with the entries separated by commas, and enclose the entire matrix within an outer set of brackets. Typically, you'll want to store it in one of the 10 matrix variables [A], . . . , [J], but as the third and fourth frames show, there is also **Ans**, which stores the most recent computational result, even if it's a matrix result:

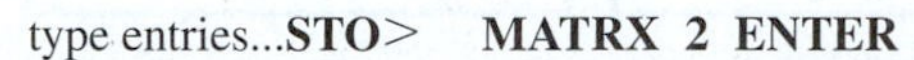

type entries...**STO>** **MATRX 2 ENTER** type entries **ENTER** **2nd ANS ENTER**

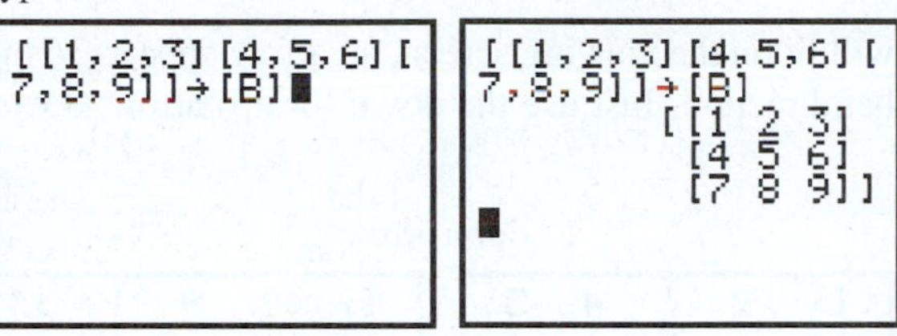

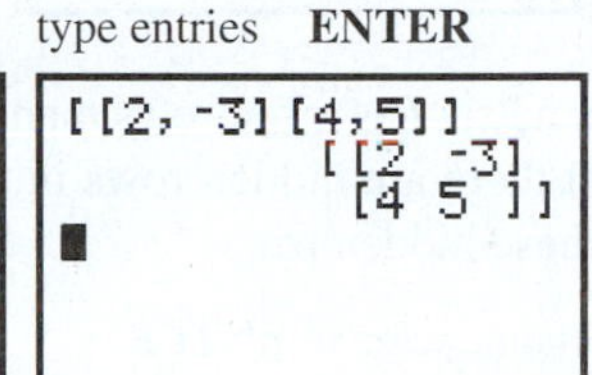

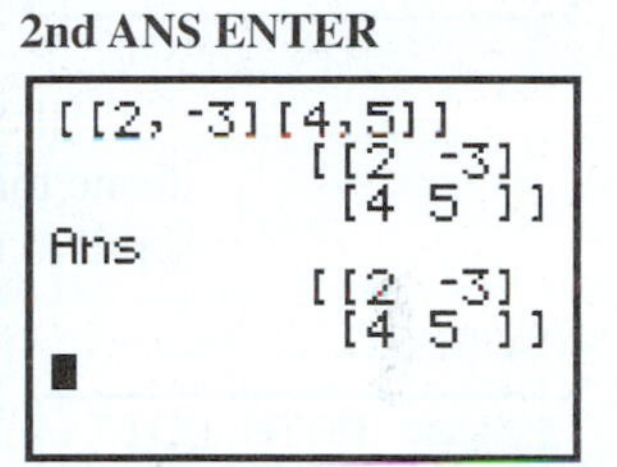

To multiply matrices, just type their names in the proper order (with or without a multiplication sign in between). You can square a matrix using the x^2 button. Multiplication by a scalar is shown in the final frame below:

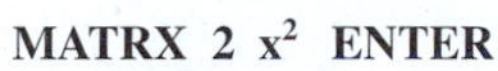

MATRX 3 MATRX 2 ENTER **MATRX 2 x^2 ENTER** **1000 MATRX 2 ENTER**

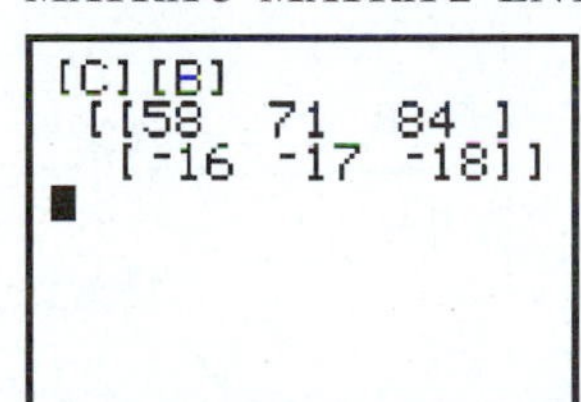

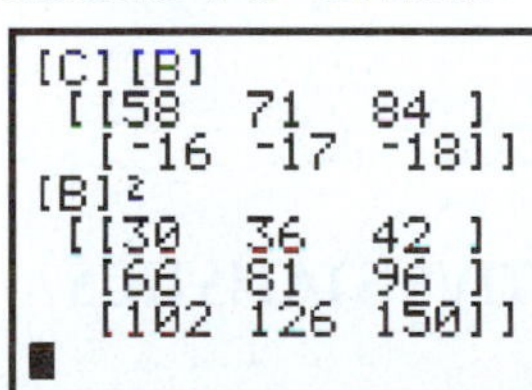

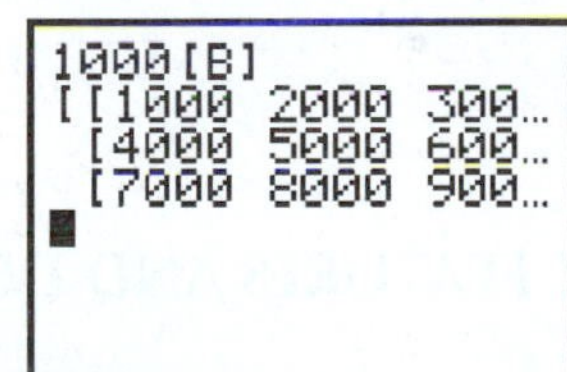

Notice that the three dots at the right of each row of the last frame indicate that there is more of the matrix in that direction; just use the right arrow key to reveal the hidden columns. To find the inverse of a (square) matrix, use the x^{-1} key following the name of the matrix. Fractional forms can be obtained by pressing **MATH 1**, then **ENTER**.

(edit [A]) **2nd Quit** **MATRX 1 ENTER** **MATRX 1 x^{-1} ENTER** **MATH 1 ENTER**

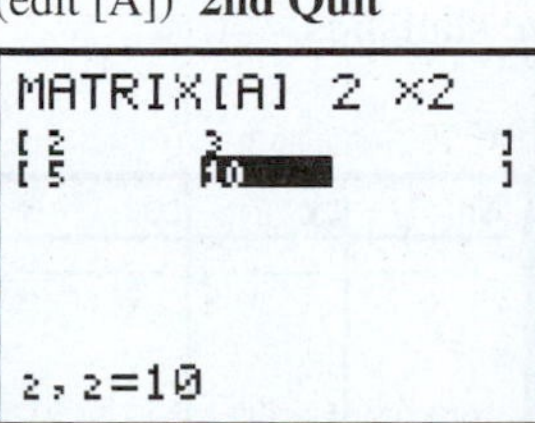

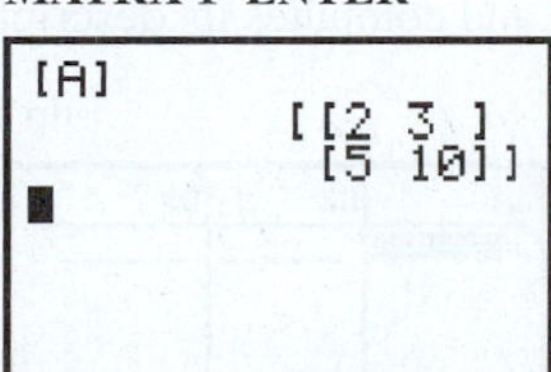

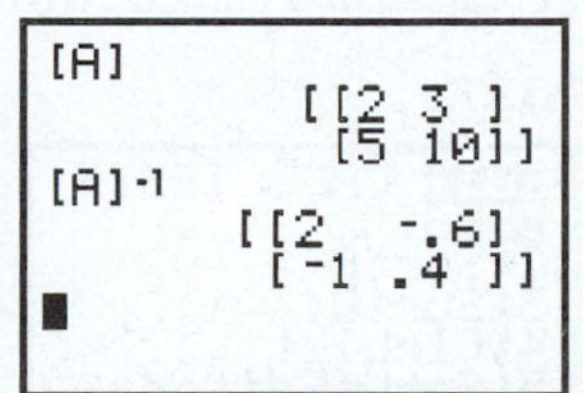

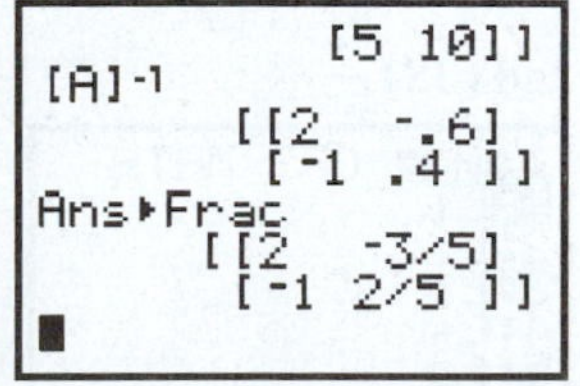

The determinant of a (square) matrix is available as a feature in the MATH submenu.

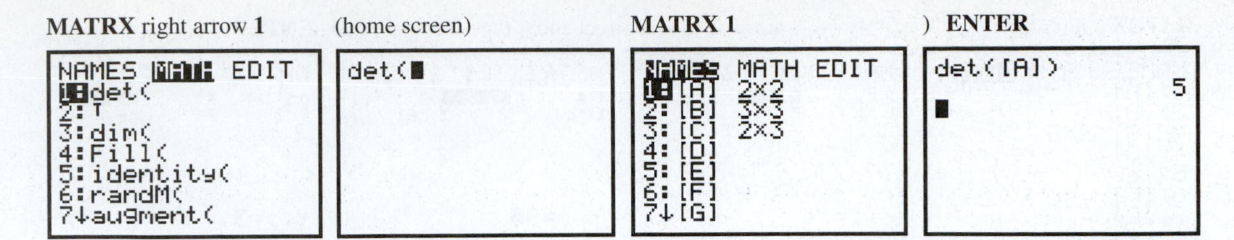

To row-reduce a matrix which might represent a system of equations, the option **rref(** is available in the MATH submenu.

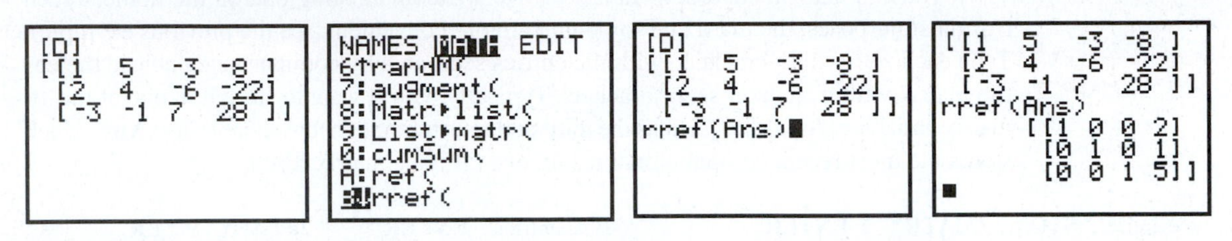

If a matrix has more rows than will fit on the viewing screen, an arrow appears to indicate that there are hidden rows in that direction. Just use the down (or up) arrow key to scroll to these hidden rows.

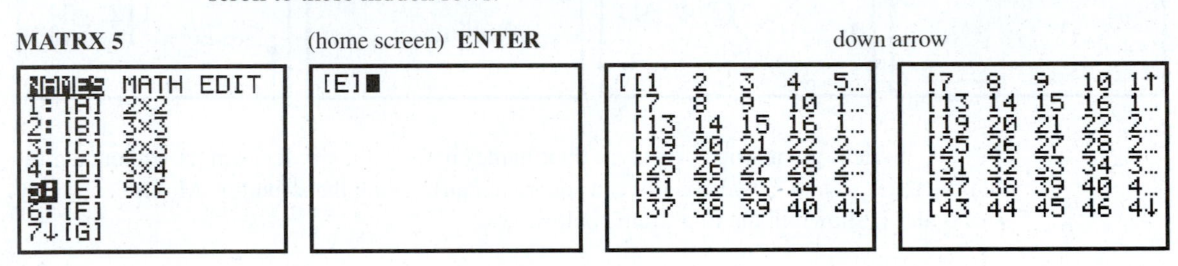

C.9 LIST FEATURES AND DESCRIPTIVE STATISTICS

The TI-83 has six built-in lists, named L_1, L_2, . . . , L_6, which can be accessed from the keyboard by pressing **2nd** followed by the subscript number of the list. You can create and name other lists (starting with a letter and using no more than five characters) and access them by pressing **2nd LIST** (**2nd**, then the **STAT** key). The TI-83 Plus and its Silver Edition will show L_1, L_2, . . . , L_6 in the **LIST** NAMES submenu, but an ordinary TI-83 will not. New list names can be created most easily in the **STAT Edit**or (press **STAT** then **1**). In this editor, go to the very top of any column and press **2nd INS** to create a new list. The example below creates a list named "RBI" and computes its descriptive statistics.

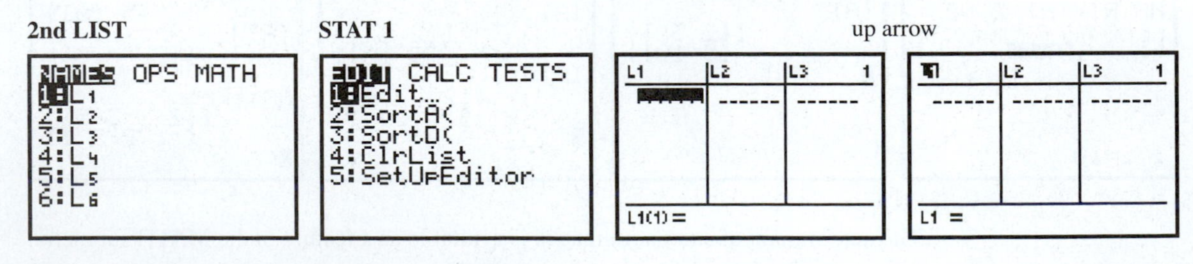

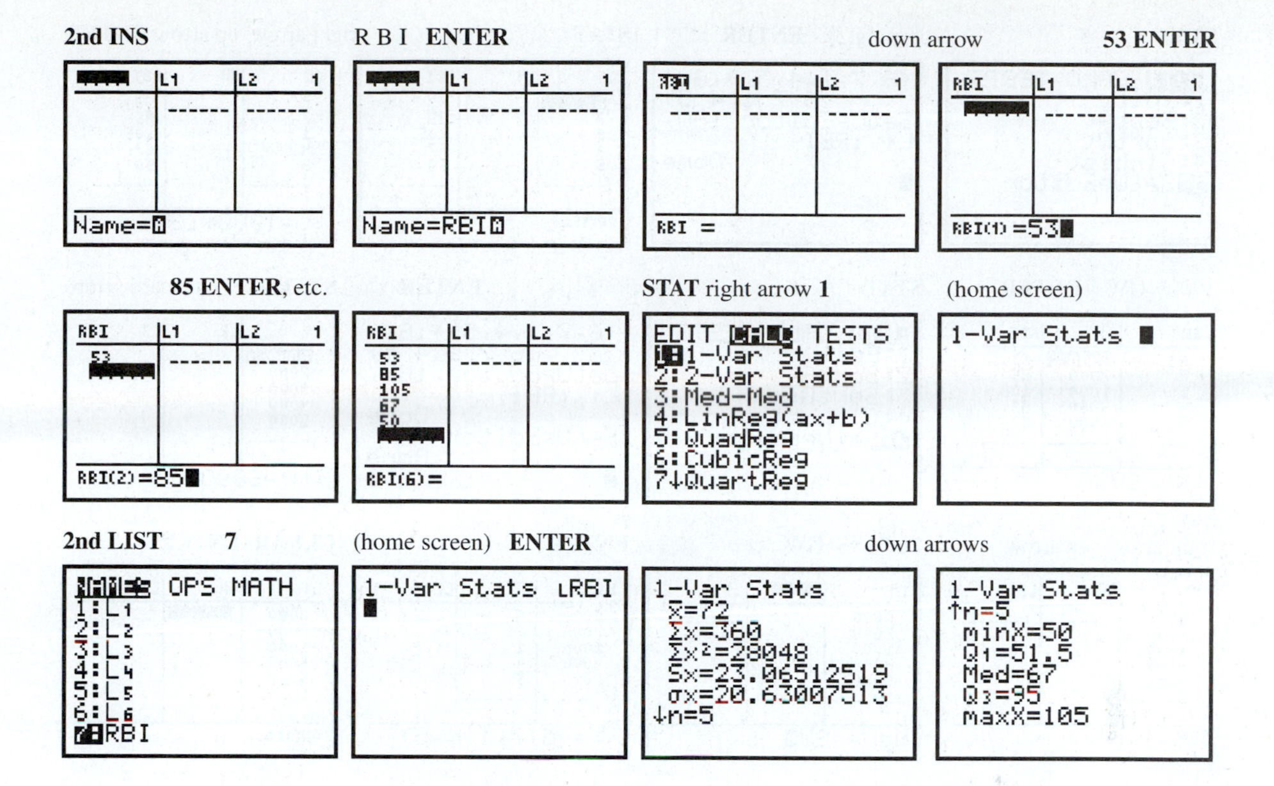

Note that when a list name is shown on the home screen, it is preceded by a small ʟ. This character is available in the **LIST OPS** menu (press **2nd LIST**, right arrow, up arrow, **ENTER**), and is also in the **CATALOG** (press **2nd CATALOG L ENTER**). You can create a list from the home screen by enclosing your data in braces { }, separating the items with commas, and **STO**ring them in a new list name (preceded by the small ʟ). However, a list created in this manner will *not* appear in the **STAT Edit**or until you include its name in a **SetUpEditor** command or **INS**ert its name along the top row of the **STAT Edit**or. To restore the standard setup, which shows just $L_1, L_2, \ldots, L_6$, execute **SetUpEditor** without specifying any list names. Deleting a list from the editor in this way (or by arrowing up to a list name at the top of the editor and pressing **DEL**) does not erase any data or delete the list from the **LIST NAMES** submenu. The list simply doesn't appear in the *editor* for the time being. To erase all data from a list (leaving it empty), arrow up to its name in the top row of the editor and press **CLEAR**, then **ENTER** (or **CLEAR**, then down arrow). To completely dispose of a list (name and all) on a TI-83 Plus, press **2nd MEM 2 4** and down arrow until the marker points at the name of the list you wish to get rid of, then press **DEL** (*not* **ENTER**). On an ordinary TI-83, press **ENTER** instead of **DEL**.

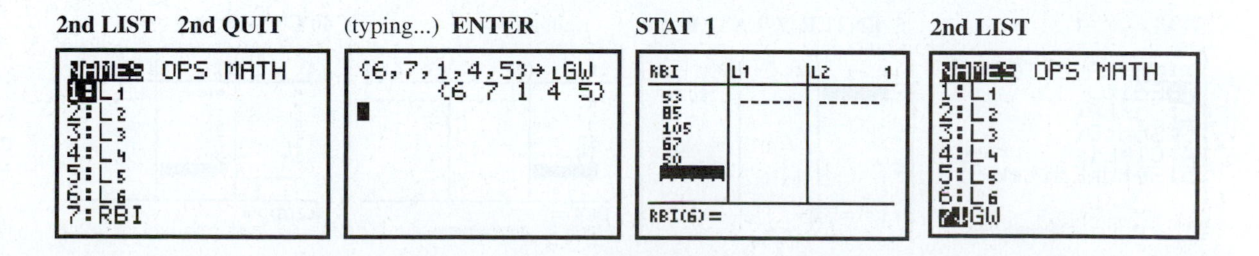

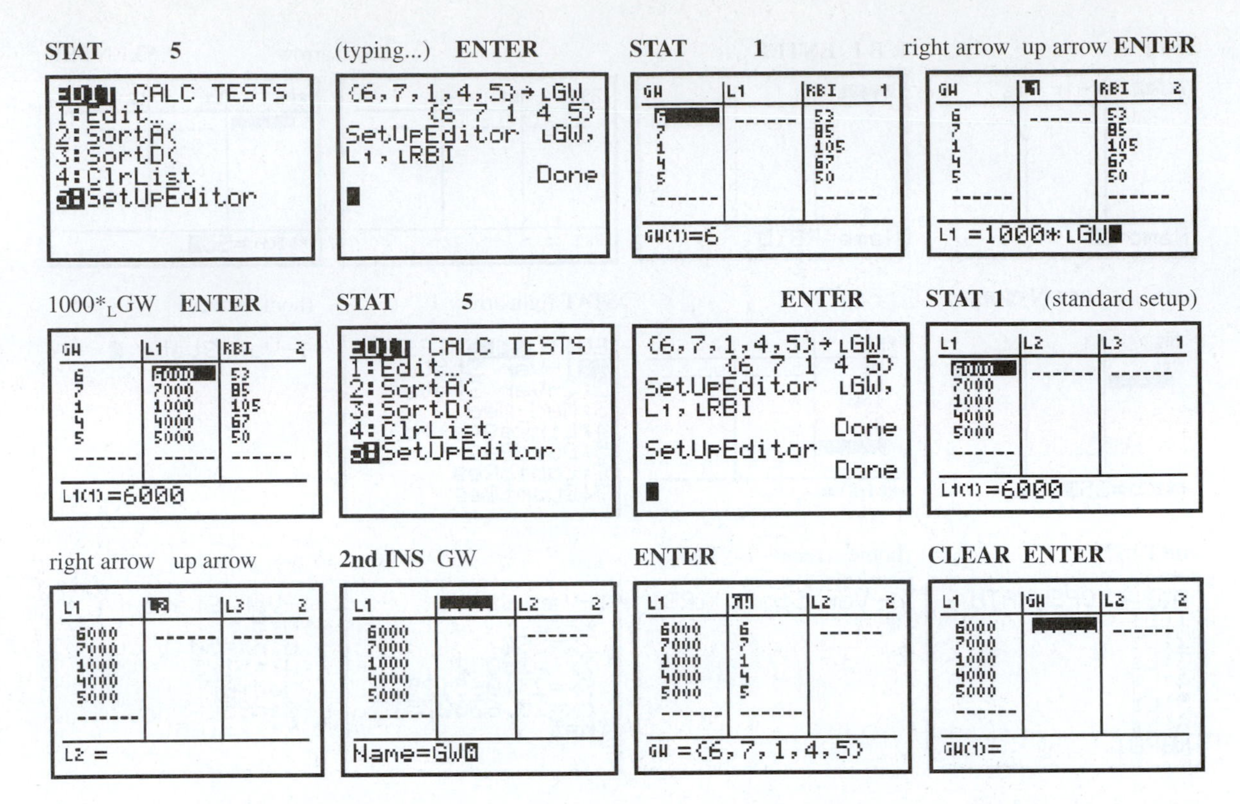

The first three frames in the bottom row show how to insert a preexisting list into the **STAT Edit**or. The last frame illustrates how to **CLEAR** a list, leaving it empty, but still named. If you highlight a list's name (in the very top row of the editor), you will be able to wrap around using the right or left arrow keys. If you wrap around using the *right* arrow, you will also notice a new blank list in the editor, ready to be named and filled with data. Just press **ENTER** or down arrow when it is highlighted, and you will be given a chance to name it. The **STAT Edit**or can hold up to 20 lists at once. Each list stored in a TI-83 or TI-83 Plus can have as many as 999 elements.

C.10 THE LINE OF BEST FIT (LINEAR REGRESSION)

EXAMPLE

Find and graph the equation of the line of best fit for the following data:

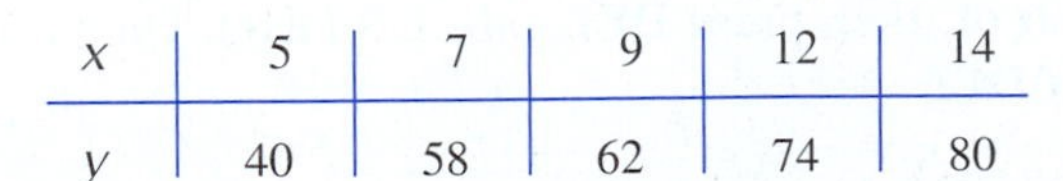

x	5	7	9	12	14
y	40	58	62	74	80

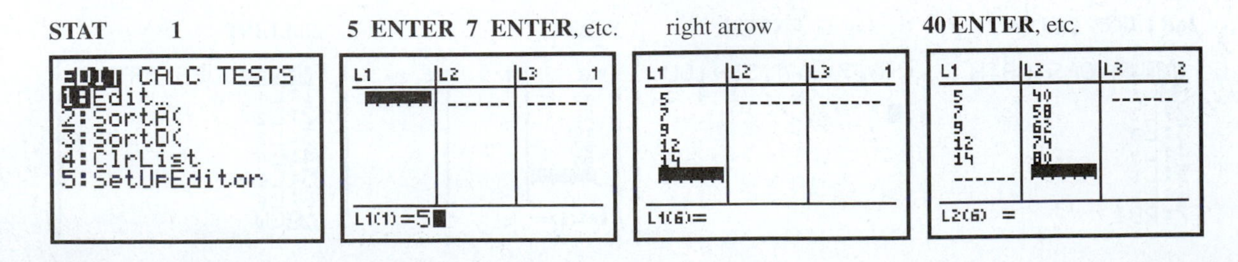

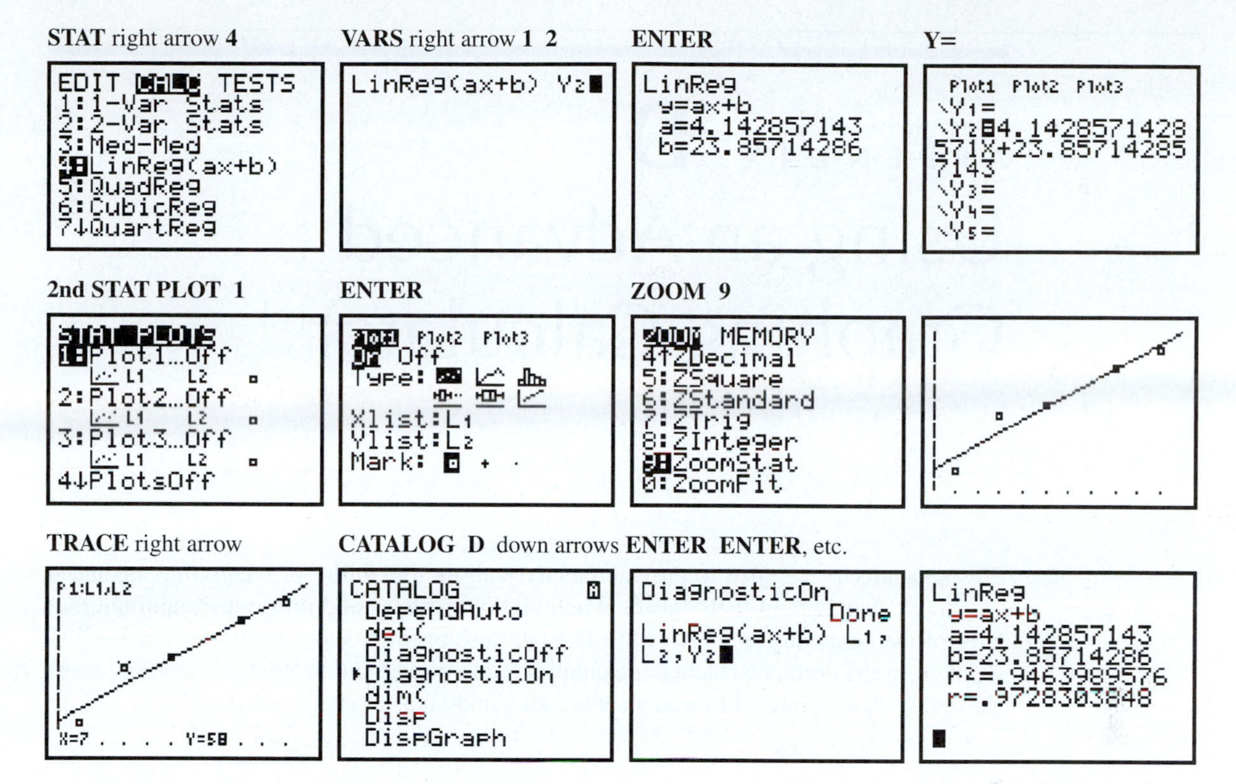

Turn **STAT PLOT 1 Off** *now* (before trying to graph anything else).

In the last three frames, this linear regression is shown in more detail. **DiagnosticOn** enables the calculator to compute and print correlation coefficients. **DiagnosticOff** is the default setting. These options appear only in the **CATALOG**. The last linear regression command shows the full syntax of the **LinReg(ax+b)** statement. All three parameters are optional. The name of the first list provides the x-values, the second list provides the y-values, and the third parameter is the name of the function where the regression equation will be stored. If you omit the two list names, the calculator will use list L_1 for the x-values and list L_2 for the y-values (see the sixth frame of the example). If you don't plan to use or graph the regression equation, you may omit the function name. To turn **PLOT 1 Off**, press **Y=**, up arrow, then **ENTER**.

APPENDIX D

Using an Advanced Graphing Calculator

This appendix is included to provide faculty with the flexibility of integrating advanced graphing calculators in their classes. Each section explains and illustrates important features of the Texas Instruments TI-89. Though this appendix was specifically designed to supplement the graphing calculator examples found throughout the text, the material is organized so that an interested student could also study it as a separate chapter.

D.1 INTRODUCTION TO THE TI-89 KEYBOARD

The Texas Instruments TI-89 has a 50-key layout, similar to earlier TI graphing calculators. The bottom six rows of keys perform the standard functions of a scientific calculator. The thin blue keys that form the top row are used to choose menus and to give quick access to specialized screens for graphing functions and calculating tables of function values. The second, third, and fourth rows of keys contain the calculator's editing and navigation features. Of special interest are the calculator's **MODE** key, which can be used to change any of its settings, and the **APPS** key, which gives access to all nine of the specialized editing and application screens. Also notice the (golden yellow) **2nd** key, the (purple) **alpha** key, and the **green diamond** key, which give additional, color-coded meanings to almost every key on the calculator. Start by pressing the **ON** key in the lower left-hand corner. There are two ways to turn the calculator **OFF**. If you want to be able to return to your work precisely as you left it, press the **green diamond** key, then **ON/OFF** (or just leave the calculator unattended for a couple of minutes). Pressing the **2nd** key, then **OFF**, exits any menus, editors, or application screens and puts the calculator back on the home screen as it turns off the power.

Look again at the scientific calculator features found in the bottom six rows of keys. The **LN** (natural logarithm), **SIN, COS,** and **TAN** functions are shown in gold as **2nd** shifted versions of the **X, Y, Z,** and **T** keys, respectively. The **green** shifted items in that row include $\mathbf{e^x}$, the inverse trig functions, and the variable θ. The square root is the **2nd** version of the multiplication key. Unlike previous models, the TI-89 does *not* include specialized keys for squaring or taking reciprocals. Another change is that the common (base 10) logarithm, **log(**, is found only in the **CATALOG**. To access any item shown in green on the keyboard, press the **green diamond** key and then press the key which has the green writing above it; to increase (or decrease) screen contrast, *hold down* the **green diamond** key while pressing the addition (or subtraction) sign repeatedly. An interesting surprise is that many of the keys have **green** shifted meanings that are not marked on the

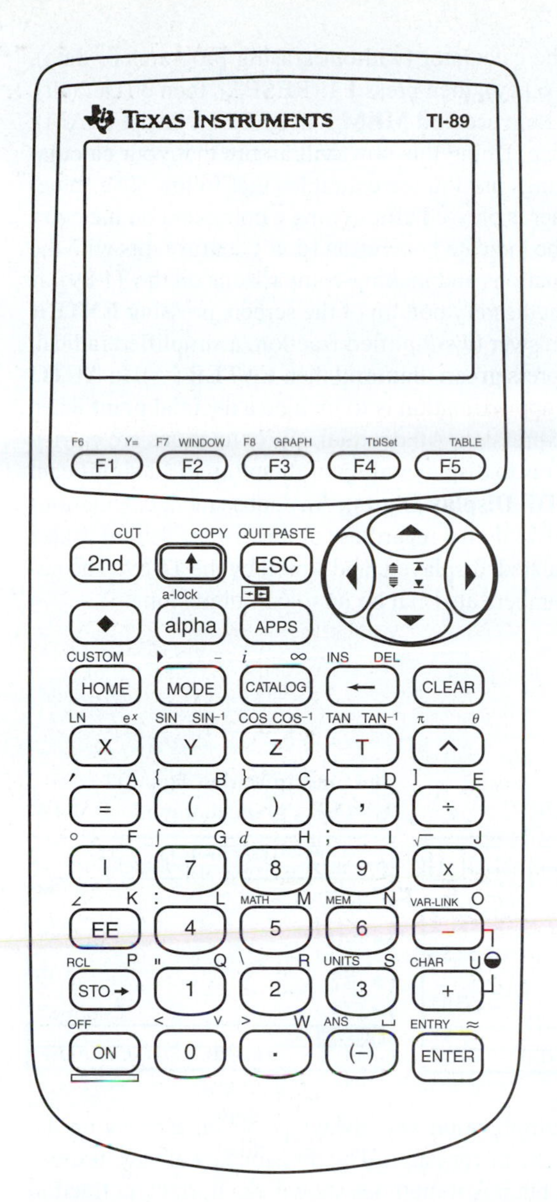

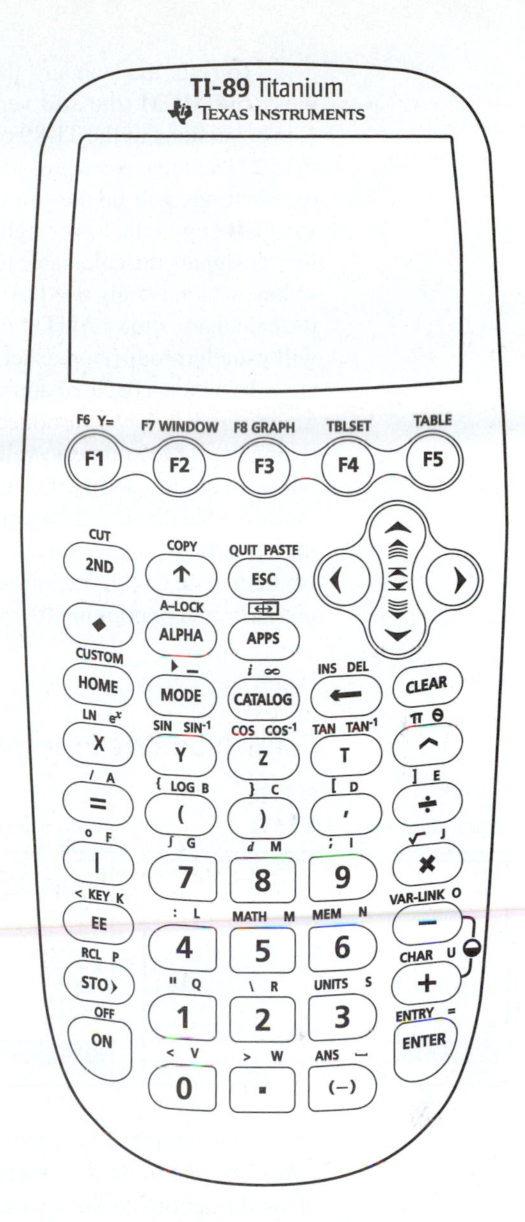

Courtesy of Texas Instruments.

face of the calculator. Among the most useful are $\leq$, $\geq$, $\neq$, and ! (the factorial symbol). They are the **green** shifted versions of zero, the decimal point, the equal sign, and the division sign, respectively.

For a map of the hidden **green** features, press **green diamond** then **EE** (just to the left of **4**):

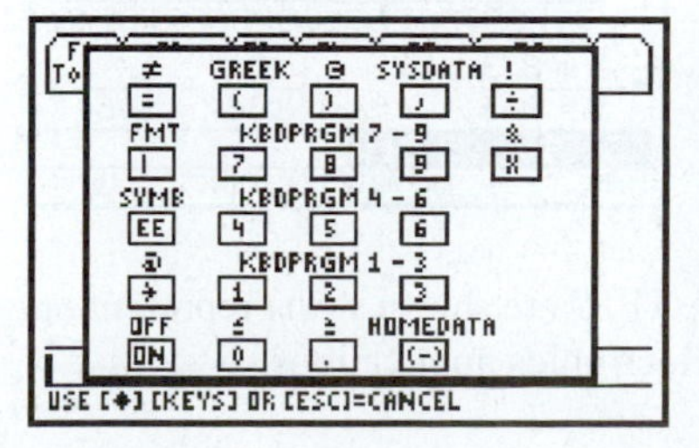

To reset the standard defaults without erasing programs or data:
(Newer TI-89s will use **2nd MEM F1 1 2**)

2nd MEM F1 3

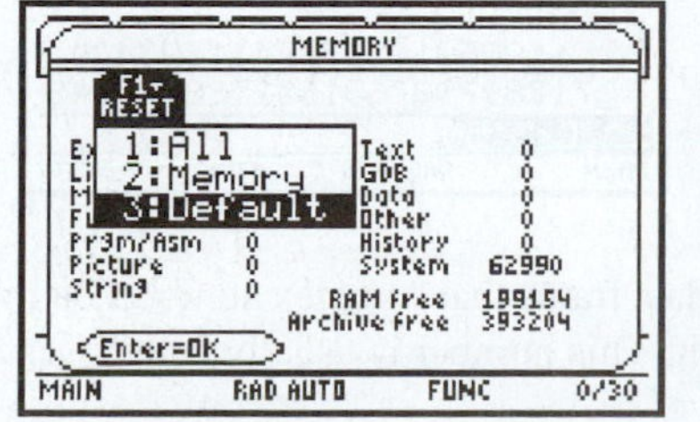

ENTER ENTER

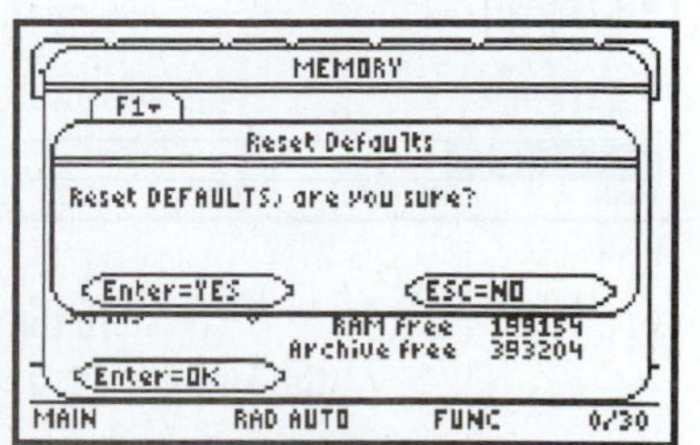

To reset the standard defaults on the calculator (without erasing programs or data), press **2nd MEM** (the **2nd** version of the **6** key), then press **F1** (RESET), then **3** (Default). Newer versions of the TI-89 operating system use **2nd MEM F1** (RESET), then **1** (RAM), then **2** (Default). Now press **ENTER** twice. Doing this now will assure that your calculator's settings will be the same as the settings used in the examples that follow. The (blue) **ENTER** key in the lower right-hand corner is pressed after typing a command on the entry line. It signals the calculator to perform the indicated operation (don't confuse this with the = key, which is only used for writing equations and making comparisons on the TI-89). If the calculator shows **AUTO** or **EXACT** at the very bottom of the screen, pressing **ENTER** will usually produce an exact algebraic answer (a simplified fraction, a simplified radical, etc.). To obtain a decimal approximation, press **green diamond** then **ENTER** (≈). In **AUTO** mode, another way to produce a decimal approximation is to include a decimal point when typing any of the numeric values in the expression. All decimal approximations are carried out to 14 significant digits, but the default is to display only 6 of them (up to 12 significant digits can be displayed by adjusting **MODE Display Digits**). Any subsequent calculations that refer to the approximation will use all 14 digits, regardless of the display setting. Exact integers up to 614 digits long can be calculated, displayed, and stored by the TI-89! The numerators and denominators of rational numbers also can be up to 614 digits long.

Some Numeric Examples

12 × 5 + 3 ÷ 7 **ENTER**
green diamond ENTER

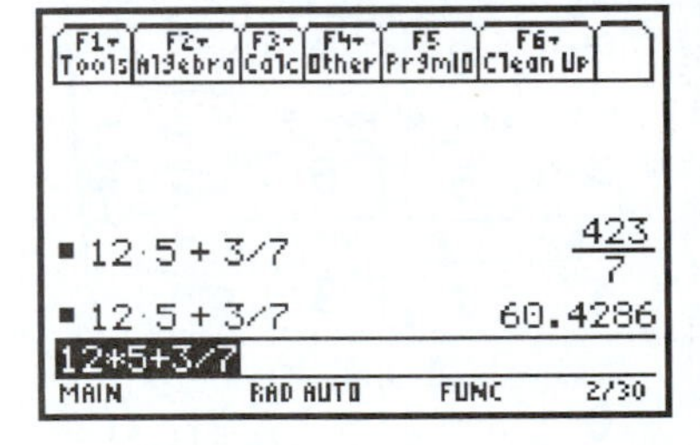

312 ÷ 816 **ENTER**
green diamond ENTER

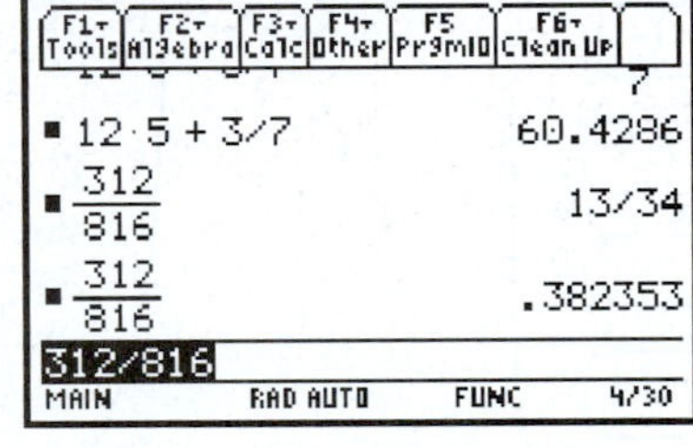

2nd multiplication sign 32)
ENTER green diamond ENTER

Note in the first frame that the multiplication key always prints an asterisk on the entry line and is shown as a raised dot in the history area. The division key always prints a slanted fraction bar line on the entry line, but it is sometimes shown as a horizontal fraction bar line in the history area.

SIN is the **2nd** version of **Y**.
π is the **2nd** version of the ^ key.

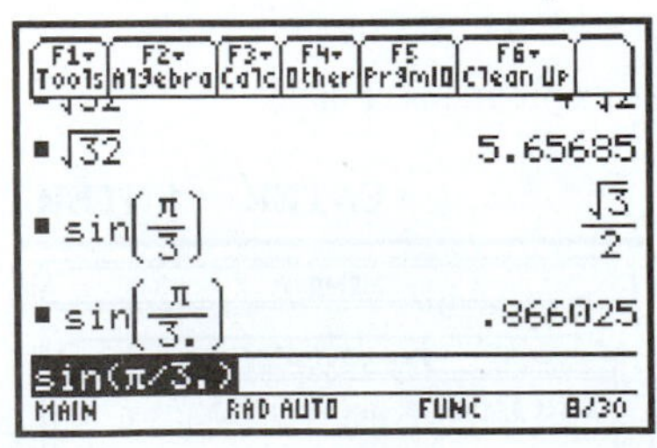

An exponent is indicated just after the ^ key is pressed.

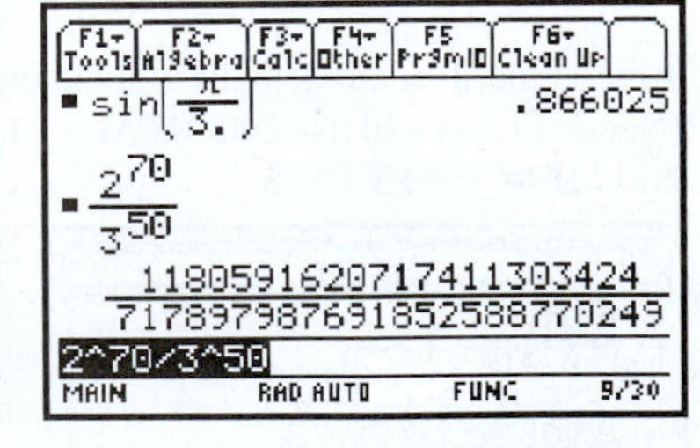

Complex numbers are available;
i is the **2nd** version of **CATALOG**.

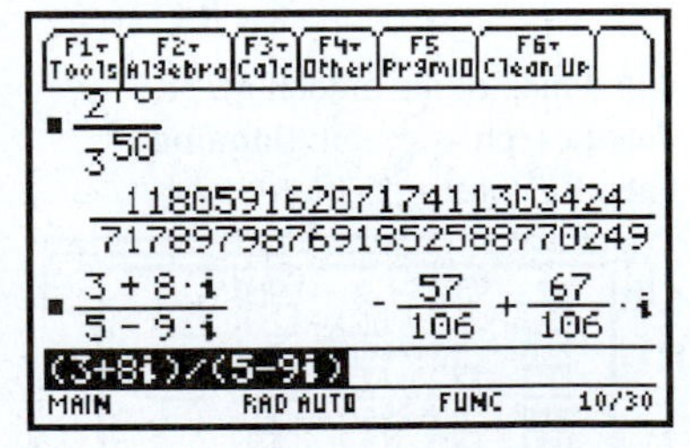

Note in the last frame that complex numbers on the TI-89 are shown with ***i*** representing the imaginary unit. This number is usually called ***j*** in electronics applications.

D.2 VARIABLES AND EDITING

The commonly used variables **X, Y, Z,** and **T** are each given their own key on the TI-89. All other letters of the alphabet are obtained by pressing the purple **alpha** key followed by the letter of the alphabet (written in purple above keys in the bottom 5 rows); a blank space can be obtained by pressing **alpha** and then the sign change key **(-)**. If you need to type several letters of the alphabet in succession, press **2nd a-lock** (the gold **2nd** key followed by the **alpha** key) or press **alpha** twice. To release alpha-lock, just press **alpha** one more time. The **backspace** key (a black key with a left arrow on it) is located just to the left of the **CLEAR** key. This key erases the character to the *left* of the cursor each time it is pressed. The **green diamond** version of this key (**DEL**) erases a character to the *right* of an insert cursor, and erases the character *highlighted* by an overstrike cursor. Both **backspace** and **DEL** will erase a block of highlighted characters on the entry line or in the history area. The right and left arrow keys are used to move the cursor one space at a time when editing the entry line. Pressing the **2nd** key just before pressing one of the (blue) arrow keys causes the cursor to travel as far as possible in that direction. Quite often, the screen is not wide enough to display the entire result of a calculation. A small arrow at the right means that there is more to view in that direction. Just arrow up to highlight that line of history and scroll using the right arrow key (or press **2nd** then right arrow to jump to the end of that line). There is also a **shift** key marked as a white up arrow on a black key (just to the right of the **2nd** key). The **shift** key enables you to type uppercase letters, but it should be noted that the symbolic algebra features of the TI-89 are *not* case sensitive. The symbols z and Z represent the *same* variable on the TI-89. To type several uppercase letters in a row, press **2nd shift alpha** or press **alpha shift alpha**. To release shift-lock, just press alpha one more time. This shift key is also used in conjunction with the arrow keys to highlight text which you may want to **cut, copy, paste,** or **delete.** *Hold down* the **shift** key while pressing the (blue) right or left arrow keys to highlight the text you want to manage, then press **green diamond CUT, COPY, PASTE,** or **DEL** (the **green** versions of the **2nd, shift, ESC,** or **backspace** keys, respectively). These text-managing options can also be found in the **F1 (Tools)** menu. Editing text or other input can be done with either an "insert" or "overstrike" cursor. The insert cursor is indicated by a blinking vertical line, which fits between the characters where the insertion will occur, while the overstrike cursor highlights a single character, which will be replaced by the one you are typing. To switch between the two cursors, just press **2nd INS** (the **2nd** version of the **backspace** key). All keystroke examples in this book will assume that the calculator currently shows an insert cursor.

Variable names for the TI-89 can be from one to eight characters in length and must start with a letter (the other characters may be alphabetic or numeric). The one-letter variables a through z can be deleted easily (press **2nd F6**, then press **ENTER** twice) and are popular for symbolic manipulation and other short-term use. Variables that store valuable long-term information should probably be named using more than one letter. Don't forget, though, that the TI-89 expression **4ac** does *not* mean **4** times **a** times **c**. Instead, it means **4** times a variable named **ac**. Note also that parentheses immediately following a variable name always indicate function notation (*not* multiplication) on the TI-89. For example $q(x/z)$ indicates a function q of the expression x/z (not q times x/z). The times sign (which prints an asterisk on the screen) is used much more often than proximity to indicate multiplication on this calculator. If multiplication is intended in the preceding examples, you could enter $4a*c$, $4*a*c$, or $(4a)c$ in the first example and $q*x/z$, $q*(x/z)$, or $(q*x)/z$ in the second example.

To erase everything to the right of the cursor on the entry line, press **CLEAR** once; pressing **CLEAR** again (with nothing to the right of the cursor) will erase the entire entry line. The **CLEAR** key (or the **backspace** key) can also be used to erase unwanted items in the history area. To erase an individual history item, just press the (blue) up arrow key until the item is highlighted, then press **CLEAR** (or **backspace**). Both the entry and its answer are erased at the same time. To scroll to the very top of the history area on the home screen, enter the history area by pressing the (blue) up arrow key, then press **green diamond**, then up arrow. Pressing **green diamond** then down arrow scrolls to the bottom of the history area. Press the **ESC** key to escape the history area and return to the entry line. Pressing **ESC** can also escape any unwanted menu without leaving the current application. The **HOME** key escapes from any menu or application and returns you to the home screen. To erase all history on the home screen, be sure that you have the standard home screen menu, then press **F1** (the blue key in the upper left-hand corner), then **8 (Clear Home)**.

D.3 THE HOME SCREEN MENUS

The home screen is used for computation and symbolic manipulation (there are specialized screens for graphing, program input/output, tables, and data editing). Two different menu systems are available for the home screen. The standard menu system (shown in the first and second frames below) allows quick access to the algebra and calculus features of the TI-89. The default custom menu system (shown in the third frame) contains useful templates, common units of measurement, and special symbols, which can save many keystrokes. As their name implies, the custom menus can also be modified to reflect an individual user's needs and preferences (see "Creating a Custom Menu" in the owner's manual that comes with your calculator). The custom menu system is activated or exited by pressing **2nd CUSTOM** (the **2nd** version of the **HOME** key). It is also exited (and the standard menu system is shown) whenever you go to another application screen and later return to the home screen.

Press **F1** then **8** to erase all history.

Options **1 Open** and **3 New** are dimmed, which indicates that they are not active on the home screen.

To switch between these two, press **2nd CUSTOM.**

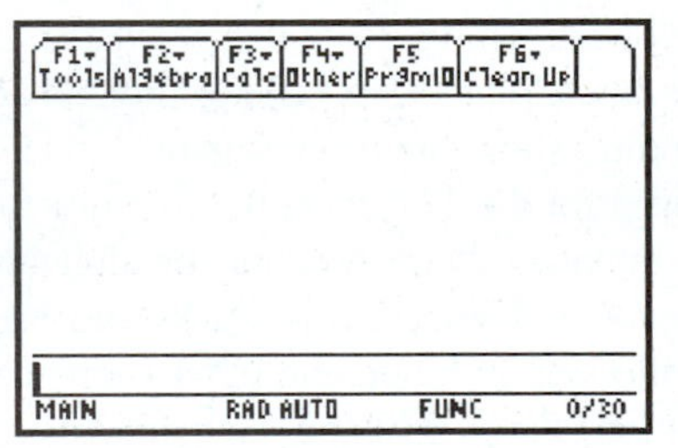

the standard menu system

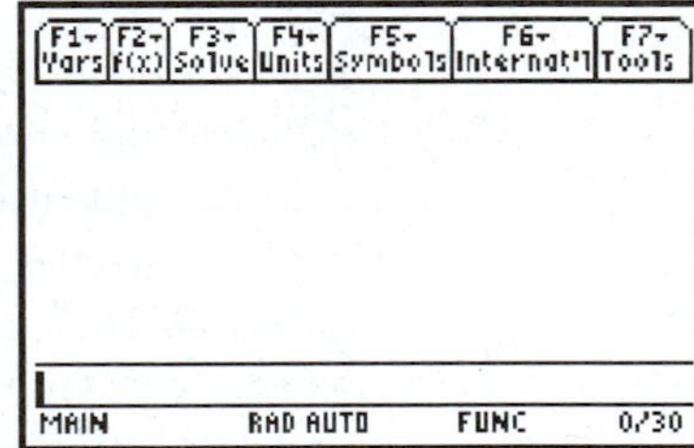

the (default) custom menu system

The Standard Home Screen Menus

To select an individual menu, press the (thin blue) key in the top row that has the same marking as the menu (to access the **Algebra** menu, press the **F2** key). Note that **F6, F7,** and **F8** are the **2nd** versions of the **F1, F2,** and **F3** keys, respectively. The top of the **Tools** menu was shown in the previous example.

the bottom of the **F1 Tools** menu

Keyboard shortcuts are shown at the right of options 7 and 9. Options 4, 5, and 6 are also on the keyboard.

the **F2 Algebra** menu (top)

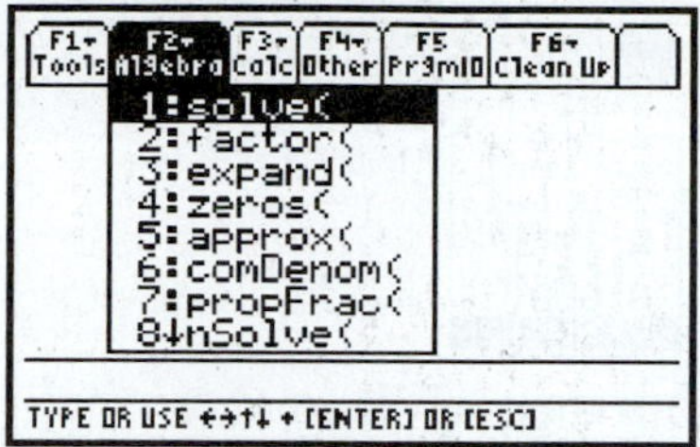

The algebra features include a wide variety of solving and simplifying commands (see Section D.14).

the **F2 Algebra** menu (bottom)

To get to the bottom of a menu quickly, press the up arrow (or **2nd**, then the down arrow key).

the **F3 Calculus** menu (top)

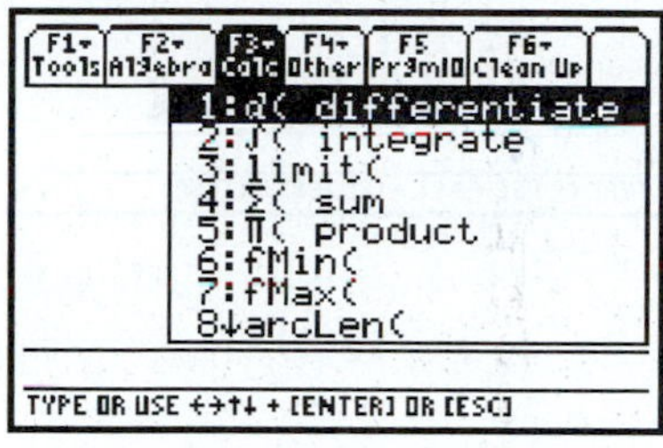

Options 1 and 2 are available on the keyboard as **2nd** versions of the **8** and **7** keys, respectively.

the **F3 Calculus** menu (bottom)

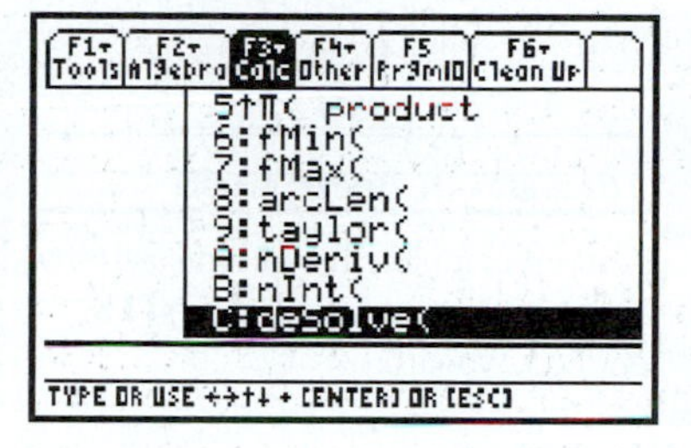

the **F4 Other** menu (top)

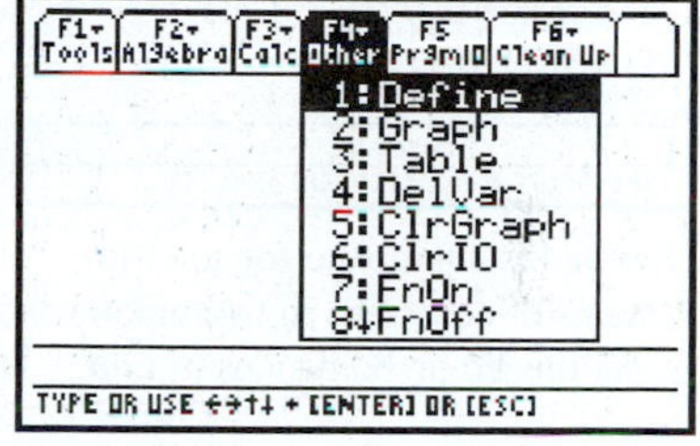

Of particular importance is option 4, which is used to free up a specific variable.

the **F4 Other** menu (bottom)

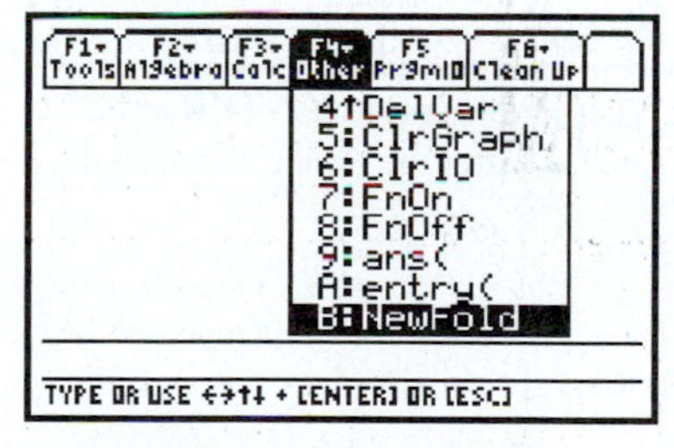

Option 9 allows reference to previous results; ans(1) means the most recent answer, etc. This feature is **2nd** (-) on the keyboard.

F5 Program Input/Output screen

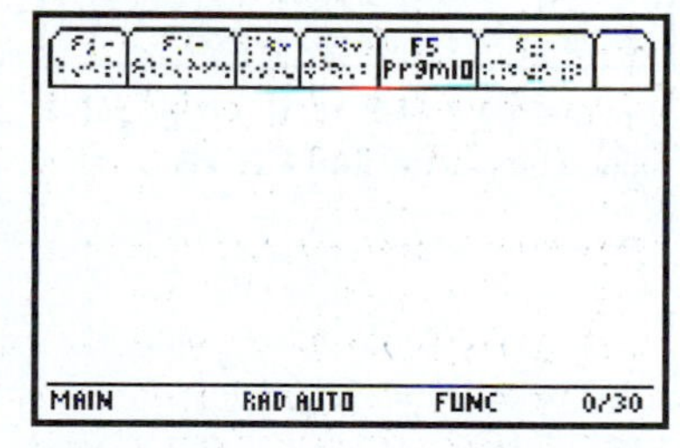

The **F5** key accesses a specialized screen for user-written programs (rather than a menu). Press **F5**, **ESC**, or **HOME** to exit.

the **F6 Clean Up** menu

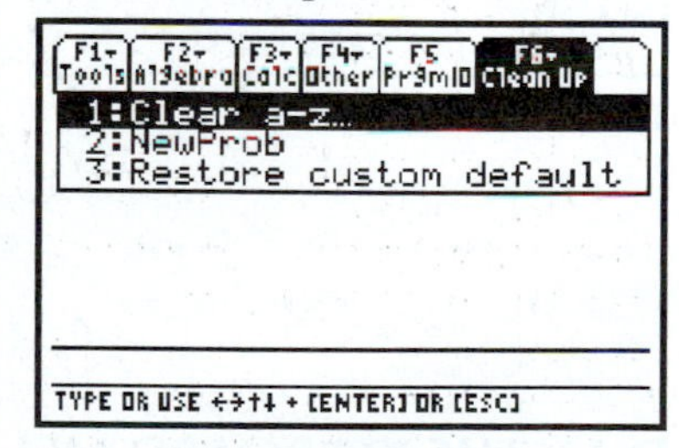

Clear a-z frees up only these variables. **NewProb** clears a-z, turns off all functions, and clears all screens.

The (Default) CUSTOM Menus for the Home Screen

Press **2nd CUSTOM** (the **2nd** version of the **HOME** key) to switch between the standard home screen menus and the custom menus for the home screen.

the **F1 Variables** menu (top)

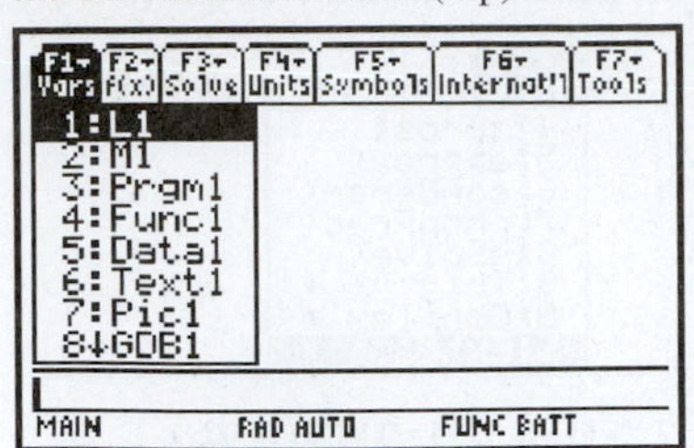

the **F1 Variables** menu (bottom)

The **F1 Variables** menu assumes that many people will name their lists L1, L2, … , their matrices M1, M2, … , and so on. These are *not* predefined variable names on the TI-89.

the **F2 f(x)** menu

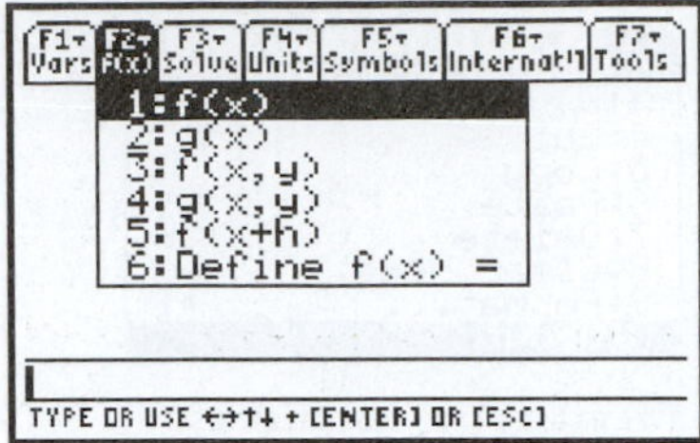

This menu saves many keystrokes. Option 6 is particularly useful in setting up function notation.

the **F3 Solve** menu

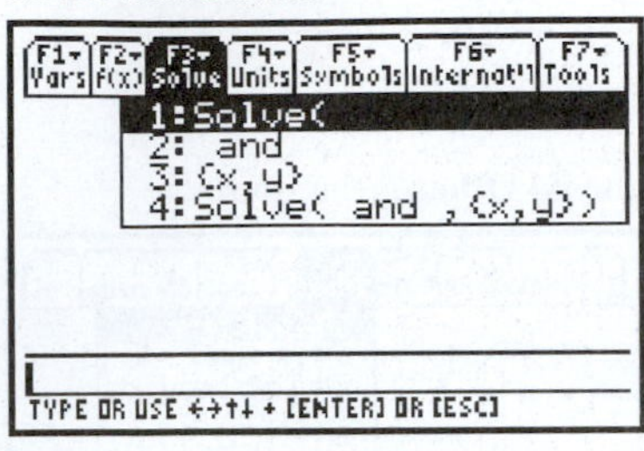

Option 4 is a template for solving systems of equations in two unknowns. Equations go on both sides of **and**.

the **F4 Units** menu (top)

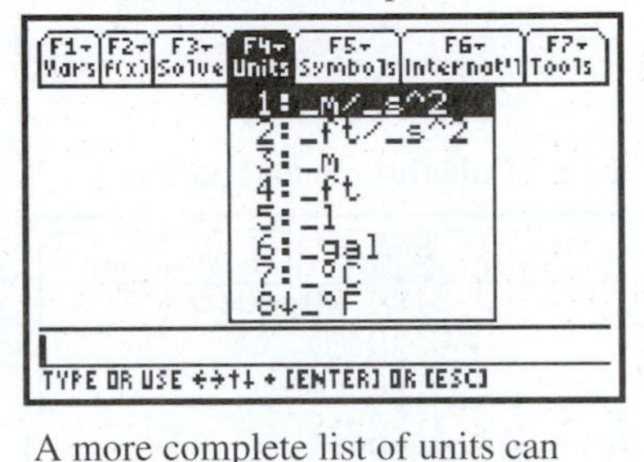

A more complete list of units can be found by pressing **2nd UNITS** on the keyboard.

the **F4 Units** menu (bottom)

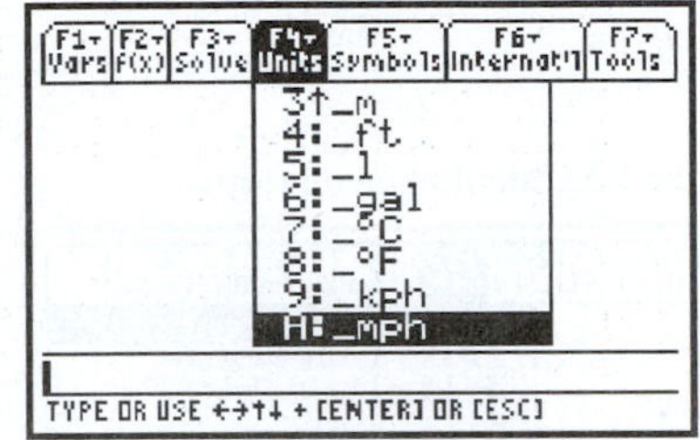

the **F5 Symbols** menu

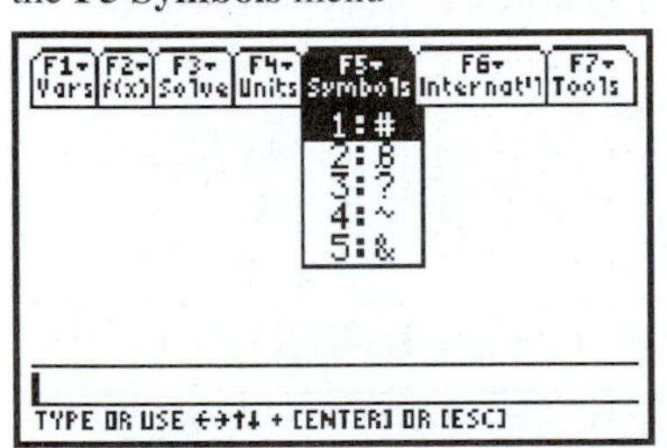

These symbols are seldom used.

the **F6 International** menu (top)

For a more complete list of special characters, press **2nd CHAR**.

the **F7 Tools** menu

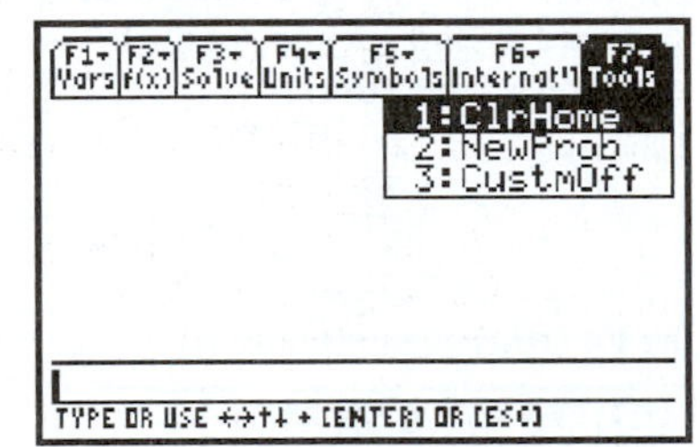

These tools require **ENTER** to be pressed after the option is selected.

D.4 THE KEYBOARD MENUS

Eight other menus are shown only when you specify them from the keyboard. Three of the keyboard menus have their own dedicated key (**APPS, MODE,** and **CATALOG**); the five others (**MATH, MEM, VAR-LINK, UNITS,** and **CHAR**) are accessed as **2nd** versions of keys in the lower right-hand portion of the keyboard. All of the keyboard menus can be accessed from any screen on the calculator (not just the home screen).

The most important navigation menu is accessed by pressing the (blue) **APPS** key (see the following two screens). Each of the options activates one of the calculator's specialized application screens. The first five options are also readily available from the keyboard (the **HOME** key and the **green diamond** versions of **F1, F2, F3,** and **F5,** respectively). Newer versions of the TI-89 show option 1 as **FlashApps** instead of **Home**.

the **APPS** menu (top)

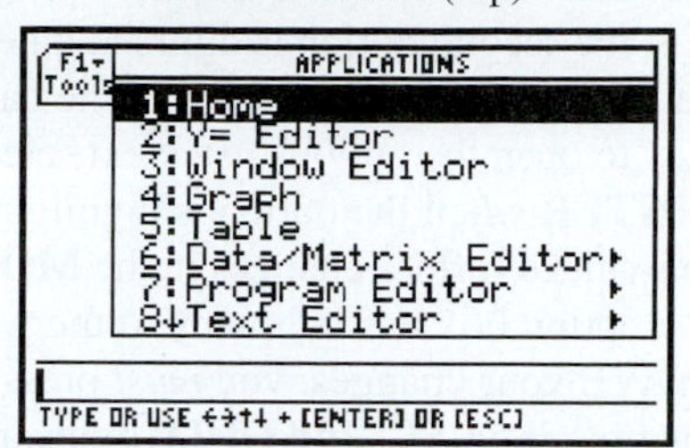

the **APPS** menu (bottom)

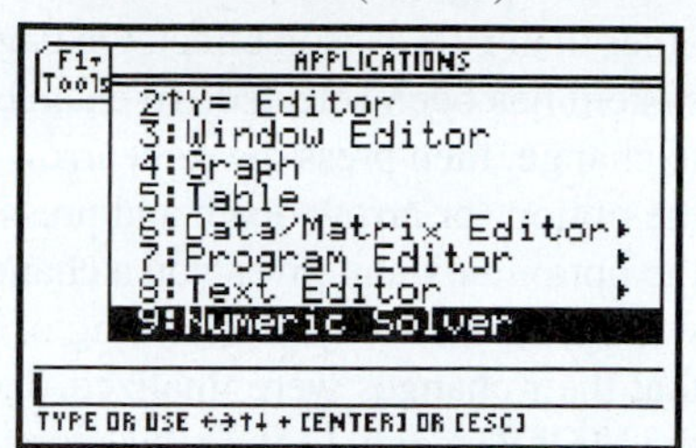

The MODE Options

Pressing the **MODE** key reveals page 1 of the three-page **MODE** menu (use **F2** and **F3** to change pages).

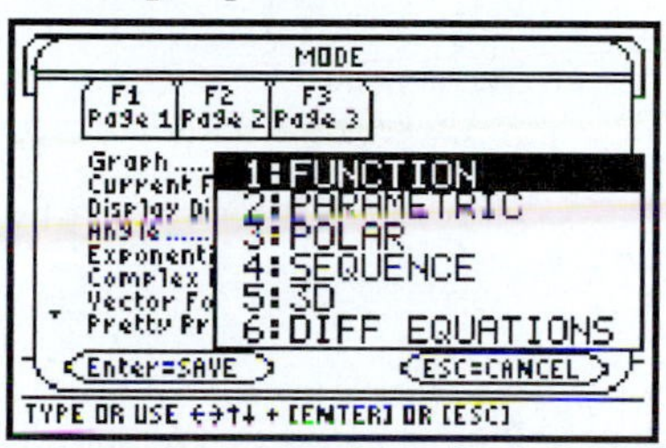

Page 1

Page 2

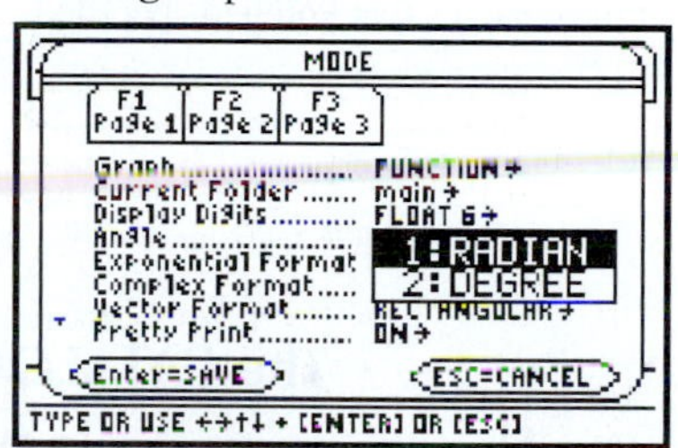

Page 3

the **Graph** options

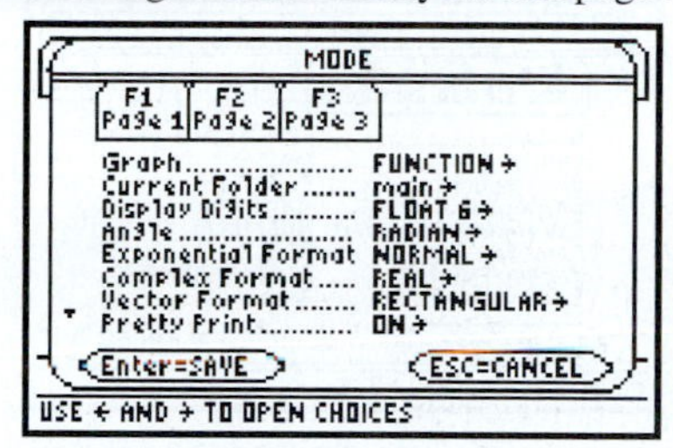

the **Display Digits** options

the **Angle** options

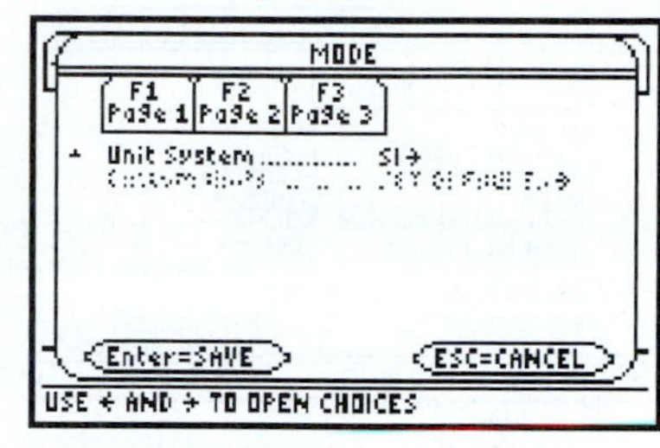

the **Exponential Format** options

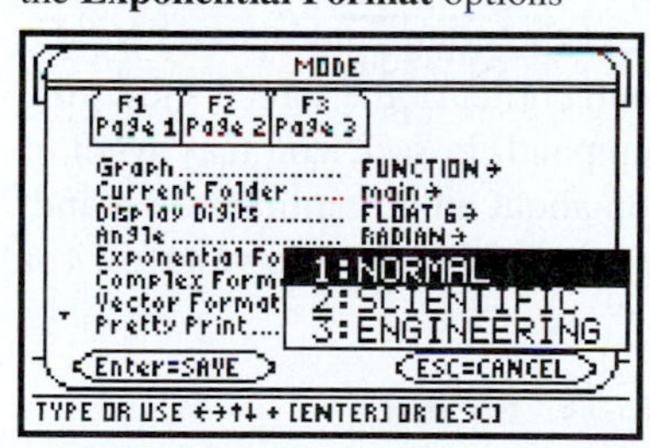

the **Complex Format** options

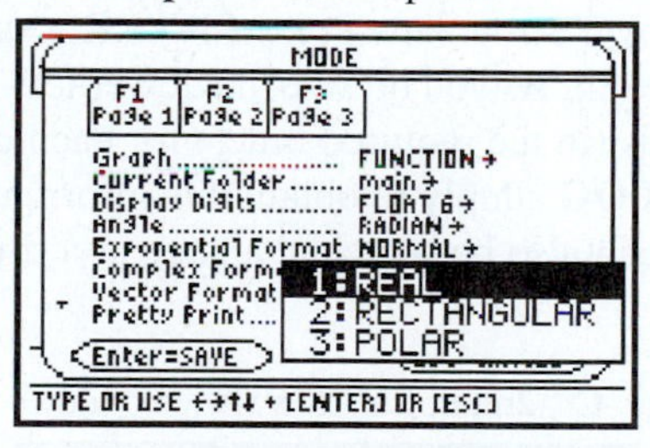

the **Vector Format** options

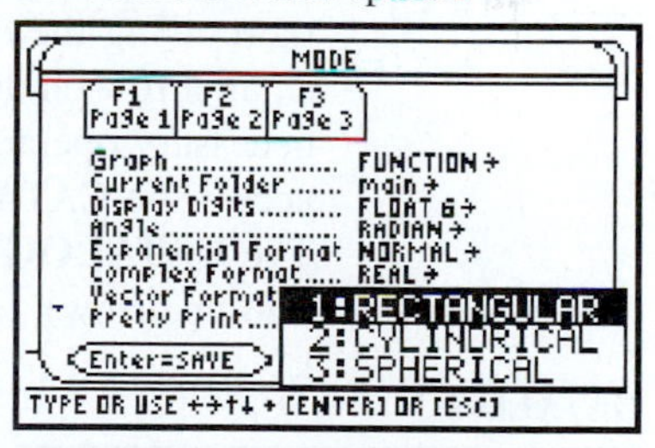

F2 the **Split Screen** options

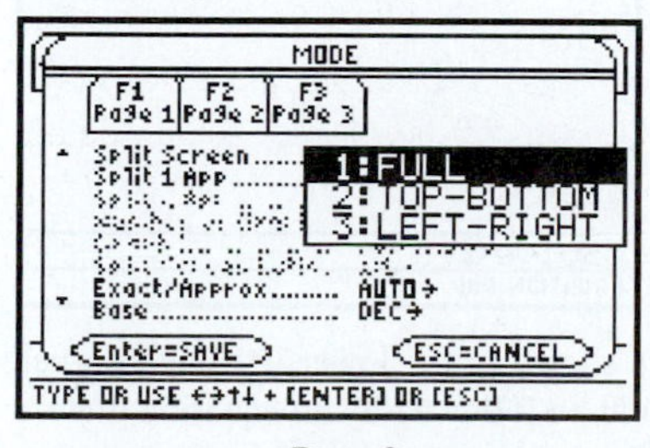

Page 2

the **Exact /Approx.** options

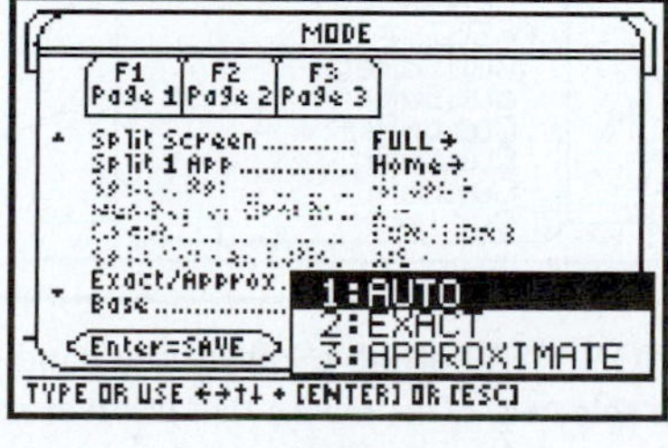

Page 2

F3 the **Unit System** options

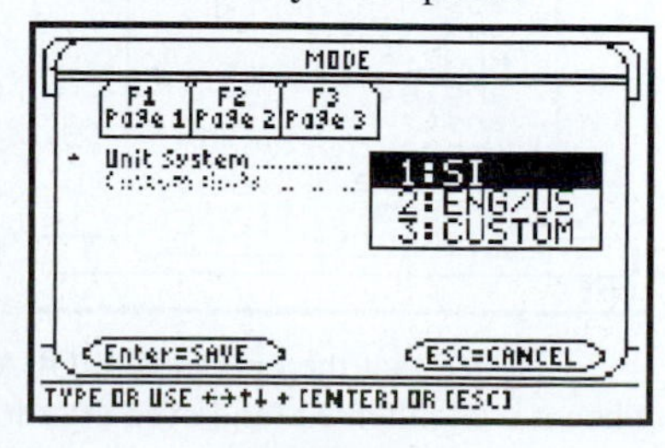

Page 3

On page 2, many of the options are dimmed (not available) unless a split screen mode other than **FULL** is in effect. On page 3, the second option is dimmed unless a custom unit system has been selected. To change an option, first arrow down to the category you wish to change, then press the right arrow key to open the options list. Next select the number of the option (or arrow down and press **ENTER** when the option is highlighted). This closes the option box and gives you a chance to make further changes in the **MODE** menu if you wish. Unfortunately, this closing of the option box has led many to the wrong conclusion that their changes were finalized. To SAVE your changes, you *must* press **ENTER** to exit the **MODE** menu. Exiting this menu by pressing **ESC, 2nd QUIT,** or an application screen key (like **green diamond Y=**, etc.) will *not* make any of the changes that you had specified! The following example shows how to set the calculator's Angle status to **DEG**rees.

MODE three down arrows right arrow down arrow **ENTER** **ENTER**

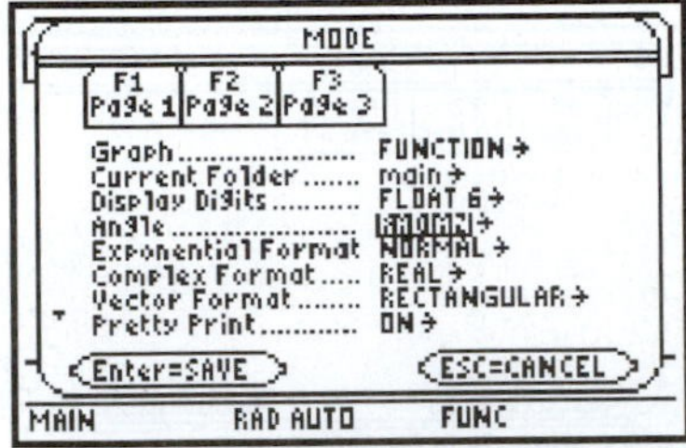

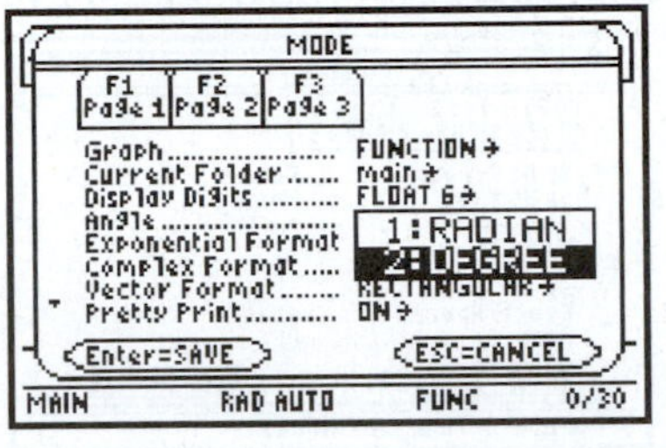

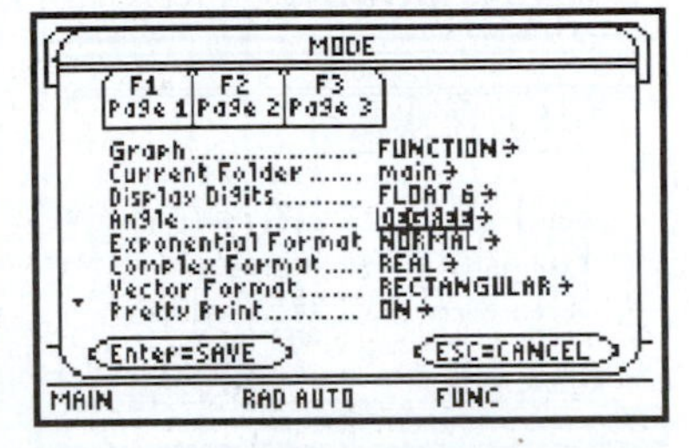

Note that even after **DEGREE** has been selected in the last frame, the calculator is still in **RAD**ian mode (as indicated in the status line at the bottom of the screen)! None of your changes are put into effect until you press **ENTER** to exit the **MODE** menu. Exiting this menu in any other manner (**ESC, 2nd Quit, green diamond Y=**, etc.) discards all of the changes that you had specified.

The CATALOG

Pressing the **CATALOG** key reveals an alphabetized listing of almost every TI-89 feature. Press the first letter of the desired command (without pressing **alpha**), then scroll from there using the down arrow key. Press **ENTER** to select the command marked by the small arrow on the left. As you browse, the lower left-hand corner of the screen shows the necessary operands (in the required order) for each command. In fact, you may want to access the **CATALOG** simply to obtain this information about an unfamiliar command. The **CATALOG** can also be searched a page at a time by pressing **2nd** down arrow (or **2nd** up arrow).

CATALOG C **2nd** down arrow (twice) down arrows **ENTER**

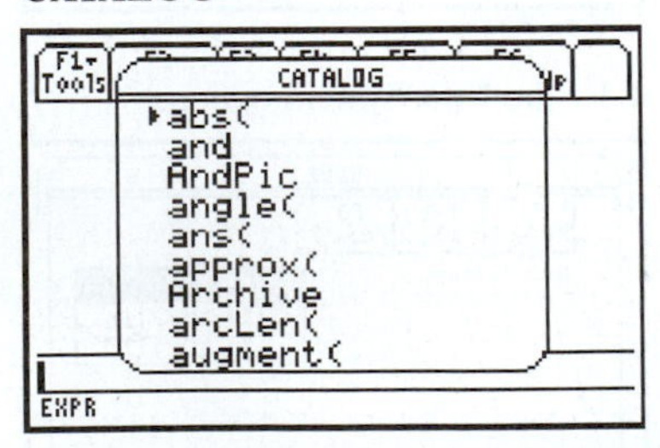

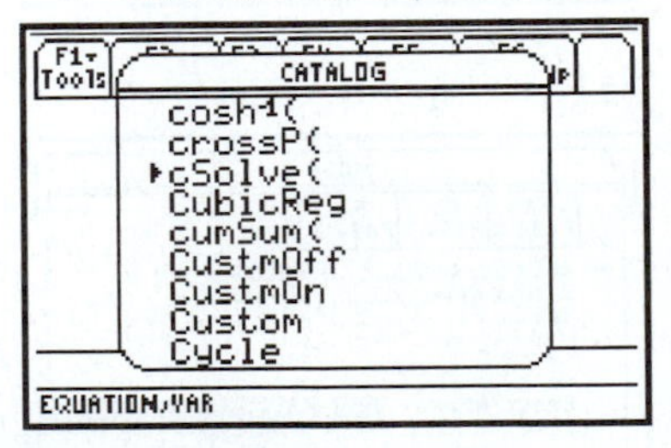

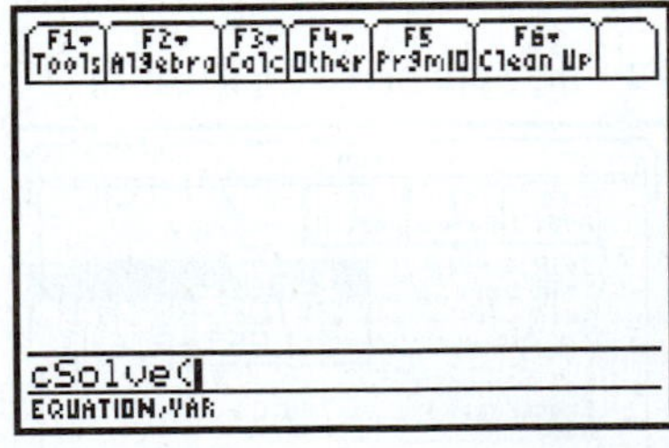

Notice that the operands "EQUATION,VAR" of the **cSolve(** command are shown in the lower left-hand corner when it is highlighted (and even after it is selected). This means that **cSolve(** should be followed by an equation, then a comma, then the name of a variable. Of course, be sure to include a closing parenthesis.

The MATH Submenus

Pressing **2nd MATH** (**2nd**, then the **5** key) gives access to more than 100 different commands. Because there are so many options available, they have been divided into 13 submenus. On the TI-89, a submenu is marked by a small arrow to the right of its name. You can open a submenu just by pressing its number or letter, or you can arrow down to the submenu you want, then press the right arrow key (or **ENTER**) to open it. Press the left arrow key (or **ESC**) if you want to close a submenu without selecting one of its options. Some of the most important **MATH** submenus are shown in the frames that follow.

2nd MATH (top of **MATH** menu)

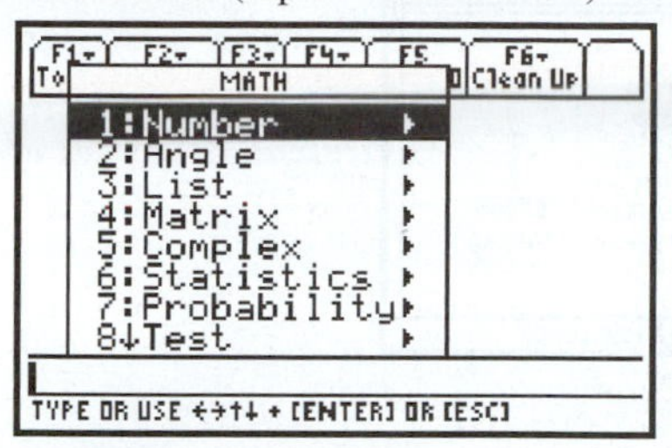

1 (top of the **Number** submenu)

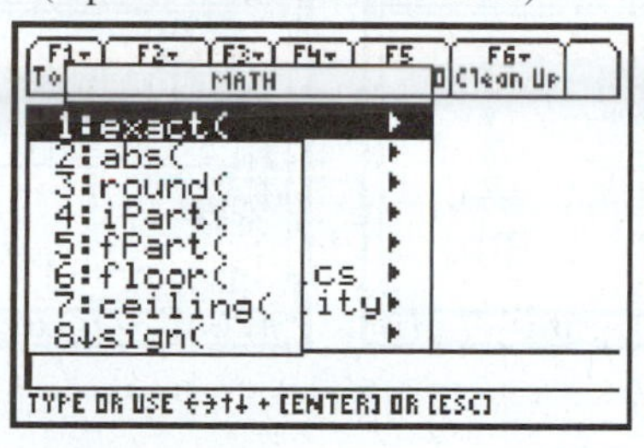

(bottom of the **Number** submenu)

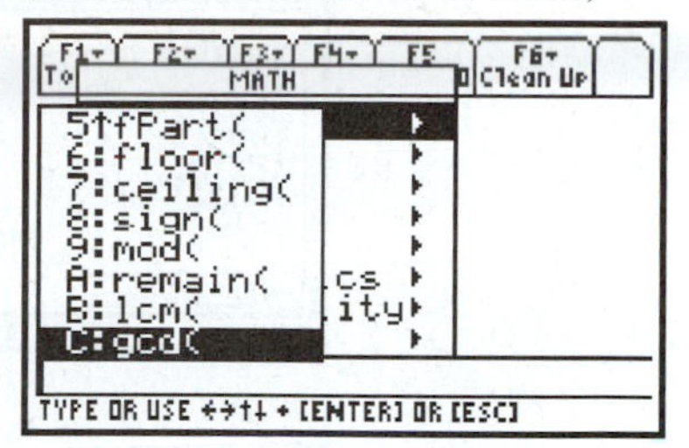

2nd MATH 2 (**Angle** submenu)

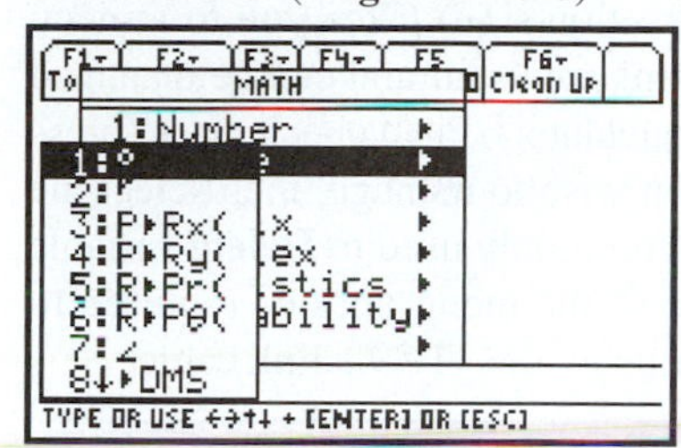

2nd MATH 3 (**List** submenu)

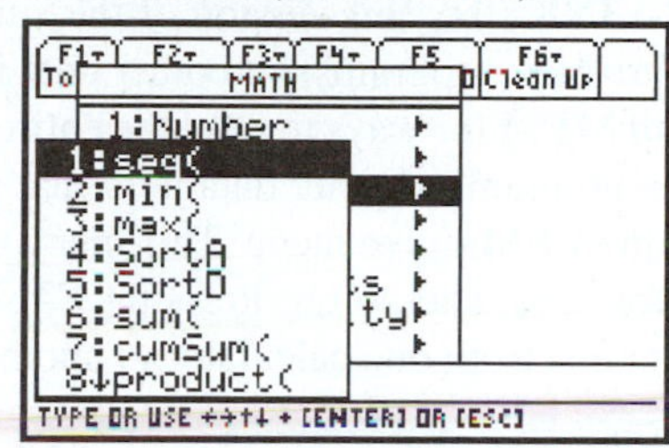

2nd MATH 4 (**Matrix** submenu)

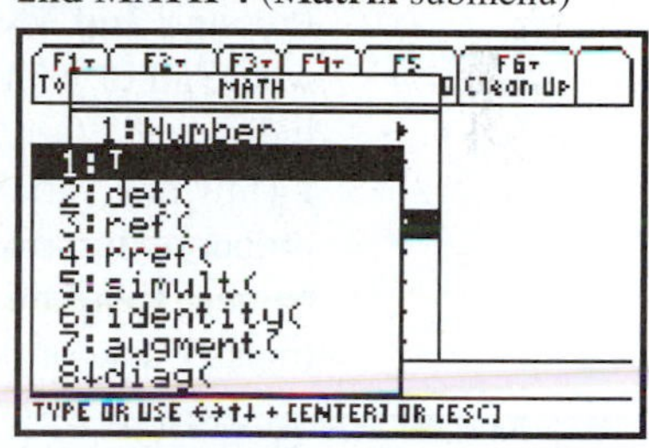

2nd MATH 5 (**Complex** submenu)

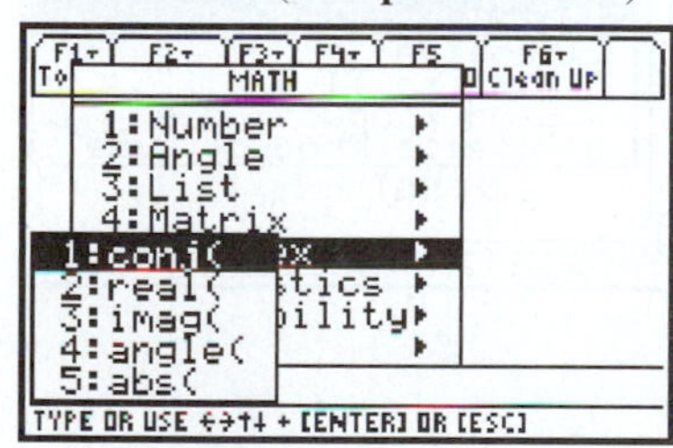

2nd MATH 6 (**Statistics** submenu)

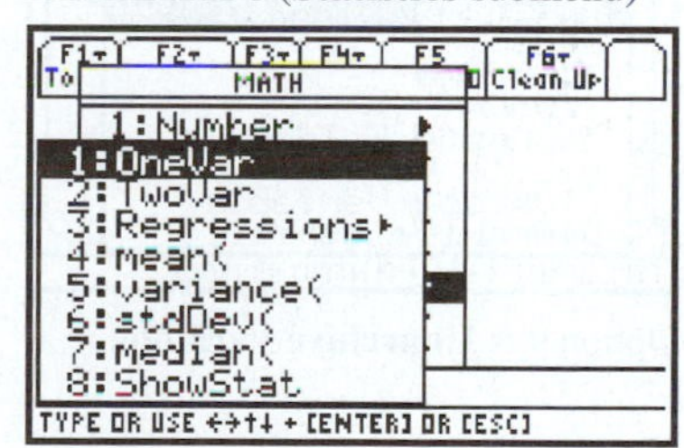

3 (**Regressions** sub-submenu)

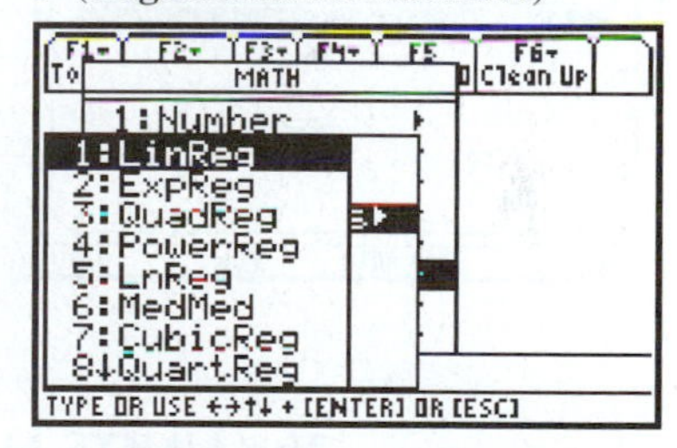

2nd MATH 7 (**Probability** submenu)

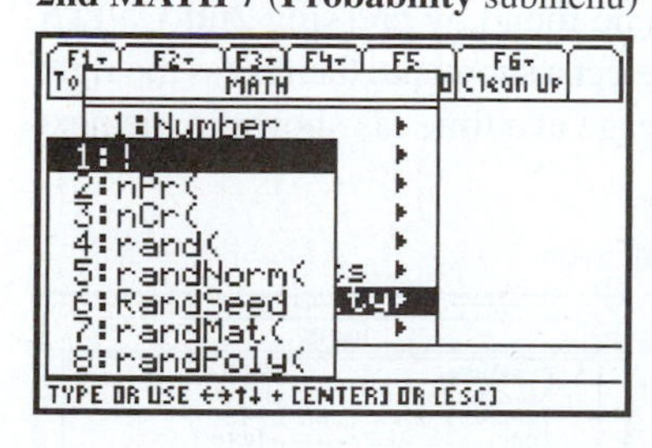

2nd MATH 8 (**Test** submenu)

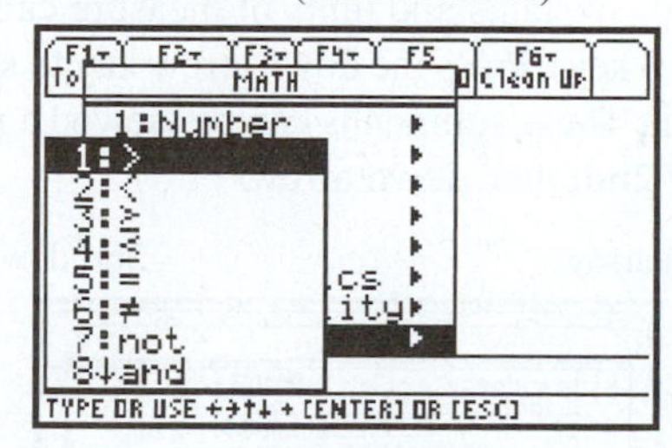

(bottom of the **MATH** menu)

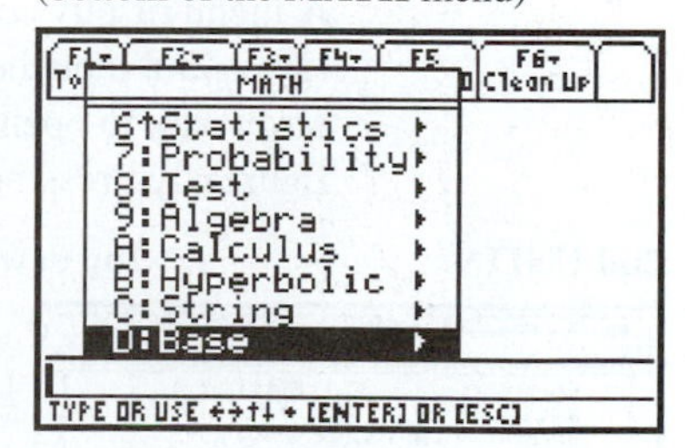

The bottom five submenus of **MATH** (shown in the last frame) include exact duplicates of the standard home screen **Algebra** (**F2**) and **Calculus** (**F3**) menus as well as access to **Hyperbolic** trig functions, **String** manipulation features, and other number **Bases**.

The MEMory Menu

Press **2nd MEM** (**2nd**, then the **6** key) to see how much memory is free and how much is allocated to the various data types in storage. The only memory management tool available in this menu is **F1 RESET**. Look for less drastic memory management options under the **VAR-LINK** menu. Of particular interest is option **3 Default**, which you used earlier to set your calculator to its original factory settings without disturbing any other data or programs.

Newer versions of the TI-89 will have a more elaborate **RESET** menu.

2nd MEM **F1**

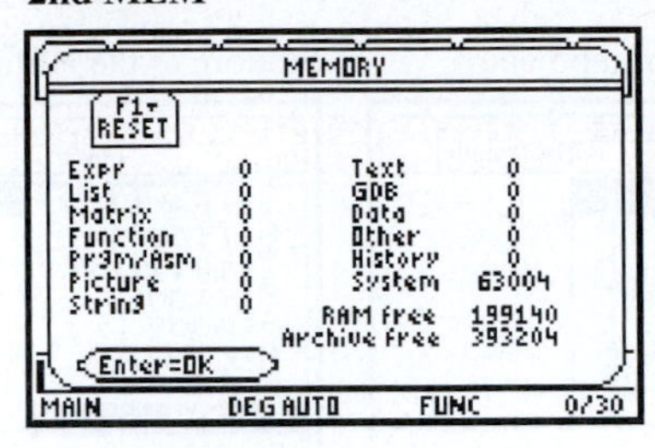

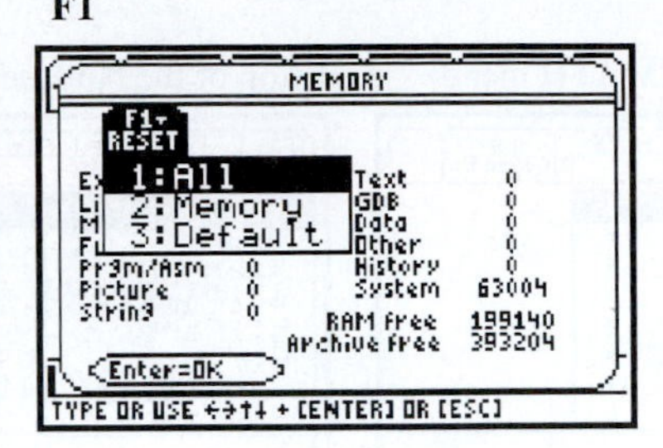

The VAR-LINK Menu

Pressing **2nd VAR-LINK** (the **2nd** version of the subtraction sign) takes you to a menu where all of your variables, programs, and other data items are listed and can be managed individually (see **2nd MEM** for ways to reset the entire calculator). Put a check mark (press **F4**) by any variable, program, or other data item that you wish to manage, then select the proper option from the **F1 Manage** menu. This menu is commonly used to **Delete** variable names, **Rename** programs, and so on. Pressing **F3** shows the menu options required to transfer data or programs from one calculator to another using the TI-89's link cable.

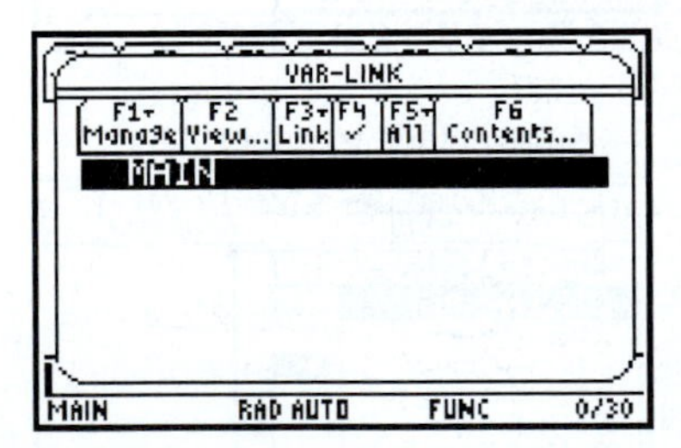

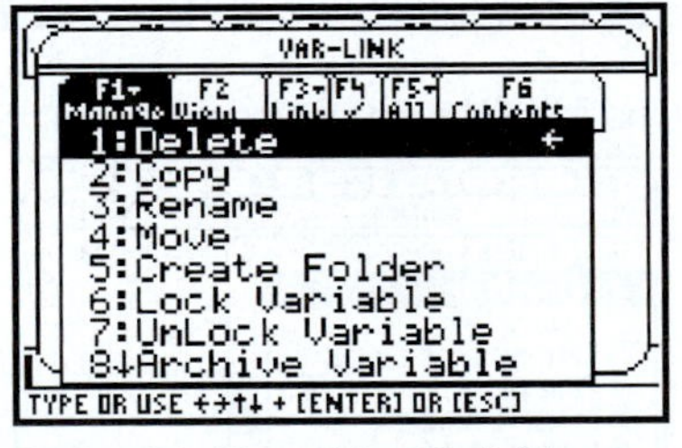

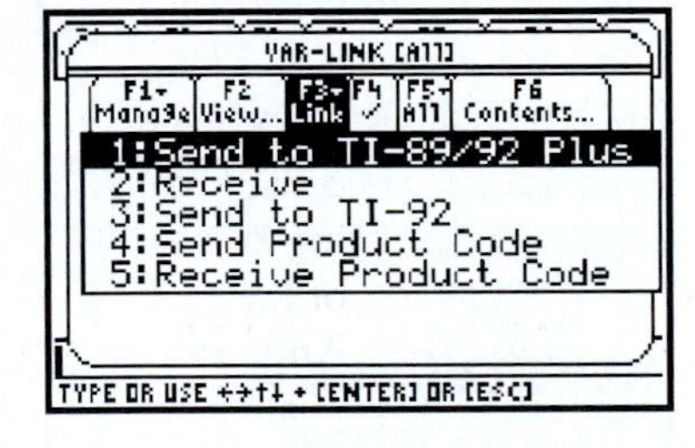

Option 9 is **Unarchive Variable.**

The UNITS Menu

A menu of physical constants and units of measure can be found by pressing **2nd UNITS** (press **2nd** then the **3** key). Press the down arrow key to select a submenu, then press the right arrow key to open it. These submenus can be viewed a page at a time (as shown in the next figure) by pressing **2nd**, then down arrow.

2nd UNITS **2nd** down arrow **2nd** down arrow

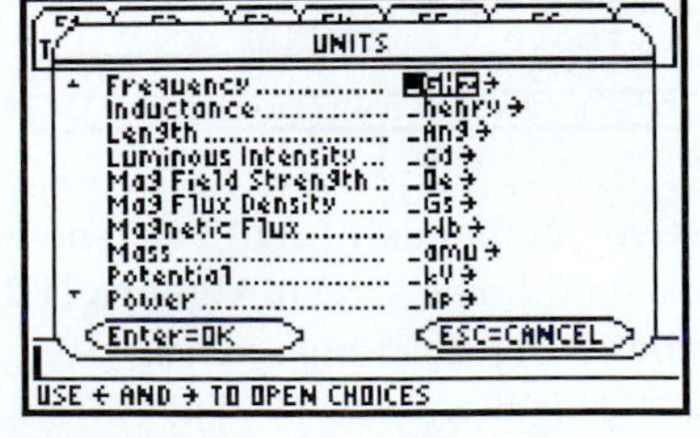

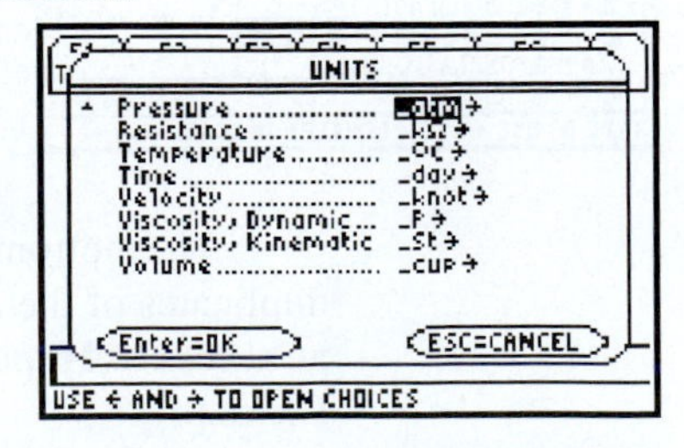

The CHARacter Menu

Press **2nd CHAR** (**2nd,** then the plus sign) to reveal a menu that contains more than 90 special characters and mathematical symbols. They are organized into five submenus.

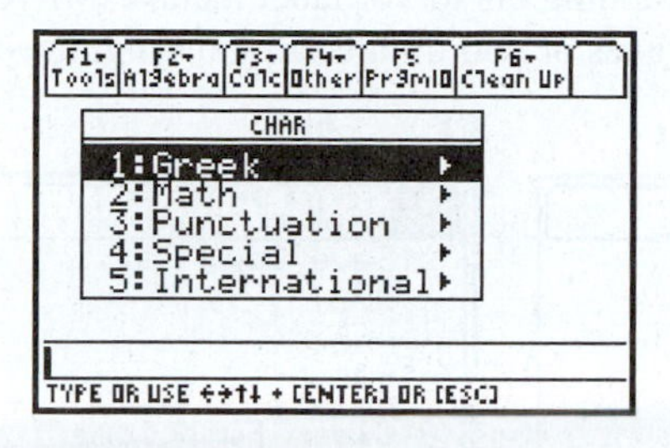

D.5 GRAPHING FUNCTIONS

The **green diamond** versions of the thin blue buttons along the top row of the calculator access most of its graphical features. The **green diamond Y=** key provides access to the calculator's list of 99 functions (assuming that the calculator's graph **MODE** is **Function**). Pressing **green diamond Y=** will most likely reveal functions y1 through y7. Press **2nd**, then the down or up arrow key to view the functions a page at a time. These functions are part of the calculator's memory, but the information stored on this screen can be easily edited, overwritten, or **CLEAR**ed. Functions are selected for graphing (turned "on") if they have a check mark immediately to the left of the function's name. The check mark appears automatically when you type in a new function or modify an old one. In other cases, to change the status of a function (from "off" to "on" or vice versa), you will need to use the up and down arrow keys to highlight the function, then press **F4** (the check mark). Unmarked functions are stored in memory, but will *not* be graphed. The default graphing style is **Line**, which means that the individually calculated points will be connected with thin line segments (giving an appearance of continuity). The optional graphing styles can be found by pressing **2nd F6**. These include options to plot only isolated **Dot**s or (larger) **Square**s for each of the calculated points, an option to plot a **Thick** continuous graph, an option that will **Animate** the curve as a path of motion and another that will mark that **Path** during the animation. There are also options to shade **Above** or **Below** the curve. To view the style setting of a particular function, highlight that function, then press **2nd F6** (**2nd**, then the **F1** key). Its current style setting is indicated by a check mark. Of course, you can also change the style setting in this menu by pressing the number of the appropriate option.

Pressing **green diamond WINDOW** allows you to *manually* specify the extents of the x- and y-values that will be visible on the calculator's graphing screen (see **ZOOM** for *automatic* ways of doing this). The **xscl** and **yscl** options specify the meaning of a mark on the x- or y-axis. For example, **xscl=5** means that each mark shown on the x-axis will mean an increment of 5 units (**xscl=1** is a common setting for algebraic functions; **xscl=**π**/2** is commonly used when graphing trigonometric functions). The last option, **xres**, allows you to control how many points will actually be calculated when a graph is drawn. Setting **xres=1** means that an accurate point will be calculated for each pixel on the x-axis (slow, but very accurate). The default setting is **xres=2**, which will calculate a point at every other pixel. The fastest (but least accurate) setting, **xres=10**, calculates a point only at every 10th pixel on the x-axis. In the examples that follow, all graphs are shown with **xres=2** unless otherwise indicated.

To see how the graphing screen is presently formatted, press **green diamond Y=** or **green diamond GRAPH**, then press **F1** followed by **9**. This menu contains options that allow you to change the way coordinates are displayed (polar instead of rectangular, or off completely, inhibiting some **Trace** features), change the way that two or more graphs will be drawn (**Seq**uentially or **Simult**aneously), provide a coordinate grid, hide the axes, provide a graphing cursor, or label the axes. If you find your graphs looking cluttered or notice that axes or coordinates are missing, the default settings are shown in the next figure.

green diamond Y= **F1 9** (These are the default graph formats.)

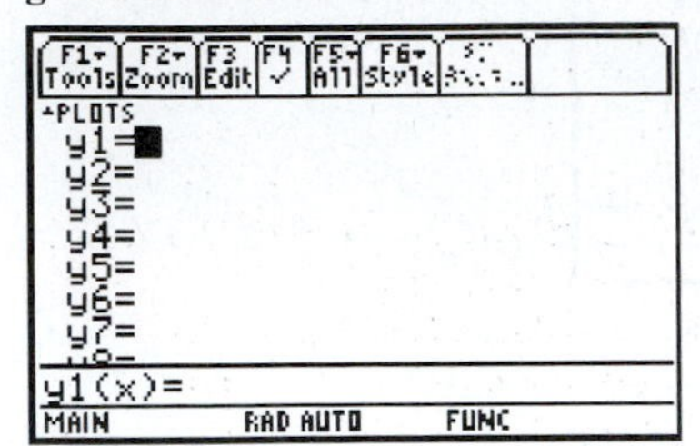

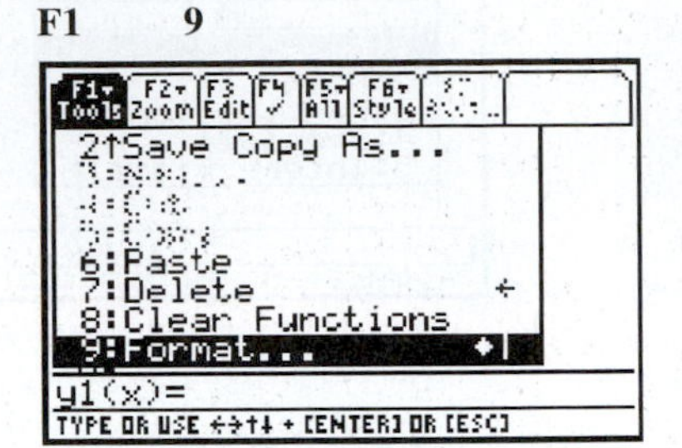

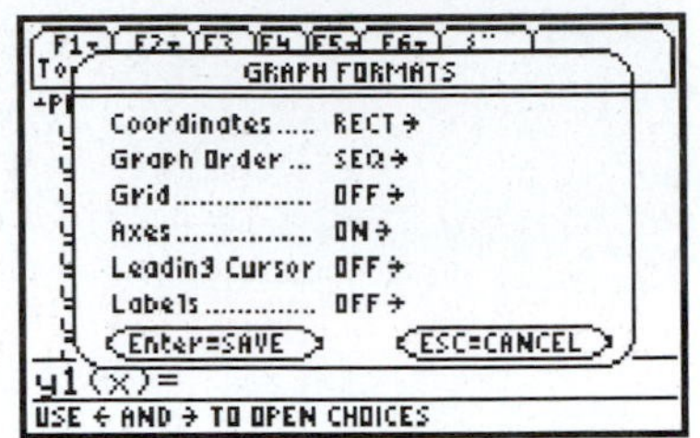

The **F2 Zoom** menu can be found on both the **Y=** editor and the graphing screen. It provides access to *automatic* ways to set the graphical viewing window. **ZoomStd** (option 6) is usually a good place to start, but you should consider option 7, **ZoomTrig,** if you're graphing trigonometric functions. **ZoomStd** shows the origin in the exact center of the screen with x- and y-values both ranging from -10 to 10. **ZoomStd** also sets **xscl=1**, **yscl=1**, and **xres=2**. From here you can **ZoomIn** or **ZoomOut** (options 2 and 3), or draw a box around a portion of the graph that you would like magnified to fit the entire screen (option 1, **ZoomBox**). There is also an option to "square up" your graph so that units along the x-axis are equal in length to units along the y-axis (option 5, **ZoomSqr**); the *smaller* unit length from the axes of the previous graph will now be used on both axes. This option makes the graph look more like it would on regular graph paper; for example, circles really look like circles. **ZoomDec** and **ZoomInt** (options 4 and 8) prepare the screen for **Traces**, which will utilize x-coordinates at exact tenths or integer values, respectively. Option 9, **ZoomData**, makes sure that all of the data in a statistical plot will fit in the viewing window. Option A, **ZoomFit**, calculates a viewing window using the present x-axis, but adjusting the y-axis so that the function fits neatly within the viewing window. All of these options work by making automatic changes to the **WINDOW** settings. Want to go back to the view you had before? The **Memory** submenu (**Zoom** option B) includes features to go back to your immediately previous view (**Memory** option 1, **ZoomPrev**) or to a window setting you saved a while ago (**Memory** option 3, **ZoomRcl**). **ZoomSto** (**Memory** option 2) is the way to save the current window setting for later (note that it can only retain one window setting, so the new information replaces whatever setting you had saved before). **Zoom** option C, **SetFactors...**, gives you the chance to control how dramatically your calculator will **ZoomIn** or **ZoomOut**. These zoom factors are set by Texas Instruments for a magnification ratio of 4 on each axis. Many people prefer smaller factors, such as 2 on each axis. It is possible to set either factor to any number greater than or equal to 1; they don't need to be whole numbers, and they don't necessarily have to be equal.

The **F3 Trace** feature is found only on the graphing screen. Its purpose is to display the x- and y-coordinates of specific points as you trace along a curve using the left and right arrow keys. Note that in **Trace**, the up and down arrow keys are used to jump from one curve to another when several curves have been drawn on the same screen. The subscript of the function you are presently tracing is shown in the upper right-hand corner of the screen. If you press **ENTER** while in **Trace**, the graph will be redrawn with the currently selected point in the exact center of the screen, even if that point is presently outside the current viewing window. This feature is a convenient way to pan up or down to see higher or lower portions of the graph. It is also the easiest way to locate a "lost" graph that doesn't appear anywhere in the current viewing window (just press **F3 Trace**, then **ENTER**). To pan left or right, just press and hold the left or right arrow key until new portions of the graph come into view. These useful features change the way the graph is centered on the screen without changing its magnification. Note also that these recentering features work *only* in **Trace**. To exit **Trace** without disturbing your view of the graph, just press **CLEAR** (or **green diamond GRAPH**). To return to the home screen, abandoning both **Trace** and the graph, just press **HOME** or **2nd QUIT** (or press **CLEAR** twice). Note that using any **F2 Zoom** feature also causes an exit from **Trace** (to resume tracing on the new zoomed version, you must press **F3 Trace** again).

Pressing **green diamond GRAPH** takes the calculator from any screen or menu to the current graph. Note that the calculator is smart enough that it will redraw the curves only if changes have been made to the function list (**green diamond Y=**). As previously mentioned, **green diamond GRAPH** can be used to turn off **Trace**. You can also hide an unwanted free cursor by pressing **green diamond GRAPH**. When you're finished with viewing a graph, press **HOME** or **CLEAR** or **2nd QUIT** to return to the home screen.

D.6 EXAMPLES OF GRAPHING

This appendix is designed to explain graphing calculator features rather than mathematics itself. Accordingly, the example format in this and subsequent sections is different from that found in the body of the text.

EXAMPLE 1

To graph $y = x^2 - 5x$, first press **green diamond Y=**, then highlight y1 (using the up arrow key, if necessary) and press **CLEAR** to erase its current formula (or use the down arrow key to find a blank function), then press **ENTER** or **F3** (**Edit**), which activates the editor line near the bottom of the screen. Type the function by pressing the **X** key, followed by the ^ button and **2**; now press the (black) minus sign key, then **5**, followed immediately by the **X** key (a multiplication sign is not needed). Your screen should look very much like the first frame shown in the following figure. Note that you are not required to press **ENTER** when you have finished typing a function's formula. To set up a good graphing window, press **F2** (**Zoom**) then **6** to choose **ZoomStd**. This causes the graph to be immediately drawn on axes that range from -10 to 10. Notice that you did not have to press **green diamond GRAPH**; the **F2 Zoom** menu items also activate the graphing screen. If you have one or more unwanted graphs drawn on top of this one, go back to your function list (**Y=**) and turn "off" the unwanted functions by highlighting them one at a time and pressing **F4** to remove their check mark. After you have turned off the unwanted functions, just press **green diamond GRAPH** and you will finally see the last screen below.

green diamond Y= (typing...) **F2 6**

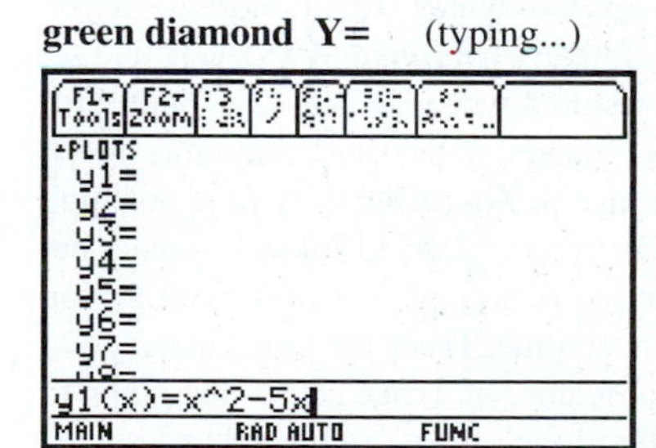

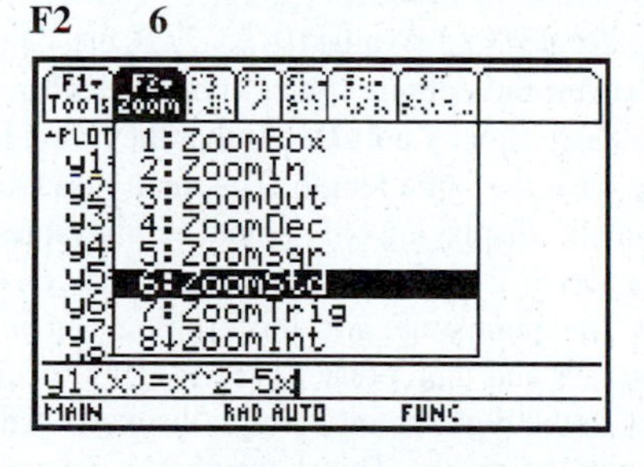

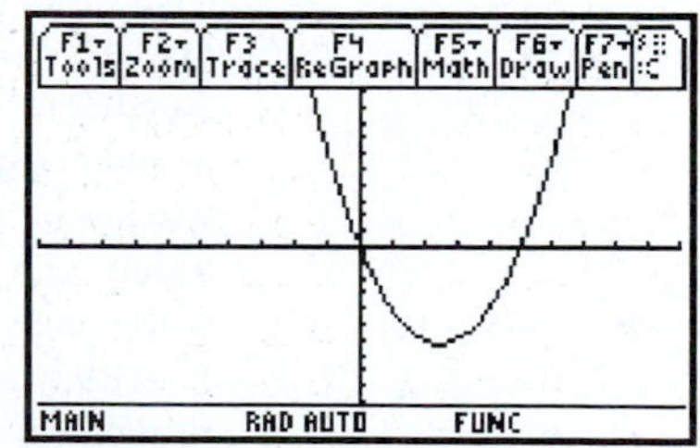

EXAMPLE 2

To modify this function to be $y = -x^2 + 6$, press **green diamond Y=**, then **ENTER** or **F3** (**Edit**). Insert the negative sign by pressing the left arrow key (which takes you to the left-hand edge of the formula), then the gray sign change key (-); now press the right arrow key three times to skip over the parts of the formula that are to be preserved. Type the plus sign and the **6**, then press **CLEAR** (only once) to delete the unwanted right-hand end of the formula. Press **green**

diamond GRAPH to plot this function, then press **F3** (**Trace**). **Trace** gives the added bonus of a highlighted point, with its coordinates shown at the bottom of the screen. Press the right or left arrow keys to highlight other points on the curve.

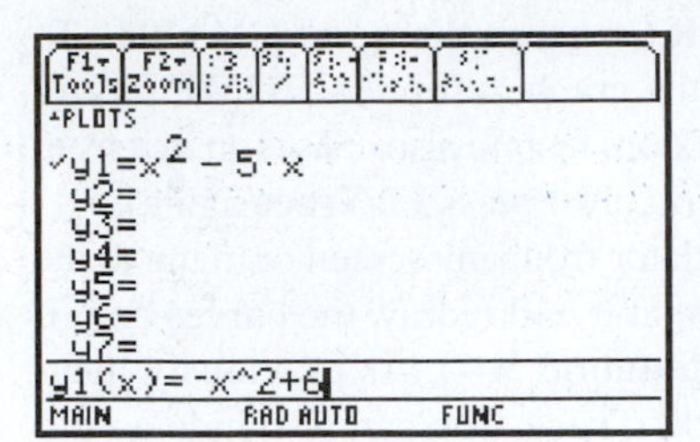

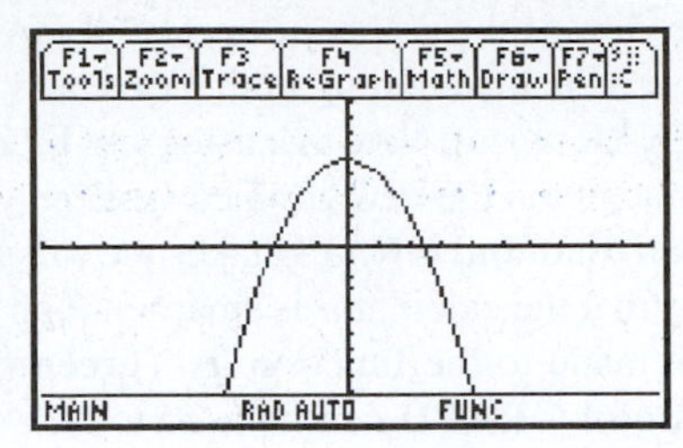

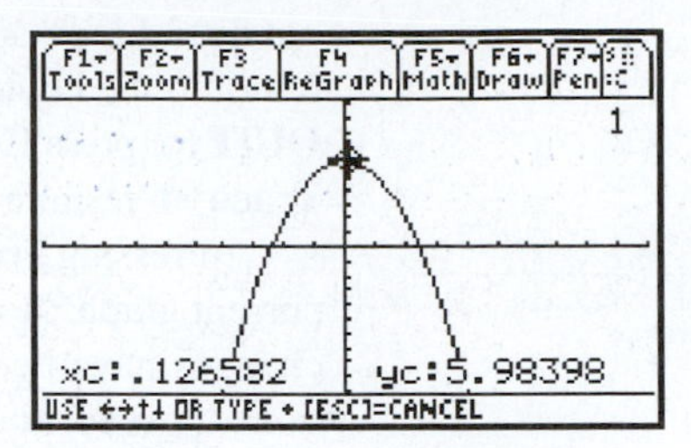

Perhaps $x = .126582$, $y = 5.98398$ was not a coordinate pair that you had expected to investigate. Two special **Zoom** features (options 4 and 8) can be used to make the **Trace** option more predictable. Press **F2 Zoom**, then **4** to select **ZoomDec**; now press **F3 Trace**.

F2 4 **F3**

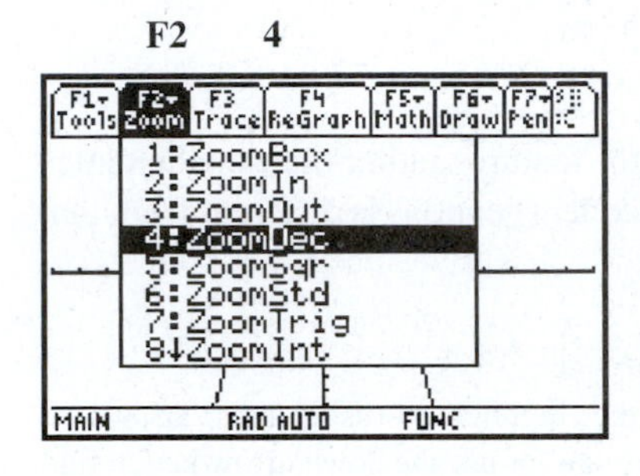

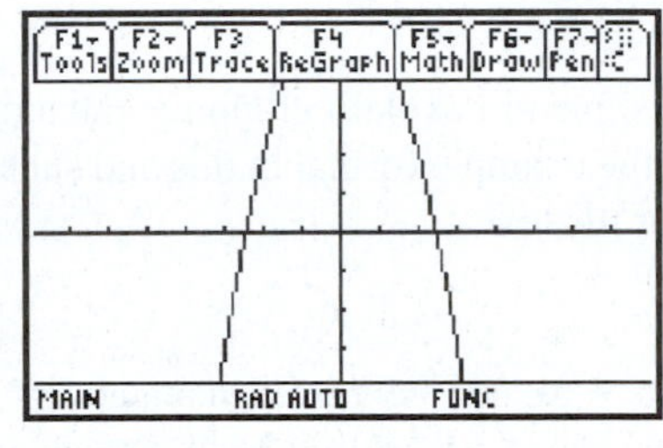

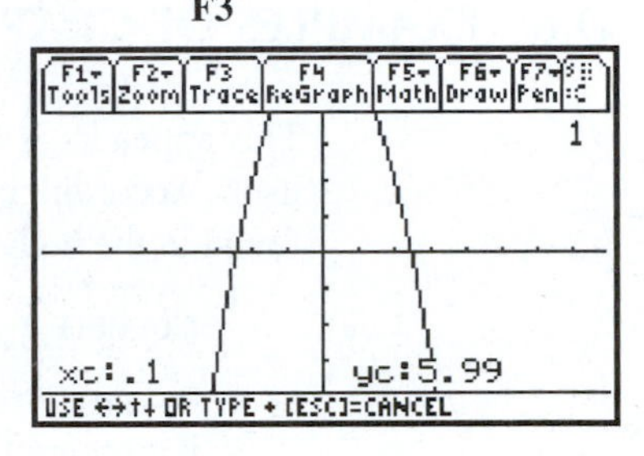

right arrows... **ENTER** (This is how to pan up or down.)

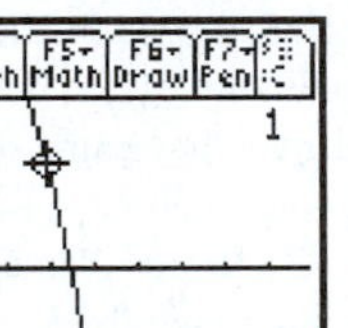

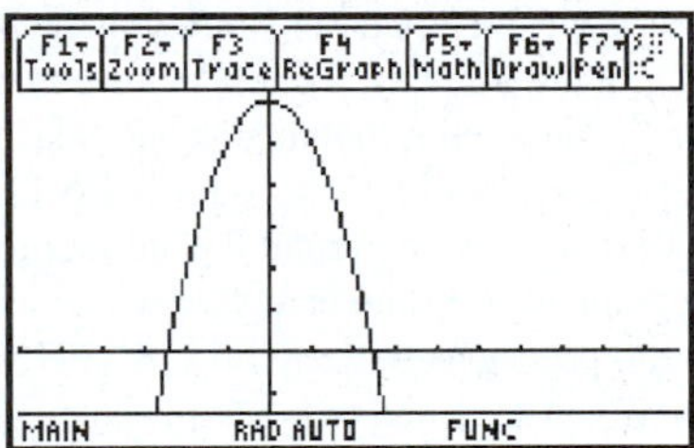

Try pressing the right or left arrow key about 10 times watching the values at the bottom of the screen. You'll quickly notice that the x-values are now all *exact tenths*, .1, .3, .5, and so on. (**Zoom** option 8, **ZoomInt**, produces **Trace**able x-values which are all integers). If you want to see every decimal x-value (0, .1, .2, .3, etc.), press **green diamond WINDOW** and set **xres=1**. However, the default value, **xres=2**, will be restored whenever you use **ZoomStd**. Another nice thing about **ZoomDec** is that the graph is "square" in the sense that units on the x- and y-axes have the same length. The main disadvantage to **ZoomDec** is that the graphing window is "small," displaying only points with x-values between -7.9 and 7.9 and y-values between -3.8 and 3.8. This disadvantage is apparent on the current graph, which runs off the top of the screen. To demonstrate how this problem can be overcome, **Trace** the graph to the point $x = 1.9$, $y = 2.39$ and then press **ENTER**. This special feature of **Trace** causes the graph to be redrawn with the highlighted point in the exact center of the screen (with no change in the magnification of the graph). This is the way to pan up or down from the current viewing window (to pan left or right, see Example 6).

EXAMPLE 3

Another way to deal with the preceding problem is to **ZoomOut**, but first you'll want to set your ZOOM FACTORS to 2 (the factory setting is 4). Press **F2** (**Zoom**), then **alpha C**. Now press **2**, down arrow, **2**, down arrow, **2**, then press **ENTER** twice.

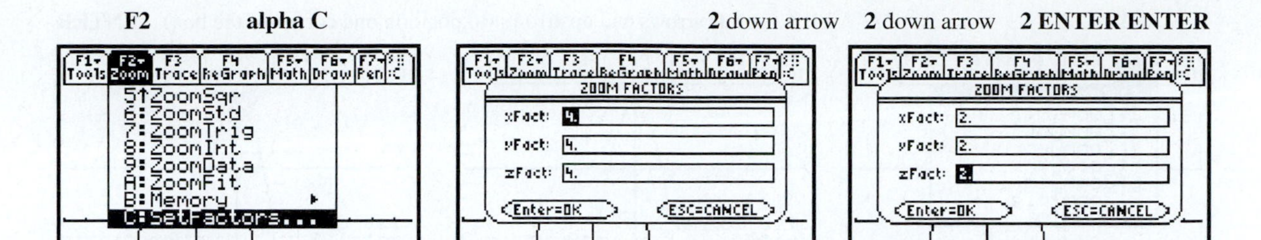

To reproduce our problem, press **F2**, then **4** (**ZoomDec**). However, this time correct it by pressing **F2** followed by **3** (**ZoomOut**). At first glance, it looks like nothing has happened, except that $xc:0$, $yc:0$, and the question "New Center?" are displayed near the bottom of the screen. The calculator is waiting for you to use your arrow keys to locate the point in the current window where you would like the exact center of the new graph to be (then press **ENTER**). Of course, if you like the way the graph is already centered, you will still have to press **ENTER** (you'll just skip pressing the arrow keys). The resulting graph is drawn on a window whose x- and y-axes represent lengths twice as long (remember the ZOOM FACTORS are set to 2) as the original axes. The scaling marks on each axis also represent twice what they originally did (**xscl** and **yscl** are now both equal to 2 instead of 1). Press **F3** (**Trace**) and notice, by pressing the left or right arrow key a few times, that the x-values are now changing by .4 (instead of .2) and the graph is still "square." Other popular square window settings can be obtained by repeating this example with both zoom factors set to 2.5 or 5. These give "larger" windows where the x-values can be changed by .25 or .5, respectively, during a **Trace** (if you also change the **WINDOW** options to **xres=1**).

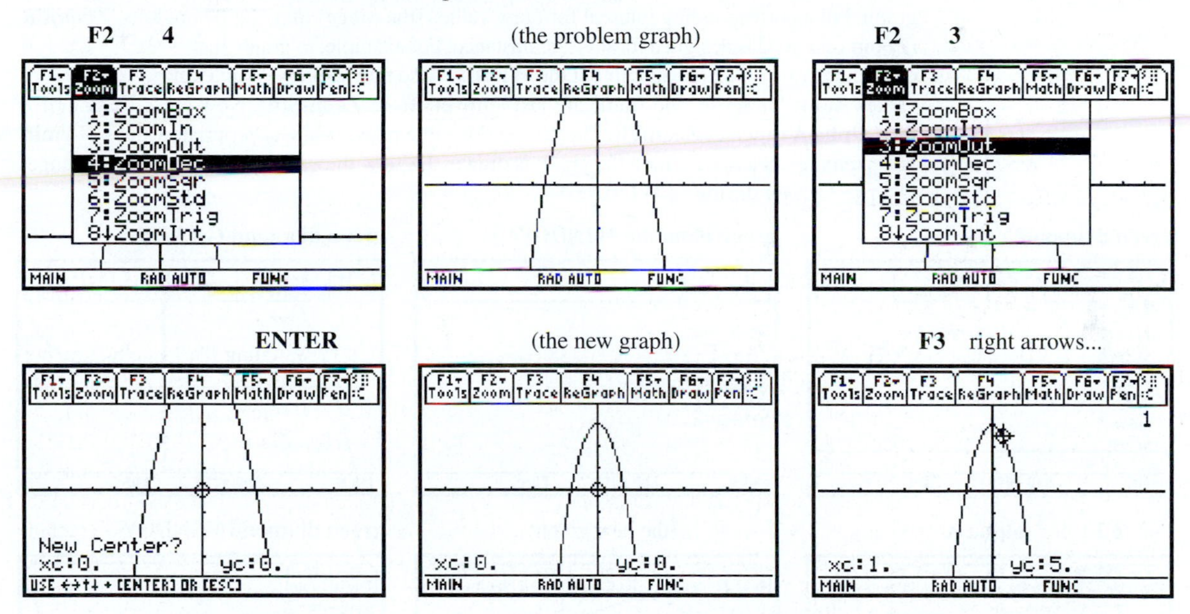

EXAMPLE 4

The extra keystroke needed to activate **ZoomOut** in the previous example has proven to be a bit confusing to beginners who think that **ZoomIn**, **ZoomOut**, and **ZoomInt** should work like the six zoom options (4, 5, 6, 7, 9, and A) that do their job without pressing **ENTER**. The only other zoom option that needs extra keystrokes is **ZoomBox** (option 1). This is a very powerful option that lets you draw a box around a part of the graph that you would like enlarged to fit the entire screen. After selecting this option, use the arrow keys to locate the position of one corner of the box and press **ENTER**. Now use the arrow keys to locate the *opposite* corner, and press **ENTER** again.

F2 1

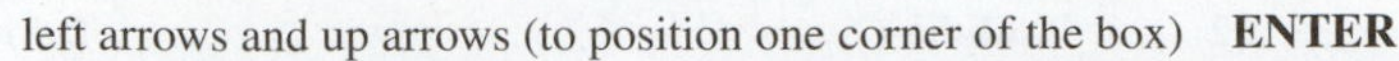
left arrows and up arrows (to position one corner of the box) **ENTER**

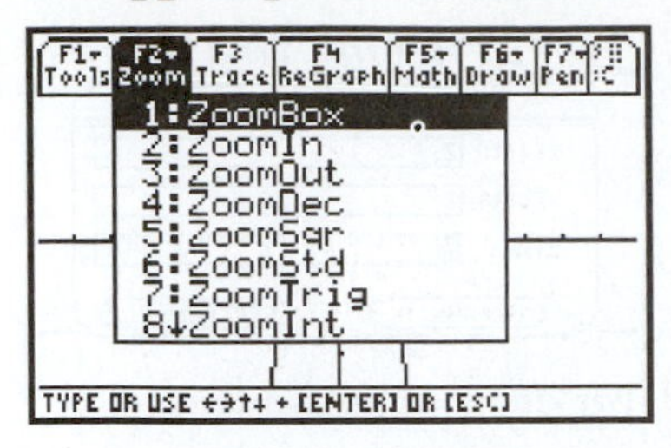

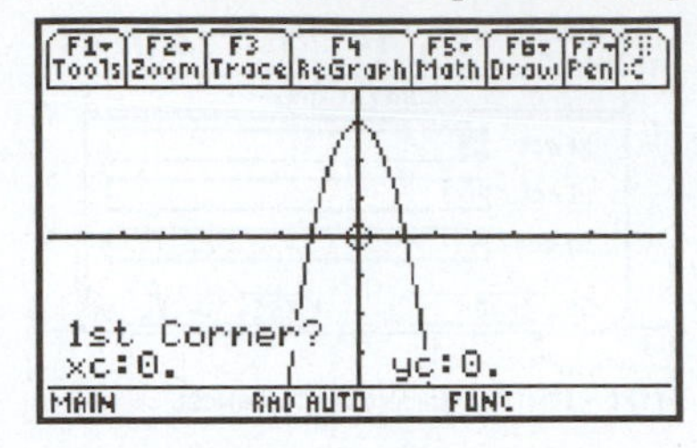

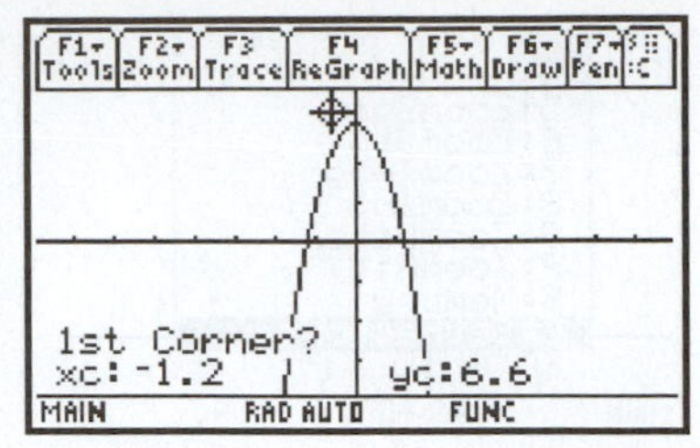

(the placement of one corner) right and down arrows (opposite corner) **ENTER**

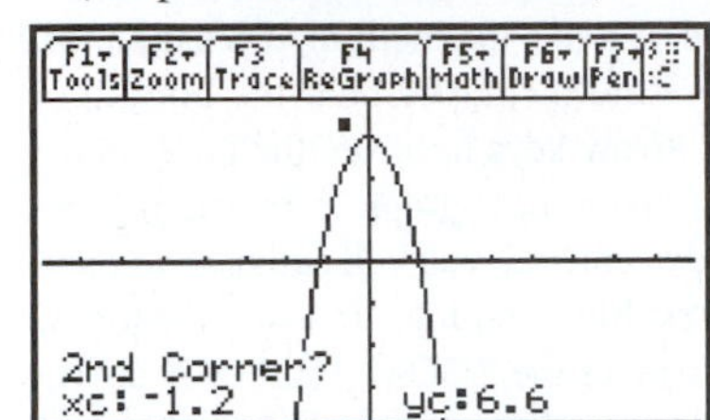

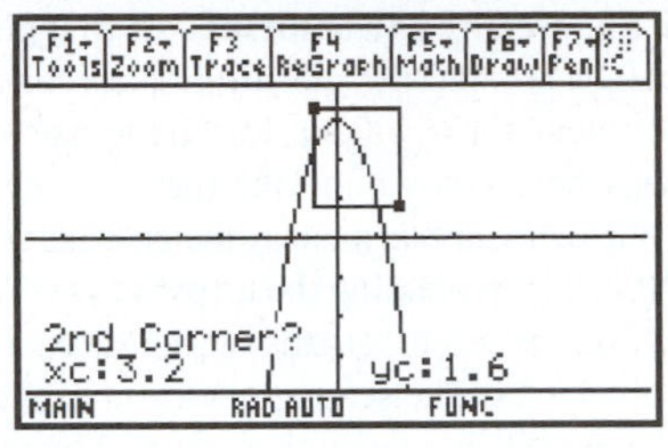

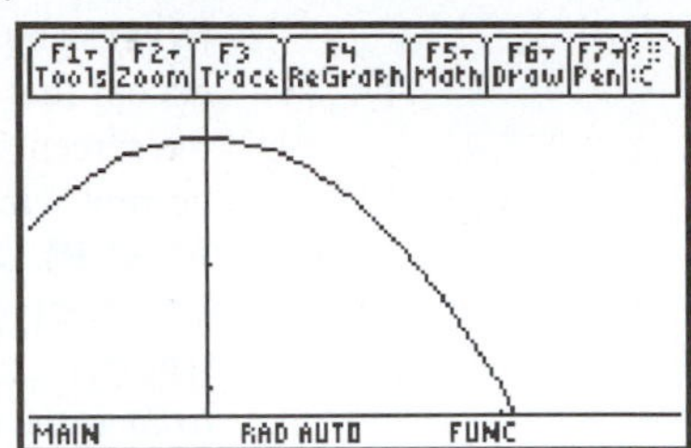

EXAMPLE 5

Sometimes, you will know the precise interval on the x-axis (the domain) that you want for a graph, but a corresponding interval for the y-values (the range) may not be obvious. **ZoomFit** (**Zoom** option A) is designed for this circumstance. For example, to graph $f(x) = 2x^3 - 8x + 9$ on the interval $[-3, 2]$, manually set the **WINDOW** so that **xmin=−3** and **xmax=2** (the other values shown in the second frame are just leftovers from **ZoomStd**). Now press **F2** (**Zoom**), then **alpha A** to select **ZoomFit**. There is a noticeable pause while appropriate values of **ymin** and **ymax** are calculated, then the graph is drawn. To view the values calculated for the range, just press **green diamond WINDOW**.

green diamond Y= **green diamond WINDOW** **green diamond GRAPH**

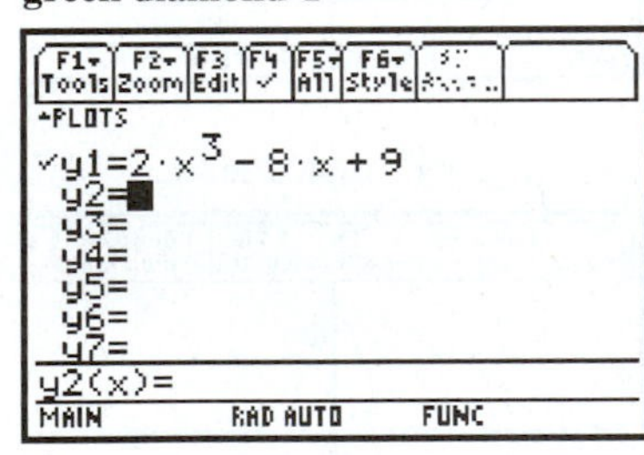

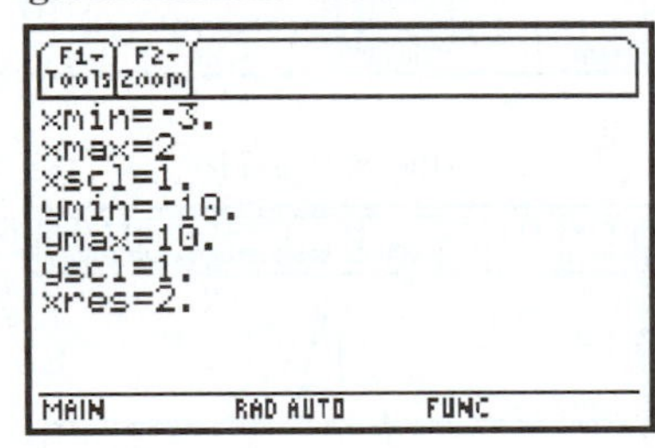

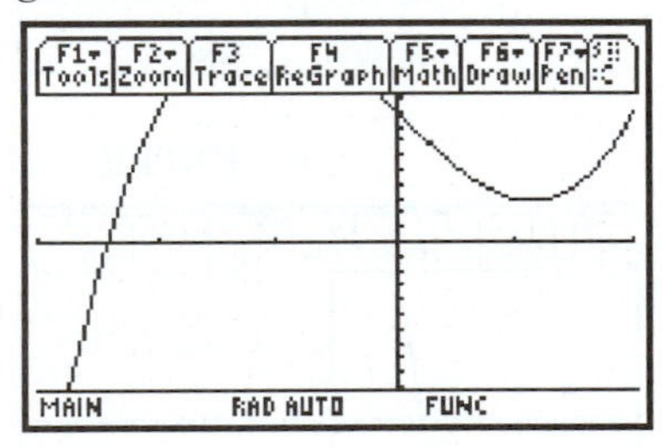

F2 **alpha A** (the new graph) **green diamond WINDOW**

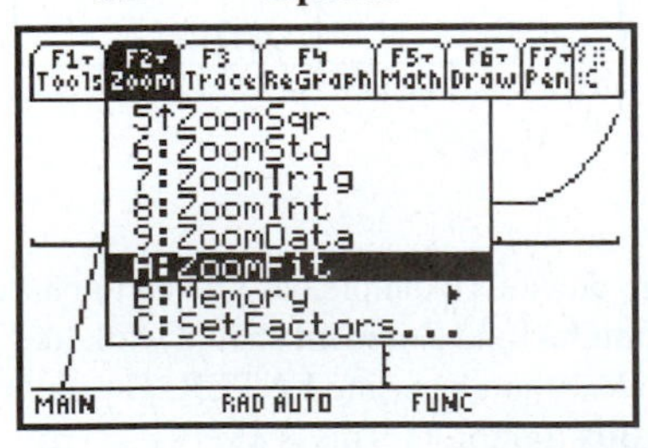

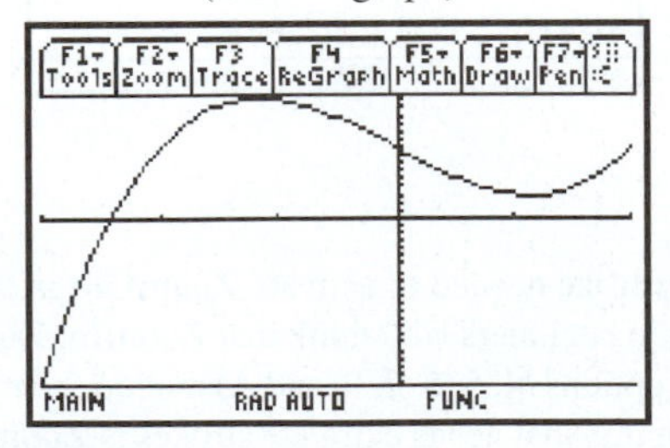

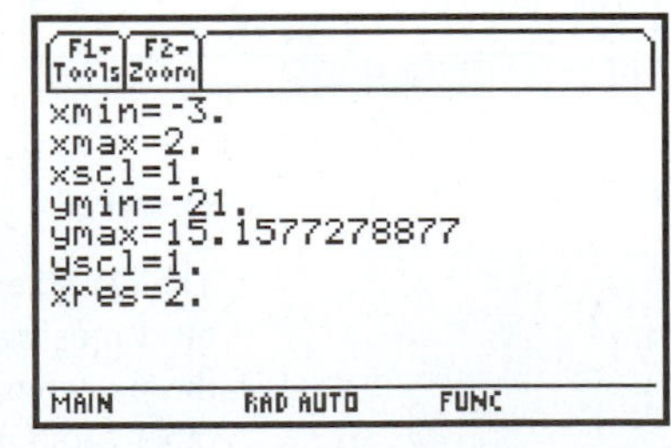

Note that **ZoomFit** changes only **ymin** and **ymax**. You may wish to change **yscl**. Note also that the maximum value of this function on $[-3, 2]$ is approximately 15.158 and that its minimum value appears to be -21. By accessing this information from the final **WINDOW** settings, **ZoomFit** can be used to find the range of a function on an interval.

EXAMPLE 6

Panning to the right is done by tracing a curve off the right-hand edge of the screen.

green diamond Y= **F2 6**

F3 (Trace) right arrows beyond the edge of the screen to pan to the right...

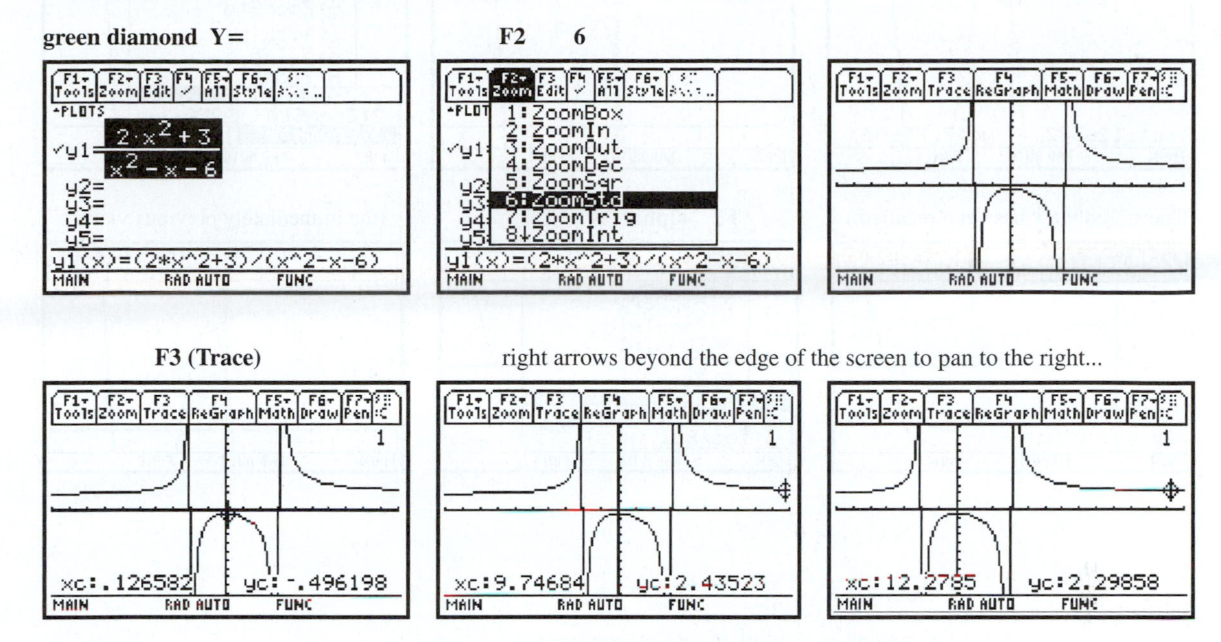

Panning to the left is done similarly. To pan *up* or *down,* see the last part of Example 2.

EXAMPLE 7

Creating, storing, and retrieving viewing windows:

green diamond Y= **green diamond WINDOW** **green diamond GRAPH**

F2 alpha B **2** (Store the current view.) **F2 6** (Let's change it.)

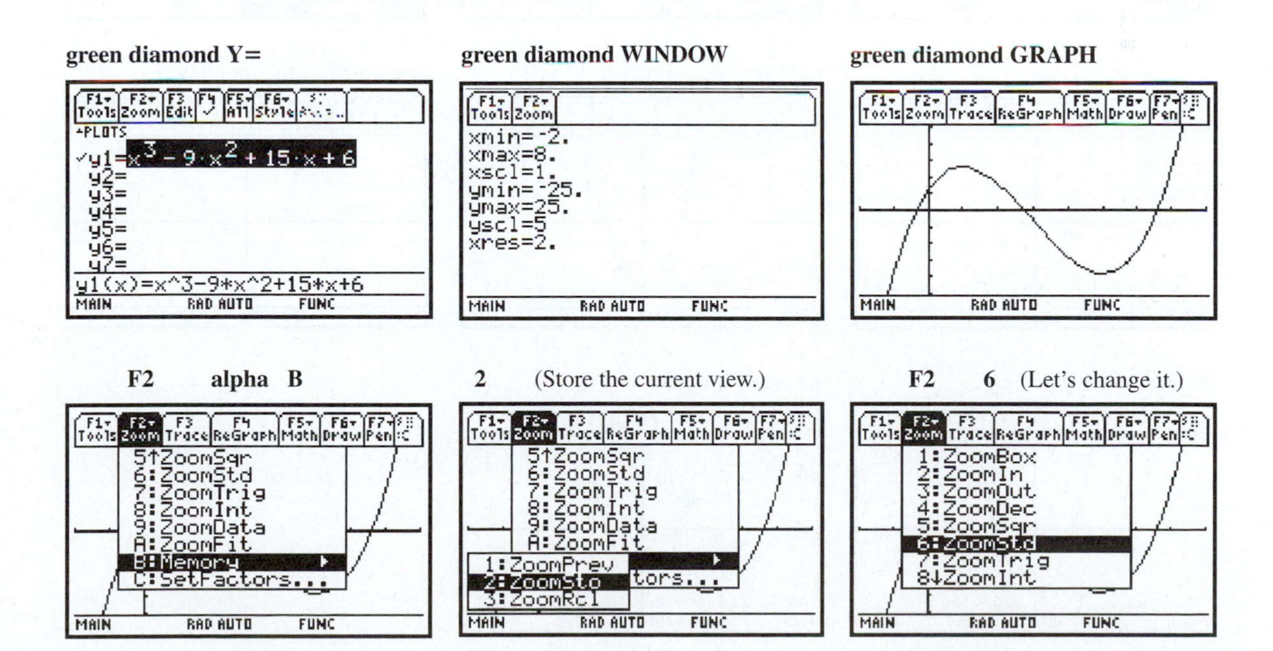

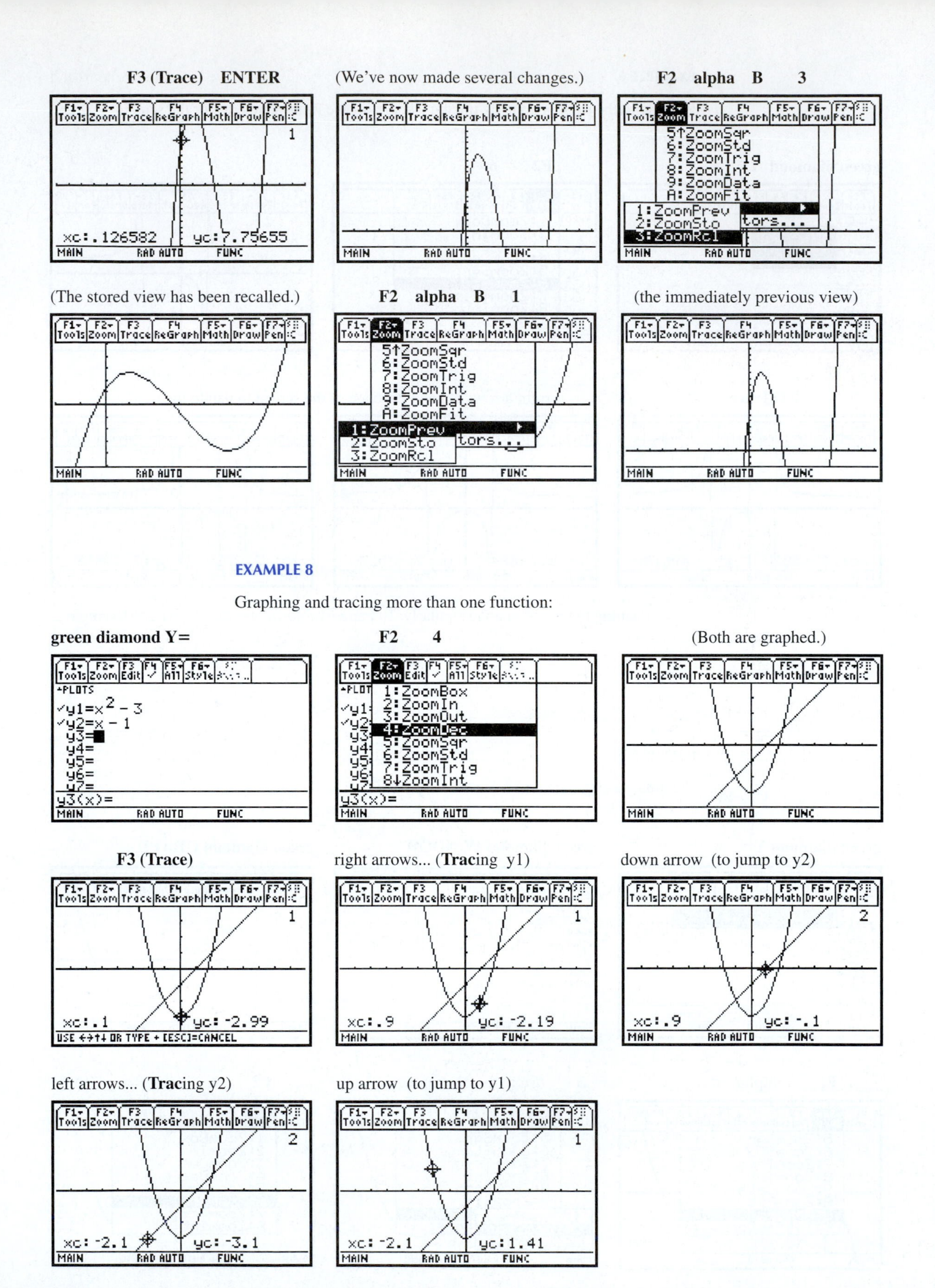
F3 (Trace) ENTER
(We've now made several changes.)
F2 alpha B 3
(The stored view has been recalled.)
F2 alpha B 1
(the immediately previous view)
EXAMPLE 8
Graphing and tracing more than one function:
green diamond Y=
F2 4
(Both are graphed.)
F3 (Trace)
right arrows... (Tracing y1)
down arrow (to jump to y2)
left arrows... (Tracing y2)
up arrow (to jump to y1)

Note that the subscript of the function being investigated is shown in the upper right-hand corner of the **Trace** screens. Pressing the down arrow increases this subscript; pressing the up arrow decreases it. Both of these arrow keys will wrap around and start over. The result has nothing to do with which graph lies above or below the other.

EXAMPLE 9

Finding a point of intersection:

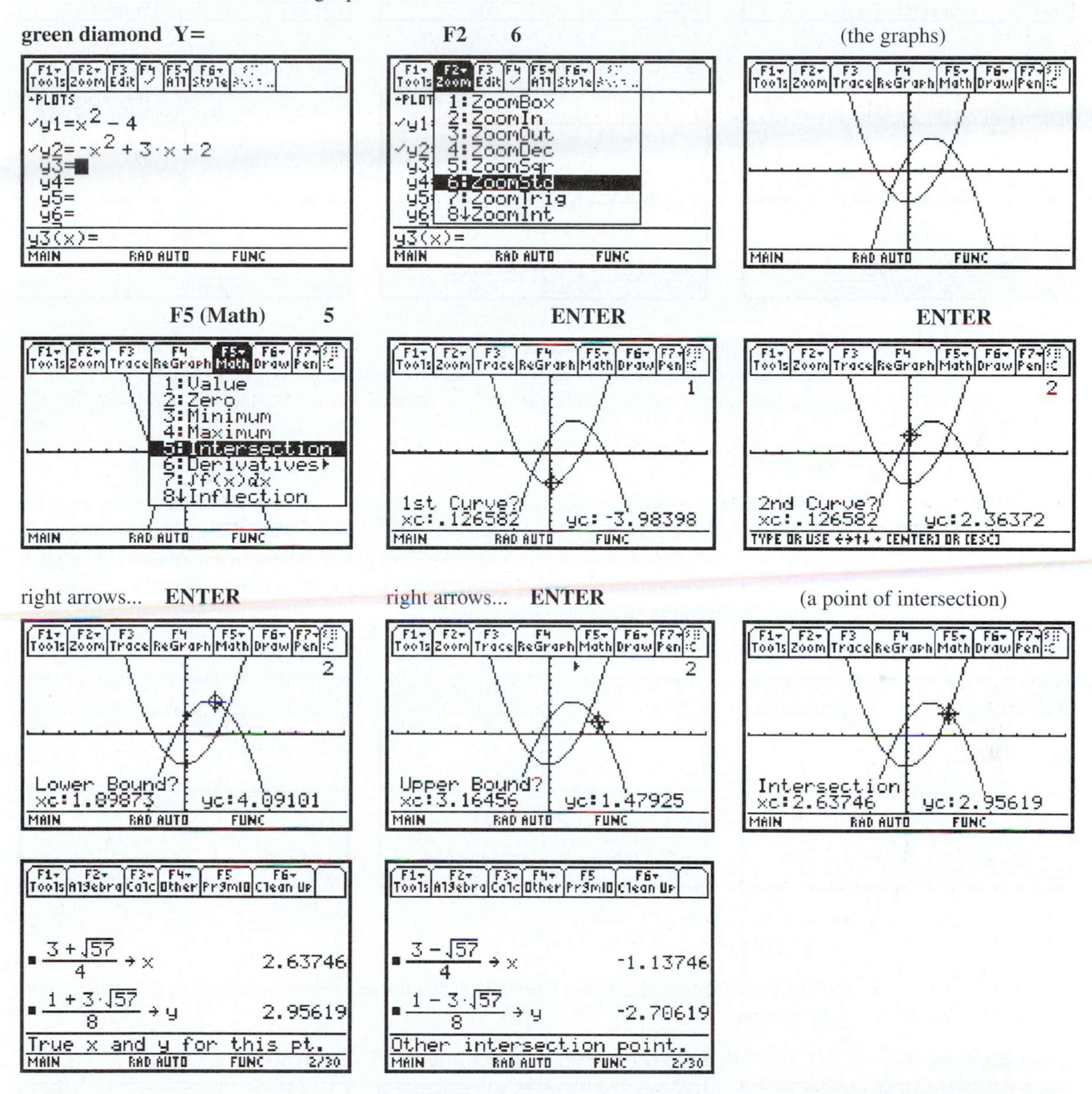

Note that the upper and lower bounds were very important in specifying *which* point of intersection would be calculated. You should now find the other point using **Intersection**.

EXAMPLE 10

Calculating y-values and locating the resulting points on your graphs:
If you have just completed Example 9, please skip to the third frame.

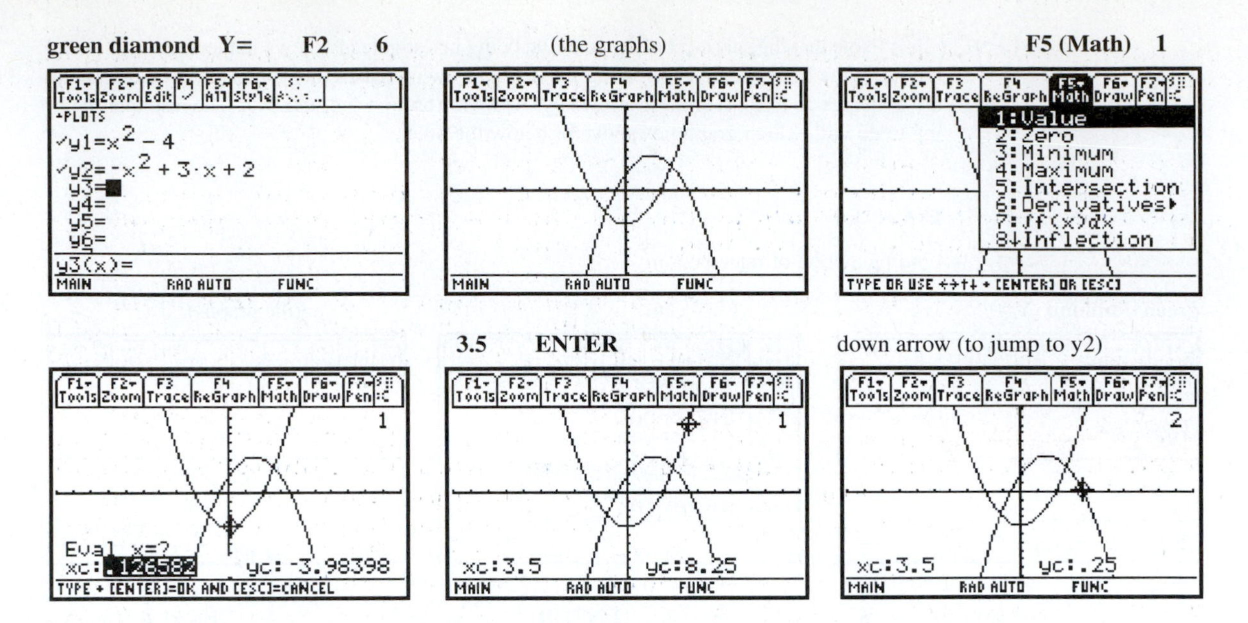

Note again that the subscript of a function (shown in the upper right-hand corner) increases as you jump from one curve to another by pressing the down arrow key. The up arrow key decreases this number, and either key can be used to wrap around and start over.

EXAMPLE 11

Another way to calculate a specific value of a function is to press **F3** (**Trace**), then type the x-value and press **ENTER**. The main difference between the **Value** feature (see Example 10) and this special **Trace** feature is that **Trace** does *not* preserve the entered x-value if you jump from one curve to another (you would usually need to retype that x-value). The following frames assume that Example 10 has just been completed.

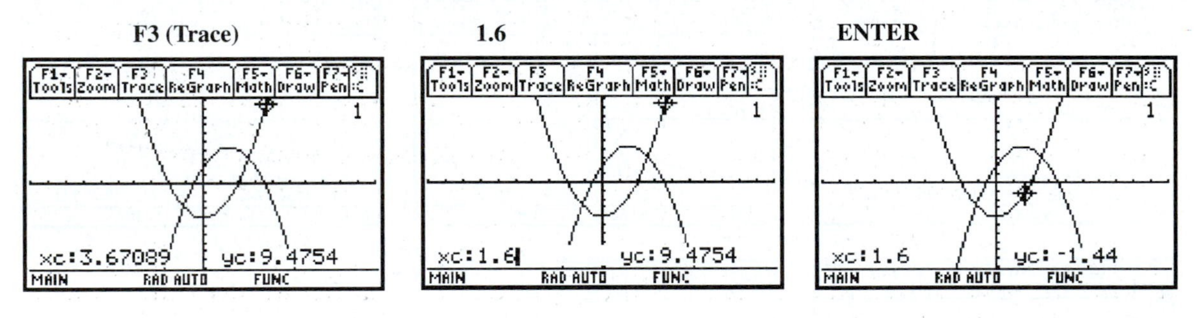

EXAMPLE 12

Using **Trace** to quickly locate a "lost" graph that doesn't appear anywhere in the current viewing window:

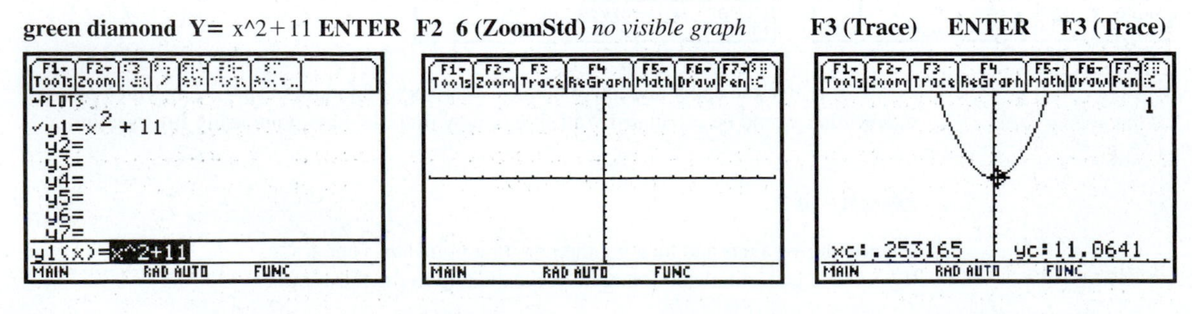

D.7 TRIG FUNCTIONS AND POLAR COORDINATES

EXAMPLE 1

Evaluating trigonometric functions using degrees, minutes, and seconds:

MODE three down arrows right arrow **2 ENTER**

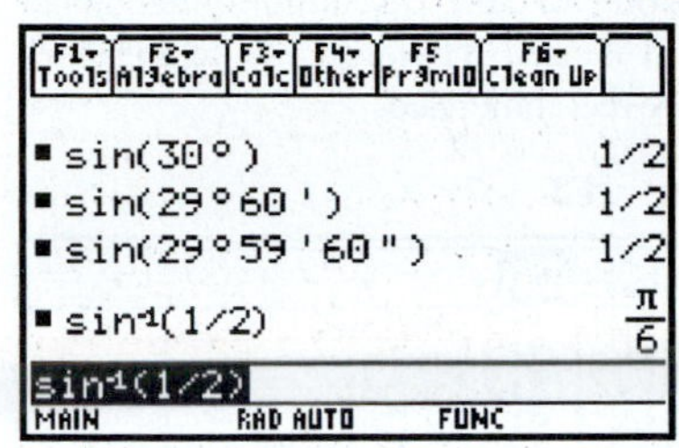

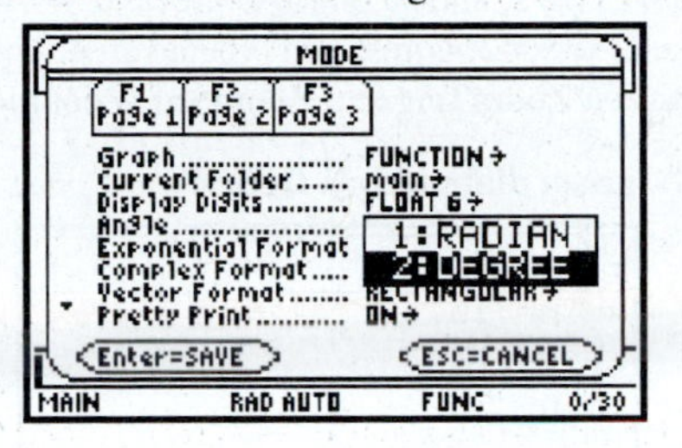

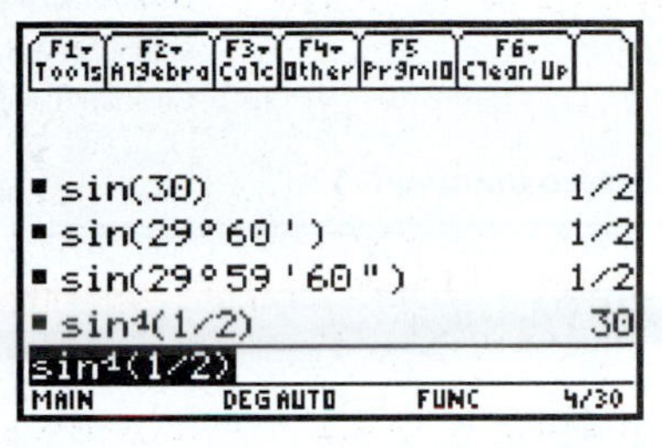

Note that degrees (**2nd** then the | key), minutes (**2nd** then the = key), and seconds (**2nd** then the **1** key) can easily be specified, even when the calculator is in **RADIAN MODE** (indicated by **RAD** in the status line). Comparing the first and third frames shows that the main difference is in the evaluation of an *inverse* trig function, as shown in the fourth calculations. Note also that a degree symbol is optional if the calculator is in **DEGREE MODE** (indicated by **DEG** in the status line) and no minutes or seconds are shown.

EXAMPLE 2

To graph $y = \sin x$, first press **green diamond Y=**, then press **CLEAR** to erase the current formula for y1 (or use the arrow keys to find a blank function), then press **2nd SIN** (**2nd** then the **Y** key) followed by **X**, and finally press the right parenthesis key. Your screen should look very much like the first one in the following figure. To set up a good graphing window, press **F2** and then press **7** to choose **ZoomTrig**. This causes the graph to be immediately drawn on axes that range from $-79\pi/24$ to $79\pi/24$ ($-592.5°$ to $592.5°$) in the x-direction and from -4 to 4 in the y-direction. Each mark along the x-axis represents a multiple of $\pi/2$ radians (90°), and each mark along the y-axis represents 0.5. The **ZoomTrig** feature is especially designed to show the same graph whether the calculator is set for **RADIAN**s or **DEGREE**s. Other automatic ways to set the viewing window (such as **ZoomStd** and **ZoomDec**) ignore the **MODE Angle** setting and should *not* be used to graph trigonometric functions in **DEGREE**s. If you have one or more unwanted graphs drawn on top of this one, go back to your function list (**green diamond Y=**) and turn "off" the unwanted functions by placing the cursor over their formulas and pressing **F4** (to erase their check mark). After you have turned off the unwanted functions, just press **green diamond GRAPH** and you will finally see the last frame in the following figure.

green diamond Y= **F2 7** (the graph)

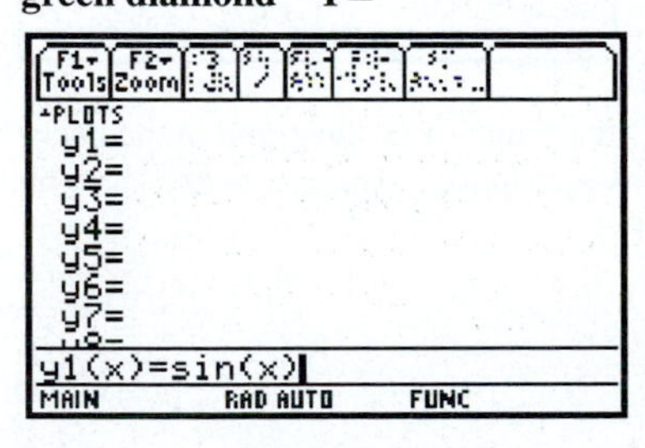

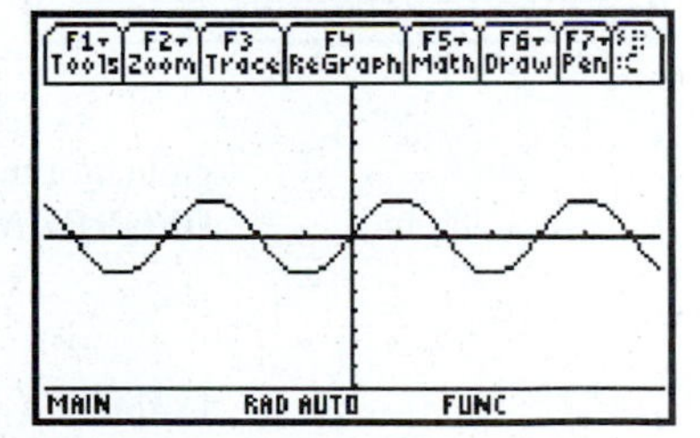

EXAMPLE 3

To modify this function to be $y = -3 \sin 2x$, press **green diamond Y=**, then **ENTER** or **F3** (to bring the current y1 to the edit line). Press the left arrow key (to start editing from the left) and type -3; now press the right arrow key four times (to skip over the part of the formula that's

OK), then press **2**. Press **F2**, then **7** (**ZoomTrig**), and **F3** (**Trace**) to graph and begin tracing this function. **Trace** gives the added bonus of a highlighted point with its coordinates shown at the bottom of the screen. Press the right or left arrow keys to highlight other points on the curve. For best results when using **Trace** with **ZoomTrig**, set **xres=1** on the **green diamond WINDOW** screen. This allows you to see a calculated point for all multiples of $\pi/24$ (7.5°), shown in decimal form. Of course, this includes all special angles values such as $0, \pi/6, \pi/4, \pi/3, \pi/2$, and so on (in degrees: 0, 30, 45, 60, 90, etc.). Be careful to remember that the default, **xres=2**, is restored whenever **ZoomStd** is used. Therefore, switching back and forth between **ZoomTrig** and **ZoomStd** is not highly recommended.

green diamond Y=

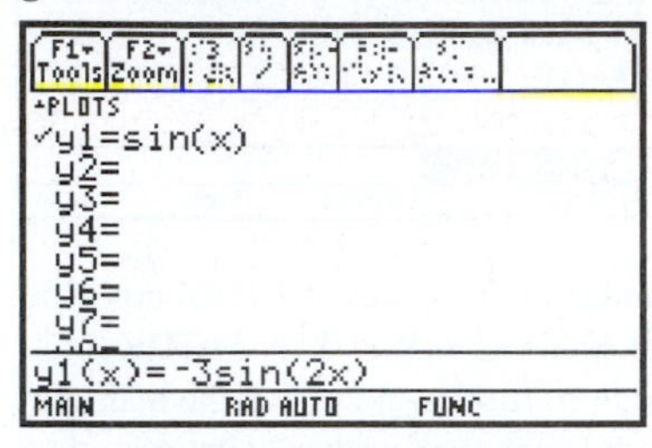

Note that the old function **sin(x)** is still shown as y1 while it is being modified.

green diamond WINDOW

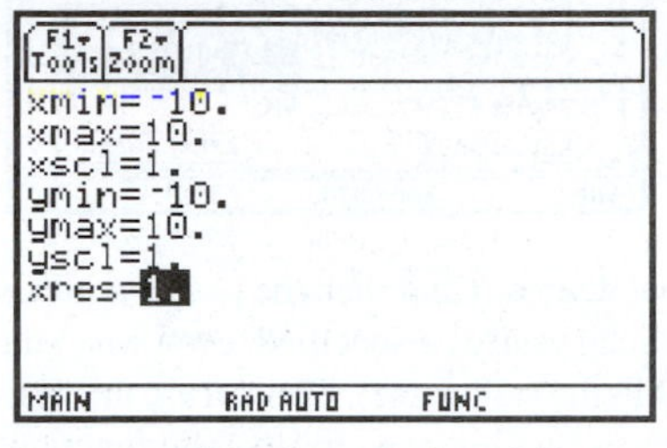

Set **xres=1** for best results with **Trace**. All other settings will be changed by **ZoomTrig**.

F2 7

(the graph)

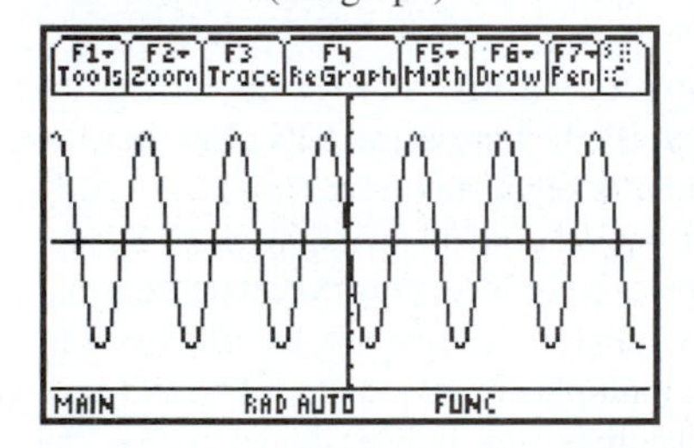

F3 (Trace)

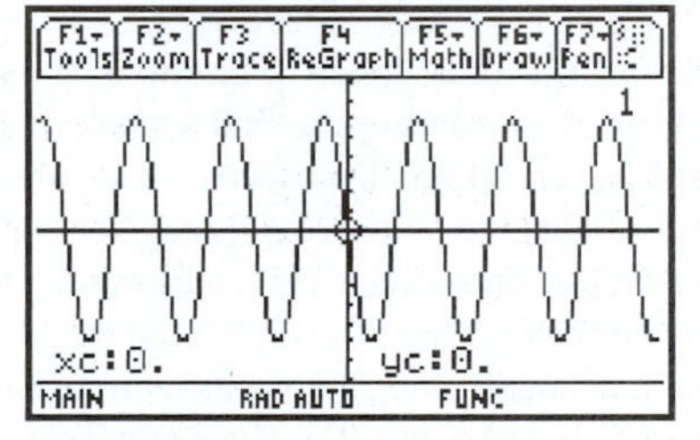

right arrows... [($5\pi/12$, -1.5)]

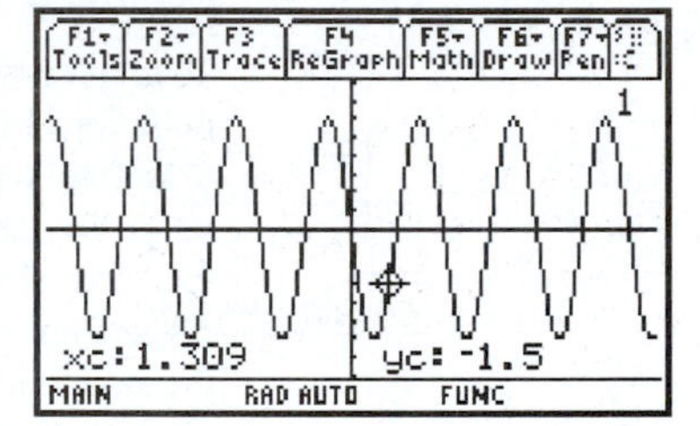

MODE three down arrows right arrow **2 ENTER F2 7**

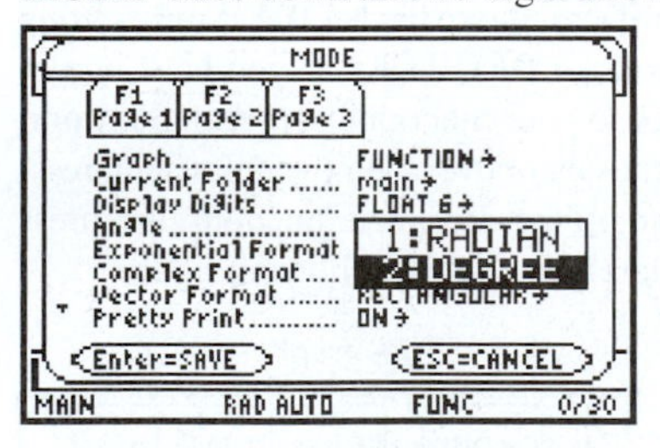

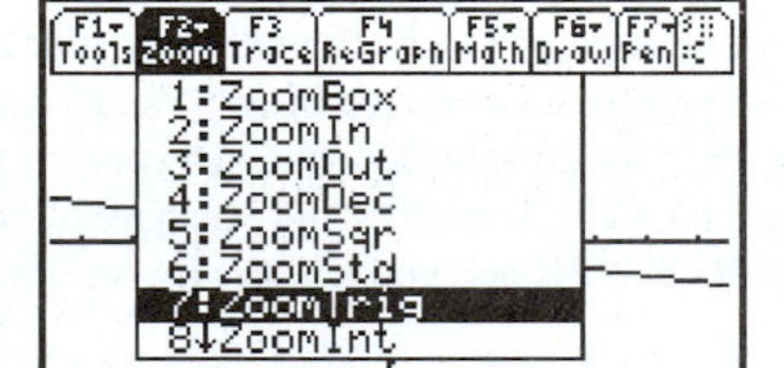

F3 right arrows... [(75°, -1.5)]

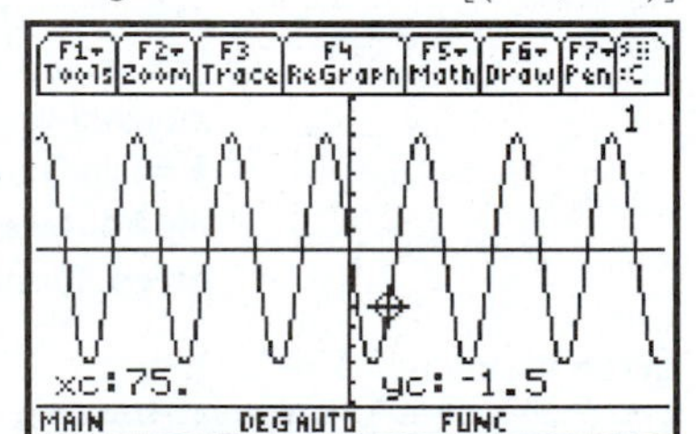

Switching to **DEGREE MODE** initially distorts the graph (note the graph in the background of the middle frame), but **ZoomTrig** restores its appearance. The last frame shows a **DEGREE MODE Trace**.

EXAMPLE 4

Several related trig functions can be drawn on the same screen, either by typing them separately in the function list **green diamond Y=** or by using a list of coefficients as shown in the following figure. A list consists of numbers separated by commas, which are enclosed by braces { }. The braces are the shifted **2nd** versions of the parentheses keys. The first row of frames indicates how to efficiently graph $y = \sin(x)$, $y = 2\sin(x)$, and $y = 4\sin(x)$ on the same screen. The last row graphs $y = 2\sin(x)$ and $y = 2\sin(3x)$.

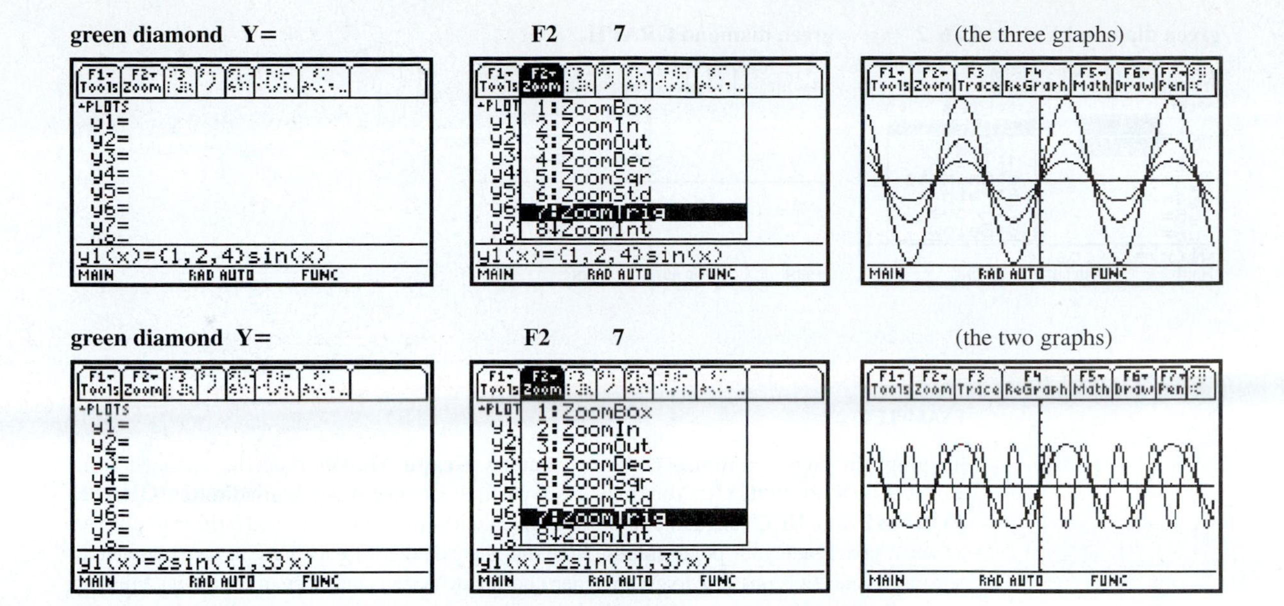

EXAMPLE 5

Multiple lists are allowed but are not highly recommended. To graph the two functions $y = 2 \sin 3x$ and $y = 4 \sin x$, you could do what's shown in the first frame or type them separately as shown in the third frame.

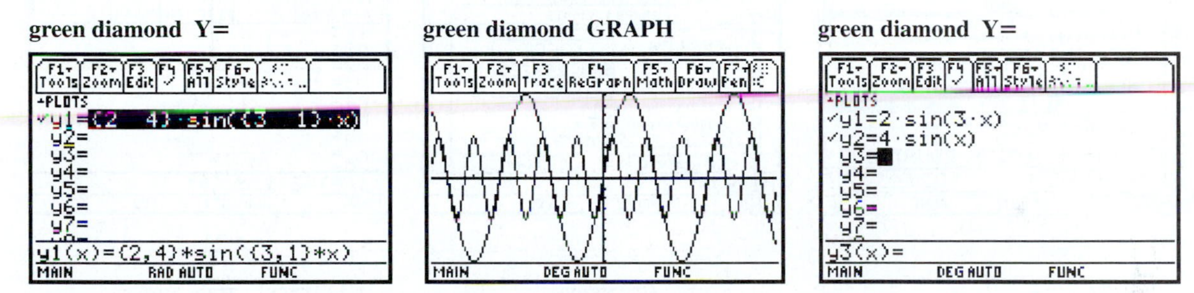

EXAMPLE 6

When graphing trig functions that have vertical asymptotes, remember that your calculator just evaluates individual points and arbitrarily assumes that it should connect those points if the function's graph style is **Line, Thick,** or **Path**. The effect is shown in the following graph of $y = \sec x$. Some people like these "vertical asymptotes" being shown on the graph (they are really just nearly vertical lines which are trying to connect two points on the curve). The last frames show the same graph in **Dot** style.

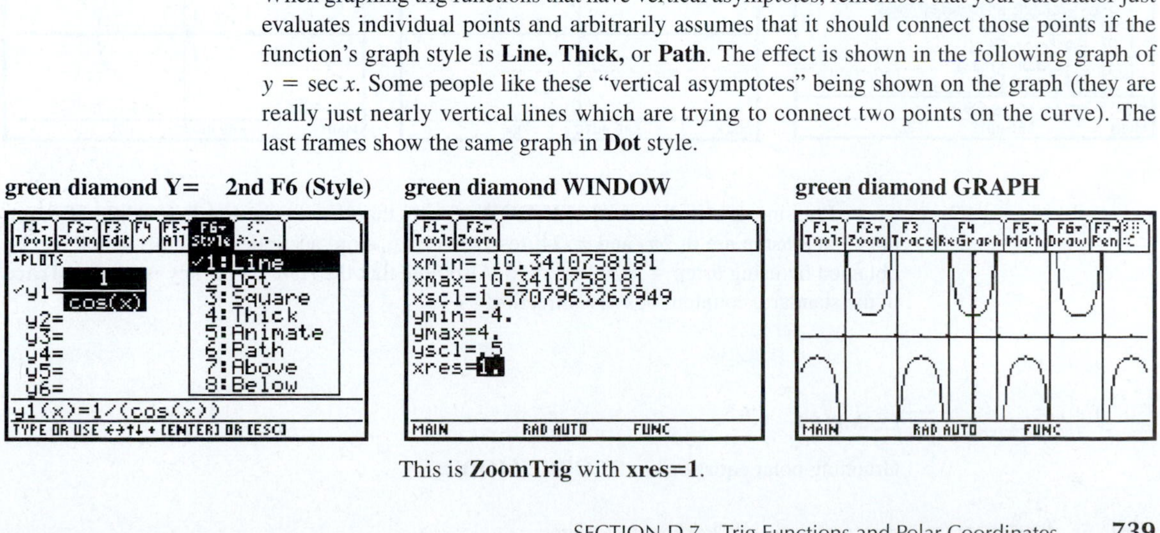

This is **ZoomTrig** with **xres=1**.

green diamond Y= 2nd F6 2 **green diamond GRAPH**

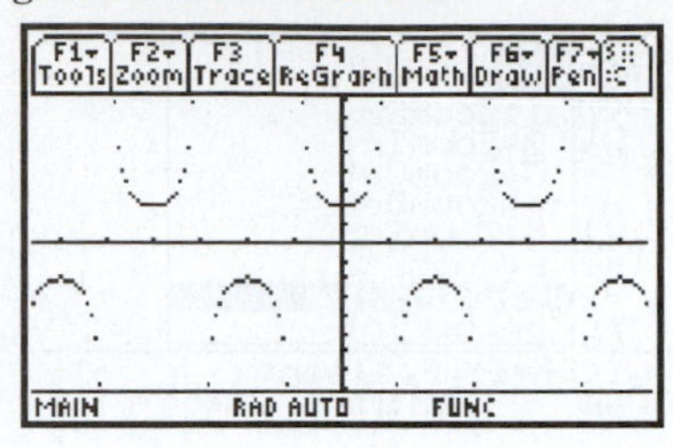

EXAMPLE 7

To graph in polar coordinates, the calculator's **Graph MODE** must be changed from **FUNCTION** to **POLAR**. You may also wish to change your **Coordinates GRAPH FORMAT** from **RECT**angular to **POLAR**, which will show values of r and θ (instead of x and y) when you **Trace** your polar graphs. Note that the calculator treats these as two completely separate issues (it is possible to graph in one coordinate system and trace in the other). Pressing **green diamond Y=** will reveal the calculator's 99 polar functions r1, r2, . . . , r99. It is important to note that θ is the **green diamond** version of the ^ key and that θ must be used as the independent variable for all **POLAR** graphs.

MODE right arrow **3** **ENTER** **green diamond Y= F1 9** right arrow **2 ENTER** etc.

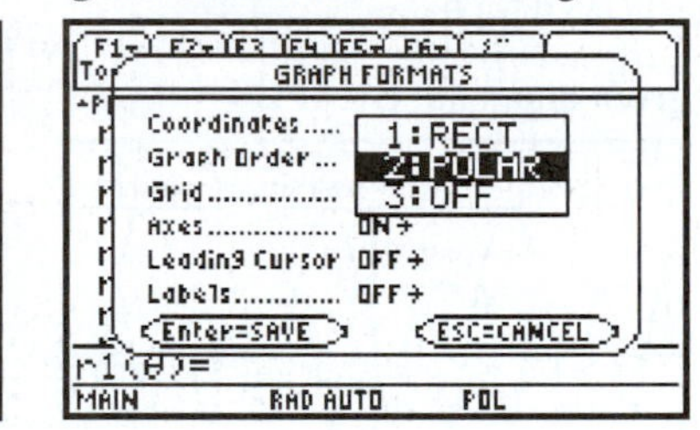

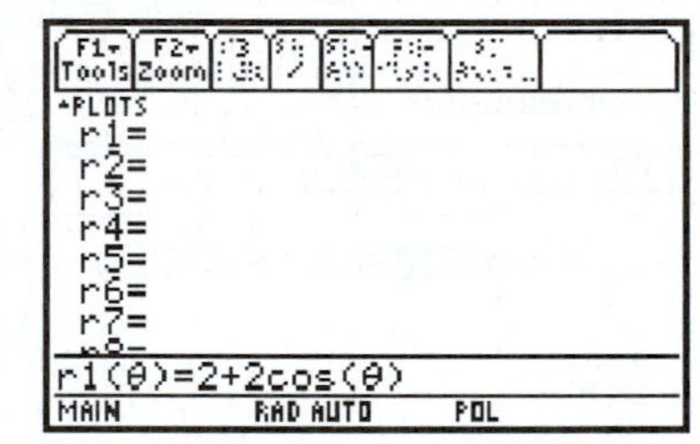

F2 **4**

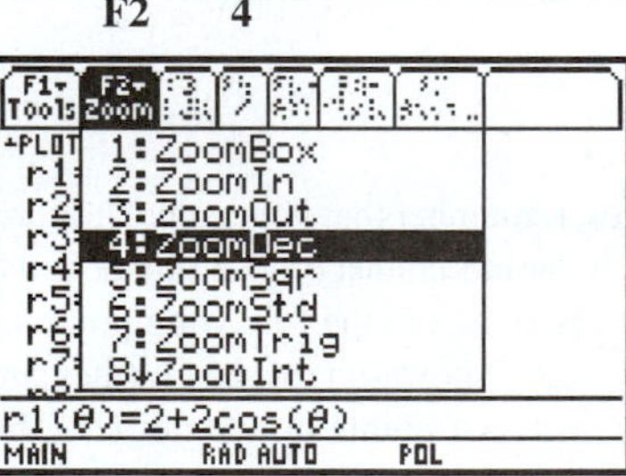

(the graph)

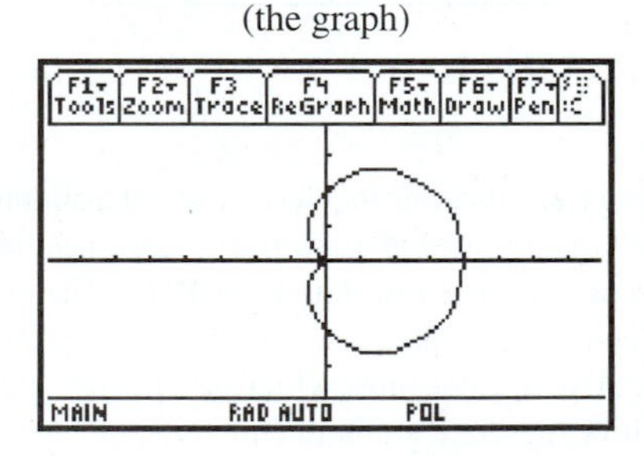

F3 (**Trace**) right arrows

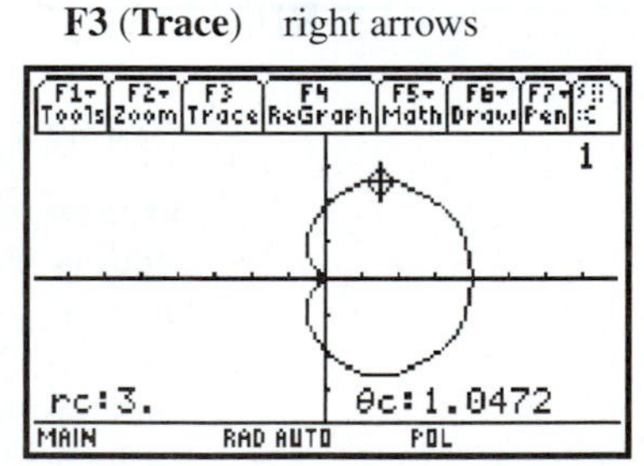

Pressing **green diamond WINDOW** reveals that the standard radian values of θmin, θmax, and θstep are 0, 2π, and $\pi/24$, respectively. A more accurate, smoother graph can be obtained by using $\theta\text{step} = \pi/48$ or $\pi/96$. Also note that the *right* arrow key is used to **Trace** in the standard, counterclockwise direction.

EXAMPLE 8

Graphing polar equations in **DEGREE MODE**:

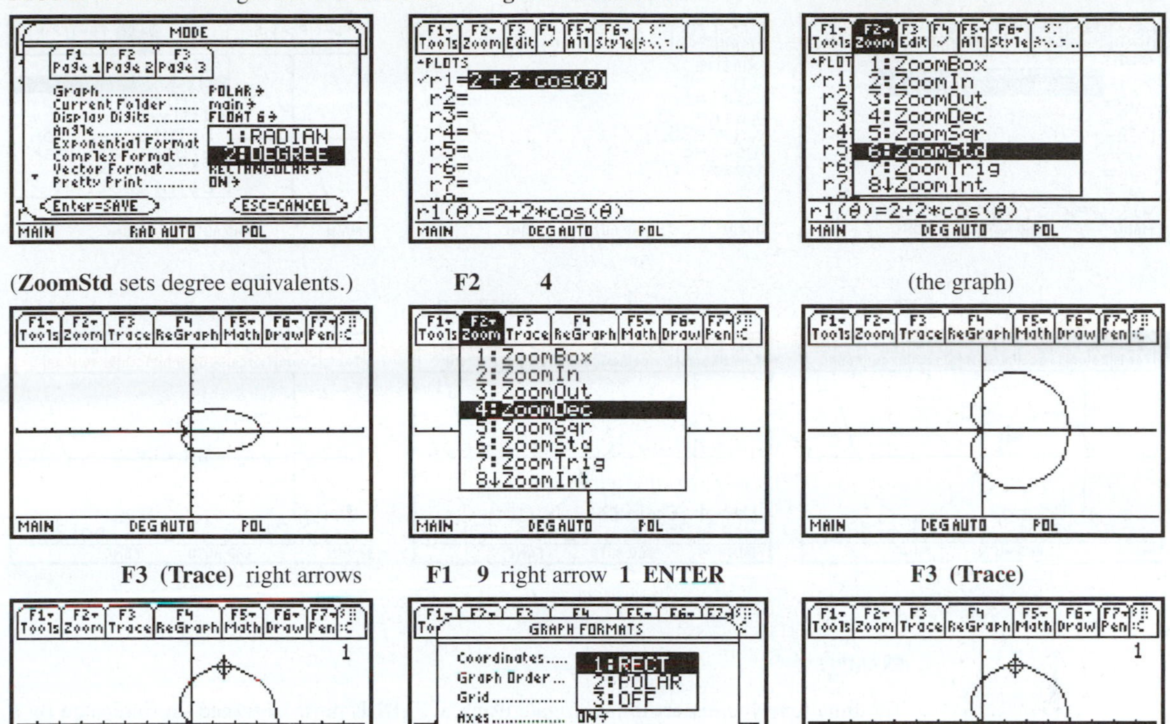

ZoomStd sets the standard degree values of θmin, θmax, and θstep, which are 0, 360, and 7.5, respectively. Smoother (but slower) graphs can be obtained by using smaller values of θstep (such as $\theta\text{step} = 2$). The last frame shows a polar graph traced in **RECT**angular **Coordinates** GRAPH FORMAT.

D.8 NUMERICAL GRAPH AND TABLE FEATURES

EXAMPLE 1

To numerically estimate solutions of equations on the graphing screen:

(a) Rewrite the equation on paper in the form $f(x) = 0$; for example, rewrite

$$x^3 + 15x = 9x^2 - 6$$

as

$$x^3 - 9x^2 + 15x + 6 = 0$$

(b) Enter this function in the **green diamond Y=** menu, and make sure that it is the only one selected for graphing.

(c) Graph it using an appropriate viewing **WINDOW** or **Zoom** option (**F2**). The zeros being estimated must appear in the viewing window.

(d) Press **F5** (**Math**) and choose option 2, **Zero**.

(e) The prompt "Lower Bound?" is asking for an x-value to the *left* of the desired zero. Either type a value from the keyboard or arrow over to it, then press **ENTER**. Likewise, "Upper Bound?" is asking for an x-value to the *right* of the desired solution. Pressing **ENTER** then produces an estimated zero in the specified interval and also shows the y-value of the point as a check (it should be zero or extremely close to zero, such as 1.E-12, which means 10^{-12}).

green diamond Y=

```
F1 Tools | F2 Zoom | F3 Edit | F4 ✓ | F5 All | F6 Style
▲PLOTS
✓y1=x^3 - 9·x^2 + 15·x + 6
 y2=
 y3=
 y4=
 y5=
 y6=
 y7=
y1(x)=x^3-9*x^2+15*x+6
MAIN   RAD AUTO   FUNC
```

green diamond WINDOW

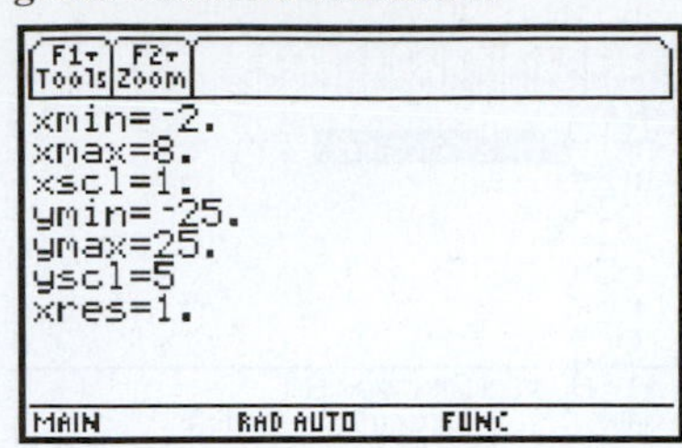

green diamond GRAPH F5 2

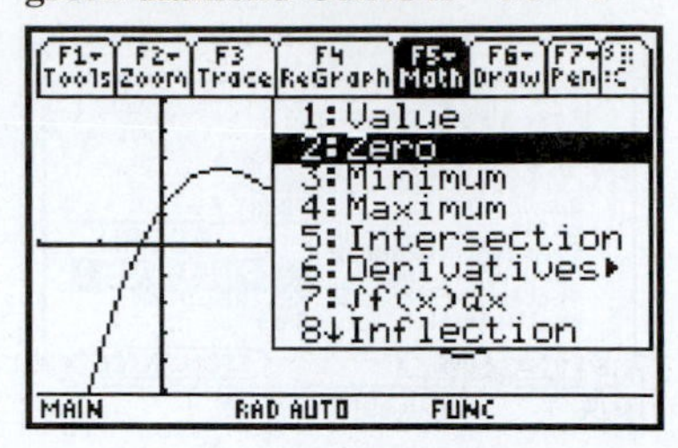

left arrows... **ENTER**

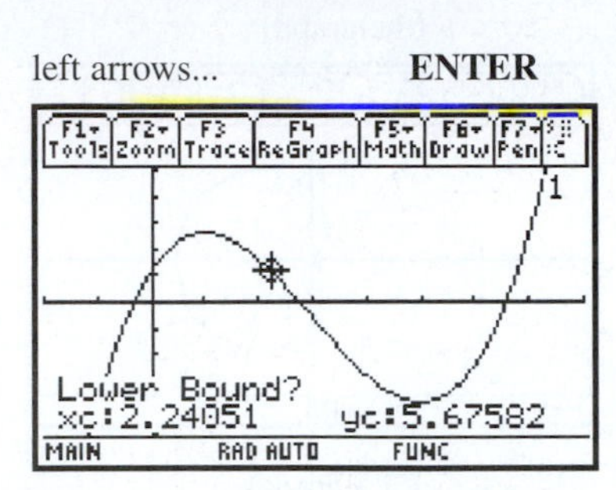

right arrows... **ENTER**

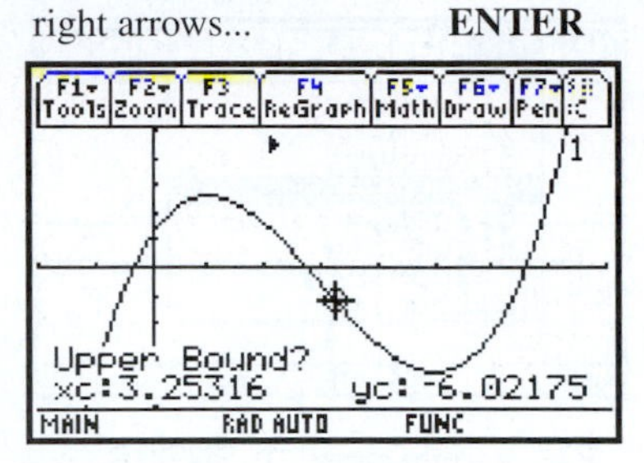

(a solution)

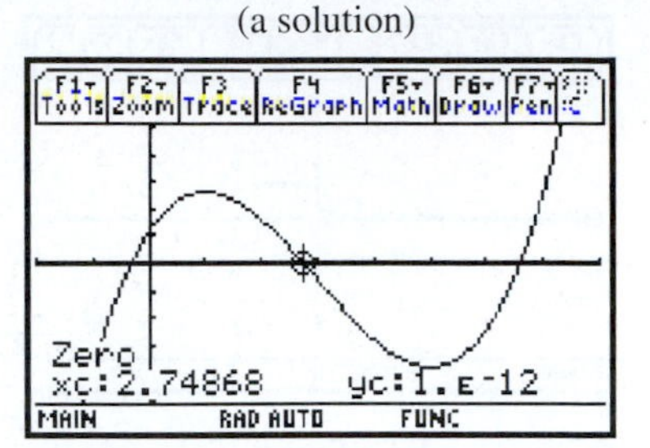

EXAMPLE 2

The functions you expect to investigate using a **TABLE** must be turned on (indicated by a check mark) in the same sense as those you want to graph. To get to the TABLE SETUP screen, press **green diamond TblSet** (**green diamond**, then the **F4** key). The specification for **tblStart** is just a beginning x-value for the table; you can scroll up or down using the arrow keys. **Δtbl** is the incremental change in x. You can give **Δtbl** the value 1 (as in the following frames) to calculate the function at consecutive integers; in calculus, you could use .001 to investigate what is happening as a function approaches a limit; or you could use a larger number like 10 or 100 to study the function's numerical behavior as x goes to ∞.

green diamond Y=

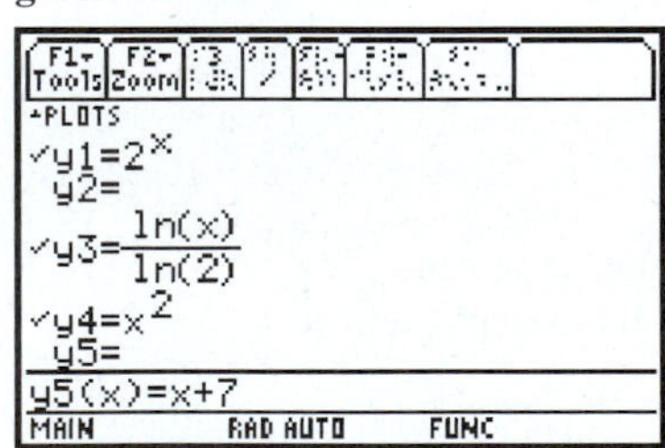

Even y5 is active for this **TABLE**, since it is defined on the edit line.

green diamond TblSet

```
TABLE SETUP
tblStart:          0.
Δtbl:              1.
Graph <-> Table:   OFF →
Independent:       AUTO →
(Enter=SAVE)        (ESC=CANCEL)
y5(x)=x+7
TYPE + [ENTER]=OK AND [ESC]=CANCEL
```

These standard values start the **TABLE** at 0 and change by 1.

green diamond TABLE

x	y1	y3	y4
0.	1.	-∞	0.
1.	2.	0.	1.
2.	4.	1.	4.
3.	8.	1.585	9.
4.	16.	2.	16.

x=0.

up arrow

x	y1	y3	y4
-1.	.5	undef	1.
0.	1.	-∞	0.
1.	2.	0.	1.
2.	4.	1.	4.
3.	8.	1.585	9.

x=-1.

down arrows right arrows

x	y1	y3	y4
3.	8.	1.585	9.
4.	16.	2.	16.
5.	32.	2.3219	25.
6.	64.	2.585	36.
7.	128.	2.8074	49.

y3(x)=2.8073549220576

two more right arrows (to reveal y5)

x	y3	y4	y5
3.	1.585	9.	10.
4.	2.	16.	11.
5.	2.3219	25.	12.
6.	2.585	36.	13.
7.	2.8074	49.	14.

y5(x)=14.

F4 (**Header**)

x	y3	y4	y5
3.	1.585	9.	10.
4.	2.	16.	11.
5.	2.3219	25.	12.
6.	2.585	36.	13.
7.	2.8074	49.	14.

y5(x)=x+7

(Change the definition of y5.) **ENTER**

x	y3	y4	y5
3.	1.585	9.	10.
4.	2.	16.	11.
5.	2.3219	25.	12.
6.	2.585	36.	13.
7.	2.8074	49.	14.

y5(x)=x^8

(**TABLE** is changed accordingly.)

x	y3	y4	y5
3.	1.585	9.	6561.
4.	2.	16.	65536.
5.	2.3219	25.	3.91E5
6.	2.585	36.	1.68E6
7.	2.8074	49.	5.76E6

y5(x)=5764801.

EXAMPLE 3

Finding local maximum (and minimum) values on the graphing screen:

green diamond Y=

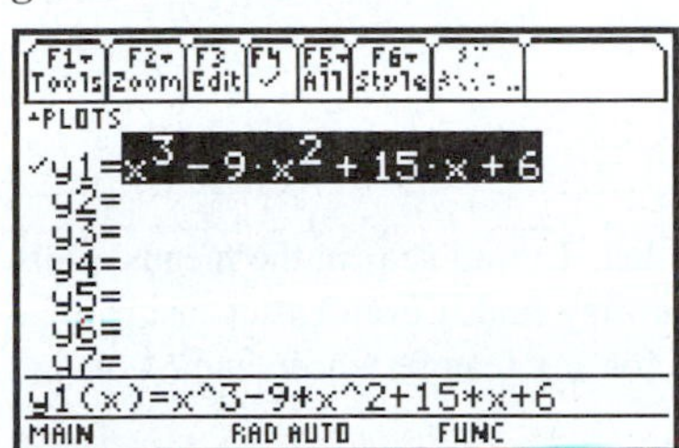

green diamond WINDOW

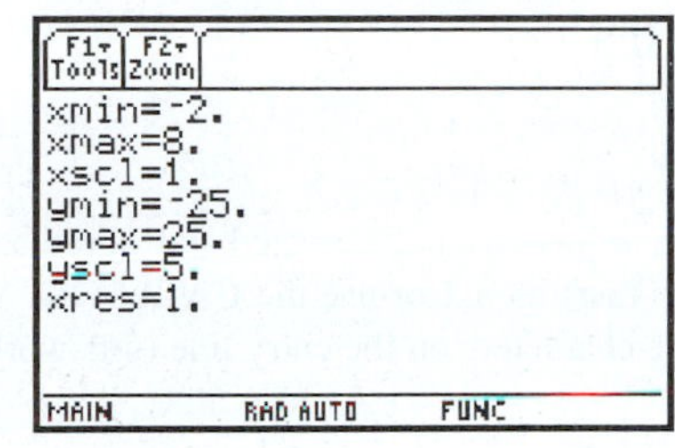

green diamond GRAPH F5 4

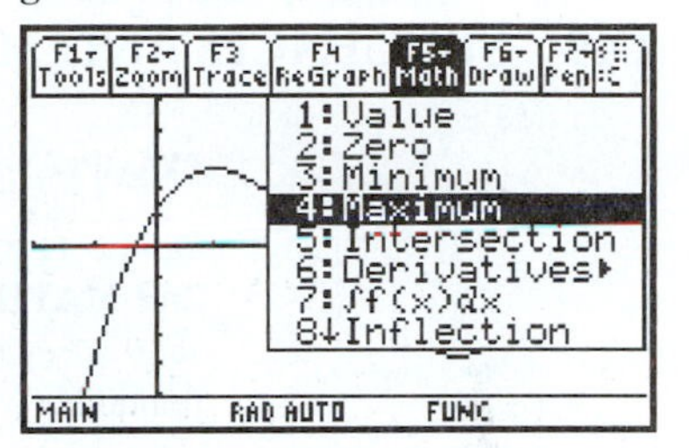

left arrows... **ENTER**

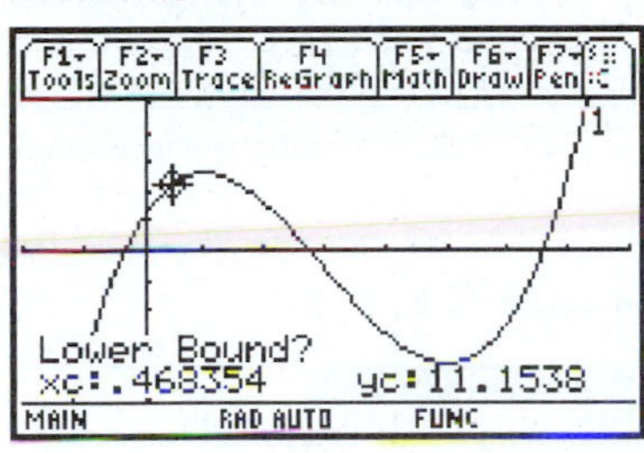

right arrows... **ENTER**

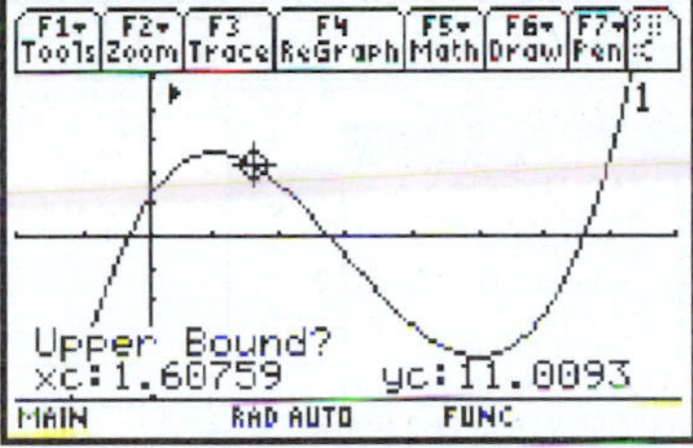

(a local maximum point)

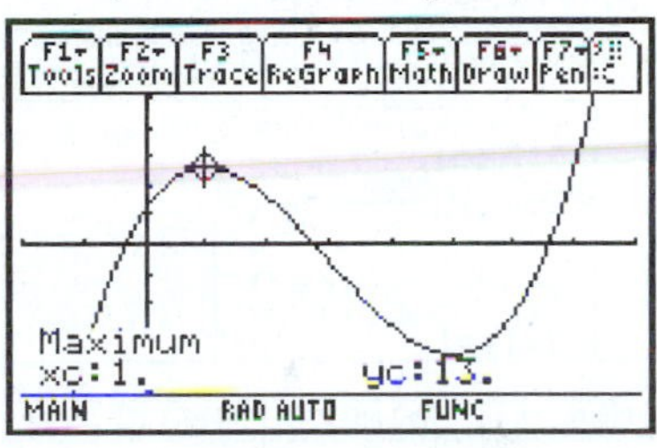

The only intended use of these graphical maximum and minimum features is to find *local* extreme values using a fairly short bracketing interval. These routines can give unpredictable results if you try to use them to compare local extrema with endpoint extrema on a given interval. **ZoomFit** and the home screen features **fMin** and **fMax** (options 6 and 7 in the **F3 Calculus** menu) are more suitable features if your problem is determining a function's overall maximum and minimum values on a specified interval.

EXAMPLE 4

Find the (straight line) distance between two points on the same curve:

green diamond Y=

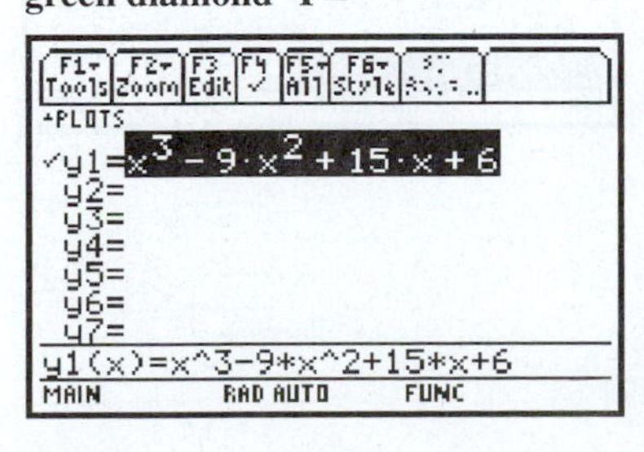

green diamond WINDOW

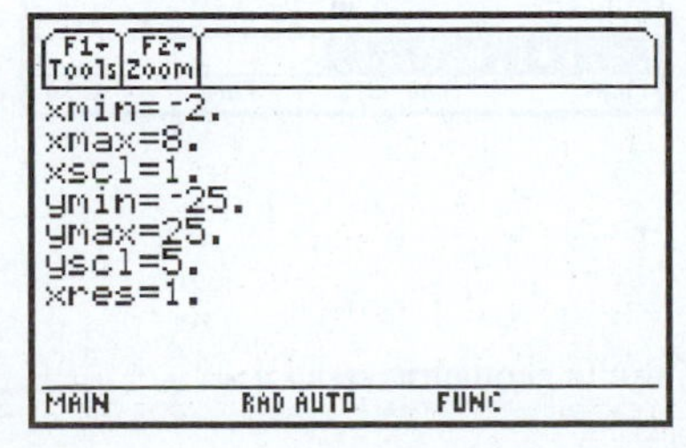

green diamond GRAPH F5 9

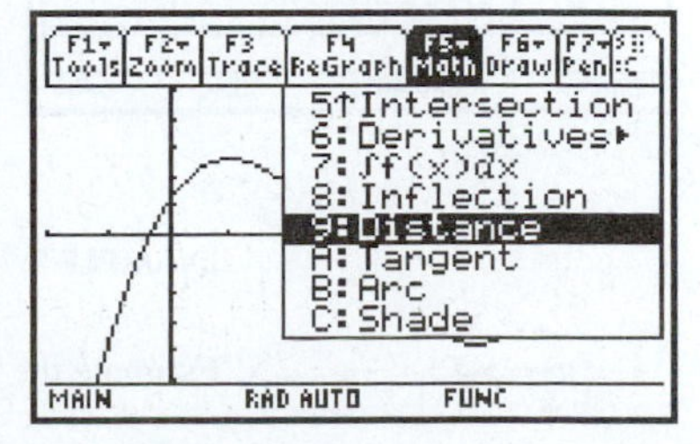

arrows... **ENTER** (the first point) arrows... **ENTER** (the second point) (the distance)

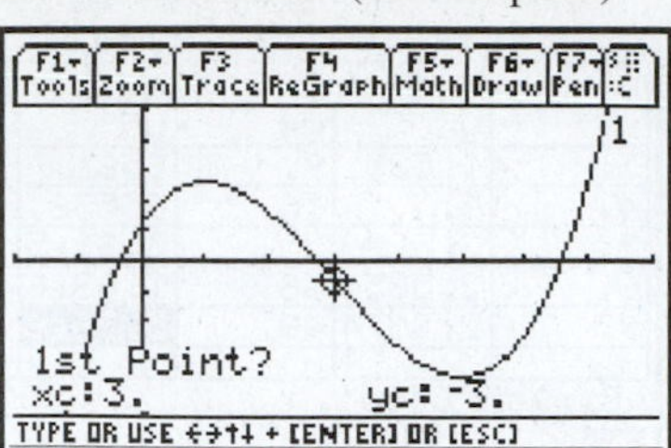

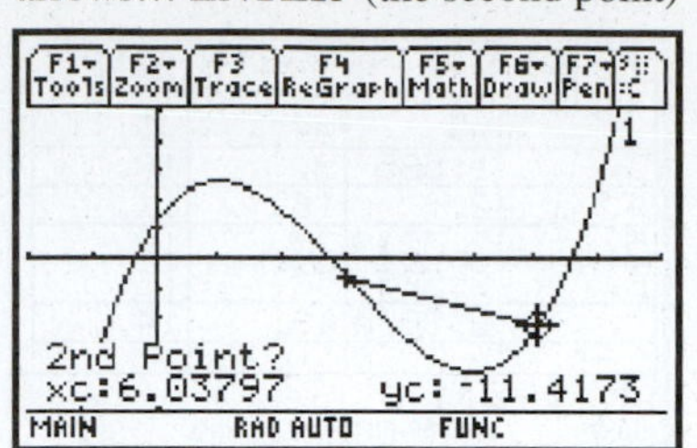

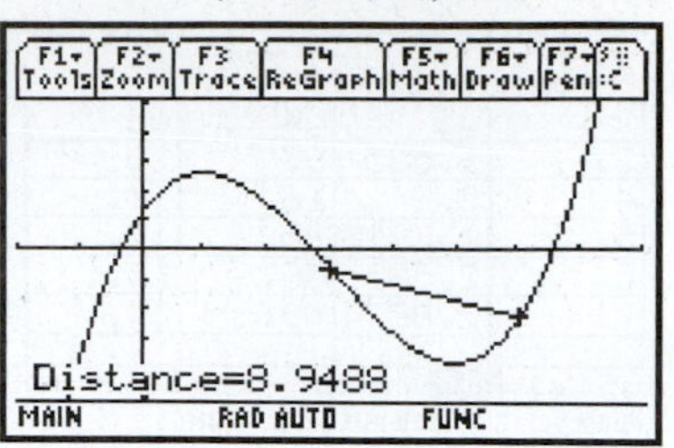

D.9 SEQUENCES AND SERIES

EXAMPLE 1

On the TI-89, a sequence can be created and stored as a list. To find **seq(** in the menus, press **2nd MATH 3** (**List**) then **1** or use the **CATALOG**. You may find it even easier just to type **seq(** as separate characters on the entry line (this works for any feature whose name you can remember).

2nd MATH **3** **1** (the squares from 1 to 5)

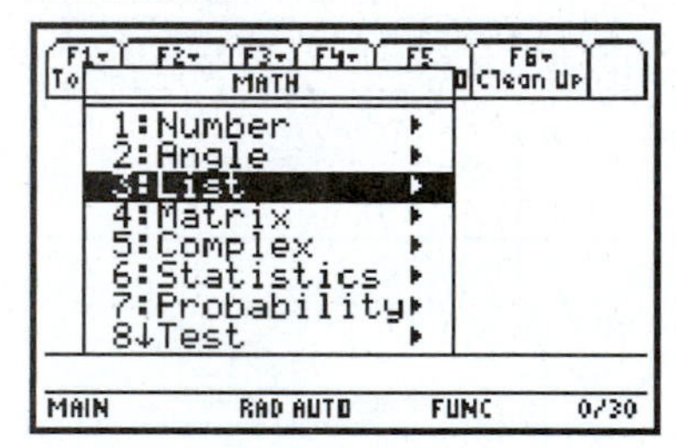

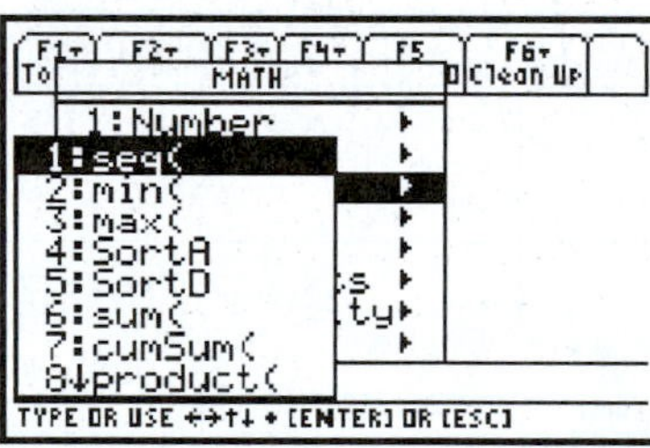

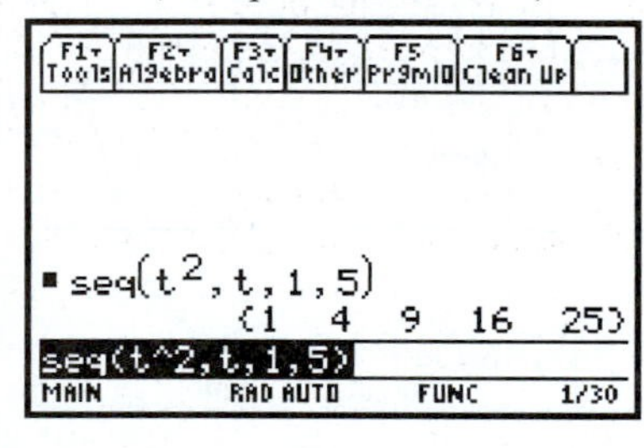

EXAMPLE 2

Evaluate the sum $7 + 9 + 11 + 13 + 15 + \cdots + 121$.

F3 **4** (There are many different ways in which this sum can be expressed.)

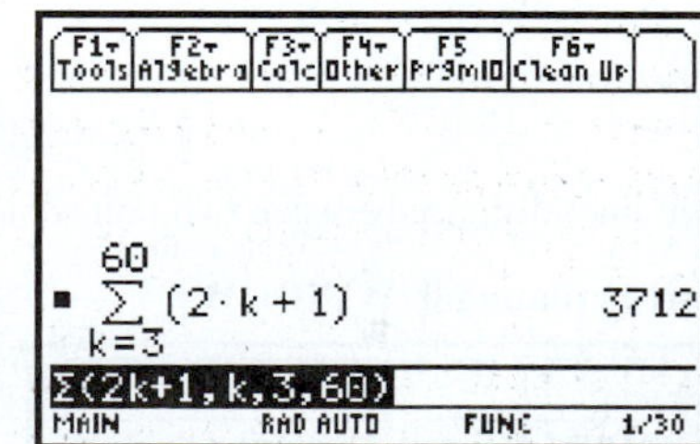

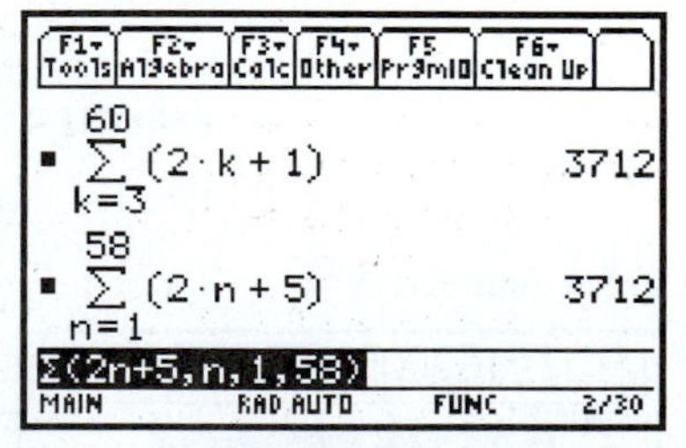

EXAMPLE 3

Estimate the infinite geometric series $1 + \frac{2}{3} + \frac{4}{9} + \frac{8}{27} + \cdots$.

F3 **4** (typing...) **ENTER** **green diamond** ≈ (∞ is **green diamond CATALOG.**)

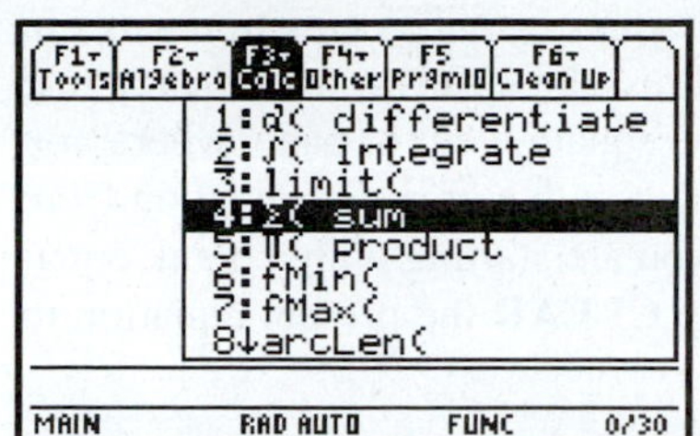

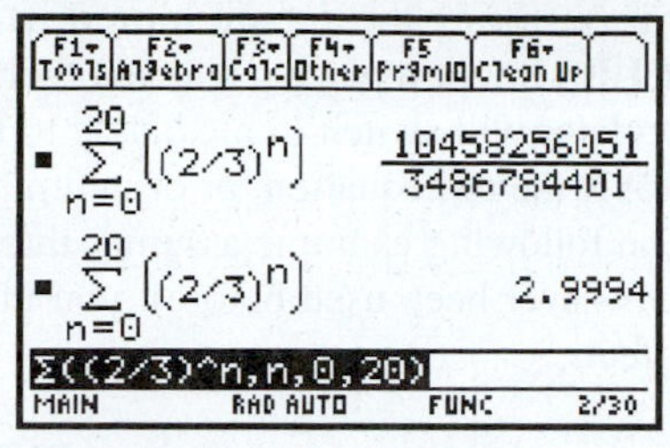

The sum of the first 21 terms is given as both a fraction and a decimal.

The exact infinite sum is given by $a_0/(1 - r)$, where $a_0 = 1$ and $r = \frac{2}{3}$.

EXAMPLE 4

Find the value of the infinite series $1/0! + 1/1! + 1/2! + 1/3! + 1/4! + \cdots$.

F3 **4** (typing...) **ENTER** **green diamond** ≈ (∞ is **green diamond CATALOG.**)

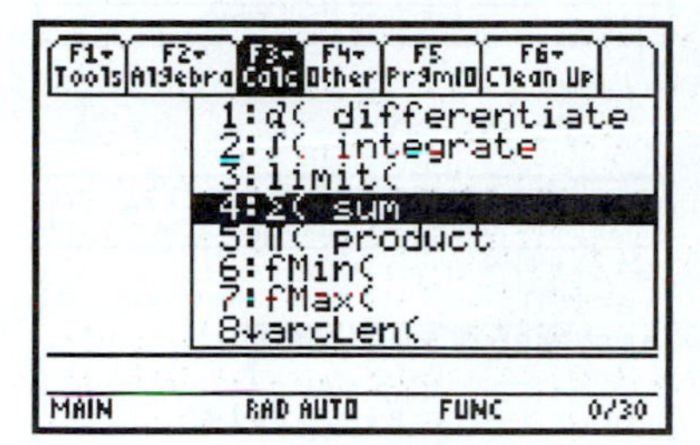

The factorial symbol is **green diamond**, then the ÷ key.

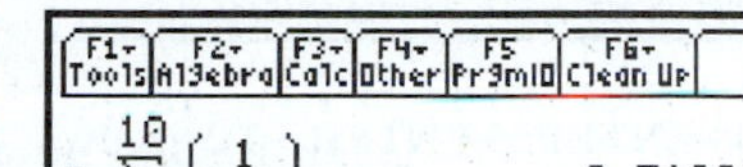

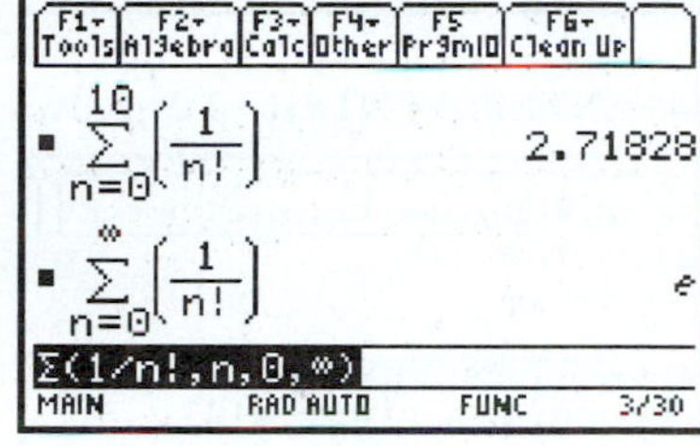

D.10 THE NUMERIC SOLVER

The **Numeric Solver** feature (press **APPS**, then **9**) is composed of two specialized screens. One is an equation editor, which shows **eqn:** on the screen, waiting for you to write the equation or formula you want to investigate. To use the formula for the total surface area of a cylinder, $A = 2\pi r(r + h)$, remember to place a times sign before the parentheses (otherwise, the TI-89 will see function notation instead of multiplication). Whether or not you capitalize the A makes no difference to the TI-89 (it's *not* case sensitive). Pressing **ENTER** takes you to the second screen, where you can enter values and solve for a variable. Enter a value for each variable name except the unknown variable you intend to solve for. If there is more than one possible solution for that variable, you can control which one will be found by typing an estimate of it. Use the up and down arrow keys to place the cursor on the line that contains the unknown variable and press **F2 (Solve)**. The solution is marked by a small square to its left. Also marked (at the bottom of the screen) is a "check" of this solution, indicating the difference between the left- and right-hand sides of the equation using all of the values shown (this should be zero or something very close to zero, such as 1.E-12). Fourteen significant digits are calculated and displayed for the solution variable.

The **bound=** option near the bottom of the screen allows you to specify an interval to be searched for the solution. The default interval {−1.E14,1.E14} is restored whenever you enter the **Numeric Solver** (any **bound** interval you specify is valid only for that session). When solving trigonometric equations, a limited bound like the interval $\{0,2\pi\}$ might be better, but in most cases, the default interval works well. If you accidentally erase this line and cannot remember the syntax, just exit the **Numeric Solver** by pressing **HOME**; then press **APPS 9** to reenter it (restoring the default **bound** in the process). The equation stays in memory until you **CLEAR** it or replace it with another. Press the up arrow until the top line is reached. The calculator will switch immediately to the equation editor page, where you can modify or **CLEAR** the old equation, or even replace it with a previous equation from the **F5 Eqns** list. The following example assumes that you are starting with a blank equation (as if **Solver** has never been used before); you may **CLEAR** the present equation to achieve the same effect.

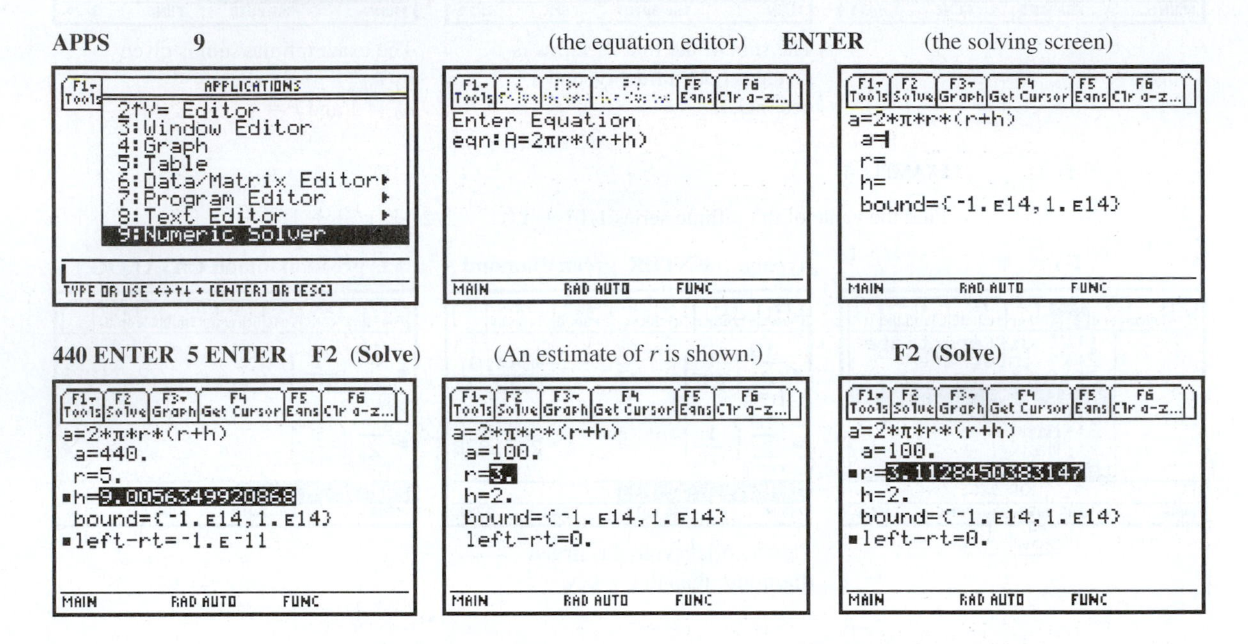

The first frame in the bottom row of the previous example found the height of a cylinder whose total surface area is 440 square units and whose radius is 5 (the answer is about 9). The last frame found the radius of a cylinder whose total surface area is 100 square units and whose height is 2. Note that the estimate $r = 3$ determined which of the two solutions would be found (the other is negative and not physically important).

The **Numeric Solver** can also be used on equations which contain only one variable, but think about the previous example (which had several variables) to understand why the **Numeric Solver** works as it does. In particular, it would seem natural to press **ENTER** after typing an estimate for the solution variable; instead, you must press **F2** (**Solve**). If you accidentally press **ENTER** first, you will have to press the up arrow key to get back to the solution variable's line (if the cursor is on **bound=** rather than a variable's line, pressing **F2** (**Solve**) does *nothing*). For direct comparison, the next example solves the same equation that was solved on the graphics screen in Section D.8. Estimates of the solutions are critical in determining which of the three solutions will be found. Solve: $x^3 + 15x = 9x^2 - 6$.

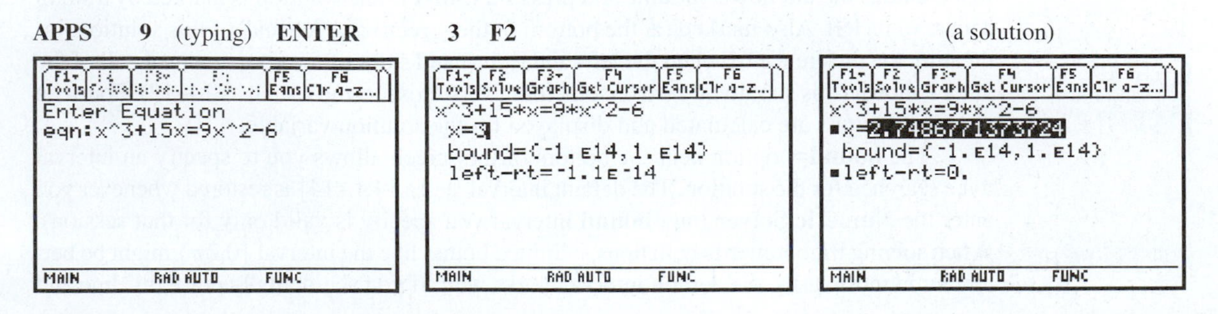

7 F2 (another solution) **-1 F2** (the third solution)

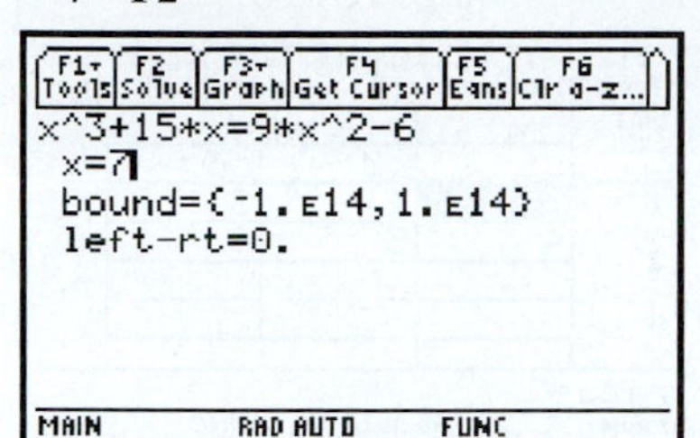

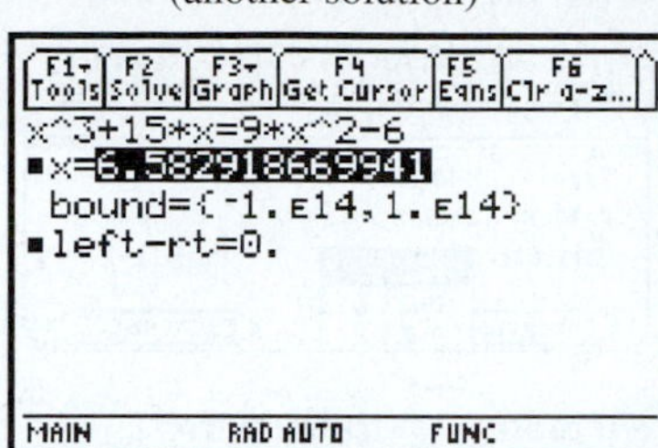

D.11 MATRIX FEATURES

Matrices on the TI-89 can easily be entered on the home screen. The entire matrix should be enclosed in brackets [ ], with rows separated by semicolons, and entries within a row separated by commas. The brackets are the **2nd** versions of the comma and ÷ keys. The semicolon is the **2nd** version of the **9** key. The effect is shown in the following frames.

matrix multiplication example **2nd MATH 4** **2**

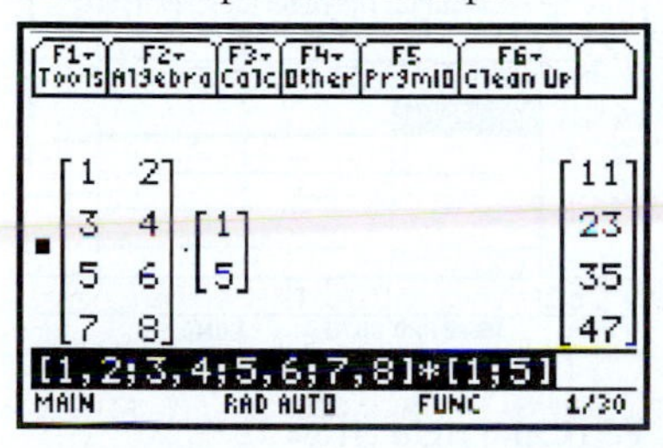

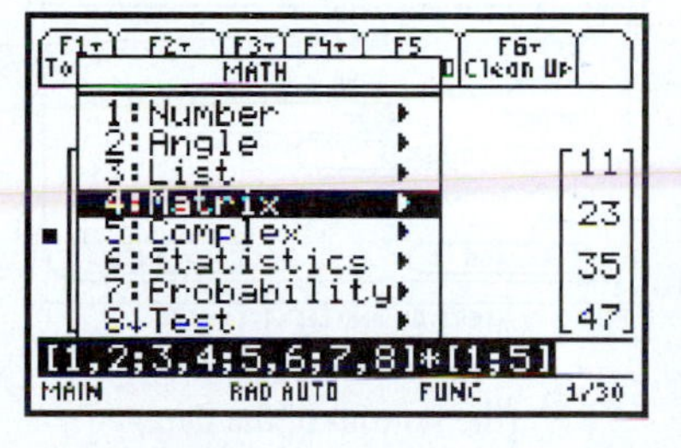

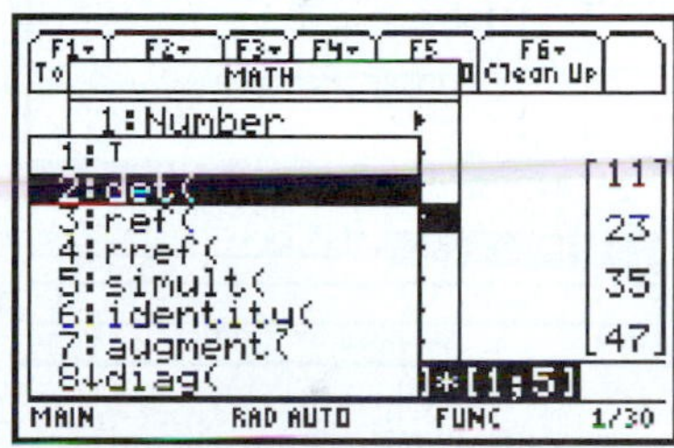

determinant example **2nd MATH 4 4** row reduction example

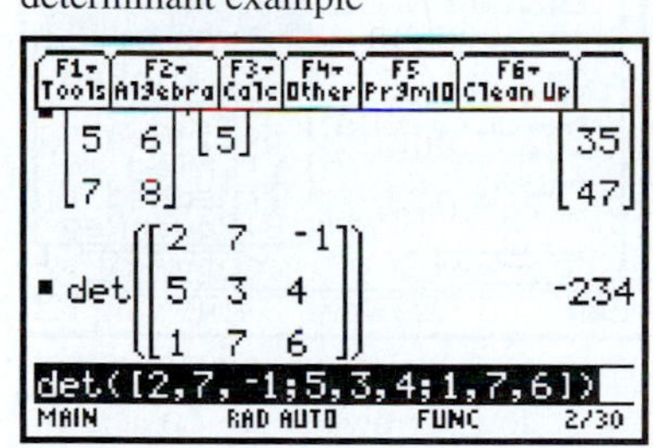

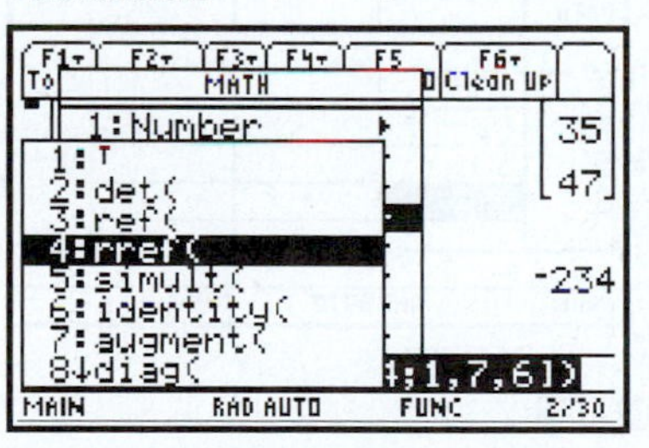

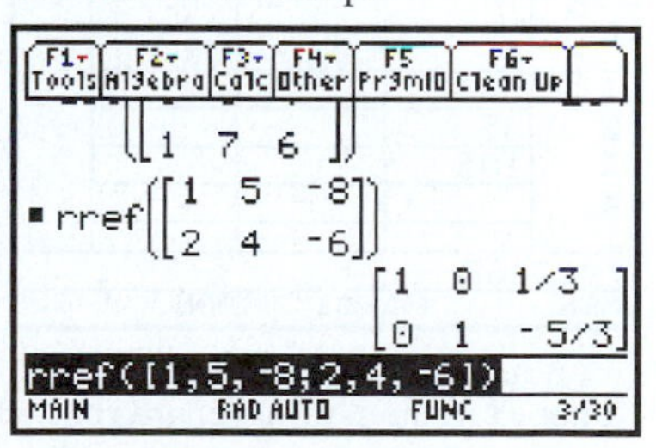

naming and storing a matrix **APPS 6** (the Matrix Editor) **2** (Open the Matrix m1.)

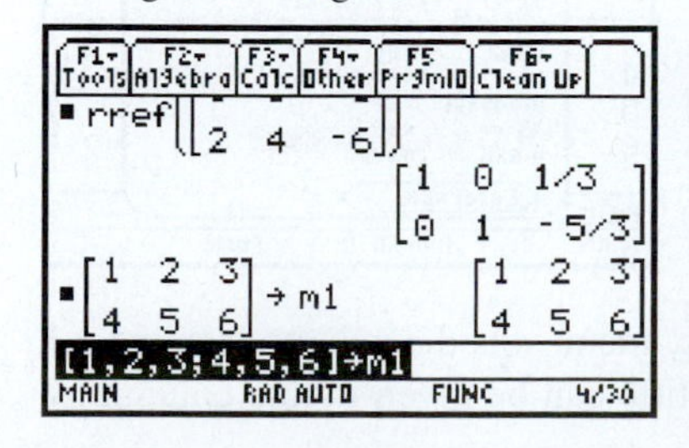

right arrow **2** (Select m1.) **ENTER** **ENTER**

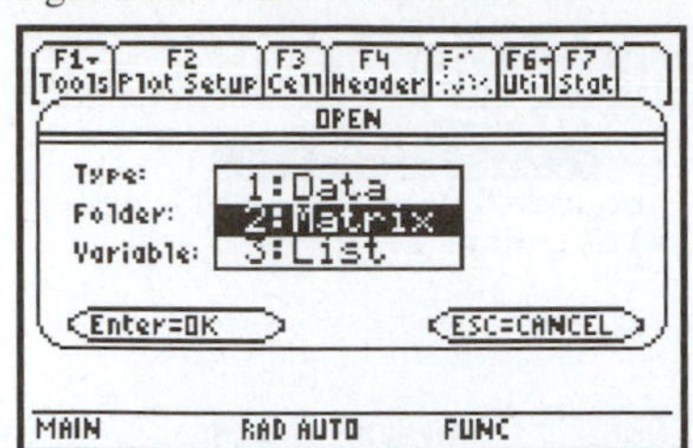

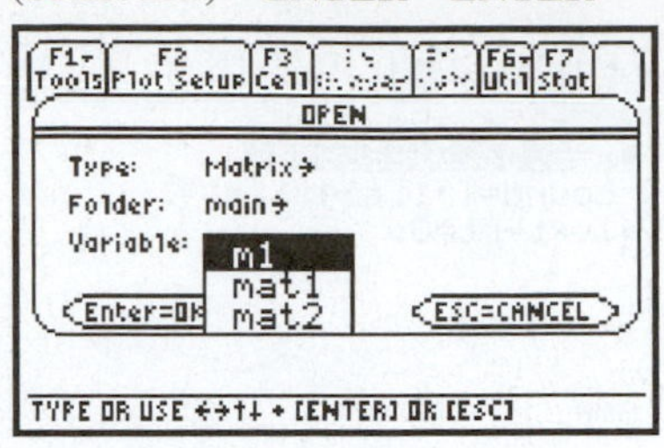

Once in the Matrix Editor, use the arrow keys to highlight any entries you may want to change and type over them. Pressing **ENTER** takes you to the next entry on the right. When the end of a row is reached, **ENTER** puts you at the beginning of the next row. The effect is very nice when you are entering a new matrix, which can thus be entered row by row by pressing **ENTER** after each matrix entry. Press **2nd QUIT** or **HOME** to return to the home screen after you finish editing a matrix.

D.12 THE DATA EDITOR AND DESCRIPTIVE STATISTICS

APPS **6** (Data/Matrix Editor) **3** (Name the Data Variable "**rbi**.") (the Data Editor) **alpha**

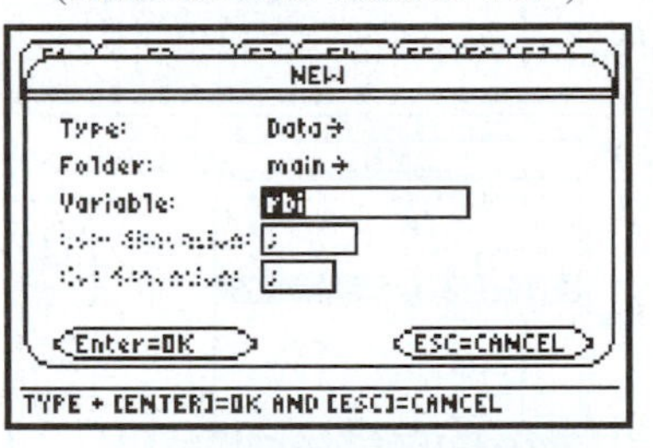

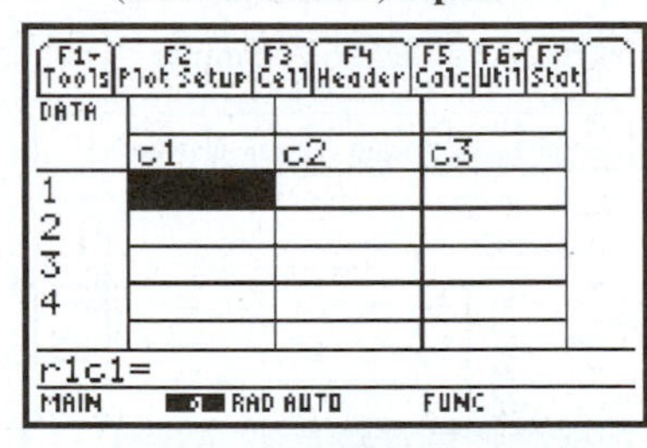

(Press **ENTER** after each entry.) (the bottom of the data) **F5** (**Calc**) right arrow **1**

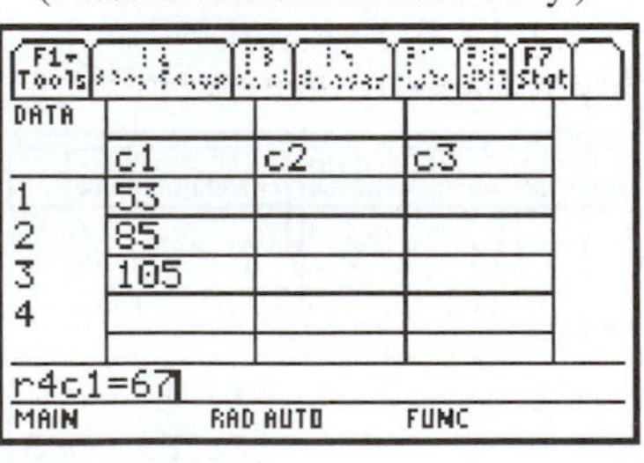

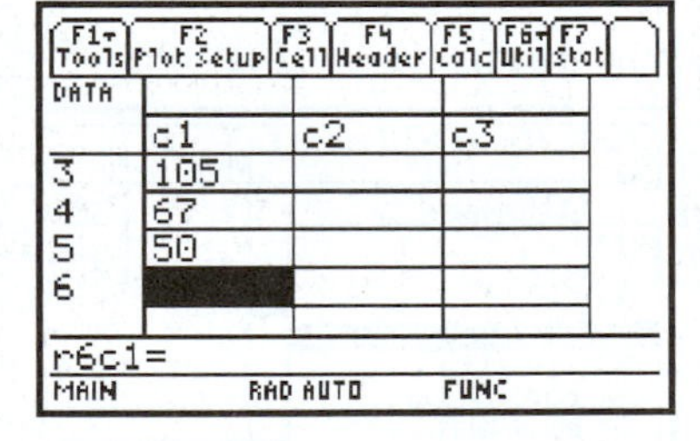

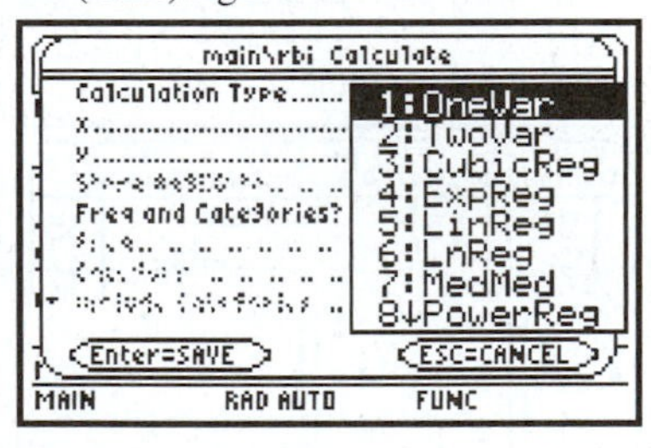

(Enter **c1** as the *x*-variable, *not* **rbi**.) **ENTER** down arrows

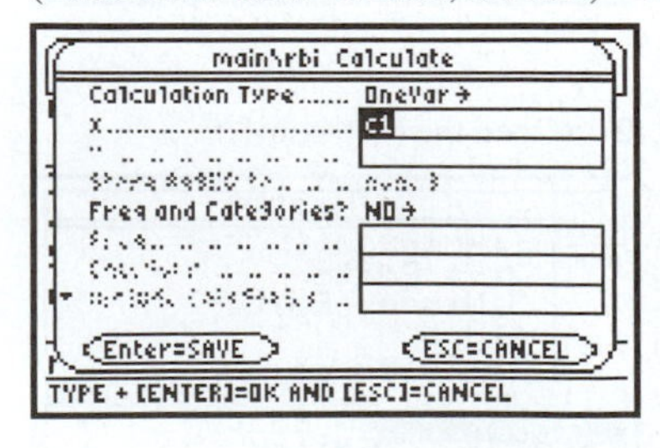

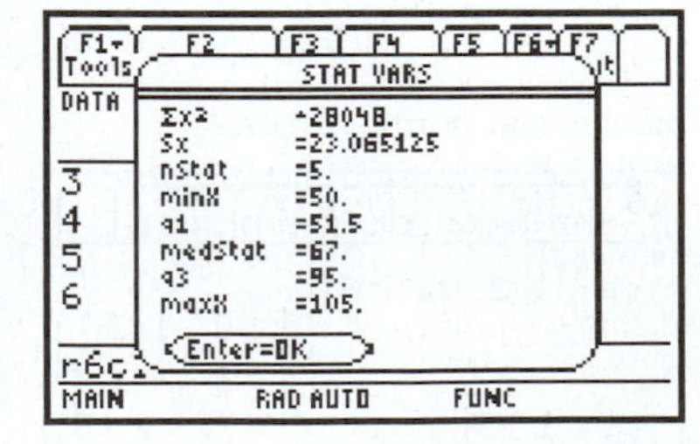

Note that the *x*-variable required in the **F5** (**Calc**) menu was the column heading, **c1** (not the data variable name **rbi**). A more descriptive title can be given to this column by

pressing the up arrow key to the blank space above **c1**; but even if you enter such a title, use **c1** as the variable name in the **F5 (Calc)** menu. Data variables can have as many as 999 rows.

D.13 THE LINE OF BEST FIT (LINEAR REGRESSION)

EXAMPLE

Find and graph the equation of the line of best fit for the following data:

x	5	7	9	12	14
y	40	58	62	74	80

APPS 6 (Data Editor) **3**

(Create the Data Variable **lin**.)

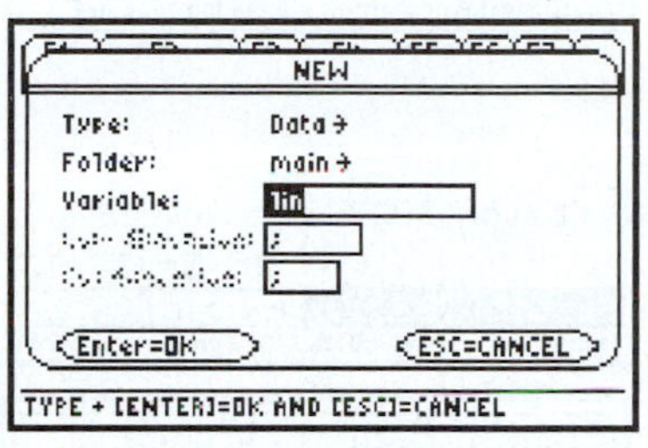

top of the data entered

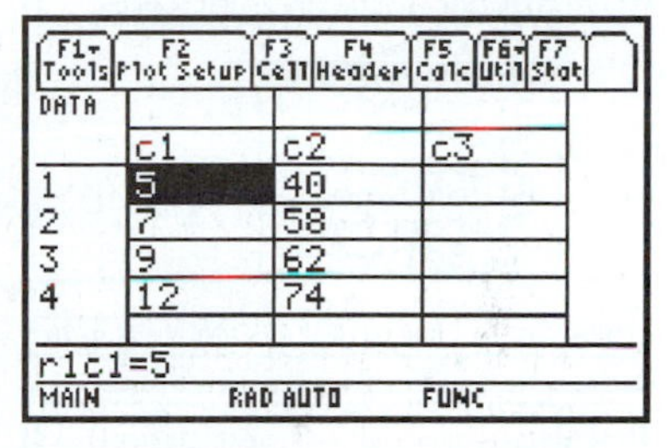

bottom of the data entered

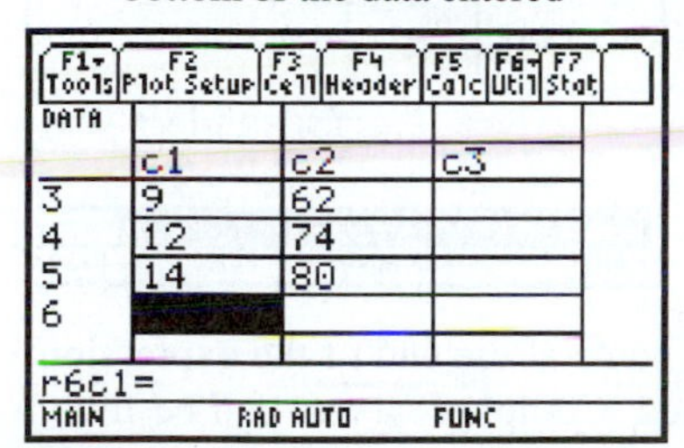

F5 (**Calc**) right arrow **5**

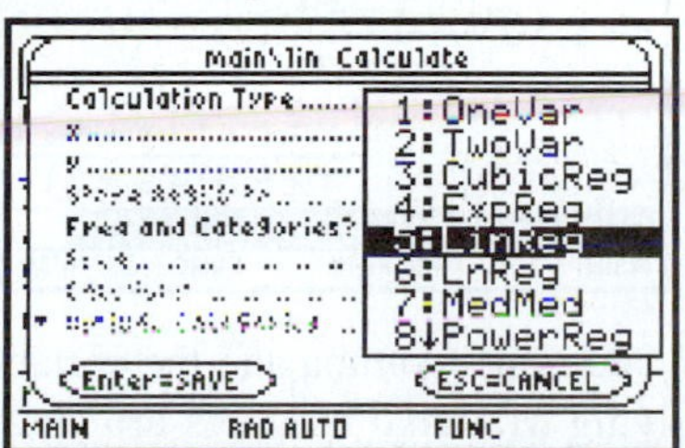

(Specify **c1** and **c2**, right arrow **y1**.)

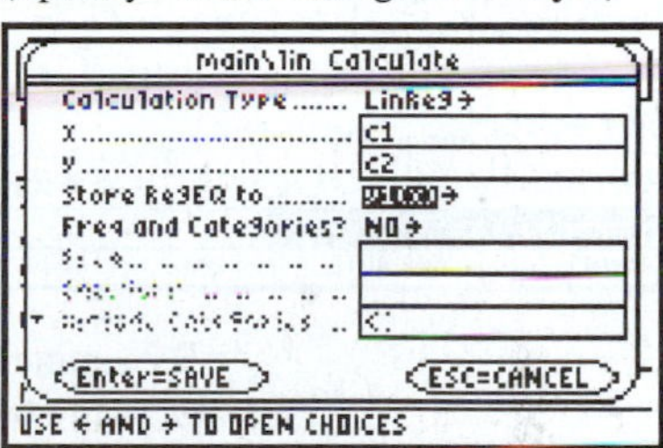

ENTER

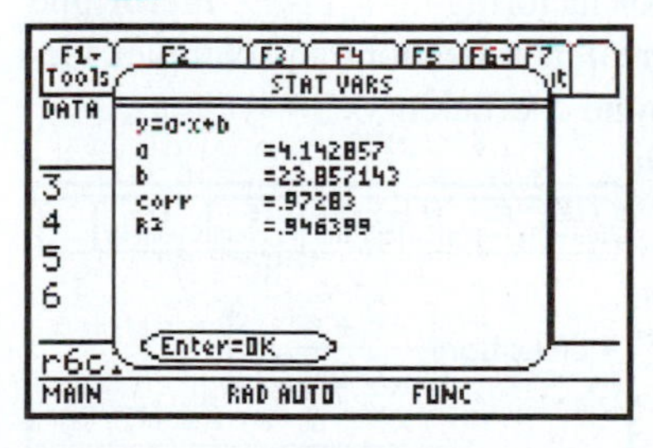

green diamond Y= up arrow **ENTER**

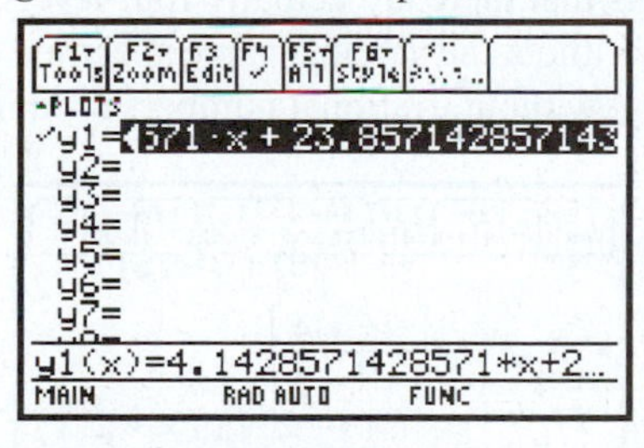

(Specify **c1** and **c2**.) **ENTER**

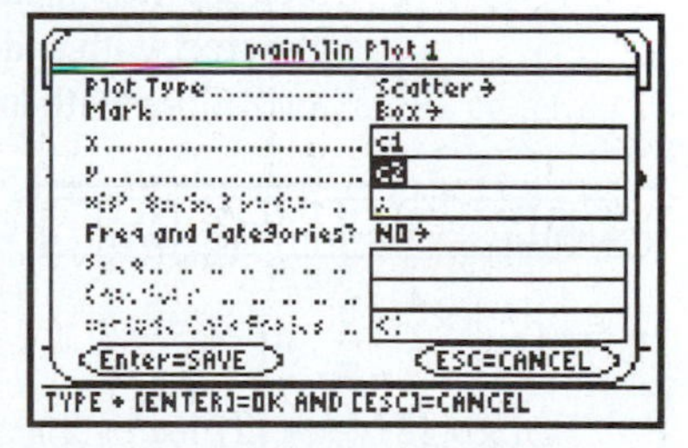

(Statistical Plot 1 is now defined.)

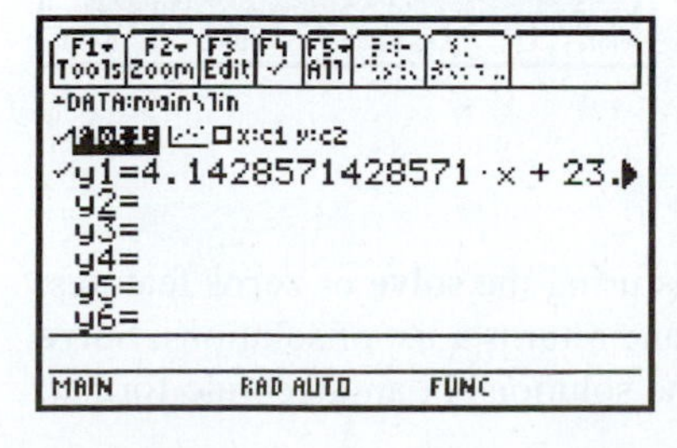

F2 **9**

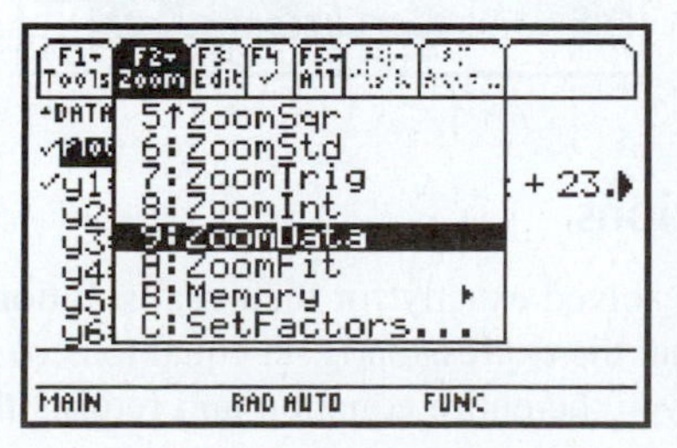

the data and the regression line

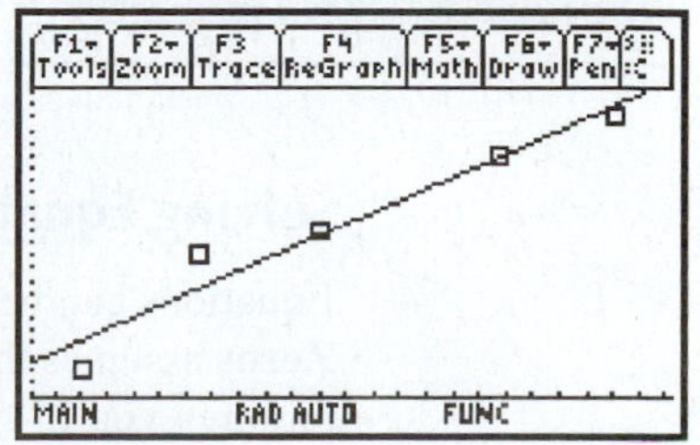

D.14 SYMBOLIC ALGEBRA FEATURES

The symbolic algebra features of the TI-89 are all found in the **F2 Algebra** menu of the home screen. If the menu **F2 f(x)** appears instead, just press **2nd CUSTOM** to restore the standard home screen menus.

Factoring Whole Numbers, Polynomials, and Fractions

The **factor** feature can factor whole numbers, fractions, polynomials, and rational expressions. If you just use **factor** followed by an expression in parentheses, the feature tries to factor using only integer or rational coefficients. In the case of fractions or rational expressions, the *simplified* versions of the numerator and denominator are shown in factored form.

Press **F2**, then **2**; finish by typing **392200**) and pressing **ENTER**.

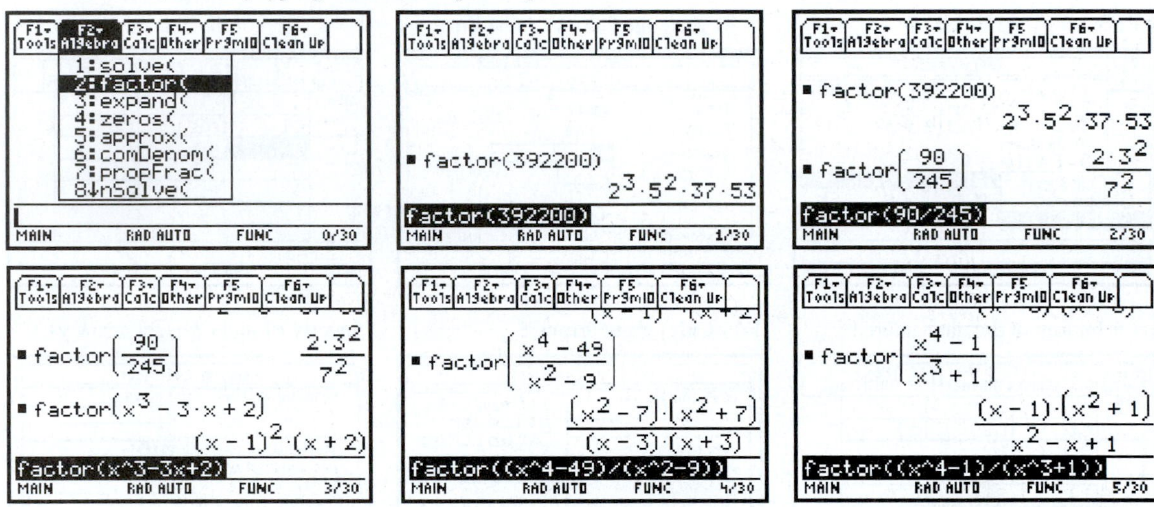

By typing an additional comma and the variable name at the end of the expression, factorizations involving irrational numbers are obtained. Complex factors can be found using **cFactor** instead of **factor** (just type a **c** in front of **factor** or press **F2,** then **alpha A** then **2**). This means that there are actually four levels of factoring on a TI-89: **factor** and **cFactor**, with and without the variable name specified (real factorizations and complex factorizations with and without irrational numbers used in the coefficients).

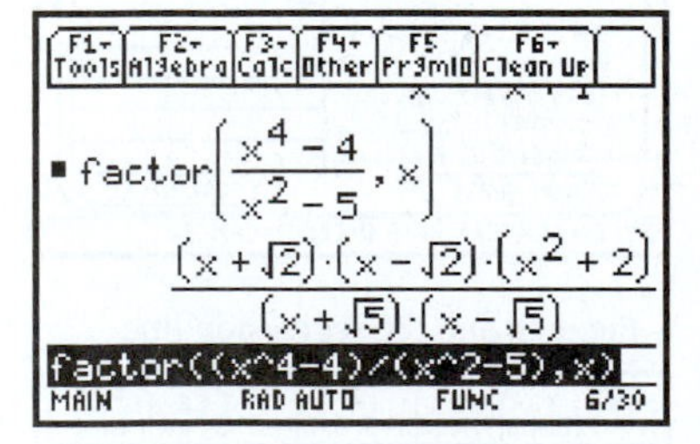

Solving Equations

Equations can be solved exactly for their real solutions using the **solve** or **zeros** features. **Zeros** assumes that the expression is set equal to zero and returns a list of solutions. **Solve** requires you to type the entire equation and returns the solutions in an algebraic format.

Both features require a comma and the variable's name near the end of the command. The commands **cSolve** and **cZeros** will find all real and complex solutions (press **F2**, then **alpha A** to see these options).

Press **F2**, then **4**; finish by typing **x^3−3x+2,x)** and pressing **ENTER**.

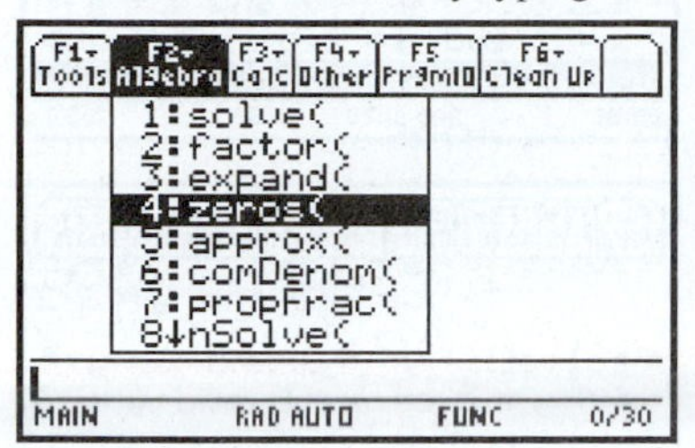

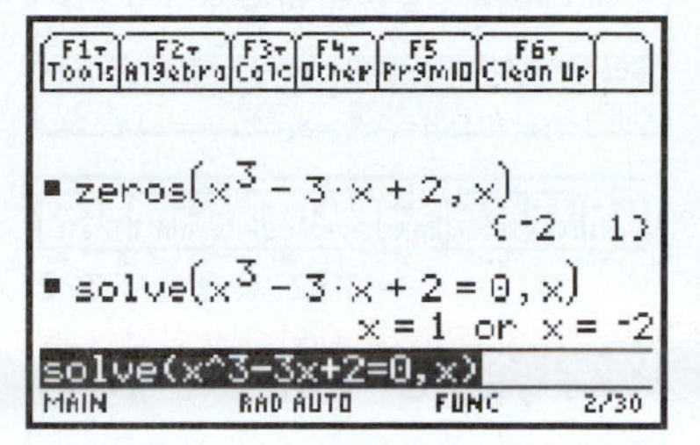

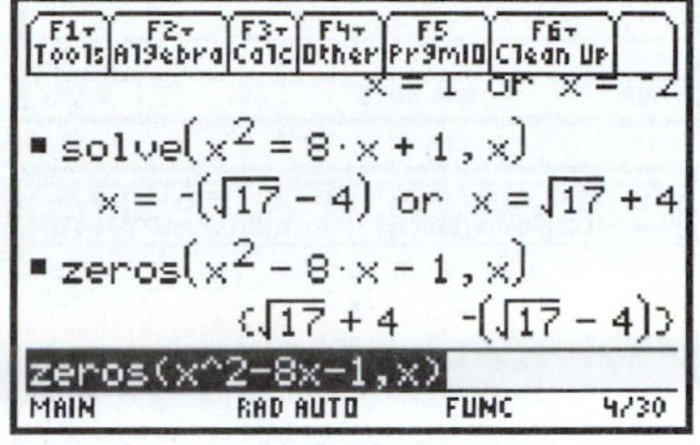

To restrict the domain of an equation, use the "with" operator (|) located just to the left of the **7** key. In the first frame below, the restriction that y is negative eliminates the well-known solutions 2 and 4. Note that decimal estimates of solutions are computed if no exact form can be found. In the second frame, the trig equation has infinitely many solutions, which are all odd multiples of π. Arbitrary integer constants are shown as @n1, @n2, and so on. In the last frame, **nSolve** is illustrated. It approximates solutions one at a time, using an estimated x-value to control which solution will be approximated.

(restricting an equation's domain)

(infinitely many periodic solutions)

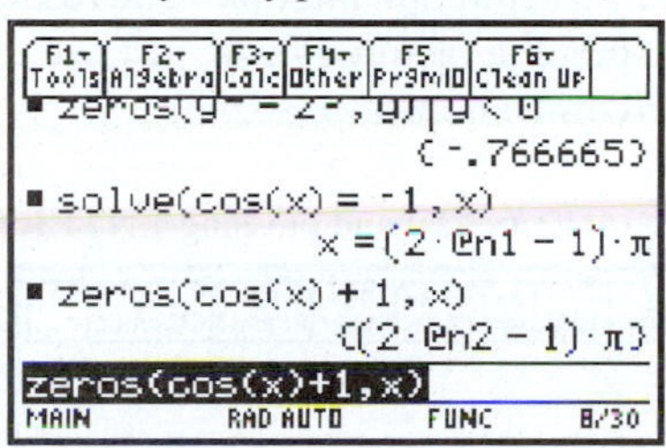

F2 alpha A 1

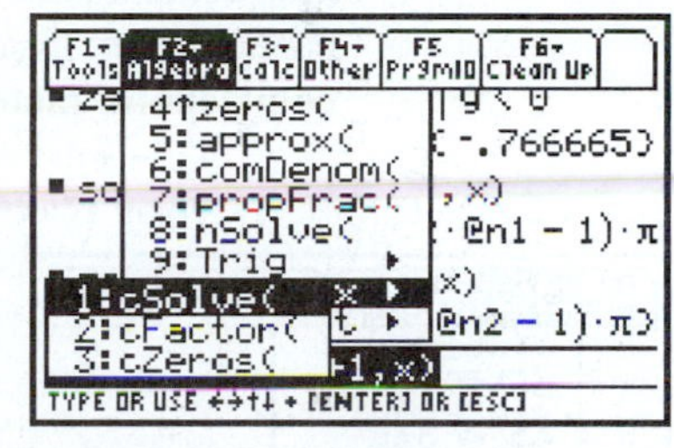

(complex-valued solutions)

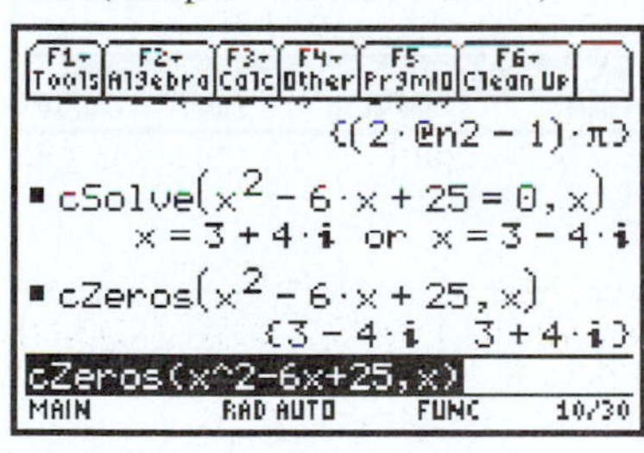

F2 8

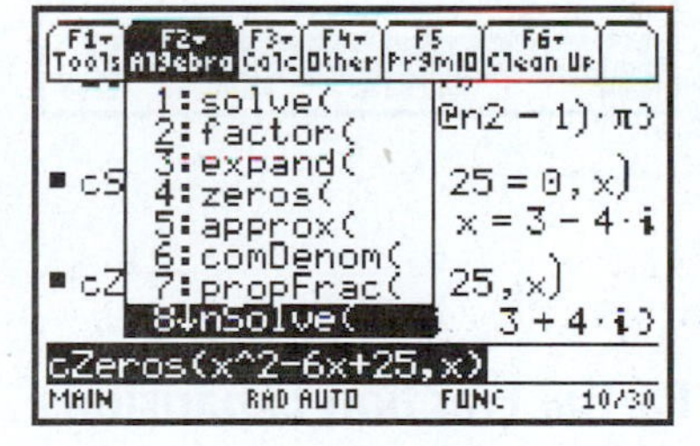

(numerically estimated solutions)

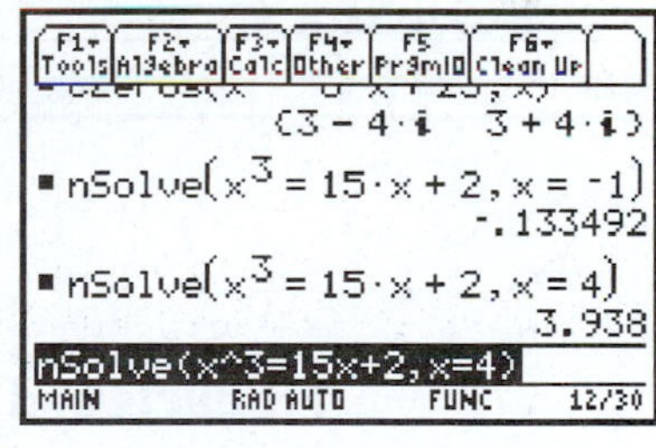

Systems of Equations

To solve systems of equations, use the solve feature with the **and** operator placed between the equations and a list of variables at the end of the command. There is a template for this purpose in the custom menu (press **2nd CUSTOM F3 4**). When you are finished with this section press **2nd CUSTOM** again to return to the standard menu. Returning to the **HOME** screen from a specialized screen also restores the standard menu.

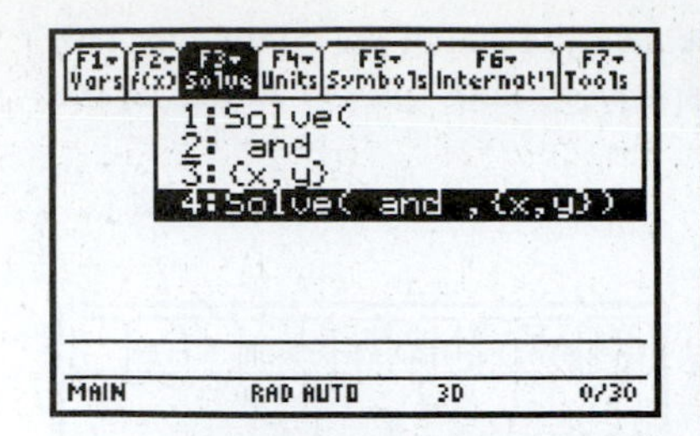

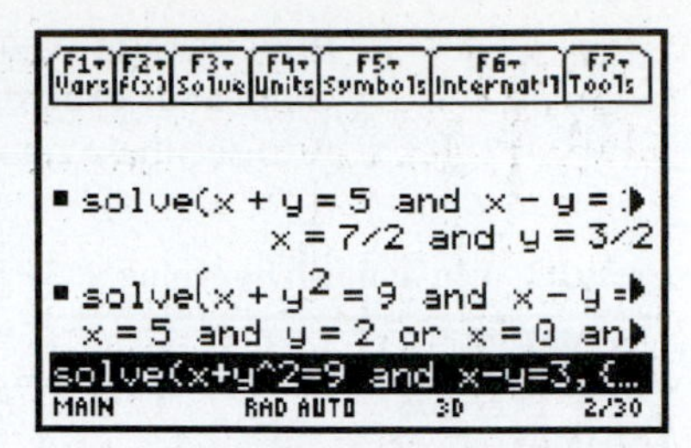

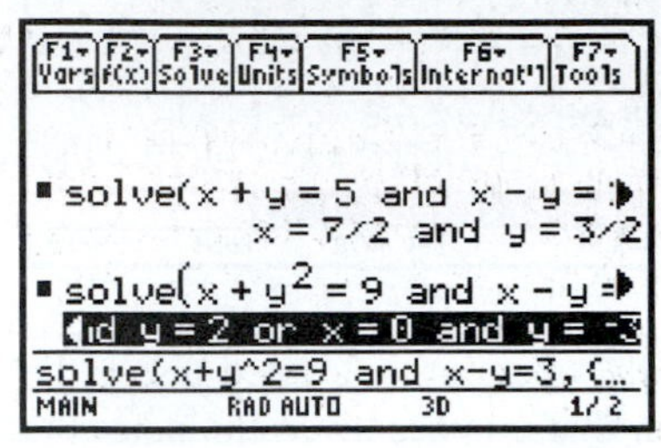

Combining Expressions into a Single Fraction

The feature that adds, subtracts, and otherwise combines several rational expressions into a single, simplified fraction is called **comDenom**. This feature is *not* for computing a common denominator or least common multiple. The result is always a simplified fraction in expanded form. If you prefer a factored form, just use the **factor** feature in conjunction with **comDenom: factor(comDenom(***expression***))**.

Press **F2**, then **6**; finish by typing **12/(x^2−4)+3/(x+2))** and pressing **ENTER**.

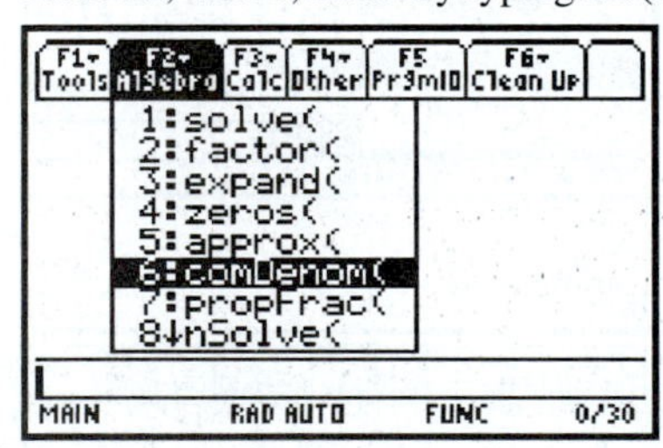

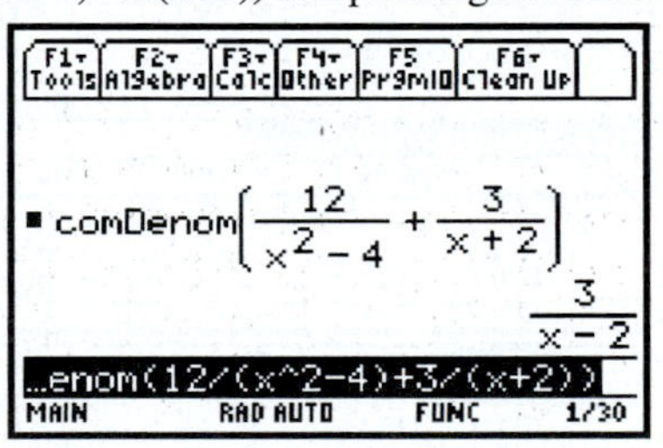

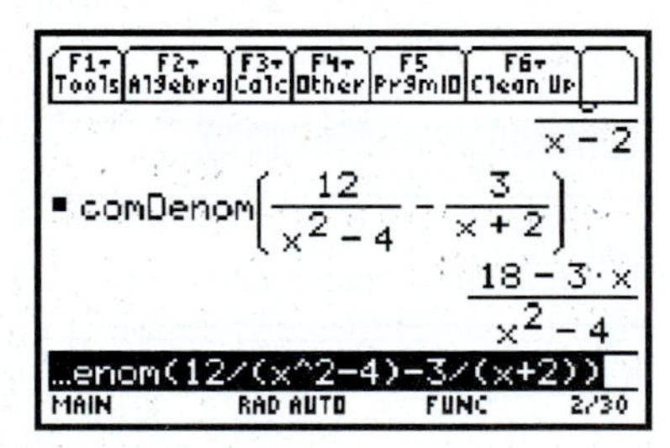

Products and Partial Fraction Expansions

The **expand** feature multiplies out indicated products and powers. For rational expressions, **expand** produces a partial fraction expansion.

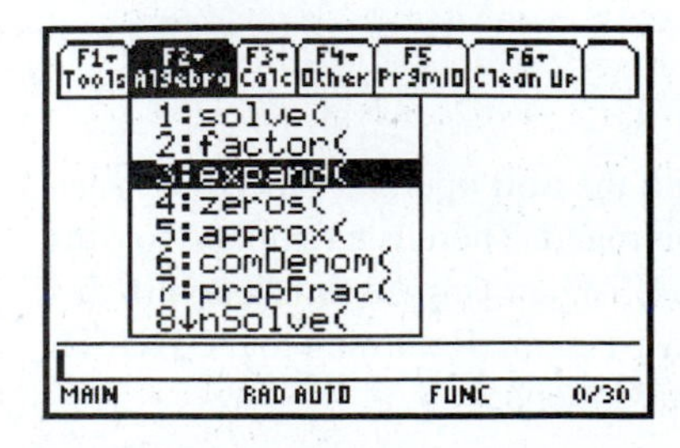

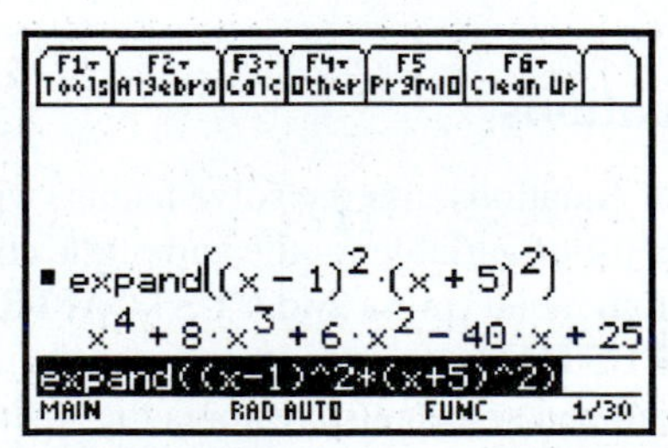

Polynomial Division

The feature that performs polynomial division is called **propFrac**. It can also be used to find a mixed number representation for improper numeric fractions.

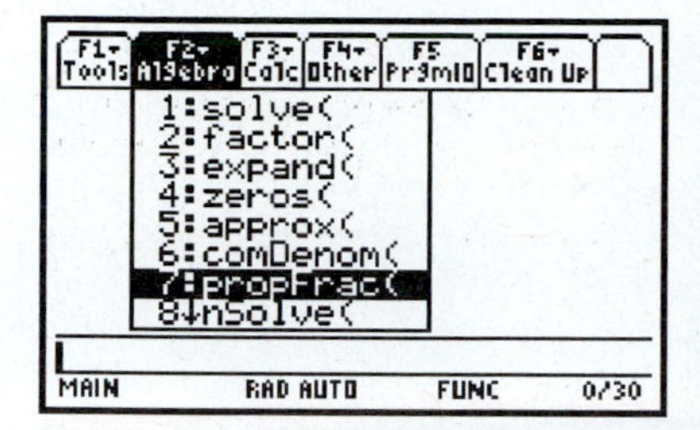

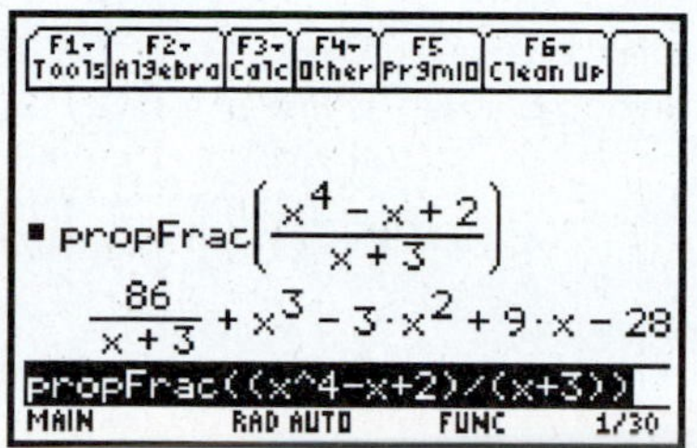

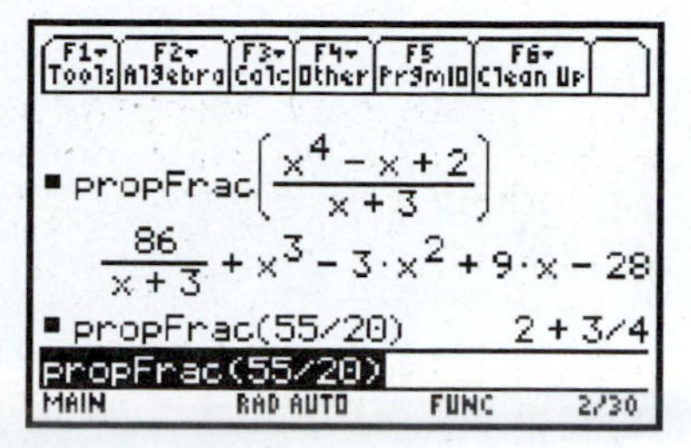

Expanding or Simplifying Trigonometric Expressions

To expand a multiple-angle trig formula in powers of sin (x) and cos (x), use the **tExpand** feature. The **tCollect** feature is the opposite of **tExpand**, taking an expression involving powers of trig functions and returning a multiple-angle equivalent.

Press **F2**, then **9**, then **1**; finish by typing **sin(3x)**) and pressing **ENTER**.

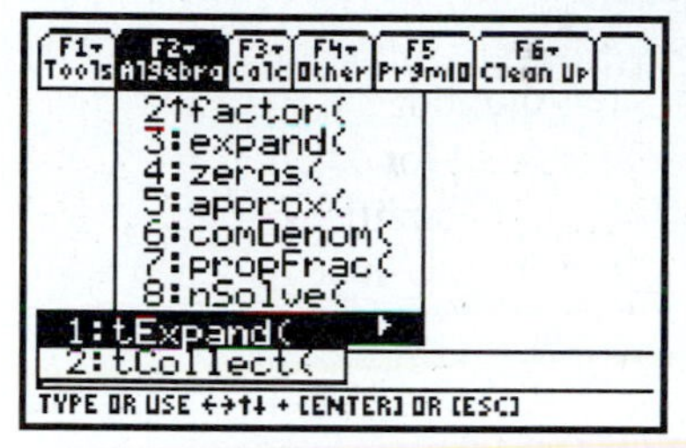

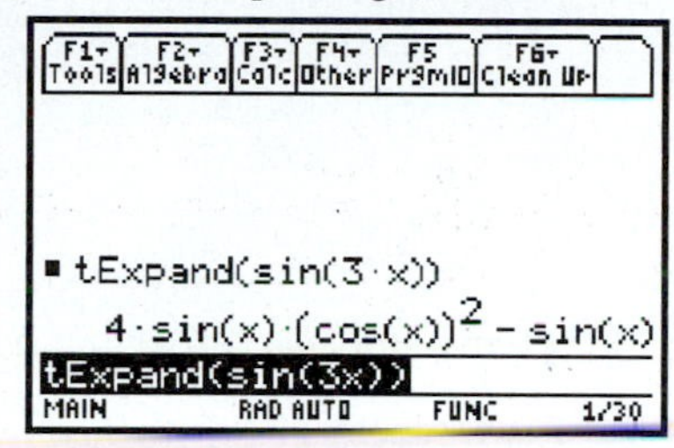

Answers to Odd-Numbered Exercises and Chapter Reviews

CHAPTER 1

Exercises 1.1, Page 5

1. 15 **3.** 11 **5.** 8 **7.** 2 **9.** -9 **11.** -2 **13.** 10 **15.** 3 **17.** $-1\frac{1}{12}$ **19.** $\frac{35}{36}$ **21.** -13
23. -7 **25.** -5 **27.** 6 **29.** 9 **31.** -14 **33.** -3 **35.** -15.7 **37.** $-\frac{5}{12}$ **39.** $-1\frac{7}{12}$ **41.** 6
43. 11 **45.** -5 **47.** -9 **49.** 30 **51.** -24 **53.** 4 **55.** $-\frac{9}{5}$ **57.** $\frac{3}{8}$ **59.** $-\frac{3}{4}$ **61.** $-\frac{2}{3}$
63. -84 **65.** 60 **67.** 7 **69.** $-\frac{27}{8}$ **71.** 20° **73.** 30° **75.** 75 ft

Exercises 1.2, Pages 8–9

1. -3 **3.** 0 **5.** 0 **7.** Meaningless **9.** 0 **11.** Indeterminate **13.** 2 **15.** $0, -\frac{4}{3}$ **17.** $-1, \frac{1}{2}$
19. 0 **21.** $\frac{5}{6}$ **23.** $3, -\frac{1}{2}$ **25.** 23 **27.** -21 **29.** 14 **31.** -29 **33.** 34 **35.** -6 **37.** -4
39. 76 **41.** 3 **43.** 21 **45.** $-\frac{12}{5}$ **47.** -2

Exercises 1.3, Pages 13–14

1. 10^{11} **3.** 10^{9} **5.** 10^{-8} **7.** 10^{3} **9.** 10^{4} **11.** 10^{-17} **13.** 10^{16} **15.** 10^{-15} **17.** 10^{4}
19. 2.07×10^{3} **21.** 9.1×10^{-2} **23.** 5.61×10^{0} **25.** 8.5×10^{6} **27.** 6×10^{-6} **29.** 1.006×10^{4}
31. 127 **33.** 0.0000614 **35.** 9,240,000 **37.** 0.00000000696 **39.** 9.66 **41.** 50,300 **43.** 3.32×10^{19}
45. -6.83×10^{-6} **47.** 8.36×10^{-11} **49.** -7.98×10^{19} **51.** -6.85×10^{1} **53.** -4.92×10^{-23}

Exercises 1.4, Pages 16–17

1. 3 **3.** 3 **5.** 4 **7.** 3 **9.** 2 **11.** 3 **13.** 0.1 cm **15.** 0.01 cm **17.** 1 mm **19.** 0.01 m
21. 10 Ω **23.** 0.0001 A **25.** **(a)** 15.2 m **(b)** 0.023 m **27.** **(a)** 14.02 cm **(b)** 0.642 cm
29. **(a)** 0.0270 A **(b)** 0.00060 A **31.** **(a)** 305,000 Ω **(b)** 305,000 Ω, 38,000 Ω **33.** **(a)** 0.08 m **(b)** 13.2 m
35. **(a)** 0.52 km **(b)** 16.8 km **37.** **(a)** 0.00009 A **(b)** 0.41 A **39.** **(a)** 500,000 Ω **(b)** 500,000 Ω

Exercises 1.5, Pages 20–21

1. 22.1 in. **3.** 84.8 cm **5.** 1.1369 g **7.** 19 V **9.** 25.09 cm **11.** 3.9 cm **13.** 2.4 mm **15.** 5.24 oz
17. 3.996 in. **19.** 2.35 in. **21.** 0.85 A **23.** 853 m^2 **25.** 0.13 in^2 **27.** 25,800 cm^3 **29.** 2.1 m
31. 73 cm/s^2 **33.** 0.078 N/m^2 **35.** 65 kg m/s^2 **37.** 110 cm^2 **39.** 614 cm^3 **41.** 1.28×10^{8} kg m^2/s^2
43. 12.48 mm **45.** 1970: 62.9 bu/acre. 2000: $14\bar{0}$ bu/acre; 77 bu/acre

Exercises 1.6, Pages 25–26

1. Binomial **3.** Trinomial **5.** Monomial **7.** Binomial **9.** Trinomial **11.** Binomial **13.** 2 **15.** 4 **17.** 10 **19.** 5 **21.** 6 **23.** 4 **25.** $3x^2 + 5x + 2$, degree 2 **27.** $9x^8 - 5x^4 + 6x^3 + 5x^2$, degree 8 **29.** $4y^5 + 3y^3 - 3y + 5$, degree 5 **31.** $4x + 2x^3 - 3x^4$, degree 4 **33.** $-7 + c + 5c^3 + 3c^4 - 8c^5$, degree 5 **35.** $2 + 2y + 5y^3 - 8y^4 - 6y^6$, degree 6 **37.** 3 **39.** 5 **41.** 8 **43.** 9 **45.** $9x^2 - 5x + 5$ **47.** $4x^2 + x - 5$ **49.** $-8x^2 - 3x + 4$ **51.** $-10x^2 - x + 10$ **53.** $4x^2 + x - 10$ **55.** $-4x^2 - 7x + 3$ **57.** $-4x^2 + 8x$ **59.** $-6x^2 - 11x + 4$ **61.** $8x^3 + 2x^2 - 5x + 3$ **63.** $-x^3 - 10x^2 + 9x - 2$ **65.** $-2y$ **67.** $x + 6y$ **69.** -5 **71.** -4 **73.** 324 **75.** -216 **77.** 3 **79.** $-30/29$ **81.** $-110{,}592$ **83.** $\frac{3}{10}$

Exercises 1.7, Pages 30–31

1. x^{12} **3.** $12a^5$ **5.** m^6 **7.** $\dfrac{1}{x^4}$ **9.** $3x^4$ **11.** $\dfrac{5}{x^3}$ **13.** a^6 **15.** c^{16} **17.** $81a^2$ **19.** $32x^{10}$ **21.** $\frac{9}{16}$ **23.** $\dfrac{16}{a^{12}}$ **25.** 1 **27.** 3 **29.** $9x^2$ **31.** t^{12} **33.** $-a^6$ **35.** $-8a^6b^3$ **37.** $9x^4y^6$ **39.** $-27x^9y^{12}z^3$ **41.** $\dfrac{4x^4}{9y^6}$ **43.** $\dfrac{16x^2}{9y^4}$ **45.** $\dfrac{1}{36y^6}$ **47.** 2 **49.** 8 **51.** 11 **53.** 5^8 **55.** 5 **57.** -6 **59.** 8 **61.** $3\sqrt{5}$ **63.** $5\sqrt{2}$ **65.** $6\sqrt{2}$ **67.** $4\sqrt{3}$ **69.** $4\sqrt{2}$ **71.** 18.1 **73.** 51.0 **75.** 0.0687 **77.** 7.21 **79.** 9.43 **81.** 1.04×10^5 **83.** 134

Exercises 1.8, Page 34

1. $32x^5$ **3.** $-24a^5b^3$ **5.** $-48a^3bc^5$ **7.** $24a^4b^9c^6$ **9.** $12a^2 - 21ab$ **11.** $6x^3 + 12x^2 - 15x$ **13.** $-15x^4 + 25x^3 - 40x^2$ **15.** $24a^3b^4 - 48a^4b^7$ **17.** $3a^6b^7 - 9a^3b^5 + 3a^2b^6$ **19.** $6x^2 + x - 35$ **21.** $48x^2 + 54x + 15$ **23.** $9x^2 - 16$ **25.** $24x^2 + 2x - 1$ **27.** $6x^2 - 23x + 21$ **29.** $18x^2 + 60xy + 32y^2$ **31.** $40s^2 - 62st - 18t^2$ **33.** $-15x^2 + 2x + 24$ **35.** $6x^4 + 19x^2 - 7$ **37.** $30x^4 - 61x^2 + 30$ **39.** $6x^3 + 23x^2 + 14x - 15$ **41.** $x^4 + 4x^2 + x + 6$ **43.** $x^2 - y^2 - 3x + 11y - 28$ **45.** $15x^4 - 2x^3 - 41x^2 + 22x + 6$ **47.** $12x^4 + 20x^3 - x^2 - 15x - 6$ **49.** $4x^2 - 20x + 25$ **51.** $9x^2 + 48x + 64$ **53.** $25x^2 - 20x + 4$ **55.** $9x^4 - 42x^3 + 73x^2 - 56x + 16$ **57.** $8x^3 - 12x^2 + 6x - 1$ **59.** $8a^3 + 60a^2b + 150ab^2 + 125b^3$

Exercises 1.9, Pages 37–38

1. $6x$ **3.** $-4ab^3$ **5.** $\dfrac{5x^3}{8y}$ **7.** $-\dfrac{5}{4x^2y^2}$ **9.** $\dfrac{b^2}{12a}$ **11.** $\dfrac{16t^6}{3}$ **13.** $3x^2 - 2x + 1$ **15.** $3x^4 - 4x^3 + 2x$ **17.** $-4x^2 + 5x - 7$ **19.** $-8x^3 + 6x + \dfrac{9}{2x} + \dfrac{3}{x^3}$ **21.** $2a - 3ab + 4$ **23.** $-3mn - 4n^2 + \dfrac{5m^2}{n}$ **25.** $8x^3z^3 - 6x^2yz^2 - 4y^2$ **27.** $2x - 5$ **29.** $x^2 - x + 3$ **31.** $x^2 - 2x + 3 + \dfrac{4}{2x - 1}$ **33.** $2x^2 + 7x - 3$ **35.** $-4x^2 - 3x + 7 - \dfrac{3}{4x - 3}$ **37.** $3x^3 - 2x^2 + 4$ **39.** $x^2 + 3x - 5$ **41.** $x^2 - 4x + 16 - \dfrac{128}{x + 4}$ **43.** $4x^2 - 2x + 1$

Exercises 1.10, Pages 41–42

1. -2 **3.** 16 **5.** 7 **7.** -72 **9.** 5 **11.** 2 **13.** -2 **15.** -5 **17.** $-\frac{17}{11}$ **19.** 14 **21.** 5 **23.** $-\frac{8}{15}$ **25.** -7 **27.** 2 **29.** 6 **31.** -2 **33.** 12 **35.** 14 **37.** $-\frac{15}{2}$ **39.** 6 **41.** $-\frac{21}{2}$ **43.** $\frac{154}{45}$ **45.** 25.6 **47.** 1.15 **49.** 6.39 **51.** 117 **53.** 1.65×10^{-4} **55.** 7.35×10^{-9}

Exercises 1.11, Pages 44–45

1. $J = \dfrac{W}{Q}$ **3.** $R_2 = R_T - R_1 - R_3$ **5.** $R = \dfrac{E}{I}$ **7.** $Q = CV$ **9.** $Q = \dfrac{W}{V}$ **11.** $R = \dfrac{JQ}{I^2t}$ **13.** $N = \dfrac{(\text{O.D.}) - 2P}{P}$ **15.** $L = \dfrac{RD^2}{k}$ **17.** $P = \dfrac{\pi}{2R}$ **19.** $T = \dfrac{VT'}{V'}$ **21.** $F = \frac{9}{5}C + 32$ **23.** $N_s = \dfrac{N_pI_p}{I_s}$ **25.** $a = \dfrac{1.22\lambda d}{\Delta d}$ **27.** $\lambda = \dfrac{2l}{n}$ **29.** $T = \dfrac{(\Delta L) + \alpha LT_0}{\alpha L}$ **31.** $v = \dfrac{f'v_s + fv_0}{f' - f}$ **33.** $q_1 = \dfrac{4\pi\epsilon_0 r^2F}{q_2}$ **35.** $R = \dfrac{E}{I} - \dfrac{q}{IC}$ or $R = \dfrac{EC - q}{IC}$ **37.** $R_2 = \dfrac{R_TR_1}{R_1 - R_T}$ **39.** $R_T = \dfrac{R_1R_2R_3}{R_1R_3 + R_1R_2 + R_2R_3}$ **41.** $s_0 = \dfrac{fs_i}{s_i - f}$ **43.** $n = \dfrac{R'R'' + fR'' - fR'}{fR'' - fR'}$ **45.** $f = \dfrac{eV + \phi}{h}$ **47.** $Z_2 = \dfrac{Z_3(I_T - I_2)}{I_2}$ **49.** $R_1 = \dfrac{R_A(R_2 + R_3)}{R_3 - R_A}$

Exercises 1.12, Pages 48–49

1. 4.00 in. **3.** 36.0 m **5.** 3.91 m **7.** 28.5 ft **9.** 131° **11.** 4280 cal **13.** 0.199 Ω/Hz **15.** 16 hp **17.** 960 cm^3 **19.** 5.22 m **21.** 94.8 m^3 **23.** 45.3 cm **25.** 396 Hz **27.** 80.0 Ω **29.** 13 Ω **31.** $6\bar{0}$ cm **33.** 21.9 m/s **35.** 0.150 mA

Exercises 1.13, Pages 53–55

1. \$54, \$162 **3.** 12 in. **5.** 12 m, 16 m **7.** 12 acres at \$650/acre, 28 acres at \$450/acre **9.** 22 **11.** **(a)** 4.5 h **(b)** 337.5 mi **13.** 85 mi/h, 105 mi/h **15.** 7.5 lb of 70%, 12.5 lb of 30% **17.** 4 qt **19.** 0.65 A, 1.40 A, 2.60 A **21.** 40 m × 80 m **23.** 150 Ω, 165 Ω, 225 Ω

Exercises 1.14, Pages 57–60

1. $\frac{9}{2}$ **3.** $\frac{1}{10}$ **5.** $\frac{72}{1}$ **7.** $\frac{500}{1}$ **9.** $\frac{16}{1}$ **11.** $\frac{30}{1}$ **13.** $\frac{1}{40{,}000}$ **15.** $\frac{4}{1}$ **17.** \$43.50/ft^2 **19.** 25 gal/acre **21.** 2.25 or $\frac{9}{4}$ **23.** $\frac{24}{5}$ **25.** 135 bu/acre **27.** $\frac{5}{3}$ **29.** 27 **31.** 168 **33.** 2 **35.** 12 **37.** $\frac{mp}{n}$ **39.** 8 **41.** 782 **43.** 8670 **45.** 0.0815 **47.** \$90, \$336 **49.** \$218.88 **51.** 900 **53.** \$132,300 **55.** 900 ft by 360 ft **57.** 20,100 and 48,240 **59.** 1500 lb and 1000 lb **61.** \$1106.56 **63.** 675 lb **65.** 62,400 bu **67.** 120 V **69.** 470,000 **71.** 450 N **73.** 51 m, 68 m **75.** Hypotenuse: 150 ft; sides: 90 ft, 120 ft

Exercises 1.15, Pages 63–65

1. Direct: $k = \frac{3}{4}$ **3.** Inverse: $k = \frac{3}{2}$ **5.** Neither **7.** $y = kz$ **9.** $a = kbc$ **11.** $r = \frac{ks}{\sqrt{t}}$ **13.** $f = \frac{kgh}{j^2}$ **15.** $k = \frac{1}{3}$; 12 **17.** $k = 54$; 3 **19.** $k = 6$; 36 **21.** 19.5 **23.** $k = 32$; $\frac{64}{3}$ **25.** 1.19×10^{18} N **27.** $k = 6.00$ cm^3/K; $18\bar{0}0$ cm^3 **29.** $k = 7\bar{5}0$ ft^3 lb/(in^2 °R); 8630 ft^3 **31.** $k = 1.0$ WΩ/V^2; 480 W **33.** 160 teeth **35.** 24 rpm **37.** **(a)** $5\bar{0}$ rpm **(b)** $1\bar{0}0$ rpm

Chapter 1 Review, Pages 70–72

1. −5 **2.** 0 **3.** −210 **4.** −2 **5.** −4 **6.** 9 **7.** 4 **8.** −1 **9.** 11 **10.** −24 **11.** $\frac{27}{13}$ **12.** $\frac{1}{6}$ **13.** 0 **14.** Indeterminate **15.** Meaningless **16.** 10^8 **17.** 10^4 **18.** 10^7 **19.** 3.42×10^6 **20.** 0.000561 **21.** 4.24×10^{-6} **22.** 3.00×10^{29} **23.** **(a)** 3 significant digits **(b)** 1 m **24.** **(a)** 2 significant digits **(b)** 0.001 A **25.** **(a)** 3 significant digits **(b)** 1000 V **26.** 57.6 L **27.** 730.9 cm **28.** $4\bar{0}0$ m^2 **29.** 3.11 m **30.** 2.5 lb/in^2 **31.** Trinomial, degree 3 **32.** $16x^2 + 4x - 21$ **33.** $7x^2 - x - 11$ **34.** $x^2 - 5x + 2$ **35.** $5x^2 + 4x$ **36.** $-5a + 8b$ **37.** $-9a - 21b$ **38.** $\frac{18}{5}$ **39.** y^9 **40.** b^9 **41.** $-15x^6$ **42.** $6m^4$ **43.** a^{12} **44.** $25a^6$ **45.** $\frac{y^3}{x^6}$ **46.** 1 **47.** $-s^9$ **48.** x^{18} **49.** $8a^9b^6$ **50.** $\frac{x^8}{4}$ **51.** $-8x^{18}$ **52.** 7 **53.** 3 **54.** $3\sqrt{7}$ **55.** $6\sqrt{3}$ **56.** 56.1 **57.** 0.143 **58.** $-24a^5b^2c^4$ **59.** $10x^2 - 30xy$ **60.** $-12x^5 + 9x^4 + 3x^3 - 12x^2$ **61.** $-15a^4b^4 + 9a^5b - 15a^2b^3$ **62.** $6x^2 + 13x - 28$ **63.** $30x^2 - 63x + 27$ **64.** $20x^2 + 38x + 12$ **65.** $6x^2 - 5x - 6$ **66.** $64x^2 - 80x + 25$ **67.** $2x^2 + xy - y^2 - 11x + 10y - 21$ **68.** $\frac{3y^2}{z}$ **69.** $8a^3$ **70.** $16x - 8 + \frac{5}{x}$ **71.** $3x - \frac{5}{x} + \frac{7}{x^2}$ **72.** $3n^2 + \frac{4}{mn^2} - \frac{4m^4}{3n}$ **73.** $3x + 4$ **74.** $3x^2 - 3x + 8 - \frac{11}{x + 1}$ **75.** $x^2 - 2x + 3 + \frac{10}{2x - 1}$ **76.** −4 **77.** −4 **78.** −30 **79.** $-\frac{2}{3}$ **80.** 23 **81.** $\frac{32}{7}$ **82.** $-\frac{5}{7}$ **83.** $-\frac{11}{42}$ **84.** 8.77 **85.** $D = \frac{12S}{\pi}$ **86.** $V = \frac{Q}{C}$ **87.** $t = \frac{v_0 - v}{g}$ **88.** $T_0 = \frac{\beta VT - (\Delta V)}{\beta V}$ **89.** 82 m/s **90.** 55.0 Ω **91.** 40.7 Ω **92.** 18 **93.** 20 m by 25 m **94.** 12 oz of 15% silver, 18 oz of 20% silver **95.** $\frac{1}{2}$ h **96.** $\frac{40}{49}$ **97.** $\frac{2}{5}$ **98.** $\frac{49}{9}$ **99.** 52 **100.** 52 **101.** 24 **102.** $\frac{a^2 - 6b}{b}$ **103.** $\frac{bc}{a}$ **104.** 56.7 **105.** 100,000 gal **106.** 711 and 1659 **107.** $y = k\sqrt{z}$ **108.** $y = kvu^2$ **109.** $y = \frac{kp}{q}$ **110.** $y = \frac{kmn}{p^2}$ **111.** 5184 **112.** 48 **113.** 160 N **114.** 18.8 lb **115.** −270 N, attractive force

Chapter 1 Special NASA Application, Page 74

Part A 1. It converts time in years to time in months. This is necessary for consistency in units since lead time and safety pad are given in months. **2.** 4 tires **3.** MSL = 4(8)(1)(9 + 12) ÷ (12) = 56 tires **4.** When the inventory level drops to 56 tires, a new order for tires should be placed.

Part B 1. **(a)** 4 tires **(b)** 8 **(c)** 12 months **(d)** 56 tires **(e)** 0 tires **(f)** 54 tires **(g)** OQ = 4(8)(1 ÷ 12)(12) + 56 − 0 − 54 = 34 tires **2.** Under the given conditions, NASA needs to order 34 more tires.

CHAPTER 2

Exercises 2.1, Pages 81–82

1. ∠2 and ∠3 **3.** ∠5 and ∠6 **5.** ∠1 and ∠4 **7.** 55° **9.** ∠2 and ∠8, ∠5 and ∠6, ∠7 and ∠8 **11.** 40° **13.** 40° **15.** 40° **17.** 40° **19.** 145°

Exercises 2.2, Pages 89–90

1. 10.0 cm **3.** 2860 m **5.** **(a)** 203 in^2 **(b)** 69.5 in. **7.** **(a)** 3,020,000 m^2 **(b)** 8660 m **9.** 6660 cm^2; 372 cm **11.** 37,800 ft^2; 939 ft **13.** The point of intersection of the three medians, the point of intersection of the three altitudes, and the point of intersection of the three angle bisectors coincide. **15.** 68.9° **17.** 11.2 ft **19.** 238 ft **21.** 254,000 ft^2 **23.** 68 ft

Exercises 2.3, Pages 93–94

1. **(a)** 776 m **(b)** 35,$\bar{0}$00 m^2 **3.** **(a)** 632 in. **(b)** 19,400 in^2 **5.** **(a)** 60.0 m **(b)** 225 m^2 **7.** **(a)** 72.4 m **(b)** 308 m^2 **9.** 16.7 m **11.** 141 cm **13.** 95.5 in. **15.** $1012.50 **17.** 44,500 ft^2 **19.** 1970 in^2

Exercises 2.4, Pages 98–100

1. **(a)** 41$\bar{0}$ m **(b)** 13,400 m^2 **3.** **(a)** 393 mi **(b)** 12,300 mi^2 **5.** **(a)** 66.0 cm **(b)** 346 cm^2 **7.** **(a)** 7040 m **(b)** 3,940,000 m^2 **9.** 3.91 ft **11.** 11.7 in. **13.** **(a)** 49.0° **(b)** 29.0° **(c)** 20.0° **(d)** 90° **15.** **(a)** 60.0° **(b)** 120.0° **(c)** 60.0° **(d)** 60.0° **(e)** 30.0° **17.** 28.7 cm **19.** 60.0° **21.** 2350 ft^2

Exercises 2.5, Pages 103–107

1. **(a)** 30,300 cm^3 **(b)** 3890 cm^2 **(c)** 6140 cm^2 **3.** 1540 in^3 **5.** **(a)** 38,200 ft^3 **(b)** 5490 ft^2 **7.** **(a)** 21,200 ft^3 **(b)** 2830 ft^2 **(c)** 4240 ft^2 **9.** **(a)** 2210 ft^3 **(b)** 723 ft^2 **(c)** 1213 ft^2 **11.** **(a)** 13$\bar{0}$0 in^3 **(b)** 531 lb **13.** **(a)** 15,100 lb **(b)** 65,400 lb **(c)** 23,600 ft^3 **15.** **(a)** 846,000 gal **(b)** 90.5 gal **17.** 407,000 cm^2 **19.** 52 truckloads

Chapter 2 Review, Pages 110–111

1. ∠1 = 150°, ∠2 = 30°, ∠3 = 150°, ∠4 = 150° **2.** **(a)** ∠1 and ∠3 **(b)** ∠3 and ∠4 **(c)** ∠1 and ∠4 **(d)** ∠1 and ∠2, ∠2 and ∠3, ∠4 and ∠5 **3.** **(a)** Triangle **(b)** Quadrilateral **(c)** Pentagon **(d)** Hexagon **(e)** Octagon **4.** **(a)** 105,000 m^2 **(b)** 1396 m **5.** **(a)** 385 ft^2 **(b)** 92.0 ft **6.** **(a)** 2.95 mi^2 **(b)** 9.74 mi **7.** **(a)** 7.05 km^2 **(b)** 11.48 km **8.** **(a)** 13$\bar{0}$0 ft^2 **(b)** 144.0 ft **9.** **(a)** 541 m^2 **(b)** 112.8 m **10.** **(a)** 661 m^2 **(b)** 91.1 m **11.** 248 ft **12.** 32.0 m **13.** 39.6 in. **14.** **(a)** 62.4 cm^2 **(b)** 36.0 cm **15.** **(a)** 15,600 ft^3 **(b)** 2160 ft^2 **(c)** 3890 ft^2 **16.** **(a)** 3190 m^3 **(b)** 1340 m^2 **(c)** 1480 m^2 **17.** **(a)** 5270 in^3 **(b)** 1490 in^2 **(c)** 1870 in^2 **18.** 6780 cm^3 **19.** **(a)** 17,200 m^3 **(b)** 3220 m^2 **20.** **(a)** 2980 ft^3 **(b)** 985 ft^2 **(c)** 113$\bar{0}$ ft^2

Chapter 2 Special NASA Application, Page 113

1. The units canceled. **2.** It would not change. **3.** The units cancel when the ratio is formed.

CHAPTER 3

Exercises 3.1, Pages 122–123

1. **3.**

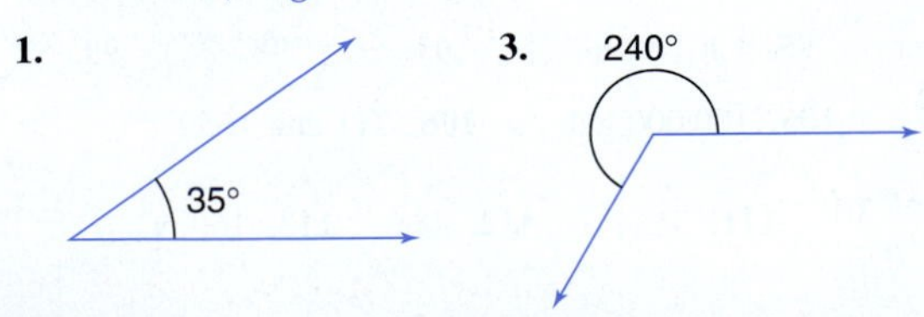

5. $\frac{1}{4}°$ **7.** 2° **9.** 30′ **11.** 24′ **13.** 37.2° **15.** 75.78° **17.** 69°20′ **19.** 23°18′ **21.** 34.4042°
23. 19.3075° **25.** 18°12′36″ **27.** 8°55′30″ **29.** a **31.** c **33.** a **35.** B **37.** B **39.** 13.0 cm
41. 173 mi **43.** 338 yd **45.** 39.5 m **47.** 3.00 cm
49. $\sin A = 0.385$, $\cos A = 0.923$, $\tan A = 0.417$, $\cot A = 2.40$, $\sec A = 1.08$, $\csc A = 2.60$
51. $\sin A = 0.489$, $\cos A = 0.872$, $\tan A = 0.561$, $\cot A = 1.78$, $\sec A = 1.15$, $\csc A = 2.04$
53. $\sin A = 0.867$, $\cos A = 0.500$, $\tan A = 1.73$, $\cot A = 0.577$, $\sec A = 2.00$, $\csc A = 1.15$
55. $\sin B = 0.923$, $\cos B = 0.385$, $\tan B = 2.40$, $\cot B = 0.417$, $\sec B = 2.60$, $\csc B = 1.08$
57. $\sin B = 0.868$, $\cos B = 0.497$, $\tan B = 1.75$, $\cot B = 0.573$, $\sec B = 2.01$, $\csc B = 1.15$
59. $\sin B = 0.500$, $\cos B = 0.866$, $\tan B = 0.578$, $\cot B = 1.73$, $\sec B = 1.16$, $\csc B = 2.00$

61.

	(a)	(b)
sin A	0.385	0.385
cos A	0.923	0.923
tan A	0.417	0.417
cot A	2.40	2.40
sec A	1.08	1.08
csc A	2.60	2.60

Exercises 3.2, Pages 129–130

1. 0.3173 **3.** 0.8816 **5.** 0.2215 **7.** 1.216 **9.** 1.483 **11.** 1.309 **13.** 0.7280 **15.** 3.443
17. 1.011 **19.** 25.5° **21.** 25.1° **23.** 81.6° **25.** 72.7° **27.** 11.5° **29.** 65.0° **31.** 59.6°
33. 26.8° **35.** 80.4° **37.** 61.08° **39.** 33.01° **41.** 24.54° **43.** 0.5934 **45.** 1.314 **47.** 0.9850
49. 1.363 **51.** 1.004 **53.** 3.237 **55.** 58°49′34″ **57.** 80°55′58″ **59.** 53°22′24″ **61.** 49°2′55″
63. 24°52′43″ **65.** 31°5′42″

Exercises 3.3, Pages 133–134

1. $A = 26.6°$, $B = 63.4°$, $c = 8.94$ ft **3.** $B = 62.7°$, $a = 9.63$ cm, $b = 18.7$ cm **5.** $A = 56.0°$, $B = 34.0°$, $a = 11.1$ m
7. $A = 58.2°$, $B = 31.8°$, $c = 14.6$ mi **9.** $A = 53.0°$, $a = 332$ km, $c = 415$ km
11. $A = 22.32°$, $B = 67.68°$, $b = 34.62$ cm **13.** $B = 21.25°$, $b = 2627$ mi, $c = 7248$ mi
15. $A = 74.20°$, $a = 43.61$ m, $b = 12.34$ m **17.** $A = 56.13°$, $B = 33.87°$, $a = 3832$ ft
19. $A = 65°$, $B = 25°$, $a = 3200$ mi **21.** $A = 4\bar{0}°$, $a = 29$ m, $b = 34$ m **23.** $B = 53°$, $b = 190$ ft, $c = 230$ ft
25. $A = 68°$, $b = 1.4$ mi, $c = 3.8$ mi **27.** $A = 52°18′30″$, $b = 1354$ m, $c = 2215$ m
29. $A = 53°25′31″$, $B = 36°34′29″$, $b = 367.7$ ft **31.** $B = 31°48′35″$, $b = 23.27$ m, $c = 44.15$ m
33. $A = 62°54′44″$, $a = 6011$ ft, $b = 3075$ ft

Exercises 3.4, Pages 137–142

1. 52 m **3.** 4° **5.** 257 ft **7.** 92.2 m **9.** 269 ft **11.** **(a)** 139 Ω, 36.4° **(b)** 270 Ω, 110 Ω
13. 2.86 cm, 36.5° **15.** $x = 2.04$ cm, $y = 5.86$ cm, $z = 5.41$ cm, $\alpha = 68.5°$, $\beta = 63.7°$, $\theta = 149.7°$, $\phi = 78.1°$
17. 0.2445 in. **19.** 12.6° **21.** $x = 31.21$ m, $\alpha = 125.22°$, $\beta = 144.78°$ **23.** 11.6 ft **25.** 9.5° **27.** 43 ft
29. 107.2 mm **31.** 92.08 mm **33.** 373.1 ft **35.** 26°33′54″; 63°26′6″

Chapter 3 Review, Pages 143–144

1. 129.5° **2.** 76.2° **3.** 35°40′ **4.** 314°18′ **5.** 16.4625° **6.** 38°24′18″ **7.** 32.2 m **8.** 31.6 mi
9. $\sin A = 0.843$, $\cos A = 0.540$, $\tan A = 1.56$, $\cot A = 0.641$, $\sec A = 1.85$, $\csc A = 1.19$ **10.** 0.9677
11. 0.7848 **12.** 0.5658 **13.** 1.044 **14.** 1.797 **15.** 1.018 **16.** 37.4° **17.** 69.4° **18.** 51.0°
19. 41.2° **20.** 47.3° **21.** 54.9° **22.** 0.6643 **23.** 3.774 **24.** 1.210 **25.** 63°26′55″
26. 13°32′21″ **27.** 16°50′12″ **28.** $A = 36.4°$, $B = 53.6°$, $c = 11.8$ m
29. $A = 53.50°$, $a = 21.28$ cm, $c = 26.48$ cm **30.** $B = 7\bar{0}°$, $a = 580$ km, $b = 1600$ km
31. $B = 54°45′28″$, $b = 347.8$ m, $c = 425.8$ m **32.** 4800 ft **33.** $5\bar{0}00$ ft **34.** 2.4°
35. $x = 22.3$ m, $y = 24.3$ m, $\alpha = 76.3°$, $\beta = 63.1°$ **36.** 10.2 ft **37.** **(a)** 86 Ω, $\phi = 61°$
(b) $R = 88$ Ω, $Z = 130$ Ω **38.** $x = 5.00$ cm, $y = 4.33$ cm **39.** 346 mi **40.** 7.6 m

CHAPTER 4

Exercises 4.1, Pages 149–150

Function	*Domain*	*Range*
1. Yes	{2, 3, 9}	{2, 4, 7}
3. No	{1, 2, 7}	{1, 3, 5}
5. Yes	{−2, 2, 3, 5}″	{2}

	Function	*Domain*	*Range*
7.	Yes	Real numbers	Real numbers
9.	Yes	Real numbers	Real numbers where $y \geq 1$
11.	No	Real numbers where $x \geq -2$	Real numbers
13.	Yes	Real numbers where $x \geq -3$	Real numbers where $y \geq 0$
15.	Yes	Real numbers where $x \geq 4$	Real numbers where $y \geq 6$

17. **(a)** 20 **(b)** -12 **(c)** -28 **19.** **(a)** 35 **(b)** 15 **(c)** -25 **21.** **(a)** 95 **(b)** 0 **(c)** 4
23. **(a)** 2 **(b)** $\frac{2}{3}$ **(c)** 0 is not in the domain of $f(t)$. **25.** **(a)** $6a + 8$ **(b)** $24a + 8$ **(c)** $6c^2 + 8$
27. **(a)** $4x^2 + 4x - 8$ **(b)** $4x^2 - 36x + 72$ **(c)** $16x^2 - 8x - 8$
29. **(a)** $x^2 - 3x$ **(b)** $-x^2 + 9x - 2$ **(c)** $3x^3 - 19x^2 + 9x - 1$ **(d)** $3x + 3h - 1$
31. All real numbers $x \neq 2$ **33.** All real numbers $t \neq 6$ and $t \neq -3$ **35.** All real numbers $x < 5$

Exercises 4.2, Pages 157–158

1.

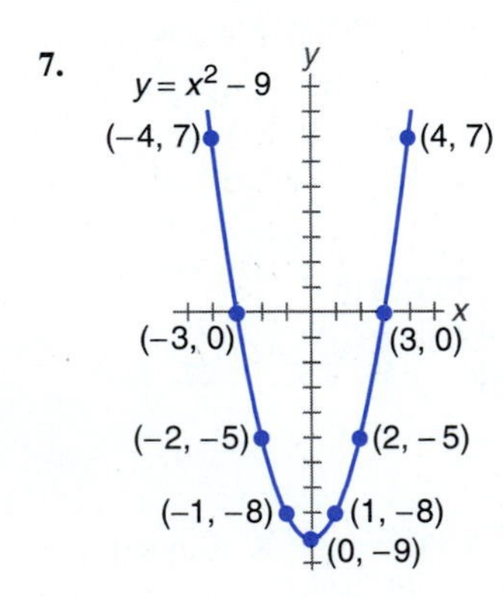

3.

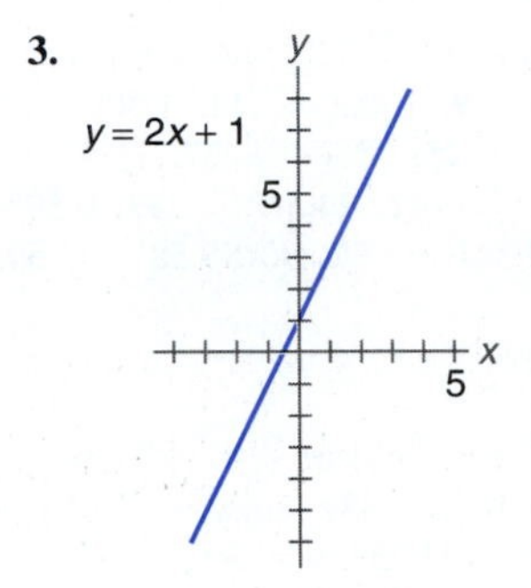

5.

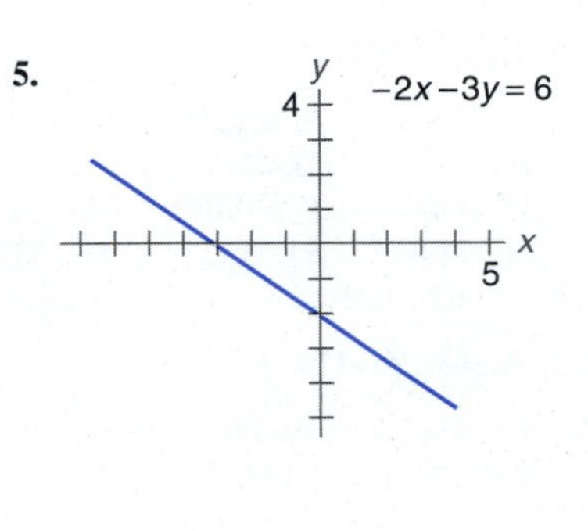

7.

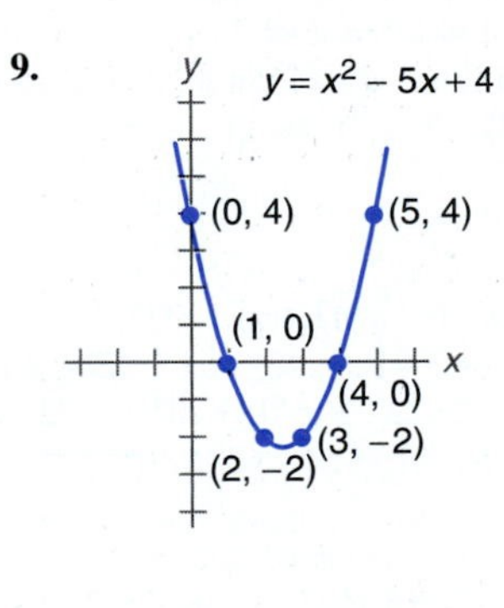

9.

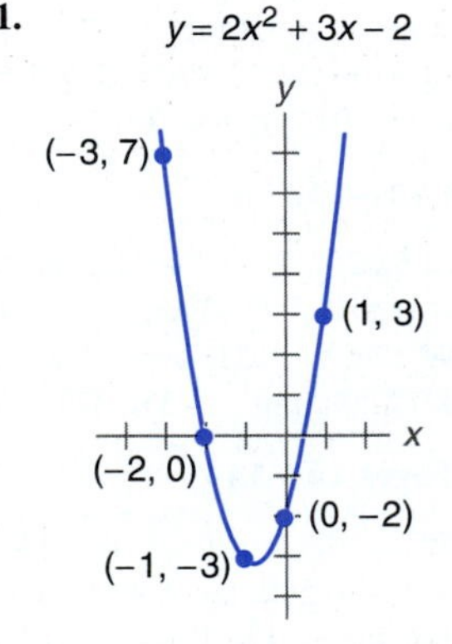

11.

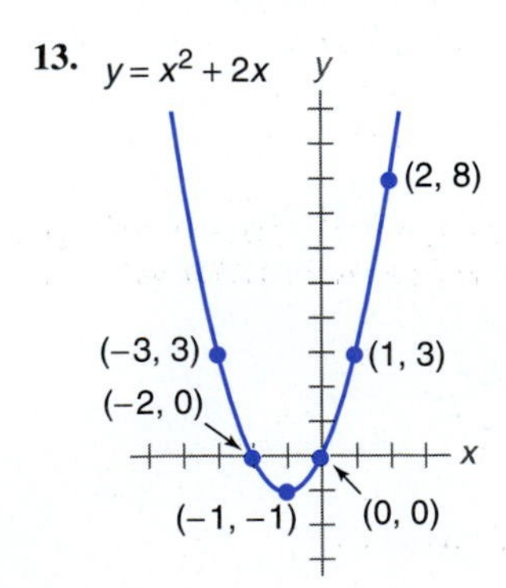

13.

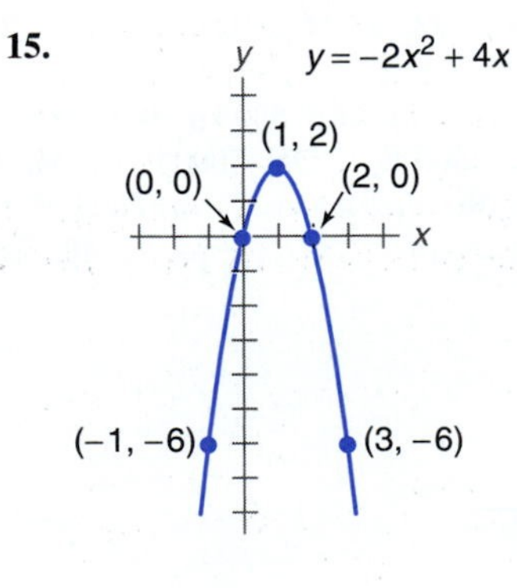

15.

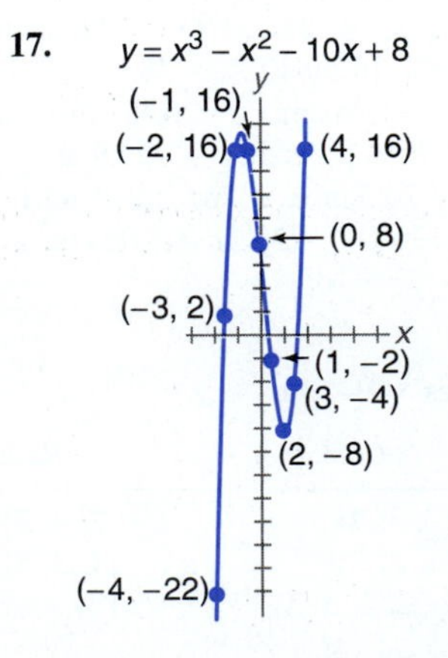

17.

19.

21.

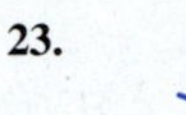

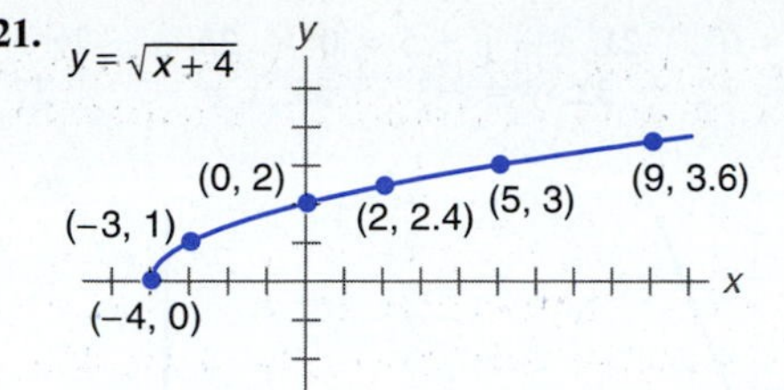

23.

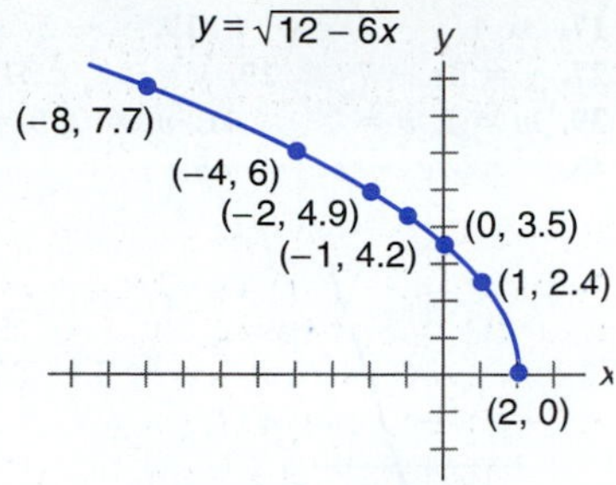

25. 3 and −3; 2 and −2; 3.3 and −3.3 **27.** 1 and 4; 4.5 and 0.5; no solution
29. −2 and $\frac{1}{2}$; 1 and −2.5; 1.3 and −2.8 **31.** 0 and −2; 1 and −3; 1.6 and −3.6
33. 0 and 2; no solution; 2.7 and −0.7; 2.3 and −0.3 **35.** 3.3, −3.1, 0.8; 3.4, −3.0, 0.6; 3.2, −3.2, 1.0
37. −4, 1; −3.8, 0, 1.8; −3.6, −0.5, 2.1

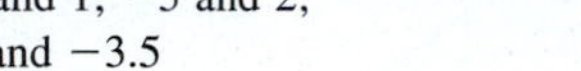

39. −4 and 1; −5 and 2; 0.5 and −3.5

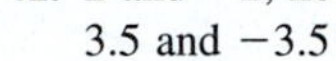

41. 2 and −2; no solution; 3.5 and −3.5

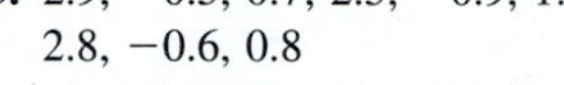

43. 2.9, −0.5, 0.7; 2.5, −0.9, 1.3; 2.8, −0.6, 0.8

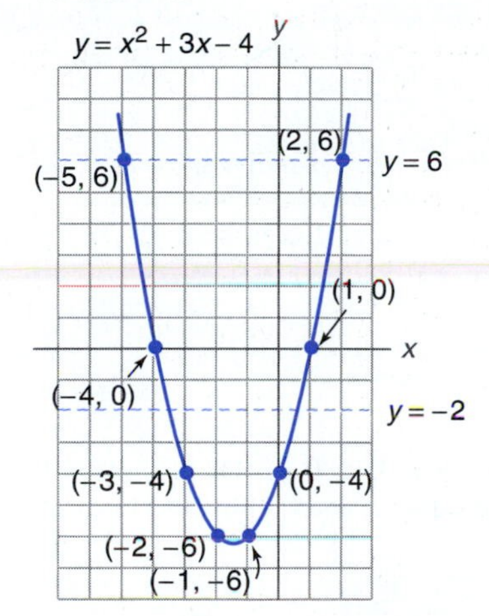

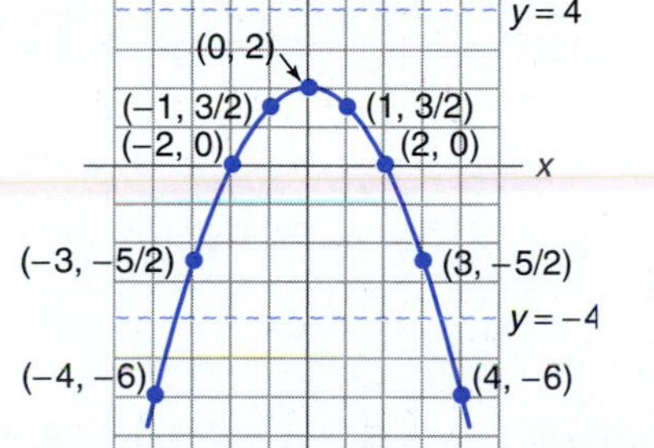

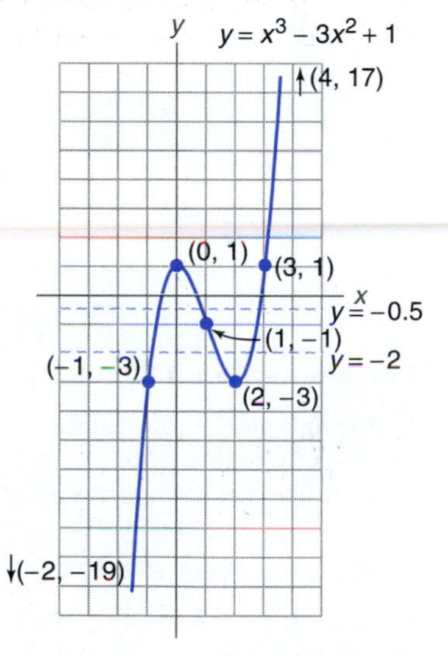

45. 2.6 ms, 4 ms, 5.5 ms **47.** 0.27 ms, 0.48 ms, 0.94 ms **49.** 2.7 s, 3.1 s
51. $A(1.18, -1.62)$, $B(2.35, -3.24)$, $C(3.53, -4.85)$

Exercises 4.3, Pages 165–166

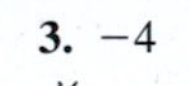

1. 1 **3.** −4 **5.** 0 **7.** $\frac{5}{8}$

9.

11.

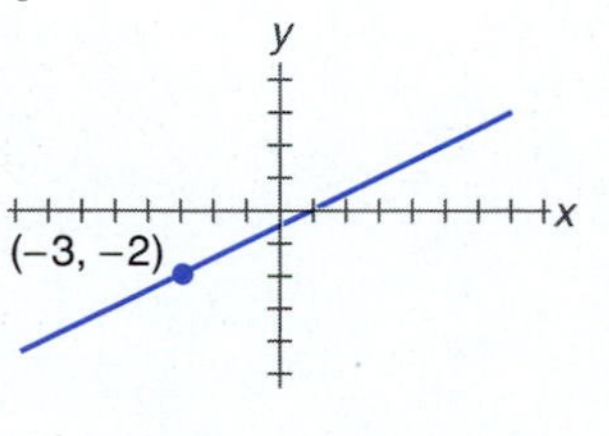

13.

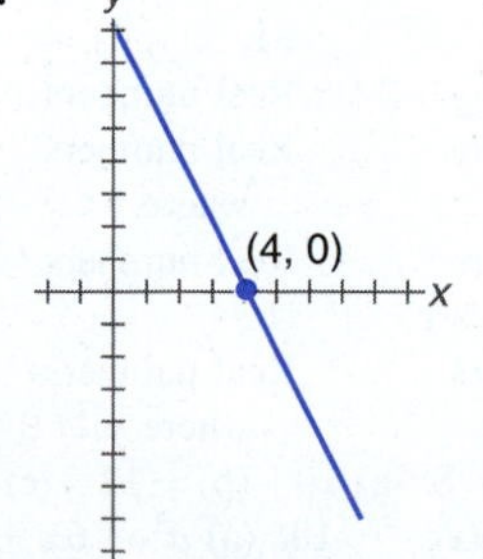

15.

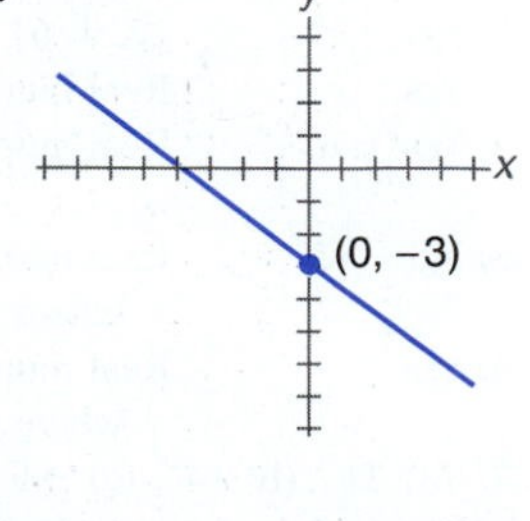

17. $3x + y - 2 = 0$ **19.** $x - 2y - 5 = 0$ **21.** $x + y - 5 = 0$ **23.** $x + 2y + 10 = 0$ **25.** $y = -5x - 2$
27. $y = 2x + 7$ **29.** $y = 5$ **31.** $x = -2$ **33.** $y = -3$ **35.** $x = -7$ **37.** $m = -\frac{1}{4}, b = 3$
39. $m = 2, b = 7$ **41.** $m = 0, b = 6$

43.

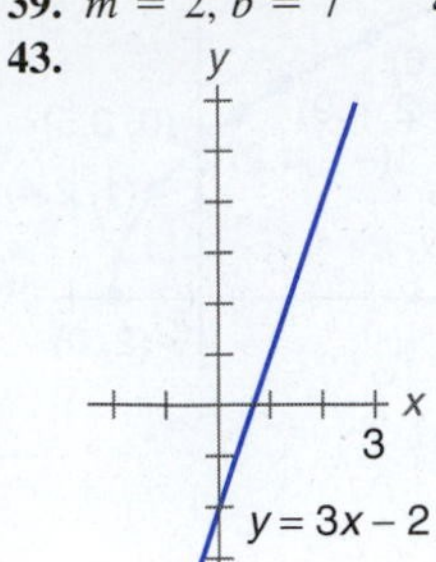

45.

5x − 2y + 4 = 0

47.

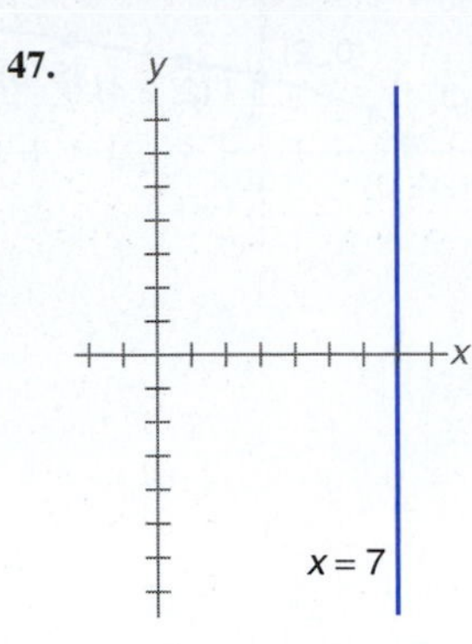

49.

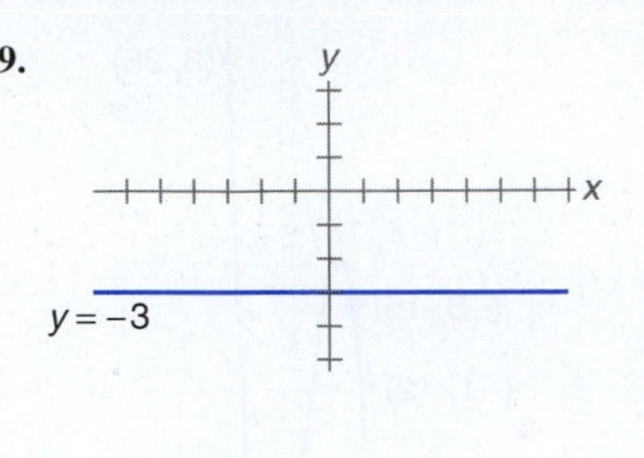

51.

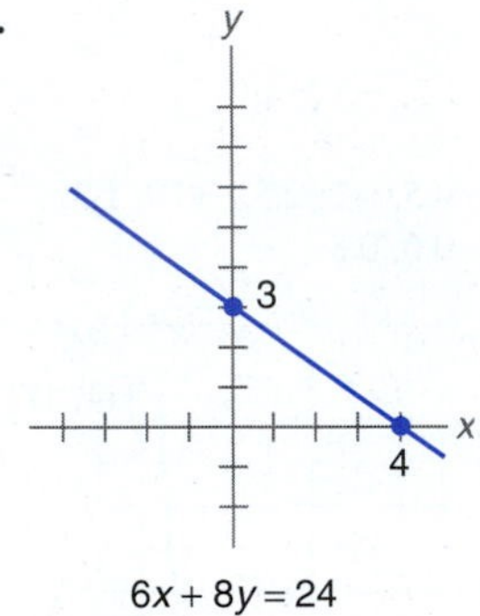

53.

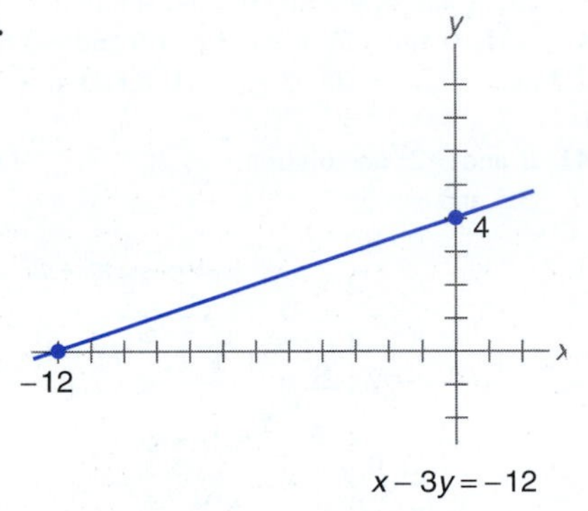

55. $\frac{1}{350}$

Exercises 4.4, Page 170

1. Perpendicular **3.** Neither **5.** Parallel **7.** $2x - y + 7 = 0$ **9.** $5x + y + 31 = 0$ **11.** $3x - 4y = 0$ **13.** $3x - 2y = 18$ **15.** $y = 8$ **17.** $x = 7$ **19. (a)** Yes, slopes of opposite sides are equal. **(b)** No, slopes of adjacent sides are not negative reciprocals.

Exercises 4.5, Pages 172–173

1. 15 **3.** 7 **5.** $4\sqrt{2}$ **7.** 7 **9.** $(3\frac{1}{2}, 5)$ **11.** $(1\frac{1}{2}, -1)$ **13.** $(0, -2\frac{1}{2})$ **15. (a)** 24 **(b)** Yes **(c)** No **(d)** 24
17. (a) $10 + \sqrt{82} + \sqrt{58}$ or 26.7 **(b)** No **(c)** No **19.** $4\sqrt{2}$ **21.** $x - 2y = 10$ **23.** $2x - y = -13$

Chapter 4 Review, Pages 174–175

	Function	*Domain*	*Range*
1.	Yes	{2, 3, 4, 5}	{3, 4, 5, 6}
2.	No	{2, 4, 6}	{1, 3, 4, 6}
3.	Yes	Real numbers	Real numbers
4.	Yes	Real numbers	Real numbers where $y \geq -5$
5.	No	Real numbers where $x \geq 4$	Real numbers
6.	Yes	Real numbers where $x \leq \frac{1}{2}$	Real numbers where $y \geq 0$

7. (a) 24 **(b)** 14 **(c)** −6 **8. (a)** 10 **(b)** −12 **(c)** 38 **9. (a)** 5 **(b)** $\frac{85}{4}$ **(c)** −15 is not in the domain of $h(x)$. **(d)** 1 is not in the domain of $h(x)$. **10. (a)** $a^2 - 6a + 4$ **(b)** $4x^2 - 12x + 4$ **(c)** $z^2 - 10z + 20$

11.

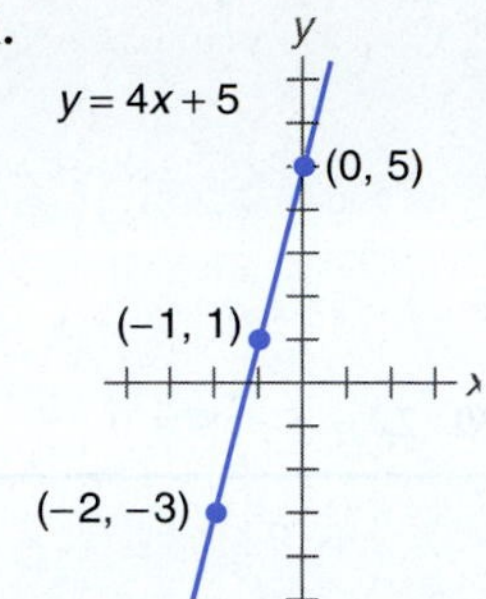

12.

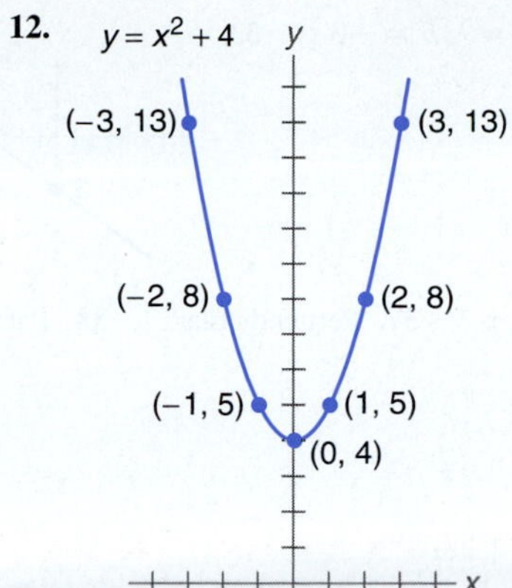

13.

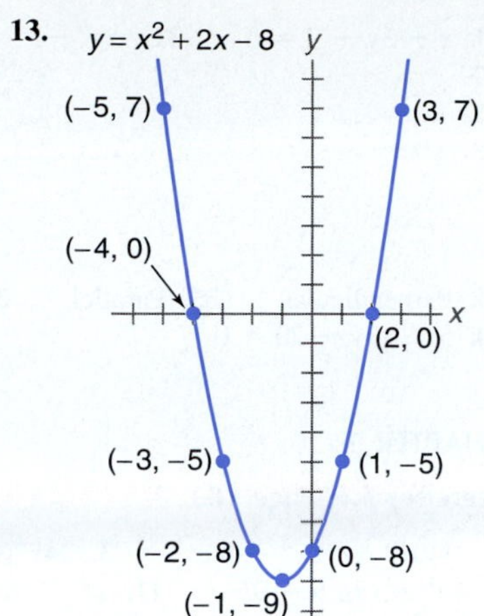

14.

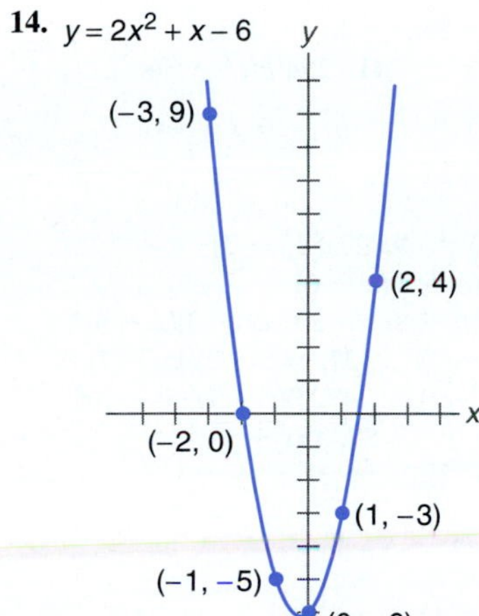

15.

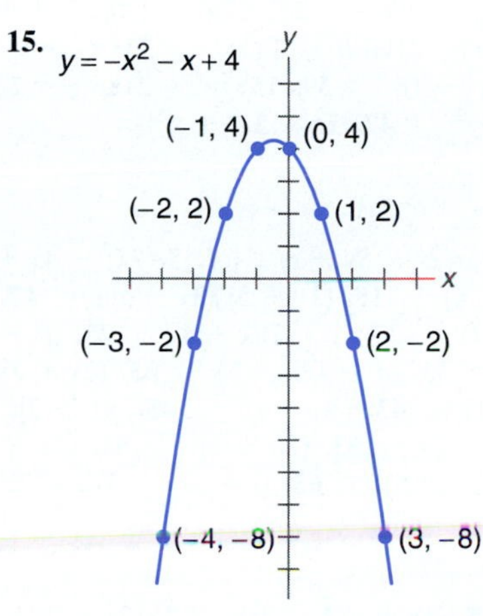

16.

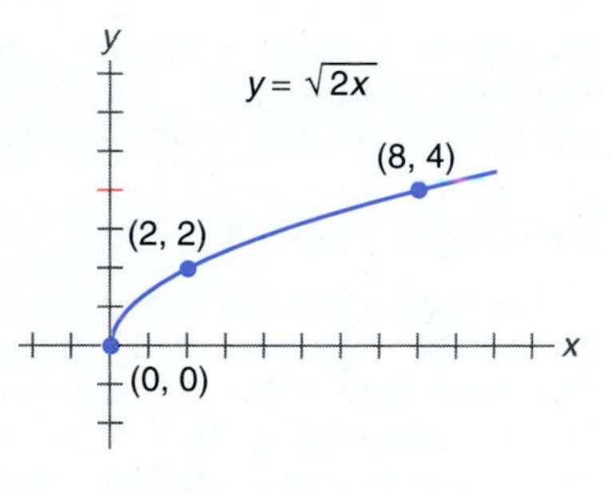

17.

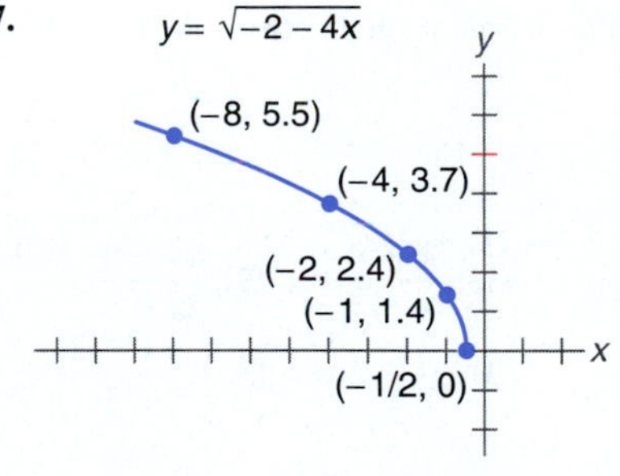

18.

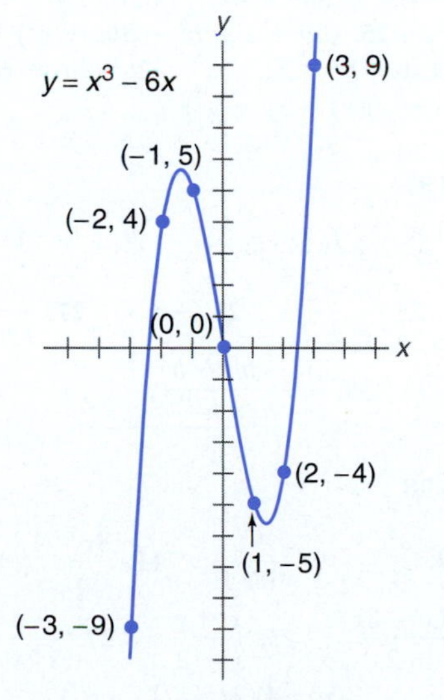

19. 1 and −1, 1.7 and −1.7, no solution **20.** −4 and 2, 1.6 and −3.6, −4.5 and 2.5
21. −2 and 1, −2.6 and 1.6, −3 and 2 **22.** 0, 2.4, −2.4; 2.6, −2.3, −0.3; 2.1, −2.7, 0.5 **23.** 1, 1.7, 2
24. 2.1, 2.4 **25.** $-\frac{2}{9}$ **26.** $\sqrt{85}$ **27.** $(-\frac{3}{2}, -3)$ **28.** $11x + 2y - 58 = 0$ **29.** $2x - 3y + 9 = 0$

30. $x + 3y + 9 = 0$ **31.** $x = -3$ **32.** $m = \frac{3}{2}; b = -3$ **33.**

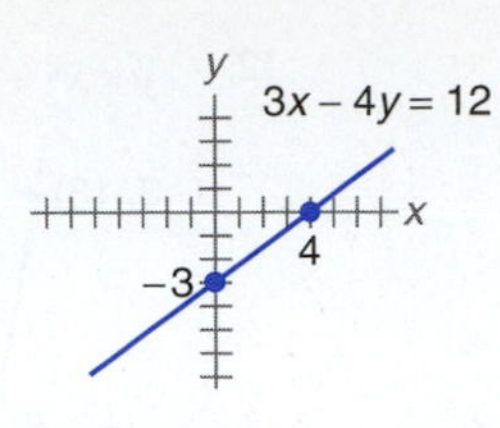

34. Perpendicular **35.** Parallel **36.** Neither **37.** Perpendicular **38.** Parallel **39.** $2x - y - 8 = 0$
40. $5x - 3y + 20 = 0$

CHAPTER 5

Exercises 5.1, Page 183

1. $-40x^5 + 72x^4 - 80x^3$ **3.** $24x^4y^6 - 42x^6y^7 + 54x^7y^5$ **5.** $4x^2 - 49$ **7.** $49x^4 + 28x^2y + 4y^2$
9. $4a^4 - 12a^2b + 9b^2$ **11.** $x^2 + 16x + 48$ **13.** $-6x^2 - 31x - 40$ **15.** $4a^2 + b^2 + 9c^2 - 4ab + 12ac - 6bc$
17. $8a^3 + 12a^2b + 6ab^2 + b^3$ **19.** $125 - 150x^2 + 60x^4 - 8x^6$ **21.** $42a^2 + 17ab - 15b^2$
23. $16x^2 + 56xy + 49y^2$ **25.** $12a^2 + 32ab - 99b^2$ **27.** $27a^9 + 270a^6b^2 + 900a^3b^4 + 1000b^6$
29. $-24a^5b^7 + 36a^7b^{10} - 80a^7b^3$ **31.** $81a^6 - 216a^3b^2 + 144b^4$ **33.** $x^2 + 9y^2 + 25z^2 - 6xy - 10xz + 30yz$
35. $54a^2 - 114ab + 40b^2$ **37.** $30x^6 - 4x^3 - 16$ **39.** $15x^2y^2 + 21xy^2z - 72y^2z^2$ **41.** $25a^2b^2c^2 - 36$
43. $3x^4 + \frac{3}{2}x^2y - \frac{1}{3}y^2$ **45.** $x^2 - \frac{4}{3}xy + \frac{4}{9}y^2$ **47.** $x^2 + 2x^4 + x^6$

Exercises 5.2, Page 188

1. $3(2x + 3y)$ **3.** $5(2x + 5y - 9z)$ **5.** $6x(2x - 5y + z)$ **7.** $3x(2x^3 - 4x + 1)$ **9.** $27x(x^2 + 2)$
11. $2xy^2(4x^2 - 3y + 6z)$ **13.** $(x + 4)(x - 4)$ **15.** $(3x + 5y)(3x - 5y)$ **17.** $2(e + 6)(e - 6)$
19. $4(2d + 5)(2d - 5)$ **21.** $4(R + r)(R - r)$ **23.** $(x + 2)(x + 4)$ **25.** $(b + 3)(b + 8)$ **27.** $(x - 3)(x - 6)$
29. $(a - 2)(a - 16)$ **31.** $(x + 5)(x - 7)$ **33.** $(a + 4)(a - 1)$ **35.** $(2x + 3)(x - 2)$ **37.** $(3x - 2)(5x - 7)$
39. $(5y + 6)(9y + 1)$ **41.** $(7a - 1)(5a + 1)$ **43.** $(3c - 4)^2$ **45.** $5(b + 2)(5b + 2)$ **47.** $(7t - 5s)(5t + 3s)$
49. $5(2k - 5t)(3k - 2t)$ **51.** $3(6x + 7)(3x - 2)$ **53.** $(2a + 5)^2$ **55.** $(3x - 8y^2)^2$ **57.** $(5x + y^3)^2$
59. $(x^2 + 9)(x + 3)(x - 3)$ **61.** $(a^2 - 8)(a^2 + 5)$ **63.** $(t + 3)(t - 3)(t + 2)(t - 2)$

Exercises 5.3, Pages 191–192

1. $(m + n)(a - b)$ **3.** $(x + y)(m + n)$ **5.** $(3x + y)(1 - 2x)$ **7.** $(3x - 2)(2x^2 + 1)$
9. $(x + y + 2z)(x + y - 2z)$ **11.** $(10 + 7x - 7y)(10 - 7x + 7y)$ **13.** $(x - 3 + 2y)(x - 3 - 2y)$
15. $(2x + 2y - 1)(2x - 2y + 1)$ **17.** $(x + y + 4)(x + y + 9)$
19. $[4(a + b) + 1][6(a + b) - 5]$, or $(4a + 4b + 1)(6a + 6b - 5)$ **21.** $(a + b)(a^2 - ab + b^2)$
23. $(x - 4)(x^2 + 4x + 16)$ **25.** $(ab + c)(a^2b^2 - abc + c^2)$ **27.** $(a - 3b)(a^2 + 3ab + 9b^2)$
29. $(3a + 4b)(9a^2 - 12ab + 16b^2)$ **31.** $(3a^2 - 2b^3)(9a^4 + 6a^2b^3 + 4b^6)$
33. $(x + y)(x^2 - xy + y^2)(x - y)(x^2 + xy + y^2)$

Exercises 5.4, Pages 194–195

1. $\frac{1}{6}$ **3.** $\frac{5}{9y}$ **5.** $\frac{4}{(x + 4)^2}$ **7.** $2(x - 3)^2$ **9.** $\frac{5}{6}$ **11.** $\frac{6}{m - 1}$ **13.** x **15.** $\frac{x + 4}{x - 7}$ **17.** $\frac{2x + 3}{2x + 9}$
19. -2 **21.** $-x - 1$ **23.** $\frac{a - 2}{a + 2}$ **25.** $-m$ **27.** $\frac{-t^2}{t + 1}$ **29.** $\frac{x}{x - 2}$ **31.** $y - 2$ **33.** $\frac{x - 2}{y^2 + 3y + 9}$
35. $\frac{1}{x - 4}$ **37.** $\frac{-3 - 2x}{2x - 3}$ **39.** $\frac{a^2 - ab + b^2}{a + b}$

Exercises 5.5, Pages 197–198

1. $\frac{3}{4}$ **3.** $\frac{5}{2}$ **5.** 2 **7.** $\frac{a^5}{3}$ **9.** $\frac{9}{10b^2y}$ **11.** $\frac{8y^3}{5a}$ **13.** $\frac{a^2b}{3}$ **15.** $\frac{1}{4y}$ **17.** $\frac{5}{27}$ **19.** $\frac{y(y + 2)}{y - 2}$
21. $-(m + 1)^2$ **23.** $\frac{-x(x + 8)}{x + 2}$ **25.** $\frac{x + y}{x - 2}$ **27.** $\frac{-x - 3}{x^3(x - 3)}$ **29.** $x + 3$ **31.** $\frac{x + 3}{x + 1}$ **33.** 1
35. $\frac{a - 4b}{5a + b}$ **37.** $\frac{y}{x + y}$ **39.** $\frac{(x - 4)(2x + 7)}{(x + 3)(3x + 1)}$ **41.** $(x - 2)(x + 2)$ **43.** $\frac{y - x}{3y}$ **45.** $\frac{(x + 2)(5x + 9)}{x}$
47. $\frac{(3x + 4)^2}{2x(x - 5)}$

Exercises 5.6, Pages 201–202

1. $24x$ **3.** $4k^2$ **5.** $30a^2b^2$ **7.** $5x(x-1)$ **9.** $6(x+2)$ **11.** $(x+5)(x-5)^2$ **13.** $(x+2)(x+4)(x-3)$ **15.** $6x^2(x+2)(x-4)$ **17.** $12c(c+2)(c-4)^2$ **19.** $\frac{4}{7}$ **21.** $\frac{7}{a}$ **23.** $\frac{15}{x+1}$ **25.** $\frac{7}{20}$ **27.** $\frac{2x+5}{30}$ **29.** $\frac{-8}{3y}$ **31.** $\frac{4p+6}{3p^2}$ **33.** $\frac{-18x^2+15x+4}{12x^3}$ **35.** $\frac{bc+ac+ab}{abc}$ **37.** $\frac{6x^2-x+4}{2x}$ **39.** $\frac{3c-1}{c(c-1)}$ **41.** $\frac{3t-18}{(t+2)(t-2)}$ **43.** $\frac{3a+2}{a(a+1)}$ **45.** $\frac{-d^2-13d}{(d+4)(d-5)}$ **47.** $\frac{6}{x-4}$ **49.** $\frac{-1}{6(a+3)}$ **51.** $\frac{-2a^2+10a-2}{(a+1)(a-1)}$ **53.** $\frac{11}{x-3}$ **55.** $\frac{3r+4}{(r-2)^2(r+3)}$ **57.** $\frac{1}{x-3}$ **59.** $\frac{5t+9}{4(t+1)}$ **61.** $\frac{5x^2+6x-9}{(x-4)(x+1)(x+3)}$ **63.** 0 **65.** $\frac{7x+21}{(x+4)(x-4)(x-3)}$ **67.** $\frac{5x^2+4x-5}{2x(4x+1)}$ **69.** $\frac{t-2}{t^2+2t+4}$ **71.** $\frac{x^2+1}{x^2-x+1}$

Exercises 5.7, Pages 204–205

1. $\frac{x+2}{3x}$ **3.** $\frac{3(t-1)}{2(t+1)}$ **5.** 1 **7.** $\frac{39}{28}$ **9.** $\frac{a}{2}$ **11.** $\frac{a-5}{a}$ **13.** $\frac{15}{xy}$ **15.** $-y$ **17.** $\frac{a+1}{a-1}$ **19.** $\frac{n+m}{n-m}$ **21.** $\frac{x-7}{x+6}$ **23.** $\frac{x-y}{(x+y)^2}$ **25.** 1 **27.** $\frac{(2a+b)^2}{(a-b)^2}$ **29.** $\frac{x}{x+1}$ **31.** $\frac{x-1}{2x-1}$ **33.** $\frac{x^2}{x^2-x+1}$ **35.** $\frac{-6}{x^2(x+5)}$ **37.** $\frac{3}{5}$

Exercises 5.8, Pages 208–209

1. 12 **3.** 10 **5.** 15 **7.** -1 **9.** -7 **11.** -1 **13.** -19 **15.** $-\frac{3}{4}$ **17.** $r=\frac{mv^2}{F}$ **19.** $R=\frac{V-I_LR_L}{I_L}$ **21.** $R_1=\frac{RR_2}{R_2-R}$ **23.** $y=\frac{x}{x-1}$ **25.** $x=-a-b$ **27.** $Q=\frac{VR_1R_2}{R_2-R_1}$ **29.** $\frac{1}{2}$ **31.** 12 **33.** 5 **35.** $-\frac{1}{3}$ **37.** No solution **39.** No solution **41.** -22 **43.** -10

Chapter 5 Review, Page 211

1. $-6a^5b^4+72a^3b^6+54a^2b^9$ **2.** $15x^2+53x-52$ **3.** $16x^2-49$ **4.** $25a^4-90a^2b^2+81b^4$ **5.** $64a^3-48a^2b^2+12ab^4-b^6$ **6.** $30x^2+23xyz^2-14y^2z^4$ **7.** $9a^2b(2a-b^3+ab)$ **8.** $(4x+3y)(4x-3y)$ **9.** $(x+4)(5x+8)$ **10.** $(7x-2)(2x+9)$ **11.** $(2a-5)^2$ **12.** $(m-n)(m^2+mn+n^2)$ **13.** $3(x+1)(x-7)$ **14.** $(x^2+9)(x+3)(x-3)$ **15.** $(3x+10)(5x-9)$ **16.** $5(x+4y)(2x-3y)$ **17.** $(x+3+5y)(x+3-5y)$ **18.** $(a+2b)(x-4y)$ **19.** $\frac{15y^6}{8x^4}$ **20.** 2 **21.** $\frac{x-8}{2x+3}$ **22.** $\frac{-(2x+1)}{x+4}$ **23.** $\frac{x}{2x+5}$ **24.** $\frac{-(2x-1)}{2x+1}$ **25.** $3y$ **26.** $\frac{-1}{6(x+2)^2}$ **27.** $\frac{16a+27-30a^2}{12a^2}$ **28.** $\frac{5x}{2x+1}$ **29.** $\frac{3x}{(x-5)(2x-1)}$ **30.** $\frac{5x^2-12x+1}{(x+1)(2x+1)(3x-4)}$ **31.** $\frac{1}{x^2-x+1}$ **32.** $\frac{14y+4}{3y+1}$ **33.** -8 **34.** $\frac{a}{a+b}$ **35.** No solution **36.** -8

CHAPTER 6

Exercises 6.1, Pages 220–222

1. $(4,-1)$ **3.** $(-3,5)$ **5.** System dependent, lines coincide **7.** $(2,7)$ **9.** $(1,-6)$ **11.** $(-1,6)$ **13.** $(-5,0)$ **15.** $(3,1)$ **17.** System inconsistent, lines parallel **19.** $(-3,5)$ **21.** $(-4,6)$ **23.** $(\frac{2}{3},-\frac{7}{3})$ **25.** $(2,8)$ **27.** $(-1,\frac{1}{3})$ **29.** $(-2,-14)$ **31.** $(-6,3)$ **33.** $(-1,-4)$ **35.** $(\frac{1}{6},\frac{2}{3})$ **37.** $(5,-\frac{1}{4})$ **39.** $(5.2, 0.6)$ **41.** $(180, 240)$ **43.** 40 μF, 15 μF **45.** Cement: 2.3 m^3; gravel: 9.2 m^3 **47.** 1600 L of 3% solution, 400 L of 8% solution **49.** 0.5 km/h, 4.5 km/h **51.** Length: 30 m; width: 18 m **53.** 120 mA; 720 mA **55.** $I_1=2$ mA; $I_2=6$ mA **57.** $\frac{7}{2}$

Exercises 6.2, Page 224

1. $(1,b-a)$ **3.** (ab,a^2) **5.** $\left(\frac{b}{a^2+b^2},\frac{a}{a^2+b^2}\right)$ **7.** $\left(\frac{c}{a+b},\frac{c}{a+b}\right)$ **9.** $\left(\frac{a-b}{a+b},\frac{a-3b}{a+b}\right)$ **11.** $(3,4)$ **13.** $(\frac{1}{2},-\frac{1}{3})$ **15.** $(\frac{2}{3},\frac{3}{2})$ **17.** $(\frac{3}{4},\frac{1}{2})$ **19.** $\left(\frac{2}{a+b},\frac{2}{a-b}\right)$

Exercises 6.3, Pages 229–230

1. (3, 2, −1) **3.** (2, 5, 0) **5.** (4, 6, −6) **7.** (13, 1, 2) **9.** (2, −3, 1) **11.** $(\frac{1}{3}, \frac{3}{4}, -2)$ **13.** (2, 4, −1, 5) **15.** $R_1 = 1200\ \Omega$; $R_2 = 150\ \Omega$, $R_3 = 600\ \Omega$ **17.** 29 cm, 24 cm, 12 cm **19.** 64 mm; 16 mm; 5 mm **21.** $f(x) = 2x^2 - 5x + 7$

Exercises 6.4, Pages 234–235

1. 14 **3.** −2 **5.** 39 **7.** −42 **9.** −12 **11.** 16 **13.** 1 **15.** 95 **17.** $mn + n^3$ **19.** 29 **21.** 78 **23.** 16 **25.** 0 **27.** 0 **29.** −138 **31.** −1139 **33.** 24 **35.** −504

Exercises 6.5, Pages 237–238

1. 0 **3.** 0 **5.** −126 **7.** −105 **9.** −30 **11.** 31 **13.** 61 **15.** −11 **17.** 450 **19.** 410 **21.** 970 **23.** 744

Exercises 6.6, Pages 243–244

1. (3, −2) **3.** (−2, 9) **5.** (0, −4) **7.** Dependent **9.** Inconsistent **11.** $(-\frac{2}{3}, \frac{5}{6})$ **13.** $(\frac{1}{2}, 2)$ **15.** $\left(\frac{2a - 4b}{a^2 - b^2}, \frac{4a - 2b}{a^2 - b^2}\right)$ **17.** $\left(\frac{a^2 + b^2}{2a + 5b}, \frac{2b - 5a}{2a + 5b}\right)$ **19.** $\left(\frac{c}{a + b^2}, \frac{bc}{a + b^2}\right)$ **21.** 75 V, 135 V **23.** (2, 1, −4) **25.** (3, 2, −4) **27.** (−1, 4, −5) **29.** (1, 2, −1) **31.** (2, −1, 5) **33.** (4, 0, −3) **35.** 6 cm, 7 cm, 8 cm **37.** (1, 3, −2, 5) **39.** (2, −2, 3, 4, −1)

Exercises 6.7, Page 251

1. $\frac{5}{x + 2} + \frac{3}{x - 7}$ **3.** $\frac{3}{2x + 3} - \frac{2}{x - 4}$ **5.** $\frac{7}{x} + \frac{2}{3x - 4} + \frac{5}{2x + 1}$ **7.** $\frac{1}{x + 1} + \frac{1}{(x + 3)^2}$ **9.** $\frac{3}{4x - 1} + \frac{1}{(4x - 1)^2} - \frac{7}{(4x - 1)^3}$ **11.** $\frac{3}{x} + \frac{8}{x - 1} - \frac{4}{(x - 1)^2}$ **13.** $\frac{x - 1}{x^2 + 1} - \frac{x}{x^2 - 3}$ **15.** $\frac{4x + 1}{x^2 + x + 1} - \frac{1}{x^2 - 5}$ **17.** $\frac{5x - 2}{x^2 + 5x + 3} + \frac{1}{x + 3} - \frac{2}{x - 3}$ **19.** $\frac{5}{x} + \frac{3x - 1}{x^2 + 1} - \frac{5}{(x^2 + 1)^2}$ **21.** $\frac{1}{x} - \frac{2}{x^2} - \frac{4x}{(x^2 + 2)^2}$ **23.** $\frac{2}{x + 3} + \frac{4}{x - 3} - \frac{6x}{x^2 + 9}$ **25.** $x + \frac{\frac{1}{2}}{x + 1} + \frac{\frac{1}{2}}{x - 1}$ **27.** $x - 1 + \frac{3}{x - 2} + \frac{1}{x + 2}$ **29.** $3x - 2 + \frac{5}{x} - \frac{8x}{x^2 + 1}$

Chapter 6 Review, Pages 255–256

1. (1, 2) **2.** (0, 2) **3.** (−1, 8) **4.** System inconsistent, lines parallel **5.** (3, 1) **6.** $(\frac{25}{8}, \frac{1}{4})$ **7.** (2, −2) **8.** $(4, \frac{7}{2})$ **9.** (4, −1) **10.** (−2, −6) **11.** (2, 3) **12.** $(-\frac{5}{2}, \frac{1}{8})$ **13.** $\left(\frac{a + 3b}{2a + 4b}, \frac{-4}{a + 2b}\right)$ **14.** $(1, \frac{1}{2})$ **15.** $(\frac{5}{2}, -\frac{1}{2}, 0)$ **16.** (2, −2, −4) **17.** 22 **18.** −25 **19.** 276 **20.** 45 **21.** (10, 2) **22.** System dependent, lines coincide **23.** (2, 0) **24.** $(\frac{8}{9}, \frac{11}{9})$ **25.** (−7, 5, −15) **26.** (0, 3, 6) **27.** $(-\frac{1}{4}, \frac{1}{2})$ **28.** (1, 5) **29.** $(\frac{7}{2}, -1)$ **30.** (−1, −24) **31.** (1, 5, 1) **32.** (2, 30, 11) **33.** 24 kg of 10% lead, 6 kg of 20% lead **34.** \$1400 at $5\frac{1}{2}$%, \$2200 at $6\frac{1}{2}$% **35.** Wind velocity: 10 km/h; plane speed: 90 km/h **36.** Jack, 13 pieces; John, 11 pieces; Bob, 8 pieces **37.** 18 **38.** 385 **39.** $\frac{4}{x - 3} + \frac{2}{x + 5}$ **40.** $\frac{1}{x} + \frac{2}{x + 1} - \frac{4}{x - 1}$ **41.** $\frac{4}{x} + \frac{1}{x^2} - \frac{1}{x + 1}$ **42.** $\frac{10}{x + 1} - \frac{6}{(x + 1)^2}$ **43.** $\frac{3x + 2}{x^2 + 1} + \frac{4}{x - 1}$ **44.** $\frac{5}{x^2 + 4} + \frac{2x + 1}{(x^2 + 4)^2}$

Chapter 6 Application, Page 260

1. $I_1 = I_5 = 0.222$ A; $I_2 = I_3 = 0.0741$ A; $I_4 = 0.148$ A **2.** $I_1 = 0.909$ A; $I_2 = 0.545$ A; $I_3 = I_4 = 0.364$ A **3.** $I_1 = I_4 = 0.319$ A; $I_2 = 0.200$ A; $I_3 = 0.120$ A **4.** $I_1 = I_2 = I_3 = 0.221$ A; $I_4 = 0.0851$ A; $I_5 = I_6 = 0.136$ A

CHAPTER 7

Exercises 7.1, Page 264

1. 7, −4 **3.** 2, 4 **5.** −5, 2 **7.** 3, $-\frac{3}{2}$ **9.** $-\frac{5}{7}, -\frac{1}{2}$ **11.** $\frac{8}{3}, \frac{7}{6}$ **13.** 0, $-\frac{1}{8}$ **15.** 0, 3 **17.** 6, −6 **19.** $\frac{5}{4}, -\frac{5}{4}$ **21.** $\pm\frac{2\sqrt{15}}{5}$ **23.** $\pm\frac{\sqrt{21}}{2}$ **25.** $\pm 2\sqrt{2}$ **27.** $\pm\sqrt{6}$ **29.** $-\frac{6}{5}, \frac{2}{3}$ **31.** −2, $-\frac{1}{2}$ **33.** −4, 3 **35.** 5, −1 **37.** 3, $-\frac{8}{5}$

Exercises 7.2, Page 266

1. 2, −8 **3.** 5, 7 **5.** $\frac{1}{2}$, −1 **7.** $-\frac{3}{5}, \frac{6}{5}$ **9.** $-2 \pm \sqrt{11}$ **11.** $\frac{-1 \pm \sqrt{19}}{2}$ **13.** $\frac{-9 \pm \sqrt{21}}{6}$ **15.** $-\frac{3}{2}$, 3 **17.** $\frac{1}{4}$, −3 **19.** $\frac{-5 \pm \sqrt{17}}{4}$

Exercises 7.3, Page 271

1. 8, −4 **3.** 3, $-\frac{3}{2}$ **5.** $\frac{4}{3}, -\frac{7}{4}$ **7.** −2, −3 **9.** $\frac{9}{4}, \frac{3}{2}$ **11.** 0, $-\frac{5}{2}$ **13.** $\frac{-5 \pm \sqrt{53}}{2}$ **15.** $2 \pm 3\sqrt{2}$ **17.** $\frac{2 \pm 3\sqrt{3}}{2}$ **19.** $\frac{7 \pm 2\sqrt{6}}{3}$ **21.** $2 \pm \sqrt{3}$ **23.** $\frac{1}{3}$, 2 **25.** 2, −12 **27.** $-\frac{3}{2}$ **29.** $\frac{5 \pm \sqrt{17}}{4}$ **31.** No solution **33.** 1.24, −0.469 **35.** 1.56, 0.578 **37.** 0.485, −1.37 **39.** 0.355, −1.88 **41.** 0, 1; 3, −2; no solution **43.** 0, $-\frac{1}{3}$; 1, $-\frac{4}{3}$; $\frac{-1 \pm \sqrt{13}}{6}$ **45.** $\frac{1}{m}$ **47.** $-\frac{2}{n}$ **49.** $-4a$, $4a - 5$ **51.** $\sqrt{\frac{A}{\pi}}$

Exercises 7.4, Pages 274–275

1. 5, 8 **3.** 3.50 cm × 10.0 cm **5.** 11.0 m, 12.0 m **7.** 2.00 s, 5.00 s; 3.00 s, 4.00 s; $\frac{7 \pm \sqrt{17}}{2}$ s, or 5.56 s and 1.44 s **9.** 1.00 μs; $\frac{2 \pm \sqrt{2}}{2}$ μs, or 1.71 μs and 0.293 μs **11.** 0.100 ms **13.** 15.0 m × 25.0 m **15.** 18.0 in. **17.** Two possible solutions: $45\bar{0}$ m × $16\bar{0}0$ m; $80\bar{0}$ m × $90\bar{0}$ m **19.** 6.97 m **21.** 6.00 in. × 18.0 in. **23.** $10\bar{0}$ m × $12\bar{0}$ m **25.** 60.0 Ω, $18\bar{0}$ Ω **27.** 50.0 m

Chapter 7 Review, Pages 276–277

1. 3, −7 **2.** $\frac{8}{3}, \frac{5}{2}$ **3.** $-\frac{1}{4}, -\frac{4}{5}$ **4.** $\frac{1}{6}, -\frac{1}{6}$ **5.** −2, $-\frac{1}{2}$ **6.** 0, 6 **7.** $-\frac{3}{4}, \frac{3}{2}$ **8.** 3, −3 **9.** −5 **10.** $\frac{4}{3}$ **11.** −4, $-\frac{5}{2}$ **12.** −1, $\frac{7}{3}$ **13.** $4 \pm \sqrt{7}$ **14.** $\frac{3 \pm \sqrt{14}}{5}$ **15.** 2, $\frac{10}{3}$ **16.** $\frac{3 \pm \sqrt{21}}{4}$ **17.** $\frac{-3 \pm \sqrt{17}}{-4}$ **18.** $\frac{-4 \pm \sqrt{46}}{6}$ **19.** $\frac{m \pm \sqrt{m^2 - 4mn^2}}{2mn}$ **20.** $\frac{m^2}{n^2}$ **21.** $t = \frac{-v \pm \sqrt{v^2 + 2as}}{a}$ **22.** 3, 12 **23.** $60\bar{0}$ ft × $90\bar{0}$ ft **24.** 10.0 cm, $90\bar{0}0$ cm^3 **25.** **(a)** 1.00 s, 3.00 s **(b)** No **(c)** 4.00 s **(d)** 64.0 ft **26.** $275 \pm 5\sqrt{385}$ Ω, or 373 Ω and 177 Ω

CHAPTER 8

Exercises 8.1, Page 281

1. x^5 **3.** 5^3 **5.** 7^6 **7.** m^3 **9.** $\frac{1}{d^7}$ **11.** $\frac{1}{2^6}$ **13.** $\frac{1}{y^8}$ **15.** s^{10} **17.** t^3 **19.** $\frac{16}{49}$ **21.** 3 **23.** $\frac{6}{5}$ **25.** 2^3 or 8 **27.** $\frac{9}{4}$ **29.** a^{10} **31.** a^2 **33.** $\frac{1}{a^{12}}$ **35.** $9a^4$ **37.** $\frac{a^4}{4b^2}$ **39.** $-\frac{48}{k^2}$ **41.** $\frac{15a}{b^2}$ **43.** $\frac{1}{w^5}$ **45.** x^6 **47.** $\frac{y^5}{x^5}$ **49.** t^8 **51.** $\frac{1}{b^2}$ **53.** $\frac{c^4}{ab^2}$ **55.** a^{18} **57.** $\frac{108}{a^{14}}$ **59.** $\frac{8a^{12}}{c^6}$ **61.** $\frac{49a^{10}}{b^8}$ **63.** $\frac{a + b}{ab}$ **65.** $\frac{b + ab}{a^2}$ **67.** $\frac{2}{9}$ **69.** $\frac{6}{5}$ **71.** $\frac{1}{2}$

Exercises 8.2, Pages 283–284

1. 6 **3.** 4 **5.** 16 **7.** 32 **9.** $\frac{1}{4}$ **11.** $\frac{1}{3}$ **13.** 9 **15.** $\frac{1}{27}$ **17.** $\frac{4}{5}$ **19.** $\frac{8}{27}$ **21.** $\frac{64}{125}$ **23.** $-\frac{1}{2}$ **25.** 9 **27.** 3 **29.** 2 **31.** x^2 **33.** $x^{1/4}$ **35.** $x^{1/2}$ **37.** $\frac{1}{x^{1/3}}$ **39.** $\frac{1}{x^{1/3}}$ **41.** $x^{1/4}$ **43.** $\frac{a^3}{b^6c^{12}}$

45. $\frac{a^{16}c^8}{b^{12}}$ **47.** $x + 4x^2$ **49.** $8t + 2$ **51.** $6c^{1/3} + \frac{8}{c}$ **53.** $x - y$ **55.** $x + 2x^{1/2}y^{1/2} + y$ **57.** 56

59. 32 **61.** 256 **63.** 640 **65.** $\frac{2}{\pi}$

Exercises 8.3, Pages 290–291

1. 7 **3.** $5\sqrt{3}$ **5.** $6\sqrt{5}$ **7.** $2a\sqrt{2}$ **9.** $6b\sqrt{2}$ **11.** $4a^2b\sqrt{5a}$ **13.** $4ab^2c^4\sqrt{2c}$ **15.** $\frac{\sqrt{3}}{2}$

17. $\frac{\sqrt{10}}{4}$ **19.** $\frac{\sqrt{15}}{5}$ **21.** $\frac{5\sqrt{6}}{12}$ **23.** $\frac{2\sqrt{15a}}{15b}$ **25.** $\frac{\sqrt{2b}}{2b}$ **27.** $\frac{a}{2b}$ **29.** 5 **31.** $2a\sqrt[3]{2a}$

33. $2a^2\sqrt[3]{5a^2}$ **35.** $3x\sqrt[3]{2x^2}$ **37.** $2x^2yz\sqrt[3]{7xy^2}$ **39.** $\frac{\sqrt[3]{3}}{2}$ **41.** $\frac{\sqrt[3]{10}}{2}$ **43.** $\frac{\sqrt[3]{90}}{6}$ **45.** $\frac{\sqrt[3]{4}}{2}$

47. $\frac{\sqrt[3]{10ab}}{5b}$ **49.** $\frac{2\sqrt[3]{147a^2}}{21ab}$ **51.** $\frac{\sqrt[3]{2a^2b}}{2b}$ **53.** $2x\sqrt[4]{5x^2}$ **55.** $ab\sqrt[4]{25a}$ **57.** $\frac{2ab^2\sqrt[5]{a^3b^2c^4}}{c^2}$ **59.** $\sqrt{a}$

61. $b\sqrt{3ab}$ **63.** $\sqrt[4]{3}$ **65.** 2 **67.** $2x\sqrt{2x}$ **69.** $2\sqrt{a^2 - b^2}$ **71.** $2(a - b)$

Exercises 8.4, Pages 292–293

1. $-\sqrt{2}$ **3.** $\sqrt{2}$ **5.** $32\sqrt{3}$ **7.** $2\sqrt{5}$ **9.** $-7\sqrt{3}$ **11.** $\sqrt{2}$ **13.** $-22\sqrt{3}$ **15.** 0 **17.** $4\sqrt{2x}$

19. $25x\sqrt{2}$ **21.** $-3\sqrt[3]{6}$ **23.** $2\sqrt[3]{2}$ **25.** $3\sqrt[3]{5}$ **27.** $6\sqrt[3]{3} - \sqrt{3}$ **29.** $-\sqrt[3]{2}$ **31.** $2\sqrt{2}$ **33.** $\frac{13\sqrt{3}}{6}$

35. $-\frac{3\sqrt{6}}{2}$ **37.** $\frac{2\sqrt{3}}{3}$ **39.** $\frac{3\sqrt{2}}{2}$ **41.** $\frac{4\sqrt{10}}{5}$ **43.** $2\sqrt[3]{3}$ **45.** $\frac{\sqrt[3]{6}}{6} - \frac{\sqrt[3]{36}}{6}$ **47.** $-\frac{17\sqrt{2}}{4}$

49. $(5x + 6ax - 2a^2)\sqrt{2ax}$ **51.** $(3x^2 + 2x + 5)\sqrt[3]{2x}$ **53.** $\left(\frac{a + 6b - 2}{6ab}\right)\sqrt{6ab}$ **55.** $\left(\frac{a + b - 1}{ab}\right)\sqrt[3]{ab}$

Exercises 8.5, Page 296

1. $3\sqrt{2}$ **3.** $2\sqrt[3]{9}$ **5.** $\sqrt{2}$ **7.** $\sqrt[6]{500}$ **9.** $2\sqrt[6]{2}$ **11.** $9\sqrt[6]{3}$ **13.** $\sqrt[6]{3}$ **15.** $\frac{\sqrt[6]{72}}{3}$ **17.** $-3 - \sqrt{5}$

19. 22 **21.** $42 + 17\sqrt{3}$ **23.** $2a^2 - 2a\sqrt{b} - 4b$ **25.** $-34 - 6\sqrt{21}$ **27.** $2a + \sqrt{ab} - 6b$ **29.** $7 - 4\sqrt{3}$

31. $17 - 4\sqrt{15}$ **33.** $a^2 + 2a\sqrt{b} + b$ **35.** $4a + 12\sqrt{ab} + 9b$ **37.** $\frac{3\sqrt{2} - 2}{7}$ **39.** $\frac{16 + 7\sqrt{5}}{-11}$

41. $\frac{6 - 16\sqrt{2}}{-17}$ **43.** 2 **45.** $\frac{3 + \sqrt{5}}{2}$ **47.** $\frac{a^2 - a\sqrt{b}}{a^2 - b}$

Exercises 8.6, Page 300

1. 53 **3.** 2 **5.** -122 **7.** 42 **9.** 9 **11.** 1 **13.** 1 **15.** 1 **17.** $\frac{3}{2}, -3$ **19.** -6

21. No solution **23.** 7 **25.** 9 **27.** $h = \frac{v_0^2 - v^2}{2g}$ **29.** $l = \frac{P^2g}{4\pi^2}$ **31.** $C = \frac{1}{4\pi^2f^2L}$ **33.** $L = \frac{\sqrt{Z^2 - R^2}}{2\pi f}$

35. $C = \frac{(R_1 + R_2)T^2}{4\pi^2R_2}$ **37.** $m = \frac{v^2Dr}{G(r - D)}$ **39.** $6, \sqrt{15}, \sqrt{21}$

Exercises 8.7, Page 302

1. $3, -3, \sqrt{2}, -\sqrt{2}$ **3.** $1, -1, \frac{1}{4}, -\frac{1}{4}$ **5.** $-3, -4$ **7.** $0, 2, 3, -1$ **9.** 9 **11.** $-\frac{1}{3}, -1$ **13.** $8, -64$

15. $\frac{7}{3}$ **17.** $\pm\sqrt{\frac{3 \pm \sqrt{5}}{2}}$ or ± 1.62; ± 0.618 **19.** 1, 64

Chapter 8 Review, Page 303

1. $\frac{3}{a^2}$ **2.** $\frac{1}{16a^4}$ **3.** a^5 **4.** a^8 **5.** $\frac{b^2c^6}{a^2}$ **6.** $\frac{a + 1}{a^2}$ **7.** 7 **8.** 64 **9.** $\frac{1}{4}$ **10.** $\frac{y^{1/5}}{x^{2/5}}$ **11.** $\frac{1}{x^{5/12}}$

12. $x^{1/2}$ **13.** $\frac{2}{9}$ **14.** $4\sqrt{5}$ **15.** $6ab\sqrt{2b}$ **16.** $4x^2y^3\sqrt{5x}$ **17.** $4x\sqrt{3xy}$ **18.** $\frac{\sqrt{30}}{18}$ **19.** $\frac{3a\sqrt{35}}{7b^2}$

20. $5\sqrt[3]{2}$ **21.** $3a\sqrt[3]{4ab^2}$ **22.** $4ab^3\sqrt[3]{4a^2b}$ **23.** $\frac{\sqrt[3]{450ab}}{10b}$ **24.** $\frac{4\sqrt[3]{25}}{5}$ **25.** $\sqrt{3a}$ **26.** $\frac{\sqrt{3a}}{a}$

27. $\frac{\sqrt[3]{18a^2}}{a}$ **28.** $\sqrt[4]{5}$ **29.** $^{2}\ \sqrt[12]{10}$ **30.** $4\sqrt{3}$ **31.** $-\frac{17\sqrt{10}}{10}$ **32.** 0 **33.** $\frac{13}{6}\sqrt[3]{6}$ **34.** $4\sqrt{10}$ **35.** 3

36. $2\sqrt[6]{2}$ **37.** $-11 - 6\sqrt{3}$ **38.** $61 - 24\sqrt{5}$ **39.** $\frac{9 + 2\sqrt{3}}{-23}$ **40.** 62 **41.** -27 **42.** 9 **43.** -9

44. $3, -3, \frac{\sqrt{5}}{2} - \frac{\sqrt{5}}{2}$ **45.** $-8, 64$

CHAPTER 9

Exercises 9.1, Pages 307–308

1.

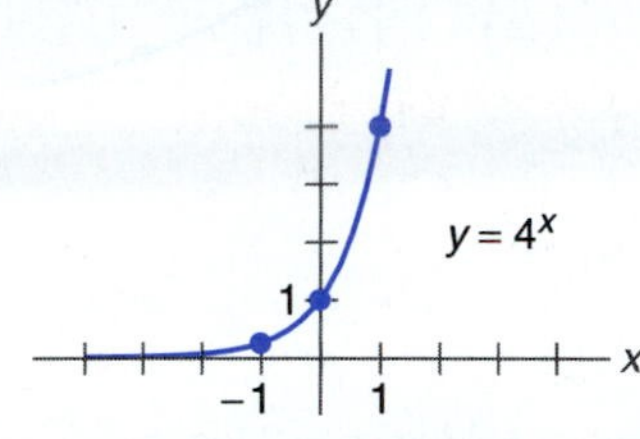

3.

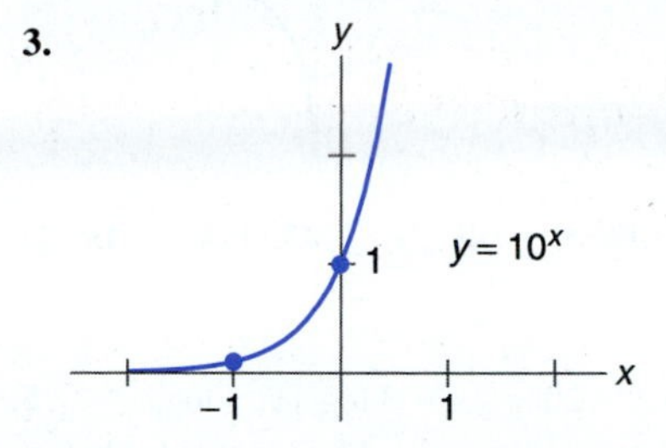

5.

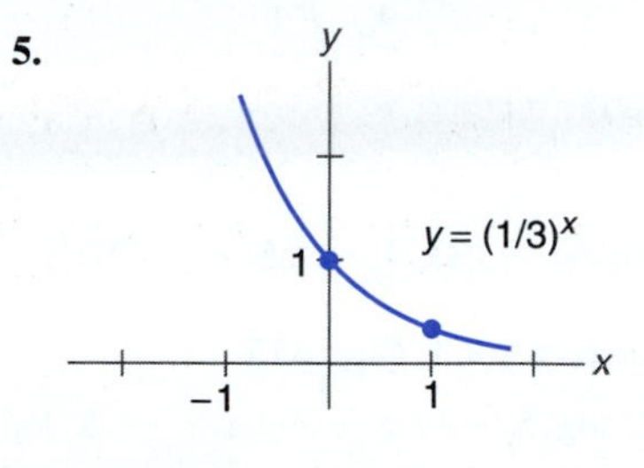

7.

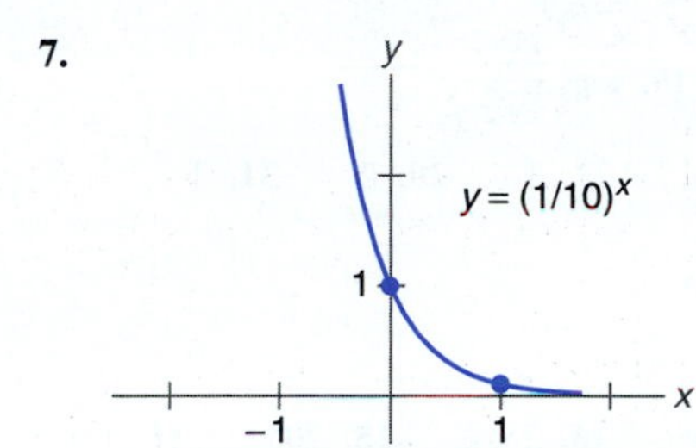

9.

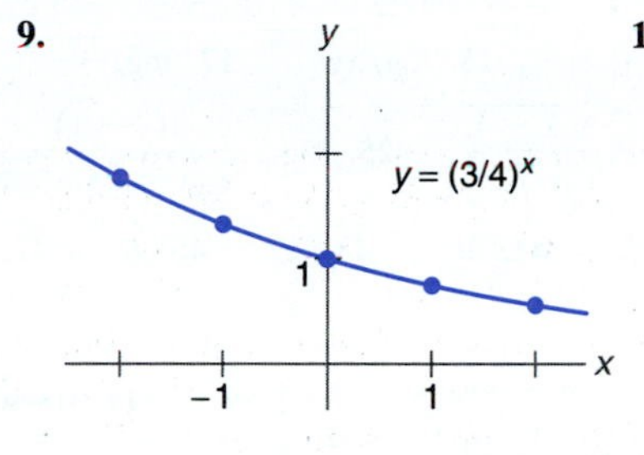

11.

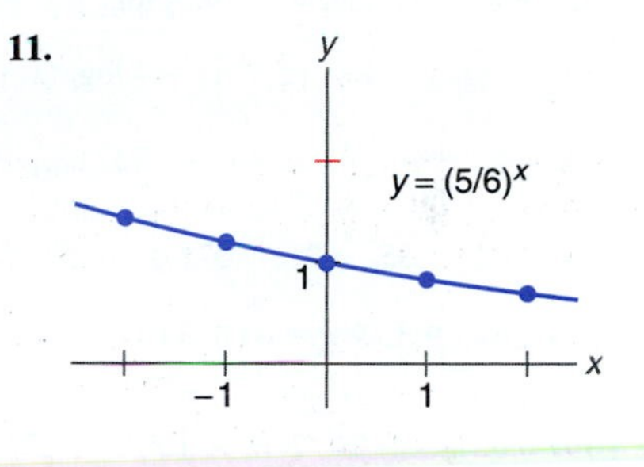

13.

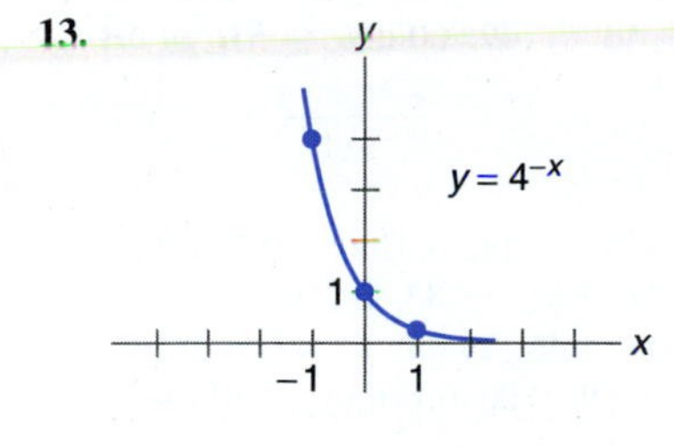

15.

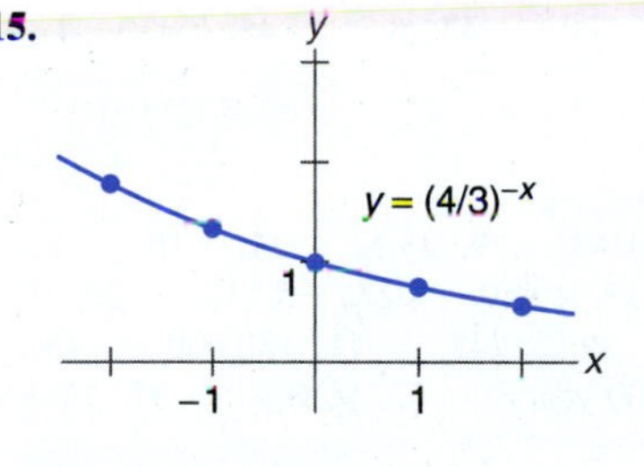

17.

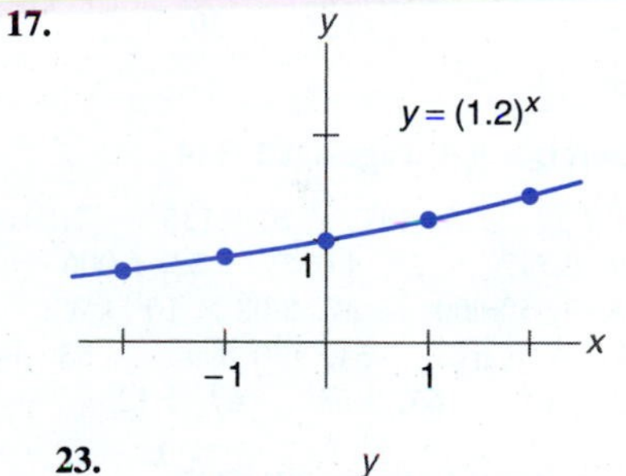

19.

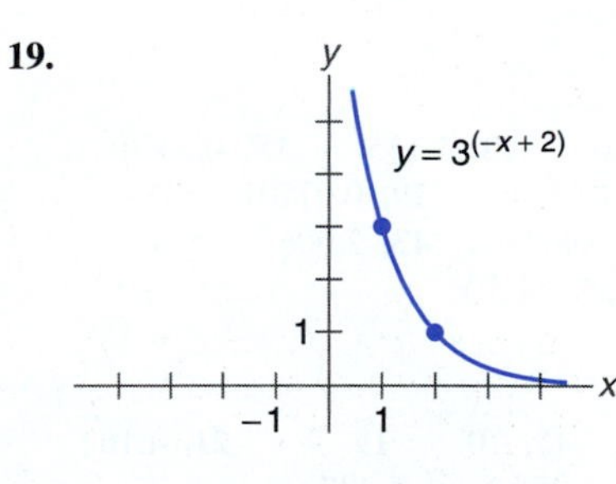

21.

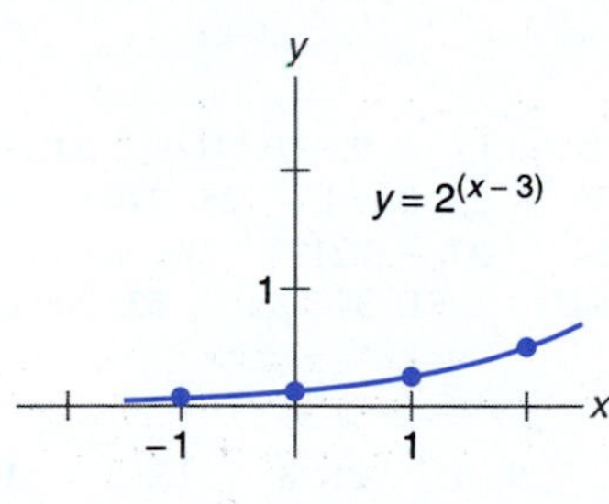

23.

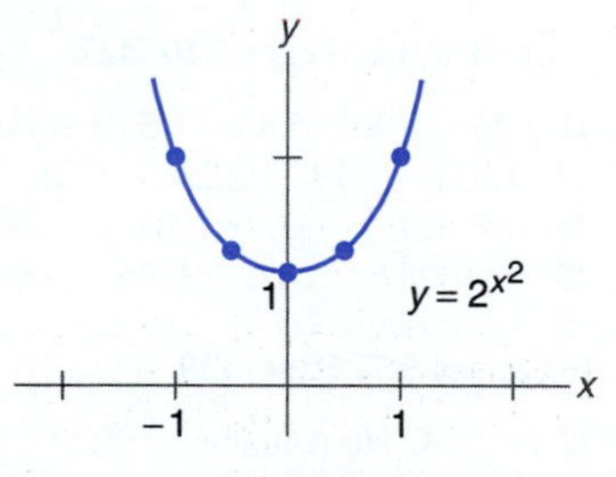

25.

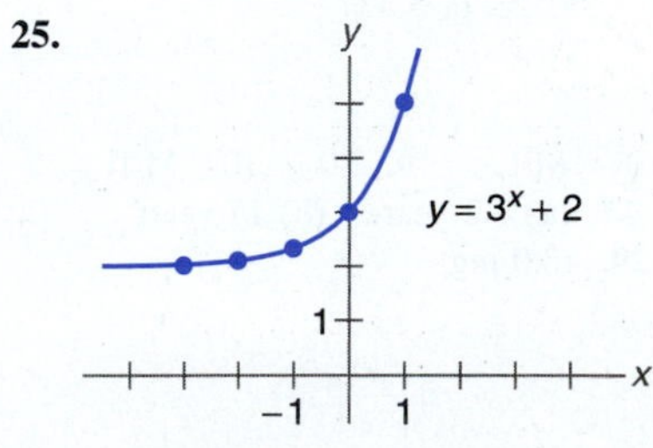

27. 1.38 **29.** 316,000 **31.** 24,600 **33.** 8.82 **35.** 5.20 **37.** 1.57 **39.** $2.\underline{2}8$
41. 7.93×10^{11} kWh **43.** **(a)** \$1260 **(b)** \$1265 **(c)** \$1268 **(d)** \$1271 **45.** $1\bar{7}00$°R

Exercises 9.2, Pages 310–311

1. $\log_3 9 = 2$ **3.** $\log_5 125 = 3$ **5.** $\log_2 32 = 5$ **7.** $\log_9 3 = \frac{1}{2}$ **9.** $\log_5 (\frac{1}{25}) = -2$ **11.** $\log_{10} 0.00001 = -5$
13. $5^2 = 25$ **15.** $2^4 = 16$ **17.** $25^{1/2} = 5$ **19.** $8^{1/3} = 2$ **21.** $2^{-2} = \frac{1}{4}$ **23.** $10^{-2} = 0.01$
25.

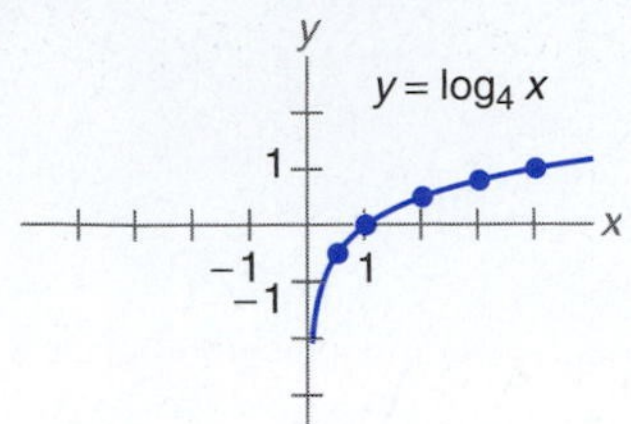

27.

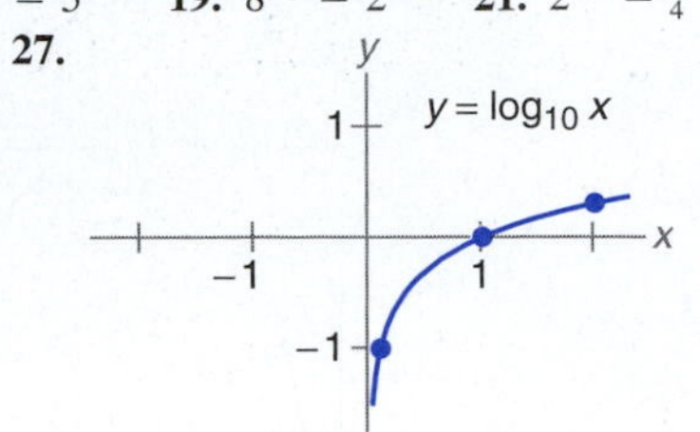

29.

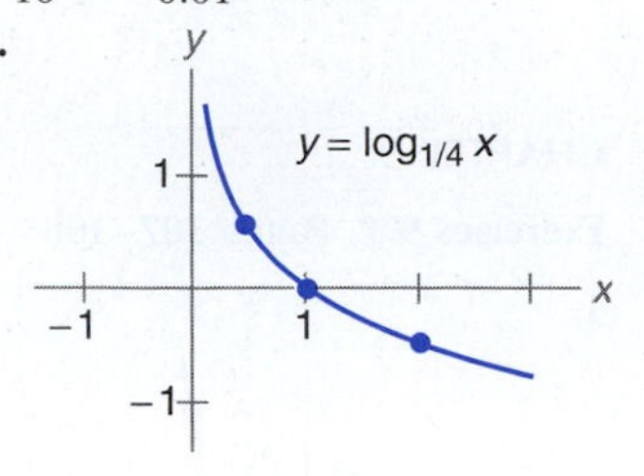

31. 64 **33.** $\frac{1}{2}$ **35.** 3 **37.** $\frac{1}{2}$ **39.** 5 **41.** 3 **43.** 144 **45.** 27 **47.** 4

Exercises 9.3, Page 315

1. $\log_2 5 + 3 \log_2 x + \log_2 y$ **3.** $\log_{10} 2 + 2 \log_{10} x - 3 \log_{10} y - \log_{10} z$ **5.** $3 \log_b y + \frac{1}{2} \log_b x - 2 \log_b z$
7. $\frac{2}{3} \log_b x - \frac{1}{3} \log_b y$ **9.** $\frac{1}{2} \log_2 y - \log_2 x - \frac{1}{2} \log_2 z$ **11.** $3 \log_b z + \frac{1}{2} \log_b x - \frac{1}{3} \log_b y$
13. $2 \log_b x + \log_b (x + 1) - \frac{1}{2} \log_b (x + 2)$ **15.** $\log_b xy^2$ **17.** $\log_b \frac{xy^2}{z^3}$ **19.** $\log_3 \frac{x\sqrt[3]{y}}{\sqrt{z}}$
21. $\log_{10} \frac{x^2}{(x + 1)\sqrt{x - 3}}$ **23.** $\log_b \frac{x^5 \sqrt[3]{x - 1}}{x + 2}$ **25.** $\log_{10} \frac{x(x - 1)^2}{\sqrt[3]{(x + 2)(x - 5)}}$ **27.** 3 **29.** 2 **31.** 3
33. −2 **35.** −3 **37.** 0 **39.** 5 **41.** 36 **43.** $\frac{1}{25}$ **45.** 6 **47.** 8

Exercises 9.4, Pages 318–319

1. 1.833 **3.** 1.653 **5.** −0.8477 **7.** −2.207 **9.** 2.906 **11.** 0.9661 **13.** 25.4 **15.** 705 **17.** 1.90
19. 0.0248 **21.** 2.32×10^{-4} **23.** 1.53×10^{-5} **25.** 3.64 **27.** 2.94 **29.** 0.389 **31.** 1.12 **33.** −0.574
35. 0.590 **37.** 3.56 **39.** 4.82 **41.** 6.0 **43.** 6.5 **45.** 7.3 **47.** 100 dB **49.** 60 dB **51.** 14 dB
53. −1.2 dB

Exercises 9.5, Pages 322–324

1. 7.39 **3.** 403 **5.** 0.135 **7.** 0.00248 **9.** 33.1 **11.** 1.16 **13.** 0.0821 **15.** 0.923 **17.** 1.95
19. 0.717 **21.** 4.025 **23.** 6.006 **25.** 1.459 **27.** −5.146 **29.** 0 **31.** 1.61 **33.** 0.236
35. 2,550,000 **37.** 8.08×10^{11} kWh **39.** \$5742 **41.** 130,000 **43.** 3700 **45.** \$4500 **47.** 6.02 g
49. 31 years **51.** $7\bar{0}0$ years **53.** 3400 years **55.** 9.18 V **57.** 17.4 V **59.** 0.0046 ft/min **61.** 4.03
63. 5.60 **65.** 3.68 **67.** 5.12

Exercises 9.6, Pages 326–327

1. 2.26 **3.** −5.45 **5.** −0.710 **7.** −3.15 **9.** −0.1411 **11.** −0.2886 **13.** 3.135 **15.** 0.8356
17. 1.851 **19.** −2.248 **21.** −0.6936 **23.** 3.164 **25.** 2.641 **27.** −0.7821 **29.** 0.07201
31. ±2.529 **33.** −1.386 **35.** 0.5754 **37.** −3.219 **39.** 4.512 **41.** 0.4621 **43.** 2.006
45. −0.3087 **47.** 0.3890 **49.** −0.5432 **51.** 34.7 μs **53.** 209 μs **55.** 51.5 V

Exercises 9.7, Page 329

1. 5 **3.** No solution **5.** 7 **7.** 7 **9.** 6 **11.** 8 **13.** 3 **15.** 6 **17.** 10 **19.** 9 **21.** 4.162
23. 899 **25.** 8.437 **27.** 0.7315 **29.** 3.25 **31.** 7.361 **33.** 0.9283 **35.** 2*e*, or 5.437

Exercises 9.8, Pages 334–335

1. **(a)** 7.9 years **(b)** 14.8 years **3.** 4.9 s **5.** 3.2×10^{-4} M/L **7.** 3.2×10^{-10} M/L **9.** 2.0×10^{-3} M/L
11. 3.2×10^{-15} W/cm² **13.** 1.0×10^{-4} W/cm² **15.** 1.0×10^{-13} W/cm² **17.** **(a)** 7.2 years **(b)** 17 years
19. $1\bar{0}$ s **21.** 1.94×10^{-3} ms **23.** 11.1 μs **25.** 2.8 **27.** 50.0 mg **29.** 48.0 mg

Exercises 9.9, Pages 340–341

1. $y = 3.5x + 15$ **3.** $y = 18.2x - 60$

5.

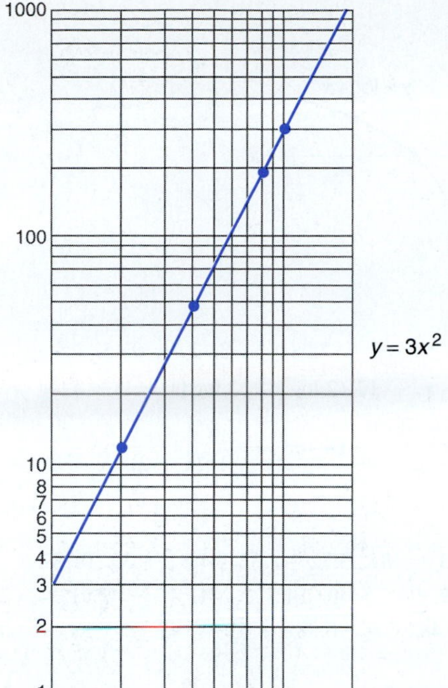

7.

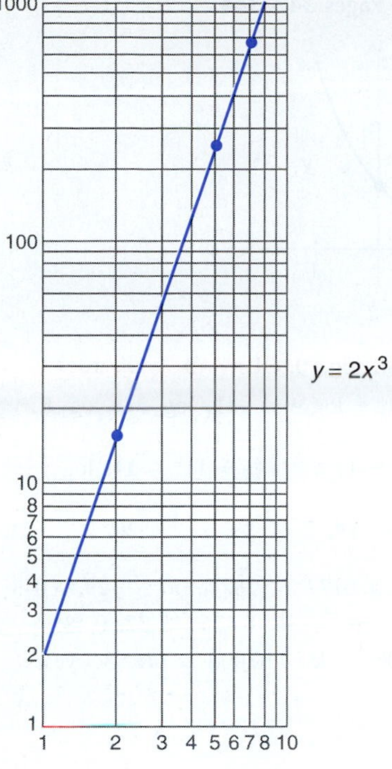

9. $y = 2x^2$ **11.** $y = 12x^2$

13.

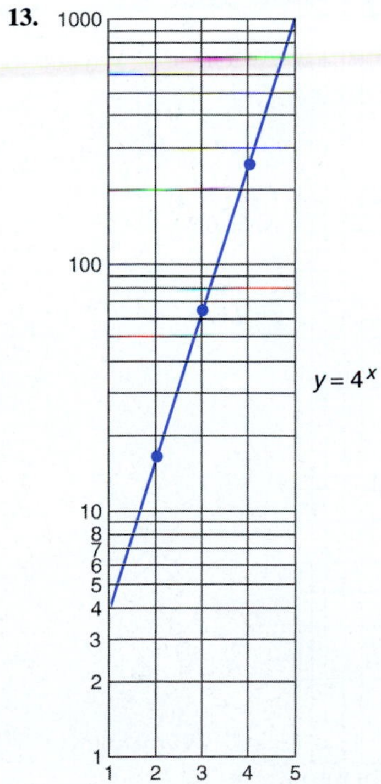

15.

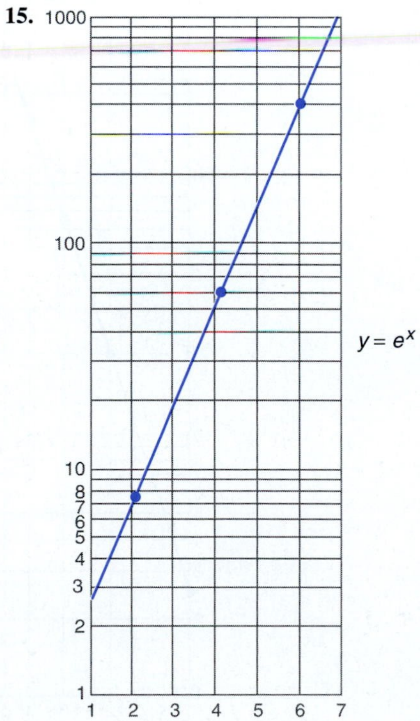

17.

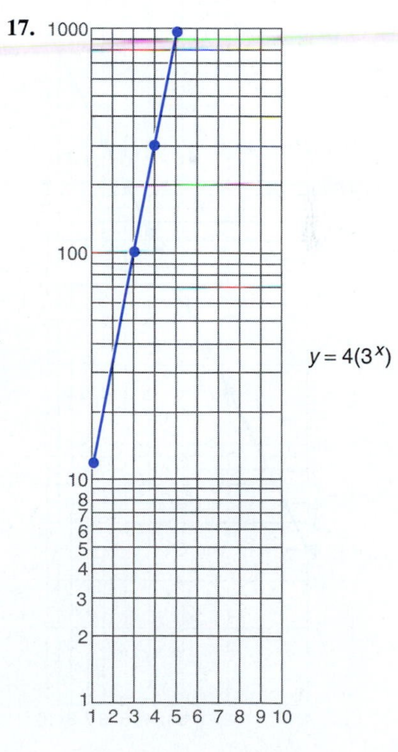

19. $y = 2.7^x$ **21.** $y = 0.9^x$ **23.** $y = 2.1(3^x)$

Chapter 9 Review, Pages 342–344

1.

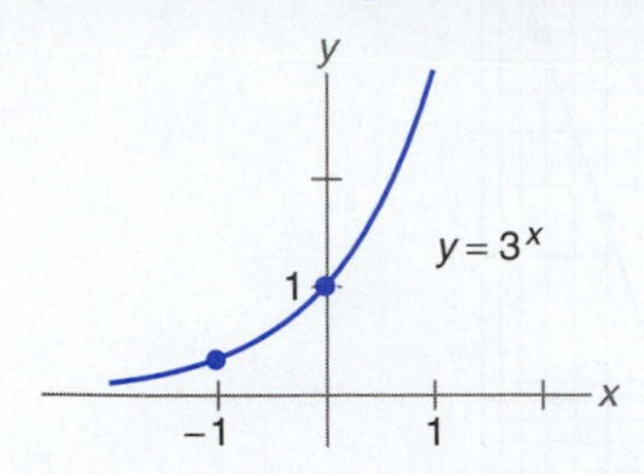

2.

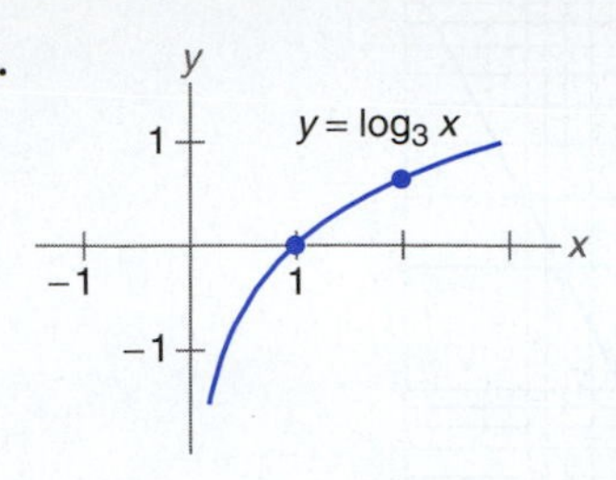

3. $\log_2 16 = 4$ **4.** $\log_{10} 0.001 = -3$ **5.** $10^{0.8686} = 7.389$ **6.** $4^{-2} = \frac{1}{16}$ **7.** 81 **8.** 2 **9.** 5
10. $\log_4 6 + 2\log_4 x + \log_4 y$ **11.** $\log_3 5 + \log_3 x + \frac{1}{2}\log_3 y - 3\log_3 z$ **12.** $2\log x + 3\log(x+1) - \frac{1}{2}\log(x-4)$
13. $3\ln x + 3\ln(x-1) - \frac{1}{2}\ln(x+1)$ **14.** $\log_2 \frac{xy^3}{z^2}$ **15.** $\log \frac{\sqrt{x+1}}{(x-2)^3}$ **16.** $\ln \frac{x^4}{(x+1)^5(x+2)}$
17. $\ln \frac{\sqrt{x(x+2)}}{(x-5)^2}$ **18.** 3 **19.** x^2 **20.** 2 **21.** x **22.** 2.823 **23.** −1.393 **24.** 4.159 **25.** 1180
26. 0.00372 **27.** 0.0177 **28.** 4.277 **29.** 6.043 **30.** −6.293 **31.** 3.72 **32.** 31.5 **33.** 0.787
34. 2.16 **35.** −1.08 **36.** 0.626 **37.** 0.485 **38.** 2.20 **39.** 96 **40.** 14.5 **41.** 4 **42.** 1
43. 0.5 **44.** 4.48 **45.** 110,000 **46.** 8.7 years **47.** 518 **48.** $y = -0.83x + 61$ **49.** $y = 3.1x + 60$

50.

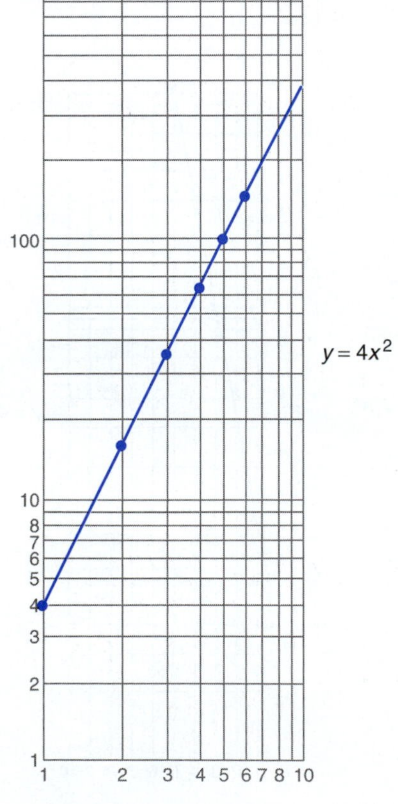

51.

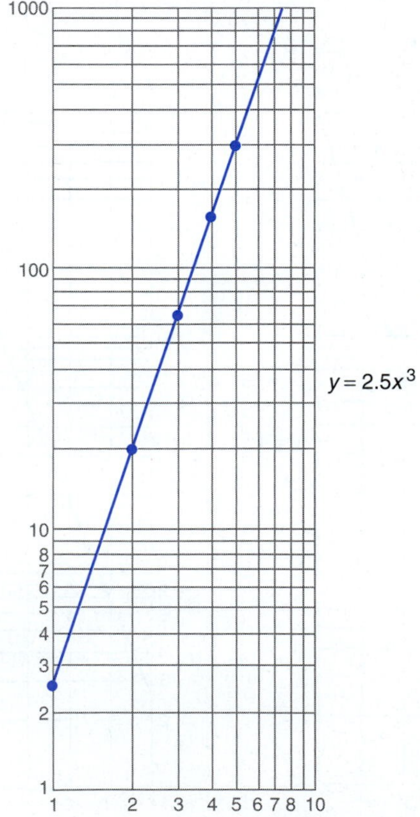

52. $y = 3.6x^2$ **53.** $y = 2.1x^2$

54.

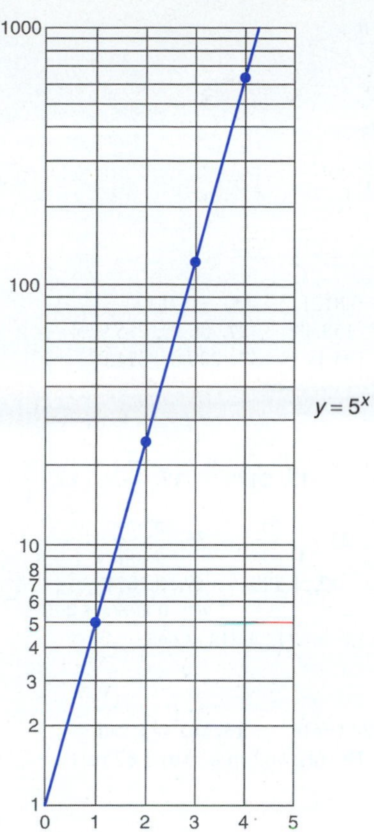

55.

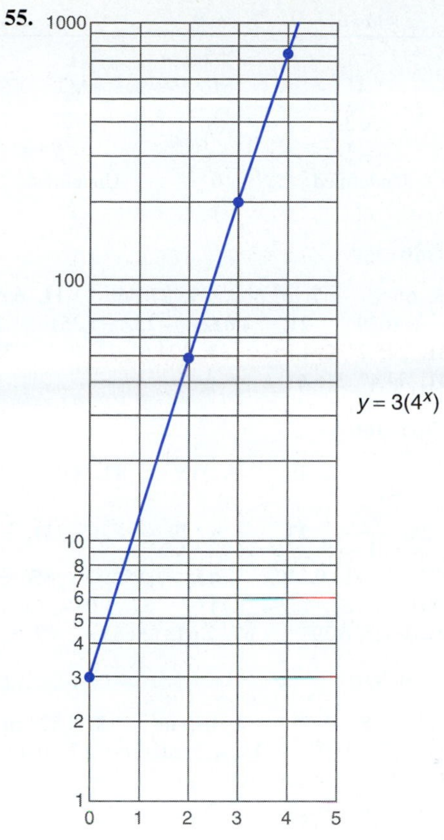

56. $y = 3.9^x$ **57.** $y = 2.5(2^x)$

CHAPTER 10

Exercises 10.1, Pages 353–354

1.

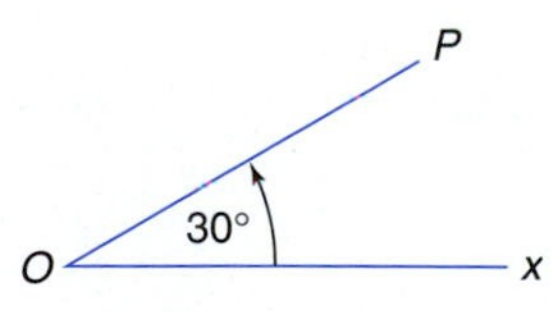

3.

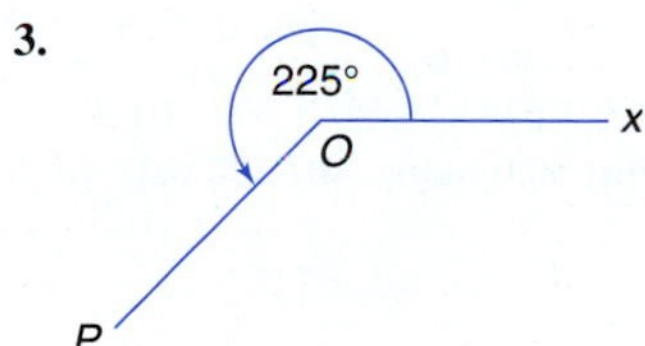

5.

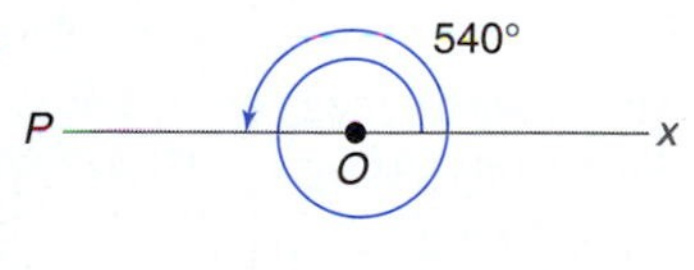

7. 420°, −300° **9.** 274°, −446° **11.** 585°, −135° **13.** 52°, −308°

15.

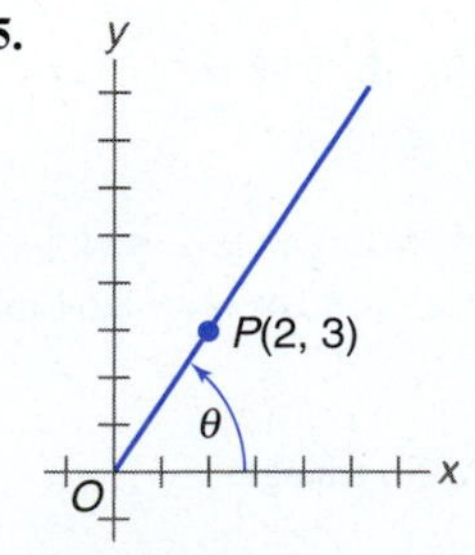

17.

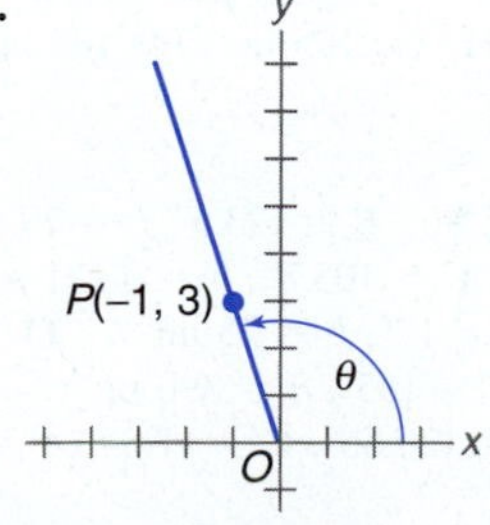

19.

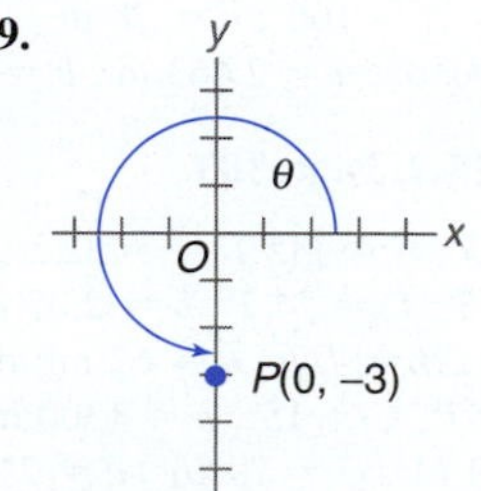

	$\sin\theta$	$\cos\theta$	$\tan\theta$	$\cot\theta$	$\sec\theta$	$\csc\theta$
21.	$-\frac{4}{5}$	$\frac{3}{5}$	$-\frac{4}{3}$	$-\frac{3}{4}$	$\frac{5}{3}$	$-\frac{5}{4}$
23.	$\frac{1}{\sqrt{2}}$	$\frac{1}{\sqrt{2}}$	1	1	$\sqrt{2}$	$\sqrt{2}$
25.	$-\frac{\sqrt{3}}{2}$	$-\frac{1}{2}$	$\sqrt{3}$	$\frac{1}{\sqrt{3}}$	-2	$-\frac{2}{\sqrt{3}}$
27.	$\frac{5}{\sqrt{41}}$	$-\frac{4}{\sqrt{41}}$	$-\frac{5}{4}$	$-\frac{4}{5}$	$-\frac{\sqrt{41}}{4}$	$\frac{\sqrt{41}}{5}$
29.	-1	0	Undefined	0	Undefined	-1
31.	$\frac{3}{5}$	$\frac{4}{5}$	$\frac{3}{4}$	$\frac{4}{3}$	$\frac{5}{4}$	$\frac{5}{3}$

Exercises 10.2, Pages 359–360

1. 60° **3.** 73° **5.** 66.6° **7.** 63.3° **9.** 77°56′ **11.** 60° **13.** 0.8121 **15.** −0.1817 **17.** −0.07498 **19.** −0.8039 **21.** −1.022 **23.** −1.351 **25.** 21.6°, 158.4° **27.** 35.9°, 215.9° **29.** 96.0°, 264.0° **31.** 245.7°, 294.3° **33.** 124.6°, 235.4° **35.** 15.1°, 195.1° **37.** 229.0°, 311.0° **39.** 65.1°, 294.9° **41.** 44.4°, 315.6°

Exercises 10.3, Pages 365–366

1. $\frac{3\pi}{4}$ **3.** $\frac{\pi}{2}$ **5.** $-\frac{5\pi}{12}$ **7.** 7π **9.** 315° **11.** 855° **13.** 1620° **15.** 212° **17.** $\frac{\pi}{4}$ **19.** $\frac{\pi}{4}$ **21.** $\frac{\pi}{12}$ **23.** $\frac{2\pi}{5}$ **25.** $\frac{\sqrt{2}}{2}$ **27.** $\sqrt{3}$ **29.** $-\sqrt{2}$ **31.** $\frac{2\pi}{3}, \frac{4\pi}{3}$ **33.** $\frac{\pi}{3}, \frac{5\pi}{3}$ **35.** $\frac{\pi}{4}, \frac{5\pi}{4}$ **37.** 0.7174 **39.** 2.572 **41.** 0.5403 **43.** −0.9969 **45.** −0.1507 **47.** 1.289 **49.** 0.7071 **51.** −3.078 **53.** 0.9914 **55.** 1.394, 1.747 **57.** 1.035, 4.177 **59.** 2.425, 3.858 **61.** 0.8060, 5.477 **63.** 1.782, 4.923 **65.** 3.628, 5.797 **67.** 2.618, 3.665 **69.** 0.5236, 2.618 **71.** 2.618, 3.665

Exercises 10.4, Pages 369–370

1. 25.1 in. **3.** 151 in^2 **5.** 47.7° **7.** 4.19 m^2 **9.** 0.524 m **11.** 344 ft-lb **13.** **(a)** 44.0 rad/s **(b)** $44\bar{0}$ rad or 70.0 rev **(c)** 77.0 ft/s **15.** 4.2 rad/s^2 **17.** 1040 mi/h **19.** **(a)** 2.62 m/s **(b)** 2.62 m/s **(c)** 32.8 rad/s

Chapter 10 Review, Pages 371–372

1. $\frac{3}{5}, \frac{4}{5}, \frac{3}{4}, \frac{4}{3}, \frac{5}{4}, \frac{5}{3}$ **2.** $-\frac{1}{2}, -\frac{\sqrt{3}}{2}, \frac{1}{\sqrt{3}}, \sqrt{3}, -\frac{2}{\sqrt{3}}, -2$ **3.** 0, −1, 0, undefined, −1, undefined **4.** 45° **5.** 28°20′ **6.** 55° **7.** 10° **8.** −0.9011 **9.** −0.4142 **10.** −6.611 **11.** 0.4602 **12.** 1.107 **13.** −4.810 **14.** 20.2°, 159.8° **15.** 123.3°, 236.7° **16.** 133.6°, 313.6° **17.** 59.3°, 300.7° **18.** 151.7°, 331.7° **19.** 47.7°, 132.3° **20.** $\frac{2\pi}{5}$ **21.** $\frac{7\pi}{4}$ **22.** 150° **23.** 135° **24.** $\frac{\pi}{3}$ **25.** $\frac{2\pi}{5}$ **26.** $-\frac{\sqrt{3}}{2}$ **27.** $-\sqrt{3}$ **28.** $\sqrt{2}$ **29.** $-\frac{1}{2}$ **30.** −1 **31.** $\frac{1}{2}$ **32.** $\frac{3\pi}{4}, \frac{5\pi}{4}$ **33.** $\frac{3\pi}{4}, \frac{7\pi}{4}$ **34.** $\frac{\pi}{3}, \frac{2\pi}{3}$ **35.** $\frac{4\pi}{3}, \frac{5\pi}{3}$ **36.** $\frac{2\pi}{3}, \frac{4\pi}{3}$ **37.** $\frac{\pi}{6}, \frac{7\pi}{6}$ **38.** 0.9975 **39.** 0.9689 **40.** 0.5774 **41.** −0.8660 **42.** 1.371, 4.912 **43.** 3.845, 5.579 **44.** 1.017, 4.159 **45.** 1.935, 4.348 **46.** 8.80 in., 39.6 in^2 **47.** 427 ft-lb **48.** **(a)** 66.0 rad/s **(b)** $33\bar{0}$ rad **(c)** 99.0 ft/s

CHAPTER 11

Exercises 11.1, Pages 378–379

1. $B = 38.0°$, $C = 73.0°$, $c = 25.6$ m **3.** $C = 52.5°$, $A = 66.1°$, $a = 129$ cm **5.** $C = 47.6°$, $a = 223$ ft, $c = 196$ ft **7.** $B = 48.5°$, $C = 16.5°$, $c = 1840$ m **9.** $A = 67.07°$, $B = 40.35°$, $a = 40.72$ cm **11.** $A = 131.94°$, $a = 89{,}460$ mi, $b = 57{,}870$ mi **13.** $B = \bar{2}0°$, $C = 135°$, $c = 84$ cm **15.** $A = 6°$, $B = 166°$, $b = 28$ m **17.** $A = 91°\,31'8''$, $C = 37°28'35''$, $a = 2506$ ft **19.** $C = 63°10''$, $a = 2.663$ mi, $b = 5.120$ mi **21.** **(a)** 195 in. **(b)** 144 in. **23.** **(a)** 16.3 ft **(b)** 9.88 ft

Exercises 11.2, Page 383

1. $B = 27.3°$, $C = 115.7°$, $c = 32.2$ cm **3.** $A = 29.6°$, $B = 123.9°$, $b = 79.4$ km; or $A = 150.4°$, $B = 3.1°$, $b = 5.18$ km **5.** $B = 34.3°$, $C = 74.2°$, $b = 2.05$ m; or $B = 2.7°$, $C = 105.8°$, $b = 0.171$ m **7.** $A = 44.6°$, $C = 30.4°$, $c = 8.64$ mi **9.** $A = 35°$, $B = 127°$, $b = 62$ mi; or $A = 145°$, $B = 17°$, $b = 23$ mi **11.** No triangle **13.** $A = 157°$, $C = 15°$, $a = 1300$ m; or $A = 7°$, $C = 165°$, $a = 390$ m **15.** $A = 59.34°$, $C = 79.16°$, $c = 21.12$ km; or $A = 120.66°$, $C = 17.84°$, $c = 6.588$ km **17.** No triangle **19.** $A = 6.06°$, $B = 165.19°$, $b = 150.1$ m

21. $A = 107°2'21''$, $C = 43°41'2''$, $a = 421.5$ m; or $A = 14°24'25''$, $C = 136°18'58''$, $a = 109.7$ m
23. $A = 27°10'25''$, $B = 127°48'50''$, $b = 907.3$ ft; or $A = 152°49'35''$, $B = 2°9'40''$, $b = 43.31$ ft

Exercises 11.3, Pages 386–387

1. $B = 47.8°$, $C = 72.2°$, $a = 22.8$ m **3.** $A = 27.0°$, $B = 44.0°$, $c = 292$ km **5.** $A = 44.6°$, $B = 51.2°$, $C = 84.2°$
7. $A = 32.1°$, $B = 104.6°$, $C = 43.3°$ **9.** $B = 85°$, $C = 5\bar{0}°$, $a = 36$ m **11.** $A = 84°$, $B = 53°$, $C = 43°$
13. $A = 19°$, $B = 26°$, $c = 78$ ft **15.** $A = 148.80°$, $C = 11.95°$, $b = 3064$ m
17. $A = 40.11°$, $B = 31.14°$, $c = 595.2$ mm **19.** $A = 113.43°$, $B = 27.84°$, $C = 38.73°$
21. $a = 1465$ m, $B = 44°17'56''$, $C = 63°23'52''$ **23.** $A = C = 37°48'52''$, $B = 104°22'16''$
25. **(a)** 76.0° **(b)** 57.6° **27.** **(a)** 17.0 ft **(b)** 20.4°

Exercises 11.4, Pages 388–392

1. $AC = 10.9$ m, $BC = 6.02$ m **3.** 63.0 m **5.** 240 m **7.** 5.13 mi from tower A, 3.34 mi from tower B
9. 98 mi **11.** 290 ft **13.** **(a)** $4\bar{0}0$ m **(b)** 140 m **15.** 321 ft or 80.2 ft **17.** 48.9 ft
19. **(a)** 3.86 m **(b)** 55.2° **(c)** 6.56 m **(d)** 35.4° **(e)** 60.4° **21.** 24.7 ft **23.** 79.4 ft **25.** 45.4 m

Exercises 11.5, Page 398

1. $1\bar{0}$ km at $9\bar{0}°$ **3.** 85 mi/h at 328° **5.** 73 mi at 281° **7.** 41 mi at 126° **9.** 4.0 km at 328°
11. 79 mi at 74° (16° east of north) **13.** 270 mi/h at $24\bar{0}°$ ($3\bar{0}°$ west of south)

Exercises 11.6, Page 400

1. 76.8 km/h at 238.2° **3.** 10.4 km at 89.6° **5.** 108 mi/h at 155.4° **7.** 1180 m at 85.4° **9.** 4.05 km at 328.2°
11. 278 km at 99.4° (9.4° west of north) **13.** 40.0 km due east **15.** 187 mi/h at 249.6° (20.4° west of south)

Exercises 11.7, Pages 405–406

1. 1.59 km, 18.1 km **3.** 0, −135 mi/h **5.** −2380 ft, 1240 ft **7.** 10.8 m at 36.3° **9.** 8.10 m/s at 305.5°
11. 14.7 km at 180.0° **13.** $|\mathbf{R}| = 85\ \Omega$, $|\mathbf{X}_L| = 31\ \Omega$ **15.** $28\bar{0}\ \Omega$, $3\bar{0}°$ **17.** 661 ft at 60.6°
19. 30.1 mi/h at 20.2° **21.** 464 km at 34.8° **23.** 10,600 m at 30.3° **25.** 36.0 mi/h at 78.2°
27. 224 mi at 211.7° (31.7° south of west)

Exercises 11.8, Pages 411–413

1. 301 mi at 258.6° (11.4° west of south) **3.** 312 km at 234.1° (54.1° south of west) **5.** 25.0 mi **7.** 127.1°
9. 45.0 km **11.** **(a)** 12 mi/h **(b)** 9 mi/h **13.** 177 mi/h at 171.9° (8.1° north of west)
15. 289 km/h at 116.9° (26.9° west of north) **17.** 175 mi/h at 148.0° (32.0° north of west) **19.** −14.3 lb, 20.5 lb
21. 3960 lb at 333.8° (26.2° south of east) **23.** 496 lb at 216.4° (36.4° south of west)
25. 193 lb at 70.9° from **R** or 65.4 lb at 109.1° **27.** 85.4 lb at 159.4° **29.** 724 lb at 45.0° **31.** 999 lb
33. $\mathbf{T}_1 = 306$ lb, $\mathbf{T}_2 = 337$ lb

Chapter 11 Review, Pages 415–417

1. $A = 59.3°$, $C = 49.3°$, $a = 371$ ft **2.** $B = 117.2°$, $c = 22.6$ m, $b = 32.9$ m
3. $B = 106.3°$, $C = 58.2°$, $a = 65.7$ cm **4.** $A = 37.3°$, $B = 86.4°$, $C = 56.3°$ **5.** No triangle
6. $B = 35°$, $C = 116°$, $c = 76$ cm; or $B = 145°$, $C = 6°$, $c = 8.8$ cm
7. $A = 32°$, $B = 112°$, $c = 2300$ ft **8.** $A = 69°$, $B = 53°$, $b = 390$ m; or $A = 111°$, $B = 11°$, $b = 92$ m
9. $A = 41.62°$, $C = 33.23°$, $b = 335.9$ m **10.** $B = 63.62°$, $C = 41.63°$, $b = 20.60$ cm
11. $A = 20.12°$, $C = 141.63°$, $c = 3023$ ft; or $A = 159.88°$, $C = 1.87°$, $c = 158.9$ ft **12.** No triangle
13. $C = 63°37'7''$, $b = 3426$ ft, $c = 3308$ ft **14.** $A = 28°23'18''$, $B = 122°11'26''$, $c = 14{,}010$ ft **15.** 4288 ft
16. **(a)** 1122 ft **(b)** 1103 ft **17.** 8.236 in. **18.** **(a)** 2.939 in. **(b)** 4.755 in.
19. **(a)** 11.5 m **(b)** 8.63 m **(c)** 5.77 m **(d)** 6.41 m **20.** 92 km/h at $1\bar{0}0°$ **21.** 215 mi at 160.4°
22. 44.7 mi/h at 142.0° **23.** −182 km, 182 km **24.** 23.0 mi/h, −35.4 mi/h **25.** 110 N, −110 N
26. $1\bar{0}0$ lb, 43 lb **27.** 36.1 N at 300.8° **28.** 74.6 mi/h at 135.9° **29.** 3870 N at 308.2°
30. 325 N at 162.0° **31.** 63.0 km at 145.4° **32.** 630 km/h at 26° (64° east of north)
33. 67.7 mi at 98.6° (8.6° west of north) **34.** 640 lb, 380 lb

Chapter 11 Special NASA Application, Page 420

(−5, 13) − (−5, −8) = (0, 21). Since the absolute value of the out-of-plane wind change is 21, which exceeds the threshold value of 20, the launch should be delayed if the launch window exceeds 3 h or scrubbed if the launch window is less than 3 h.

CHAPTER 12

Exercises 12.1, Page 426

1. 2; 2π

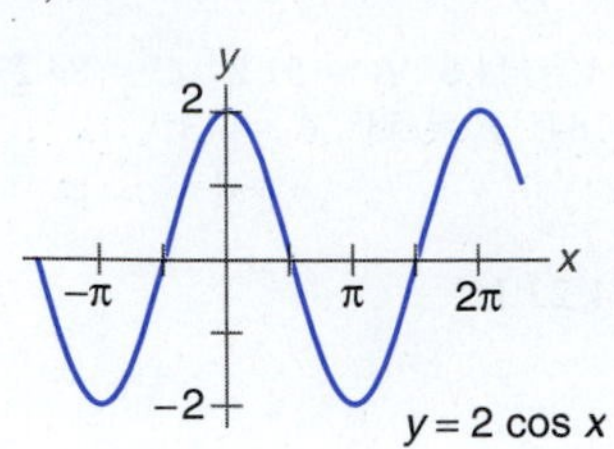

3. 3; 2π

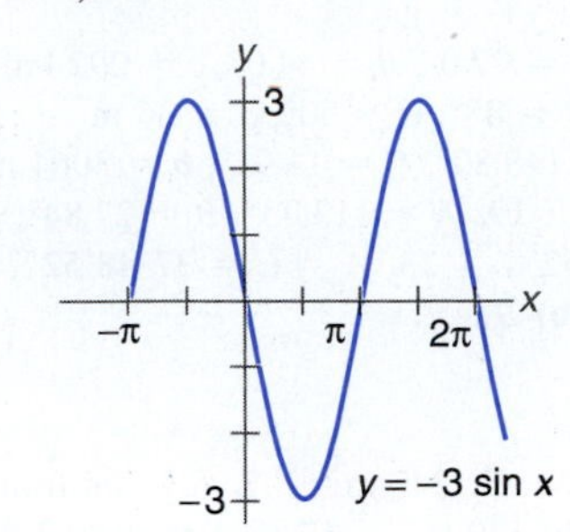

5. 1; $\frac{2\pi}{3}$

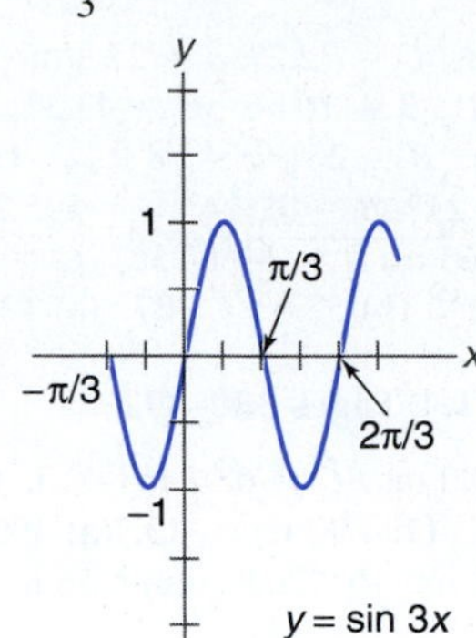

7. 1; π

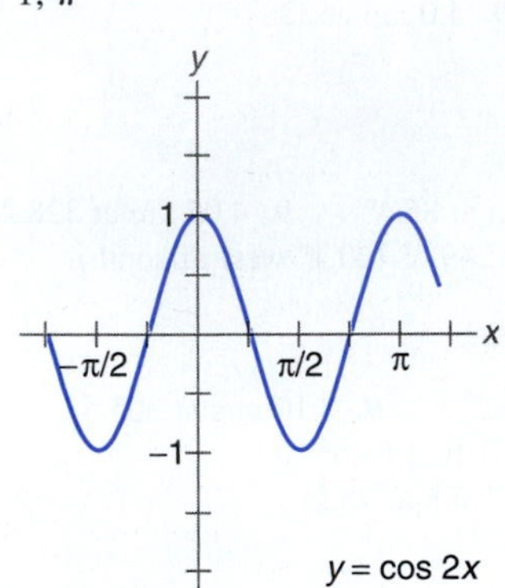

9. 2; $\frac{\pi}{2}$

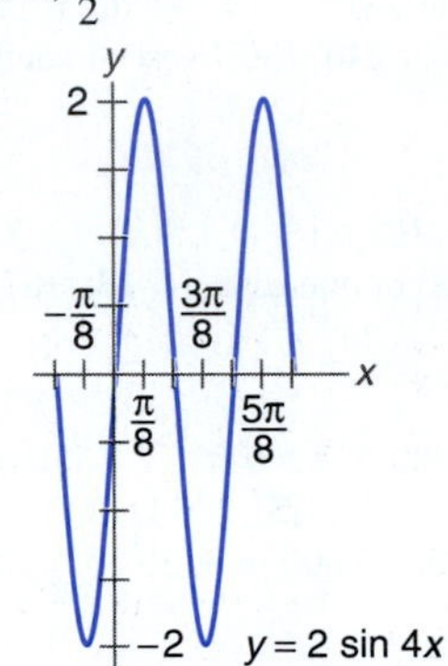

11. $\frac{5}{2}$; 4π

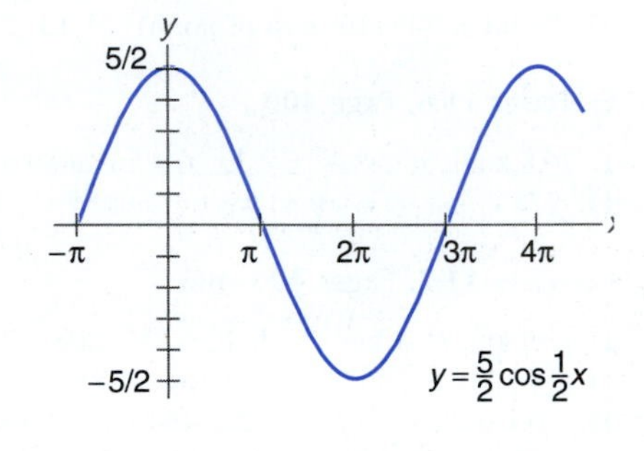

13. $\frac{1}{2}$; 3π

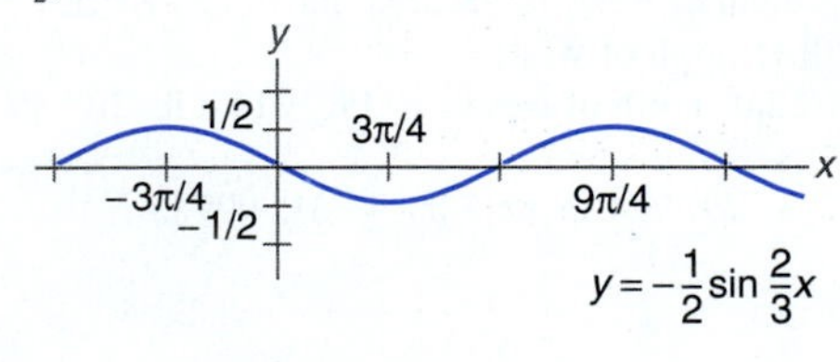

15. 2; $\frac{2}{3}$

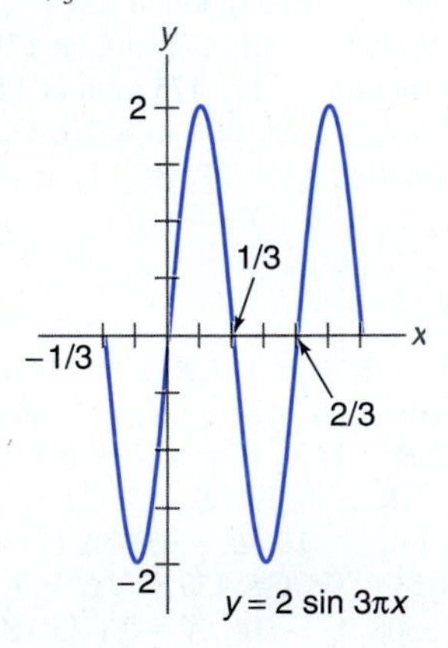

17. 3; 2

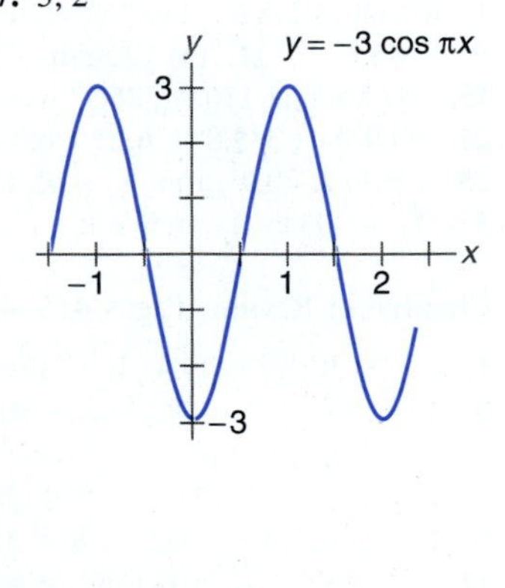

19. 6.5; $\frac{1}{60}$

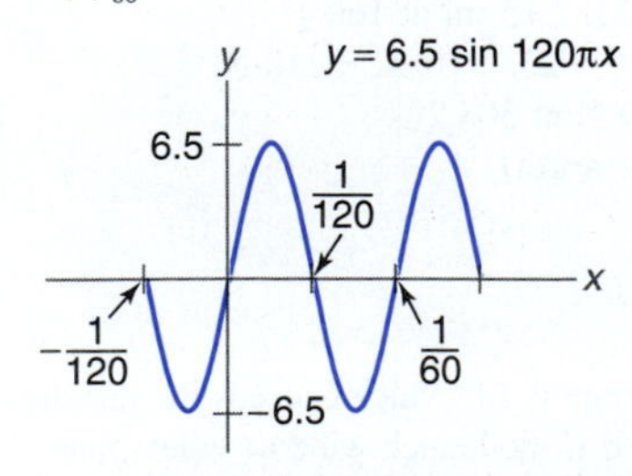

21. 40; $\frac{\pi}{30}$

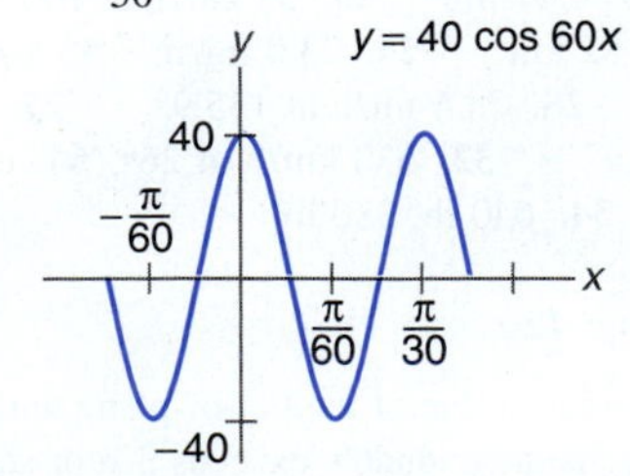

23. 60; $\frac{1}{40}$

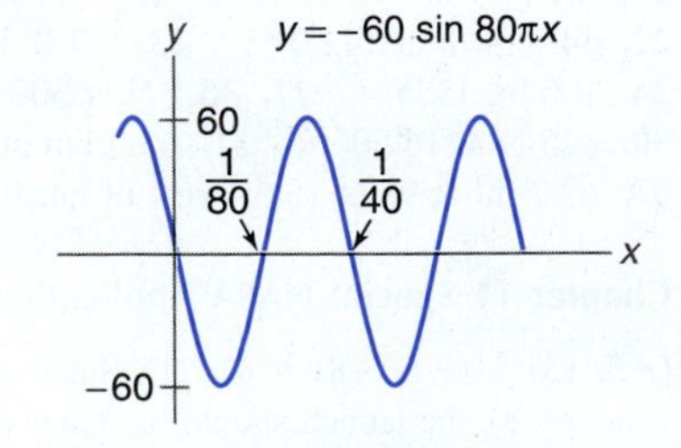

Exercises 12.2, Page 430

1. 1; 2π; $\frac{\pi}{3}$

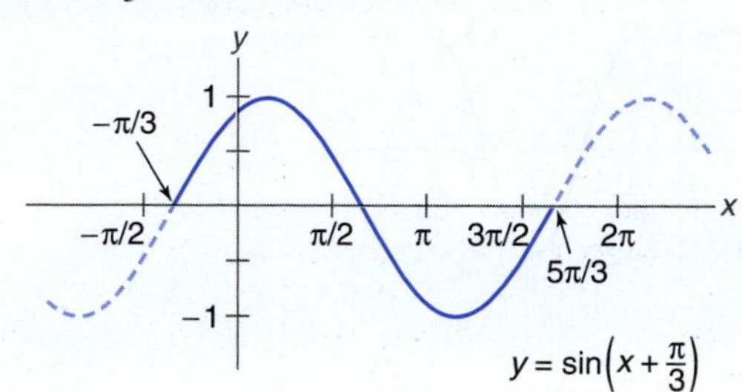

3. 2; 2π; $-\frac{\pi}{6}$

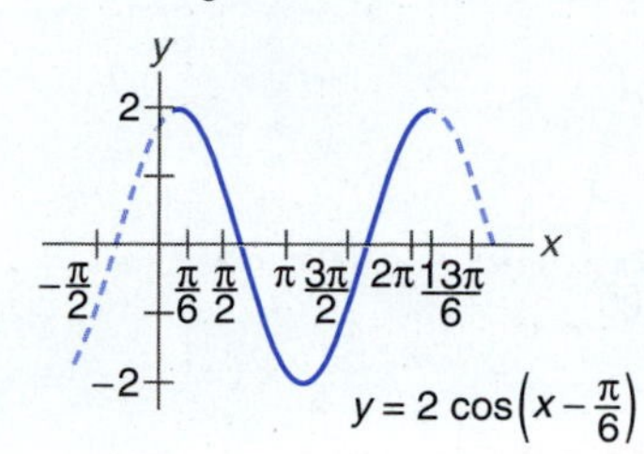

5. 1; $\frac{2\pi}{3}$; $-\frac{\pi}{3}$

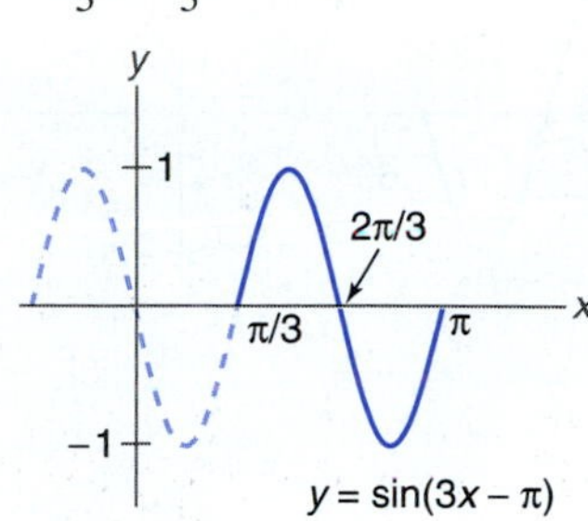

7. 1; $\frac{\pi}{2}$; $\frac{\pi}{4}$

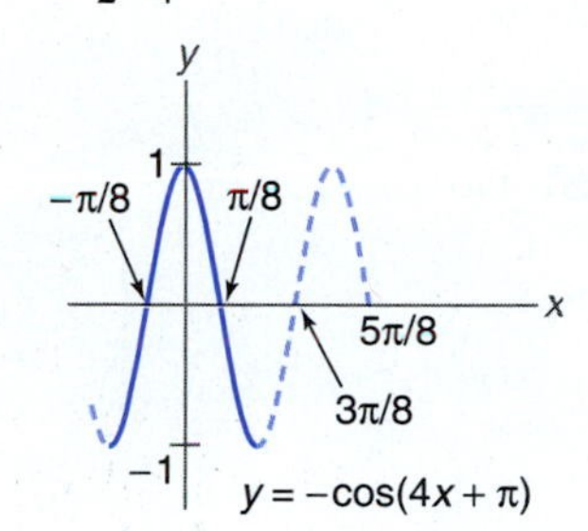

9. 3; 4π; $-\frac{\pi}{2}$

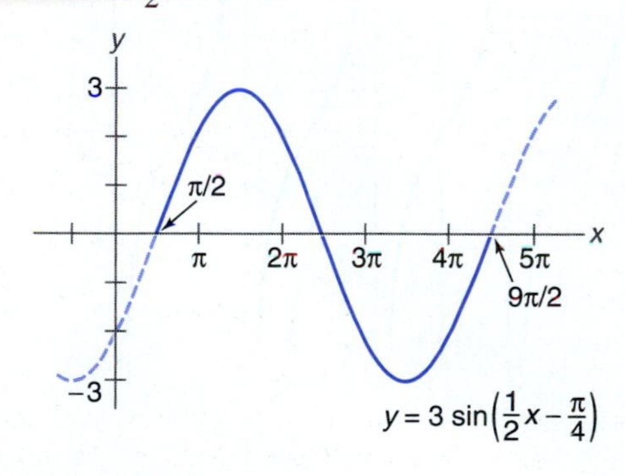

11. 2; $\frac{3\pi}{2}$; $\frac{\pi}{4}$

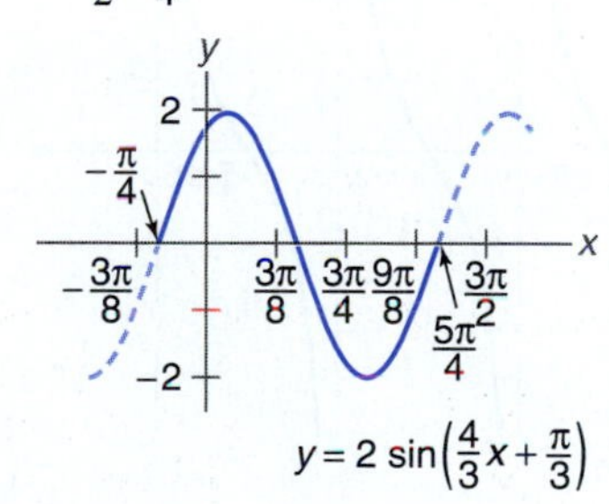

13. 3; 2; 1

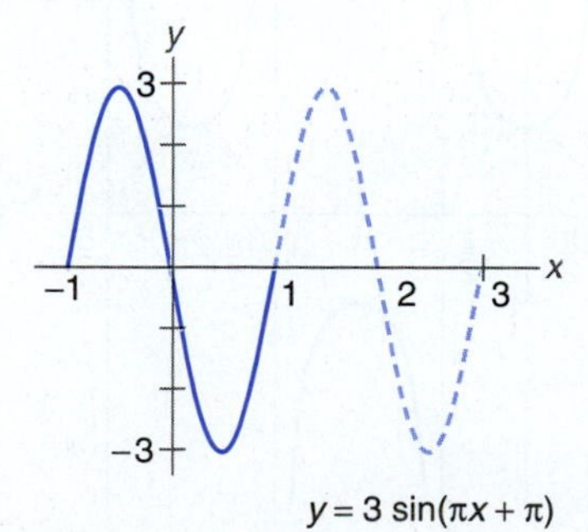

15. 2; 4; −1

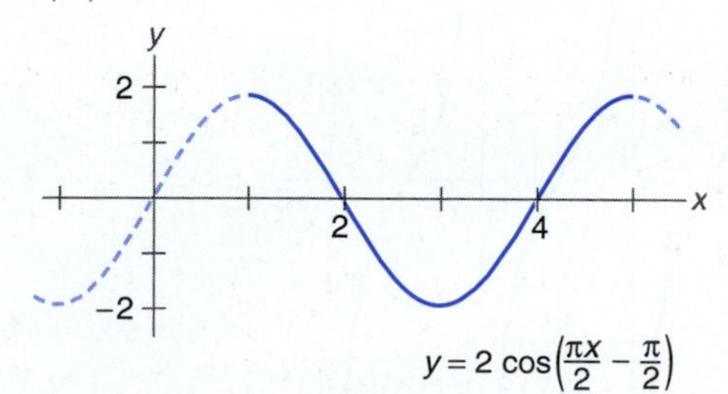

17. $40; \frac{\pi}{30}; -\frac{\pi}{180}$

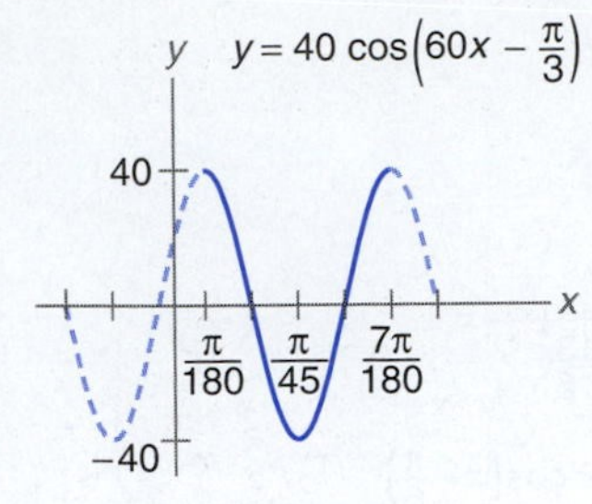

19. $120; \frac{1}{20}; -\frac{1}{80}$

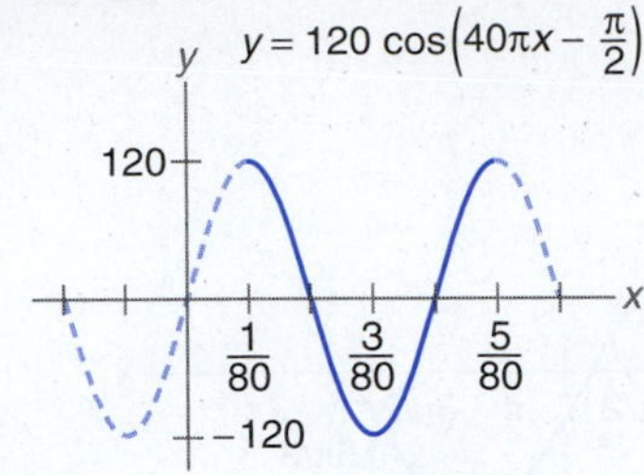

21. $7.5; \frac{\pi}{110}; \frac{\pi}{220}$

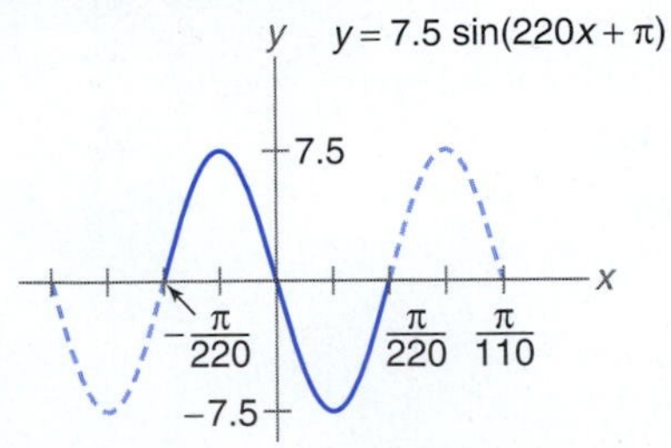

23. $20; \frac{1}{60}; \frac{1}{30}$

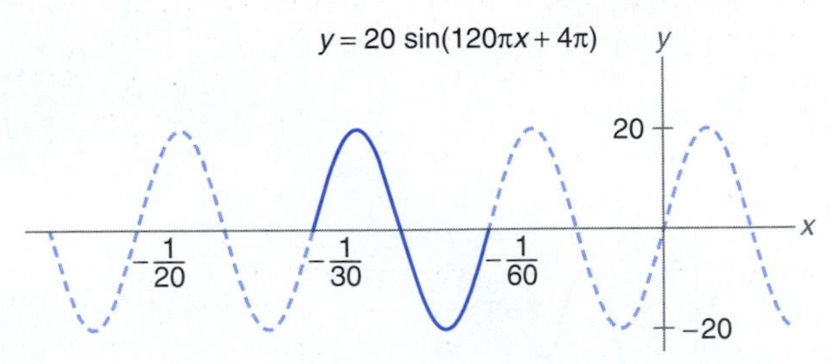

Exercises 12.3, Page 433

1. $\frac{\pi}{3}$

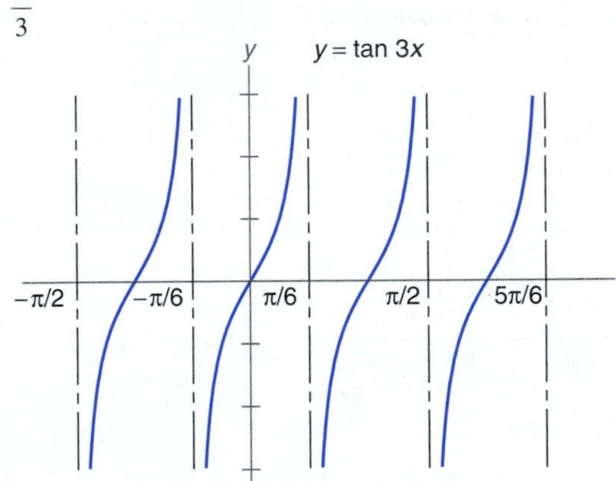

3. 2π

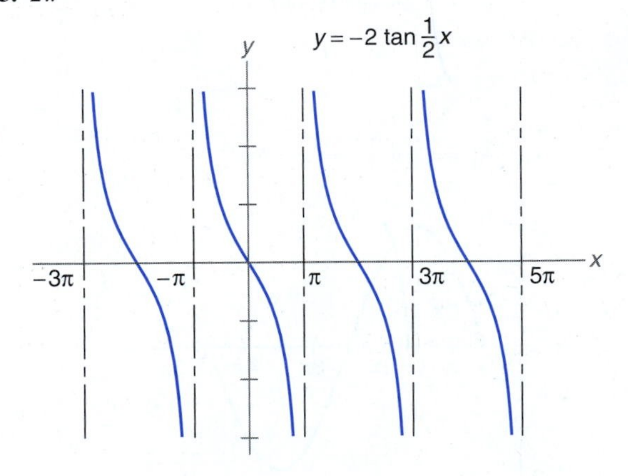

5. $\frac{\pi}{6}$

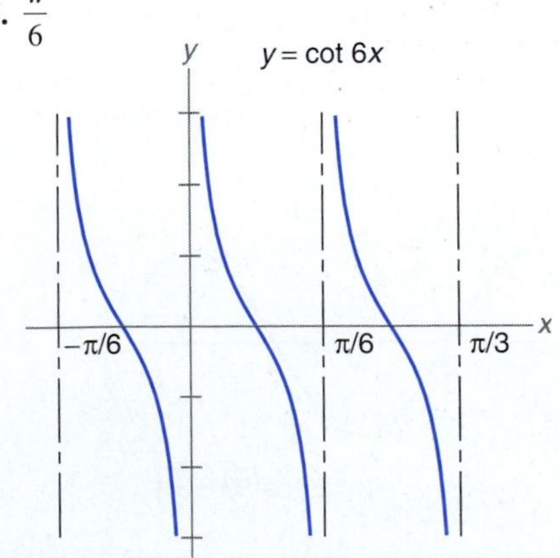

7. 2π

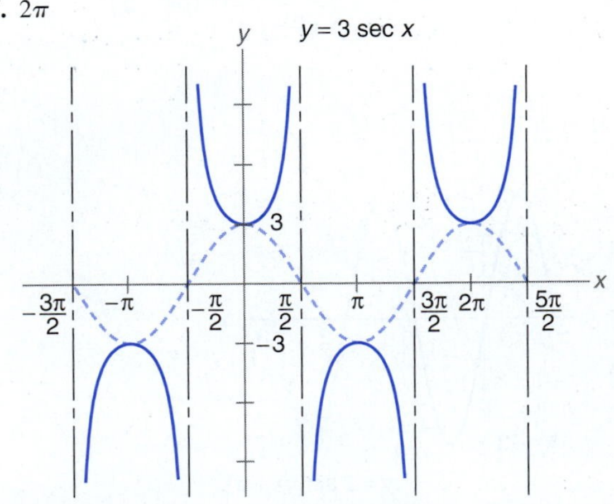

9. 2π

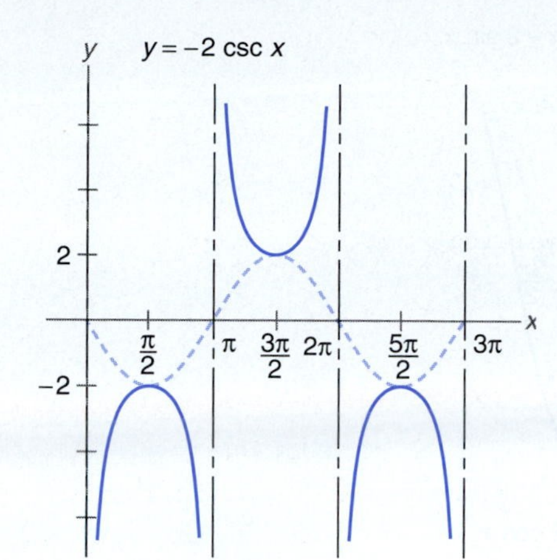

11. $\frac{\pi}{3}$

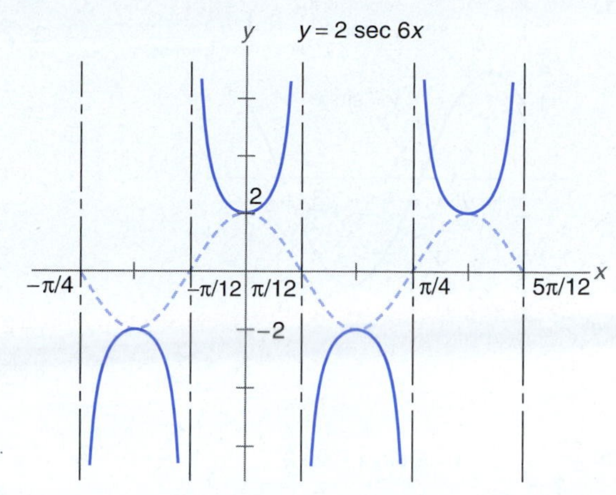

13. 4π

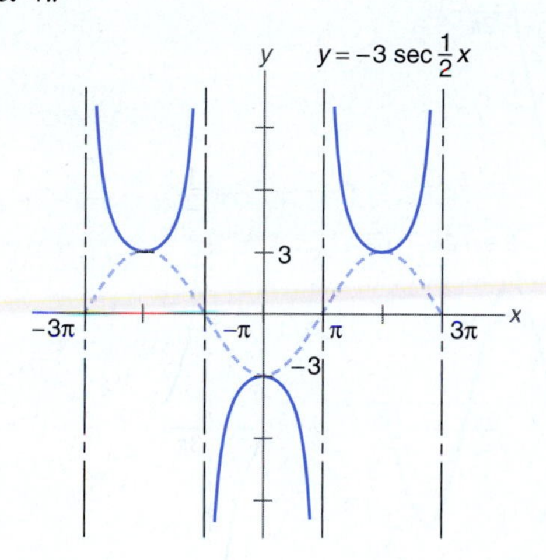

15. π

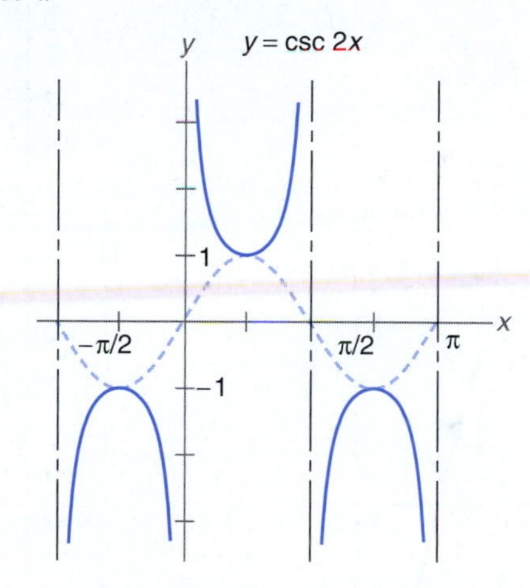

17.

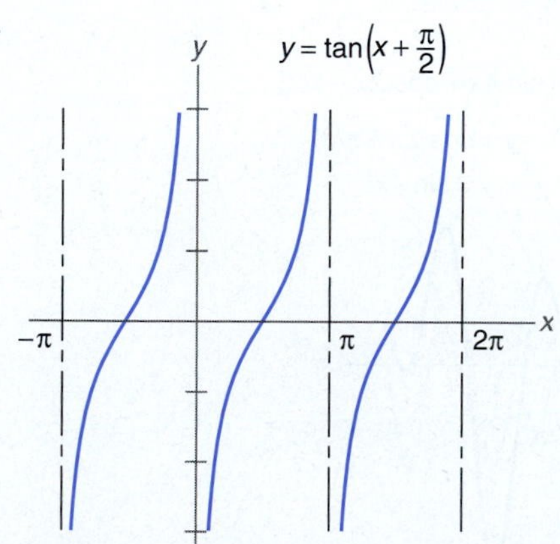

19.

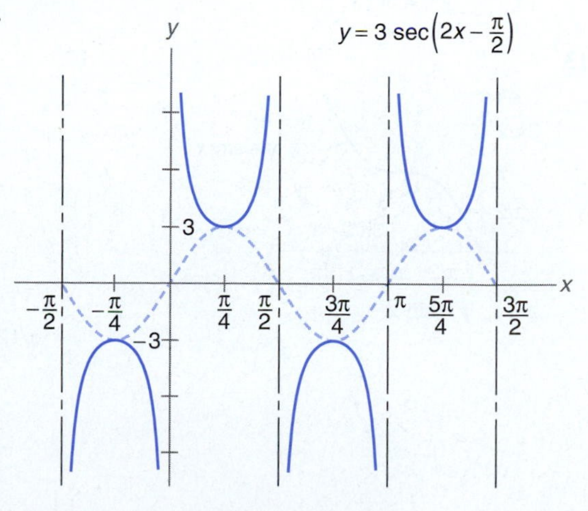

Exercises 12.4, Page 435

1.

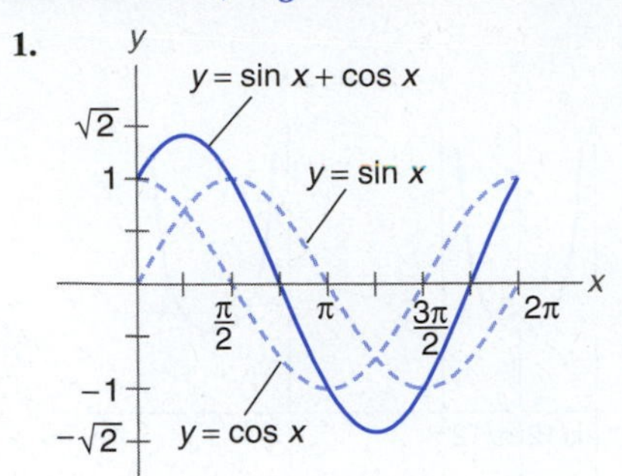

3.

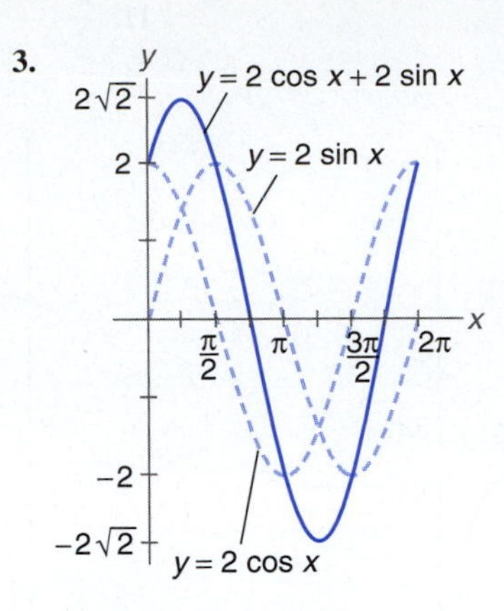

5.

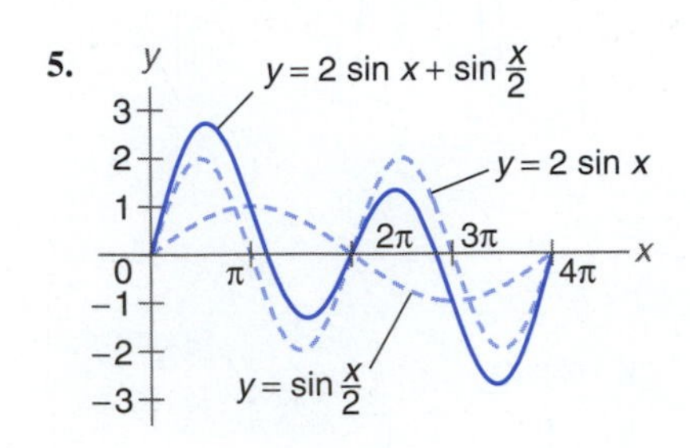

7.

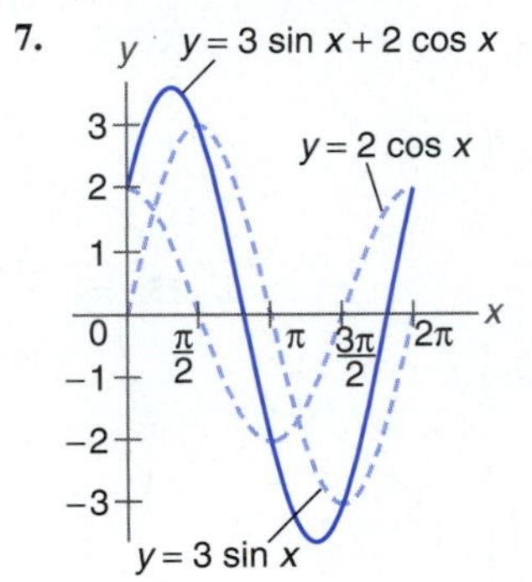

9.

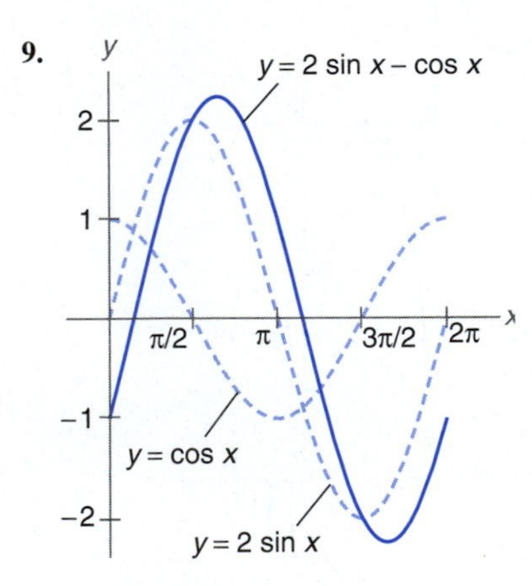

11.

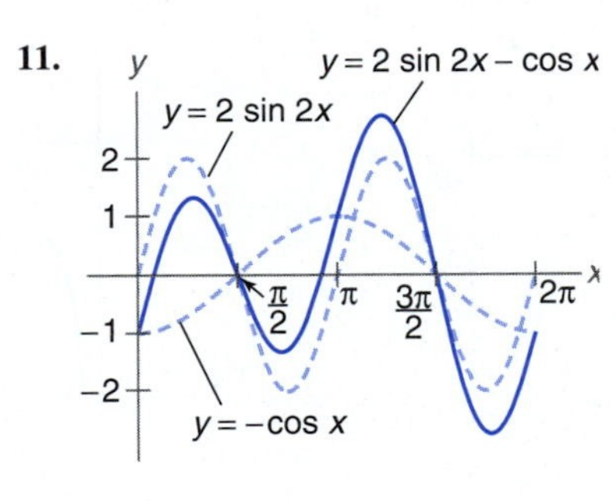

13.

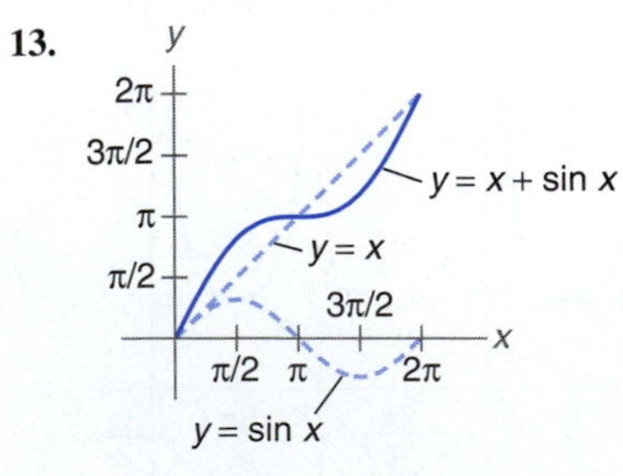

15.

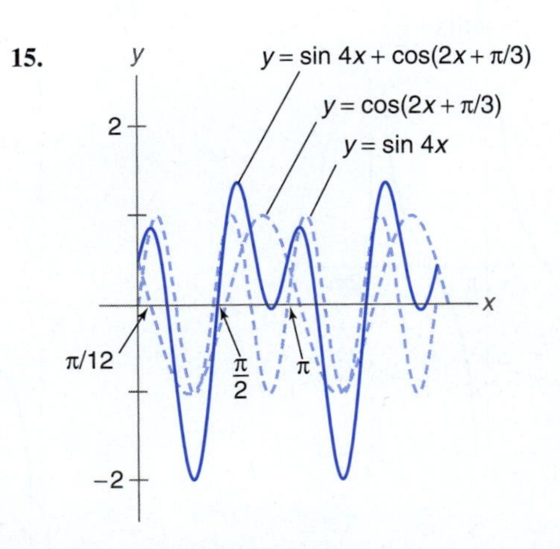

Exercises 12.5, Page 440

1.

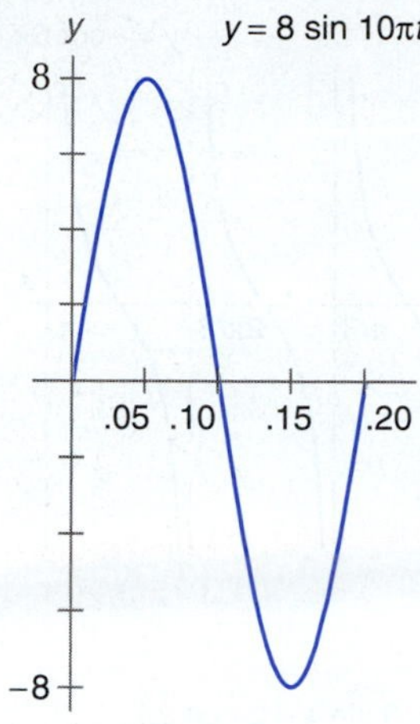

3.

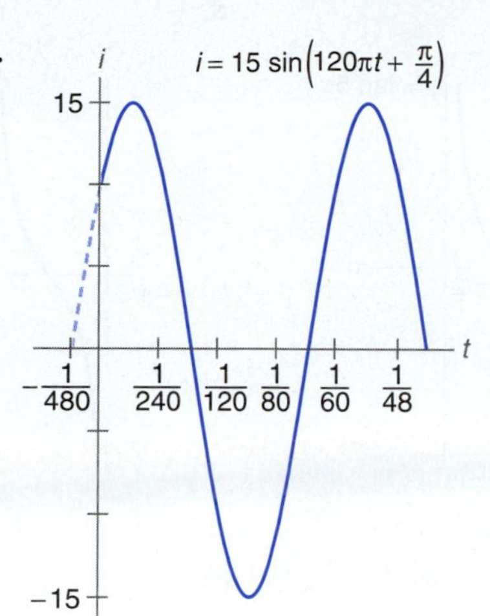

5.

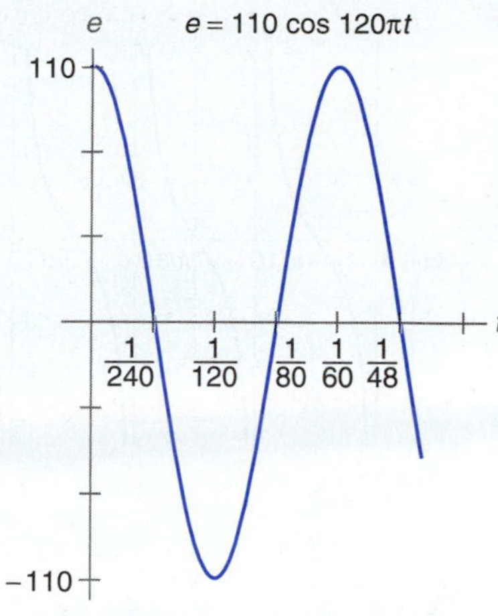

7.

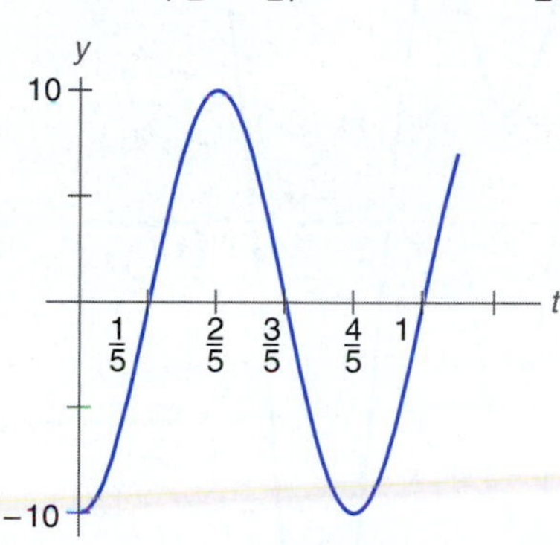

9.

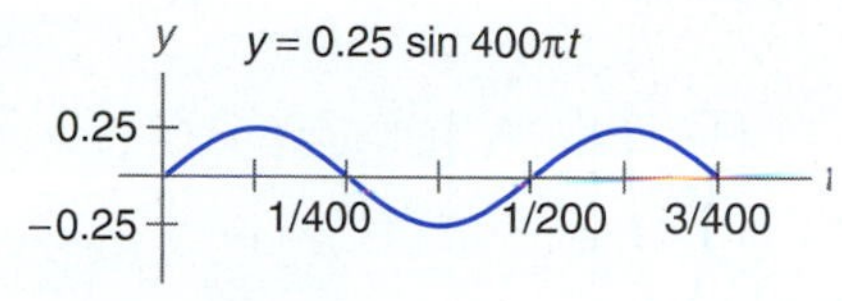

Chapter 12 Review, Pages 441–442

1. 4; $\frac{\pi}{3}$

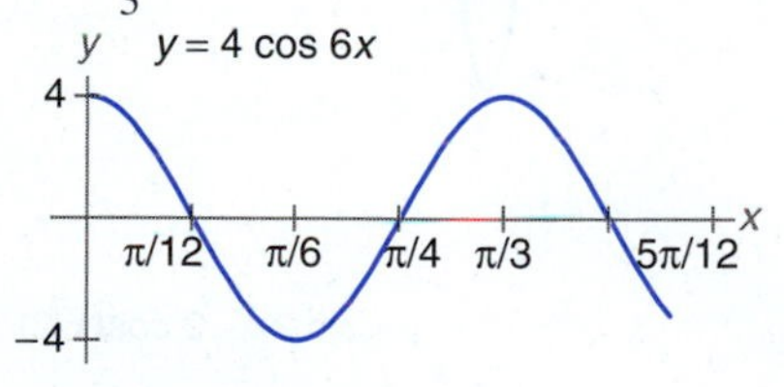

2. 2; 6π

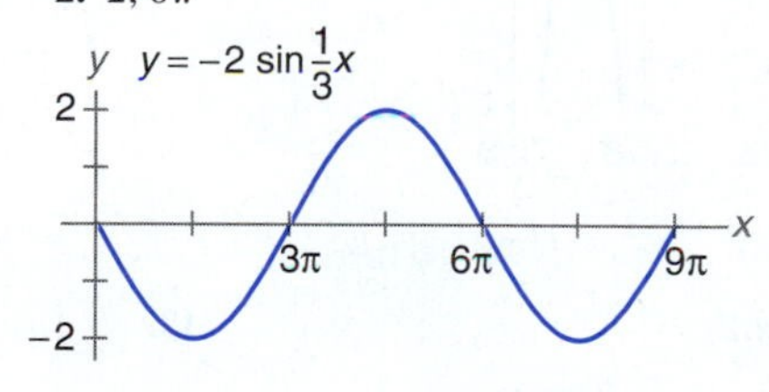

3. 3; 1

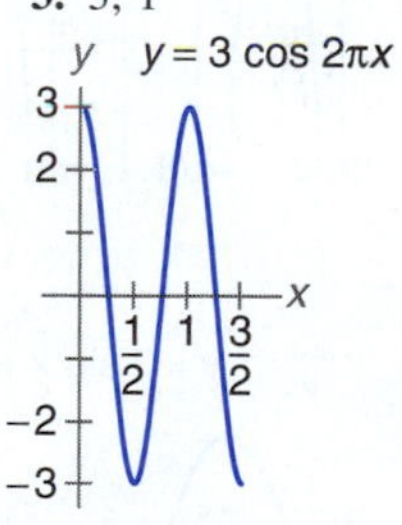

4. 3; 2π; $-\frac{\pi}{4}$

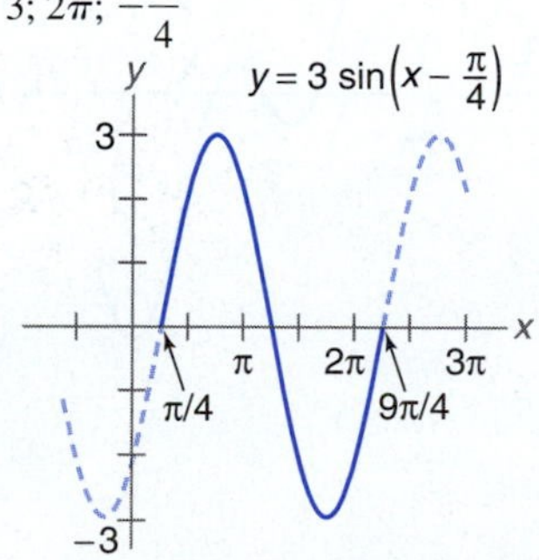

5. 1; π; $\frac{\pi}{3}$

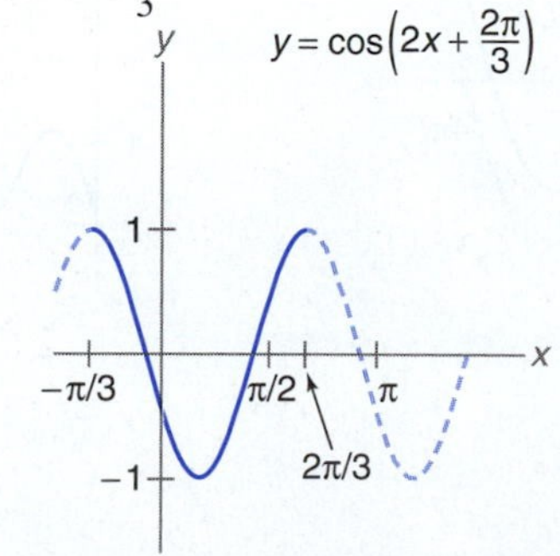

6. 4; 2; $\frac{1}{2}$

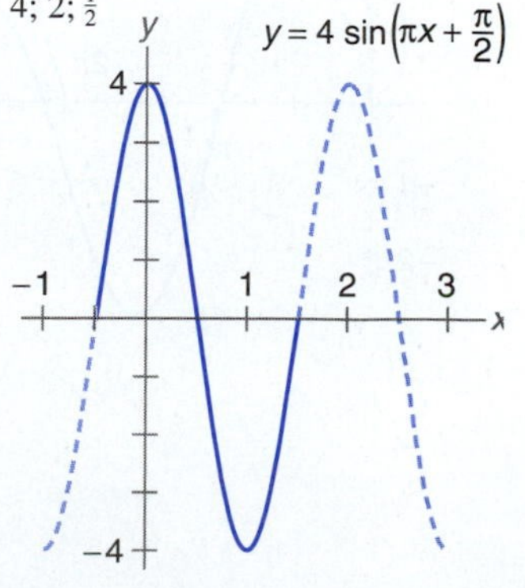

7. $\frac{\pi}{5}$

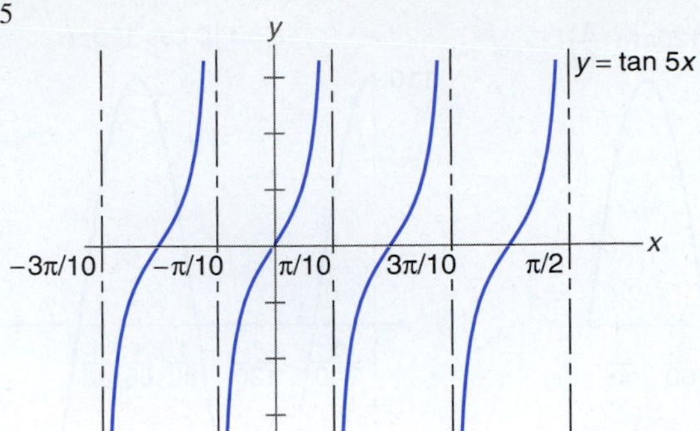

8. $\frac{\pi}{3}$

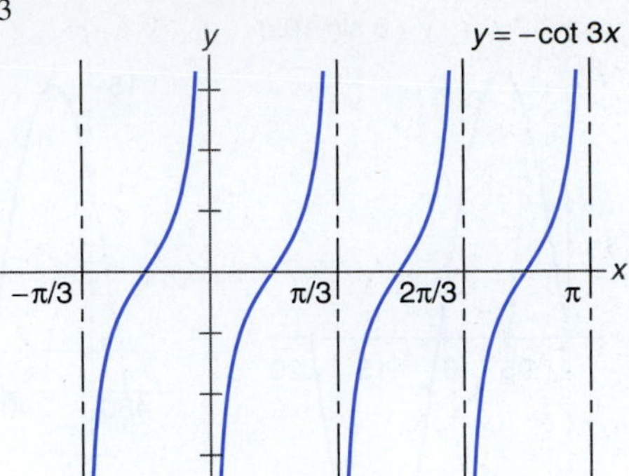

9. $\frac{\pi}{2}$

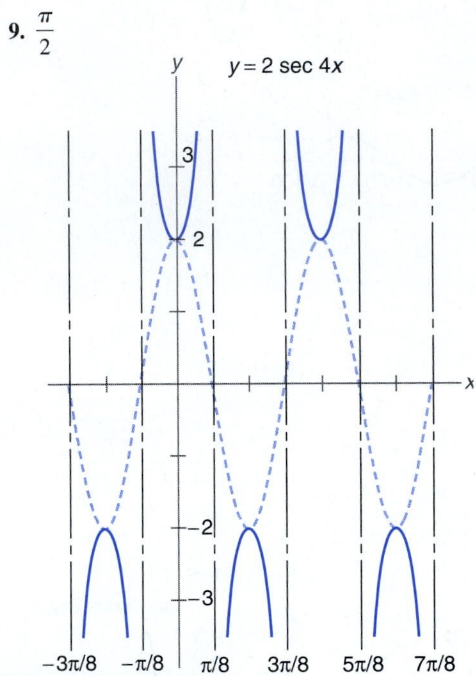

10.

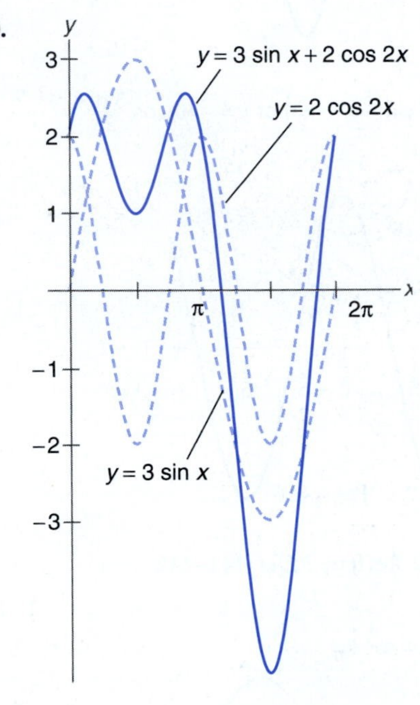

11.

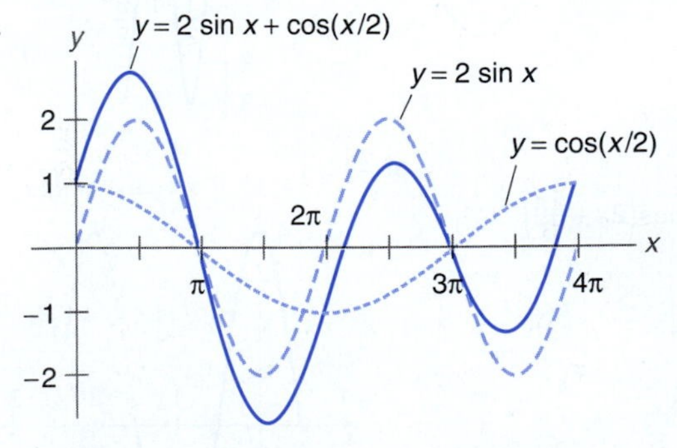

12.

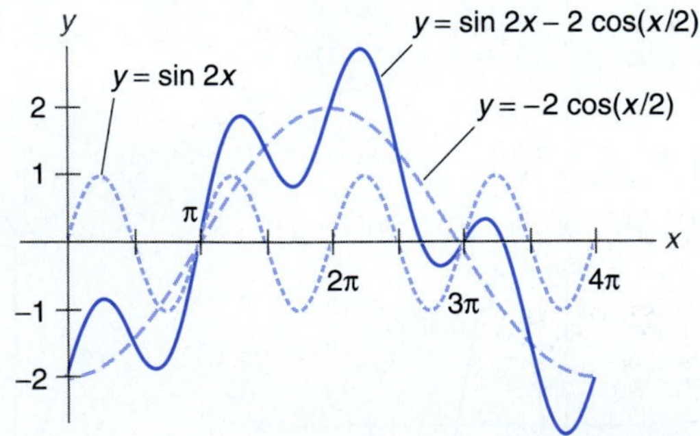

13.

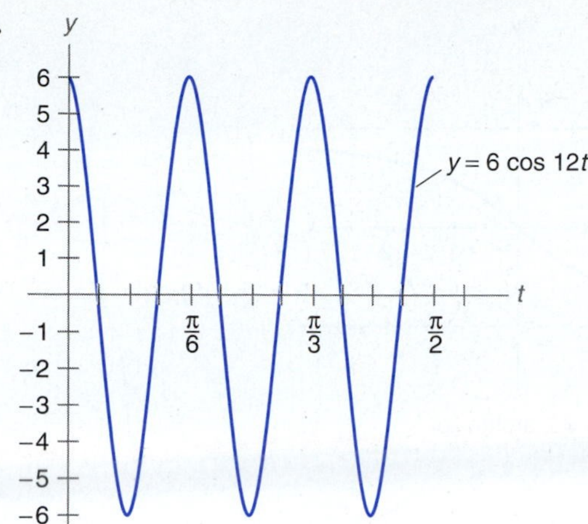

14.

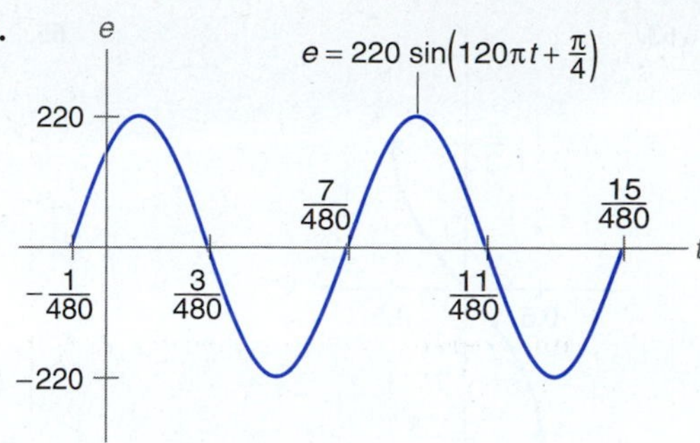

CHAPTER 13

Exercises 13.2, Pages 455–456

21. $\cos(M + N)$ **23.** $\sin 4\theta$ **25.** $\cos\theta$ **27.** $\tan 5\theta$ **29.** $2\sin\theta\cos\phi$ **31.** $\sin(A + 2B)$

Exercises 13.3, Pages 461–462

1. $\sin\frac{x}{2}$ **3.** $\cos 6x$ **5.** $\cos\frac{\theta}{8}$ **7.** $\cos\frac{x}{3}$ **9.** $10\sin 8\theta$ **11.** $\cos 125°$ **13.** $4\cos 2\theta$ **15.** $50\sin 60t$

17. $\frac{24}{25}$ **19.** $\frac{120}{119}$ **21.** $-\sqrt{\frac{3-\sqrt{5}}{6}}$

Exercises 13.4, Page 465

1. 0° **3.** 45°, 225° **5.** 30°, 150°, 210°, 330° **7.** 45°, 225°
9. 10°, 50°, 70°, 110°, 130°, 170°, 190°, 230°, 250°, 290°, 310°, 350° **11.** 90°, 270°, 210°, 330°
13. 0°, 180°, 270° **15.** 21.5°, 158.5° **17.** $\frac{\pi}{2}, \frac{7\pi}{6}, \frac{11\pi}{6}$ **19.** $0, \pi$ **21.** $\frac{\pi}{4}, \frac{3\pi}{4}, \frac{5\pi}{4}, \frac{7\pi}{4}$
23. $\frac{\pi}{4}, \frac{5\pi}{4}, \frac{7\pi}{12}, \frac{11\pi}{12}, \frac{19\pi}{12}, \frac{23\pi}{12}$ **25.** $\frac{\pi}{3}, \frac{2\pi}{3}, \frac{4\pi}{3}, \frac{5\pi}{3}$ **27.** $\frac{\pi}{24}, \frac{5\pi}{24}, \frac{13\pi}{24}, \frac{17\pi}{24}, \frac{25\pi}{24}, \frac{29\pi}{24}, \frac{37\pi}{24}, \frac{41\pi}{24}$
29. $0, \frac{4\pi}{3}$ **31.** $\frac{2\pi}{3}, \pi$ **33.** 0

Exercises 13.5, Pages 471–472

1. y is the angle whose sine is x. **3.** y is the angle whose cotangent is $4x$.
5. y is three times the angle whose cosecant is $\frac{1}{2}x$. **7.** $x = \frac{1}{3}\arcsin y$ **9.** $x = \arccos\frac{y}{4}$ **11.** $x = 2\arctan\frac{y}{5}$
13. $x = 4\,\text{arccot}\,\frac{2y}{3}$ **15.** $x = 1 + \arcsin\frac{y}{3}$ **17.** $x = -\frac{1}{3} + \frac{1}{3}\arccos 2y$ **19.** $\frac{\pi}{3}$ **21.** $-\frac{\pi}{6}$ **23.** $\frac{5\pi}{6}$
25. $\frac{\pi}{4}$ **27.** $\frac{\pi}{3}$ **29.** $\frac{\pi}{4}$ **31.** $-\frac{\pi}{3}$ **33.** 2.554 **35.** 1.889 **37.** −1.054 **39.** $\frac{1}{2}$ **41.** $\frac{1}{\sqrt{2}}$
43. 0 **45.** $\frac{\sqrt{3}}{2}$ **47.** 0.8 **49.** −0.1579 **51.** $\sqrt{1-x^2}$ **53.** $\frac{\sqrt{x^2-1}}{x}$ **55.** $\frac{1}{x}$ **57.** x
59. $\sqrt{1-4x^2}$ **61.** $2x\sqrt{1-x^2}$

63.

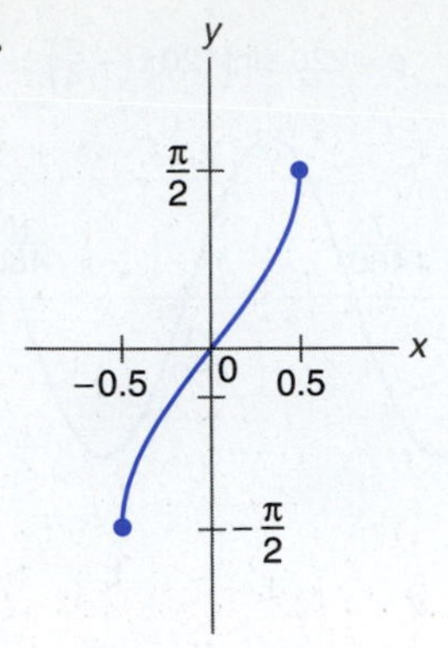

y = arcsin 2x

65.

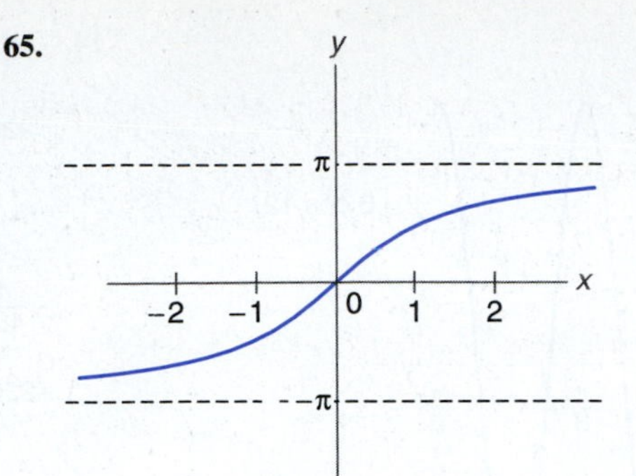

y = 2 arctan 3x

Chapter 13 Review, Pages 474–475

11. $\frac{1}{2}\sin 2\theta$ **12.** $\cos 6\theta$ **13.** $\cos^2 2\theta$ **14.** $\cos\frac{2\theta}{3}$ **15.** $\cos 5x$ **16.** $\sin x$ **17.** $-\frac{120}{169}$ **18.** $-\frac{\sqrt{5}}{5}$

19. $\frac{\pi}{2}, \frac{3\pi}{2}, \frac{\pi}{3}, \frac{5\pi}{3}$ **20.** $\frac{\pi}{6}, \frac{5\pi}{6}, \frac{7\pi}{6}, \frac{11\pi}{6}$ **21.** $\frac{\pi}{6}, \frac{5\pi}{6}, \frac{3\pi}{2}, \frac{7\pi}{18}, \frac{11\pi}{18}, \frac{19\pi}{18}, \frac{23\pi}{18}, \frac{31\pi}{18}, \frac{35\pi}{18}$ **22.** $0, \pi$ **23.** $\frac{3\pi}{2}$

24. $\frac{\pi}{8}, \frac{5\pi}{8}, \frac{9\pi}{8}, \frac{13\pi}{8}$ **25.** $x = \frac{4}{3}\arcsin 2y$ **26.** $\frac{\pi}{4}$ **27.** $-\frac{\pi}{6}$ **28.** π **29.** $\frac{2\pi}{3}$ **30.** $\frac{\sqrt{3}}{2}$ **31.** $\sqrt{3}$

32. 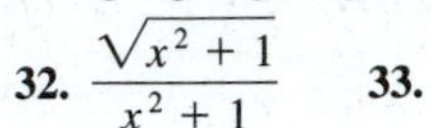 $\frac{\sqrt{x^2+1}}{x^2+1}$ **33.**

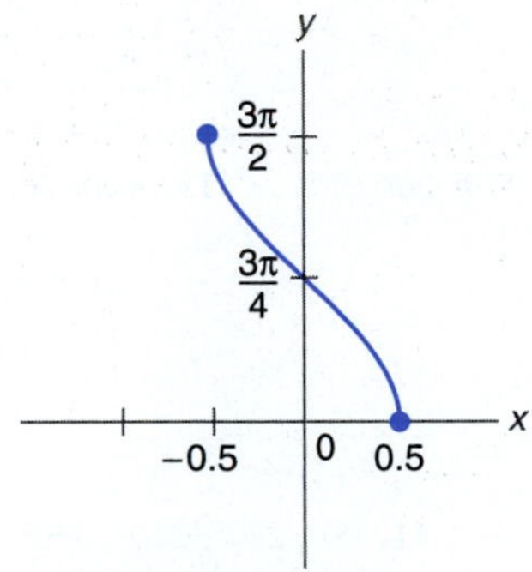

y = 1.5 arccos 2x

CHAPTER 14

Exercises 14.1, Pages 481–482

1. $7j$ **3.** $8j$ **5.** $2j\sqrt{3}$ **7.** $3j\sqrt{6}$ **9.** $-j$ **11.** -1 **13.** j **15.** -1 **17.** $12 + 6j$ **19.** $2 - 8j$

21. $-j$ **23.** $19 + 2j$ **25.** $-22 + 7j$ **27.** 29 **29.** $-7 - 24j$ **31.** $\frac{5 + 31j}{17}$ **33.** $-\frac{3j}{4}$ **35.** $\frac{-7 + 3j}{4}$

37. $2j, -2j$ **39.** $\frac{-2 \pm j\sqrt{23}}{3}$ **41.** $\frac{1 \pm 2j\sqrt{6}}{5}$ **43.** $\frac{1 \pm j\sqrt{2}}{3}$ **45.** $-1, \frac{1 \pm j\sqrt{3}}{2}$ **47.** $1, -1, j, -j$

49. $0, \pm 4j\sqrt{5}$ **51.** $0, -3, \frac{3 \pm 3j\sqrt{3}}{2}$ **53.** $0, 1, \frac{-1 \pm j\sqrt{3}}{2}$ **57.**

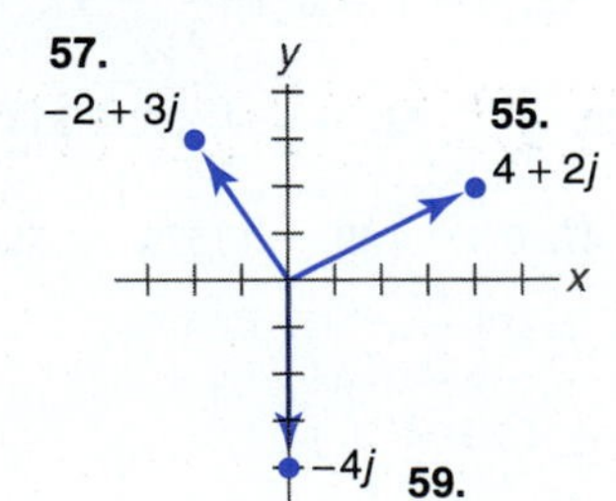

61. $1 + 3j$ **63.** $4 - 8j$ **65.** $3 - 6j$ **67.** $7j$

Exercises 14.2, Pages 485–486

1. $2\sqrt{2}(\cos 45° + j \sin 45°)$ **3.** $2(\cos 240° + j \sin 240°)$ **5.** $4(\cos 90° + j \sin 90°)$ **7.** $6\sqrt{2}(\cos 225° + j \sin 225°)$ **9.** $\sqrt{13}(\cos 124° + j \sin 124°)$ **11.** $2 + 2j\sqrt{3}$ **13.** $\sqrt{3} - j$ **15.** $-3 + 3j$ **17.** $-3j$ **19.** $7.00 - 2.01j$ **21.** $2e^{5.76j}$ **23.** $2e^{3.93j}$ **25.** $2\sqrt{13}e^{0.983j}$ or $7.21e^{0.983j}$ **27.** $3(\cos 77° + j \sin 77°) = 0.67 + 2.92j$ **29.** $4(\cos 330° + j \sin 330°) = 3.46 - 2j$ **31.** $2(\cos 57° + j \sin 57°) = 1.09 + 1.68j$

Exercises 14.3, Pages 489–490

1. $28e^{4j}$ **3.** $27e^{2j}$ **5.** $24e^{9.3j}$ or $24e^{3.0j}$ **7.** $-12 - 12\sqrt{3}j$ **9.** $-74.7 - 28.7j$ **11.** $-39.5 + 36.8j$ **13.** $12(\cos 113° + j \sin 113°)$ **15.** $9(\cos 300° + j \sin 300°)$ **17.** $7(\cos 135° + j \sin 135°)$ **19.** $\frac{25}{2} + \frac{25\sqrt{3}}{2}j$ **21.** -12 **23.** $54.1 + 14.5j$ **25.** $\frac{e^{4j}}{3}$ **27.** $4e^{-7j}$ or $4e^{5.57j}$ **29.** $\frac{e^{-2.2j}}{3}$ or $\frac{e^{4.1j}}{3}$ **31.** $-\sqrt{3} - j$ **33.** $-6.91 - 1.10j$ **35.** $3.94 - 3.08j$ **37.** $5(\cos 70° + j \sin 70°)$ **39.** $6(\cos 119° + j \sin 119°)$ **41.** $8(\cos 231° + j \sin 231°)$ **43.** $-\frac{9}{2} + \frac{9\sqrt{3}}{2}j$ **45.** $-\frac{\sqrt{3}}{10} + \frac{1}{10}j$ **47.** $3 - 3\sqrt{3}j$ **49.** -8 **51.** $-81j$

Exercises 14.4, Page 495

1. $243e^{7j}$ or $243e^{0.72j}$ **3.** $25e^{9.2j}$ or $25e^{2.9j}$ **5.** $81(\cos 80° + j \sin 80°)$ **7.** $32(\cos 30° + j \sin 30°)$ **9.** $\frac{1}{8}(\cos 0° + j \sin 0°)$ or $\frac{1}{8}$ **11.** 16 **13.** $-64j$ **15.** $16 - 16\sqrt{3}j$ **17.** $-\frac{1}{324}$ **19.** $1, -\frac{1}{2} + \frac{\sqrt{3}}{2}j, -\frac{1}{2} - \frac{\sqrt{3}}{2}j$ **21.** $0.951 + 0.309j, j, -0.951 + 0.309j, -0.588 - 0.809j, 0.588 - 0.809j$ **23.** $3(\cos 135° + j \sin 135°) = -2.12 + 2.12j, 3(\cos 255° + j \sin 255°) = -0.776 - 2.90j,$ $3(\cos 15° + j \sin 15°) = 2.90 + 0.776j$ **25.** $\cos 45° + j \sin 45° = \frac{\sqrt{2}}{2} + \frac{\sqrt{2}}{2}j, \cos 225° + j \sin 225° = -\frac{\sqrt{2}}{2} - \frac{\sqrt{2}}{2}j$ **27.** $\cos 0° + j \sin 0° = 1, \cos 72° + j \sin 72° = 0.309 + 0.951j, \cos 144° + j \sin 144° = -0.809 + 0.588j,$ $\cos 216° + j \sin 216° = -0.809 - 0.588j, \cos 288° + j \sin 288° = 0.309 - 0.951j$ **29.** $\cos 36° + j \sin 36° = 0.809 + 0.588j, \cos 108° + j \sin 108° = -0.309 + 0.951j, \cos 180° + j \sin 180° = -1,$ $\cos 252° + j \sin 252° = -0.309 - 0.951j, \cos 324° + j \sin 324° = 0.809 - 0.588j$ **31.** $2(\cos 45° + j \sin 45°) = \sqrt{2} + \sqrt{2}j, 2(\cos 135° + j \sin 135°) = -\sqrt{2} + \sqrt{2}j,$ $2(\cos 225° + j \sin 225°) = -\sqrt{2} - \sqrt{2}j, 2(\cos 315° + j \sin 315°) = \sqrt{2} - \sqrt{2}j$ **33.** $\sqrt{5}(\cos 33° + j \sin 33°) = 1.88 + 1.22j, \sqrt{5}(\cos 153° + j \sin 153°) = -1.99 + 1.02j,$ $\sqrt{5}(\cos 273° + j \sin 273°) = 0.117 - 2.23j$

Chapter 14 Review, Page 497

1. $9j$ **2.** $3\sqrt{2}j$ **3.** -1 **4.** $-j$ **5.** 1 **6.** j **7.** $5 + 10j$ **8.** $3 - 4j$ **9.** $44 - 23j$ **10.** $5 - 12j$ **11.** $\frac{2}{17} - \frac{9}{17}j$ **12.** $\frac{4}{29} + \frac{19}{29}j$ **13.** $-\frac{2}{5} + \frac{11}{5}j$ **14.** $\frac{19}{10} - \frac{13}{10}j$ **15.** $6j, -6j$ **16.** $\frac{-3 \pm \sqrt{7}j}{4}$ **17.** $\frac{5}{6} \pm \frac{\sqrt{23}}{6}j$ **18.** $\pm 5j$ **19.** $\sqrt{2}(\cos 135° + j \sin 135°), \sqrt{2}e^{2.36j}$ **20.** $2(\cos 300° + j \sin 300°), 2e^{5.24j}$ **21.** $\sqrt{34}(\cos 301° + j \sin 301°), \sqrt{34}e^{5.25j}$ **22.** $2\sqrt{17}(\cos 256° + j \sin 256°), 2\sqrt{17}e^{4.47j}$ **23.** $3\sqrt{2} - 3\sqrt{2}j, 6e^{5.50j}$ **24.** $-2\sqrt{3} - 2j, 4e^{3.67j}$ **25.** $-1.43 + 5.33j, 5.52e^{1.83j}$ **26.** $-1.90 - 6.22j, 6.50e^{4.42j}$ **27.** $2(\cos 28° + j \sin 28°) = 1.77 + 0.939j$ **28.** $3(\cos 141° + j \sin 141°) = -2.33 + 1.89j$ **29.** $7(\cos 84° + j \sin 84°), 0.732 + 6.96j$ **30.** $8(\cos 132° + j \sin 132°), -5.35 + 5.95j$ **31.** $15e^{5j}$ **32.** $8(\cos 90° + j \sin 90°)$ **33.** $4e^{5j}$ **34.** $3(\cos 235° + j \sin 235°)$ **35.** $64e^{6j}$ **36.** $128(\cos 60° + j \sin 60°)$ **37.** -64 **38.** 64 **39.** $-\frac{1}{4}$ **40.** $\cos 30° + j \sin 30° = \frac{\sqrt{3}}{2} + \frac{1}{2}j, \cos 150° + j \sin 150° = -\frac{\sqrt{3}}{2} + \frac{1}{2}j, \cos 270° + j \sin 270° = -j$ **41.** $\cos 45° + j \sin 45° = \frac{\sqrt{2}}{2} + \frac{\sqrt{2}}{2}j, \cos 135° + j \sin 135° = -\frac{\sqrt{2}}{2} + \frac{\sqrt{2}}{2}j, \cos 225° + j \sin 225° = -\frac{\sqrt{2}}{2} - \frac{\sqrt{2}}{2}j,$ $\cos 315° + j \sin 315° = \frac{\sqrt{2}}{2} - \frac{\sqrt{2}}{2}j$ **42.** $2(\cos 0° + j \sin 0°) = 2, 2(\cos 90° + j \sin 90°) = 2j, 2(\cos 180° + j \sin 180°) = -2, 2(\cos 270° + j \sin 270°) = -2j$ **43.** $\sqrt{2}(\cos 63° + j \sin 63°) = 0.642 + 1.26j, \sqrt{2}(\cos 135° + j \sin 135°) = -1 + j,$ $\sqrt{2}(\cos 207° + j \sin 207°) = -1.26 - 0.642j, \sqrt{2}(\cos 279° + j \sin 279°) = 0.221 - 1.40j,$ $\sqrt{2}(\cos 351° + j \sin 351°) = 1.40 - 0.221j$

CHAPTER 15

Exercises 15.1, Pages 501–503

1. 2×2 **3.** 2×4 **5.** 4×1 **7.** 1×2 **9.** $x = 2, y = 4, z = -5$
11. $x = 1, y = 3, z = -4, a = 4, b = 12, c = -12$ **13.** $x = 4, y = -6$
15. $x = 3, y = 5, w = 6, z = 10$

17. $\begin{bmatrix} 8 & 4 & -3 \\ -4 & 8 & 16 \end{bmatrix}$ **19.** $\begin{bmatrix} 5 & -10 & 8 \\ 4 & -7 & 1 \\ 10 & 1 & -14 \end{bmatrix}$ **21.** $\begin{bmatrix} -1 & -10 & -6 \\ -5 & 2 & 0 \end{bmatrix}$ **23.** $\begin{bmatrix} 3 & 30 & 18 \\ 15 & -6 & 0 \end{bmatrix}$ **25.** $\begin{bmatrix} -2 & 18 \\ -8 & -6 \end{bmatrix}$

27. $\begin{bmatrix} 1 & 8 & -2 \\ 9 & 5 & 5 \end{bmatrix}$ **29.** $\begin{bmatrix} -1 & -12 & -14 \\ -1 & 9 & 5 \end{bmatrix}$ **31.** Not defined **33.** $\begin{bmatrix} 3 & 38 & 50 \\ -1 & -34 & -20 \end{bmatrix}$ **35.** Not defined

43. $\begin{bmatrix} 170 & 130 & 96 \\ 140 & 120 & 50 \\ 68 & 50 & 30 \end{bmatrix}$ **45.** $\begin{bmatrix} 3 & 8 & 6 & 6 \\ 2 & 5 & 1 & 10 \\ 4 & 16 & 8 & 2 \\ 7 & 5 & 2 & 1 \end{bmatrix}$

Exercises 15.2, Pages 507–508

1. [15] **3.** [0 15] **5.** $\begin{bmatrix} -3 & 6 \\ -2 & 9 \end{bmatrix}$ **7.** $\begin{bmatrix} 7 & 9 & -5 \\ 20 & 0 & 8 \end{bmatrix}$ **9.** $\begin{bmatrix} 25 \\ 57 \end{bmatrix}$ **11.** Not defined **13.** $\begin{bmatrix} 6 & -11 & 8 \\ 2 & -11 & -12 \\ 10 & -15 & 20 \end{bmatrix}$

15. $\begin{bmatrix} -10 & 41 & 3 & 21 \\ 30 & 21 & -27 & -9 \\ -8 & 20 & 4 & 12 \\ -18 & 13 & 13 & 15 \end{bmatrix}$ **17.** Not defined **19.** $\begin{bmatrix} -9 & -8 & -18 \\ 27 & 3 & -22 \\ -9 & -1 & -10 \end{bmatrix}$ **21.** $\begin{bmatrix} -7 \\ 31 \end{bmatrix}$

23. **(a)** $\begin{bmatrix} 10 & 7 \\ 5 & -2 \end{bmatrix}$ **(b)** $\begin{bmatrix} -4 & 1 \\ 7 & 12 \end{bmatrix}$ **(c)** $\begin{bmatrix} 1 & 8 \\ -16 & -7 \end{bmatrix}$

25. **(a)** $\begin{bmatrix} 0 & -9 & 33 \\ -4 & 7 & 24 \\ 12 & -36 & 45 \end{bmatrix}$ **(b)** $\begin{bmatrix} 7 & 2 & 12 \\ 19 & 2 & 24 \\ 40 & -36 & 43 \end{bmatrix}$ **(c)** $\begin{bmatrix} 16 & 4 & 23 \\ 5 & 14 & 10 \\ 35 & -10 & 49 \end{bmatrix}$

27. **(a)** $\begin{bmatrix} 8 & 7 \\ -13 & 0 \end{bmatrix}$ **(b)** Not defined **(c)** Not defined

Exercises 15.3, Pages 515–516

1. $\begin{bmatrix} 2 & -3 \\ -\frac{1}{2} & 1 \end{bmatrix}$ **3.** $\begin{bmatrix} -2 & 1 \\ -\frac{11}{2} & \frac{5}{2} \end{bmatrix}$ **5.** $\begin{bmatrix} -\frac{5}{18} & -\frac{2}{9} \\ \frac{1}{6} & \frac{1}{3} \end{bmatrix}$ **7.** $\begin{bmatrix} 1 & 1 \\ \frac{5}{2} & 2 \end{bmatrix}$ **9.** $\begin{bmatrix} -3 & 4 \\ 1 & -1 \end{bmatrix}$ **11.** $\begin{bmatrix} \frac{1}{2} & -\frac{1}{4} \\ \frac{1}{6} & \frac{1}{12} \end{bmatrix}$ **13.** $\begin{bmatrix} \frac{5}{2} & 2 \\ \frac{3}{2} & 1 \end{bmatrix}$

15. $\begin{bmatrix} -2\frac{1}{3} & 2 \\ 1\frac{1}{3} & -1 \end{bmatrix}$ **17.** $\begin{bmatrix} 0 & -\frac{3}{4} & \frac{1}{2} \\ -\frac{1}{2} & \frac{1}{8} & \frac{1}{4} \\ \frac{1}{4} & \frac{3}{16} & -\frac{1}{8} \end{bmatrix}$ **19.** $\begin{bmatrix} 2 & -3 & 1 \\ -6 & 9 & -2 \\ -3 & 5 & -1 \end{bmatrix}$ **21.** $\begin{bmatrix} \frac{1}{2} & \frac{1}{4} & -\frac{1}{4} \\ 4\frac{1}{2} & -\frac{1}{4} & -1\frac{3}{4} \\ -2 & 0 & 1 \end{bmatrix}$ **23.** $\begin{bmatrix} \frac{1}{2} & 1 & 1\frac{1}{2} \\ 1 & 2\frac{1}{2} & 3\frac{1}{2} \\ 1\frac{1}{2} & 3\frac{1}{2} & 4 \end{bmatrix}$

25. $\begin{bmatrix} -\frac{1}{3} & \frac{1}{3} & -\frac{2}{3} & 0 \\ -3 & \frac{1}{2} & -2 & 1 \\ \frac{2}{3} & -\frac{1}{6} & \frac{1}{3} & 0 \\ \frac{1}{3} & \frac{1}{6} & \frac{2}{3} & 0 \end{bmatrix}$ **27.** $\begin{bmatrix} 1 & 0 & 0 & -\frac{1}{2} \\ \frac{7}{10} & -\frac{2}{5} & \frac{1}{2} & -\frac{1}{2} \\ \frac{1}{4} & 0 & \frac{1}{4} & -\frac{1}{4} \\ -\frac{3}{5} & \frac{1}{5} & 0 & \frac{1}{2} \end{bmatrix}$ **29.** (2, 5) **31.** (−4, −1) **33.** (1, 8) **35.** (−3, 5)

37. (−3, 4, −6) **39.** (3, −2, 12) **41.** (8, −5, 10, 3)

Exercises 15.4, Pages 519–520

1. (3, −4) **3.** (17, −2, 5) **5.** (3, 15, −6, 10) **7.** (2, 3) **9.** (−4, 8) **11.** (3, −5) **13.** (7, 7)
15. (1, 2, −4) **17.** (3, −2, 5) **19.** Dependent system **21.** (3, 4, 0, −1) **23.** (3, −2, 4, 5)
25. 7 in., 13 in., 14 in.

Chapter 15 Review, Pages 521–523

1. 2×3 **2.** 3×1 **3.** 3×4 **4.** $x = 12, y = -3, z = 10$ **5.** $x = 1, y = 3, z = 4$

6. $\begin{bmatrix} 7 & 7 & 7 \\ 0 & 15 & -1 \end{bmatrix}$ 7. $\begin{bmatrix} 0 & 9 \\ -1 & -13 \\ 2 & 17 \end{bmatrix}$ 8. $\begin{bmatrix} -2 & 4 \\ -8 & -11 \end{bmatrix}$ 9. $\begin{bmatrix} 6 & -9 & 15 \\ 21 & 0 & -3 \end{bmatrix}$ 10. $\begin{bmatrix} 8 & -5 & 0 \\ 17 & 4 & 7 \end{bmatrix}$

11. Not defined 12. $\begin{bmatrix} 4 & 1 & -10 \\ 3 & 4 & 9 \end{bmatrix}$ 13. $\begin{bmatrix} 22 & -12 & -5 \\ 44 & 12 & 22 \end{bmatrix}$ 14. $\begin{bmatrix} -2 & -11 & 35 \\ 15 & -8 & -21 \end{bmatrix}$ 15. Not defined

16. $\begin{bmatrix} -5 & 2 \\ -32 & 18 \end{bmatrix}$ 17. $\begin{bmatrix} -13 & -1 & 8 \\ -23 & 10 & -12 \\ -6 & 0 & 2 \end{bmatrix}$ 18. Not defined 19. $\begin{bmatrix} 5 & 3 & -7 \\ -1 & 12 & 9 \end{bmatrix}$ 20. $\begin{bmatrix} -2 & 6 \\ -3 & 7 \end{bmatrix}$

21. $\begin{bmatrix} 5 & 2 \\ -8 & -3 \end{bmatrix}$ 22. $\begin{bmatrix} \frac{3}{4} & -\frac{5}{4} \\ 1 & -2 \end{bmatrix}$ 23. $\begin{bmatrix} \frac{1}{3} & -1 & \frac{1}{3} \\ -\frac{1}{6} & 1 & -\frac{1}{6} \\ -\frac{1}{2} & 1 & 0 \end{bmatrix}$ 24. $\begin{bmatrix} 0 & -\frac{5}{3} & \frac{2}{3} \\ 0 & \frac{4}{3} & -\frac{1}{3} \\ -1 & \frac{2}{3} & \frac{1}{3} \end{bmatrix}$ 25. $\begin{bmatrix} 2 & 1 & -1 & -2 \\ -8 & -5 & 6 & 12 \\ 1 & 1 & -1 & -2 \\ -5 & -3 & 3 & 7 \end{bmatrix}$

26. (3, 4) 27. (5, −2) 28. (1, −2, 4) 29. (3, 2, 2) 30. (3, 1, −2, 4) 31. (−6, 10) 32. (4, −3, 15)
33. (4, −11, 8, 0) 34. (3, 3) 35. (4, −2) 36. (1, 1, −4) 37. (3, −6, 8) 38. (1, 0, −2, 3)

CHAPTER 16

Exercises 16.1, Page 531

1. x: 0, 2, 5; y: 0 3. x: 1, 2.5, 4; y: −20 5. x: −3, 0, 3; y: 0 7. $f(x) = (x + 4)(x + 3)(x - 2)$
9. $f(x) = (x + 3)(x - 1)(x - 3)$ 11. $f(x) = x(x + 4)(x - 1)$ 13. $f(x) = (x - 1)(x - 3)^2$
15. $f(x) = (x + 2)(x - 4)^2$ 17. $f(x) = (x + 3)(x + 1)(x - 4)(x - 2)$ 19. $f(x) = (x + 2)(x - 1)(x - 3)^2$
21. $f(x) = (x + 3)(x - 3)(x - 1)^2$ 23. $f(x) = (x + 3)^2(x - 1)^2$

Exercises 16.2, Page 535

1. Estimates: −3, 0.5, 1.5; better: −2.93543, 0.462598, 1.47283
3. Estimates: −0.3, 0.6, 2; better: −0.340665, 0.678963, 2.16170 5. Estimate: 0.7; better: 0.738984
7. Estimates: −3, −2, 1, 3; better: −2.84100, −2.23078, 1.12412, 2.94766
9. Estimates: −1.9, 2.5; better: −1.64891, 2.59720

Exercises 16.3, Page 539

1. 5 3. 4 5. 10 7. $-2, 4 \pm j$ 9. $\pm j, 1 \pm j$ 11. $3, \pm j$ (multiplicity 2) 13. $3 \pm 4j, -3, 1$
15. $f(x) = x^3 - 4x^2 + 4x - 16 = 0$ 17. $f(x) = x^4 - 3x^3 + 2x^2 + 2x - 4 = 0$
19. $f(x) = x^4 - 4x^3 + 11x^2 - 14x + 10 = 0$ 21. $f(x) = x^5 - 2x^4 - 14x^3 - 8x^2 + 40x = 0$
23. Since complex roots occur in conjugate pairs, there would be five roots for a fourth degree equation.
25. Since complex roots occur in conjugate pairs, there would be four roots for a third degree equation.

Chapter 16 Review, Page 541

1. 7 2. 5 3. 12 4. x: 0, −7, 0.5; y: 0 5. x: −3, 2; y: −18 6. x: 0, 6; y: 0
7. x: 5, −3, −3.5; y: −105 8. x: 6, −4, 1; y: 24 9. $f(x) = (x - 1)(x - 2)(x - 4)$
10. $f(x) = (x + 2)(x - 1)(x + 3)$ 11. $f(x) = (x + 6)(x + 5)(x - 1)^2$ 12. $f(x) = (x + 3)(x - 1)(x - 2)(x - 4)$
13. $f(x) = (x + 1)(x - 4)(x + 3)^2$ 14. $f(x) = (x - 6)(2x + 5)(x - 7)^2$
15. 4 with multiplicity 3, −4.5 with multiplicity 2 16. 6 with multiplicity 1, −10 with multiplicity 3
17. 0 with multiplicity 4, 1 with multiplicity 2 18. 3 with multiplicity 1, −5 with multiplicity 2
19. 0.302776, −3.30278 20. 3.54682 21. −0.592366, 1.27047, 3.32190 22. 1.20754
23. 1.28514, 0.221876, −3.50702 24. 1.14129, 0.393780, −1.33507 25. 8 26. 5 27. $3 \pm j, 5$
28. $2 \pm 3j, 3$ 29. −4 with multiplicity of 2, $\pm 3j$ 30. $2 \pm j, 5, -7$ 31. $f(x) = x^3 - 8x^2 + 49x - 392 = 0$
32. $f(x) = x^3 - 2x^2 - 11x + 52 = 0$ 33. $f(x) = x^4 - 2x^3 + 6x^2 - 8x + 8 = 0$
34. $f(x) = 50x^5 - 85x^4 + 98x^3 - 94x^2 + 48x - 9 = 0$

CHAPTER 17

Exercises 17.1, Pages 549–550

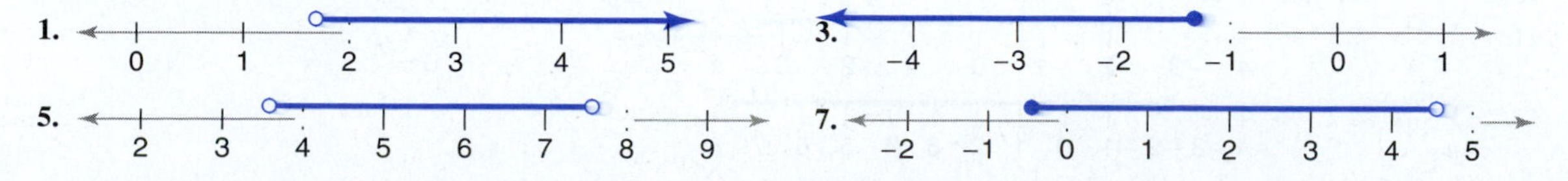

9.

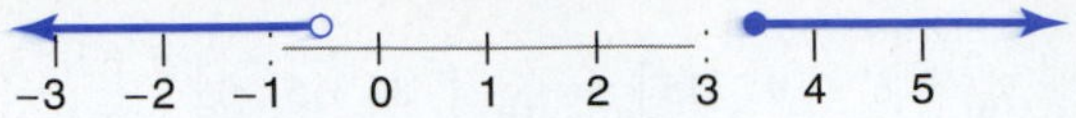

11. $x \geq 3$ **13.** $x < -1$ **15.** $4 < x < 7$ **17.** $-8 \leq x < -5$ **19.** $x < -1$ or $x \geq 1$

21. $x < 3$

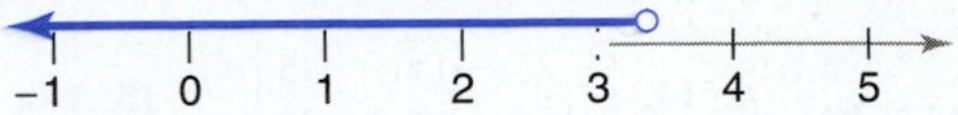

23. $x \leq -3$

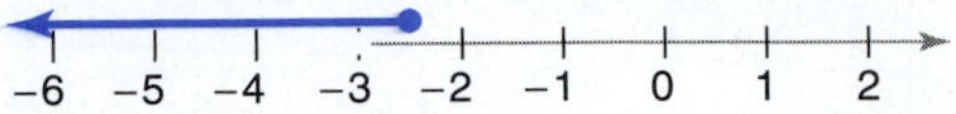

25. $x < 6$

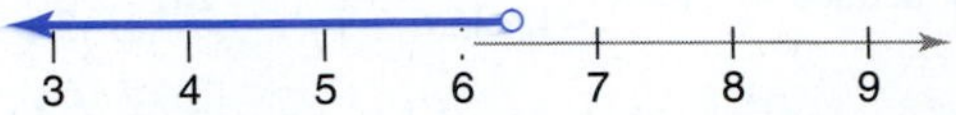

27. $x \leq -6$

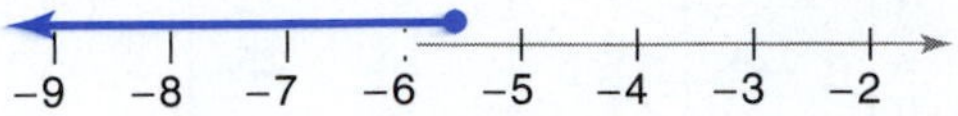

29. $x \leq 3$

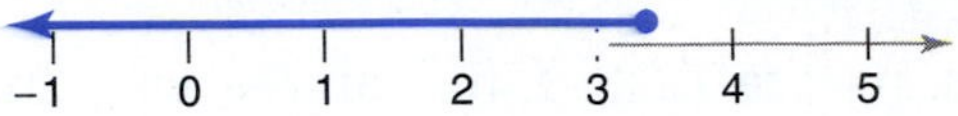

31. $x < 3$

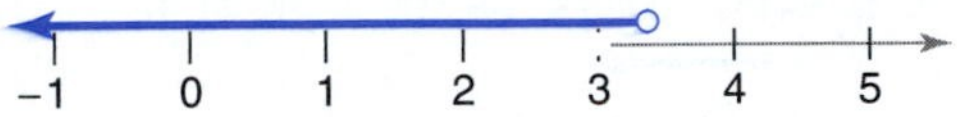

33. $x \geq 8$

35. $x < -\frac{5}{14}$

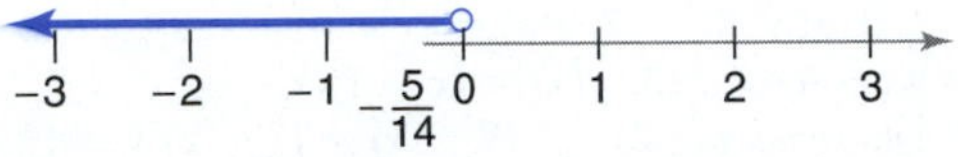

37. $x > \frac{1}{6}$

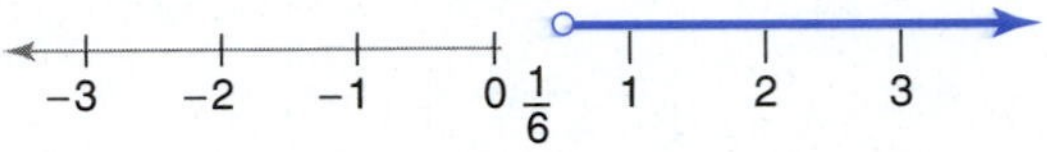

39. $x \leq -\frac{32}{9}$

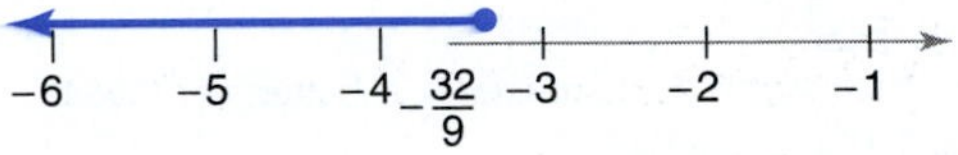

41. $x \geq 10$

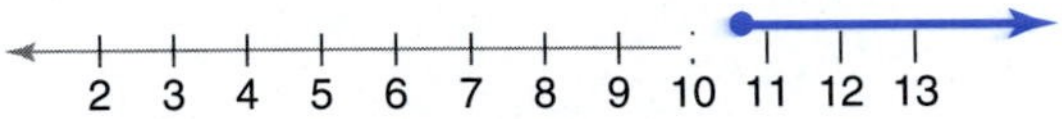

43. $x < 4$

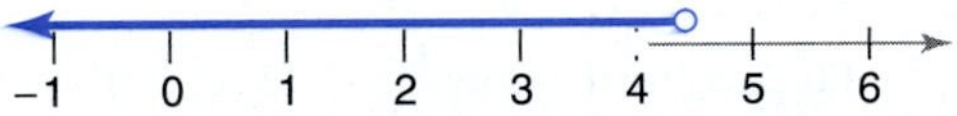

45. $7 < x \leq 10$

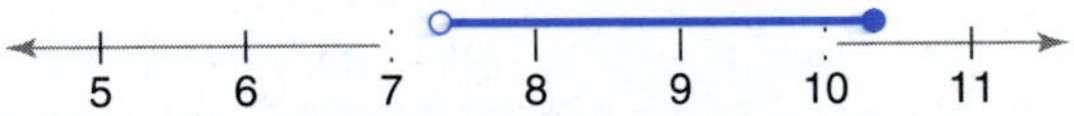

47. Set of real numbers

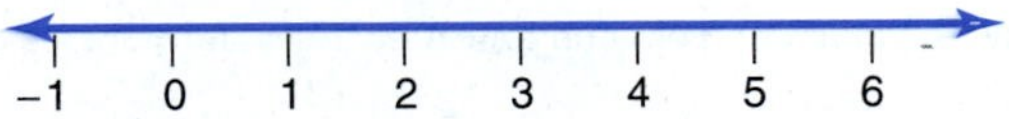

49. $-3 \leq x < 4$

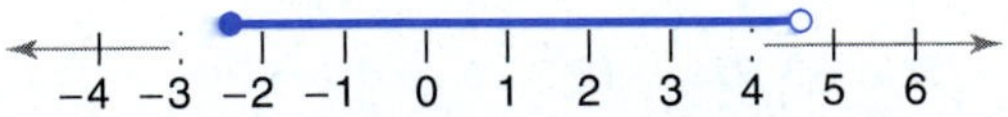

51. $-6 \leq x \leq 2$

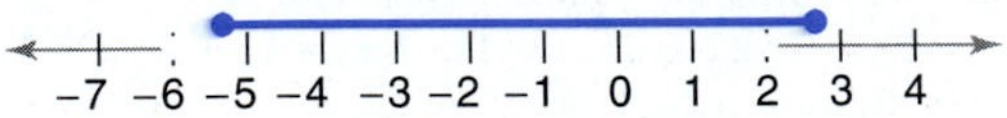

53. $0 < x \leq 5$

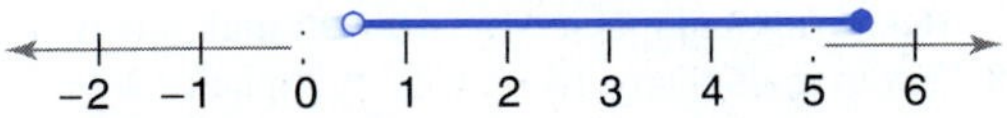

55. $1 < x < 5$

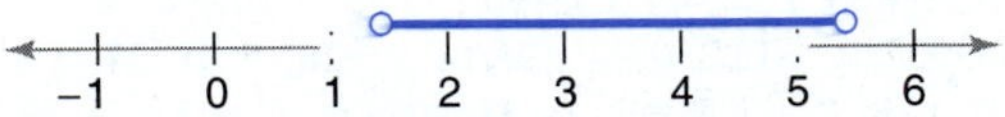

57. $-7 \leq x < -\frac{10}{3}$

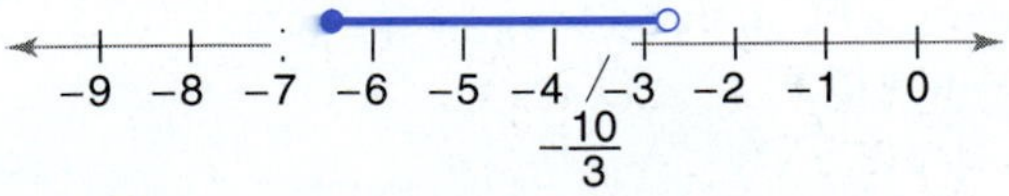

Exercises 17.2, Page 552

1. 6, −14 **3.** $\frac{2}{5}$, 2 **5.** $-\frac{2}{3}$, −6 **7.** $-\frac{1}{3}$ **9.** 5, $\frac{29}{9}$

11. $-3 < x < 3$

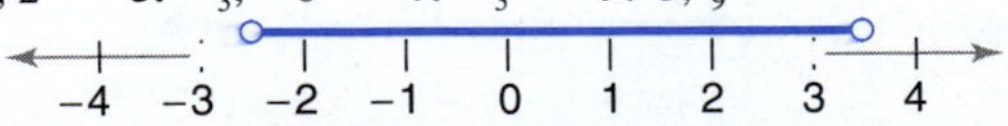

13. $-3 < x < 7$

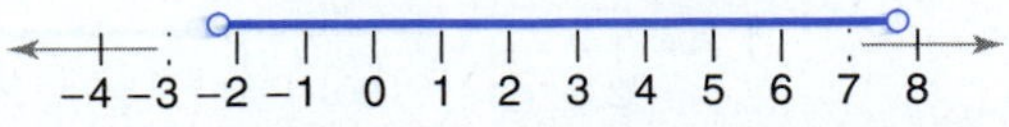

15. $x < -2$ or $x > -1$

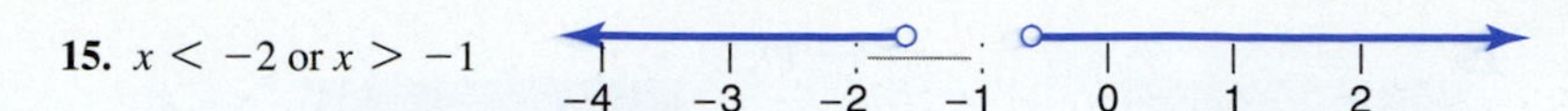

17. $-4 \le x \le \frac{11}{2}$

19. $x \le -\frac{17}{3}$ or $x \ge 5$

21. $\frac{13}{2} < x < 9$

23. $x \le -1$ or $x \ge \frac{4}{3}$

25. $x < -12$ or $x > 6$

27. No solution

Exercises 17.3, Page 556

1. $x < -3$ or $x > 5$ **3.** $-\frac{1}{4} \le x \le \frac{7}{3}$ **5.** $x < -7$ or $x > -2$ **7.** $\frac{1}{2} < x < 4$ **9.** $x \le -\frac{2}{5}$ or $x \ge \frac{1}{3}$
11. $-\frac{1}{4} \le x \le 3$ **13.** All real numbers $x \ne -\frac{3}{2}$ **15.** $x < -2$ or $x > 2$ **17.** $-\sqrt{5} \le x \le \sqrt{5}$
19. $x < 0$ or $x > 3$ **21.** $-4 \le x \le 0$ or $x \ge 3$ **23.** $x > 6$ **25.** $-2 < x < -1$ or $1 < x < 2$
27. $-4 < x < 2$ **29.** $-3 < x \le -2$ or $x \ge -1$ **31.** $-4 < x \le -3$ or $3 \le x < 4$ **33.** $x < -2$ or $x > 5$
35. $-1 \le x \le 0$ or $1 < x \le 2$ **37.** $2 < x \le 3$ **39.** $x < 0$ or $x > \frac{1}{2}$

Exercises 17.4, Page 558

1.

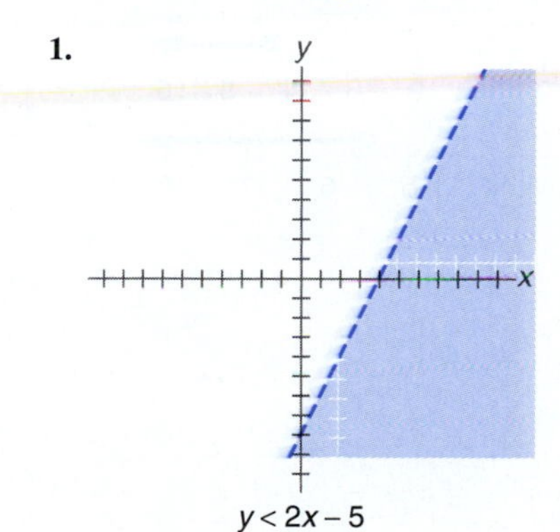

$y < 2x - 5$

3.

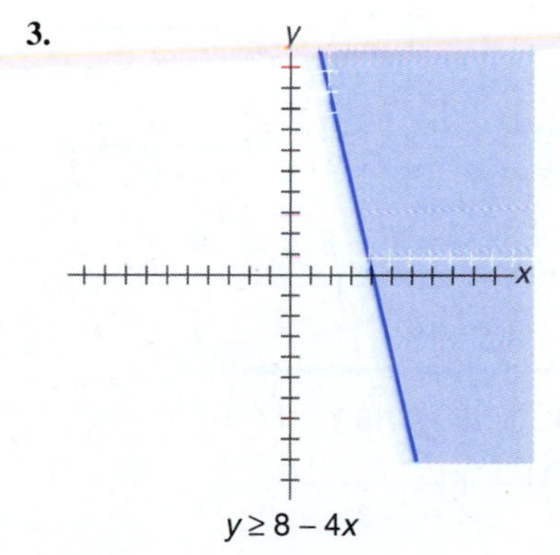

$y \ge 8 - 4x$

5.

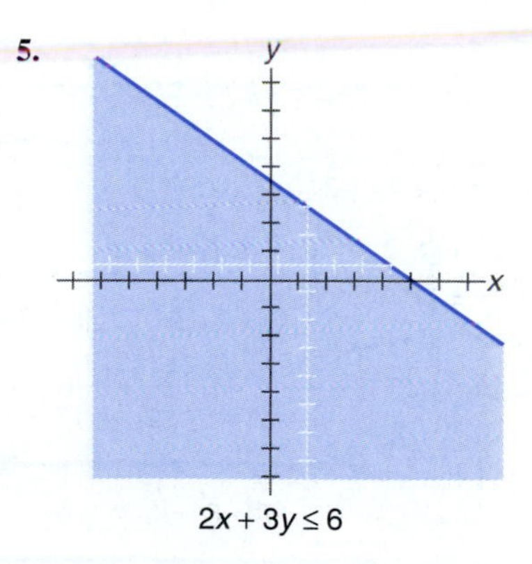

$2x + 3y \le 6$

7.

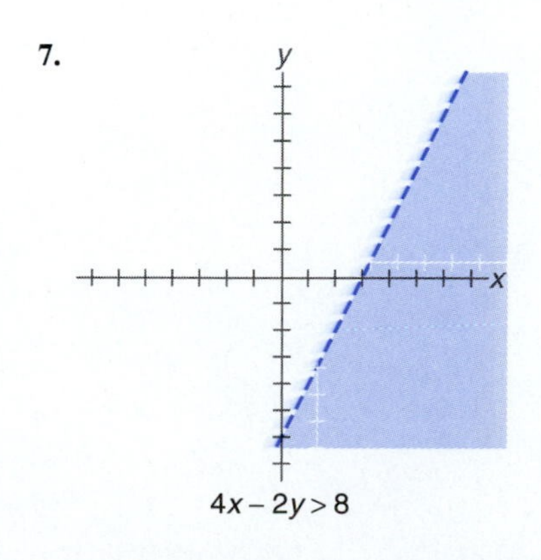

$4x - 2y > 8$

9.

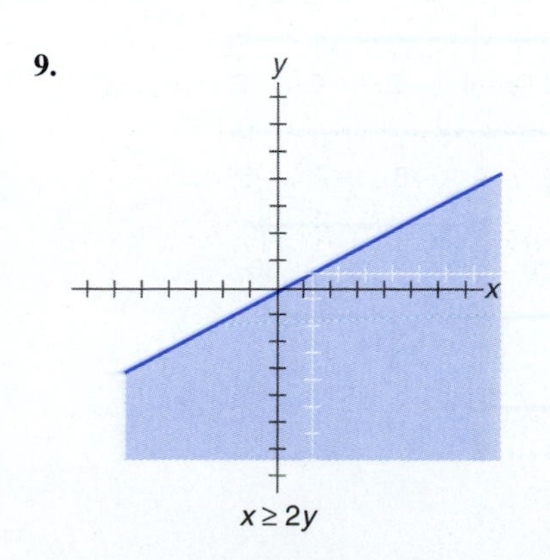

$x \ge 2y$

11.

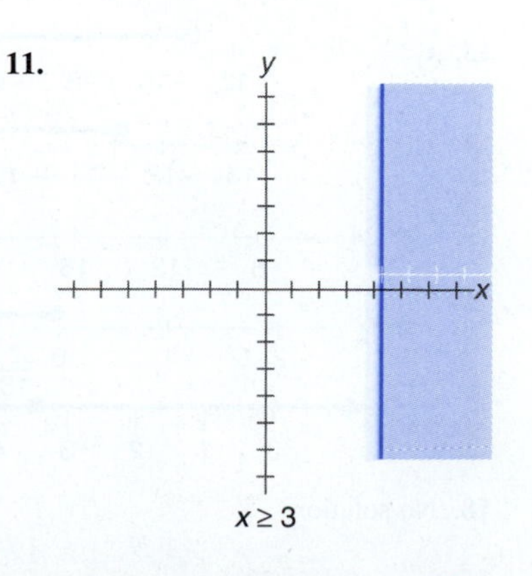

$x \ge 3$

13.

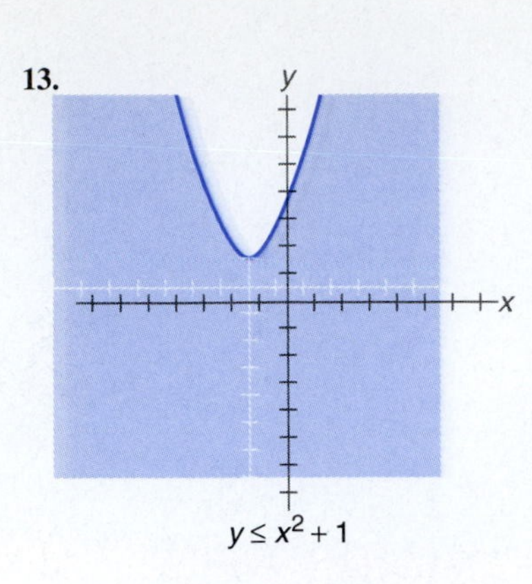

15.

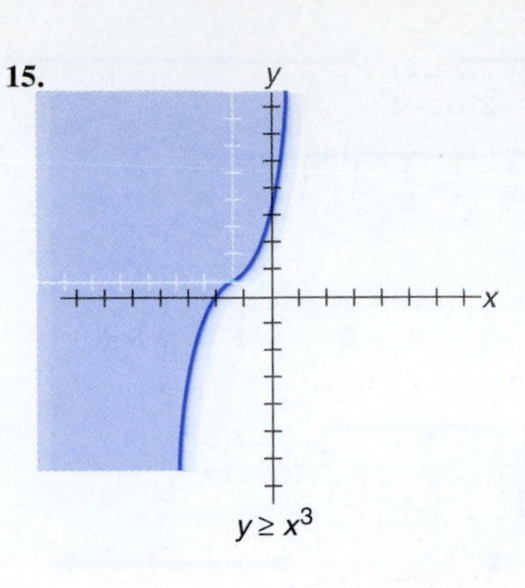

17.

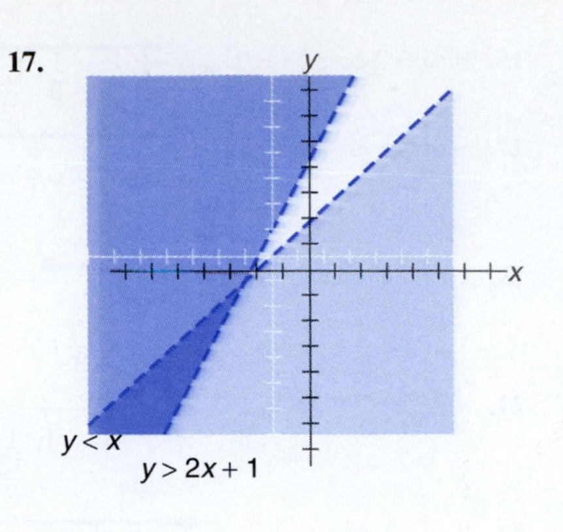

19.

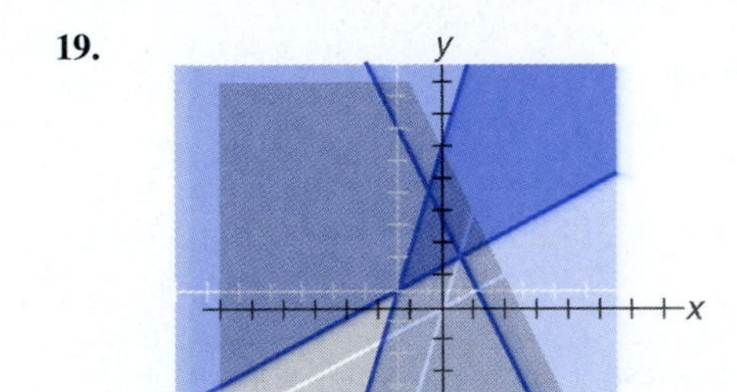

21.

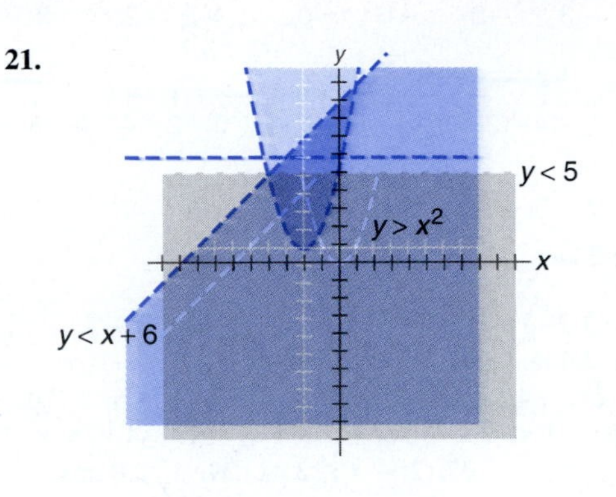

Chapter 17 Review, Pages 559–560

1.

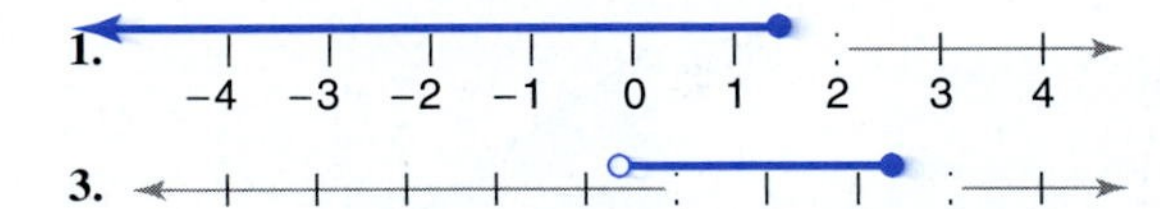

2.

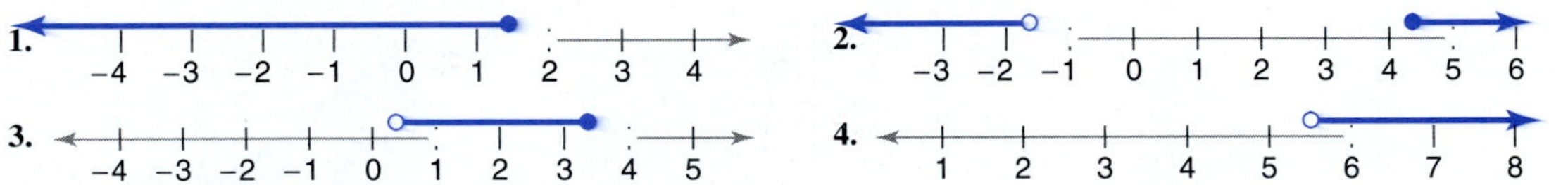

3. −4 −3 −2 −1 0 1 2 3 4 5

4. 1 2 3 4 5 6 7 8

5. $x > 2$ **6.** $-1 < x \le 2$ **7.** $x \le 4$ or $x \ge 7$ **8.** $x \le -3$

9. $x \le -4$ −6 −5 −4 −3 −2 −1 0

10. $x > 14$ 11 12 13 14 15 16 17

11. $x \le -5$ −8 −7 −6 −5 −4 −3 −2

12. $x < 3$ −2 −1 0 1 2 3 4 5

13. $x > -10$ −12 −10 −8 −6 −4 −2 0 2

14. $x \ge -11$ −13 −12 −11 −10 −9 −8 −7 −6

15. $x \ge 24$ 8 12 16 20 24 28 30

16. $x \ge \frac{7}{18}$ −2 −1 0 $\frac{7}{18}$ 1 2 3

17. $x \le 6$ 0 1 2 3 4 5 6 7 8

18. No solution

19. $x \le 0$ or $x > 3$

20. $x \le -1$

21. $-6 \le x \le 2$

22. $-\frac{7}{3} \le x < 1$

23. 4, −5 **24.** $\frac{13}{3}$, −3 **25.** $\frac{5}{7}$, 9 **26.** $-\frac{16}{7}$, −4 **27.** $-12 < x < 4$ **28.** $x < -6$ or $x > 12$
29. $x \le -11$ or $x \ge 6$ **30.** $-6 \le x \le 9$ **31.** $x < -5$ or $x > \frac{17}{3}$ **32.** $-\frac{5}{2} \le x \le 6$ **33.** $-\frac{1}{2} \le x \le 4$
34. $x < \frac{2}{3}$ or $x > 6$ **35.** $-7 \le x \le 7$ **36.** $-6 \le x \le 0$ or $x \ge 6$ **37.** $x > 0$ **38.** $-4 < x < 9$
39. $x \le -3$ or $\frac{1}{3} \le x < 3$ **40.** $-4 < x < \frac{4}{3}$ **41.** $x < -5$ or $x \ge 4.9$ **42.** $x < 4$ or $x \ge 13$

43.

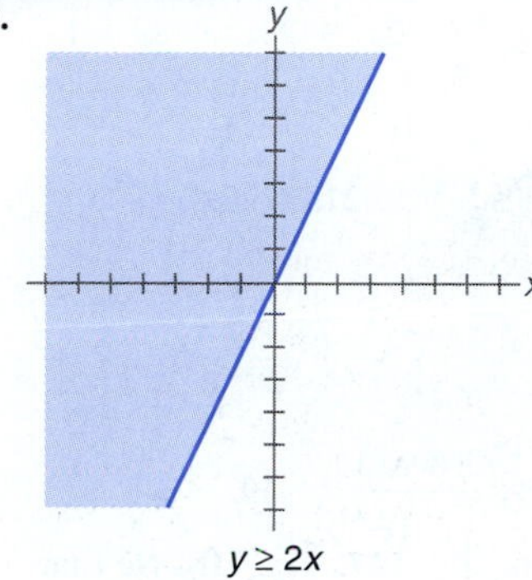

44.

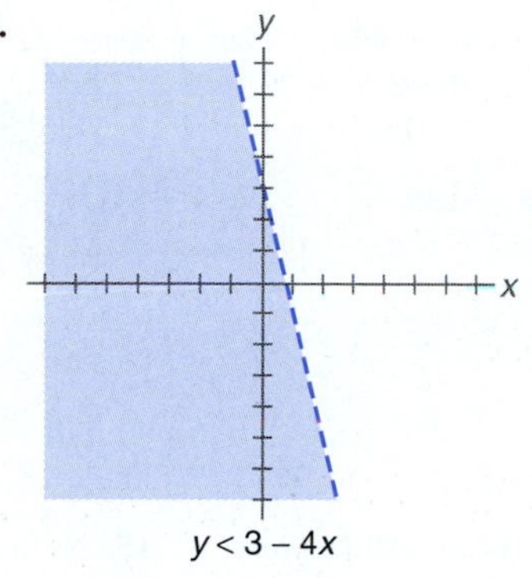

45.

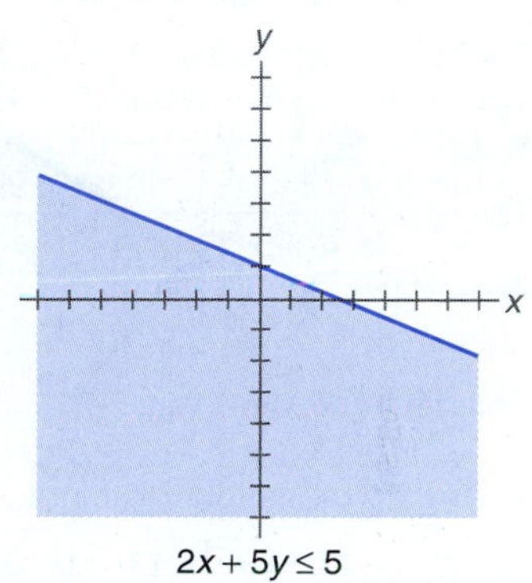

46.

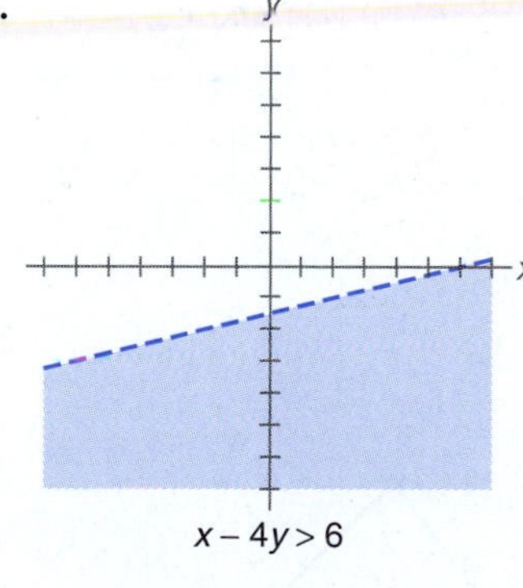

47.

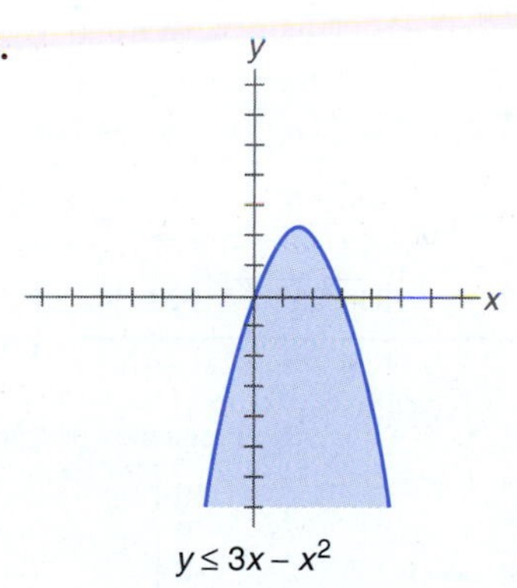

48.

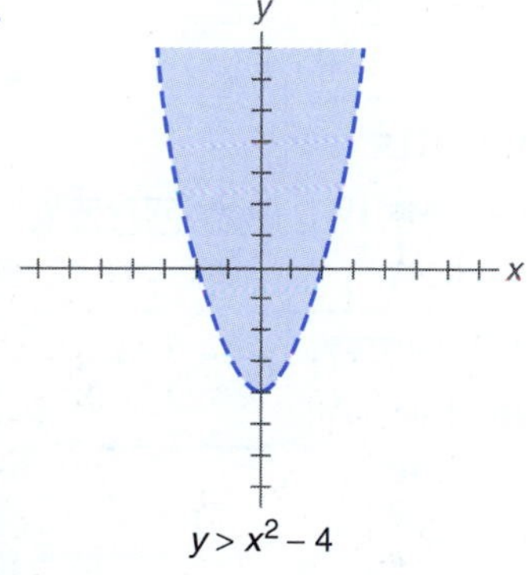

49.

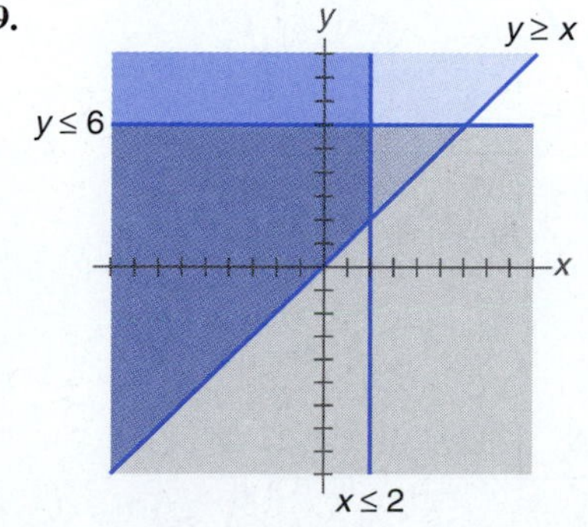

50.

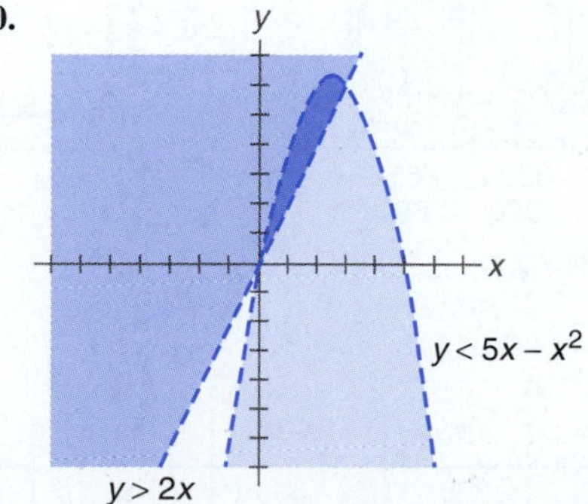

CHAPTER 18

Exercises 18.1, Pages 563–564

1. 17 **3.** 24 **5.** -95 **7.** 57 **9.** 202.5 **11.** -546 **13.** $2, -1, -4, -7, -10$ **15.** $5, 5\frac{2}{3}, 6\frac{1}{3}, 7, 7\frac{2}{3}$
17. 4 **19.** 1,000,000 **21.** \$31,200

Exercises 18.2, Pages 567–568

1. $\frac{20}{2187}$ **3.** 8 **5.** $-\frac{1}{64}$ **7.** $\frac{65{,}600}{2187}$ **9.** $7\sqrt{2} + 14$ **11.** $\frac{341}{64}$ **13.** $3, \frac{3}{2}, \frac{3}{4}, \frac{3}{8}, \frac{3}{16}$ **15.** $5, -\frac{5}{4}, \frac{5}{16}, -\frac{5}{64}, \frac{5}{256}$
17. $-4, -12, -36, -108, -324$ **19.** $\frac{1}{2}$ **21.** \$17,531 **23.** $\frac{3}{8}$ ft **25.** 15.1°C **27.** $\frac{14}{3}$ **29.** $\frac{8}{3}$
31. No sum **33.** 6.25 **35.** $\frac{1}{3}$ **37.** $\frac{2}{165}$ **39.** $\frac{13}{15}$

Exercises 18.3, Pages 571–572

1. $27x^3 + 27x^2y + 9xy^2 + y^3$ **3.** $a^5 - 10a^4 + 40a^3 - 80a^2 + 80a - 32$ **5.** $16x^4 - 32x^3 + 24x^2 - 8x + 1$
7. $64a^6 + 576a^5b + 2160a^4b^2 + 4320a^3b^3 + 4860a^2b^4 + 2916ab^5 + 729b^6$
9. $\frac{32}{243}x^5 - \frac{160}{81}x^4 + \frac{320}{27}x^3 - \frac{320}{9}x^2 + \frac{160}{3}x - 32$ **11.** $a^2 + 12a^{3/2}b^2 + 54ab^4 + 108a^{1/2}b^6 + 81b^8$
13. $\frac{x^4}{y^4} - \frac{8x^3}{y^3z} + \frac{24x^2}{y^2z^2} - \frac{32x}{yz^3} + \frac{16}{z^4}$ **15.** $-126x^4y^5$ **17.** $41{,}184x^5y^8$ **19.** $280x^3y^8$ **21.** $4320x^3y^3$
23. $-8064x^5$

Chapter 18 Review, Page 573

1. 47 **2.** $\frac{1}{16}$ **3.** -81 **4.** -62 **5.** $\frac{2}{81}$ **6.** 95 **7.** 300 **8.** $\frac{127}{16}$ **9.** $\frac{-80\sqrt{3}}{1 + \sqrt{3}}$ **10.** -348
11. $\frac{728}{81}$ **12.** 500 **13.** 1,001,000 **14.** \$2988 (approx.) **15.** No sum **16.** $\frac{35}{6}$ **17.** $\frac{3}{2}$ **18.** No sum
19. $\frac{5}{11}$ **20.** $\frac{152}{165}$ **21.** $a^6 - 6a^5b + 15a^4b^2 - 20a^3b^3 + 15a^2b^4 - 6ab^5 + b^6$
22. $32x^{10} - 80x^8 + 80x^6 - 40x^4 + 10x^2 - 1$ **23.** $16x^4 + 96x^3y + 216x^2y^2 + 216xy^3 + 81y^4$
24. $1 + 8x + 28x^2 + 56x^3 + 70x^4 + 56x^5 + 28x^6 + 8x^7 + x^8$ **25.** $90x^2$ **26.** $1280a^3b^3$ **27.** $8064x^5b^{10}$
28. $3{,}247{,}695x^{16}$

CHAPTER 19

Exercises 19.1, Pages 577–579

1.

Hours	205–243	244–282	283–321	322–360	361–399
Frequency	1	1	6	10	7

1b. Histogram **1c.** Frequency polygon

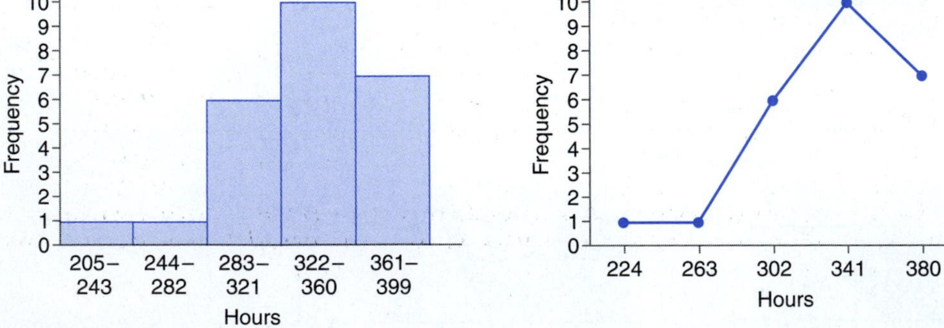

3.

Bars	1–3	4–6	7–9	10–12	13–15	16–18	19–21	22–24
Frequency	5	3	2	4	4	5	4	3

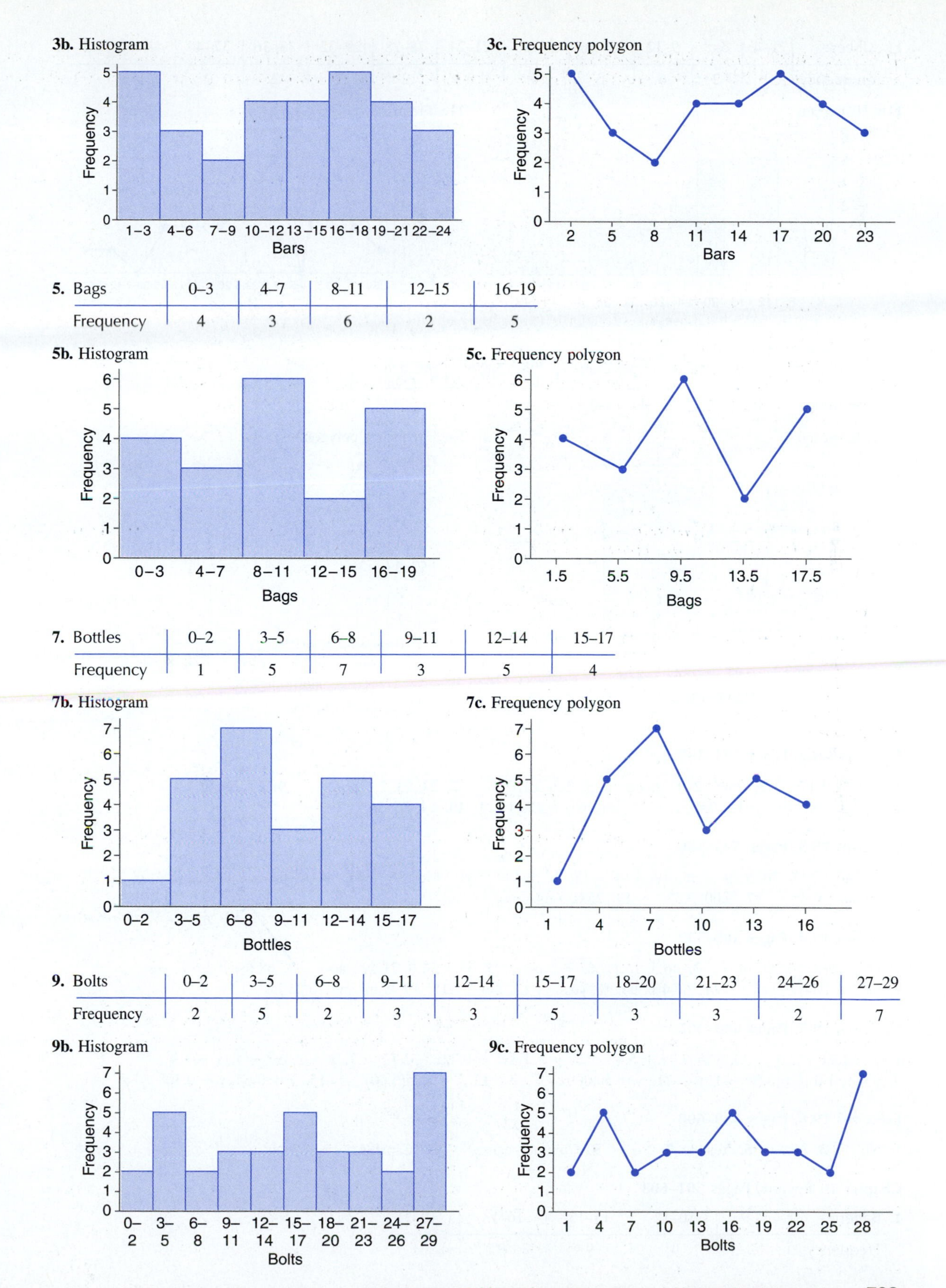

5.

Bags	0–3	4–7	8–11	12–15	16–19
Frequency	4	3	6	2	5

7.

Bottles	0–2	3–5	6–8	9–11	12–14	15–17
Frequency	1	5	7	3	5	4

9.

Bolts	0–2	3–5	6–8	9–11	12–14	15–17	18–20	21–23	24–26	27–29
Frequency	2	5	2	3	3	5	3	3	2	7

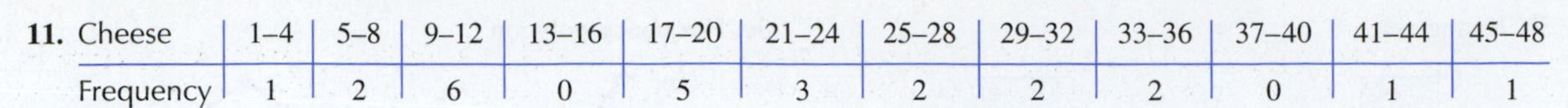

11. Cheese	1–4	5–8	9–12	13–16	17–20	21–24	25–28	29–32	33–36	37–40	41–44	45–48
Frequency	1	2	6	0	5	3	2	2	2	0	1	1

11b. Histogram

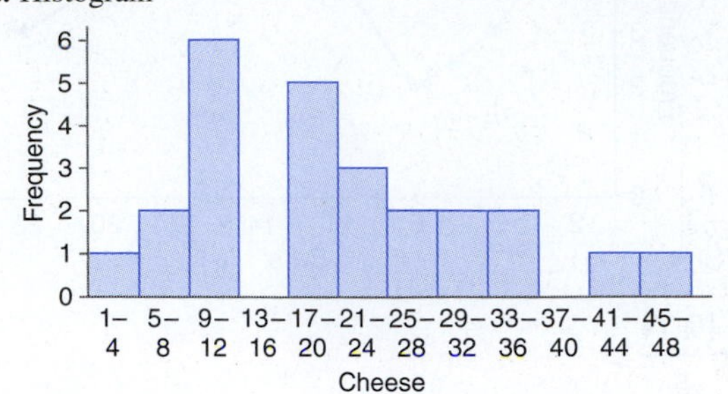

11c. Frequency polygon

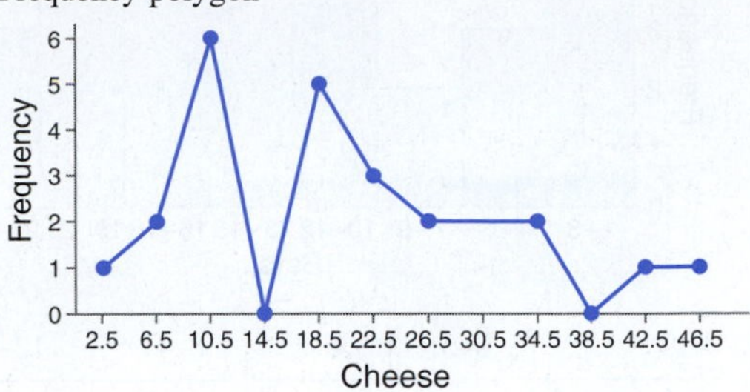

13. Hamburgers	220–234	235–249	250–264	265–279
Frequency	7	4	7	7

13b. Histogram

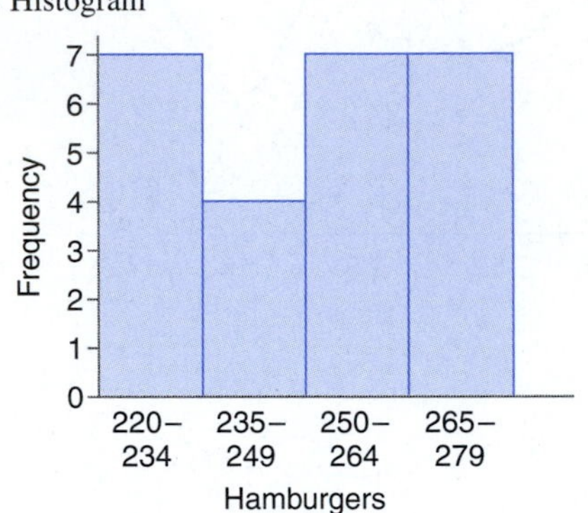

13c. Frequency polygon

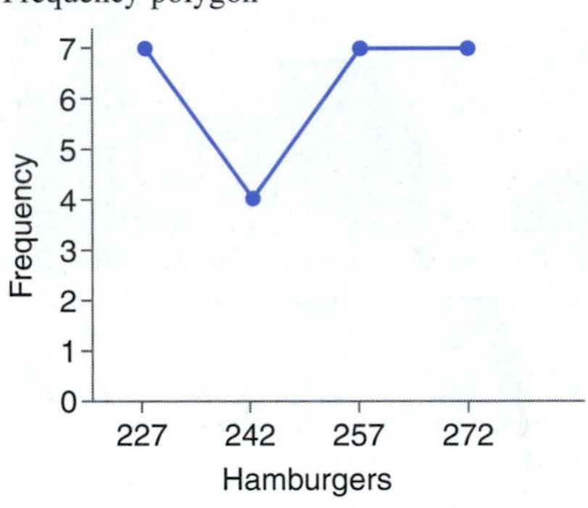

Exercises 19.2, Pages 581–582

1. 21.4, 21, 17 **3.** 2526, 2497, none **5.** 2.3, 2.1, 2.5 **7.** 23, 23, 23 and 27 **9.** 17, 3 and 27 **11.** 19, 11 and 19 **13.** 250, 268 **15.** 16 **17.** $1\overline{0}$ **19.** 54, 56, 56

Exercises 19.3, Pages 585–586

1. 9, 2.83 **3.** 16, 5.29 **5.** 19, 6.14 **7.** 16, 6.40 **9.** 17, 6.78 **11.** 44.0, 16.1 **13.** 83.0, 31.6 **15.** 2.50, 0.969 **17.** 9.00, 3.32 **19.** 25.0, 7.64

Exercises 19.4, Pages 588–589

1. 49.3, 20.8, 50.0%, no **3.** 46.3, 10.1, 66.7%, yes **5.** 34.7, 11.0, 71.4%, no **7.** 68.6%, 93.6%, yes **9.** 68.2%, 94.0%, yes **11.** 66.4%, 94.2%, yes **13.** No **15.** 19.75 in. and 20.25 in.

Exercises 19.5, Pages 593–595

1. $y = 1.9x - 2.2$ **3.** $y = 2.7x + 5.4$ **5.** $y = 1.1x^2 + 0.72x - 2.7$ **7.** $y = 2.0x^2 - 4.1x + 4.9$ **9.** $y = -1.0x^2 + 6.5x + 11$ **11.** $y = 50{,}\overline{0}00x^{-0.982}$ **13.** $y = 325(5.00)^x$ **15.** $y = 0.641x + 0.308$

Exercises 19.6, Pages 599–600

1. No **3.** No **5.** Yes **7.** No **9.** Common cause **11.** Capable

Chapter 19 Review, Pages 601–603

1. Batteries	7–9	10–12	13–15	16–18	19–21
Frequency	7	4	9	6	4

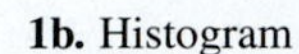

1b. Histogram

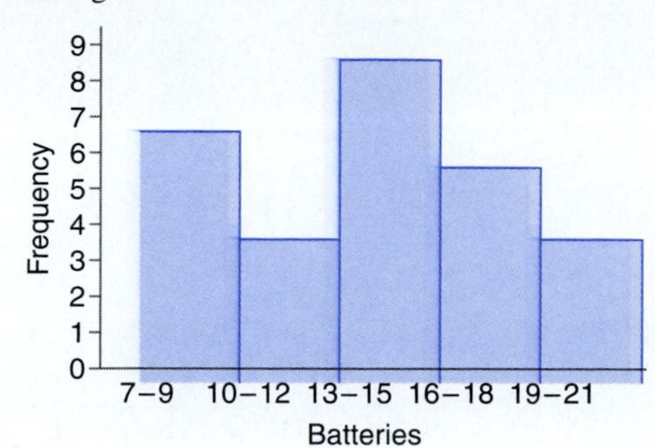

1c. Frequency polygon

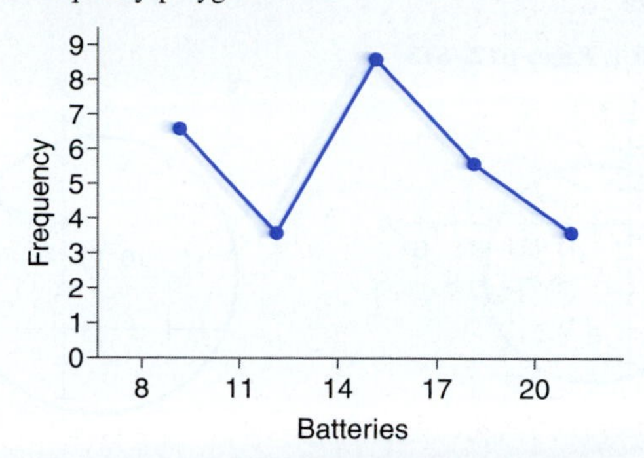

2.

psi	101–105	106–110	111–115	116–120	121–125	126–130
Frequency	4	2	6	6	3	4

2b. Histogram

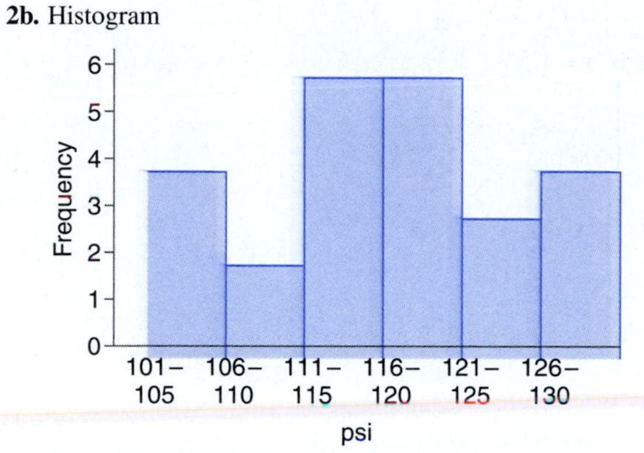

2c. Frequency polygon

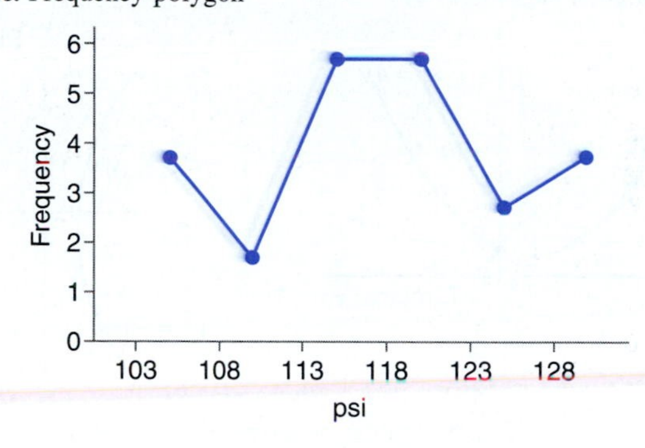

3.

Bottles	53–59	60–66	67–73	74–80	81–87
Frequency	5	4	2	3	6

3b. Histogram

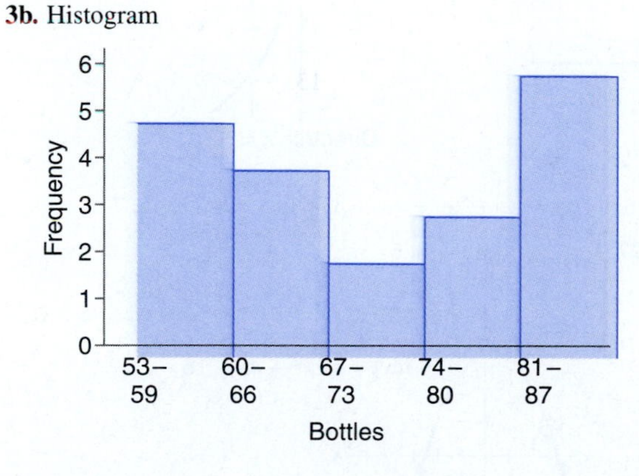

3c. Frequency polygon

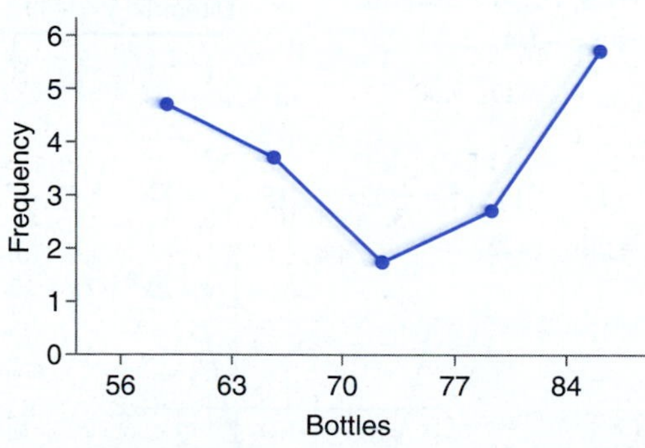

4. 22.6, 23.0, 18.0 **5.** 13.1, 14.0, 15.0 **6.** 36.4, 36.5, 38.0 **7.** 22.0, 21.0, 21.0 **8.** 19.8, 20.0, 20.0 **9.** 6.00, 2.28 **10.** 10.0, 4.03 **11.** 14.0, 5.34 **12.** 11.0, 3.92 **13.** 68.5%, 94.6%, yes **14.** 67.7%, 93.9%, yes **15.** $y = 5.4x - 3.7$ **16.** $y = 3.8x + 2.9$ **17.** $y = 3.1x - 0.97$ **18.** $y = 2.1x - 0.43$ **19.** $y = 0.84x^2 + 4.5x + 1.1$ **20.** $y = 0.57x^2 + 5.7x - 10$ **21.** $y = 2.1\,(3.0)^x$ **22.** $y = 2.0\,(0.63)^x$ **23.** $y = 51{,}500x^{-1.02}$ **24.** Yes **25.** No

CHAPTER 20

Exercises 20.1, Pages 612–613

1.

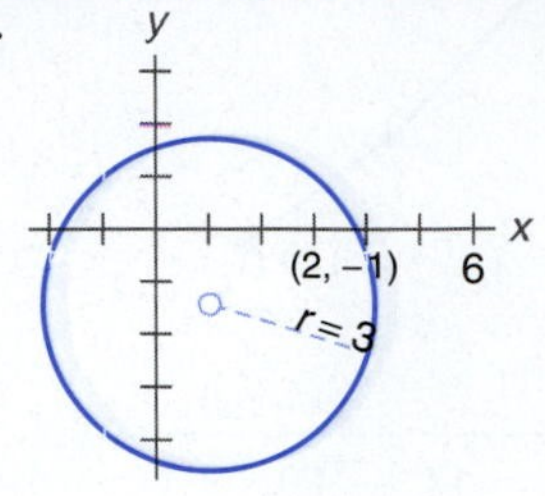

3.

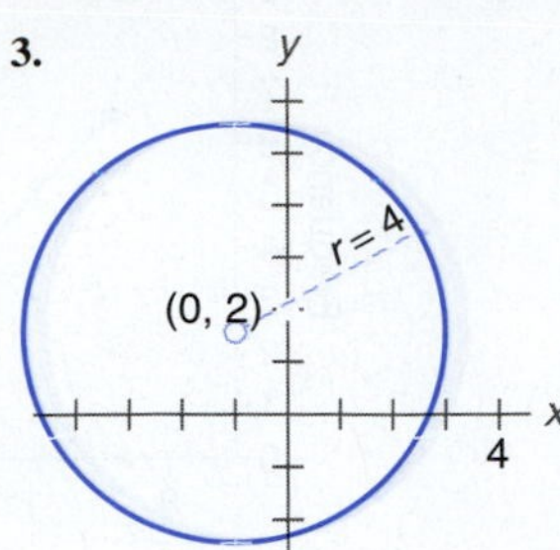

5. $(x - 1)^2 + (y + 1)^2 = 16$ **7.** $(x + 2)^2 + (y + 4)^2 = 34$ **9.** $x^2 + y^2 = 36$ **11.** $(0, 0); r = 4$
13. $(-3, 4); r = 8$ **15.** $(4, -6); r = 2\sqrt{15}$ **17.** $(6, 1); r = 7$ **19.** $(-\frac{7}{2}, -\frac{3}{2}); r = \sqrt{94}/2$
21. $x^2 + y^2 - 2y - 9 = 0; (0, 1); r = \sqrt{10}$ **23.** $x^2 + y^2 + 10x - 40y = 0; (-5, 20); r = 5\sqrt{17}$

Exercises 20.2, Pages 620–621

1.

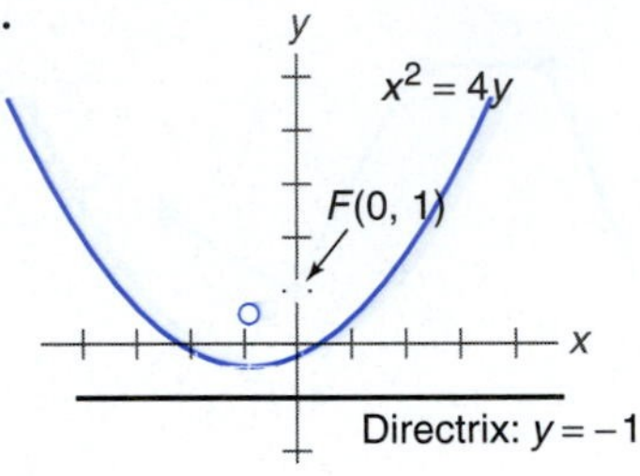

3.

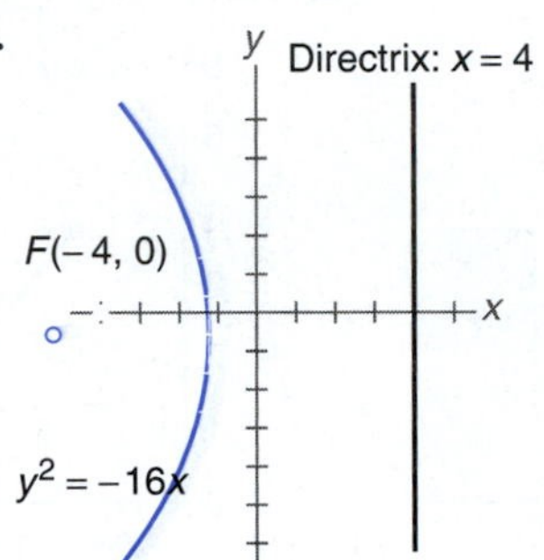

5.

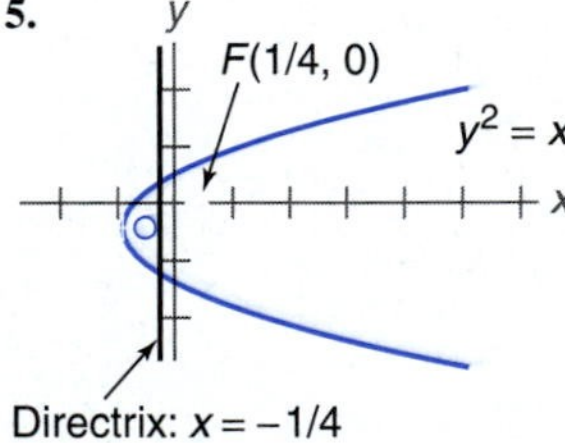

7.

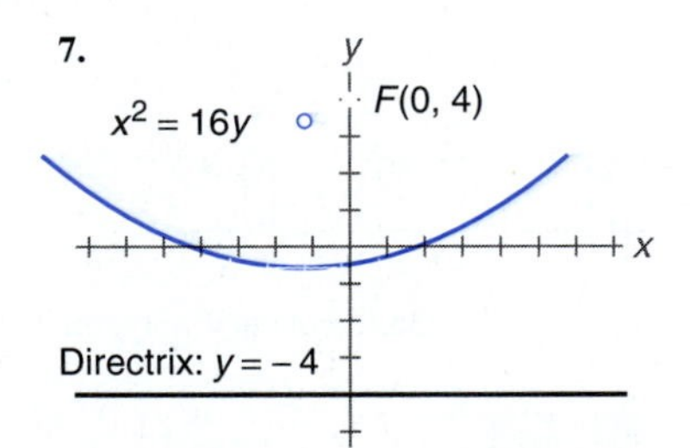

9.

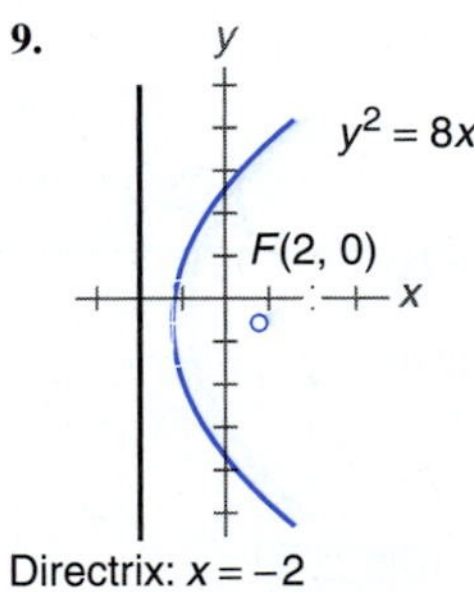

11. $y^2 = 8x$ **13.** $y^2 = -32x$ **15.** $x^2 = 24y$ **17.** $y^2 = -16x$ **19.** $y^2 - 6y + 8x + 1 = 0$ **21.** 15 m, 7 m
23. $x^2 = 32y$ **25.**

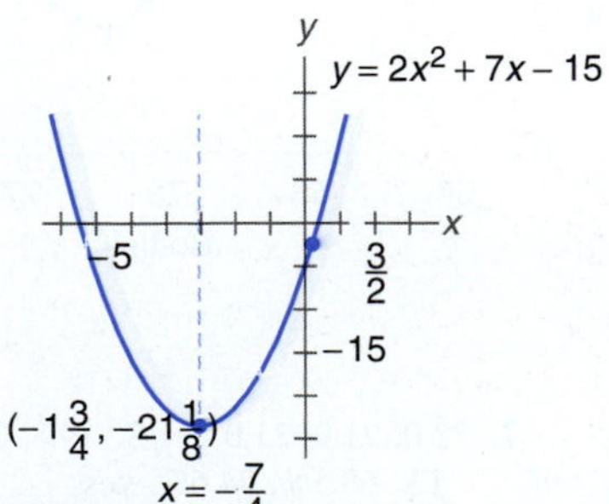

27.

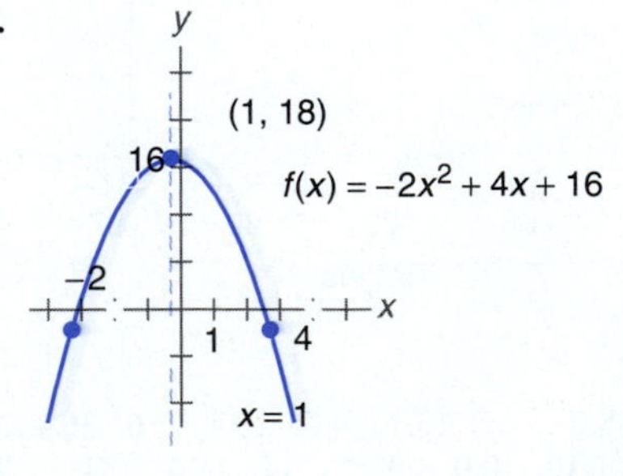

29. **(a)** 1024 m; **(b)** 1024 m **31.** 3600 m^2

Exercises 20.3, Page 625

	Vertices	Foci	Major axis	Minor axis
1.	(5, 0)(−5, 0)	(3, 0)(−3, 0)	10	8
3.	(4, 0) (−4, 0)	$(\sqrt{7}, 0)$ $(-\sqrt{7}, 0)$	8	6
5.	(0, 6)(0, −6)	$(0, \sqrt{35})$ $(0, -\sqrt{35})$	12	2
7.	(0, 4) (0, −4)	$(0, \sqrt{7})$ $(0, -\sqrt{7})$	8	6

1.

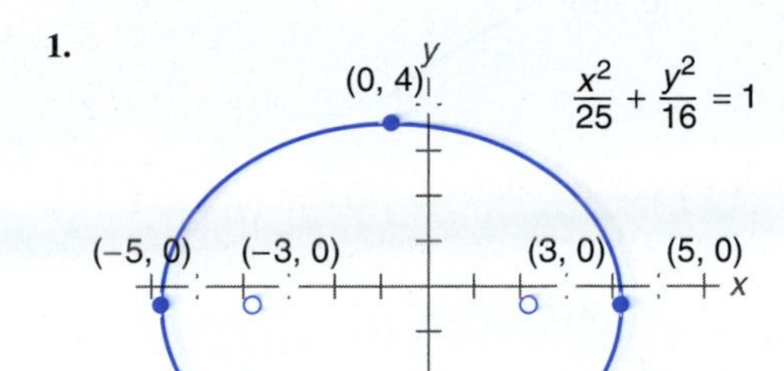

3.

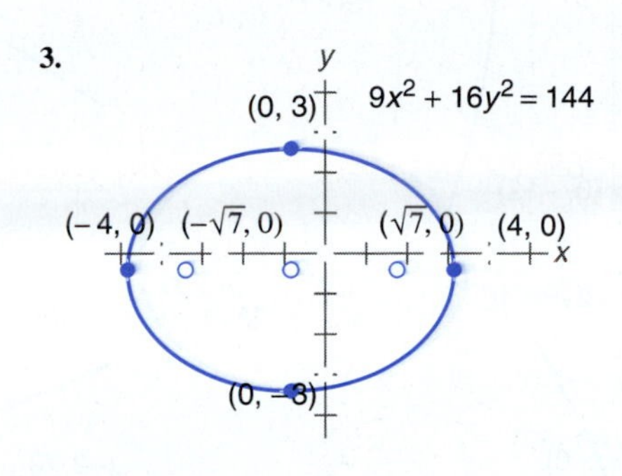

5.

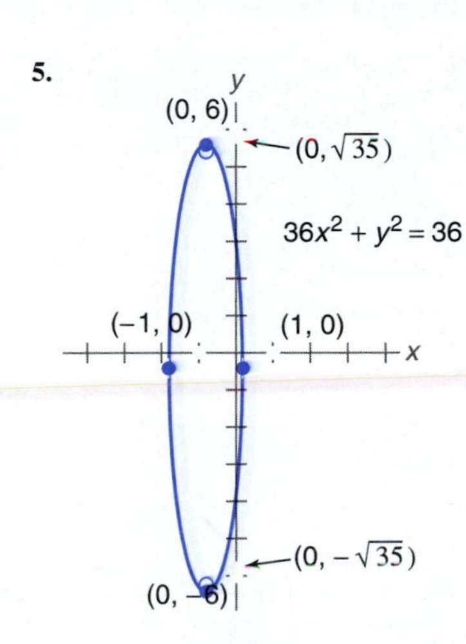

7.

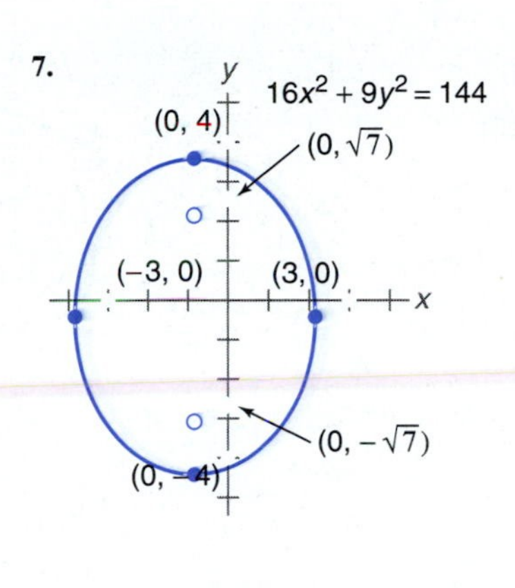

9. $\dfrac{x^2}{16} + \dfrac{y^2}{12} = 1$ or $3x^2 + 4y^2 = 48$ **11.** $\dfrac{x^2}{45} + \dfrac{y^2}{81} = 1$ or $9x^2 + 5y^2 = 405$

13. $\dfrac{x^2}{36} + \dfrac{y^2}{25} = 1$ or $25x^2 + 36y^2 = 900$ **15.** $\dfrac{x^2}{39} + \dfrac{y^2}{64} = 1$ or $64x^2 + 39y^2 = 2496$

17. $\dfrac{x^2}{5600^2} + \dfrac{y^2}{5000^2} = 1$ or $625x^2 + 784y^2 = 1.96 \times 10^{10}$

Exercises 20.4, Pages 630–631

	Vertices	Foci	Transverse axis	Conjugate axis	Asymptotes
1.	(5, 0) (−5, 0)	(13, 0) (−13, 0)	10	24	$y = \pm\dfrac{12}{5}x$
3.	(0, 3) (0, −3)	(0, 5) (0, −5)	6	8	$y = \pm\dfrac{3}{4}x$
5.	$(\sqrt{2}, 0)$ $(-\sqrt{2}, 0)$	$(\sqrt{7}, 0)$ $(-\sqrt{7}, 0)$	$2\sqrt{2}$	$2\sqrt{5}$	$y = \pm\sqrt{\dfrac{5}{2}}x$
7.	(0, 1) (0, −1)	$(0, \sqrt{5})$ $(0, -\sqrt{5})$	2	4	$y = \pm\dfrac{1}{2}x$

1.

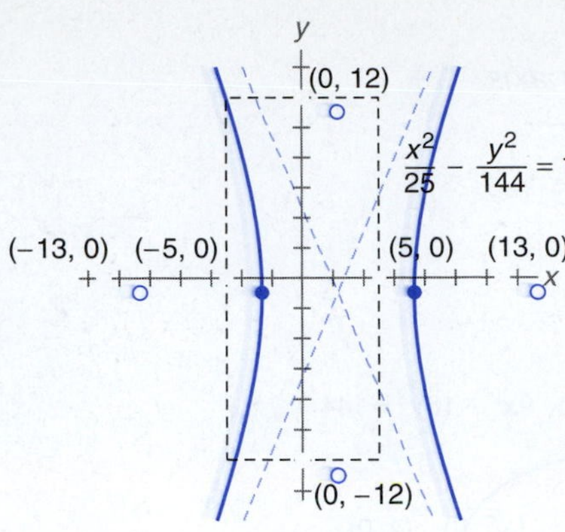

3.

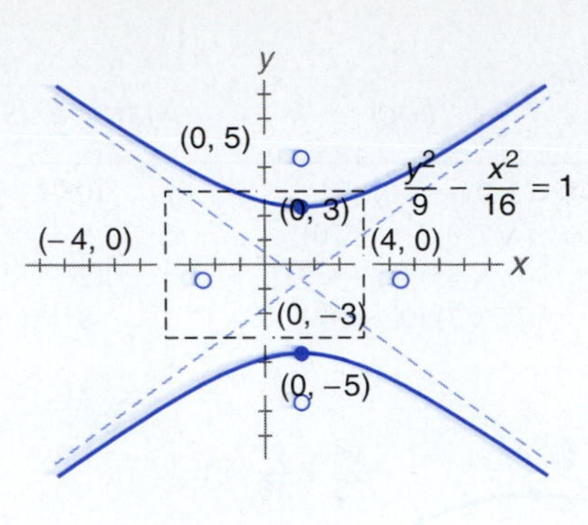

5.

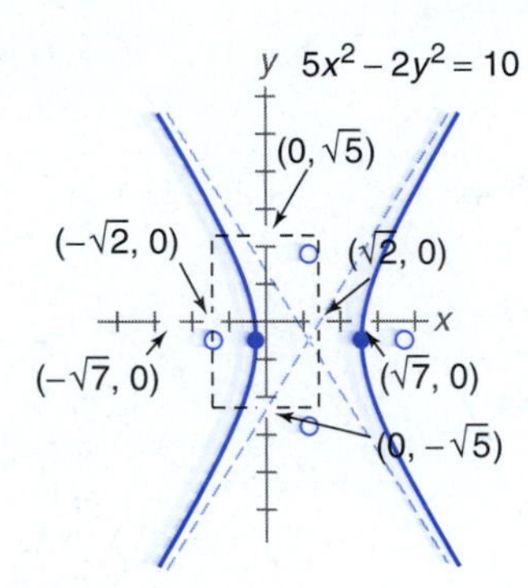

7.

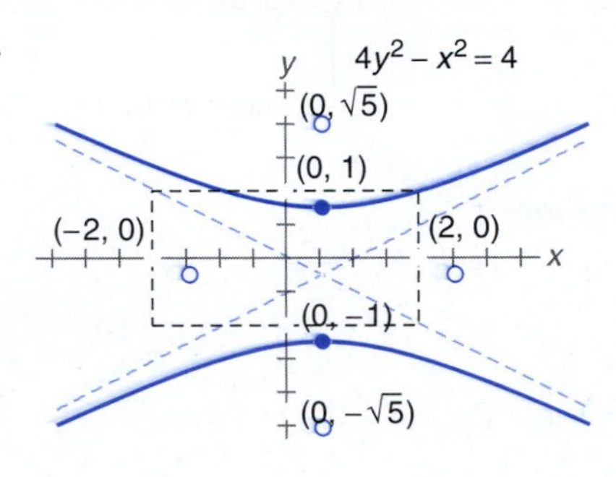

9. $\frac{x^2}{16} - \frac{y^2}{20} = 1$ or $5x^2 - 4y^2 = 80$ **11.** $\frac{y^2}{36} - \frac{x^2}{28} = 1$ or $7y^2 - 9x^2 = 252$

13. $\frac{x^2}{9} - \frac{y^2}{25} = 1$ or $25x^2 - 9y^2 = 225$ **15.** $\frac{x^2}{25} - \frac{y^2}{11} = 1$ or $11x^2 - 25y^2 = 275$

17.

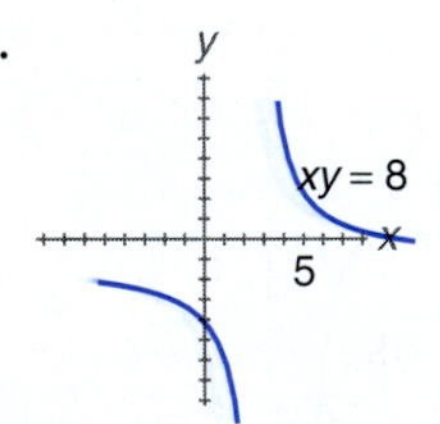

Exercises 20.5, Pages 635–636

1. $\frac{(x-1)^2}{16} + \frac{(y+1)^2}{12} = 1$ **3.** $\frac{(y-1)^2}{36} - \frac{(x-1)^2}{28} = 1$ **5.** $(y+1)^2 = 8(x-3)$

7. Parabola; vertex: $(2, -3)$ **9.** Hyperbola; center: $(-2, 0)$ **11.** Ellipse; center: $(2, 0)$

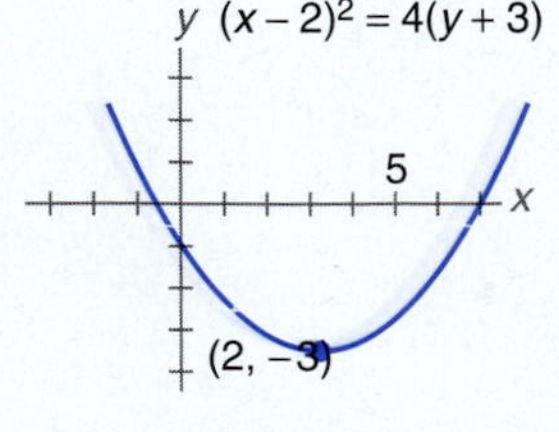

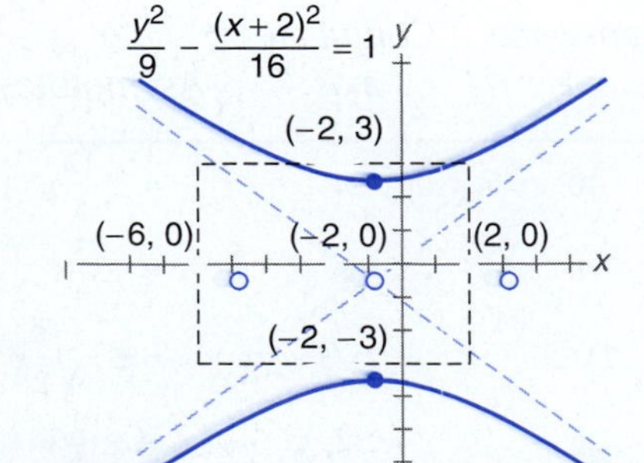

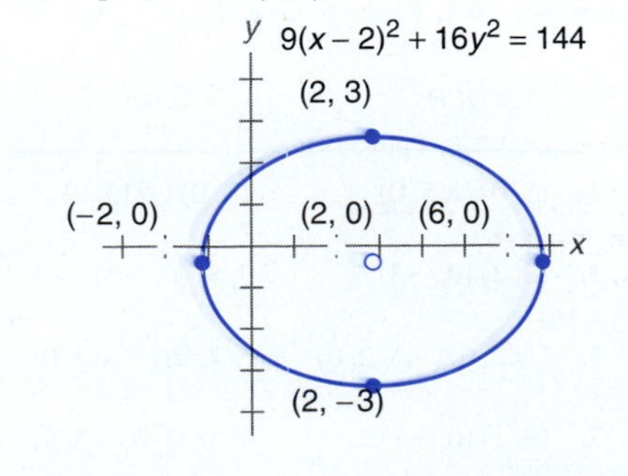

13. Ellipse; center: (3, 1)

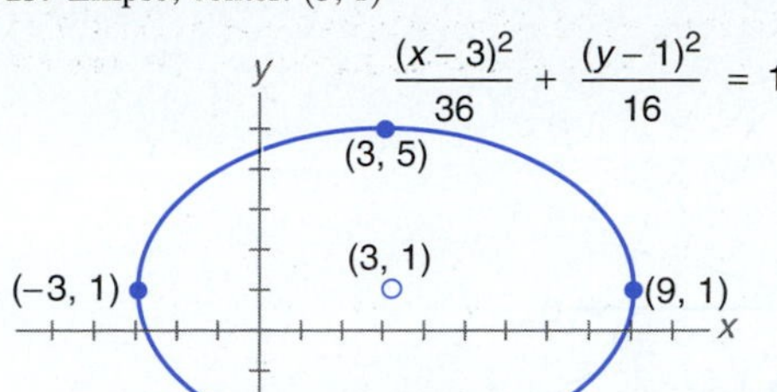

15. Parabola; vertex: (1, −3)

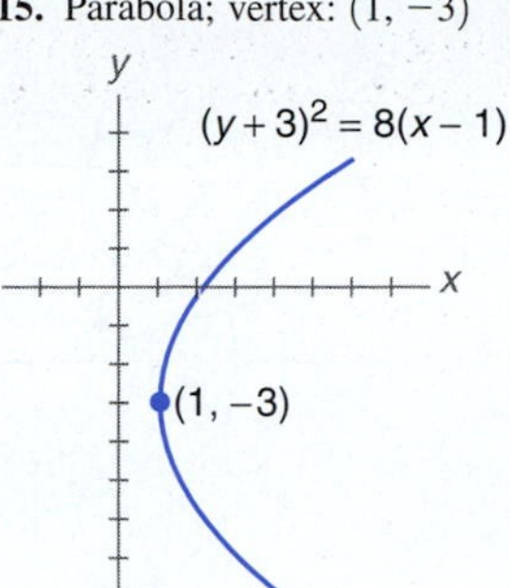

17. Hyperbola; center: (−1, −1)

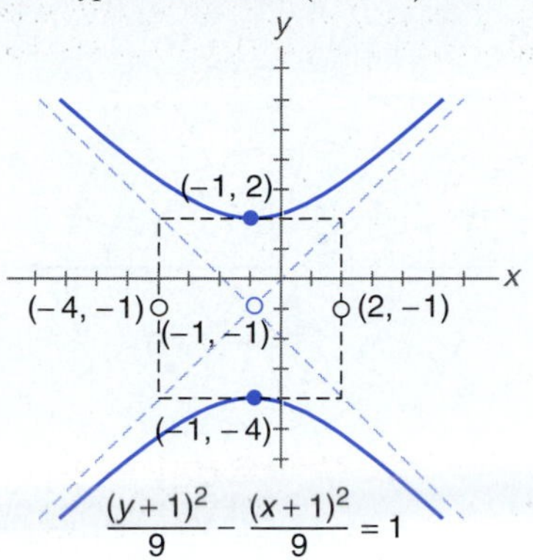

19. Parabola; vertex: (2, −1)

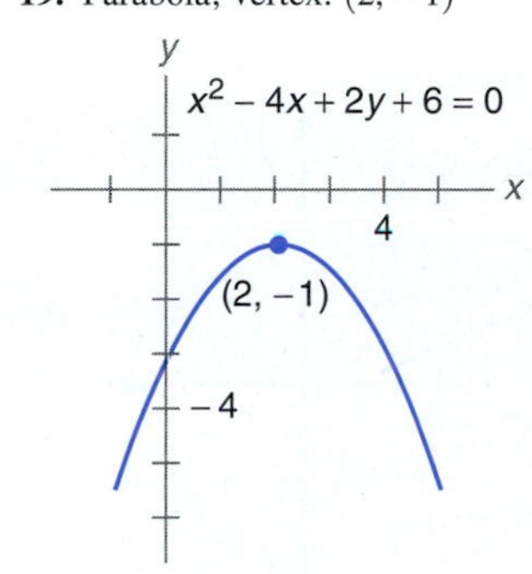

21. Ellipse; center: (−2, 1)

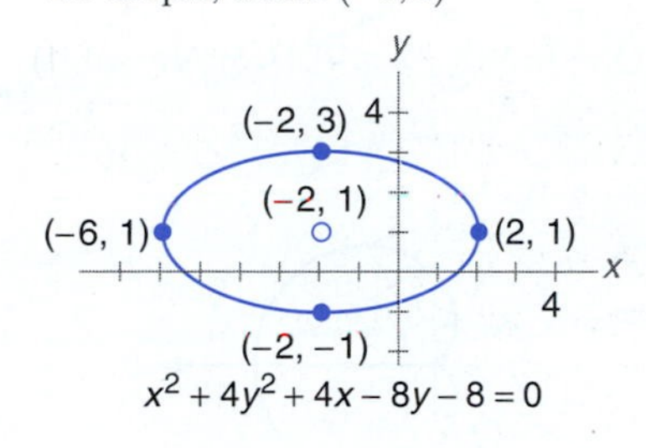

23. Hyperbola; center: (1, 1)

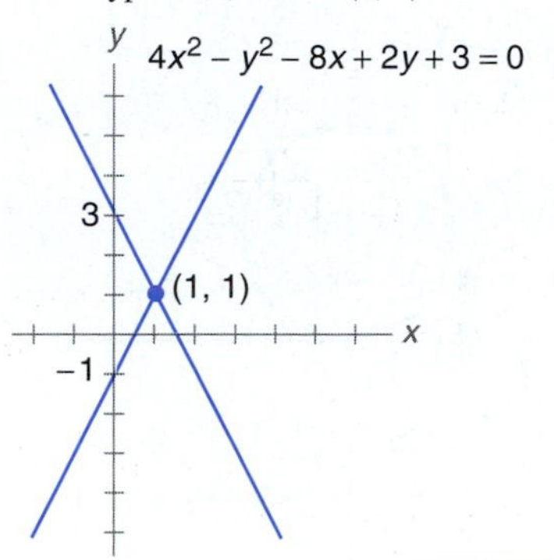

25. Hyperbola; center: (−3, 3)

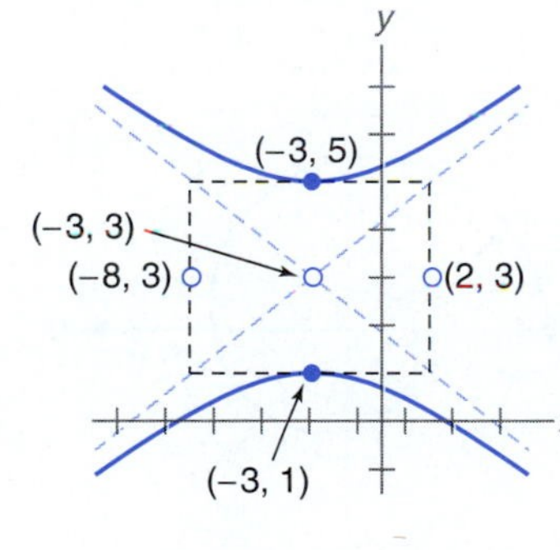

27. Parabola; vertex: (−8, −2)

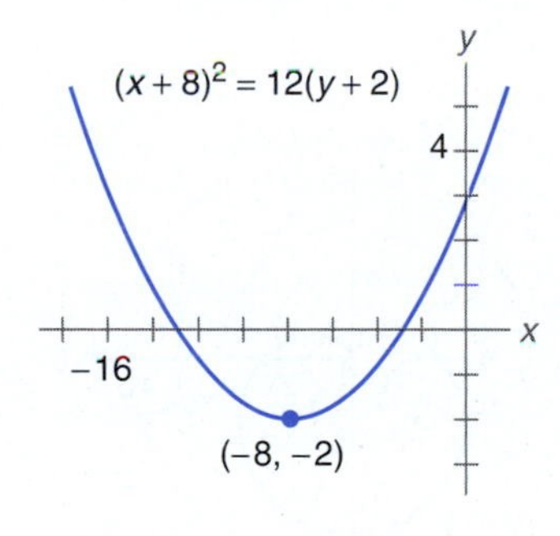

29. Ellipse; center: (−6, −2)

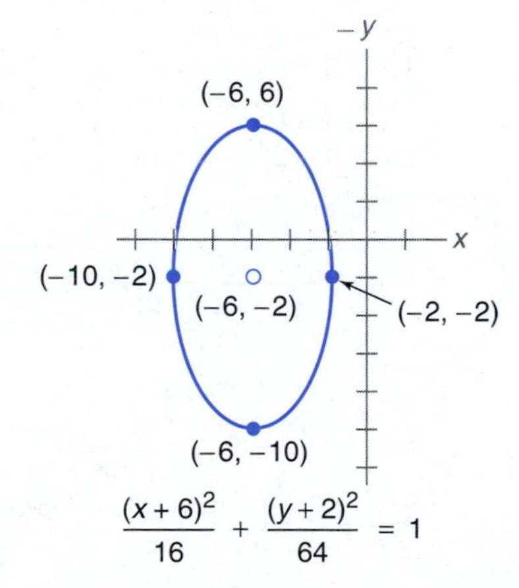

Exercises 20.6, Page 637

1. Ellipse **3.** Parabola **5.** Hyperbola **7.** Circle **9.** Circle **11.** Ellipse **13.** Hyperbola **15.** Parabola

Exercises 20.7, Page 641

1. (3, 3) **3.** $(1, \sqrt{3}), (1, -\sqrt{3})$ **5.** $(2\sqrt{6}, 6), (-2\sqrt{6}, 6)$ **7.** (−2, 0), (2, 0) **9.** (−6, 6), (6, 6) **11.** (−2, 2), (−2, −2) **13.** (2.3, 5.5), (−2.3, 5.5) **15.** (5, 4), (5, −4), (−5, 4), (−5, −4) **17.** (1, 4), (−1, −4), (4, 1), (−4, −1)

Exercises 20.8, Pages 647–648

1.

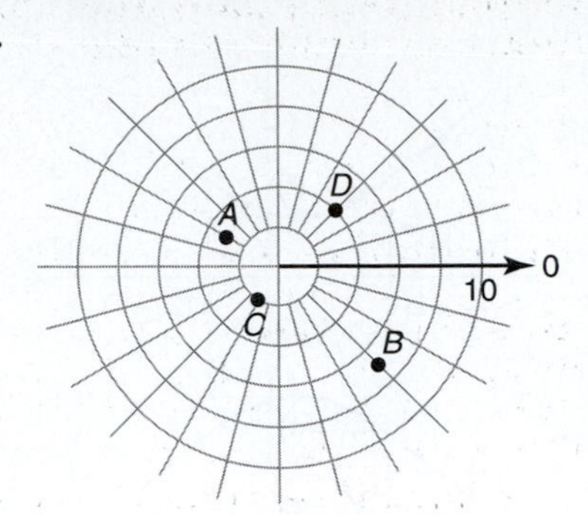

3.

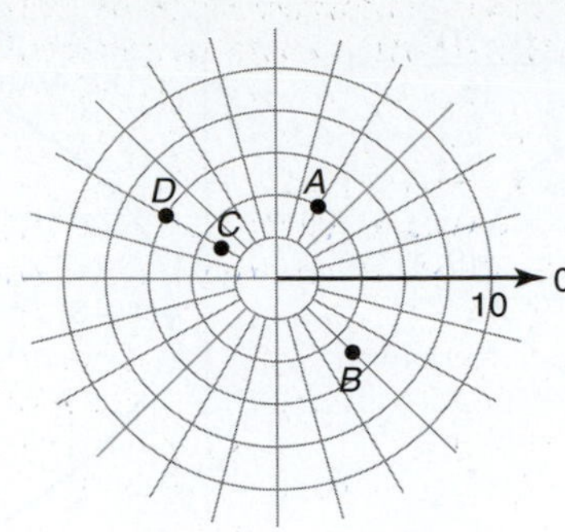

5. $(-3, 240°), (-3, -120°), (3, -300°)$ **7.** $(5, 135°), (-5, -45°), (5, -225°)$
9. $(-4, -315°), (-4, 45°), (4, 225°)$ **11.** $(-3, 7\pi/6), (-3, -5\pi/6), (3, -11\pi/6)$
13. $(9, 5\pi/3), (9, -\pi/3), (-9, -4\pi/3)$ **15.** $(4, -3\pi/4), (-4, \pi/4), (4, 5\pi/4)$

17.

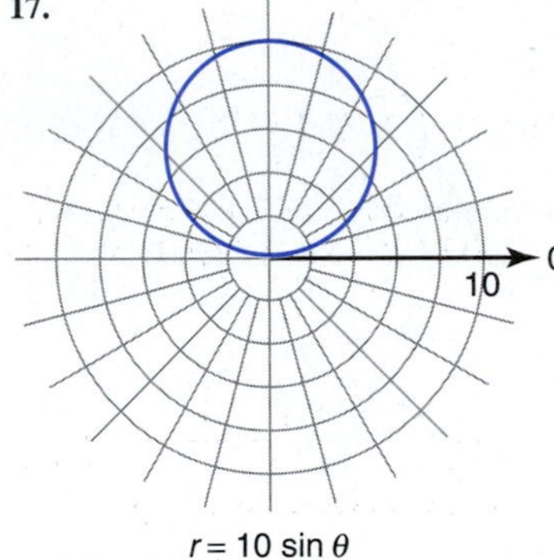

$r = 10 \sin \theta$

19.

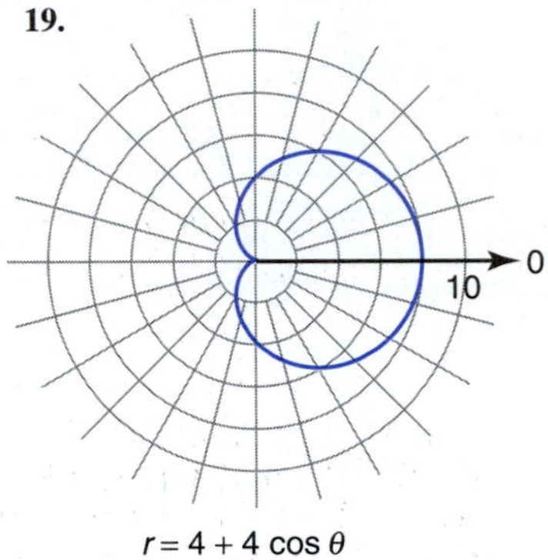

$r = 4 + 4 \cos \theta$

21.

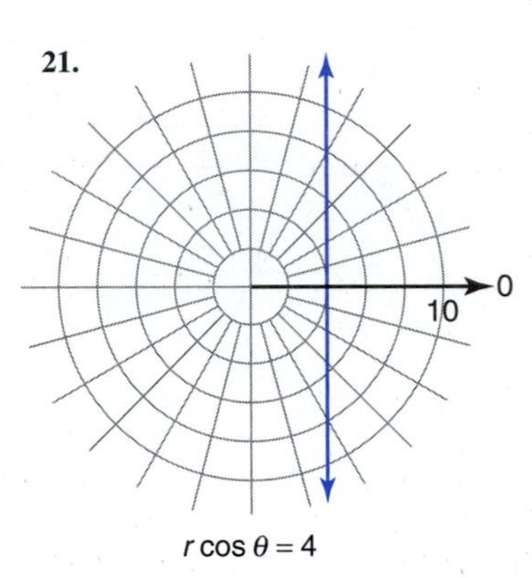

$r \cos \theta = 4$

23.

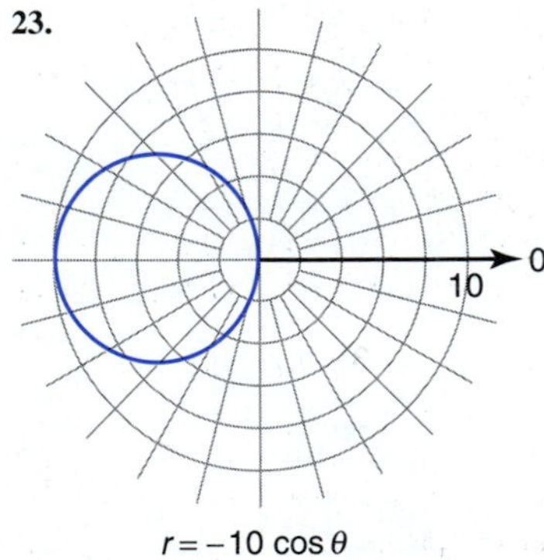

$r = -10 \cos \theta$

25.

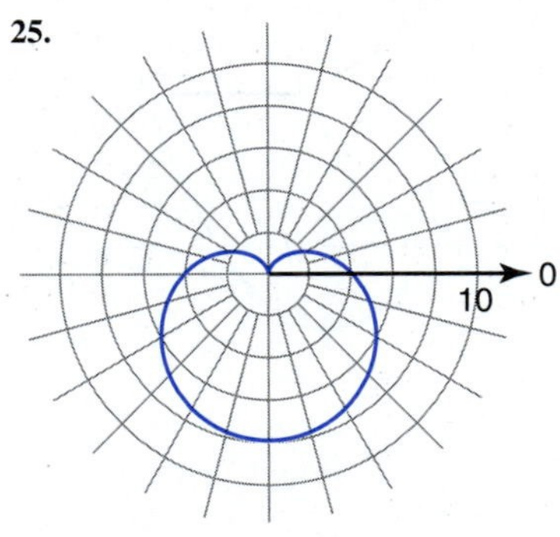

$r = 4 - 4 \sin \theta$

27.

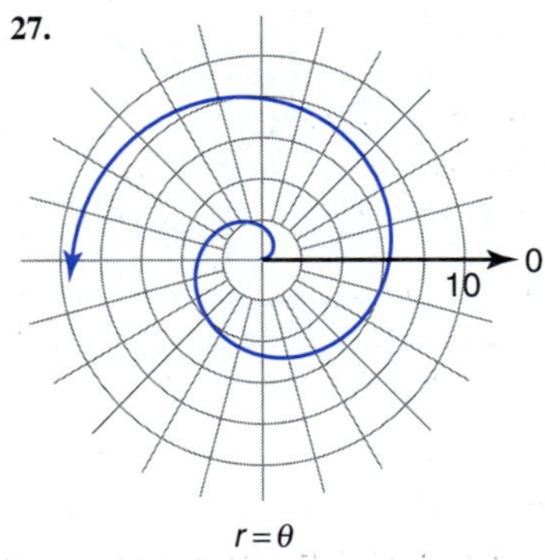

$r = \theta$

29. $(3\sqrt{3}/2, 3/2)$ **31.** $(1, \sqrt{3})$ **33.** $(2\sqrt{3}, -2)$ **35.** $(0, 6)$ **37.** $(2.5, -4.33)$ **39.** $(1.4, 1.4)$
41. $(7.1, 45°)$ **43.** $(4, 90°)$ **45.** $(4, 240°)$ **47.** $(4\sqrt{2}, 3\pi/4)$ **49.** $(2\sqrt{2}, 5\pi/6)$ **51.** $(4, 3\pi/2)$
53. $r \cos \theta = 3$ **55.** $r = 6$ **57.** $r + 2 \cos \theta + 5 \sin \theta = 0$ **59.** $r = 12/(4 \cos \theta - 3 \sin \theta)$
61. $r^2 = 36/(9 - 5 \sin^2 \theta)$ **63.** $r = 4 \sec \theta \tan^2 \theta$ **65.** $y = -3$ **67.** $x^2 + y^2 = 25$ **69.** $y = x$
71. $x^2 + y^2 - 5x = 0$ **73.** $x^2 + y^2 - 3x + 3\sqrt{3}y = 0$ **75.** $y^2 = 3x$ **77.** $xy = 1$
79. $x^4 + 2x^2y^2 + y^4 - 2xy = 0$ **81.** $y^2 = x^2(x^2 + y^2)$ **83.** $x^2 + 6y - 9 = 0$
85. $x^4 + 2x^2y^2 + y^4 + 4y^3 - 12x^2y = 0$ **87.** $x^4 + 2x^2y^2 + y^4 - 8x^2y - 8y^3 - 4x^2 + 12y^2 = 0$ **89.** $\sqrt{13}$
91. $d = \sqrt{r_1^2 + r_2^2 - 2r_1r_2 \cos(\theta_1 - \theta_2)}$

Exercises 20.9, Pages 654–655

1.

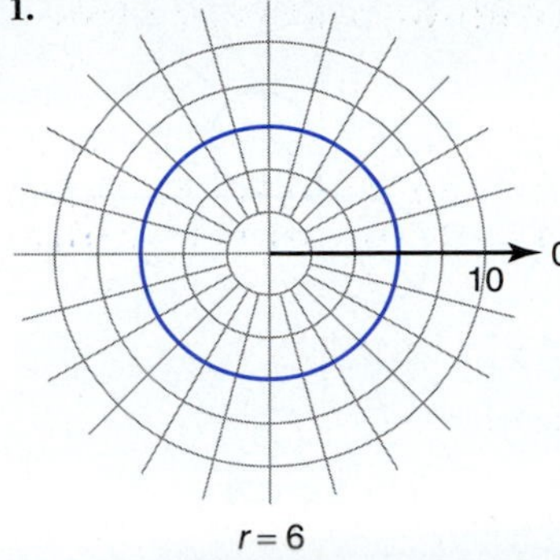

$r = 6$

3.

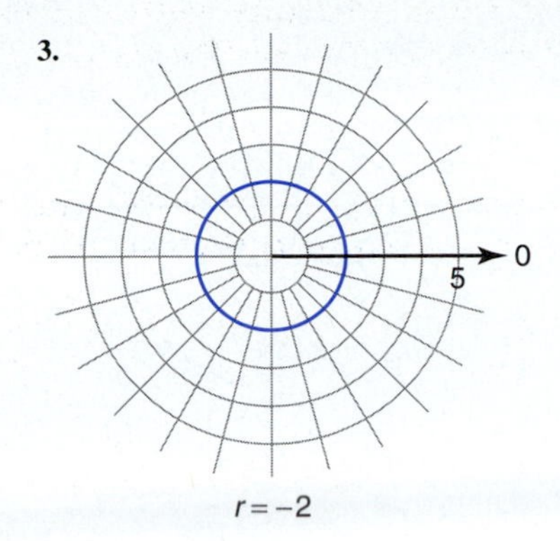

$r = -2$

5.

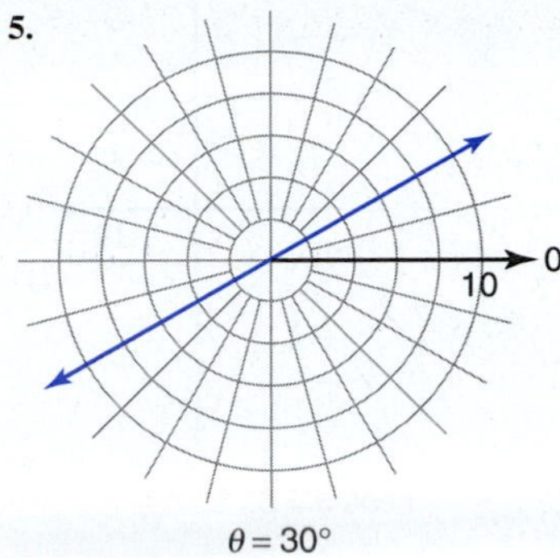

$\theta = 30°$

7.

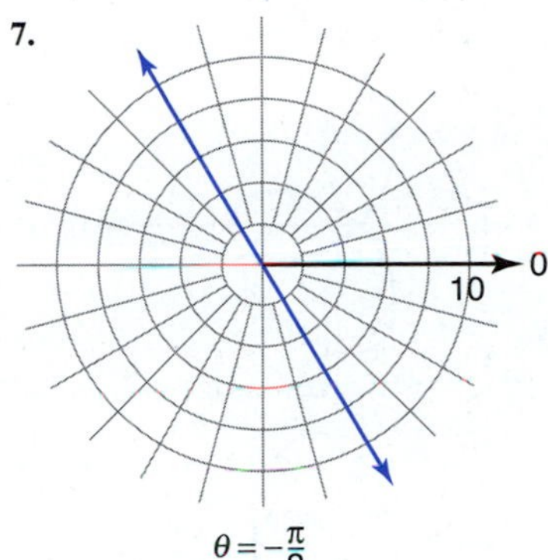

$\theta = -\frac{\pi}{3}$

9.

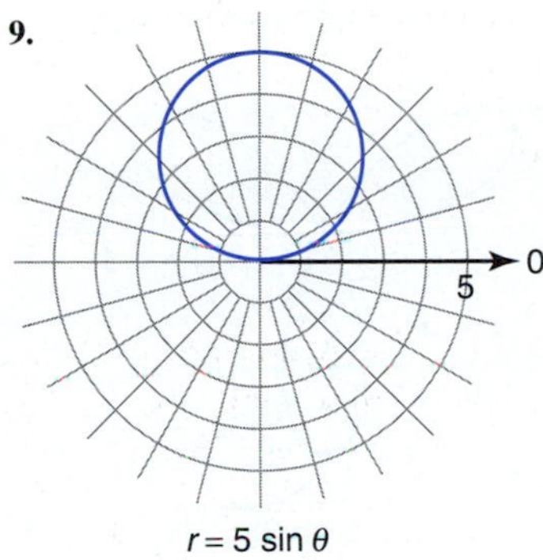

$r = 5 \sin \theta$

11.

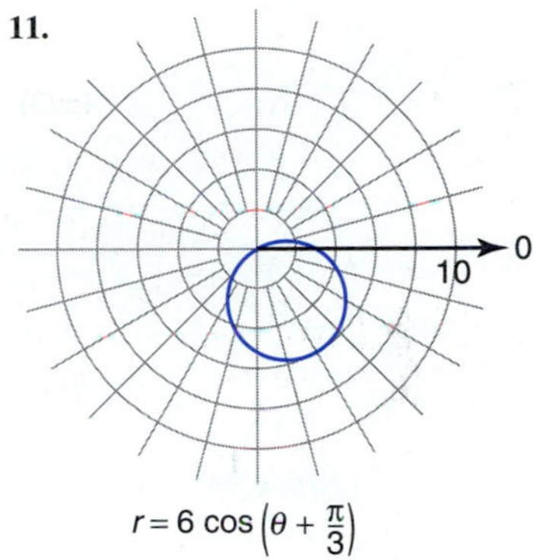

$r = 6 \cos\left(\theta + \frac{\pi}{3}\right)$

13.

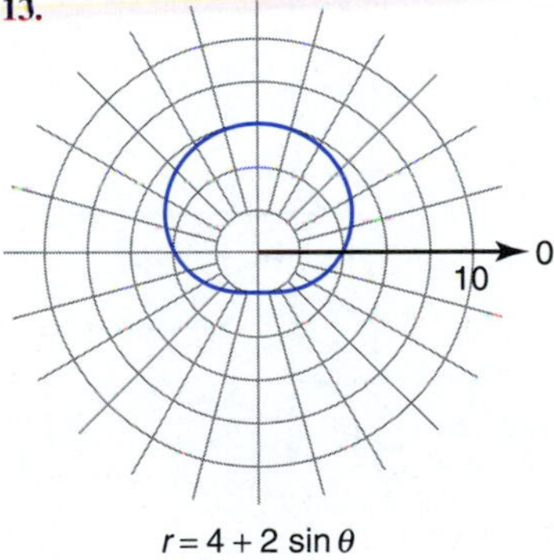

$r = 4 + 2 \sin \theta$

15.

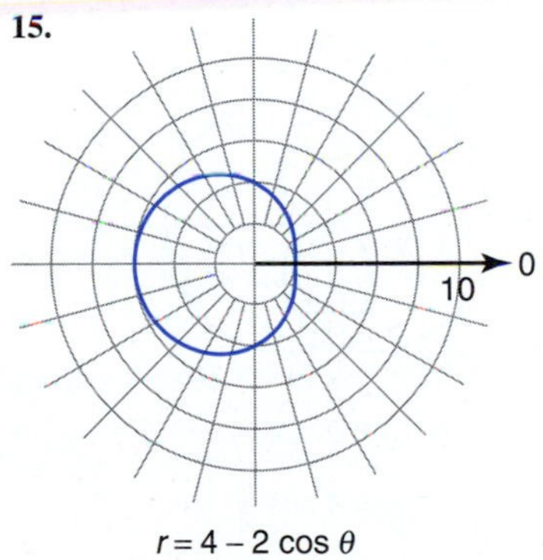

$r = 4 - 2 \cos \theta$

17.

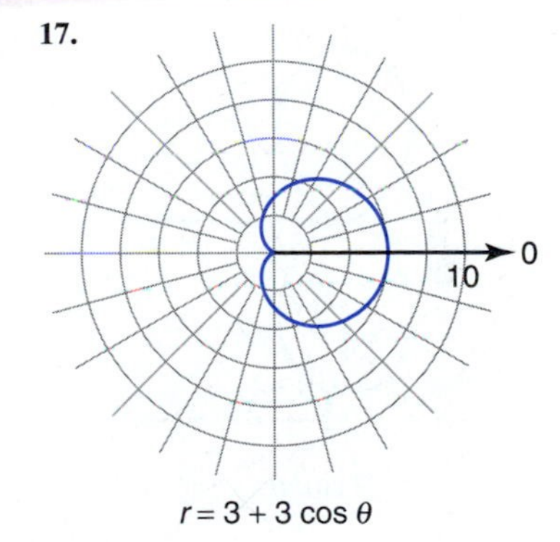

$r = 3 + 3 \cos \theta$

19.

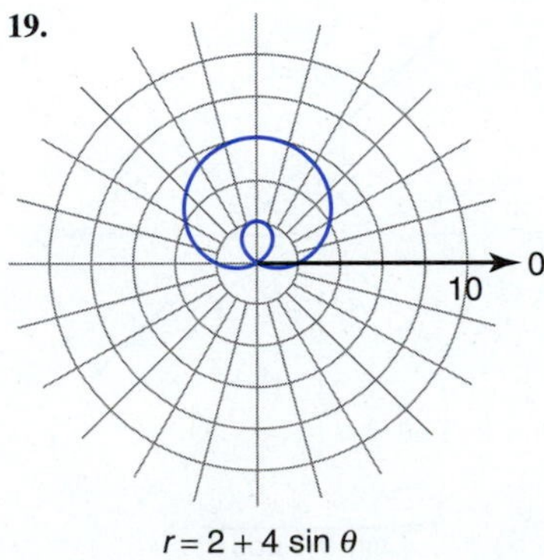

$r = 2 + 4 \sin \theta$

21.

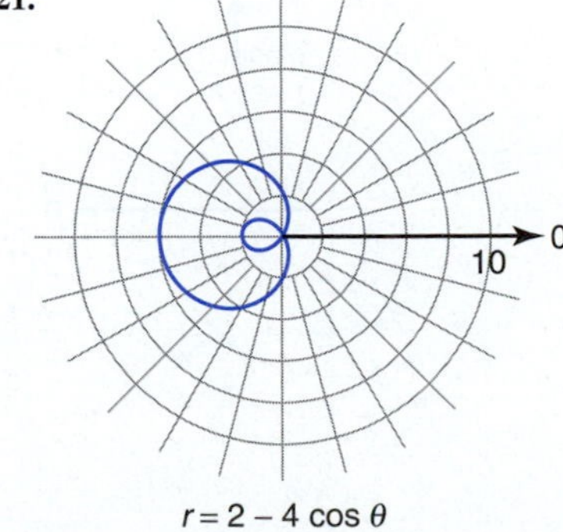

$r = 2 - 4 \cos \theta$

23.

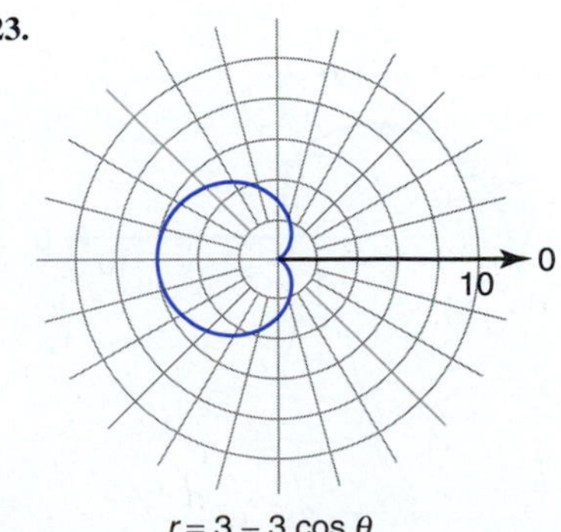

$r = 3 - 3 \cos \theta$

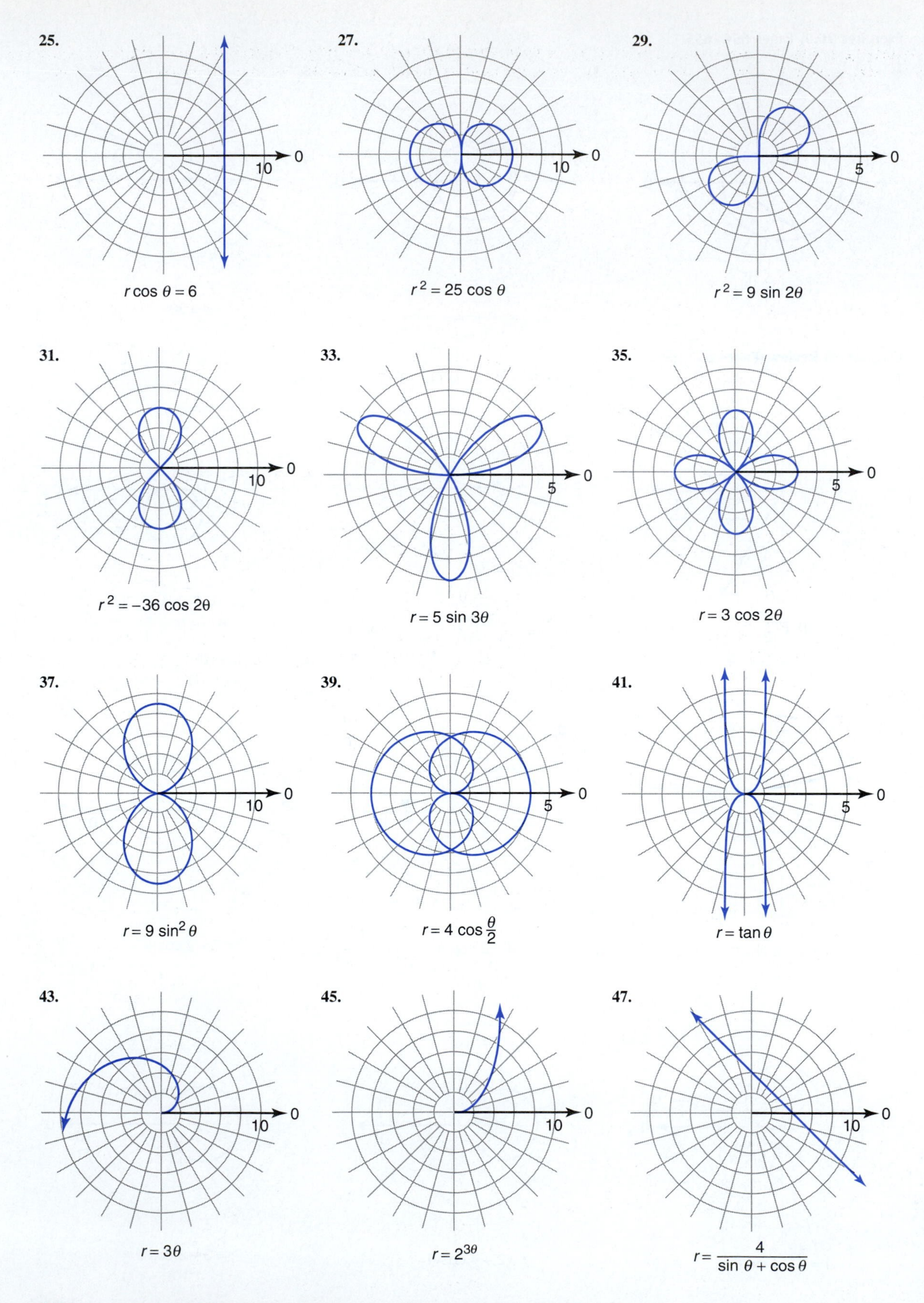
25.
0
10
$r \cos \theta = 6$
27.
0
10
$r^2 = 25 \cos \theta$
29.
0
5
$r^2 = 9 \sin 2\theta$
31.
0
10
$r^2 = -36 \cos 2\theta$
33.
0
5
$r = 5 \sin 3\theta$
35.
0
5
$r = 3 \cos 2\theta$
37.
0
10
$r = 9 \sin^2 \theta$
39.
0
5
$r = 4 \cos \frac{\theta}{2}$
41.
0
5
$r = \tan \theta$
43.
0
10
$r = 3\theta$
45.
0
10
$r = 2^{3\theta}$
47.
0
10
$r = \frac{4}{\sin \theta + \cos \theta}$

49.

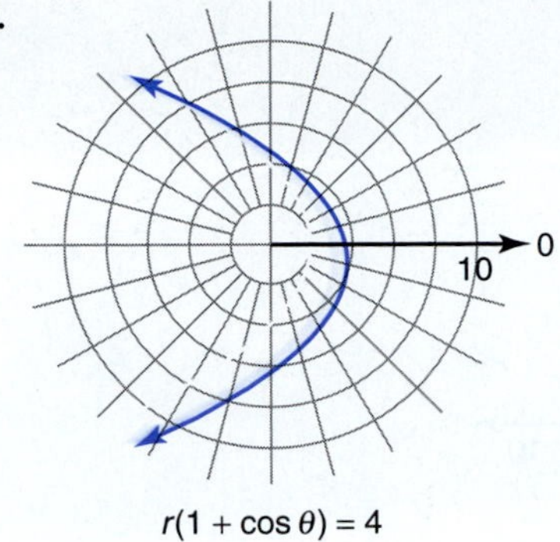

Chapter 20 Review, Pages 657–658

1. $(x-5)^2 + (y+7)^2 = 36$ or $x^2 + y^2 - 10x + 14y + 38 = 0$ **2.** $(4, -3)$; 7

3. $(0, \frac{3}{2})$; $y = -\frac{3}{2}$

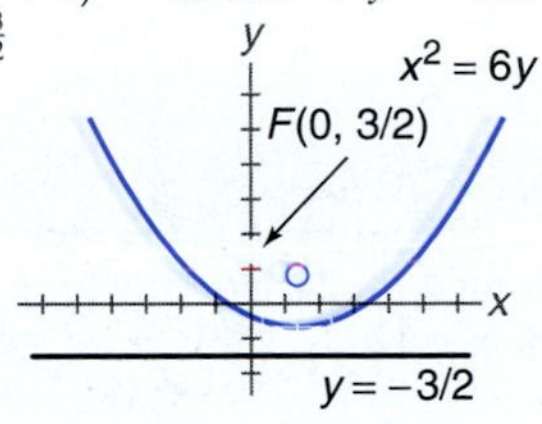

4. $y^2 = -16x$ **5.** $(y-3)^2 = 8(x-2)$ or $y^2 - 6y - 8x + 25 = 0$

6. $V(7, 0), (-7, 0)$; $F(3\sqrt{5}, 0), (-3\sqrt{5}, 0)$

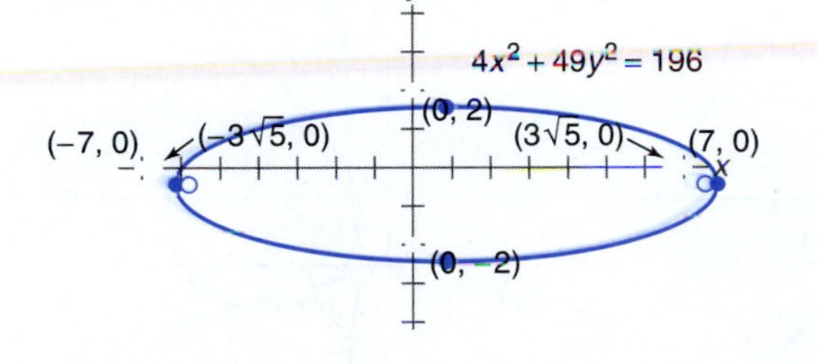

7. $\frac{x^2}{4} + \frac{y^2}{16} = 1$ or $4x^2 + y^2 = 16$

8. $V(6, 0), (-6, 0)$; $F(2\sqrt{13}, 0), (-2\sqrt{13}, 0)$

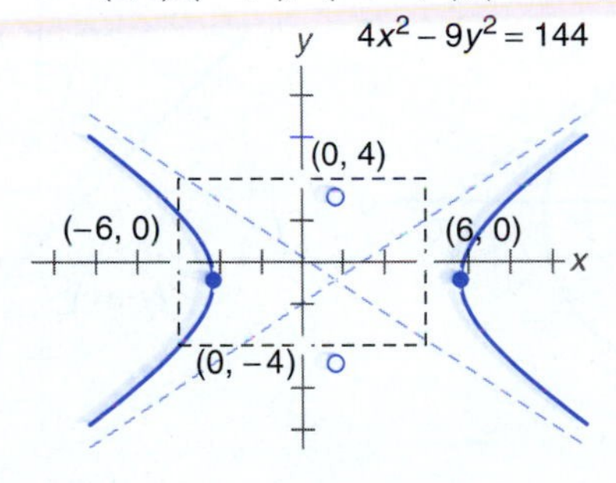

9. $\frac{y^2}{25} - \frac{x^2}{16} = 1$ or $16y^2 - 25x^2 = 400$ **10.** $\frac{(x-3)^2}{9} + \frac{(y+4)^2}{25} = 1$ **11.** $\frac{(x+7)^2}{81} - \frac{(y-4)^2}{9} = 1$

12. Hyperbola

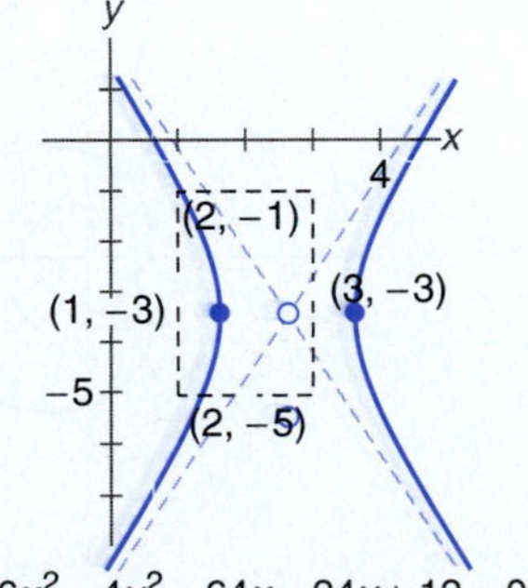

13. $(0, 0), (-12, -6)$ **14.** $(4, \sqrt{3}), (4, -\sqrt{3}), (-4, \sqrt{3}), (-4, -\sqrt{3})$

15.

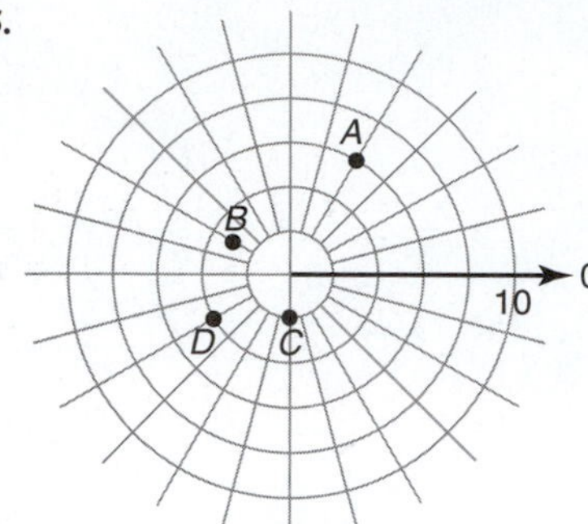

16.

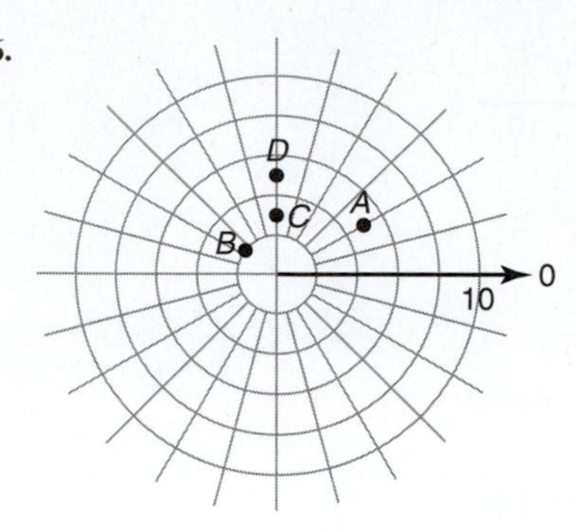

17. $(5, -225°), (-5, -45°), (-5, 315°)$ **18.** $(2, -11\pi/6), (-2, -5\pi/6), (2, \pi/6)$
19. (a) $(-2.6, -1.5)$ **(b)** $(-1, -1.7)$ **(c)** $(-4.3, 2.5)$ **(d)** $(0, 6)$ **20. (a)** $(4.2, 135°)$ **(b)** $(6, 270°)$ **(c)** $(2, 120°)$
21. (a) $(5, \pi)$ **(b)** $(12, 5\pi/6)$ **(c)** $(\sqrt{2}, 7\pi/4)$ **22.** $r = 7$ **23.** $r \sin^2 \theta = 9 \cos \theta$ **24.** $r = 8/(5 \cos \theta + 2 \sin \theta)$
25. $r^2 = 12/(1 - 5 \sin^2 \theta)$ **26.** $r = 6 \csc \theta \cot^2 \theta$ **27.** $r = \cos \theta \cot \theta$ **28.** $x = 12$ **29.** $x^2 + y^2 = 81$
30. $y = -\sqrt{3}x$ **31.** $x^2 + y^2 - 8x = 0$ **32.** $y^2 = 5x$ **33.** $xy = 4$ **34.** $x^4 + 2x^2y^2 + y^4 + 4y^2 - 4x^2 = 0$
35. $y = 1$ **36.** $x^4 + y^4 + 2x^2y^2 - 2x^2y - 2y^3 - x^2 = 0$ **37.** $x^2 = 4(y + 1)$

38.

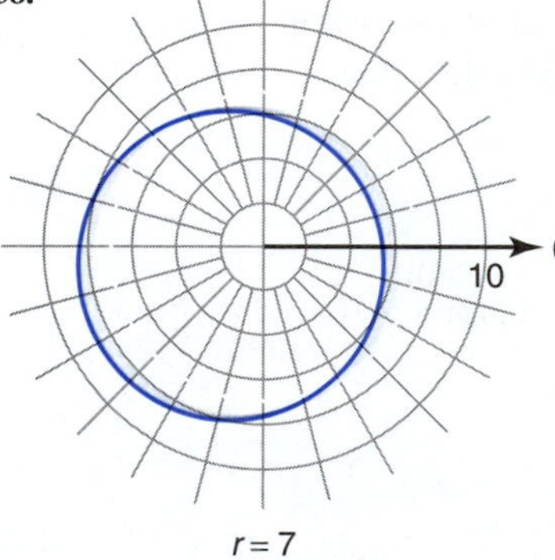

$r = 7$

39.

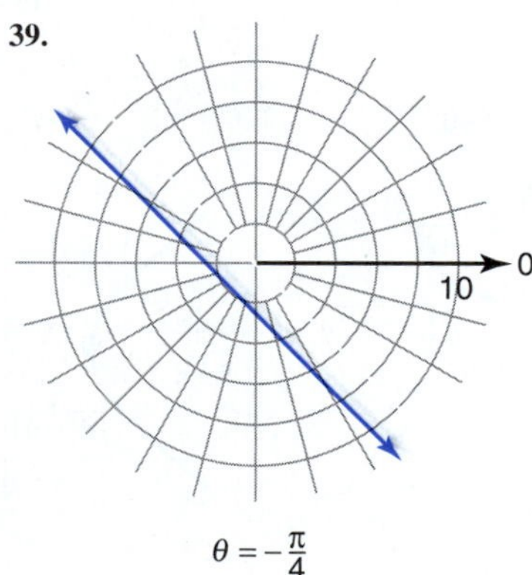

$\theta = -\frac{\pi}{4}$

40.

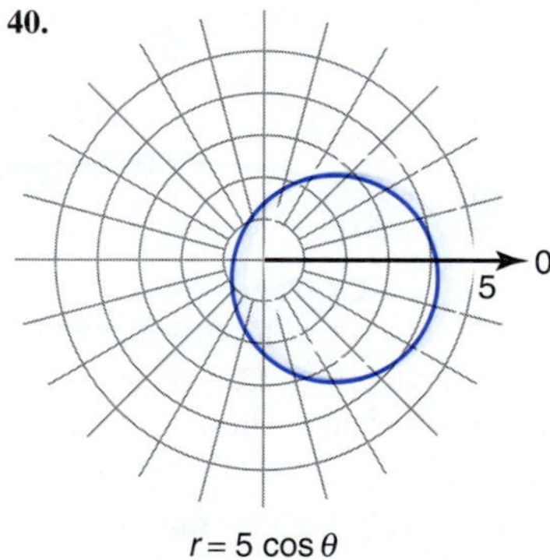

$r = 5 \cos \theta$

41.

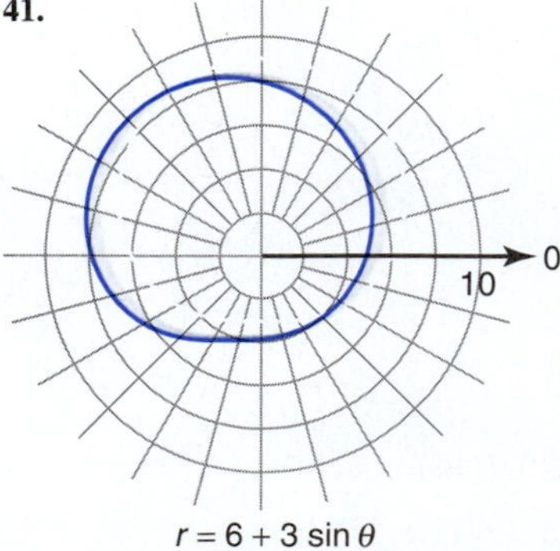

$r = 6 + 3 \sin \theta$

42.

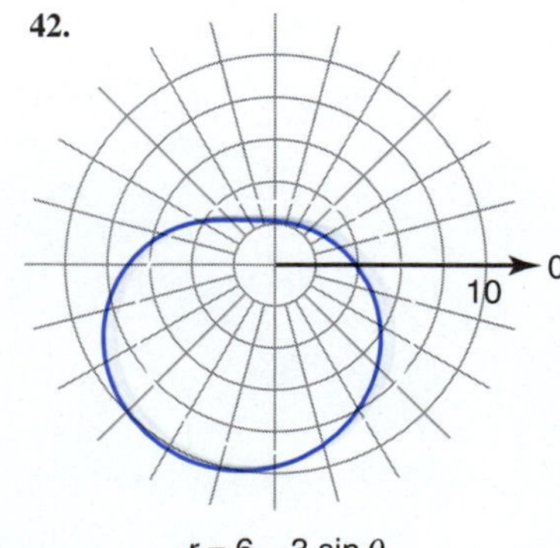

$r = 6 - 3 \sin \theta$

43.

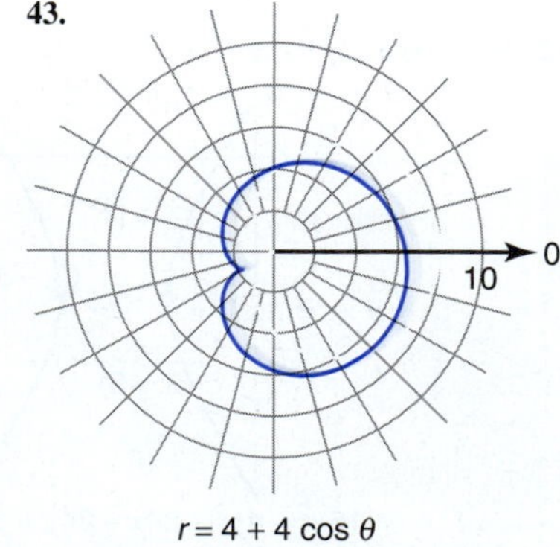

$r = 4 + 4 \cos \theta$

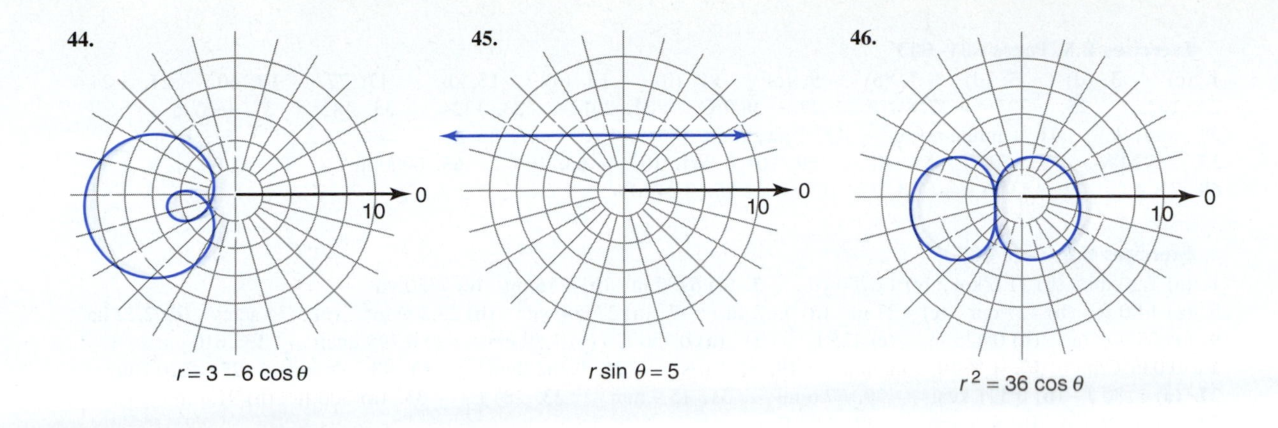

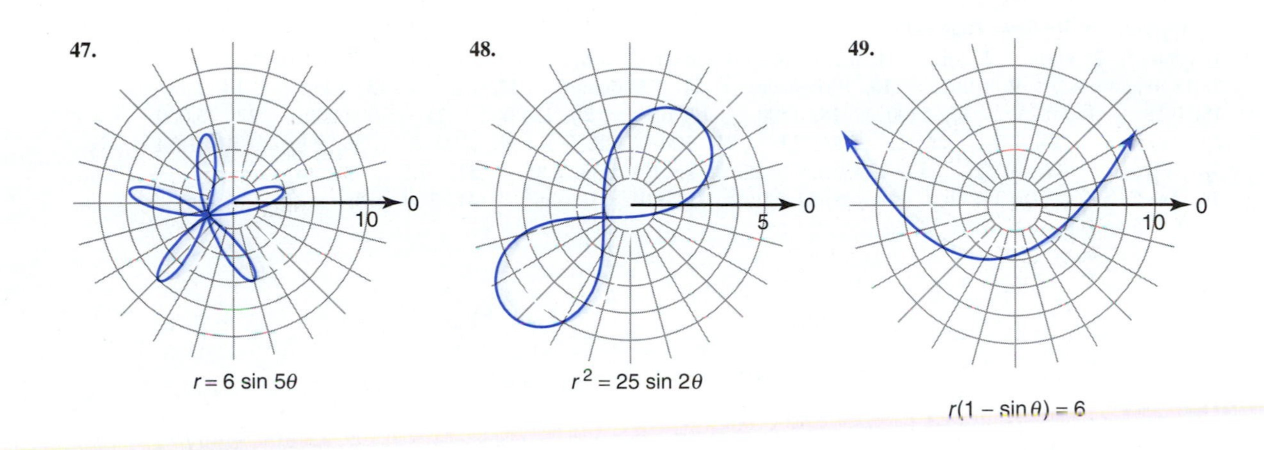

APPENDIX B

Exercises B.1, Page 666

1. kilo **3.** hecto **5.** milli **7.** mega **9.** h **11.** m **13.** M **15.** c **17.** 133 mm **19.** 18 kL **21.** 19 cg **23.** 72 hm **25.** 14 metres **27.** 19 grams **29.** 17 millimetres **31.** 25 dekametres **33.** 16 megametres **35.** metre **37.** litre and cubic metre **39.** second

Exercises B.2, Pages 669–670

1. 1 metre **3.** 1 kilometre **5.** 1 kilometre **7.** cm **9.** m **11.** mm **13.** km **15.** mm **17.** m **19.** km **21.** cm **23.** km **25.** cm **27.** km **29.** mm **31.** cm **33.** 1000 **35.** 100 **37.** 0.1 **39.** 1000 **41.** 100 **43.** 0.001 **45.** 10 **47.** 0.23 km **49.** 198,000 m **51.** 8.4 m **53.** 4750 mm **55.** 4,000,000 μm

Exercises B.3, Page 671

1. 1 gram **3.** 1 kilogram **5.** 1 kilogram **7.** kg **9.** kg **11.** metric ton **13.** g **15.** mg **17.** kg **19.** g **21.** kg **23.** g **25.** g **27.** kg **29.** kg **31.** metric ton **33.** kg **35.** 1000 **37.** 100 **39.** 0.1 **41.** 1000 **43.** 100 **45.** 0.001 **47.** 1,000,000 **49.** 565,000 mg **51.** 0.85 g **53.** 5 g **55.** 80,000,000 mg **57.** 1.5 kg **59.** 0.5 mg

Exercises B.4, Pages 676–677

1. 1 litre **3.** 1 cubic centimetre **5.** 1 square kilometre **7.** L **9.** m^2 **11.** m^3 **13.** ha **15.** mL **17.** m^3 **19.** L **21.** mL **23.** L **25.** m^2 **27.** L **29.** ha **31.** m^3 **33.** m^2 **35.** L **37.** 1000 **39.** 0.1 **41.** 0.01 **43.** 10 **45.** 1 **47.** 1,000,000 **49.** 0.001 **51.** 10,000 **53.** 100 **55.** 0.01 **57.** 100 **59.** 6.5 L **61.** 1400 mL **63.** 225,000 mm^3 **65.** 2×10^9 mm^3 **67.** 175 mL **69.** 1000 L **71.** 7500 cm^3 **73.** 50 cm^2 **75.** 50,000 cm^2 **77.** 400 km^2 **79.** 750 g

Exercises B.5, Pages 681–683

1. (c) **3.** (d) **5.** (d) **7.** (b) **9.** (c) **11.** (b) **13.** (d) **15.** (b) **17.** 77 **19.** 60 **21.** −24.4 **23.** 1022 **25.** −223 **27.** 41°F **29.** −38.9°C **31.** 2012 **33.** 1324 **35.** 482 **37.** second; s **39.** watt; W **41.** 1 milliampere **43.** 1 megawatt **45.** 1 A **47.** 2.7 μA **49.** 9.5 kW **51.** 15 ns **53.** 135 MW **55.** 10^3 **57.** 10^9 **59.** 10^6 **61.** 10^3 **63.** 10^{12} **65.** 6000 mA **67.** 42,000 μW **69.** 7.8 A **71.** 15,915 s **73.** 3×10^9 ns **75.** 0.04 MW

Exercises B.6, Page 685

1. **(a)** 7.20 mi **(b)** 11.6 km **(c)** 12,700 yd **3.** **(a)** 6700 m **(b)** 4.16 mi **(c)** 7330 yd **5.** **(a)** 14.0 ft **(b)** 427 cm **(c)** 4.27 m **(d)** 4.67 yd **7.** **(a)** 27,800 yd^2 **(b)** 23,200 m^2 **(c)** 5.74 acres **(d)** 2.32 ha **9.** **(a)** 25,900 in^3 **(b)** 0.425 m^3 **(c)** 425 L **11.** **(a)** 0.750 L **(b)** 1.59 pints **(c)** 0.795 quart **13.** 610 cm/s **15.** 0.036°C/h **17.** 1×10^{-4} in./min **19.** 98.4 ft/s^2 **21.** 0.362 lb/in^2 **23.** 43.3 oz/in^3 **25.** 2220 Btu **27.** **(a)** 2180 J **(b)** 0.521 kcal **29.** 77.6 mi **31.** 15.9 mm **33.** 284 L **35.** **(a)** 300 ft **(b)** 91.5 m **37.** **(a)** 0.826 cm **(b)** 8.26 mm **39.** 4.27 fl oz

Appendix B Review, Page 685

1. centi **2.** kilo **3.** mL **4.** μg **5.** 16 kilometres **6.** 250 milliamperes **7.** 1.1 hectolitres **8.** 18 megawatts **9.** 1 litre **10.** 1 kilometre **11.** 1 kilogram **12.** 1 m^3 **13.** 1 km^2 **14.** 1 ns **15.** 0.18 **16.** 0.25 **17.** 5700 **18.** 1500 **19.** 650 **20.** 15,000 **21.** 15,000,000 **22.** 75,000 **23.** 750,000 **24.** 1.8 **25.** 21 **26.** 23 **27.** 100 **28.** 0 **29.** 1200 m **30.** 1 kg **31.** 1.5 L **32.** 170 cm **33.** 50 kg **34.** 12 km/L **35.** 80 km/h **36.** 8.85 **37.** 77.5 **38.** **(a)** $12\bar{0}0$ yd **(b)** $11\bar{0}0$ m **39.** **(a)** 53,500 g **(b)** 118 lb **40.** **(a)** 32,400 ft^2 **(b)** 3.01×10^7 cm^2 **41.** 31.1 mi/h **42.** 15.6 lb/ft^3

Index